土木工程师资料手册

[美]罗伯特·O·帕姆利　著
杨　军　译

土木工程师资料手册

[美]罗伯特·O·帕姆利　著
杨　军　译

中国建筑工业出版社

著作权合同登记图字：01－2004－1183号

图书在版编目(CIP)数据

土木工程师资料手册/(美)帕姆利著；杨军译．—北京：中国建筑工业出版社，2004
ISBN 7－112－06743－X

I．土… II．①帕… ②杨… III．土木工程－技术手册 IV．TU－62

中国版本图书馆CIP数据核字(2004)第067543号

Civil Engineer's Illustrated Sourcebook/Robert O. Parmley, P.E.
ISBN 0－07－137607－0

责任编辑：董苏华 丁洪良
责任设计：郑秋菊
责任校对：赵明霞

土木工程师资料手册
[美]罗伯特·O·帕姆利 著
杨 军 译
*
中国建筑工业出版社出版、发行(北京西郊百万庄)
新 华 书 店 经 销
北京建筑工业印刷厂印刷
*
开本：880×1230毫米 1/16 印张：45¼ 字数：1400千字
2004年11月第一版 2004年11月第一次印刷
定价：115.00元
ISBN 7－112－06743－X
TU·5891 (12697)

(邮政编码 100037)
本社网址：http://www.china-abp.com.cn
网上书店：http://www.china-building.com.cn.

献给鼓励我从事专业工程
职业的哈罗德(吉姆)·克利里

总目录

招标程序

施工

补充材料

前　言

任何在私营实践领域寻求事业发展的工程师进入到这个行业后，都会由一个经验丰富的专家进行指导。但是在日益专业化的当今世界，土木工程领域很少有人能够精通工程咨询的所有方面。即便真的找到这么一个人，这个人也往往由于过于繁忙而没有足够时间来训练入门水平的从业人员。这样，新入行的工程师基本上处于一种很不利的状态。

良好的训练和教育对于从事顾问工程事业并不够，还必须拥有丰富的实践经验。每个州和有立法权地区的职业注册法律都规定在发放注册专业工程师（P.E.，Professional Engineer）注册证之前对工作经验有强制要求。然而这些要求是有广泛基础的。本手册并不打算取代任何注册要求，而是设计为扩展工程师训练中的这个环节。

市面上有大量关于某个专业主题的优秀图书，包括规划、设计、监理、预算、报告书写、施工管理和散布在整个工程文献领域中的相关主题。不幸的是，这些材料都过于详尽而且是针对特定主题的，以至于包含项目工程概览或指南的单卷本并未见到。这正是本手册打算填补的文献空白。

《土木工程师资料手册》被设计为引导读者历经项目的所有过程；从最初的创意到项目竣工。项目的里程碑式事件按照时间顺序依次介绍，主要焦点集中在典型土木工程项目的施工设计图准备上。这些设计图（包括相关文件），是从最近40年的成功项目中精选的。

很明显，本手册在有限的篇幅内不可能覆盖一个项目的所有方面，然而我们有足够的理由说这本手册给出的内容是丰富的，而且对读者来说是有用的指导和参考。

必须注意的是，在承接任何工程服务委托之前，工程师和客户都应对将要履行的服务范围、执行这些工作的合理时间表以及对上述服务的付费等协议有一个清楚的了解。通常在执行任何工作之前，这些协议都被整理成正式的书面文件，并由所有各方在合适的位置签名。

不同的专业组织拥有大量得到应用的不同的工程服务标准合同格式。这些协议标准格式中第一流的原始资料出自：工程师联合合同文本委员会（由来自美国顾问工程师委员会、美国土木工程师学会、施工规范协会和美国职业工程师协会的代表组成）。另外，很多出资/贷款机构和一些管理机构备有他们自己的典型协议格式或者重要的特殊要求。所以，不管是签署的建议书、协议信函或者正式的服务合同，本手册假定正确执行了工程合同，这样工程师就可以开始工作了。在理解这些之后，读者现在就应已准备好浏览本书的主体。我希望这样的阅读是有益的，获取的价值是持久的。

所有资料和文件的来源都在引用处做了说明。特别感谢 Lana 和 Ethne 在文字处理方面的帮助。如同以前一样，Wayne 在最终格式的版面和设计上的工作非常出色，与他在一个大项目中再次合作非常愉快。

罗伯特·O·帕姆利
注册专业工程师、主编
莱迪史密斯，威斯康星州

导 言

本手册的目的是为用户提供有关土木工程的通用样板的全面概述，这些样板正是被私营咨询公司普遍使用的。

手册的范围

通过已竣工的实际项目的图解说明，本手册总的目标是为那些对私营土木工程实践感兴趣的人提供相关规则的概要介绍，不论他们是学生、教师、初级工程师、地方官员、技术员、规划师、设计师或者普通大众。尽管这些信息并不新鲜，但它分散在不同的工程文献中，到目前为止还没有一本按逻辑顺序囊括所有这些信息的单卷。我们的目标就是要填补这一空白，为用户提供一本有益的有持久专业价值的手册。

目前，关于这些主题的大多数文献都是以文字形式存在的。因为本手册有意成为包容一切的和有教益的，所以在接下来的篇幅中包含了大量的插图。只有需要对图纸、图表、表格和草图进行补充时，才有必要包括进文字材料。

准备书稿的一个首要问题是精选已有的材料，让它们能够装进一本单卷的手册中。对未来的读者的需求预期更加增大了这项工作的难度。他们感兴趣的必然要涵盖从最基本的到非常复杂的内容。为了解决这些难题，选择能够均衡表达的材料就显得尤为重要。这样，将从40余年的项目数据库中做出选择，重点放在基本的或通用的实践领域。

主编提醒读者注意，本书给出的材料包括通用的应用、基本的样板和一般的范例。敏锐的读者会注意到，书中给出的很多示例设计图是按照过去的规范设计的，现在已不能使用了。我们请求读者许可我们收录这些设计，因为它们被成功地用于建造至今仍发挥着初始功能的设备、结构和体系。然而，对于特定项目，读者必须参考现行的规范、规程和条例，同时满足对项目参数有管辖权的特定管理机构的最新要求。

大多数土木工程项目实际上并不大，也不为公众关注或者在历史上有名。相反，它们通常是相对较小的、日常的、实用的，持续不断地推进现代文明。大部分社区的人口都少于10000，但仍然存在大都市的基本需求：例如排污管道、饮用水、交通网络、消防、废物处理和处置、市政建筑和相关设施。考虑到这些情况，我们广泛收集了规模相对较小却很基本的设计图来说明较大范畴的项目，它们是大量常见的市政工程，而且对于土木工程实践是很基本的。

如前所述，本手册的材料是从数十年的档案中摘取出来的。这样，读者将会注意到设计图格式和绘制技术的变化。潮流在变，技术方法也随之演变。本书后续大多数的设计图都是用传统工具（例如墨笔或者铅笔）手绘在牛皮纸或硫酸纸上的。只是在最近几年，这种绘图方法才很快被计算机辅助制图（Computer Aided Drafting，CAD）所取代。我很担心因为完全依赖这种电子过程，一些有价值的技术从此失传。由此，我们倾向于收录大量手绘的设计图。读者将学习这些早期设计图的技巧和组织，并能考虑集成其中可用的概念到他们将来的图纸中。然而必须注意的是，我们并不是想说明 CAD 系统处于劣势或者使用范围不广泛，而是想警告 CAD 系统存在潜在的程序约束和图形极限。让 CAD 系统为我们服务，而不是让设计师适应其局限性。为了对照比较，本书收录了几套完全由 CAD 绘制的图纸，它们展示了经过适当编程的 CAD 完成指定程序的能力。

本手册的体例遵从典型项目的一般逻辑顺序，也就是说，规划、设计、投标程序和施工。下面对每个范畴的介绍将充实它们的相应章节。

规　　划

所有的项目都开始于一个好的创意。随着这个创意扎根并开始发育，紧跟着必须进行合理的规划，否则项目可能会遭遇无数的延迟、技术上的陷阱、可能的冲突、缺乏支持、公众的误解、预算问题、环境矛盾和规章限制。

每个项目都有其自身特定的必须强调的要素。本手册显然不可能预见到所有的这一切。然而，土木工程实践中的大部分要素对所有项目都是共同的。所以，本手册致力于把这些最基本的东西按照规则的、逻辑合理的、时间发生先后的顺序包含进来，这样读者就能够看到一个项目是如何走向圆满成功的。

因为项目的构想来自个人的思想或者团体的集体智慧，所以必须及时做记录并整理出一个概要，为工程或技术报告做准备。这份报告在描述项目和项目子项时必须完整和精确。第 1 节介绍了不同类型的技术报告，并提供了概要示例。

第 2 节介绍了项目进度计划，并给出了摘自过去已经竣工的项目的概要和格式示例。

随着项目的继续，通常需要进行现场勘察、测量和绘图。地图、数据和信息示例于第 3 节，让读者熟悉项目发展的这个阶段需要的不同材料。

第 4 节介绍了一个项目的公开会议部分。这个阶段将极为关键。很多项目因为没有很好地进行公开会议而被终止或者延期。不管是信息通报会、评估听证会、日常会议还是正式会议，工程师必须进行充分准备并了解其目的以防止犯错。

因为管理部门的审查通过不可能先于提交最终设计图和规范，所以了解哪些部门对待建项目有管辖权就是非常明智的。在项目开展早期要及时了解现行规范和规程的全部完整知识，以避免潜在的设计冲突。这一部分在第 5 节中重点强调。

理智的做法总是首先计算一个项目的成本，再进行财务评价。第 6 节详细介绍一些土木工程项目常用的估算方式以及一些基本的表格来帮助读者进行成本计算。需要记住的是，除了保护公众的健康和安全，土木工程师应该首先正直地为其客户提供最经济和环境友好的项目。

设　计

在概念规划、财务安排和客户的开工通知后，工程师可以着手开始设计阶段。项目的这个阶段无疑是最细致的和技术密集型的。

施工设计图以及与之伴随的规范的准备必须由注册的专业设计师控制和进行持续的监管。尽管有大量的技术人员参与到设计图和技术规范的准备工作中，但监管工程师负责提供最终的设计文件。最终设计图和技术规范通常需要由监管专业设计师封缄、注明日期并签名，随后送给相关的管理机构审查。

设计图的一般格式和图纸安排随项目类型、管理机构的要求和每个设计公司的偏好不同而不同。然而，大多数设计图的标题页包含一些基本参数。本手册的第 7 节将介绍一些典型示例。

第 8 节到第 20 节包含了实际设计图示例，为读者提供了较大范围的项目，从中可以获得图纸安排的感性认识。经过缩小的设计图来自多年前的档案，大多数是手绘图，但有一些是用 CAD 绘制的。这里并不打算标准化图纸，因为主编想为读者提供尽可能大范围的图纸和绘图技术。CAD 理所当然成为绘图的首选方法，但是已建项目的原始图纸仍然必须参考，而它们的绘制是在使用计算机之前。这样，年轻的工程师需要在一定程度上适应以前的实践。我们鼓励读者审视这些实例，希望能够发现能用于今后设计图绘制的技术和图形体例。

第 21 节给出了一些经常用于相似项目的典型标准详图示例。

第 22 节简要介绍了施工项目技术说明的编制过程。这些书面文件为进一步描述待建项目，以一种图纸不能完全详细表达的方式。虽然一张图抵得过千言万语，然而有时一个词或者短语能够用来描述绘图不能充分表达的细节。这样，设计图和技术说明必须互相补充，确保对待建项目的完整理解。

招、投标程序

施工设计图和技术说明完成后，管理部门开始审查工作。因为管理部门众多而且监管领域不同，本手册不打算将它们一一罗列或者分类。有充足的理由说，专业设计师有责任完全了解每个特定项目的现行规范、条例、规程和需要的审查。

一般来说，在收到所有必需的批准和经过签署的土地使用权之后，客户就可授权工程师启动招投标程序。每个待建项目都有其特别参数，一些项目有州或联邦的工资率要求，特别的少数民族法令，联邦反歧视条款等。所有这些特殊的规定和条例必须加入到招标文件中，作为设计图和技术规范的补充。第 23 节介绍了这个阶段，不但提供了常用文件示例，还讨论了招投标过程。

在招标套件（即招标文件、设计图和技术规程）汇编完成后，就可以准备招标广告并在当地报纸和其他合适的出版物（例如商业杂志，建筑商交易所等）上刊登（见第 24 节）。在此时，通常将招标套件送给客户，建筑商交易所和合适的承包商、分包商和供应商。在从首次发布招标广告到开标的周期内，工程师必须能够回复可能的投标人（承包商）提出的问题，这些问题可能涉及招标文件、设计图和技术规范中的任何部分。有时需要召开招标预备会议来答复承包商的问题。如果某个问题或利害关系重大，工程师应准备一份书面的补遗并在开标前发放给所有设计图拥有者。这份文件的目的是在开标前澄清一些误解或者纠正所有已发现的错误，这样所有的可能投标人位于同一水平并进行公平竞争。

当到了开标的时候，通常由客户或者工程师宣布标底。这是一个正式的过程，应该是一个公开事件，附有一份签名的出席名单。所有的标书都必须在截止日期之前封装在信封中提交，并由接收工作人员用姓名的首字母签名。招标截止后，工程师和客户按照提交顺序打开每份标书并大声朗读每个投标者的出价。

唱标完成后，观众回避，工程师负责详细地审查每一份标书，需要检查数字、投标保单和其他可用的文件。通常在开标数天后，工程师给所有设计图拥有者和客户发布一份正式标书表格。在这个时刻，工程师向客户提供建议。如果较低价投标者拥有一份真诚的投标保单、合理的平衡出价、优秀的介绍人而且满足了所有特殊招标要求，那么工程师将向客户推荐他。如果客户同意，中标通知书通常通过工程师发送给成功的承包商。在 10 - 15 天的时间内，承包商向工程师提供一份履约保单、支付保单、保险证明书和其他相关合同文件，工程师将它们整合进合同供客户的律师审查。按照律师的证明或者合同规定，工程师通知承包商并安排好开工前会议的日期。施工合同请参见第 25 节。

施　工

施工阶段使得项目能够最终实际成形。在经过漫长的几个星期、几个月，通常是数年的努力后，项目最终完成。

第 26 节介绍了开工前会议的一般形式。在这次会议上，所有的参与者正式集合在一起，包括业主代表、工程师、监理、承包商、分包商、公共事业代表和其他利益相关团体。施工承包合同正式授予承包商，并附有一份开工通知。审查计划和规程，提出施工进度计划。讨论权利体系并规划好各项的时间表。这时就可以安置项目的标识牌了，择日举行破土动工典礼。

两三周后，施工图开始陆续抵达工程师办公室供审查、讨论和通过。生产商和制造车间往往比建造者更早地需要施工图纸。工程师有责任审查这些图纸以确保它们符合项目设计图和技术说明。这些构件不是监理能够在项目工地能够看到其施工的。这些构件的例子包括：检修孔、水泵、电路板、结构钢材、特殊设备、控制模块、预应力混凝土、钢制水箱、细木家具、门、窗、屋面桁架、特殊硬件等。一些实际的施工图可参见第 27 节。

建筑现场的安全是一个重大问题。它通常会在开工前会议上讨论，但应受到与施工阶段相关的所有人员的持续不断关注。一些承包商每周召开会议，在员工中强调安全第一的思想。第 28 节提到了施工安全的一些方面。

项目监理是工程师在现场的代表，负责在现场直接监理施工过程。监理保持精确的、详细的、每天书面记录的项目日志是非常重要的。关键单元的施工草图和照片对于历史记录也是极为重要的。监理并不向施工企业负责，然而工程中的任何缺陷和质量较差的工艺都会被尽可能快地汇报到工程师那里，这样就能及时地采取补救措施。优秀的承包商将与公正的监理协作，以保证设计图和技术说明得到遵循。第 29 节介绍了监理的一些职责。

对材料、土体、混凝土和项目系统的测试通常都应在监理的监督下进行。混凝土试样和其他材料的收集可由监理来协调和编目。对给排水系统和类似管道网络的测试通常是工程师在监理的协助下完成的。

项目施工的立界桩是由工程师和监理或者工程的测量队完成的。

监理通常要陪伴测量队，这样他就能知道立界桩的方法并能控制地坪和/或切片给工长。进一步的讨论请参阅第 30 节。

项目的结束阶段特别耗费时间而且强度很大。一个项目的最后 5% 是最艰巨的。承包商和分包商急于转移到它们下一个项目，在现场只留有非常少的工作人员。在这个时间点，工程师在驻现场监理的帮助下需要完成从头到尾验收已完工项目的任务，向客户（业主）确保施工实际上符合设计图和技术说明，而且满足管理部门许可的需求。这个过程衍生出的文件即是承包工程项目清单（Punch List）。所有在项目验收过程中发现的未完成项目、缺陷、问题等都将罗列在承包工程项目清单中，它将被正式地送达给主承包商，便于其执行完善。在工程师签署项目竣工报告前，所有在承包工程项目清单上的问题都必须加以解决或改善。另外，在尾款支付前，承包商必须向业主提交留置权弃权书。

附　　录

本手册的最后一节名为技术参考。它的内容包括了大量的技术数据信息，它们通常并未包含在普通工程手册中。这些材料经过浓缩，便于使用，并且避免进行过于冗长的讨论。

本节的最后 8 页是公制计量。描述了国际单位制的基本单位，其后紧跟的是对它的详细描述和转换系数。

希望读者将此参考材料加入他们个人的技术档案或者利用这些数据着手创建个人技术档案。

规　划

第 1 节　技术报告

例　1

报告在词典中定义为关于某个研究课题的说明或者某些事实的正式叙述。

准备好的书面技术报告是为读者提供关于某个专门的或特定的主题的信息。

本手册的这一节将简要地介绍多种技术报告并给出典型范例的要点。

目　录

概　览

市面上有多种优秀的书籍详尽地讨论了技术报告写作的多个方面和阶段。本手册不打算与它们竞争。我们想要简要地总结整个过程并在合适的地方提供一些有用的提示和来自实际项目的图表示例。

有研究表明，人类作为交流者是迟钝的，效率非常低。虽然生来就被赋予了一套高度复杂的交流系统，而我们却常常未能正确地使用它。这套生物交流系统由传送器和接收器耦合成一个由计算机控制的单元；所有单元都装在我们每个人的头部。传播以三种方式进行：身体动作、言语和书面记录（以多种方式存在）。本节将简要介绍的是最后一种方式。

无论技术报告的主题是什么，对读者的考虑是非常重要的。读者必须能理解报告的内容及其想达到的目标。所有的事实都必须是精确的并备有相应的证明文件。结论必须严格地基于这样的事实和建议——它们是依据可靠的专业审查得到的而且由那些相关事实的合理评价所支持。

准备一份有用的、能很好实施的技术报告包括很多方面。虽然要将这些所有方面完整地罗列出来是不可能的，作者还是愿意提到那些在数年内都有价值的关键因素。它们是：

1－了解你的主题

2－准备一个提纲

3－用文件记录所有的事实，并将它们按照合理的顺序组织起来

4－绘制一份草图

5－简短、扼要

6－收集可用的文件作为支持证据

7－获得高质量的图片、图表、地图和照片

8－协调

9－校对

10－坚持要求良好的文字处理，也就是字体、页面设计和标题的强化

11－终读和高层审查

12－专业的封面和装订

必须注意的是，技术报告作为一种记录必须保证使用正确的语法而且遵循公认的书写规范。然而，本书篇幅和目标不允许将这些材料包含进来，主编在这里强烈建议读者获取一本优秀的技术报告写作教科书作为参考。

工程报告

工程报告是启动项目的首要文件。这样，就必须很好地将它们组织起来，技术上要精确而且有一个合理的顺序来完整地展现它们的项目建议。然而，文件在格式上必须是有弹性的，允许随着项目逐渐演进到成熟对其进行合理的调整。因为在大多数情况下，工程报告是公共会议或者官方听证会的主要焦点，所以它必须是经过有资格的合伙人审查后的最终版本。另外，工程报告也是提交给管理机构作规划审查并获得许可的关键材料。

下面给出一个典型案例的工程报告目录。

M & P Project No. 01-115

-ENGINEERING REPORT-
for
MUNICIPAL UTILITY REHABILITATION PROJECT
W. THIRD AVENUE

Village
of
Sheldon, Wisconsin
February, 2002

Prepared by:

MORGAN & PARMLEY, LTD.
115 West 2nd Street, South
Ladysmith, Wisconsin 54848

Source: Morgan & Parmley, Ltd.

-TABLE OF CONTENTS-

-i-

-LIST OF ILLUSTRATIONS-

-LIST OF EXHIBITS-

-ii-

M & P Project No. 93-159

ENGINEERING REPORT
for
PROPOSED
MUNICIPAL WATER SYSTEM
Village
of
Tony, Wisconsin
February, 1994

prepared by:

MORGAN & PARMLEY LTD.
Professional Consulting Engineers
115 West 2nd Street South
Ladysmith, Wisconsin 54848

-TABLE OF CONTENTS-

-i-

Source: Morgan & Parmley, Ltd.

-ii-

-LIST OF ILLUSTRATIONS-

-LIST OF TABLES-

-iii-

-LIST OF REFERENCE EXHIBITS-

-ABBREVIATIONS USED IN REPORT-

SYMBOL		DESCRIPTION
DNR	-	Department of Natural Resources
Fe	-	Iron
FmHA	-	Farmers Home Administration
gpcd	-	gallons per capita per day
gpd	-	gallons per day
gpm	-	gallons per minute
ISO	-	Insurance Services Office of Wisconsin
MG	-	million gallons
MGD	-	million gallons per day
mg/l	-	milligrams per liter
Mn	-	Manganense
ppm	-	Parts per Million
PSC	-	Public Service Commission of Wisconsin
psi	-	pounds per square inch (gauge)
USGS	-	U.S. Geological Survey (All elevations referred to in this report are USGS)
VOCs	-	Volatile Organic Compounds

初步报告

所谓初步报告就是指它们还是“初步的”。因为所有的事实、测试和（或）参数暂时还无法全部获得，所以在最终报告完成之前初步报告可以扮演重要的角色。这样，虽然最终的结论和建议被暂且搁置，但相关主题可以被讨论到所需的深度。一般来说，通过向相关的审阅者提供概念和细节，初步报告用来确保建议项目的平顺进展，而不被耽误。后面所附的目录可作为指南。值得注意的是，初步报告总体格式与最终工程报告类似，只是缺少一些支持性的证据和文件。

M & P Project No. 90-139

PRELIMINARY
-ENGINEERING REPORT-

MUNICIPAL STREET IMPROVEMENT PROJECT
(PHASE ONE)

Village
of
Gilman, Wisconsin

JAN 8 1992

prepared by:

MORGAN & PARMLEY LTD.
Professional Consulting Engineers
115 West 2nd Street South
Ladysmith, Wisconsin 54848

Source: Morgan & Parmley, Ltd.

TABLE OF CONTENTS

LIST OF ILLUSTRATIONS

LIST OF TABLES

LIST OF EXHIBITS

可行性研究

可行性研究（或者评估报告）首先是创意和概念，然后进一步发展，最后通过分析来评价项目在技术上和（或）经济上是否合理可行。

可行性研究报告可以只是一封信函，也可以是一份成熟的正式报告。当选择信函形式时，项目通常较小而且审阅者（个人或者委员会）拥有充分的背景知识来快速地彻底了解提交的材料。正式的可行性研究报告用于更加复杂或者综合的项目。后面给出的是一份正式报告的目录示例。

M & P Project No. 90-123

ENGINEERING REPORT

-PRELIMINARY FEASIBILE CONCEPT-
for
COMPOSTING FACILITY

City
of
Ladysmith, Wisconsin
October 15, 1990

prepared by:

MORGAN & PARMLEY, LTD.
Professional Consulting Engineers
115 West 2nd Street South
Ladysmith, Wisconsin 54848

Source: Morgan & Parmley, Ltd.

-TABLE OF CONTENTS-

-i-

-LISTING OF ILLUSTRATIONS-

-LISTINGS OF EXHIBITS-

-ii-

MAJOR TECHNICAL REFERENCES

1-Applicable sections of Wisconsin Administrative Code.

2-Final Report: Wisconsin Co-Composting Demonstration Study, by Dr. Aga S. Razvi (March, 1987)

3-The Utilization of Solid Waste Composts, Co-Composts, and Shredded Refuse on Agricultural Lands (Literature Review)

4-Solid Waste Composting in the U.S., by Nora Goldstein, Bio-Cycle-Nov., 1989

5-Solid Waste Composting Facilities, by Goldstein & Spencer, BioCycle-Jan. 1990

6-Fillmore County, Minn. Resource Recovery Center, by Tim. L. Goodman, Sept. 29, 1989

7-Portions of: Pollution Control Agency Solid Waste Management Rules-Printed Jan., 1989

8-Wisconsin Act 335-The Recycling Law-DNR Publ-IE-041 Rev 6/90

9-Grants Provided by the Recycling Law-Published by DNR, Publ-IE-046-90 Rev.

-iii-

10-Accounting for and Spending Recycling Grants-Published by DNR, Publ-IE-048-90 REV

11-Forming Responsible Units-Published by DNR, PUBL-IE 044-90 REV 8/24/90

12-How New Units Can Receive Recycling Grants, Published by DNR, Publ-IE-047-90 REV

13-Small Town Designs for Big Impact, by Mark Selby & Joe Carruth, BioCycle, Aug., 1989

14-Solid Waste Management, by D. Joseph Hagerty, et al, Van Norstrand Reinhold 1973

15-Environmental Engineers' Handbook, by Liptak, editor, Chilton Book Co., 1974

-iv-

废水处理设施初步设计文件

对于大型项目，例如新建市政污水处理设施，废水处理设施初步设计是一份包括一切的、通常也是大部头的文件。废水处理设施初步设计极为重要的目标是，通过建议一种不仅在环境上有效而且最经济的方案，来为现存的环境问题提供解决方案。

毫无疑问，这是需要花费最多费用、耗费大量时间和技术需求来准备的报告。这份报告的多数领域都必须被恰当地提及，而且由于向公开听证会的最终汇报、管理部门的审查以及来自特定利益集团的质询，都使得它成为技术报告写作中最重要的部分。一般来说，这一份文件需要 18 – 24 个月才能最终定稿。

下面的目录摘自一份实际的废水处理设施初步设计，揭示了需要包含进能为各方认可的最终文件的很多方面。下面几页目录显示了成功准备文件所需要完成的复杂多样的任务。

M & P Project No. 96-113

-AMENDMENT-
to
FACILITIES PLAN
for
MUNICIPAL WASTEWATER TREATMENT FACILITY

Village
of
Sheldon, Wisconsin
July, 1997

Prepared by:

MORGAN & PARMLEY, LTD.
Professional Consulting Engineers
115 West 2nd Street, South
Ladysmith, Wisconsin 54848

Source: Morgan & Parmley, Ltd.

-TABLE OF CONTENTS-

-iii-

-iv-

-LIST OF ILLUSTRATIONS-

-LIST OF TABLES-

下水道使用条例

虽然下水道使用条例并不是真正意义上的技术报告，但是本手册包含这部分内容以使我们的覆盖范围更完整。

下水道使用条例（Sewer Use Ordinance，SUO）是大多数州级管理部门要求的必需文件，这些管理部门有权管理市政污水收集和处理设施（WWCTF）。这份文件包括管理当地生活污水管道设施的规范和规程。另外，这份文件（或者条例）还包括用户收费系统（User Charge System，UCS），它建立了财务运行体系，也就是用户费用、运行和维护成本以及相关的开销。下水道使用条例的典型格式请参考下面的目录。

M & P Project No. 98-109

-MUNICIPAL WWCTF-
SEWER USE ORDINANCE
and
USER CHARGE SYSTEM

Village
of
Sheldon, Wisconsin
May, 2001

Prepared by:

MORGAN & PARMLEY, LTD.
Professional Consulting Engineers
115 West 2nd Street, South
Ladysmith, Wisconsin 54848

Source: Morgan & Parmley, Ltd.

TABLE OF CONTENTS

水资源保护计划

水资源保护计划（Wellhead Protection Plan，WHPP）是最近由美国环境保护局(US/EPA)推动并被大多数州委托管理的文件。文件的目标是，通过规范在用水井周边区域内的环境保护活动来保护公共水井的饮用水源。

考虑到前述的几个主题，主编认为关于 WHPP 文件的简要介绍是恰当的。优质饮用水的短缺是对我们居民的一个非常严峻的威胁，它必须受到保护。不断增加的需求（既有生活用水也有工业用水）当然使得这成为需要优先考虑的问题。这样，读者可以通过下面给出的最近一份水资源保护计划(WHPP)文件的目录来对这个主题有一个简要的了解。

M & P Project No. 96-172

-WELLHEAD PROTECTION PLAN-

Village
of
Gilman, Wisconsin
April, 1998

Prepared by:

MORGAN & PARMLEY, LTD.
Professional Consulting Engineers
115 West 2nd Street, South
Ladysmith, Wisconsin 54848

Source: Morgan & Parmley, Ltd.

TABLE OF CONTENTS

-ILLUSTRATIONS-

-APPENDIX-

规　划

第2节　项目进度计划

例　2

为了项目获得成功，必须很好地将其组织。总体控制计划是每个项目都必须遵循的，否则会使项目受到损害，并导致各种努力大打折扣。

目　录

项目工程师

如前所述，项目工程师们必须很好地组织起来才能确保成功。另外，他们还必须在技术上有能力而且非常自律。大多数情况下，错误的决策和不当的行动会造成大量浪费并对项目进展造成负面影响，而且还会危及到预算。

一个优秀的项目工程师永远不会失去自制力，并能一直对项目负责。缺少设计完美的进度计划，项目工程师将不能掌握一个合理的过程。这样，从一开始，项目工程师就应该着手勾画出项目进度计划并规划关键过程。

当然，项目工程师通过项目要与大量各种人员打交道，包括技术人员、建筑零售商、公共事业代表、管理人员、环境组织、选举产生的官员、普通市民、媒体、法律部门、公众听证会、管理监察人员、会计师事务所、政府补贴管理部门、政治官员和咨询工程公司的工程人员。当然，为了成为一个高效率的项目工程师，他必须拥有良好的交流技能和领导素质以成功地完成一个项目。时间不允许，而且这也不是本书的主要目的来罗列成为一个顶级优秀的项目工程师所需要的全部非工程方面的特质。然而值得注意的是，这些素质是一种混合体，来自对商业的理解、职业道德、人格魅力、经验、服务意识、目标取向和对专业人员的真实尊重。

总而言之，一个专业的项目工程师必须拥有广泛范围的技能，但是如果不遵循一个设计良好的进度计划，项目不可能得到成功的运作。

事件项目日程表—基本任务一览

Event Calendar–Summary of Basic Tasks

Item No.	Description	Projected Dates	Notes
i	Define Project Scope		
ii	Municipality Authorizes Engineer to Begin		
1	Initial Conference: Municipality/Engineer		
2	Research Records & Files		
3	Field Survey & Reconnaissance		
4	Video Photograph Project Site (prior to Commencing Construction)		
5	Buried Utilities Located & Flagged		
6	Research Property Survey Records		
7	Soil Borings, Test Drilling & Subsurface Investigation		
8	Public Works Committee Meeting-Refine Project Scope		
9	Reduce Field Notes		
10	Preliminary Budget-Construction Cost Estimate		
11	Utility Committee Meeting-Construction Cost Estimate		
12	Finalize Budget-Construction Cost Estimate		
13	Preliminary Plans & Specifications		
14	Municipality Meeting – Review & Approve Prel. P & S		
15	Easement Survey & Legal Documents (If Req'd)		
16	Final Plans & Specifications		
17	Engineering Report (Summary of Work)		
18	Proposal & Bidding Documents		
19	Obtain Concurrence from Municipality for Final P & S		
20	Submit Plans & Specifications to Regulatory Agencies		
21	Obtain Approvals from Regulatory Agencies:		
22	DNR (If Req'd.)		
23	PSC (If Req'd.)		
24	DOT (If Req'd.)		
25	Corps of Engineers (If Req'd.)		
26	Private Utility Company Approvals:		
27	Cable TV		
28	Electrical & Gas		
29	Telephone		
30	Public Hearing (Assessments to Property Owners) (If Req'd.)		

Source: Morgan & Parmley, Ltd.

Item No.	Description	Projected Dates	Notes
31	Organize Bidding Process (Commence)		
32	Obtain Wage Rates (State and/or Federal)		
33	Advertisement for Bids		
34	Supply Plans & Specs to Builders Exchanges		
35	Send Plans & Specs to Contractors		
36	Pre-Bid Meeting (If Req'd.)		
37	Bid Opening:		
38	Minutes, Bid Tabulation & Analysis of Proposals		
39	Recommendation to Municipality		
40	Municipality gives Notice of Award		
41	Prepare Construction Contract Documents		
42	Secure Municipality Atty's Certification of Construction Contract		
43	Obtain Municipality's Resolution to Sign Construction Contract		
44	Supervise Signing of Construction Contract		
45	Pre-Construction Conference		
46	Municipality to Sign Notice to Proceed		
47	Shop Drawing Review		
48	Construction Staking		
49	General Inspection Services, Daily Log, etc.		
50	Construction Records		
51	Contractor Payment Requests Review & Certification		
52	Status Reports (Periodic)		
53	Inspect Testing of Installation		
54	Final Inspection & Certification		
55	Preparation of Construction Record Drawings		
56	Final Payment Request Review & Close-Out Certification		

Note: The projected dates are targets and may vary, due to conditions beyond the control of the Municipality.

工程服务一览（项目获得批准之后）
Summary of Engineering Services (Post Plan Approval)

1. Apply for State Wage Rates
2. Prepare Advertisement for Bids
3. Submit Ad to Ladysmith News & Western Builder
4. Prepare Bidding Documents
5. Print & Bind Plans, Specifications & Bidding Documents
6. Send Plans, Specs. & Bidding Documents to Builders Exchanges
7. Distribute Plans, Specs & Bidding Documents to Prospective Bidders
8. Tabulate Plan Holders List
9. Prepare Addendas (if necessary)
10. Interpret Plans & Specs; i.e. answer technical questions from Prospective Bidders
11. Supervise Bid Opening
12. Prepare Bid Tabulation & Distribute to Plan Holders
13. Review Bids & Investigate Qualifications of Low Bidder
14. Prepare Summary for Village Board
15. Prepare Notice of Award for Village to Execute
16. Prepare Construction Contract
17. Review Construction Contract w/Village Attorney for Certification
18. Coordinate & Manage Pre-Construction Conference
19. Attend Groundbreaking Ceremony
20. Review Shop Drawings
21. Coordinate Plan Submittal for Roof Trusses and Precast Concrete
22. Provide Construction Staking
23. Periodic Inspection Visits to Site @ Key Events
24. Perform Soil & Concrete Sampling & Lab Services
25. Process Change Orders (if required)
26. Review & Certify Pay Requests from Contractor
27. Perform Final Inspection
28. Prepare Punch List
29. Supervise Close-Out Process
30. Obtain Lien Waivers & Consent of Surety for Final Payment
31. Verify that Contractor demonstrate operation of HVAC equipment to Village Personnel
32. Verify that O & M Manuals for Equipment are Supplied
33. Brief Status Reports to Village Board at Regular Meetings

Please Note: This does not include any resident inspector services.

Source: Morgan & Parmley, Ltd.

规　划

第3节　现场勘察、测量和绘图

例　3

最初，建议的场地必须经过调查研究以确定它对将来使用的总体适应性；环境问题必须得到解决；考古和历史文物必须得到保护；必须完整地提出对野生动植物和濒危物种的影响；安全因素和市民的关心必须经过评估，敏感环境区域必须得到保护；如果适当，应进行植被测量调查；必须完成对现有场地土和地下水的勘察，以确定对建议项目的适应性。

在完成上述各项工作以及文物得到处理后，就可以开始进行最终的现场勘察、测量和绘图。本节提供了大量的图纸示例，包括项目场地图、地形图、子区图、分块地图、土地资格测量图、复合图纸、经核准的测量图、所有权调查、调查鉴定图、市政图、通行权图、污水管道的通行编号图和服务区域图。

目　录

Project Site Mapping 项目场地图

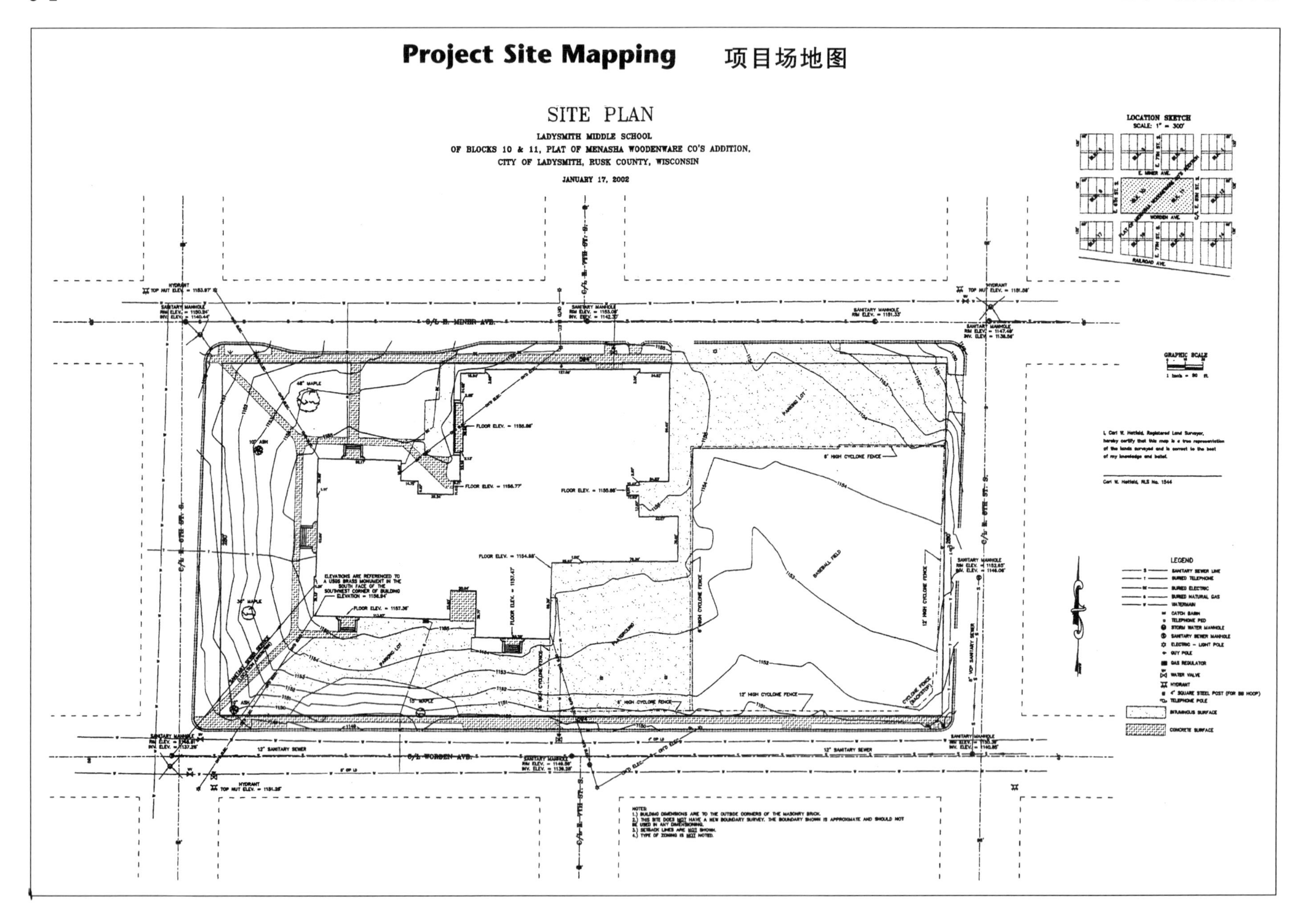

Source: Polk County Land Surveying

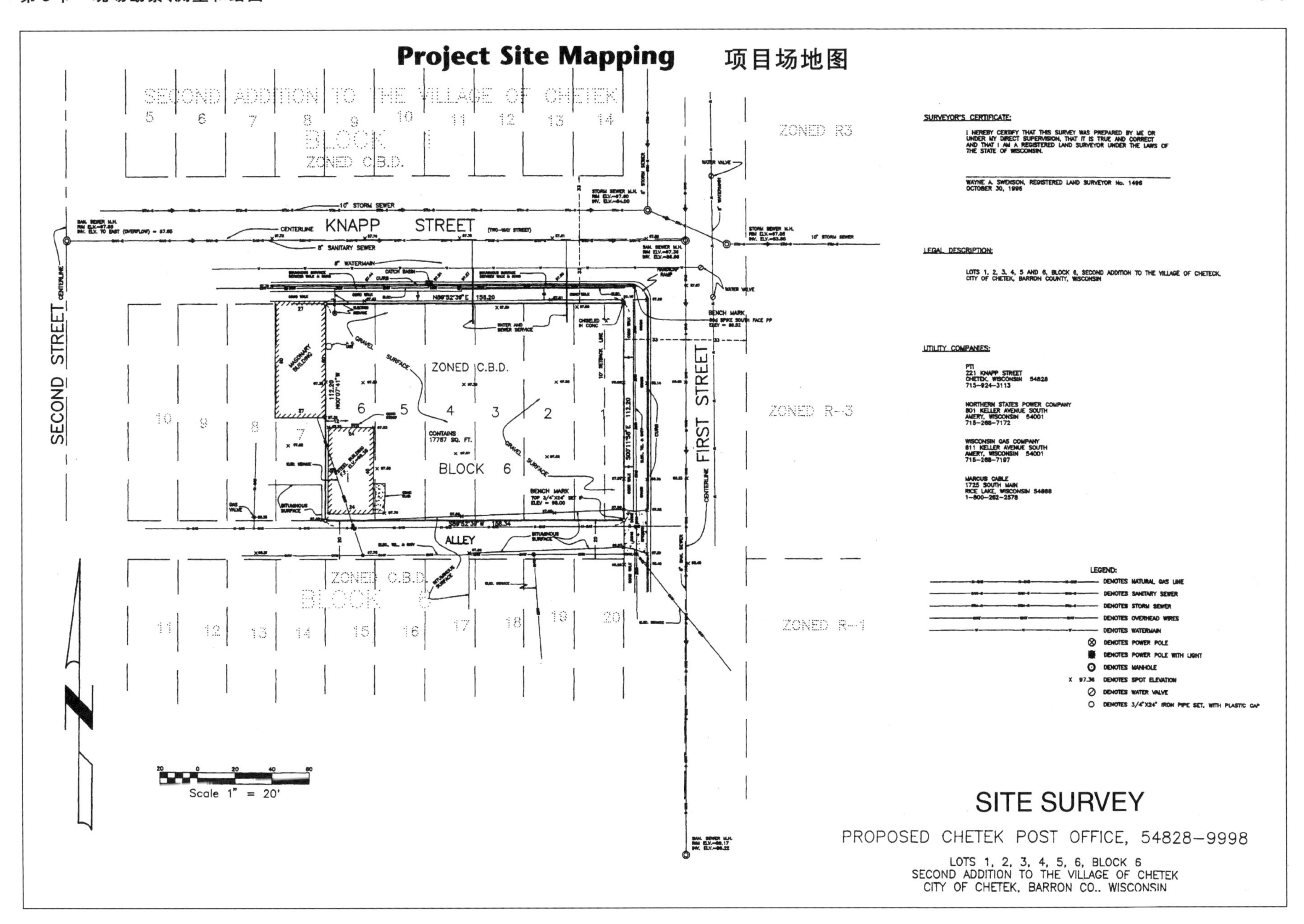

Source: Polk County Land Surveying

Topographic Mapping 地形图

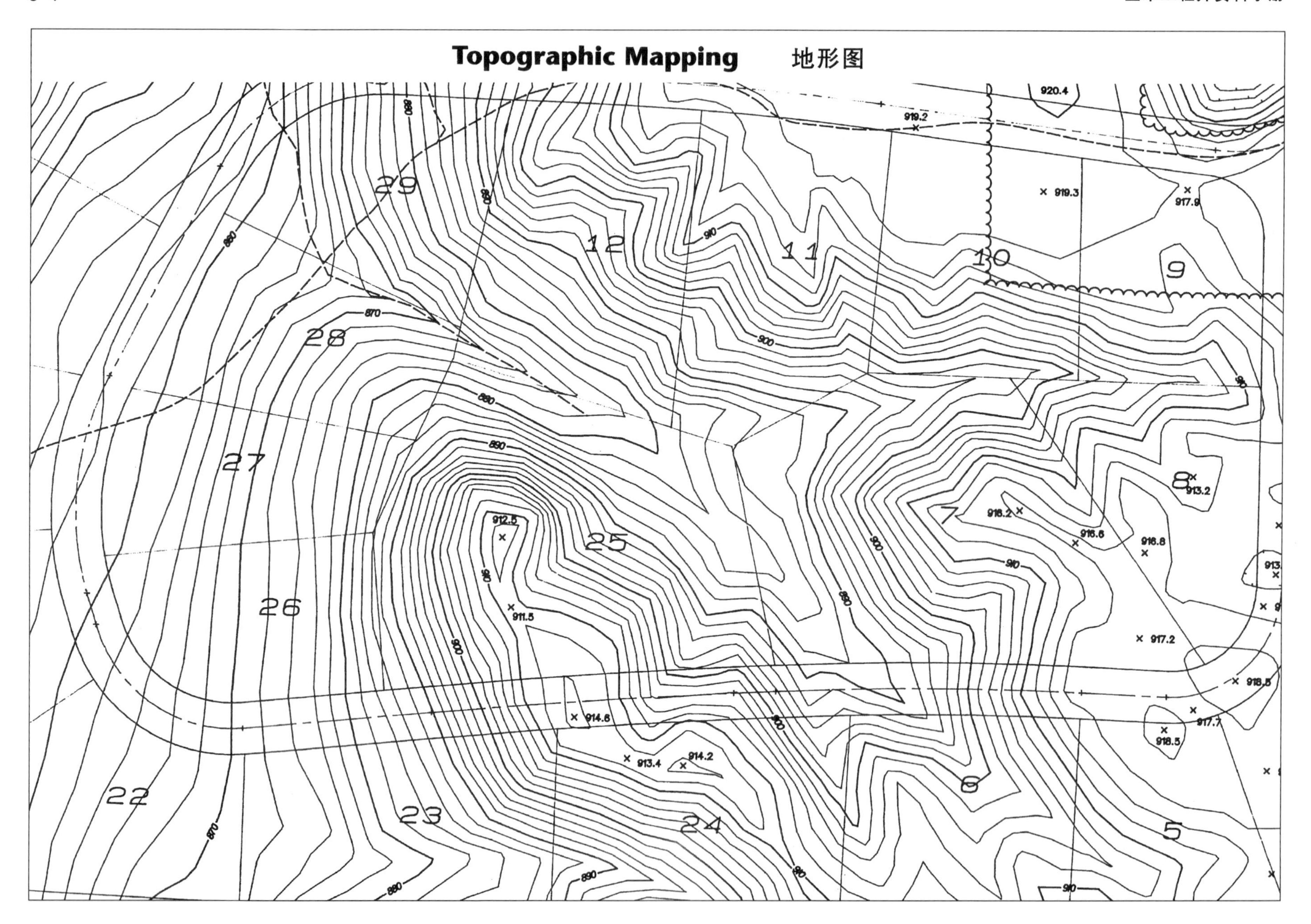

Source: Morgan & Parmley, Ltd.

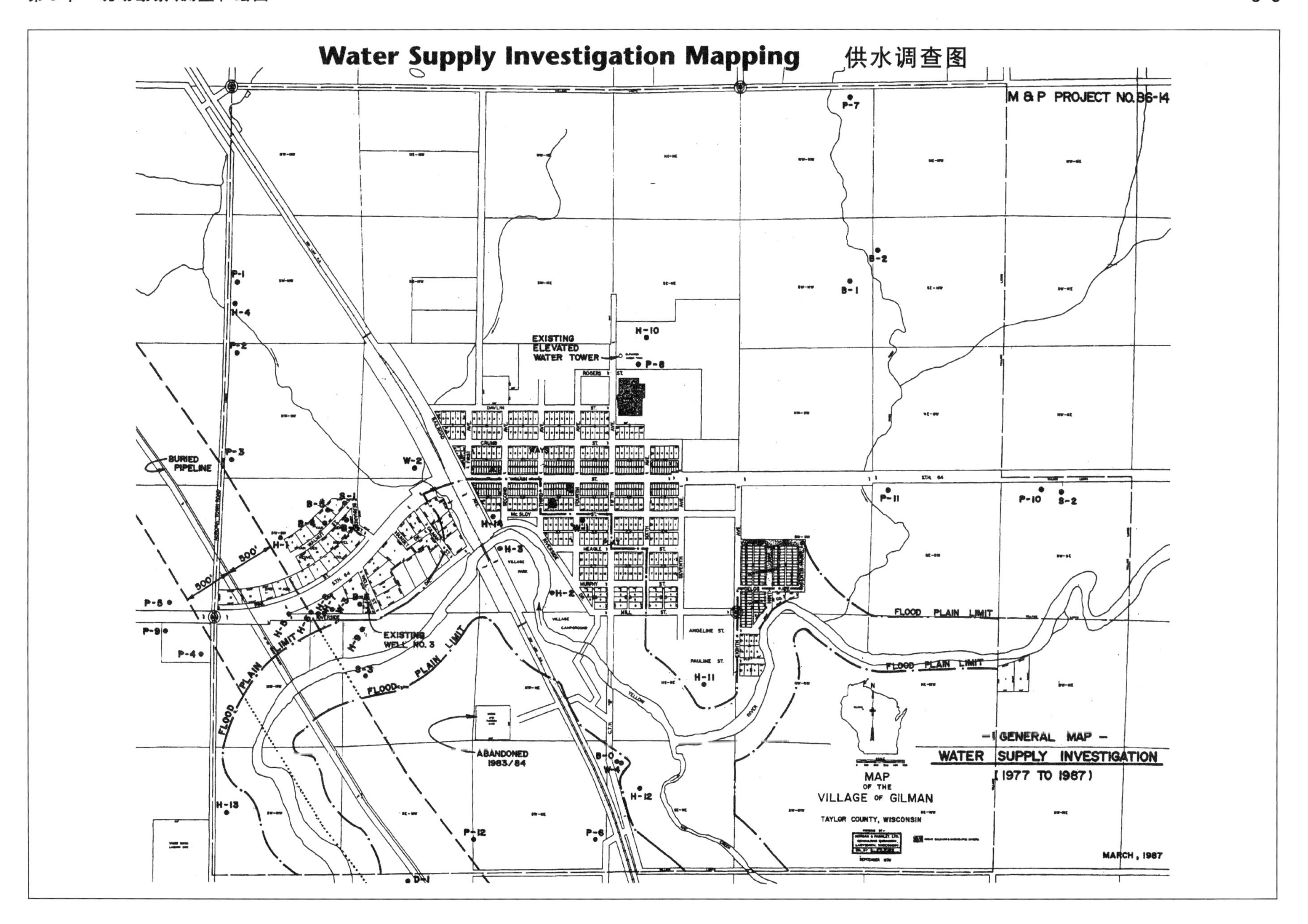

Source: Morgan & Parmley, Ltd.

Municipal Mapping 市政图

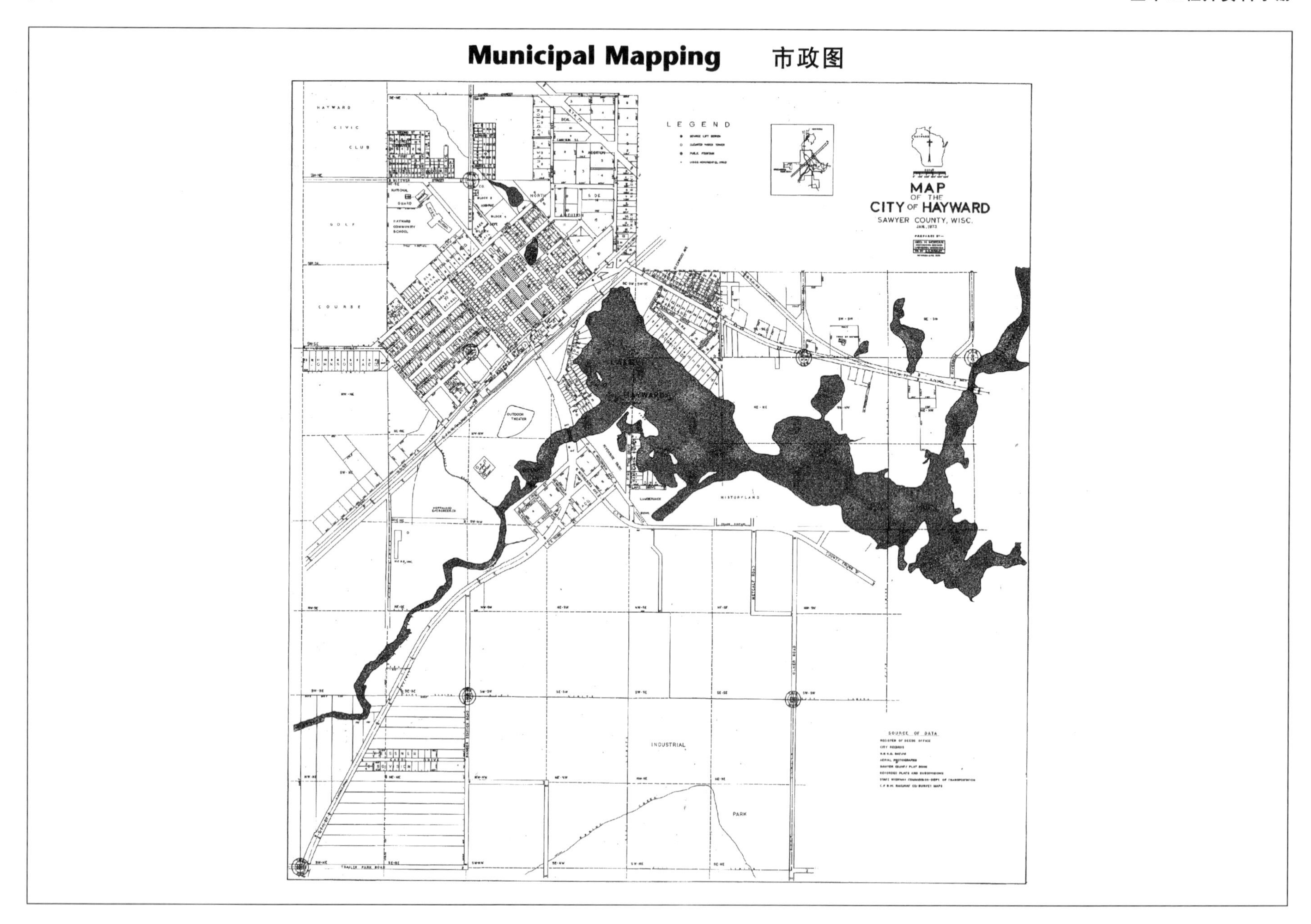

Source: Morgan & Parmley, Ltd.

Sub-Division Mapping　子区图

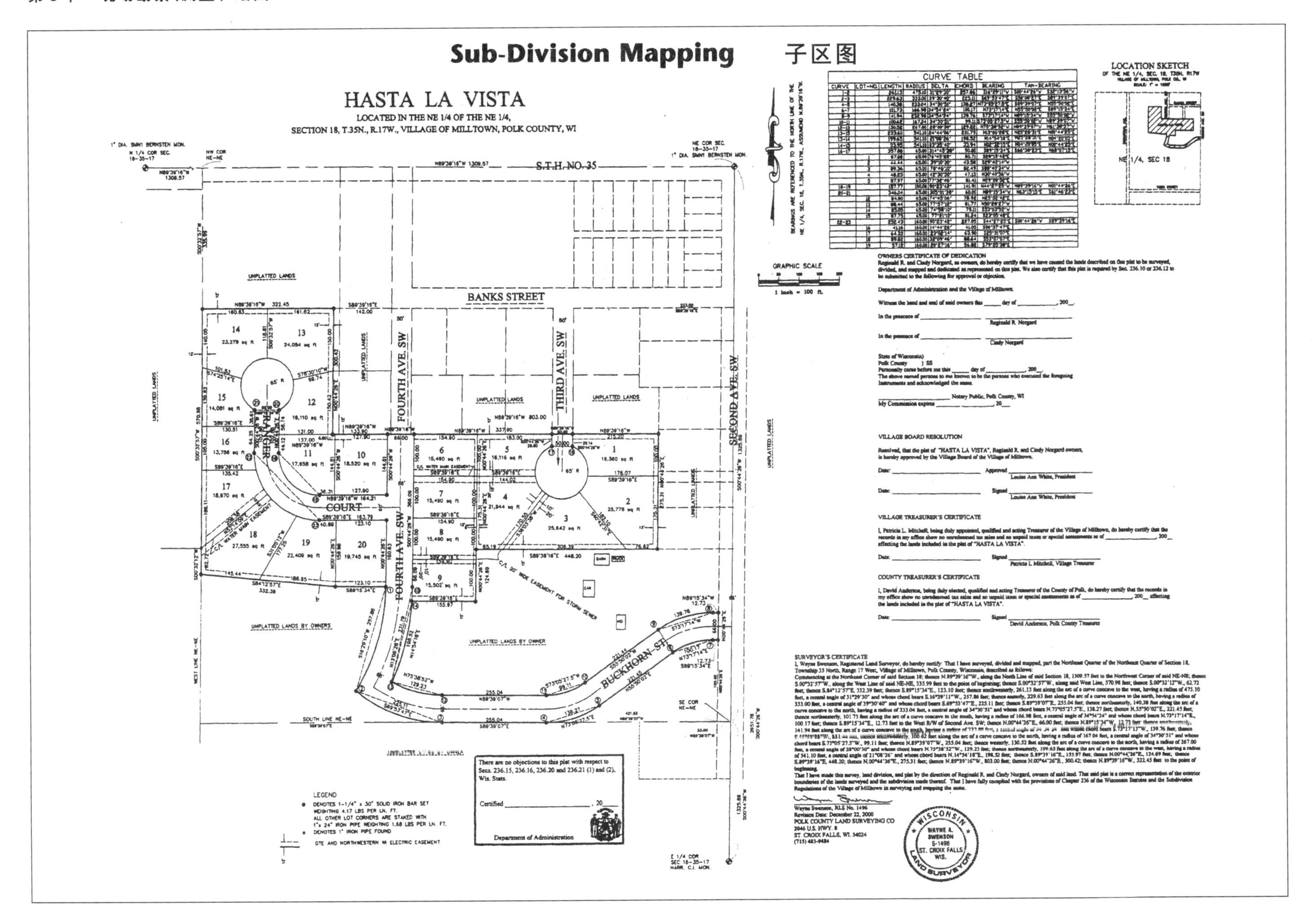

Source: Polk County Land Surveying

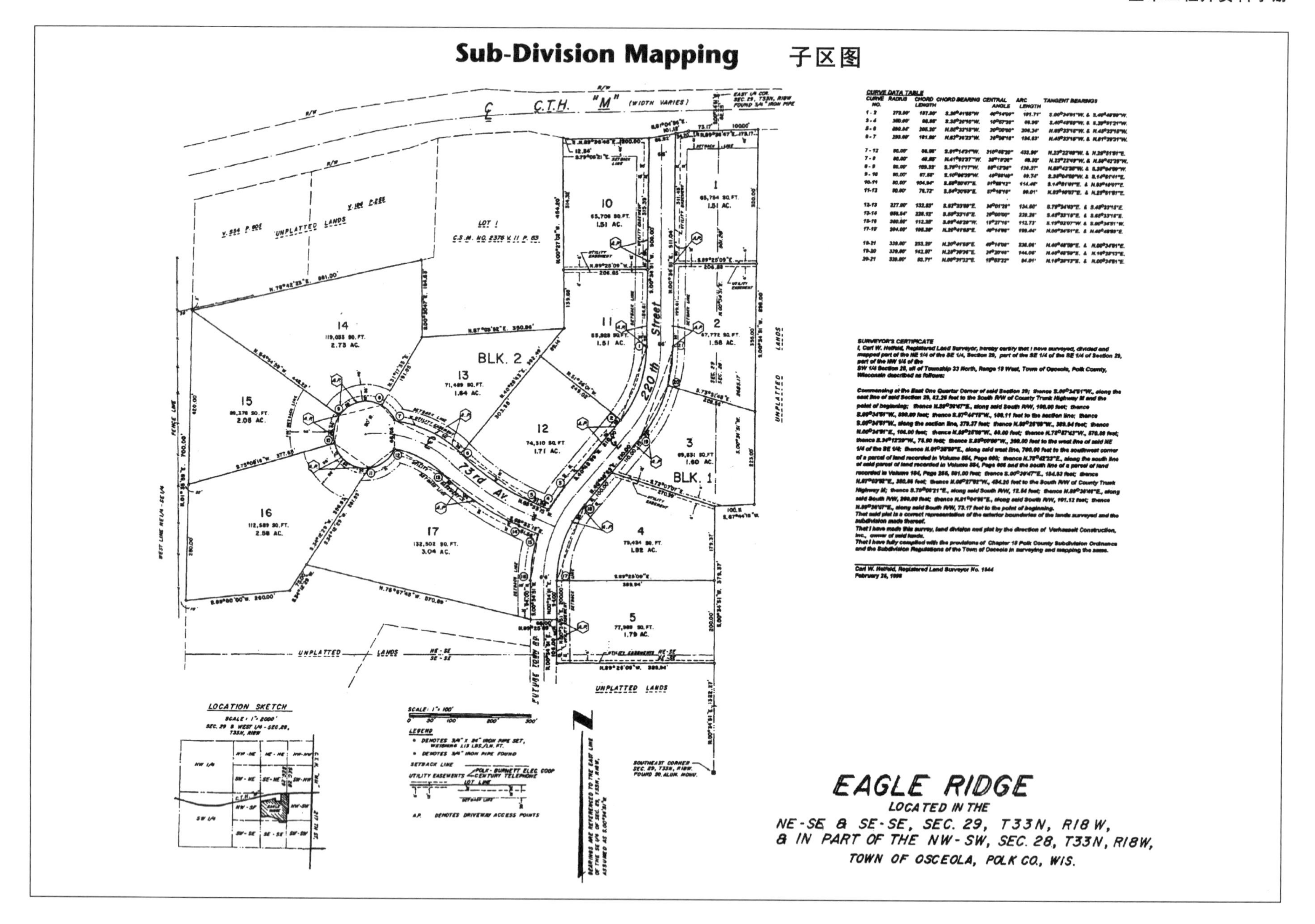

Source: Polk County Land Surveying

Map of Survey　测量图

OF PART OF THE S 1/2 – SE 1/4
SEC. 17, T.35N., R.17W.,
TOWN OF MILLTOWN, POLK CO., WI.

DATE: JULY 31, 2001
CLIENT: TOWN OF MILLTOWN

GRAPHIC SCALE
100 50 0 150
1 inch = 250 ft.

SW-SE | SE-SE

S89°51'44"W 314.00

STEEL BLD.

S00°05'05"W 300.00 267.00 33.00

N00°05'05"E 300.00 267.00 33.00

EAST LINE SECTION 17 S00°05'05"W

N89°51'44"E 314.00

N89°51'44"E 314.00

R/W R/W

S89°51'44"W 1110.58

S89°51'44"W 1200.00

C/L 210TH AVE.

S 1/4 COR SEC. 17–35–17

SE COR SEC. 17–35–17

S89°51'44"W 2624.58

○ DENOTES 1"x24" IRON PIPE SET

⊕ DENOTES POLK CO. SURVEYORS MONUMENT

LEGAL DESCRIPTION

That part of the S 1/2 of SE 1/4, Section 17, T.35N., R.17W., described as follows: Beginning at a point on the South Line of said Section 17 that is 1200 feet West of the Southeast Corner of said Section 17; running thence North parallel to the East Section Line 300 feet; running thence West parallel to the South Section Line 314 feet; running thence South parallel to the East Section Line 300 feet; running thence East on the South Section Line 314 feet to the point of beginning.

SURVEYOR'S CERTIFICATE

I, Wayne Swenson, Registered Land Surveyor, hereby certify that this map is a true representation of the lands surveyed and is correct to the best of my knowledge and belief.

Wayne Swenson, RLS No. 1496
JULY 31, 2001

Source: Polk County Land Surveying

Map of Survey　测量图

OF LOTS 9 AND 10, PLAT OF OASE HIGH ACRES,
BEING LOCATED IN THE NE 1/4, SEC. 12, T.32N., R.18W.,
TOWN OF ALDEN, POLK COUNTY, WISCONSIN

DATE: SEPTEMBER 25, 2001
CLIENT: ROGER HOIBY

LEGEND
○ 1" X 24" IRON PIPE SET WEIGHING 1.68 LBS./LINEAL FT.
◉ 1" I.P. FND.

LOT 7

LOT 8

LOT 9
25,989 SQ. FT.
0.60 ACRES

LOT 10
37,747 SQ. FT.
0.87 ACRES

LOT 11

LOT 12

CHURCH PINE LAKE

C/L 184TH ST. (PLATTED AS OASE ROAD)

HOUSE GAR. HOUSE GAR.

(REC. AS N71°49'E, 258.60') N71°47'32"E 258.49' 131.22' 127.27'

(REC. AS N41°17'W) N41°16'55"W 100.00' 43'

(REC. AS N76°45'E, 303.75') N76°44'16"E 303.65' 145.36' 158.29'

(REC. AS N20°6'W) N20°16'54"W 100.00' 45'

S14°00'29"E 168.00' (REC. AS S14°00'E) 68.00' 100.00'

N76°36'10"E 314.57' (REC. AS N76°37'E, 314.64') 198.26' 116.31' 90'

66' 33' 33'

SCALE: 1" = 80'

SURVEYOR'S CERTIFICATE:

I, Carl W. Hetfeld, Registered Land Surveyor, hereby certify that this map is a true representation of the lands surveyed and is correct to the best of my knowledge and belief.

Carl W. Hetfeld 9/25/01

Carl W. Hetfeld, R.L.S. No. 1544

WISCONSIN
CARL W.
HETFELD
S-1544
ST. CROIX FALLS,
WIS.
LAND SURVEYOR

Source: Polk County Land Surveying

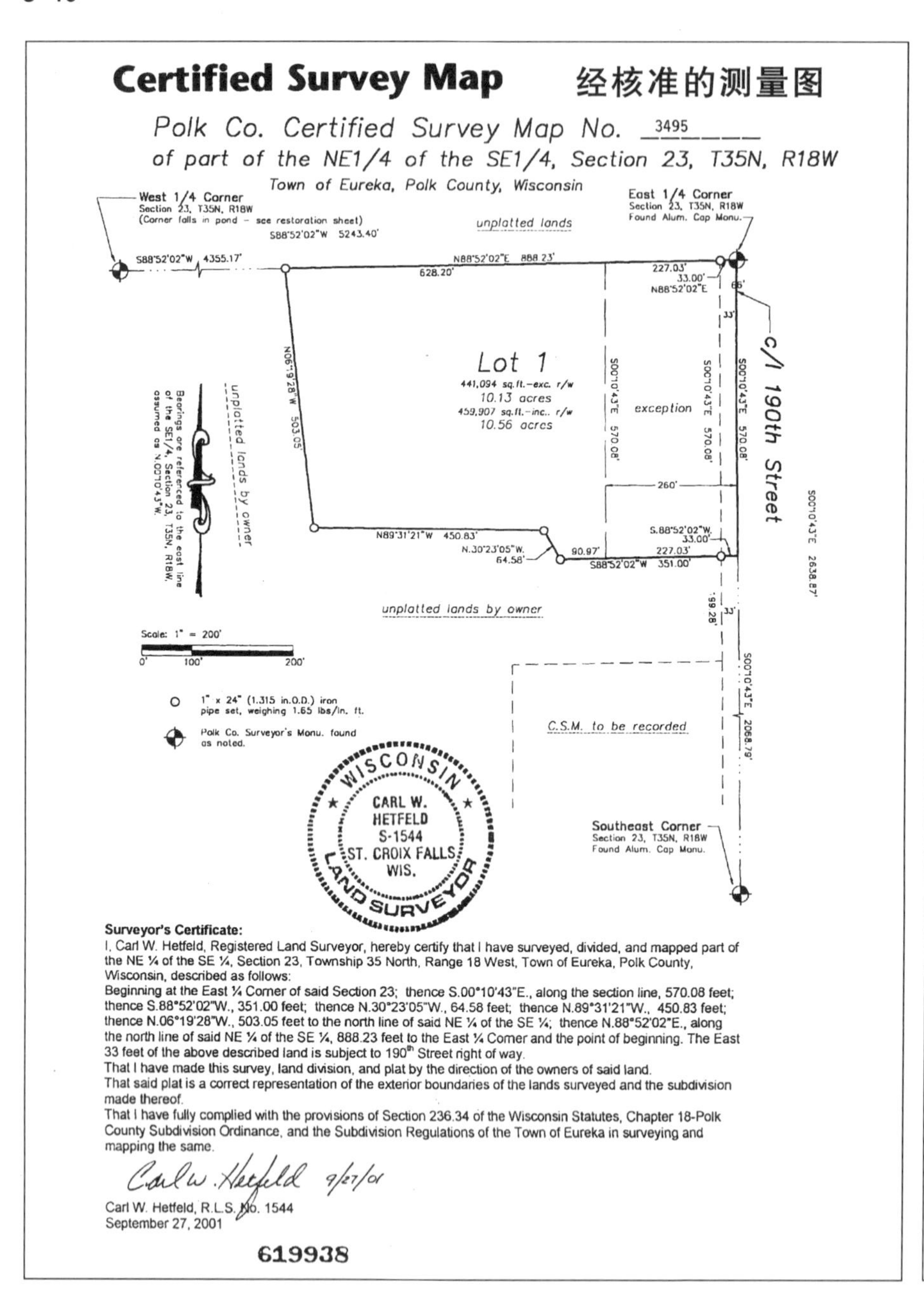

Source: Polk County Land Surveying

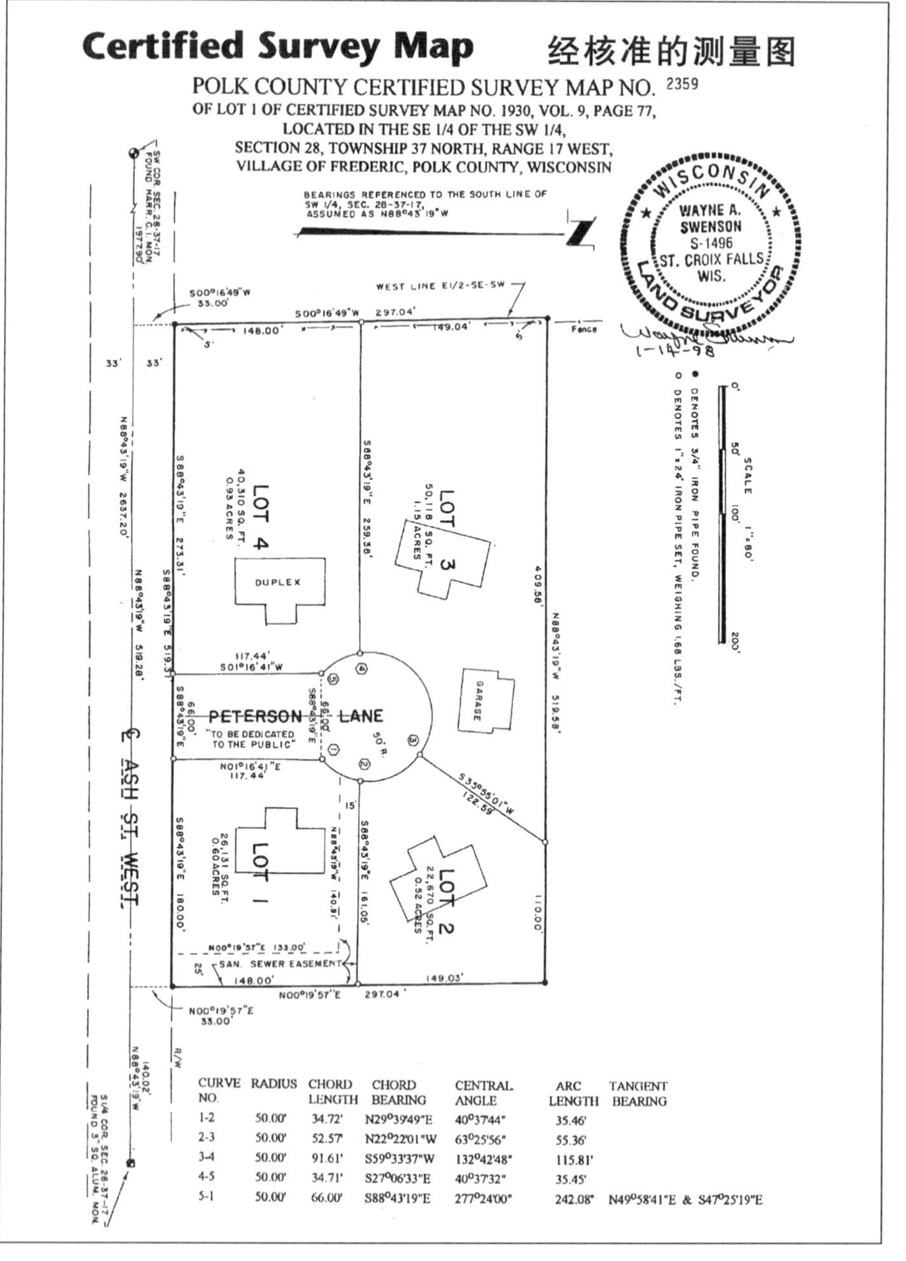

CURVE NO.	RADIUS	CHORD LENGTH	CHORD BEARING	CENTRAL ANGLE	ARC LENGTH	TANGENT BEARING
1-2	50.00'	34.72'	N29°39'49"E	40°37'44"	35.46'	
2-3	50.00'	52.57'	N22°22'01"W	63°25'56"	55.36'	
3-4	50.00'	91.61'	S59°33'37"W	132°42'48"	115.81'	
4-5	50.00'	34.71'	S27°06'33"E	40°37'32"	35.45'	
5-1	50.00'	66.00'	S88°43'19"E	277°24'00"	242.08'	N49°58'41"E & S47°25'19"E

Source: Polk County Land Surveying

土地资格测量图

Land Title Survey Mapping

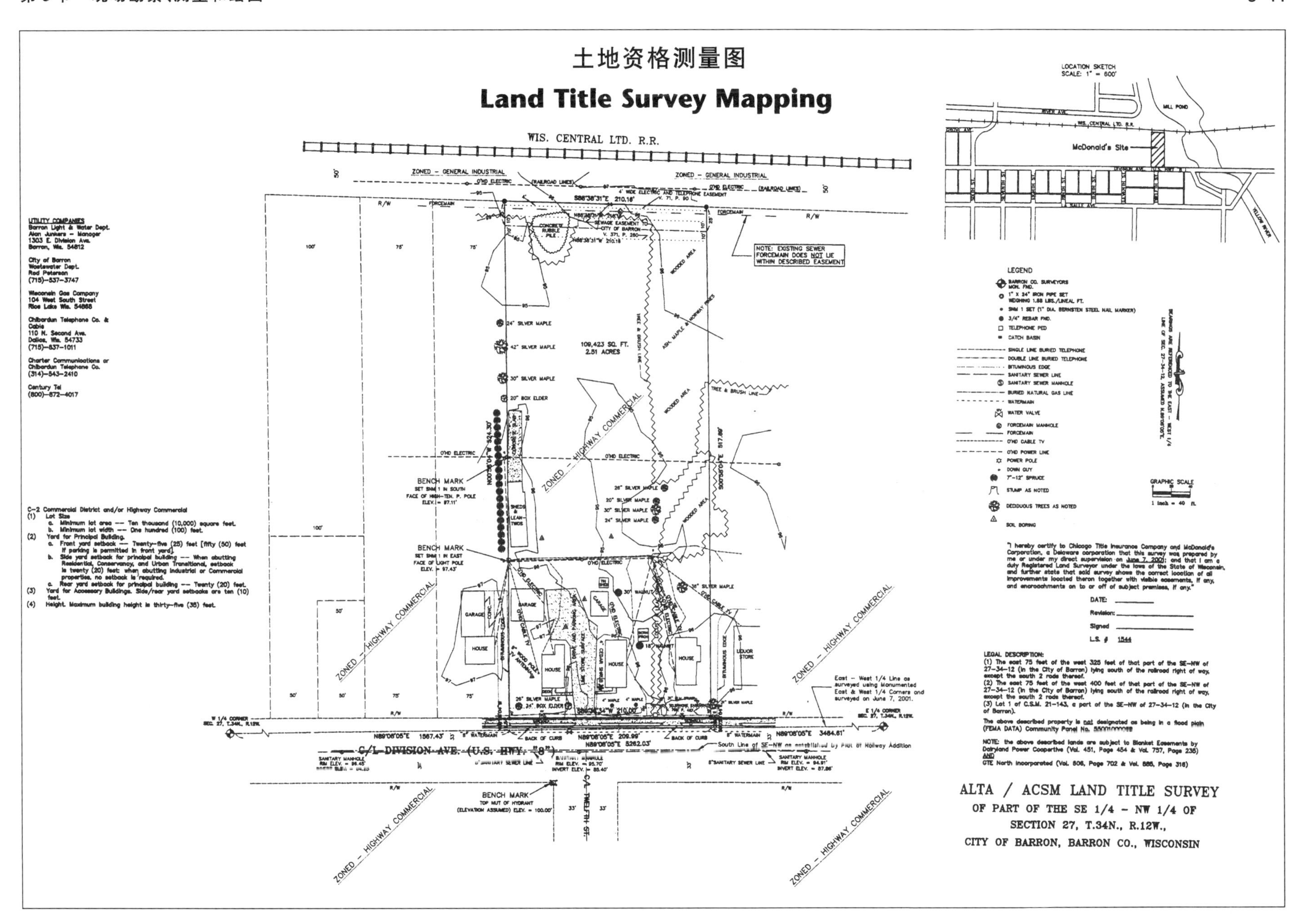

Source: Polk County Land Surveying

Important Facts About Land Descriptions 关于土地描述的重要事实

LAND MEASUREMENTS, TOWNSHIPS, SECTIONS, MEANDERED WATER, GOVERNMENT LOTS, ETC.

WHAT IS A LAND DESCRIPTION?

A land description is a description of a tract of land in legally acceptable terms, so as to show exactly where it is located and how many acres it contains.

TABLE OF LAND MEASUREMENTS

LINEAR MEASURE				SQUARE MEASURE			
1 inch	.0833 foot	16½ feet	1 rod	144 sq. in.	1 sq. ft.	43560 sq. ft.	1 acre
7.92 inches	1 link	5½ yards	1 rod	9 sq. ft.	1 sq. yd.	640 acres	1 sq. mile
12 inches	1 foot	4 rods	100 links	30¼ sq. yds.	1 sq. rod	1 sq. mile	1 section
1 vara	33 inches	66 feet	1 chain	16 sq. rods	1 sq. chain	36 sq. miles	1 township
2¾ feet	1 vara	80 chains	1 mile	1 sq. rod	272¼ sq. ft.	6 miles sq.	1 township
3 feet	1 yard	320 rods	1 mile	1 sq. chain	4356 sq. ft.	208 ft. 8 in. sq.	1 acre
25 links	16½ feet	8000 links	1 mile	10 sq. chains	1 acre	80 rods sq.	40 acres
25 links	1 rod	5280 feet	1 mile	160 sq. rods	1 acre	160 rods sq.	160 acres
100 links	1 chain	1760 yards	1 mile	4840 sq. yds.	1 acre		

In non-rectangular land descriptions, distance is usually described in terms of either feet or rods (this is especially true in surveying today), and square measure in terms of acres. Such descriptions are called Metes and Bounds descriptions and will be explained in detail later.

In rectangular land descriptions, square measure is again in terms of acres, and the location of the land in such terms as N½ (north one-half), SE¼ (south east one-fourth or quarter), etc. as shown in Figures 2, 3, 4 and 5.

MEANDERED WATER & GOVERNMENT LOTS

A meandered lake or stream is water, next to which the adjoining landowner pays taxes on the land only. Such land is divided into divisions of land called government lots. The location, acreage and lot number of each such a tract of land, was determined, surveyed and platted by the original government surveyors.

The original survey of your county (complete maps of each township, meandered lakes, government lots, etc.) is in your courthouse, and this original survey is the basis for all land descriptions in your county (see figure 1).

IMPORTANT
THE GOVERNMENT LOT NUMBER GIVEN TO A PIECE OF LAND, IS THE LEGAL DESCRIPTION OF THAT TRACT OF LAND.

HOW CAN YOU TELL WHETHER WATER IS MEANDERED OR PRIVATELY OWNED?

On our township maps, if you find government lots adjoining a body of water or stream, those waters are meandered. If there are no government lots surrounding water, that water is privately owned, the owner is paying taxes on the land under the water, and the owner controls the hunting, fishing, trapping rights, etc., on that water, within the regulations of the State and Federal laws, EXCEPT where such water is deemed navigable, other rulings may sometimes pertain.

As a generality (but not always), meandered water is public water which the public may use for recreational purposes, fishing, hunting, trapping, etc., provided that there is legal access to the water, or in other words, if the public can get to such waters without trespassing. There still is much litigation concerning the same to be decided by the courts.

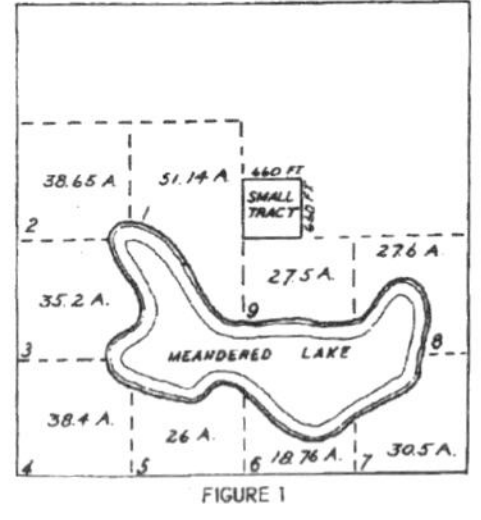

FIGURE 1

SAMPLE SECTIONS SHOWING RECTANGULAR LAND DESCRIPTIONS, ACREAGES AND DISTANCES

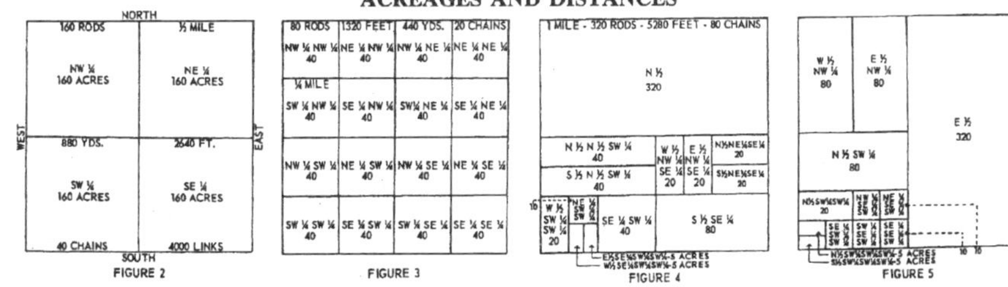

FIGURE 2 FIGURE 3 FIGURE 4 FIGURE 5

THE BEST WAY TO READ LAND DESCRIPTIONS IS FROM THE REAR OR BACKWARDS

Descriptions of land always read FIRST from either the North or the South. In figures 2, 3, 4 and 5, notice that they all start with N (north), S (south), such as NW, SE, etc. They are never WN (west north), ES (east south) etc.
IMPORTANT: It is comparatively simple for anyone to understand a description, that is, determine where a tract of land is located, from even a long description. The SECRET is to read or analyze the description from the rear or backwards.

EXAMPLE: Under figure 4, the first description reads E½, SE¼, SW¼, SW¼. The last part of the description reads SW¼, which means that the tract of land we are looking for is somewhere in that quarter (as shown in figure 2). Next back we find SW¼, which means the tract we are after is somewhere in the SW¼ SW¼ (as shown in figure 3). Next back, we find the SE¼, which means that the tract is in the SE¼ SW¼ SW¼ (as shown in figure 5). Next back and our last part to look up, is the E½ of the above, which is the location of the tract described by the whole description (as shown in figure 4).

TO INTERPRET A LAND DESCRIPTION - LOCATE THE AREA ON YOUR TOWNSHIP PLAT, THEN ANALYZE THE DESCRIPTION & FOLLOW IT ON THE PLAT MAP.

TOWNSHIP SURVEY INFORMATION

A CONGRESSIONAL TOWNSHIP CONTAINS 36 SECTIONS OF LAND 1 MILE SQUARE

A CIVIL OR POLITICAL TOWN MAY BE LARGER OR SMALLER THAN A CONGRESSIONAL TOWNSHIP.

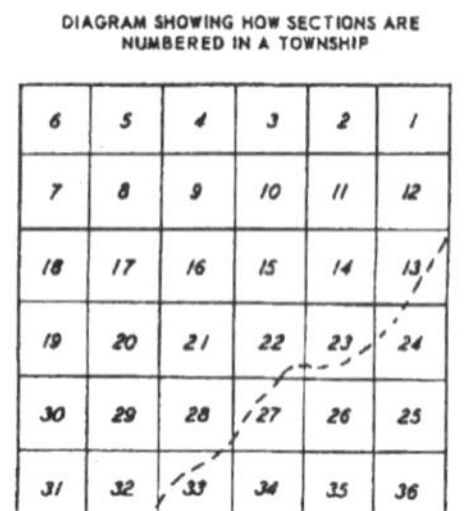

FIGURE 6

FIGURE 7

TOWNSHIPS

Theoretically, a township is a square tract of land with sides of six miles each, and containing 36 sections of land. Actually this is not the case. Years ago, when the original survey of this state was made by the government engineers, they knew that it was impossible to keep a true north and south direction of township lines, and still keep getting township squares of 36 square miles. As they surveyed toward the north pole, they were constantly running out of land, because the township lines were converging toward the north pole.

If you will turn to one of the township maps in this plat book, you will notice that on the north and on the west of each township, there are divisions of land which show odd acreages. In some townships, these odd acreages are called government lots (because they were given a lot number), and at other times left as FRACTIONAL FORTIES OR EIGHTIES. It was at the option of the original government surveyors as to whether they would call these odd acreages government lots, or fractional forties and eighties.

The reason for these odd acreages is that the government surveyors adjusted for shortages of land which developed as they went north, by making fractional forties, eighties or government lots out of the land on the west side of a township, and the same for the land on the north side of a township to keep east and west lines running parallel. In other words it was impossible to fit full squares into a circle.

Townships sometimes vary in size from the regularly laid-out township (see figure 6). Suppose that the dotted line in figure 6 is a river separating two counties. The land north and west of the river could be a township in one county, the land south and east could be a township in another county. Whichever county the land is in, it still retains the same section, township and range numbers for purposes of land descriptions.

Each township has a township number and also a range number (sometimes more than one of each if the township is oversized, or a combination of more than one township and range).

Government surveying of townships is run from starting lines called base lines and principal meridians. Each township has a township number. This number is the number of rows or tiers of townships that a township is either north or south of the base line. Also each township has a range number. This number is the number of rows or tiers of townships that a township is either east or west of the principal meridian (see figure 7). EVERY DESCRIPTION OF LAND SHOULD SHOW THE SECTION, TOWNSHIP AND RANGE IT IS LOCATED IN.

TOWNSHIPS MAY BE EITHER NORTH OR SOUTH OF THE BASE LINE RANGES MAY BE EITHER EAST OR WEST OF THE PRINCIPAL MERIDIAN.

METES AND BOUNDS DESCRIPTIONS

AND EXPLANATION OF DIRECTION IN TERMS OF DEGREES

WHAT IS A METES AND BOUNDS DESCRIPTION? It is a description of a tract of land by starting at a given point, running so many feet a certain direction, so many feet another direction etc., back to the point of beginning. EXAMPLE: In figure 1 notice the small tract of land outlined. The following would be a typical metes and bounds description of that tract of land. "Begin at the center of the section, thence north 660 feet, thence east 660 feet, thence south 660 feet, thence west 660 feet, back to the point of beginning, and containing 10 acres, being a part of Sec. No. etc."

IMPORTANT: To locate a tract of land from a metes and bounds description, start from the point of beginning, and follow it out (do not read it backwards as in the case of a rectangular description).

The small tract of land just located by the above metes and bounds description could also be described as the SW¼ SW¼ NE¼ of the section. In most cases, the same tract of land may be described in different ways. The rectangular system of describing and locating land as shown in figures 2, 3, 4 and 5 is the most simple and almost always used when possible.

A circle contains 360 degrees. Explanation: If you start at the center of a circle and run 360 straight lines an equal angle apart to the edge of the circle, so as to divide the circle into 360 equal parts, THE DIFFERENCE OF DIRECTION BETWEEN EACH LINE IS ONE DEGREE.

In land descriptions, degree readings are not a measure of distance. They are combined with either North or South, to show the direction a line runs from a given point.

HOW TO READ DESCRIPTIONS WHICH SHOW DIRECTIONS IN TERMS OF DEGREES

In figure 8, the north-south line, and the east-west line divide the circle into 4 equal parts, which means that each part contains 90 degrees as shown. Several different direction lines are shown in this diagram, with the number of degrees each varies east or west from the north and south starting points (remember again that all descriptions read from the north or south).

We all know what north-west is. It is a direction which is half-way between North and West. In terms of degrees the direction north-west would read, north 45 degrees west (see figure 8).

EXAMPLE OF A LAND DESCRIPTION IN TERMS OF DEGREES

At this time, study figure 8 for a minute or two.

In figure 8, notice the small tract. The following metes and bounds description will locate this small tract. "Begin at the beginning point, thence N 20 degrees west — 200 feet, thence N 75 degrees east — 1190 feet, thence S 30 degrees east — 240 feet, thence S 45 degrees west — 420 feet, thence west — 900 feet back to the point of beginning, containing so many acres, etc."

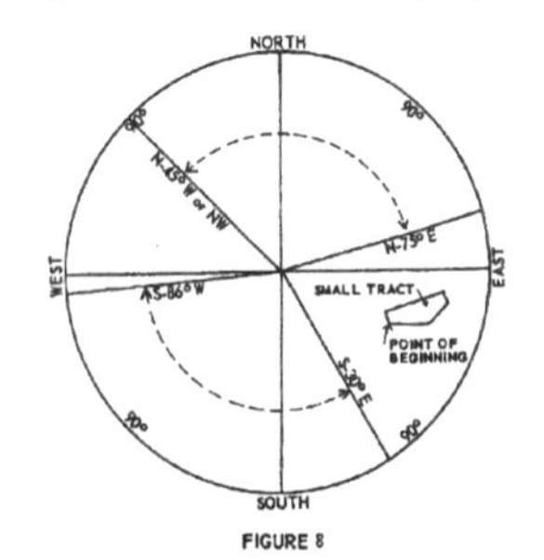

FIGURE 8

Source: Rockford Map Publishing, Inc.

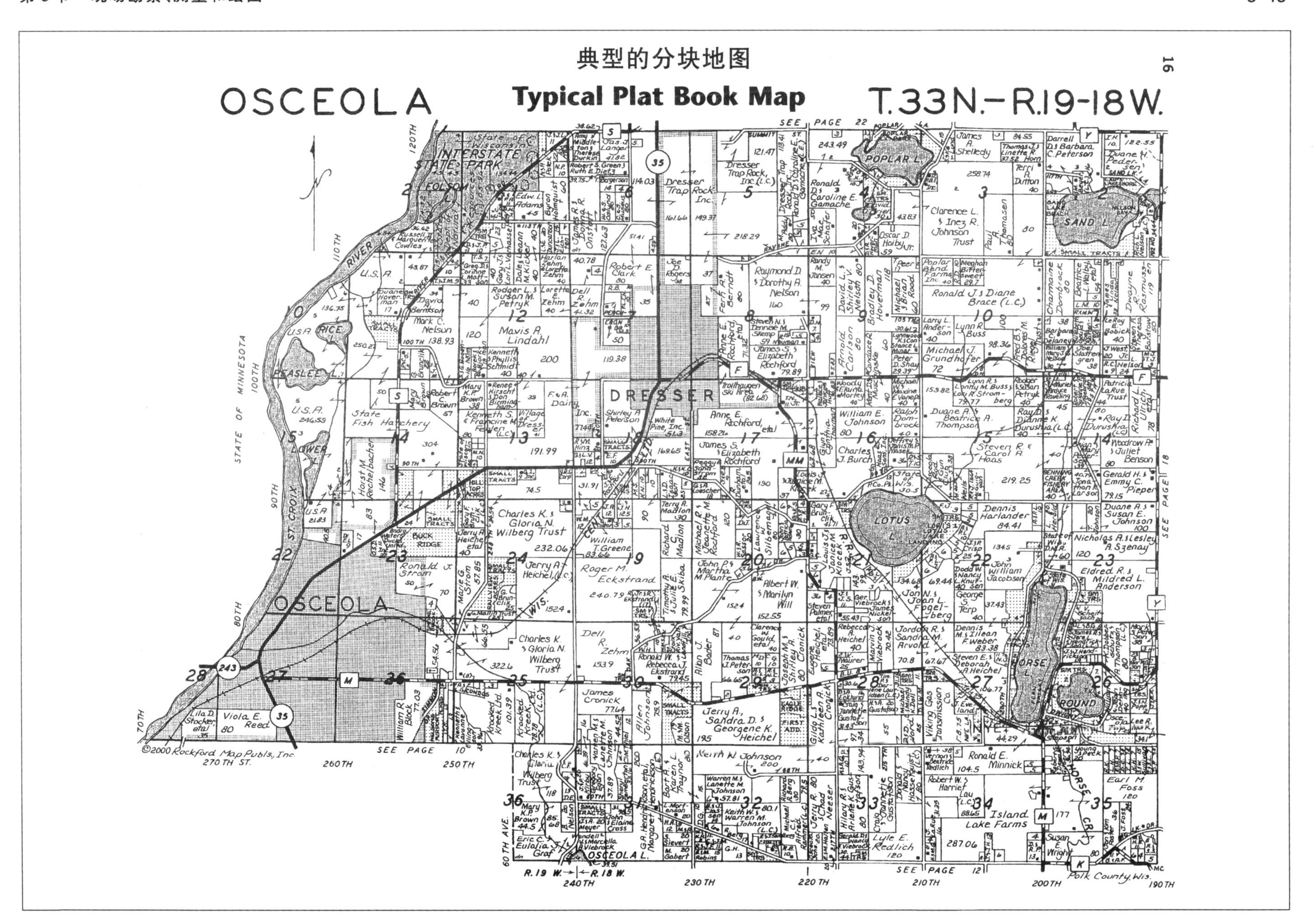

Source: © 2004 Rockford Map Publishing, Inc.

Map of Right-of-Way Easement 通行权图

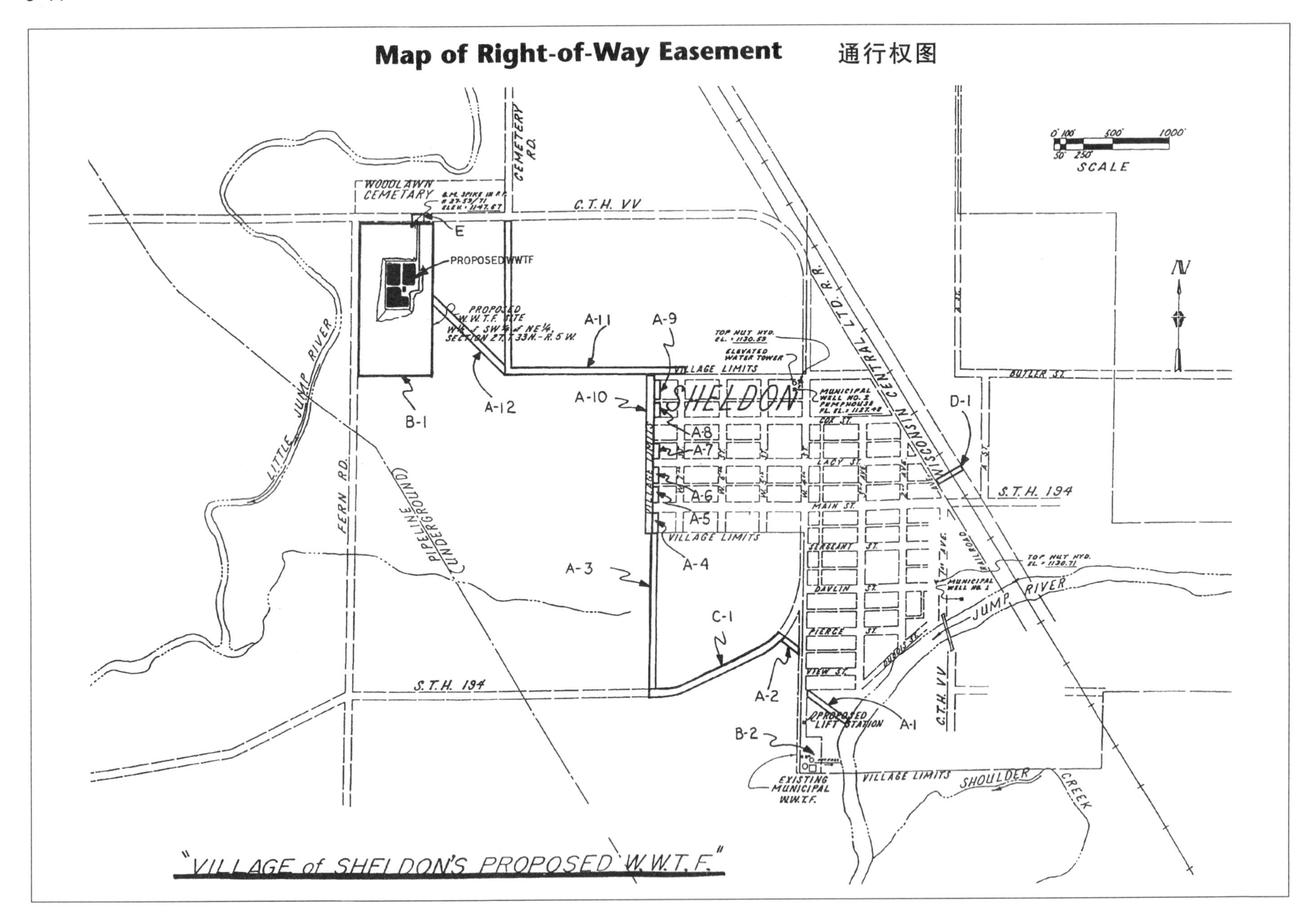

Source: Morgan & Parmley, Ltd.

Easement I.D. Mapping for Sanitary Sewer Connections　污水管道的通行编号图

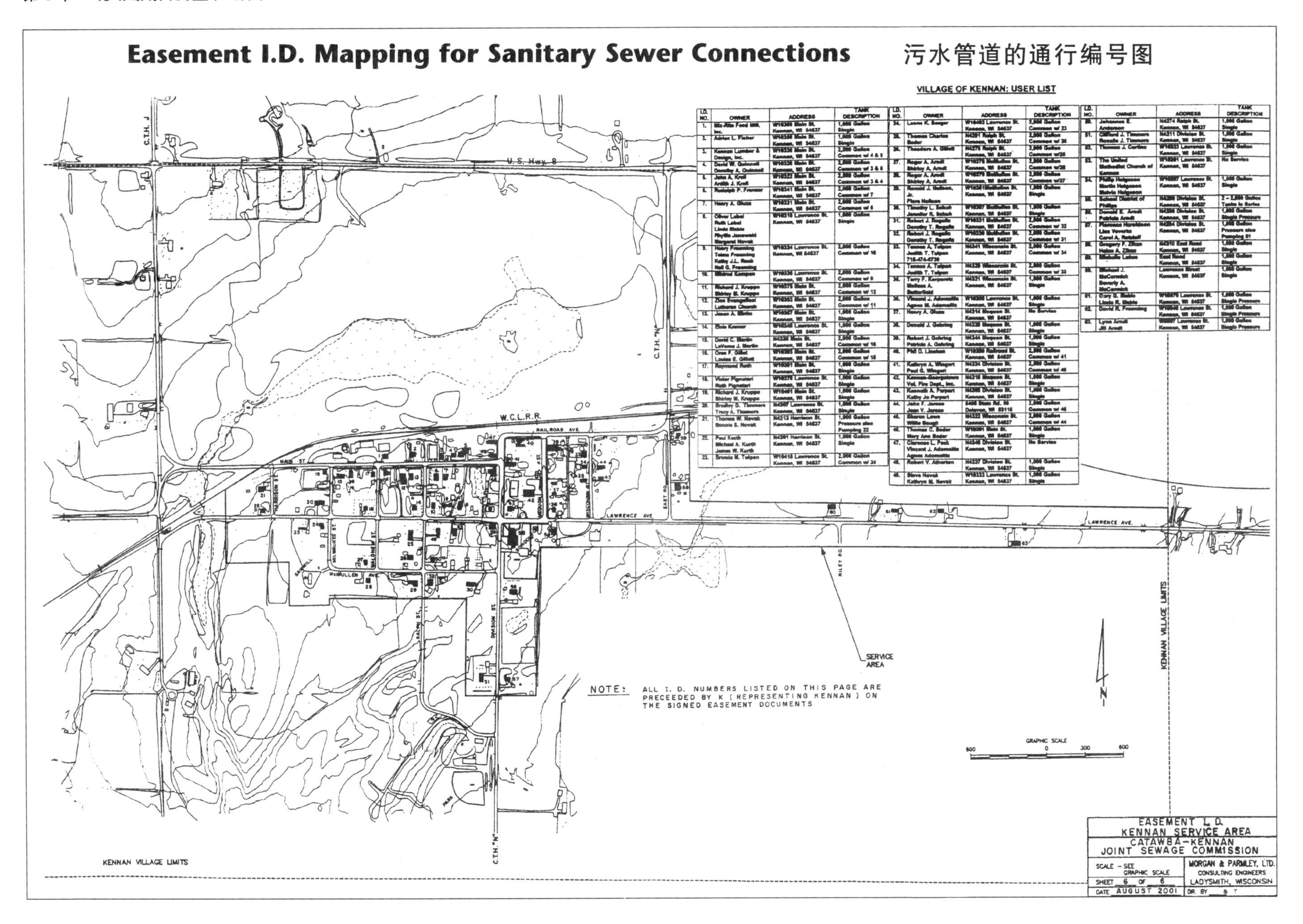

VILLAGE OF KENNAN: USER LIST

I.D. NO.	OWNER	ADDRESS	TANK DESCRIPTION
1.	Mix-Rite Feed Mill, Inc.	W16326 Main St. Kennan, WI 54637	1,000 Gallon Single
2.	Adrian L. Fisher	W16326 Main St. Kennan, WI 54637	1,000 Gallon Single
3.	Kennan Lumber & Design, Inc.	W16336 Main St. Kennan, WI 54637	2,500 Gallon Common w/ 4 & 5
4.	David W. Quinnell Dorothy A. Quinnell	W16326 Main St. Kennan, WI 54637	2,500 Gallon Common w/ 3 & 5
5.	John A. Kroll Ardith J. Kroll	W16322 Main St. Kennan, WI 54637	2,500 Gallon Common w/ 3 & 4
6.	Rudolph P. Fromm	W16341 Main St. Kennan, WI 54637	2,000 Gallon Common w/ 7
7.	Henry A. Glaza	W16331 Main St. Kennan, WI 54637	2,000 Gallon Common w/ 6
8.	Oliver Lebel Ruth Lebel Linda Mable Phyllis Jaszewski Margaret Novak	W16318 Lawrence St. Kennan, WI 54637	1,000 Gallon Single
9.	Henry Froemming Telma Froemming Kathy J.L. Raab Neil G. Froemming	W16334 Lawrence St. Kennan, WI 54637	2,000 Gallon Common w/ 10
10.	Mildred Kempen	W16336 Lawrence St. Kennan, WI 54637	2,000 Gallon Common w/ 9
11.	Richard J. Kruppe Shirley M. Kruppe	W16375 Main St. Kennan, WI 54637	2,000 Gallon Common w/ 12
12.	Zion Evangelical Lutheran Church	W16383 Main St. Kennan, WI 54637	2,000 Gallon Common w/ 11
13.	Jason A. Mirko	W16387 Main St. Kennan, WI 54637	1,000 Gallon Single
14.	Elsie Kramer	W16346 Lawrence St. Kennan, WI 54637	1,000 Gallon Single
15.	David C. Martin LaVerne J. Martin	N4336 Main St. Kennan, WI 54637	2,000 Gallon Common w/ 16
16.	Oren F. Gillett Louise E. Gillett	W16393 Main St. Kennan, WI 54637	2,000 Gallon Common w/ 15
17.	Raymond Roth	W16391 Main St. Kennan, WI 54637	1,000 Gallon Single
18.	Victor Pignatari Ruth Pignatari	W16378 Lawrence St. Kennan, WI 54637	1,000 Gallon Single
19.	Richard J. Kruppe Shirley M. Kruppe	W16401 Main St. Kennan, WI 54637	1,000 Gallon Single
20.	Bradley D. Timmers Tracy A. Timmers	N4367 Lawrence St. Kennan, WI 54637	1,000 Gallon Single
21.	Thomas W. Novak Bonnie S. Novak	N4313 Harrison St. Kennan, WI 54637	1,000 Gallon Pressure also Pumping 22
22.	Paul Kurth Michael A. Kurth James W. Kurth	N4301 Harrison St. Kennan, WI 54637	1,000 Gallon Single
23.	Brenda M. Tulpan	W16415 Lawrence St. Kennan, WI 54637	2,000 Gallon Common w/ 24
24.	Leona K. Seeger	W16403 Lawrence St. Kennan, WI 54637	2,000 Gallon Common w/ 23
25.	Thomas Charles Bader	N4291 Ralph St. Kennan, WI 54637	2,000 Gallon Common w/ 26
26.	Theodore A. Gillett	N4275 Ralph St. Kennan, WI 54637	2,000 Gallon Common w/25
27.	Roger A. Arndt Shirley A. Arndt	W16378 McMullen St. Kennan, WI 54637	2,000 Gallon Common w/28
28.	Roger A. Arndt Shirley A. Arndt	W16376 McMullen St. Kennan, WI 54637	2,000 Gallon Common w/27
29.	Ronald J. Neilson, Jr. Flora Neilson	W16361McMullen St. Kennan, WI 54637	1,000 Gallon Single
30.	Timothy L. Schuh Jennifer R. Schuh	W16387 McMullen St. Kennan, WI 54637	1,000 Gallon Single
31.	Robert J. Regalia Dorothy T. Regalia	W16331 McMullen St. Kennan, WI 54637	2,000 Gallon Common w/ 32
32.	Robert J. Regalia Dorothy T. Regalia	W16330 McMullen St. Kennan, WI 54637	2,000 Gallon Common w/ 31
33.	Tennes A. Tulpan Judith T. Tulpan 715-474-6730	N4341 Wisconsin St. Kennan, WI 54637	2,000 Gallon Common w/ 34
34.	Tennes A. Tulpan Judith T. Tulpan	N4325 Wisconsin St. Kennan, WI 54637	2,000 Gallon Common w/ 33
35.	Terry F. Korpowitz Melissa A. Butterfield	N4321 Wisconsin St. Kennan, WI 54637	1,000 Gallon Single
36.	Vincent J. Adomaitis Agnes M. Adomaitis	W16298 Lawrence St. Kennan, WI 54637	1,000 Gallon Single
37.	Henry A. Glaza	N4314 Bloquon St. Kennan, WI 54637	No Service
38.	Donald J. Gehring	N4328 Bloquon St. Kennan, WI 54637	1,000 Gallon Single
39.	Robert J. Gehring Patricia A. Gehring	N4344 Bloquon St. Kennan, WI 54637	1,000 Gallon Single
40.	Phil D. Linehan	W16299 Railroad St. Kennan, WI 54637	2,000 Gallon Common w/ 41
41.	Kathryn A. Wingert Paul G. Wingert	N4334 Division St. Kennan, WI 54637	2,000 Gallon Common w/ 40
42.	Kennan-Georgetown Vol. Fire Dept., Inc.	N4318 Bloquon St. Kennan, WI 54637	1,000 Gallon Single
43.	Kenneth A. Parpart Kathy Jo Parpart	N4386 Division St. Kennan, WI 54637	1,000 Gallon Single
44.	John F. Jarosz Jean V. Jarosz	5406 State Rd. 50 Delavan, WI 53115	2,000 Gallon Common w/ 45
45.	Sharon Lowe Willie Baugh	N4322 Wisconsin St. Kennan, WI 54637	2,000 Gallon Common w/ 44
46.	Thomas C. Bader Mary Ann Bader	W16391 Main St. Kennan, WI 54637	1,000 Gallon Single
47.	Clarence L. Peck Vincent J. Adomaitis Agnes Adomaitis	N4348 Division St. Kennan, WI 54637	No Service
48.	Robert V. Atherton	N4337 Division St. Kennan, WI 54637	1,000 Gallon Single
49.	Steve Novak Kathryn M. Novak	W16333 Lawrence St. Kennan, WI 54637	1,000 Gallon Single
50.	Johannes E. Anderson	N4274 Ralph St. Kennan, WI 54637	1,000 Gallon Single
51.	Clifford J. Timmers Rosalie J. Timmers	N4211 Division St. Kennan, WI 54637	1,000 Gallon Single
52.	Thomas J. Cortise	W16323 Lawrence St. Kennan, WI 54637	1,000 Gallon Single
53.	The United Methodist Church of Kennan	W16361 Lawrence St. Kennan, WI 54637	No Service
54.	Philip Helgeson Martin Helgeson Melvin Helgeson	W16287 Lawrence St. Kennan, WI 54637	1,000 Gallon Single
55.	School District of Phillips	N4288 Division St. Kennan, WI 54637	2 – 2,000 Gallon Tanks in Series
56.	Donald E. Arndt Patricia Arndt	N4288 Division St. Kennan, WI 54637	1,000 Gallon Single Pressure
57.	Florence Hendrickson Lisa Yovertz Carol A. Rotsdorf	N4284 Division St. Kennan, WI 54637	1,000 Gallon Pressure also Pumping 51
58.	Gregory F. Zilkan Helen A. Zilkan	N4318 East Road Kennan, WI 54637	1,000 Gallon Single
59.	Michelle Lakes	East Road Kennan, WI 54637	1,000 Gallon Single
60.	Michael J. McCormick Beverly A. McCormick	Lawrence Street Kennan, WI 54637	1,000 Gallon Single
61.	Gary G. Mable Linda R. Mable	W16076 Lawrence St. Kennan, WI 54637	1,000 Gallon Single Pressure
62.	David R. Froemming	W16040 Lawrence St. Kennan, WI 54637	1,000 Gallon Single Pressure
63.	Lynn Arndt Jill Arndt	W16097 Lawrence St. Kennan, WI 54637	1,000 Gallon Single Pressure

Source: Morgan & Parmley, Ltd.

Easement I.D. Mapping for Sanitary Sewer Connections　污水管道的通行编号图

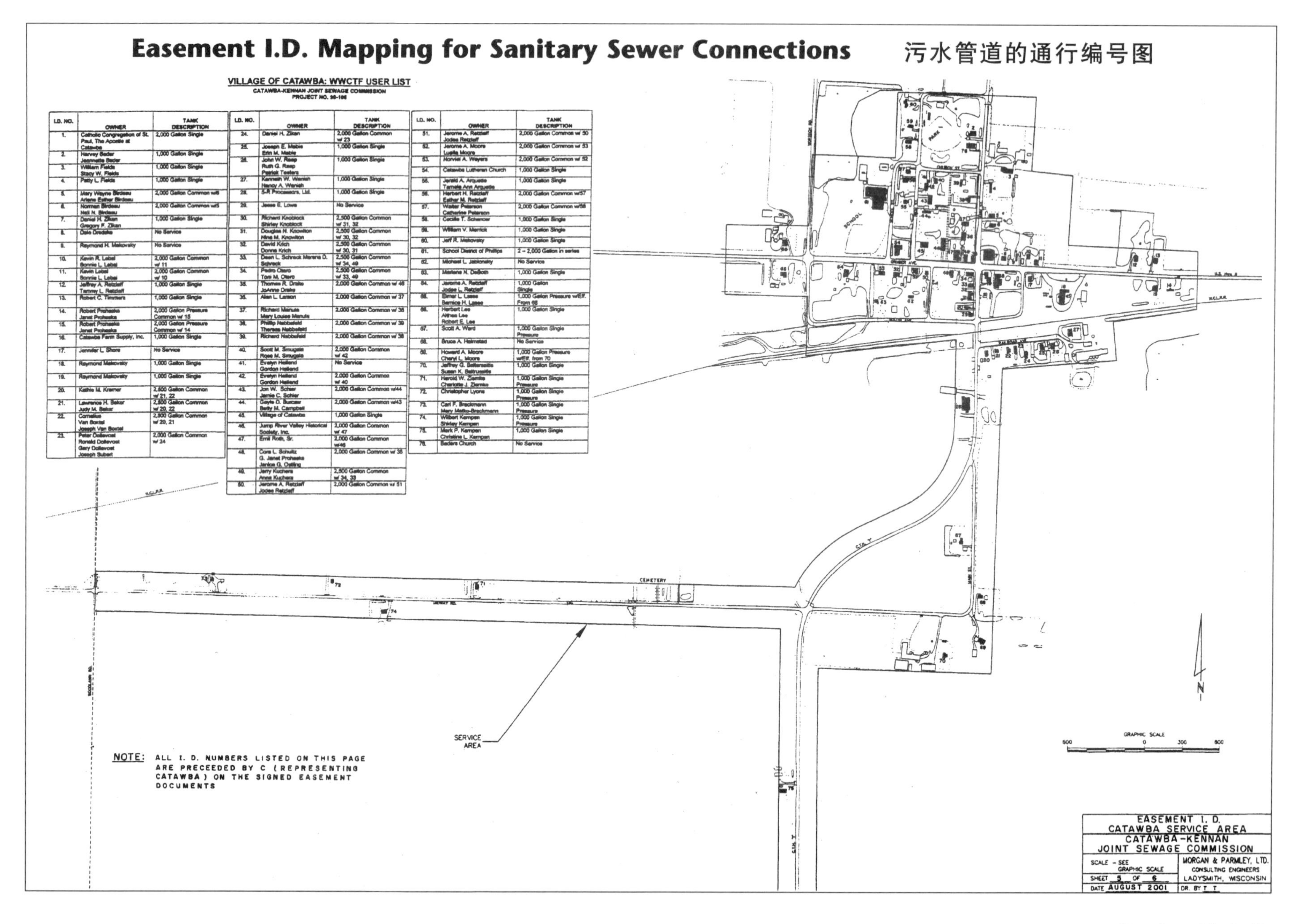

VILLAGE OF CATAWBA: WWCTF USER LIST

CATAWBA-KENNAN JOINT SEWAGE COMMISSION
PROJECT NO. 98-106

I.D. NO.	OWNER	TANK DESCRIPTION
1.	Catholic Congregation of St. Paul, The Apostle at Catawba	2,000 Gallon Single
2.	Harvey Bader Jeannette Bader	1,000 Gallon Single
3.	William Fields Stacy W. Fields	1,000 Gallon Single
4.	Patty L. Fields	1,000 Gallon Single
5.	Mary Wayne Birdeau Arlene Esther Birdeau	2,000 Gallon Common w/6
6.	Norman Birdeau Neil N. Birdeau	2,000 Gallon Common w/5
7.	Daniel H. Zilkan Gregory F. Zilkan	1,000 Gallon Single
8.	Dale Dredske	No Service
9.	Raymond H. Makovsky	No Service
10.	Kevin R. Lebel Bonnie L. Lebel	2,000 Gallon Common w/ 11
11.	Kevin Lebel Bonnie L. Lebel	2,000 Gallon Common w/ 10
12.	Jeffrey A. Retzlaff Tammy L. Retzlaff	1,000 Gallon Single
13.	Robert C. Timmers	1,000 Gallon Single
14.	Robert Prohaska Janet Prohaska	2,000 Gallon Pressure Common w/ 15
15.	Robert Prohaska Janet Prohaska	2,000 Gallon Pressure Common w/ 14
16.	Catawba Farm Supply, Inc.	1,000 Gallon Single
17.	Jennifer L. Shore	No Service
18.	Raymond Makovsky	1,000 Gallon Single
19.	Raymond Makovsky	1,000 Gallon Single
20.	Kathie M. Kramer	2,500 Gallon Common w/ 21, 22
21.	Lawrence H. Baker Judy M. Baker	2,500 Gallon Common w/ 20, 22
22.	Cornelius Van Boxtel Joseph Van Boxtel	2,500 Gallon Common w/ 20, 21
23.	Peter Dallevost Ronald Dallevost Gary Dallevost Joseph Subert	2,000 Gallon Common w/ 24
24.	Daniel H. Zilkan	2,000 Gallon Common w/ 23
25.	Joseph E. Mabie Erin M. Mabie	1,000 Gallon Single
26.	John W. Rasp Ruth G. Rasp Patrick Testers	1,000 Gallon Single
27.	Kenneth W. Wanish Nancy A. Wanish	1,000 Gallon Single
28.	S-R Processors, Ltd.	1,000 Gallon Single
29.	Jesse E. Lowe	No Service
30.	Richard Knoblock Shirley Knoblock	2,500 Gallon Common w/ 31, 32
31.	Douglas N. Knowlton Nina M. Knowlton	2,500 Gallon Common w/ 30, 32
32.	David Krich Donna Krich	2,500 Gallon Common w/ 30, 31
33.	Dean L. Schreck Marena D. Schreck	2,500 Gallon Common w/ 34, 49
34.	Pedro Otero Toni M. Otero	2,500 Gallon Common w/ 33, 49
35.	Thomas R. Drake JoAnne Drake	2,000 Gallon Common w/ 46
36.	Alan L. Larson	2,000 Gallon Common w/ 37
37.	Richard Manula Mary Louise Manula	2,000 Gallon Common w/ 36
38.	Phillip Nebbefeld Theresa Nebbefeld	2,000 Gallon Common w/ 39
39.	Richard Nebbefeld	2,000 Gallon Common w/ 38
40.	Scott M. Smugala Rose M. Smugala	2,000 Gallon Common w/ 42
41.	Evelyn Helland Gordon Helland	No Service
42.	Evelyn Helland Gordon Helland	2,000 Gallon Common w/ 40
43.	Jon W. Schier Jamie C. Schier	2,000 Gallon Common w/44
44.	Gayle D. Burcaw Betty M. Campbell	2,000 Gallon Common w/43
45.	Village of Catawba	1,000 Gallon Single
46.	Jump River Valley Historical Society, Inc.	2,000 Gallon Common w/ 47
47.	Emil Roth, Sr.	2,000 Gallon Common w/46
48.	Cora L. Schultz G. Janet Prohaska Janice G. Ostling	2,000 Gallon Common w/ 35
49.	Jerry Kuchera Anna Kuchera	2,500 Gallon Common w/ 34, 33
50.	Jerome A. Retzlaff Jodee Retzlaff	2,000 Gallon Common w/ 51
51.	Jerome A. Retzlaff Jodee Retzlaff	2,000 Gallon Common w/ 50
52.	Jerome A. Moore Luella Moore	2,000 Gallon Common w/ 53
53.	Norviel A. Weyers	2,000 Gallon Common w/ 52
54.	Catawba Lutheran Church	1,000 Gallon Single
55.	Jerald A. Arquette Tamela Ann Arquette	1,000 Gallon Single
56.	Herbert H. Retzlaff Esther M. Retzlaff	2,000 Gallon Common w/57
57.	Walter Peterson Catherine Peterson	2,000 Gallon Common w/56
58.	Cecilia T. Schenow	1,000 Gallon Single
59.	William V. Merrick	1,000 Gallon Single
60.	Jeff R. Makovsky	1,000 Gallon Single
61.	School District of Phillips	2 – 2,000 Gallon in series
62.	Michael L. Jablonsky	No Service
63.	Marlene N. DeBoth	1,000 Gallon Single
64.	Jerome A. Retzlaff Jodee L. Retzlaff	1,000 Gallon Single
65.	Elmer L. Lasee Bernice H. Lasee	1,000 Gallon Pressure w/Eff. From 66
66.	Herbert Lee Althea Lee Robert E. Lee	1,000 Gallon Single
67.	Scott A. Ward	1,000 Gallon Single Pressure
68.	Bruce A. Halmstad	No Service
69.	Howard A. Moore Cheryl L. Moore	1,000 Gallon Pressure w/Eff. from 70
70.	Jeffrey G. Baltrusaitis Susan K. Baltrusaitis	1,000 Gallon Single
71.	Harold W. Ziemke Charlotte J. Ziemke	1,000 Gallon Single Pressure
72.	Christopher Lyons	1,000 Gallon Single Pressure
73.	Carl F. Brackmann Mary Matke-Brackmann	1,000 Gallon Single Pressure
74.	Wilbert Kempen Shirley Kempen	1,000 Gallon Single Pressure
75.	Mark P. Kempen Christine L. Kempen	1,000 Gallon Single
76.	Baders Church	No Service

Source: Morgan & Parmley, Ltd.

洪泛区图

Flood Plain Mapping

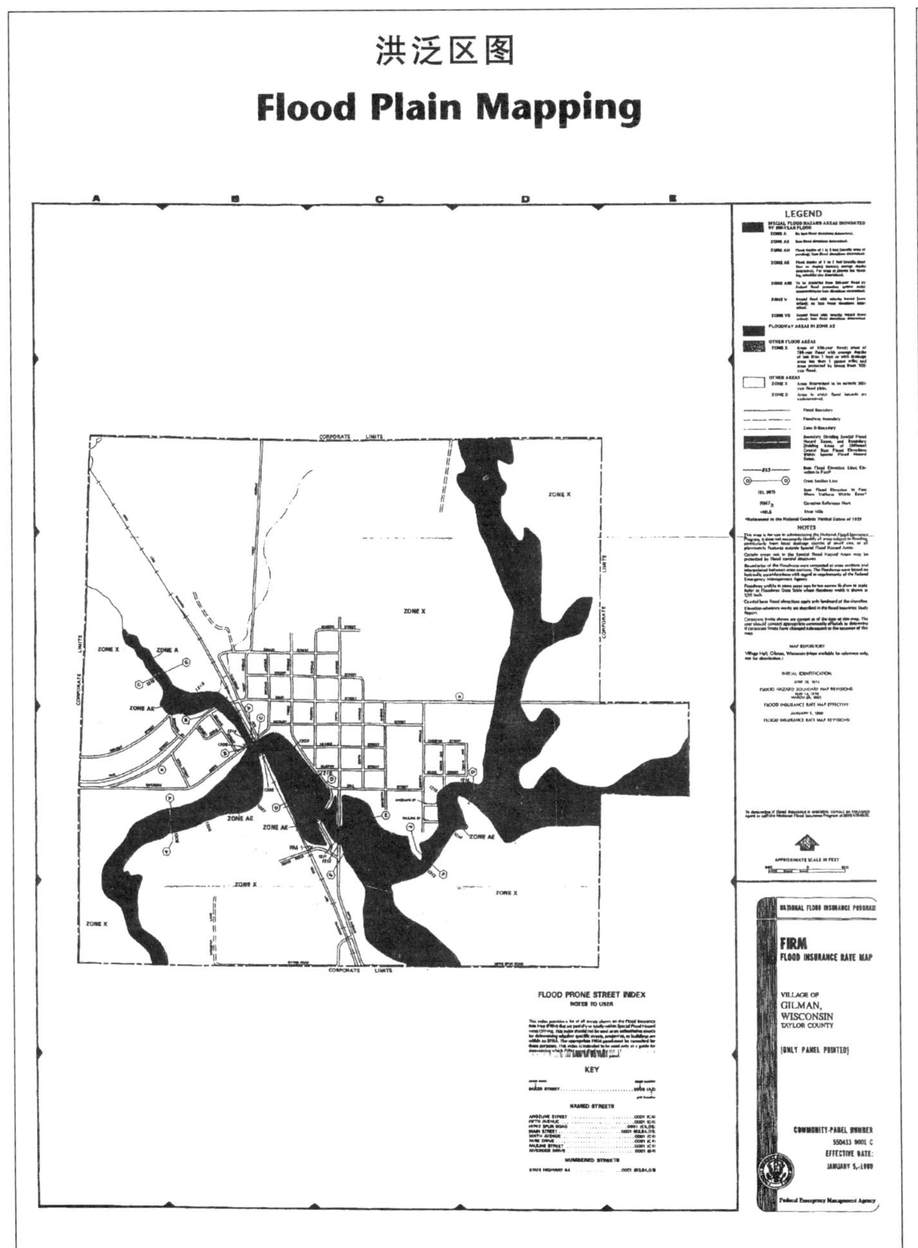

Source: FEMA

规划用地红线图

Planning Area Boundary Mapping

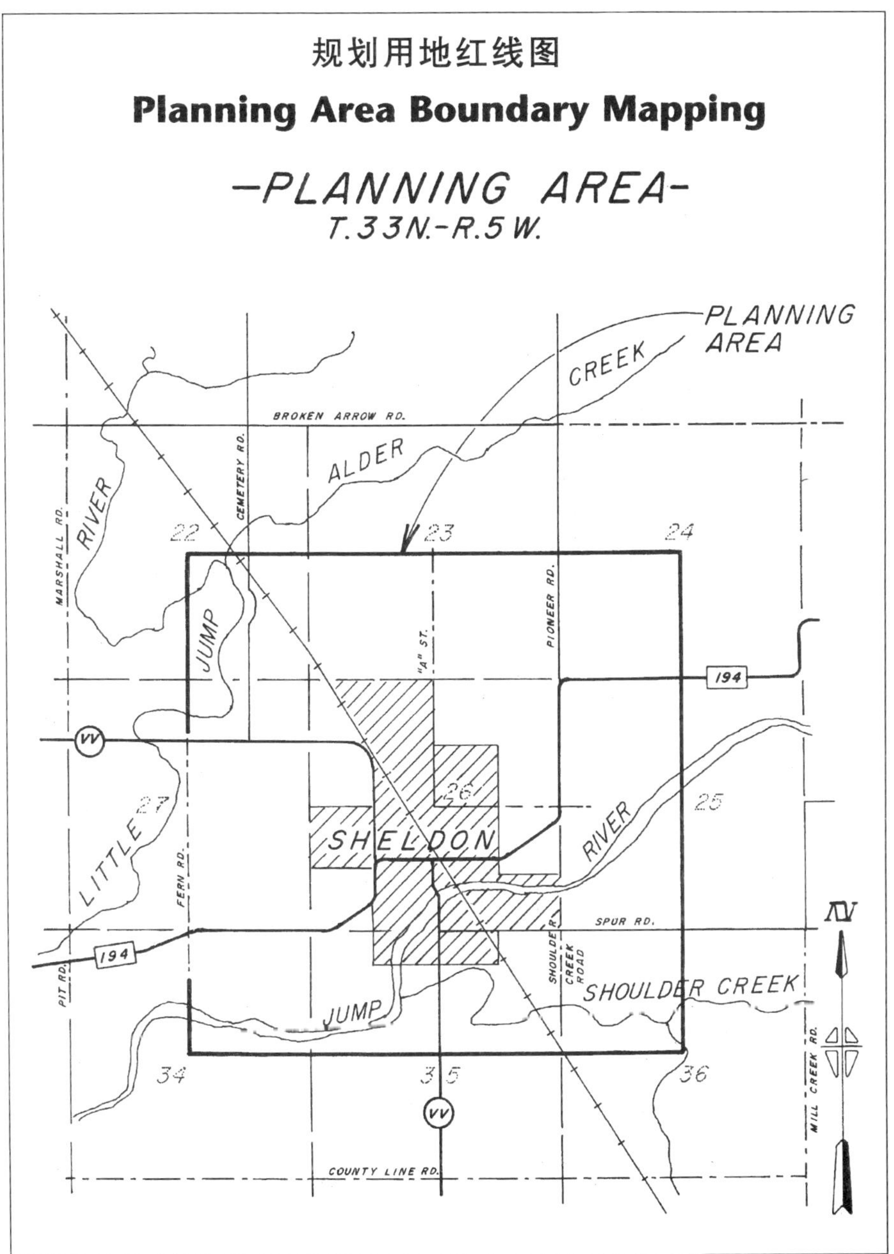

Source: Morgan & Parmley, Ltd.

Sanitary Sewer Service Area Mapping 污水管道服务区域图

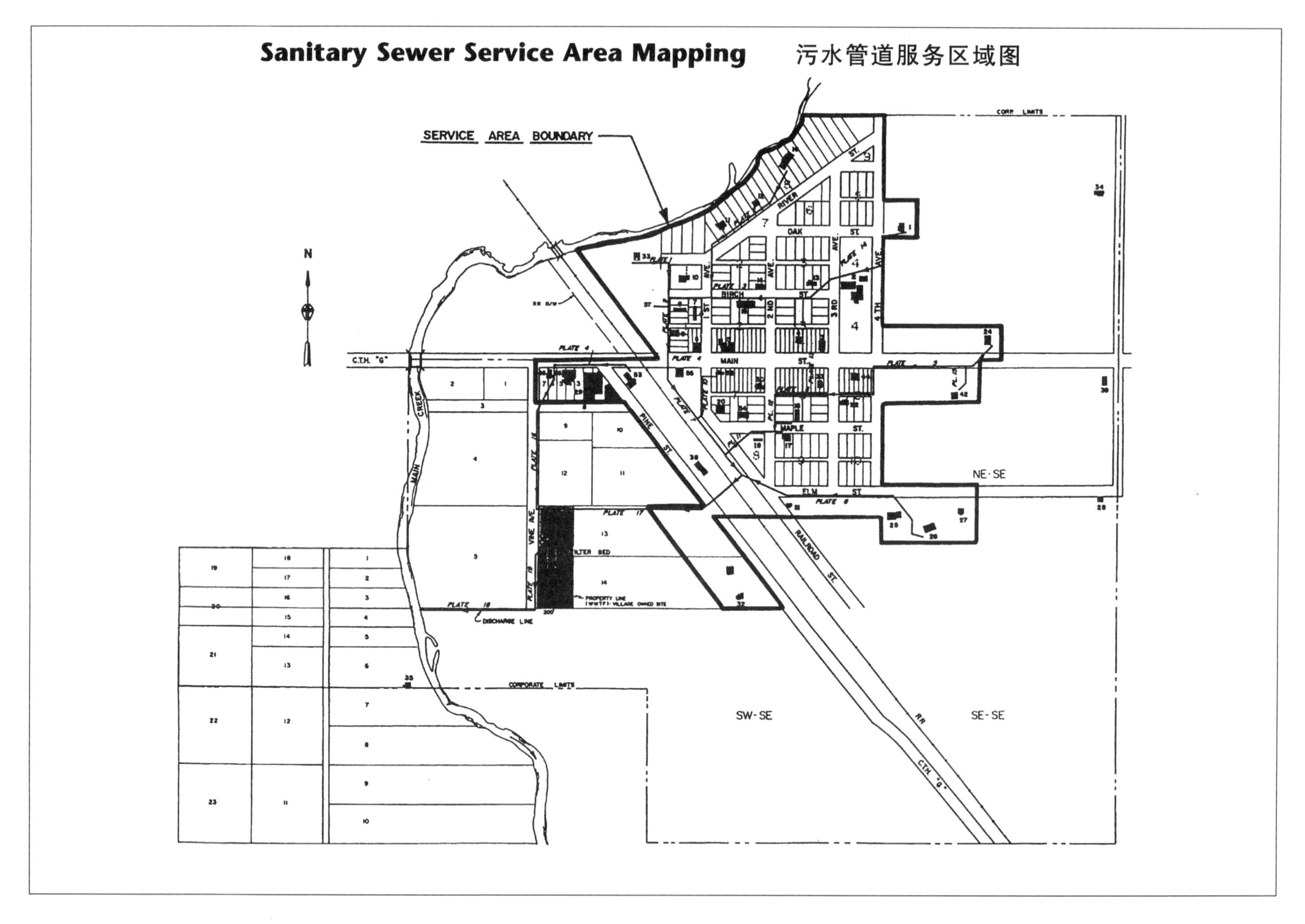

Source: Morgan & Parmley, Ltd.

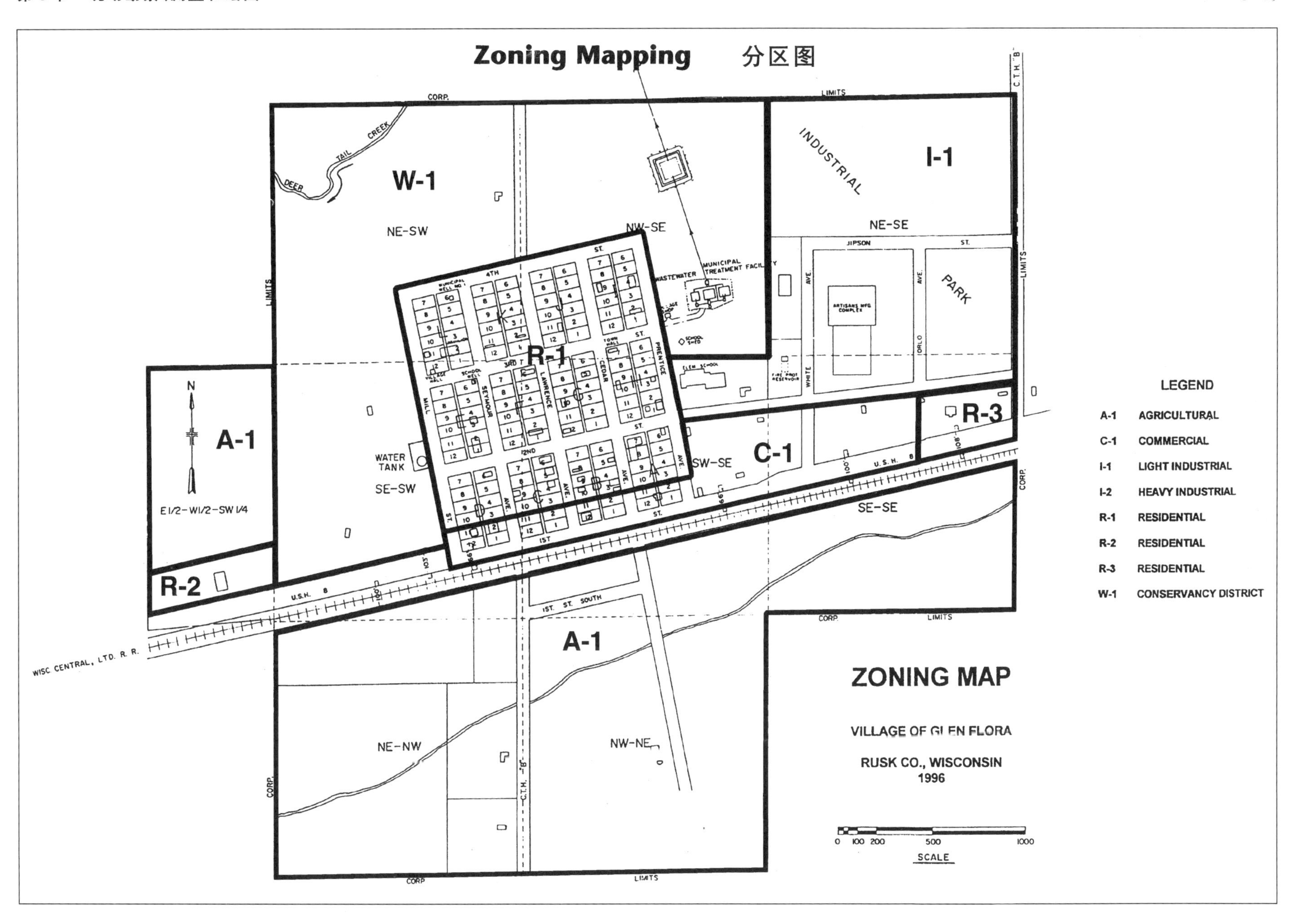

Source: Morgan & Parmley, Ltd.

规　划

第4节　公开会议

例　4

顾问工程师在执行他们的职业任务期间，需要不断地参加大量的公开会议。大多数项目要面向至少一个公开会议，而且顾问工程师通常是一个关键的与会者。这样，项目工程师能够了解并掌握各种公开会议的结构和类型就显得尤为重要。

目　录

常规的市政会议

选举产生的市政官员以一种向公众公开的方式召开会议，来履行他们的政府监管职能。这些会议一般是相当公开的，而且按照一份详细的议程进行。当要讨论有争议的问题或者大型项目时，通常会有新闻媒体出席，而且还有不同的市民组织与会。

无论这些会议是市议会、农村理事会、污水委员会、卫生管区、县议会、水资源委员会、镇议会、学校管区或者类似的市政当局，它们在概念和运作形式上基本上是相似的。所以，强烈建议读者获取一本名为《罗伯特秩序规则》的经典著作并研习其内容。大多数的会议都遵循这些规则而且所有的会议都应避免混乱和不正确的程序。

当工程师被要求或者需要出席一次市政会议，它通常直接关系到一项活跃的项目。这种对特定项目的关注要求工程师做好充分准备，并能应付各种问题和关注。建议准备分发给观众的恰当的直观辅助材料和分发的印刷材料，以备陈述时辅助使用。工程师必须牢记将项目描述得在本质上更宽泛一些，要避免使用可能引起公众困惑的专业术语。然而，如果有人提出了特定的技术问题，工程师必须能够详细地回答而且有充足的文件来支持关键结论。尽管这很少见，但工程师不能假设具体的技术问题不会被问及。

另外，工程师应该评估观众和市政当局官员来决定陈述应该到何种详细程度以及持续多长时间。最后，工程师必须一直保持一种放松的、沉着的和专业的风格，以免危及项目的进展。

委员会会议

委员会的成员人数通常少于他们所服务的市政理事会，他们开会对特定主题进行讨论并提出建议。这些委员会包括：分区、林业、财务、停车场和地面、公共工程、保卫和消防、人力、福利、交通、给排水、设施等。

如前所述，这些委员会通常成员较少，这样就因为更宽松的气氛而变得更易于沟通。这些团体的议程较小，可以集中关注主题而不受冲突干扰。委员会收集并评估各种信息来向他的上级组织提出建议；也就是向它们相应的市政理事会提出建议。工程师必须设法赢得委员会的专业敬重并建立良好的工作关系，否则项目将开始出现问题，有可能会危害项目本身。

信息通报会议

信息通报会是因为多种目的而召开的。当然最主要的原因还是在建议项目仍然还是处于概念阶段时就向公众通报其相关信息。基本上，在项目或进度进行得过远之前，市政当局、公共事业公司、委员会或者管理机构正在从市民中获取反馈意见。

这些特别会议范围很宽泛而且也可以很不正式。气氛更加宽松，而且通常这也被反映在听众的热情参与上。基层对特定项目的反馈无疑将帮助工程师开发出一个更可行的方案。

这种类型的会议需要一份广泛的日程安排。然后，需要一个优秀的会议主持来很好地安排会议的进程并控制参与者有良好的秩序。秘书或者职员应该做会议纪要，而且项目工程师必须保留一份记录。

评估听证会

评估听证会能够引起一些最热烈的评论和观众的反对，因为将为特定私人财产进行估价。这些听证会非常正式，通常都有录像，录音和/或有资质的法庭书记员做记录。通常这些听证会参会热烈，新闻媒体将做事件的全面报道。

在评估听证会之前，整个项目被记述在一份工程报告中，它详细地介绍了项目需求、概念设计和详细的估计成本。市政当局展示估价表，它总结了对每项受影响的财产进行的估价。在举行公开评估听证会之前，这份估价表向公众公布，被鉴定的复印件邮寄给所有受影响的财产的业主。在听证会召开之前，所有市民都可以查询工程报告和估价表以便审查。听证会的通知将被印刷，日程也将被安排。

在公开听证会上，项目工程师将被要求叙述该项目，回顾评估的成本并回答问题。在工程师的陈述之前，官方的会议主持人将宣布听证会的目的并设置会议程序。通常为了避免混乱，任何希望发言或者提问的个人都被要求填写一份简短的表格，登记姓名、地址和陈述，这样官方记录才够准确。

工程师应有合适的直观辅助材料和分发的印刷品来辅助发言。记住，主持人负责而且必须控制听证会；而这不是工程师的责任。所有的问题都应在工程师的陈述结束之后才被提出来，工程师只应回答提出的那些问题。一种非常罕见的情形是，如果你不知道答案，那么请说明你将尽快找到答案并通过书面材料（复印件提交给合适的市政官员）单独回答。在所有的时间里，保持你的镇定和机敏，留意所有的与会者并使得你的回答简短而精要。千万不要做出那些你并没有坚实和可靠信息支持的陈述。

废水处理设施初步设计听证会

与评估听证会一样，废水处理设施初步设计听证会也可能会引起强烈的反对意见。污水处理设施、电力输送线路、核电站、采矿和类似的项目的计划施工通常会对环境敏感区域造成冲击，还会引起不同环境保护组织的反应。

如果废水处理设施初步设计过程在敏感区域潜在冲击评估和确保恰当的保护措施方面做了充足的工作，被建议的项目通常将被允许继续建设。然而，总有一些问题是存在争议的，这样就需要举行公开听证会。另外，也有必要保证公众的知情权。

通常，听证会在形式上与前面介绍的评估听证会非常相似。废水处理设施初步设计文件描述待建项目，评估文化的、历史的、环境的、考古的、社会的和项目的可选方案，并选择成本最有效/环境最良好的方案。后面再附上项目时间表。在听证会之前，该文件应向公众和适当的管理机构公布并接受审查。工程师和其他支持人员被要求在听证会上做一个正式的陈述。陈述应该是详尽的，并且通常也是较长的，因为听众中可能会有专业人士准备给出技术上的回应。

通常，这些听证会不包括提问和回答问题的时间。但是，评论和陈述可以允许作为记录。很多这种听证会是由管理机构来召开的，这个管理机构将允许或者否决建设被建议的项目。这样，最主要的目的是收集额外的数据以便帮助他们润色最终决定。这些听证会非常正式，而且一份正式的笔记将作为永久记录。通常，听证会将延续一定的时间来接受书面的评论和陈述，加到记录中。

开工前会议

在开始实施任何实际的活动之前，每个项目都会召开一次开工前会议。这个会议是非常详尽而且明确的。项目的各个方面都会涉及到，也就是说业主、承包商、分包商、公共事业、管理机构、工程师、监理等都必须出席。这是全部过程的一个重要里程碑。有关详细论述，读者可参考本手册的第 26 节。

典　　礼

项目工程师有时会被邀请出席参加一些施工项目的典礼。一般来说，市政当局会为一项大型或者特殊项目公开举行破土典礼。新闻媒体通常也受到邀请，这样项目就会受到应有的重视。这些破土仪式的形式可以是官员们的合影，也可以是包括项目历史和嘉宾们致词的正式议程。项目工程师非常偶然的也会被邀请对人群发表演说，这应当被视为一种荣誉，演说辞应该保持尽量短小。

研讨会和讲座

工程师经常受邀参加技术研讨会。这是一种良好的公共关系，而且能让你所在的咨询公司得以亮相。记住一点，如果你不被认为是专家你是不会受到邀请的。这样，明智的做法是多花一些时间来做好充分的准备。

顾问工程师偶尔也会被邀请做一次讲座，通常是面向高中生或者大学生。这既是一种非常的荣誉，同时也是一场挑战。取决于讲座的主题，这为工程师提供了很好的机会向专心的听众传授特定的技术信息。也许某一场这样的讲座会为某些人点亮对工程的兴趣，而他们可能本来根本就不可能接受这项职业。无论如何，作者认为顾问工程师有责任用最高的方式来描述这个职业并愿意宣传它的优势和机会。

接受新闻媒体采访

顾问工程师有时也会接受新闻媒体关于公众当前关注的项目的采访。不论它是负面的问题或者仅仅是一般的好奇，工程师都应该以一种积极正面的方式回答问题。工程师必须牢记它的首要任务是保护公众的健康和安全并以一种完善和专业的方式为客户服务。

建议给新闻报道者一份关于项目的书面描述，确保每个事项都得到准确的沟通。同时，也建议提供精细的、可复印的场地平面图和待建项目的典型说明或示意图表。如果是房屋建筑项目，楼层平面图、立面图或者透视渲染图将为发表的文章添彩不少。

如果采访者的主要本质是反对意见，工程师在回答任何问题时都必须非常小心。有些回答可能会引起误解或者被错误地报道。这样就要特别小心落入用词的陷阱。当必要时应谢绝采访，特别是存在可能的法律诉讼时。

规　划

第5节　管理机构审核批准

例　5

项目的一个关键要素是获得相关管理机构的适当审核批准和同意。

目　录

一般介绍

在施工之前，确保获得所有相关管理机构的审核批准和许可是设计师的专业职责。

设计公司在项目运作的早期就争取获得所有有用的审核批准和许可无疑是良好的策略，这样就能将审核批准和许可的复印件包含在招标文件中。这将提供给投标者与这些审核批准和许可相联系的所有特殊情况的清楚认识。有时候，一个管理机构的审核批准将包括一些附加的特殊事项，可能会影响造价。这样，谨慎的做法是将这些事项在开标之前提前罗列出来，这样潜在的承包商能够将这些费用考虑进它们的报价中。这种做法可以避免一些潜在的费用高昂的变更，这些变更是承包商在获得合同后提出的，有关工程范围的修改。

这些审核批准和许可对项目而言是特别重要的。它们的条件必须得到遵循，而且不能做任何修改除非事先征得相关部门的书面同意。这些文件与设计图和技术说明是相互关联的，并且是施工合同的一部分。它们从不被理所当然的接受，尽管它们的要求可能是强制性的，而确保所有要求得到执行是项目工程师的职责。

管理机构

土木工程是致力于市政、公共或私人设施的规划、设计和施工的工程学科。这些领域包括：污水管道系统、水井、水坝、配水网络、消防、房屋建筑、结构、环境控制、公路、道路、桥梁、隧道、机场、海洋设施和相关的专业。所有这些领域都直接与公众的健康和安全相关。这样，适用的指南和规范经过较长时间逐步发展，对每类项目建立了最低标准。

适当的政府立法机关制定特定的法律，目的在于监管、规范并强制实施这些标准和规范。反过来，通过建立不同的管理机构来确保这些法律得到参与这些类型结构和设施的设计和施工人员的遵循。

本节并不打算列出所有这些管理机构或者提到需要的每一种类型的审核批准和许可。然而，下表总结了需要管理机构审核批准（或许可）的大部分主要领域；不管是联邦、州、县或当地市政府。任何遗漏都不是故意的。

废弃物规划（填埋场、水坝、水井等）/Abandonment Plans (Landfills, Dams, Wells, etc.)
空气排放控制设施（Air Emission Control Facilities）
机场候机楼（Air Terminals）
机场和跑道（Airports & Runways）
体育设施（Athletic Facilities）
桥梁（Bridges）
建筑（结构，供暖通风和空调、管道和防火）/Buildings (Structural, HVAC, Plumbing & Fire Protection)
露营场所（Campgrounds）
运河（Canals）
化学过程设施（Chemical Processing Facilities）
交通运输系统（Communication Transmission Systems）
水坝（水电和水资源控制）/Dams (Hydroelectric & Water Control)
电力系统（Electrical Systems）
电力传输网络（Electrical Transmission Networks）
教学综合楼和设施（Educational Complex & Facilities）
有害废物填埋场（Hazardous Waste Landfills）
公路（Highways）
医院（Hospitals）
供暖通风和空调系统（HVAC Systems）
工业综合建筑（Industrial Complexes）
人造湖和湿地（Man-Made Lakes & Wetlands）

生产设施（Manufacturing Facilities）
海洋设施（Marine Facilities）
采矿（Mining Operations）
核电站（Nuclear Power Plants）
疗养院（Nursing Homes）
停车场、坡道和结构（Parking Lots，Ramps & Structures）
管道（Pipeline）
发电设施（Power Generating Facilities）
私有土地开发（Private Land Development）
水泵系统（Pumping Systems）
铁路网络（Railroad Networks）
水库（Reservoirs）
道路（Roads）
污水收集系统（Sanitary Collection System）
垃圾填埋场（Sanitary Landfills）
污水管道系统（Sanitary Sewage Systems）
学校设施（School Facilities）
雨水排水系统（Storm Sewer Systems）
街道（Streets）
结构（Structures）
通信系统和设施（Telecommunication Systems & Facilities）
活动住屋场地（Trailer Courts）
隧道（Tunnels）
废水收集系统（Wastewater Collection System）
废水处理设施（Wastewater Treatment Facilities）
供水网络（Water Distribution Networks）
水箱和水塔（Water Tanks & Towers）
水井和供水设施（Wells & Water Supply Facilities）
塘地（施工中）/Wetlands （Constructed）

规　划

第6节　成本估算

一个项目如果没有充足的资金是不可能顺利完成的。这样，项目在准备好成本估算和拟定出良好的财务计划之前是不能走得太远的。

目　录

项目估算

估算一项待建项目的造价是一个非常复杂的过程，包括很多可变因素。造价估算不是轻易就能掌握的技能，在精通这个工程领域之前需要经过适当的学习、训练和实践。

本节没有足够的篇幅而且作者也不打算全面地讨论成本估算，本节主要目的在于强调这项基础工作对每个项目的重要性，并列出有助于读者精细调整他们的估算的几项要领。

一些因素对项目估算有显著影响，估算师应在完成成本估算之前考虑它们并准确地计入它们的影响。参考以下几项：

1）类似项目：最好的参照就是类似项目，参考它们的最终成本项目和相关费用可以作为可靠的估算基础。类似项目的经验是无价之宝。

2）材料价格：在开始估算制表之前，获取可靠的材料和物资价格，计入运输费用。

3）工资率：确定该项目是遵从美国的州或联邦工资率。另外，还需确认是否需要执行当地工资率。在估算中考虑这个因素是强制性的。

4）场地条件：可能增加项目造价的场地条件包括：不利的地基条件，湿地，受污染的材料，设备之间的冲突（地下管道、电缆、高架管线等），环境敏感区域，地下水，河流横越，交通繁忙，地下储库，文物考古现场，濒危生物栖息地，以及其他类似的现状。

5）通货膨胀：通货膨胀总是最多变的因素。在采用以前类似项目做为估价基础时，需要考虑“工程新闻报道”上刊登的造价指数。这一全美范围的建设行业列表已经连续记录了数十年。

6）投标时机：开标的时机对于获得低价投标有显著影响，其中的关键因素是建设活动的季节性变化以及与其他开标的冲突。

7）项目进度：施工进度当然会影响造价。如果项目对时间的要求过于苛刻，一般来说价格会上涨，特别当存在未在指定日期之前完成的巨额违约赔偿金时。相反，如果中标通知超过合理期限以及开工通知不确定，承包商担心材料涨价或者有其他优先项目。这样，大多数的竞标者都将提高报价以保护出现这些情况。任何超过60天的期限都会导致更高的报价。

8）设计图和技术说明的质量：没有什么能够取代精心准备的设计图和技术说明。设计中的每个细节和组成部分都被准确地完成并充分描述是非常重要的。任何含糊的措辞或者拙劣的图纸不但会造成误解，而且还会在承包商的头脑中造成怀疑，这通常会引起较高的报价。

9）出资机构：如果出资机构涉足为项目投资部分资金，承包商在准备投标时应考虑这一情况。一些出资机构需要大量可观的文书工作，而这在非投资项目一般是不需要的。有时这些预期的额外文书工作将抬高报价。

10）工程师的声誉：如果项目工程师或工程公司在承包商中的声誉很好，它将反映在合理的投标报价中。如果一个承包商与一个特定的工程师或工程公司能够很好合作，那么项目将能够顺利进展，从而节省费用。

11）管理要求：有时管理机构的许可是需要花费一定费用的。这样，为了对潜在竞标者完全公开，强烈建议将所有管理许可的复印件都包括进招标文件中。

12）保险要求：一般保险要求，例如履约保证金、付款保证金和承包商普通债务是开展业务的正常费用。然而，一些特殊项目需要额外的保证金，铁路交叉口就是一个很好的例子。必须事先考虑这些补充规定的保险费对项目成本的贡献。

13）项目的大小：项目的大小和复杂程度决定了本地承包商是否

有能力完成这项工程。待建项目越大越复杂，就越能吸引更多潜在竞标者的关注。这有利于竞争，但可能会增加动员成本。

14）工程地点：待建工程的地点是进行合理成本估算需要考虑的一个重要因素。如果工地在农村，那么建筑市场上有经验的劳动力将非常有限。这样，承包商必须引入工人，而且通常还需要支付出差津贴，也就是不在城里的住宿及相关费用。另外，偏远地区会增加材料运输费用。

15）价值工程：一些机构规定数百万美元的项目在完成设计或者投标之前需要进行价值工程审查。这样，估价师应该提前考虑这个因素。

16）意外费用：历史经验表明，需要增加 10% 的意外费用来考虑那些在项目实施过程中突然出现的不可预见费用。当通货膨胀率很高或者关键建筑材料和物资短缺时，最好将意外费用增加至 15% 或 20%，这样的估算更合理并能提供一定的安全系数。

17）补充研究和调查：如在因素 4 中所述，一些项目场地需要特别的研究和调查。这些特殊工作的费用应包括在最初的成本估算中，以免将来被动。

18）评审：在最后的分析中，一份优秀估价的最关键部分是进行合理的技术评审，这是由经验并在高级经理的指导下完成的。

公共给水系统成本估算

Cost Estimate

Public Water Supply System

for Glen Flora, Wisconsin

DESCRIPTION	TOTAL
WATER SUPPLY:	
Well w/Concrete Pump Base	$ 31,580
Pumphouse Structure	28,500
Turbine Pump w/Controls	12,500
Interior Piping	5,000
Electrical Service W/Main Panel & Controls	6,000
Electrical System & Lighting	3,520
Telemetry System & Electrical Retrofit	5,000
Chemical Feed Equipment	3,000
Auxiliary Power Mechanism	5,500
Surge Eliminator & Pressure Switch	750
Yard Piping	3,500
Yard Hydrant	1,500
Site Restoration	1,000
STORAGE:	
Tower Foundation	$22,000
Elevated Water Tank w/Recirculation Pump	121,000
Painting	19,800
AUXILIARY WELL:	
Purchase Existing School Well	10,000
Upgrade Well & Pump	15,000
Upgrade Pumphouse	8,500
Chemical Feed Equipment	3,000
Right Angle Gear Drive Unit	5,500
Yard Piping	3,500
Controls (Interconnect to Telemetry)	2,500
Interior Piping Retrofit	4,000
Electrical Upgrading	3,500
DISTRIBUTION SYSTEM:	
Watermain	142,450
Main Valves	11,550
Hydrants	25,725
Water Services w/Curb Stops	20,000
Water Meters	11 825
Street Repair & Restoration	47,115
SUB-TOTAL (CONSTRUCTION)	$584,315
Construction Contingency @ 12%	$ 57,141
Basic Engineering Services	49,700
Construction Staking	6,000
Inspection and Related Services	25,500
Testing Fees & Soil Borings	6,000
PSC Coordination	7,500
Wellhead Protection Plan	6,500
Legal Services	2,500
Administrative Costs	2,500
Accounting Fees	1,500
TOTAL ESTIMATED PROJECT	$749,156

NOTE: Does not include cost of grant administration, publication fees, property surveys, tower site purchase, audit fees, bond council services and related tasks.

Source: Morgan & Parmley, Ltd.

Present Worth Analysis

of Aerated Lagoon Facility

通风蓄污池现值分析

SHELDON, WISCONSIN

DESIGN YEAR 2016

COMPONENT DESCRIPTION	Initial Cost ($)	Service Life (Years)	Future Cost 15th Yr. ($)	Cost Per Yr. ($)	Salvage Value 20th Yr. ($)
MOBILIZATION	5,000	N/A	-0-	-0-	-0-
INTERCEPTOR:					
INFLUENT PIPING	4,000	50	-0-	80	2,400
MANHOLE (FINAL)	3,000	30	-0-	100	1,000
INFLUENT FLOW METER & SAMPLER	6,000	15	6,000	400	4,000
RAW SEWAGE PUMPING:					
ELECTRICAL SERVICE & CONTROLS	6,000	40	-0-	200	4,000
STRUCTURAL	25,000	40	-0-	625	12,500
MECHANICAL	15,000	20	-0-	750	-0-
TELEMETRY	15,000	30	-0-	500	5,000
FORCE MAIN:					
PIPING	75,000	50	-0-	1,500	45,000
AIR RELIEF M.H. W/ VALVING	10,000	30	-0-	333	3,333
HIGHWAY CROSSING (BORED/CASED)	15,000	50	-0-	300	9,000
EROSION CONTROL	5,000	N/A	-0-	-0-	-0-
TRAFFIC FLAGGING & SIGNAGE	2,500	N/A	-0-	-0-	-0-
RESTORATION (SEED, FER. & MULCH)	5,000	N/A	-0-	-0-	-0-
AERATED LAGOON SYSTEM:					
CLEARING & GRUBBING	5,000	N/A	-0-	-0-	-0-
CELL CONSTRUCTION	71,500	50	-0-	1,430	42,900
PVC LINER	38,000	50	-0-	760	22,800
RIP RAP	12,000	40	-0-	300	6,000
AIR SYSTEM (PIPING)	17,000	40	-0-	425	8,500
CELL PIPING (WET)	40,000	40	-0-	1,000	20,000
SECURITY FENCING & SIGNAGE	4,200	30	-0-	140	1,400
EROSION CONTROL	7,500	N/A	-0-	-0-	-0-
RESTORATION (SEED, FERT. & MULCH)	10,000	N/A	-0-	-0-	-0-
OUTFALL (EFFLUENT DISCHARGE)	66,000	50	-0-	1,320	39,600
CONTROL BUILDING:					
GENERAL CONSTRUCTION	45,000	40	-0-	1,125	22,500
OFFICE & LAB EQUIPMENT	4,000	20	-0-	200	-0-
ELECTRICAL (SERVICE)	5,000	30	-0-	167	1,667
ELECTRICAL / H & V (BUILDING)	12,000	30	-0-	400	4,000
ELECTRICAL (BLOWER SYSTEM)	18,000	20	-0-	900	-0-
MECHANICAL (BLOWER SYSTEM)	21,500	15	21,500	1,433	14,333
EFF. FLOW METER & SAMPLER	6,000	15	6,000	400	4,000
WATER SERVICE (POTABLE)	17,000	40	-0-	425	8,500
PLUMBING SYSTEM	6,000	30	-0-	200	2,000
UV DISINFECTION SYSTEM	18,000	30	-0-	600	6,000
AUXILIARY POWER: (DIESEL) L.S.:					
PORTABLE TRASH PUMP W / TRAILER	12,000	20	-0-	600	-0-
SAFETY EQUIPMENT	4,500	20	-0-	225	-0-
MAINTENANCE EQUIPMENT:					
SHOP TOOLS & DUCK BOAT	4,250	20	-0-	212	-0-
UTILITY TRACTOR W / MOWER	18,125	15	18,125	1,208	12,083
ACCESS DRIVE PAVING	10,000	20	-0-	500	-0-
SEPTAGE RECEIVING STATION (BASIC)	9,500	30	-0-	317	3,167
CONSTRUCTION SIGN	750	N/A	-0-	-0-	-0-
DEMOLITION OF EXISTING STP	5,000	N/A	-0-	-0-	-0-
CONTINGENCY @ 10%	68,132	N/A	-0-	-0-	-0-
ESTIMATED CONSTRUCTION	**$749,457**	**TOTALS**	**$51,625**	**$19,075**	**▲$305,683**
ENGINEERING SERVICES	89,935	N/A	-0-	-0-	-0-
LEGAL	5,000	N/A	-0-	-0-	-0-
ADMINISTRATION	3,500	N/A	-0-	-0-	-0-
CONSTRUCTION INSPECTION & STAKING	45,000	N/A	-0-	-0-	-0-
TESTING & LAB FEES	4,000	N/A	-0-	-0-	-0-
LAND PURCHASE	50,000	N/A	-0-	-0-	-0-
INITIAL ESTIMATED COST (TOTAL)	**$946,892**	—	—	—	**▲$73,656**

▲ DOES NOT INCLUDE LAND VALUE

Source: Morgan & Parmley, Ltd.

贷款偿还计划

Loan Payment Schedule

Monthly Payment Necessary on a $1,000 Loan

Years	5%	5½%	6%	6½%	7%	7½%	8%	8½%	9%	9½%
1	$85.61	$85.84	$86.07	$86.30	$86.53	$86.76	$86.99	$87.22	$87.45	$87.68
2	43.87	44.10	44.32	44.55	44.77	45.00	45.22	45.45	45.67	45.90
3	29.97	30.20	30.42	30.65	30.88	31.11	31.34	31.57	31.80	32.02
4	23.03	23.26	23.49	23.72	23.95	24.18	24.42	24.65	24.89	25.11
5	18.87	19.10	19.33	19.57	19.80	20.04	20.27	20.50	20.74	20.97
6	16.11	16.34	16.57	16.81	17.05	17.29	17.52	17.76	18.00	18.23
7	14.13	14.37	14.61	14.85	15.09	15.34	15.59	15.83	16.08	16.32
8	12.66	12.90	13.14	13.39	13.64	13.89	14.14	14.39	14.63	14.88
9	11.52	11.76	12.01	12.26	12.51	12.76	13.01	13.27	13.52	13.77
10	10.61	10.85	11.10	11.36	11.60	11.87	12.13	12.39	12.64	12.90
11	9.86	10.11	10.37	10.62	10.89	11.15	11.41	11.66	11.92	12.18
12	9.25	9.50	9.76	10.02	10.28	10.55	10.80	11.06	11.32	11.59
13	8.73	8.99	9.24	9.51	9.78	10.05	10.32	10.58	10.85	11.11
14	8.29	8.55	8.81	9.08	9.35	9.63	9.89	10.16	10.43	10.70
15	7.91	8.17	8.44	8.71	8.99	9.27	9.56	9.84	10.10	10.38
16	7.58	7.84	8.11	8.39	8.67	8.96	9.24	9.51	9.78	10.06
17	7.29	7.56	7.83	8.11	8.40	8.69	8.96	9.24	9.51	9.79
18	7.03	7.30	7.58	7.87	8.16	8.45	8.74	9.02	9.30	9.49
19	6.80	7.08	7.36	7.65	7.94	8.24	8.50	8.78	9.06	9.35
20	6.60	6.88	7.16	7.46	7.75	8.06	8.34	8.63	8.92	9.22
25	5.85	6.14	6.44	6.75	7.07	7.39	7.69	8.00	8.30	8.61
30	5.37	5.68	6.00	6.32	6.65	6.99	7.29	7.60	7.93	8.24
35	5.05	5.37	5.70	6.04	6.39	6.74	7.06	7.40	7.74	8.08
40	4.82	5.16	5.50	5.85	6.21	6.58	6.90	7.25	7.60	7.96

Example: $4,100 Loan @ 7% for 10 years

Multiply $\frac{4,100}{1,000}$ X 11.60 = $47.56 per month

给定本金的复利

Compound Interest Amount of a Given Principal

The amount A at the end of n years of a given principal P placed at compound interest to-day is A = P × x, the interest (at the rate of r percent per annum) is compounded annually, the factor x being taken from the following tables. Values of x.

Years	r = 4	5	6	7	8	9	10	11	12
2	1.082	1.102	1.124	1.145	1.166	1.188	1.210	1.232	1.254
3	1.125	1.158	1.191	1.225	1.260	1.295	1.331	1.368	1.405
4	1.170	1.216	1.262	1.311	1.360	1.412	1.464	1.518	1.574
5	1.217	1.276	1.338	1.403	1.469	1.539	1.611	1.685	1.762
6	1.265	1.340	1.419	1.501	1.587	1.677	1.772	1.870	1.974
7	1.316	1.407	1.504	1.606	1.714	1.828	1.949	2.076	2.211
8	1.369	1.477	1.594	1.718	1.851	1.993	2.144	2.305	2.476
9	1.423	1.551	1.689	1.838	1.999	2.172	2.358	2.558	2.773
10	1.480	1.629	1.791	1.967	2.159	2.367	2.594	2.839	3.106
11	1.539	1.710	1.898	2.105	2.332	2.580	2.853	3.152	3.479
12	1.601	1.796	2.012	2.252	2.518	2.813	3.138	3.498	3.896
13	1.665	1.886	2.133	2.410	2.720	3.066	3.452	3.883	4.363
14	1.732	1.980	2.261	2.579	2.937	3.342	3.797	4.310	4.887
15	1.801	2.079	2.397	2.759	3.172	3.642	4.177	4.785	5.474
16	1.873	2.183	2.540	2.952	3.426	3.970	4.595	5.311	6.130
17	1.948	2.292	2.693	3.159	3.700	4.328	5.054	5.895	6.866
18	2.026	2.407	2.854	3.380	3.996	4.717	5.560	6.543	7.690
19	2.107	2.527	3.026	3.616	4.316	5.142	6.116	7.263	8.613
20	2.191	2.653	3.207	3.870	4.661	5.604	6.727	8.062	9.646
25	2.666	3.386	4.292	5.427	6.848	8.623	10.834	13.585	17.000
30	3.243	4.322	5.743	7.612	10.062	13.267	17.449	22.892	29.960
40	4.801	7.040	10.285	14.974	21.724	31.408	45.258	64.999	93.049

This table computed from the formula

$$x = [1 + (r/100)]^n$$

年金总计

Amount of an Annuity

The amount S accumulated at the end of n years by a given annual payment Y set aside at the end of each year is S=Y × v, where the factor v is to be taken from the following table. (Interest at r percent per annum, compounded annually.) Values of v.

Years	r = 4	5	6	7	8	9	10	11	12
2	2.040	2.050	2.060	2.070	2.080	2.090	2.100	2.110	2.120
3	3.122	3.152	3.184	3.215	3.246	3.278	3.310	3.342	3.374
4	4.246	4.310	4.375	4.440	4.506	4.573	4.641	4.710	4.779
5	5.416	5.526	5.637	5.751	5.867	5.985	6.105	6.228	6.353
6	6.633	6.802	6.975	7.153	7.336	7.523	7.716	7.913	8.115
7	7.898	8.142	8.394	8.654	8.923	9.200	9.487	9.783	10.089
8	9.214	9.549	9.897	10.260	10.637	11.028	11.436	11.859	12.300
9	10.583	11.026	11.491	11.978	12.487	13.021	13.579	14.164	14.776
10	12.006	12.578	13.181	13.816	14.486	15.193	15.937	16.722	17.549
11	13.486	14.206	14.971	15.783	16.645	17.560	18.531	19.561	20.654
12	15.026	15.917	16.870	17.888	18.977	20.140	21.384	22.713	24.133
13	16.627	17.712	18.882	20.140	21.495	22.953	24.522	26.211	28.029
14	18.292	19.598	21.015	22.550	24.215	26.019	27.975	30.095	32.392
15	20.023	21.578	23.275	25.129	27.152	29.360	31.772	34.405	37.279
16	21.824	23.657	25.672	27.887	30.324	33.003	35.949	39.189	42.753
17	23.697	25.840	28.212	30.840	33.750	36.973	40.544	44.500	48.883
18	25.645	28.132	30.905	33.998	37.450	41.300	45.598	50.395	55.749
19	27.671	30.538	33.759	37.378	41.446	46.017	51.158	56.939	63.439
20	29.777	33.065	36.785	40.995	45.761	51.159	57.274	64.202	72.052
25	41.645	47.725	54.863	63.247	73.105	84.699	98.345	114.411	133.332
30	56.083	66.436	79.055	94.458	113.281	136.303	164.489	199.017	241.330
40	95.023	20.794	154.755	199.628	259.050	337.869	442.576	581.810	767.079

Formula:

$$v = [\{1 + (r/100)\}^{n} - 1] \div (r/100)$$
$$= (x - 1) \div (r/100)$$

总计为给定值的本金

Principal Which Will Amount to a Given Sum

The principal P, which, if placed at compound interest to-day, will amount to a given sum A at the end of n years is P=A × x′, the interest (at the rate of r percent per annum) is compounded annually, the factor x′ being taken from the following table. Values of x′.

Years	r = 4	5	6	7	8	9	10	11	12
2	0.925	0.907	0.890	0.873	0.857	0.842	0.826	0.812	0.797
3	0.889	0.864	0.840	0.816	0.794	0.772	0.751	0.731	0.712
4	0.855	0.823	0.792	0.763	0.735	0.708	0.683	0.659	0.636
5	0.822	0.784	0.747	0.713	0.681	0.650	0.621	0.593	0.567
6	0.790	0.746	0.705	0.666	0.630	0.596	0.564	0.535	0.507
7	0.760	0.711	0.665	0.623	0.583	0.547	0.513	0.482	0.452
8	0.731	0.677	0.627	0.582	0.540	0.502	0.467	0.434	0.404
9	0.703	0.645	0.592	0.544	0.500	0.460	0.424	0.391	0.361
10	0.676	0.614	0.558	0.508	0.463	0.422	0.386	0.352	0.322
11	0.650	0.585	0.527	0.475	0.429	0.388	0.350	0.317	0.287
12	0.625	0.557	0.497	0.444	0.397	0.356	0.319	0.286	0.257
13	0.601	0.530	0.469	0.415	0.368	0.326	0.290	0.258	0.229
14	0.577	0.505	0.442	0.388	0.340	0.299	0.263	0.232	0.205
15	0.555	0.481	0.417	0.362	0.315	0.275	0.239	0.209	0.183
16	0.534	0.458	0.394	0.339	0.292	0.252	0.218	0.188	0.163
17	0.513	0.436	0.371	0.317	0.270	0.231	0.198	0.170	0.146
18	0.494	0.416	0.350	0.296	0.250	0.212	0.180	0.153	0.130
19	0.475	0.396	0.331	0.277	0.232	0.194	0.164	0.138	0.116
20	0.456	0.377	0.312	0.258	0.215	0.178	0.149	0.124	0.104
25	0.375	0.295	0.233	0.184	0.146	0.116	0.092	0.074	0.059
30	0.308	0.231	0.174	0.131	0.099	0.075	0.057	0.044	0.033
40	0.208	0.142	0.097	0.067	0.046	0.032	0.022	0.015	0.011

Formula:

$$x' = [1 + (r/100)]^{-n} = 1/x$$

总计为给定值的年金(偿债基金)

Annuity Which Will Amount to a Given Sum (Sinking Fund)

The annual payment, Y, which, if set aside at the end of each year, will amount with accumulated interest to a given sum S at the end of n years is Y=S×v′, where the factor v′ is given below. (Interest at r percent per annum, compounded annually.)　Values of v′.

Years	r = 4	5	6	7	8	9	10	11	12
2	0.490	0.488	0.485	0.483	0.481	0.478	0.476	0.474	0.472
3	0.320	0.317	0.314	0.311	0.308	0.305	0.302	0.299	0.296
4	**0.235**	**0.232**	**0.229**	**0.225**	**0.222**	**0.219**	**0.215**	**0.212**	**0.209**
5	0.185	0.181	0.177	0.174	0.170	0.167	0.164	0.161	0.157
6	0.151	0.147	0.143	0.140	0.136	0.133	0.130	0.126	0.123
7	**0.127**	**0.123**	**0.119**	**0.116**	**0.112**	**0.109**	**0.105**	**0.102**	**0.099**
8	0.109	0.105	0.101	0.097	0.094	0.091	0.087	0.084	0.081
9	0.094	0.091	0.087	0.083	0.080	0.077	0.074	0.071	0.068
10	**0.083**	**0.080**	**0.076**	**0.072**	**0.069**	**0.066**	**0.063**	**0.060**	**0.057**
11	0.074	0.070	0.067	0.063	0.060	0.057	0.054	0.051	0.048
12	0.067	0.063	0.059	0.056	0.053	0.050	0.047	0.044	0.041
13	0.060	0.056	0.053	0.050	0.047	0.044	0.041	0.038	0.036
14	0.055	0.051	0.048	0.044	0.041	0.038	0.036	0.033	0.031
15	**0.050**	**0.046**	**0.043**	**0.040**	**0.037**	**0.034**	**0.031**	**0.029**	**0.027**
16	0.046	0.042	0.039	0.036	0.033	0.030	0.028	0.026	0.023
17	0.042	0.039	0.035	0.032	0.030	0.027	0.025	0.022	0.020
18	**0.039**	**0.036**	**0.032**	**0.029**	**0.027**	**0.024**	**0.022**	**0.020**	**0.018**
19	0.036	0.033	0.030	0.027	0.024	0.022	0.020	0.018	0.016
20	0.034	0.030	0.027	0.024	0.022	0.020	0.017	0.016	0.014
25	**0.024**	**0.021**	**0.018**	**0.016**	**0.014**	**0.012**	**0.010**	**0.009**	**0.008**
30	0.018	0.015	0.013	0.011	0.009	0.007	0.006	0.005	0.004
40	0.011	0.008	0.006	0.005	0.004	0.003	0.002	0.002	0.001

Formula:

$$v' = (r/100) + [\{1 + (r/100)\}^{n} - 1] = 1/v$$

年金的现值

Present Worth of an Annuity

The Capital C, which, if placed at interest to-day, will provide for a given annual payment Y for a term of n years before it is exhausted is C=Y×w, where the factor w is given below. (Interest at r percent per annum, compounded annually.)　Values of w.

Years	r = 4	5	6	7	8	9	10	11	12
2	1.886	1.859	1.833	1.808	1.783	1.759	1.736	1.713	1.690
3	2.775	2.723	2.673	2.624	2.577	2.531	2.487	2.444	2.402
4	**3.630**	**3.546**	**3.465**	**3.387**	**3.312**	**3.240**	**3.170**	**3.102**	**3.037**
5	4.452	4.329	4.212	4.100	3.993	3.890	3.791	3.696	3.605
6	5.242	5.076	4.917	4.766	4.623	4.486	4.355	4.231	4.111
7	**6.002**	**5.786**	**5.582**	**5.389**	**5.206**	**5.033**	**4.868**	**4.712**	**4.564**
8	6.733	6.463	6.210	5.971	5.747	5.535	5.335	5.146	4.968
9	7.435	7.108	6.802	6.515	6.247	5.995	5.759	5.537	5.328
10	**8.111**	**7.722**	**7.360**	**7.024**	**6.710**	**6.418**	**6.145**	**5.889**	**5.650**
11	8.760	8.306	7.887	7.499	7.139	6.805	6.495	6.206	5.938
12	9.385	8.863	8.384	7.943	7.536	7.161	6.814	6.492	6.194
13	9.986	9.393	8.853	8.358	7.904	7.487	7.103	6.750	6.424
14	10.563	9.899	9.295	8.745	8.244	7.786	7.367	6.982	6.628
15	**11.118**	**10.380**	**9.712**	**9.108**	**8.559**	**8.061**	**7.606**	**7.191**	**6.811**
16	11.652	10.838	10.106	9.447	8.851	8.313	7.824	7.379	6.974
17	12.166	11.274	10.477	9.763	9.122	8.544	8.022	7.549	7.120
18	**12.659**	**11.689**	**10.828**	**10.059**	**9.372**	**8.756**	**8.201**	**7.702**	**7.250**
19	13.134	12.085	11.158	10.336	9.604	8.950	8.365	7.839	7.366
20	13.590	12.462	11.470	10.594	9.818	9.129	8.514	7.963	7.469
25	**15.622**	**14.094**	**12.783**	**11.654**	**10.675**	**9.823**	**9.077**	**8.422**	**7.843**
30	17.292	15.372	13.765	12.409	11.258	10.274	9.427	8.694	8.055
40	19.793	17.159	15.046	13.332	11.925	10.757	9.779	8.951	8.244

Formula:

$$w = [1 - \{1 + (r/100)]^{-n} \div [r/100] = v/x$$

给定资本金提供的年金

Annuity Provided for by a Given Capital

The annual payment Y provided for a term of n years by a given capital C placed at interest to-day is $Y = C \times w'$. (Interest at r percent per annum, compounded annually; the fund supposed to be exhausted at the end of the term.) Values of w'.

Years	r = 4	5	6	7	8	9	10	11	12
2	0.530	0.538	0.545	0.553	0.561	0.568	0.576	0.584	0.592
3	0.360	0.367	0.374	0.381	0.388	0.395	0.402	0.409	0.416
4	**0.275**	**0.282**	**0.289**	**0.295**	**0.302**	**0.309**	**0.315**	**0.322**	**0.329**
5	0.225	0.231	0.237	0.244	0.250	0.257	0.264	0.271	0.277
6	0.191	0.197	0.203	0.210	0.216	0.223	0.230	0.236	0.243
7	**0.167**	**0.173**	**0.179**	**0.186**	**0.192**	**0.199**	**0.205**	**0.212**	**0.219**
8	0.149	0.155	0.161	0.167	0.174	0.181	0.187	0.194	0.201
9	0.134	0.141	0.147	0.153	0.160	0.167	0.174	0.181	0.188
10	**0.123**	**0.130**	**0.136**	**0.142**	**0.149**	**0.156**	**0.163**	**0.170**	**0.177**
11	0.114	0.120	0.127	0.133	0.140	0.147	0.154	0.161	0.168
12	0.107	0.113	0.119	0.126	0.133	0.140	0.147	0.154	0.161
13	0.100	0.106	0.113	0.120	0.127	0.134	0.141	0.148	0.156
14	0.095	0.101	0.108	0.114	0.121	0.128	0.136	0.143	0.151
15	**0.090**	**0.096**	**0.103**	**0.110**	**0.117**	**0.124**	**0.131**	**0.139**	**0.147**
16	0.086	0.092	0.099	0.106	0.113	0.120	0.128	0.136	0.143
17	0.082	0.089	0.095	0.102	0.110	0.117	0.125	0.132	0.140
18	**0.079**	**0.086**	**0.092**	**0.099**	**0.107**	**0.114**	**0.122**	**0.130**	**0.138**
19	0.076	0.083	0.090	0.097	0.104	0.112	0.120	0.128	0.136
20	0.074	0.080	0.087	0.094	0.102	0.110	0.117	0.126	0.134
25	**0.064**	**0.071**	**0.078**	**0.086**	**0.094**	**0.102**	**0.110**	**0.119**	**0.127**
30	0.058	0.065	0.073	0.081	0.089	0.097	0.106	0.115	0.124
40	0.051	0.058	0.066	0.075	0.084	0.093	0.102	0.112	0.121

Formula:

$$w' = [r/100] \div [1 - \{1 + (r/100)\}^{-n}] = 1/w = v' + (r/100)$$

设　计

第 7 节　图纸封面的组织

例 7

项目设计图的首页是图纸封面，它应当正确地显示待建工程的关键要素。尽管每个专业设计师都有自己安排数据形式的偏好，但是一些特定的内容是应该显示的。记住这一点后，下面的几页用于展现常规的概览。除了本节讨论的基本要素外，在最终出封面图纸时还应了解相应管理机构的特殊要求。

随着读者继续阅读本手册，他们将注意到图纸封面设计的多样性。后文包括大量范围的实际项目设计图将给读者提供丰富的图纸封面形式的资源。

下面提供的典型空白平面图和剖面图纸用于一般参考。

目 录

General Title Sheet Arrangement 图纸封面的一般安排

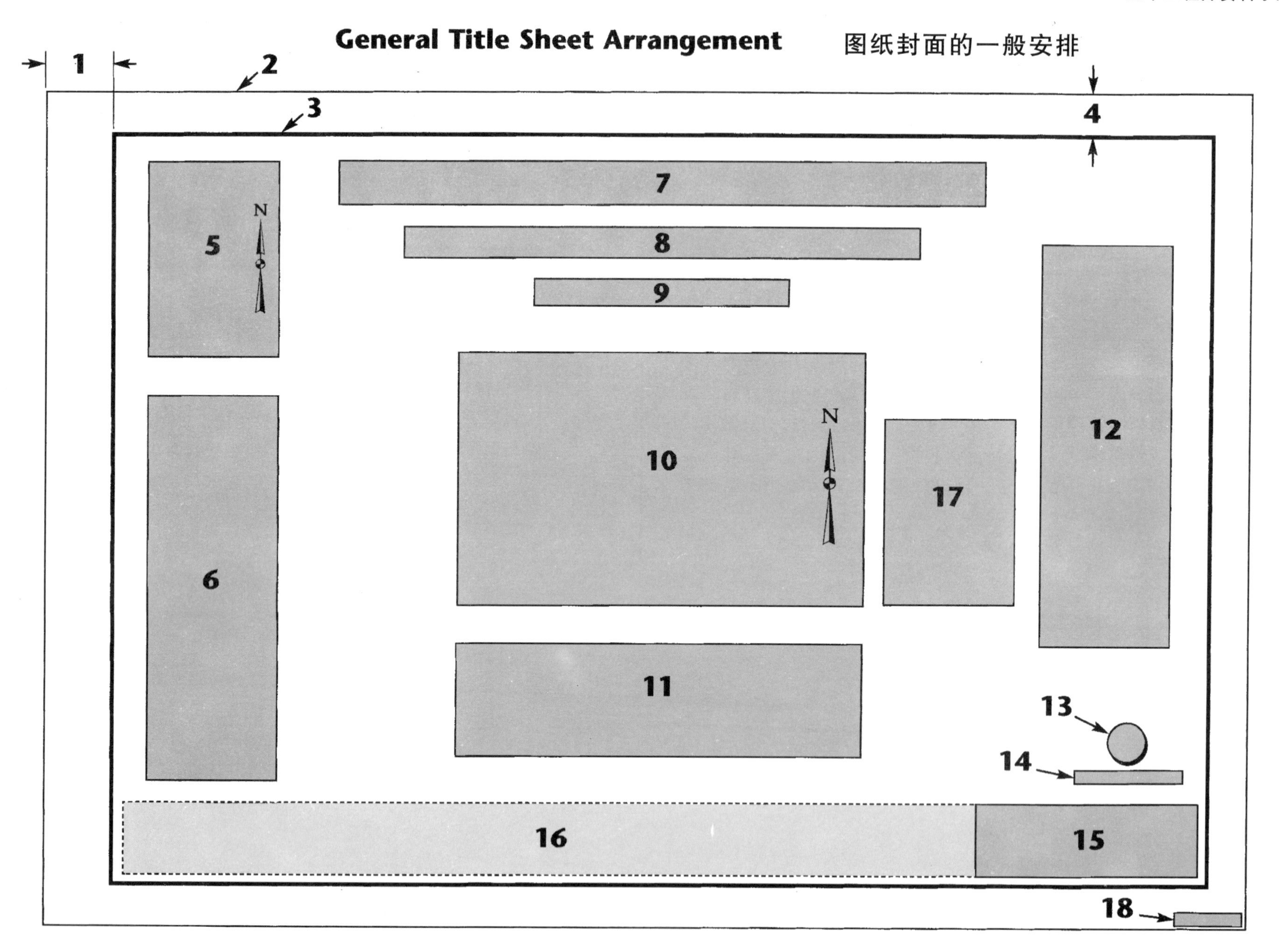

图纸封面组成要素的说明

1 – 装订空间
2 – 图纸边
3 – 边框（通常是粗实线）
4 – 图纸裁剪边界
5 – 总体位置图，突出项目所在场地
6 – 总说明
7 – 项目名称
8 – 项目业主和地址
9 – 项目编号和日期
10 – 总场地平面图，突出建议的工程区域
11 – 特别说明和影响到的公用事业
12 – 图纸目录和页码
13 – 专业封缄处
14 – 专业设计师签名处
15 – 设计公司名称和地址
16 – 罗列补充信息的扩展区域
17 – 图例
18 – 内部编号（如果有）

Typical Blank Plan Sheet 典型空白平面图纸

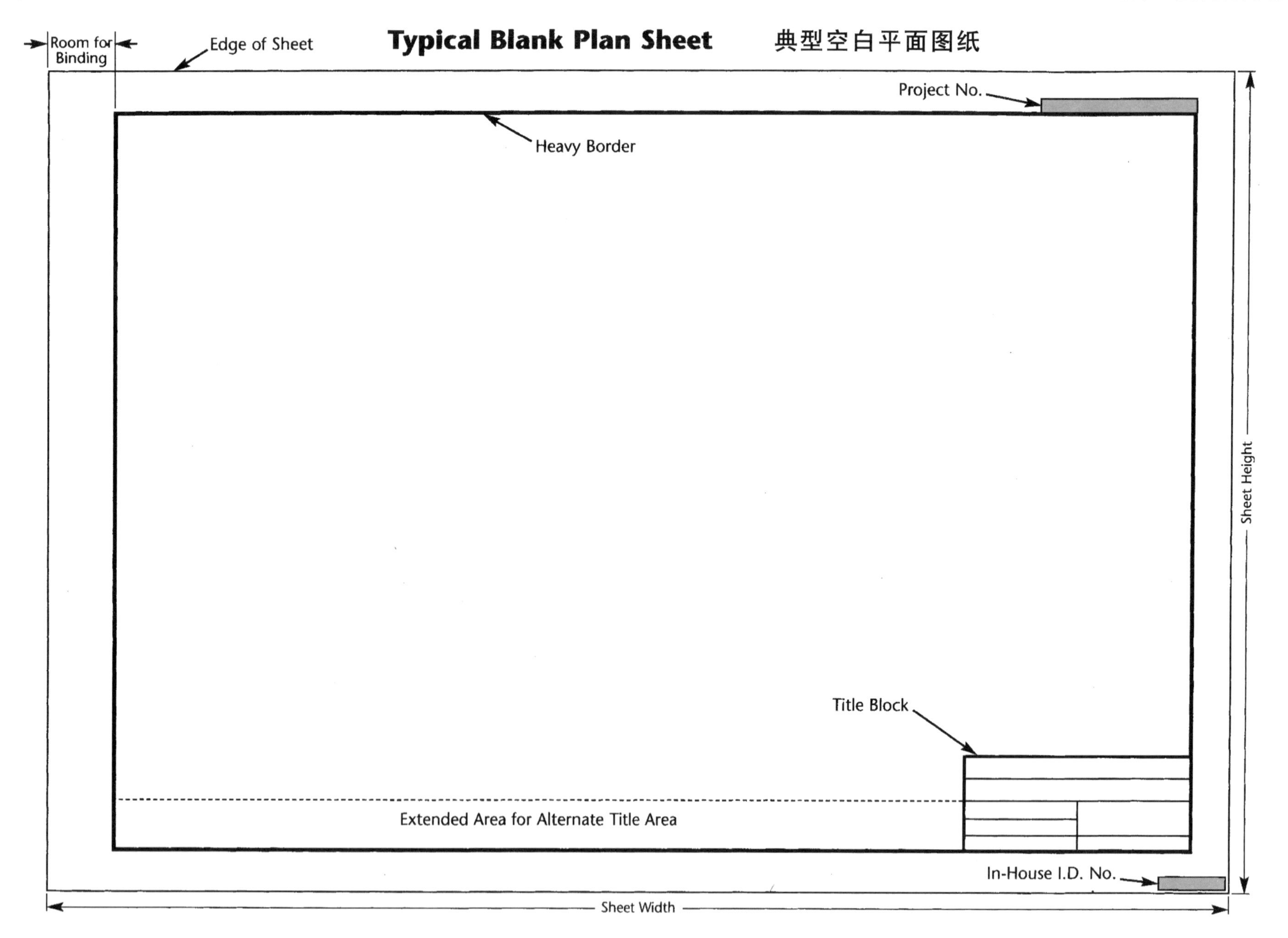

Generic Blank Plan & Profile Sheet　一般平面图和剖面图纸

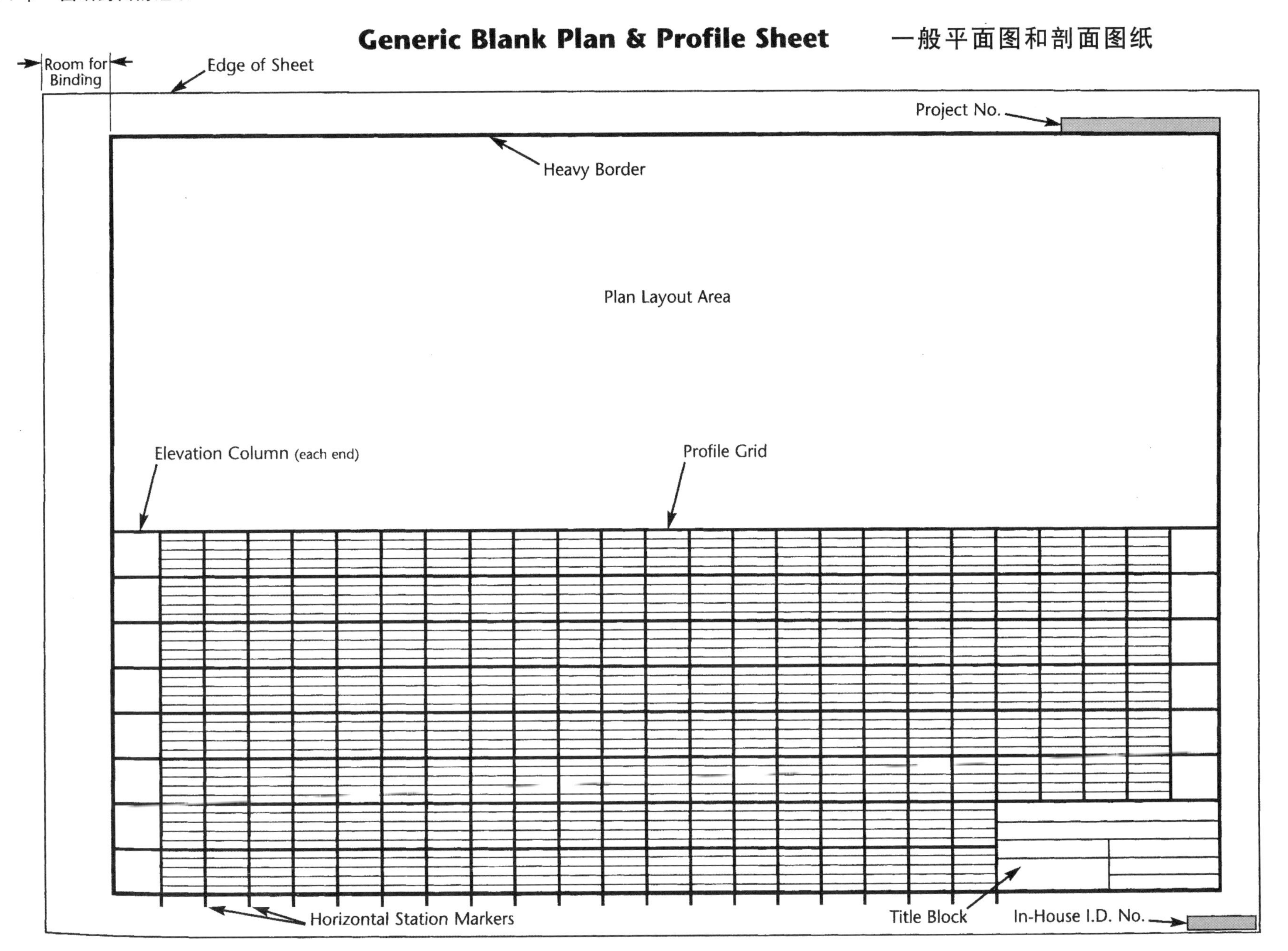

设　计

第8节 房屋建筑设计

例 8

房屋建筑设计是土木工程师工作的主要部分。这些结构包括：城市和乡村会堂、图书馆、服务站、仓库、教堂、学校、消防站、停车结构、泵房、车库、维护建筑、制造综合建筑、商业建筑和私人结构等。

建筑设计图包括配套系统的设计，也就是供暖、通风和空调系统、电力系统、管道、遥测装置和喷淋装置。这些系统的图纸通常都整合在全套的设计图纸中。

下面的设计图示例给出的是随机选取的由土木工程师创作的几类建筑。读者应当注意到，建筑物通常是一个更大型项目的一个较小的组成部分，例如污水处理设施的服务建筑物就是这样，只起到从属的作用。

目 录

乡村会堂和图书馆（布鲁斯乡）

VILLAGE HALL & LIBRARY for the VILLAGE OF BRUCE BRUCE, WISCONSIN

BUILDING VOLUME= 109,200 CU. FT.
BUILDING AREA= 7,360 SQ. FT.

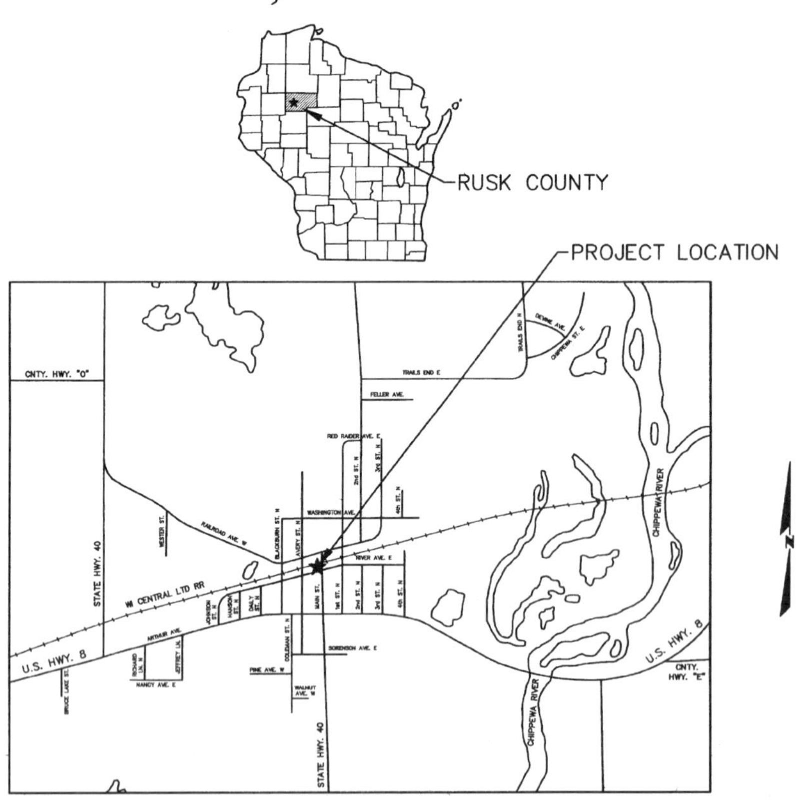

OWNER:

VILLAGE OF BRUCE, RUSK COUNTY
VILLAGE HALL
BRUCE, WI 54819

INDEX OF SHEETS

SHEET NO.	DESCRIPTION
T1	TITLE PAGE
AP1	SITE PLAN
AP2	WEST SITE GRADING PLAN
AP3	EAST SITE GRADING PLAN
AP4	SITE MISCELLANEOUS DETAILS
B1	FLOOR PLAN
B2	BUILDING SECTIONS
B3	SECTIONS
B4	SECTIONS AND DETAILS
B5	RESTROOM DETAILS
B6	BUILDING ELEVATIONS
B7	ROOM FINISH & DOOR SCHEDULE AND DETAILS
B8	SECTIONS AND DETAILS
B9	CEILING PLAN
S1	FOUNDATION PLAN
S2	FOUNDATION DETAILS
S3	ROOF FRAMING PLAN
S4	ROOF FRAMING DETAILS & STRUCTURAL NOTES
P1	PLUMBING SITE PLAN
P2	UNDERGROUND PLUMBING PLAN
P3	PLUMBING FLOOR PLAN
P4	PLUMBING ISOMETRIC DIAGRAMS
M1	HVAC SCHEDULES
M2	HVAC FLOOR PLANS
M3	HVAC PARTIAL PLAN AND DETAILS
E1	ELECTRIC SYMBOLS
E2	ELECTRIC SITE PLAN
E3	LIGHTING PLAN
E4	POWER AND SYSTEMS PLAN
E5	ELECTRIC SCHEDULES

BRUCE VILLAGE HALL & LIBRARY		
BRUCE, WISCONSIN		
SCALE	NONE	MORGAN & PARMLEY, LTD. CONSULTING ENGINEERS LADYSMITH, WISCONSIN
SHEET	T1	
DATE	DEC. 31, 2001	DR. BY J.T.D.

Courtesy: Morgan & Parmley, Ltd.

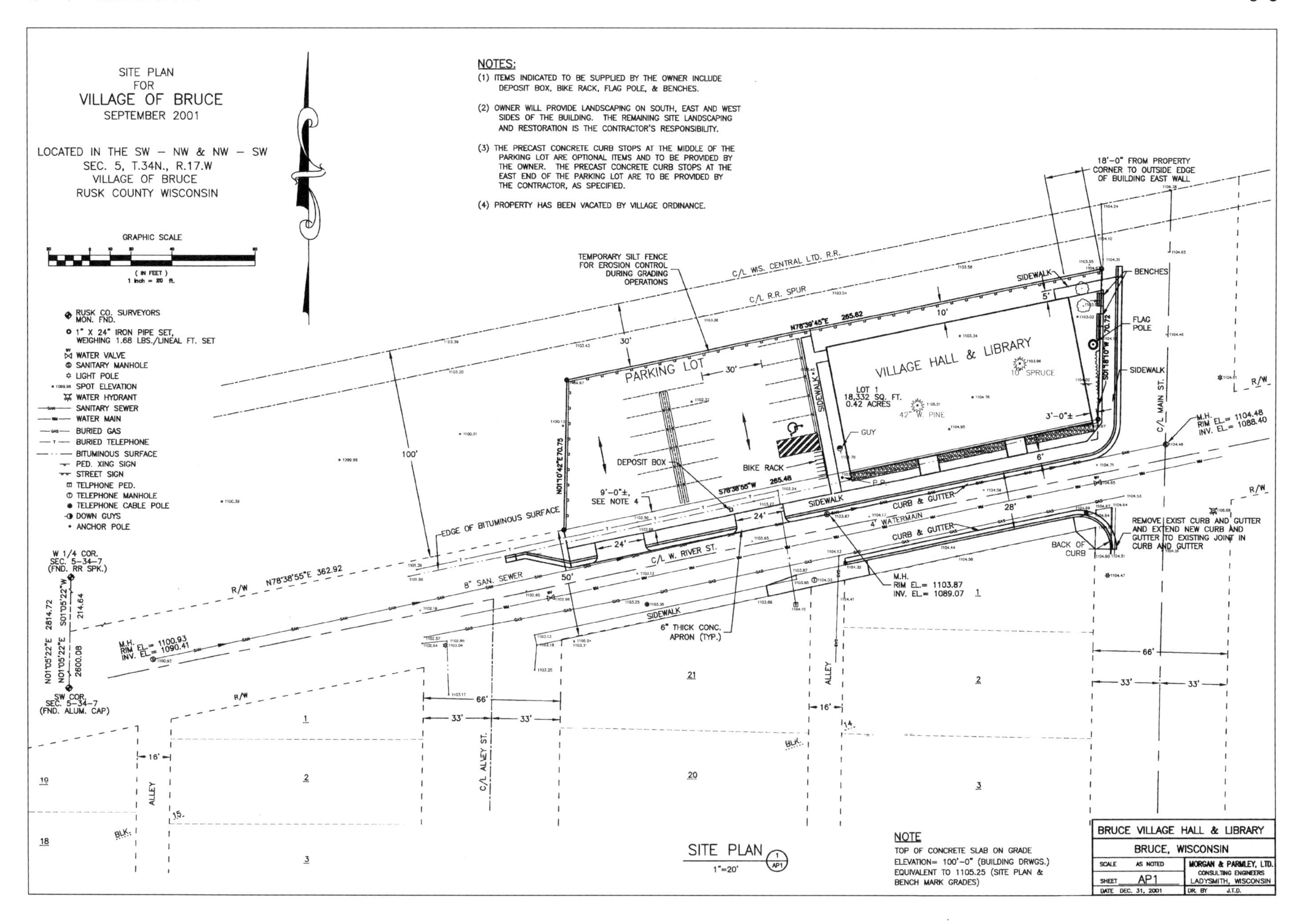
SITE PLAN
FOR
VILLAGE OF BRUCE
SEPTEMBER 2001
LOCATED IN THE SW – NW & NW – SW
SEC. 5, T.34N., R.17.W
VILLAGE OF BRUCE
RUSK COUNTY WISCONSIN
GRAPHIC SCALE
(IN FEET)
1 inch = 20 ft.
NOTES:
(1) ITEMS INDICATED TO BE SUPPLIED BY THE OWNER INCLUDE DEPOSIT BOX, BIKE RACK, FLAG POLE, & BENCHES.
(2) OWNER WILL PROVIDE LANDSCAPING ON SOUTH, EAST AND WEST SIDES OF THE BUILDING. THE REMAINING SITE LANDSCAPING AND RESTORATION IS THE CONTRACTOR'S RESPONSIBILITY.
(3) THE PRECAST CONCRETE CURB STOPS AT THE MIDDLE OF THE PARKING LOT ARE OPTIONAL ITEMS AND TO BE PROVIDED BY THE OWNER. THE PRECAST CONCRETE CURB STOPS AT THE EAST END OF THE PARKING LOT ARE TO BE PROVIDED BY THE CONTRACTOR, AS SPECIFIED.
(4) PROPERTY HAS BEEN VACATED BY VILLAGE ORDINANCE.
RUSK CO. SURVEYORS MON. FND.
1" X 24" IRON PIPE SET, WEIGHING 1.68 LBS./LINEAL FT. SET
WATER VALVE
SANITARY MANHOLE
LIGHT POLE
SPOT ELEVATION
WATER HYDRANT
SANITARY SEWER
WATER MAIN
BURIED GAS
BURIED TELEPHONE
BITUMINOUS SURFACE
PED. XING SIGN
STREET SIGN
TELPHONE PED.
TELEPHONE MANHOLE
TELEPHONE CABLE POLE
DOWN GUYS
ANCHOR POLE
TEMPORARY SILT FENCE FOR EROSION CONTROL DURING GRADING OPERATIONS
C/L WIS. CENTRAL LTD. R.R.
C/L R.R. SPUR
18'-0" FROM PROPERTY CORNER TO OUTSIDE EDGE OF BUILDING EAST WALL
SIDEWALK
BENCHES
FLAG POLE
SIDEWALK
PARKING LOT
VILLAGE HALL & LIBRARY
LOT 1
18,332 SQ. FT.
0.42 ACRES
42" W. PINE
10" SPRUCE
N78°39'45"E 265.62
S01°18'10"W 70.72
N01°10'42"E 70.75
S78°38'55"W 265.48
DEPOSIT BOX
BIKE RACK
GUY
9'-0"±, SEE NOTE 4
EDGE OF BITUMINOUS SURFACE
C/L MAIN ST.
M.H. RIM EL= 1104.48 INV. EL= 1088.40
R/W
CURB & GUTTER
4" WATERMAIN
REMOVE EXIST CURB AND GUTTER AND EXTEND NEW CURB AND GUTTER TO EXISTING JOINT IN CURB AND GUTTER
BACK OF CURB
C/L W. RIVER ST.
8" SAN. SEWER
R/W N78°38'55"E 362.92
M.H. RIM EL= 1103.87 INV. EL= 1089.07
6" THICK CONC. APRON (TYP.)
W 1/4 COR. SEC. 5-34-7 (FND. RR SPK.)
N01°05'22"E 2814.72
S01°05'22"W 214.64
N01°05'22"E 2600.08
M.H. RIM EL= 1100.93 INV. EL= 1090.41
SW COR. SEC. 5-34-7 (FND. ALUM. CAP)
C/L ALLEY ST.
ALLEY
BLK.
NOTE
TOP OF CONCRETE SLAB ON GRADE ELEVATION= 100'-0" (BUILDING DRWGS.) EQUIVALENT TO 1105.25 (SITE PLAN & BENCH MARK GRADES)
SITE PLAN 1"=20'
BRUCE VILLAGE HALL & LIBRARY
BRUCE, WISCONSIN
SCALE AS NOTED
SHEET AP1
DATE DEC. 31, 2001
MORGAN & PARMLEY, LTD. CONSULTING ENGINEERS LADYSMITH, WISCONSIN
DR. BY J.T.D.

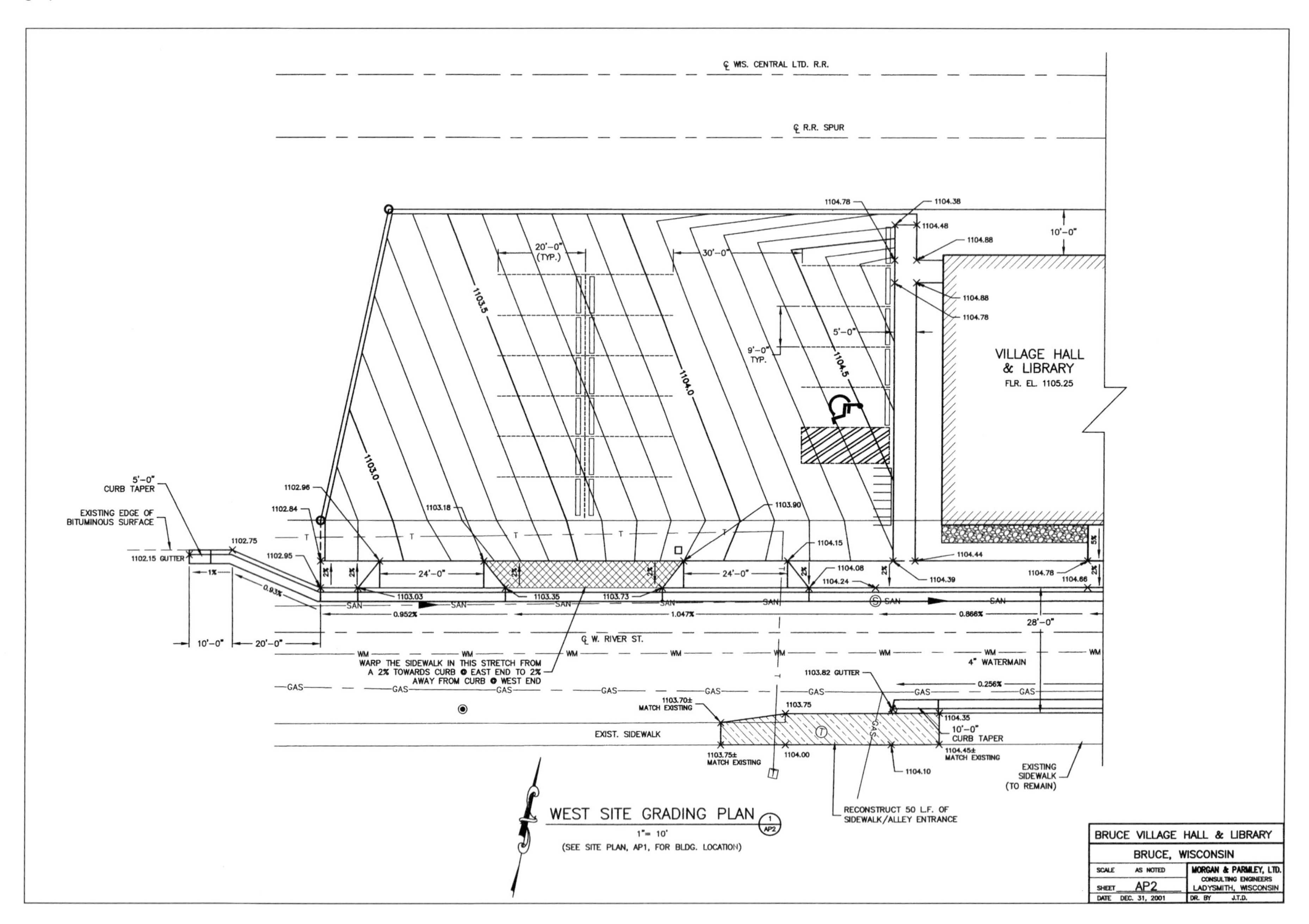

℄ WIS. CENTRAL LTD. R.R.
℄ R.R. SPUR
VILLAGE HALL & LIBRARY
FLR. EL. 1105.25
℄ W. RIVER ST.
EXIST. SIDEWALK
WARP THE SIDEWALK IN THIS STRETCH FROM A 2% TOWARDS CURB @ EAST END TO 2% AWAY FROM CURB @ WEST END
4" WATERMAIN
5'-0" CURB TAPER
EXISTING EDGE OF BITUMINOUS SURFACE
1102.15 GUTTER
10'-0" CURB TAPER
EXISTING SIDEWALK (TO REMAIN)
RECONSTRUCT 50 L.F. OF SIDEWALK/ALLEY ENTRANCE
1103.70± MATCH EXISTING
1103.75± MATCH EXISTING
1104.45± MATCH EXISTING
1103.82 GUTTER
WEST SITE GRADING PLAN
1"= 10'
(SEE SITE PLAN, AP1, FOR BLDG. LOCATION)
BRUCE VILLAGE HALL & LIBRARY
BRUCE, WISCONSIN
SCALE AS NOTED
SHEET AP2
DATE DEC. 31, 2001
MORGAN & PARMLEY, LTD.
CONSULTING ENGINEERS
LADYSMITH, WISCONSIN
DR. BY J.T.D.

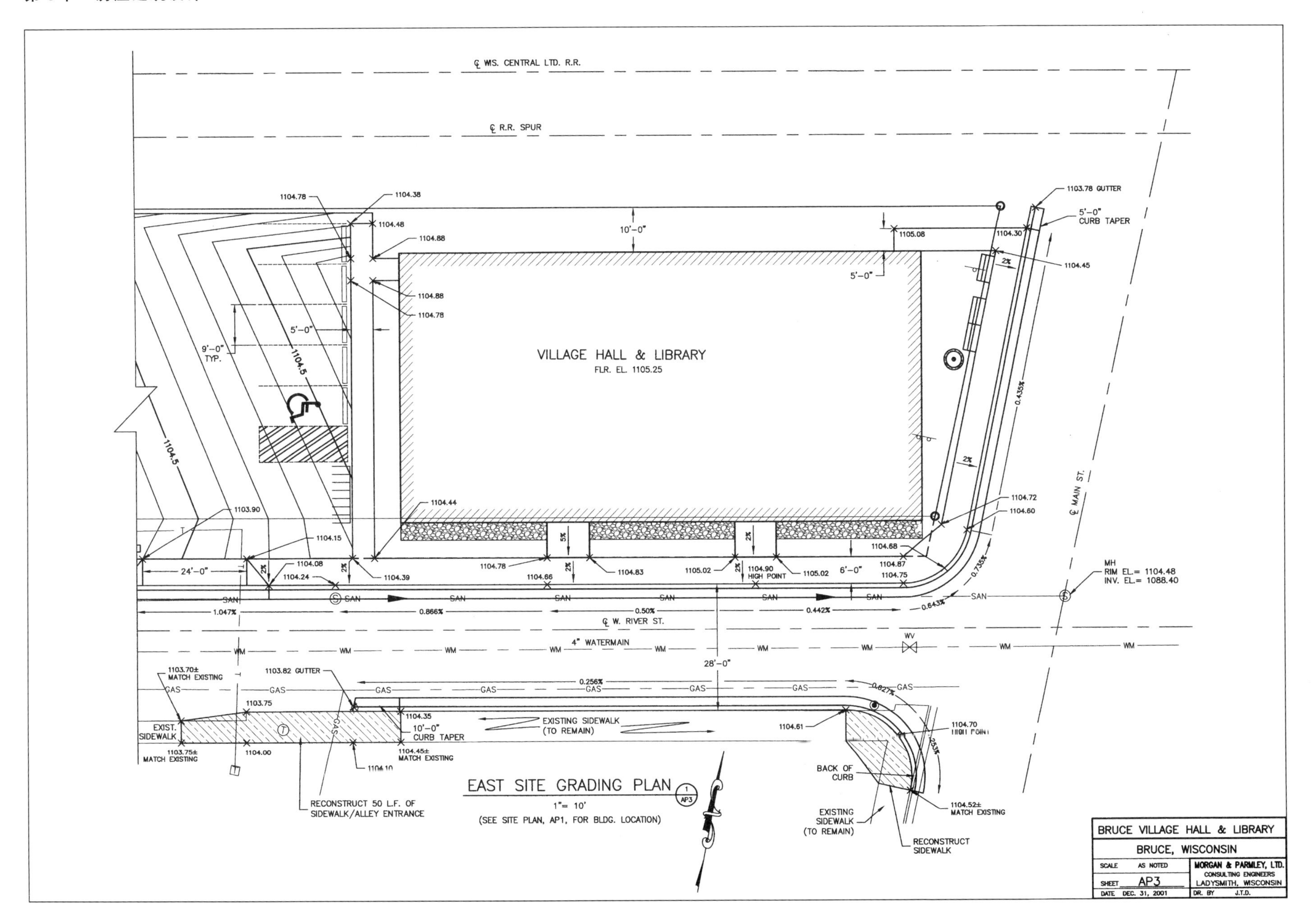

℄ WIS. CENTRAL LTD. R.R.
℄ R.R. SPUR
VILLAGE HALL & LIBRARY
FLR. EL. 1105.25
℄ W. RIVER ST.
4" WATERMAIN
℄ MAIN ST.
MH
RIM EL.= 1104.48
INV. EL.= 1088.40
1103.78 GUTTER
5'-0"
CURB TAPER
EXISTING SIDEWALK
(TO REMAIN)
10'-0"
CURB TAPER
EXIST.
SIDEWALK
1103.70±
MATCH EXISTING
1103.75±
MATCH EXISTING
1104.45±
MATCH EXISTING
1104.52±
MATCH EXISTING
1103.82 GUTTER
1104.70
HIGH POINT
1104.90
HIGH POINT
BACK OF
CURB
EXISTING
SIDEWALK
(TO REMAIN)
RECONSTRUCT
SIDEWALK
RECONSTRUCT 50 L.F. OF
SIDEWALK/ALLEY ENTRANCE
9'-0"
TYP.
EAST SITE GRADING PLAN
1"= 10'
(SEE SITE PLAN, AP1, FOR BLDG. LOCATION)
BRUCE VILLAGE HALL & LIBRARY
BRUCE, WISCONSIN
SCALE AS NOTED
SHEET AP3
DATE DEC. 31, 2001
MORGAN & PARMLEY, LTD.
CONSULTING ENGINEERS
LADYSMITH, WISCONSIN
DR. BY J.T.D.

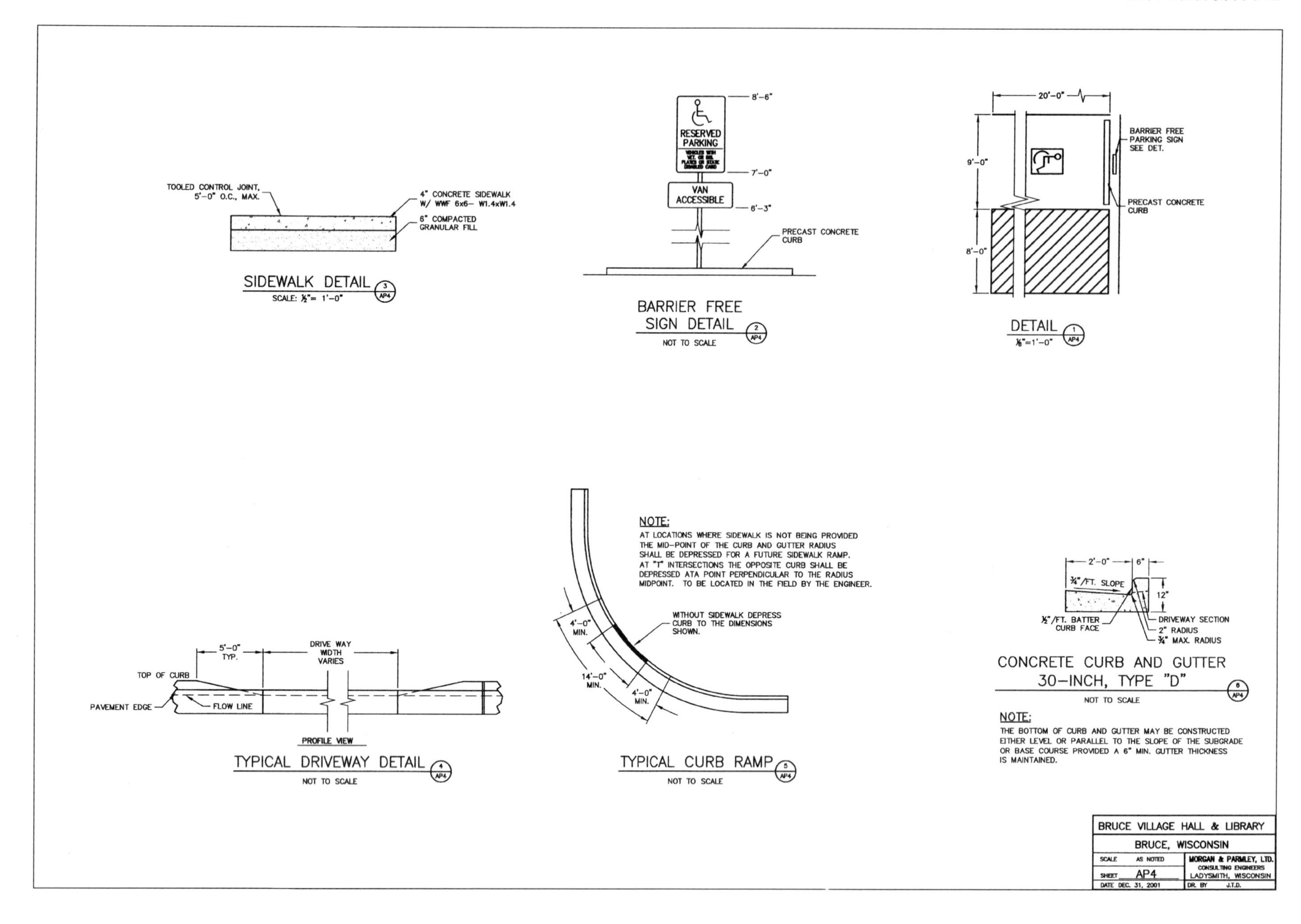

TOOLED CONTROL JOINT, 5'-0" O.C., MAX.
4" CONCRETE SIDEWALK W/ WWF 6x6— W1.4xW1.4
6" COMPACTED GRANULAR FILL
SIDEWALK DETAIL
SCALE: ½"= 1'-0"
3 AP4
RESERVED PARKING
VAN ACCESSIBLE
8'-6"
7'-0"
6'-3"
PRECAST CONCRETE CURB
BARRIER FREE SIGN DETAIL
2 AP4
NOT TO SCALE
20'-0"
9'-0"
8'-0"
BARRIER FREE PARKING SIGN SEE DET.
PRECAST CONCRETE CURB
DETAIL
⅛"=1'-0"
1 AP4
NOTE:
AT LOCATIONS WHERE SIDEWALK IS NOT BEING PROVIDED THE MID-POINT OF THE CURB AND GUTTER RADIUS SHALL BE DEPRESSED FOR A FUTURE SIDEWALK RAMP. AT "T" INTERSECTIONS THE OPPOSITE CURB SHALL BE DEPRESSED ATA POINT PERPENDICULAR TO THE RADIUS MIDPOINT. TO BE LOCATED IN THE FIELD BY THE ENGINEER.
WITHOUT SIDEWALK DEPRESS CURB TO THE DIMENSIONS SHOWN.
4'-0" MIN.
14'-0" MIN.
4'-0" MIN.
TYPICAL CURB RAMP
5 AP4
NOT TO SCALE
5'-0" TYP.
DRIVE WAY WIDTH VARIES
TOP OF CURB
PAVEMENT EDGE
FLOW LINE
PROFILE VIEW
TYPICAL DRIVEWAY DETAIL
4 AP4
NOT TO SCALE
2'-0"
6"
¾"/FT. SLOPE
12"
½"/FT. BATTER CURB FACE
DRIVEWAY SECTION
2" RADIUS
¾" MAX. RADIUS
CONCRETE CURB AND GUTTER 30-INCH, TYPE "D"
6 AP4
NOT TO SCALE
NOTE:
THE BOTTOM OF CURB AND GUTTER MAY BE CONSTRUCTED EITHER LEVEL OR PARALLEL TO THE SLOPE OF THE SUBGRADE OR BASE COURSE PROVIDED A 6" MIN. GUTTER THICKNESS IS MAINTAINED.
BRUCE VILLAGE HALL & LIBRARY
BRUCE, WISCONSIN
SCALE AS NOTED
SHEET AP4
DATE DEC. 31, 2001
MORGAN & PARMLEY, LTD.
CONSULTING ENGINEERS
LADYSMITH, WISCONSIN
DR. BY J.T.D.

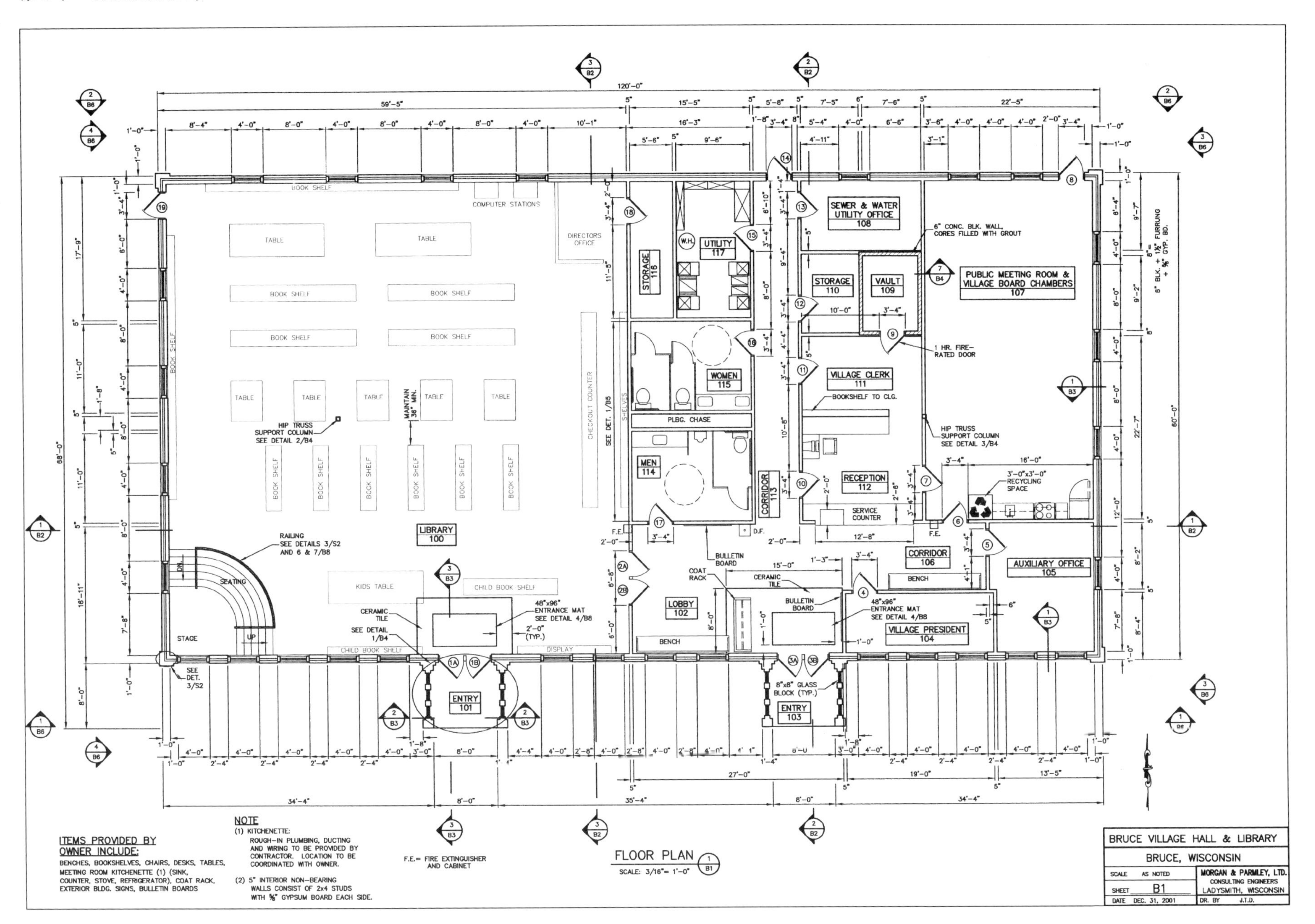
LIBRARY 100
STAGE
SEATING
COMPUTER STATIONS
DIRECTORS OFFICE
CHECKOUT COUNTER
STORAGE 116
UTILITY 117
WOMEN 115
MEN 114
PLBG. CHASE
CORRIDOR 113
SEWER & WATER UTILITY OFFICE 108
STORAGE 110
VAULT 109
PUBLIC MEETING ROOM & VILLAGE BOARD CHAMBERS 107
VILLAGE CLERK 111
RECEPTION 112
CORRIDOR 106
AUXILIARY OFFICE 105
VILLAGE PRESIDENT 104
LOBBY 102
ENTRY 101
ENTRY 103
6" CONC. BLK. WALL, CORES FILLED WITH GROUT
1 HR. FIRE-RATED DOOR
HIP TRUSS SUPPORT COLUMN SEE DETAIL 2/B4
HIP TRUSS SUPPORT COLUMN SEE DETAIL 3/B4
RAILING SEE DETAILS 3/S2 AND 6 & 7/B8
48"x96" ENTRANCE MAT SEE DETAIL 4/B8
8"x8" GLASS BLOCK (TYP.)
3'-0"x3'-0" RECYCLING SPACE
SERVICE COUNTER
BULLETIN BOARD
COAT RACK
CERAMIC TILE
BENCH
120'-0"
60'-0"
68'-0"
ITEMS PROVIDED BY OWNER INCLUDE:
BENCHES, BOOKSHELVES, CHAIRS, DESKS, TABLES, MEETING ROOM KITCHENETTE (1) (SINK, COUNTER, STOVE, REFRIGERATOR), COAT RACK, EXTERIOR BLDG. SIGNS, BULLETIN BOARDS
NOTE
(1) KITCHENETTE: ROUGH-IN PLUMBING, DUCTING AND WIRING TO BE PROVIDED BY CONTRACTOR. LOCATION TO BE COORDINATED WITH OWNER.
(2) 5" INTERIOR NON-BEARING WALLS CONSIST OF 2x4 STUDS WITH 5/8" GYPSUM BOARD EACH SIDE.
F.E.= FIRE EXTINGUISHER AND CABINET
FLOOR PLAN
SCALE: 3/16"= 1'-0"
BRUCE VILLAGE HALL & LIBRARY
BRUCE, WISCONSIN
SCALE AS NOTED
SHEET B1
DATE DEC. 31, 2001
MORGAN & PARMLEY, LTD.
CONSULTING ENGINEERS
LADYSMITH, WISCONSIN
DR. BY J.T.D.

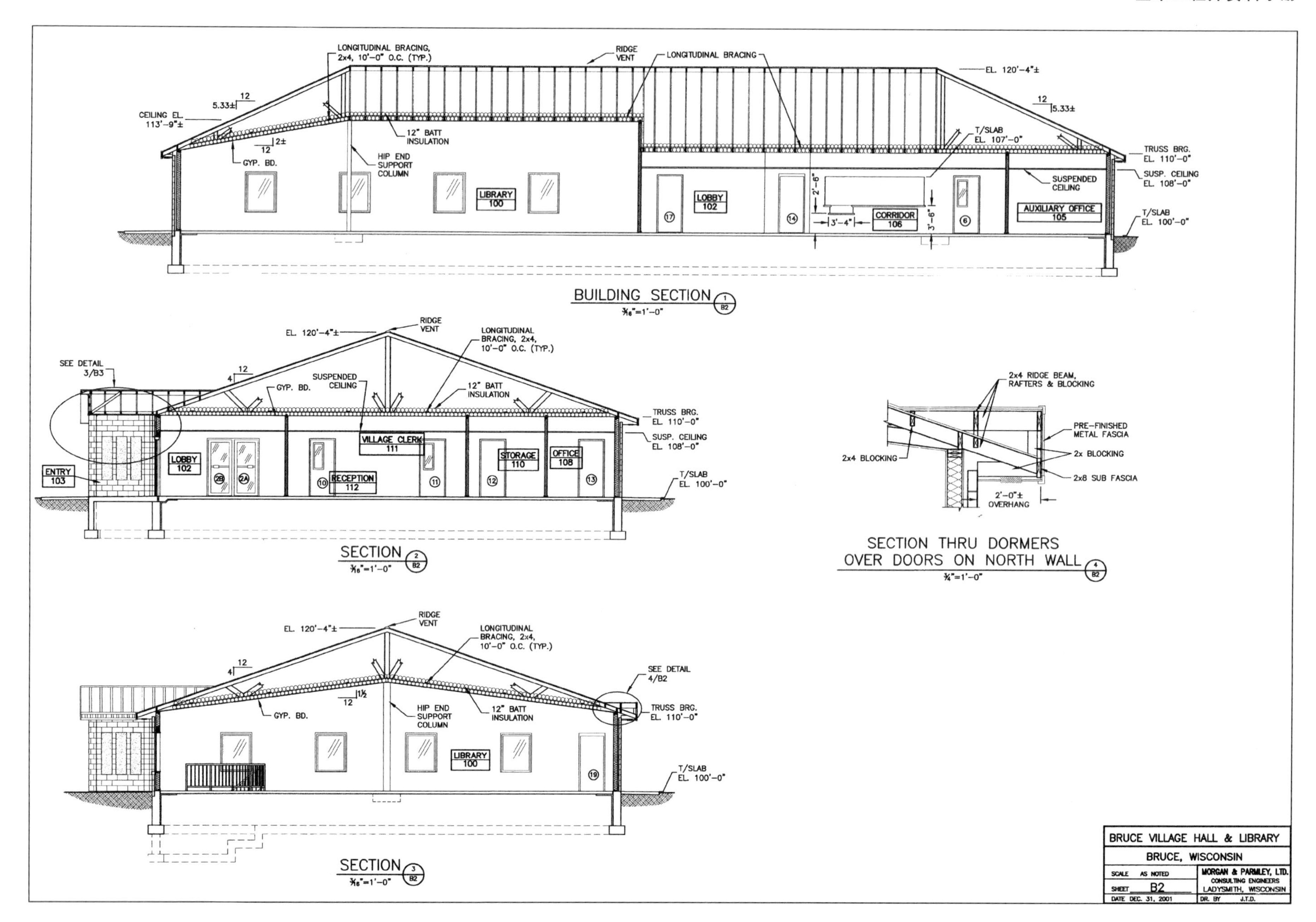
LONGITUDINAL BRACING, 2x4, 10'-0" O.C. (TYP.)
RIDGE VENT
LONGITUDINAL BRACING
EL. 120'-4"±
CEILING EL. 113'-9"±
12" BATT INSULATION
GYP. BD.
HIP END SUPPORT COLUMN
T/SLAB EL. 107'-0"
TRUSS BRG. EL. 110'-0"
SUSPENDED CEILING
SUSP. CEILING EL. 108'-0"
T/SLAB EL. 100'-0"
LIBRARY 100
LOBBY 102
CORRIDOR 106
AUXILIARY OFFICE 105
BUILDING SECTION
3/16"=1'-0"
SEE DETAIL 3/B3
SUSPENDED CEILING
ENTRY 103
VILLAGE CLERK 111
RECEPTION 112
STORAGE 110
OFFICE 108
SECTION
2x4 RIDGE BEAM, RAFTERS & BLOCKING
PRE-FINISHED METAL FASCIA
2x BLOCKING
2x4 BLOCKING
2x8 SUB FASCIA
2'-0"± OVERHANG
SECTION THRU DORMERS OVER DOORS ON NORTH WALL
3/4"=1'-0"
SEE DETAIL 4/B2
SECTION
BRUCE VILLAGE HALL & LIBRARY
BRUCE, WISCONSIN
SCALE AS NOTED
SHEET B2
DATE DEC. 31, 2001
MORGAN & PARMLEY, LTD.
CONSULTING ENGINEERS
LADYSMITH, WISCONSIN
DR. BY J.T.D.

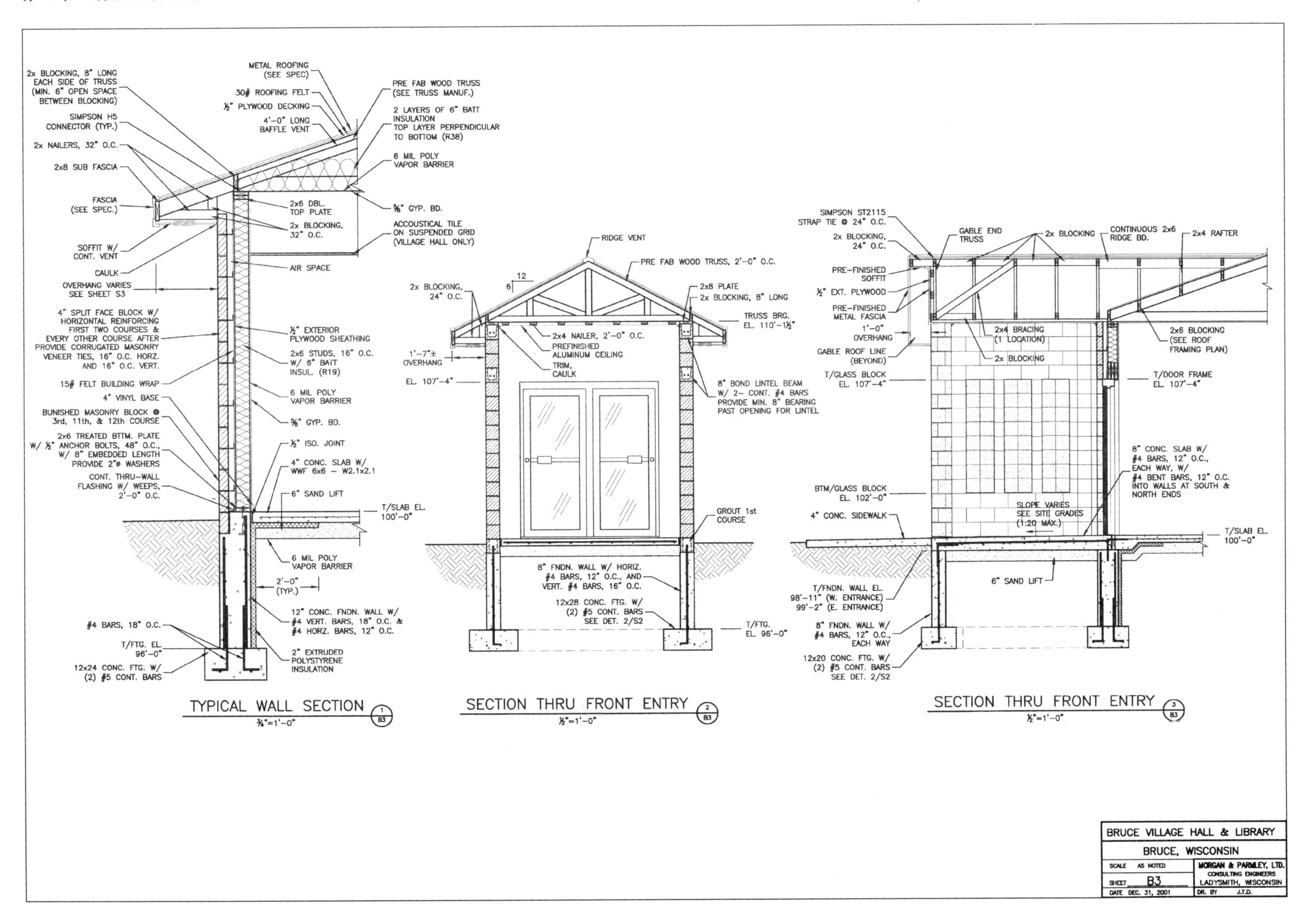
TYPICAL WALL SECTION
¾"=1'-0"
SECTION THRU FRONT ENTRY
½"=1'-0"
SECTION THRU FRONT ENTRY
½"=1'-0"
BRUCE VILLAGE HALL & LIBRARY
BRUCE, WISCONSIN
MORGAN & PARMLEY, LTD.
CONSULTING ENGINEERS
LADYSMITH, WISCONSIN
SHEET B3
DATE DEC. 31, 2001

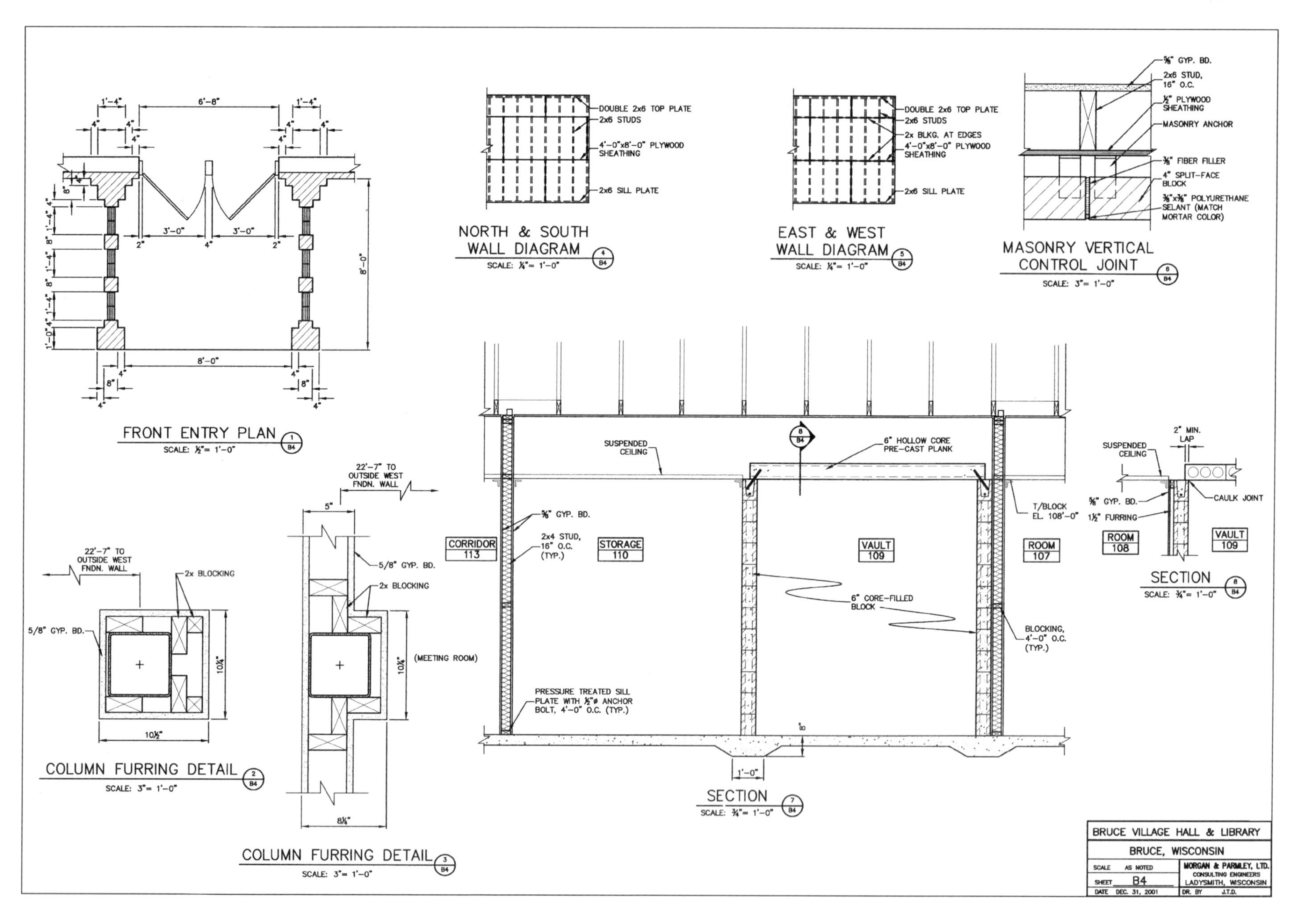
FRONT ENTRY PLAN
SCALE: ½"= 1'-0"
NORTH & SOUTH WALL DIAGRAM
SCALE: ¼"= 1'-0"
DOUBLE 2x6 TOP PLATE
2x6 STUDS
4'-0"x8'-0" PLYWOOD SHEATHING
2x6 SILL PLATE
EAST & WEST WALL DIAGRAM
2x BLKG. AT EDGES
MASONRY VERTICAL CONTROL JOINT
SCALE: 3"= 1'-0"
⅝" GYP. BD.
2x6 STUD, 16" O.C.
½" PLYWOOD SHEATHING
MASONRY ANCHOR
⅜" FIBER FILLER
4" SPLIT-FACE BLOCK
⅜"x⅜" POLYURETHANE SELANT (MATCH MORTAR COLOR)
COLUMN FURRING DETAIL
SCALE: 3"= 1'-0"
22'-7" TO OUTSIDE WEST FNDN. WALL
2x BLOCKING
5/8" GYP. BD.
(MEETING ROOM)
SUSPENDED CEILING
6" HOLLOW CORE PRE-CAST PLANK
CORRIDOR 113
STORAGE 110
VAULT 109
ROOM 107
ROOM 108
2x4 STUD, 16" O.C. (TYP.)
6" CORE-FILLED BLOCK
T/BLOCK EL. 108'-0"
BLOCKING, 4'-0" O.C. (TYP.)
PRESSURE TREATED SILL PLATE WITH ½"ø ANCHOR BOLT, 4'-0" O.C. (TYP.)
SECTION
SCALE: ¾"= 1'-0"
2" MIN. LAP
CAULK JOINT
1½" FURRING
BRUCE VILLAGE HALL & LIBRARY
BRUCE, WISCONSIN
SCALE AS NOTED
SHEET B4
DATE DEC. 31, 2001
MORGAN & PARMLEY, LTD. CONSULTING ENGINEERS LADYSMITH, WISCONSIN
DR. BY J.T.D.

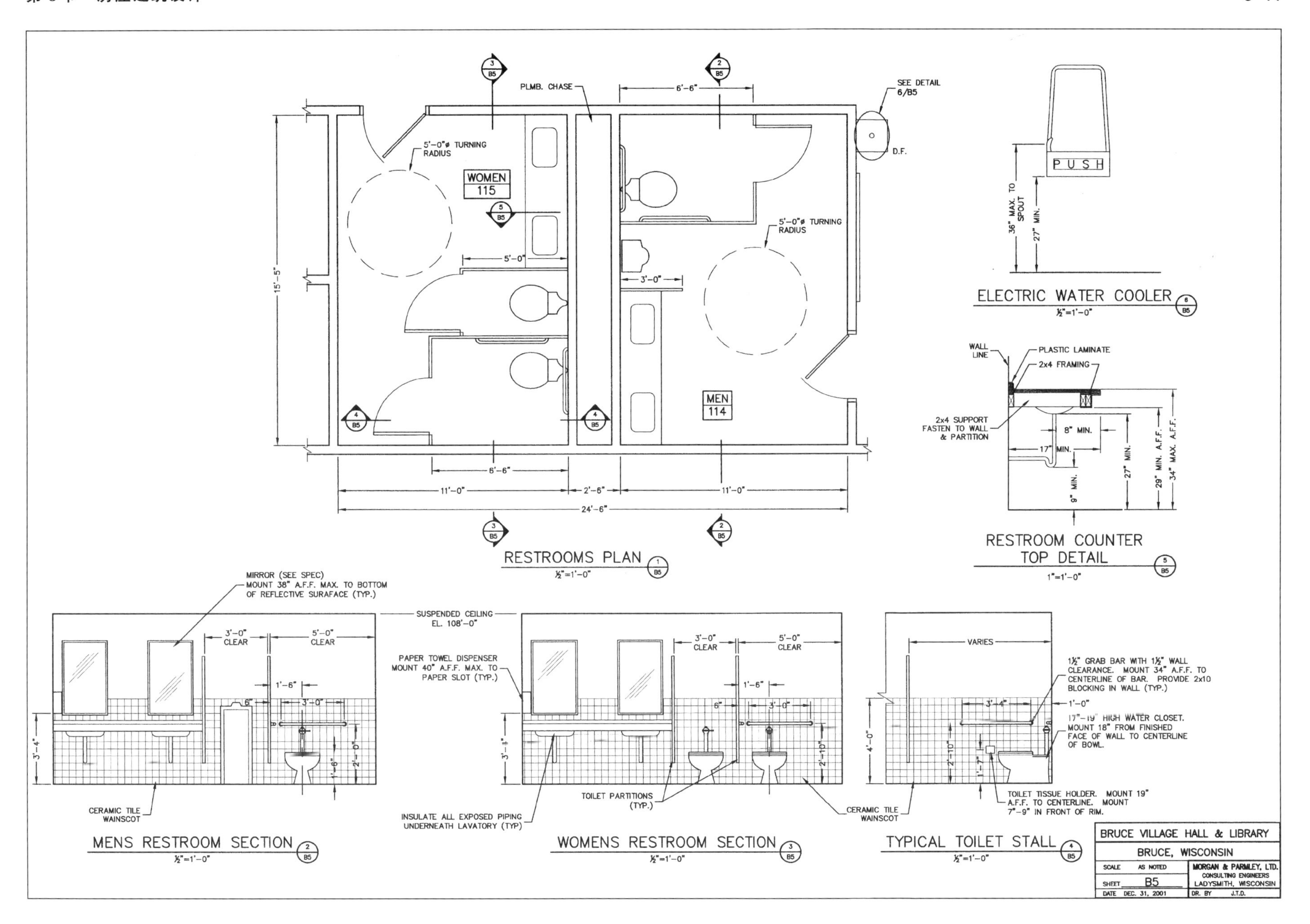
PLMB. CHASE
6'-6"
SEE DETAIL 6/B5
D.F.
5'-0"ø TURNING RADIUS
WOMEN 115
MEN 114
5'-0"
3'-0"
15'-5"
11'-0"
2'-6"
24'-6"
RESTROOMS PLAN
½"=1'-0"
ELECTRIC WATER COOLER
PUSH
36" MAX. TO SPOUT
27" MIN.
WALL LINE
PLASTIC LAMINATE
2x4 FRAMING
2x4 SUPPORT FASTEN TO WALL & PARTITION
8" MIN.
17" MIN.
9" MIN.
29" MIN. A.F.F.
34" MAX. A.F.F.
RESTROOM COUNTER TOP DETAIL
1"=1'-0"
MIRROR (SEE SPEC) MOUNT 38" A.F.F. MAX. TO BOTTOM OF REFLECTIVE SURAFACE (TYP.)
SUSPENDED CEILING EL. 108'-0"
PAPER TOWEL DISPENSER MOUNT 40" A.F.F. MAX. TO PAPER SLOT (TYP.)
3'-0" CLEAR
5'-0" CLEAR
1'-6"
VARIES
1½" GRAB BAR WITH 1½" WALL CLEARANCE. MOUNT 34" A.F.F. TO CENTERLINE OF BAR. PROVIDE 2x10 BLOCKING IN WALL (TYP.)
17"-19" HIGH WATER CLOSET. MOUNT 18" FROM FINISHED FACE OF WALL TO CENTERLINE OF BOWL.
TOILET TISSUE HOLDER. MOUNT 19" A.F.F. TO CENTERLINE. MOUNT 7"-9" IN FRONT OF RIM.
CERAMIC TILE WAINSCOT
TOILET PARTITIONS (TYP.)
INSULATE ALL EXPOSED PIPING UNDERNEATH LAVATORY (TYP)
MENS RESTROOM SECTION
WOMENS RESTROOM SECTION
TYPICAL TOILET STALL
BRUCE VILLAGE HALL & LIBRARY
BRUCE, WISCONSIN
SCALE AS NOTED
SHEET B5
DATE DEC. 31, 2001
MORGAN & PARMLEY, LTD.
CONSULTING ENGINEERS
LADYSMITH, WISCONSIN
DR. BY J.T.D.

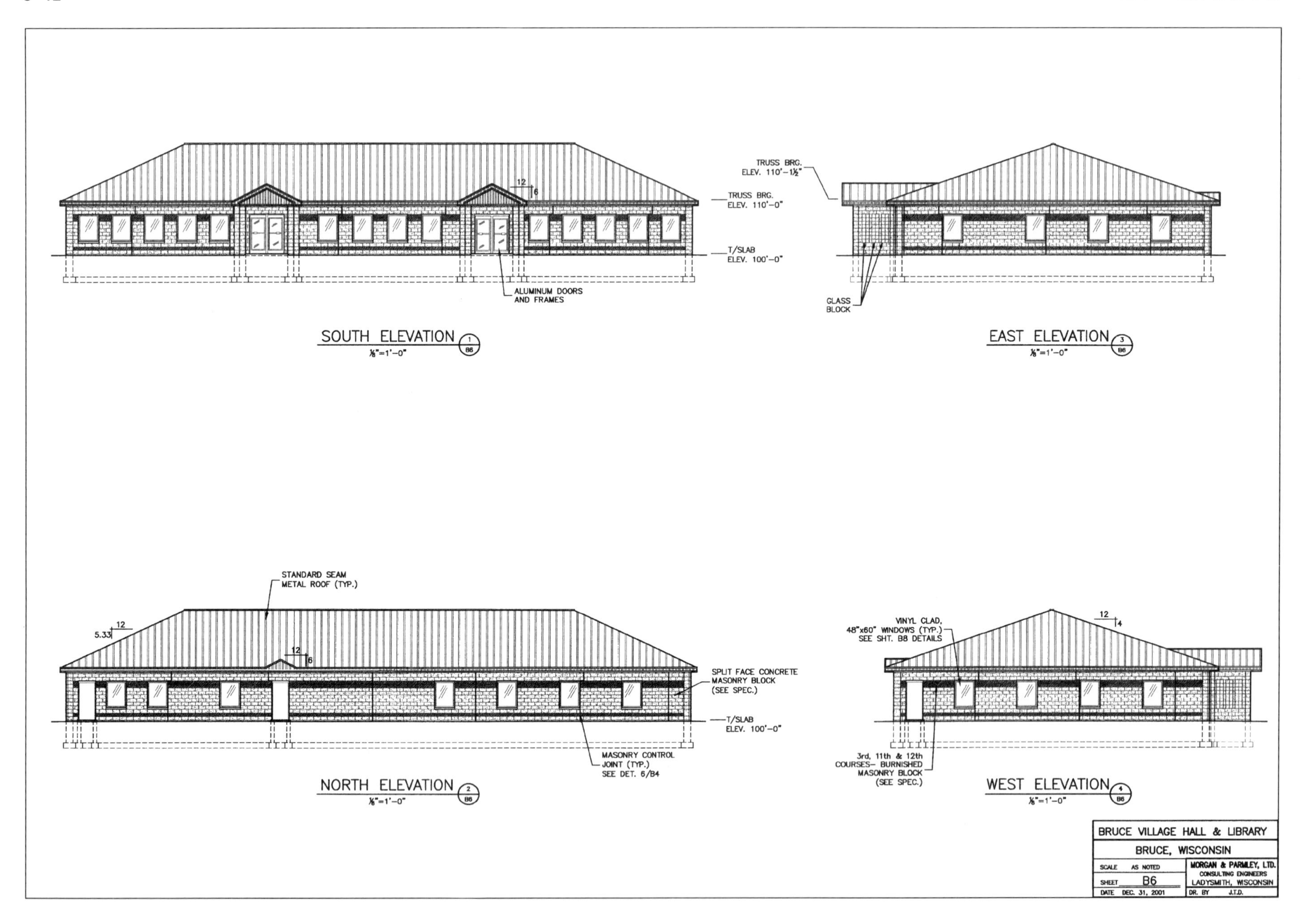

TRUSS BRG. ELEV. 110'-1½"
TRUSS BRG. ELEV. 110'-0"
T/SLAB ELEV. 100'-0"
ALUMINUM DOORS AND FRAMES
GLASS BLOCK
SOUTH ELEVATION
⅛"=1'-0"
EAST ELEVATION
⅛"=1'-0"
STANDARD SEAM METAL ROOF (TYP.)
SPLIT FACE CONCRETE MASONRY BLOCK (SEE SPEC.)
T/SLAB ELEV. 100'-0"
MASONRY CONTROL JOINT (TYP.) SEE DET. 6/B4
NORTH ELEVATION
⅛"=1'-0"
VINYL CLAD, 48"x60" WINDOWS (TYP.) SEE SHT. B8 DETAILS
3rd, 11th & 12th COURSES– BURNISHED MASONRY BLOCK (SEE SPEC.)
WEST ELEVATION
⅛"=1'-0"
BRUCE VILLAGE HALL & LIBRARY
BRUCE, WISCONSIN
SCALE AS NOTED
SHEET B6
DATE DEC. 31, 2001
MORGAN & PARMLEY, LTD.
CONSULTING ENGINEERS
LADYSMITH, WISCONSIN
DR. BY J.T.D.

DOOR SCHEDULE								
DOOR NO.	DOOR SIZE			DOOR		FRAME		NOTES
	WIDTH	HEIGHT	THICKNESS	MATERIAL	TYPE	MATERIAL	TYPE	
1A & 1B	3'-0"	7'-0"	THK.	ALUM./GLASS	C	ALUM.	B	FRONT ENTRANCE, EXT.
2A & 2B	3'-0"	7'-0"	THK.	ALUM./GLASS	C	ALUM.	B	LIBRARY, INT.
3A & 3B	3'-0"	7'-0"	THK.	ALUM./GLASS	C	ALUM.	B	FRONT ENTRANCE, EXT.
4	3'-0"	7'-0"	THK.	WOOD/GLASS	B	STEEL	A	VILLAGE PRESIDENT, INT.
5	3'-0"	7'-0"	THK.	WOOD/GLASS	B	STEEL	A	ASSESSOR & TOWN CLERK OFFICE, INT.
6	3'-0"	7'-0"	THK.	WOOD/GLASS	B	STEEL	A	PUBLIC MEETING ROOM & VILLAGE BOARD CHAMBERS, INT.
7	3'-0"	7'-0"	THK.	WOOD/GLASS	B	STEEL	A	PUBLIC MEETING ROOM & VILLAGE BOARD CHAMBERS, INT.
8	3'-0"	7'-0"	THK.	STEEL	A	STEEL	C	PUBLIC MEETING ROOM & VILLAGE BOARD CHAMBERS, EXT.
9	3'-0"	7'-0"	THK.	STEEL	A	STEEL	C	VAULT, INT., 1 HR. FIRE RATED
10	3'-0"	7'-0"	THK.	WOOD/GLASS	B	STEEL	A	RECEPTION, INT.
11	3'-0"	7'-0"	THK.	WOOD/GLASS	B	STEEL	A	VILLAGE CLERK, INT.
12	3'-0"	7'-0"	THK.	STEEL	A	STEEL	A	STORAGE, INT.
13	3'-0"	7'-0"	THK.	STEEL	B	STEEL	A	SEWER & WATER UTILITY OFFICE, INT.
14	3'-0"	7'-0"	THK.	STEEL	B	STEEL	C	CORRIDOR, EXT.
15	3'-0"	7'-0"	THK.	STEEL	A	STEEL	A	UTILITY, INT.
16	3'-0"	7'-0"	THK.	WOOD	A	STEEL	A	WOMEN'S BATHROOM, INT., PROVIDE 12"x12" LOUVER
17	3'-0"	7'-0"	THK.	WOOD	A	STEEL	A	MEN'S BATHROOM, INT., PROVIDE 12"x12" LOUVER
18	3'-0"	7'-0"	THK.	STEEL	A	STEEL	A	STORAGE, INT.
19	3'-0"	7'-0"	THK.	STEEL	A	STEEL	C	LIBRARY, EXT.

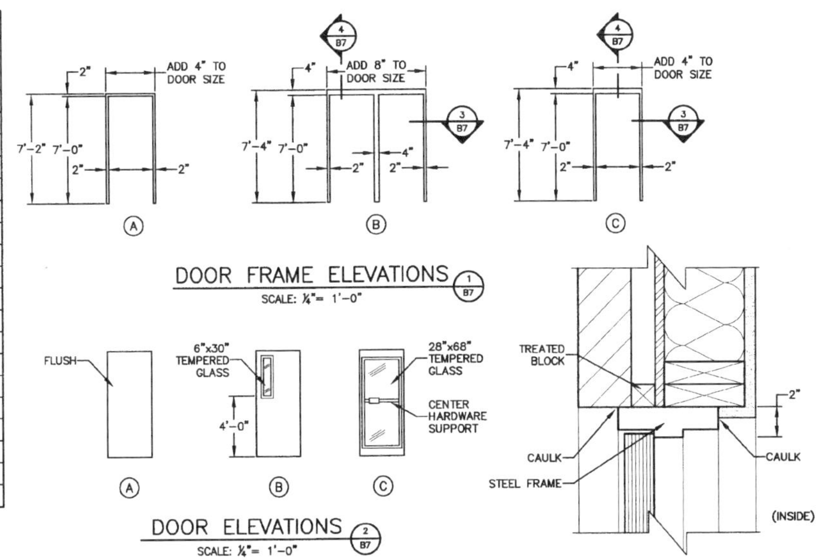

ROOM FINISH SCHEDULE					
ROOM NAME	FLOOR/BASE	WALLS	CEILING	CEILING HEIGHT	NOTES
LIBRARY (100)	CARPET/VINYL COVING	PAINTED GYP. BD.	PAINTED GYP. BD.	10'-0" - 13'-9"	-
ENTRY (101)	SEALED CONC.	CONC. BLK.	METAL SOFFIT	10'-0 1/2"	-
LOBBY (102)	CARPET & CERAMIC TILE/VINYL COVING	PAINTED GYP. BD.	SUSPENDED TILE	8'-0"	-
ENTRY (103)	SEALED CONC.	CONC. BLK.	METAL SOFFIT	10'-0 1/2"	-
VILLAGE PRESIDENT (104)	CARPET/VINYL COVING	PAINTED GYP. BD.	SUSPENDED TILE	8'-0"	-
AUXILIARY OFFICE (105)	CARPET/VINYL COVING	PAINTED GYP. BD.	SUSPENDED TILE	8'-0"	-
CORRIDOR (106)	CARPET/VINYL COVING	PAINTED GYP. BD.	SUSPENDED TILE	8'-0"	-
PUBLIC MEETING RM. & VILLAGE BD. CHAMBERS (107)	CARPET/VINYL COVING	PAINTED GYP. BD.	SUSPENDED TILE	8'-0"	-
SEWER & WATER UTILITY OFFICE (108)	SEALED CONC./VINYL COVING	PAINTED GYP. BD.	SUSPENDED TILE	8'-0"	-
VAULT (109)	SEALED CONC./VINYL COVING	CONC. BLK.	CONCRETE	8'-0"	-
STORAGE (110)	SEALED CONC./VINYL COVING	CONC. BLK./PAINTED GYP. BD.	SUSPENDED TILE	8'-0"	-
VILLAGE CLERK (111)	CARPET/VINYL COVING	PAINTED GYP. BD.	SUSPENDED TILE	8'-0"	-
RECEPTION (112)	CARPET/VINYL COVING	PAINTED GYP. BD.	SUSPENDED TILE	8'-0"	-
CORRIDOR (113)	CARPET/VINYL COVING	PAINTED GYP. BD.	SUSPENDED TILE	8'-0"	
MEN'S BATHROOM (114)	CERAMIC TILE & COVING	PAINTED GYP. BD.	SUSPENDED TILE	8'-0"	4'-0" CERAMIC WAINSCOAT
WOMEN'S BATHROOM (115)	CERAMIC TILE & COVING	PAINTED GYP. BD.	SUSPENDED TILE	8'-0"	4'-0" CERAMIC WAINSCOAT
STORAGE (116)	SEALED CONC./VINYL COVING	PAINTED GYP. BD.	SUSPENDED TILE	8'-0"	-
UTILITY (117)	SEALED CONC./VINYL COVING	PAINTED GYP. BD.	PAINTED GYP. BD.	10'-0"	-

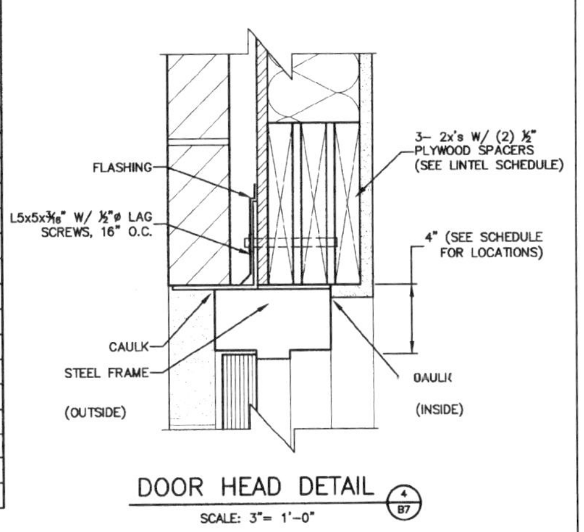

NOTES
CARPET WILL BE PROVIDED BY THE OWNER.
WALLS TO BE PAINTED IN ACCORDANCE WITH SPECIFICATIONS.

BRUCE VILLAGE HALL & LIBRARY
BRUCE, WISCONSIN
SCALE AS NOTED
SHEET B7
DATE DEC. 31, 2001
MORGAN & PARMLEY, LTD.
CONSULTING ENGINEERS
LADYSMITH, WISCONSIN
DR. BY J.T.D.

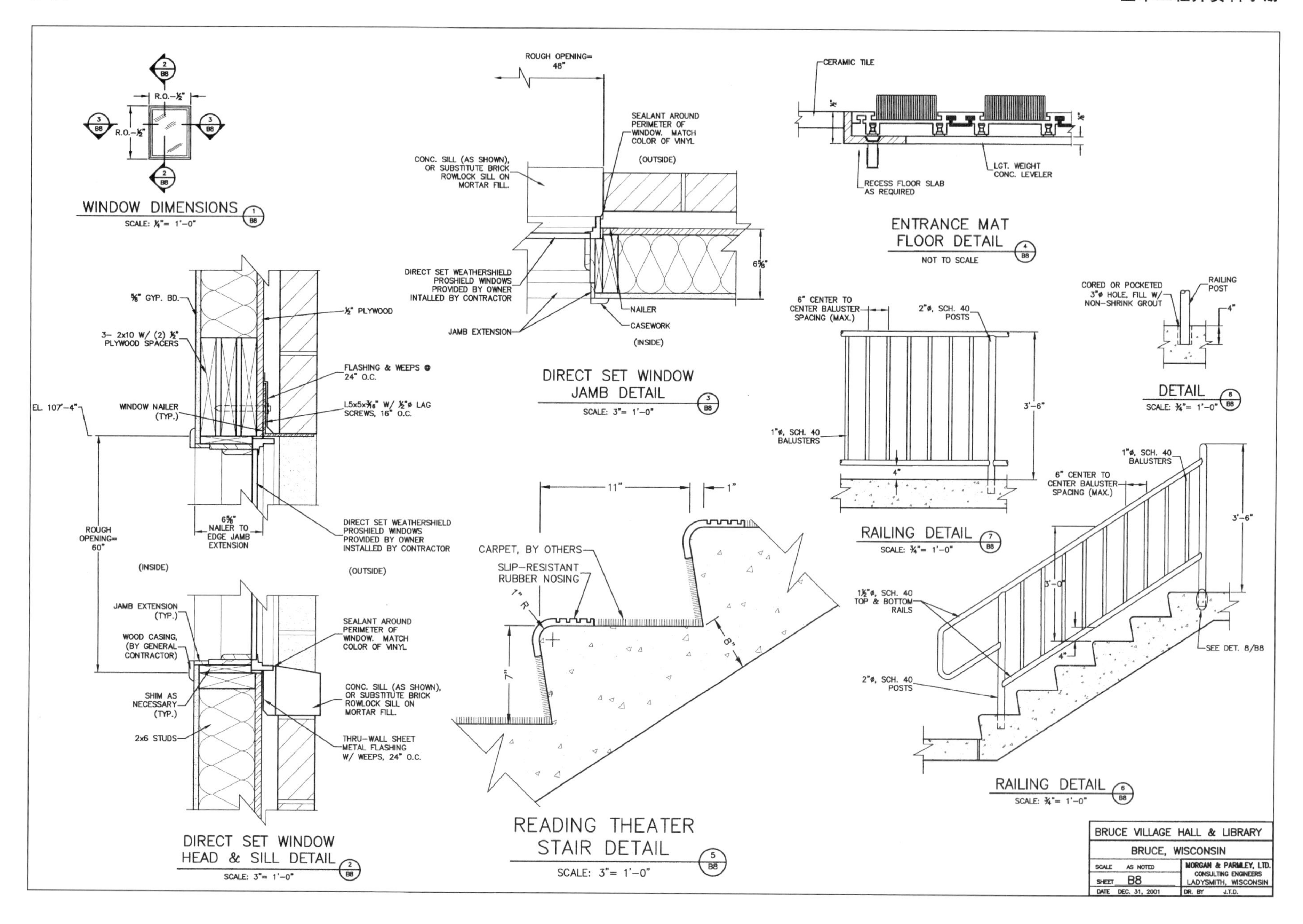
WINDOW DIMENSIONS
SCALE: ¼"= 1'-0"
ROUGH OPENING= 48"
SEALANT AROUND PERIMETER OF WINDOW. MATCH COLOR OF VINYL
(OUTSIDE)
CONC. SILL (AS SHOWN), OR SUBSTITUTE BRICK ROWLOCK SILL ON MORTAR FILL.
DIRECT SET WEATHERSHIELD PROSHIELD WINDOWS PROVIDED BY OWNER INTALLED BY CONTRACTOR
JAMB EXTENSION
NAILER
CASEWORK
(INSIDE)
DIRECT SET WINDOW JAMB DETAIL
SCALE: 3"= 1'-0"
CERAMIC TILE
LGT. WEIGHT CONC. LEVELER
RECESS FLOOR SLAB AS REQUIRED
ENTRANCE MAT FLOOR DETAIL
NOT TO SCALE
⅝" GYP. BD.
½" PLYWOOD
3- 2x10 W/ (2) ½" PLYWOOD SPACERS
FLASHING & WEEPS @ 24" O.C.
WINDOW NAILER (TYP.)
L5x5x⁵⁄₁₆" W/ ½"ø LAG SCREWS, 16" O.C.
EL. 107'-4"
ROUGH OPENING= 60"
NAILER TO EDGE JAMB EXTENSION
DIRECT SET WEATHERSHIELD PROSHIELD WINDOWS PROVIDED BY OWNER INSTALLED BY CONTRACTOR
(INSIDE)
(OUTSIDE)
JAMB EXTENSION (TYP.)
WOOD CASING, (BY GENERAL CONTRACTOR)
SHIM AS NECESSARY (TYP.)
2x6 STUDS
SEALANT AROUND PERIMETER OF WINDOW. MATCH COLOR OF VINYL
CONC. SILL (AS SHOWN), OR SUBSTITUTE BRICK ROWLOCK SILL ON MORTAR FILL.
THRU-WALL SHEET METAL FLASHING W/ WEEPS, 24" O.C.
DIRECT SET WINDOW HEAD & SILL DETAIL
SCALE: 3"= 1'-0"
6" CENTER TO CENTER BALUSTER SPACING (MAX.)
2"ø, SCH. 40 POSTS
1"ø, SCH. 40 BALUSTERS
3'-6"
RAILING DETAIL
SCALE: ¾"= 1'-0"
CORED OR POCKETED 3"ø HOLE, FILL W/ NON-SHRINK GROUT
RAILING POST
DETAIL
SCALE: ¾"= 1'-0"
CARPET, BY OTHERS
SLIP-RESISTANT RUBBER NOSING
READING THEATER STAIR DETAIL
SCALE: 3"= 1'-0"
1½"ø, SCH. 40 TOP & BOTTOM RAILS
SEE DET. 8/B8
RAILING DETAIL
SCALE: ¾"= 1'-0"
BRUCE VILLAGE HALL & LIBRARY
BRUCE, WISCONSIN
SCALE AS NOTED
SHEET B8
DATE DEC. 31, 2001
MORGAN & PARMLEY, LTD. CONSULTING ENGINEERS LADYSMITH, WISCONSIN
DR. BY J.T.D.

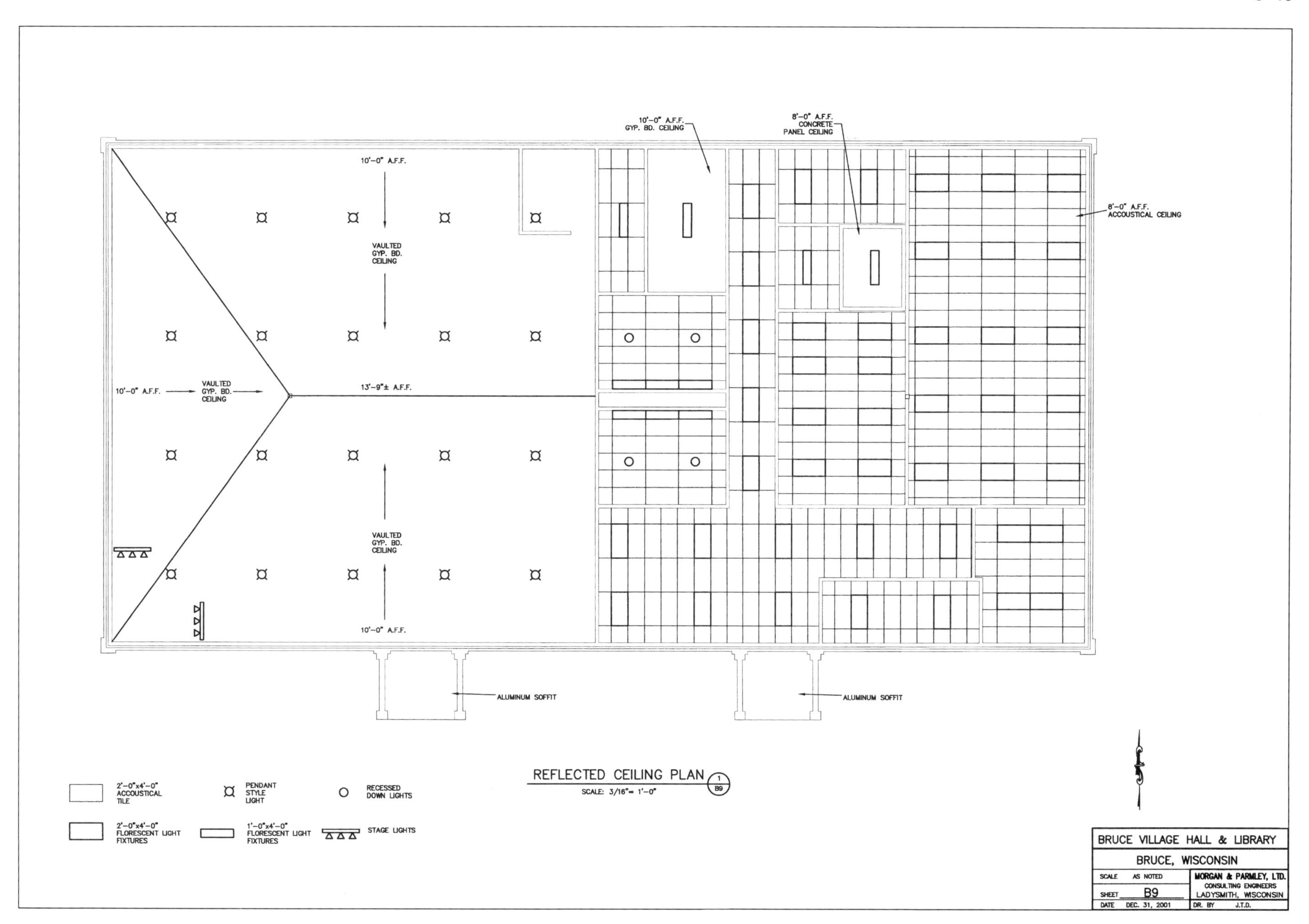
10'-0" A.F.F.
GYP. BD. CEILING
8'-0" A.F.F.
CONCRETE
PANEL CEILING
10'-0" A.F.F.
VAULTED
GYP. BD.
CEILING
8'-0" A.F.F.
ACCOUSTICAL CEILING
10'-0" A.F.F.
VAULTED
GYP. BD.
CEILING
13'-9"± A.F.F.
VAULTED
GYP. BD.
CEILING
10'-0" A.F.F.
ALUMINUM SOFFIT
ALUMINUM SOFFIT
2'-0"x4'-0"
ACCOUSTICAL
TILE
PENDANT
STYLE
LIGHT
RECESSED
DOWN LIGHTS
2'-0"x4'-0"
FLORESCENT LIGHT
FIXTURES
1'-0"x4'-0"
FLORESCENT LIGHT
FIXTURES
STAGE LIGHTS
REFLECTED CEILING PLAN
SCALE: 3/16"= 1'-0"
1
B9
BRUCE VILLAGE HALL & LIBRARY
BRUCE, WISCONSIN
SCALE AS NOTED
SHEET B9
DATE DEC. 31, 2001
MORGAN & PARMLEY, LTD.
CONSULTING ENGINEERS
LADYSMITH, WISCONSIN
DR. BY J.T.D.

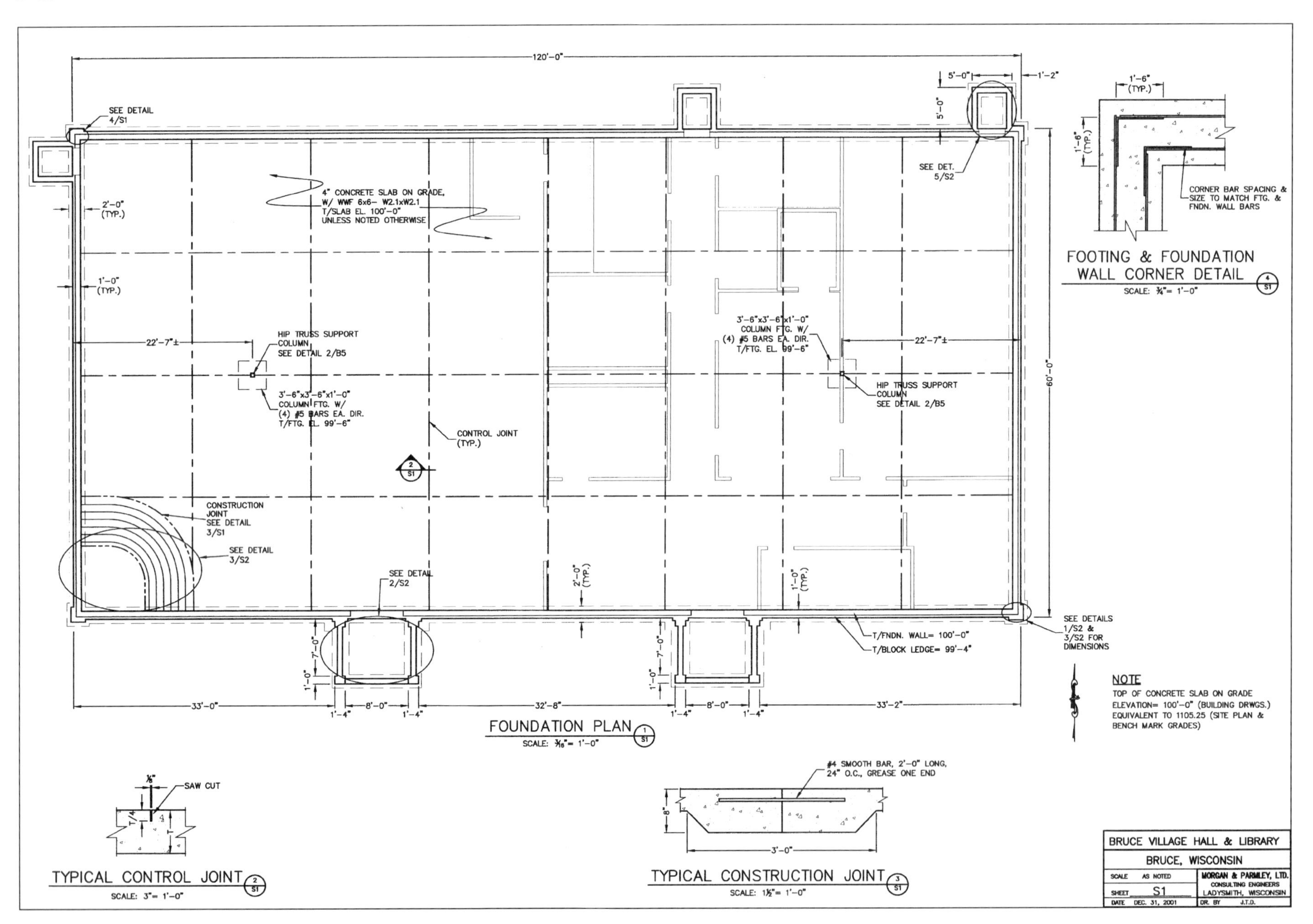

120'-0"
SEE DETAIL 4/S1
2'-0" (TYP.)
4" CONCRETE SLAB ON GRADE, W/ WWF 6x6- W2.1xW2.1 T/SLAB EL. 100'-0" UNLESS NOTED OTHERWISE
SEE DET. 5/S2
1'-0" (TYP.)
HIP TRUSS SUPPORT COLUMN SEE DETAIL 2/B5
3'-6"x3'-6"x1'-0" COLUMN FTG. W/ (4) #5 BARS EA. DIR. T/FTG. EL. 99'-6"
22'-7"±
CONTROL JOINT (TYP.)
CONSTRUCTION JOINT SEE DETAIL 3/S1
SEE DETAIL 3/S2
SEE DETAIL 2/S2
T/FNDN. WALL= 100'-0"
T/BLOCK LEDGE= 99'-4"
SEE DETAILS 1/S2 & 3/S2 FOR DIMENSIONS
60'-0"
33'-0" 8'-0" 32'-8" 8'-0" 33'-2"
FOUNDATION PLAN
SCALE: 3/16"= 1'-0"
CORNER BAR SPACING & SIZE TO MATCH FTG. & FNDN. WALL BARS
FOOTING & FOUNDATION WALL CORNER DETAIL
SCALE: 3/4"= 1'-0"
NOTE
TOP OF CONCRETE SLAB ON GRADE ELEVATION= 100'-0" (BUILDING DRWGS.) EQUIVALENT TO 1105.25 (SITE PLAN & BENCH MARK GRADES)
SAW CUT
TYPICAL CONTROL JOINT
SCALE: 3"= 1'-0"
#4 SMOOTH BAR, 2'-0" LONG, 24" O.C., GREASE ONE END
TYPICAL CONSTRUCTION JOINT
SCALE: 1½"= 1'-0"
BRUCE VILLAGE HALL & LIBRARY
BRUCE, WISCONSIN
SCALE AS NOTED
SHEET S1
DATE DEC. 31, 2001
MORGAN & PARMLEY, LTD. CONSULTING ENGINEERS LADYSMITH, WISCONSIN
DR. BY J.T.D.

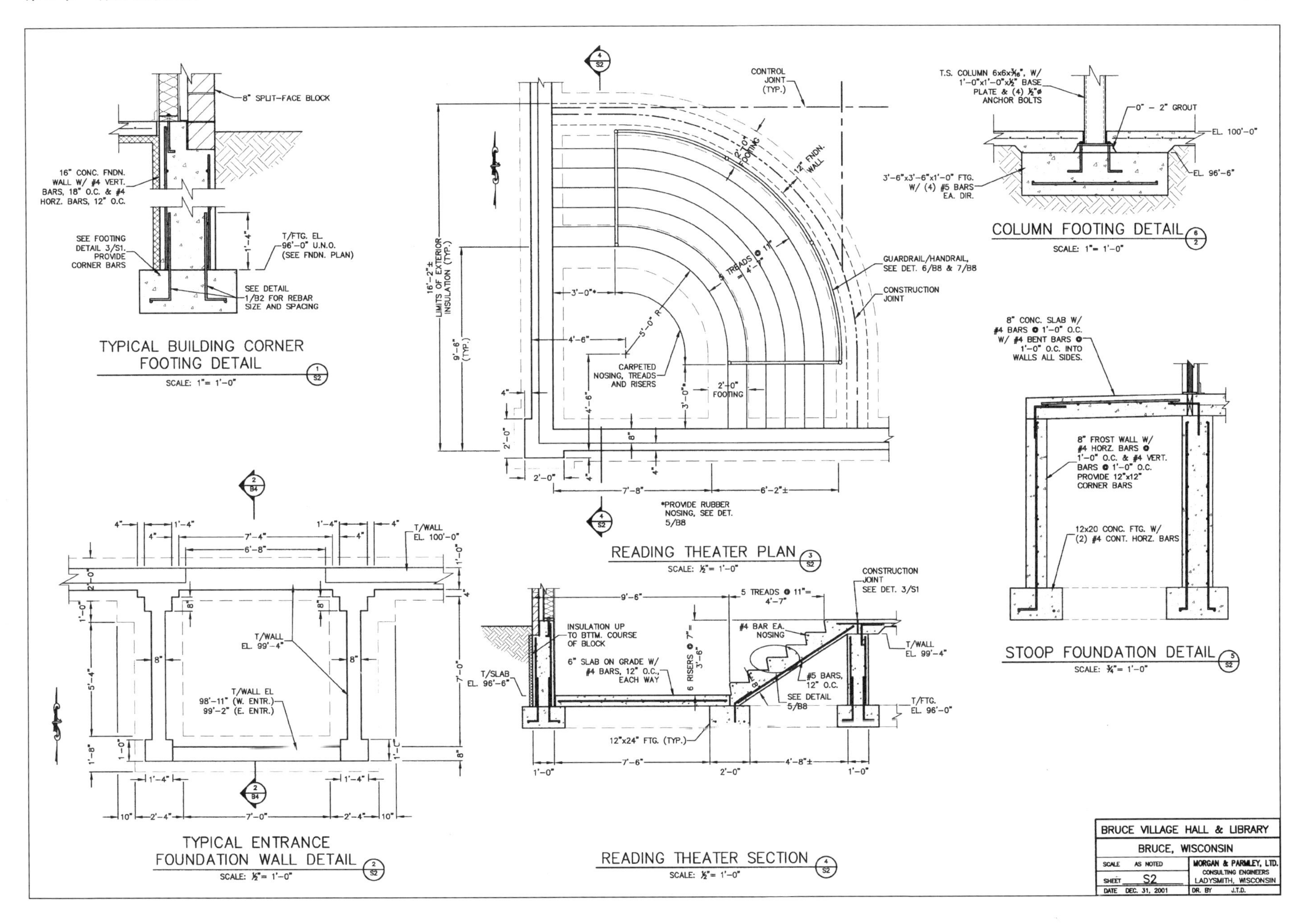
8" SPLIT-FACE BLOCK
16" CONC. FNDN. WALL W/ #4 VERT. BARS, 18" O.C. & #4 HORZ. BARS, 12" O.C.
SEE FOOTING DETAIL 3/S1. PROVIDE CORNER BARS
T/FTG. EL. 96'-0" U.N.O. (SEE FNDN. PLAN)
SEE DETAIL 1/B2 FOR REBAR SIZE AND SPACING
TYPICAL BUILDING CORNER FOOTING DETAIL
SCALE: 1"= 1'-0"
CONTROL JOINT (TYP.)
2'-0" FOOTING
12" FNDN. WALL
5 TREADS @ 11" = 4'-7"
GUARDRAIL/HANDRAIL, SEE DET. 6/B8 & 7/B8
CONSTRUCTION JOINT
LIMITS OF EXTERIOR INSULATION (TYP.)
CARPETED NOSING, TREADS AND RISERS
*PROVIDE RUBBER NOSING, SEE DET. 5/B8
READING THEATER PLAN
SCALE: ½"= 1'-0"
T.S. COLUMN 6x6x3/16", W/ 1'-0"x1'-0"x½" BASE PLATE & (4) ½"ø ANCHOR BOLTS
0" - 2" GROUT
EL. 100'-0"
EL. 96'-6"
3'-6"x3'-6"x1'-0" FTG. W/ (4) #5 BARS EA. DIR.
COLUMN FOOTING DETAIL
SCALE: 1"= 1'-0"
8" CONC. SLAB W/ #4 BARS @ 1'-0" O.C. W/ #4 BENT BARS @ 1'-0" O.C. INTO WALLS ALL SIDES.
8" FROST WALL W/ #4 HORZ. BARS @ 1'-0" O.C. & #4 VERT. BARS @ 1'-0" O.C. PROVIDE 12"x12" CORNER BARS
12x20 CONC. FTG. W/ (2) #4 CONT. HORZ. BARS
STOOP FOUNDATION DETAIL
SCALE: ¾"= 1'-0"
T/WALL EL. 100'-0"
T/WALL EL. 99'-4"
T/WALL EL 98'-11" (W. ENTR.) 99'-2" (E. ENTR.)
TYPICAL ENTRANCE FOUNDATION WALL DETAIL
SCALE: ½"= 1'-0"
CONSTRUCTION JOINT SEE DET. 3/S1
INSULATION UP TO BTTM. COURSE OF BLOCK
#4 BAR EA. NOSING
6" SLAB ON GRADE W/ #4 BARS, 12" O.C., EACH WAY
6 RISERS @ 7" = 3'-6"
T/SLAB EL. 96'-6"
#5 BARS, 12" O.C.
SEE DETAIL 5/B8
T/WALL EL. 99'-4"
T/FTG. EL. 96'-0"
12"x24" FTG. (TYP.)
READING THEATER SECTION
SCALE: ½"= 1'-0"
BRUCE VILLAGE HALL & LIBRARY
BRUCE, WISCONSIN
SCALE AS NOTED
SHEET S2
DATE DEC. 31, 2001
MORGAN & PARMLEY, LTD. CONSULTING ENGINEERS LADYSMITH, WISCONSIN
DR. BY J.T.D.

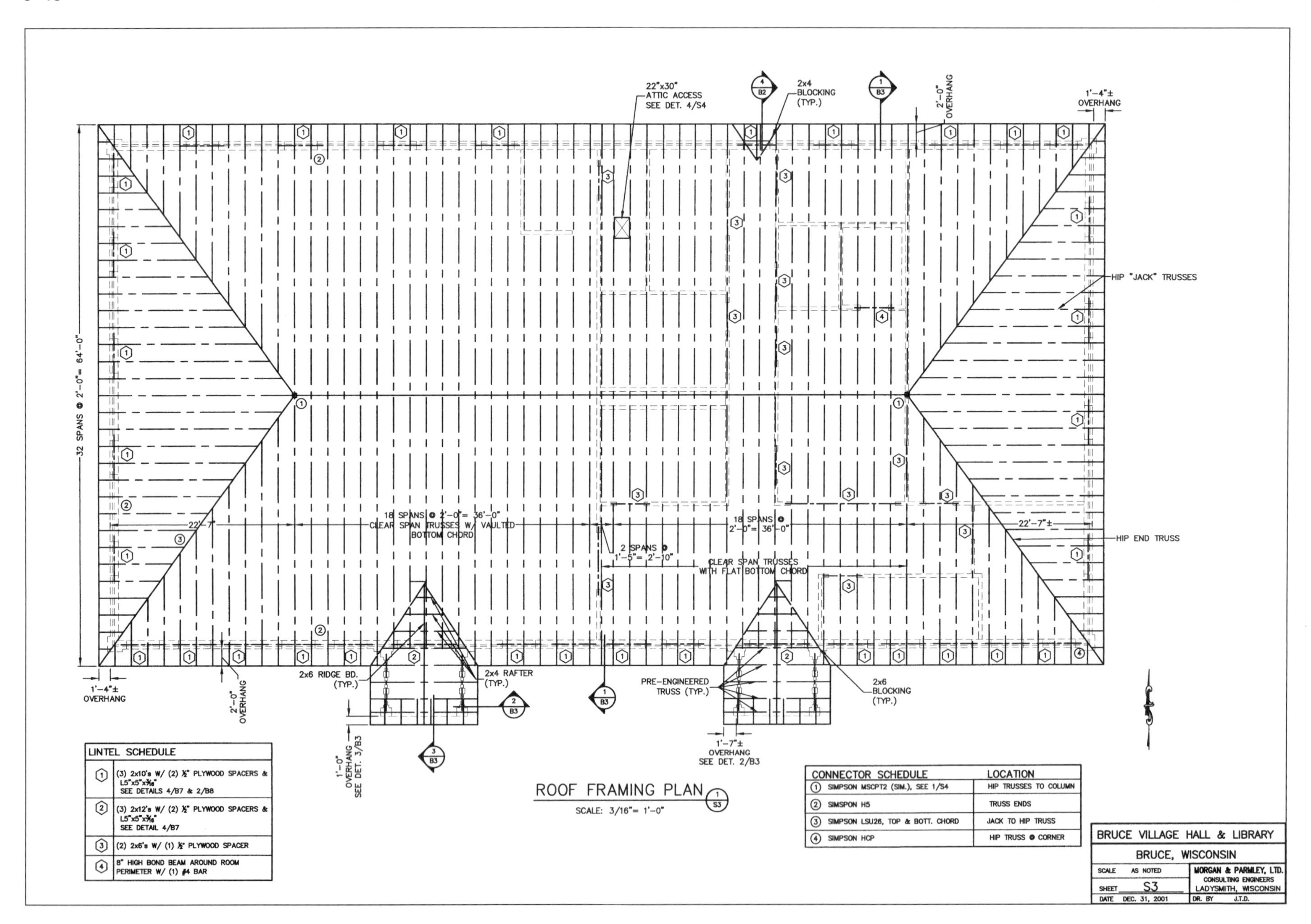
22"x30"
ATTIC ACCESS
SEE DET. 4/S4
2x4
BLOCKING
(TYP.)
2'-0"
OVERHANG
1'-4"±
OVERHANG
HIP "JACK" TRUSSES
32 SPANS @ 2'-0"= 64'-0"
18 SPANS @ 2'-0"= 36'-0"
CLEAR SPAN TRUSSES W/ VAULTED BOTTOM CHORD
22'-7"
18 SPANS @ 2'-0"= 36'-0"
22'-7"±
HIP END TRUSS
2 SPANS @ 1'-5"= 2'-10"
CLEAR SPAN TRUSSES WITH FLAT BOTTOM CHORD
2x6 RIDGE BD. (TYP.)
2x4 RAFTER (TYP.)
PRE-ENGINEERED TRUSS (TYP.)
2x6 BLOCKING (TYP.)
1'-4"± OVERHANG
2'-0" OVERHANG
1'-0" OVERHANG SEE DET. 3/B3
1'-7"± OVERHANG SEE DET. 2/B3
ROOF FRAMING PLAN
SCALE: 3/16"= 1'-0"
LINTEL SCHEDULE
1 (3) 2x10's W/ (2) ½" PLYWOOD SPACERS & L5"x5"x5/16" SEE DETAILS 4/B7 & 2/B8
2 (3) 2x12's W/ (2) ½" PLYWOOD SPACERS & L5"x5"x5/16" SEE DETAIL 4/B7
3 (2) 2x6's W/ (1) ½" PLYWOOD SPACER
4 8" HIGH BOND BEAM AROUND ROOM PERIMETER W/ (1) #4 BAR
CONNECTOR SCHEDULE — LOCATION
1 SIMPSON MSCPT2 (SIM.), SEE 1/S4 — HIP TRUSSES TO COLUMN
2 SIMPSON H5 — TRUSS ENDS
3 SIMPSON LSU26, TOP & BOTT. CHORD — JACK TO HIP TRUSS
4 SIMPSON HCP — HIP TRUSS @ CORNER
BRUCE VILLAGE HALL & LIBRARY
BRUCE, WISCONSIN
SCALE AS NOTED
SHEET S3
DATE DEC. 31, 2001
MORGAN & PARMLEY, LTD.
CONSULTING ENGINEERS
LADYSMITH, WISCONSIN
DR. BY J.T.D.

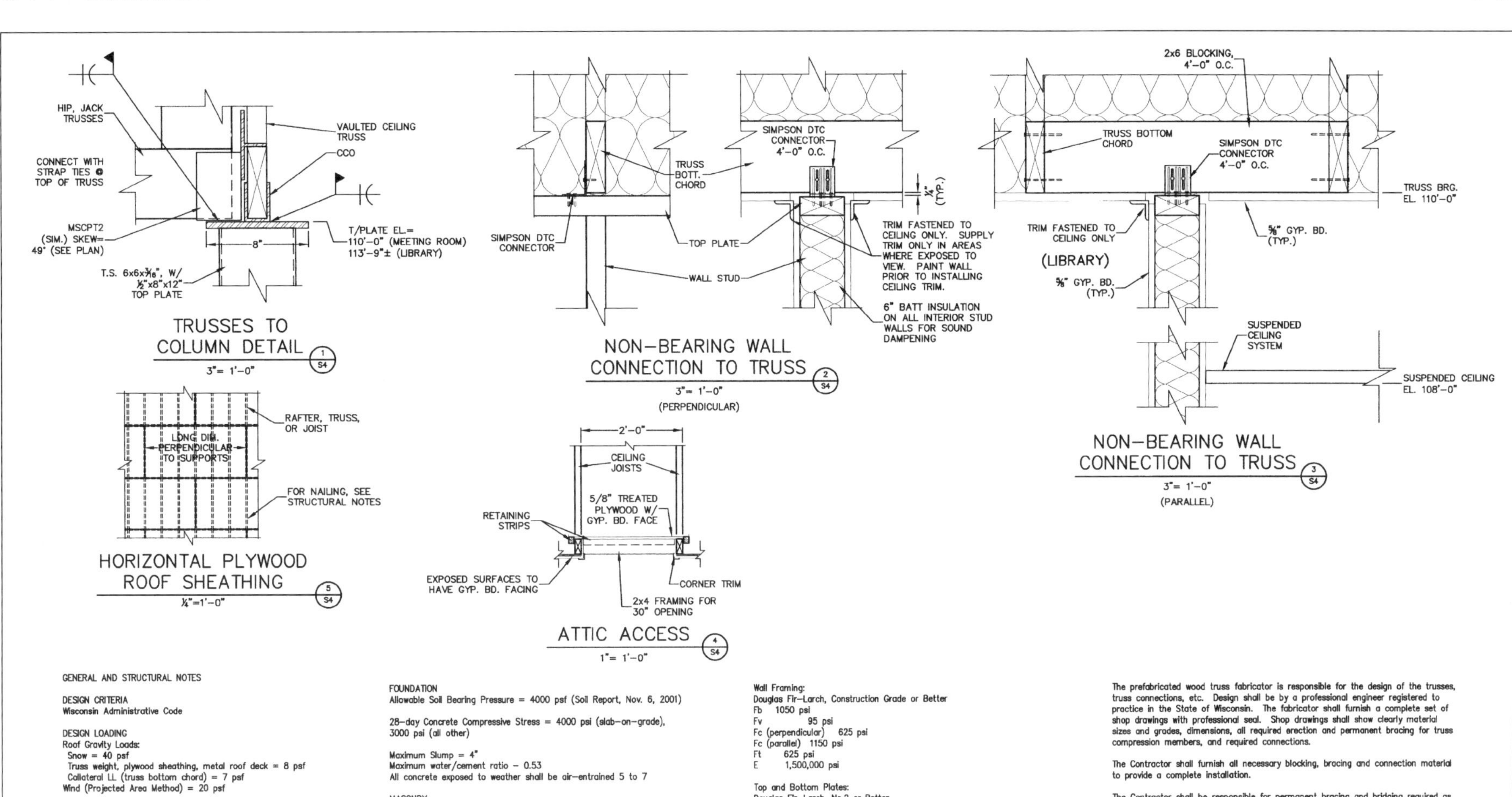

GENERAL AND STRUCTURAL NOTES

DESIGN CRITERIA
Wisconsin Administrative Code

DESIGN LOADING
Roof Gravity Loads:
Snow = 40 psf
Truss weight, plywood sheathing, metal roof deck = 8 psf
Collateral LL (truss bottom chord) = 7 psf
Wind (Projected Area Method) = 20 psf

OCCUPANCY AND SANITARY FIXTURE REQUIREMENTS
Village Offices Net Occupied Area = 2750 sq ft
No. Employees = 2750/75 sq ft/person = 37 people (19 males, 19 females)

Library Net Occupied Space = 2616 sq ft
No. Public = 2616/20 sq ft/person = 131 people (66 males, 66 females)

Total = 85 males, 85 females

Table 54.12-A;
Males - 1 WC, 1 LAV
Females - 1 WC, 1 LAV
1 DF
Table 54.12-B:
Males - 1 WC, 1 Urinal, 1 LAV
Females - 2 WC, 1 LAV
1 DF

RECYCLING SPACE (Comm 52.24, Appendix C)
Library - 2.2 cu ft / 1000 sq ft
Office (Assembly Occupancy) - 2.2 cu ft / 1000 sq ft
Required Recycling Space = 120 x 60 x 2.2 = 15.8 cubic feet
Provide 3' x 3' floor area x 2 ft high = 18 cubic feet OK

FOUNDATION
Allowable Soil Bearing Pressure = 4000 psf (Soil Report, Nov. 6, 2001)

28-day Concrete Compressive Stress = 4000 psi (slab-on-grade), 3000 psi (all other)

Maximum Slump = 4"
Maximum water/cement ratio - 0.53
All concrete exposed to weather shall be air-entrained 5 to 7

MASONRY
Units shall be standard weight, Grade N-1, 4" concrete block, set in running bond. Mortar to be Type M or S. See specifications for unit finishes and coloring, horizontal reinforcing and veneer tie requirements.

STRUCTURAL WOOD
Dimension Lumber: General construction requirements and provisions per Wisconsin Administrative Code Chapter 53.

Reference: American Forest and Paper Association; National Design Specification for Wood Construction; plus Supplement, "Design Values for Wood Construction"

Unless noted otherwise, dimension lumber shall have the following base design values:

Structural Framing, Lintels:
Douglas Fir-Larch, No.2 or Better
Fb 1450 psi
Fv 95 psi
Fc (perpendicular) 625 psi
Fc (parallel) 1000 psi
Ft 850 psi
E 1,700,000 psi

Wall Framing:
Douglas Fir-Larch, Construction Grade or Better
Fb 1050 psi
Fv 95 psi
Fc (perpendicular) 625 psi
Fc (parallel) 1150 psi
Ft 625 psi
E 1,500,000 psi

Top and Bottom Plates:
Douglas Fir-Larch, No.2 or Better
Fb 1450 psi
Fv 95 psi
Fc (perpendicular) 625 psi
Fc (parallel) 1000 psi
Ft 850 psi
E 1,700,000 psi

All member sizes indicated on the plans are nominal dimensions.

All nailing shall be in accordance with Table 53-XIX of the Wisconsin Administrative Code, unless noted otherwise.

All wood bearing on concrete or masonry shall be preservative treated.

PREFABRICATED ROOF TRUSSES
Trusses shall be designed to meet all loading and spans as indicated on the plans. Design trusses for concentrated loads and cantilevers as indicated on the architectural and structural plans.

Trusses shall comply with all provisions of the latest edition of the design specification for light metal plate wood trusses of the Truss Plate Institute.

The prefabricated wood truss fabricator is responsible for the design of the trusses, truss connections, etc. Design shall be by a professional engineer registered to practice in the State of Wisconsin. The fabricator shall furnish a complete set of shop drawings with professional seal. Shop drawings shall show clearly material sizes and grades, dimensions, all required erection and permanent bracing for truss compression members, and required connections.

The Contractor shall furnish all necessary blocking, bracing and connection material to provide a complete installation.

The Contractor shall be responsible for permanent bracing and bridging required as shown by the plans.

ROOF AND WALL FRAMING NOTES
Roof sheathing shall be as follows: ½ inch APA rated sheathing 32/16 EXP 1 plywood.

Plywood Roof and Wall Diaphragms: Shall be fastened with 10d (roof) and 8d (wall) nails, spaced 6" o/c along panel edges and 12" o/c at intermediate supports.

All nailing shall be accomplished with common nails and shall be driven flush but not fracture the sheathing surface. Box or sinker nails must be approved prior to use. to use.

As an alternate to hand nailing, the contractor shall submit for approval the size and type of nail used for automatic nailing with approved technical data for its use in nailing horizontal diaphragms.

BRUCE VILLAGE HALL & LIBRARY	
BRUCE, WISCONSIN	
SCALE AS NOTED	MORGAN & PARMLEY, LTD. CONSULTING ENGINEERS LADYSMITH, WISCONSIN
SHEET S4	
DATE DEC. 31, 2001	DR. BY J.T.D.

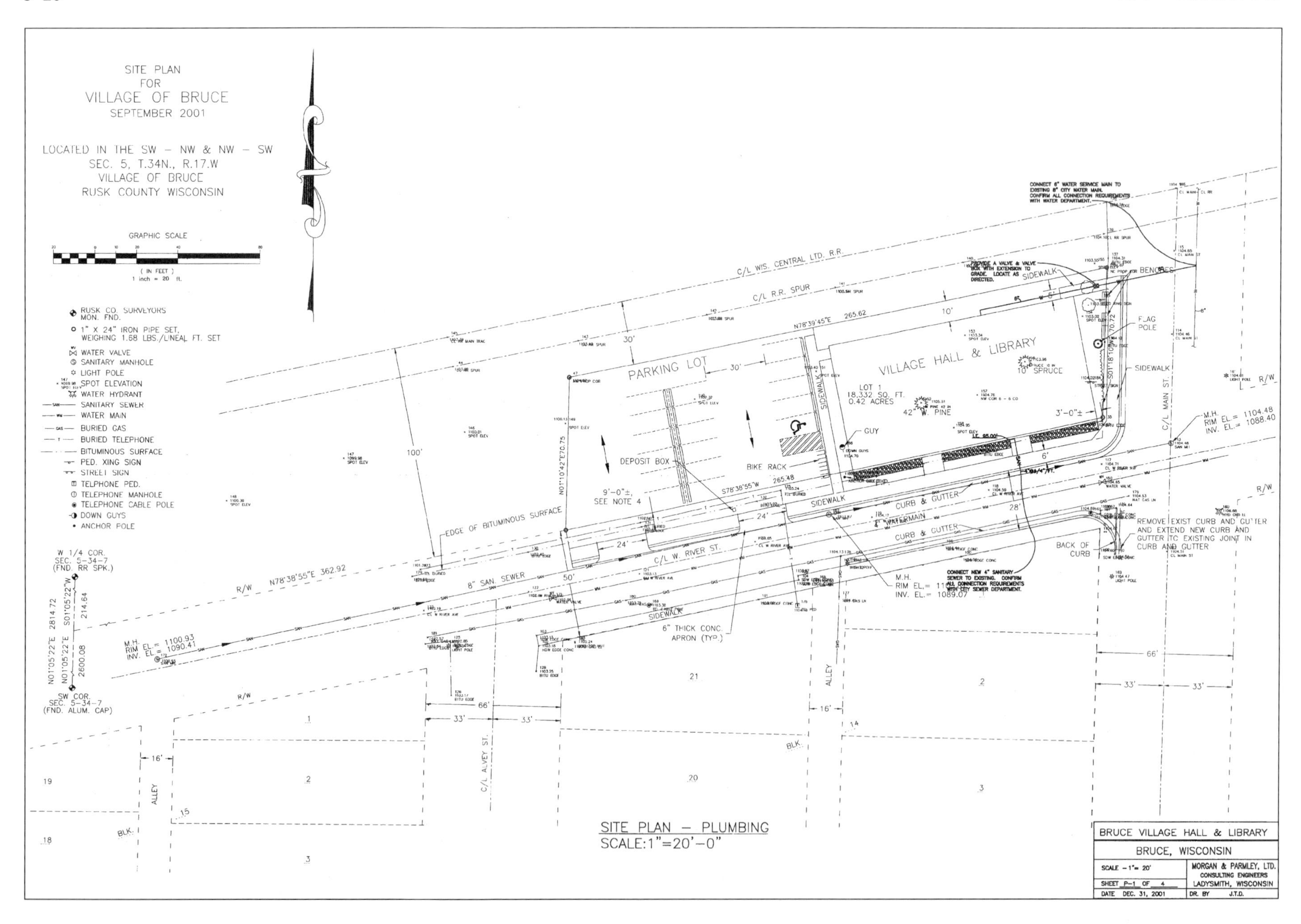
SITE PLAN
FOR
VILLAGE OF BRUCE
SEPTEMBER 2001
LOCATED IN THE SW – NW & NW – SW
SEC. 5, T.34N., R.17.W
VILLAGE OF BRUCE
RUSK COUNTY WISCONSIN
GRAPHIC SCALE
(IN FEET)
1 inch = 20 ft.
RUSK CO. SURVEYORS MON. FND.
1" X 24" IRON PIPE SET, WEIGHING 1.68 LBS./LINEAL FT. SET
WATER VALVE
SANITARY MANHOLE
LIGHT POLE
SPOT ELEVATION
WATER HYDRANT
SANITARY SEWER
WATER MAIN
BURIED GAS
BURIED TELEPHONE
BITUMINOUS SURFACE
PED. XING SIGN
STREET SIGN
TELPHONE PED.
TELEPHONE MANHOLE
TELEPHONE CABLE POLE
DOWN GUYS
ANCHOR POLE
W 1/4 COR. SEC. 5-34-7 (FND. RR SPK.)
SW COR. SEC. 5-34-7 (FND. ALUM. CAP)
C/L WIS. CENTRAL LTD. R.R.
C/L R.R. SPUR
PARKING LOT
VILLAGE HALL & LIBRARY
LOT 1
18,332 SQ. FT.
0.42 ACRES
10" SPRUCE
42" W. PINE
FLAG POLE
SIDEWALK
BENCHES
DEPOSIT BOX
BIKE RACK
GUY
EDGE OF BITUMINOUS SURFACE
9'-0"± SEE NOTE 4
CURB & GUTTER
WATERMAIN
C/L W. RIVER ST.
8" SAN. SEWER
6" THICK CONC. APRON (TYP.)
M.H. RIM EL.= 1100.93 INV. EL.= 1090.41
M.H. RIM EL.= 1104.48 INV. EL.= 1088.40
M.H. INV. EL.= 1089.07
CONNECT 6" WATER SERVICE MAIN TO EXISTING 8" CITY WATER MAIN. CONFIRM ALL CONNECTION REQUIREMENTS WITH WATER DEPARTMENT.
PROVIDE A VALVE & VALVE BOX WITH EXTENSION TO GRADE. LOCATE AS DIRECTED.
CONNECT NEW 4" SANITARY SEWER TO EXISTING. CONFIRM ALL CONNECTION REQUIREMENTS WITH CITY SEWER DEPARTMENT.
REMOVE EXIST CURB AND GUTTER AND EXTEND NEW CURB AND GUTTER TO EXISTING JOINT IN CURB AND GUTTER
BACK OF CURB
C/L MAIN ST.
C/L ALVEY ST.
ALLEY
BLK.
R/W
N78°39'45"E 265.62
S78°38'55"W 265.48
N78°38'55"E 362.92
N01°10'42"E 70.75
S01°18'10"W 70.72
N01°05'22"E 2814.72
S01°05'22"W 214.64
N01°05'22"E 2600.08
SITE PLAN – PLUMBING
SCALE:1"=20'-0"
BRUCE VILLAGE HALL & LIBRARY
BRUCE, WISCONSIN
SCALE – 1"= 20'
SHEET P-1 OF 4
DATE DEC. 31, 2001
MORGAN & PARMLEY, LTD. CONSULTING ENGINEERS LADYSMITH, WISCONSIN
DR. BY J.T.D.

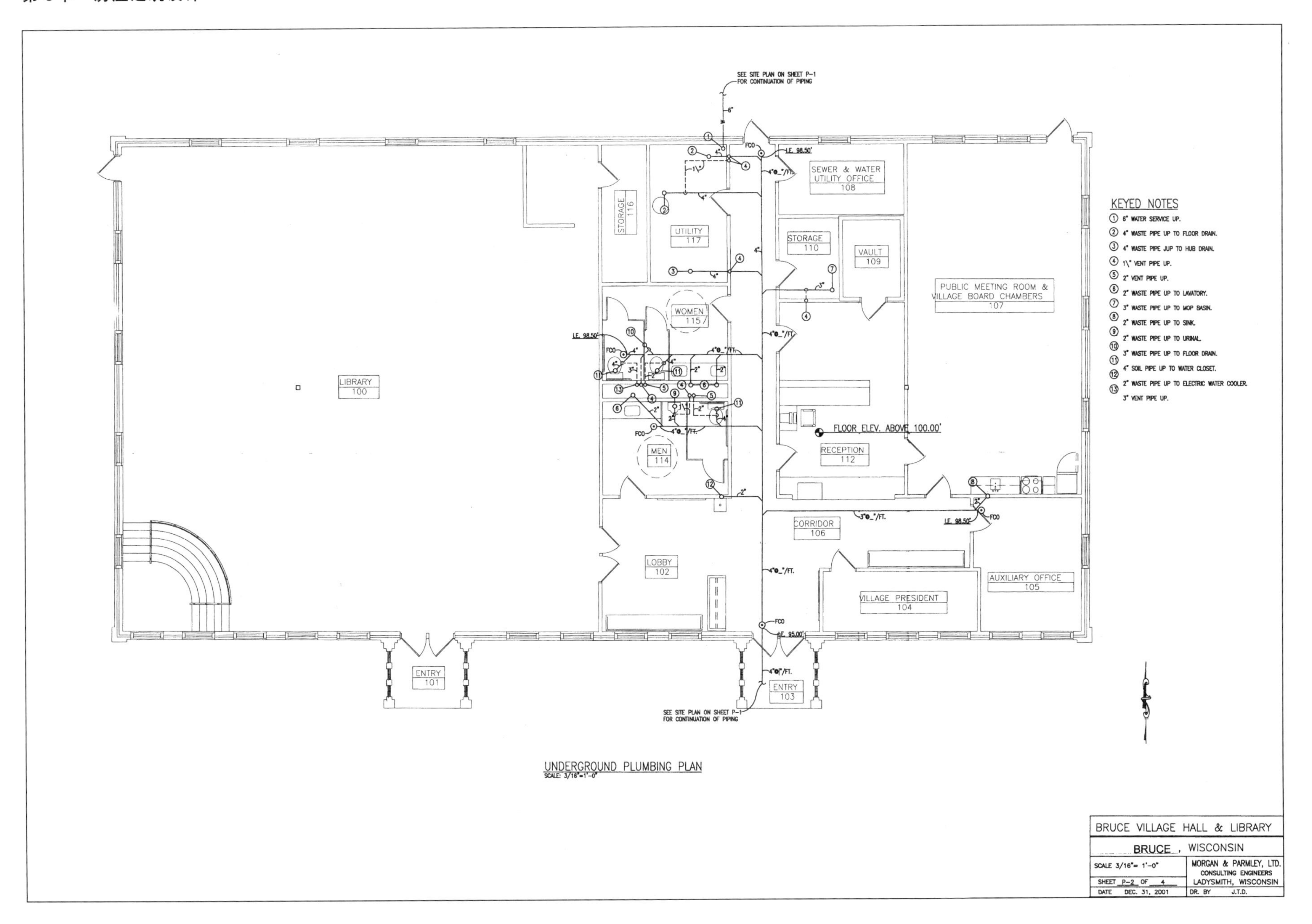
SEE SITE PLAN ON SHEET P-1
FOR CONTINUATION OF PIPING
LIBRARY 100
STORAGE 116
UTILITY 117
WOMEN 115
MEN 114
SEWER & WATER UTILITY OFFICE 108
STORAGE 110
VAULT 109
PUBLIC MEETING ROOM & VILLAGE BOARD CHAMBERS 107
FLOOR ELEV. ABOVE 100.00'
RECEPTION 112
CORRIDOR 106
LOBBY 102
VILLAGE PRESIDENT 104
AUXILIARY OFFICE 105
ENTRY 101
ENTRY 103
KEYED NOTES
① 6" WATER SERVICE UP.
② 4" WASTE PIPE UP TO FLOOR DRAIN.
③ 4" WASTE PIPE JUP TO HUB DRAIN.
④ 1\" VENT PIPE UP.
⑤ 2" VENT PIPE UP.
⑥ 2" WASTE PIPE UP TO LAVATORY.
⑦ 3" WASTE PIPE UP TO MOP BASIN.
⑧ 2" WASTE PIPE UP TO SINK.
⑨ 2" WASTE PIPE UP TO URINAL.
⑩ 3" WASTE PIPE UP TO FLOOR DRAIN.
⑪ 4" SOIL PIPE UP TO WATER CLOSET.
⑫ 2" WASTE PIPE UP TO ELECTRIC WATER COOLER.
⑬ 3" VENT PIPE UP.
UNDERGROUND PLUMBING PLAN
SCALE: 3/16"=1'-0"
BRUCE VILLAGE HALL & LIBRARY
BRUCE, WISCONSIN
SCALE 3/16"= 1'-0"
SHEET P-2 OF 4
DATE DEC. 31, 2001
MORGAN & PARMLEY, LTD.
CONSULTING ENGINEERS
LADYSMITH, WISCONSIN
DR. BY J.T.D.

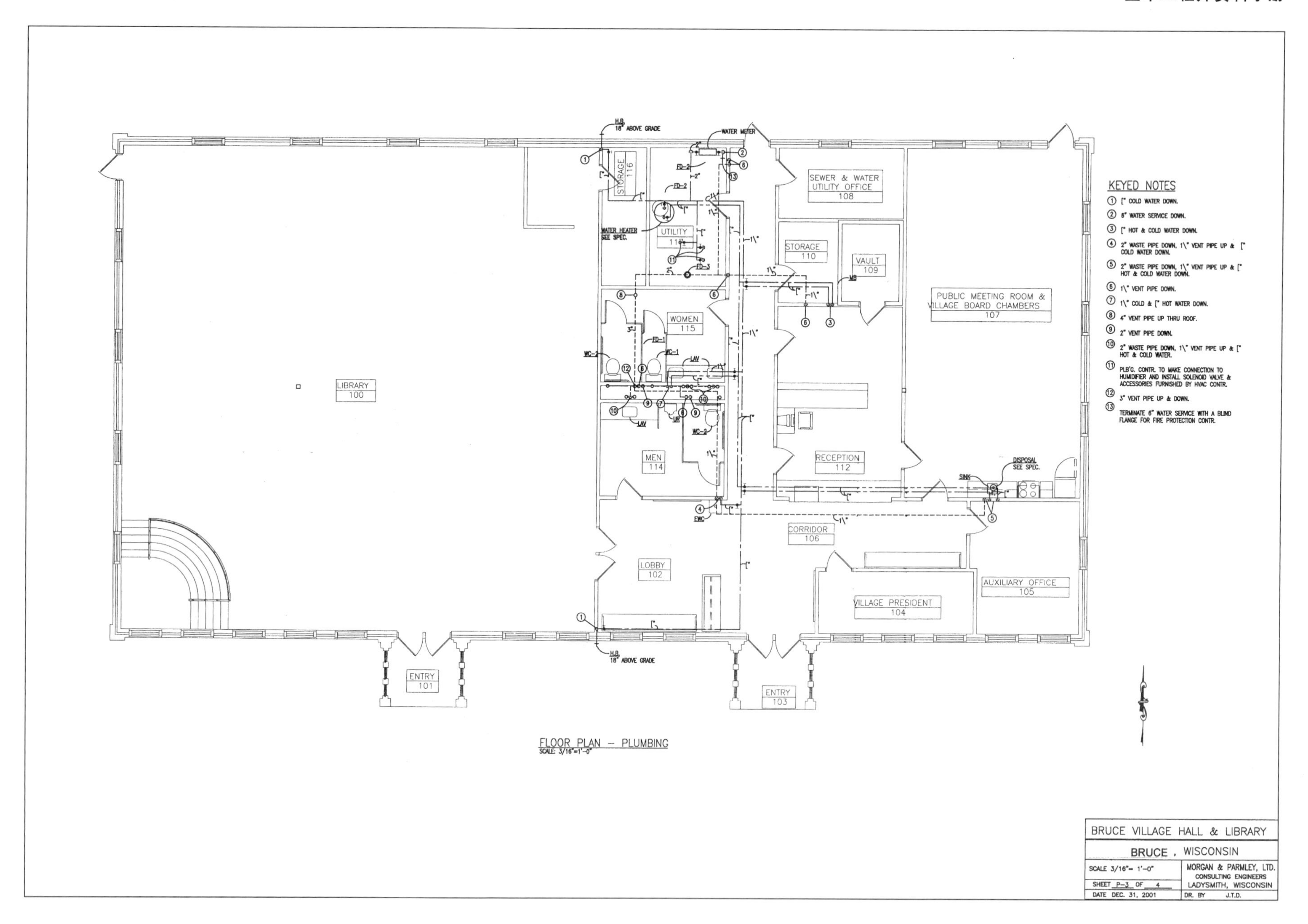
H.B.
18" ABOVE GRADE
WATER METER
STORAGE 116
WATER HEATER SEE SPEC.
UTILITY 118
SEWER & WATER UTILITY OFFICE 108
STORAGE 110
VAULT 109
PUBLIC MEETING ROOM & VILLAGE BOARD CHAMBERS 107
WOMEN 115
LIBRARY 100
MEN 114
RECEPTION 112
DISPOSAL SEE SPEC.
SINK
CORRIDOR 106
LOBBY 102
AUXILIARY OFFICE 105
VILLAGE PRESIDENT 104
ENTRY 101
ENTRY 103
FLOOR PLAN – PLUMBING
SCALE: 3/16"=1'-0"
KEYED NOTES
① [" COLD WATER DOWN.
② 6" WATER SERVICE DOWN.
③ [" HOT & COLD WATER DOWN.
④ 2" WASTE PIPE DOWN, 1\" VENT PIPE UP & [" COLD WATER DOWN.
⑤ 2" WASTE PIPE DOWN, 1\" VENT PIPE UP & [" HOT & COLD WATER DOWN.
⑥ 1\" VENT PIPE DOWN.
⑦ 1\" COLD & [" HOT WATER DOWN.
⑧ 4" VENT PIPE UP THRU ROOF.
⑨ 2" VENT PIPE DOWN.
⑩ 2" WASTE PIPE DOWN, 1\" VENT PIPE UP & [" HOT & COLD WATER.
⑪ PLB'G. CONTR. TO MAKE CONNECTION TO HUMIDIFIER AND INSTALL SOLENOID VALVE & ACCESSORIES FURNISHED BY HVAC CONTR.
⑫ 3" VENT PIPE UP & DOWN.
⑬ TERMINATE 6" WATER SERVICE WITH A BLIND FLANGE FOR FIRE PROTECTION CONTR.
BRUCE VILLAGE HALL & LIBRARY
BRUCE , WISCONSIN
SCALE 3/16"= 1'-0"
SHEET P-3 OF 4
DATE DEC. 31, 2001
MORGAN & PARMLEY, LTD.
CONSULTING ENGINEERS
LADYSMITH, WISCONSIN
DR. BY J.T.D.

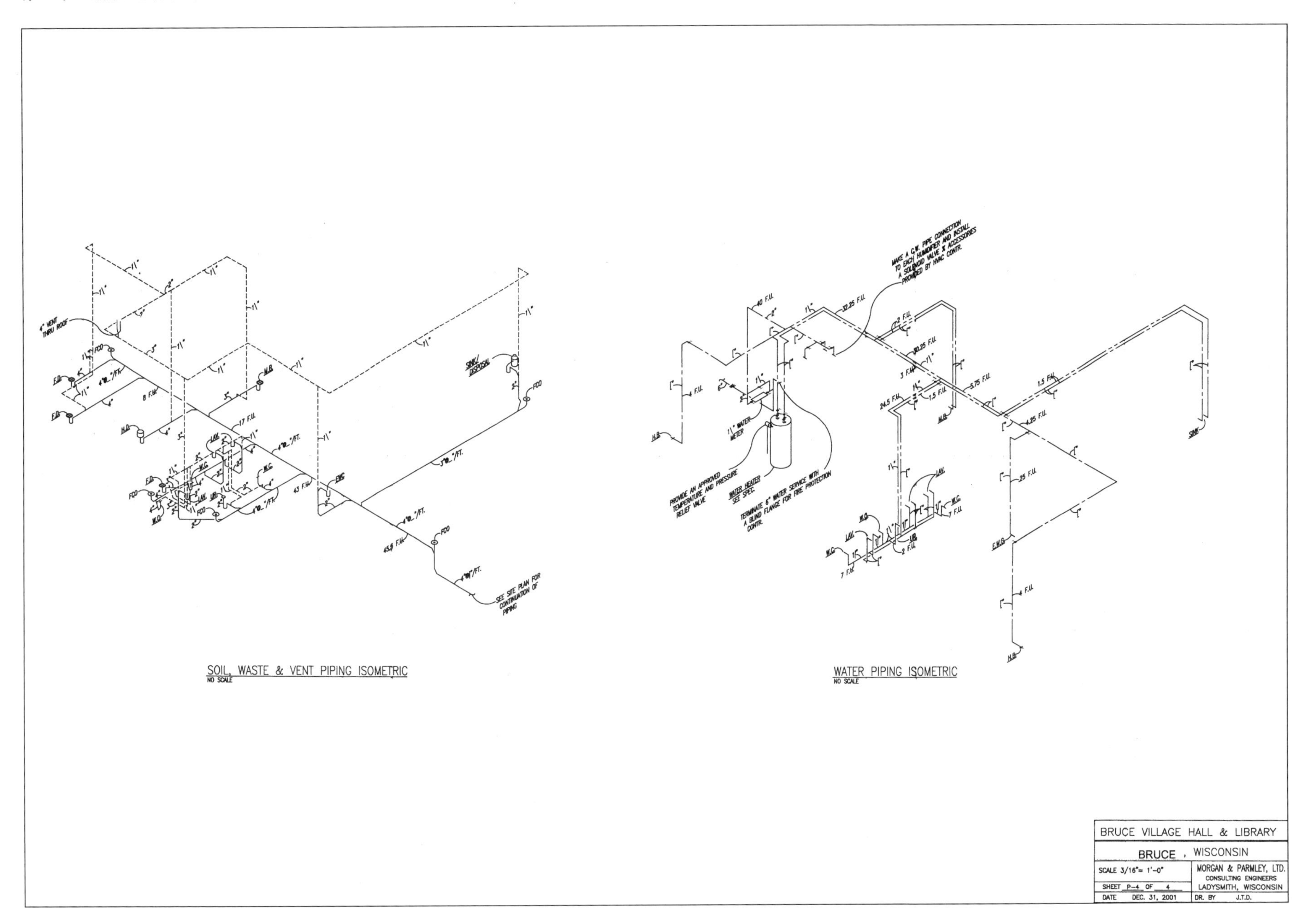
4" VENT THRU ROOF
SINK/ DISPOSAL
SEE SITE PLAN FOR CONTINUATION OF PIPING
SOIL, WASTE & VENT PIPING ISOMETRIC
NO SCALE
MAKE A C.W. PIPE CONNECTION TO EACH HUMIDIFIER AND INSTALL A SOLENOID VALVE & ACCESSORIES PROVIDED BY HVAC CONTR.
PROVIDE AN APPROVED TEMPERATURE AND PRESSURE RELIEF VALVE
WATER HEATER SEE SPEC.
TERMINATE 6" WATER SERVICE WITH A BLIND FLANGE FOR FIRE PROTECTION CONTR.
WATER PIPING ISOMETRIC
NO SCALE
BRUCE VILLAGE HALL & LIBRARY
BRUCE , WISCONSIN
SCALE 3/16"= 1'-0"
MORGAN & PARMLEY, LTD.
CONSULTING ENGINEERS
LADYSMITH, WISCONSIN
SHEET P-4 OF 4
DATE DEC. 31, 2001
DR. BY J.T.D.

GAS FIRED FURNACE SCHEDULE

UNIT NO.	AREA SERVED	MFR.	MODEL NO.	AIR HANDLING SECTION: SUP. CFM	OA CFM	E.S.P.	FAN RPM	FAN HP	HEATING SECTION: MBH INPUT	MBH OUTPUT	% EFF.	GAS PIPE SIZE	INTAKE/VENT SIZE	ELECT. RQMTS.: VOLTS	PHASE	REMARKS
F-1	LIBRARY-NORTH	CARRIER	58MCAØ80-20	16ØØ	16Ø	.5	HIGH	¾	8Ø	74	91	¾"	2"	12ØV	1Φ	①
F-2	LIBRARY-SOUTH	CARRIER	58MCAØ80-20	16ØØ	16Ø	.5	HIGH	¾	8Ø	74	91	¾"	2"	12ØV	1Φ	①
F-3	MEETING ROOM	CARRIER	58MCAØ80-16	135Ø	3ØØ	.5	HIGH	½	8Ø	74	91	¾"	2"	12ØV	1Φ	①
F-4	VILLAGE OFFICES	CARRIER	58MCAØ80-20	16ØØ	16Ø	.5	HIGH	¾	8Ø	74	91	¾"	2"	12ØV	1Φ	①

① CONCENTRIC COMB. AIR INTAKE/VENT KIT.
SIDE FILTER RACK W/ 1" THICK WASHABLE FILTER.
7 DAY ELECTRONIC PROGRAMMABLE THERMOSTAT.

FURNACE COOLING COIL SCHEDULE

UNIT NO.	UNIT SERVED	MFR.	MODEL NO.	TYPE	CFM	CLG. CAP. (MBH): TC	SC	EAT (°F): DB/WB	APD (IN H_2O)	SST COIL (°F)	REFRIG. PIPE SIZES: SUCT.	LIQ.
CC-1	F-1	CARRIER	CKBAØ48	CASED COIL	16ØØ	46.Ø	32.7	8Ø/67	.31	45	1⅛" O.D.	½" O.D.
CC-2	F-2	CARRIER	CKBAØ48	CASED COIL	16ØØ	46.Ø	32.7	8Ø/67	.31	45	1⅛" O.D.	½" O.D.
CC-3	F-3	CARRIER	CKBAØ36	CASED COIL	135Ø	34.Ø	25.7	8Ø/67	.3Ø	45	⅞" O.D.	½" O.D.
CC-4	F-4	CARRIER	CKBAØ48	CASED COIL	16ØØ	46.Ø	32.7	8Ø/67	.31	45	1⅛" O.D.	½" O.D.

AIR COOLED CONDENSING UNIT SCHEDULE

CU — EQUIPMENT NAME — UNIT NUMBER

UNIT NO.	UNIT SERVED	UNIT LOCATION	MFR.	MODEL NO.	NOM. CAP. AT 95°F (TONS)	ELECTRICAL REQUIREMENTS: VOLTS/PHASE	MIN. CIRC. AMPS	MAX. FUSE SIZE	EER	UNIT WEIGHT	REMARKS
CU-1	F-1	ON GRADE	CARRIER	38 TUA-Ø48	4	24ØV/1Φ	3Ø.5 A	5Ø A	11.Ø	2ØØ	①
CU-2	F-2	ON GRADE	CARRIER	38 TUA-Ø48	4	24ØV/1Φ	3Ø.5 A	5Ø A	11.Ø	2ØØ	①
CU-3	F-3	ON GRADE	CARRIER	38 TUA-Ø36	3	24ØV/1Φ	23.4 A	4Ø A	11.Ø	14Ø	①
CU-4	F-4	ON GRADE	CARRIER	38 TUA-Ø48	4	24ØV/1Φ	3Ø.5 A	5Ø A	11.Ø	2ØØ	①

① BALL BEARING COND. FAN MOTOR.
COMPRESSOR START ASSIST CAPACI[illegible] % RELAY.
CRANKCASE HEATER.
CYCLE PROTECTOR.
EVAPORATOR FREEZE STAT.
FILTER DRIERS.
HIGH PRESSURE SWITCH.
LOW PRESSURE SWITCH.
CONDENSER FAN SPEED LOW AMBIENT CONTROL.
WINTER START CONTROL.

FAN SCHEDULE

EF — EQUIPMENT NAME — UNIT NUMBER

UNIT NO.	AREA SERVED	MFR.	MODEL NO.	CFM	S.P.	RPM	H.P.	VOLTS	PHASE	SONES	NO. REQ'D	NOTES
EF-1	WOMENS TOILET	CARNES	VCDDØ3ØC	296	.375	9Ø5	2.7 A	12ØV	1Φ	3.5	1	①
EF-2	MENS TOILET	CARNES	VCDDØ3ØC	296	.375	9Ø5	2.7 A	12ØV	1Φ	3.5	1	①
EF-3	JANITORS CLOSET	CARNES	VCDDØ2ØC	15Ø	.375	74Ø	1.8 A	12ØV	1Φ	2.9	1	①

① SPEED SWITCH.
SLOPING ROOF CAP.

SCHEDULE OF GRILLES, REGISTERS AND DIFFUSERS

TYPE	DESCRIPTION	MANUF.	MODEL NO.	MATERIAL	DAMPER	FINISH	REMARKS
SR-1	SUPPLY REGISTER	CARNES	RNDAH	ALUMINUM	YES	PRIME COAT	DOUBLE DEFLECTION, FLANGE FRAME
CD-1	CEILING DIFFUSER	CARNES	SETA	ALUMINUM	YES	OFF WHITE	LOUVER FACE, LAY-IN T-BAR FRAME
CD-2	CEILING DIFFUSER	CARNES	SEFA	ALUMINUM	YES	OFF WHITE	LOUVER FACE, FLANGE FRAME
CD-3	CEILING DIFFUSER	ACUTHERM THERMAUSER	TF-HC	STEEL	VAV DIFFUSER	OFF WHITE	VAV DIFFUSER, LAY-IN T-BAR FRAME
RG/EG/TG-1	RETURN/TRANSFER/ EXHAUST GRILLE	CARNES	R[illegible]AF	ALUMINUM	NO	OFF WHITE	EGG CRATE, FLANGE FRAME
RG/EG/TG-2	RETURN/TRANSFER/ EXHAUST GRILLE	CARNES	RA[illegible]AH	ALUMINUM	NO	OFF WHITE	EGG CRATE, LAY-IN T-BAR FRAME
RG/EG/TG-3	RETURN/TRANSFER/ EXHAUST GRILLE	CARNES	RALA[illegible]	ALUMINUM	NO	PRIME COAT	4Ø° DEFLECT., LVR. FACE, FLANGE FRAME

NOTES: 1. SEE DRAWINGS FOR AIR QUANTITIES, NECK SIZES AND THR[illegible] PATTERN.
2. COORDINATE EXACT LOCATION OF CEILING OUTLETS W[illegible] REFLECTED CEILING PLAN AND LIGHTING LAYOUT.

ELECTRIC HEATER SCHEDULE

EW — EQUIPMENT NAME — UNIT NUMBER

UNIT NO.	AREA SERVED	MFR.	MODEL NO.	HEAT. CAP. (MBH)	HEAT. CAP. (KW)	AMPS	VOLTS/PHASE	MOUNT. HEIGHT (FT.)	REMARKS
EW-1	SEE DRAWINGS	Q-MARK	CWH-21Ø1	3.4	1.Ø	8.4 A	12ØV/1Φ	12'	①
EW-2	SEE DRAWINGS	Q-MARK	CWH-2151	5.1	1.5	12.6 A	12ØV/1Φ	12'	①
EB-1	SEE DRAWINGS	Q-MARK	HBB-1257	3.4	1.Ø	8.3 A	12ØV/1Φ	-	②

① RECESS MOUNTING BOX.
INTEGRAL TAMPERPROOF THERMOSTAT.
POWER DISCONNECT SWITCH.

② INTEGRAL THERMOSTAT SECTION.

SYMBOL SCHEDULE

SYMBOL	DESCRIPTION	SYMBOL	DESCRIPTION
—— RS ——	REFRIGERANT SUCTION PIPING	F	FURNACE
—— RL ——	REFRIGERANT LIQUID PIPING	H	HUMIDIFIER
—— G ——	GAS PIPING	CU	AIR COOLED CONDENSING UNIT
—— D ——	DRAIN PIPING	EF	EXHAUST FAN
—— V ——	COMBUSTION AIR VENT PIPING	RH	RANGE HOOD
—— I ——	COMBUSTION AIR INTAKE PIPING	EW	ELECTRIC WALL HEATER
Ⓣ	THERMOSTAT/TEMPERATURE SENSOR	SR	SUPPLY REGISTER
Ⓢ	STARTER/SPEED SWITCH	CD	CEILING DIFFUSER
V	VOLUME DAMPER	TG	TRANSFER GRILLE
F	FIRE DAMPER	EG	EXHAUST GRILLE
A	AUTOMATIC DAMPER	RG	RETURN GRILLE
B	BACKDRAFT DAMPER	DG	DOOR GRILLE

MOTOR AND MOTOR STARTER SCHEDULE

ITEM NO.	EQUIP. DESCRIPTION	EQUIP. LOCATION	HP	VOLTS	PHASE	STARTER MOD. NO (ALLEN BRADLEY)	NO. REQ'D	STARTER LOCATION	ATC INTRLK	EMERG. POWER	REMARKS
1	F-1	UTILITY ROOM	¾	12ØV	1Φ	BY UNIT MFR.	1	IN UNIT	-	-	①
2	F-2	UTILITY ROOM	¾	12ØV	1Φ	BY UNIT MFR.	1	IN UNIT	-	-	①
3	F-3	UTILITY ROOM	½	12ØV	1Φ	BY UNIT MFR.	1	IN UNIT	-	-	①
4	F-4	UTILITY ROOM	¾	12ØV	1Φ	BY UNIT MFR.	1	IN UNIT	-	-	①
5	H-1	UTILITY ROOM	.6 A	12ØV	1Φ	BY UNIT MFR.	1	IN UNIT	1 & 2	-	①
6	H-2	UTILITY ROOM	.6 A	12ØV	1Φ	BY UNIT MFR.	1	IN UNIT	3	-	①
7	H-3	UTILITY ROOM	.6 A	12ØV	1Φ	BY UNIT MFR.	1	IN UNIT	4	-	①
8	CU-1	OUTSIDE ON GRADE	3Ø.5 A	24ØV	1Φ	BY UNIT MFR.	1	IN UNIT	1	-	②
9	CU-2	OUTSIDE ON GRADE	3Ø.5 A	24ØV	1Φ	BY UNIT MFR.	1	IN UNIT	2	-	②
1Ø	CU-3	OUTSIDE ON GRADE	23.4 A	24ØV	1Φ	BY UNIT MFR.	1	IN UNIT	3	-	②
11	CU-4	OUTSIDE ON GRADE	3Ø.5 A	24ØV	1Φ	BY UNIT MFR.	1	IN UNIT	4	-	②
12	EF-1	WOMEN'S TOILET	2.7 A	12ØV	1Φ	6ØØ TAX 9	1	UTILITY ROOM	1,2,3 & 4	-	③
13	EF-2	MEN'S TOILET	2.7 A	12ØV	1Φ	6ØØ TAX 9	1	UTILITY ROOM	1,2,3 & 4	-	③
14	EF-3	STORAGE	1.8 A	12ØV	1Φ	6ØØ TAX 9	1	UTILITY ROOM	1,2,3 & 4	-	③
15	EB-1	LIBRARY-STAGE	8.3 A	12ØV	1Φ	BY UNIT MFR.	1	IN UNIT	-	-	⑤
16	EW-1	SEE DRAWING	8.4 A	12ØV	1Φ	BY UNIT MFR.	2	IN UNIT	-	-	④⑤
17	EW-2	SEE DRAWING	12.6 A	12ØV	1Φ	BY UNIT MFR.	2	IN UNIT	-	-	④⑤

NOTES: ① E.C. TO FURNISH AND INSTALL DISCONNECT SWITCH.
② E.C. TO FURNISH AND INSTALL WEATHERPROOF SWITCH.
③ E.C. TO INSTALL AND WIRE MFR. SUPPLIED SPEED SWITCH ON FAN HOUSING.
④ E.C. TO INSTALL AND WIRE MFR. SUPPLIED DISCONNECT SWITCH.
⑤ E.C. TO INSTALL AND WIRE ELECTRIC HEATER.

HUMIDIFIER SCHEDULE

UNIT NO.	UNIT SERVED	MFR.	MODEL NO.	EVAP. RATE AT 12Ø°F PLENUM	ELECT. REQRMTS.	SUPPLY PLENUM CONNECTION	BYPASS DUCT SIZE	REMARKS
H-1	F-1/2	APRILAIRE	448	.75 GPH	.6 A/12ØV/1Φ	13¾"Wx11¾"H	6"Φ	①
H-2	F-3	APRILAIRE	448	.75 GPH	.6 A/12ØV/1Φ	13¾"Wx11¾"H	6"Φ	①
H-3	F-4	APRILAIRE	448	.75 GPH	.6 A/12ØV/1Φ	13¾"Wx11¾"H	6"Φ	①

① LOW VOLTAGE WALL MOUNTED HUMIDISTAT.
MAKE-UP WATER CONTROL VALVE.

BRUCE VILLAGE HALL & LIBRARY	
BRUCE, WISCONSIN	
SCALE: NONE	MORGAN & PARMLEY, LTD. CONSULTING ENGINEERS LADYSMITH, WISCONSIN
SHEET: M-1 OF 3	
DATE: DEC. 31, 2001	DR. BY: FLM, INC.

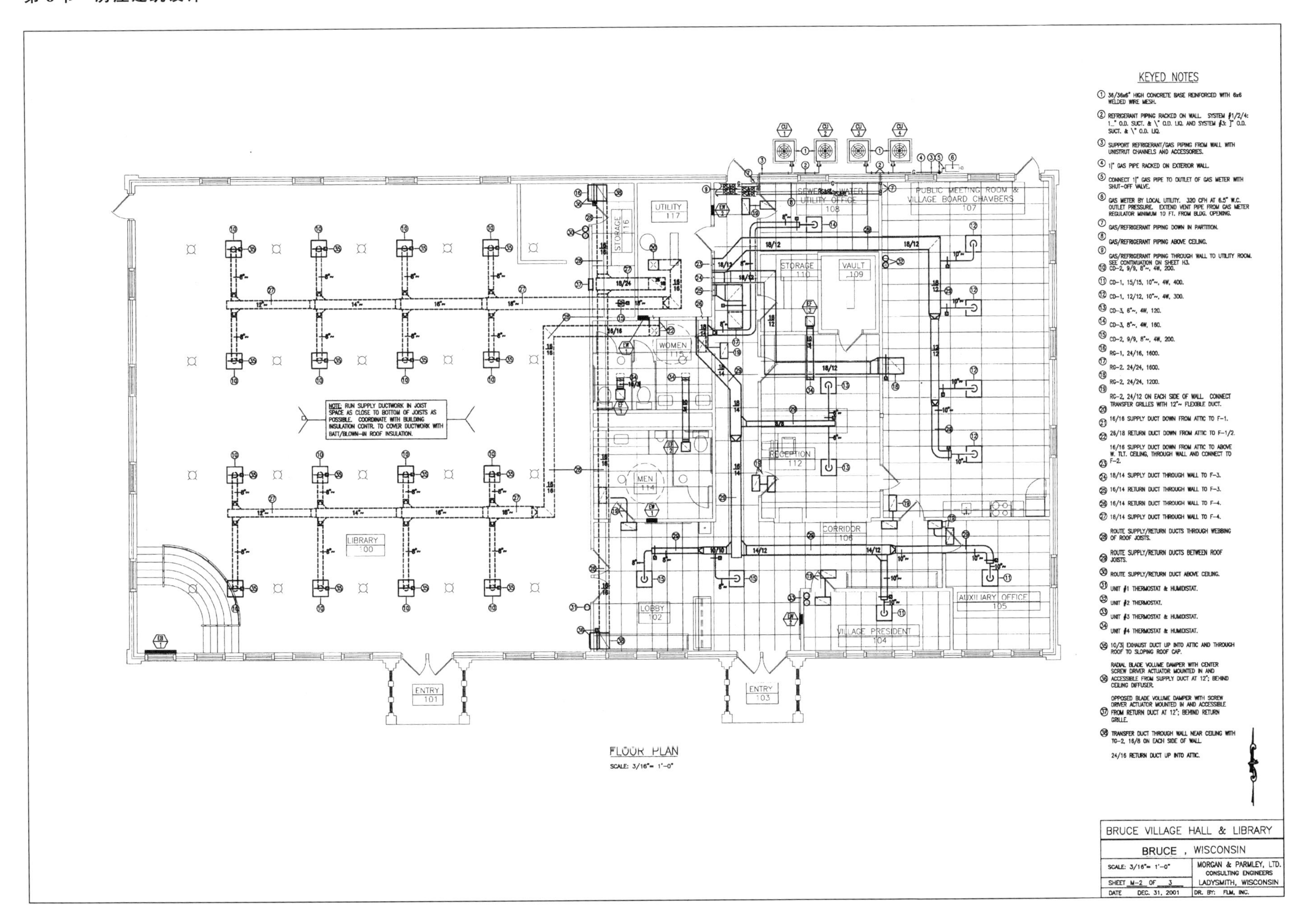
KEYED NOTES
① 36/36x6" HIGH CONCRETE BASE REINFORCED WITH 6x6 WELDED WIRE MESH.
② REFRIGERANT PIPING RACKED ON WALL. SYSTEM #1/2/4: 1_" O.D. SUCT. & \" O.D. LIQ. AND SYSTEM #3:]" O.D. SUCT. & \" O.D. LIQ.
③ SUPPORT REFRIGERANT/GAS PIPING FROM WALL WITH UNISTRUT CHANNELS AND ACCESSORIES.
④ 1|" GAS PIPE RACKED ON EXTERIOR WALL.
⑤ CONNECT 1|" GAS PIPE TO OUTLET OF GAS METER WITH SHUT-OFF VALVE.
⑥ GAS METER BY LOCAL UTILITY. 320 CFH AT 6.5" W.C. OUTLET PRESSURE. EXTEND VENT PIPE FROM GAS METER REGULATOR MINIMUM 10 FT. FROM BLDG. OPENING.
⑦ GAS/REFRIGERANT PIPING DOWN IN PARTITION.
⑧ GAS/REFRIGERANT PIPING ABOVE CEILING.
⑨ GAS/REFRIGERANT PIPING THROUGH WALL TO UTILITY ROOM. SEE CONTINUATION ON SHEET H3.
⑩ CD-2, 9/9, 8"~, 4W, 200.
⑪ CD-1, 15/15, 10"~, 4W, 400.
⑫ CD-1, 12/12, 10"~, 4W, 300.
⑬ CD-3, 6"~, 4W, 120.
⑭ CD-3, 8"~, 4W, 160.
⑮ CD-2, 9/9, 8"~, 4W, 200.
⑯ RG-1, 24/16, 1600.
⑰ RG-2, 24/24, 1600.
⑱ RG-2, 24/24, 1200.
⑲ RG-2, 24/12 ON EACH SIDE OF WALL. CONNECT TRANSFER GRILLES WITH 12"~ FLEXIBLE DUCT.
⑳ 16/16 SUPPLY DUCT DOWN FROM ATTIC TO F-1.
㉑ 26/18 RETURN DUCT DOWN FROM ATTIC TO F-1/2.
㉒ 16/16 SUPPLY DUCT DOWN FROM ATTIC TO ABOVE W. TLT. CEILING, THROUGH WALL AND CONNECT TO F-2.
㉓ 18/14 SUPPLY DUCT THROUGH WALL TO F-3.
㉔ 16/14 RETURN DUCT THROUGH WALL TO F-3.
㉕ 16/14 RETURN DUCT THROUGH WALL TO F-4.
㉖ 18/14 SUPPLY DUCT THROUGH WALL TO F-4.
㉗ ROUTE SUPPLY/RETURN DUCTS THROUGH WEBBING OF ROOF JOISTS.
㉘ ROUTE SUPPLY/RETURN DUCTS BETWEEN ROOF JOISTS.
㉙ ROUTE SUPPLY/RETURN DUCT ABOVE CEILING.
㉚ UNIT #1 THERMOSTAT & HUMIDISTAT.
㉛ UNIT #2 THERMOSTAT.
㉜ UNIT #3 THERMOSTAT & HUMIDISTAT.
㉝ UNIT #4 THERMOSTAT & HUMIDISTAT.
㉞ 10/3| EXHAUST DUCT UP INTO ATTIC AND THROUGH ROOF TO SLOPING ROOF CAP.
㉟ RADIAL BLADE VOLUME DAMPER WITH CENTER SCREW DRIVER ACTUATOR MOUNTED IN AND ACCESSIBLE FROM SUPPLY DUCT AT 12"; BEHIND CEILING DIFFUSER.
㊱ OPPOSED BLADE VOLUME DAMPER WITH SCREW DRIVER ACTUATOR MOUNTED IN AND ACCESSIBLE FROM RETURN DUCT AT 12"; BEHIND RETURN GRILLE.
㊲ TRANSFER DUCT THROUGH WALL NEAR CEILING WITH TG-2, 16/8 ON EACH SIDE OF WALL.
㊳ 24/16 RETURN DUCT UP INTO ATTIC.
NOTE: RUN SUPPLY DUCTWORK IN JOIST SPACE AS CLOSE TO BOTTOM OF JOISTS AS POSSIBLE. COORDINATE WITH BUILDING INSULATION CONTR. TO COVER DUCTWORK WITH BATT/BLOWN-IN ROOF INSULATION.
LIBRARY 100
ENTRY 101
LOBBY 102
ENTRY 103
VILLAGE PRESIDENT 104
AUXILIARY OFFICE 105
CORRIDOR 106
PUBLIC MEETING ROOM & VILLAGE BOARD CHAMBERS 107
SEWER & WATER UTILITY OFFICE 108
VAULT 109
STORAGE 110
RECEPTION 112
MEN 114
WOMEN 115
STORAGE 116
UTILITY 117
FLOOR PLAN
SCALE: 3/16"= 1'-0"
BRUCE VILLAGE HALL & LIBRARY
BRUCE , WISCONSIN
SCALE: 3/16"= 1'-0"
MORGAN & PARMLEY, LTD. CONSULTING ENGINEERS LADYSMITH, WISCONSIN
SHEET M-2 OF 3
DATE DEC. 31, 2001
DR. BY: FLM, INC.

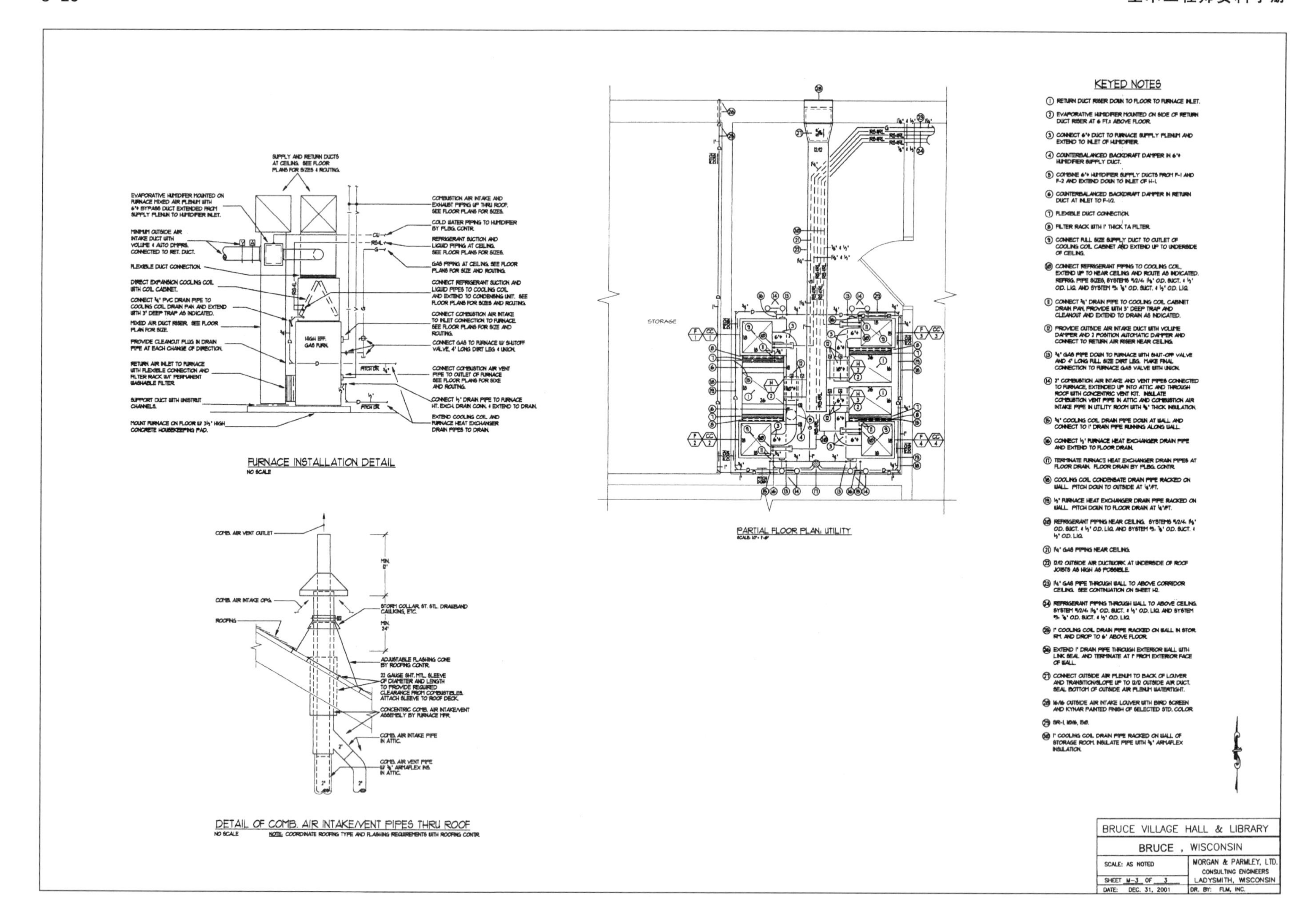
KEYED NOTES
FURNACE INSTALLATION DETAIL
NO SCALE
PARTIAL FLOOR PLAN: UTILITY
STORAGE
DETAIL OF COMB. AIR INTAKE/VENT PIPES THRU ROOF
NO SCALE
BRUCE VILLAGE HALL & LIBRARY
BRUCE , WISCONSIN
SCALE: AS NOTED
MORGAN & PARMLEY, LTD.
CONSULTING ENGINEERS
LADYSMITH, WISCONSIN
SHEET M-3 OF 3
DATE: DEC. 31, 2001
DR. BY: FLM, INC.

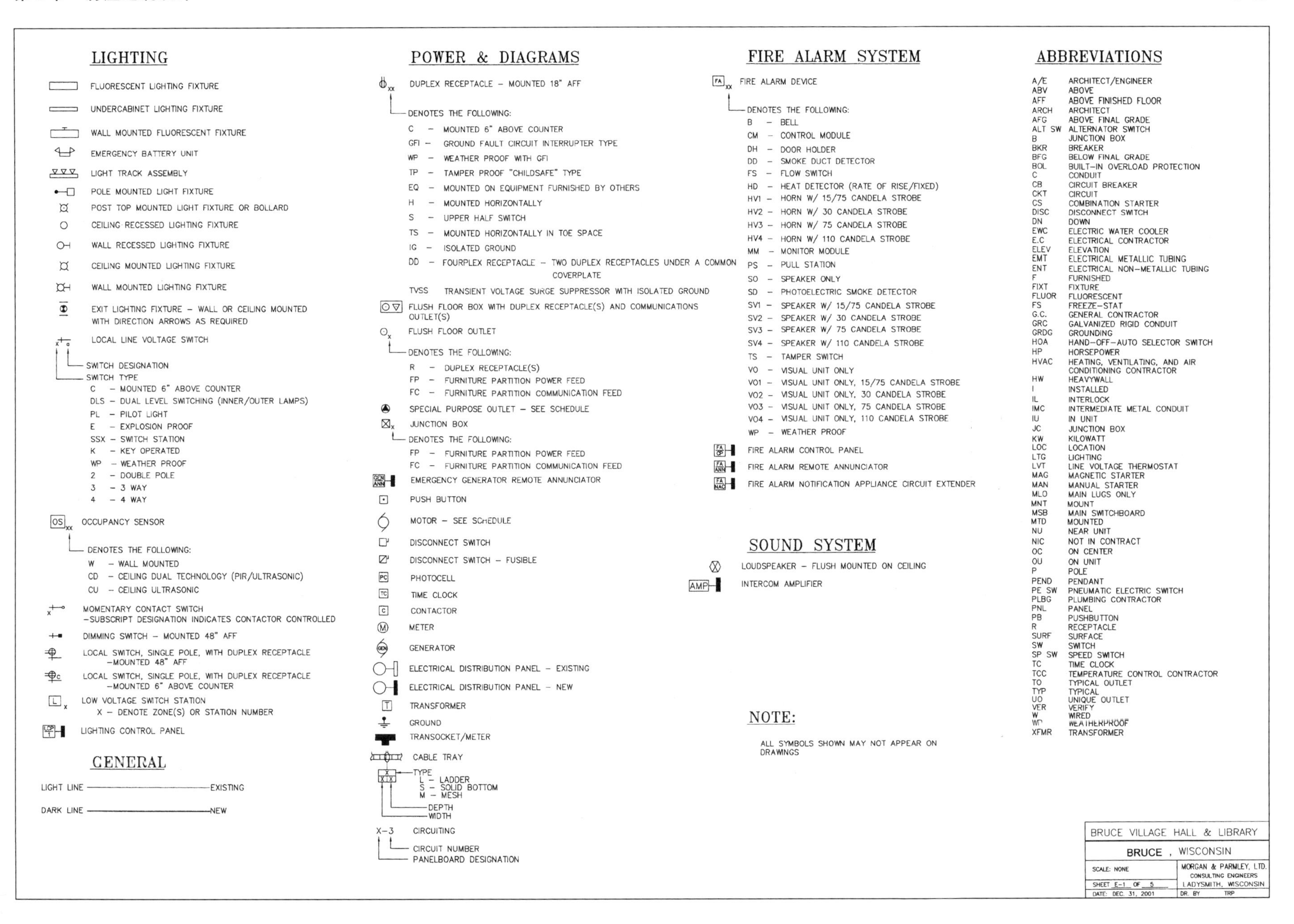
LIGHTING
FLUORESCENT LIGHTING FIXTURE
UNDERCABINET LIGHTING FIXTURE
WALL MOUNTED FLUORESCENT FIXTURE
EMERGENCY BATTERY UNIT
LIGHT TRACK ASSEMBLY
POLE MOUNTED LIGHT FIXTURE
POST TOP MOUNTED LIGHT FIXTURE OR BOLLARD
CEILING RECESSED LIGHTING FIXTURE
WALL RECESSED LIGHTING FIXTURE
CEILING MOUNTED LIGHTING FIXTURE
WALL MOUNTED LIGHTING FIXTURE
EXIT LIGHTING FIXTURE – WALL OR CEILING MOUNTED WITH DIRECTION ARROWS AS REQUIRED
LOCAL LINE VOLTAGE SWITCH
SWITCH DESIGNATION
SWITCH TYPE
C – MOUNTED 6" ABOVE COUNTER
DLS – DUAL LEVEL SWITCHING (INNER/OUTER LAMPS)
PL – PILOT LIGHT
E – EXPLOSION PROOF
SSX – SWITCH STATION
K – KEY OPERATED
WP – WEATHER PROOF
2 – DOUBLE POLE
3 – 3 WAY
4 – 4 WAY
OCCUPANCY SENSOR
DENOTES THE FOLLOWING:
W – WALL MOUNTED
CD – CEILING DUAL TECHNOLOGY (PIR/ULTRASONIC)
CU – CEILING ULTRASONIC
MOMENTARY CONTACT SWITCH
–SUBSCRIPT DESIGNATION INDICATES CONTACTOR CONTROLLED
DIMMING SWITCH – MOUNTED 48" AFF
LOCAL SWITCH, SINGLE POLE, WITH DUPLEX RECEPTACLE –MOUNTED 48" AFF
LOCAL SWITCH, SINGLE POLE, WITH DUPLEX RECEPTACLE –MOUNTED 6" ABOVE COUNTER
LOW VOLTAGE SWITCH STATION
X – DENOTE ZONE(S) OR STATION NUMBER
LIGHTING CONTROL PANEL
GENERAL
LIGHT LINE — EXISTING
DARK LINE — NEW
POWER & DIAGRAMS
DUPLEX RECEPTACLE – MOUNTED 18" AFF
DENOTES THE FOLLOWING:
C – MOUNTED 6" ABOVE COUNTER
GFI – GROUND FAULT CIRCUIT INTERRUPTER TYPE
WP – WEATHER PROOF WITH GFI
TP – TAMPER PROOF "CHILDSAFE" TYPE
EQ – MOUNTED ON EQUIPMENT FURNISHED BY OTHERS
H – MOUNTED HORIZONTALLY
S – UPPER HALF SWITCH
TS – MOUNTED HORIZONTALLY IN TOE SPACE
IG – ISOLATED GROUND
DD – FOURPLEX RECEPTACLE – TWO DUPLEX RECEPTACLES UNDER A COMMON COVERPLATE
TVSS TRANSIENT VOLTAGE SURGE SUPPRESSOR WITH ISOLATED GROUND
FLUSH FLOOR BOX WITH DUPLEX RECEPTACLE(S) AND COMMUNICATIONS OUTLET(S)
FLUSH FLOOR OUTLET
DENOTES THE FOLLOWING:
R – DUPLEX RECEPTACLE(S)
FP – FURNITURE PARTITION POWER FEED
FC – FURNITURE PARTITION COMMUNICATION FEED
SPECIAL PURPOSE OUTLET – SEE SCHEDULE
JUNCTION BOX
DENOTES THE FOLLOWING:
FP – FURNITURE PARTITION POWER FEED
FC – FURNITURE PARTITION COMMUNICATION FEED
EMERGENCY GENERATOR REMOTE ANNUNCIATOR
PUSH BUTTON
MOTOR – SEE SCHEDULE
DISCONNECT SWITCH
DISCONNECT SWITCH – FUSIBLE
PHOTOCELL
TIME CLOCK
CONTACTOR
METER
GENERATOR
ELECTRICAL DISTRIBUTION PANEL – EXISTING
ELECTRICAL DISTRIBUTION PANEL – NEW
TRANSFORMER
GROUND
TRANSOCKET/METER
CABLE TRAY
TYPE
L – LADDER
S – SOLID BOTTOM
M – MESH
DEPTH
WIDTH
X-3 CIRCUITING
CIRCUIT NUMBER
PANELBOARD DESIGNATION
FIRE ALARM SYSTEM
FIRE ALARM DEVICE
DENOTES THE FOLLOWING:
B – BELL
CM – CONTROL MODULE
DH – DOOR HOLDER
DD – SMOKE DUCT DETECTOR
FS – FLOW SWITCH
HD – HEAT DETECTOR (RATE OF RISE/FIXED)
HV1 – HORN W/ 15/75 CANDELA STROBE
HV2 – HORN W/ 30 CANDELA STROBE
HV3 – HORN W/ 75 CANDELA STROBE
HV4 – HORN W/ 110 CANDELA STROBE
MM – MONITOR MODULE
PS – PULL STATION
SO – SPEAKER ONLY
SD – PHOTOELECTRIC SMOKE DETECTOR
SV1 – SPEAKER W/ 15/75 CANDELA STROBE
SV2 – SPEAKER W/ 30 CANDELA STROBE
SV3 – SPEAKER W/ 75 CANDELA STROBE
SV4 – SPEAKER W/ 110 CANDELA STROBE
TS – TAMPER SWITCH
VO – VISUAL UNIT ONLY
VO1 – VISUAL UNIT ONLY, 15/75 CANDELA STROBE
VO2 – VISUAL UNIT ONLY, 30 CANDELA STROBE
VO3 – VISUAL UNIT ONLY, 75 CANDELA STROBE
VO4 – VISUAL UNIT ONLY, 110 CANDELA STROBE
WP – WEATHER PROOF
FIRE ALARM CONTROL PANEL
FIRE ALARM REMOTE ANNUNCIATOR
FIRE ALARM NOTIFICATION APPLIANCE CIRCUIT EXTENDER
SOUND SYSTEM
LOUDSPEAKER – FLUSH MOUNTED ON CEILING
AMP
INTERCOM AMPLIFIER
NOTE:
ALL SYMBOLS SHOWN MAY NOT APPEAR ON DRAWINGS
ABBREVIATIONS
A/E ARCHITECT/ENGINEER
ABV ABOVE
AFF ABOVE FINISHED FLOOR
ARCH ARCHITECT
AFG ABOVE FINAL GRADE
ALT SW ALTERNATOR SWITCH
B JUNCTION BOX
BKR BREAKER
BFG BELOW FINAL GRADE
BOL BUILT-IN OVERLOAD PROTECTION
C CONDUIT
CB CIRCUIT BREAKER
CKT CIRCUIT
CS COMBINATION STARTER
DISC DISCONNECT SWITCH
DN DOWN
EWC ELECTRIC WATER COOLER
E.C ELECTRICAL CONTRACTOR
ELEV ELEVATION
EMT ELECTRICAL METALLIC TUBING
ENT ELECTRICAL NON-METALLIC TUBING
F FURNISHED
FIXT FIXTURE
FLUOR FLUORESCENT
FS FREEZE-STAT
G.C. GENERAL CONTRACTOR
GRC GALVANIZED RIGID CONDUIT
GRDG GROUNDING
HOA HAND-OFF-AUTO SELECTOR SWITCH
HP HORSEPOWER
HVAC HEATING, VENTILATING, AND AIR CONDITIONING CONTRACTOR
HW HEAVYWALL
I INSTALLED
IL INTERLOCK
IMC INTERMEDIATE METAL CONDUIT
IU IN UNIT
JC JUNCTION BOX
KW KILOWATT
LOC LOCATION
LTG LIGHTING
LVT LINE VOLTAGE THERMOSTAT
MAG MAGNETIC STARTER
MAN MANUAL STARTER
MLO MAIN LUGS ONLY
MNT MOUNT
MSB MAIN SWITCHBOARD
MTD MOUNTED
NU NEAR UNIT
NIC NOT IN CONTRACT
OC ON CENTER
OU ON UNIT
P POLE
PEND PENDANT
PE SW PNEUMATIC ELECTRIC SWITCH
PLBG PLUMBING CONTRACTOR
PNL PANEL
PB PUSHBUTTON
R RECEPTACLE
SURF SURFACE
SW SWITCH
SP SW SPEED SWITCH
TC TIME CLOCK
TCC TEMPERATURE CONTROL CONTRACTOR
TO TYPICAL OUTLET
TYP TYPICAL
UO UNIQUE OUTLET
VER VERIFY
W WIRED
WP WEATHERPROOF
XFMR TRANSFORMER
BRUCE VILLAGE HALL & LIBRARY
BRUCE , WISCONSIN
SCALE: NONE
MORGAN & PARMLEY, LTD.
CONSULTING ENGINEERS
LADYSMITH, WISCONSIN
SHEET E-1 OF 5
DATE: DEC. 31, 2001
DR. BY TRP

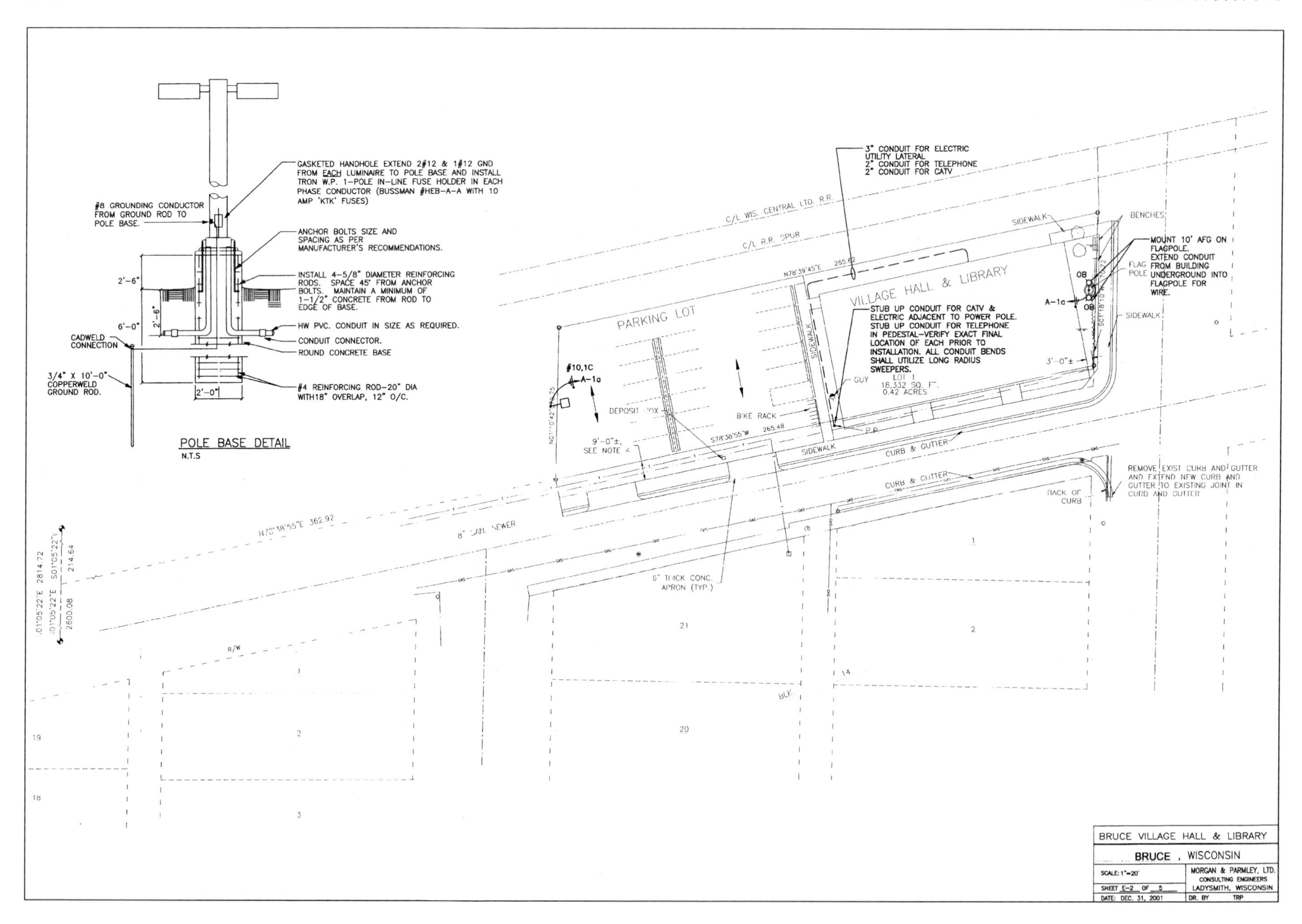
GASKETED HANDHOLE EXTEND 2#12 & 1#12 GND FROM EACH LUMINAIRE TO POLE BASE AND INSTALL TRON W.P. 1-POLE IN-LINE FUSE HOLDER IN EACH PHASE CONDUCTOR (BUSSMAN #HEB-A-A WITH 10 AMP 'KTK' FUSES)
#8 GROUNDING CONDUCTOR FROM GROUND ROD TO POLE BASE.
ANCHOR BOLTS SIZE AND SPACING AS PER MANUFACTURER'S RECOMMENDATIONS.
INSTALL 4-5/8" DIAMETER REINFORCING RODS. SPACE 45° FROM ANCHOR BOLTS. MAINTAIN A MINIMUM OF 1-1/2" CONCRETE FROM ROD TO EDGE OF BASE.
HW PVC. CONDUIT IN SIZE AS REQUIRED.
CONDUIT CONNECTOR.
ROUND CONCRETE BASE
CADWELD CONNECTION
3/4" X 10'-0" COPPERWELD GROUND ROD.
#4 REINFORCING ROD-20" DIA WITH18" OVERLAP, 12" O/C.
2'-6"
6'-0"
2'-0"
POLE BASE DETAIL
N.T.S
3" CONDUIT FOR ELECTRIC UTILITY LATERAL
2" CONDUIT FOR TELEPHONE
2" CONDUIT FOR CATV
C/L WIS. CENTRAL LTD. R.R.
C/L R.R. SPUR
SIDEWALK
BENCHES
MOUNT 10' AFG ON FLAGPOLE. EXTEND CONDUIT FROM BUILDING UNDERGROUND INTO FLAGPOLE FOR WIRE.
FLAG POLE
OB
A-1a
VILLAGE HALL & LIBRARY
PARKING LOT
STUB UP CONDUIT FOR CATV & ELECTRIC ADJACENT TO POWER POLE. STUB UP CONDUIT FOR TELEPHONE IN PEDESTAL-VERIFY EXACT FINAL LOCATION OF EACH PRIOR TO INSTALLATION. ALL CONDUIT BENDS SHALL UTILIZE LONG RADIUS SWEEPERS.
GUY
LOT 1
18,332 SQ. FT.
0.42 ACRES
3'-0"±
#10,1C
DEPOSIT BOX
BIKE RACK
N78°39'45"E 265.62
S78°38'55"W 265.48
9'-0"±, SEE NOTE 4
SIDEWALK
CURB & GUTTER
REMOVE EXIST CURB AND GUTTER AND EXTEND NEW CURB AND GUTTER TO EXISTING JOINT IN CURB AND GUTTER
BACK OF CURB
N78°38'55"E 362.92
8" SAN. SEWER
6" THICK CONC. APRON (TYP.)
R/W
BLK
BRUCE VILLAGE HALL & LIBRARY
BRUCE, WISCONSIN
SCALE: 1"=20'
SHEET E-2 OF 5
DATE: DEC. 31, 2001
MORGAN & PARMLEY, LTD.
CONSULTING ENGINEERS
LADYSMITH, WISCONSIN
DR. BY TRP

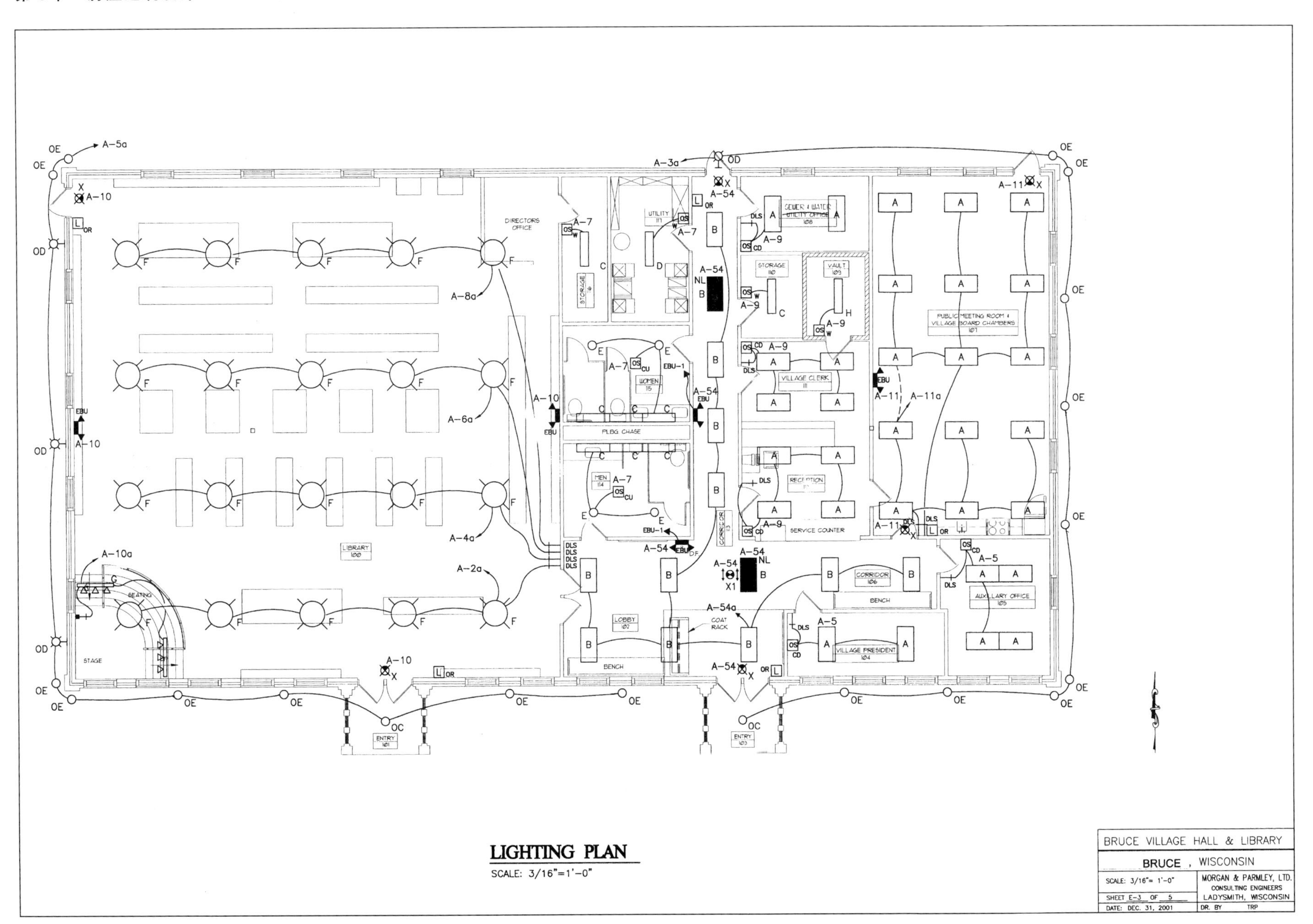
LIGHTING PLAN
SCALE: 3/16"=1'-0"
BRUCE VILLAGE HALL & LIBRARY
BRUCE , WISCONSIN
SCALE: 3/16"= 1'-0"
MORGAN & PARMLEY, LTD.
CONSULTING ENGINEERS
LADYSMITH, WISCONSIN
SHEET E-3 OF 5
DATE: DEC. 31, 2001
DR. BY TRP
LIBRARY 100
DIRECTORS OFFICE
STAGE
SEATING
LOBBY 102
CORRIDOR 106
AUXILLARY OFFICE 105
VILLAGE PRESIDENT 104
VILLAGE CLERK 111
RECEPTION
SERVICE COUNTER
STORAGE 110
VAULT 109
SEWER & WATER UTILITY OFFICE 108
PUBLIC MEETING ROOM & VILLAGE BOARD CHAMBERS 107
UTILITY 117
WOMEN 115
MEN 114
PLBG CHASE
BENCH
COAT RACK
ENTRY 101
ENTRY 103

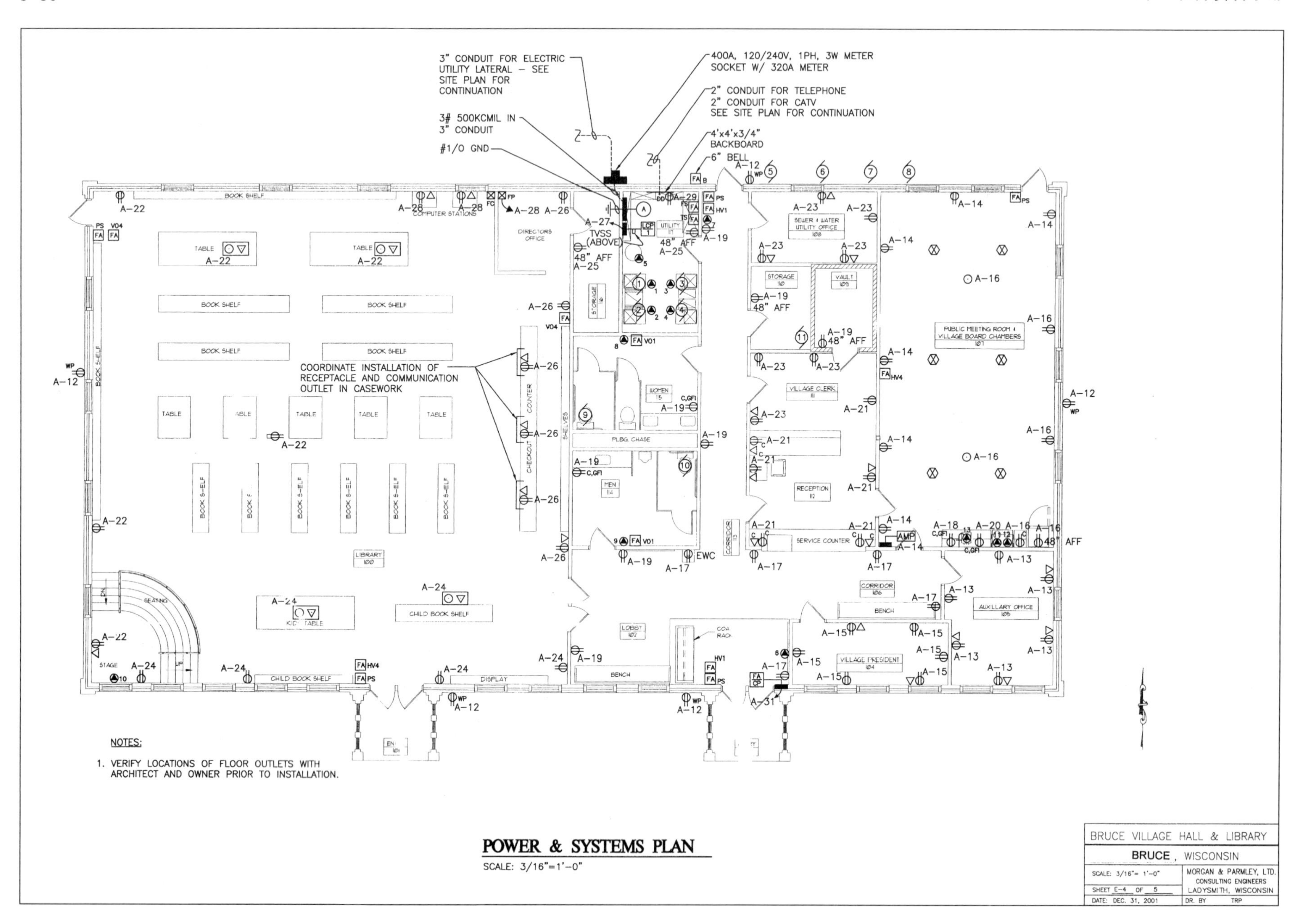

3" CONDUIT FOR ELECTRIC UTILITY LATERAL – SEE SITE PLAN FOR CONTINUATION
400A, 120/240V, 1PH, 3W METER SOCKET W/ 320A METER
2" CONDUIT FOR TELEPHONE
2" CONDUIT FOR CATV
SEE SITE PLAN FOR CONTINUATION
3# 500KCMIL IN 3" CONDUIT
4'x4'x3/4" BACKBOARD
#1/0 GND
6" BELL
COORDINATE INSTALLATION OF RECEPTACLE AND COMMUNICATION OUTLET IN CASEWORK
NOTES:
1. VERIFY LOCATIONS OF FLOOR OUTLETS WITH ARCHITECT AND OWNER PRIOR TO INSTALLATION.
POWER & SYSTEMS PLAN
SCALE: 3/16"=1'-0"
BRUCE VILLAGE HALL & LIBRARY
BRUCE, WISCONSIN
SCALE: 3/16"= 1'-0"
SHEET E-4 OF 5
DATE: DEC. 31, 2001
MORGAN & PARMLEY, LTD.
CONSULTING ENGINEERS
LADYSMITH, WISCONSIN
DR. BY TRP

SPECIAL OUTLET SCHEDULE

TERMINAL: R – RECEPTACLE　D – DISCONNECT SWITCH CONNECTION　B – JUNCTION BOX CONNECTION

NO.	TO FEED	LOC.	FEED FROM		BREAKER		WIRING				TERMINAL			VOLT	ø	LOAD	SEE NOTE
			PANEL	CKT.	SIZE	POLE	NO.	SIZE	GND	COND.	R	D	B				
1	HUMIDIFIER	UTILITY	A	33	20	1	2	12	12	1/2		X		120	1	.6A	
2	HUMIDIFIER	UTILITY	A	35	20	1	2	12	12	1/2		X		120	1	.6A	
3	HUMIDIFIER	UTILITY	A	37	20	1	2	12	12	1/2		X		120	1	.6A	
4	HUMIDIFIER	UTILITY	A	39	20	1	2	12	12	1/2		X		120	1	.6A	
5	WATER HEATER	UTILITY	A	57,59	25	2	2	10	10	1/2		X		240	1	4500W MAX	
6	EW-2	CORRIDOR	A	44	20	1	2	12	12	1/2		X		120	1	12.6A	1,2
7	EW-2	CORRIDOR	A	46	20	1	2	12	12	1/2		X		120	1	12.6A	1,2
8	EW-1	WOMEN	A	48	20	1	2	12	12	1/2		X		120	1	8.4A	1,2
9	EW-1	MEN	A	50	20	1	2	12	12	1/2		X		120	1	8.4A	1,2
10	EB-1	LIBRARY	A	47	20	1	2	12	12	1/2			X	120	1	8.3A	2
11	STOVE	MTG	A	53,55	50	2	3	8	10	3/4	X			120/240	1	8KW	
12	HOOD	MTG	A	42	20	1	2	12	12	1/2			X	120	1	5A	
13	DISPOSER	MTG	A	52	20	1	2	12	12	1/2		X		120	1	6A	

NOTES:
1. INSTALL AND WIRE MFR. SUPPLIED DISCONNECT SWITCH.
2. INSTALL AND WIRE HEATER.

LIGHTING FIXTURE SCHEDULE

ABBREVIATIONS

ES – EXPOSED STRUCTURE　LG – LAYIN GRID　V – VARIES　F – FLUSH　C – CONCRETE
G – GYP BOARD　S – SURFACE　W – WALL　U – UNIVERSAL　R – RECESSED

DES.	DESCRIPTION	LAMP DATA		VOLT	DEPTH	LIGHTING FIXTURE		MTG.	MTG. SURF	SEE NOTE
		NO.	TYPE			MFR.	CAT. NO.			
OA	PARKING LOT LIGHTING UNIT	2	250W MH	120	–	LITHONIA	(2) KAD250MR4120SPD04 (1) SSA25SG	POLE	CONC BASE	
OB	FLAGPOLE LIGHT	1	100W MH/PAR 38	120	–	DALTOR	FP/818-100MH-RAD	POLE	METAL	2
OC	CANOPY LIGHT	1	70W MH	120	7 3/8"	LITHONIA	LP6H70M120/6L4	R	V	
OD	WALL LIGHT	1	70W MH	120	–	LITHONIA	TWA70M120	S	CONC	3
OE	SOFFIT DOWNLIGHT	1	50W MH	120	7 3/8"	LITHONIA	LP6H50M120/6L4	R	V	4
A	2x4 PARABOLIC	3	F32T8/SP35	120	5 15/16"	LITHONIA	2PM3GB33218LD120GEB10	R	LG	
B	2x4 PARABOLIC	2	F32T8/SP35	120	5 15/16"	LITHONIA	2PM3GB23212LD120GEB10	R	LG	
C	1x4 LENSED	2	F32T8/SP35	120	4 7/8"	LITHONIA	SPG232A12125120GEB10	R	LG	
D	INDUSTRIAL	2	F32T8/SP35	120	–	LITHONIA	AF10232120GEB10	S	ES	
E	DOWNLIGHT	2	PL18/SP35	120	5 1/2"	LITHONIA	AF218DTT6WR120	R	LG	
F	PENDENT	6	BX50/SP35	120	–	VISA	CP36696F50120BA	P	GYP	1
G	TRACK	3	75W PAR30/FL	120	–	LITHONIA	(3) LTCBLSPPAR30WH (1) LT4WH/LTAWH/LTA11	S	GYP	
H	SURFACE	2	F32T8/SP35	120	–	LITHONIA	LB232120 GEB10	S	C	
X	EXIT SIGN – 1 FACE	–	LED	120	–	LITHONIA	LES1G120 ELN SD	UNIV	V	
X1	EXIT SIGN – 2 FACES	–	LED	120	–	LITHONIA	LES2G120 ELN SD	UNIV	V	
EBU	BATTERY UNIT	2	5W MR16(INCL)	120	–	DUAL-LITE	LZ15PI	S	V	
EBU-1	REMOTE HEAD	1	5W MR16(INCL)	6	–	DUAL-LITE	LZR6V5W	S	V	

NOTES:
1. MOUNTING HEIGHT IS 9' AFF – PROVIDE STEM LENGTH AS REQUIRED.
2. FIXTURE DESIGNED TO BE INSTALLED FROM OUTSIDE OF FLAGPOLE. DRILL 1/2" HOLE AND PROVIDE GROMMET FOR WIRING IN POLE. FIXTURE TO HAVE FINISH TO MATCH POLE.
3. MOUNT 1'-0" BELOW SOFFIT.
4. LOCATE CENTERED IN SOFFIT. ADD AND/OR ADJUST 2x4's IN SOFFIT TO ACCOMMODATE FIXTURES AS NEEDED.

LIGHTING RELAY CABINET SCHEDULE

DESIGNATION: LCP　　LOCATION: UTILITY

RELAY NO(S)	CIRCUIT NO(S)	LOAD DESCRIPTION	PROG. HOURS OF OPERATION		LOW VOLTAGE SWITCH LOCATION		NOTES
			OPEN	DUSK/DAWN	ROOM	QUANTITY	
1	A-1a	EXTERIOR LIGHTING (SITE)		x			
2,3	A-3a,5a	EXTERIOR LIGHTING (BLDG)		x			
4	A-54a	CORRIDOR	x		CORRIDOR	2	
5,6,7,8,9	A-2a,4a,6a,8a,10a	LIBRARY	x		LIBRARY	2	
10	A-11a	MEETING ROOM	x		MEETING ROOM	1	
11-16	–	SPARE					

MOTOR WIRING SCHEDULE ◊

ABBREVIATIONS

IU – IN UNIT　OU – ON UNIT　NU – NEAR UNIT
WP – WEATHER PROOF　MCC – MOTOR CONTROL CENTER　MFR – MANUFACTURER
EC – ELECTRICAL CONTRACTOR　MC – MECHANICAL CONTRACTOR　MAN – MANUAL STARTER
PC – PLUMBING CONTRACTOR　ES – EQUIPMENT SUPPLIER　TS – TOGGLE SWITCH
CS – COMBINATION STARTER　MAG – MAGNETIC STARTER　OS – OCCUPANCY SENSOR
NF – NON FUSED　F – FUSED

NO	HP	VOLTS	ø	LOC	DRIVING	FED FROM		STARTER					DISCONNECT					BKR		BR. WIRING				SEE NOTE
						PNL.	NO.	TYPE	F	I	W	LOC	TYPE	F	I	W	LOC	SIZE	POLE	NO	SIZE	GND	COND	
1	3/4	120	1	UTILITY	FURNACE	A	33	MAG	MFR	MFR	EC	IU	NF	EC	EC	EC	NU	20	1	2	12	12	1/2	
2	3/4	120	1	UTILITY	FURNACE	A	35	MAG	MFR	MFR	EC	IU	NF	EC	EC	EC	NU	20	1	2	12	12	1/2	
3	1/2	120	1	UTILITY	FURNACE	A	37	MAG	MFR	MFR	EC	IU	NF	EC	EC	EC	NU	20	1	2	12	12	1/2	
4	3/4	120	1	UTILITY	FURNACE	A	39	MAG	MFR	MFR	EC	IU	NF	EC	EC	EC	NU	20	1	2	12	12	1/2	
5	30.5A	240	1	EXT	CONDENSER	A	30,32	MAG	MFR	MFR	EC	IU	F/WP	EC	EC	EC	NU	50	2	2	8	10	3/4	
6	30.5A	240	1	EXT	CONDENSER	A	34,36	MAG	MFR	MFR	EC	IU	F/WP	EC	EC	EC	NU	50	2	2	8	10	3/4	
7	23.4A	240	1	EXT	CONDENSER	A	38,40	MAG	MFR	MFR	EC	IU	F/WP	EC	EC	EC	NU	40	2	2	8	10	3/4	
8	30.5A	240	1	EXT	CONDENSER	A	43,45	MAG	MFR	MFR	EC	IU	F/WP	EC	EC	EC	NU	50	2	2	8	10	3/4	
9	2.7A	120	1	WOMEN	EF-1	A	36	MAN	EC	EC	EC	117	–	–	–	–	–	20	1	2	12	12	1/2	1,2
10	2.7A	120	1	MEN	EF-2	A	36	MAN	EC	EC	EC	117	–	–	–	–	–	20	1	2	12	12	1/2	1,2
11	1.8A	120	1	STORAGE	EF-3	A	36	MAN	EC	EC	EC	117	–	–	–	–	–	20	1	2	12	12	1/2	1,2

NOTES:
1. WIRE STARTER THROUGH RELAY SUPPLIED BY CONTROL CONTRACTOR.
2. WIRE THROUGH MFR. SUPPLIED SPEED SWITCH INSTALLED ON SIDE OF FAN HOUSING (SPEED SWITCH HAS OFF POSITION).

PANEL A

PANEL: A	BUS AMPS: 400	MAIN: CB 400A MCB	MOUNTING		REMARKS:
	VOLTAGE: 120/240		SURFACE	X	SERVICE ENTRANCE RATED, 22K AIC
	PHASE: 1		FLUSH		(2) 42CKT PANELS
	WIRE: 3				

CIRCUIT DESCRIPTION	AMPS A	AMPS B	CB	CCT	CCT	CB	AMPS A	AMPS B	CIRCUIT DESCRIPTION
EXT LTG	5		20	1	2	20	12.5		LIBRARY LTG
EXT LTG		7.6	20	3	4	20		12.5	LIBRARY LTG
HALL/VP/OFFICE EXT. LTG	8.6		20	5	6	20	12.5		LIBRARY LTG
BATHRM/UTIL LTG		11.2	20	7	8	20		12.5	LIBRARY LTG
OFFICE LTG	8.5		20	9	10	20	8		LIBRARY LTG
MTG RM LTG		11.3	20	11	12	20		7.5	EXT REC
REC	9		20	13	14	20	7.5		REC
REC		9	20	15	16	20		10	REC
REC	6		20	17	18	20	10		REC
REC		7.5	20	19	20	20		10	REC
REC	7.5		20	21	22	20	6		REC
REC		7.5	20	23	24	20		6	REC
REC	6		20	25	26	20	6		REC
LCP-1		5	20	27	28	20		6	REC
BACKBOARD REC	6		20	29	30	50	30.5		MTR#5
FACP		10	20	31	32	2		30.5	MTR#5
MTR#1,SO#1	10		20	33	34	50	30.5		MTR#6
MTR#2,SO#2		10	20	35	36	2		30.5	MTR#6
MTR#3,SO#3	10		20	37	38	40	23.4		MTR#7
MTR#4,SO#4		10	20	39	40	2		23.4	MTR#7
SPARE	–		20	41	42	20	5		SO#12
MTR#8		30.5	50	43	44	20		8.4	SO#6
MTR#8	30.5		2	45	46	20	8.4		SO#7
SO#10		7.9	20	47	48	20		12.5	SO#8
TVSS	–		60	49	50	20	12.5		SO#9
TVSS		–	2	51	52	20		6	SO#13
SO#11	30		50	53	54	20	11		HALL/VP/OFFICE EXT. LTG
SO#11		30	2	55	56	20		7.2	MTR#9,#10,#11
SO#5	19		25	57	58	20	–		SPARE
SO#5		19	2	59	60	20		–	SPARE
	–			61	62		–		
		–		63	64			–	
	–			65	66		–		
		–		67	68			–	
	–			69	70		–		
		–		71	72			–	
	–			73	74		–		
		–		75	76			–	
	–			77	78		–		
		–		79	80			–	
	–			81	82		–		
		–		83	84			–	

BRUCE VILLAGE HALL & LIBRARY	
BRUCE, WISCONSIN	
SCALE: NONE	MORGAN & PARMLEY, LTD. CONSULTING ENGINEERS LADYSMITH, WISCONSIN
SHEET E-5 OF 5	
DATE: DEC. 31, 2001	DR. BY TRP

斯波肯·任德教堂

SPOKEN WORD CHURCH

LADYSMITH , WISCONSIN

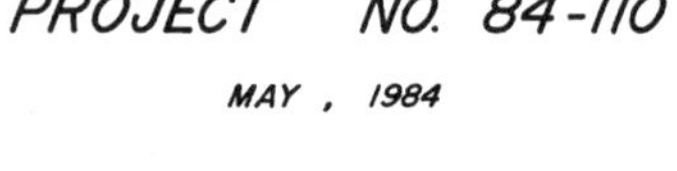

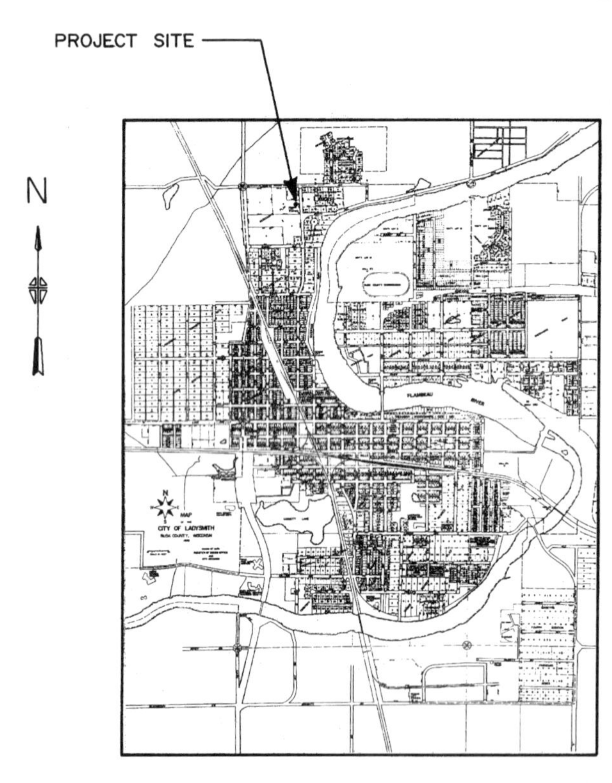

GENERAL LOCATION MAP

INDEX OF SHEETS

SHEET NO.	DESCRIPTION
1	TITLE SHEET W/ GENERAL LOCATION MAP
2	SITE PLAN
3	FLOOR PLAN
4	ELEVATION VIEWS
5	BUILDING CROSS-SECTION
6	ROOF FRAMING PLAN W/ CONSTRUCTION DETAILS
7	ELECTRICAL, HEATING & VENTILATING
8	PLUMBING

PREPARED BY:

MORGAN & PARMLEY, LTD.
CONSULTING ENGINEERS
LADYSMITH, WISCONSIN

Courtesy: Morgan & Parmley, Ltd.

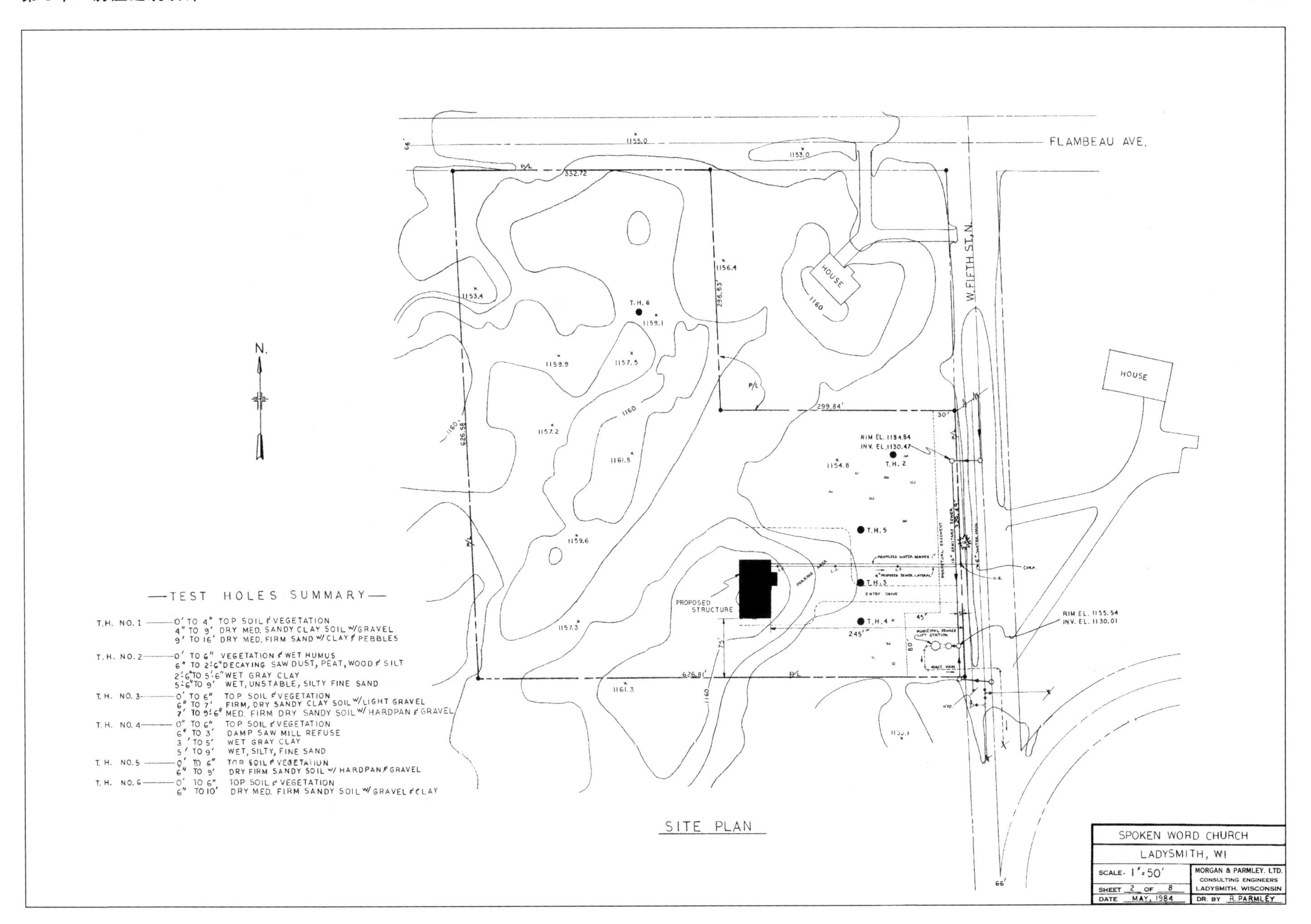
FLAMBEAU AVE.
W. FIFTH ST. N.
HOUSE
HOUSE
N.
T.H. 6
T.H. 2
T.H. 5
T.H. 3
T.H. 4
RIM EL. 1154.84
INV. EL. 1130.47
RIM EL. 1155.54
INV. EL. 1130.01
PROPOSED STRUCTURE
ENTRY DRIVE
PARKING AREA
—TEST HOLES SUMMARY—
T.H. NO. 1 — 0' TO 4" TOP SOIL & VEGETATION
4" TO 9' DRY MED. SANDY CLAY SOIL W/GRAVEL
9' TO 16' DRY MED. FIRM SAND W/CLAY & PEBBLES
T.H. NO. 2 — 0' TO 6" VEGETATION & WET HUMUS
6" TO 2'-6" DECAYING SAW DUST, PEAT, WOOD & SILT
2'-6" TO 5'-6" WET GRAY CLAY
5'-6" TO 9' WET, UNSTABLE, SILTY FINE SAND
T.H. NO. 3 — 0' TO 6" TOP SOIL & VEGETATION
6" TO 7' FIRM, DRY SANDY CLAY SOIL W/LIGHT GRAVEL
7' TO 9'-6" MED. FIRM DRY SANDY SOIL W/ HARDPAN & GRAVEL
T.H. NO. 4 — 0" TO 6" TOP SOIL & VEGETATION
6" TO 3' DAMP SAW MILL REFUSE
3' TO 5' WET GRAY CLAY
5' TO 9' WET, SILTY, FINE SAND
T.H. NO. 5 — 0' TO 6" TOP SOIL & VEGETATION
6" TO 9' DRY FIRM SANDY SOIL W/ HARDPAN & GRAVEL
T.H. NO. 6 — 0' TO 6" TOP SOIL & VEGETATION
6" TO 10' DRY MED. FIRM SANDY SOIL W/ GRAVEL & CLAY
SITE PLAN
SPOKEN WORD CHURCH
LADYSMITH, WI
SCALE- 1"=50'
SHEET 2 OF 8
DATE MAY, 1984
MORGAN & PARMLEY. LTD.
CONSULTING ENGINEERS
LADYSMITH, WISCONSIN
DR. BY R. PARMLEY

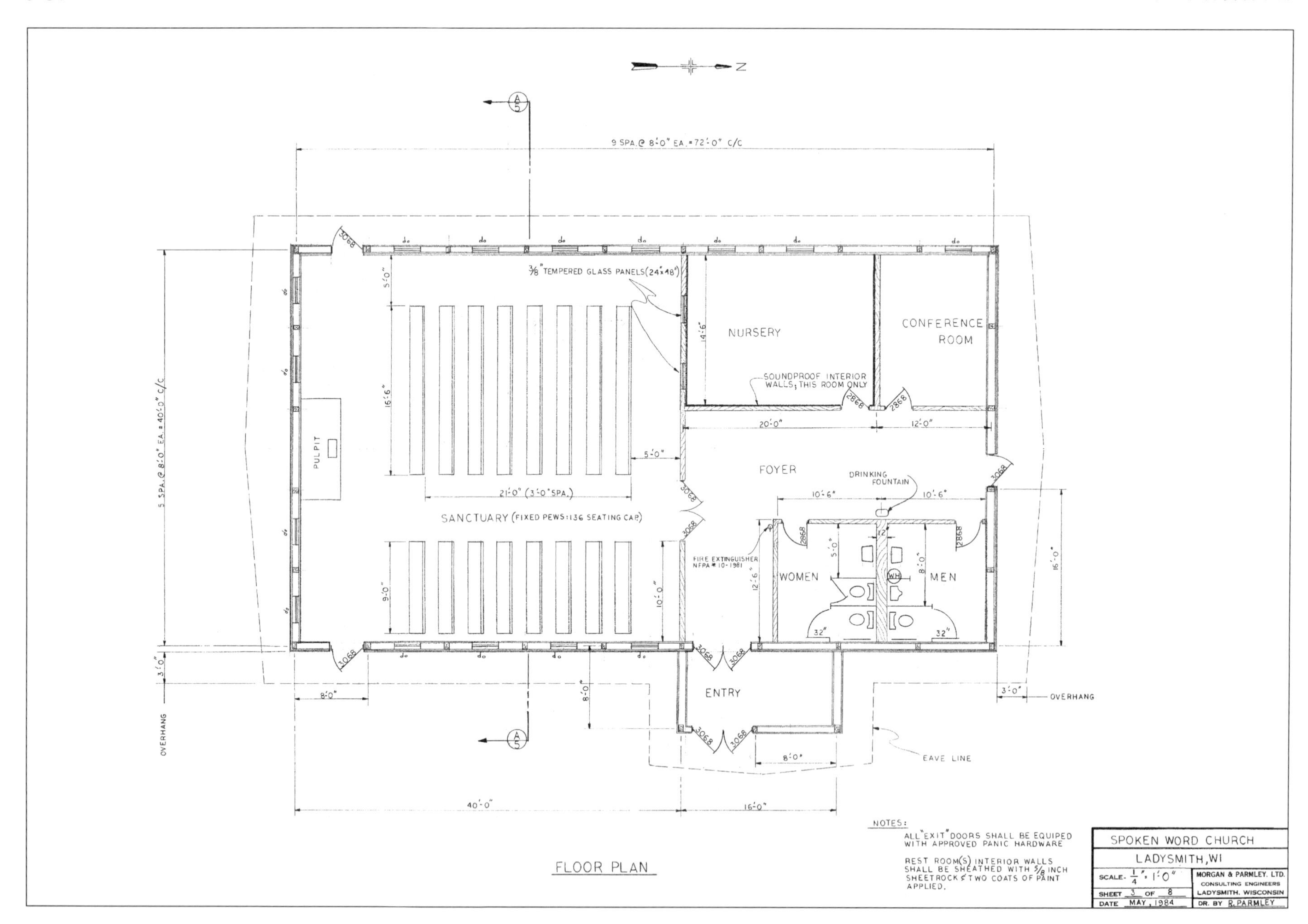
Z
9 SPA.@ 8'-0" EA.=72'-0" C/C
5 SPA.@ 8'-0" EA.=40'-0" C/C
3/8" TEMPERED GLASS PANELS (24"x48")
NURSERY
CONFERENCE ROOM
SOUNDPROOF INTERIOR WALLS, THIS ROOM ONLY
PULPIT
FOYER
DRINKING FOUNTAIN
SANCTUARY (FIXED PEWS: 136 SEATING CAP.)
21'-0" (3'-0" SPA.)
FIRE EXTINGUISHER NFPA # 10-1981
WOMEN
MEN
ENTRY
OVERHANG
EAVE LINE
40'-0"
16'-0"
NOTES:
ALL "EXIT" DOORS SHALL BE EQUIPED WITH APPROVED PANIC HARDWARE
REST ROOM(S) INTERIOR WALLS SHALL BE SHEATHED WITH 5/8 INCH SHEETROCK & TWO COATS OF PAINT APPLIED.
FLOOR PLAN
SPOKEN WORD CHURCH
LADYSMITH, WI
SCALE 1/4"=1'-0"
SHEET 3 OF 8
DATE MAY, 1984
MORGAN & PARMLEY, LTD. CONSULTING ENGINEERS LADYSMITH, WISCONSIN
DR. BY R. PARMLEY

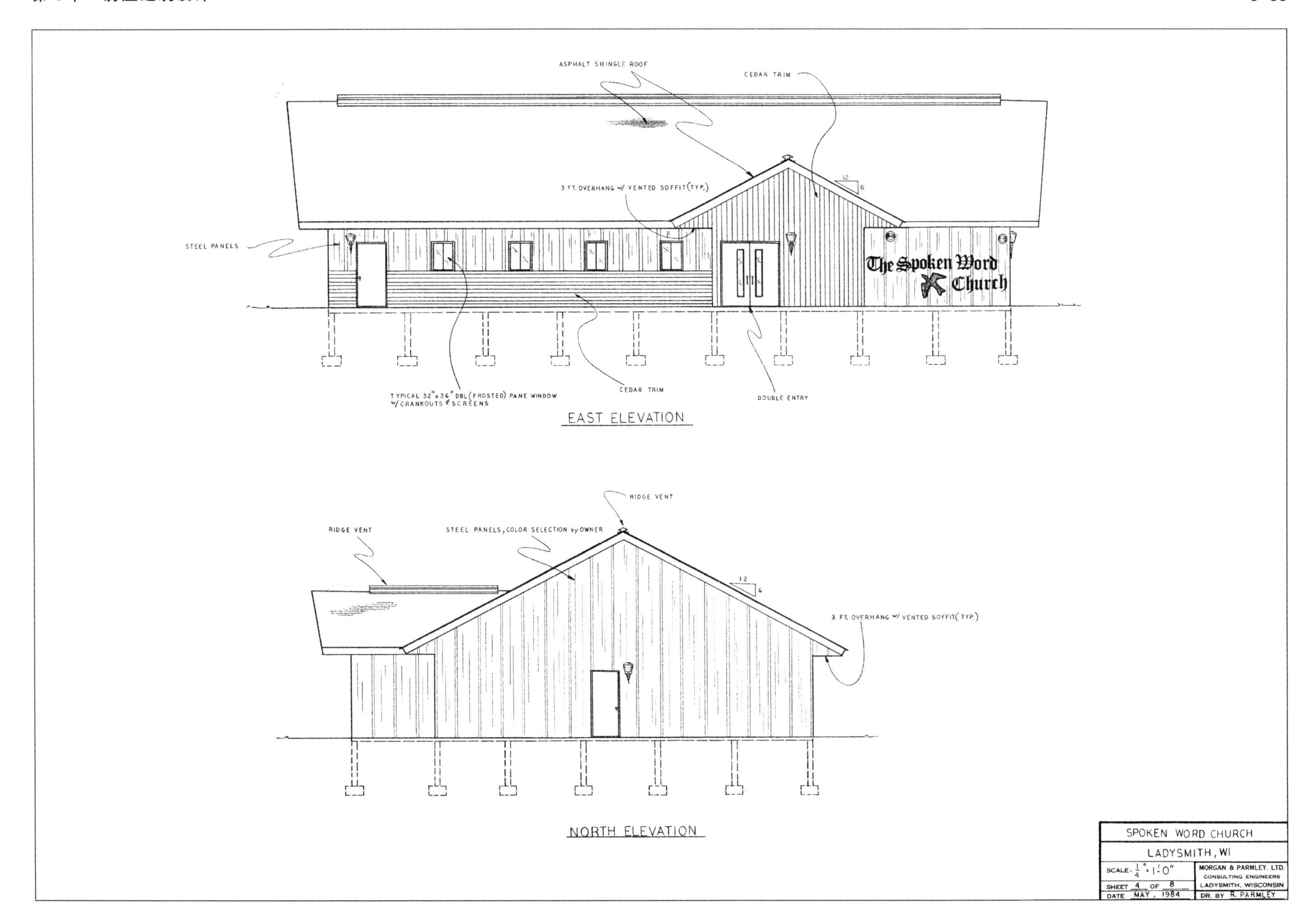
ASPHALT SHINGLE ROOF
CEDAR TRIM
3 FT. OVERHANG w/ VENTED SOFFIT (TYP.)
12
6
STEEL PANELS
The Spoken Word Church
TYPICAL 32"x36" DBL (FROSTED) PANE WINDOW w/ CRANKOUTS & SCREENS
CEDAR TRIM
DOUBLE ENTRY
EAST ELEVATION
RIDGE VENT
RIDGE VENT
STEEL PANELS, COLOR SELECTION by OWNER
12
6
3 FT. OVERHANG w/ VENTED SOFFIT (TYP.)
NORTH ELEVATION
SPOKEN WORD CHURCH
LADYSMITH, WI
SCALE - 1/4" = 1'-0"
SHEET 4 OF 8
DATE MAY, 1984
MORGAN & PARMLEY, LTD.
CONSULTING ENGINEERS
LADYSMITH, WISCONSIN
DR. BY R. PARMLEY

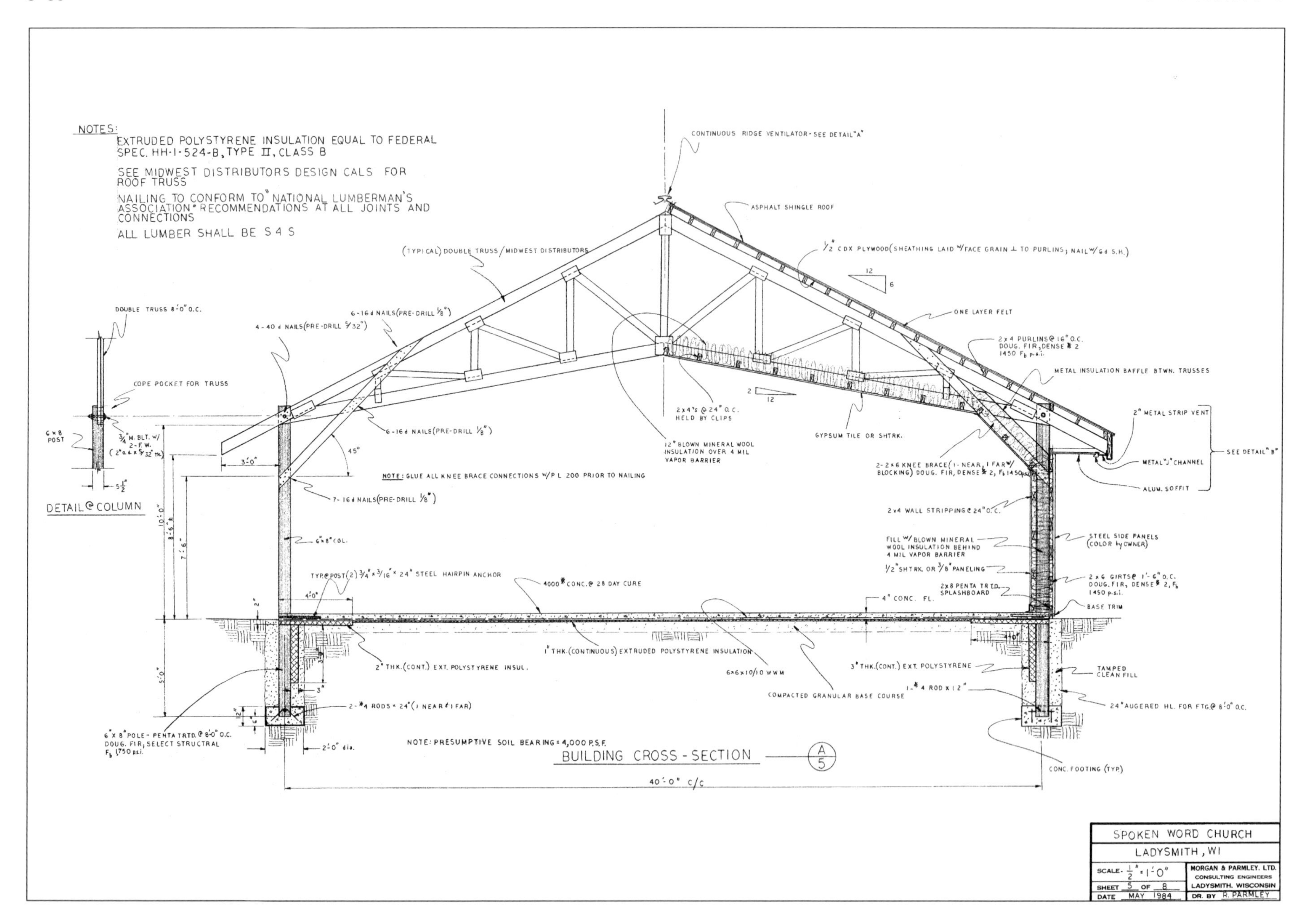
NOTES:
EXTRUDED POLYSTYRENE INSULATION EQUAL TO FEDERAL SPEC. HH-I-524-B, TYPE II, CLASS B
SEE MIDWEST DISTRIBUTORS DESIGN CALS FOR ROOF TRUSS
NAILING TO CONFORM TO "NATIONAL LUMBERMAN'S ASSOCIATION" RECOMMENDATIONS AT ALL JOINTS AND CONNECTIONS
ALL LUMBER SHALL BE S 4 S
CONTINUOUS RIDGE VENTILATOR-SEE DETAIL "A"
ASPHALT SHINGLE ROOF
(TYPICAL) DOUBLE TRUSS/MIDWEST DISTRIBUTORS
DETAIL @ COLUMN
NOTE: GLUE ALL KNEE BRACE CONNECTIONS W/PL 200 PRIOR TO NAILING
BUILDING CROSS-SECTION
NOTE: PRESUMPTIVE SOIL BEARING = 4,000 P.S.F.
40'-0" c/c
SPOKEN WORD CHURCH
LADYSMITH, WI
MORGAN & PARMLEY, LTD.
CONSULTING ENGINEERS
LADYSMITH, WISCONSIN
DATE MAY 1984
DR. BY R. PARMLEY

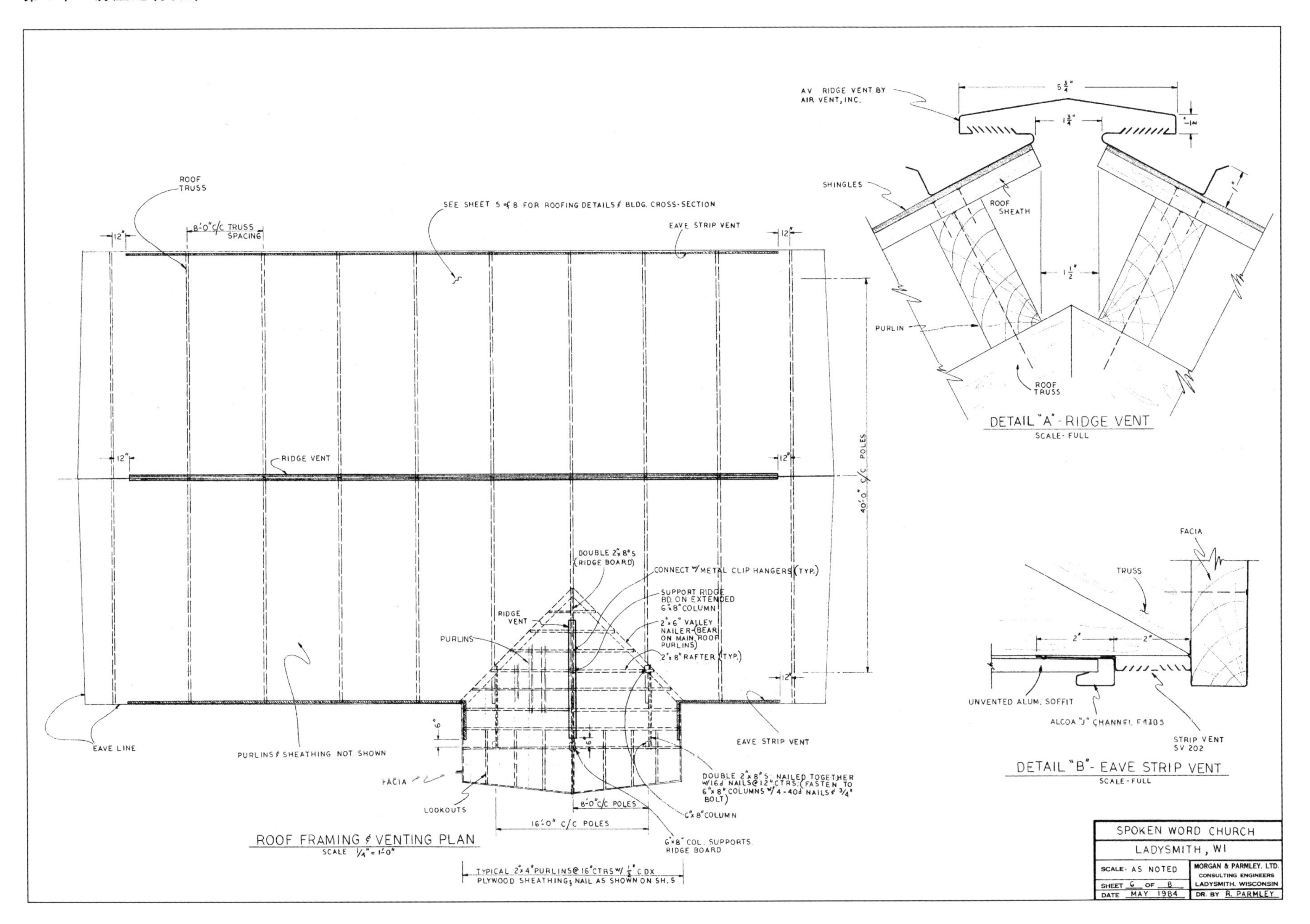
AV RIDGE VENT BY AIR VENT, INC.
SHINGLES
ROOF SHEATH
PURLIN
ROOF TRUSS
DETAIL "A" - RIDGE VENT
SCALE - FULL
ROOF TRUSS
SEE SHEET 5 of 8 FOR ROOFING DETAILS & BLDG. CROSS-SECTION
8'-0" C/C TRUSS SPACING
EAVE STRIP VENT
RIDGE VENT
40'-0" C/C POLES
DOUBLE 2"x8"S (RIDGE BOARD)
CONNECT W/ METAL CLIP HANGERS (TYP.)
SUPPORT RIDGE BD. ON EXTENDED 6"x8" COLUMN
2"x6" VALLEY NAILER (BEAR ON MAIN ROOF PURLINS)
2"x8" RAFTER (TYP.)
PURLINS
EAVE LINE
PURLINS & SHEATHING NOT SHOWN
FACIA
LOOKOUTS
8'-0" C/C POLES
16'-0" C/C POLES
6"x8" COLUMN
6"x8" COL. SUPPORTS RIDGE BOARD
DOUBLE 2"x8"S NAILED TOGETHER W/16d NAILS @ 12" CTRS. (FASTEN TO 6"x8" COLUMNS W/ 4-40d NAILS & 3/4" BOLT)
EAVE STRIP VENT
ROOF FRAMING & VENTING PLAN
SCALE 1/4" = 1'-0"
TYPICAL 2"x4" PURLINS @ 16" CTRS W/ 1/2" CDX PLYWOOD SHEATHING; NAIL AS SHOWN ON SH. 5
FACIA
TRUSS
UNVENTED ALUM. SOFFIT
ALCOA "J" CHANNEL F4305
STRIP VENT SV 202
DETAIL "B" - EAVE STRIP VENT
SCALE - FULL
SPOKEN WORD CHURCH
LADYSMITH, WI
SCALE - AS NOTED
SHEET 6 OF 8
DATE MAY 1984
MORGAN & PARMLEY, LTD.
CONSULTING ENGINEERS
LADYSMITH, WISCONSIN
DR. BY R. PARMLEY

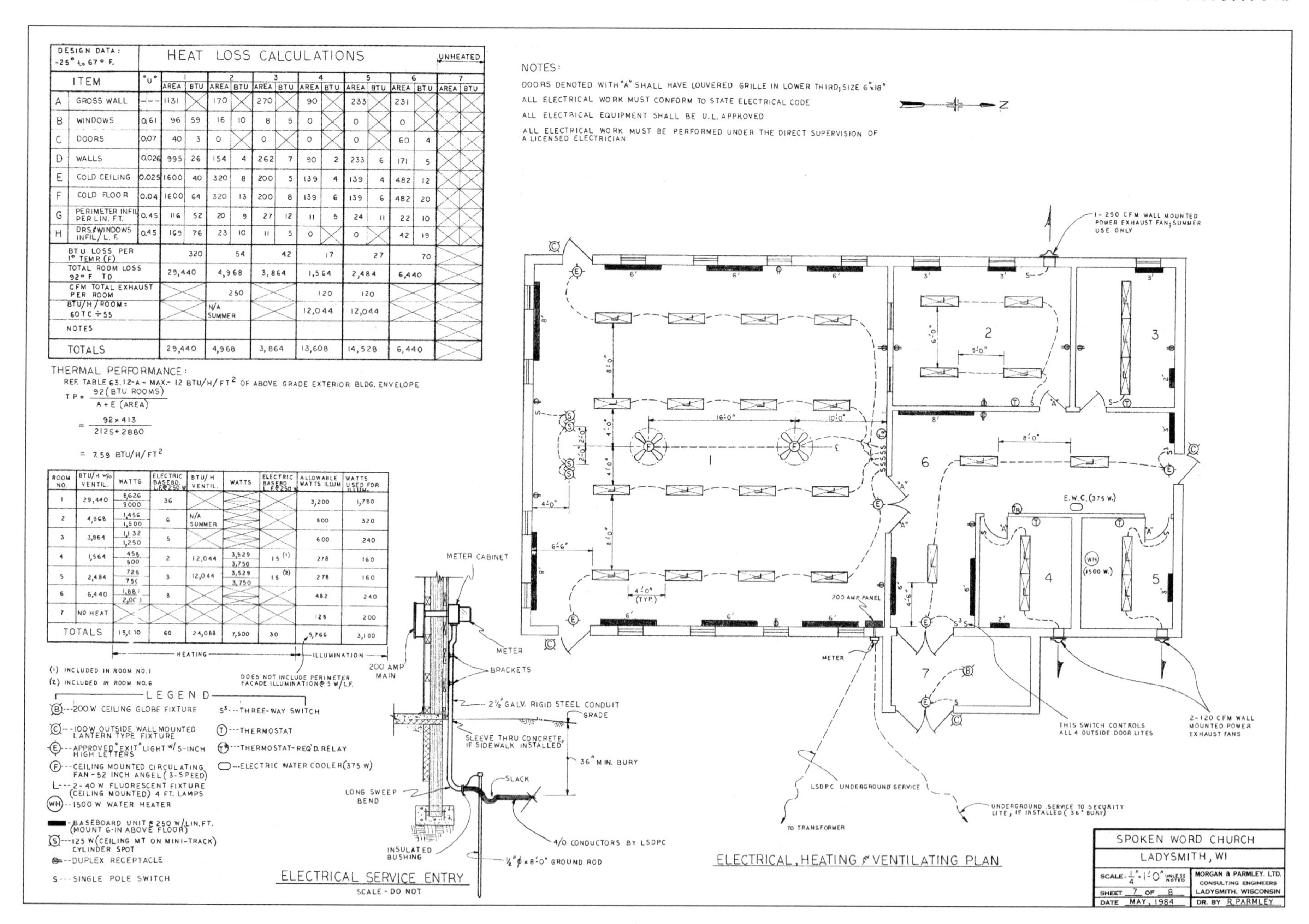

DESIGN DATA: -25° to 67° F.	HEAT LOSS CALCULATIONS													UNHEATED	
ITEM	"U"	1		2		3		4		5		6		7	
		AREA	BTU	AREA	BTU	AREA	BTU	AREA	BTU	AREA	BTU	AREA	BTU	AREA	BTU
A GROSS WALL	---	1131		170		270		90		233		231			
B WINDOWS	0.61	96	59	16	10	8	5	0		0		0			
C DOORS	0.07	40	3	0		0		0		0		60	4		
D WALLS	0.026	995	26	154	4	262	7	90	2	233	6	171	5		
E COLD CEILING	0.025	1600	40	320	8	200	5	139	4	139	4	482	12		
F COLD FLOOR	0.04	1600	64	320	13	200	8	139	6	139	6	482	20		
G PERIMETER INFIL PER LIN. FT.	0.45	116	52	20	9	27	12	11	5	24	11	22	10		
H DRS.&WINDOWS INFIL/L.F.	0.45	169	76	23	10	11	5	0		0		42	19		
BTU LOSS PER 1° TEMP. (F)		320		54		42		17		27		70			
TOTAL ROOM LOSS 92° F TD		29,440		4,968		3,864		1,564		2,484		6,440			
CFM TOTAL EXHAUST PER ROOM				250				120		120					
BTU/H/ROOM= 60TC÷55				N/A SUMMER				12,044		12,044					
NOTES															
TOTALS		29,440		4,968		3,864		13,608		14,528		6,440			

THERMAL PERFORMANCE:

REF. TABLE 63.12-A - MAX.- 12 $BTU/H/FT^2$ OF ABOVE GRADE EXTERIOR BLDG. ENVELOPE

$$TP=\frac{92(\text{BTU ROOMS})}{A+E(\text{AREA})}=\frac{92\times413}{2125+2880}=7.59\ BTU/H/FT^2$$

ROOM NO.	BTU/H w/o VENTIL.	WATTS	ELECTRIC BASEBD. L.F.@250 W	BTU/H VENTIL.	WATTS	ELECTRIC BASEBD L.F.@250 W	ALLOWABLE WATTS ILLUM	WATTS USED FOR ILLUM.
1	29,440	8,626 / 9000	36				3,200	1,780
2	4,968	1,456 / 1,500	6	N/A SUMMER			800	320
3	3,864	1,132 / 1,250	5				600	240
4	1,564	458 / 500	2	12,044	3,529 / 3,750	15 (1)	278	160
5	2,484	728 / 75[illegible]	3	12,044	3,529 / 3,750	15 (2)	278	160
6	6,440	1,88[illegible] / 2,0[illegible]	8				482	240
7	NO HEAT						128	200
TOTALS		15,[illegible]0	60	24,088	7,500	30	5,766	3,100

HEATING — ILLUMINATION

(1) INCLUDED IN ROOM NO.1
(2) INCLUDED IN ROOM NO.6

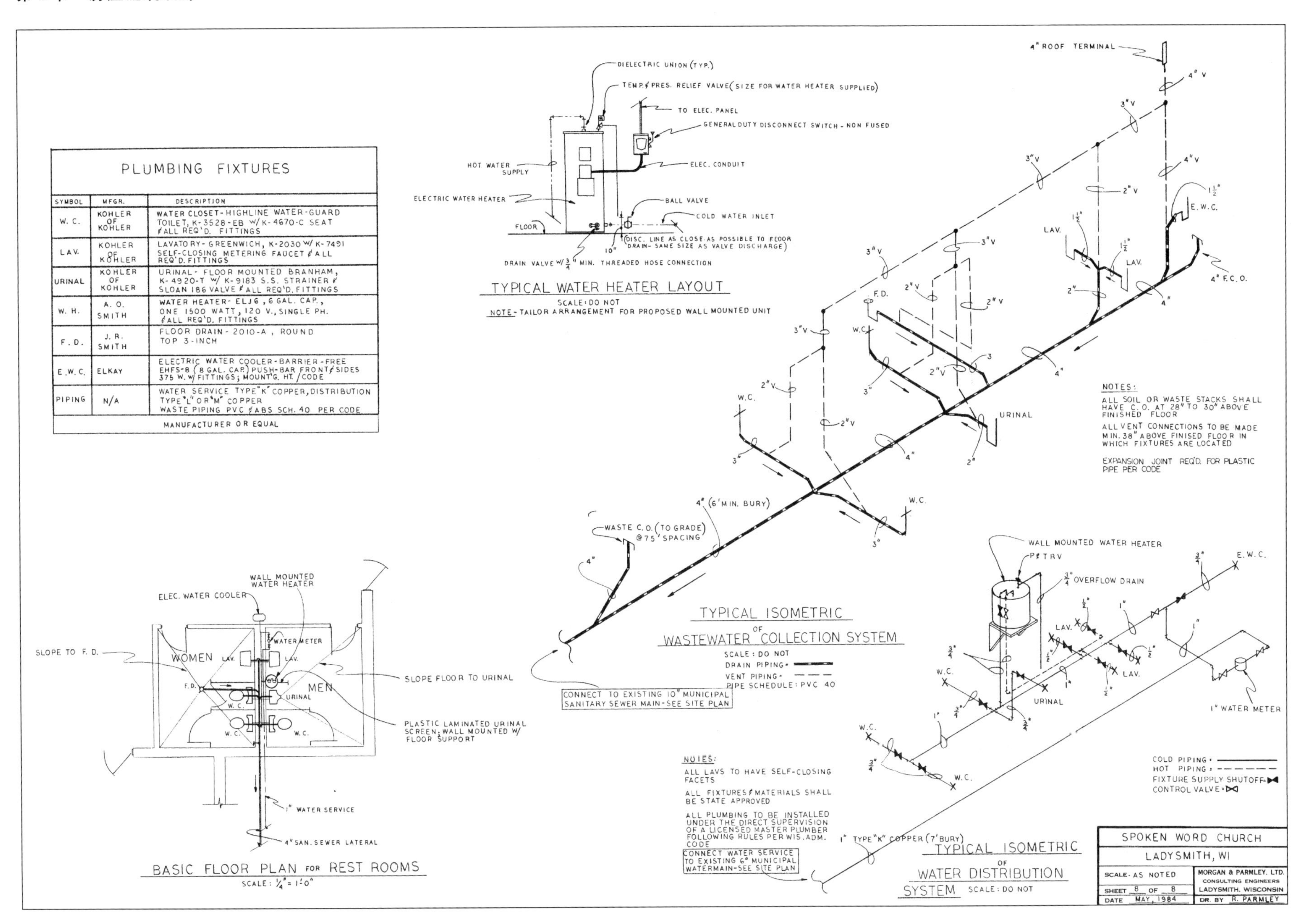
PLUMBING FIXTURES
SYMBOL | MFGR. | DESCRIPTION
W. C. | KOHLER OF KOHLER | WATER CLOSET-HIGHLINE WATER-GUARD TOILET, K-3528-EB W/K-4670-C SEAT & ALL REQ'D. FITTINGS
LAV. | KOHLER OF KOHLER | LAVATORY-GREENWICH, K-2030 W/ K-7491 SELF-CLOSING METERING FAUCET & ALL REQ'D. FITTINGS
URINAL | KOHLER OF KOHLER | URINAL-FLOOR MOUNTED BRANHAM, K-4920-T W/ K-9183 S.S. STRAINER & SLOAN 186 VALVE & ALL REQ'D. FITTINGS
W. H. | A. O. SMITH | WATER HEATER-ELJ6, 6 GAL. CAP., ONE 1500 WATT, 120 V., SINGLE PH. & ALL REQ'D. FITTINGS
F. D. | J. R. SMITH | FLOOR DRAIN-2010-A, ROUND TOP 3-INCH
E.W.C. | ELKAY | ELECTRIC WATER COOLER-BARRIER-FREE EHFS-8 (8 GAL. CAP.) PUSH-BAR FRONT & SIDES 375 W. W/ FITTINGS; MOUNT'G. HT. / CODE
PIPING | N/A | WATER SERVICE TYPE "K" COPPER, DISTRIBUTION TYPE "L" OR "M" COPPER WASTE PIPING PVC & ABS SCH. 40 PER CODE
MANUFACTURER OR EQUAL
DIELECTRIC UNION (TYP.)
TEMP. & PRES. RELIEF VALVE (SIZE FOR WATER HEATER SUPPLIED)
TO ELEC. PANEL
GENERAL DUTY DISCONNECT SWITCH-NON FUSED
HOT WATER SUPPLY
ELEC. CONDUIT
ELECTRIC WATER HEATER
BALL VALVE
COLD WATER INLET
FLOOR
(DISC. LINE AS CLOSE AS POSSIBLE TO FLOOR DRAIN-SAME SIZE AS VALVE DISCHARGE)
DRAIN VALVE W/ 3/4" MIN. THREADED HOSE CONNECTION
TYPICAL WATER HEATER LAYOUT
SCALE: DO NOT
NOTE-TAILOR ARRANGEMENT FOR PROPOSED WALL MOUNTED UNIT
4" ROOF TERMINAL
E.W.C.
LAV.
F.D.
W.C.
URINAL
4" F.C.O.
NOTES:
ALL SOIL OR WASTE STACKS SHALL HAVE C.O. AT 28" TO 30" ABOVE FINISHED FLOOR
ALL VENT CONNECTIONS TO BE MADE MIN. 38" ABOVE FINISED FLOOR IN WHICH FIXTURES ARE LOCATED
EXPANSION JOINT REQ'D. FOR PLASTIC PIPE PER CODE
4" (6' MIN. BURY)
WASTE C.O. (TO GRADE) @75' SPACING
TYPICAL ISOMETRIC OF WASTEWATER COLLECTION SYSTEM
SCALE: DO NOT
DRAIN PIPING
VENT PIPING
PIPE SCHEDULE: PVC 40
CONNECT TO EXISTING 10" MUNICIPAL SANITARY SEWER MAIN-SEE SITE PLAN
WALL MOUNTED WATER HEATER
P & T R V
3/4" OVERFLOW DRAIN
1" WATER METER
COLD PIPING
HOT PIPING
FIXTURE SUPPLY SHUTOFF
CONTROL VALVE
NOTES:
ALL LAVS TO HAVE SELF-CLOSING FACETS
ALL FIXTURES & MATERIALS SHALL BE STATE APPROVED
ALL PLUMBING TO BE INSTALLED UNDER THE DIRECT SUPERVISION OF A LICENSED MASTER PLUMBER FOLLOWING RULES PER WIS. ADM. CODE
CONNECT WATER SERVICE TO EXISTING 6" MUNICIPAL WATERMAIN-SEE SITE PLAN
1" TYPE "K" COPPER (7' BURY)
TYPICAL ISOMETRIC OF WATER DISTRIBUTION SYSTEM
SCALE: DO NOT
WALL MOUNTED WATER HEATER
ELEC. WATER COOLER
SLOPE TO F. D.
WOMEN
MEN
WATER METER
SLOPE FLOOR TO URINAL
PLASTIC LAMINATED URINAL SCREEN; WALL MOUNTED W/ FLOOR SUPPORT
1" WATER SERVICE
4" SAN. SEWER LATERAL
BASIC FLOOR PLAN FOR REST ROOMS
SCALE: 1/4" = 1'-0"
SPOKEN WORD CHURCH
LADYSMITH, WI
SCALE- AS NOTED
SHEET 8 OF 8
DATE MAY, 1984
MORGAN & PARMLEY, LTD.
CONSULTING ENGINEERS
LADYSMITH, WISCONSIN
DR. BY R. PARMLEY

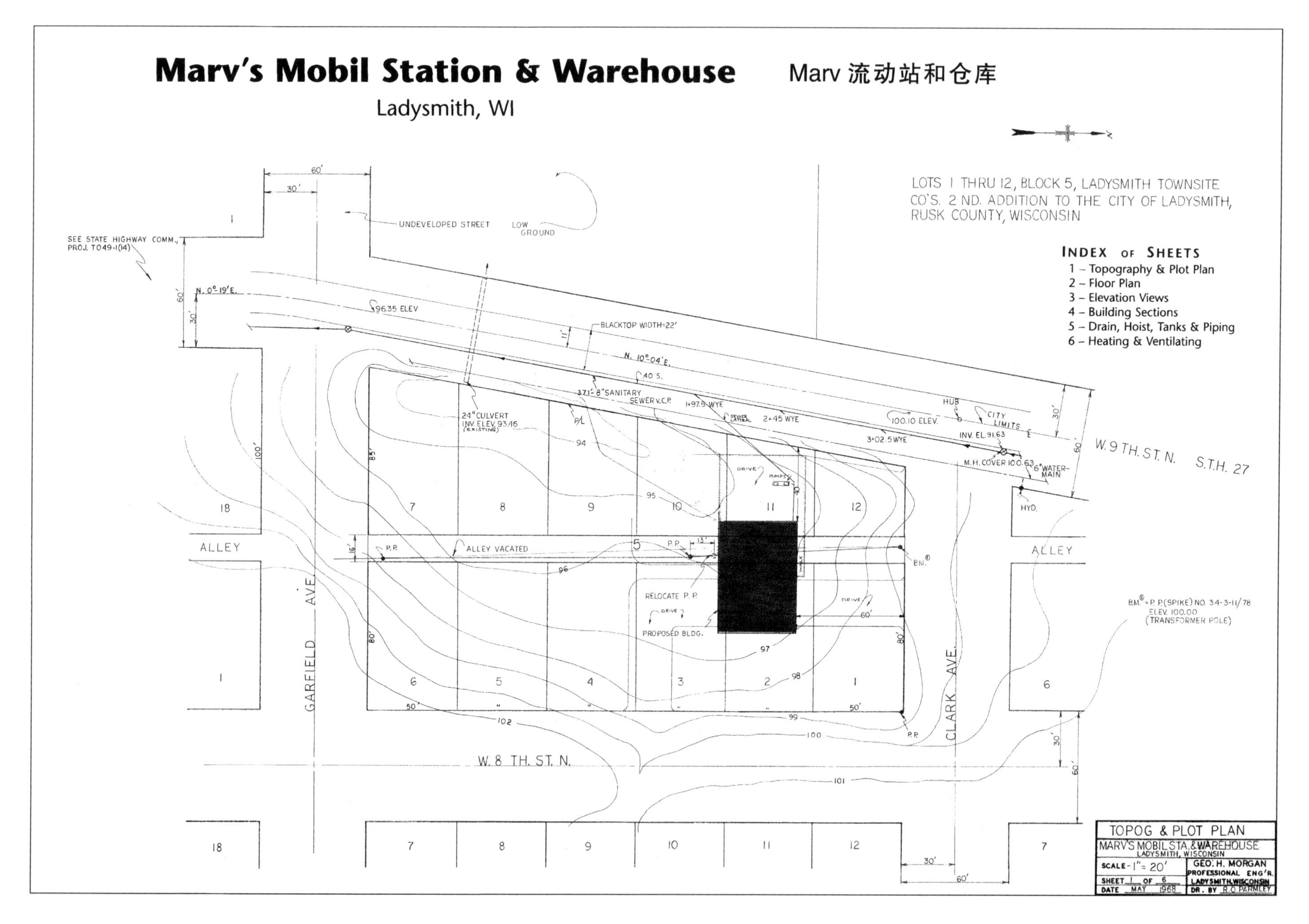

Courtesy: Morgan & Parmley, Ltd.

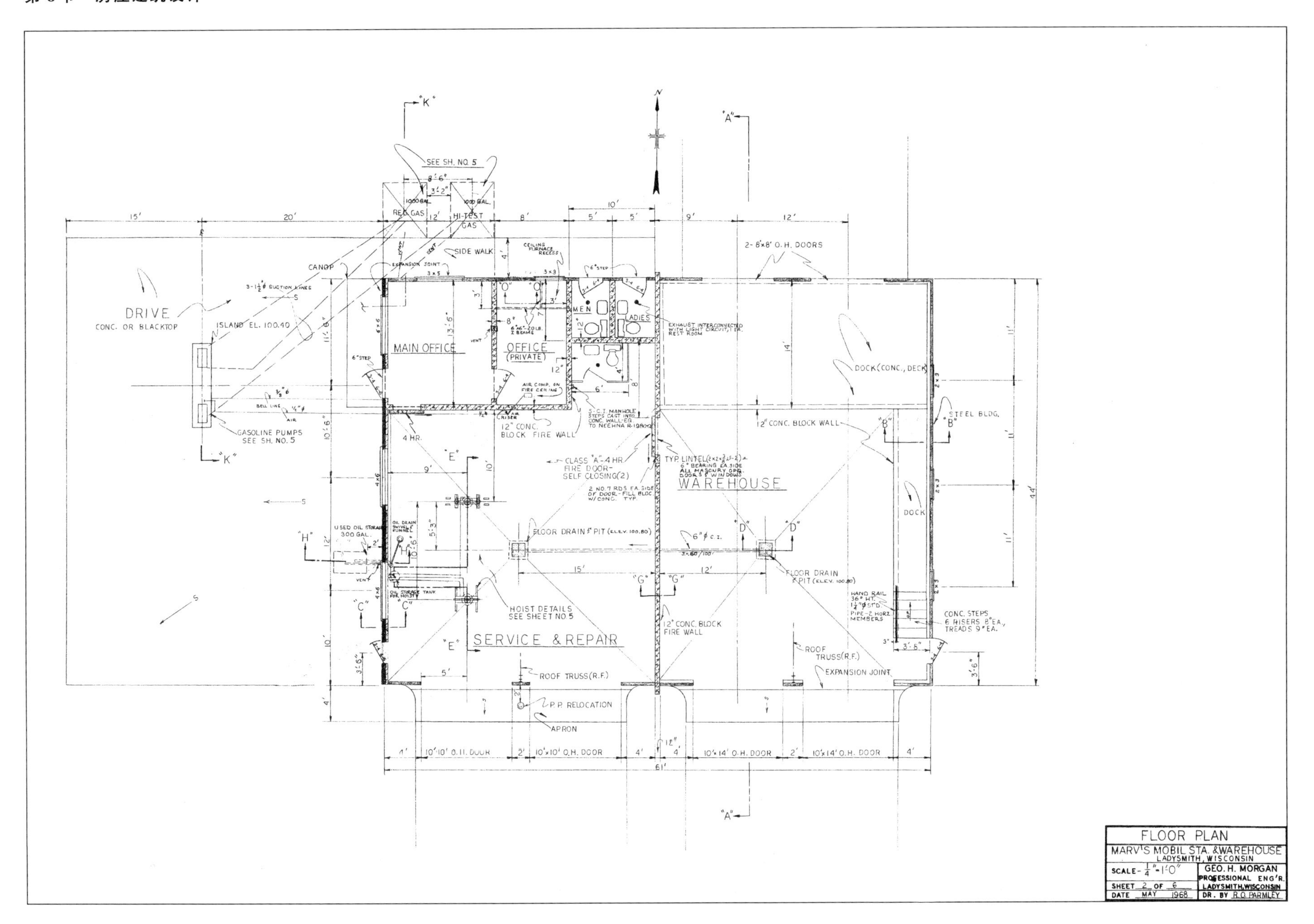
FLOOR PLAN
MARV'S MOBIL STA. & WAREHOUSE
LADYSMITH, WISCONSIN
SCALE - 1/4" = 1'-0"
GEO. H. MORGAN
PROFESSIONAL ENG'R.
LADYSMITH, WISCONSIN
SHEET 2 OF 6
DATE MAY 1968
DR. BY R.O. PARMLEY
DRIVE
CONC. OR BLACKTOP
MAIN OFFICE
OFFICE (PRIVATE)
MEN
LADIES
WAREHOUSE
SERVICE & REPAIR
DOCK (CONC., DECK)
DOCK
STEEL BLDG.
2-8'×8' O.H. DOORS
12" CONC. BLOCK FIRE WALL
12" CONC. BLOCK WALL
GASOLINE PUMPS SEE SH. NO. 5
HOIST DETAILS SEE SHEET NO. 5
ROOF TRUSS (R.F.)
EXPANSION JOINT
P.P. RELOCATION
APRON
CONC. STEPS 6 RISERS 8" EA., TREADS 9" EA.
10'×10' O.H. DOOR
10'×14' O.H. DOOR
61'

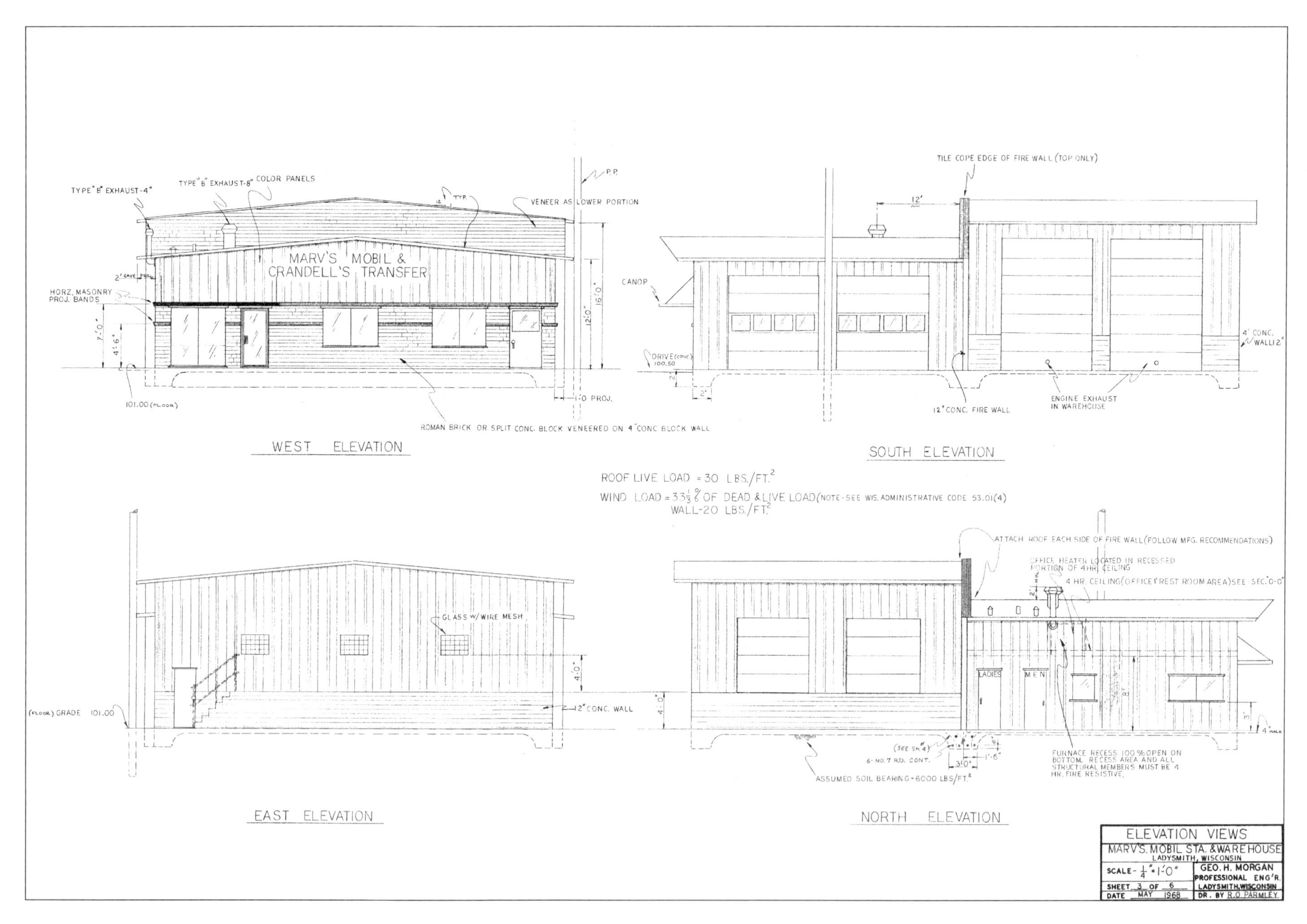

WEST ELEVATION
SOUTH ELEVATION
EAST ELEVATION
NORTH ELEVATION
MARV'S MOBIL & CRANDELL'S TRANSFER
ROOF LIVE LOAD = 30 LBS./FT.²
WIND LOAD = 33⅓% OF DEAD & LIVE LOAD (NOTE-SEE WIS. ADMINISTRATIVE CODE 53.01(4)
WALL-20 LBS./FT.²
ROMAN BRICK OR SPLIT CONC. BLOCK VENEERED ON 4" CONC BLOCK WALL
TYPE "B" EXHAUST-4"
TYPE "B" EXHAUST-8"
COLOR PANELS
VENEER AS LOWER PORTION
HORZ. MASONRY PROJ. BANDS
101.00 (FLOOR)
1'-0 PROJ.
TILE COPE EDGE OF FIRE WALL (TOP ONLY)
CANOPY
DRIVE (CONC.) 100.50
12" CONC. FIRE WALL
ENGINE EXHAUST IN WAREHOUSE
4" CONC. WALL 12"
GLASS W/WIRE MESH
(FLOOR) GRADE 101.00
12" CONC. WALL
ATTACH ROOF EACH SIDE OF FIRE WALL (FOLLOW MFG. RECOMMENDATIONS)
OFFICE HEATER LOCATED IN RECESSED PORTION OF 4HR. CEILING
4 HR. CEILING (OFFICE & REST ROOM AREA) SEE SEC. 0-0
LADIES
MEN
6-NO.7 RD. CONT.
ASSUMED SOIL BEARING-6000 LBS/FT.²
FURNACE RECESS 100% OPEN ON BOTTOM. RECESS AREA AND ALL STRUCTURAL MEMBERS MUST BE 4 HR. FIRE RESISTIVE.
4" WALK
ELEVATION VIEWS
MARV'S MOBIL STA. & WAREHOUSE
LADYSMITH, WISCONSIN
SCALE - 1/4" = 1'-0"
GEO. H. MORGAN
PROFESSIONAL ENG'R.
LADYSMITH, WISCONSIN
SHEET 3 OF 6
DATE MAY 1968
DR. BY R.O. PARMLEY

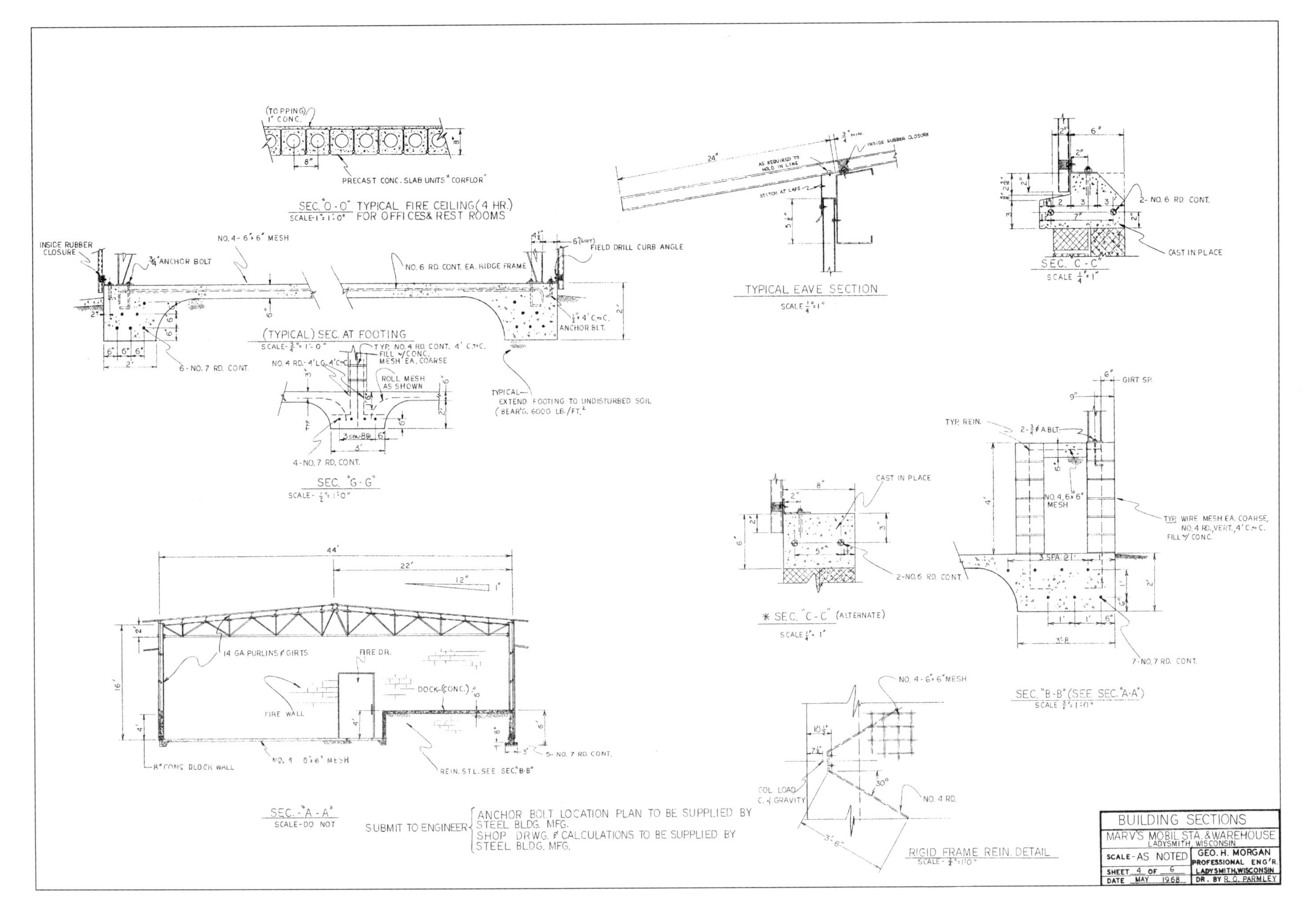
SEC. "O-O" TYPICAL FIRE CEILING (4 HR.) FOR OFFICES & REST ROOMS
PRECAST CONC. SLAB UNITS "CORFLOR"
TYPICAL EAVE SECTION
SEC. "C-C"
CAST IN PLACE
2-NO. 6 RD. CONT.
(TYPICAL) SEC. AT FOOTING
INSIDE RUBBER CLOSURE
3/4" ANCHOR BOLT
NO. 4-6"×6" MESH
NO. 6 RD. CONT. EA. RIDGE FRAME
FIELD DRILL CURB ANGLE
6-NO. 7 RD. CONT.
TYPICAL — EXTEND FOOTING TO UNDISTURBED SOIL (BEAR'G. 6000 LB./FT.²)
SEC. "G-G"
4-NO. 7 RD. CONT.
ROLL MESH AS SHOWN
* SEC. "C-C" (ALTERNATE)
SEC. "B-B" (SEE SEC. "A-A")
7-NO. 7 RD. CONT.
TYP. WIRE MESH EA. COARSE, NO. 4 RD., VERT., 4' C. to C. FILL w/ CONC.
14 GA PURLINS & GIRTS
FIRE DR.
FIRE WALL
DOCK-(CONC.)
8" CONC. BLOCK WALL
REIN. STL. SEE SEC. "B-B"
5-NO. 7 RD. CONT.
SEC. "A-A"
SCALE-DO NOT
SUBMIT TO ENGINEER
ANCHOR BOLT LOCATION PLAN TO BE SUPPLIED BY STEEL BLDG. MFG.
SHOP DRWG. & CALCULATIONS TO BE SUPPLIED BY STEEL BLDG. MFG.
RIGID FRAME REIN. DETAIL
COL. LOAD C. of GRAVITY
NO. 4 RD.
BUILDING SECTIONS
MARV'S MOBIL STA. & WAREHOUSE
LADYSMITH, WISCONSIN
SCALE-AS NOTED
GEO. H. MORGAN PROFESSIONAL ENG'R. LADYSMITH, WISCONSIN
SHEET 4 OF 6
DATE MAY 1968
DR. BY R. C. PARMLEY

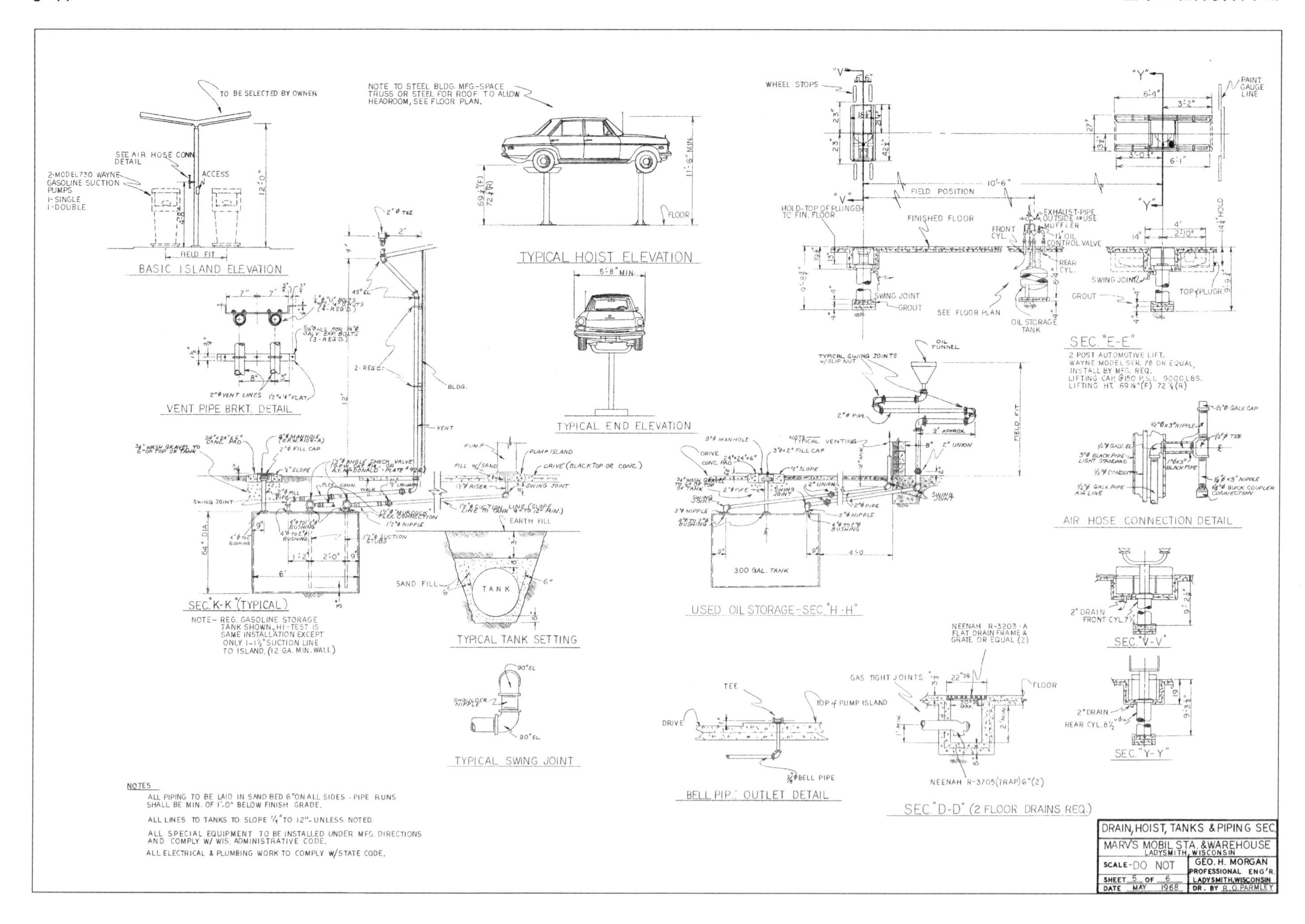

BASIC ISLAND ELEVATION
VENT PIPE BRKT. DETAIL
TYPICAL HOIST ELEVATION
TYPICAL END ELEVATION
SEC."E-E"
2 POST AUTOMOTIVE LIFT, WAYNE MODEL SER. 78 OR EQUAL, INSTALL BY MFG. REQ.
AIR HOSE CONNECTION DETAIL
SEC."K-K"(TYPICAL)
NOTE-REG. GASOLINE STORAGE TANK SHOWN, HI-TEST IS SAME INSTALLATION EXCEPT ONLY 1-1½" SUCTION LINE TO ISLAND. (12 GA. MIN. WALL)
TYPICAL TANK SETTING
USED OIL STORAGE-SEC."H-H"
SEC."V-V"
SEC."Y-Y"
TYPICAL SWING JOINT
BELL PIPE OUTLET DETAIL
SEC."D-D"(2 FLOOR DRAINS REQ.)
NOTES
ALL PIPING TO BE LAID IN SAND BED 6" ON ALL SIDES - PIPE RUNS SHALL BE MIN. OF 1'-0" BELOW FINISH GRADE.
ALL LINES TO TANKS TO SLOPE ¼" TO 12"-UNLESS NOTED.
ALL SPECIAL EQUIPMENT TO BE INSTALLED UNDER MFG. DIRECTIONS AND COMPLY W/ WIS. ADMINISTRATIVE CODE.
ALL ELECTRICAL & PLUMBING WORK TO COMPLY W/STATE CODE.
DRAIN, HOIST, TANKS & PIPING SEC.
MARV'S MOBIL STA. & WAREHOUSE
LADYSMITH, WISCONSIN
SCALE-DO NOT
GEO. H. MORGAN PROFESSIONAL ENG'R. LADYSMITH, WISCONSIN
SHEET 5 OF 6
DATE MAY 1968
DR. BY R.O. PARMLEY

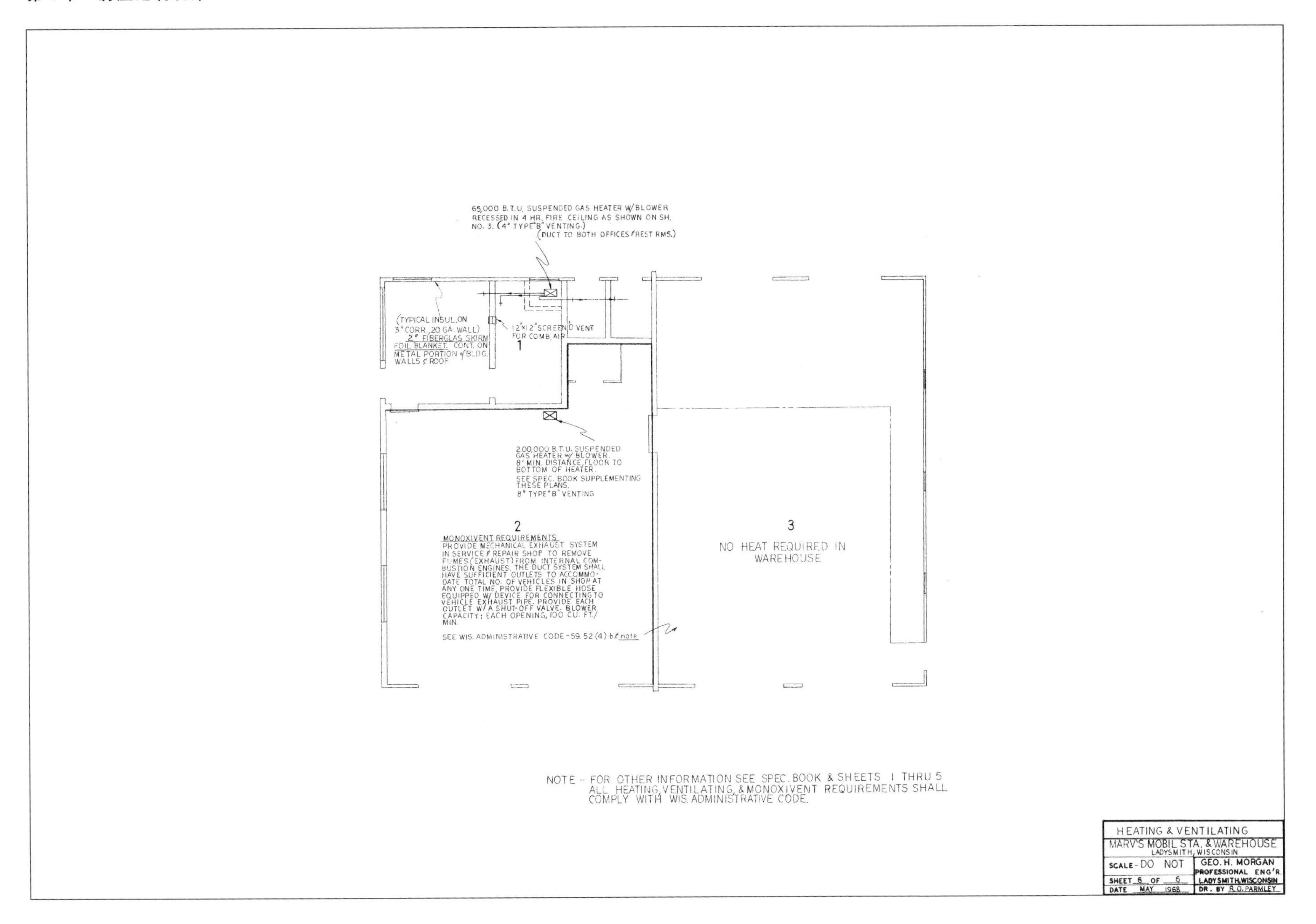
65,000 B.T.U. SUSPENDED GAS HEATER W/BLOWER
RECESSED IN 4 HR. FIRE CEILING AS SHOWN ON SH.
NO. 3. (4" TYPE "B" VENTING.)
(DUCT TO BOTH OFFICES & REST RMS.)
(TYPICAL INSUL. ON
3" CORR., 20 GA. WALL)
2" FIBERGLAS SKIRM
FOIL BLANKET. CONT. ON
METAL PORTION of BLDG.
WALLS & ROOF
12"×12" SCREEN'D VENT
FOR COMB. AIR
1
200,000 B.T.U. SUSPENDED
GAS HEATER w/ BLOWER.
8' MIN. DISTANCE, FLOOR TO
BOTTOM OF HEATER.
SEE SPEC. BOOK SUPPLEMENTING
THESE PLANS.
8" TYPE "B" VENTING
2
MONOXIVENT REQUIREMENTS
PROVIDE MECHANICAL EXHAUST SYSTEM
IN SERVICE & REPAIR SHOP TO REMOVE
FUMES (EXHAUST) FROM INTERNAL COM-
BUSTION ENGINES. THE DUCT SYSTEM SHALL
HAVE SUFFICIENT OUTLETS TO ACCOMMO-
DATE TOTAL NO. OF VEHICLES IN SHOP AT
ANY ONE TIME. PROVIDE FLEXIBLE HOSE
EQUIPPED W/ DEVICE FOR CONNECTING TO
VEHICLE EXHAUST PIPE. PROVIDE EACH
OUTLET W/ A SHUT-OFF VALVE. BLOWER
CAPACITY: EACH OPENING, 100 CU. FT./
MIN.
SEE WIS. ADMINISTRATIVE CODE-59.52 (4) b & note
3
NO HEAT REQUIRED IN
WAREHOUSE
NOTE -- FOR OTHER INFORMATION SEE SPEC. BOOK & SHEETS 1 THRU 5
ALL HEATING, VENTILATING, & MONOXIVENT REQUIREMENTS SHALL
COMPLY WITH WIS. ADMINISTRATIVE CODE.
HEATING & VENTILATING
MARV'S MOBIL STA. & WAREHOUSE
LADYSMITH, WISCONSIN
SCALE - DO NOT
SHEET 6 OF 6
DATE MAY 1968
GEO. H. MORGAN
PROFESSIONAL ENG'R.
LADYSMITH, WISCONSIN
DR. BY R.O. PARMLEY

Warehouse Addition 扩建仓库（Rusk 县公路局）

Rusk County Highway Department

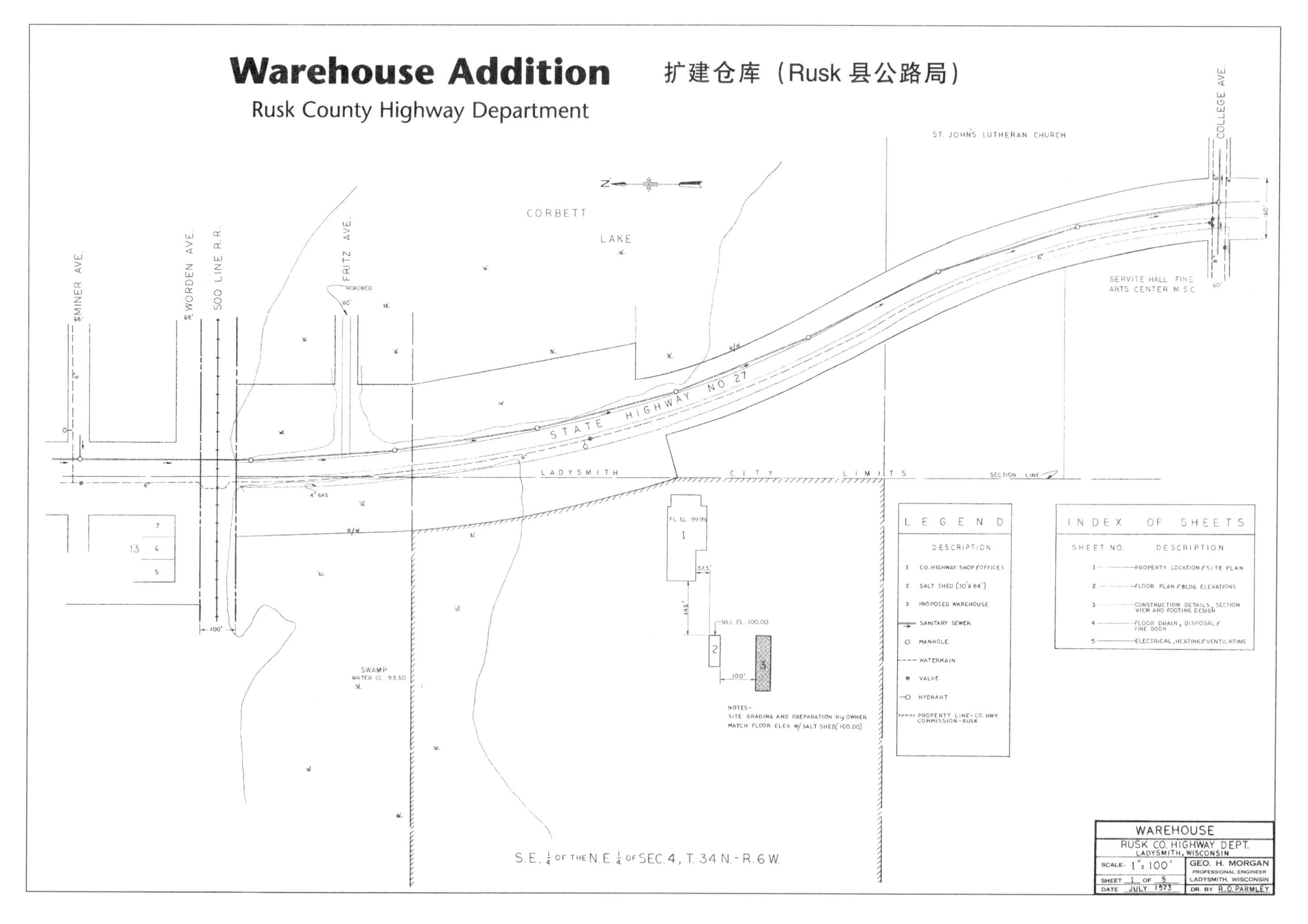

Courtesy: Morgan & Parmley, Ltd.

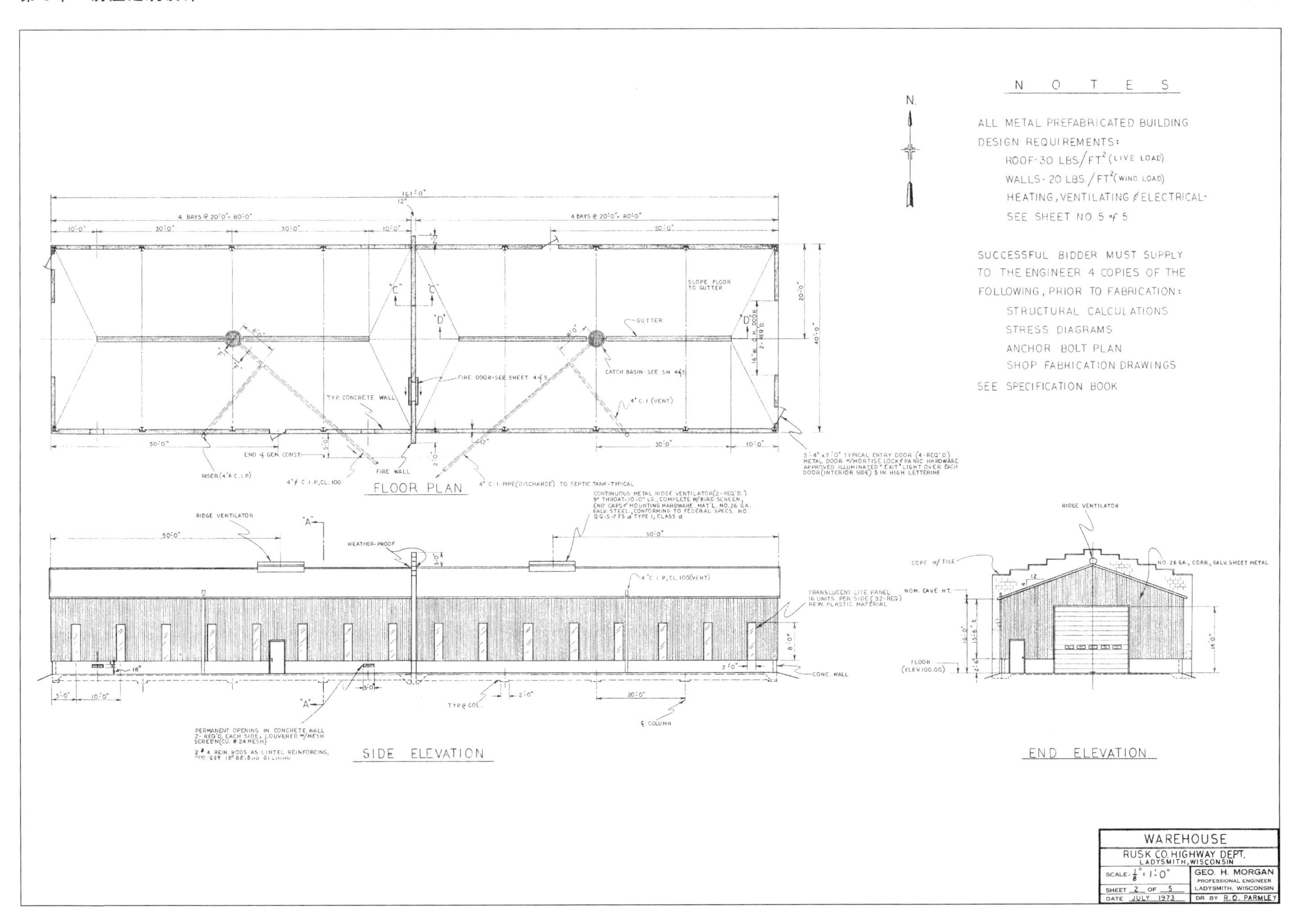
N O T E S
ALL METAL PREFABRICATED BUILDING
DESIGN REQUIREMENTS:
ROOF-30 LBS./FT.^{2} (LIVE LOAD)
WALLS-20 LBS./FT.^{2} (WIND LOAD)
HEATING, VENTILATING & ELECTRICAL-
SEE SHEET NO. 5 of 5
SUCCESSFUL BIDDER MUST SUPPLY
TO THE ENGINEER 4 COPIES OF THE
FOLLOWING, PRIOR TO FABRICATION:
STRUCTURAL CALCULATIONS
STRESS DIAGRAMS
ANCHOR BOLT PLAN
SHOP FABRICATION DRAWINGS
SEE SPECIFICATION BOOK
N.
161'-0"
4 BAYS @ 20'-0" = 80'-0"
10'-0"
30'-0"
50'-0"
20'-0"
40'-0"
SLOPE FLOOR TO GUTTER
GUTTER
FIRE DOOR-SEE SHEET 4 of 5
CATCH BASIN-SEE SH. 4 of 5
TYP. CONCRETE WALL
4" C.I. (VENT)
END of GEN. CONST.
RISER (4" C.I.P.)
4" φ C.I.P., CL. 100
FIRE WALL
4" C.I. PIPE (DISCHARGE) TO SEPTIC TANK-TYPICAL
3'-4" x 7'-0" TYPICAL ENTRY DOOR (4-REQ'D.)
FLOOR PLAN
RIDGE VENTILATOR
WEATHER-PROOF
4" C.I.P., CL. 100 (VENT)
TRANSLUCENT LITE PANEL
CONC. WALL
TYP. COL.
℄ COLUMN
SIDE ELEVATION
COPE w/ TILE
NOM. EAVE HT.
NO. 26 GA., CORR., GALV. SHEET METAL
FLOOR (ELEV. 100.00)
END ELEVATION
WAREHOUSE
RUSK CO. HIGHWAY DEPT.
LADYSMITH, WISCONSIN
SCALE-1/8" = 1'-0"
SHEET 2 OF 5
DATE JULY 1973
GEO. H. MORGAN
PROFESSIONAL ENGINEER
LADYSMITH, WISCONSIN
DR BY R.O. PARMLEY

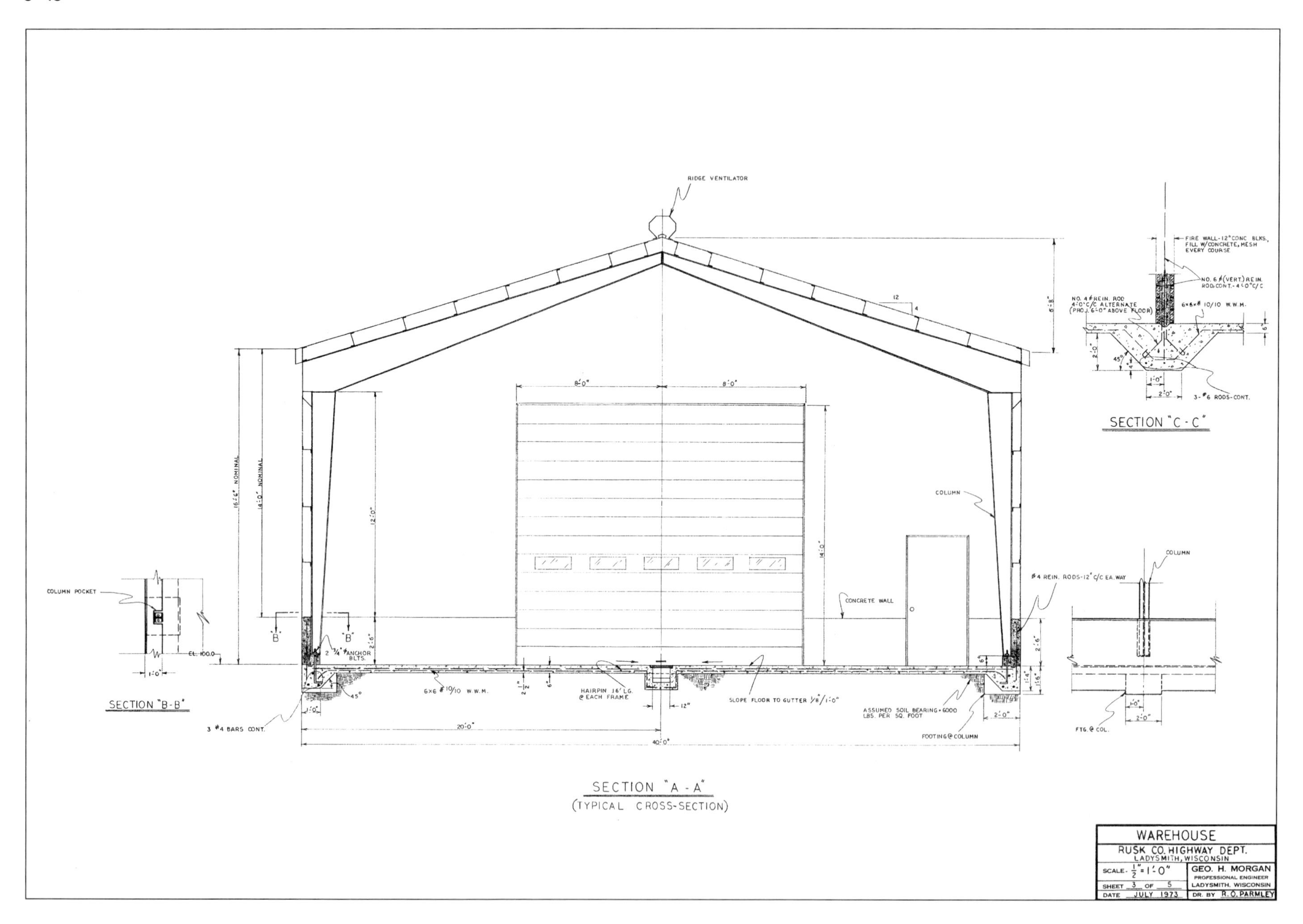
RIDGE VENTILATOR
COLUMN
COLUMN POCKET
CONCRETE WALL
SECTION "A - A"
(TYPICAL CROSS-SECTION)
SECTION "B-B"
SECTION "C - C"
WAREHOUSE
RUSK CO. HIGHWAY DEPT.
LADYSMITH, WISCONSIN
GEO. H. MORGAN
PROFESSIONAL ENGINEER
LADYSMITH, WISCONSIN
SHEET 3 OF 5
DATE JULY 1973
DR. BY R.O. PARMLEY

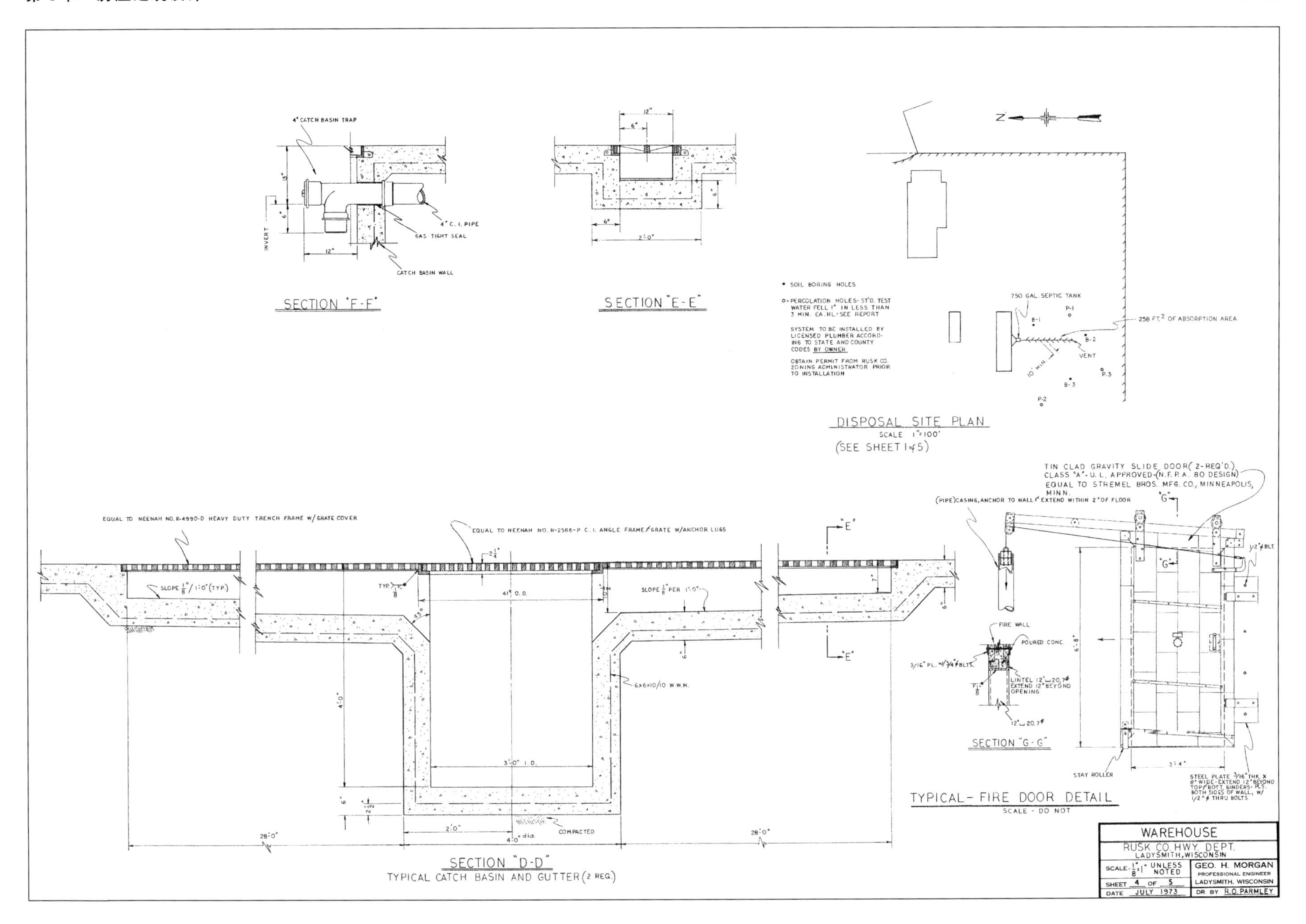

4" CATCH BASIN TRAP
4" C. I. PIPE
GAS TIGHT SEAL
CATCH BASIN WALL
SECTION "F-F"
SECTION "E-E"
DISPOSAL SITE PLAN
SCALE 1"=100'
(SEE SHEET 1 of 5)
• SOIL BORING HOLES
o PERCOLATION HOLES-51"D. TEST WATER FELL 1" IN LESS THAN 3 MIN. EA. HL.-SEE REPORT
SYSTEM TO BE INSTALLED BY LICENSED PLUMBER ACCORDING TO STATE AND COUNTY CODES BY OWNER
OBTAIN PERMIT FROM RUSK CO. ZONING ADMINISTRATOR PRIOR TO INSTALLATION
750 GAL. SEPTIC TANK
258 FT.2 OF ABSORPTION AREA
VENT
EQUAL TO NEENAH NO. R-4990-D HEAVY DUTY TRENCH FRAME w/ GRATE COVER
EQUAL TO NEENAH NO. R-2586-P C. I. ANGLE FRAME/GRATE W/ANCHOR LUGS
SLOPE 1/8"/1'-0" (TYP)
SLOPE 1/8" PER 1'-0"
6x6x10/10 W.W.M.
COMPACTED
SECTION "D-D"
TYPICAL CATCH BASIN AND GUTTER (2 REQ.)
TIN CLAD GRAVITY SLIDE DOOR (2-REQ'D.) CLASS "A"-U.L. APPROVED-(N.F.P.A. 80 DESIGN) EQUAL TO STREMEL BROS. MFG. CO., MINNEAPOLIS, MINN.
(PIPE) CASING, ANCHOR TO WALL & EXTEND WITHIN 2" OF FLOOR
FIRE WALL
POURED CONC.
LINTEL 12" ⊔ 20.7# EXTEND 12" BEYOND OPENING
SECTION "G-G"
STAY ROLLER
STEEL PLATE 3/16" THK. X 8" WIDE-EXTEND 12" BEYOND TOP & BOTT. BINDERS-PLS. BOTH SIDES OF WALL, W/ 1/2" ϕ THRU BOLTS
TYPICAL-FIRE DOOR DETAIL
SCALE - DO NOT
WAREHOUSE
RUSK CO. HWY. DEPT.
LADYSMITH, WISCONSIN
SCALE 1/8"=1' UNLESS NOTED
GEO. H. MORGAN
PROFESSIONAL ENGINEER
LADYSMITH, WISCONSIN
SHEET 4 OF 5
DATE JULY 1973
DR. BY R.O. PARMLEY

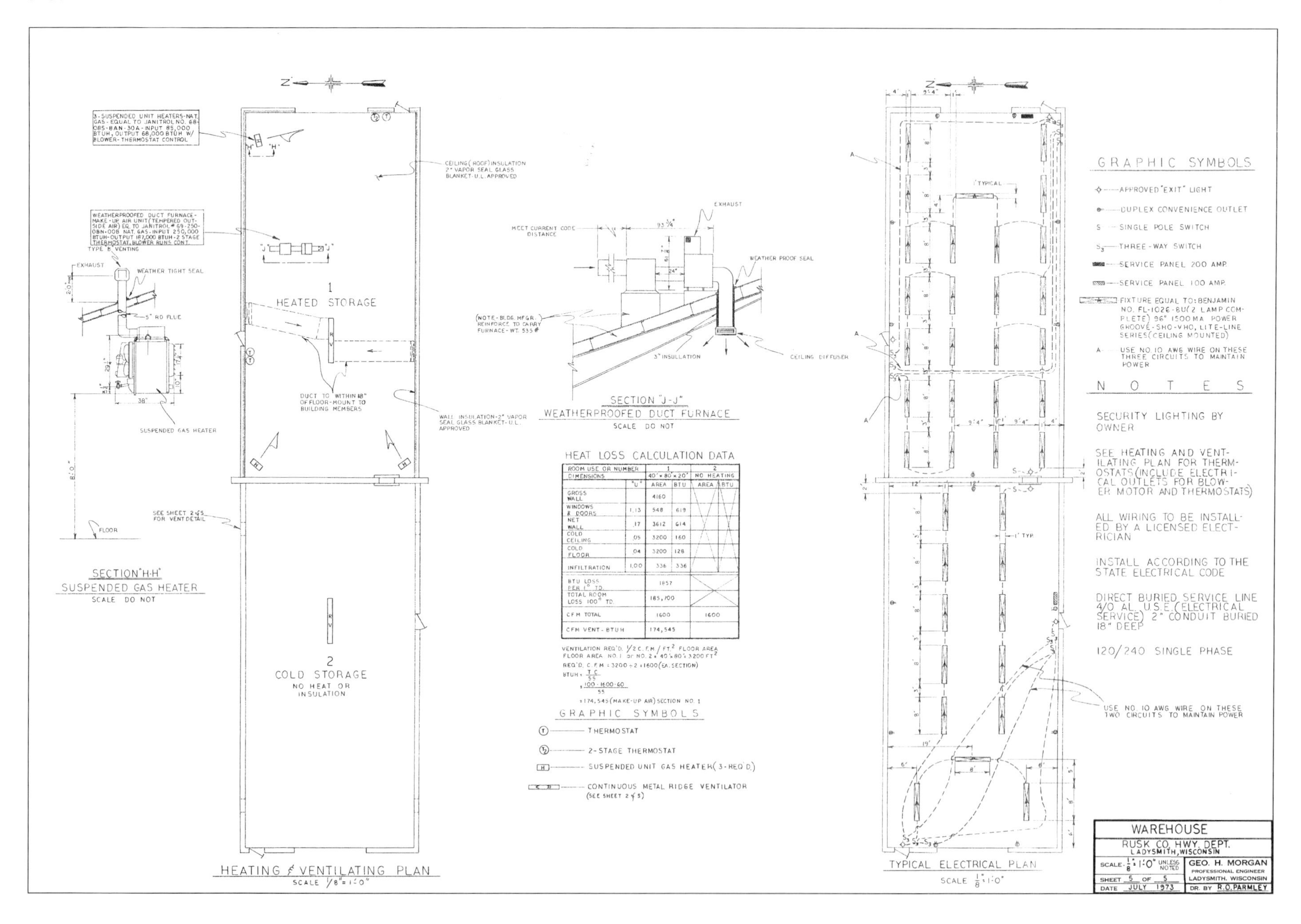
3-SUSPENDED UNIT HEATERS-NAT. GAS-EQUAL TO JANITROL NO. 68-085-BAN-30A-INPUT 85,000 BTUH, OUTPUT 68,000 BTUH W/ BLOWER-THERMOSTAT CONTROL
CEILING (ROOF) INSULATION 2" VAPOR SEAL GLASS BLANKET-U.L. APPROVED
WEATHERPROOFED DUCT FURNACE-MAKE-UP AIR UNIT (TEMPERED OUTSIDE AIR) EQ. TO JANITROL # 69-250-OBN-00B NAT. GAS-INPUT 250,000 BTUH-OUTPUT 187,000 BTUH-2 STAGE THERMOSTAT, BLOWER RUNS CONT.
TYPE 'B' VENTING
EXHAUST
WEATHER TIGHT SEAL
5" RD FLUE
38"
SUSPENDED GAS HEATER
FLOOR
SEE SHEET 2 of 5 FOR VENT DETAIL
SECTION "H-H"
SUSPENDED GAS HEATER
SCALE DO NOT
1
HEATED STORAGE
DUCT TO WITHIN 18" OF FLOOR-MOUNT TO BUILDING MEMBERS
WALL INSULATION-2" VAPOR SEAL GLASS BLANKET-U.L. APPROVED
2
COLD STORAGE
NO HEAT OR INSULATION
HEATING & VENTILATING PLAN
SCALE 1/8"=1'-0"
MEET CURRENT CODE DISTANCE
EXHAUST
WEATHER PROOF SEAL
(NOTE-BLDG. MFGR. REINFORCE TO CARRY FURNACE-WT. 535#)
3" INSULATION
CEILING DIFFUSER
SECTION "J-J"
WEATHERPROOFED DUCT FURNACE
SCALE DO NOT
HEAT LOSS CALCULATION DATA
ROOM USE OR NUMBER | 1 | 2
DIMENSIONS | 40'×80'×20' | NO HEATING
"U" | AREA | BTU | AREA | BTU
GROSS WALL | | 4160 |
WINDOWS & DOORS | 1.13 | 548 | 619
NET WALL | .17 | 3612 | 614
COLD CEILING | .05 | 3200 | 160
COLD FLOOR | .04 | 3200 | 128
INFILTRATION | 1.00 | 336 | 336
BTU LOSS PER 1° TD. | 1857
TOTAL ROOM LOSS 100° TD. | 185,700
CFM TOTAL | 1600 | 1600
CFM VENT-BTUH | 174,545
VENTILATION REQ'D. 1/2 C.F.M./FT.² FLOOR AREA
FLOOR AREA NO. 1 or NO. 2 = 40'×80' = 3200 FT²
REQ'D. C.F.M = 3200 ÷ 2 = 1600 (EA. SECTION)
BTUH = T C / 55
= 100·1600·60 / 55
= 174,545 (MAKE-UP AIR) SECTION NO. 1
GRAPHIC SYMBOLS
THERMOSTAT
2-STAGE THERMOSTAT
SUSPENDED UNIT GAS HEATER (3-REQ'D.)
CONTINUOUS METAL RIDGE VENTILATOR (SEE SHEET 2 of 5)
TYPICAL ELECTRICAL PLAN
SCALE 1/8"=1'-0"
USE NO. 10 AWG WIRE ON THESE TWO CIRCUITS TO MAINTAIN POWER
GRAPHIC SYMBOLS
APPROVED "EXIT" LIGHT
DUPLEX CONVENIENCE OUTLET
S — SINGLE POLE SWITCH
S3 — THREE-WAY SWITCH
SERVICE PANEL 200 AMP.
SERVICE PANEL 100 AMP.
FIXTURE EQUAL TO: BENJAMIN NO. FL-1026-8U (2 LAMP COMPLETE) 96" 1500 MA POWER GROOVE-SHO-VHO, LITE-LINE SERIES (CEILING MOUNTED)
A — USE NO. 10 AWG WIRE ON THESE THREE CIRCUITS TO MAINTAIN POWER
NOTES
SECURITY LIGHTING BY OWNER
SEE HEATING AND VENTILATING PLAN FOR THERMOSTATS (INCLUDE ELECTRICAL OUTLETS FOR BLOWER MOTOR AND THERMOSTATS)
ALL WIRING TO BE INSTALLED BY A LICENSED ELECTRICIAN
INSTALL ACCORDING TO THE STATE ELECTRICAL CODE
DIRECT BURIED SERVICE LINE 4/0 AL. U.S.E. (ELECTRICAL SERVICE) 2" CONDUIT BURIED 18" DEEP
120/240 SINGLE PHASE
WAREHOUSE
RUSK CO. HWY. DEPT.
LADYSMITH, WISCONSIN
SCALE 1/8"=1'-0" UNLESS NOTED
GEO. H. MORGAN
PROFESSIONAL ENGINEER
LADYSMITH, WISCONSIN
SHEET 5 OF 5
DATE JULY 1973
DR. BY R.O. PARMLEY

Warehouse Addition　扩建仓库（马歇尔干酪厂）

Marshall Cheese Factory

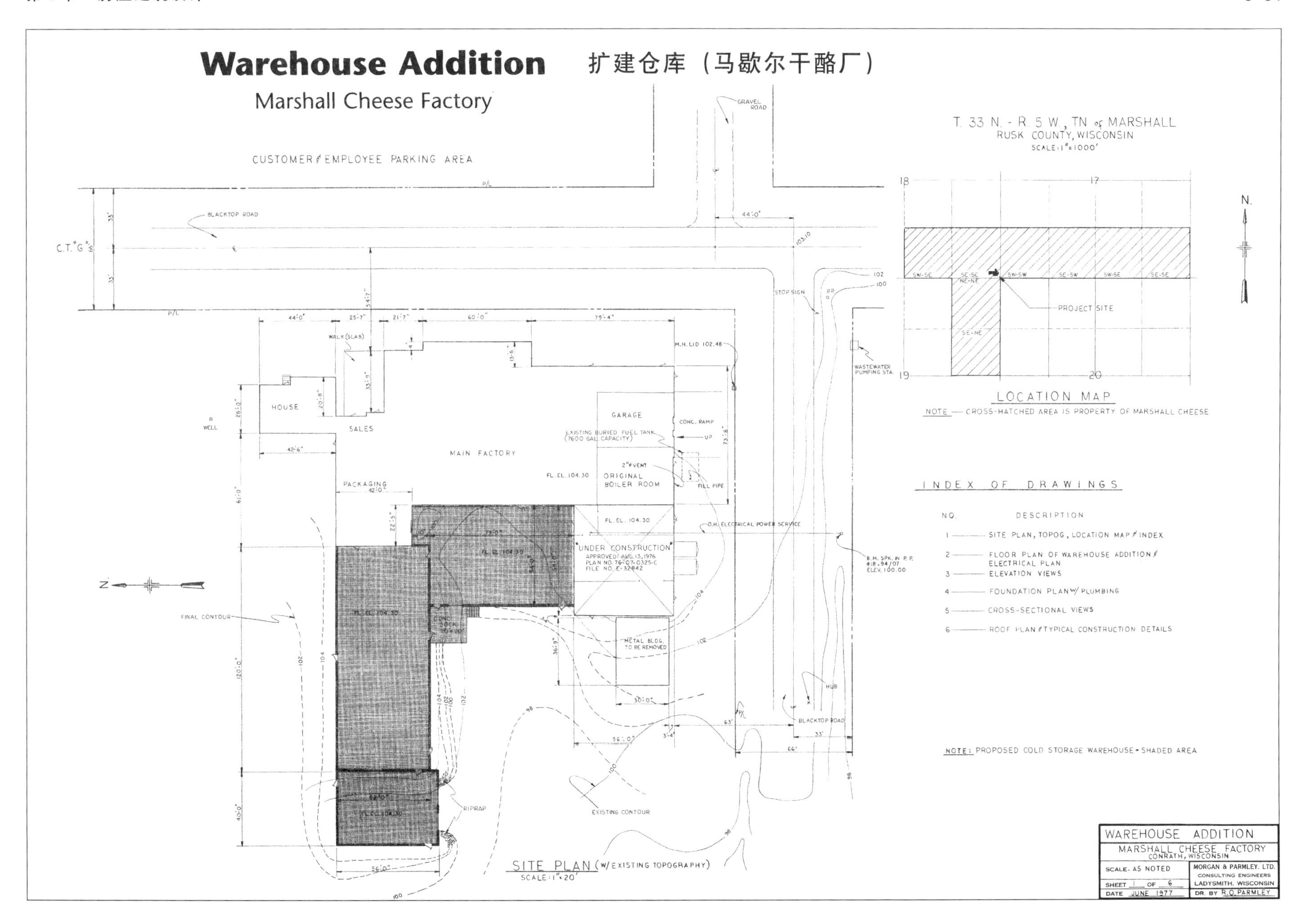

Courtesy: Morgan & Parmley, Ltd.

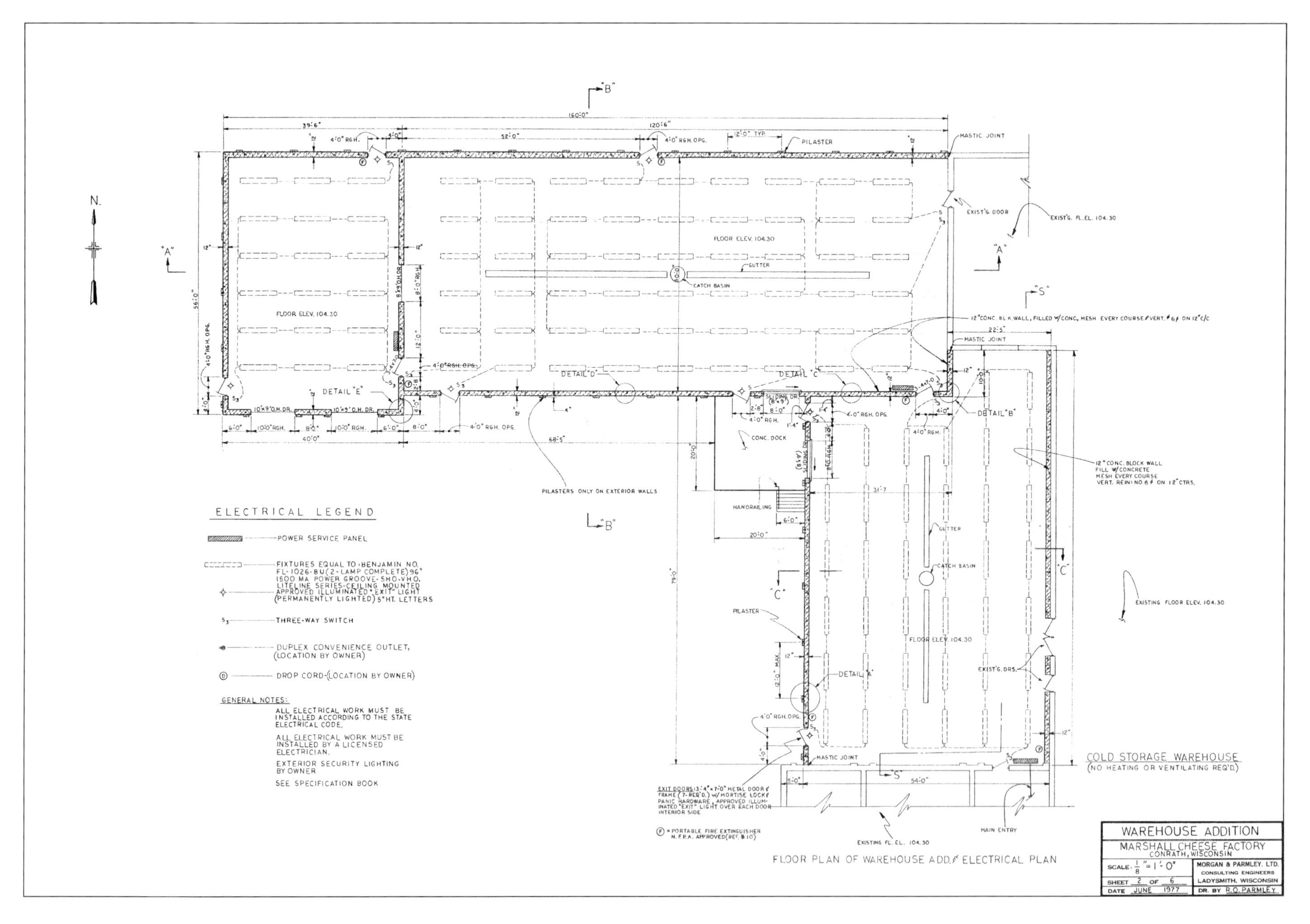
ELECTRICAL LEGEND
POWER SERVICE PANEL
FIXTURES EQUAL TO BENJAMIN NO. FL-1026-BU (2-LAMP COMPLETE) 96" 1500 MA POWER GROOVE-SHO-VHO. LITELINE SERIES-CEILING MOUNTED
APPROVED ILLUMINATED "EXIT" LIGHT (PERMANENTLY LIGHTED) 5" HT. LETTERS
THREE-WAY SWITCH
DUPLEX CONVENIENCE OUTLET, (LOCATION BY OWNER)
DROP CORD-(LOCATION BY OWNER)
GENERAL NOTES:
ALL ELECTRICAL WORK MUST BE INSTALLED ACCORDING TO THE STATE ELECTRICAL CODE.
ALL ELECTRICAL WORK MUST BE INSTALLED BY A LICENSED ELECTRICIAN.
EXTERIOR SECURITY LIGHTING BY OWNER
SEE SPECIFICATION BOOK
FLOOR ELEV. 104.30
CATCH BASIN
GUTTER
PILASTER
MASTIC JOINT
EXIST'G DOOR
EXIST'G. FL. EL. 104.30
12" CONC. BLK. WALL, FILLED W/CONC. MESH EVERY COURSE & VERT. #6 ON 12" C/C
DETAIL "A"
DETAIL "B"
DETAIL "C"
DETAIL "D"
DETAIL "E"
CONC. DOCK
HANDRAILING
PILASTERS ONLY ON EXTERIOR WALLS
12" CONC. BLOCK WALL FILL W/CONCRETE MESH EVERY COURSE VERT. REIN. NO. 6 ON 12" CTRS.
EXISTING FLOOR ELEV. 104.30
EXIST'G. DRS.
COLD STORAGE WAREHOUSE
(NO HEATING OR VENTILATING REQ'D.)
EXIT DOORS: 3'-4" × 7'-0" METAL DOOR & FRAME (7-REQ'D.) W/ MORTISE LOCK & PANIC HARDWARE, APPROVED ILLUMINATED "EXIT" LIGHT OVER EACH DOOR INTERIOR SIDE
(F) = PORTABLE FIRE EXTINGUISHER N.F.P.A. APPROVED (REF. B 10)
EXISTING FL. EL. 104.30
MAIN ENTRY
FLOOR PLAN OF WAREHOUSE ADD. & ELECTRICAL PLAN
WAREHOUSE ADDITION
MARSHALL CHEESE FACTORY
CONRATH, WISCONSIN
SCALE 1/8" = 1'-0"
SHEET 2 OF 6
DATE JUNE 1977
MORGAN & PARMLEY, LTD.
CONSULTING ENGINEERS
LADYSMITH, WISCONSIN
DR. BY R.O. PARMLEY

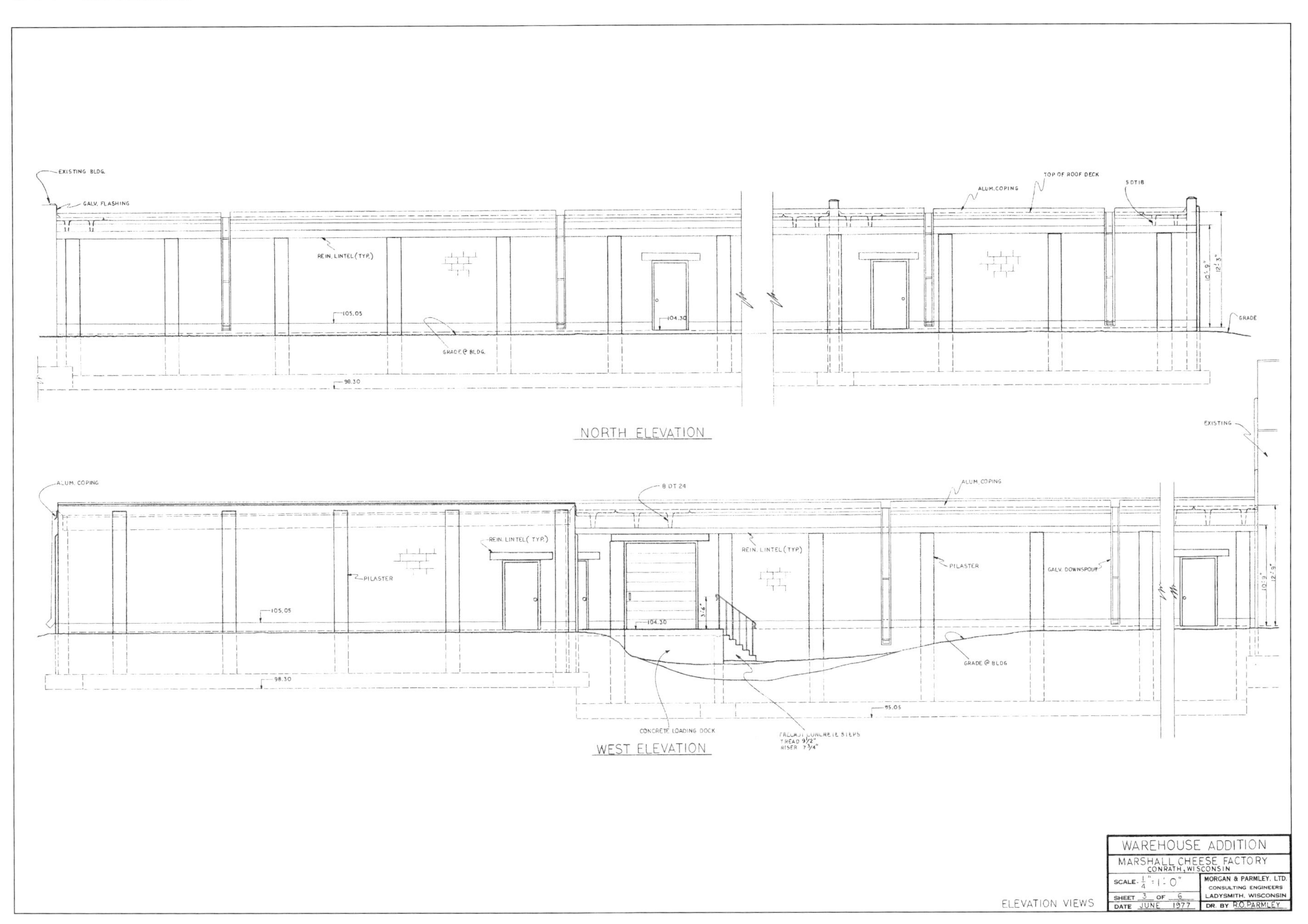
EXISTING BLDG.
GALV. FLASHING
REIN. LINTEL (TYP.)
105.05
104.30
GRADE @ BLDG.
98.30
ALUM. COPING
TOP OF ROOF DECK
5 DT18
10'-9"
12'-3"
GRADE
NORTH ELEVATION
EXISTING
ALUM. COPING
8 DT 24
REIN. LINTEL (TYP.)
PILASTER
GALV. DOWNSPOUT
3'-6"
95.05
CONCRETE LOADING DOCK
PRECAST CONCRETE STEPS
TREAD 9 1/2"
RISER 7 3/4"
WEST ELEVATION
ELEVATION VIEWS
WAREHOUSE ADDITION
MARSHALL CHEESE FACTORY
CONRATH, WISCONSIN
SCALE: 1/4" = 1'-0"
SHEET 3 OF 6
DATE JUNE 1977
MORGAN & PARMLEY, LTD.
CONSULTING ENGINEERS
LADYSMITH, WISCONSIN
DR. BY R.O. PARMLEY

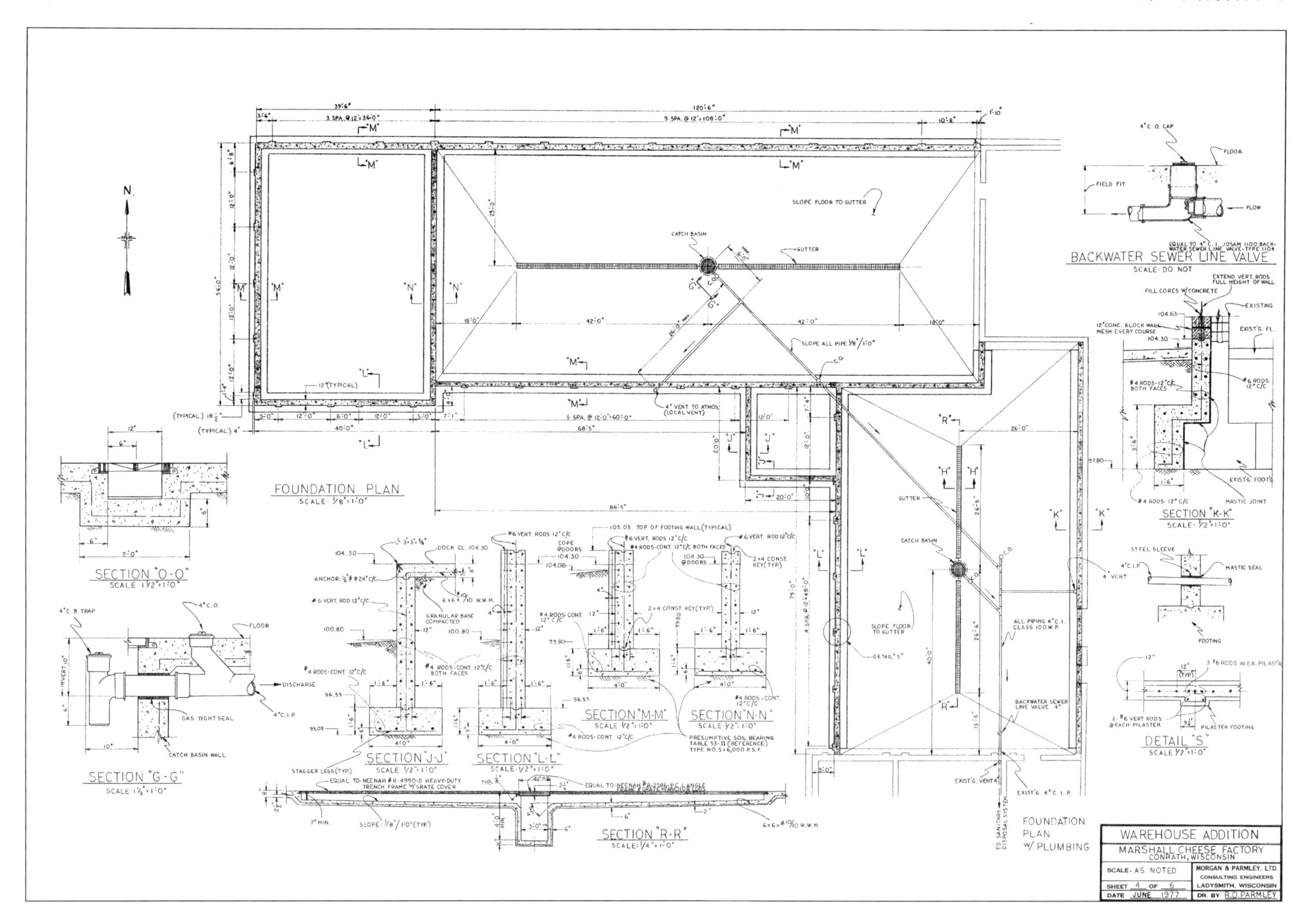
FOUNDATION PLAN
SCALE: 1/8"=1'-0"
BACKWATER SEWER LINE VALVE
SCALE: DO NOT
SECTION "O-O"
SCALE 1 1/2"=1'-0"
SECTION "G-G"
SCALE 1 1/2"=1'-0"
SECTION "J-J"
SCALE 1/2"=1'-0"
SECTION "L-L"
SCALE 1/2"=1'-0"
SECTION "M-M"
SCALE 1/2"=1'-0"
SECTION "N-N"
SCALE 1/2"=1'-0"
SECTION "K-K"
SCALE 1/2"=1'-0"
DETAIL "S"
SCALE 1/2"=1'-0"
SECTION "R-R"
SCALE: 1/4"=1'-0"
FOUNDATION PLAN W/ PLUMBING
SLOPE FLOOR TO GUTTER
CATCH BASIN
GUTTER
SLOPE ALL PIPE 1/8"/1'-0"
4" VENT TO ATMOS. (LOCAL VENT)
ALL PIPING 4" C.I. CLASS 100 W.P.
BACKWATER SEWER LINE VALVE 4"
PRESUMPTIVE SOIL BEARING TABLE 53-II (REFERENCE) TYPE NO.5=6,000 P.S.F.
WAREHOUSE ADDITION
MARSHALL CHEESE FACTORY
CONRATH, WISCONSIN
SCALE- AS NOTED
SHEET 4 OF 6
DATE JUNE 1977
MORGAN & PARMLEY, LTD.
CONSULTING ENGINEERS
LADYSMITH, WISCONSIN
DR. BY R.O. PARMLEY

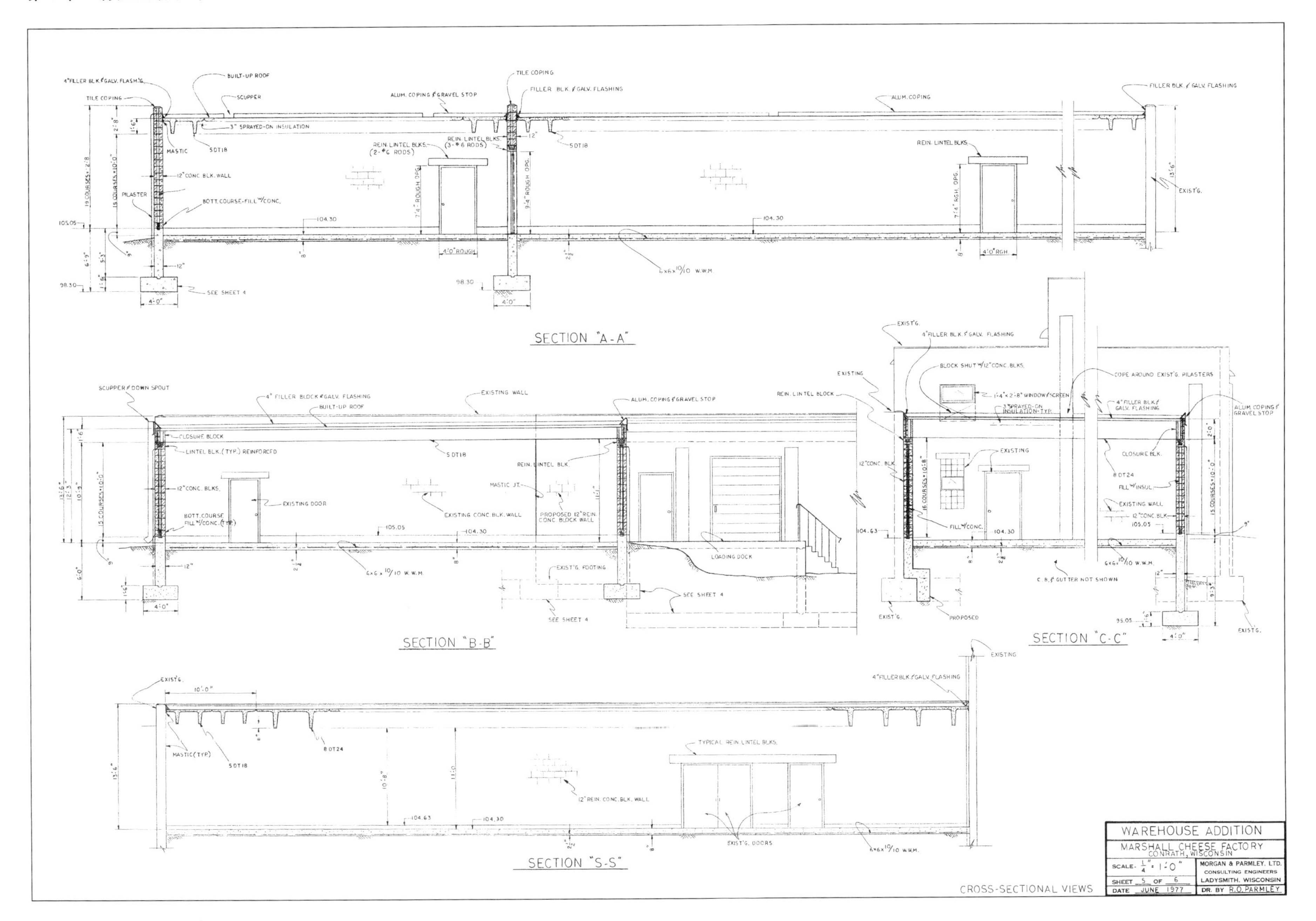

SECTION "A-A"
SECTION "B-B"
SECTION "C-C"
SECTION "S-S"
BUILT-UP ROOF
TILE COPING
SCUPPER
3" SPRAYED-ON INSULATION
MASTIC
5DT18
12" CONC. BLK. WALL
PILASTER
BOTT. COURSE-FILL W/CONC.
ALUM. COPING & GRAVEL STOP
FILLER BLK. & GALV. FLASHING
REIN. LINTEL BLKS.
ALUM. COPING
104.30
105.05
98.30
SEE SHEET 4
4'-0" ROUGH
6×6×10/10 W.W.M.
EXIST'G.
SCUPPER & DOWN SPOUT
4" FILLER BLOCK & GALV. FLASHING
EXISTING WALL
CLOSURE BLOCK
LINTEL BLK. (TYP.) REINFORCED
12" CONC. BLKS.
EXISTING DOOR
EXISTING CONC. BLK. WALL
PROPOSED 12" REIN. CONC. BLOCK WALL
LOADING DOCK
EXIST'G. FOOTING
EXISTING
REIN. LINTEL BLOCK
BLOCK SHUT W/12" CONC. BLKS.
COPE AROUND EXIST'G. PILASTERS
8 DT24
FILL W/CONC.
C. B. & GUTTER NOT SHOWN
PROPOSED
MASTIC (TYP)
TYPICAL REIN. LINTEL BLKS.
12" REIN. CONC. BLK. WALL
EXIST'G. DOORS
104.63
WAREHOUSE ADDITION
MARSHALL CHEESE FACTORY
CONRATH, WISCONSIN
SCALE 1/4" = 1'-0"
SHEET 5 OF 6
DATE JUNE 1977
MORGAN & PARMLEY, LTD.
CONSULTING ENGINEERS
LADYSMITH, WISCONSIN
DR. BY R.O. PARMLEY
CROSS-SECTIONAL VIEWS

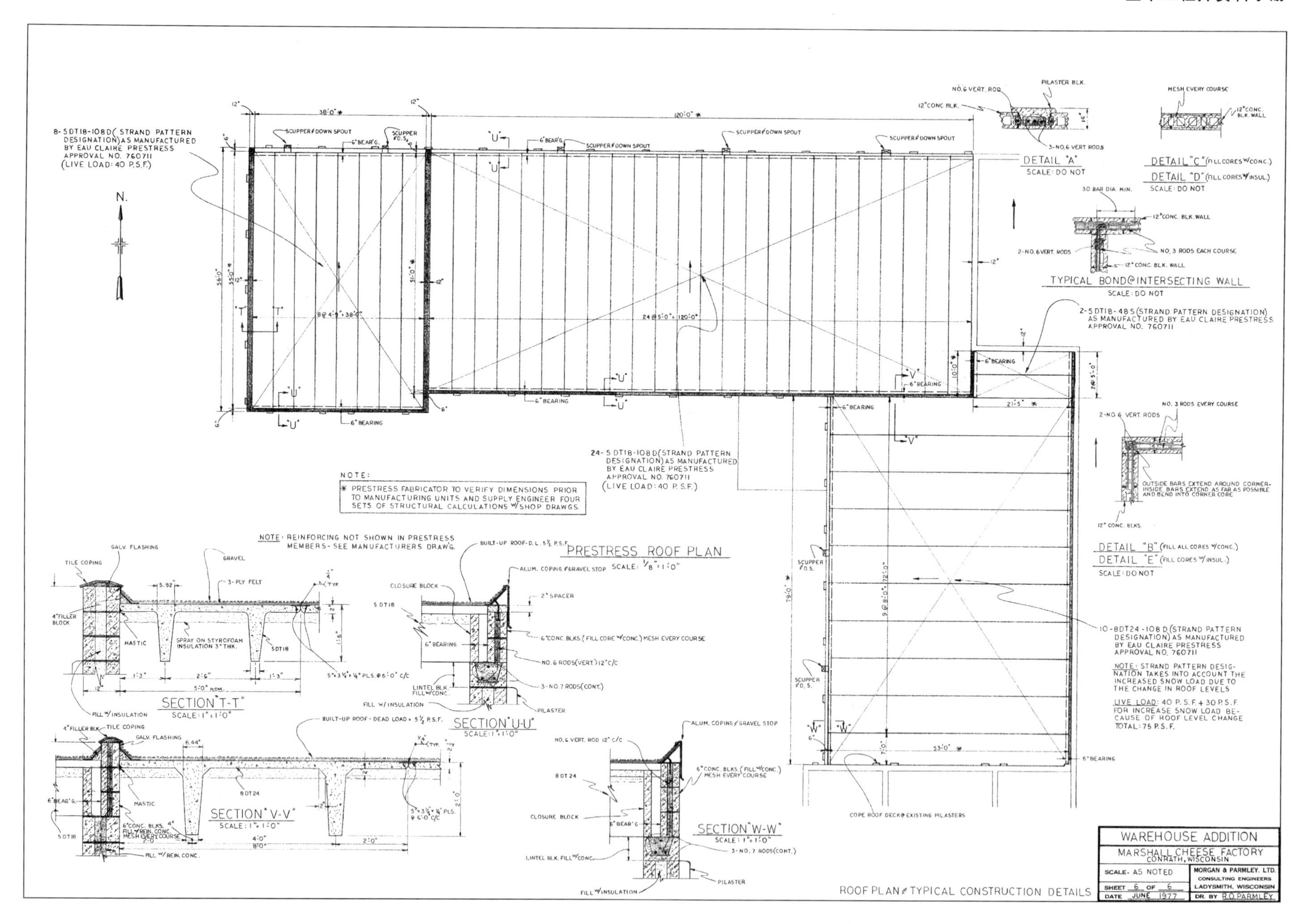

PRESTRESS ROOF PLAN
SCALE: 1/8"=1'-0"
NOTE:
* PRESTRESS FABRICATOR TO VERIFY DIMENSIONS PRIOR TO MANUFACTURING UNITS AND SUPPLY ENGINEER FOUR SETS OF STRUCTURAL CALCULATIONS W/SHOP DRAWGS.
NOTE: REINFORCING NOT SHOWN IN PRESTRESS MEMBERS-SEE MANUFACTURERS DRAW'G.
8-5DT18-108D (STRAND PATTERN DESIGNATION) AS MANUFACTURED BY EAU CLAIRE PRESTRESS APPROVAL NO. 760711 (LIVE LOAD: 40 P.S.F.)
24-5DT18-108D (STRAND PATTERN DESIGNATION) AS MANUFACTURED BY EAU CLAIRE PRESTRESS APPROVAL NO. 760711 (LIVE LOAD: 40 P.S.F.)
2-5DT18-48S (STRAND PATTERN DESIGNATION) AS MANUFACTURED BY EAU CLAIRE PRESTRESS APPROVAL NO. 760711
10-8DT24-108D (STRAND PATTERN DESIGNATION) AS MANUFACTURED BY EAU CLAIRE PRESTRESS APPROVAL NO. 760711
NOTE: STRAND PATTERN DESIGNATION TAKES INTO ACCOUNT THE INCREASED SNOW LOAD DUE TO THE CHANGE IN ROOF LEVELS
LIVE LOAD: 40 P.S.F + 30 P.S.F FOR INCREASE SNOW LOAD BECAUSE OF ROOF LEVEL CHANGE TOTAL: 75 P.S.F.
DETAIL "A"
SCALE: DO NOT
DETAIL "C" (FILL CORES W/CONC.)
DETAIL "D" (FILL CORES W/INSUL.)
SCALE: DO NOT
TYPICAL BOND@INTERSECTING WALL
SCALE: DO NOT
DETAIL "B" (FILL ALL CORES W/CONC.)
DETAIL "E" (FILL CORES W/INSUL.)
SCALE: DO NOT
OUTSIDE BARS EXTEND AROUND CORNER-INSIDE BARS EXTEND AS FAR AS POSSIBLE AND BEND INTO CORNER CORE
SECTION "T-T"
SCALE: 1"=1'-0"
SECTION "U-U"
SCALE: 1"=1'-0"
SECTION "V-V"
SCALE: 1"=1'-0"
SECTION "W-W"
SCALE: 1"=1'-0"
COPE ROOF DECK@EXISTING PILASTERS
WAREHOUSE ADDITION
MARSHALL CHEESE FACTORY
CONRATH, WISCONSIN
SCALE: AS NOTED
SHEET 6 OF 6
DATE JUNE 1977
MORGAN & PARMLEY, LTD.
CONSULTING ENGINEERS
LADYSMITH, WISCONSIN
DR. BY R.O. PARMLEY
ROOF PLAN & TYPICAL CONSTRUCTION DETAILS

南部小学扩建建筑

SOUTH SIDE ELEMENTARY SCHOOL ADDITION

LADYSMITH—HAWKINS SCHOOL DISTRICT

LADYSMITH, WISCONSIN

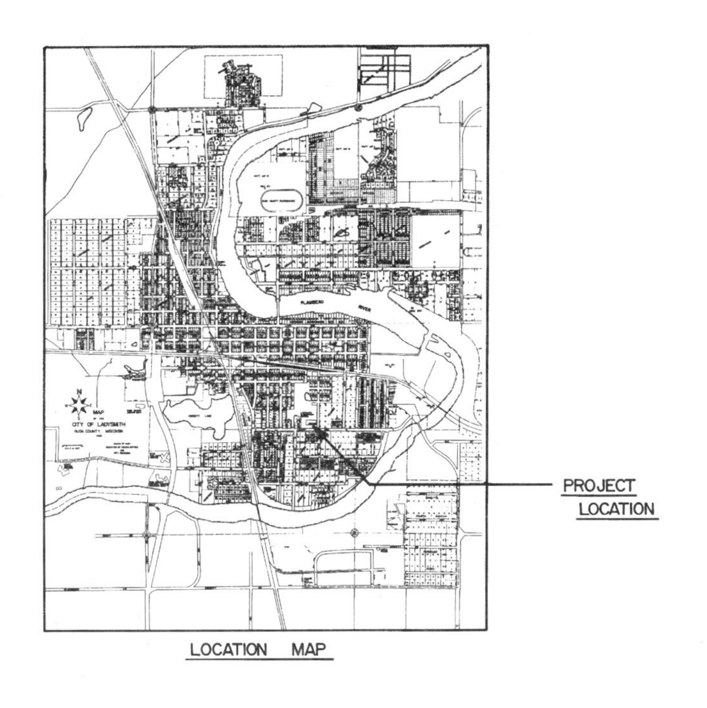

LOCATION MAP

INDEX OF SHEETS

SHEET NO.	*DESCRIPTION*
1	TITLE SHEET W/MAP & INDEX.
2	SITE PLAN & EXISTING ELEVATIONS.
3	EXISTING & PROPOSED FLOOR PLANS, SECTIONS & CONSTRUCTION DETAILS.
4	SECTIONS & CONSTRUCTION DETAILS.
5	ELECTRICAL, HEATING & VENTILATING PLANS.

PREPARED BY: MORGAN & PARMLEY LTD. LADYSMITH, WISCONSIN

Courtesy: Morgan & Parmley, Ltd.

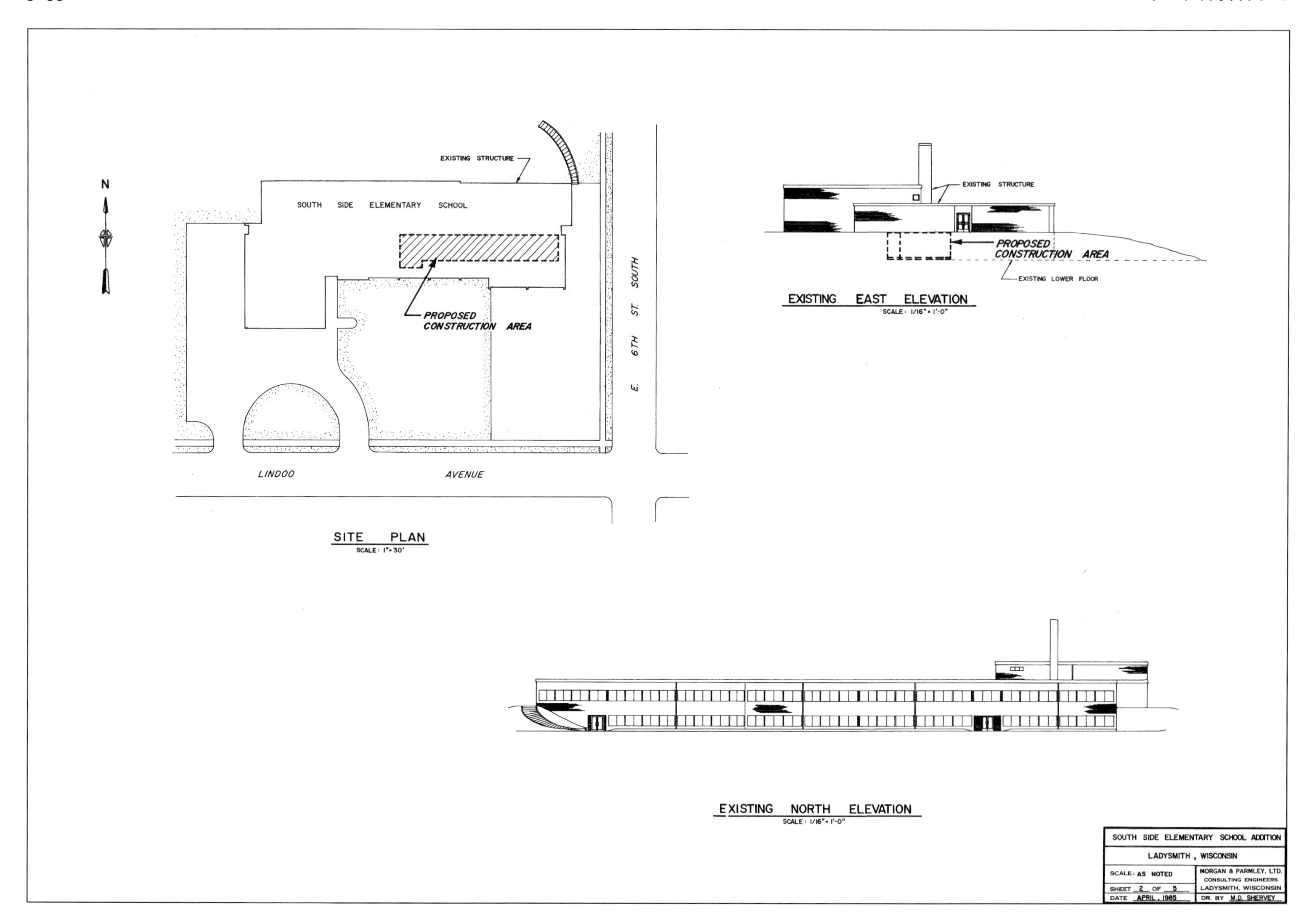
N
EXISTING STRUCTURE
SOUTH SIDE ELEMENTARY SCHOOL
PROPOSED CONSTRUCTION AREA
E. 6TH ST. SOUTH
LINDOO AVENUE
SITE PLAN
SCALE: 1"=30'
EXISTING STRUCTURE
PROPOSED CONSTRUCTION AREA
EXISTING LOWER FLOOR
EXISTING EAST ELEVATION
SCALE: 1/16"=1'-0"
EXISTING NORTH ELEVATION
SCALE: 1/16"=1'-0"
SOUTH SIDE ELEMENTARY SCHOOL ADDITION
LADYSMITH, WISCONSIN
SCALE- AS NOTED
SHEET 2 OF 5
DATE APRIL, 1985
MORGAN & PARMLEY, LTD.
CONSULTING ENGINEERS
LADYSMITH, WISCONSIN
DR. BY M.D. SHERVEY

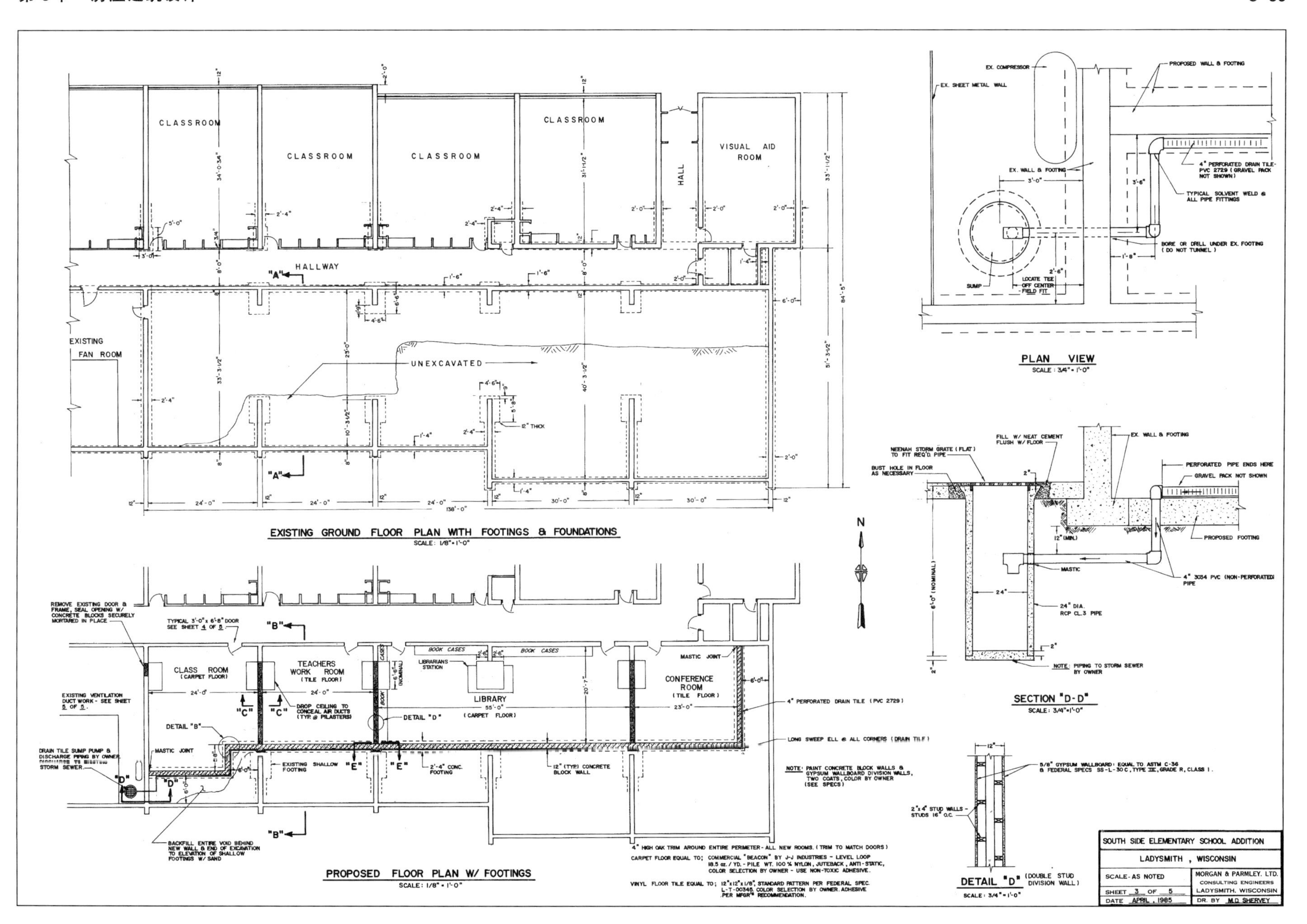
CLASSROOM
CLASSROOM
CLASSROOM
CLASSROOM
VISUAL AID ROOM
HALL
HALLWAY
EXISTING FAN ROOM
UNEXCAVATED
EXISTING GROUND FLOOR PLAN WITH FOOTINGS & FOUNDATIONS
SCALE: 1/8"=1'-0"
PLAN VIEW
SCALE: 3/4"=1'-0"
SECTION "D-D"
SCALE: 3/4"=1'-0"
PROPOSED FLOOR PLAN W/ FOOTINGS
SCALE: 1/8"=1'-0"
DETAIL "D" (DOUBLE STUD DIVISION WALL)
SCALE: 3/4"=1'-0"
CLASS ROOM (CARPET FLOOR)
TEACHERS WORK ROOM (TILE FLOOR)
LIBRARY (CARPET FLOOR)
CONFERENCE ROOM (TILE FLOOR)
4" HIGH OAK TRIM AROUND ENTIRE PERIMETER - ALL NEW ROOMS. (TRIM TO MATCH DOORS)
SOUTH SIDE ELEMENTARY SCHOOL ADDITION
LADYSMITH, WISCONSIN
SCALE - AS NOTED
SHEET 3 OF 5
DATE APRIL, 1985
MORGAN & PARMLEY, LTD.
CONSULTING ENGINEERS
LADYSMITH, WISCONSIN
DR. BY M.D. SHERVEY

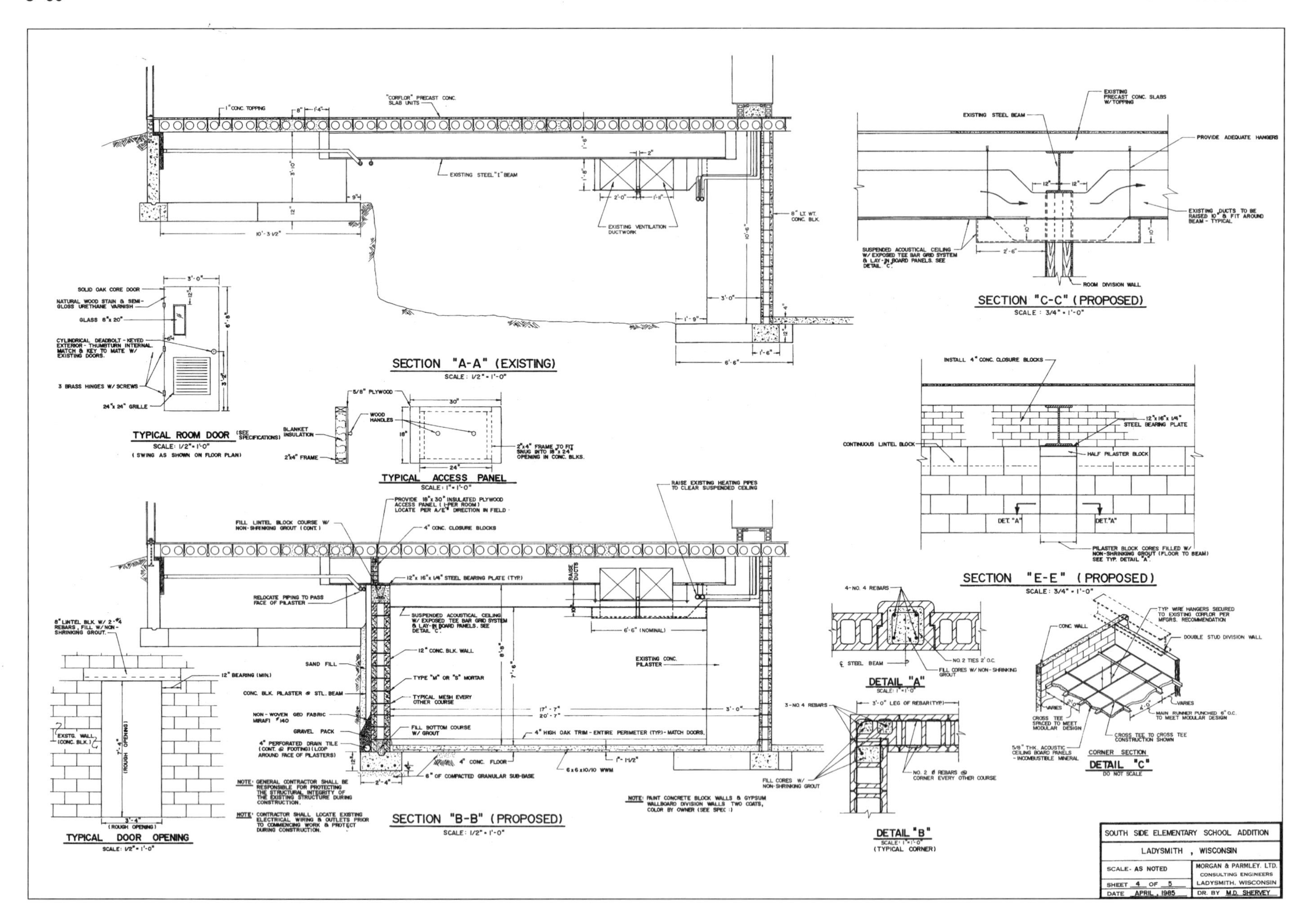
1" CONC. TOPPING
"CORFLOR" PRECAST CONC. SLAB UNITS
EXISTING STEEL "I" BEAM
EXISTING VENTILATION DUCTWORK
8" LT. WT. CONC. BLK.
10'-3 1/2"
SECTION "A-A" (EXISTING)
SCALE: 1/2" = 1'-0"
SOLID OAK CORE DOOR
NATURAL WOOD STAIN & SEMI-GLOSS URETHANE VARNISH
GLASS 8"x 20"
CYLINDRICAL DEADBOLT - KEYED EXTERIOR - THUMBTURN INTERNAL. MATCH & KEY TO MATE W/ EXISTING DOORS.
3 BRASS HINGES W/ SCREWS
24"x 24" GRILLE
TYPICAL ROOM DOOR (SEE SPECIFICATIONS)
SCALE: 1/2" = 1'-0"
(SWING AS SHOWN ON FLOOR PLAN)
5/8" PLYWOOD
WOOD HANDLES
BLANKET INSULATION
2"x4" FRAME
2"x4" FRAME TO FIT SNUG INTO 18"x 24" OPENING IN CONC. BLKS.
TYPICAL ACCESS PANEL
SCALE: 1" = 1'-0"
EXISTING PRECAST CONC. SLABS W/ TOPPING
EXISTING STEEL BEAM
PROVIDE ADEQUATE HANGERS
EXISTING DUCTS TO BE RAISED 10" & FIT AROUND BEAM - TYPICAL
SUSPENDED ACOUSTICAL CEILING W/ EXPOSED TEE BAR GRID SYSTEM & LAY-IN BOARD PANELS. SEE DETAIL "C".
ROOM DIVISION WALL
SECTION "C-C" (PROPOSED)
SCALE: 3/4" = 1'-0"
INSTALL 4" CONC. CLOSURE BLOCKS
12"x 16"x 1/4" STEEL BEARING PLATE
CONTINUOUS LINTEL BLOCK
HALF PILASTER BLOCK
DET. "A"
PILASTER BLOCK CORES FILLED W/ NON-SHRINKING GROUT (FLOOR TO BEAM) SEE TYP. DETAIL "A".
SECTION "E-E" (PROPOSED)
SCALE: 3/4" = 1'-0"
RAISE EXISTING HEATING PIPES TO CLEAR SUSPENDED CEILING
PROVIDE 18"x 30" INSULATED PLYWOOD ACCESS PANEL (1-PER ROOM) LOCATE PER A/E'S DIRECTION IN FIELD
FILL LINTEL BLOCK COURSE W/ NON-SHRINKING GROUT (CONT.)
4" CONC. CLOSURE BLOCKS
12"x 16"x 1/4" STEEL BEARING PLATE (TYP.)
RELOCATE PIPING TO PASS FACE OF PILASTER
RAISE DUCTS
SUSPENDED ACOUSTICAL CEILING W/ EXPOSED TEE BAR GRID SYSTEM & LAY-IN BOARD PANELS. SEE DETAIL "C".
6'-6" (NOMINAL)
12" CONC. BLK. WALL
EXISTING CONC. PILASTER
SAND FILL
TYPE "M" OR "S" MORTAR
CONC. BLK. PILASTER @ STL. BEAM
TYPICAL MESH EVERY OTHER COURSE
NON-WOVEN GEO FABRIC MIRAFI #140
GRAVEL PACK
FILL BOTTOM COURSE W/ GROUT
4" HIGH OAK TRIM - ENTIRE PERIMETER (TYP.) - MATCH DOORS.
4" PERFORATED DRAIN TILE (CONT. @ FOOTING) (LOOP AROUND FACE OF PILASTERS)
4" CONC. FLOOR
6" OF COMPACTED GRANULAR SUB-BASE
6x6x10/10 WWM
NOTE: GENERAL CONTRACTOR SHALL BE RESPONSIBLE FOR PROTECTING THE STRUCTURAL INTEGRITY OF THE EXISTING STRUCTURE DURING CONSTRUCTION.
NOTE: CONTRACTOR SHALL LOCATE EXISTING ELECTRICAL WIRING & OUTLETS PRIOR TO COMMENCING WORK & PROTECT DURING CONSTRUCTION.
NOTE: PAINT CONCRETE BLOCK WALLS & GYPSUM WALLBOARD DIVISION WALLS TWO COATS, COLOR BY OWNER (SEE SPEC.)
SECTION "B-B" (PROPOSED)
SCALE: 1/2" = 1'-0"
8" LINTEL BLK. W/ 2-#4 REBARS, FILL W/ NON-SHRINKING GROUT.
12" BEARING (MIN.)
EXSTG. WALL (CONC. BLK.)
7'-4" (ROUGH OPENING)
3'-4" (ROUGH OPENING)
TYPICAL DOOR OPENING
SCALE: 1/2" = 1'-0"
4-NO. 4 REBARS
NO. 2 TIES 2' O.C.
℄ STEEL BEAM
FILL CORES W/ NON-SHRINKING GROUT
DETAIL "A"
SCALE: 1" = 1'-0"
3-NO. 4 REBARS
3'-0" LEG OF REBAR (TYP.)
NO. 2 Ø REBARS @ CORNER EVERY OTHER COURSE
FILL CORES W/ NON-SHRINKING GROUT
DETAIL "B"
SCALE: 1" = 1'-0"
(TYPICAL CORNER)
TYP. WIRE HANGERS SECURED TO EXISTING CORFLOR PER MFGRS. RECOMMENDATION
CONC. WALL
DOUBLE STUD DIVISION WALL
VARIES
CROSS TEE SPACED TO MEET MODULAR DESIGN
MAIN RUNNER PUNCHED 6" O.C. TO MEET MODULAR DESIGN
CROSS TEE TO CROSS TEE CONSTRUCTION SHOWN
5/8" THK. ACOUSTIC CEILING BOARD PANELS - INCOMBUSTIBLE MINERAL
CORNER SECTION
DETAIL "C"
DO NOT SCALE
SOUTH SIDE ELEMENTARY SCHOOL ADDITION
LADYSMITH, WISCONSIN
SCALE - AS NOTED
SHEET 4 OF 5
DATE APRIL, 1985
MORGAN & PARMLEY, LTD. CONSULTING ENGINEERS LADYSMITH, WISCONSIN
DR. BY M.D. SHERVEY

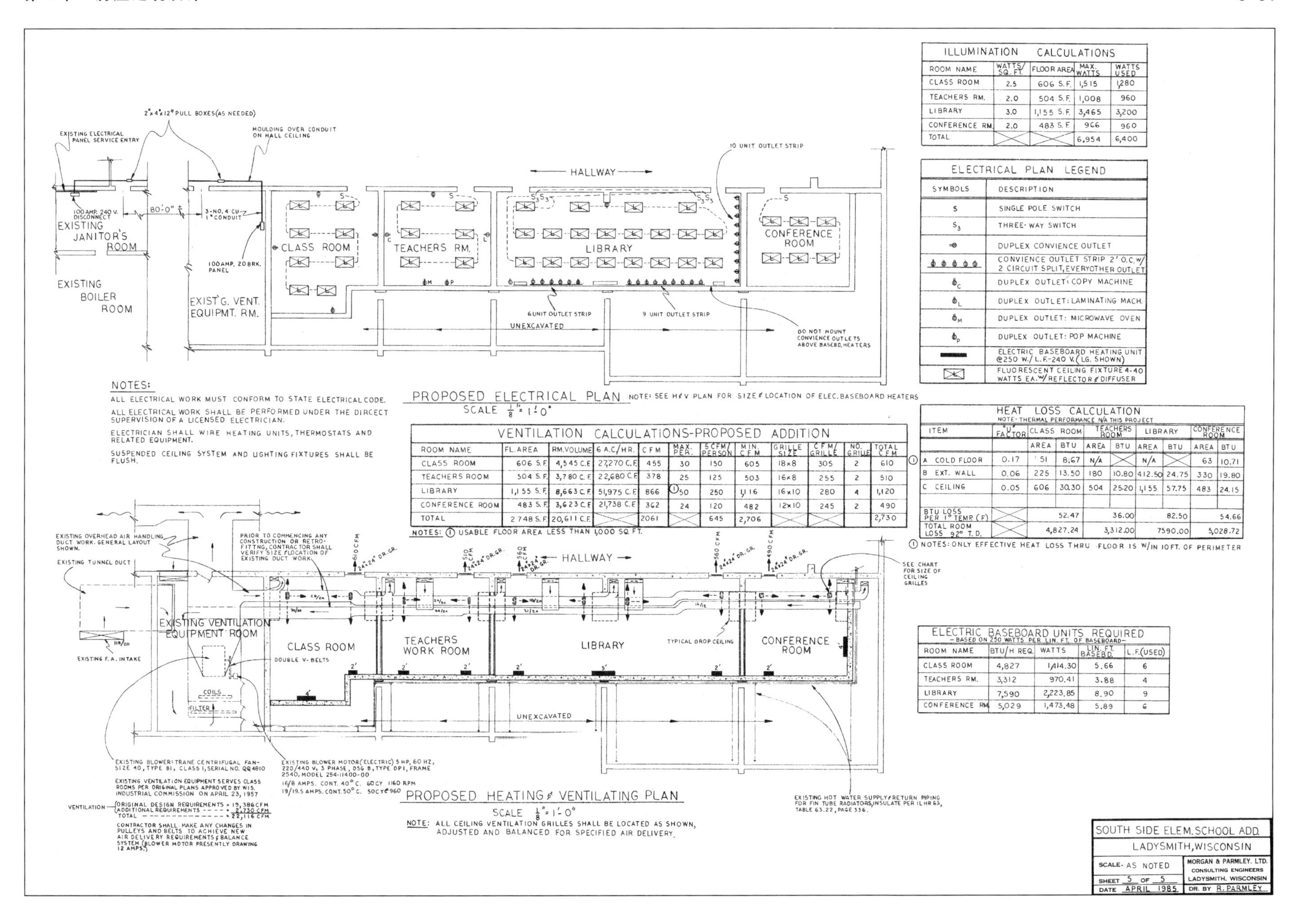

ILLUMINATION CALCULATIONS

ROOM NAME	WATTS/SQ.FT.	FLOOR AREA	MAX. WATTS	WATTS USED
CLASS ROOM	2.5	606 S.F.	1,515	1,280
TEACHERS RM.	2.0	504 S.F.	1,008	960
LIBRARY	3.0	1,155 S.F.	3,465	3,200
CONFERENCE RM.	2.0	483 S.F.	966	960
TOTAL			6,954	6,400

ELECTRICAL PLAN LEGEND

SYMBOLS	DESCRIPTION
S	SINGLE POLE SWITCH
S_3	THREE-WAY SWITCH
	DUPLEX CONVIENCE OUTLET
	CONVIENCE OUTLET STRIP 2' O.C. W/ 2 CIRCUIT SPLIT, EVERYOTHER OUTLET
C	DUPLEX OUTLET: COPY MACHINE
L	DUPLEX OUTLET: LAMINATING MACH.
M	DUPLEX OUTLET: MICROWAVE OVEN
P	DUPLEX OUTLET: POP MACHINE
	ELECTRIC BASEBOARD HEATING UNIT @250 W./L.F.-240 V. (LG. SHOWN)
	FLUORESCENT CEILING FIXTURE 4-40 WATTS EA. W/ REFLECTOR & DIFFUSER

VENTILATION CALCULATIONS-PROPOSED ADDITION

ROOM NAME	FL. AREA	RM. VOLUME	6 A.C./HR.	CFM	MAX. PER.	5 CFM/PERSON	MIN. CFM	GRILLE SIZE	CFM/GRILLE	NO. GRILLE	TOTAL CFM
CLASS ROOM	606 S.F.	4,545 C.F.	27,270 C.F.	455	30	150	605	18×8	305	2	610
TEACHERS ROOM	504 S.F.	3,780 C.F.	22,680 C.F.	378	25	125	503	16×8	255	2	510
LIBRARY	1,155 S.F.	8,663 C.F.	51,975 C.F.	866	(1) 50	250	1,116	16×10	280	4	1,120
CONFERENCE ROOM	483 S.F.	3,623 C.F.	21,738 C.F.	362	24	120	482	12×10	245	2	490
TOTAL	2748 S.F.	20,611 C.F.		2061		645	2,706				2,730

NOTES: (1) USABLE FLOOR AREA LESS THAN 1,000 SQ.FT.

HEAT LOSS CALCULATION

NOTE-THERMAL PERFORMANCE N/A THIS PROJECT

ITEM	"U" FACTOR	CLASS ROOM AREA	CLASS ROOM BTU	TEACHERS ROOM AREA	TEACHERS ROOM BTU	LIBRARY AREA	LIBRARY BTU	CONFERENCE ROOM AREA	CONFERENCE ROOM BTU
(1) A COLD FLOOR	0.17	51	8.67	N/A		N/A		63	10.71
B EXT. WALL	0.06	225	13.50	180	10.80	412.50	24.75	330	19.80
C CEILING	0.05	606	30.30	504	25.20	1,155	57.75	483	24.15
BTU LOSS PER 1° TEMP. (F)			52.47		36.00		82.50		54.66
TOTAL ROOM LOSS 92° T.D.			4,827.24		3,312.00		7590.00		5,028.72

(1) NOTES: ONLY EFFECTIVE HEAT LOSS THRU FLOOR IS W/IN 10 FT. OF PERIMETER

ELECTRIC BASEBOARD UNITS REQUIRED

—BASED ON 250 WATTS PER LIN. FT. OF BASEBOARD—

ROOM NAME	BTU/H REQ.	WATTS	LIN. FT. BASEBD.	L.F. (USED)
CLASS ROOM	4,827	1,414.30	5.66	6
TEACHERS RM.	3,312	970.41	3.88	4
LIBRARY	7,590	2,223.85	8.90	9
CONFERENCE RM.	5,029	1,473.48	5.89	6

设　计

第9节　供水和配水设计

例 9

市政供水和配水系统的设计是土木工程师专业活动的一个重要部分。持续的、充足的饮用水供应是现代城市的生命线。安全的饮用水水源、蓄水和配水由相关的管理部门来管理和监督。在施工之前，这些设施的所有设计图和技术说明必须得到监控部门的正式核准。

目　录

第4号市政井

MUNICIPAL WELL NO. 4

VILLAGE OF
GILMAN, WISCONSIN

APRIL, 1987

WELL PROJECT SITE

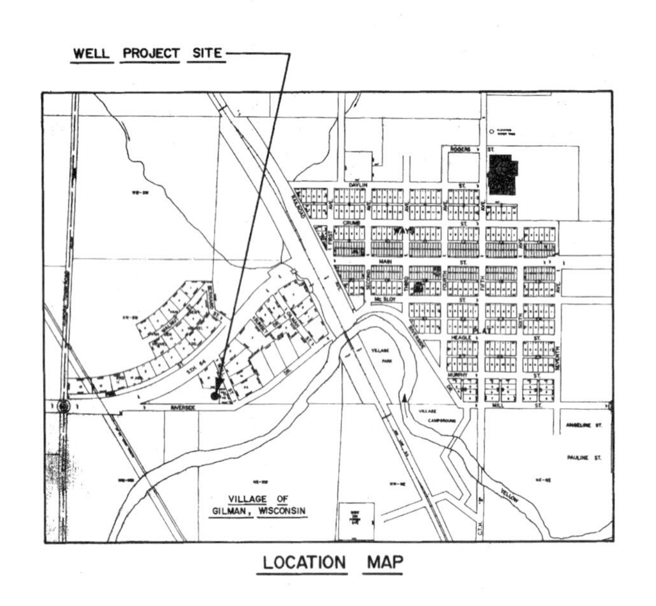

LOCATION MAP

INDEX OF SHEETS

PREPARED BY: MORGAN & PARMLEY, LTD. LADYSMITH, WISCONSIN

Courtesy: Morgan & Parmley, Ltd.

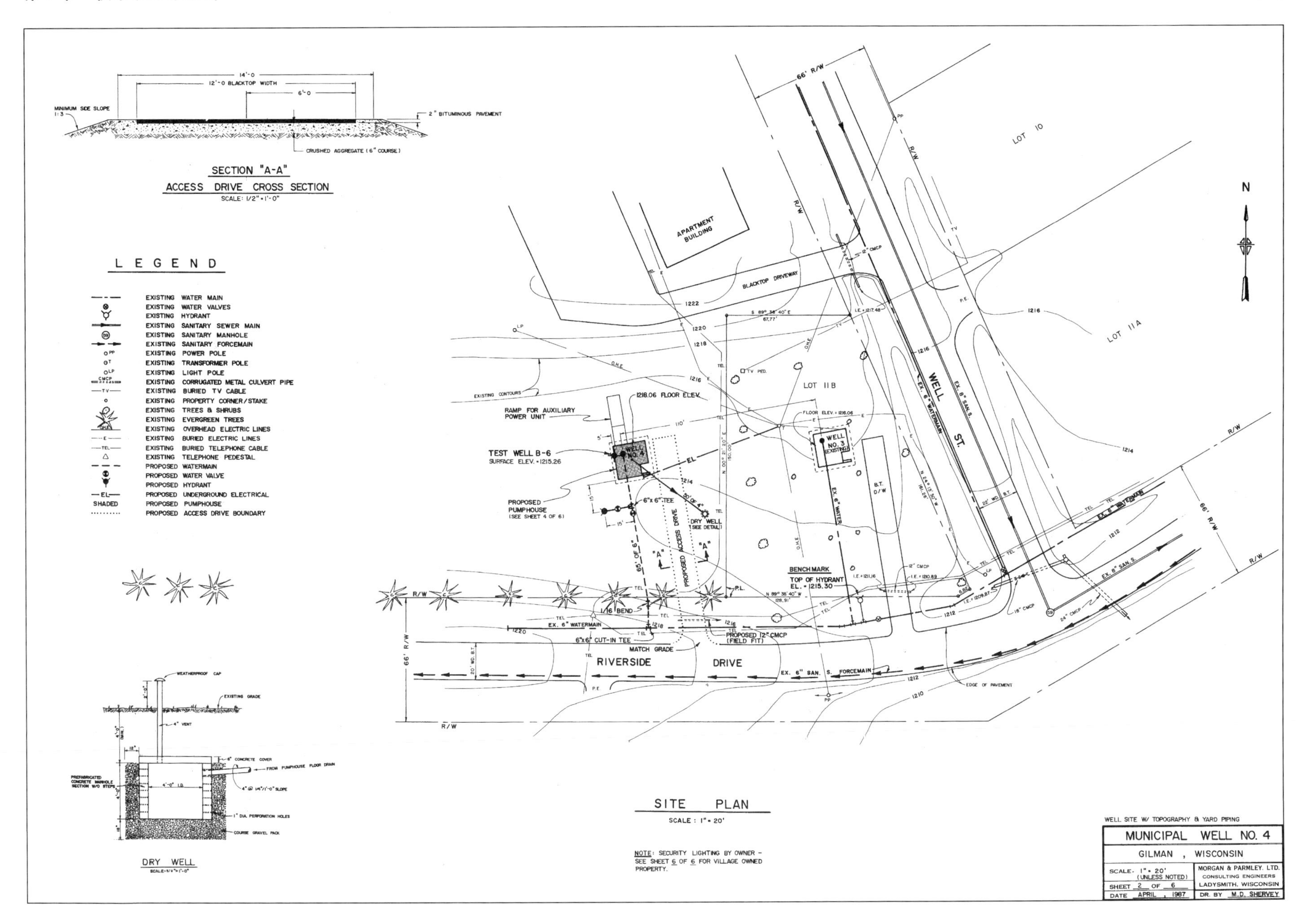
SECTION "A-A"
ACCESS DRIVE CROSS SECTION
SCALE: 1/2" = 1'-0"
LEGEND
EXISTING WATER MAIN
EXISTING WATER VALVES
EXISTING HYDRANT
EXISTING SANITARY SEWER MAIN
EXISTING SANITARY MANHOLE
EXISTING SANITARY FORCEMAIN
EXISTING POWER POLE
EXISTING TRANSFORMER POLE
EXISTING LIGHT POLE
EXISTING CORRUGATED METAL CULVERT PIPE
EXISTING BURIED TV CABLE
EXISTING PROPERTY CORNER/STAKE
EXISTING TREES & SHRUBS
EXISTING EVERGREEN TREES
EXISTING OVERHEAD ELECTRIC LINES
EXISTING BURIED ELECTRIC LINES
EXISTING BURIED TELEPHONE CABLE
EXISTING TELEPHONE PEDESTAL
PROPOSED WATERMAIN
PROPOSED WATER VALVE
PROPOSED HYDRANT
PROPOSED UNDERGROUND ELECTRICAL
SHADED PROPOSED PUMPHOUSE
PROPOSED ACCESS DRIVE BOUNDARY
DRY WELL
SITE PLAN
SCALE: 1" = 20'
NOTE: SECURITY LIGHTING BY OWNER - SEE SHEET 6 OF 6 FOR VILLAGE OWNED PROPERTY.
WELL SITE W/ TOPOGRAPHY & YARD PIPING
MUNICIPAL WELL NO. 4
GILMAN, WISCONSIN
SCALE- 1" = 20' (UNLESS NOTED)
SHEET 2 OF 6
DATE APRIL, 1987
MORGAN & PARMLEY, LTD. CONSULTING ENGINEERS LADYSMITH, WISCONSIN
DR. BY M.D. SHERVEY
RIVERSIDE DRIVE
WELL ST.
APARTMENT BUILDING
LOT 10
LOT 11A
LOT 11B
TEST WELL B-6
SURFACE ELEV. = 1215.26
PROPOSED PUMPHOUSE
WELL NO. 3 (EXISTING)
BENCHMARK
TOP OF HYDRANT EL. = 1215.30
N

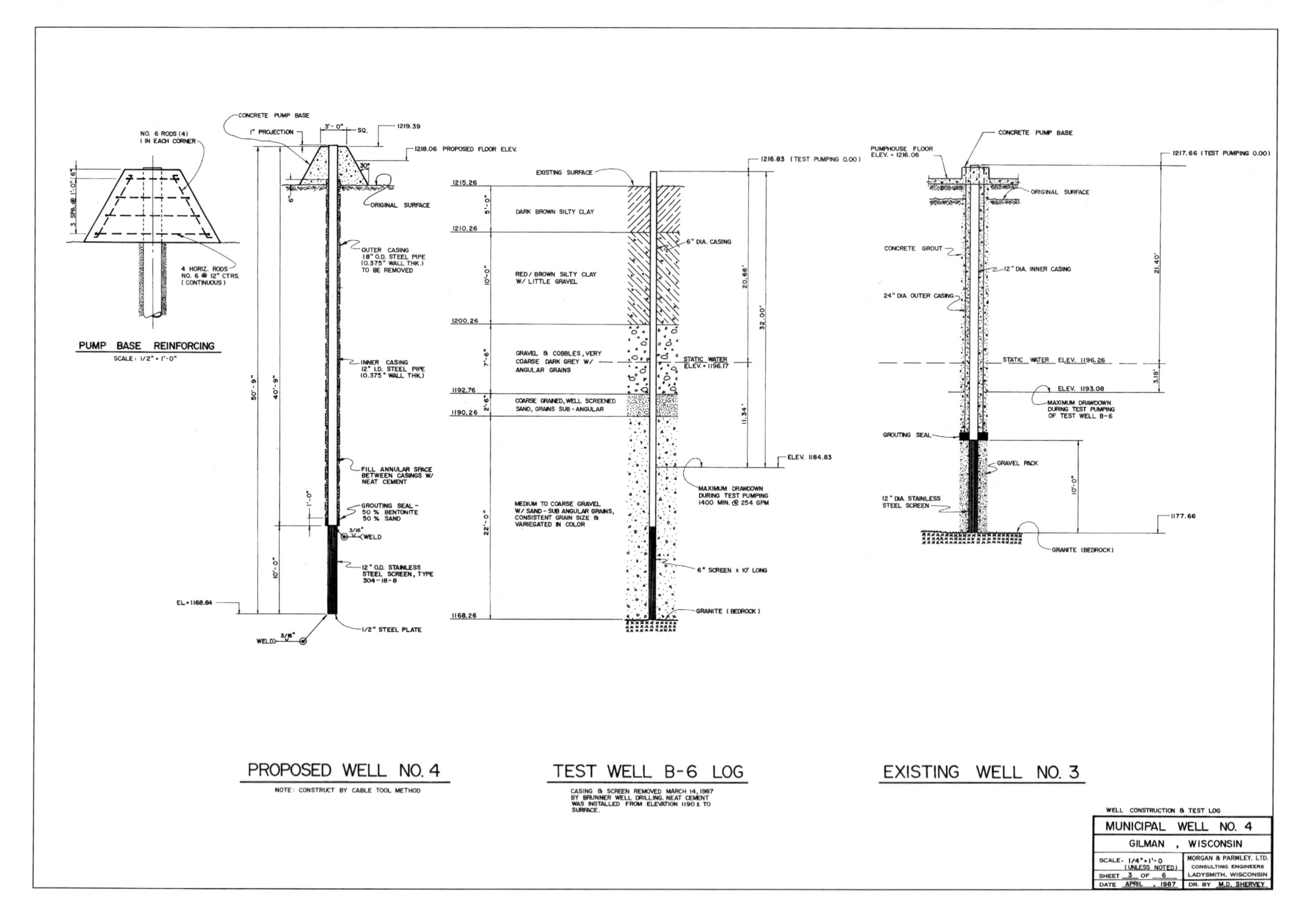
NO. 6 RODS (4)
1 IN EACH CORNER
3 SPA. @ 1'-0"
6"
4 HORIZ. RODS
NO. 6 @ 12" CTRS.
(CONTINUOUS)
PUMP BASE REINFORCING
SCALE: 1/2" = 1'-0"
CONCRETE PUMP BASE
1" PROJECTION
3'-0" SQ.
1219.39
1218.06 PROPOSED FLOOR ELEV.
30°
6"
ORIGINAL SURFACE
OUTER CASING
18" O.D. STEEL PIPE
(0.375" WALL THK.)
TO BE REMOVED
INNER CASING
12" I.D. STEEL PIPE
(0.375" WALL THK.)
50'-9"
40'-9"
FILL ANNULAR SPACE
BETWEEN CASINGS W/
NEAT CEMENT
1'-0"
GROUTING SEAL -
50 % BENTONITE
50 % SAND
3/16" WELD
10'-0"
12" O.D. STAINLESS
STEEL SCREEN, TYPE
304-18-8
EL. = 1168.64
1/2" STEEL PLATE
WELD 3/16"
PROPOSED WELL NO. 4
NOTE: CONSTRUCT BY CABLE TOOL METHOD
EXISTING SURFACE
1216.83 (TEST PUMPING 0.00)
1215.26
5'-0"
DARK BROWN SILTY CLAY
1210.26
6" DIA. CASING
10'-0"
RED/BROWN SILTY CLAY
W/ LITTLE GRAVEL
20.66'
32.00'
1200.26
7'-6"
GRAVEL & COBBLES, VERY
COARSE DARK GREY W/
ANGULAR GRAINS
STATIC WATER
ELEV. = 1196.17
1192.76
2'-6"
COARSE GRAINED, WELL SCREENED
SAND, GRAINS SUB-ANGULAR
1190.26
11.34'
ELEV. 1184.83
MAXIMUM DRAWDOWN
DURING TEST PUMPING
1400 MIN. @ 254 GPM
22'-0"
MEDIUM TO COARSE GRAVEL
W/ SAND - SUB ANGULAR GRAINS,
CONSISTENT GRAIN SIZE &
VARIEGATED IN COLOR
6" SCREEN x 10' LONG
GRANITE (BEDROCK)
1168.26
TEST WELL B-6 LOG
CASING & SCREEN REMOVED MARCH 14, 1987
BY BRUNNER WELL DRILLING. NEAT CEMENT
WAS INSTALLED FROM ELEVATION 1190 ± TO
SURFACE.
CONCRETE PUMP BASE
PUMPHOUSE FLOOR
ELEV. = 1216.06
1217.66 (TEST PUMPING 0.00)
ORIGINAL SURFACE
CONCRETE GROUT
12" DIA. INNER CASING
24" DIA. OUTER CASING
21.40'
STATIC WATER ELEV. 1196.26
3.18'
ELEV. 1193.08
MAXIMUM DRAWDOWN
DURING TEST PUMPING
OF TEST WELL B-6
GROUTING SEAL
GRAVEL PACK
10'-0"
12" DIA. STAINLESS
STEEL SCREEN
1177.66
GRANITE (BEDROCK)
EXISTING WELL NO. 3
WELL CONSTRUCTION & TEST LOG
MUNICIPAL WELL NO. 4
GILMAN , WISCONSIN
SCALE- 1/4" = 1'-0
(UNLESS NOTED)
SHEET 3 OF 6
DATE APRIL , 1987
MORGAN & PARMLEY, LTD.
CONSULTING ENGINEERS
LADYSMITH, WISCONSIN
DR. BY M.D. SHERVEY

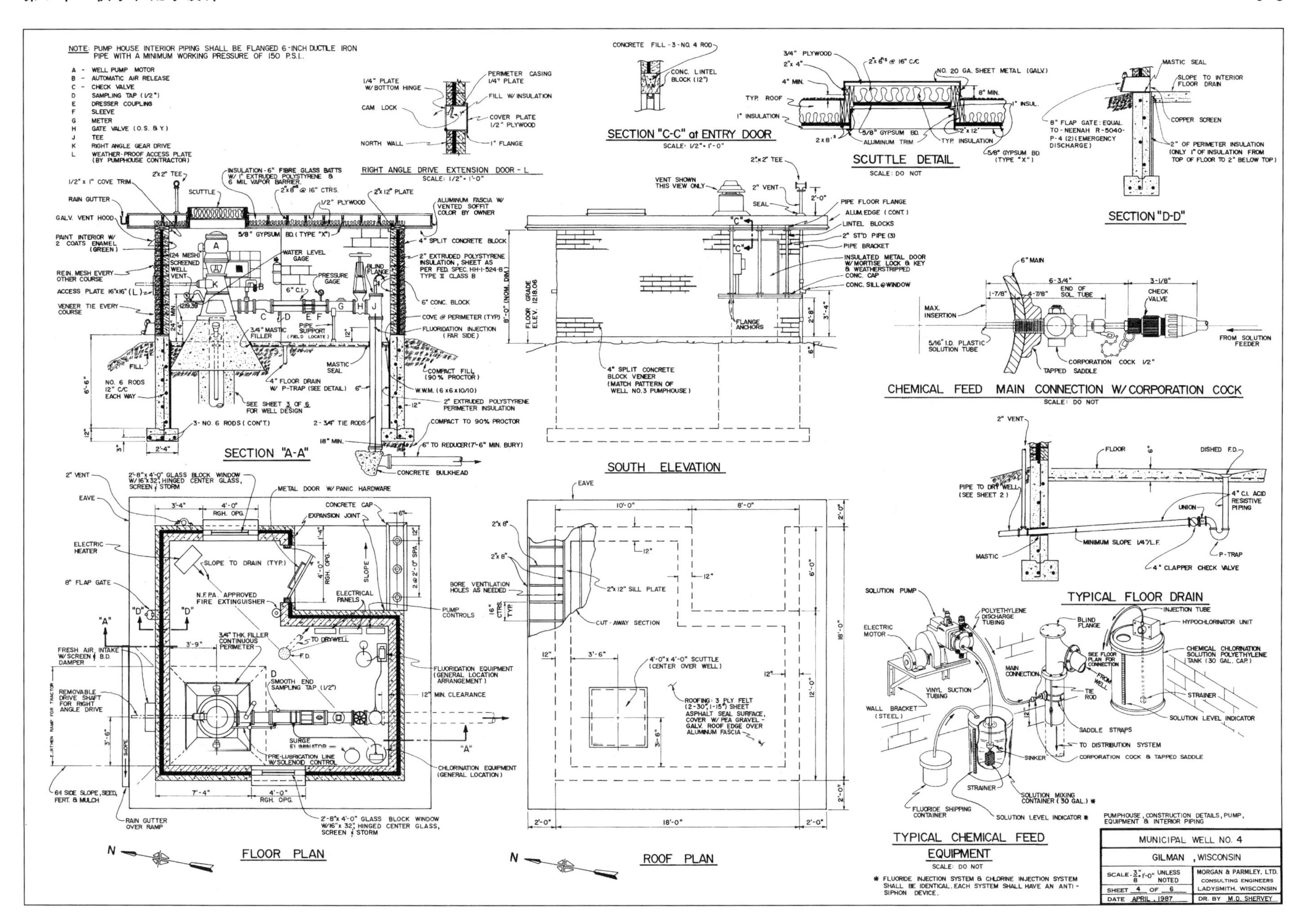

NOTE: PUMP HOUSE INTERIOR PIPING SHALL BE FLANGED 6-INCH DUCTILE IRON PIPE WITH A MINIMUM WORKING PRESSURE OF 150 P.S.I..
A - WELL PUMP MOTOR
B - AUTOMATIC AIR RELEASE
C - CHECK VALVE
D - SAMPLING TAP (1/2")
E - DRESSER COUPLING
F - SLEEVE
G - METER
H - GATE VALVE (O.S. & Y.)
J - TEE
K - RIGHT ANGLE GEAR DRIVE
L - WEATHER-PROOF ACCESS PLATE (BY PUMPHOUSE CONTRACTOR)
1/4" PLATE W/ BOTTOM HINGE
CAM LOCK
NORTH WALL
PERIMETER CASING 1/4" PLATE
FILL W/ INSULATION
COVER PLATE 1/2" PLYWOOD
1" FLANGE
RIGHT ANGLE DRIVE EXTENSION DOOR - L
SCALE: 1/2" = 1'-0"
1/2" x 1" COVE TRIM
RAIN GUTTER
GALV. VENT HOOD
PAINT INTERIOR W/ 2 COATS ENAMEL (GREEN)
REIN. MESH EVERY OTHER COURSE
ACCESS PLATE 16"x16" (L)
VENEER TIE EVERY COURSE
2"x2" TEE
SCUTTLE
INSULATION - 6" FIBRE GLASS BATTS W/ 1" EXTRUDED POLYSTYRENE & 6 MIL VAPOR BARRIER
2"x8" @ 16" CTRS.
2"x12" PLATE
1/2" PLYWOOD
5/8" GYPSUM BD. (TYPE "X")
ALUMINUM FASCIA W/ VENTED SOFFIT COLOR BY OWNER
4" SPLIT CONCRETE BLOCK
(24 MESH) SCREENED WELL VENT
WATER LEVEL GAGE
PRESSURE GAGE
BLIND FLANGE
6" C.I.
2" EXTRUDED POLYSTYRENE INSULATION, SHEET AS PER FED. SPEC. HH-I-524-B TYPE II CLASS B
6" CONC. BLOCK
COVE @ PERIMETER (TYP)
FLUORIDATION INJECTION (FAR SIDE)
3/4" MASTIC FILLER
PIPE SUPPORT (FIELD LOCATE)
MASTIC SEAL
COMPACT FILL (90% PROCTOR)
W.W.M. (6x6x10/10)
4" FLOOR DRAIN W/ P-TRAP (SEE DETAIL)
2" EXTRUDED POLYSTYRENE PERIMETER INSULATION
FILL
NO. 6 RODS 12" C/C EACH WAY
SEE SHEET 3 OF 6 FOR WELL DESIGN
3-NO. 6 RODS (CON'T.)
2-3/4" TIE RODS
COMPACT TO 90% PROCTOR
18" MIN.
6" TO REDUCER (7'-6" MIN. BURY)
CONCRETE BULKHEAD
SECTION "A-A"
CONCRETE FILL - 3 - NO. 4 ROD
CONC. LINTEL BLOCK (12")
SECTION "C-C" at ENTRY DOOR
SCALE: 1/2" = 1'-0"
3/4" PLYWOOD
2"x4"
2"x6"s @ 16" C/C
NO. 20 GA. SHEET METAL (GALV.)
4" MIN.
8" MIN.
TYP. ROOF
1" INSUL.
1" INSULATION
5/8" GYPSUM BD.
ALUMINUM TRIM
2"x8"s
2"x12"
TYP. INSULATION
5/8" GYPSUM BD. (TYPE "X")
SCUTTLE DETAIL
SCALE: DO NOT
MASTIC SEAL
SLOPE TO INTERIOR FLOOR DRAIN
8" FLAP GATE: EQUAL TO - NEENAH R-5040-P-4 (2) (EMERGENCY DISCHARGE)
COPPER SCREEN
2" OF PERIMETER INSULATION (ONLY 1" OF INSULATION FROM TOP OF FLOOR TO 2" BELOW TOP)
SECTION "D-D"
VENT SHOWN THIS VIEW ONLY
2"x2" TEE
2" VENT
SEAL
PIPE FLOOR FLANGE
ALUM. EDGE (CONT.)
LINTEL BLOCKS
2" STD PIPE (3)
PIPE BRACKET
INSULATED METAL DOOR W/ MORTISE LOCK & KEY & WEATHERSTRIPPED
CONC. CAP
CONC. SILL @ WINDOW
FLOOR GRADE ELEV. 1218.06
FLANGE ANCHORS
4" SPLIT CONCRETE BLOCK VENEER (MATCH PATTERN OF WELL NO. 3 PUMPHOUSE)
SOUTH ELEVATION
6" MAIN
MAX. INSERTION
END OF SOL. TUBE
CHECK VALVE
5/16" I.D. PLASTIC SOLUTION TUBE
CORPORATION COCK 1/2"
TAPPED SADDLE
FROM SOLUTION FEEDER
CHEMICAL FEED MAIN CONNECTION W/ CORPORATION COCK
SCALE: DO NOT
2" VENT
FLOOR
DISHED F.D.
PIPE TO DRY WELL (SEE SHEET 2)
4" C.I. ACID RESISTIVE PIPING
UNION
MINIMUM SLOPE 1/4"/L.F.
P-TRAP
MASTIC
4" CLAPPER CHECK VALVE
TYPICAL FLOOR DRAIN
2" VENT
2'-8" x 4'-0" GLASS BLOCK WINDOW W/ 16"x 32" HINGED CENTER GLASS, SCREEN & STORM
METAL DOOR W/ PANIC HARDWARE
EAVE
CONCRETE CAP
EXPANSION JOINT
ELECTRIC HEATER
SLOPE TO DRAIN (TYP.)
8" FLAP GATE
N.F.P.A. APPROVED FIRE EXTINGUISHER
ELECTRICAL PANELS
PUMP CONTROLS
3/4" THK. FILLER CONTINUOUS PERIMETER
TO DRYWELL
F.D.
FRESH AIR INTAKE W/ SCREEN & B.D. DAMPER
REMOVABLE DRIVE SHAFT FOR RIGHT ANGLE DRIVE
SMOOTH END SAMPLING TAP (1/2")
FLUORIDATION EQUIPMENT (GENERAL LOCATION ARRANGEMENT)
12" MIN. CLEARANCE
SURGE ELIMINATOR
PRE-LUBRICATION LINE W/ SOLENOID CONTROL
CHLORINATION EQUIPMENT (GENERAL LOCATION)
6:1 SIDE SLOPE, SEED, FERT. & MULCH
RAIN GUTTER OVER RAMP
2'-8" x 4'-0" GLASS BLOCK WINDOW W/ 16" x 32" HINGED CENTER GLASS, SCREEN & STORM
RGH. OPG.
FLOOR PLAN
2"x8"
BORE VENTILATION HOLES AS NEEDED
2"x12" SILL PLATE
CUT-AWAY SECTION
4'-0" x 4'-0" SCUTTLE (CENTER OVER WELL)
ROOFING: 3 PLY FELT (2-30", 1-15") SHEET ASPHALT SEAL SURFACE, COVER W/ PEA GRAVEL - GALV. ROOF EDGE OVER ALUMINUM FASCIA
ROOF PLAN
SOLUTION PUMP
ELECTRIC MOTOR
POLYETHYLENE DISCHARGE TUBING
BLIND FLANGE
INJECTION TUBE
HYPOCHLORINATOR UNIT
SEE FLOOR PLAN FOR CONNECTION
MAIN CONNECTION
FROM WELL
CHEMICAL CHLORINATION SOLUTION POLYETHYLENE TANK (30 GAL. CAP.)
VINYL SUCTION TUBING
TIE ROD
STRAINER
WALL BRACKET (STEEL)
SADDLE STRAPS
SOLUTION LEVEL INDICATOR
TO DISTRIBUTION SYSTEM
SINKER
CORPORATION COCK & TAPPED SADDLE
STRAINER
SOLUTION MIXING CONTAINER (30 GAL.) *
FLUORIDE SHIPPING CONTAINER
SOLUTION LEVEL INDICATOR *
TYPICAL CHEMICAL FEED EQUIPMENT
SCALE: DO NOT
* FLUORIDE INJECTION SYSTEM & CHLORINE INJECTION SYSTEM SHALL BE IDENTICAL. EACH SYSTEM SHALL HAVE AN ANTI-SIPHON DEVICE.
PUMPHOUSE, CONSTRUCTION DETAILS, PUMP, EQUIPMENT & INTERIOR PIPING
MUNICIPAL WELL NO. 4
GILMAN, WISCONSIN
SCALE - 3/8" = 1'-0" UNLESS NOTED
MORGAN & PARMLEY, LTD. CONSULTING ENGINEERS LADYSMITH, WISCONSIN
SHEET 4 OF 6
DATE APRIL, 1987
DR. BY M.D. SHERVEY

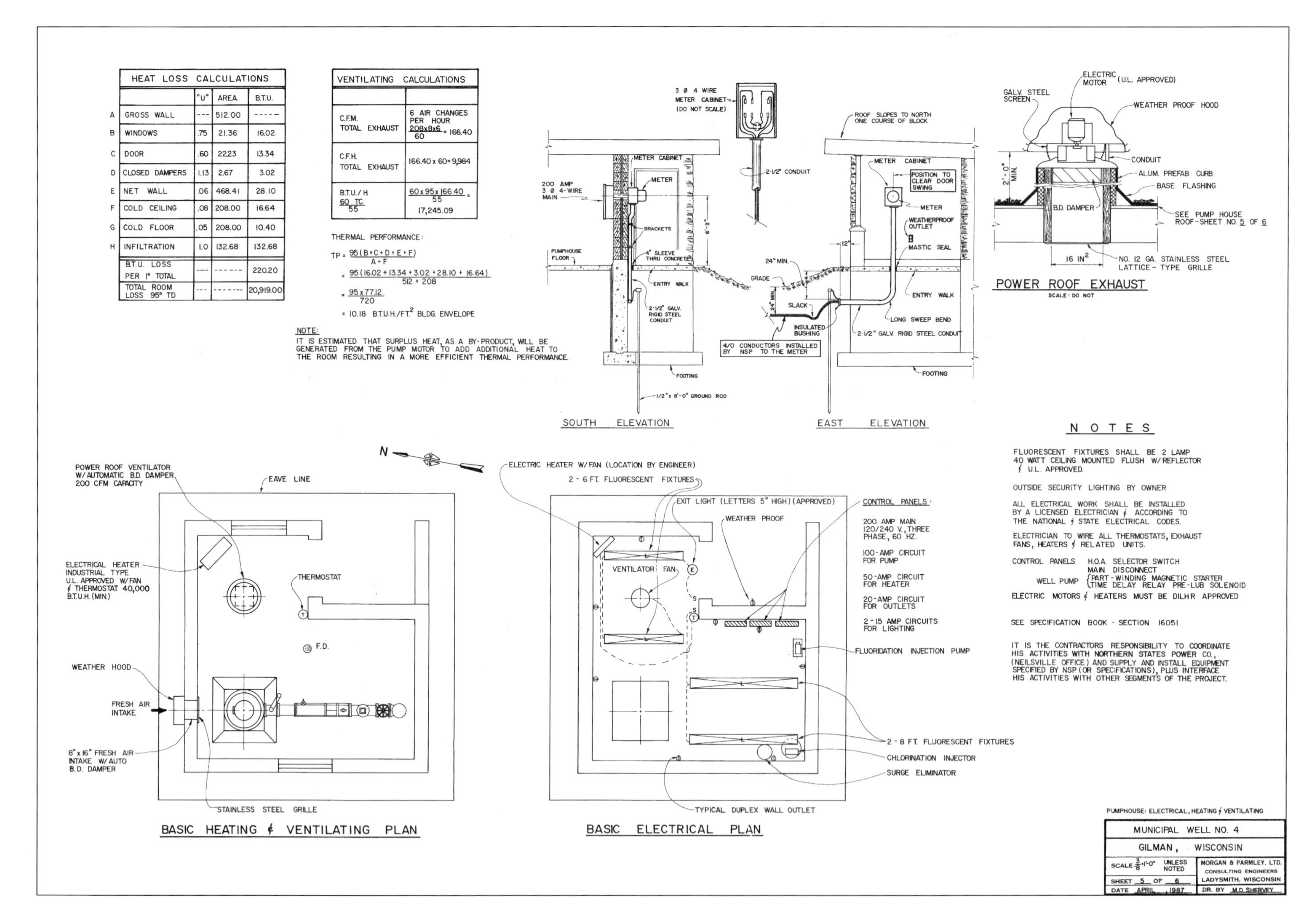

HEAT LOSS CALCULATIONS
VENTILATING CALCULATIONS
THERMAL PERFORMANCE:
NOTE:
IT IS ESTIMATED THAT SURPLUS HEAT, AS A BY-PRODUCT, WILL BE GENERATED FROM THE PUMP MOTOR TO ADD ADDITIONAL HEAT TO THE ROOM RESULTING IN A MORE EFFICIENT THERMAL PERFORMANCE.
SOUTH ELEVATION
EAST ELEVATION
POWER ROOF EXHAUST
N O T E S
FLUORESCENT FIXTURES SHALL BE 2 LAMP 40 WATT CEILING MOUNTED FLUSH W/ REFLECTOR & U.L. APPROVED.
OUTSIDE SECURITY LIGHTING BY OWNER
ALL ELECTRICAL WORK SHALL BE INSTALLED BY A LICENSED ELECTRICIAN & ACCORDING TO THE NATIONAL & STATE ELECTRICAL CODES.
ELECTRICIAN TO WIRE ALL THERMOSTATS, EXHAUST FANS, HEATERS & RELATED UNITS.
CONTROL PANELS H.O.A. SELECTOR SWITCH MAIN DISCONNECT
WELL PUMP PART-WINDING MAGNETIC STARTER TIME DELAY RELAY PRE-LUB SOLENOID
ELECTRIC MOTORS & HEATERS MUST BE DILHR APPROVED
SEE SPECIFICATION BOOK - SECTION 16051
IT IS THE CONTRACTORS RESPONSIBILITY TO COORDINATE HIS ACTIVITIES WITH NORTHERN STATES POWER CO., (NEILSVILLE OFFICE) AND SUPPLY AND INSTALL EQUIPMENT SPECIFIED BY NSP (OR SPECIFICATIONS), PLUS INTERFACE HIS ACTIVITIES WITH OTHER SEGMENTS OF THE PROJECT.
BASIC HEATING & VENTILATING PLAN
BASIC ELECTRICAL PLAN
PUMPHOUSE: ELECTRICAL, HEATING & VENTILATING
MUNICIPAL WELL NO. 4
GILMAN, WISCONSIN
MORGAN & PARMLEY, LTD. CONSULTING ENGINEERS LADYSMITH, WISCONSIN
SHEET 5 OF 6
DATE APRIL, 1987
DR. BY M.D. SHERVEY

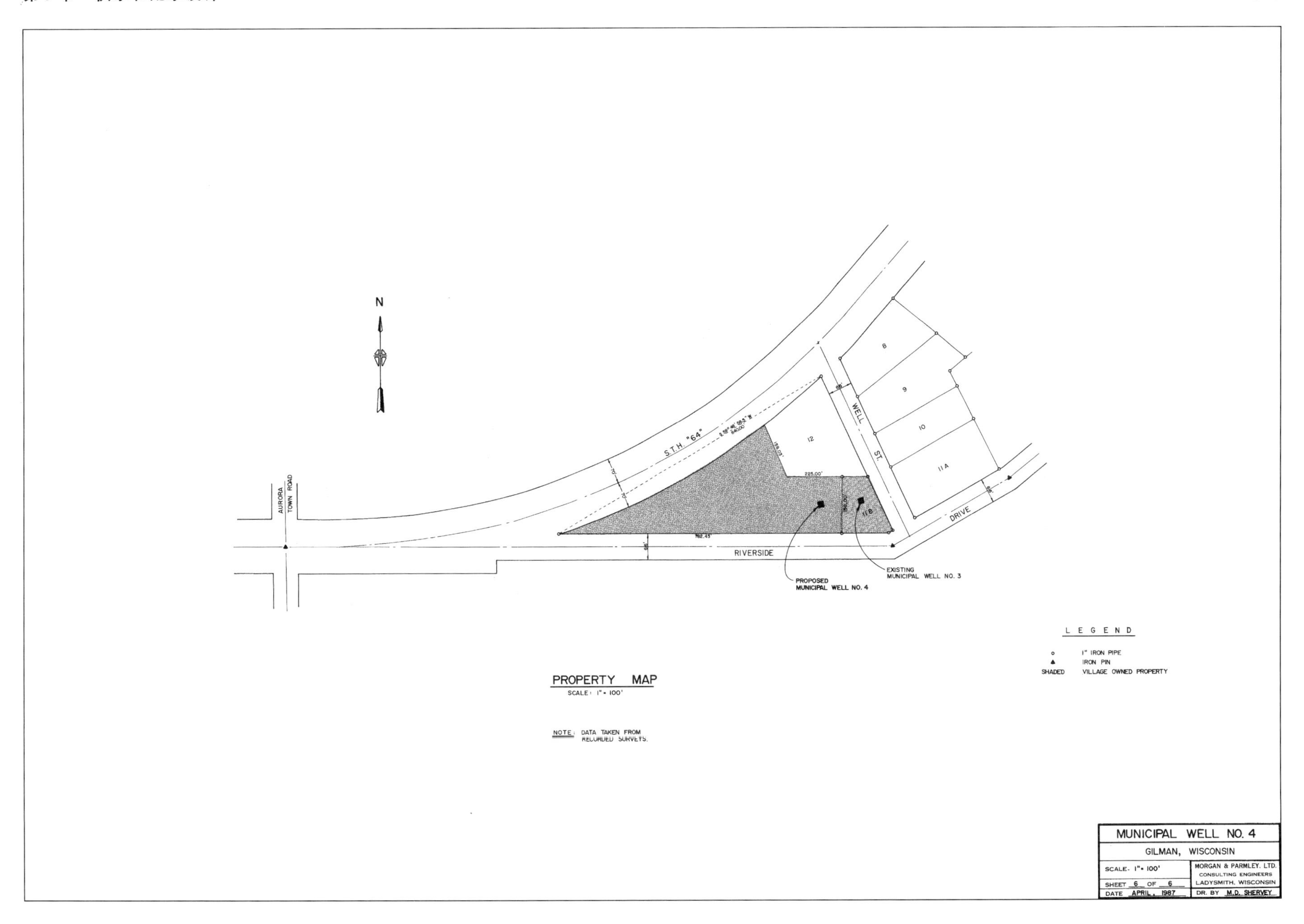
N
S.T.H. "64"
WELL ST.
AURORA TOWN ROAD
RIVERSIDE
DRIVE
225.00'
762.45'
8
9
10
11A
11B
12
PROPOSED MUNICIPAL WELL NO. 4
EXISTING MUNICIPAL WELL NO. 3
LEGEND
1" IRON PIPE
IRON PIN
SHADED VILLAGE OWNED PROPERTY
PROPERTY MAP
SCALE: 1" = 100'
NOTE: DATA TAKEN FROM RECORDED SURVEYS.
MUNICIPAL WELL NO. 4
GILMAN, WISCONSIN
SCALE- 1" = 100'
SHEET 6 OF 6
DATE APRIL, 1987
MORGAN & PARMLEY, LTD.
CONSULTING ENGINEERS
LADYSMITH, WISCONSIN
DR. BY M.D. SHERVEY

供水、泵房和输水干管

WATER SUPPLY, PUMPHOUSE & TRANSMISSION MAIN

FOR

INDUSTRIAL PARK COMPLEX

E.D.A. PROJECT NO. 06-01-02571

VILLAGE OF GLEN FLORA, WISCONSIN

NOV., 1992

UTILITIES

CONTACT - NORTHERN STATES POWER CO.
711 W. 9th Street, North
Ladysmith, WI. 54848

Attn: John Rymarkiewicz
715-532-6226

CONTACT - UNIVERSAL TELEPHONE CO.
OF NORTHERN WISCONSIN
Highway "8"
Hawkins, WI 54530

Attn: Jim Arquette
715-585-7707

CONTACT - DIGGERS HOTLINE

1-800 242-8511

CONTACT - VILLAGE OF GLEN FLORA WWCTF

Les Evjen, Operator
P.O. Box 253
Glen Flora, WI. 54526

715-322-5511

CONTACT - GLEN FLORA ELEMENTARY SCHOOL
SCHOOL WELL & PIPELINE
Larry Johnson, Custodian

715-322-5271

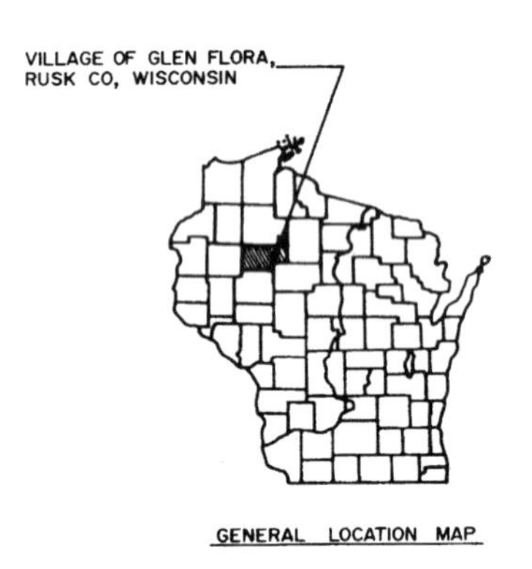

GENERAL LOCATION MAP

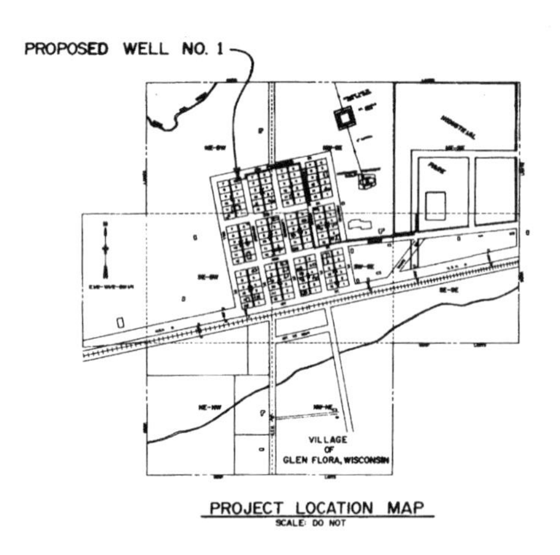

PROJECT LOCATION MAP
SCALE: DO NOT

INDEX OF SHEETS

THESE PLANS REPRESENT ONLY 1 SECTION OF THE TOTAL PROJECT

THE TOTAL PROJECT EDA NO. 06-01-02571 IS PARTIALLY FUNDED BY A 60% GRANT FROM THE UNITED STATES ECONOMIC DEVELOPMENT ADMINISTRATION IN THE TOTAL GRANT AMOUNT OF $590,520

PREPARED BY:
MORGAN & PARMLEY LTD.
CONSULTING ENGINEERS
LADYSMITH, WISCONSIN

Courtesy: Morgan & Parmley, Ltd.

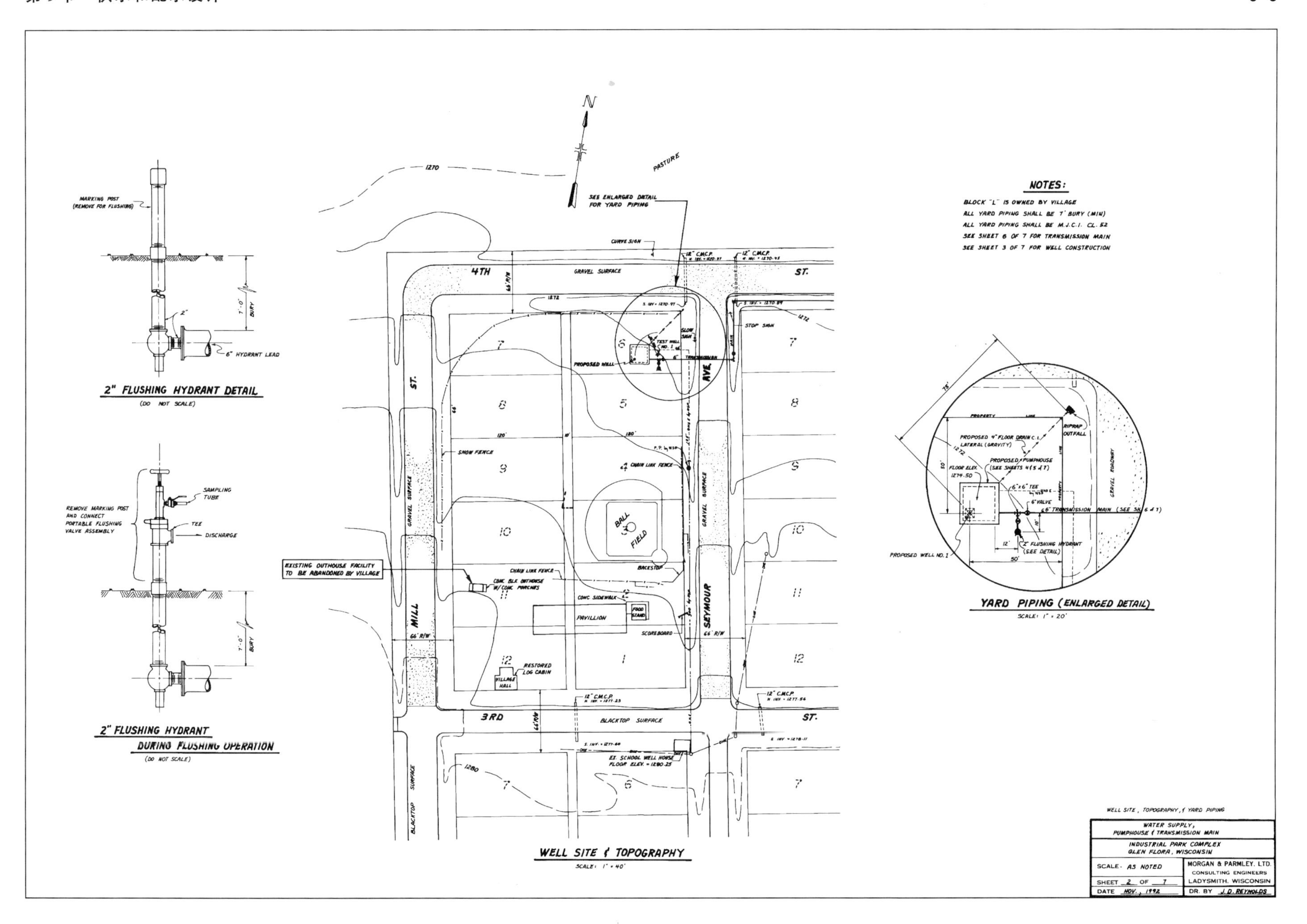
NOTES:
BLOCK "L" IS OWNED BY VILLAGE
ALL YARD PIPING SHALL BE 7' BURY (MIN)
ALL YARD PIPING SHALL BE M.J.C.I. CL. 52
SEE SHEET 6 OF 7 FOR TRANSMISSION MAIN
SEE SHEET 3 OF 7 FOR WELL CONSTRUCTION
2" FLUSHING HYDRANT DETAIL
(DO NOT SCALE)
2" FLUSHING HYDRANT DURING FLUSHING OPERATION
(DO NOT SCALE)
WELL SITE & TOPOGRAPHY
SCALE: 1" = 40'
YARD PIPING (ENLARGED DETAIL)
SCALE: 1" = 20'
WATER SUPPLY, PUMPHOUSE & TRANSMISSION MAIN
INDUSTRIAL PARK COMPLEX
GLEN FLORA, WISCONSIN
SCALE- AS NOTED
SHEET 2 OF 7
DATE NOV., 1992
MORGAN & PARMLEY, LTD.
CONSULTING ENGINEERS
LADYSMITH, WISCONSIN
DR. BY J.D. REYNOLDS

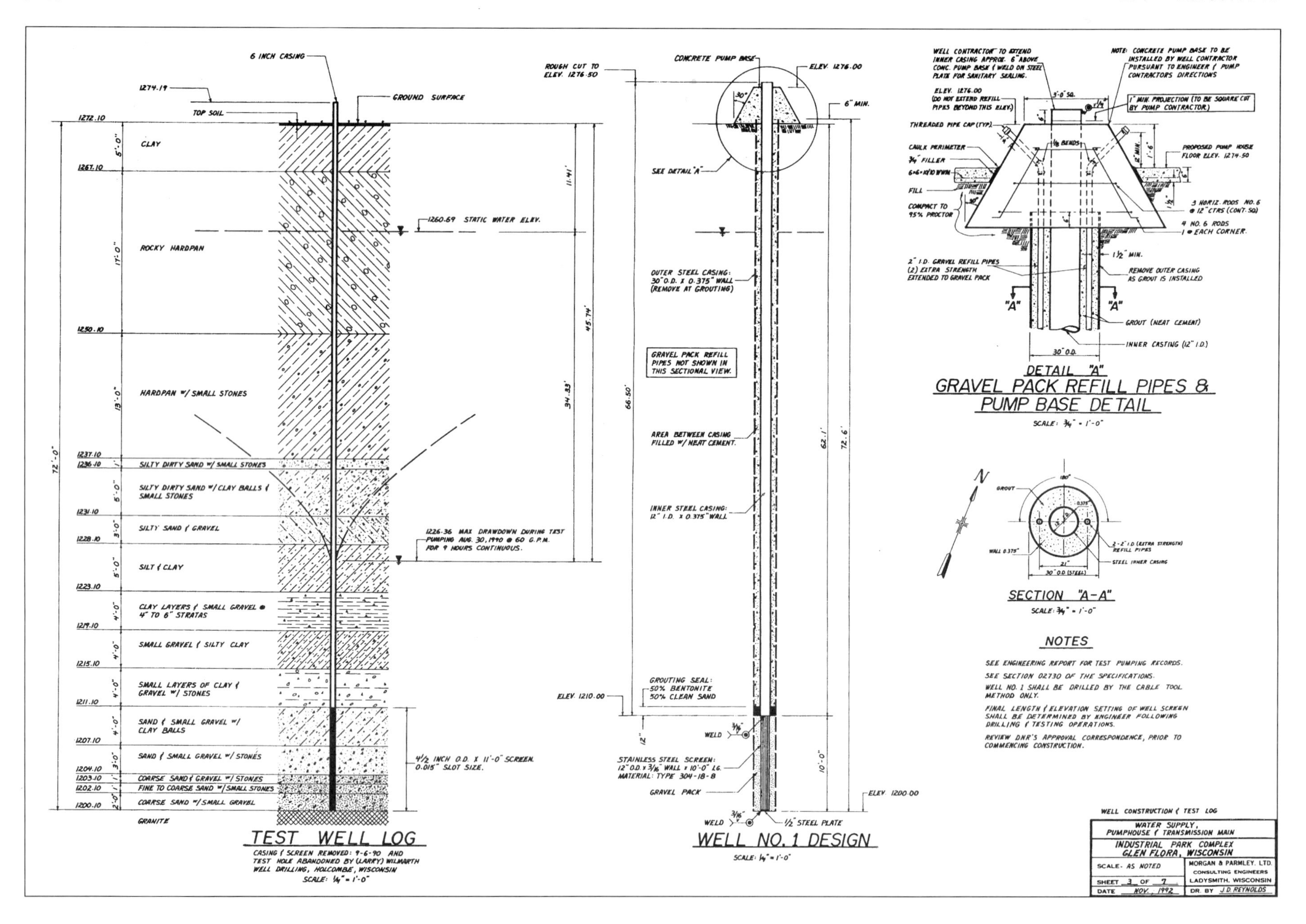
TEST WELL LOG
CASING & SCREEN REMOVED: 9-6-90 AND TEST HOLE ABANDONED BY (LARRY) WILMARTH WELL DRILLING, HOLCOMBE, WISCONSIN
SCALE: 1/4" = 1'-0"
WELL NO.1 DESIGN
SCALE: 1/4" = 1'-0"
DETAIL "A"
GRAVEL PACK REFILL PIPES & PUMP BASE DETAIL
SCALE: 3/4" = 1'-0"
SECTION "A-A"
SCALE: 3/4" = 1'-0"
NOTES
SEE ENGINEERING REPORT FOR TEST PUMPING RECORDS.
SEE SECTION 02730 OF THE SPECIFICATIONS.
WELL NO. 1 SHALL BE DRILLED BY THE CABLE TOOL METHOD ONLY.
FINAL LENGTH & ELEVATION SETTING OF WELL SCREEN SHALL BE DETERMINED BY ENGINEER FOLLOWING DRILLING & TESTING OPERATIONS.
REVIEW DNR'S APPROVAL CORRESPONDENCE, PRIOR TO COMMENCING CONSTRUCTION.
WELL CONSTRUCTION & TEST LOG
WATER SUPPLY, PUMPHOUSE & TRANSMISSION MAIN
INDUSTRIAL PARK COMPLEX GLEN FLORA, WISCONSIN
SCALE- AS NOTED
SHEET 3 OF 7
DATE NOV. 1992
MORGAN & PARMLEY, LTD. CONSULTING ENGINEERS LADYSMITH, WISCONSIN
DR. BY J.D. REYNOLDS

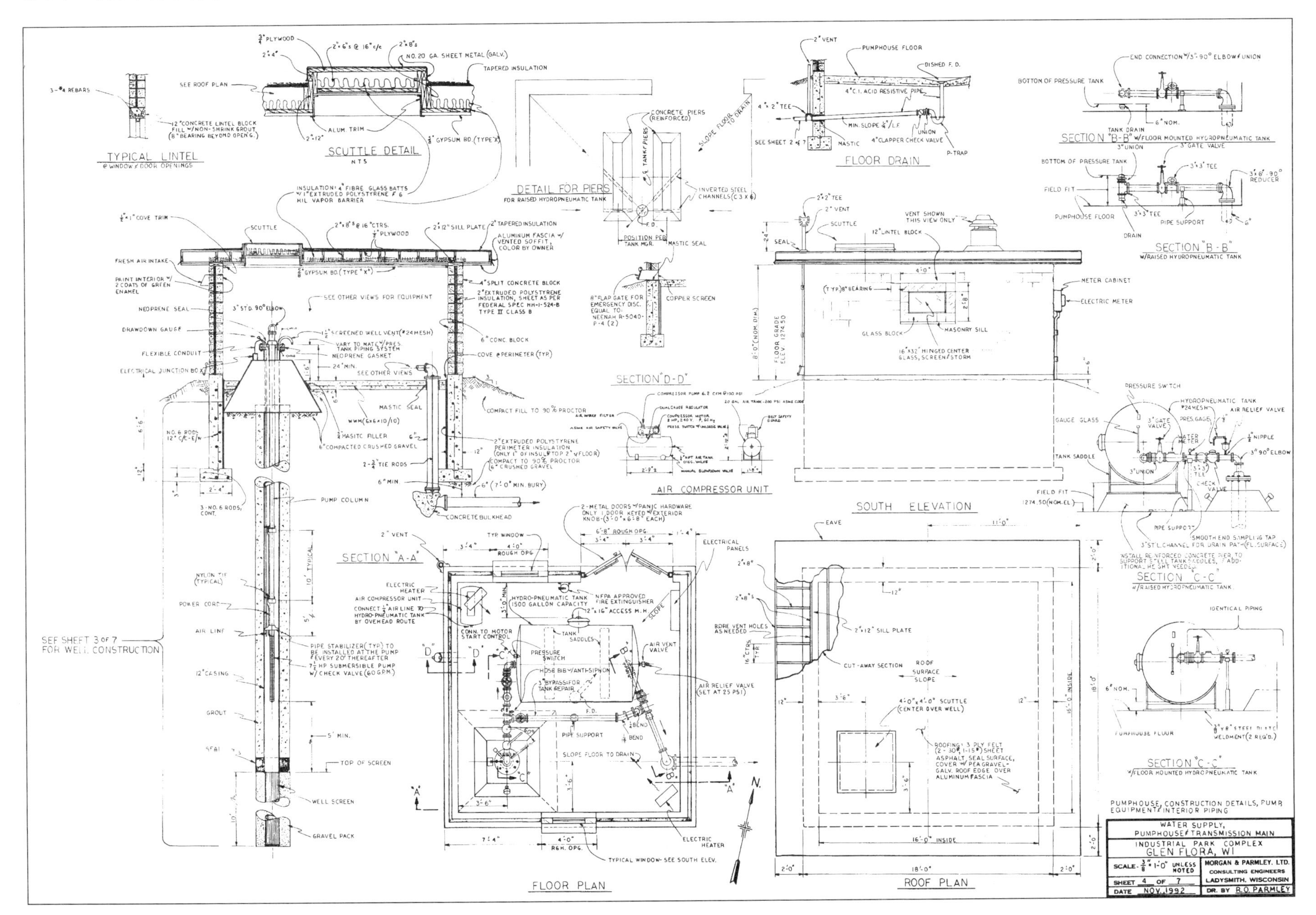
TYPICAL LINTEL
SCUTTLE DETAIL
DETAIL FOR PIERS
FLOOR DRAIN
SECTION "B-B" W/FLOOR MOUNTED HYDROPNEUMATIC TANK
SECTION "B-B" W/RAISED HYDROPNEUMATIC TANK
SECTION "D-D"
AIR COMPRESSOR UNIT
SOUTH ELEVATION
SECTION "A-A"
SECTION "C-C" W/RAISED HYDROPNEUMATIC TANK
SECTION "C-C" W/FLOOR MOUNTED HYDROPNEUMATIC TANK
FLOOR PLAN
ROOF PLAN
PUMPHOUSE, CONSTRUCTION DETAILS, PUMP EQUIPMENT & INTERIOR PIPING
WATER SUPPLY, PUMPHOUSE & TRANSMISSION MAIN
INDUSTRIAL PARK COMPLEX
GLEN FLORA, WI
MORGAN & PARMLEY, LTD.
CONSULTING ENGINEERS
LADYSMITH, WISCONSIN
SHEET 4 OF 7
DATE NOV. 1992
DR. BY R.O. PARMLEY

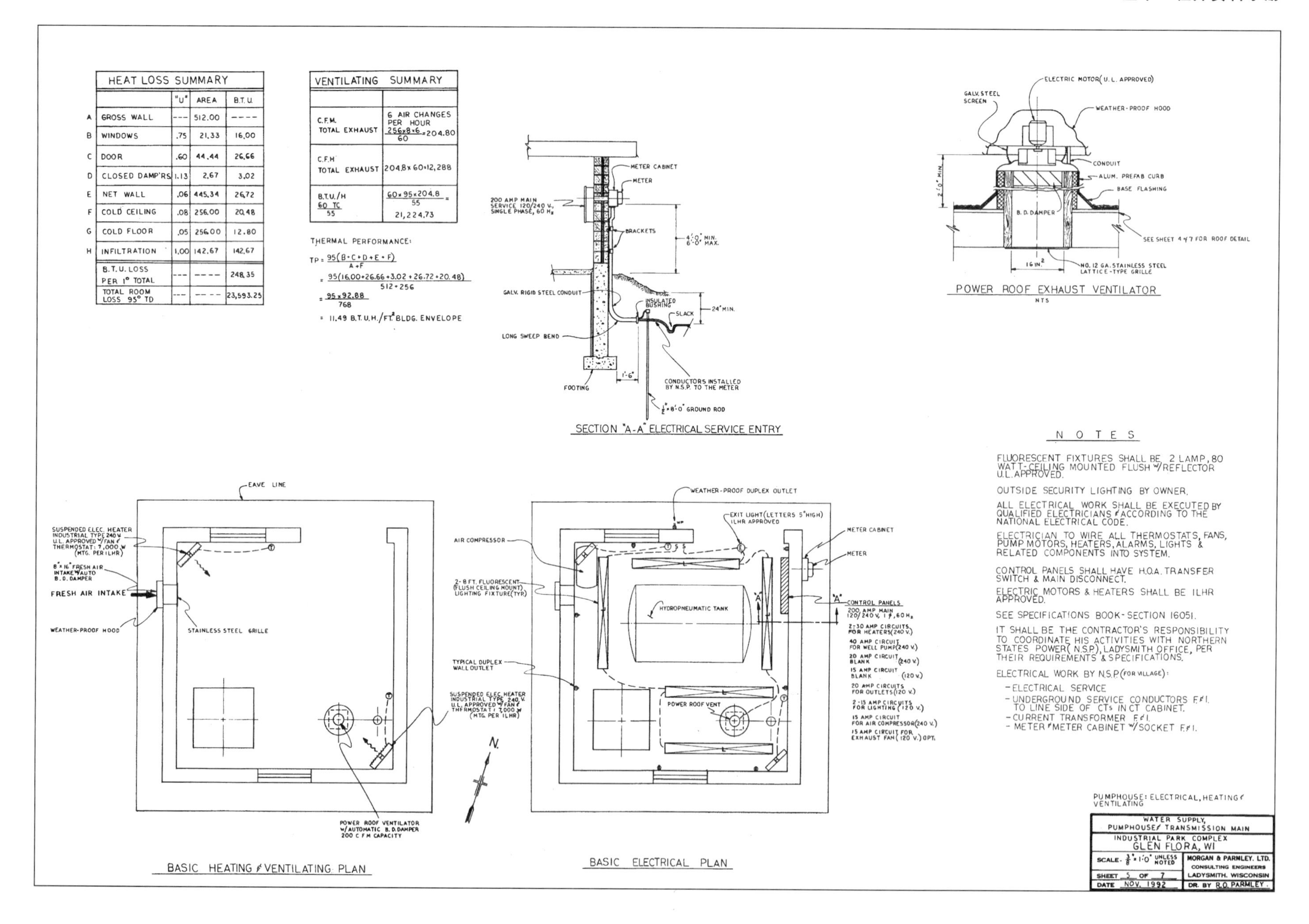

	HEAT LOSS SUMMARY	"U"	AREA	B.T.U.
A	GROSS WALL	---	512.00	----
B	WINDOWS	.75	21.33	16.00
C	DOOR	.60	44.44	26.66
D	CLOSED DAMP'RS	1.13	2.67	3.02
E	NET WALL	.06	445.34	26.72
F	COLD CEILING	.08	256.00	20.48
G	COLD FLOOR	.05	256.00	12.80
H	INFILTRATION	1.00	142.67	142.67
	B.T.U. LOSS PER 1° TOTAL	---	----	248.35
	TOTAL ROOM LOSS 95° TD	---	----	23,593.25

VENTILATING SUMMARY	
C.F.M. TOTAL EXHAUST	6 AIR CHANGES PER HOUR $\frac{256\times8\times6}{60}=204.80$
C.F.H. TOTAL EXHAUST	204.8×60=12,288
B.T.U./H $\frac{60\ TC}{55}$	$\frac{60\times95\times204.8}{55}=$ 21,224.73

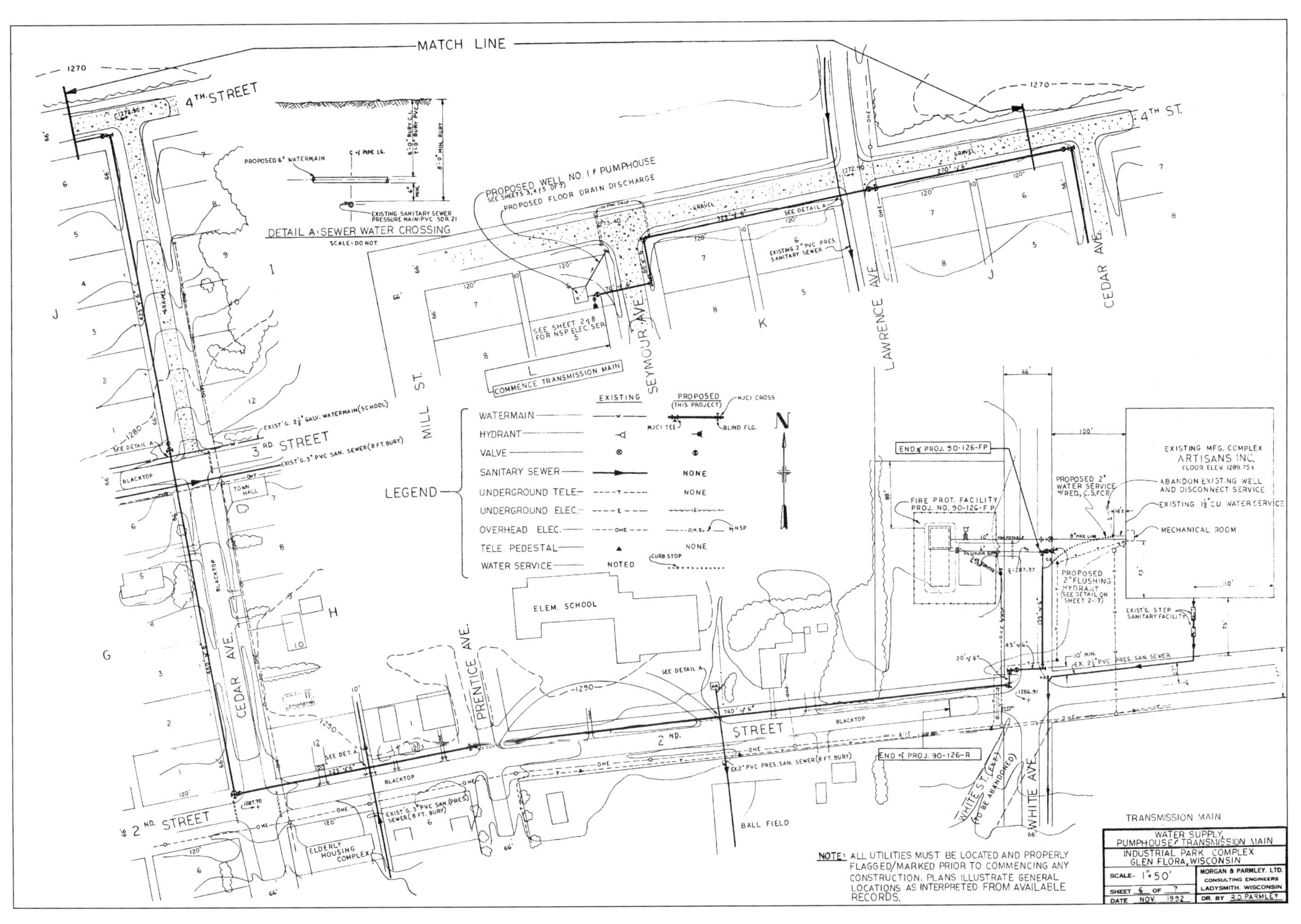
MATCH LINE
4TH STREET
4TH ST.
PROPOSED 6" WATERMAIN
C-ℓ PIPE LG.
8'-0" MIN. BURY
EXISTING SANITARY SEWER PRESSURE MAIN: PVC SDR 21
DETAIL A: SEWER WATER CROSSING
SCALE: DO NOT
PROPOSED WELL NO. 1 & PUMPHOUSE
SEE SHEETS 3,4 & 5 OF 7
PROPOSED FLOOR DRAIN DISCHARGE
GRAVEL
SEE DETAIL A
EXISTING 2" PVC PRES. SANITARY SEWER
SEE SHEET 2 & 8 FOR NSP ELEC. SER.
COMMENCE TRANSMISSION MAIN
MILL ST.
SEYMOUR AVE.
LAWRENCE AVE.
CEDAR AVE.
EXIST'G 2½" GALV. WATERMAIN (SCHOOL)
3RD. STREET
EXIST'G 3" PVC SAN. SEWER (8 FT. BURY)
BLACKTOP
TOWN HALL
LEGEND
EXISTING
PROPOSED (THIS PROJECT)
MJCI CROSS
MJCI TEE
BLIND FLG.
WATERMAIN
HYDRANT
VALVE
SANITARY SEWER
UNDERGROUND TELE.
UNDERGROUND ELEC.
OVERHEAD ELEC.
TELE. PEDESTAL
WATER SERVICE
NONE
NOTED
CURB STOP
NSP
N
END ℓ PROJ. 90-126-FP
FIRE PROT. FACILITY PROJ. NO. 90-126-F.P.
EXISTING MFG. COMPLEX
ARTISANS INC.
FLOOR ELEV. 1289.75±
PROPOSED 2" WATER SERVICE W/RED. C.S. & CR
ABANDON EXISTING WELL AND DISCONNECT SERVICE
EXISTING 1½" CU WATER SERVICE
MECHANICAL ROOM
PROPOSED 2" FLUSHING HYDRANT (SEE DETAIL ON SHEET 2-7)
EXIST'G. STEP SANITARY FACILITY
EX. 2½" PVC PRES. SAN. SEWER
ELEM. SCHOOL
SEE DETAIL A
PRENTICE AVE.
2ND. STREET
END ℓ PROJ. 90-126-R
EX. 2" PVC PRES. SAN. SEWER (8 FT. BURY)
EXIST'G. 3" PVC SAN. (PRES.) SEWER (8 FT. BURY)
ELDERLY HOUSING COMPLEX
BALL FIELD
WHITE ST. (EXT.) (TO BE ABANDONED)
WHITE AVE.
TRANSMISSION MAIN
NOTE: ALL UTILITIES MUST BE LOCATED AND PROPERLY FLAGGED/MARKED PRIOR TO COMMENCING ANY CONSTRUCTION. PLANS ILLUSTRATE GENERAL LOCATIONS AS INTERPRETED FROM AVAILABLE RECORDS.
WATER SUPPLY, PUMPHOUSE, TRANSMISSION MAIN
INDUSTRIAL PARK COMPLEX
GLEN FLORA, WISCONSIN
SCALE: 1"=50'
SHEET 6 OF 7
DATE NOV. 1992
MORGAN & PARMLEY, LTD.
CONSULTING ENGINEERS
LADYSMITH, WISCONSIN
DR. BY R.O. PARMLEY

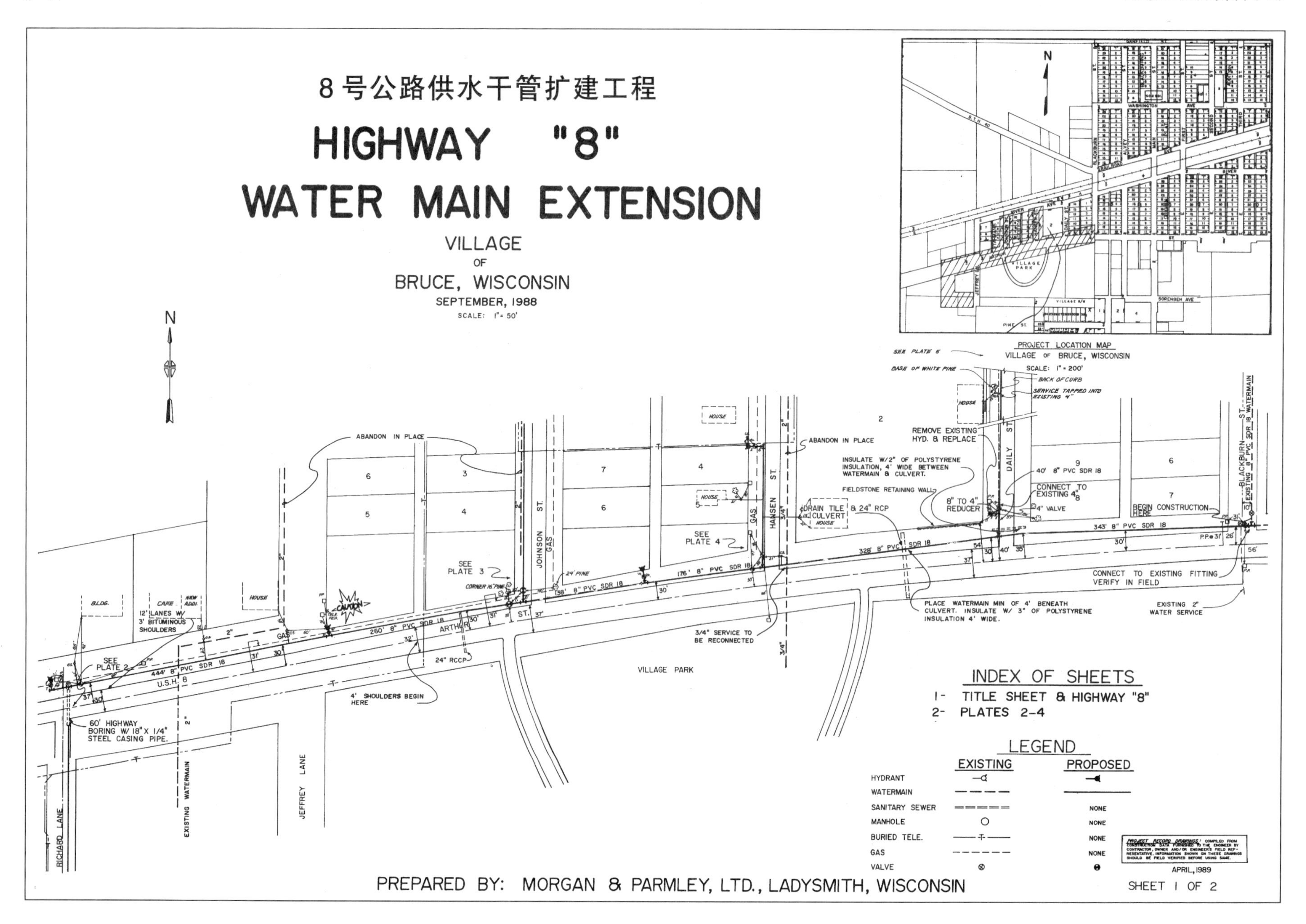

Courtesy: Morgan & Parmley, Ltd.

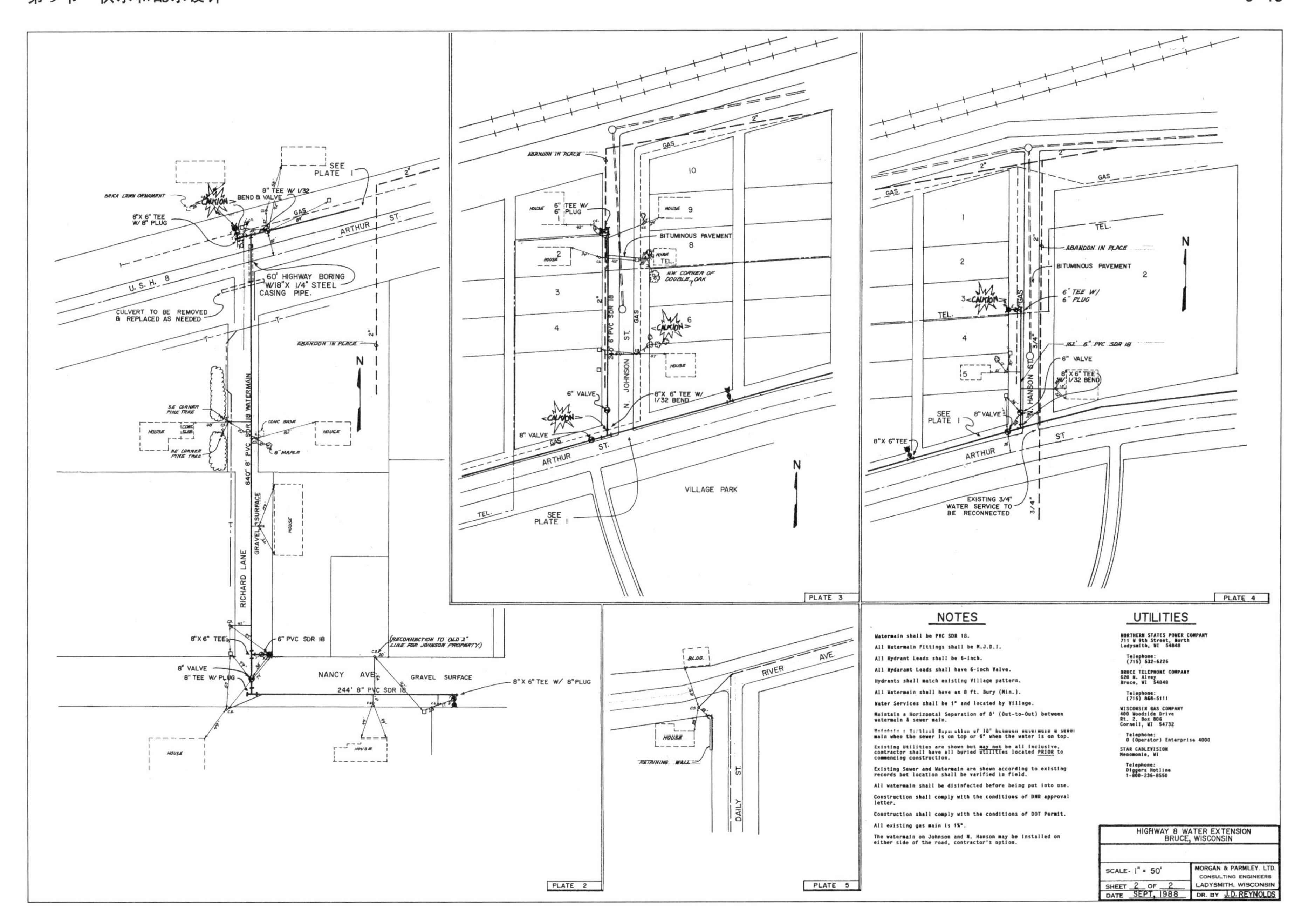
SEE PLATE 1
BRICK LAWN ORNAMENT
CAUTION
BEND & VALVE
8" TEE W/ 1/32
8"X 6" TEE W/ 8" PLUG
GAS
ARTHUR ST.
U. S. H. 8
60' HIGHWAY BORING W/18"X 1/4" STEEL CASING PIPE.
CULVERT TO BE REMOVED & REPLACED AS NEEDED
ABANDON IN PLACE
N
SE CORNER PINE TREE
HOUSE
CONC SLAB
CONC BASE
NE CORNER PINE TREE
8" MAPLE
640' 8" PVC SDR 18 WATERMAIN
GRAVEL SURFACE
RICHARD LANE
8"X 6" TEE
6" PVC SDR 18
(RECONNECTION TO OLD 2" LINE FOR JOHNSON PROPERTY)
8" VALVE
8" TEE W/ PLUG
NANCY AVE
GRAVEL SURFACE
244' 8" PVC SDR 18
8"X 6" TEE W/ 8" PLUG
PLATE 2
ABANDON IN PLACE
6" TEE W/ 6" PLUG
BITUMINOUS PAVEMENT
NW CORNER OF DOUBLE OAK
240' 6" PVC SDR 18
N. JOHNSON ST.
6" VALVE
8"X 6" TEE W/ 1/32 BEND
8" VALVE
VILLAGE PARK
TEL.
SEE PLATE 1
PLATE 3
ABANDON IN PLACE
BITUMINOUS PAVEMENT
6" TEE W/ 6" PLUG
162' 6" PVC SDR 18
6" VALVE
8"X 6" TEE W/ 1/32 BEND
N. HANSON ST.
SEE PLATE 1
8" VALVE
8"X 6" TEE
EXISTING 3/4" WATER SERVICE TO BE RECONNECTED
PLATE 4
BLDG.
RIVER AVE.
HOUSE
RETAINING WALL
DAILY ST.
PLATE 5
NOTES
Watermain shall be PVC SDR 18.
All Watermain Fittings shall be M.J.D.I.
All Hydrant Leads shall be 6-inch.
All Hydarant Leads shall have 6-inch Valve.
Hydrants shall match existing Village pattern.
All Watermain shall have an 8 ft. Bury (Min.).
Water Services shall be 1" and located by Village.
Maintain a Horizontal Separation of 8' (Out-to-Out) between watermain & sewer main.
main when the sewer is on top or 6" when the water is on top.
Existing Utilities are shown but may not be all inclusive, contractor shall have all buried utilities located PRIOR to commencing construction.
Existing Sewer and Watermain are shown according to existing records but location shall be varified in field.
All watermain shall be disinfected before being put into use.
Construction shall comply with the conditions of DNR approval letter.
Construction shall comply with the conditions of DOT Permit.
All existing gas main is 1½".
The watermain on Johnson and N. Hanson may be installed on either side of the road, contractor's option.
UTILITIES
NORTHERN STATES POWER COMPANY
711 W 9th Street, North
Ladysmith, WI 54848
Telephone:
(715) 532-6226
BRUCE TELEPHONE COMPANY
620 N. Alvey
Bruce, WI 54848
Telephone:
(715) 868-5111
WISCONSIN GAS COMPANY
400 Woodside Drive
Rt. 2, Box 806
Cornell, WI 54732
Telephone:
0 (Operator) Enterprise 4000
STAR CABLEVISION
Menomonie, WI
Telephone:
Diggers Hotline
1-800-236-8550
HIGHWAY 8 WATER EXTENSION
BRUCE, WISCONSIN
SCALE- 1" = 50'
SHEET 2 OF 2
DATE SEPT, 1988
MORGAN & PARMLEY, LTD.
CONSULTING ENGINEERS
LADYSMITH, WISCONSIN
DR. BY J.D. REYNOLDS

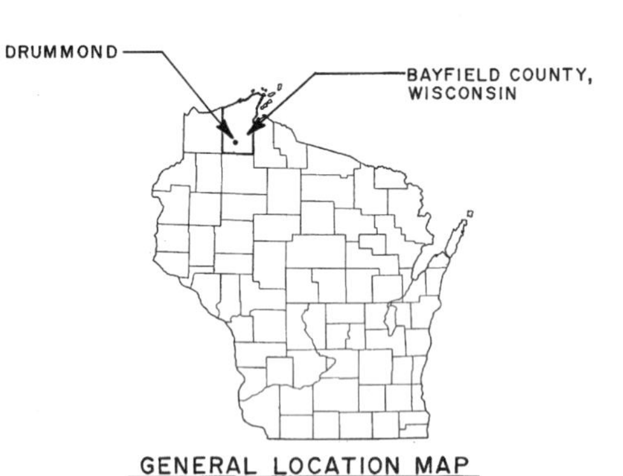

GENERAL LOCATION MAP

供水干管扩建工程

WATERMAIN EXTENSION

DRUMMOND SANITARY DISTRICT

DRUMMOND, WISCONSIN

M. & P. PROJECT NO. 99-157-S

JULY, 2002

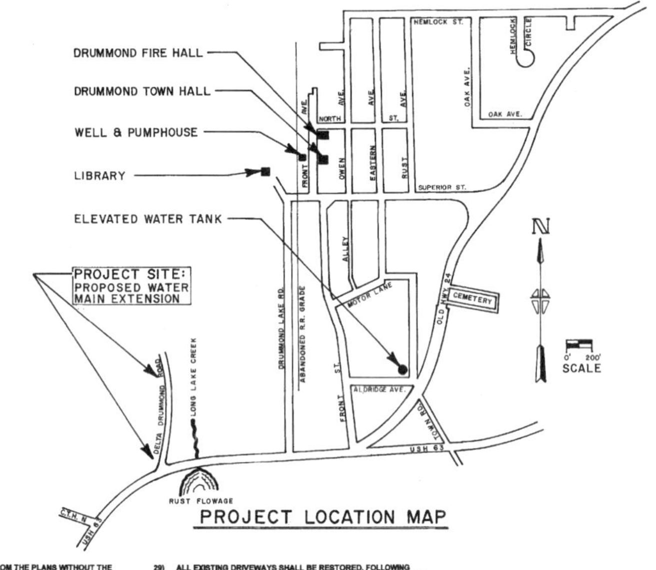

PROJECT LOCATION MAP

-NOTES-

1) ALL ELEVATIONS BASED ON DATUM OBTAINED FROM PRIOR PLANS AND SANITARY DISTRICT UTILITY RECORDS.
2) CONSTRUCTION SHALL COMPLY WITH THE CONDITIONS OF DNR APPROVAL LETTER, AND OTHER REGULATORY APPROVALS AS LISTED IN THE SPECIFICATION BOOK.
3) ALL WATERMAIN SHALL BE PVC DR18 C900, SIZED AS SHOWN ON PLANS.
4) ALL WATERMAIN FITTINGS SHALL BE M.J.D.I.
5) HYDRANT LEAD SHALL BE 6-INCH D.I., CL. 52.
6) HYDRANT LEAD SHALL HAVE A 6-INCH GATE VALVE W/RISER.
7) HYDRANT SHALL MATCH THE VILLAGE OF DRUMMOND'S PATTERN.
8) ALL WATERMAIN SHALL HAVE A 8 FT. BURY (MIN.)
9) 8 FT. (MINIMUM) HORIZONTAL DISTANCE (℄ TO ℄) BETWEEN WATERMAIN AND SANITARY SEWER OR STORM SEWER.
10) 18" (MINIMUM) VERTICAL DISTANCE (OUT TO OUT) BETWEEN WATERMAIN UNDER SANITARY SEWER OR STORM SEWER @ THEIR INTERSECTION.
11) 6" (MINIMUM) VERTICAL DISTANCE (OUT TO OUT) BETWEEN WATERMAIN OVER SANITARY SEWER OR STORM SEWER @ THEIR INTERSECTION.
12) VALVES SHALL BE SET TO GRADE SHOWN ON PLANS, UNLESS DIRECTED OTHERWISE.
13) WATERMAIN SHALL BE INSULATED AT ALL STORM SEWER, CULVERT, STREET AND DITCH CROSSINGS WITH 2" THICK EXTRUDED POLYSTYRENE INSULATION BOARD AND JOINTS LAPPED A MINIMUM OF 2 FT.
14) ALL WATERMAIN SHALL BE DISINFECTED AND RECEIVE A "SAFE" LAB CERTIFICATION BEFORE BEING PUT INTO SERVICE.
15) EXISTING UTILITIES ARE SHOWN BUT MAY NOT BE ALL INCLUSIVE, CONTRACTOR SHALL HAVE ALL BURIED UTILITIES LOCATED PRIOR TO COMMENCING CONSTRUCTION.
16) EXISTING SANITARY SEWER IS SHOWN ACCORDING TO EXISTING RECORDS, BUT LOCATION SHALL BE VERIFIED IN FIELD.
17) EROSION CONTROL MEASURES SHALL BE IN PLACE BEFORE CONSTRUCTION BEGINS.
18) THE CONTRACTOR SHALL MINIMIZE EROSION DURING CONSTRUCTION USING GOOD CONSTRUCTION TECHNIQUES AND UTILIZING SILT FENCES AND BALE DITCH CHECKS. THE CONTRACTOR SHALL RELEASE RUNOFF FROM THE SITE IN A NUISANCE FREE MANNER.
19) THE CONTRACTOR IS RESPONSIBLE TO MAINTAIN THE SITE IN A SAFE CONDITION. THE SOLE RESPONSIBILITY FOR WARNING SIGNS, BARRICADES, FLAGGING PERSONNEL AND ALL ASPECTS OF SAFETY LIE WITH THE CONTRACTOR.
20) ALL DISTRUBED YARD & ROADWAY DITCHES SHALL BE RESTORED W /SEEDING AS FOLLOWS:
   A) REPLACE ALL SALVAGED TOPSOIL
   B) SEED MIXUTRE: PERMANENT SEEDING – (LAWN TYPE TURF)
      35% KENTUCKY BLUEGRASS
      25% IMPROVED FINE PERENNIAL RYEGRASS
      15% CREEPING RED FESCUE
      15% IMPROVED HARD FESCUE
      10% WHITE CLOVER
   C) SEED RATE: 2# / 1000 SQ. FT.
   D) MULCH RATE: 3 TON / ACRE
   E) PERMANENT SEEDING: ALLOWED TO SEPT. 7
      TEMPORARY SEEDING: SEPT. 8 THROUGH Nov. 10 – (4 BU. OATS / ACRE)
   F) SEEDING MAY BE BROADCAST BUT MUST BE COVERED WITH A MAXIMUM OF ½" SOIL. MULCH MUST BE ANCHORED WITH A MULCH TILLER OR BLOWN W / 75 – 100 GAL. OF ASPHALT / TON OF MULCH.
   G) FERTILIZER: 500# / ACRE 20-10-10
21) THERE SHALL BE NO DEVIATION FROM THE PLANS WITHOUT THE DESIGN ENGINEER'S APPROVAL.
22) CONSTRUCTION STAKES DESTROYED BY THE CONTRACTOR THAT NEED TO BE REPLACED, WILL BE CHARGED TO THE CONTRACTOR.
23) PROPERTY CORNERS KNOWN TO EXIST SHALL NOT BE DISTURBED. DAMAGED CORNERS WILL BE REPLACED AT THE CONTRACTOR'S EXPENSE. PRESERVE ALL SURVEY MONUMENTS.
24) SQUARE CUT ALL ASPHALT PAVEMENT PRIOR TO TRENCH EXCAVATION. THOROUGHLY COMPACT BACKFILL TO 90% PROCTOR. INSTALL A MINIMUM OF 9 INCHES OF GRANULAR SUB-BASE AND 6 INCHES OF BASE COURSE. PAVE WITH 2 – 1½ INCH LIFTS OF ASPHALT.
25) SEE SPECIFICATION BOOK AND FOLLOWING PLAN SHEETS FOR ADDITIONAL DETAILS AND NOTES.
26) IT SHALL BE THE CONTRACTOR'S RESPONSIBILITY TO PROPERLY NOTIFY ALL UTILITIES & DIGGERS HOTLINE PRIOR TO COMMENCING ANY CONSTRUCTION ON THIS PROJECT.
27) DO NOT CAST, DEPOSIT OR STOCKPILE ANY MATERIAL ON EXISTING ASPHALT PAVEMENT.
28) NO HEAVY CONSTRUCTION EQUIPMENT ALLOWED ON ASPHALT PAVEMENT.
29) ALL EXISTING DRIVEWAYS SHALL BE RESTORED, FOLLOWING WATERMAIN INSTALLATION AND NEW CULVERTS (12") INSTALLED WITH ENDWALLS.
30) AN (APPROVED) TRENCH BOX SHALL BE USED FOR WATERMAIN INSTALLATION.

UTILITIES

MUNICIPAL WATER SYSTEM:

DRUMMOND SANITARY DISTRICT
P.O. BOX 43 — FRONT AVE.
DRUMMOND, WISCONSIN 54832
MARK JEROME: WATERWORKS OPERATOR
TELE. NO. 715/ 739-6741

TELEPHONE COMPANY:

CHEQUAMEGON TELEPHONE CO.
1st AVE. — P.O. BOX 67
CABLE, WISCONSIN 54821
TELE. NO. 800/ 250-8927

ELECTRICAL SERVICE:

XCEL ENERGY
(A.K.A. NSP)
16048 ELECTRIC AVE.
HAYWARD, WISCONSIN 54843
KEN DISHER: FIELD ENGINEER
TELE. NO. 800/ 895-4999

FIRE DEPARTMENT

DRUMMOND FIRE & RESCUE
MARK JEROME: CHIEF
TELE. NO. 715 / 739-6696

DIGGERS HOTLINE

TELE. NO. 800 / 242-8511

-INDEX OF SHEETS-

SHEET NO.	DESCRIPTION
1	TITLE SHEET W/LOCATION MAPS AND NOTES
2	WATERMAIN PLAN & PROFILE
3 & 4	TYPICAL EROSION CONTROL DETAILS & NOTES This Sheet Not Shown

GENERAL SPECIFICATIONS: STANDARD SPECIFICATIONS FOR SEWER AND WATER CONSTRUCTION IN WISCONSIN (5TH EDITION)

WISCONSIN CONSTRUCTION SITE BEST MANAGEMENT PRACTIC HANDBOOK

SPECIFIC SPECIFICATIONS: MORGAN & PARMLEY, LTD. SPEC. BOOK

BIDDING DOCUMENTS: MORGAN & PARMLEY, LTD. SPEC. BOOK

PREPARED BY:
MORGAN & PARMLEY, Ltd.
PROFESSIONAL CONSULTING ENGINEERS
115 W. 2nd STREET SOUTH
LADYSMITH, WISCONSIN 54848

Courtesy: Morgan & Parmley, Ltd.

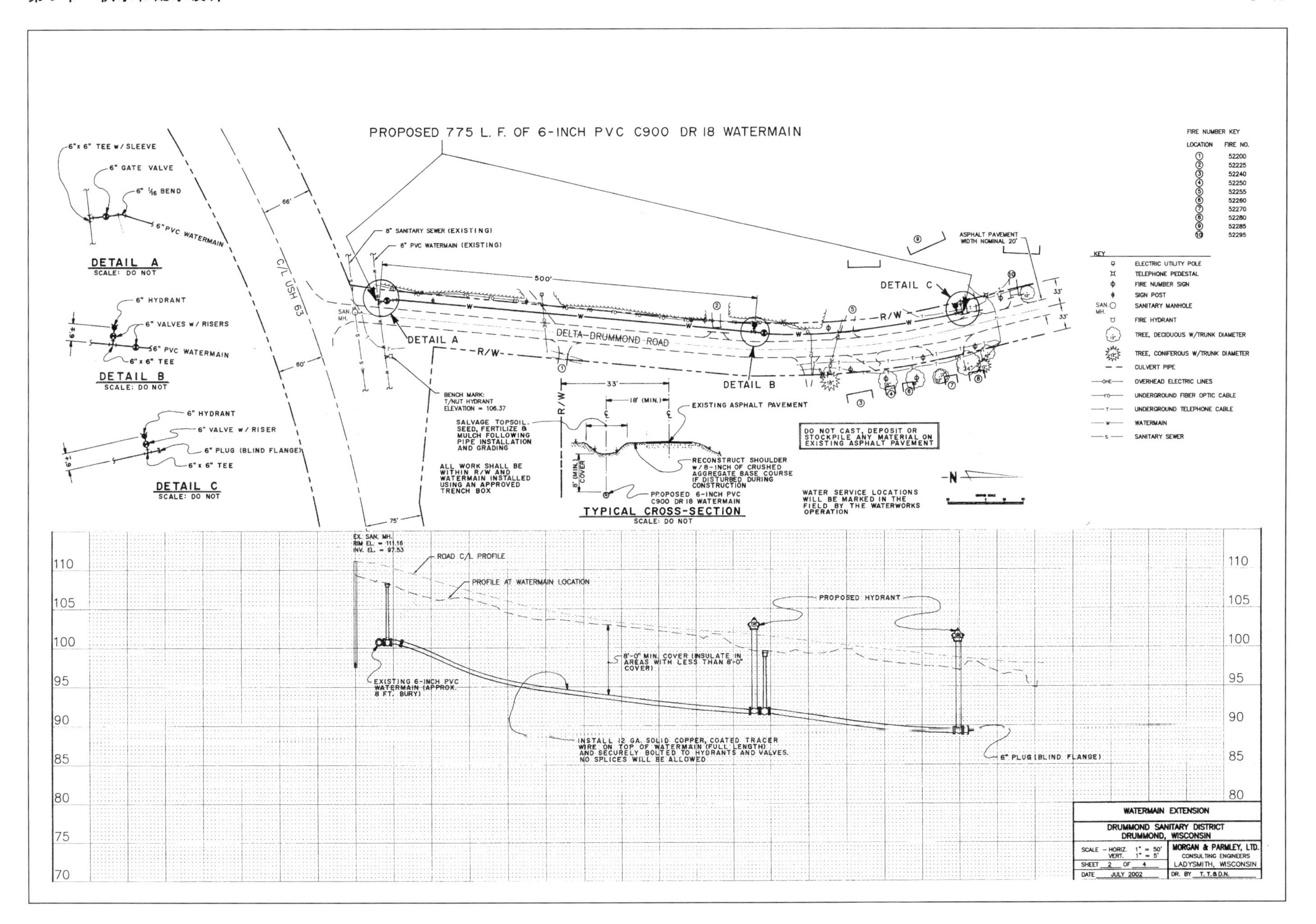
PROPOSED 775 L. F. OF 6-INCH PVC C900 DR 18 WATERMAIN
DETAIL A
SCALE: DO NOT
DETAIL B
SCALE: DO NOT
DETAIL C
SCALE: DO NOT
DELTA-DRUMMOND ROAD
TYPICAL CROSS-SECTION
SCALE: DO NOT
DO NOT CAST, DEPOSIT OR STOCKPILE ANY MATERIAL ON EXISTING ASPHALT PAVEMENT
WATER SERVICE LOCATIONS WILL BE MARKED IN THE FIELD BY THE WATERWORKS OPERATION
ALL WORK SHALL BE WITHIN R/W AND WATERMAIN INSTALLED USING AN APPROVED TRENCH BOX
PROPOSED HYDRANT
INSTALL 12 GA. SOLID COPPER, COATED TRACER WIRE ON TOP OF WATERMAIN (FULL LENGTH) AND SECURELY BOLTED TO HYDRANTS AND VALVES. NO SPLICES WILL BE ALLOWED
6" PLUG (BLIND FLANGE)
WATERMAIN EXTENSION
DRUMMOND SANITARY DISTRICT
DRUMMOND, WISCONSIN
MORGAN & PARMLEY, LTD.
CONSULTING ENGINEERS
LADYSMITH, WISCONSIN

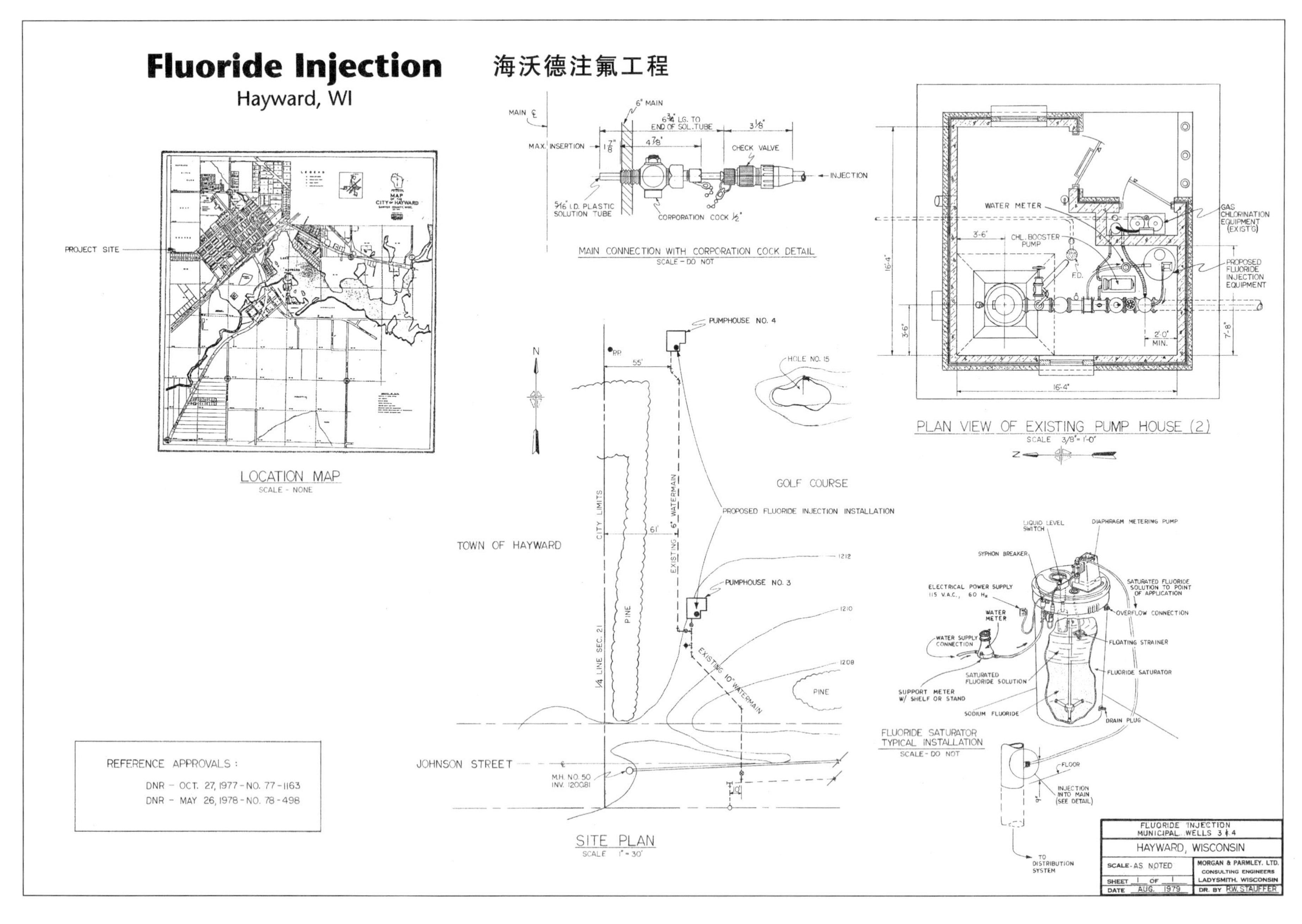

Courtesy: Morgan & Parmley, Ltd.

设　计

第 10 节 消防设计

例 10

目 录

无论是由公共供水系统或者非饮用水设备提供消防用水，消防是保护我们社会安全并远离火灾带来的毁坏和高昂代价的关键因素。

本节的第一套设计图是在饮用水配水系统中蓄水并保持足够水头压力的水塔。可以从分布在管道网络上的消火栓获得用于灭火的水源。读者可以参考前一节了解配水系统的相关内容。

当供水有限时，非饮用水可以用于消防服务设施。考虑到这一点，书中包含了设计用于乡村的工业综合建筑的设施设计图。厂房结构拥有一套完全的“湿”喷淋系统，这套系统由应急泵驱动。主泵是电动的，但是备用泵是由柴油机驱动的并能在电力瘫痪时自动启动。

水塔

-PROPOSED-
ELEVATED WATER RESERVOIR
VILLAGE of GILMAN, WISCONSIN

JUNE, 1998

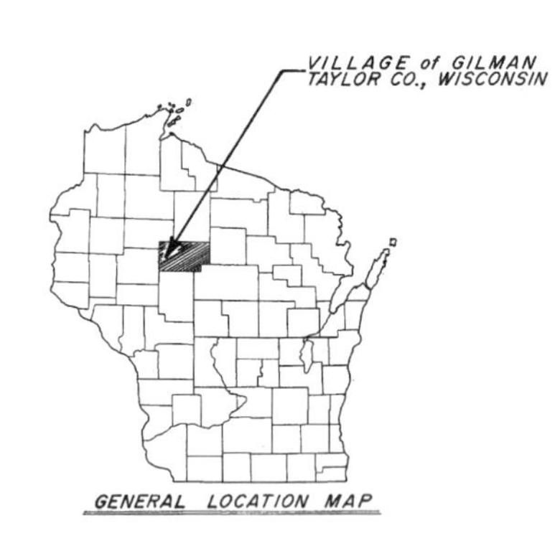

-NOTES-

1 ------ ALL ELEVATIONS BASED ON USGS DATUM.

2 ------ CONSTRUCTION SHALL COMPLY WITH THE CONDITIONS OF DNR APPROVAL LETTER, PSC AUTHORIZATION TO CONSTRUCT AND OTHER REGULATORY APPROVALS AS LISTED IN THE SPECIFICATION BOOK.

3 ------ ALL WATERMAIN SHALL BE M.J.D.I., CL. 52; SIZED AS SHOWN ON PLANS.

4 ------ ALL WATERMAIN FITTINGS SHALL BE M.J.D.I.

5 ------ ALL HYDRANT LEADS SHALL BE 6-INCH M.J.D.I., CL. 52.

6 ------ ALL HYDRANT LEADS SHALL HAVE A 6-INCH GATE VALVE.

7 ------ HYDRANTS SHALL MATCH THE VILLAGE OF GILMAN'S PATTERN.

8 ------ ALL WATER MAIN SHALL HAVE AN 8 FT. BURY (MIN.).

9 ------ 8 FT. (MINIMUM) HORIZONTAL DISTANCE (℄ TO ℄) BETWEEN WATERMAIN AND SANITARY SEWER OR STORM SEWER.

10 ----- 18" (MINIMUM) VERTICAL DISTANCE (OUT TO OUT) BETWEEN WATERMAIN UNDER SANITARY SEWER OR STORM SEWER @ THEIR INTERSECTION.

11 ----- 6" (MINIMUM) VERTICAL DISTANCE (OUT TO OUT) BETWEEN WATERMAIN OVER SANITARY SEWER OR STORM SEWER @ THEIR INTERSECTION.

12 ----- VALVES SHALL BE SET TO GRADE SHOWN ON PLANS, UNLESS DIRECTED OTHERWISE.

13 ----- WATERMAIN SHALL BE INSULATED AT ALL STORM SEWER, CULVERT AND DITCH CROSSINGS WITH 2" THICK EXTRUDED POLYSTYRENE INSULATION BOARD.

14 ----- ALL WATERMAIN SHALL BE DISINFECTED BEFORE BEING PUT INTO SERVICE.

15 ----- EXISTING UTILITIES ARE SHOWN BUT MAY NOT BE ALL INCLUSIVE, CONTRACTOR SHALL HAVE ALL BURIED UTILITIES LOCATED PRIOR TO COMMENCING CONSTRUCTION.

16 ----- EXISTING SANITARY SEWER IS SHOWN ACCORDING TO EXISTING RECORDS, BUT LOCATION SHALL BE VERIFIED IN FIELD.

17 ----- EROSION CONTROL MEASURES SHALL BE IN PLACE BEFORE CONSTRUCTION BEGINS.

18 ----- THE CONTRACTORS SHALL MINIMIZE EROSION DURING CONSTRUCTION USING GOOD CONSTRUCTION TECHNIQUES AND UTILIZING SILT FENCES AND BALE DITCH CHECKS. THE CONTRACTOR SHALL RELEASE RUNOFF FROM THE SITE IN A NUISANCE FREE MANNER.

19 ----- THE CONTRACTOR IS RESPONSIBLE TO MAINTAIN THE SITE IN A SAFE CONDITION. THE SOLE RESPONSIBILITY FOR WARNING SIGNS, BARRICADES, FLAGGING PERSONNEL AND ALL ASPECTS OF SAFETY LIE WITH THE CONTRACTOR.

20 ----- ALL DISTURBED AREAS SHALL BE SEEDED, FERTILIZED AND MULCHED, OR COVERED W/BASE COURSE AFTER CONSTRUCTION, OR RESTORED WITH BITUMINOUS PAVING WHERE APPLICABLE.

21 ----- THERE SHALL BE NO DEVIATION FROM THE PLANS WITHOUT THE DESIGN ENGINEER'S APPROVAL.

22 ----- CONSTRUCTION STAKES DESTROYED BY THE CONTRACTOR THAT NEED TO BE REPLACED, WILL BE CHARGED TO THE CONTRACTOR.

23 ----- PROPERTY CORNERS KNOWN TO EXIST SHALL NOT BE DISTURBED. DAMAGED CORNERS WILL BE REPLACED AT THE CONTRACTOR'S EXPENSE. PRESERVE ALL SURVEY MONUMENTS.

24 ----- SQUARE CUT ALL BLACKTOP PAVEMENT PRIOR TO TRENCH EXCAVATION.

25 ----- SEE SPECIFICATION BOOK AND FOLLOWING PLAN SHEETS FOR ADDITIONAL DETAILS AND NOTES.

26 ----- SEE SPECIFICATION BOOK AND PLANS FOR ELEVATED TANK & TOWER.

PROJECT LOCATION
SEE PROJECT SITE LOCATION

EXISTING ELEVATED TANK

VILLAGE OF GILMAN, WISCONSIN

LOCATION MAP

GENERAL SPECIFICATIONS: STANDARD SPECIFICATIONS FOR SEWER AND WATER CONSTRUCTION IN WISCONSIN (5TH EDITION)

WISCONSIN CONSTRUCTION SITE BEST BEST MANAGEMENT PRACTICE HANDBOOK

SPECIFIC SPECIFICATIONS: MORGAN & PARMLEY, LTD. SPEC. BOOK

BIDDING DOCUMENTS: MORGAN & PARMLEY, LTD. SPEC. BOOK

UTILITIES

S & K TV SYSTEMS
508 W. MINER
LADYSMITH, WI 54848
ATTN: RANDY SCOTT

LAKEHEAD PIPELINE
803 HIGHLAND AVENUE
P.O. BOX 308
FT. ATKINSON, WI 53538
ATTN: MARK KINBLOM

NORTHERN STATES POWER CO.
500 N. 5TH STREET
ABBOTSFORD, WI 54405
ATTN: DAVE PEPPER

DIGGER'S HOTLINE

CENTURY TELEPHONE OF NORTHERN WISCONSIN, INC.
425 ELLINGSON AVE.
P.O. BOX 78
HAWKINS, WISCONSIN 54530
ATTN: JAMES ARQUETTE

INDEX OF DRAWINGS

SHEET NO.	DESCRIPTION
1	TITLE SHEET, LOCATION MAP, INDEX & NOTES
2	PROJECT SITE PLAN
3	BASIC GEOMETRY of ELEVATED WATER RESERVOIR
4	DETAILS of ELEVATED WATER RESERVOIR
5	TYPICAL EROSION CONTROL DETAILS　This Sheet Not Shown

PROJECT SITE LOCATION

NOTE-SEE TAYLOR CO. REGISTER OF DEEDS' RECORDS; VOL. 35 PAGES 92 & 93 FOR COMPLETE DESCRIPTION OF THIS C.S.M. DOCUMENT.

TAYLOR COUNTY CERTIFIED SURVEY MAP NO. 693

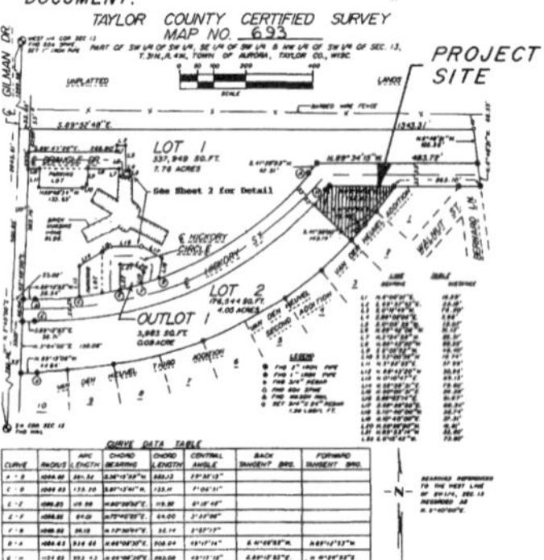

PREPARED BY:
MORGAN & PARMLEY, Ltd.
PROFESSIONAL CONSULTING ENGINEERS
LADYSMITH, WISCONSIN

Courtesy: Morgan & Parmley, Ltd.

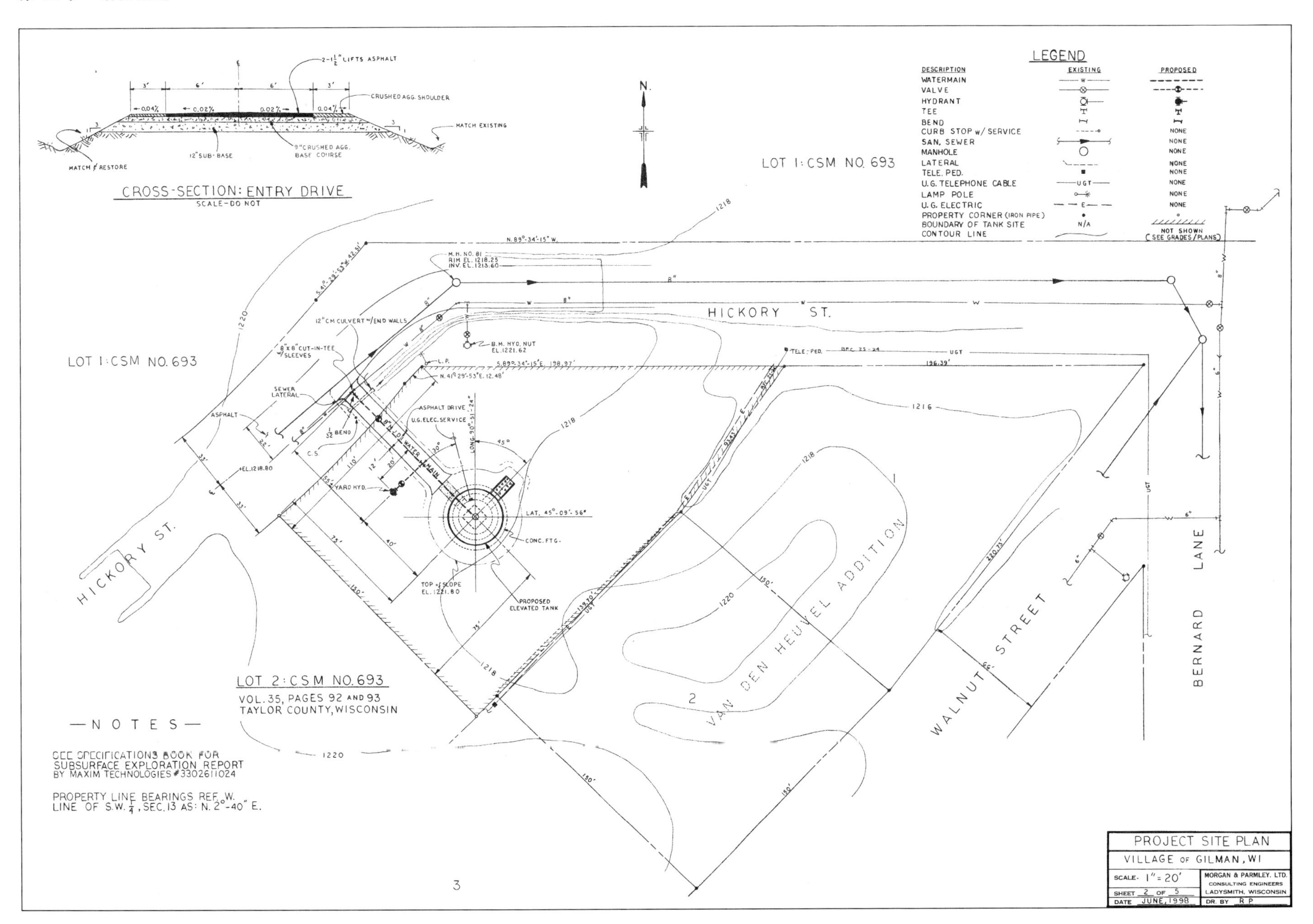

CROSS-SECTION: ENTRY DRIVE
SCALE-DO NOT
LEGEND
LOT 1: CSM NO. 693
HICKORY ST.
LOT 2: CSM NO. 693
VOL. 35, PAGES 92 AND 93
TAYLOR COUNTY, WISCONSIN
VAN DEN HEUVEL ADDITION
WALNUT STREET
BERNARD LANE
PROPOSED ELEVATED TANK
— NOTES —
SEE SPECIFICATIONS BOOK FOR SUBSURFACE EXPLORATION REPORT BY MAXIM TECHNOLOGIES #3302611024
PROPERTY LINE BEARINGS REF. W. LINE OF S.W. 1/4, SEC. 13 AS: N. 2°-40" E.
PROJECT SITE PLAN
VILLAGE OF GILMAN, WI
SCALE- 1" = 20'
SHEET 2 OF 5
DATE JUNE, 1998
MORGAN & PARMLEY, LTD.
CONSULTING ENGINEERS
LADYSMITH, WISCONSIN
DR. BY R P

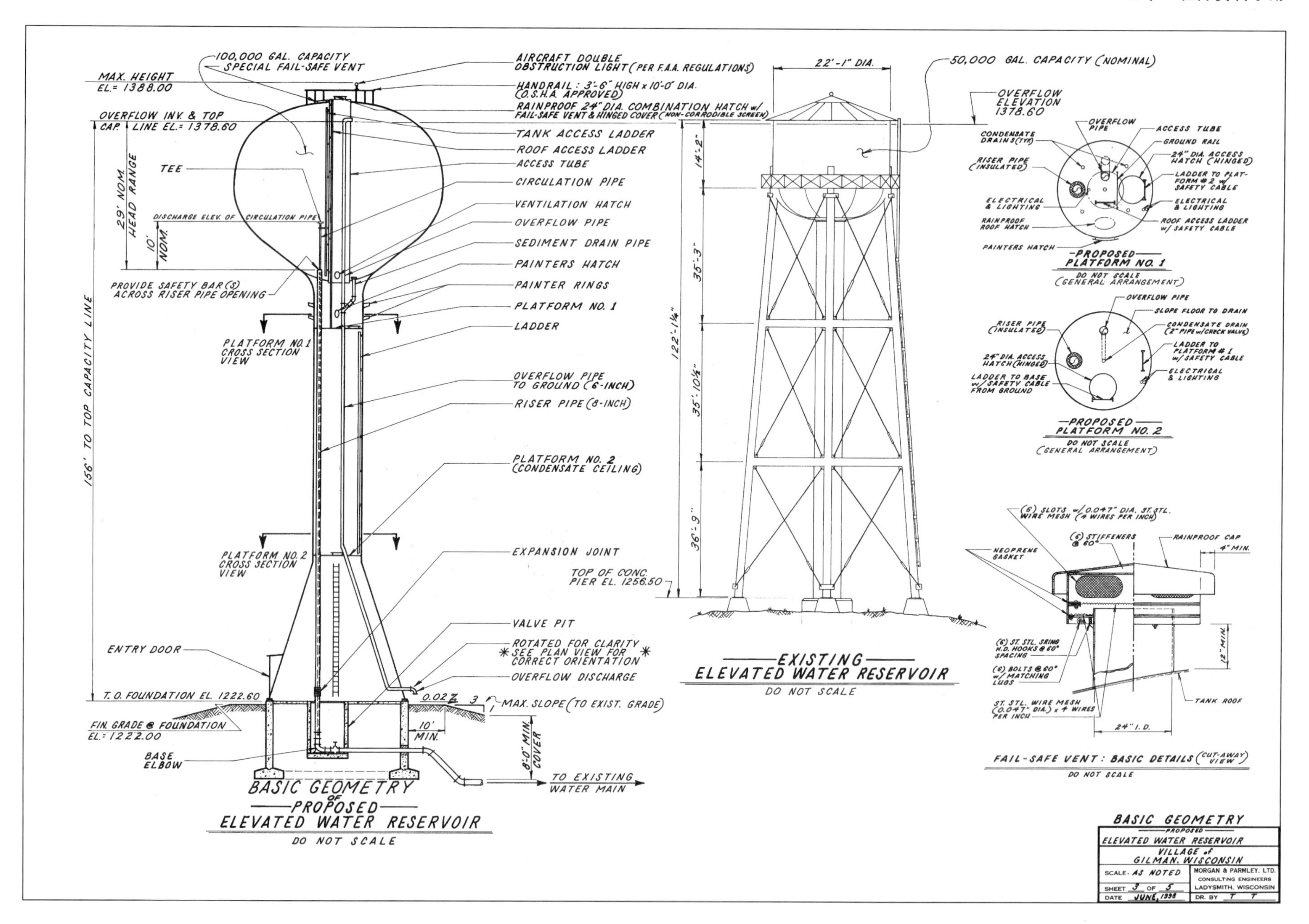
100,000 GAL. CAPACITY
SPECIAL FAIL-SAFE VENT
MAX. HEIGHT EL.= 1388.00
OVERFLOW INV. & TOP CAP. LINE EL.= 1378.60
AIRCRAFT DOUBLE OBSTRUCTION LIGHT (PER F.A.A. REGULATIONS)
HANDRAIL: 3'-6" HIGH x 10'-0" DIA. (O.S.H.A. APPROVED)
RAINPROOF 24" DIA. COMBINATION HATCH w/ FAIL-SAFE VENT & HINGED COVER (NON-CORRODIBLE SCREEN)
TANK ACCESS LADDER
ROOF ACCESS LADDER
ACCESS TUBE
CIRCULATION PIPE
VENTILATION HATCH
OVERFLOW PIPE
SEDIMENT DRAIN PIPE
PAINTERS HATCH
PAINTER RINGS
PLATFORM NO. 1
LADDER
TEE
29' NOM. HEAD RANGE
DISCHARGE ELEV. OF CIRCULATION PIPE
10' NOM.
PROVIDE SAFETY BAR(S) ACROSS RISER PIPE OPENING
PLATFORM NO. 1 CROSS SECTION VIEW
156' TO TOP CAPACITY LINE
OVERFLOW PIPE TO GROUND (6-INCH)
RISER PIPE (8-INCH)
PLATFORM NO. 2 (CONDENSATE CEILING)
PLATFORM NO. 2 CROSS SECTION VIEW
EXPANSION JOINT
VALVE PIT
ROTATED FOR CLARITY * SEE PLAN VIEW FOR CORRECT ORIENTATION *
OVERFLOW DISCHARGE
ENTRY DOOR
T. O. FOUNDATION EL. 1222.60
0.02 Z 3 1 MAX. SLOPE (TO EXIST. GRADE)
FIN. GRADE @ FOUNDATION EL.= 1222.00
10' MIN.
8'-0" MIN. COVER
BASE ELBOW
TO EXISTING WATER MAIN
BASIC GEOMETRY of PROPOSED ELEVATED WATER RESERVOIR
DO NOT SCALE
22'-1" DIA.
50,000 GAL. CAPACITY (NOMINAL)
OVERFLOW ELEVATION 1378.60
14'-2"
35'-3"
122'-1¼"
35'-10¼"
36'-9"
TOP OF CONC. PIER EL. 1256.50
EXISTING ELEVATED WATER RESERVOIR
DO NOT SCALE
OVERFLOW PIPE
ACCESS TUBE
GROUND RAIL
CONDENSATE DRAINS (TYP.)
RISER PIPE (INSULATED)
24" DIA. ACCESS HATCH (HINGED)
LADDER TO PLATFORM #2 w/ SAFETY CABLE
ELECTRICAL & LIGHTING
RAINPROOF ROOF HATCH
ROOF ACCESS LADDER w/ SAFETY CABLE
PAINTERS HATCH
PROPOSED PLATFORM NO. 1
DO NOT SCALE (GENERAL ARRANGEMENT)
SLOPE FLOOR TO DRAIN
CONDENSATE DRAIN (2" PIPE w/CHECK VALVE)
LADDER TO PLATFORM # 1 w/ SAFETY CABLE
LADDER TO BASE w/ SAFETY CABLE FROM GROUND
PROPOSED PLATFORM NO. 2
DO NOT SCALE (GENERAL ARRANGEMENT)
(6) SLOTS w/ 0.047" DIA. ST. STL. WIRE MESH (4 WIRES PER INCH)
(6) STIFFENERS @ 60°
RAINPROOF CAP
4" MIN.
NEOPRENE GASKET
(6) ST. STL. SPRING H.D. HOOKS @ 60° SPACING
(6) BOLTS @ 60° w/ MATCHING LUGS
ST. STL. WIRE MESH (0.047" DIA.) x 4 WIRES PER INCH
12" MIN.
TANK ROOF
24" I. D.
FAIL-SAFE VENT: BASIC DETAILS (CUT-AWAY VIEW)
DO NOT SCALE
BASIC GEOMETRY
PROPOSED
ELEVATED WATER RESERVOIR
VILLAGE of GILMAN, WISCONSIN
SCALE: AS NOTED
SHEET 3 OF 5
DATE JUNE, 1998
MORGAN & PARMLEY, LTD. CONSULTING ENGINEERS LADYSMITH, WISCONSIN
DR. BY T T

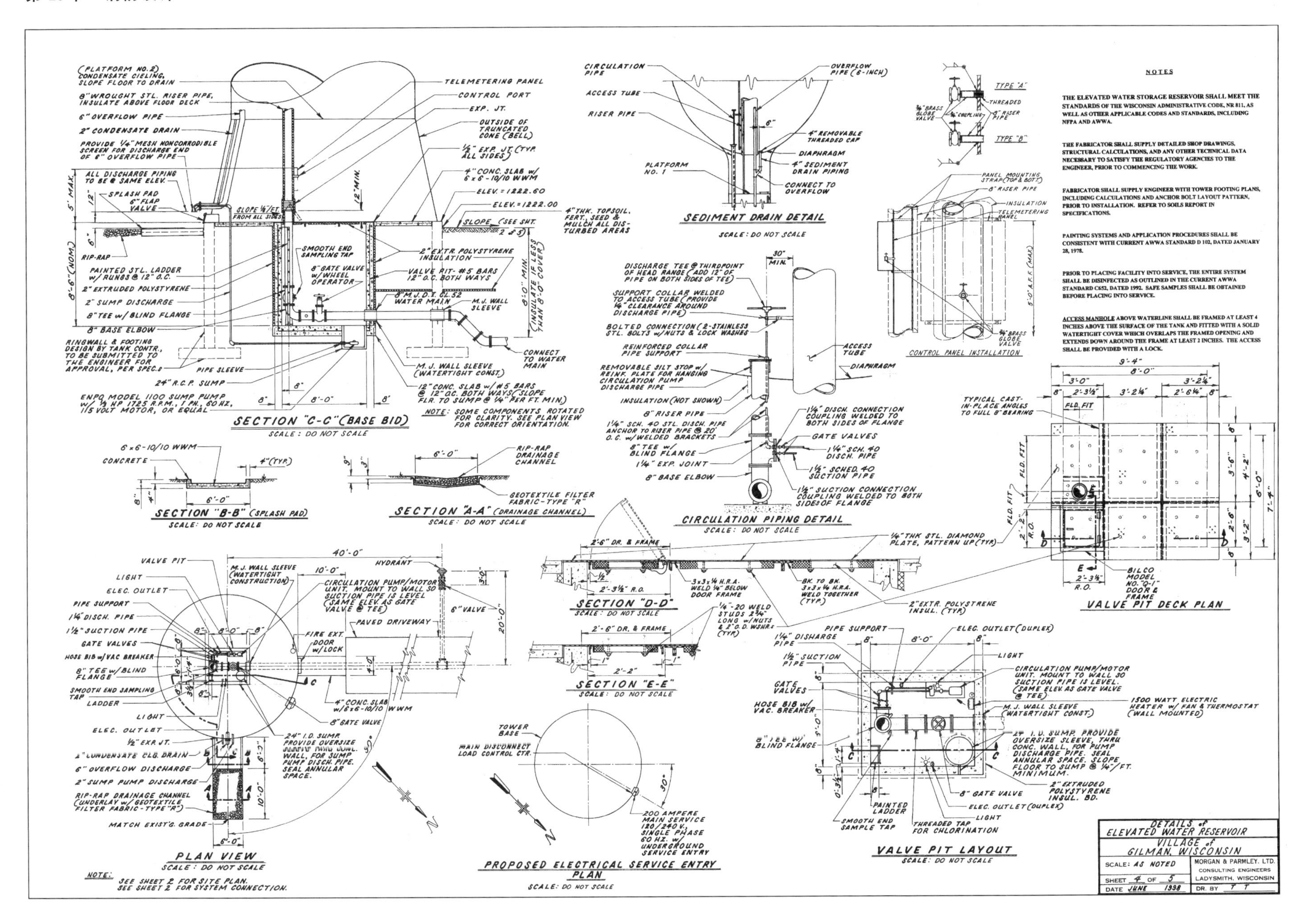

NOTES
THE ELEVATED WATER STORAGE RESERVOIR SHALL MEET THE STANDARDS OF THE WISCONSIN ADMINISTRATIVE CODE, NR 811, AS WELL AS OTHER APPLICABLE CODES AND STANDARDS, INCLUDING NFPA AND AWWA.
THE FABRICATOR SHALL SUPPLY DETAILED SHOP DRAWINGS, STRUCTURAL CALCULATIONS, AND ANY OTHER TECHNICAL DATA NECESSARY TO SATISFY THE REGULATORY AGENCIES TO THE ENGINEER, PRIOR TO COMMENCING THE WORK.
FABRICATOR SHALL SUPPLY ENGINEER WITH TOWER FOOTING PLANS, INCLUDING CALCULATIONS AND ANCHOR BOLT LAYOUT PATTERN, PRIOR TO INSTALLATION. REFER TO SOILS REPORT IN SPECIFICATIONS.
PAINTING SYSTEMS AND APPLICATION PROCEDURES SHALL BE CONSISTENT WITH CURRENT AWWA STANDARD D 102, DATED JANUARY 28, 1978.
PRIOR TO PLACING FACILITY INTO SERVICE, THE ENTIRE SYSTEM SHALL BE DISINFECTED AS OUTLINED IN THE CURRENT AWWA STANDARD C652, DATED 1992. SAFE SAMPLES SHALL BE OBTAINED BEFORE PLACING INTO SERVICE.
ACCESS MANHOLE ABOVE WATERLINE SHALL BE FRAMED AT LEAST 4 INCHES ABOVE THE SURFACE OF THE TANK AND FITTED WITH A SOLID WATERTIGHT COVER WHICH OVERLAPS THE FRAMED OPENING AND EXTENDS DOWN AROUND THE FRAME AT LEAST 2 INCHES. THE ACCESS SHALL BE PROVIDED WITH A LOCK.
SECTION "C-C" (BASE BID)
SCALE: DO NOT SCALE
NOTE: SOME COMPONENTS ROTATED FOR CLARITY. SEE PLAN VIEW FOR CORRECT ORIENTATION.
SEDIMENT DRAIN DETAIL
SCALE: DO NOT SCALE
CONTROL PANEL INSTALLATION
CIRCULATION PIPING DETAIL
SCALE: DO NOT SCALE
SECTION "B-B" (SPLASH PAD)
SCALE: DO NOT SCALE
SECTION "A-A" (DRAINAGE CHANNEL)
SCALE: DO NOT SCALE
SECTION "D-D"
SCALE: DO NOT SCALE
SECTION "E-E"
SCALE: DO NOT SCALE
VALVE PIT DECK PLAN
PLAN VIEW
SCALE: DO NOT SCALE
NOTE: SEE SHEET 2 FOR SITE PLAN. SEE SHEET 2 FOR SYSTEM CONNECTION.
PROPOSED ELECTRICAL SERVICE ENTRY PLAN
SCALE: DO NOT SCALE
VALVE PIT LAYOUT
SCALE: DO NOT SCALE
DETAILS of ELEVATED WATER RESERVOIR
VILLAGE of GILMAN, WISCONSIN
SCALE: AS NOTED
SHEET 4 OF 5
DATE JUNE 1998
MORGAN & PARMLEY, LTD.
CONSULTING ENGINEERS
LADYSMITH, WISCONSIN
DR. BY T T

消防设施

FIRE PROTECTION FACILITY

FOR

INDUSTRIAL PARK COMPLEX

E.D.A. PROJECT NO. 06-01-02571

VILLAGE
OF
GLEN FLORA, WISCONSIN
SEPT., 1992

UTILITIES

CONTACT – NORTHERN STATES POWER CO.
711 W. 9th Street, North
Ladysmith, WI 54848

Attn: John Rymarkiewicz
715-532-6226

CONTACT – UNIVERSAL TELEPHONE CO.
OF NORTHERN WISCONSIN
Highway "8"
Hawkins, WI 54530

Attn: Jim Arquette
715-585-7707

CONTACT – DIGGER'S HOTLINE

1-800-242-8511

Ticket's Nos.: 1598953
1598962

CONTACT – VILLAGE OF GLEN FLORA WWCTF

Les Evjen, Operator
P.O. Box 253
Glen Flora, WI 54526

715-322-5511

VILLAGE OF GLEN FLORA,
RUSK CO, WISCONSIN

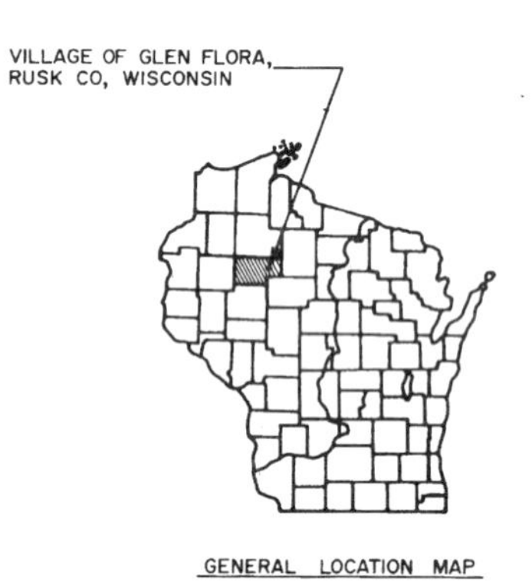

GENERAL LOCATION MAP

FIRE PROTECTION FACILITY

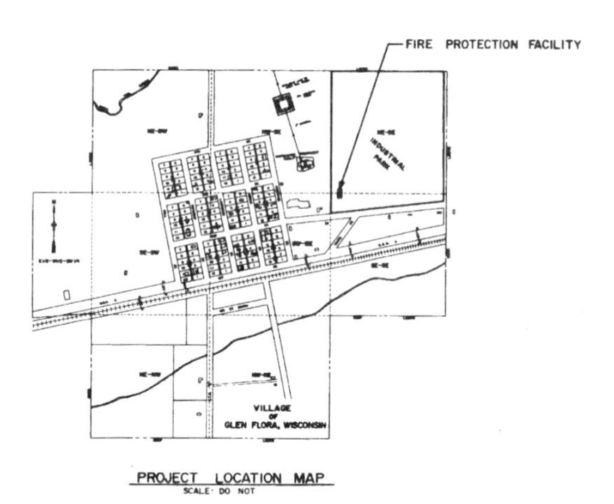

PROJECT LOCATION MAP
SCALE: DO NOT

INDEX

NO.	DESCRIPTION
1	TITLE SHEET, GENERAL LOCATION & INDEX
2	GENERAL SITE PLAN W/ SOIL BORINGS & TOPOG.
3	RESERVOIR PLAN, LONGITUDINAL-SECTION & YARD PIPING
4	RESERVOIR ROOF PLAN & PUMPHOUSE FLOOR
5	RESERVOIR DETAILS
6	PUMPHOUSE EQUIPMENT ARRANGEMENT
7	EQUIPMENT DETAILS & CROSS-SECTIONS
8	PUMPHOUSE LAYOUT & DETAILS
9	PUMPHOUSE HEATING, VENTILATING & ELEC.
10	MISC. DIAGRAMS, LAYOUTS, FUEL TANK, PUMPHOUSE SPRINKLER SYSTEM & DETAILS
11	TYPICAL EROSION CONTROL DETAILS

PREPARED BY:
MORGAN & PARMLEY LTD.
CONSULTING ENGINEERS
LADYSMITH, WISCONSIN

Courtesy: Morgan & Parmley, Ltd.

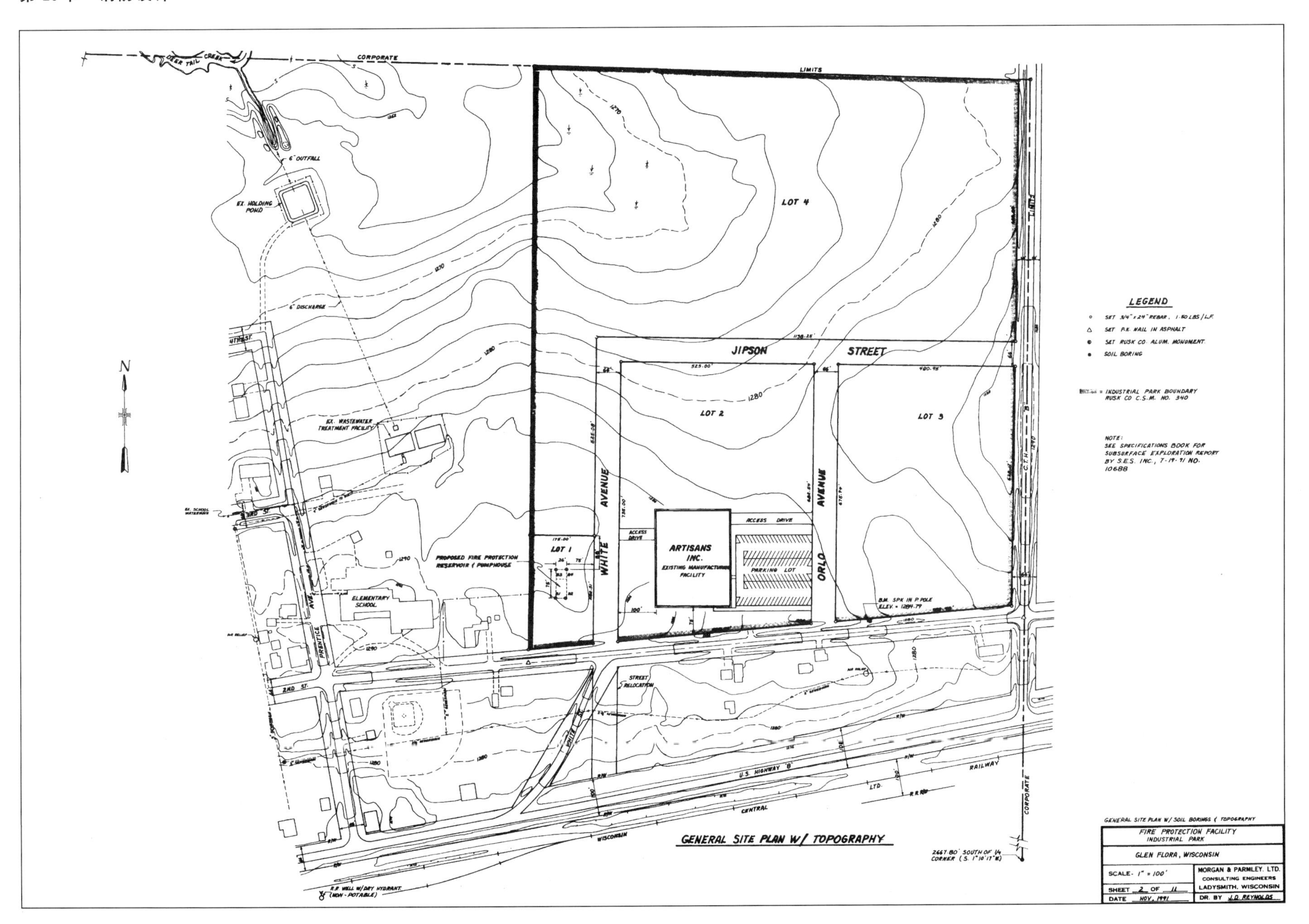
GENERAL SITE PLAN W/ TOPOGRAPHY
LEGEND
SET 3/4" x 24" REBAR, 1.50 LBS/L.F.
SET P.K. NAIL IN ASPHALT
SET RUSK CO. ALUM. MONUMENT
SOIL BORING
= INDUSTRIAL PARK BOUNDARY RUSK CO C.S.M. NO. 340
NOTE: SEE SPECIFICATIONS BOOK FOR SUBSURFACE EXPLORATION REPORT BY S.E.S. INC., 7-19-91 NO. 10688
LOT 1
LOT 2
LOT 3
LOT 4
JIPSON STREET
WHITE AVENUE
ORLO AVENUE
ARTISANS INC. EXISTING MANUFACTURING FACILITY
ACCESS DRIVE
PARKING LOT
PROPOSED FIRE PROTECTION RESERVOIR & PUMPHOUSE
ELEMENTARY SCHOOL
EX. WASTEWATER TREATMENT FACILITY
EX. HOLDING POND
6" OUTFALL
6" DISCHARGE
CORPORATE LIMITS
U.S. HIGHWAY "B"
RAILWAY
STREET RELOCATION
GENERAL SITE PLAN W/ SOIL BORINGS & TOPOGRAPHY
FIRE PROTECTION FACILITY
INDUSTRIAL PARK
GLEN FLORA, WISCONSIN
SCALE: 1" = 100'
SHEET 2 OF 11
DATE NOV. 1991
MORGAN & PARMLEY, LTD.
CONSULTING ENGINEERS
LADYSMITH, WISCONSIN
DR. BY J.D. REYNOLDS

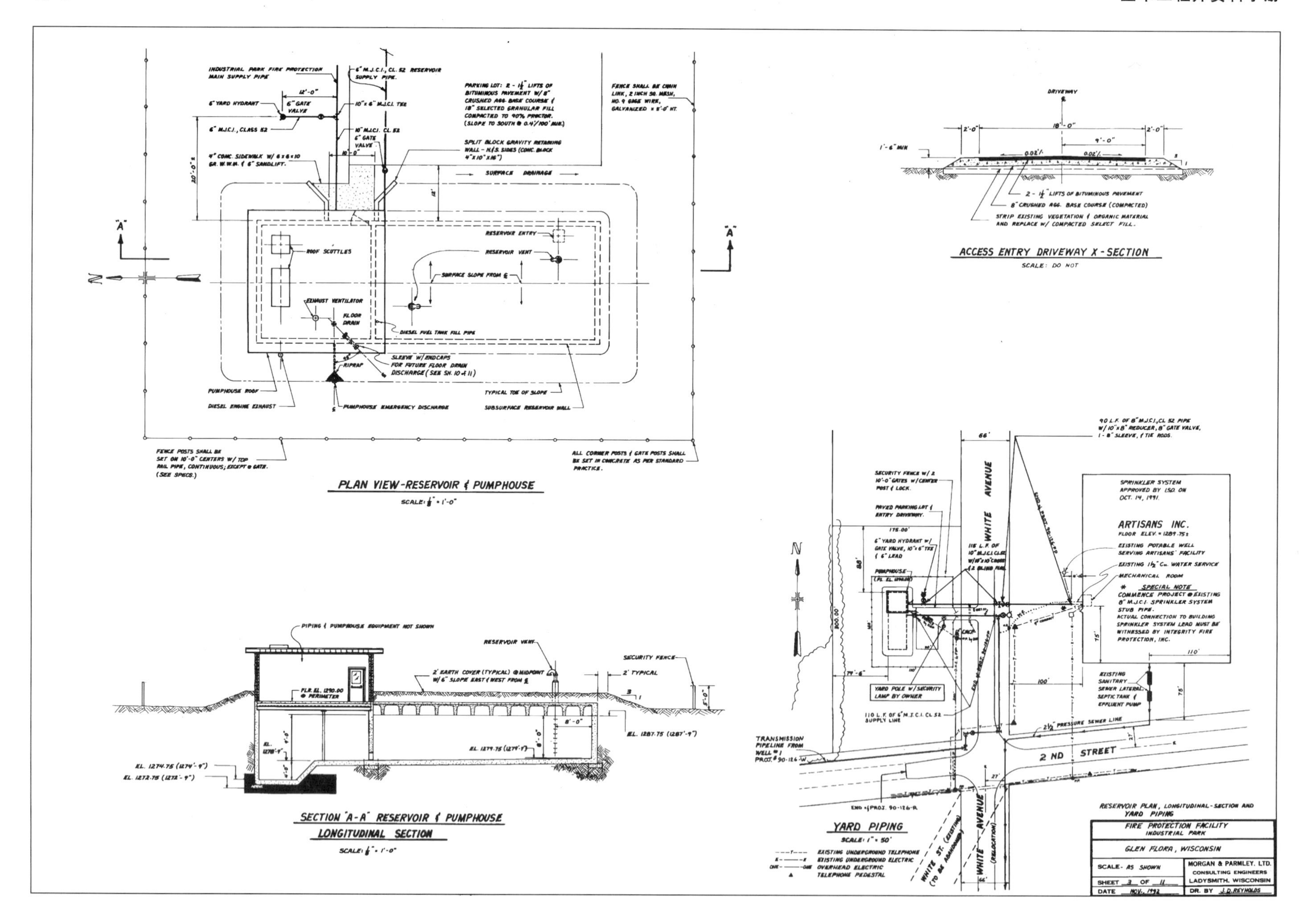
PLAN VIEW-RESERVOIR & PUMPHOUSE
SCALE: 1/8" = 1'-0"
INDUSTRIAL PARK FIRE PROTECTION MAIN SUPPLY PIPE
6" M.J.C.I., CL. 52 RESERVOIR SUPPLY PIPE.
6" YARD HYDRANT
6" GATE VALVE
10" x 6" M.J.C.I. TEE
6" M.J.C.I., CLASS 52
10" M.J.C.I. CL. 52
6" GATE VALVE
4" CONC. SIDEWALK W/ 6 x 6 x 10 GA. W.W.M. & 6" SANDLIFT.
PARKING LOT: 2 - 1 1/2" LIFTS OF BITUMINOUS PAVEMENT W/ 8" CRUSHED AGG. BASE COURSE & 18" SELECTED GRANULAR FILL COMPACTED TO 90% PROCTOR. (SLOPE TO SOUTH @ 0.4'/100' MIN.)
SPLIT BLOCK GRAVITY RETAINING WALL - N/S. SIDES (CONC. BLOCK 4"x10"x16")
FENCE SHALL BE CHAIN LINK, 2 INCH SQ. MESH, NO. 9 GAGE WIRE, GALVANIZED x 5'-0" HT.
SURFACE DRAINAGE
ROOF SCUTTLES
RESERVOIR ENTRY
RESERVOIR VENT
SURFACE SLOPE FROM ℄
EXHAUST VENTILATOR
FLOOR DRAIN
DIESEL FUEL TANK FILL PIPE
RIPRAP
SLEEVE W/ ENDCAPS FOR FUTURE FLOOR DRAIN DISCHARGE (SEE SH. 10 & 11)
PUMPHOUSE ROOF
DIESEL ENGINE EXHAUST
PUMPHOUSE EMERGENCY DISCHARGE
TYPICAL TOE OF SLOPE
SUBSURFACE RESERVOIR WALL
FENCE POSTS SHALL BE SET ON 10'-0" CENTERS W/ TOP RAIL PIPE, CONTINUOUS; EXCEPT @ GATE. (SEE SPECS.)
ALL CORNER POSTS & GATE POSTS SHALL BE SET IN CONCRETE AS PER STANDARD PRACTICE.
ACCESS ENTRY DRIVEWAY X-SECTION
SCALE: DO NOT
DRIVEWAY
18'-0"
9'-0"
2'-0"
1'-6" MIN
0.02'/'
2 - 1 1/2" LIFTS OF BITUMINOUS PAVEMENT
8" CRUSHED AGG. BASE COURSE (COMPACTED)
STRIP EXISTING VEGETATION & ORGANIC MATERIAL AND REPLACE W/ COMPACTED SELECT FILL.
SECTION "A-A" RESERVOIR & PUMPHOUSE LONGITUDINAL SECTION
SCALE: 1/8" = 1'-0"
PIPING & PUMPHOUSE EQUIPMENT NOT SHOWN
RESERVOIR VENT
SECURITY FENCE
2' TYPICAL
2' EARTH COVER (TYPICAL) @ MIDPOINT W/ 6" SLOPE EAST & WEST FROM ℄
FLR. EL. 1290.00 @ PERIMETER
EL. 1287.75 (1287'-9")
EL. 1279.75 (1279'-9")
EL. 1274.75 (1274'-9")
EL. 1272.75 (1272'-9")
YARD PIPING
SCALE: 1" = 50'
90 L.F. OF 8" M.J.C.I. CL 52 PIPE W/ 10" x 8" REDUCER, 8" GATE VALVE, 1 - 8" SLEEVE, & TIE RODS.
SECURITY FENCE W/ 2 10'-0" GATES W/ CENTER POST & LOCK.
PAVED PARKING LOT & ENTRY DRIVEWAY.
6" YARD HYDRANT W/ GATE VALVE, 10" x 6" TEE & 6" LEAD
PUMPHOUSE
115 L.F. OF 10" M.J.C.I. CL52
WHITE AVENUE
SPRINKLER SYSTEM APPROVED BY I.S.O. ON OCT. 14, 1991.
ARTISANS INC.
FLOOR ELEV. = 1289.75±
EXISTING POTABLE WELL SERVING ARTISANS' FACILITY
EXISTING 1 1/2" Cu. WATER SERVICE
MECHANICAL ROOM
* SPECIAL NOTE
COMMENCE PROJECT @ EXISTING 8" M.J.C.I. SPRINKLER SYSTEM STUB PIPE.
ACTUAL CONNECTION TO BUILDING SPRINKLER SYSTEM LEAD MUST BE WITNESSED BY INTEGRITY FIRE PROTECTION, INC.
EXISTING SANITARY SEWER LATERAL, SEPTIC TANK & EFFLUENT PUMP
YARD POLE W/ SECURITY LAMP BY OWNER
110 L.F. OF 6" M.J.C.I. CL 52 SUPPLY LINE
TRANSMISSION PIPELINE FROM WELL #1 PROJ. # 90-126-W
2 1/2" PRESSURE SEWER LINE
2 ND STREET
END of PROJ. 90-126-R
WHITE ST. (EXISTING) (TO BE ABANDONED)
WHITE AVENUE (RELOCATION)
EXISTING UNDERGROUND TELEPHONE
EXISTING UNDERGROUND ELECTRIC
OVERHEAD ELECTRIC
TELEPHONE PEDESTAL
RESERVOIR PLAN, LONGITUDINAL-SECTION AND YARD PIPING
FIRE PROTECTION FACILITY INDUSTRIAL PARK
GLEN FLORA, WISCONSIN
SCALE- AS SHOWN
SHEET 3 OF 11
DATE NOV., 1992
MORGAN & PARMLEY, LTD. CONSULTING ENGINEERS LADYSMITH, WISCONSIN
DR. BY J.D. REYNOLDS

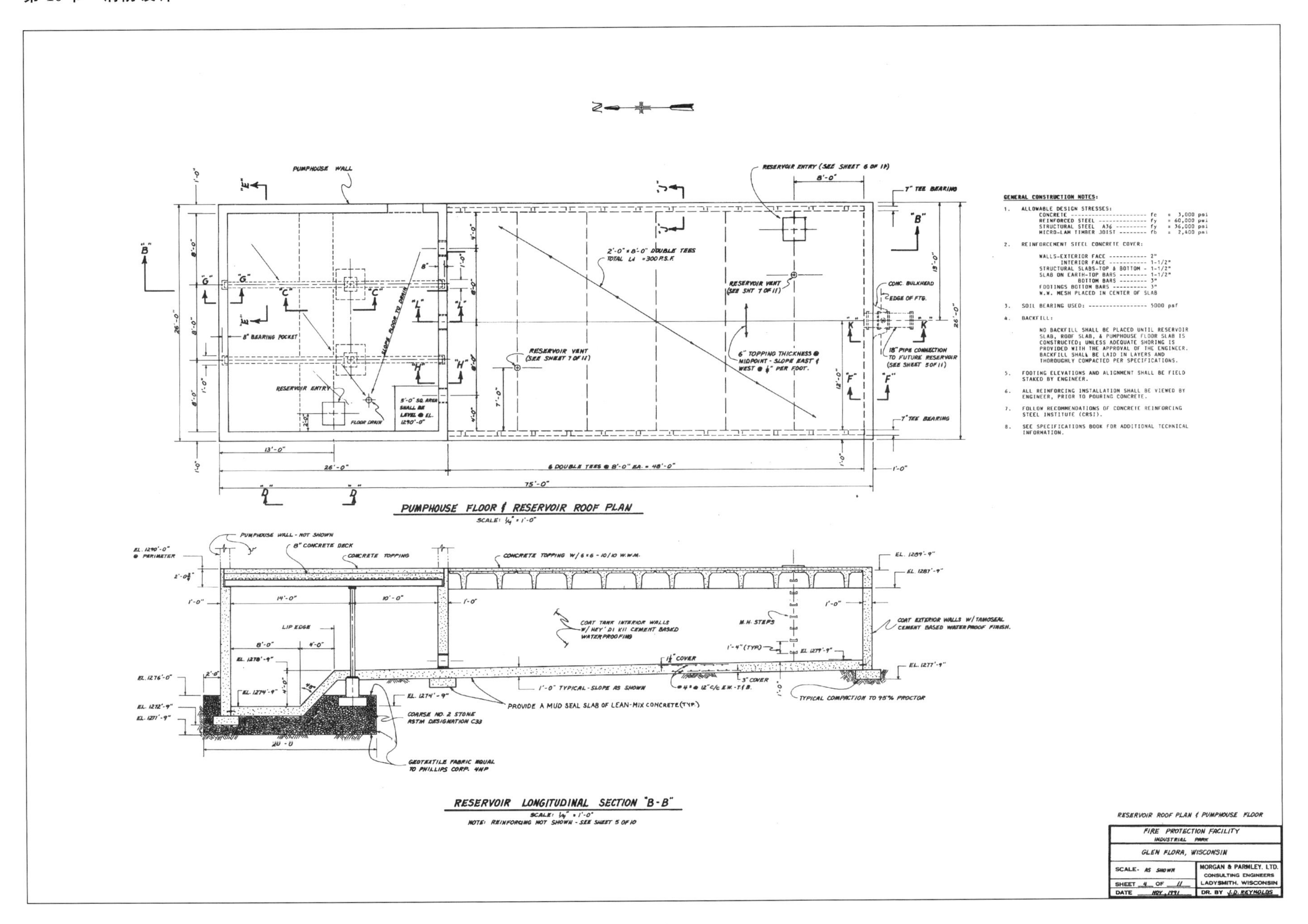
PUMPHOUSE WALL
RESERVOIR ENTRY (SEE SHEET 6 OF 11)
7" TEE BEARING
2'-0" x 8'-0" DOUBLE TEES
TOTAL LL = 300 P.S.F.
RESERVOIR VENT (SEE SHT 7 OF 11)
CONC. BULKHEAD
EDGE OF FTG.
8" BEARING POCKET
6" TOPPING THICKNESS @ MIDPOINT - SLOPE EAST & WEST @ 1/4" PER FOOT.
RESERVOIR VENT (SEE SHEET 7 OF 11)
18" PIPE CONNECTION TO FUTURE RESERVOIR (SEE SHEET 5 OF 11)
RESERVOIR ENTRY
FLOOR DRAIN
5'-0" SQ. AREA SHALL BE LEVEL @ EL. 1290'-0"
6 DOUBLE TEES @ 8'-0" EA. = 48'-0"
75'-0"
PUMPHOUSE FLOOR & RESERVOIR ROOF PLAN
SCALE: 1/4" = 1'-0"
GENERAL CONSTRUCTION NOTES:
1. ALLOWABLE DESIGN STRESSES:
CONCRETE ---------------------- fc = 3,000 psi
REINFORCED STEEL -------------- fy = 60,000 psi
STRUCTURAL STEEL A36 ---------- fy = 36,000 psi
MICRO-LAM TIMBER JOIST -------- fb = 2,400 psi
2. REINFORCEMENT STEEL CONCRETE COVER:
WALLS-EXTERIOR FACE ----------- 2"
INTERIOR FACE ----------- 1-1/2"
STRUCTURAL SLABS-TOP & BOTTOM - 1-1/2"
SLAB ON EARTH-TOP BARS -------- 1-1/2"
BOTTOM BARS -------- 3"
FOOTINGS BOTTOM BARS ---------- 3"
W.W. MESH PLACED IN CENTER OF SLAB
3. SOIL BEARING USED: ----------------- 5000 psf
4. BACKFILL:
NO BACKFILL SHALL BE PLACED UNTIL RESERVOIR SLAB, ROOF SLAB, & PUMPHOUSE FLOOR SLAB IS CONSTRUCTED; UNLESS ADEQUATE SHORING IS PROVIDED WITH THE APPROVAL OF THE ENGINEER. BACKFILL SHALL BE LAID IN LAYERS AND THOROUGHLY COMPACTED PER SPECIFICATIONS.
5. FOOTING ELEVATIONS AND ALIGNMENT SHALL BE FIELD STAKED BY ENGINEER.
6. ALL REINFORCING INSTALLATION SHALL BE VIEWED BY ENGINEER, PRIOR TO POURING CONCRETE.
7. FOLLOW RECOMMENDATIONS OF CONCRETE REINFORCING STEEL INSTITUTE (CRSI).
8. SEE SPECIFICATIONS BOOK FOR ADDITIONAL TECHNICAL INFORMATION.
PUMPHOUSE WALL - NOT SHOWN
8" CONCRETE DECK
CONCRETE TOPPING
CONCRETE TOPPING W/ 6x6 - 10/10 W.W.M.
EL. 1290'-0" @ PERIMETER
EL. 1289'-9"
EL. 1287'-9"
LIP EDGE
COAT TANK INTERIOR WALLS W/ HEY' DI KII CEMENT BASED WATERPROOFING
M.H. STEPS
COAT EXTERIOR WALLS W/ TAMOSEAL CEMENT BASED WATERPROOF FINISH.
1 1/2" COVER
3" COVER
EL. 1277'-9"
EL. 1276'-0"
EL. 1278'-9"
EL. 1274'-9"
EL. 1272'-9"
EL. 1271'-9"
1'-0" TYPICAL - SLOPE AS SHOWN
PROVIDE A MUD SEAL SLAB OF LEAN-MIX CONCRETE (TYP.)
TYPICAL COMPACTION TO 95% PROCTOR
COARSE NO. 2 STONE ASTM DESIGNATION C33
GEOTEXTILE FABRIC EQUAL TO PHILLIPS CORP. 4NP
RESERVOIR LONGITUDINAL SECTION "B-B"
SCALE: 1/4" = 1'-0"
NOTE: REINFORCING NOT SHOWN - SEE SHEET 5 OF 10
RESERVOIR ROOF PLAN & PUMPHOUSE FLOOR
FIRE PROTECTION FACILITY
INDUSTRIAL PARK
GLEN FLORA, WISCONSIN
SCALE- AS SHOWN
SHEET 4 OF 11
DATE NOV. 1991
MORGAN & PARMLEY, LTD.
CONSULTING ENGINEERS
LADYSMITH, WISCONSIN
DR. BY J.D. REYNOLDS

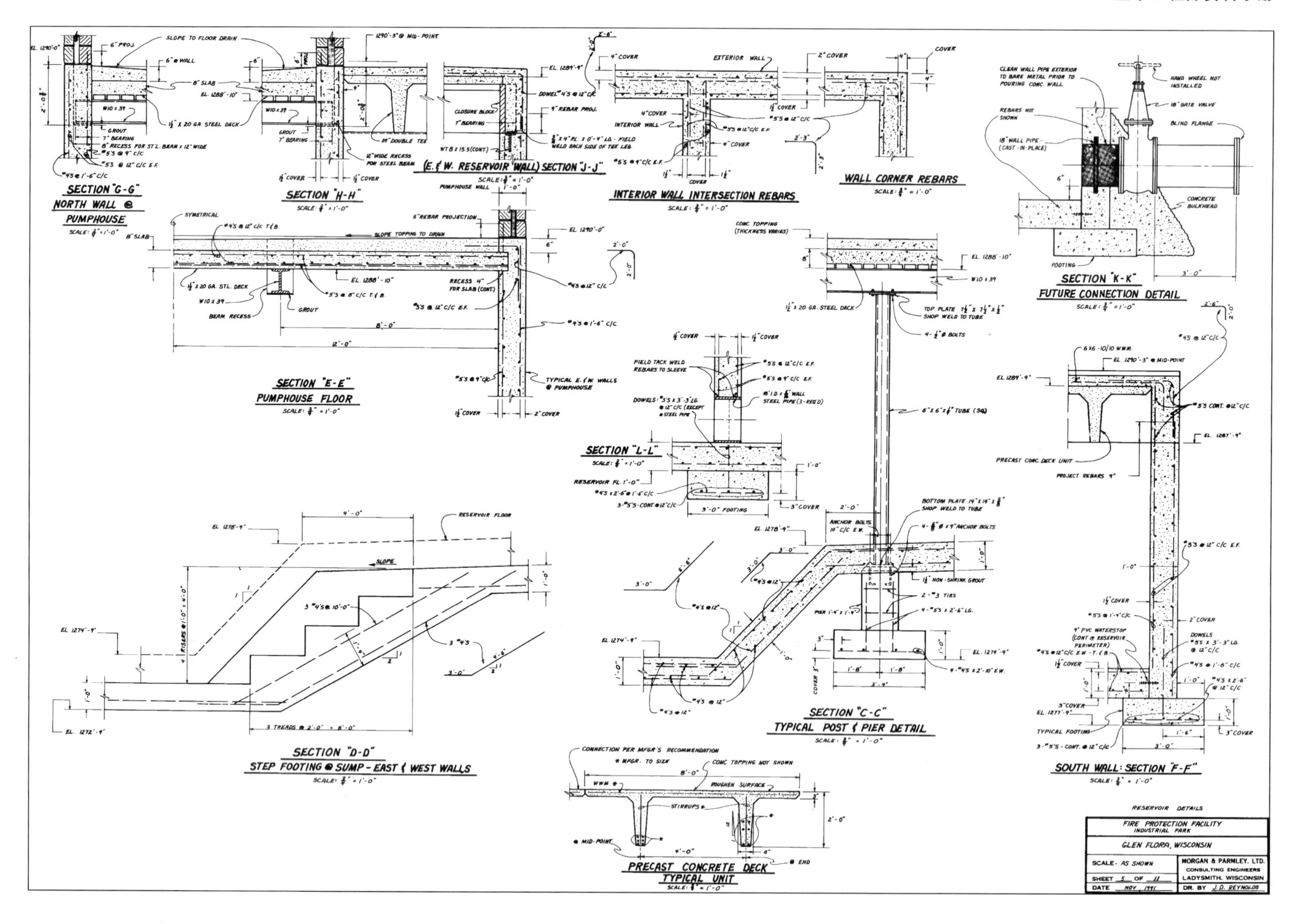
SECTION "G-G"
NORTH WALL @ PUMPHOUSE
SECTION "H-H"
(E. & W. RESERVOIR WALL) SECTION "J-J"
INTERIOR WALL INTERSECTION REBARS
WALL CORNER REBARS
SECTION "K-K"
FUTURE CONNECTION DETAIL
SECTION "E-E"
PUMPHOUSE FLOOR
SECTION "L-L"
SECTION "C-C"
TYPICAL POST & PIER DETAIL
SECTION "D-D"
STEP FOOTING @ SUMP - EAST & WEST WALLS
SOUTH WALL: SECTION "F-F"
PRECAST CONCRETE DECK
TYPICAL UNIT
RESERVOIR DETAILS
FIRE PROTECTION FACILITY
INDUSTRIAL PARK
GLEN FLORA, WISCONSIN
SCALE: AS SHOWN
SHEET 5 OF 11
DATE NOV 1991
MORGAN & PARMLEY, LTD.
CONSULTING ENGINEERS
LADYSMITH, WISCONSIN
DR. BY J.D. REYNOLDS

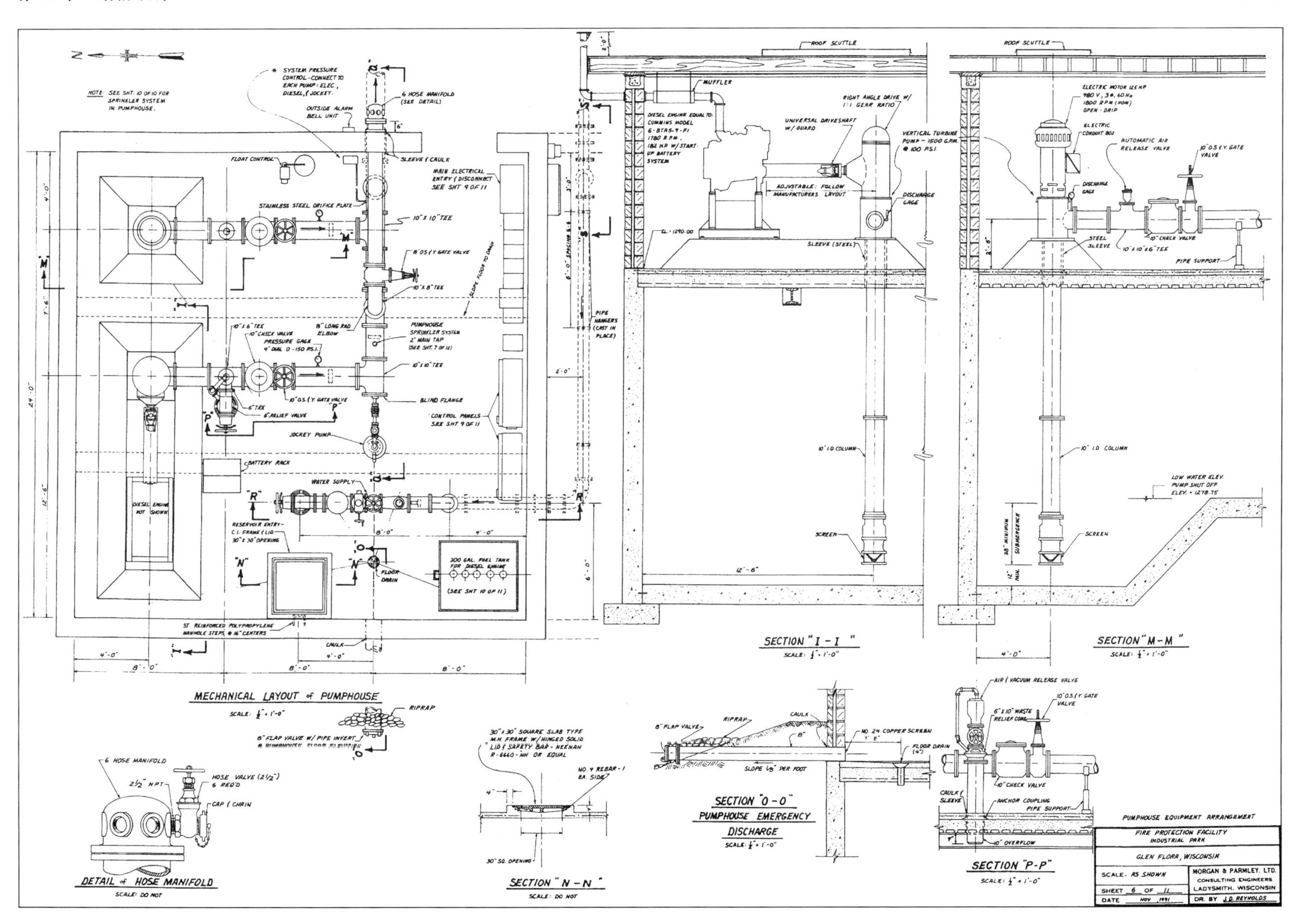
MECHANICAL LAYOUT of PUMPHOUSE
SCALE: 1/2" = 1'-0"
SECTION "I - I"
SCALE: 1/4" = 1'-0"
SECTION "M-M"
SCALE: 1/4" = 1'-0"
DETAIL of HOSE MANIFOLD
SCALE: DO NOT
SECTION "N - N"
SCALE: DO NOT
SECTION "O - O"
PUMPHOUSE EMERGENCY
DISCHARGE
SCALE: 1/4" = 1'-0"
SECTION "P-P"
SCALE: 1/4" = 1'-0"
PUMPHOUSE EQUIPMENT ARRANGEMENT
FIRE PROTECTION FACILITY
INDUSTRIAL PARK
GLEN FLORA, WISCONSIN
MORGAN & PARMLEY, LTD.
CONSULTING ENGINEERS
LADYSMITH, WISCONSIN
ROOF SCUTTLE
MUFFLER
RIGHT ANGLE DRIVE W/ 1:1 GEAR RATIO
UNIVERSAL DRIVESHAFT W/ GUARD
VERTICAL TURBINE PUMP - 1500 G.P.M. @ 100 P.S.I.
DISCHARGE GAGE
10" I.D. COLUMN
SCREEN
FLOAT CONTROL
STAINLESS STEEL ORIFICE PLATE
10" x 10" TEE
8" O.S.& Y. GATE VALVE
JOCKEY PUMP
BATTERY RACK
WATER SUPPLY
BLIND FLANGE
RIPRAP
6 HOSE MANIFOLD
HOSE VALVE (2 1/2") 6 REQ'D
CAP & CHAIN
10" CHECK VALVE
AIR & VACUUM RELEASE VALVE
ANCHOR COUPLING
PIPE SUPPORT
10" OVERFLOW

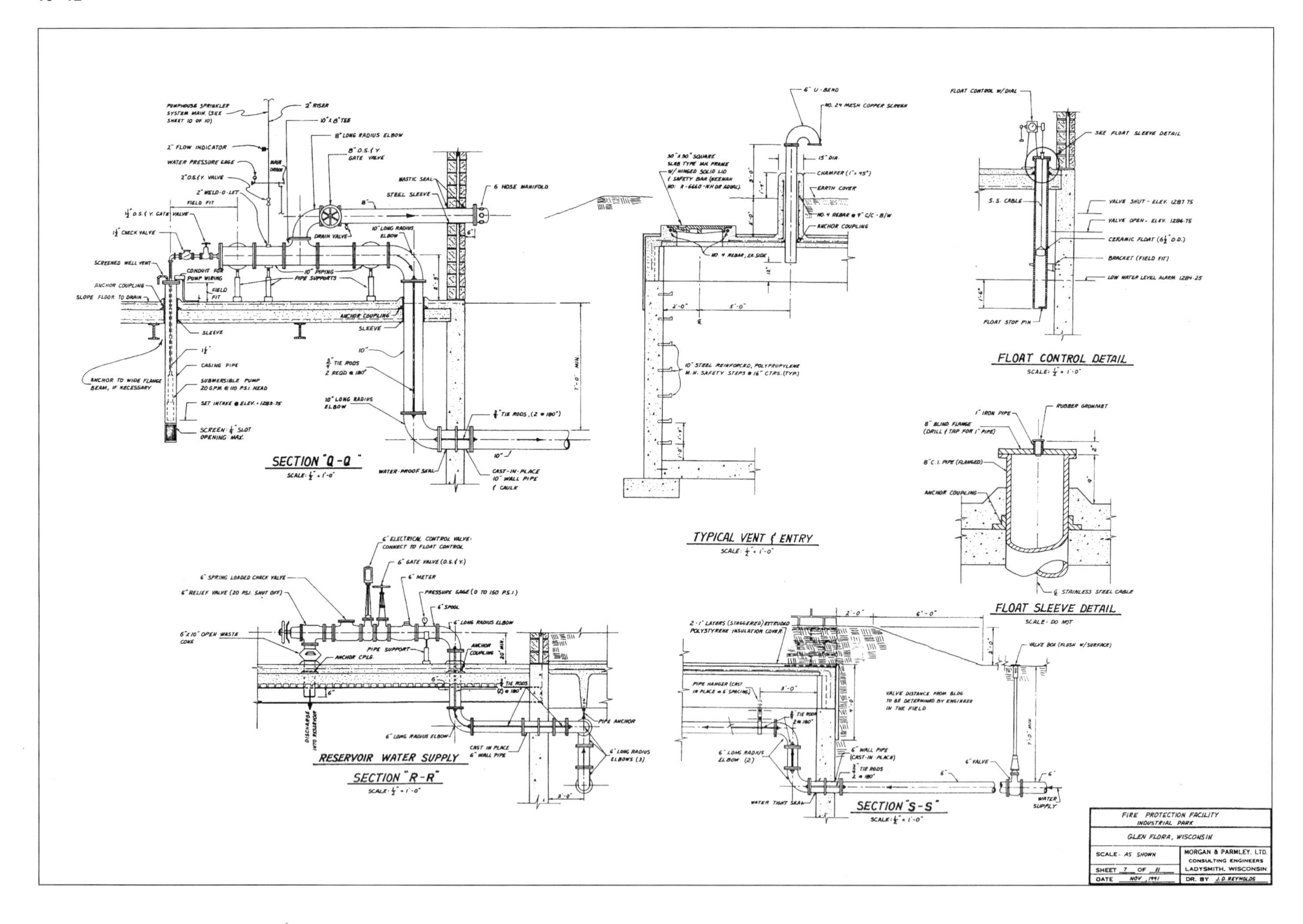
SECTION "Q-Q"
SCALE: 1/2" = 1'-0"
TYPICAL VENT & ENTRY
SCALE: 1/2" = 1'-0"
FLOAT CONTROL DETAIL
SCALE: 1/2" = 1'-0"
FLOAT SLEEVE DETAIL
SCALE: DO NOT
RESERVOIR WATER SUPPLY
SECTION "R-R"
SCALE: 1/2" = 1'-0"
SECTION "S-S"
SCALE: 1/2" = 1'-0"
FIRE PROTECTION FACILITY
INDUSTRIAL PARK
GLEN FLORA, WISCONSIN
SCALE - AS SHOWN
SHEET 7 OF 11
DATE NOV, 1991
MORGAN & PARMLEY, LTD.
CONSULTING ENGINEERS
LADYSMITH, WISCONSIN
DR. BY J.D. REYNOLDS

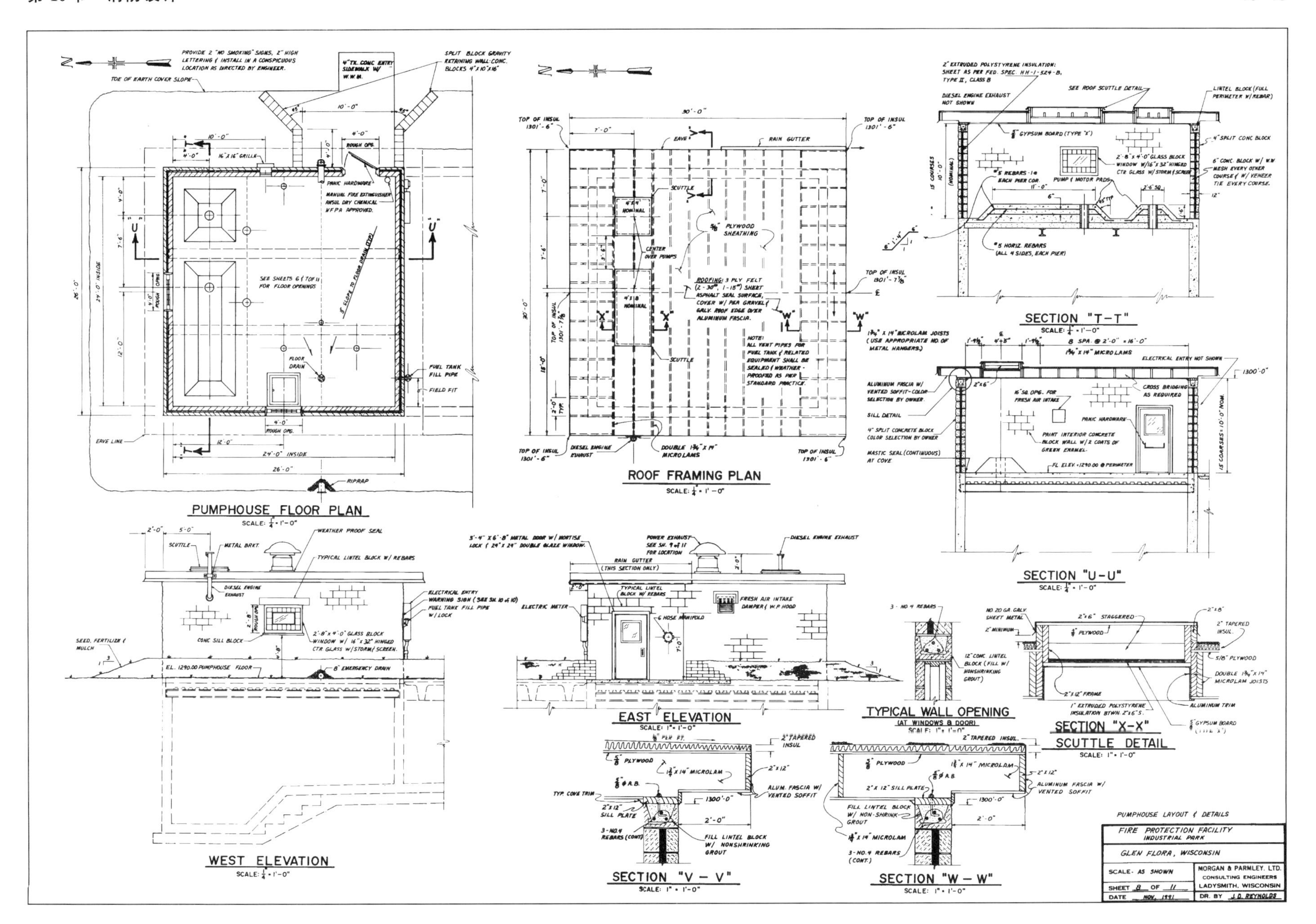
PUMPHOUSE FLOOR PLAN
SCALE: 1/4" = 1'-0"
ROOF FRAMING PLAN
SCALE: 1/4" = 1'-0"
SECTION "T-T"
SCALE: 1/4" = 1'-0"
SECTION "U-U"
SCALE: 1/4" = 1'-0"
WEST ELEVATION
SCALE: 1/4" = 1'-0"
EAST ELEVATION
SCALE: 1" = 1'-0"
TYPICAL WALL OPENING
(AT WINDOWS & DOOR)
SECTION "X-X"
SCUTTLE DETAIL
SCALE: 1" = 1'-0"
SECTION "V-V"
SCALE: 1" = 1'-0"
SECTION "W-W"
SCALE: 1" = 1'-0"
PUMPHOUSE LAYOUT & DETAILS
FIRE PROTECTION FACILITY
INDUSTRIAL PARK
GLEN FLORA, WISCONSIN
SCALE- AS SHOWN
MORGAN & PARMLEY, LTD.
CONSULTING ENGINEERS
LADYSMITH, WISCONSIN
SHEET 8 OF 11
DATE NOV. 1991
DR. BY J.D. REYNOLDS

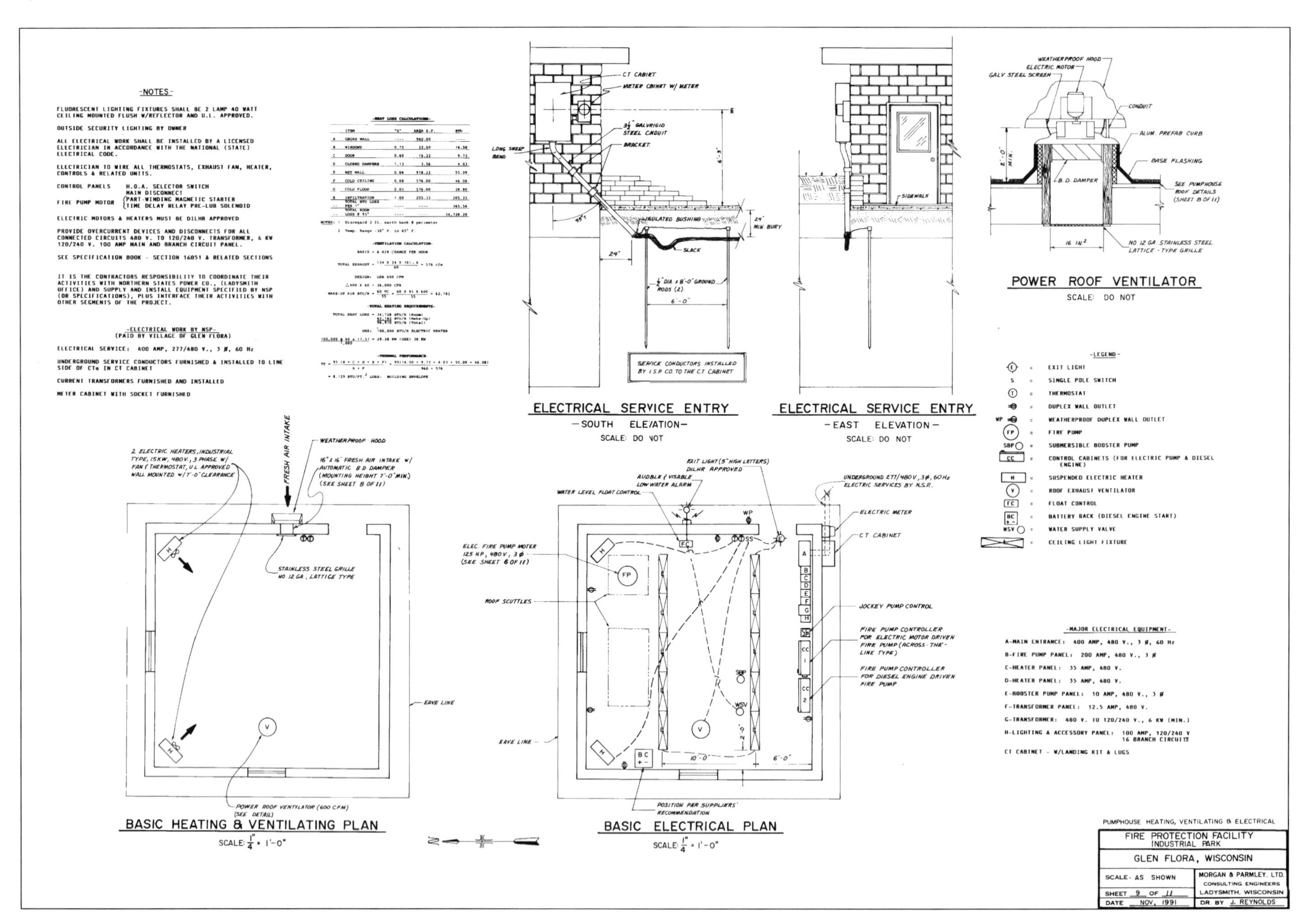
-NOTES-
FLUORESCENT LIGHTING FIXTURES SHALL BE 2 LAMP 40 WATT CEILING MOUNTED FLUSH W/REFLECTOR AND U.L. APPROVED.
OUTSIDE SECURITY LIGHTING BY OWNER
ALL ELECTRICAL WORK SHALL BE INSTALLED BY A LICENSED ELECTRICIAN IN ACCORDANCE WITH THE NATIONAL (STATE) ELECTRICAL CODE.
ELECTRICIAN TO WIRE ALL THERMOSTATS, EXHAUST FAN, HEATER, CONTROLS & RELATED UNITS.
CONTROL PANELS H.O.A. SELECTOR SWITCH
MAIN DISCONNECT
FIRE PUMP MOTOR PART-WINDING MAGNETIC STARTER
TIME DELAY RELAY PRE-LUB SOLENOID
ELECTRIC MOTORS & HEATERS MUST BE DILHR APPROVED
PROVIDE OVERCURRENT DEVICES AND DISCONNECTS FOR ALL CONNECTED CIRCUITS 480 V. TO 120/240 V. TRANSFORMER, 6 KW 120/240 V. 100 AMP MAIN AND BRANCH CIRCUIT PANEL.
SEE SPECIFICATION BOOK - SECTION 16051 & RELATED SECTIONS
IT IS THE CONTRACTORS RESPONSIBILITY TO COORDINATE THEIR ACTIVITIES WITH NORTHERN STATES POWER CO., (LADYSMITH OFFICE) AND SUPPLY AND INSTALL EQUIPMENT SPECIFIED BY NSP (OR SPECIFICATIONS), PLUS INTERFACE THEIR ACTIVITIES WITH OTHER SEGMENTS OF THE PROJECT.
-ELECTRICAL WORK BY NSP-
(PAID BY VILLAGE OF GLEN FLORA)
ELECTRICAL SERVICE: 400 AMP, 277/480 V., 3 Ø, 60 Hz
UNDERGROUND SERVICE CONDUCTORS FURNISHED & INSTALLED TO LINE SIDE OF CTs IN CT CABINET
CURRENT TRANSFORMERS FURNISHED AND INSTALLED
METER CABINET WITH SOCKET FURNISHED
CT CABIET
METER CBINKT W/ METER
3½" GALVRIGID STEEL CNDUIT
BRACKET
LONG SWEEP BEND
INSULATED BUSHING
SLACK
24" MIN BURY
½" DIA x 8'-0" GROUND RODS (2)
6'-0"
SERVICE CONDUCTORS INSTALLED BY I.S.P. CO. TO THE C.T. CABINET
ELECTRICAL SERVICE ENTRY
-SOUTH ELEVATION-
SCALE: DO NOT
SIDEWALK
ELECTRICAL SERVICE ENTRY
-EAST ELEVATION-
SCALE: DO NOT
WEATHERPROOF HOOD
ELECTRIC MOTOR
GALV. STEEL SCREEN
CONDUIT
ALUM. PREFAB CURB
BASE FLASHING
B.D. DAMPER
SEE PUMPHOUSE ROOF DETAILS (SHEET 8 OF 11)
16 IN²
NO. 12 GA. STAINLESS STEEL LATTICE-TYPE GRILLE
POWER ROOF VENTILATOR
SCALE: DO NOT
-LEGEND-
EXIT LIGHT
SINGLE POLE SWITCH
THERMOSTAT
DUPLEX WALL OUTLET
WEATHERPROOF DUPLEX WALL OUTLET
FIRE PUMP
SUBMERSIBLE BOOSTER PUMP
CONTROL CABINETS (FOR ELECTRIC PUMP & DIESEL ENGINE)
SUSPENDED ELECTRIC HEATER
ROOF EXHAUST VENTILATOR
FLOAT CONTROL
BATTERY BACK (DIESEL ENGINE START)
WATER SUPPLY VALVE
CEILING LIGHT FIXTURE
FRESH AIR INTAKE
WEATHERPROOF HOOD
2 ELECTRIC HEATERS, INDUSTRIAL TYPE, 15KW, 480V., 3 PHASE W/ FAN, THERMOSTAT, U.L. APPROVED WALL MOUNTED W/7'-0" CLEARANCE
16"x16" FRESH AIR INTAKE W/ AUTOMATIC B.D. DAMPER (MOUNTING HEIGHT 7'-0" MIN.) (SEE SHEET 8 OF 11)
STAINLESS STEEL GRILLE NO. 12 GA., LATTICE TYPE
EAVE LINE
POWER ROOF VENTILATOR (600 CFM) (SEE DETAIL)
BASIC HEATING & VENTILATING PLAN
SCALE: 1/4" = 1'-0"
EXIT LIGHT (5" HIGH LETTERS) DILHR APPROVED
AUDIBLE & VISABLE LOW WATER ALARM
WATER LEVEL FLOAT CONTROL
UNDERGROUND 277/480V, 3Ø, 60Hz ELECTRIC SERVICES BY N.S.P.
ELECTRIC METER
CT CABINET
ELEC. FIRE PUMP MOTER 125 HP, 480V, 3Ø (SEE SHEET 6 OF 11)
ROOF SCUTTLES
JOCKEY PUMP CONTROL
FIRE PUMP CONTROLLER FOR ELECTRIC MOTOR DRIVEN FIRE PUMP (ACROSS-THE-LINE TYPE)
FIRE PUMP CONTROLLER FOR DIESEL ENGINE DRIVEN FIRE PUMP
EAVE LINE
POSITION PER SUPPLIERS' RECOMMENDATION
BASIC ELECTRICAL PLAN
SCALE: 1/4" = 1'-0"
-MAJOR ELECTRICAL EQUIPMENT-
A-MAIN ENTRANCE: 400 AMP, 480 V., 3 Ø, 60 Hz
B-FIRE PUMP PANEL: 200 AMP, 480 V., 3 Ø
C-HEATER PANEL: 35 AMP, 480 V.
D-HEATER PANEL: 35 AMP, 480 V.
E-BOOSTER PUMP PANEL: 10 AMP, 480 V., 3 Ø
F-TRANSFORMER PANEL: 12.5 AMP, 480 V.
G-TRANSFORMER: 480 V. TO 120/240 V., 6 KW (MIN.)
H-LIGHTING & ACCESSORY PANEL: 100 AMP, 120/240 V 16 BRANCH CIRCUITS
CT CABINET - W/LANDING KIT & LUGS
PUMPHOUSE HEATING, VENTILATING & ELECTRICAL
FIRE PROTECTION FACILITY
INDUSTRIAL PARK
GLEN FLORA, WISCONSIN
SCALE- AS SHOWN
SHEET 9 OF 11
DATE NOV. 1991
MORGAN & PARMLEY, LTD.
CONSULTING ENGINEERS
LADYSMITH, WISCONSIN
DR. BY J. REYNOLDS

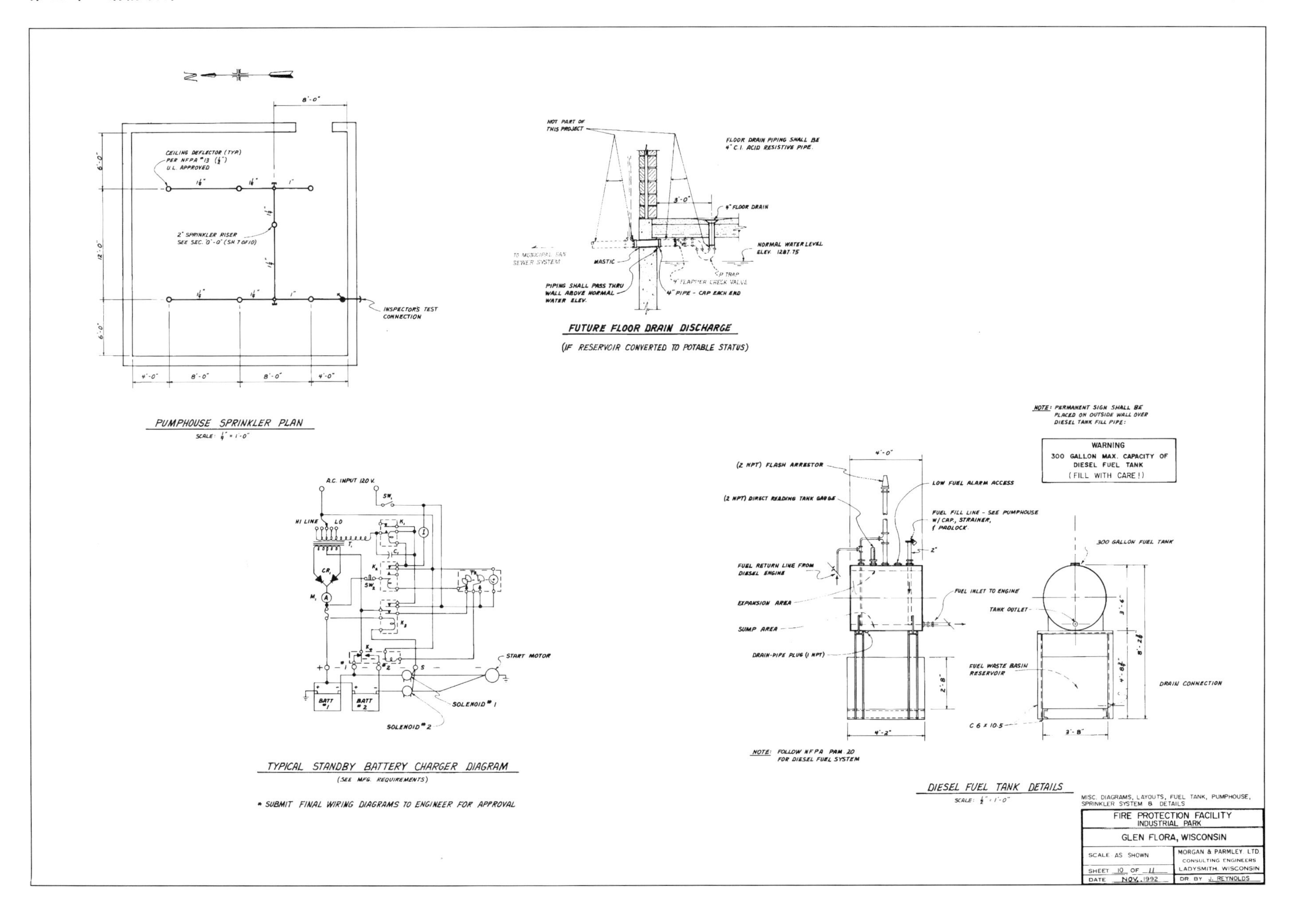
CEILING DEFLECTOR (TYP)
PER NFPA #13 (1/2")
U.L. APPROVED
2" SPRINKLER RISER
SEE SEC. 'O'-'O' (SH. 7 OF 10)
INSPECTORS TEST
CONNECTION
PUMPHOUSE SPRINKLER PLAN
SCALE: 1/4" = 1'-0"
NOT PART OF
THIS PROJECT
FLOOR DRAIN PIPING SHALL BE
4" C.I. ACID RESISTIVE PIPE.
4" FLOOR DRAIN
NORMAL WATER LEVEL
ELEV. 1287.75
TO MUNICIPAL SAN
SEWER SYSTEM
MASTIC
P TRAP
4" FLAPPER CHECK VALVE
PIPING SHALL PASS THRU
WALL ABOVE NORMAL
WATER ELEV.
4" PIPE - CAP EACH END
FUTURE FLOOR DRAIN DISCHARGE
(IF RESERVOIR CONVERTED TO POTABLE STATUS)
A.C. INPUT 120 V.
HI LINE
LO
START MOTOR
SOLENOID # 1
SOLENOID # 2
BATT #1
BATT #2
TYPICAL STANDBY BATTERY CHARGER DIAGRAM
(SEE MFG. REQUIREMENTS)
* SUBMIT FINAL WIRING DIAGRAMS TO ENGINEER FOR APPROVAL
NOTE: PERMANENT SIGN SHALL BE
PLACED ON OUTSIDE WALL OVER
DIESEL TANK FILL PIPE:
WARNING
300 GALLON MAX. CAPACITY OF
DIESEL FUEL TANK
(FILL WITH CARE!)
(2 NPT) FLASH ARRESTOR
(2 NPT) DIRECT READING TANK GAUGE
LOW FUEL ALARM ACCESS
FUEL FILL LINE - SEE PUMPHOUSE
W/ CAP., STRAINER,
& PADLOCK.
300 GALLON FUEL TANK
FUEL RETURN LINE FROM
DIESEL ENGINE
EXPANSION AREA
SUMP AREA
DRAIN-PIPE PLUG (1 NPT)
FUEL INLET TO ENGINE
TANK OUTLET
FUEL WASTE BASIN
RESERVOIR
DRAIN CONNECTION
C 6 x 10.5
NOTE: FOLLOW NFPA PAM. 20
FOR DIESEL FUEL SYSTEM
DIESEL FUEL TANK DETAILS
SCALE: 1/2" = 1'-0"
MISC. DIAGRAMS, LAYOUTS, FUEL TANK, PUMPHOUSE,
SPRINKLER SYSTEM & DETAILS
FIRE PROTECTION FACILITY
INDUSTRIAL PARK
GLEN FLORA, WISCONSIN
SCALE AS SHOWN
SHEET 10 OF 11
DATE NOV., 1992
MORGAN & PARMLEY LTD.
CONSULTING ENGINEERS
LADYSMITH, WISCONSIN
DR. BY J. REYNOLDS

设　计

第 11 节　污水收集和处理

例　11

为了保证公共健康，安全地收集并恰当地处理市政产生的污水是非常重要的。最后的排出物必须经过处理达到能排放到地表水或者地下水而不污染环境的标准。

所有新建的污水处理设施和对已有污水处理厂的改造都必须遵循严格的管理制度和规范。初期规划通常称为第一步或者废水处理设施初步设计过程。此多个方面研究的目的是在开始着手准备施工设计图之前，调查所有可行的处理方案并选择成本效益最高和环境最完好的设计。

本节的第一套设计图是服务于 5000 人社区得带紫外线消毒的三箱曝气蓄污池。主抽水站和压力干管将最初的污水从市政管道中抽运到偏远的污水处理厂，这些都包含在设计图中。第二个例子是一个服务于少于 60 栋建筑的农村的小型可选/创新污水收集和处理设施。这套系统开始于单个独立的污水处理池，其污泥通过小直径重力/压力收集系统被运到下层的快速砂滤。第三套设计图给出了对一现有工业污水处理厂的建议改造。最后两套图纸示例是对现有污水收集系统的扩建工程。读者应当注意到，这些设计图还包括了供水干管的扩建，这在市政服务设施扩建工程中是常见的。

目　录

M & P PROJECT NO.
(DESIGN) 87-110
(CONSTRUCTION) 88-161

污水处理设施、抽水站和压力干管

WASTEWATER TREATMENT FACILITY, LIFT STATION & FORCEMAIN

LADYSMITH, WISCONSIN
APRIL, 1988

PROJECT NO. 0551098-02 (DESIGN)
0551098-03 (CONSTRUCTION)

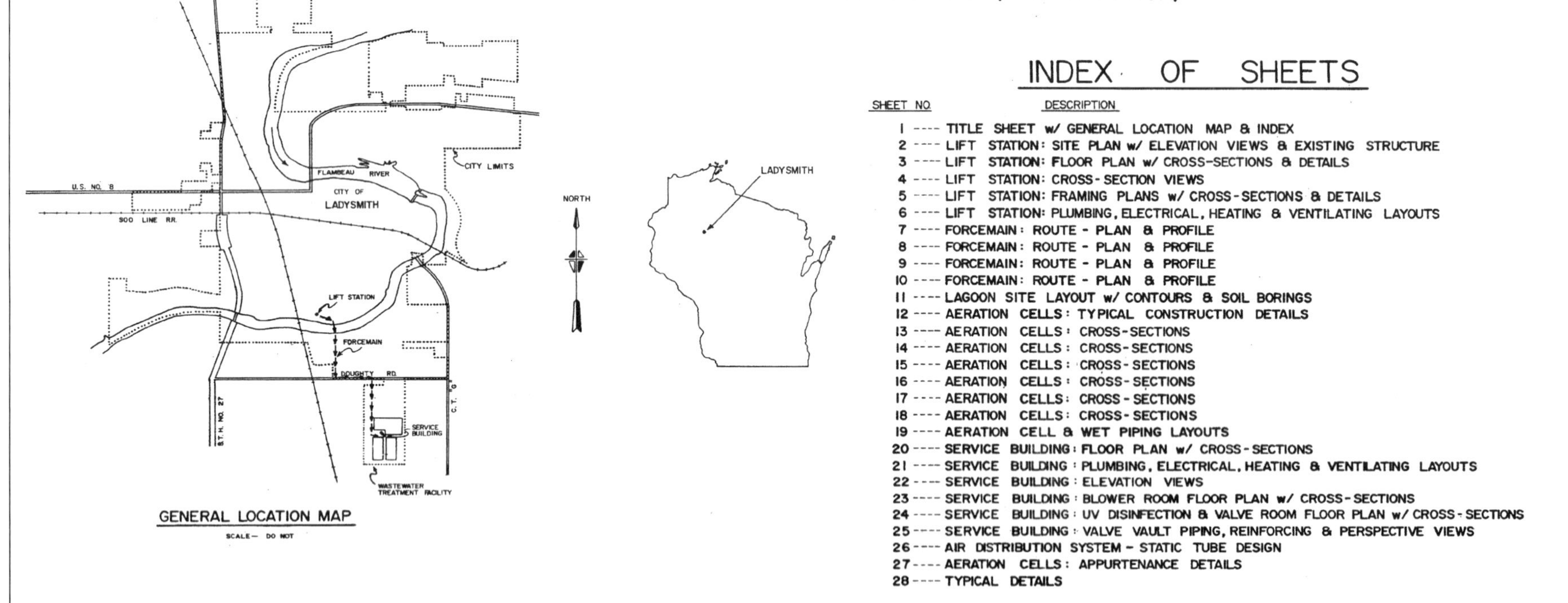

INDEX OF SHEETS

SHEET NO.	DESCRIPTION
1	TITLE SHEET w/ GENERAL LOCATION MAP & INDEX
2	LIFT STATION: SITE PLAN w/ ELEVATION VIEWS & EXISTING STRUCTURE
3	LIFT STATION: FLOOR PLAN w/ CROSS-SECTIONS & DETAILS
4	LIFT STATION: CROSS-SECTION VIEWS
5	LIFT STATION: FRAMING PLANS w/ CROSS-SECTIONS & DETAILS
6	LIFT STATION: PLUMBING, ELECTRICAL, HEATING & VENTILATING LAYOUTS
7	FORCEMAIN: ROUTE - PLAN & PROFILE
8	FORCEMAIN: ROUTE - PLAN & PROFILE
9	FORCEMAIN: ROUTE - PLAN & PROFILE
10	FORCEMAIN: ROUTE - PLAN & PROFILE
11	LAGOON SITE LAYOUT w/ CONTOURS & SOIL BORINGS
12	AERATION CELLS: TYPICAL CONSTRUCTION DETAILS
13	AERATION CELLS: CROSS-SECTIONS
14	AERATION CELLS: CROSS-SECTIONS
15	AERATION CELLS: CROSS-SECTIONS
16	AERATION CELLS: CROSS-SECTIONS
17	AERATION CELLS: CROSS-SECTIONS
18	AERATION CELLS: CROSS-SECTIONS
19	AERATION CELL & WET PIPING LAYOUTS
20	SERVICE BUILDING: FLOOR PLAN w/ CROSS-SECTIONS
21	SERVICE BUILDING: PLUMBING, ELECTRICAL, HEATING & VENTILATING LAYOUTS
22	SERVICE BUILDING: ELEVATION VIEWS
23	SERVICE BUILDING: BLOWER ROOM FLOOR PLAN w/ CROSS-SECTIONS
24	SERVICE BUILDING: UV DISINFECTION & VALVE ROOM FLOOR PLAN w/ CROSS-SECTIONS
25	SERVICE BUILDING: VALVE VAULT PIPING, REINFORCING & PERSPECTIVE VIEWS
26	AIR DISTRIBUTION SYSTEM - STATIC TUBE DESIGN
27	AERATION CELLS: APPURTENANCE DETAILS
28	TYPICAL DETAILS

PREPARED BY: MORGAN & PARMLEY LTD. LADYSMITH, WISCONSIN

Source: Morgan & Parmley, Ltd.

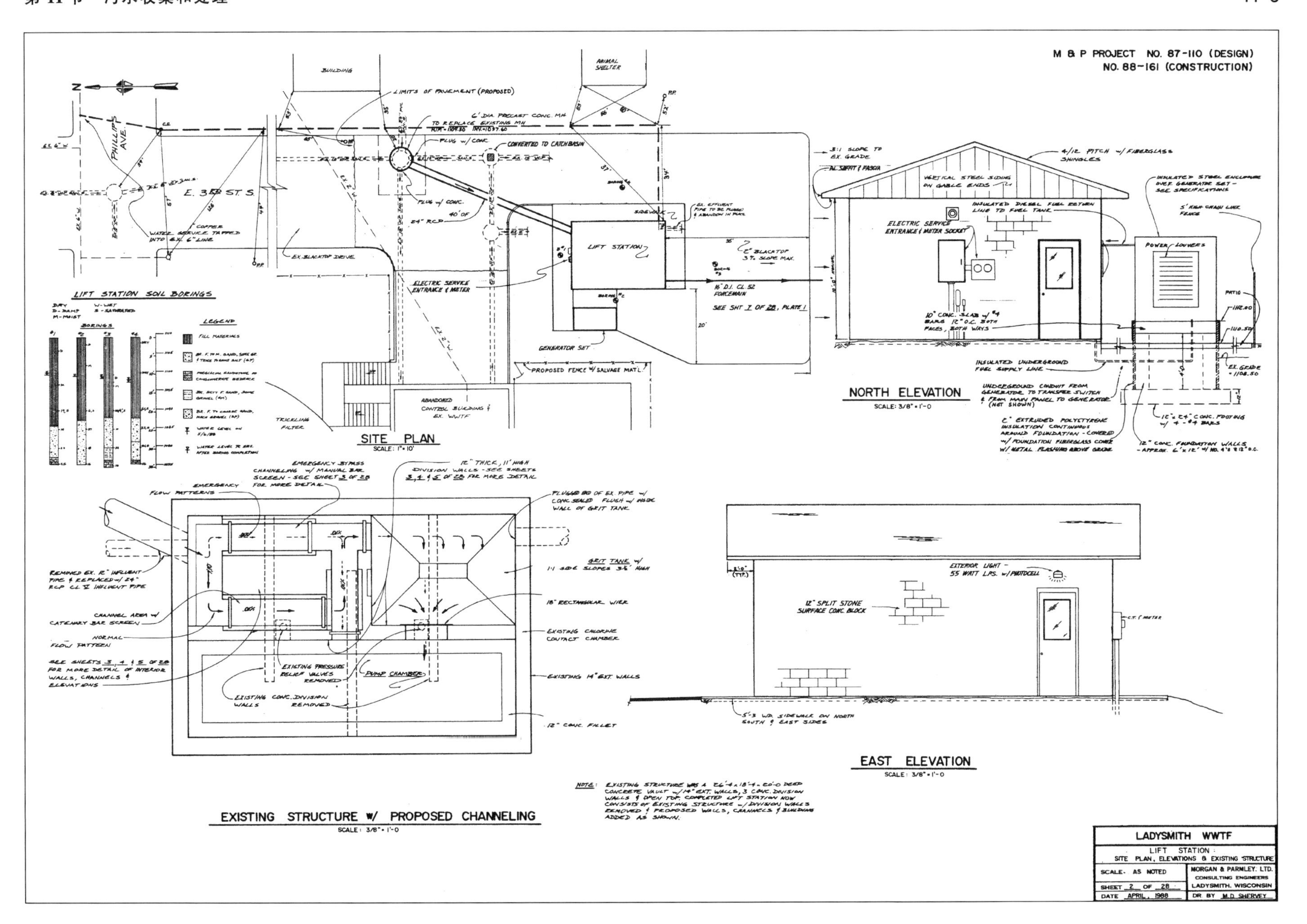
M & P PROJECT NO. 87-110 (DESIGN)
NO. 88-161 (CONSTRUCTION)
SITE PLAN
SCALE: 1"=10'
NORTH ELEVATION
SCALE: 3/8"=1'-0
EAST ELEVATION
SCALE: 3/8"=1'-0
EXISTING STRUCTURE W/ PROPOSED CHANNELING
SCALE: 3/8"=1'-0
LIFT STATION SOIL BORINGS
LIFT STATION
GENERATOR SET
PROPOSED FENCE W/ SALVAGE MATL
CONVERTED TO CATCH BASIN
LADYSMITH WWTF
LIFT STATION:
SITE PLAN, ELEVATIONS & EXISTING STRUCTURE
SCALE- AS NOTED
MORGAN & PARMLEY, LTD.
CONSULTING ENGINEERS
LADYSMITH, WISCONSIN
SHEET 2 OF 28
DATE APRIL, 1988
DR BY M.D. SHERVEY

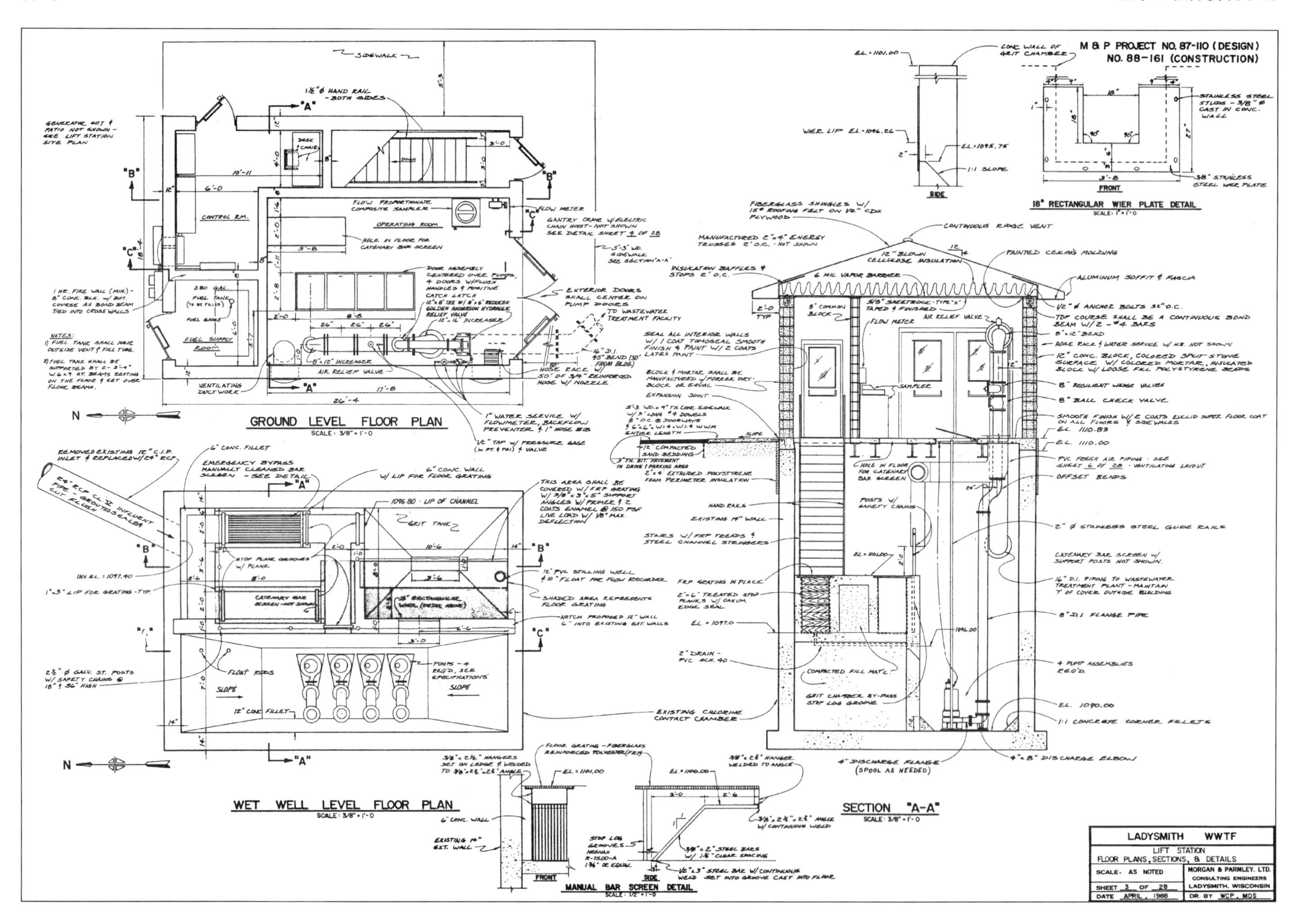

M & P PROJECT NO. 87-110 (DESIGN)
NO. 88-161 (CONSTRUCTION)
GROUND LEVEL FLOOR PLAN
SCALE: 3/8" = 1'-0
WET WELL LEVEL FLOOR PLAN
SCALE: 3/8" = 1'-0
MANUAL BAR SCREEN DETAIL
SCALE: 1/2" = 1'-0
SECTION "A-A"
SCALE: 3/8" = 1'-0
18" RECTANGULAR WIER PLATE DETAIL
SCALE: 1" = 1'-0
SIDE
FRONT
CONTROL RM.
OPERATING ROOM
VENTILATING DUCT WORK
AIR RELIEF VALVE
EXISTING CHLORINE CONTACT CHAMBER
LADYSMITH WWTF
LIFT STATION
FLOOR PLANS, SECTIONS, & DETAILS
SCALE- AS NOTED
MORGAN & PARMLEY, LTD.
CONSULTING ENGINEERS
LADYSMITH, WISCONSIN
SHEET 3 OF 28
DATE APRIL, 1988
DR. BY WCP, MDS

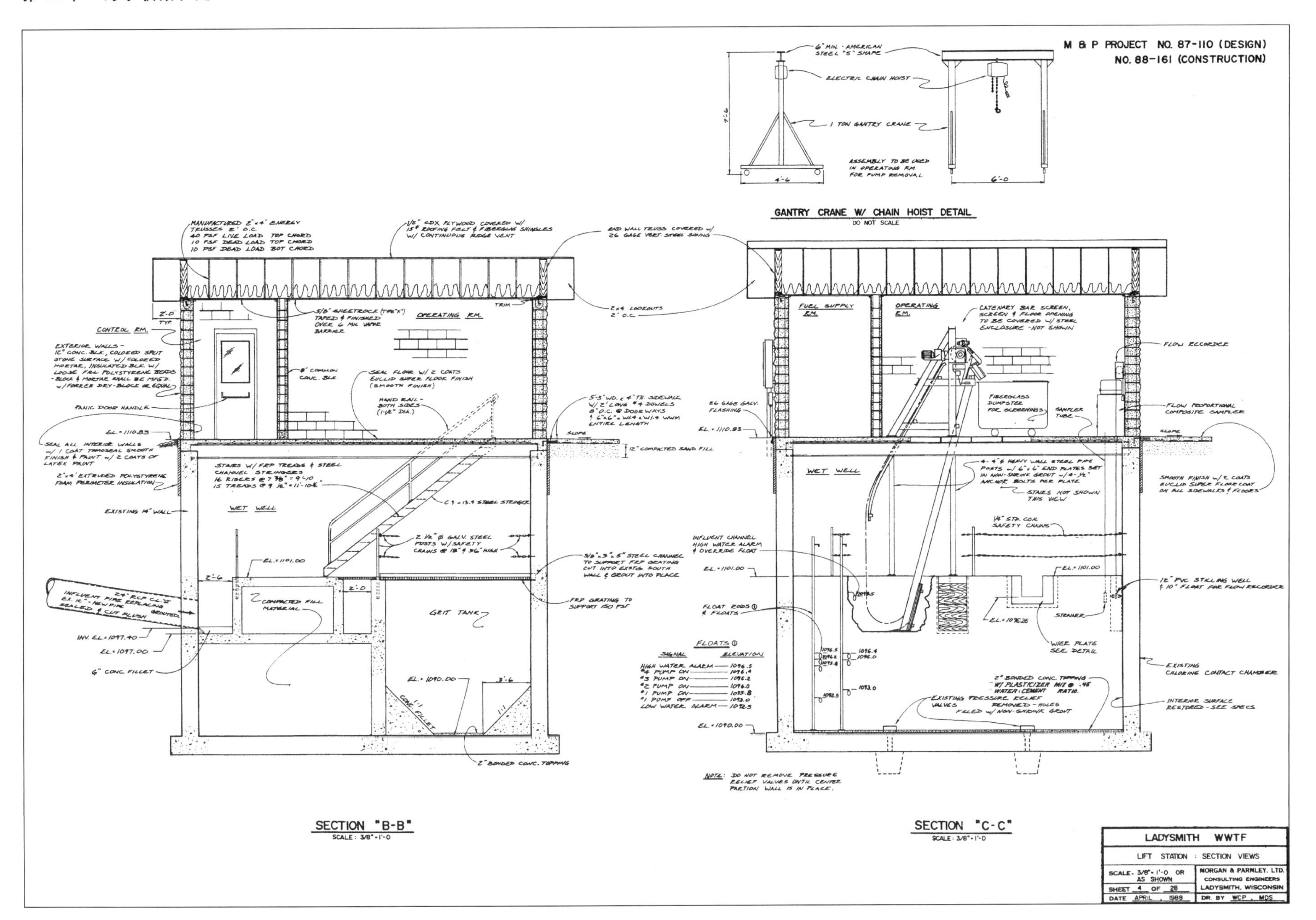

M & P PROJECT NO. 87-110 (DESIGN)
NO. 88-161 (CONSTRUCTION)
GANTRY CRANE W/ CHAIN HOIST DETAIL
DO NOT SCALE
SECTION "B-B"
SCALE: 3/8"=1'-0
SECTION "C-C"
SCALE: 3/8"=1'-0
LADYSMITH WWTF
LIFT STATION : SECTION VIEWS
MORGAN & PARMLEY, LTD.
CONSULTING ENGINEERS
LADYSMITH, WISCONSIN

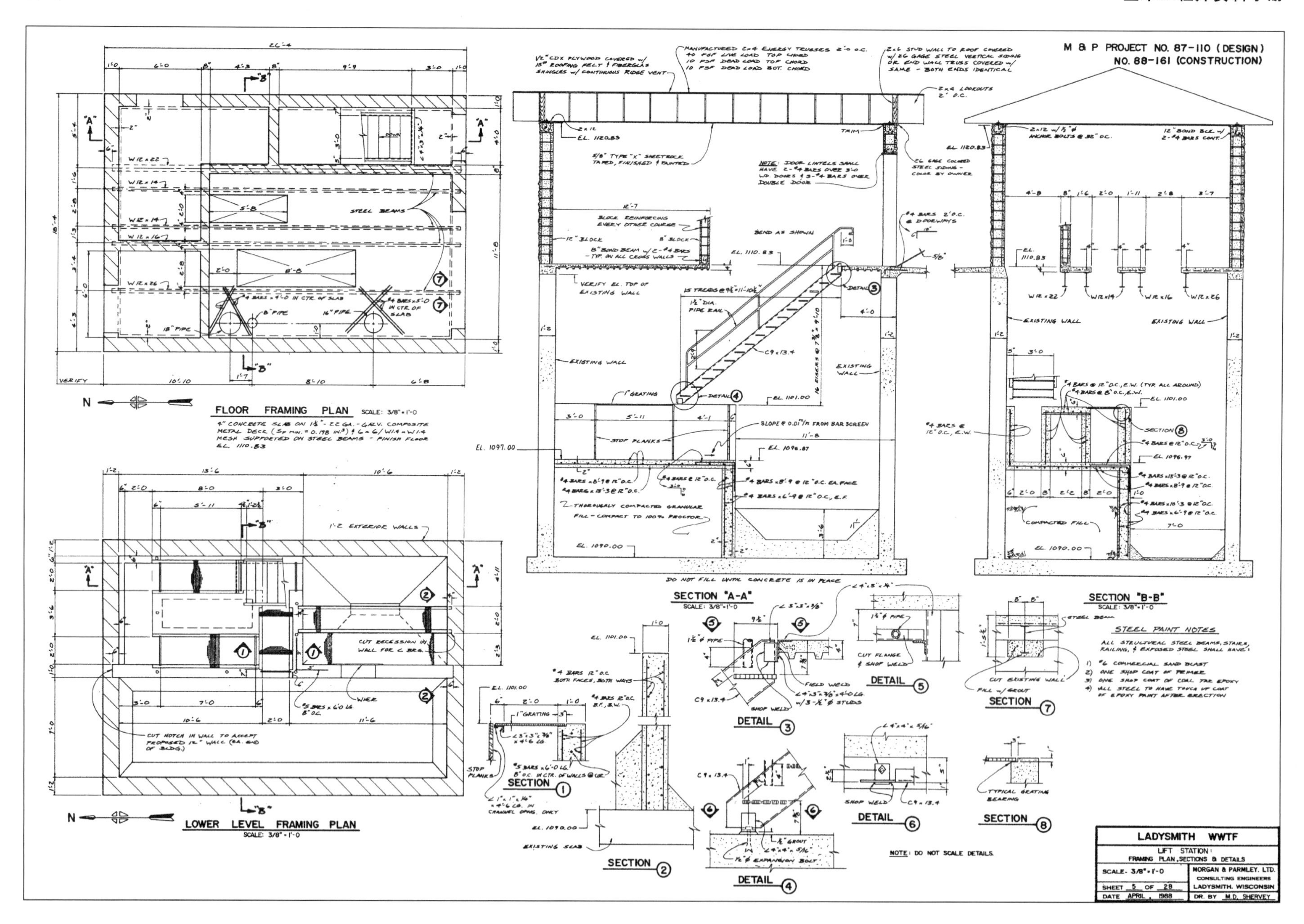
M & P PROJECT NO. 87-110 (DESIGN)
NO. 88-161 (CONSTRUCTION)
FLOOR FRAMING PLAN SCALE: 3/8"=1'-0
LOWER LEVEL FRAMING PLAN SCALE: 3/8"=1'-0
SECTION "A-A" SCALE: 3/8"=1'-0
SECTION "B-B" SCALE: 3/8"=1'-0
STEEL PAINT NOTES
DETAIL 3
DETAIL 4
DETAIL 5
DETAIL 6
SECTION 1
SECTION 2
SECTION 7
SECTION 8
NOTE: DO NOT SCALE DETAILS.
LADYSMITH WWTF
LIFT STATION:
FRAMING PLAN, SECTIONS & DETAILS
SCALE: 3/8"=1'-0
SHEET 5 OF 28
DATE APRIL, 1988
MORGAN & PARMLEY, LTD.
CONSULTING ENGINEERS
LADYSMITH, WISCONSIN
DR. BY M.D. SHERVEY

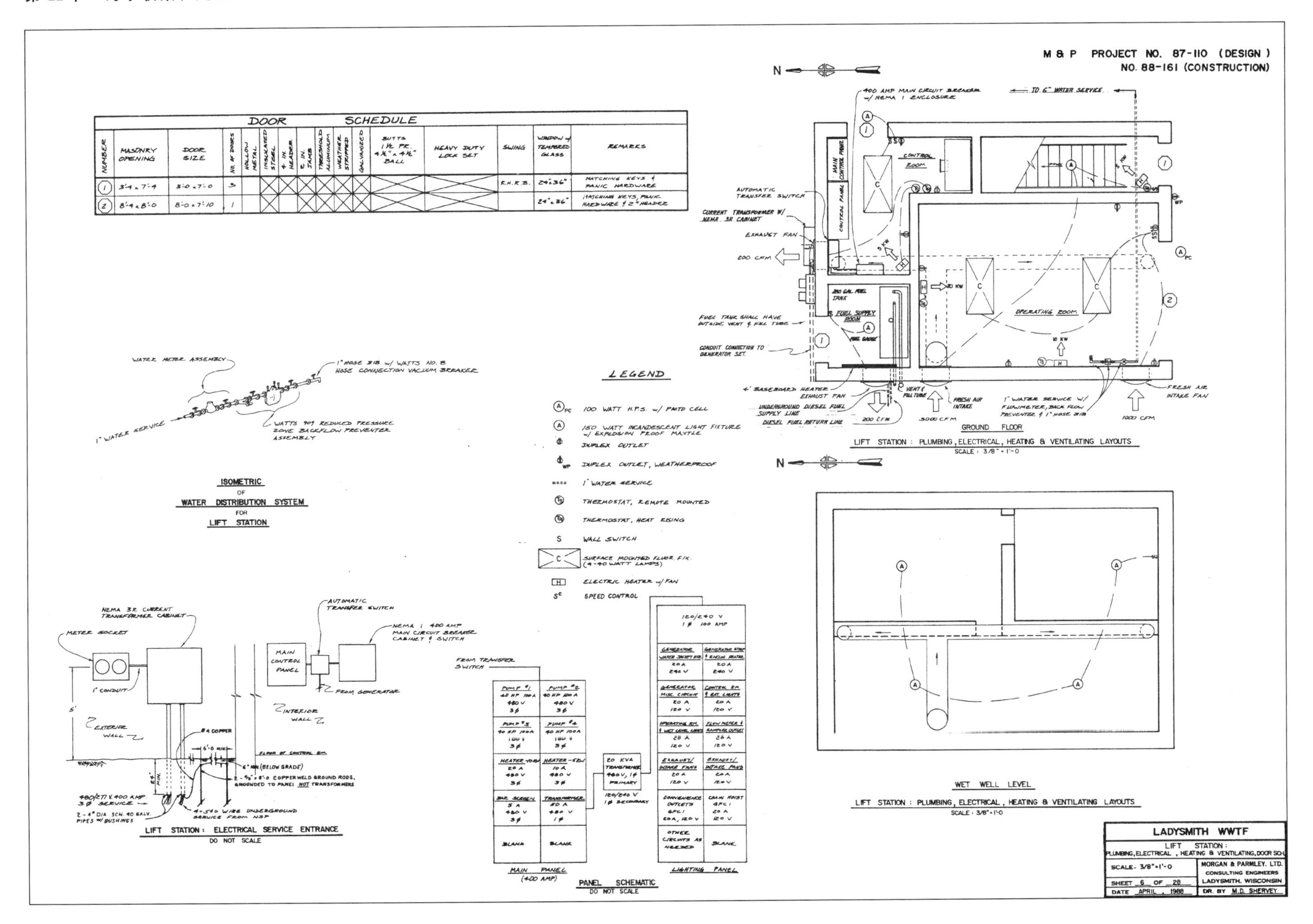
M & P PROJECT NO. 87-110 (DESIGN)
NO. 88-161 (CONSTRUCTION)
DOOR SCHEDULE
ISOMETRIC OF WATER DISTRIBUTION SYSTEM FOR LIFT STATION
LEGEND
100 WATT H.P.S. w/ PHOTO CELL
DUPLEX OUTLET
DUPLEX OUTLET, WEATHERPROOF
1" WATER SERVICE
THERMOSTAT, REMOTE MOUNTED
THERMOSTAT, HEAT RISING
WALL SWITCH
ELECTRIC HEATER w/ FAN
SPEED CONTROL
GROUND FLOOR
LIFT STATION : PLUMBING, ELECTRICAL, HEATING & VENTILATING LAYOUTS
WET WELL LEVEL
LIFT STATION : PLUMBING, ELECTRICAL, HEATING & VENTILATING LAYOUTS
LIFT STATION : ELECTRICAL SERVICE ENTRANCE
DO NOT SCALE
MAIN PANEL
PANEL SCHEMATIC
DO NOT SCALE
LIGHTING PANEL
LADYSMITH WWTF
LIFT STATION :
MORGAN & PARMLEY, LTD.
CONSULTING ENGINEERS
LADYSMITH, WISCONSIN

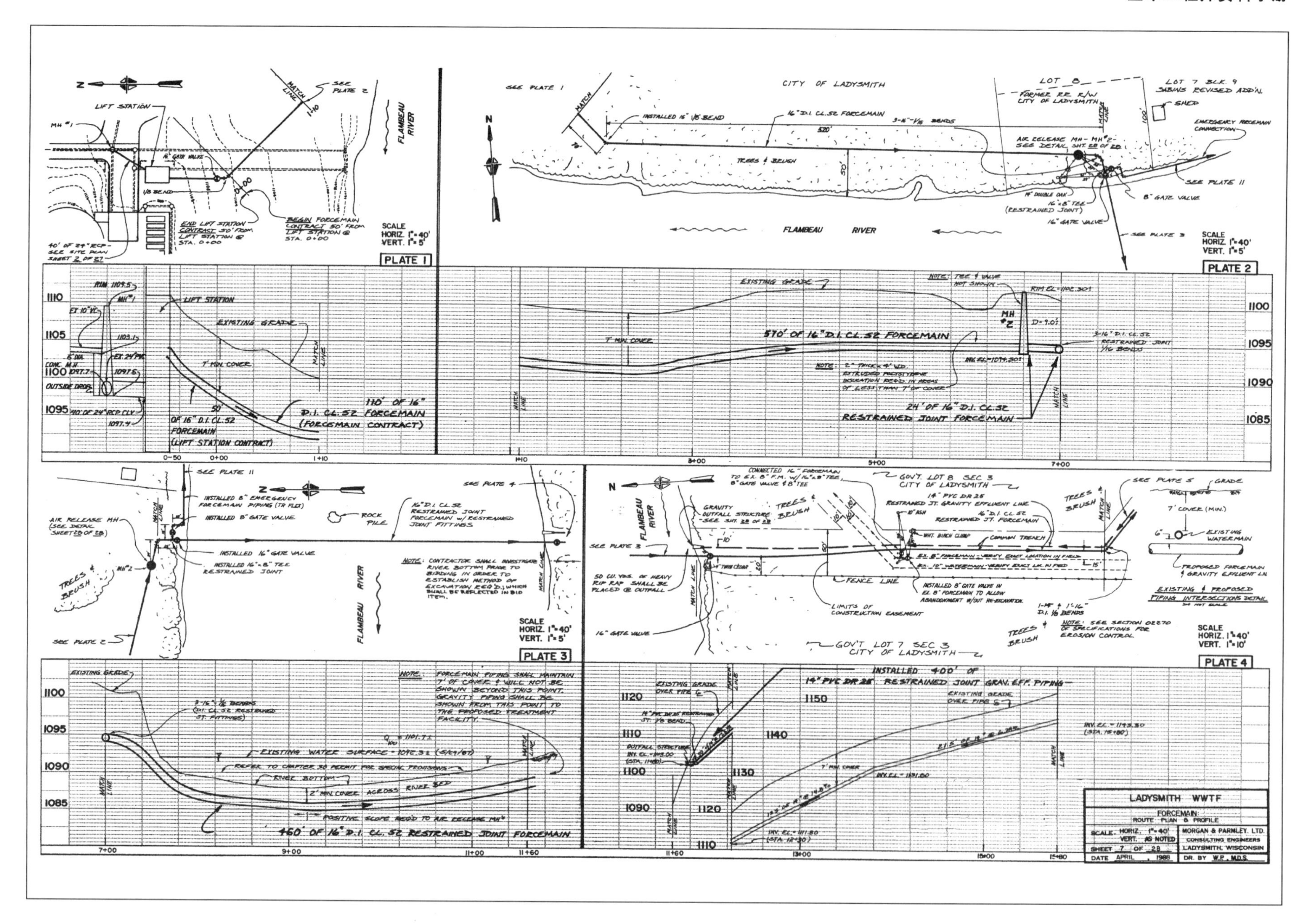
PLATE 1
PLATE 2
PLATE 3
PLATE 4
FLAMBEAU RIVER
CITY OF LADYSMITH
LIFT STATION
EXISTING GRADE
570' OF 16" D.I. CL.52 FORCEMAIN
110' OF 16" D.I. CL.52 FORCEMAIN (FORCEMAIN CONTRACT)
24' OF 16" D.I. CL.52 RESTRAINED JOINT FORCEMAIN
460' OF 16" D.I. CL.52 RESTRAINED JOINT FORCEMAIN
INSTALLED 400' OF 14" PVC DR 25 RESTRAINED JOINT GRAV. EFF. PIPING
EXISTING & PROPOSED PIPING INTERSECTIONS DETAIL
SCALE HORIZ. 1"=40' VERT. 1"=5'
SCALE HORIZ. 1"=40' VERT. 1"=10'
LADYSMITH WWTF
FORCEMAIN: ROUTE PLAN & PROFILE
MORGAN & PARMLEY, LTD. CONSULTING ENGINEERS LADYSMITH, WISCONSIN
SHEET 7 OF 28
DATE APRIL 1988
DR. BY W.P., M.D.S.

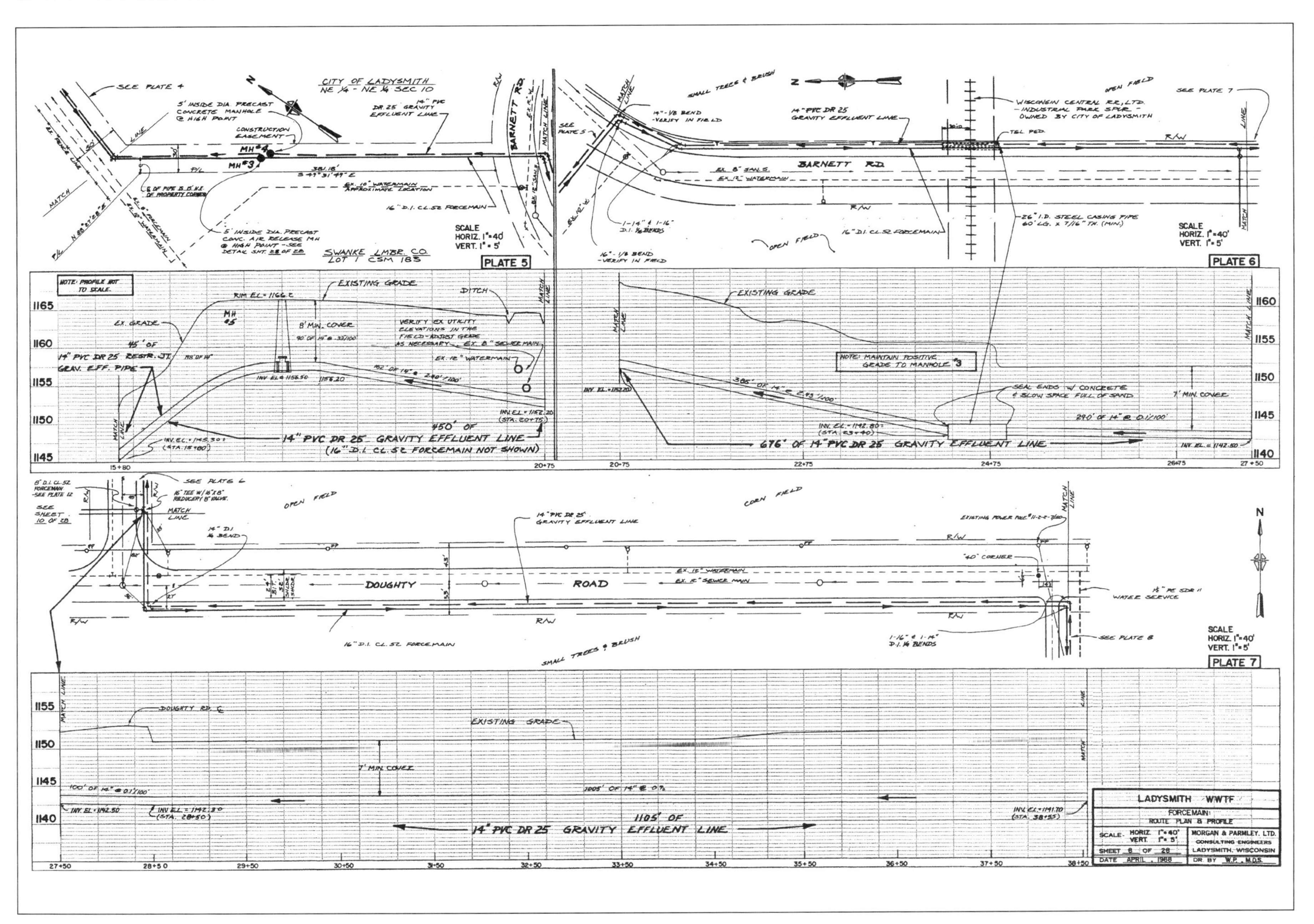
LADYSMITH WWTF
FORCEMAIN: ROUTE PLAN & PROFILE
MORGAN & PARMLEY, LTD.
CONSULTING ENGINEERS
LADYSMITH, WISCONSIN
SHEET 6 OF 28
DATE APRIL, 1988
PLATE 5
PLATE 6
PLATE 7
BARNETT RD.
DOUGHTY ROAD
14" PVC DR 25 GRAVITY EFFLUENT LINE
676' OF 14" PVC DR 25 GRAVITY EFFLUENT LINE
1105' OF 14" PVC DR 25 GRAVITY EFFLUENT LINE
EXISTING GRADE
SCALE HORIZ. 1"=40' VERT. 1"=5'

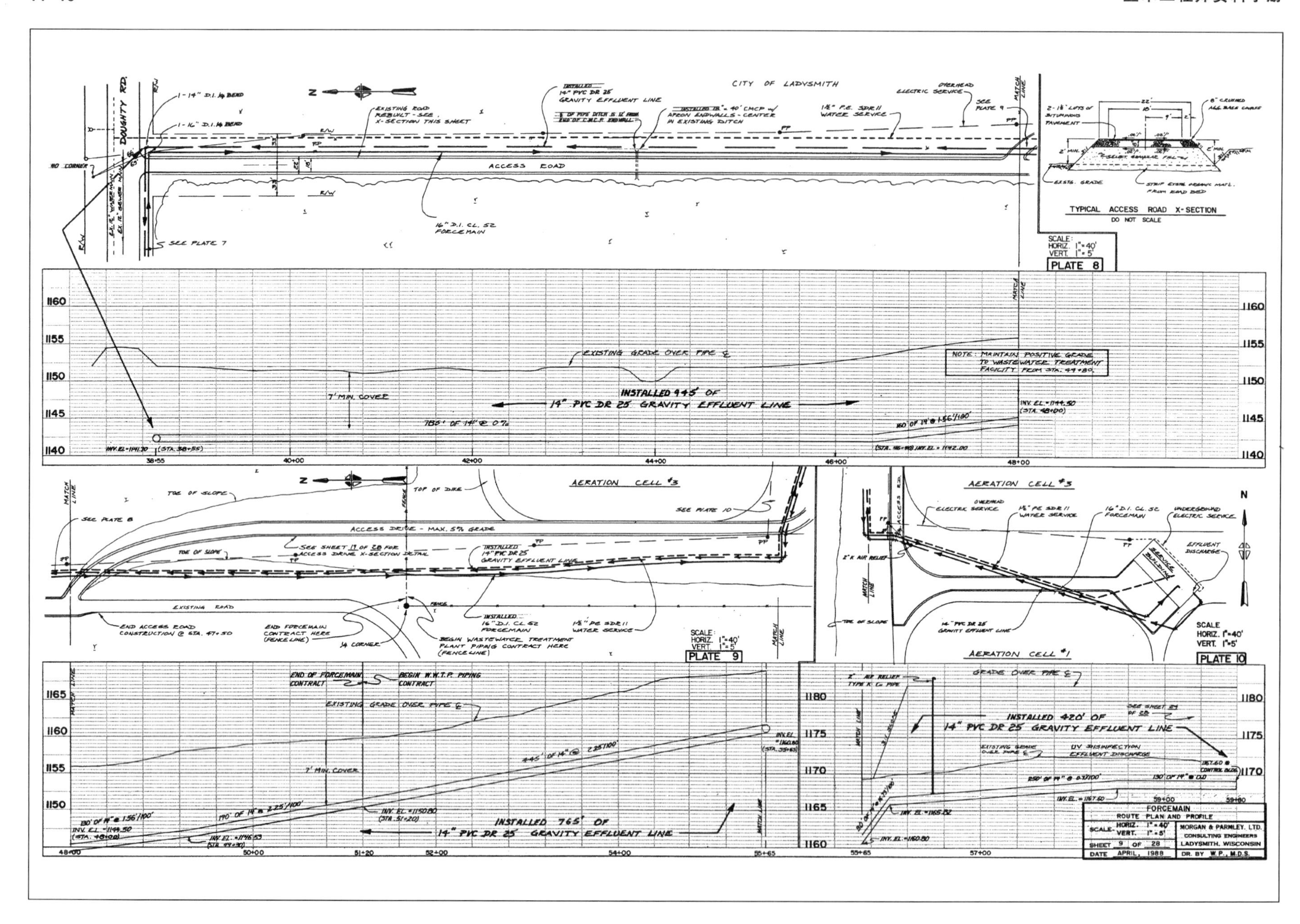

CITY OF LADYSMITH
ACCESS ROAD
TYPICAL ACCESS ROAD X-SECTION
DO NOT SCALE
PLATE 8
INSTALLED 445' OF
14" PVC DR 25 GRAVITY EFFLUENT LINE
EXISTING GRADE OVER PIPE ℄
7' MIN. COVER
AERATION CELL #3
ACCESS DRIVE - MAX. 5% GRADE
EXISTING ROAD
PLATE 9
PLATE 10
AERATION CELL #1
INSTALLED 765' OF
14" PVC DR 25 GRAVITY EFFLUENT LINE
INSTALLED 420' OF
14" PVC DR 25 GRAVITY EFFLUENT LINE
FORCEMAIN
ROUTE PLAN AND PROFILE
MORGAN & PARMLEY, LTD.
CONSULTING ENGINEERS
LADYSMITH, WISCONSIN
SHEET 9 OF 28
DATE APRIL, 1988

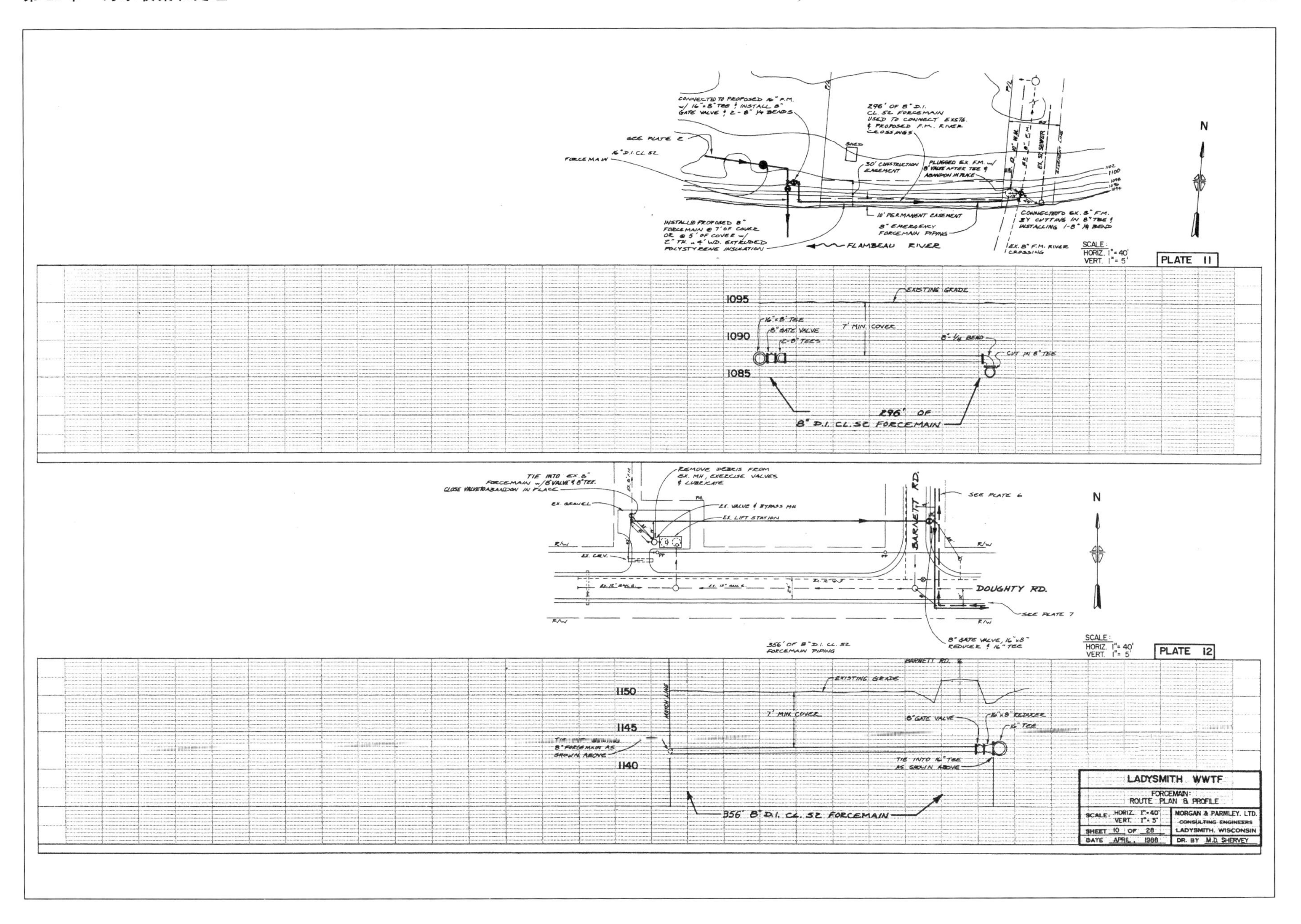
PLATE 11
PLATE 12
FLAMBEAU RIVER
DOUGHTY RD.
BARNETT RD.
296' OF 8" D.I. CL.52 FORCEMAIN
356' 8" D.I. CL.52 FORCEMAIN
LADYSMITH WWTF
FORCEMAIN: ROUTE PLAN & PROFILE
MORGAN & PARMLEY, LTD.
CONSULTING ENGINEERS
LADYSMITH, WISCONSIN
SHEET 10 OF 28
DATE APRIL, 1988
DR. BY M.D. SHERVEY

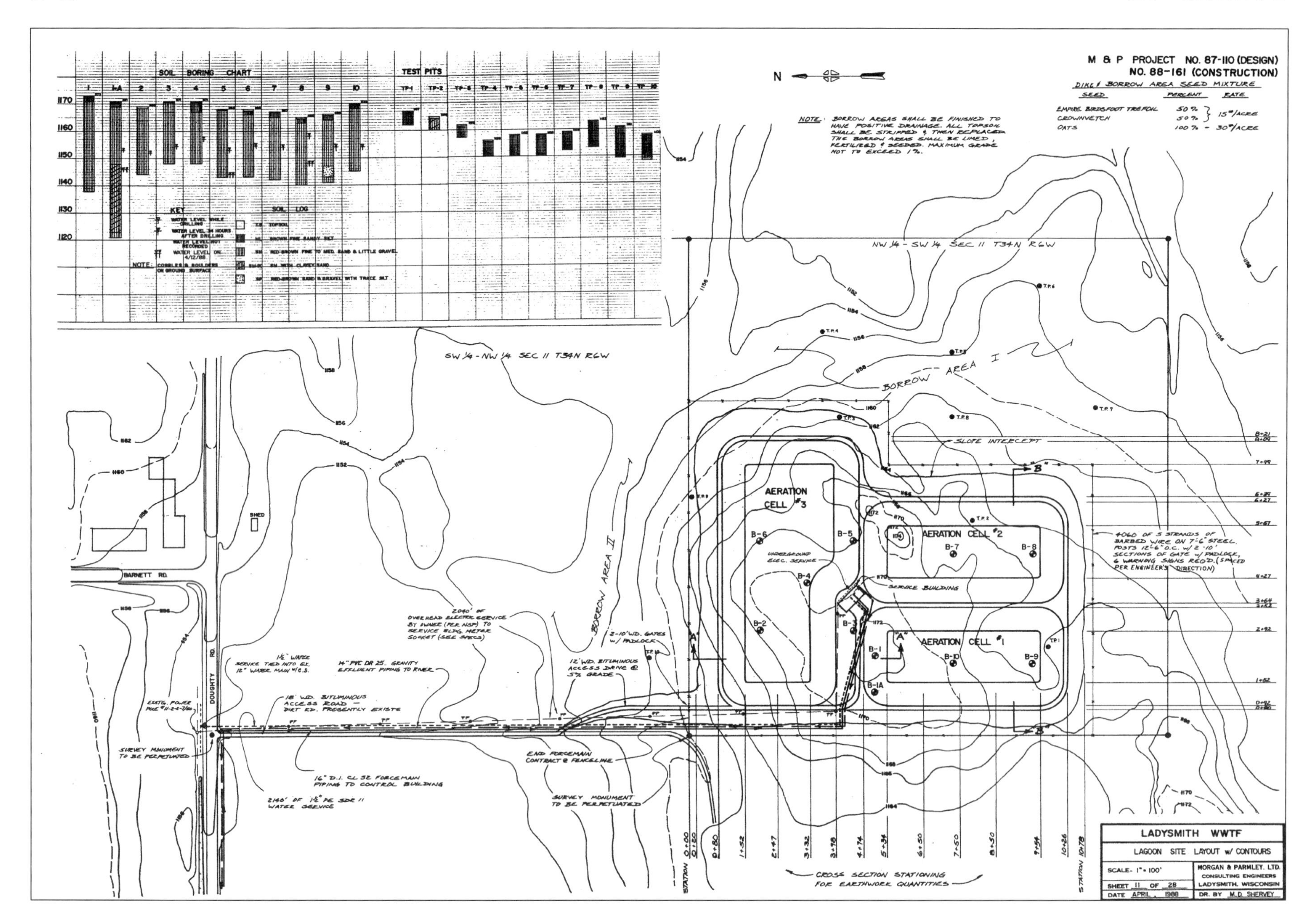

M & P PROJECT NO. 87-110 (DESIGN)
NO. 88-161 (CONSTRUCTION)
DIKE & BORROW AREA SEED MIXTURE
SOIL BORING CHART
TEST PITS
KEY
SOIL LOG
NOTE: BORROW AREAS SHALL BE FINISHED TO HAVE POSITIVE DRAINAGE. ALL TOPSOIL SHALL BE STRIPPED & THEN REPLACED. THE BORROW AREAS SHALL BE LIMED, FERTILIZED & SEEDED. MAXIMUM GRADE NOT TO EXCEED 1%.
NW 1/4 - SW 1/4 SEC 11 T34N R6W
SW 1/4 - NW 1/4 SEC 11 T34N R6W
BORROW AREA I
BORROW AREA II
SLOPE INTERCEPT
AERATION CELL #3
AERATION CELL #2
AERATION CELL #1
SERVICE BUILDING
BARNETT RD.
DOUGHTY RD
SHED
CROSS SECTION STATIONING FOR EARTHWORK QUANTITIES
LADYSMITH WWTF
LAGOON SITE LAYOUT w/ CONTOURS
SCALE - 1" = 100'
SHEET 11 OF 28
DATE APRIL, 1988
MORGAN & PARMLEY, LTD.
CONSULTING ENGINEERS
LADYSMITH, WISCONSIN
DR. BY M.D. SHERVEY

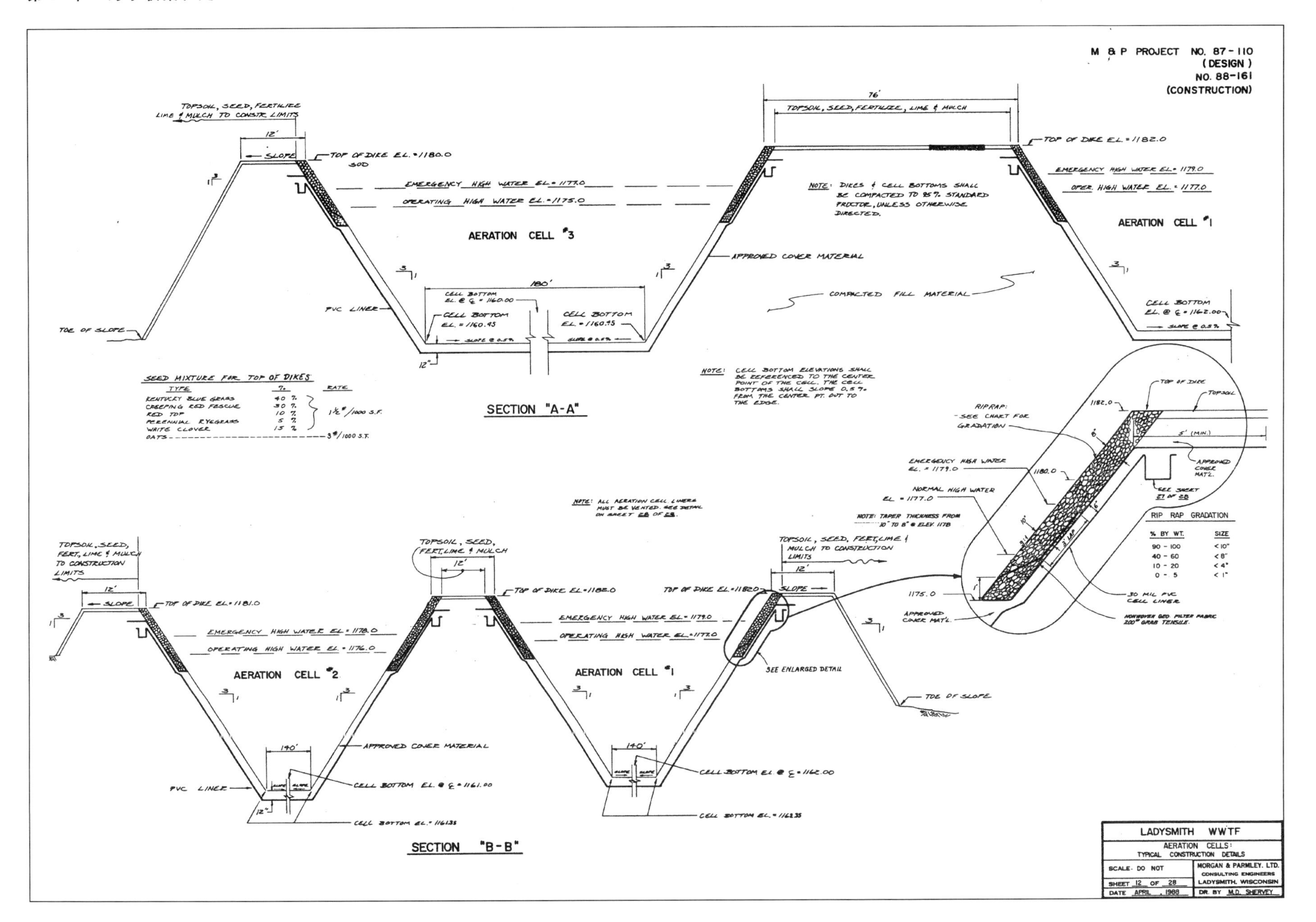
M & P PROJECT NO. 87-110
(DESIGN)
NO. 88-161
(CONSTRUCTION)
AERATION CELL #3
AERATION CELL #1
AERATION CELL #2
SECTION "A-A"
SECTION "B-B"
SEED MIXTURE FOR TOP OF DIKES
RIP RAP GRADATION
SEE ENLARGED DETAIL
LADYSMITH WWTF
AERATION CELLS:
TYPICAL CONSTRUCTION DETAILS
MORGAN & PARMLEY, LTD.
CONSULTING ENGINEERS
LADYSMITH, WISCONSIN

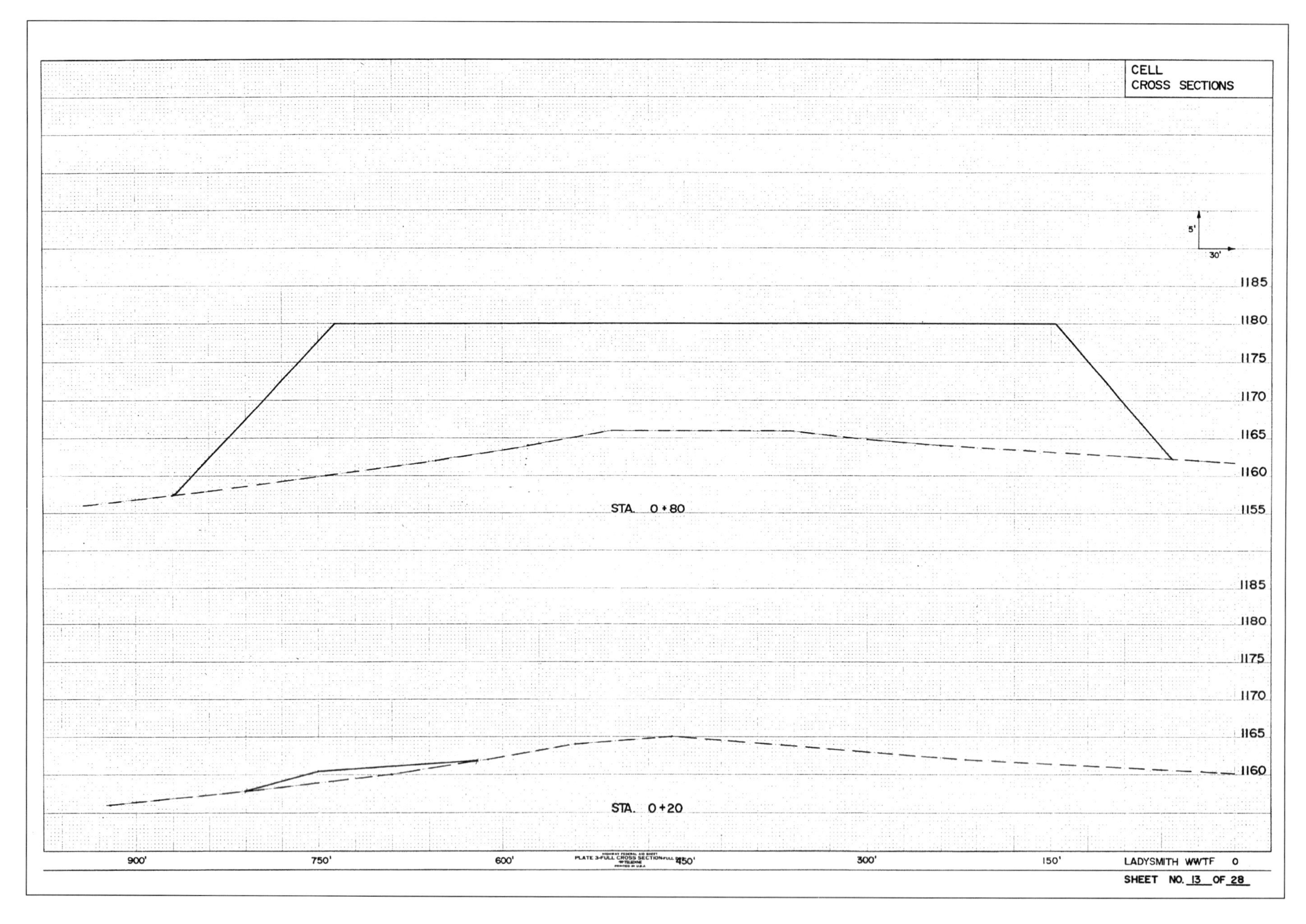
CELL
CROSS SECTIONS
5'
30'
1185
1180
1175
1170
1165
1160
1155
STA. 0+80
1185
1180
1175
1170
1165
1160
STA. 0+20
900'
750'
600'
450'
300'
150'
LADYSMITH WWTF 0
SHEET NO. 13 OF 28

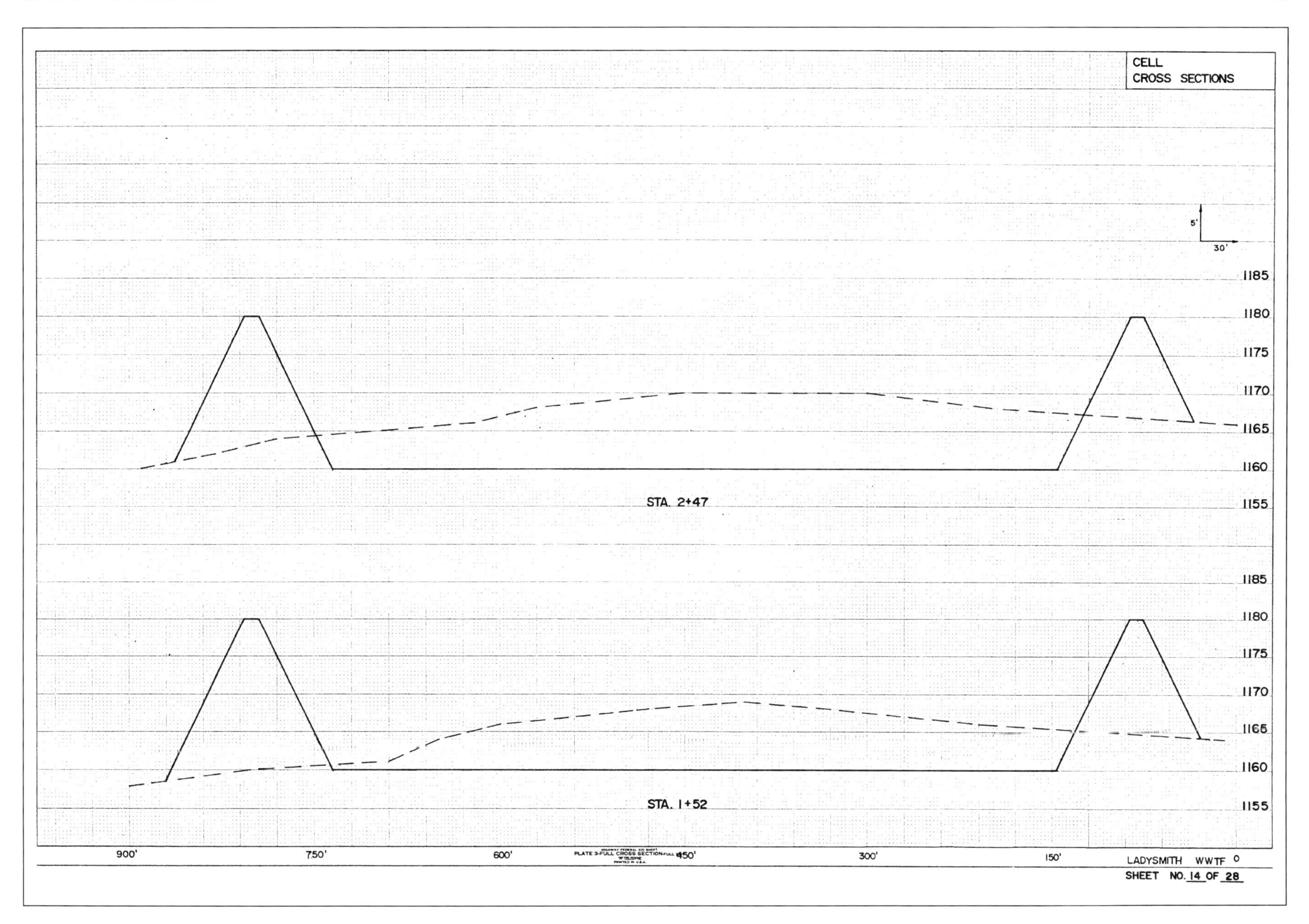
CELL
CROSS SECTIONS
5'
30'
1185
1180
1175
1170
1165
1160
1155
STA. 2+47
1185
1180
1175
1170
1165
1160
1155
STA. 1+52
900'
750'
600'
450'
300'
150'
0
LADYSMITH WWTF
SHEET NO. 14 OF 28

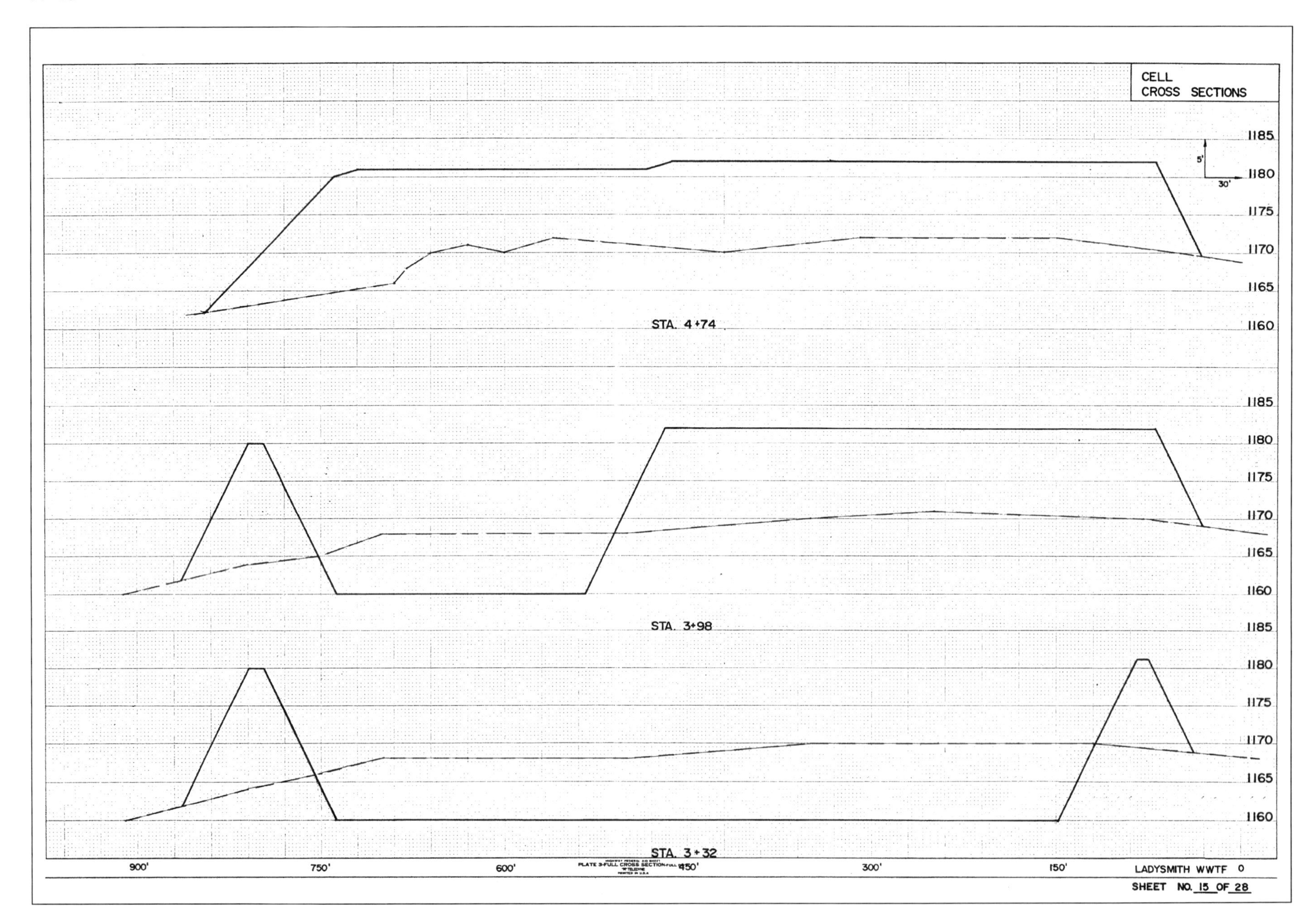
CELL
CROSS SECTIONS
1185
1180
1175
1170
1165
1160
5'
30'
STA. 4+74
STA. 3+98
STA. 3+32
900'
750'
600'
450'
300'
150'
0
LADYSMITH WWTF
SHEET NO. 15 OF 28

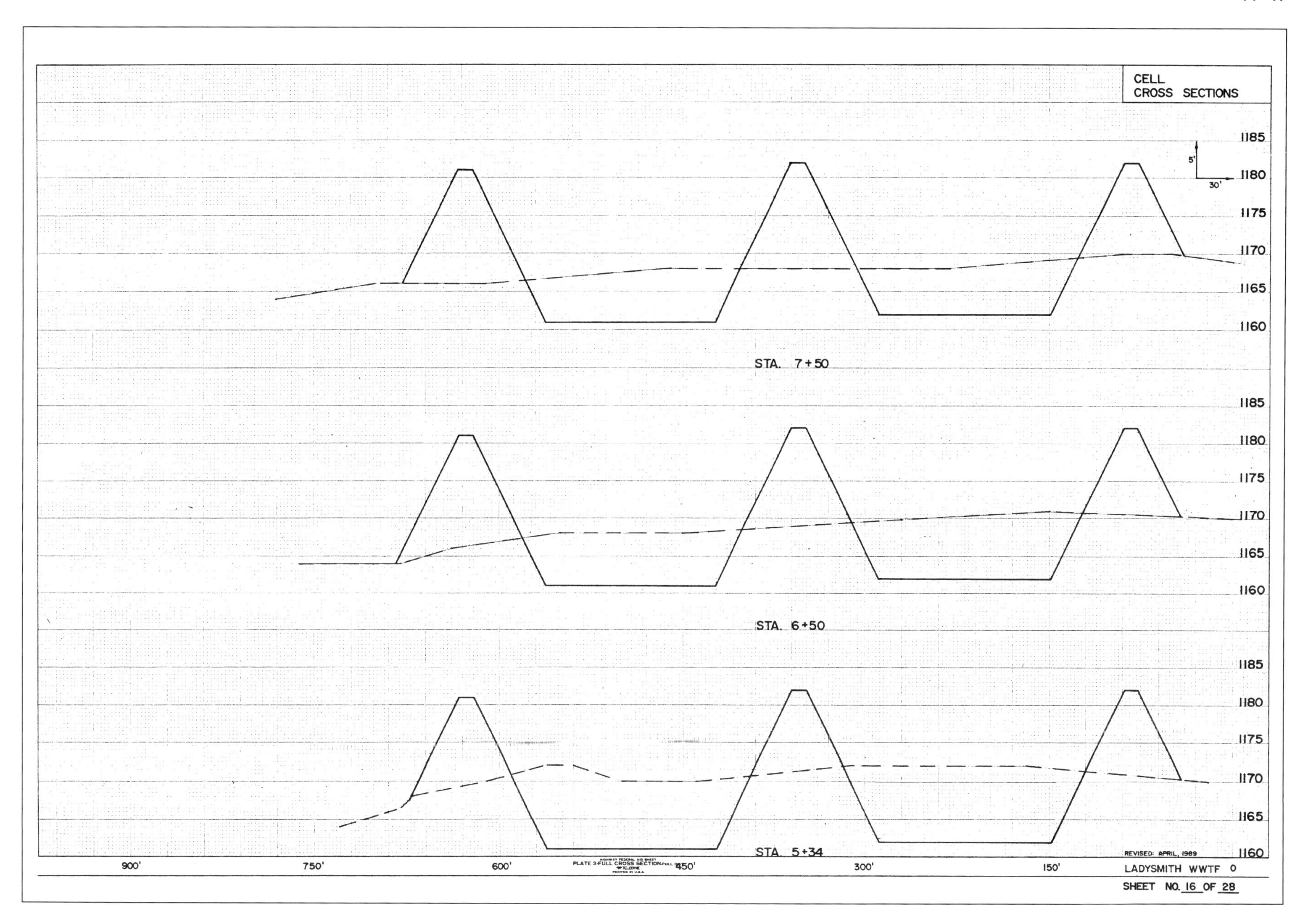
CELL
CROSS SECTIONS
1185
1180
1175
1170
1165
1160
5'
30'
STA. 7+50
STA. 6+50
STA. 5+34
900'
750'
600'
450'
300'
150'
REVISED: APRIL, 1989
LADYSMITH WWTF
SHEET NO. 16 OF 28

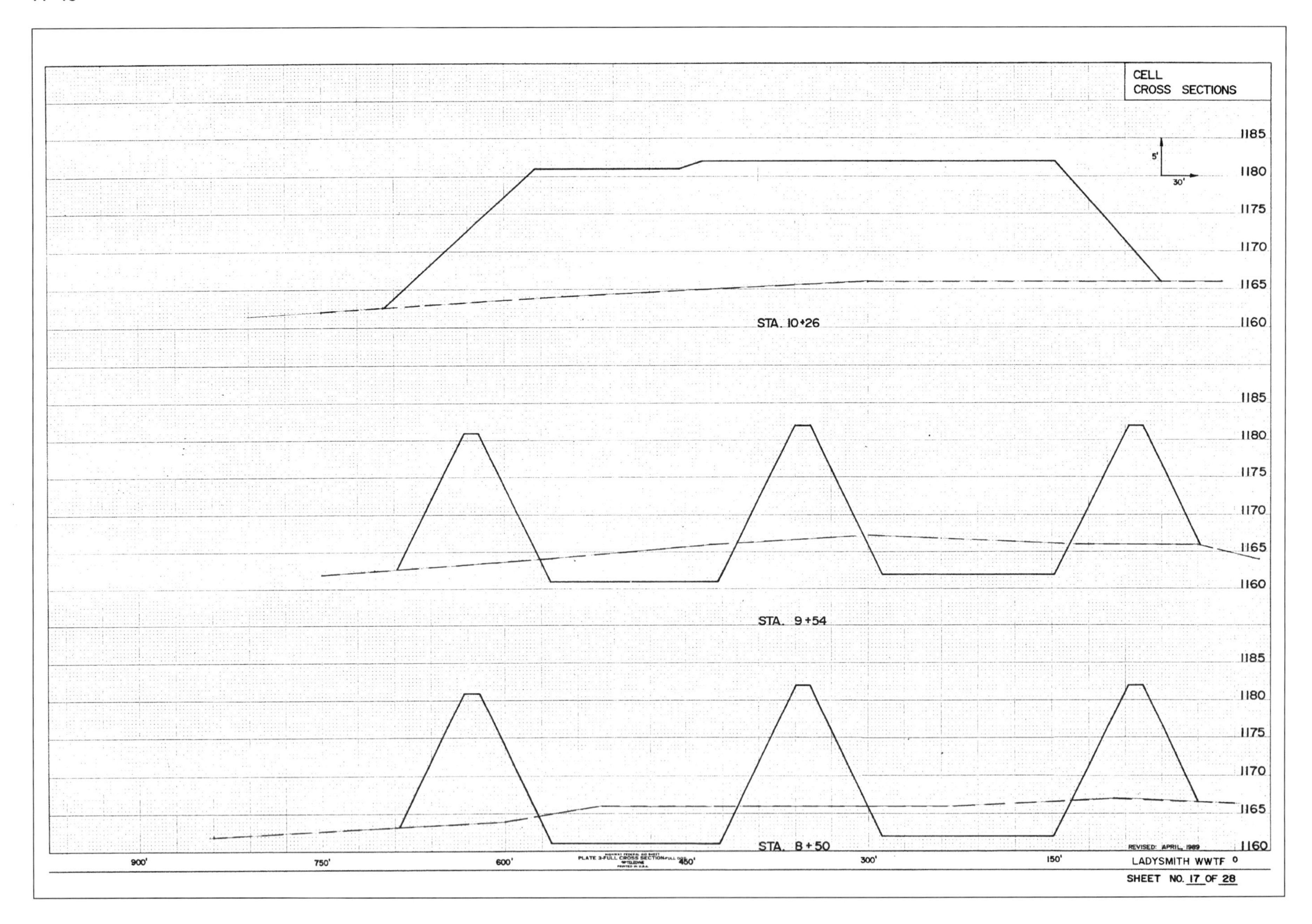
CELL
CROSS SECTIONS
1185
1180
1175
1170
1165
1160
5'
30'
STA. 10+26
STA. 9+54
STA. 8+50
900'
750'
600'
450'
300'
150'
REVISED: APRIL, 1989
LADYSMITH WWTF
SHEET NO. 17 OF 28

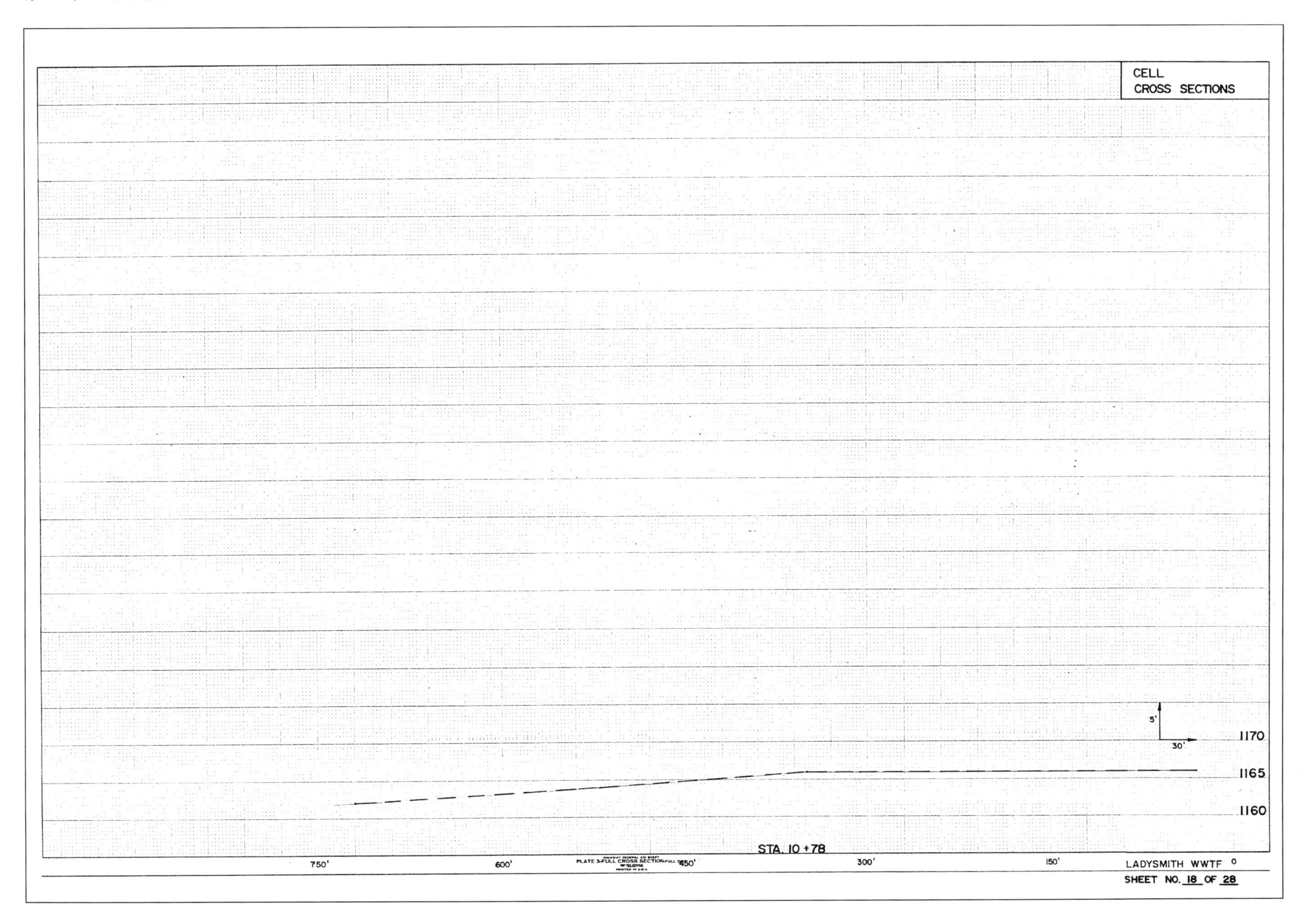
CELL
CROSS SECTIONS
5'
30'
1170
1165
1160
STA. 10 +78
750'
600'
450'
300'
150'
LADYSMITH WWTF 0
SHEET NO. 18 OF 28

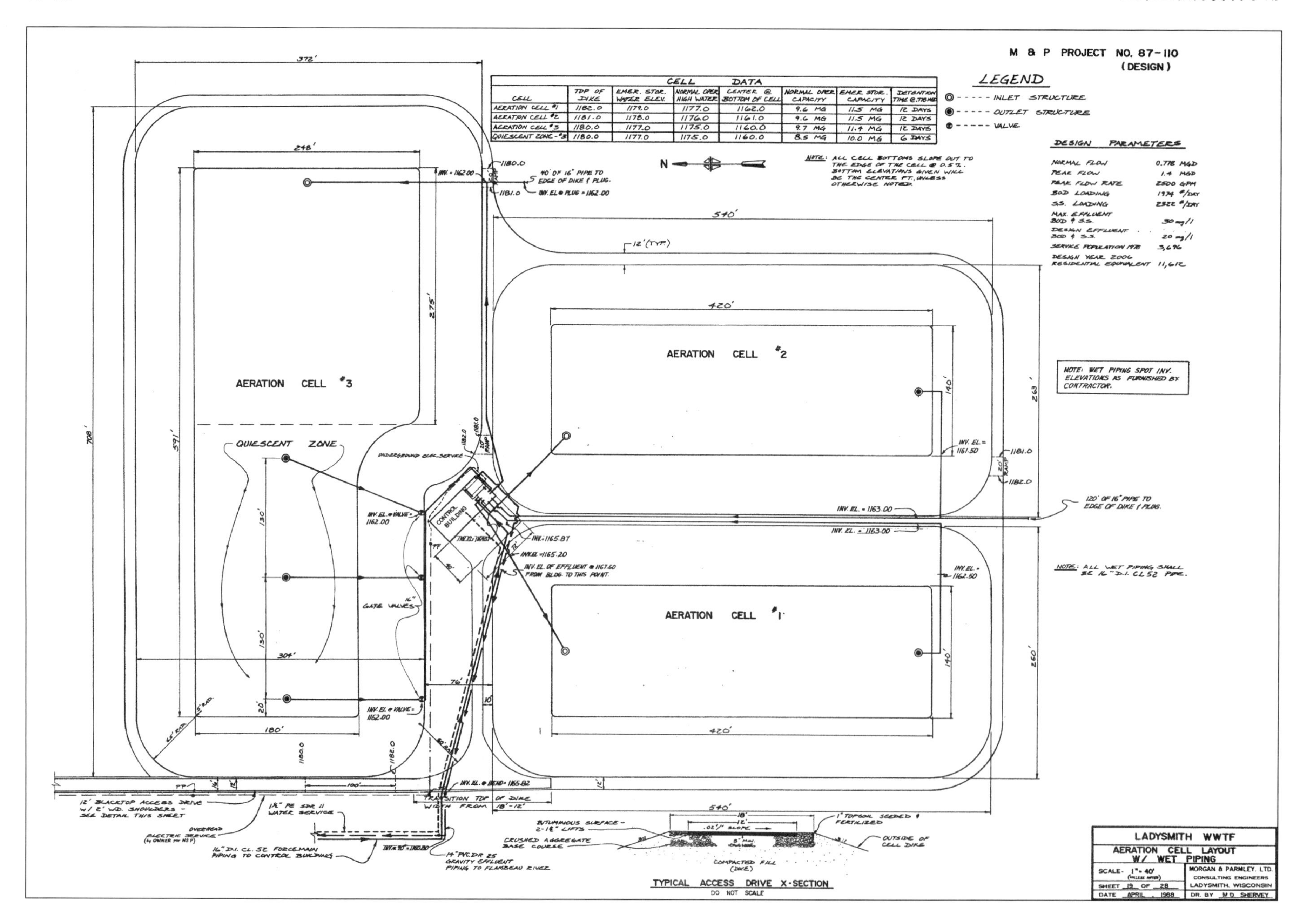

CELL	TOP OF DIKE	EMER. STOR. WATER ELEV.	NORMAL OPER. HIGH WATER	CENTER @ BOTTOM OF CELL	NORMAL OPER. CAPACITY	EMER. STOR. CAPACITY	DETENTION TIME @.778 MGD
AERATION CELL #1	1182.0	1179.0	1177.0	1162.0	9.6 MG	11.5 MG	12 DAYS
AERATION CELL #2	1181.0	1178.0	1176.0	1161.0	9.6 MG	11.5 MG	12 DAYS
AERATION CELL #3	1180.0	1177.0	1175.0	1160.0	9.7 MG	11.4 MG	12 DAYS
QUIESCENT ZONE-#3	1180.0	1177.0	1175.0	1160.0	8.5 MG	10.0 MG	6 DAYS

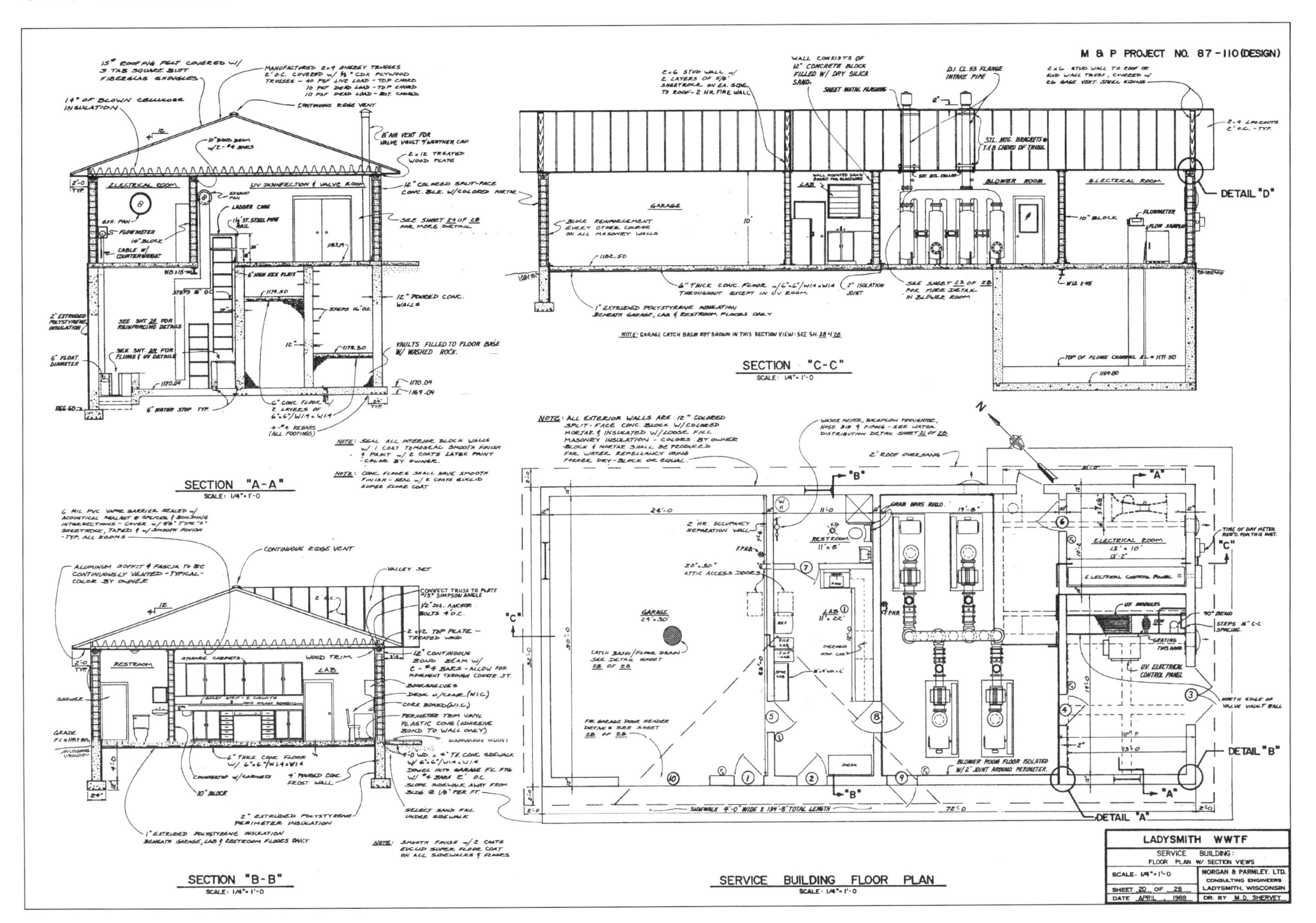
M & P PROJECT NO. 87-110(DESIGN)
SECTION "A-A"
SCALE: 1/4"=1'-0
SECTION "B-B"
SCALE: 1/4"=1'-0
SECTION "C-C"
SCALE: 1/4"=1'-0
SERVICE BUILDING FLOOR PLAN
SCALE: 1/4"=1'-0
DETAIL "A"
DETAIL "B"
DETAIL "D"
ELECTRICAL ROOM
UV DISINFECTION & VALVE ROOM
GARAGE
LAB
BLOWER ROOM
RESTROOM
ELECTRICAL CONTROL PANEL
UV MODULES
UV ELECTRICAL CONTROL PANEL
NOTE: ALL EXTERIOR WALLS ARE 12" COLORED SPLIT-FACE CONC. BLOCK W/COLORED MORTAR & INSULATED W/LOOSE FILL MASONRY INSULATION - COLORS BY OWNER. BLOCK & MORTAR SHALL BE PRODUCED FOR WATER REPELLANCY USING FORRER DRY-BLOCK OR EQUAL.
NOTE: CONC. FLOORS SHALL HAVE SMOOTH FINISH - SEAL W/ 2 COATS EUCLID SUPER FLOOR COAT
NOTE: SMOOTH FINISH W/ 2 COATS EUCLID SUPER FLOOR COAT ON ALL SIDEWALKS & FLOORS
NOTE: GARAGE CATCH BASIN NOT SHOWN IN THIS SECTION VIEW: SEE SH. 28 of 28
SIDEWALK 4'-0" WIDE X 134'-8" TOTAL LENGTH
LADYSMITH WWTF
SERVICE BUILDING: FLOOR PLAN W/ SECTION VIEWS
SCALE: 1/4"=1'-0
SHEET 20 OF 28
DATE APRIL, 1988
MORGAN & PARMLEY, LTD.
CONSULTING ENGINEERS
LADYSMITH, WISCONSIN
DR. BY M.D. SHERVEY

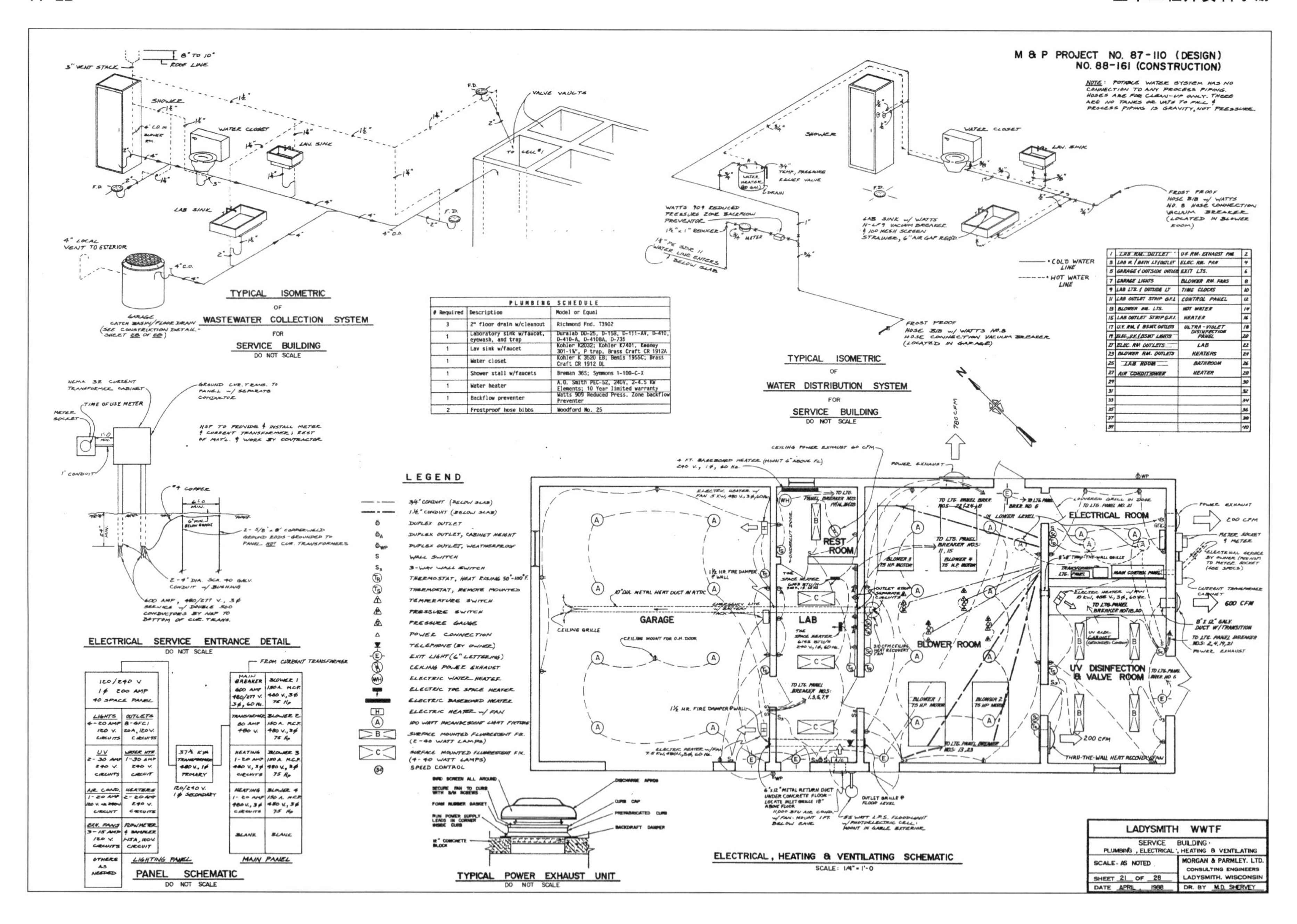

PLUMBING SCHEDULE		
# Required	Description	Model or Equal
3	2" floor drain w/cleanout	Richmond Fnd. T3902
1	Laboratory sink w/faucet, eyewash, and trap	Duralab DD-25, D-158, D-111-AV, D-410, D-410-A, D-410BA, D-735
1	Lav sink w/faucet	Kohler K2032; Kohler K7401, Keeney 301-1¼", P trap, Brass Craft CR 1912A
1	Water closet	Kohler K 3520 EB; Bemis 1955C; Brass Craft CR 1912 DL
1	Shower stall w/faucets	Breman 36S; Symmons 1-100-C-X
1	Water heater	A.O. Smith PEC-52, 240V, 2-4.5 KW Elements; 10 Year limited warranty
1	Backflow preventer	Watts 909 Reduced Press. Zone backflow Preventer
2	Frostproof hose bibbs	Woodford No. 25

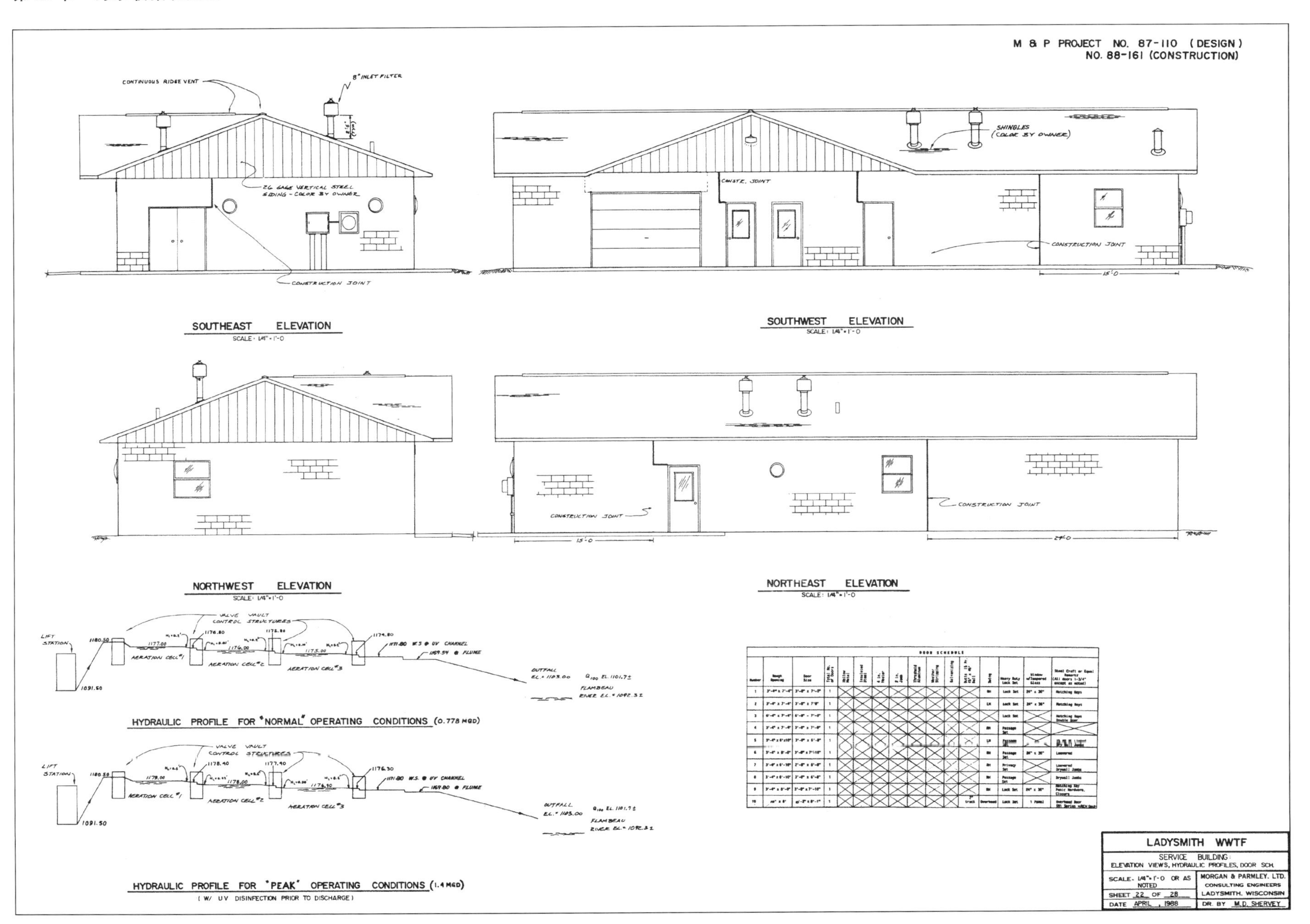

M & P PROJECT NO. 87-110 (DESIGN)
NO. 88-161 (CONSTRUCTION)
CONTINUOUS RIDGE VENT
8" INLET FILTER
26 GAGE VERTICAL STEEL SIDING - COLOR BY OWNER
CONSTRUCTION JOINT
SHINGLES (COLOR BY OWNER)
CONSTR. JOINT
CONSTRUCTION JOINT
15'-0
SOUTHEAST ELEVATION
SCALE: 1/4"=1'-0
SOUTHWEST ELEVATION
SCALE: 1/4"=1'-0
CONSTRUCTION JOINT
CONSTRUCTION JOINT
15'-0
24'-0
NORTHWEST ELEVATION
SCALE: 1/4"=1'-0
NORTHEAST ELEVATION
SCALE: 1/4"=1'-0
VALVE VAULT
CONTROL STRUCTURES
LIFT STATION
1180.50
1177.00
1176.80
1176.00
1175.80
1175.00
1174.80
1091.50
AERATION CELL #1
AERATION CELL #2
AERATION CELL #3
1171.80 W.S. @ UV CHANNEL
1169.54 @ FLUME
OUTFALL EL.= 1103.00
FLAMBEAU RIVER EL.= 1092.3±
HYDRAULIC PROFILE FOR "NORMAL" OPERATING CONDITIONS (0.778 MGD)
1178.40
1178.00
1177.40
1176.30
1176.30
1171.80 W.S. @ UV CHANNEL
1169.80 @ FLUME
OUTFALL EL.= 1103.00
FLAMBEAU RIVER EL.= 1092.3±
HYDRAULIC PROFILE FOR "PEAK" OPERATING CONDITIONS (1.4 MGD)
(W/ UV DISINFECTION PRIOR TO DISCHARGE)
DOOR SCHEDULE
LADYSMITH WWTF
SERVICE BUILDING:
ELEVATION VIEWS, HYDRAULIC PROFILES, DOOR SCH.
SCALE: 1/4"=1'-0 OR AS NOTED
SHEET 22 OF 28
DATE APRIL, 1988
MORGAN & PARMLEY, LTD.
CONSULTING ENGINEERS
LADYSMITH, WISCONSIN
DR. BY M.D. SHERVEY

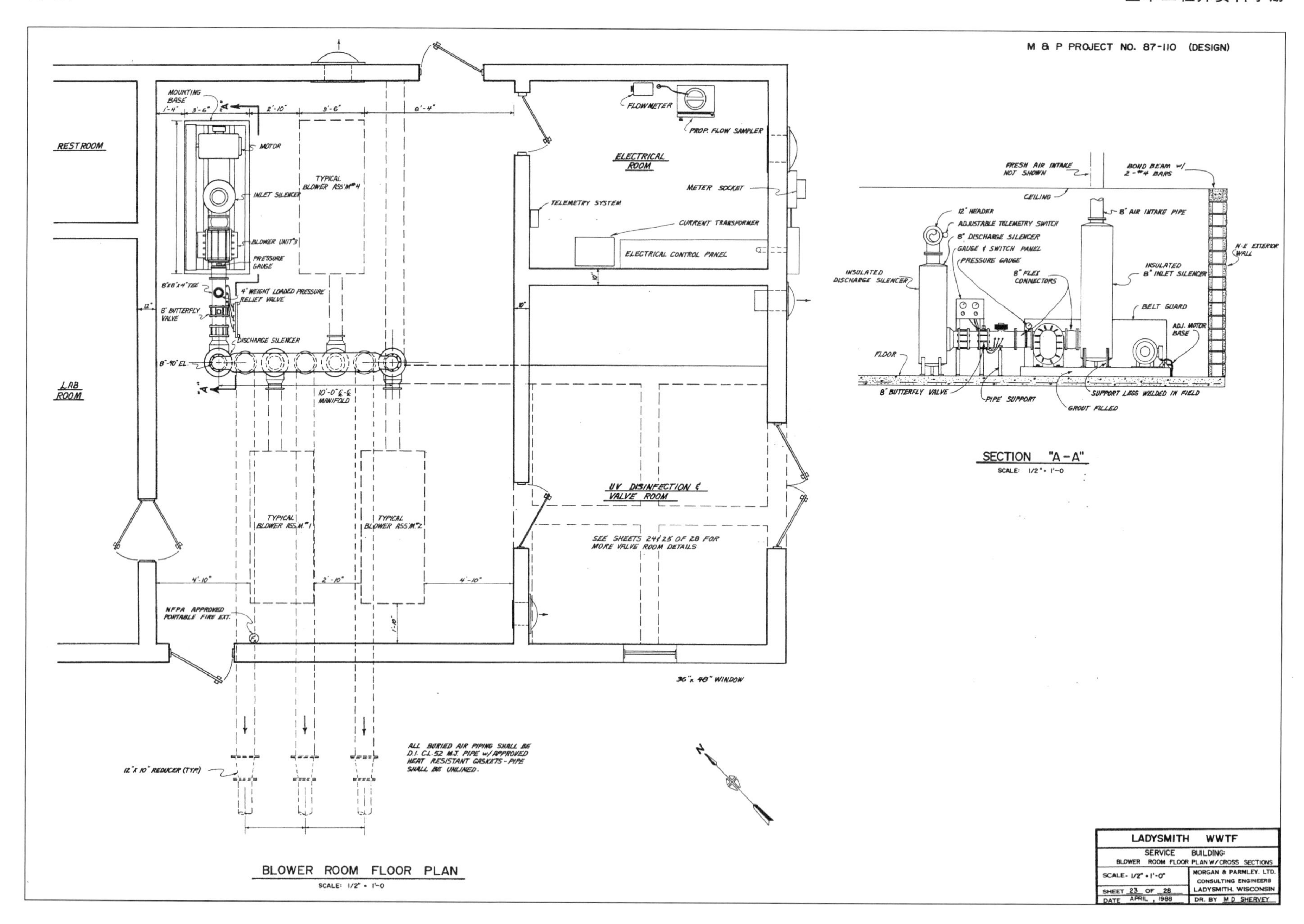
M & P PROJECT NO. 87-110 (DESIGN)
RESTROOM
LAB ROOM
MOUNTING BASE
MOTOR
INLET SILENCER
BLOWER UNIT#3
PRESSURE GAUGE
8"x8"x4" TEE
4" WEIGHT LOADED PRESSURE RELIEF VALVE
8" BUTTERFLY VALVE
DISCHARGE SILENCER
8"-90° EL.
10'-0" 6-6 MANIFOLD
TYPICAL BLOWER ASS'M #4
TYPICAL BLOWER ASS'M #1
TYPICAL BLOWER ASS'M #2
NFPA APPROVED PORTABLE FIRE EXT.
FLOWMETER
PROP. FLOW SAMPLER
ELECTRICAL ROOM
METER SOCKET
TELEMETRY SYSTEM
CURRENT TRANSFORMER
ELECTRICAL CONTROL PANEL
UV DISINFECTION & VALVE ROOM
SEE SHEETS 24 & 25 OF 28 FOR MORE VALVE ROOM DETAILS
36"x 48" WINDOW
12"x 10" REDUCER (TYP)
ALL BURIED AIR PIPING SHALL BE D.I. CL.52 MJ. PIPE w/ APPROVED HEAT RESISTANT GASKETS - PIPE SHALL BE UNLINED.
N
BLOWER ROOM FLOOR PLAN
SCALE: 1/2" = 1'-0
FRESH AIR INTAKE NOT SHOWN
BOND BEAM w/ 2-#4 BARS
CEILING
12" HEADER
ADJUSTABLE TELEMETRY SWITCH
8" DISCHARGE SILENCER
GAUGE & SWITCH PANEL
PRESSURE GAUGE
8" AIR INTAKE PIPE
N-E EXTERIOR WALL
INSULATED DISCHARGE SILENCER
8" FLEX CONNECTORS
INSULATED 8" INLET SILENCER
BELT GUARD
ADJ. MOTOR BASE
FLOOR
8" BUTTERFLY VALVE
PIPE SUPPORT
SUPPORT LEGS WELDED IN FIELD
GROUT FILLED
SECTION "A-A"
SCALE: 1/2" = 1'-0
LADYSMITH WWTF
SERVICE BUILDING
BLOWER ROOM FLOOR PLAN W/CROSS SECTIONS
SCALE- 1/2" = 1'-0"
SHEET 23 OF 28
DATE APRIL, 1988
MORGAN & PARMLEY, LTD. CONSULTING ENGINEERS LADYSMITH, WISCONSIN
DR. BY M.D. SHERVEY

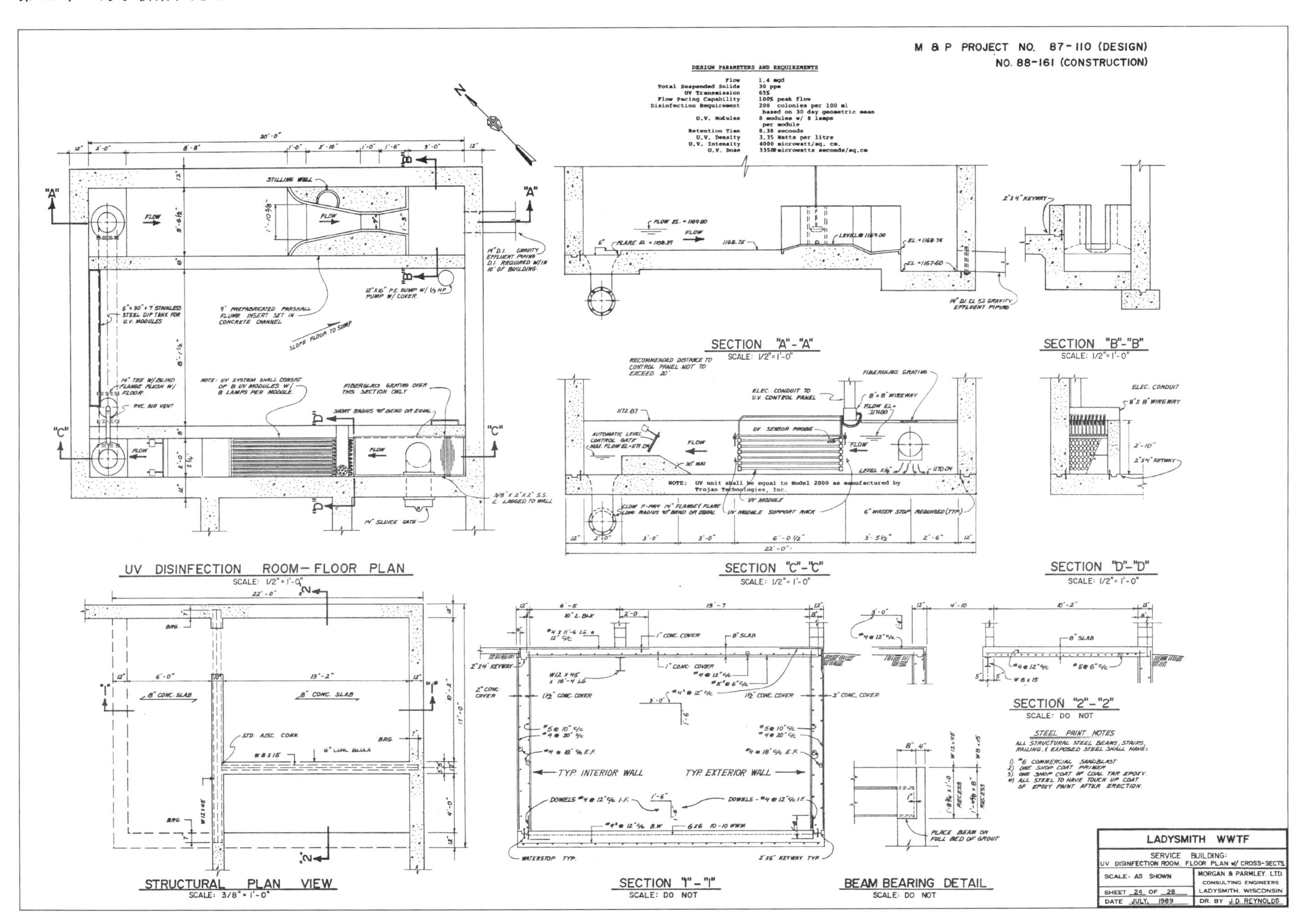
M & P PROJECT NO. 87-110 (DESIGN)
NO. 88-161 (CONSTRUCTION)
DESIGN PARAMETERS AND REQUIREMENTS
Flow 1.4 mgd
Total Suspended Solids 30 ppm
UV Transmission 65%
Flow Pacing Capability 100% peak flow
Disinfection Requirement 200 colonies per 100 ml based on 30 day geometric mean
U.V. Modules 8 modules w/ 8 lamps per module
Retention Time 8.38 seconds
U.V. Density 3.35 Watts per litre
U.V. Intensity 4000 microwatt/sq. cm.
U.V. Dose 33500 microwatts seconds/sq.cm
STILLING WELL
FLOW
9" PREFABRICATED PARSHALL FLUME INSERT SET IN CONCRETE CHANNEL
SLOPE FLOOR TO SUMP
NOTE: UV SYSTEM SHALL CONSIST OF 8 UV MODULES W/ 8 LAMPS PER MODULE.
FIBERGLASS GRATING OVER THIS SECTION ONLY
14" SLUICE GATE
UV DISINFECTION ROOM- FLOOR PLAN
SCALE: 1/2" = 1'-0"
SECTION "A"-"A"
SCALE: 1/2" = 1'-0"
SECTION "B"-"B"
SCALE: 1/2" = 1'-0"
NOTE: UV unit shall be equal to Model 2000 as manufactured by Trojan Technologies, Inc.
SECTION "C"-"C"
SCALE: 1/2" = 1'-0"
SECTION "D"-"D"
SCALE: 1/2" = 1'-0"
STRUCTURAL PLAN VIEW
SCALE: 3/8" = 1'-0"
TYP. INTERIOR WALL
TYP. EXTERIOR WALL
SECTION "1"-"1"
SCALE: DO NOT
BEAM BEARING DETAIL
SCALE: DO NOT
SECTION "2"-"2"
SCALE: DO NOT
STEEL PAINT NOTES
LADYSMITH WWTF
SERVICE BUILDING:
UV DISINFECTION ROOM, FLOOR PLAN w/ CROSS-SECTS.
SCALE- AS SHOWN
SHEET 24 OF 28
DATE JULY, 1989
MORGAN & PARMLEY, LTD.
CONSULTING ENGINEERS
LADYSMITH, WISCONSIN
DR. BY J.D. REYNOLDS

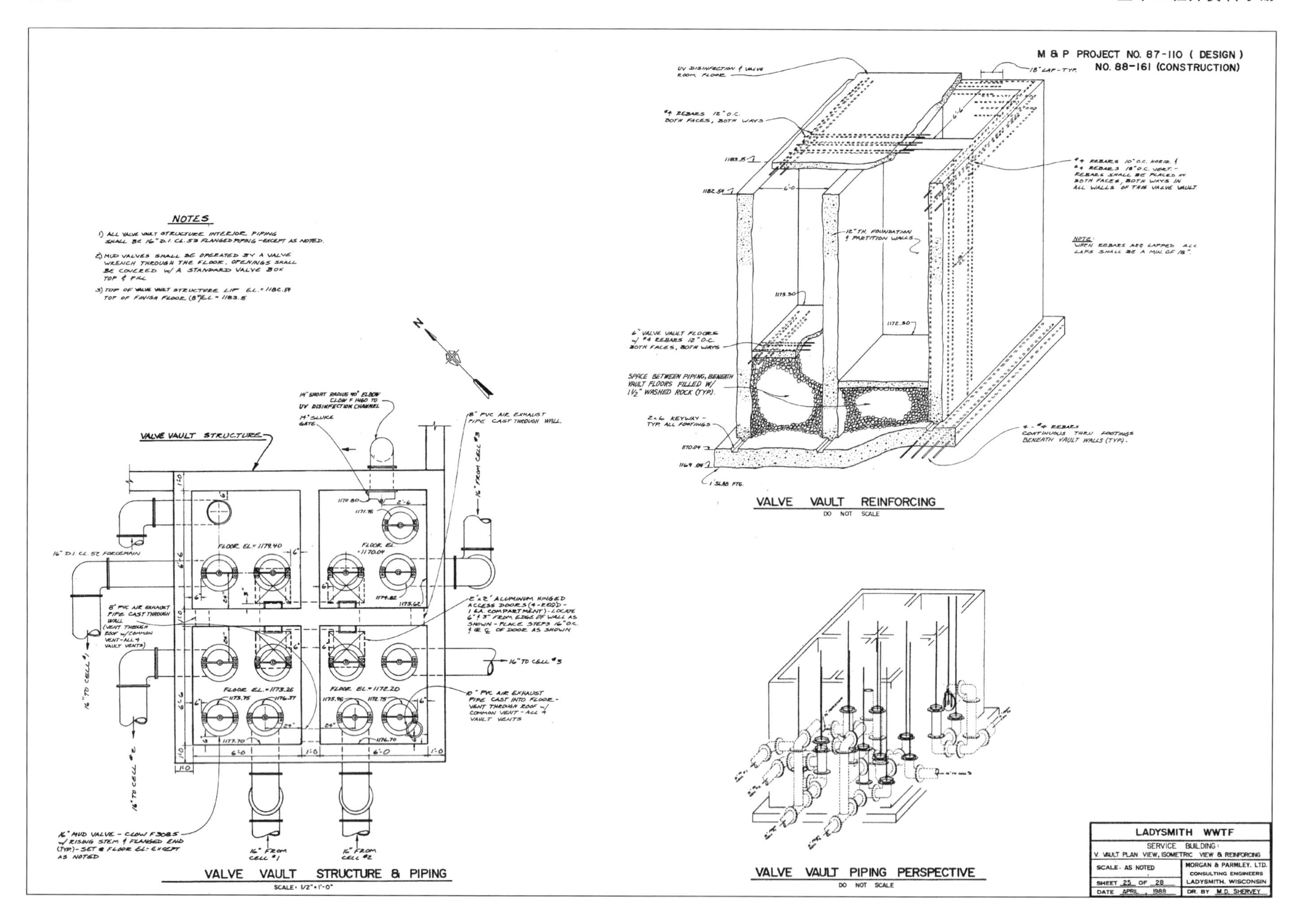
M & P PROJECT NO. 87-110 (DESIGN)
NO. 88-161 (CONSTRUCTION)
NOTES
VALVE VAULT STRUCTURE & PIPING
SCALE: 1/2"=1'-0"
VALVE VAULT REINFORCING
DO NOT SCALE
VALVE VAULT PIPING PERSPECTIVE
DO NOT SCALE
LADYSMITH WWTF
SERVICE BUILDING:
V. VAULT PLAN VIEW, ISOMETRIC VIEW & REINFORCING
SCALE- AS NOTED
SHEET 25 OF 28
DATE APRIL, 1988
MORGAN & PARMLEY, LTD.
CONSULTING ENGINEERS
LADYSMITH, WISCONSIN
DR. BY M.D. SHERVEY

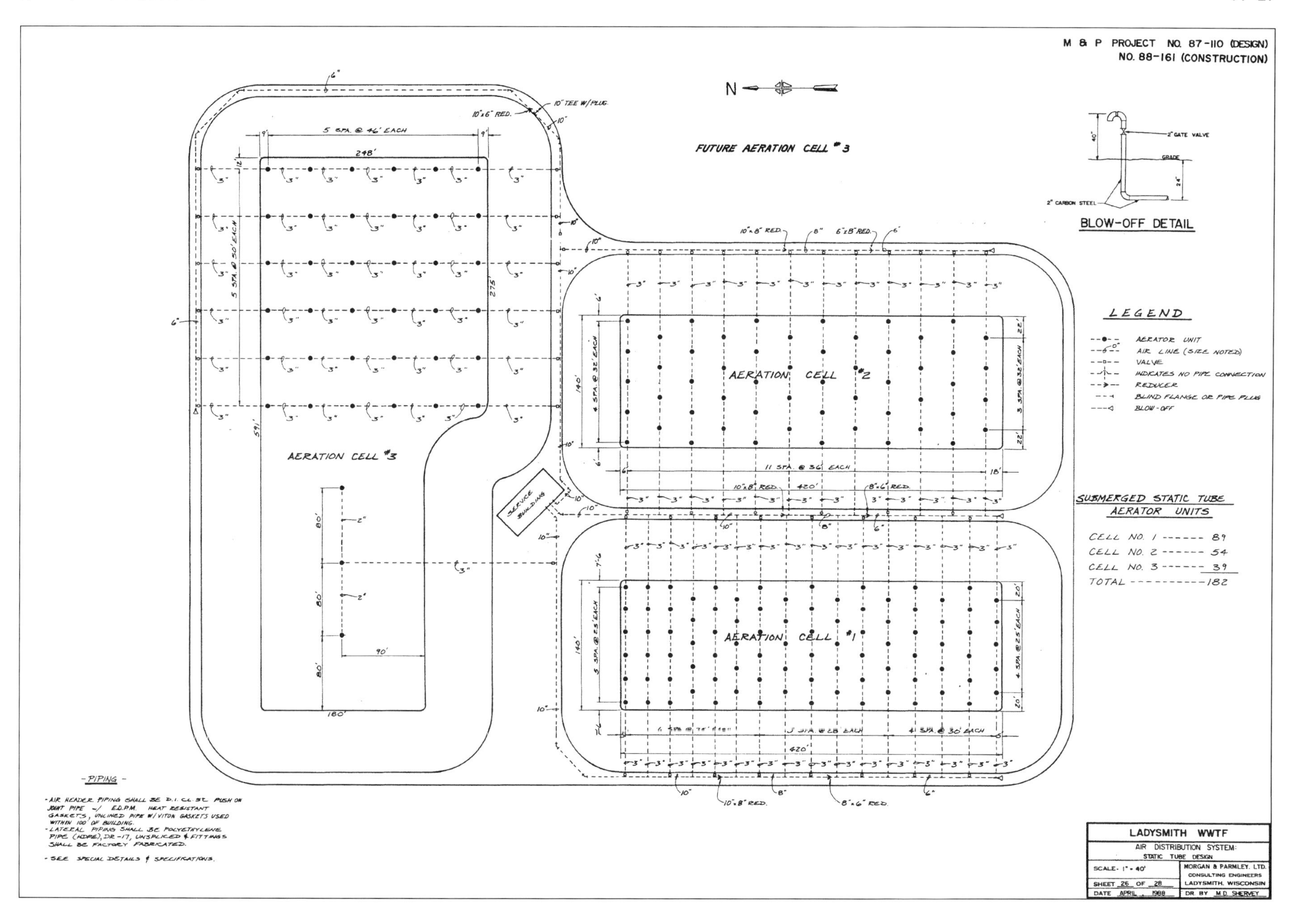
M & P PROJECT NO. 87-110 (DESIGN)
NO. 88-161 (CONSTRUCTION)
FUTURE AERATION CELL #3
AERATION CELL #3
AERATION CELL #2
AERATION CELL #1
SERVICE BUILDING
BLOW-OFF DETAIL
2" GATE VALVE
GRADE
2" CARBON STEEL
LEGEND
AERATOR UNIT
AIR LINE (SIZE NOTED)
VALVE
INDICATES NO PIPE CONNECTION
REDUCER
BLIND FLANGE OR PIPE PLUG
BLOW-OFF
SUBMERGED STATIC TUBE AERATOR UNITS
CELL NO. 1 ------ 89
CELL NO. 2 ------ 54
CELL NO. 3 ------ 39
TOTAL ---------- 182
-PIPING-
• AIR HEADER PIPING SHALL BE D.I. CL 52 PUSH ON JOINT PIPE W/ E.D.P.M. HEAT RESISTANT GASKETS, UNLINED PIPE W/VITON GASKETS USED WITHIN 100' OF BUILDING.
• LATERAL PIPING SHALL BE POLYETHYLENE PIPE (HDPE), DR-17, UNSPLICED & FITTINGS SHALL BE FACTORY FABRICATED.
• SEE SPECIAL DETAILS & SPECIFICATIONS.
LADYSMITH WWTF
AIR DISTRIBUTION SYSTEM: STATIC TUBE DESIGN
SCALE: 1" = 40'
SHEET 26 OF 28
DATE APRIL, 1988
MORGAN & PARMLEY, LTD. CONSULTING ENGINEERS LADYSMITH, WISCONSIN
DR. BY M.D. SHERVEY

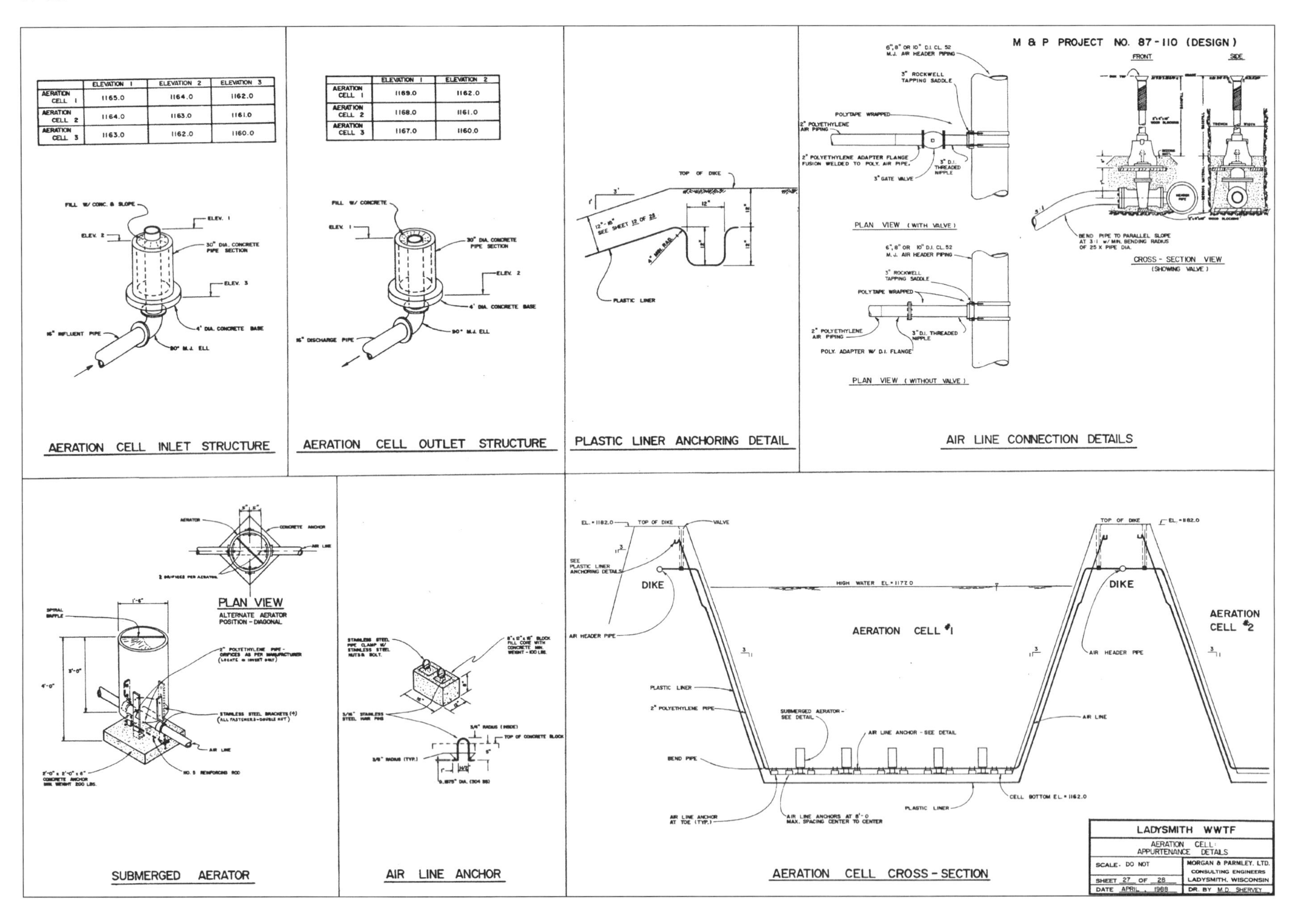

ELEVATION 1
ELEVATION 2
ELEVATION 3
AERATION CELL 1 1165.0 1164.0 1162.0
AERATION CELL 2 1164.0 1163.0 1161.0
AERATION CELL 3 1163.0 1162.0 1160.0
FILL W/ CONC. & SLOPE
ELEV. 1
ELEV. 2
30" DIA. CONCRETE PIPE SECTION
ELEV. 3
4' DIA. CONCRETE BASE
16" INFLUENT PIPE
90° M.J. ELL
AERATION CELL INLET STRUCTURE
AERATION CELL 1 1169.0 1162.0
AERATION CELL 2 1168.0 1161.0
AERATION CELL 3 1167.0 1160.0
FILL W/ CONCRETE
16" DISCHARGE PIPE
AERATION CELL OUTLET STRUCTURE
TOP OF DIKE
PLASTIC LINER
PLASTIC LINER ANCHORING DETAIL
M & P PROJECT NO. 87-110 (DESIGN)
FRONT
SIDE
6", 8" OR 10" D.I. CL. 52 M.J. AIR HEADER PIPING
3" ROCKWELL TAPPING SADDLE
POLYTAPE WRAPPED
2" POLYETHYLENE AIR PIPING
2" POLYETHYLENE ADAPTER FLANGE FUSION WELDED TO POLY. AIR PIPE.
3" D.I. THREADED NIPPLE
3" GATE VALVE
PLAN VIEW (WITH VALVE)
BEND PIPE TO PARALLEL SLOPE AT 3:1 w/ MIN. BENDING RADIUS OF 25 X PIPE DIA.
CROSS-SECTION VIEW (SHOWING VALVE)
POLY. ADAPTER W/ D.I. FLANGE
PLAN VIEW (WITHOUT VALVE)
AIR LINE CONNECTION DETAILS
AERATOR
CONCRETE ANCHOR
AIR LINE
PLAN VIEW
ALTERNATE AERATOR POSITION - DIAGONAL
SPIRAL BAFFLE
2" POLYETHYLENE PIPE - ORIFICES AS PER MANUFACTURER
STAINLESS STEEL BRACKETS (4)
NO. 5 REINFORCING ROD
SUBMERGED AERATOR
STAINLESS STEEL PIPE CLAMP W/ STAINLESS STEEL NUTS & BOLT.
TOP OF CONCRETE BLOCK
AIR LINE ANCHOR
EL. = 1182.0
TOP OF DIKE
VALVE
SEE PLASTIC LINER ANCHORING DETAILS
DIKE
HIGH WATER EL. = 1177.0
AIR HEADER PIPE
AERATION CELL #1
AERATION CELL #2
PLASTIC LINER
2" POLYETHYLENE PIPE
SUBMERGED AERATOR - SEE DETAIL
AIR LINE ANCHOR - SEE DETAIL
AIR LINE
BEND PIPE
CELL BOTTOM EL. = 1162.0
AIR LINE ANCHOR AT TOE (TYP.)
AIR LINE ANCHORS AT 8'-0 MAX. SPACING CENTER TO CENTER
AERATION CELL CROSS-SECTION
LADYSMITH WWTF
AERATION CELL APPURTENANCE DETAILS
SCALE - DO NOT
SHEET 27 OF 28
DATE APRIL, 1988
MORGAN & PARMLEY, LTD.
CONSULTING ENGINEERS
LADYSMITH, WISCONSIN
DR. BY M.D. SHERVEY

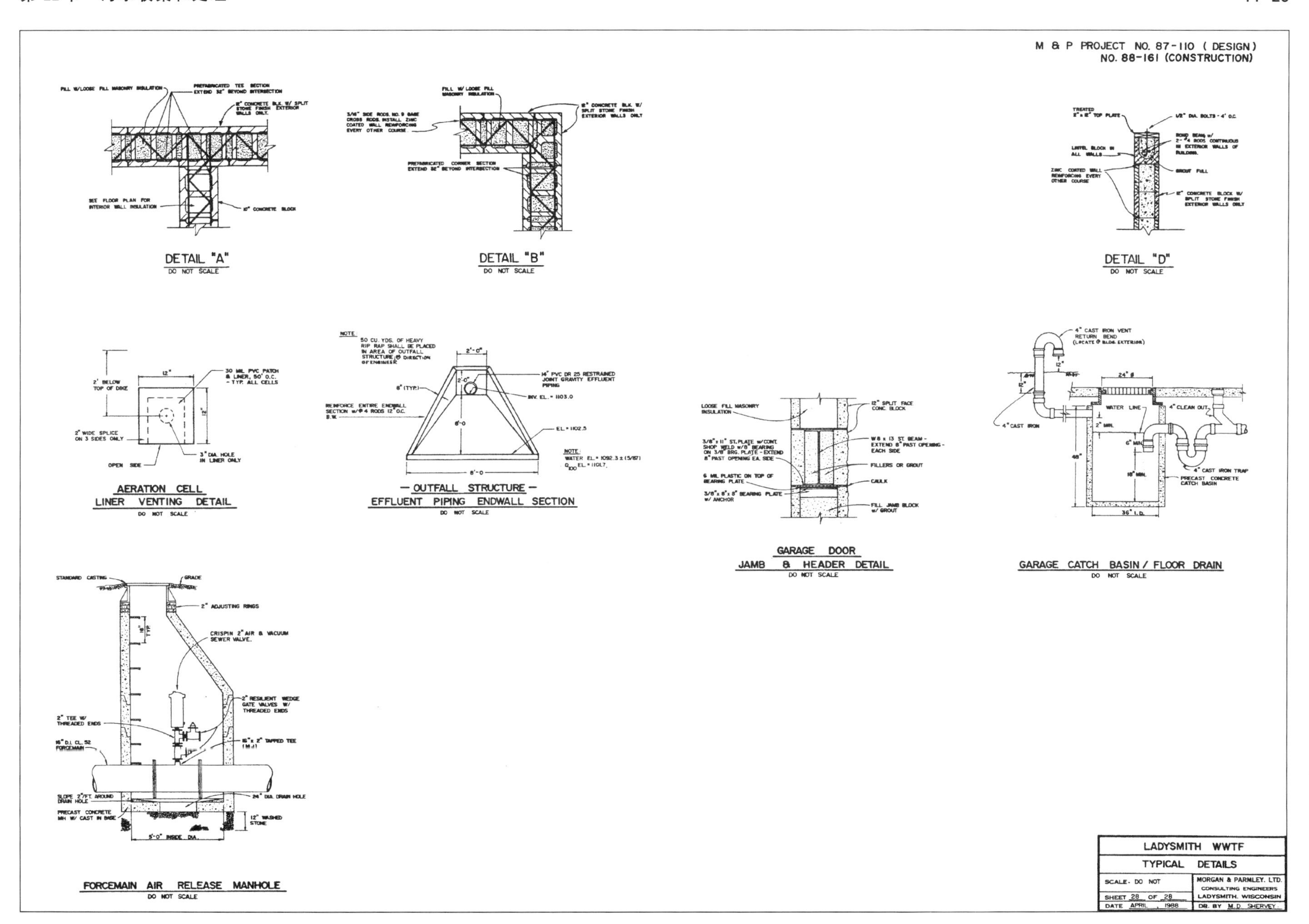

M & P PROJECT NO. 87-110 (DESIGN)
NO. 88-161 (CONSTRUCTION)
DETAIL "A"
DO NOT SCALE
DETAIL "B"
DO NOT SCALE
DETAIL "D"
DO NOT SCALE
AERATION CELL
LINER VENTING DETAIL
DO NOT SCALE
30 MIL PVC PATCH & LINER, 50' O.C. - TYP. ALL CELLS
2' BELOW TOP OF DIKE
2" WIDE SPLICE ON 3 SIDES ONLY
OPEN SIDE
3" DIA. HOLE IN LINER ONLY
NOTE
50 CU. YDS. OF HEAVY RIP RAP SHALL BE PLACED IN AREA OF OUTFALL STRUCTURE @ DIRECTION OF ENGINEER
14" PVC DR 25 RESTRAINED JOINT GRAVITY EFFLUENT PIPING
INV. EL. = 1103.0
EL. = 1102.5
NOTE:
WATER EL. = 1092.3 ± (5/87)
- OUTFALL STRUCTURE -
EFFLUENT PIPING ENDWALL SECTION
DO NOT SCALE
LOOSE FILL MASONRY INSULATION
12" SPLIT FACE CONC. BLOCK
FILLERS OR GROUT
CAULK
FILL JAMB BLOCK w/ GROUT
GARAGE DOOR
JAMB & HEADER DETAIL
DO NOT SCALE
4" CAST IRON VENT RETURN BEND
WATER LINE
4" CLEAN OUT
4" CAST IRON
4" CAST IRON TRAP
PRECAST CONCRETE CATCH BASIN
GARAGE CATCH BASIN / FLOOR DRAIN
DO NOT SCALE
STANDARD CASTING
GRADE
2" ADJUSTING RINGS
CRISPIN 2" AIR & VACUUM SEWER VALVE.
2" RESILIENT WEDGE GATE VALVES W/ THREADED ENDS
2" TEE W/ THREADED ENDS
16" D.I. CL. 52 FORCEMAIN
16" x 2" TAPPED TEE
SLOPE 2"/FT. AROUND DRAIN HOLE
24" DIA. DRAIN HOLE
PRECAST CONCRETE MH W/ CAST IN BASE
12" WASHED STONE
5'-0" INSIDE DIA.
FORCEMAIN AIR RELEASE MANHOLE
DO NOT SCALE
LADYSMITH WWTF
TYPICAL DETAILS
SCALE- DO NOT
MORGAN & PARMLEY, LTD.
CONSULTING ENGINEERS
LADYSMITH, WISCONSIN
SHEET 28 OF 28
DATE APRIL , 1988
DR. BY M.D. SHERVEY

M & P PROJECT NO. 85-182
NO. 88-116

污水收集和处理设施

WASTEWATER COLLECTION & TREATMENT FACILITY

VILLAGE
OF
CONRATH, WISCONSIN

PROJECT NO. C551225-02 (DESIGN)
C551225-03 (CONSTRUCTION)

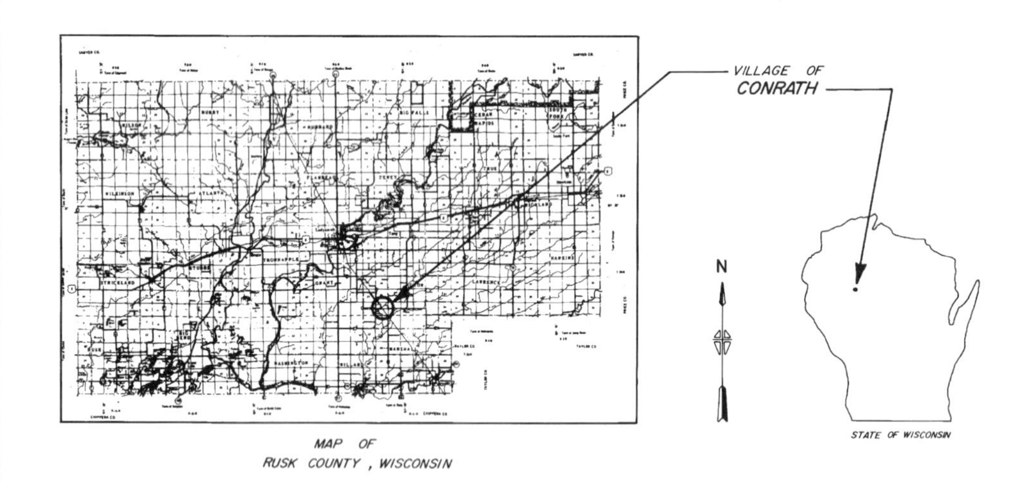

MAP OF
RUSK COUNTY, WISCONSIN

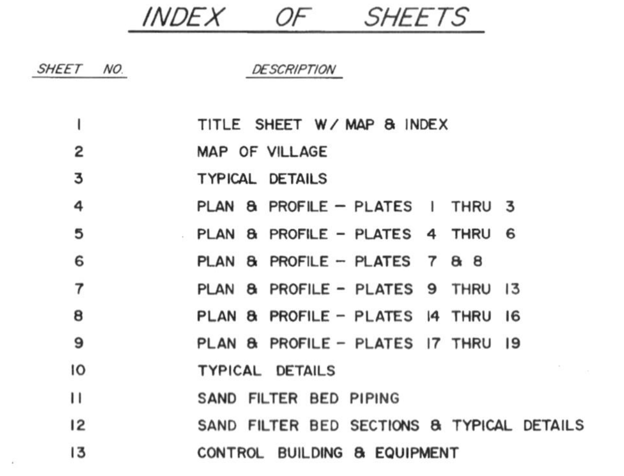

INDEX OF SHEETS

SHEET NO.	DESCRIPTION
1	TITLE SHEET W/ MAP & INDEX
2	MAP OF VILLAGE
3	TYPICAL DETAILS
4	PLAN & PROFILE – PLATES 1 THRU 3
5	PLAN & PROFILE – PLATES 4 THRU 6
6	PLAN & PROFILE – PLATES 7 & 8
7	PLAN & PROFILE – PLATES 9 THRU 13
8	PLAN & PROFILE – PLATES 14 THRU 16
9	PLAN & PROFILE – PLATES 17 THRU 19
10	TYPICAL DETAILS
11	SAND FILTER BED PIPING
12	SAND FILTER BED SECTIONS & TYPICAL DETAILS
13	CONTROL BUILDING & EQUIPMENT

PREPARED BY: MORGAN & PARMLEY, LTD. LADYSMITH, WISCONSIN

Courtesy: Morgan & Parmley, Ltd.

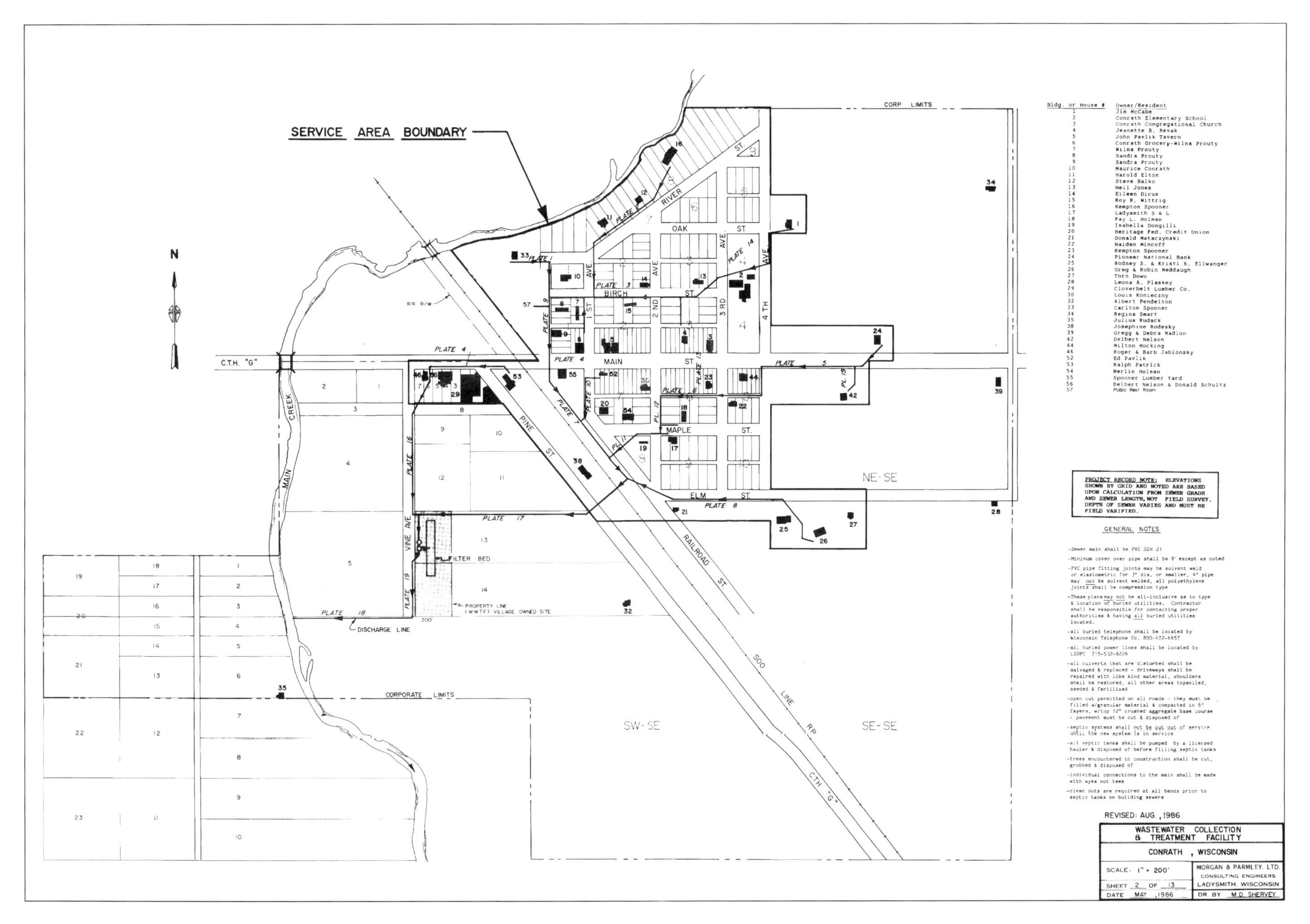
SERVICE AREA BOUNDARY
CORP LIMITS
N
C.T.H. "G"
MAIN CREEK
RR R/W
OAK ST.
BIRCH ST.
MAIN ST.
MAPLE ST.
ELM ST.
RIVER ST.
1 ST AVE
2 ND AVE
3 RD AVE
4 TH AVE
PINE ST
VINE AVE
RAILROAD ST.
SOO LINE RR
CTH "G"
FILTER BED
PROPERTY LINE (WWTF) VILLAGE OWNED SITE
DISCHARGE LINE
CORPORATE LIMITS
NE-SE
SW-SE
SE-SE
PLATE 1
PLATE 3
PLATE 4
PLATE 5
PLATE 6
PLATE 7
PLATE 8
PLATE 14
PLATE 16
PLATE 17
PLATE 18
PLATE 19
Bldg. or House #　Owner/Resident
1　Jim McCabe
2　Conrath Elementary School
3　Conrath Congregational Church
4　Jeanette B. Revak
5　John Pavlik Tavern
6　Conrath Grocery-Wilna Prouty
7　Wilna Prouty
8　Sandra Prouty
9　Sandra Prouty
10　Maurice Conrath
11　Harold Elton
12　Steve Balko
13　Neil Jones
14　Eileen Dicus
15　Roy R. Wittrig
16　Kempton Spooner
17　Ladysmith S & L
18　Fay L. Holman
19　Isabella Dongilli
20　Heritage Fed. Credit Union
21　Donald Mataczynski
22　Naiden Mincoff
23　Kempton Spooner
24　Pioneer National Bank
25　Rodney S. & Kristi K. Ellwanger
26　Greg & Robin Meddaugh
27　Torn Down
28　Leona A. Plaskey
29　Cloverbelt Lumber Co.
30　Louis Konieczny
32　Albert Pendelton
33　Carlton Spooner
34　Regina Smart
35　Julius Rudack
38　Josephine Rodesky
39　Gregg & Debra Madlon
42　Delbert Nelson
44　Milton Hocking
46　Roger & Barb Jablonsky
52　Ed Pavlik
53　Ralph Patrick
54　Merlin Holman
55　Spooner Lumber Yard
56　Delbert Nelson & Donald Schultz
57　Public Rest Room
PROJECT RECORD NOTE: ELEVATIONS SHOWN BY GRID AND NOTED ARE BASED UPON CALCULATION FROM SEWER GRADE AND SEWER LENGTH, NOT FIELD SURVEY. DEPTH OF SEWER VARIES AND MUST BE FIELD VARIFIED.
GENERAL NOTES
-Sewer main shall be PVC SDR 21
-Minimum cover over pipe shall be 8' except as noted
-PVC pipe fitting joints may be solvent weld or elastometric for 3" dia. or smaller, 4" pipe may not be solvent welded, all polyethylene joints shall be compression type
-These plans may not be all-inclusive as to type & location of buried utilities. Contractor shall be responsible for contacting proper authorities & having all buried utilities located.
-all buried telephone shall be located by Wisconsin Telephone Co. 800-472-6657
-all buried power lines shall be located by LSDPC 715-532-6226
-all culverts that are disturbed shall be salvaged & replaced - driveways shall be repaired with like kind material, shoulders shall be restored, all other areas topsoiled, seeded & fertilized
-open cut permitted on all roads - they must be filled w/granular material & compacted in 6" layers, w/top 12" crushed aggregate base course - pavement must be cut & disposed of
-septic systems shall not be put out of service until the new system is in service
-all septic tanks shall be pumped by a licensed hauler & disposed of before filling septic tanks
-trees encountered in construction shall be cut, grubbed & disposed of
-individual connections to the main shall be made with wyes not tees
-clean outs are required at all bends prior to septic tanks on building sewers
REVISED: AUG., 1986
WASTEWATER COLLECTION & TREATMENT FACILITY
CONRATH, WISCONSIN
SCALE- 1" = 200'
SHEET 2 OF 13
DATE MAY, 1986
MORGAN & PARMLEY, LTD.
CONSULTING ENGINEERS
LADYSMITH, WISCONSIN
DR BY M.D. SHERVEY

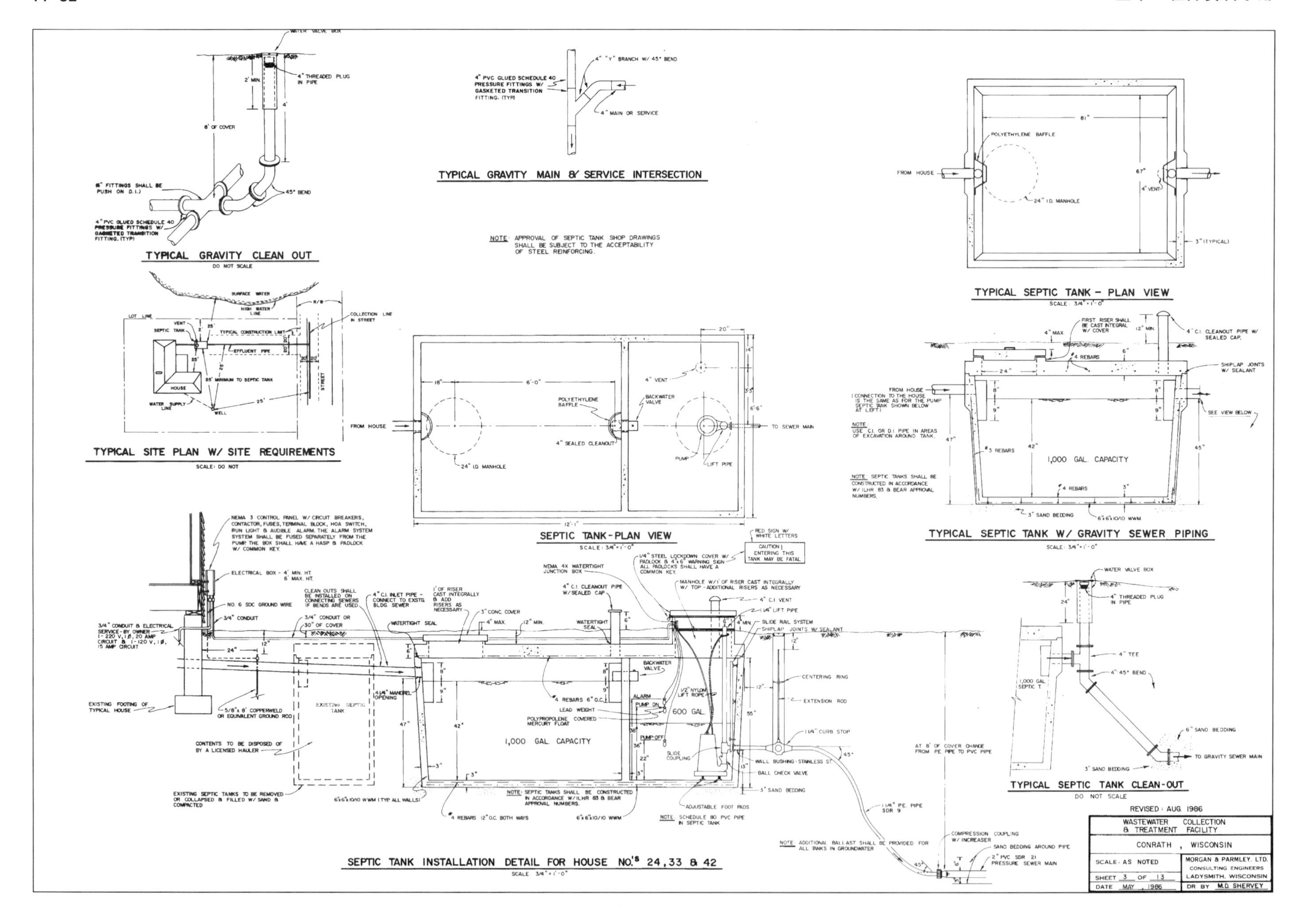
TYPICAL GRAVITY CLEAN OUT
DO NOT SCALE
TYPICAL GRAVITY MAIN & SERVICE INTERSECTION
NOTE: APPROVAL OF SEPTIC TANK SHOP DRAWINGS SHALL BE SUBJECT TO THE ACCEPTABILITY OF STEEL REINFORCING.
TYPICAL SEPTIC TANK - PLAN VIEW
SCALE: 3/4"=1'-0"
TYPICAL SITE PLAN W/ SITE REQUIREMENTS
SCALE: DO NOT
SEPTIC TANK-PLAN VIEW
SCALE: 3/4"=1'-0"
TYPICAL SEPTIC TANK W/ GRAVITY SEWER PIPING
SCALE: 3/4"=1'-0"
1,000 GAL. CAPACITY
600 GAL.
TYPICAL SEPTIC TANK CLEAN-OUT
DO NOT SCALE
SEPTIC TANK INSTALLATION DETAIL FOR HOUSE NO.'S 24, 33 & 42
SCALE 3/4"=1'-0"
REVISED: AUG. 1986
WASTEWATER COLLECTION & TREATMENT FACILITY
CONRATH, WISCONSIN
SCALE: AS NOTED
SHEET 3 OF 13
DATE MAY, 1986
MORGAN & PARMLEY, LTD.
CONSULTING ENGINEERS
LADYSMITH, WISCONSIN
DR BY M.D. SHERVEY

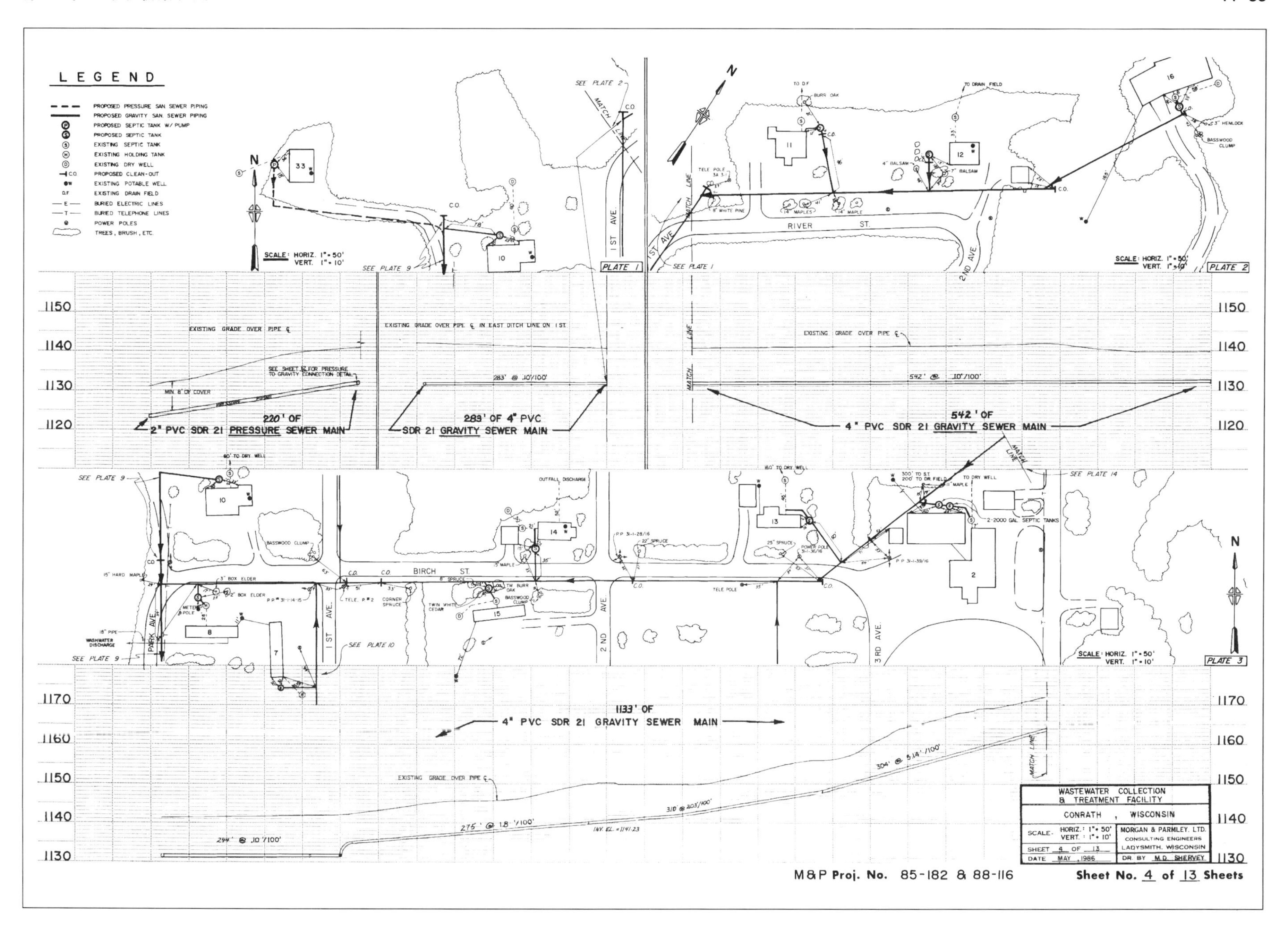
LEGEND
PROPOSED PRESSURE SAN. SEWER PIPING
PROPOSED GRAVITY SAN. SEWER PIPING
PROPOSED SEPTIC TANK W/ PUMP
PROPOSED SEPTIC TANK
EXISTING SEPTIC TANK
EXISTING HOLDING TANK
EXISTING DRY WELL
PROPOSED CLEAN-OUT
EXISTING POTABLE WELL
EXISTING DRAIN FIELD
BURIED ELECTRIC LINES
BURIED TELEPHONE LINES
POWER POLES
TREES, BRUSH, ETC.
220' OF 2" PVC SDR 21 PRESSURE SEWER MAIN
283' OF 4" PVC SDR 21 GRAVITY SEWER MAIN
542' OF 4" PVC SDR 21 GRAVITY SEWER MAIN
1133' OF 4" PVC SDR 21 GRAVITY SEWER MAIN
EXISTING GRADE OVER PIPE ℄
RIVER ST.
BIRCH ST.
PLATE 1
PLATE 2
PLATE 3
SCALE: HORIZ. 1" = 50' VERT. 1" = 10'
WASTEWATER COLLECTION & TREATMENT FACILITY
CONRATH, WISCONSIN
MORGAN & PARMLEY, LTD. CONSULTING ENGINEERS LADYSMITH, WISCONSIN
SHEET 4 OF 13
DATE MAY, 1986
DR. BY M.D. SHERVEY
M&P Proj. No. 85-182 & 88-116
Sheet No. 4 of 13 Sheets

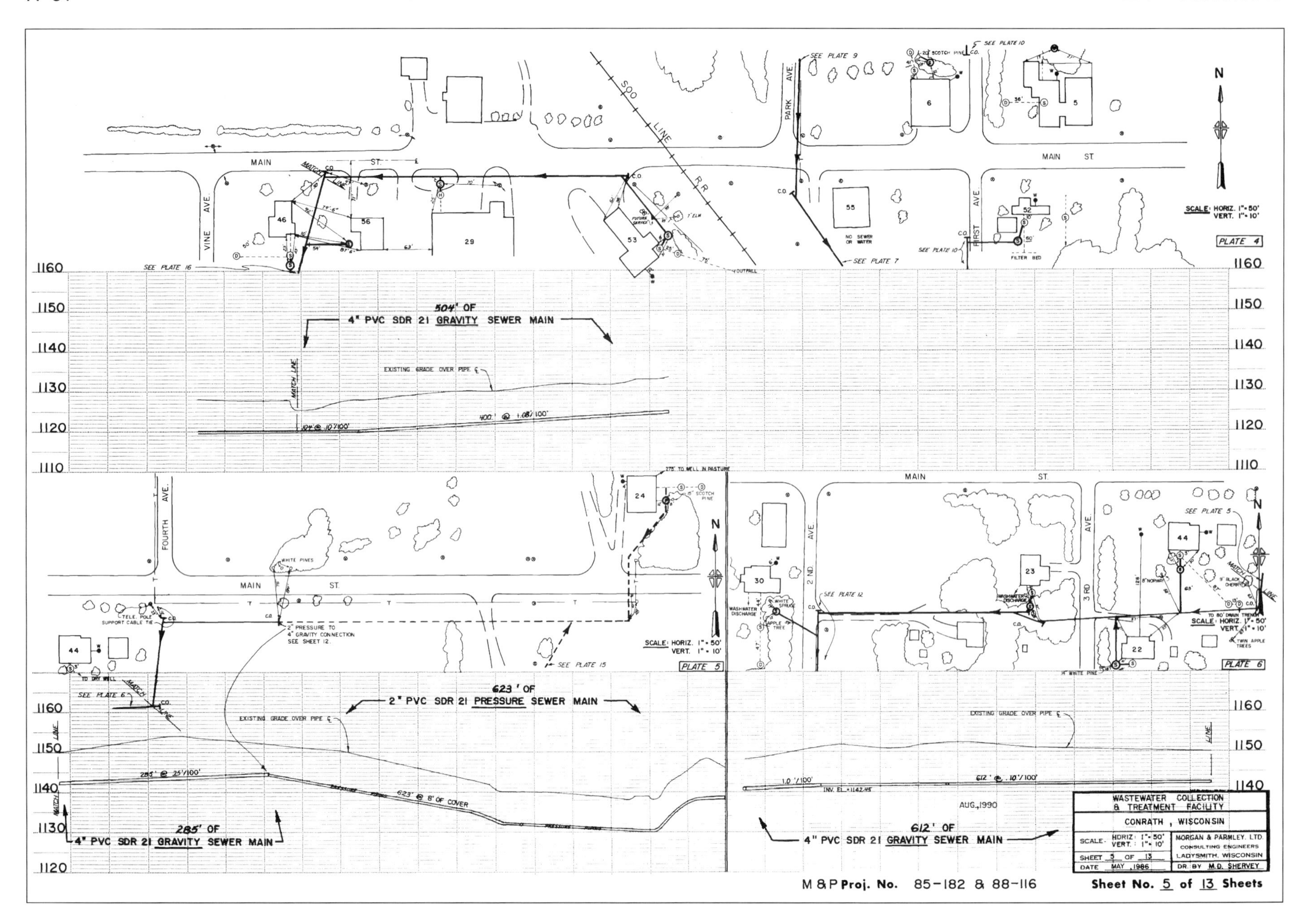

MAIN ST.
VINE AVE.
PARK AVE.
FIRST AVE.
SOO LINE R.R.
SEE PLATE 16
SEE PLATE 9
SEE PLATE 7
SEE PLATE 10
PLATE 4
SCALE: HORIZ. 1"=50' VERT. 1"=10'
504' OF 4" PVC SDR 21 GRAVITY SEWER MAIN
EXISTING GRADE OVER PIPE ℄
400' @ 1.08'/100'
104' @ 1.0'/100'
FOURTH AVE.
2ND AVE.
3RD AVE.
WHITE PINES
2" PRESSURE TO 4" GRAVITY CONNECTION SEE SHEET 12.
SEE PLATE 15
SEE PLATE 6
SEE PLATE 5
SEE PLATE 12
PLATE 5
PLATE 6
623' OF 2" PVC SDR 21 PRESSURE SEWER MAIN
285' @ .25'/100'
623' @ 8' OF COVER
285' OF 4" PVC SDR 21 GRAVITY SEWER MAIN
612' @ .10'/100'
612' OF 4" PVC SDR 21 GRAVITY SEWER MAIN
AUG,1990
WASTEWATER COLLECTION & TREATMENT FACILITY
CONRATH, WISCONSIN
MORGAN & PARMLEY, LTD. CONSULTING ENGINEERS LADYSMITH, WISCONSIN
SCALE - HORIZ. 1"=50' VERT. 1"=10'
SHEET 5 OF 13
DATE MAY, 1986
DR. BY M.D. SHERVEY
M&P Proj. No. 85-182 & 88-116
Sheet No. 5 of 13 Sheets

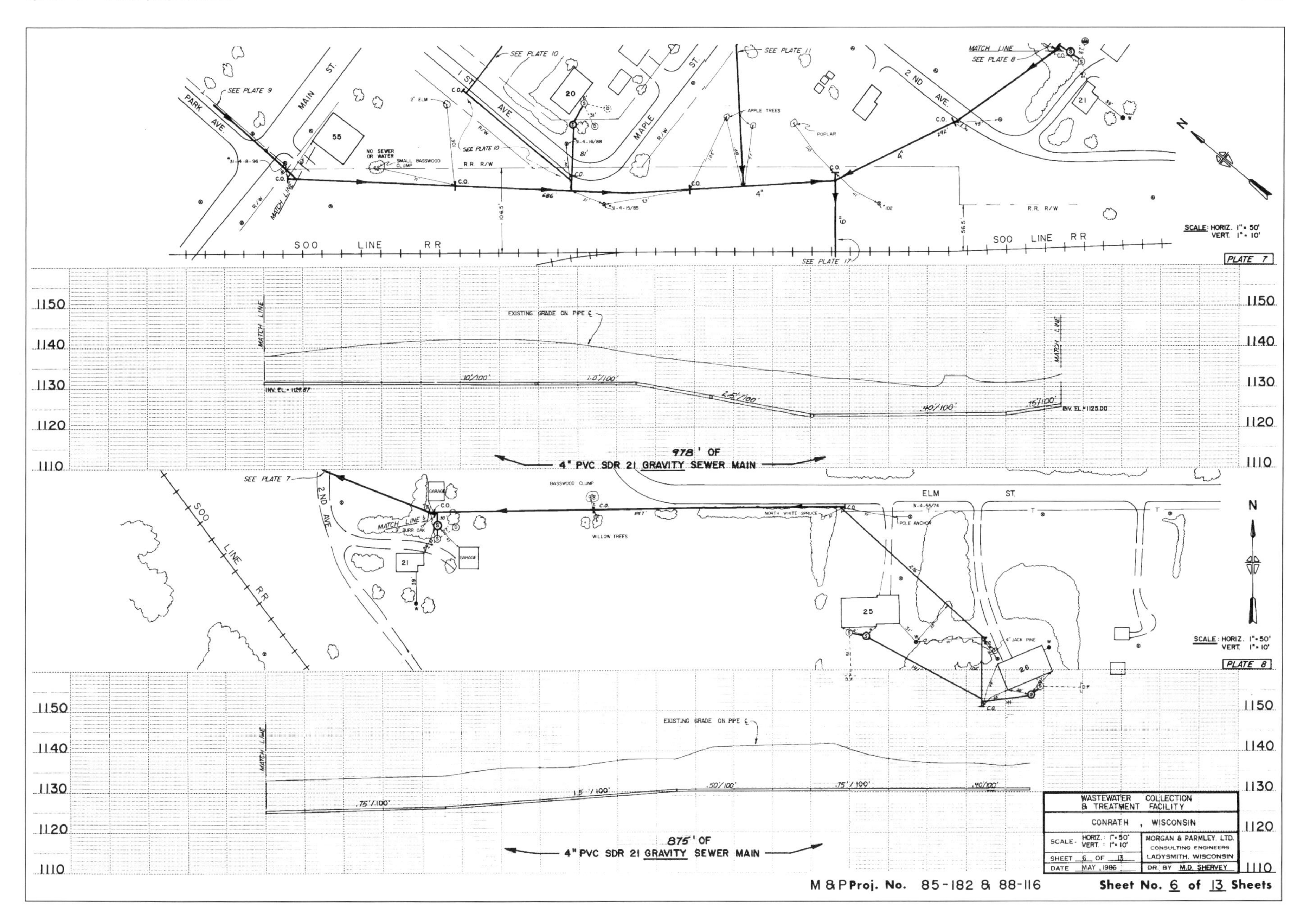

PLATE 7
SOO LINE RR
SCALE: HORIZ. 1" = 50' VERT. 1" = 10'
978' OF 4" PVC SDR 21 GRAVITY SEWER MAIN
EXISTING GRADE ON PIPE ℄
PLATE 8
ELM ST.
875' OF 4" PVC SDR 21 GRAVITY SEWER MAIN
WASTEWATER COLLECTION & TREATMENT FACILITY
CONRATH, WISCONSIN
MORGAN & PARMLEY, LTD. CONSULTING ENGINEERS LADYSMITH, WISCONSIN
SHEET 6 OF 13
DATE MAY, 1986
DR. BY M.D. SHERVEY
M & P Proj. No. 85-182 & 88-116
Sheet No. 6 of 13 Sheets

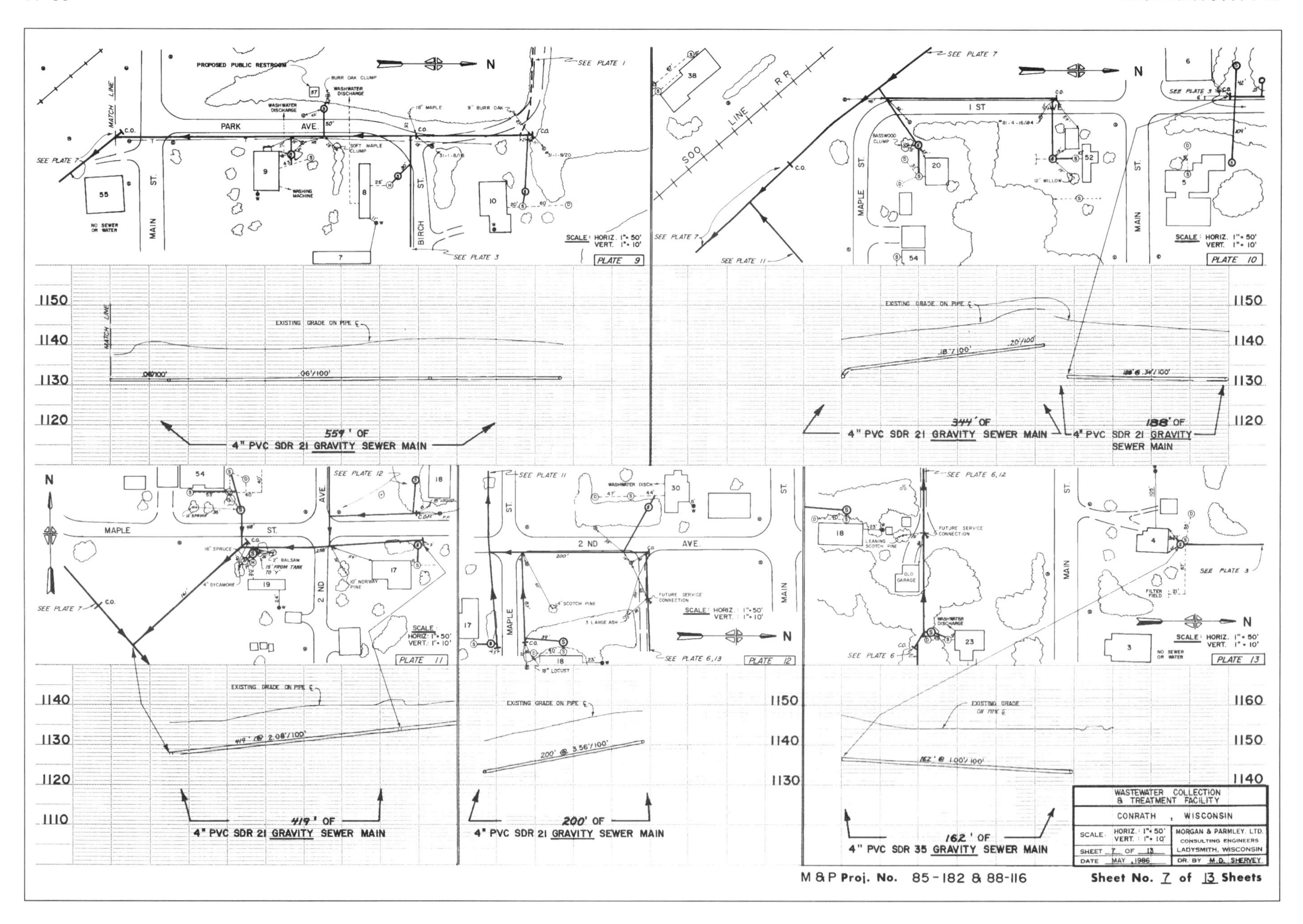
PLATE 9
PLATE 10
PLATE 11
PLATE 12
PLATE 13
559' OF 4" PVC SDR 21 GRAVITY SEWER MAIN
344' OF 4" PVC SDR 21 GRAVITY SEWER MAIN
188' OF 4" PVC SDR 21 GRAVITY SEWER MAIN
419' OF 4" PVC SDR 21 GRAVITY SEWER MAIN
200' OF 4" PVC SDR 21 GRAVITY SEWER MAIN
162' OF 4" PVC SDR 35 GRAVITY SEWER MAIN
EXISTING GRADE ON PIPE ℄
SCALE: HORIZ. 1"=50' VERT. 1"=10'
WASTEWATER COLLECTION & TREATMENT FACILITY
CONRATH, WISCONSIN
MORGAN & PARMLEY LTD. CONSULTING ENGINEERS LADYSMITH, WISCONSIN
SHEET 7 OF 13
DATE MAY, 1986
DR. BY M.D. SHERVEY
M & P Proj. No. 85-182 & 88-116
Sheet No. 7 of 13 Sheets

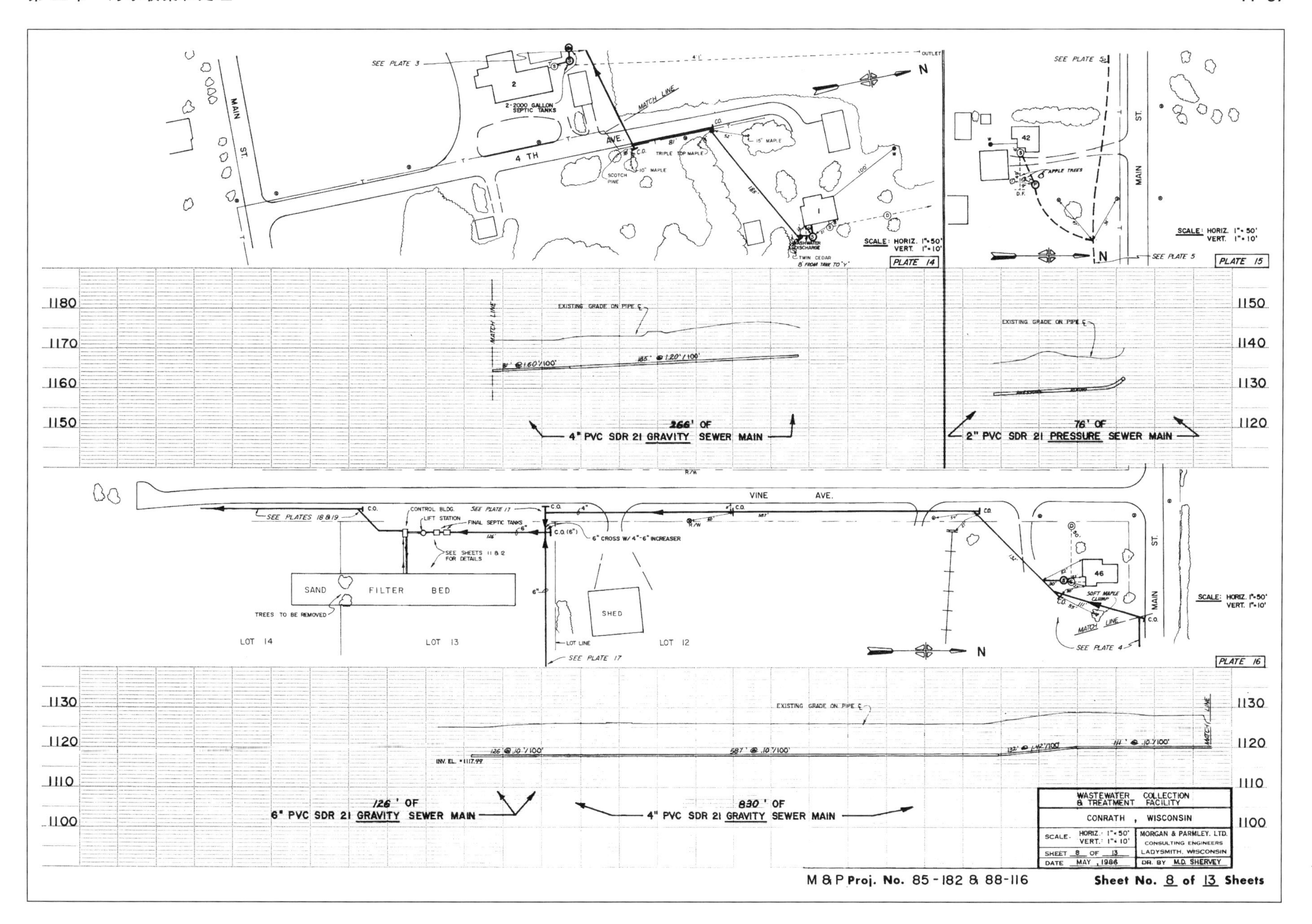

266' OF
4" PVC SDR 21 GRAVITY SEWER MAIN
76' OF
2" PVC SDR 21 PRESSURE SEWER MAIN
126' OF
6" PVC SDR 21 GRAVITY SEWER MAIN
830' OF
4" PVC SDR 21 GRAVITY SEWER MAIN
PLATE 14
PLATE 15
PLATE 16
WASTEWATER COLLECTION & TREATMENT FACILITY
CONRATH, WISCONSIN
MORGAN & PARMLEY, LTD. CONSULTING ENGINEERS LADYSMITH, WISCONSIN
SHEET 8 OF 13
DATE MAY, 1986
DR. BY M.D. SHERVEY
M & P Proj. No. 85-182 & 88-116
Sheet No. 8 of 13 Sheets

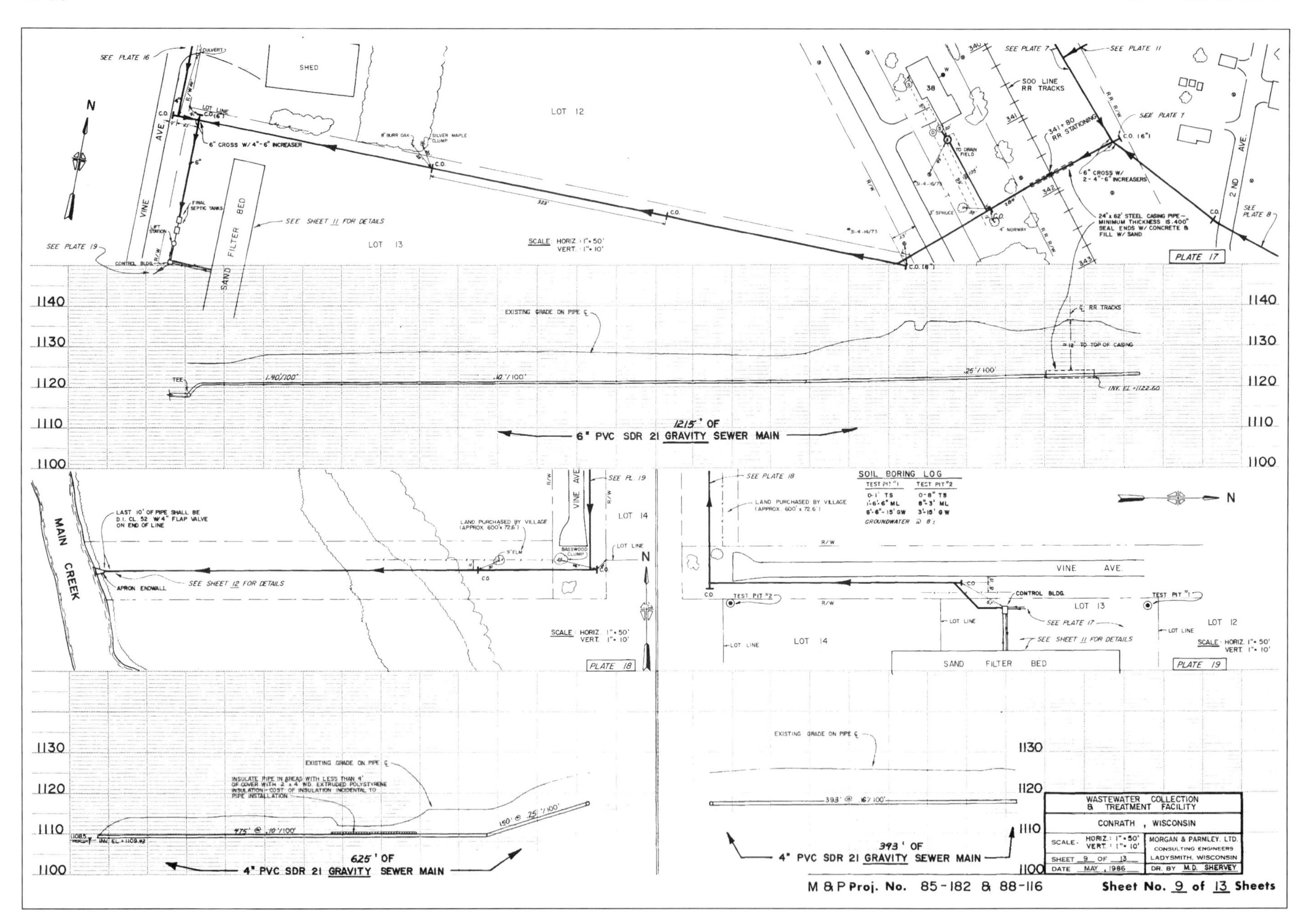
SEE PLATE 16
SHED
LOT 12
VINE AVE.
6" CROSS W/ 4"-6" INCREASER
FINAL SEPTIC TANKS
LIFT STATION
CONTROL BLDG.
SAND FILTER BED
SEE SHEET 11 FOR DETAILS
SEE PLATE 19
LOT 13
SCALE: HORIZ.: 1"=50' VERT.: 1"=10'
SOO LINE RR TRACKS
SEE PLATE 7
SEE PLATE 11
RR STATIONING
6" CROSS W/ 2-4"-6" INCREASERS
24" x 62' STEEL CASING PIPE— MINIMUM THICKNESS IS .400" SEAL ENDS W/ CONCRETE & FILL W/ SAND
2 ND AVE.
SEE PLATE 8
PLATE 17
EXISTING GRADE ON PIPE ℄
℄ RR TRACKS
TO TOP OF CASING
INV. EL.=1122.60
1215' OF
6" PVC SDR 21 GRAVITY SEWER MAIN
MAIN CREEK
LAST 10' OF PIPE SHALL BE D.I. CL. 52 W/ 4" FLAP VALVE ON END OF LINE
APRON ENDWALL
SEE SHEET 12 FOR DETAILS
LAND PURCHASED BY VILLAGE (APPROX. 600'x 72.6')
LOT 14
SEE PL. 19
PLATE 18
SEE PLATE 18
SOIL BORING LOG
TEST PIT #1
TEST PIT #2
GROUNDWATER @ 8'±
VINE AVE.
CONTROL BLDG.
TEST PIT #2
TEST PIT #1
LOT 13
LOT 12
LOT LINE
SEE PLATE 17
SEE SHEET 11 FOR DETAILS
SAND FILTER BED
PLATE 19
INSULATE PIPE IN AREAS WITH LESS THAN 4' OF COVER WITH 2" x 4' WD. EXTRUDED POLYSTYRENE INSULATION - COST OF INSULATION INCIDENTAL TO PIPE INSTALLATION
475' @ .10'/100'
150' @ .25'/100'
625' OF
4" PVC SDR 21 GRAVITY SEWER MAIN
393' @ .16'/100'
393' OF
4" PVC SDR 21 GRAVITY SEWER MAIN
WASTEWATER COLLECTION & TREATMENT FACILITY
CONRATH, WISCONSIN
MORGAN & PARMLEY, LTD. CONSULTING ENGINEERS LADYSMITH, WISCONSIN
SHEET 9 OF 13
DATE MAY, 1986
DR. BY M.D. SHERVEY
M & P Proj. No. 85-182 & 88-116
Sheet No. 9 of 13 Sheets

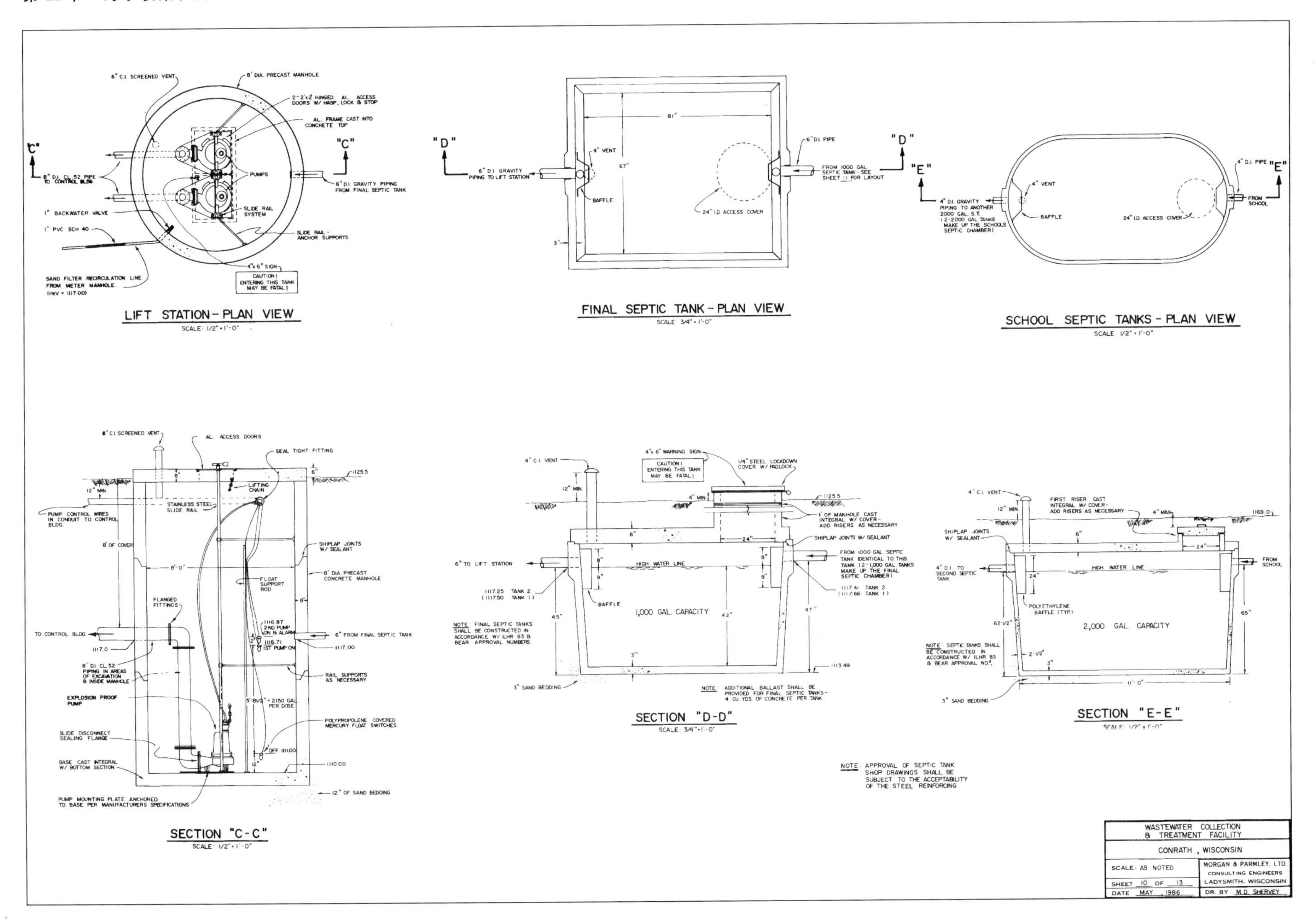
LIFT STATION - PLAN VIEW
SCALE: 1/2" = 1'-0"
8' DIA. PRECAST MANHOLE
6" C.I. SCREENED VENT
2-2'x2' HINGED AL. ACCESS DOORS W/ HASP, LOCK & STOP
AL. FRAME CAST INTO CONCRETE TOP
PUMPS
SLIDE RAIL SYSTEM
SLIDE RAIL - ANCHOR SUPPORTS
1" BACKWATER VALVE
1" PVC SCH 40
6" D.I. GRAVITY PIPING FROM FINAL SEPTIC TANK
SAND FILTER RECIRCULATION LINE FROM METER MANHOLE. (INV = 1117.00)
4'x6" SIGN
CAUTION! ENTERING THIS TANK MAY BE FATAL!
FINAL SEPTIC TANK - PLAN VIEW
SCALE: 3/4" = 1'-0"
6" D.I. GRAVITY PIPING TO LIFT STATION
4" VENT
BAFFLE
6" D.I. PIPE
FROM 1000 GAL. SEPTIC TANK - SEE SHEET 11 FOR LAYOUT
24" I.D. ACCESS COVER
SCHOOL SEPTIC TANKS - PLAN VIEW
SCALE: 1/2" = 1'-0"
4" D.I. GRAVITY PIPING TO ANOTHER 2000 GAL. S.T. (2-2000 GAL TANKS MAKE UP THE SCHOOLS SEPTIC CHAMBER)
4" D.I. PIPE
FROM SCHOOL
SECTION "C-C"
SCALE: 1/2" = 1'-0"
AL. ACCESS DOORS
SEAL TIGHT FITTING
LIFTING CHAIN
STAINLESS STEEL SLIDE RAIL
PUMP CONTROL WIRES IN CONDUIT TO CONTROL BLDG.
8' OF COVER
SHIPLAP JOINTS W/ SEALANT
8' DIA. PRECAST CONCRETE MANHOLE
FLOAT SUPPORT ROD
FLANGED FITTINGS
TO CONTROL BLDG.
6" FROM FINAL SEPTIC TANK
8" D.I. CL.52 PIPING IN AREAS OF EXCAVATION & INSIDE MANHOLE
RAIL SUPPORTS AS NECESSARY
EXPLOSION PROOF PUMP
POLYPROPOLENE COVERED MERCURY FLOAT SWITCHES
SLIDE DISCONNECT SEALING FLANGE
BASE CAST INTEGRAL W/ BOTTOM SECTION
PUMP MOUNTING PLATE ANCHORED TO BASE PER MANUFACTURERS SPECIFICATIONS
12" OF SAND BEDDING
SECTION "D-D"
SCALE: 3/4" = 1'-0"
4" C.I. VENT
4'x6" WARNING SIGN
1/4" STEEL LOCKDOWN COVER W/ PADLOCK
1' OF MANHOLE CAST INTEGRAL W/ COVER - ADD RISERS AS NECESSARY
SHIPLAP JOINTS W/ SEALANT
6" TO LIFT STATION
HIGH WATER LINE
FROM 1000 GAL. SEPTIC TANK IDENTICAL TO THIS TANK (2-1,000 GAL. TANKS MAKE UP THE FINAL SEPTIC CHAMBER)
1,000 GAL. CAPACITY
NOTE: FINAL SEPTIC TANKS SHALL BE CONSTRUCTED IN ACCORDANCE W/ ILHR 83 & BEAR APPROVAL NUMBERS.
3" SAND BEDDING
NOTE: ADDITIONAL BALLAST SHALL BE PROVIDED FOR FINAL SEPTIC TANKS - 4 CU. YDS. OF CONCRETE PER TANK.
SECTION "E-E"
4" C.I. VENT
FIRST RISER CAST INTEGRAL W/ COVER - ADD RISERS AS NECESSARY
SHIPLAP JOINTS W/ SEALANT
4" D.I. TO SECOND SEPTIC TANK
FROM SCHOOL
POLYETHYLENE BAFFLE (TYP.)
2,000 GAL. CAPACITY
NOTE: SEPTIC TANKS SHALL BE CONSTRUCTED IN ACCORDANCE W/ ILHR 83 & BEAR APPROVAL NO.S.
NOTE: APPROVAL OF SEPTIC TANK SHOP DRAWINGS SHALL BE SUBJECT TO THE ACCEPTABILITY OF THE STEEL REINFORCING
WASTEWATER COLLECTION & TREATMENT FACILITY
CONRATH, WISCONSIN
SCALE: AS NOTED
SHEET 10 OF 13
DATE MAY 1986
MORGAN & PARMLEY, LTD.
CONSULTING ENGINEERS
LADYSMITH, WISCONSIN
DR. BY M.D. SHERVEY

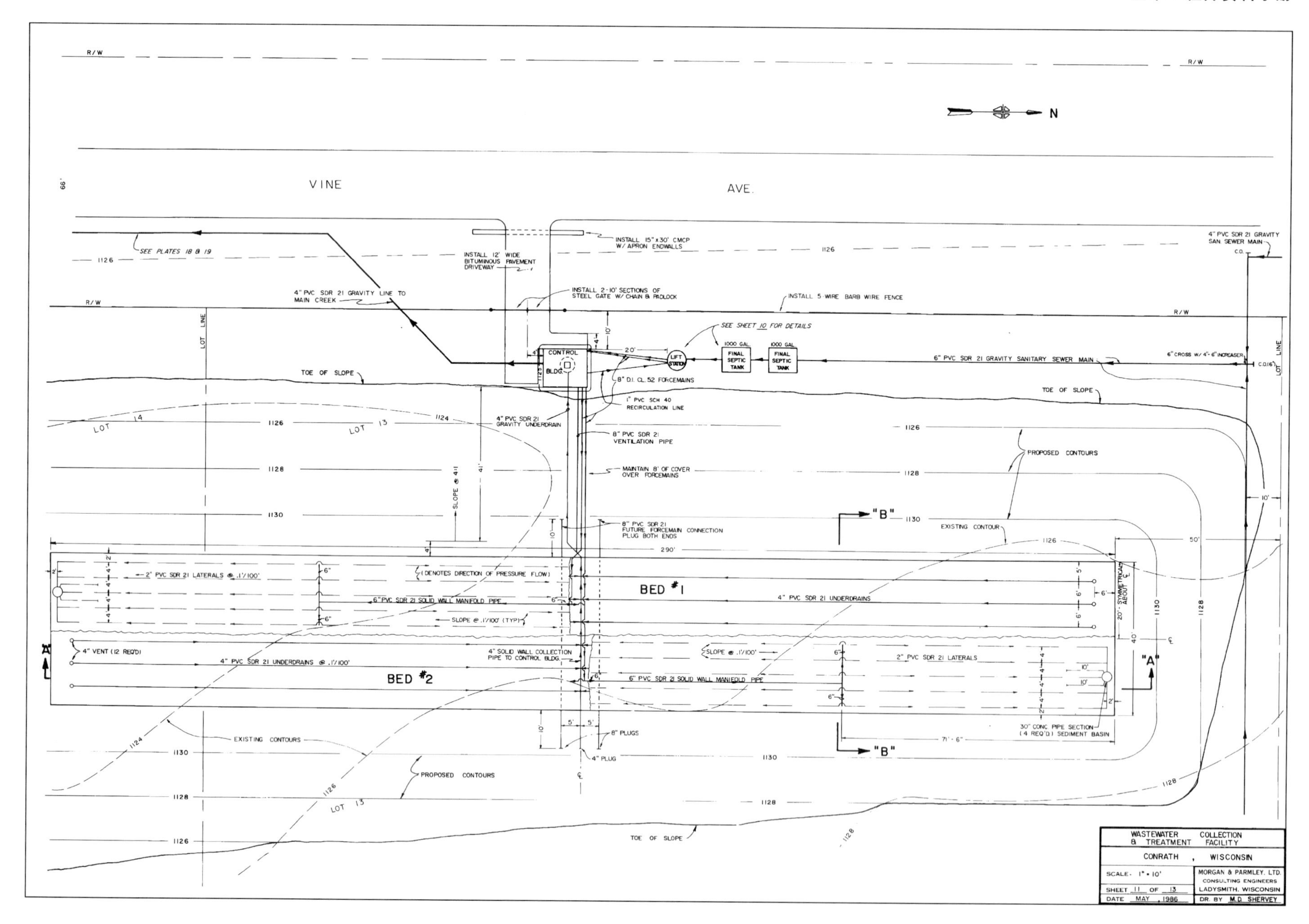
VINE AVE.
INSTALL 15"x30' CMCP W/ APRON ENDWALLS
INSTALL 12' WIDE BITUMINOUS PAVEMENT DRIVEWAY
SEE PLATES 18 & 19
4" PVC SDR 21 GRAVITY LINE TO MAIN CREEK
INSTALL 2-10' SECTIONS OF STEEL GATE W/ CHAIN & PADLOCK
INSTALL 5-WIRE BARB WIRE FENCE
SEE SHEET 10 FOR DETAILS
CONTROL BLDG.
LIFT STATION
1000 GAL FINAL SEPTIC TANK
6" PVC SDR 21 GRAVITY SANITARY SEWER MAIN
4" PVC SDR 21 GRAVITY SAN. SEWER MAIN
TOE OF SLOPE
8" D.I. CL. 52 FORCEMAINS
1" PVC SCH 40 RECIRCULATION LINE
4" PVC SDR 21 GRAVITY UNDERDRAIN
8" PVC SDR 21 VENTILATION PIPE
MAINTAIN 8' OF COVER OVER FORCEMAINS
PROPOSED CONTOURS
EXISTING CONTOUR
8" PVC SDR 21 FUTURE FORCEMAIN CONNECTION PLUG BOTH ENDS
BED #1
BED #2
2" PVC SDR 21 LATERALS @ .1'/100'
6" PVC SDR 21 SOLID WALL MANIFOLD PIPE
4" PVC SDR 21 UNDERDRAINS
4" VENT (12 REQ'D)
4" SOLID WALL COLLECTION PIPE TO CONTROL BLDG.
30" CONC PIPE SECTION (4 REQ'D) SEDIMENT BASIN
8" PLUGS
4" PLUG
EXISTING CONTOURS
WASTEWATER COLLECTION & TREATMENT FACILITY
CONRATH, WISCONSIN
SCALE: 1" = 10'
SHEET 11 OF 13
DATE MAY, 1986
MORGAN & PARMLEY, LTD. CONSULTING ENGINEERS LADYSMITH, WISCONSIN
DR. BY M.D. SHERVEY

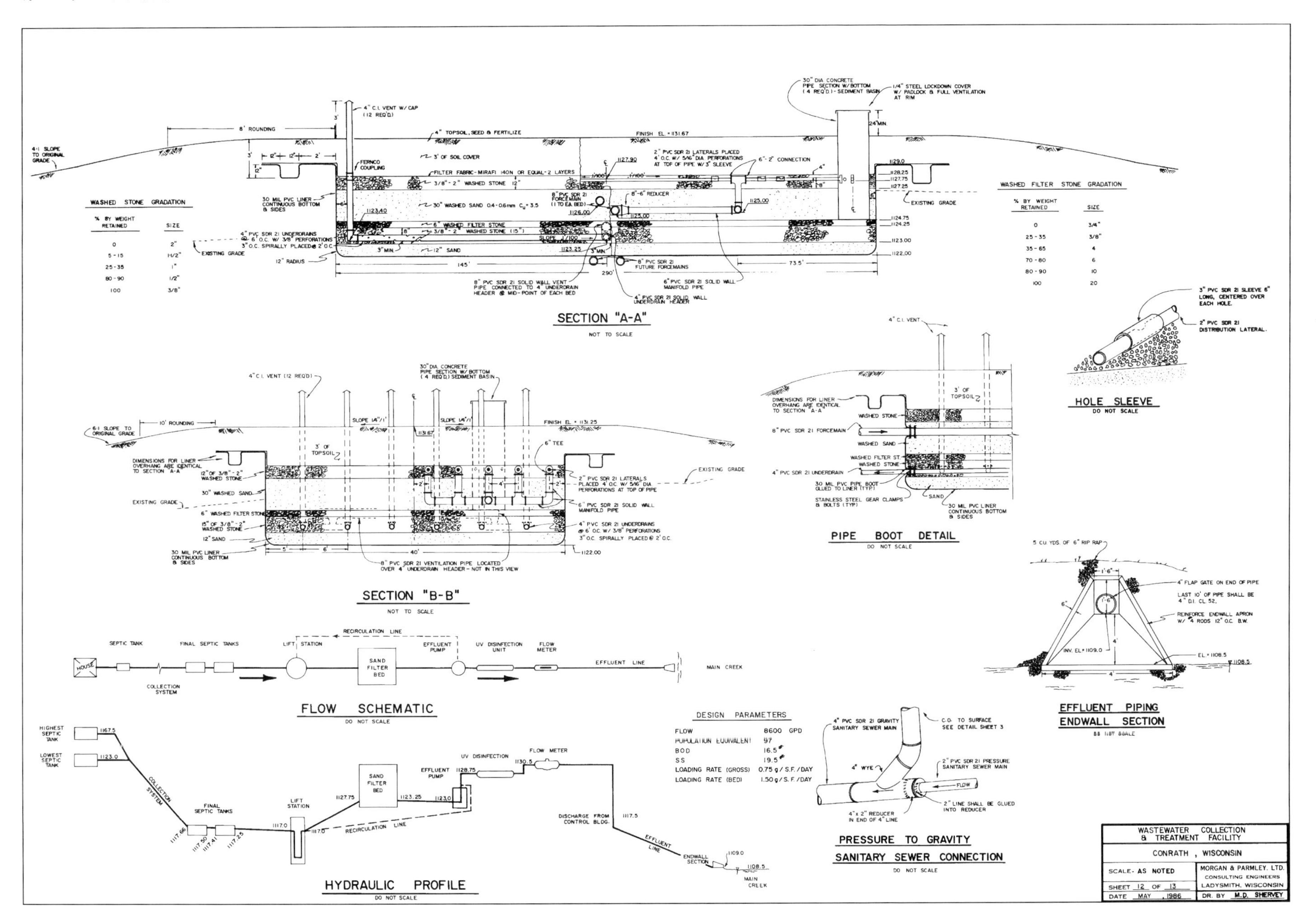
SECTION "A-A"
NOT TO SCALE
4" C.I. VENT W/ CAP (12 REQ'D)
8' ROUNDING
4:1 SLOPE TO ORIGINAL GRADE
4" TOPSOIL, SEED & FERTILIZE
3' OF SOIL COVER
FERNCO COUPLING
FILTER FABRIC-MIRAFI 140N OR EQUAL-2 LAYERS
3/8"-2" WASHED STONE 12"
30 MIL PVC LINER - CONTINUOUS BOTTOM & SIDES
30" WASHED SAND 0.4-0.6mm Cu = 3.5
6" WASHED FILTER STONE
3/8"-2" WASHED STONE (15")
12" SAND
12" RADIUS
4" PVC SDR 21 UNDERDRAINS @ 6" O.C. W/ 3/8" PERFORATIONS 3" O.C. SPIRALLY PLACED @ 2' O.C.
EXISTING GRADE
FINISH EL. = 1131.67
2" PVC SDR 21 LATERALS PLACED 4' O.C. W/ 5/16" DIA PERFORATIONS AT TOP OF PIPE W/ 3" SLEEVE
6"-2" CONNECTION
8" PVC SDR 21 FORCEMAIN (1 TO EA. BED)
8"-6" REDUCER
8" PVC SDR 21 FUTURE FORCEMAINS
8" PVC SDR 21 SOLID WALL VENT PIPE CONNECTED TO 4" UNDERDRAIN HEADER @ MID-POINT OF EACH BED
6" PVC SDR 21 SOLID WALL MANIFOLD PIPE
4" PVC SDR 21 SOLID WALL UNDERDRAIN HEADER
30" DIA CONCRETE PIPE SECTION W/ BOTTOM (4 REQ'D) - SEDIMENT BASIN
1/4" STEEL LOCKDOWN COVER W/ PADLOCK & FULL VENTILATION AT RIM
24" MIN.
SLOPE 1/100
145'
290'
73.5'
WASHED STONE GRADATION
% BY WEIGHT RETAINED | SIZE
0 | 2"
5-15 | 1 1/2"
25-35 | 1"
80-90 | 1/2"
100 | 3/8"
WASHED FILTER STONE GRADATION
% BY WEIGHT RETAINED | SIZE
0 | 3/4"
25-35 | 3/8"
35-65 | 4
70-80 | 6
80-90 | 10
100 | 20
HOLE SLEEVE
DO NOT SCALE
3" PVC SDR 21 SLEEVE 6" LONG, CENTERED OVER EACH HOLE.
2" PVC SDR 21 DISTRIBUTION LATERAL.
SECTION "B-B"
NOT TO SCALE
4" C.I. VENT (12 REQ'D)
30" DIA. CONCRETE PIPE SECTION W/ BOTTOM (4 REQ'D) SEDIMENT BASIN
6:1 SLOPE TO ORIGINAL GRADE
10' ROUNDING
SLOPE 1/4"/1'
FINISH EL. = 1131.25
3' OF TOPSOIL
6" TEE
DIMENSIONS FOR LINER OVERHANG ARE IDENTICAL TO SECTION "A-A"
12" OF 3/8"-2" WASHED STONE
30" WASHED SAND
6" WASHED FILTER STONE
15" OF 3/8"-2" WASHED STONE
12" SAND
30 MIL PVC LINER CONTINUOUS BOTTOM & SIDES
2" PVC SDR 21 LATERALS PLACED 4' O.C. W/ 5/16" DIA PERFORATIONS AT TOP OF PIPE
6" PVC SDR 21 SOLID WALL MANIFOLD PIPE
4" PVC SDR 21 UNDERDRAINS @ 6' O.C. W/ 3/8" PERFORATIONS 3" O.C. SPIRALLY PLACED @ 2' O.C.
8" PVC SDR 21 VENTILATION PIPE LOCATED OVER 4" UNDERDRAIN HEADER - NOT IN THIS VIEW
40'
1122.00
PIPE BOOT DETAIL
DO NOT SCALE
4" C.I. VENT
3' OF TOPSOIL
WASHED STONE
8" PVC SDR 21 FORCEMAIN
WASHED SAND
WASHED FILTER ST.
WASHED STONE
4" PVC SDR 21 UNDERDRAIN
30 MIL PVC PIPE BOOT GLUED TO LINER (TYP)
STAINLESS STEEL GEAR CLAMPS & BOLTS (TYP)
SAND
30 MIL PVC LINER CONTINUOUS BOTTOM & SIDES
FLOW SCHEMATIC
DO NOT SCALE
HOUSE
SEPTIC TANK
COLLECTION SYSTEM
FINAL SEPTIC TANKS
RECIRCULATION LINE
LIFT STATION
SAND FILTER BED
EFFLUENT PUMP
UV DISINFECTION UNIT
FLOW METER
EFFLUENT LINE
MAIN CREEK
HYDRAULIC PROFILE
DO NOT SCALE
HIGHEST SEPTIC TANK
LOWEST SEPTIC TANK
COLLECTION SYSTEM
FINAL SEPTIC TANKS
LIFT STATION
SAND FILTER BED
EFFLUENT PUMP
UV DISINFECTION
FLOW METER
RECIRCULATION LINE
DISCHARGE FROM CONTROL BLDG.
EFFLUENT LINE
ENDWALL SECTION
MAIN CREEK
DESIGN PARAMETERS
FLOW 8600 GPD
POPULATION EQUIVALENT 97
BOD 16.5#
SS 19.5#
LOADING RATE (GROSS) 0.75 g/S.F./DAY
LOADING RATE (BED) 1.50 g/S.F./DAY
PRESSURE TO GRAVITY SANITARY SEWER CONNECTION
DO NOT SCALE
4" PVC SDR 21 GRAVITY SANITARY SEWER MAIN
C.O. TO SURFACE SEE DETAIL SHEET 3
4" WYE
2" PVC SDR 21 PRESSURE SANITARY SEWER MAIN
FLOW
2" LINE SHALL BE GLUED INTO REDUCER
4"x2" REDUCER IN END OF 4" LINE
EFFLUENT PIPING ENDWALL SECTION
5 CU. YDS. OF 6" RIP RAP
4" FLAP GATE ON END OF PIPE
LAST 10' OF PIPE SHALL BE 4" D.I. CL 52.
REINFORCE ENDWALL APRON W/ #4 RODS 12" O.C. B.W.
INV. EL.= 1109.0
EL.= 1108.5
WASTEWATER COLLECTION & TREATMENT FACILITY
CONRATH, WISCONSIN
SCALE- AS NOTED
SHEET 12 OF 13
DATE MAY, 1986
MORGAN & PARMLEY, LTD. CONSULTING ENGINEERS LADYSMITH, WISCONSIN
DR. BY M.D. SHERVEY

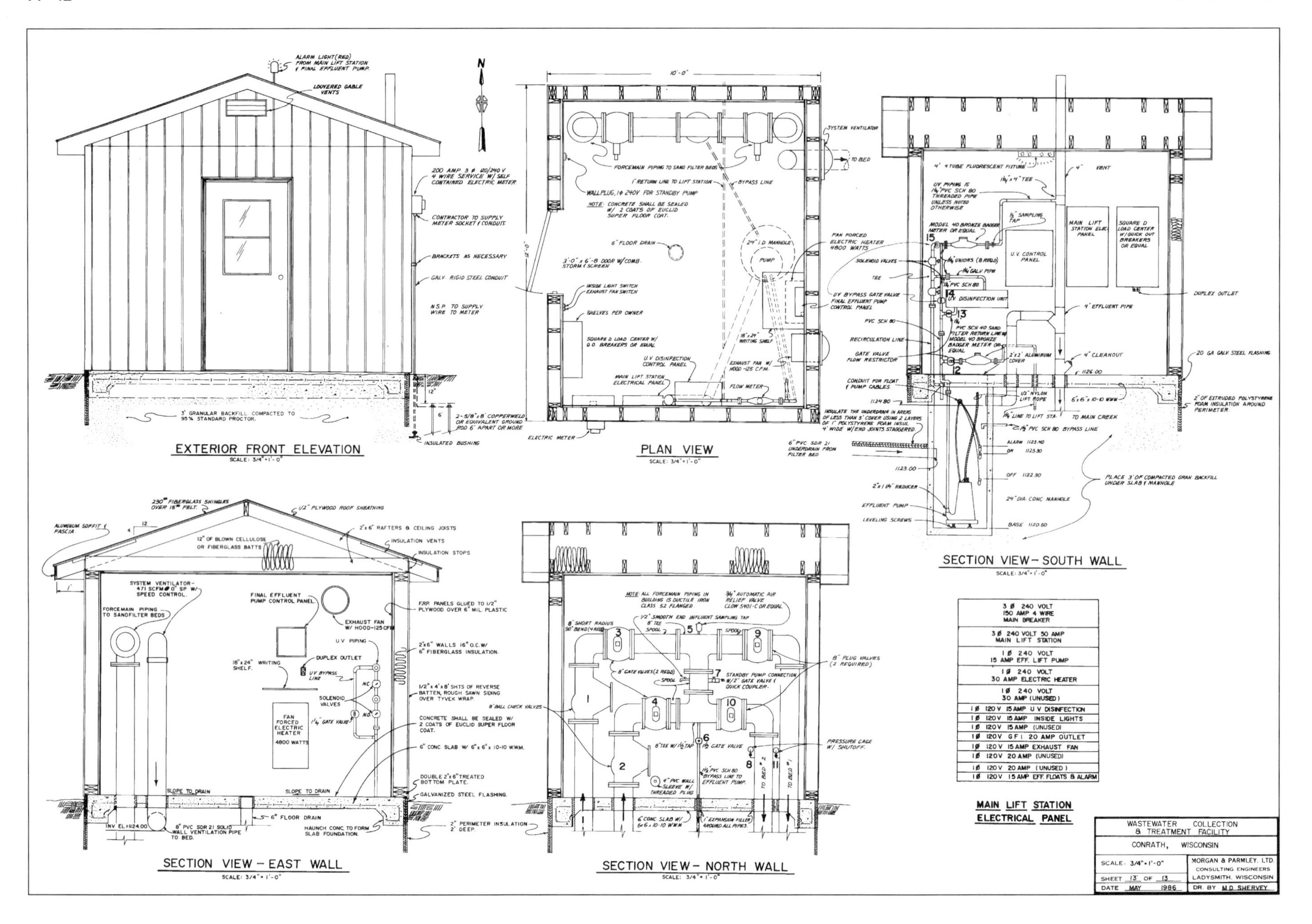

ALARM LIGHT (RED) FROM MAIN LIFT STATION & FINAL EFFLUENT PUMP.
LOUVERED GABLE VENTS
200 AMP 3 Ø 120/240 V 4 WIRE SERVICE W/ SELF CONTAINED ELECTRIC METER
CONTRACTOR TO SUPPLY METER SOCKET & CONDUIT.
BRACKETS AS NECESSARY
GALV. RIGID STEEL CONDUIT
N.S.P. TO SUPPLY WIRE TO METER
3' GRANULAR BACKFILL COMPACTED TO 95% STANDARD PROCTOR.
2-5/8"x8' COPPERWELD OR EQUIVALENT GROUND ROD 6' APART OR MORE
INSULATED BUSHING
EXTERIOR FRONT ELEVATION
SCALE: 3/4"=1'-0"
N
10'-0"
12'-0"
FORCEMAIN PIPING TO SAND FILTER BEDS
1" RETURN LINE TO LIFT STATION
BYPASS LINE
SYSTEM VENTILATOR
TO BED
WALLPLUG, 1Ø 240V FOR STANDBY PUMP
NOTE: CONCRETE SHALL BE SEALED W/ 2 COATS OF EUCLID SUPER FLOOR COAT.
6" FLOOR DRAIN
24" I.D. MANHOLE
PUMP
3'-0" x 6'-8 DOOR W/COMB. STORM & SCREEN
INSIDE LIGHT SWITCH EXHAUST FAN SWITCH
SHELVES PER OWNER
SQUARE D LOAD CENTER W/ Q.O. BREAKERS OR EQUAL
18"x24" WRITING SHELF
U.V DISINFECTION CONTROL PANEL
EXHAUST FAN W/ HOOD-125 C.F.M.
MAIN LIFT STATION ELECTRICAL PANEL
FLOW METER
ELECTRIC METER
PLAN VIEW
SCALE: 3/4"=1'-0"
FAN FORCED ELECTRIC HEATER 4800 WATTS
4" 4 TUBE FLUORESCENT FIXTURE
4" VENT
1¼"x4" TEE
UV PIPING IS 1¼" PVC SCH 80 THREADED PIPE UNLESS NOTED OTHERWISE
½" SAMPLING TAP
MODEL 40 BRONZE BADGER METER OR EQUAL
MAIN LIFT STATION ELEC. PANEL
SQUARE D LOAD CENTER W/ QUICK OUT BREAKERS OR EQUAL
U.V. CONTROL PANEL
SOLENOID VALVES
1¼" UNIONS (8 REQ'D)
TEE
1¼" GALV PIPE
1¼" PVC SCH 80
UV BYPASS GATE VALVE
UV DISINFECTION UNIT
FINAL EFFLUENT PUMP CONTROL PANEL
DUPLEX OUTLET
4" EFFLUENT PIPE
PVC SCH 80
PVC SCH 40 SAND FILTER RETURN LINE
RECIRCULATION LINE
MODEL 40 BRONZE BADGER METER OR EQUAL
GATE VALVE FLOW RESTRICTOR
2'x2' ALUMINUM COVER
4" CLEANOUT
20 GA GALV. STEEL FLASHING
CONDUIT FOR FLOAT & PUMP CABLES
1126.00
½" NYLON LIFT ROPE
6"x6" x 10-10 WWM
2" OF EXTRUDED POLYSTYRENE FOAM INSULATION AROUND PERIMETER.
1124.80
INSULATE THE UNDERDRAIN IN AREAS OF LESS THAN 5' COVER USING 2 LAYERS OF 1" POLYSTYRENE FOAM INSUL. 4' WIDE W/END JOINTS STAGGERED.
1¼" LINE TO LIFT STA.
TO MAIN CREEK
1½" PVC SCH 80 BYPASS LINE
6" PVC SDR 21 UNDERDRAIN FROM FILTER BED
ALARM 1123.40
ON 1123.30
1123.00
OFF 1122.30
PLACE 3' OF COMPACTED GRAN BACKFILL UNDER SLAB & MANHOLE
2"x1¼" REDUCER
24" DIA. CONC MANHOLE
EFFLUENT PUMP
LEVELING SCREWS
BASE 1120.50
SECTION VIEW-SOUTH WALL
SCALE: 3/4"=1'-0"
230# FIBERGLASS SHINGLES OVER 15# FELT.
1/2" PLYWOOD ROOF SHEATHING
ALUMINUM SOFFIT & FASCIA
12" OF BLOWN CELLULOSE OR FIBERGLASS BATTS
2"x6" RAFTERS & CEILING JOISTS
INSULATION VENTS
INSULATION STOPS
SYSTEM VENTILATOR-471 SCFM@0" SP W/ SPEED CONTROL.
FINAL EFFLUENT PUMP CONTROL PANEL
FORCEMAIN PIPING TO SANDFILTER BEDS
EXHAUST FAN W/ HOOD-125CFM
F.R.P. PANELS GLUED TO 1/2" PLYWOOD OVER 6" MIL. PLASTIC
U.V. PIPING
DUPLEX OUTLET
18"x24" WRITING SHELF.
UV BYPASS LINE
2"x6" WALLS 16" O.C. W/ 6" FIBERGLASS INSULATION.
SOLENOID VALVES
1/2"x4'x8' SHTS OF REVERSE BATTEN, ROUGH SAWN SIDING OVER TYVEK WRAP.
FAN FORCED ELECTRIC HEATER 4800 WATTS
1¼" GATE VALVE
CONCRETE SHALL BE SEALED W/ 2 COATS OF EUCLID SUPER FLOOR COAT.
6" CONC SLAB W/ 6"x6"x 10-10 W.W.M.
DOUBLE 2"x6" TREATED BOTTOM PLATE.
SLOPE TO DRAIN
GALVANIZED STEEL FLASHING.
6" FLOOR DRAIN
INV EL.1124.00
8" PVC SDR 21 SOLID WALL VENTILATION PIPE TO BED.
HAUNCH CONC. TO FORM SLAB FOUNDATION.
2" PERIMETER INSULATION 2' DEEP.
SECTION VIEW-EAST WALL
SCALE: 3/4"=1'-0"
NOTE: ALL FORCEMAIN PIPING IN BUILDING IS DUCTILE IRON CLASS 52 FLANGED.
3/4" AUTOMATIC AIR RELIEF VALVE CLOW 5401-C OR EQUAL
1/2" SMOOTH END INFLUENT SAMPLING TAP
8" SHORT RADIUS 90° BEND (4 REQ'D)
8" TEE
SPOOL
8" GATE VALVES (2 REQ'D)
STANDBY PUMP CONNECTION W/ 1/2" GATE VALVE & QUICK COUPLER.
8" PLUG VALVES (2 REQUIRED)
8" BALL CHECK VALVES
8" TEE W/ 1½" TAP
1½" GATE VALVE
PRESSURE GAGE W/ SHUTOFF.
4" PVC WALL SLEEVE W/ THREADED PLUG
1½" PVC SCH 80 BYPASS LINE TO EFFLUENT PUMP.
TO BED #2
TO BED #1
6" CONC SLAB W/ 6x6 x 10-10 WWM
1" EXPANSION FILLER AROUND ALL PIPES.
SECTION VIEW-NORTH WALL
SCALE: 3/4"=1'-0"
3 Ø 240 VOLT 150 AMP 4 WIRE MAIN BREAKER
3 Ø 240 VOLT 50 AMP MAIN LIFT STATION
1 Ø 240 VOLT 15 AMP EFF. LIFT PUMP
1 Ø 240 VOLT 30 AMP ELECTRIC HEATER
1 Ø 240 VOLT 30 AMP (UNUSED)
1 Ø 120V 15 AMP U V DISINFECTION
1 Ø 120V 15 AMP INSIDE LIGHTS
1 Ø 120V 15 AMP (UNUSED)
1 Ø 120V G F I 20 AMP OUTLET
1 Ø 120 V 15 AMP EXHAUST FAN
1 Ø 120V 20 AMP (UNUSED)
1 Ø 120V 20 AMP (UNUSED)
1 Ø 120V 15 AMP EFF. FLOATS & ALARM
MAIN LIFT STATION ELECTRICAL PANEL
WASTEWATER COLLECTION & TREATMENT FACILITY
CONRATH, WISCONSIN
SCALE 3/4"=1'-0"
MORGAN & PARMLEY, LTD. CONSULTING ENGINEERS LADYSMITH, WISCONSIN
SHEET 13 OF 13
DATE MAY 1986
DR. BY M.D SHERVEY

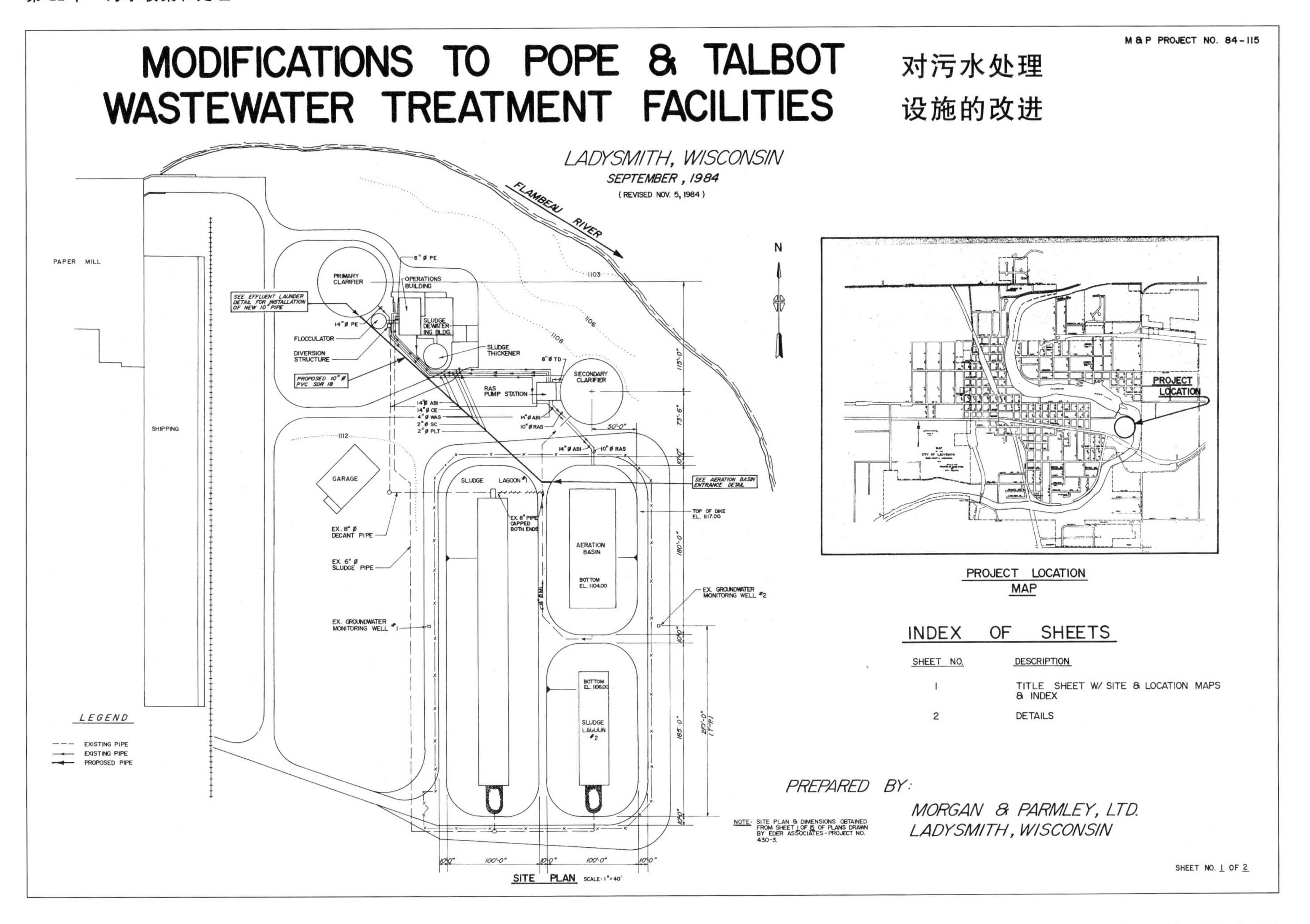

Courtesy: Morgan & Parmley, Ltd.

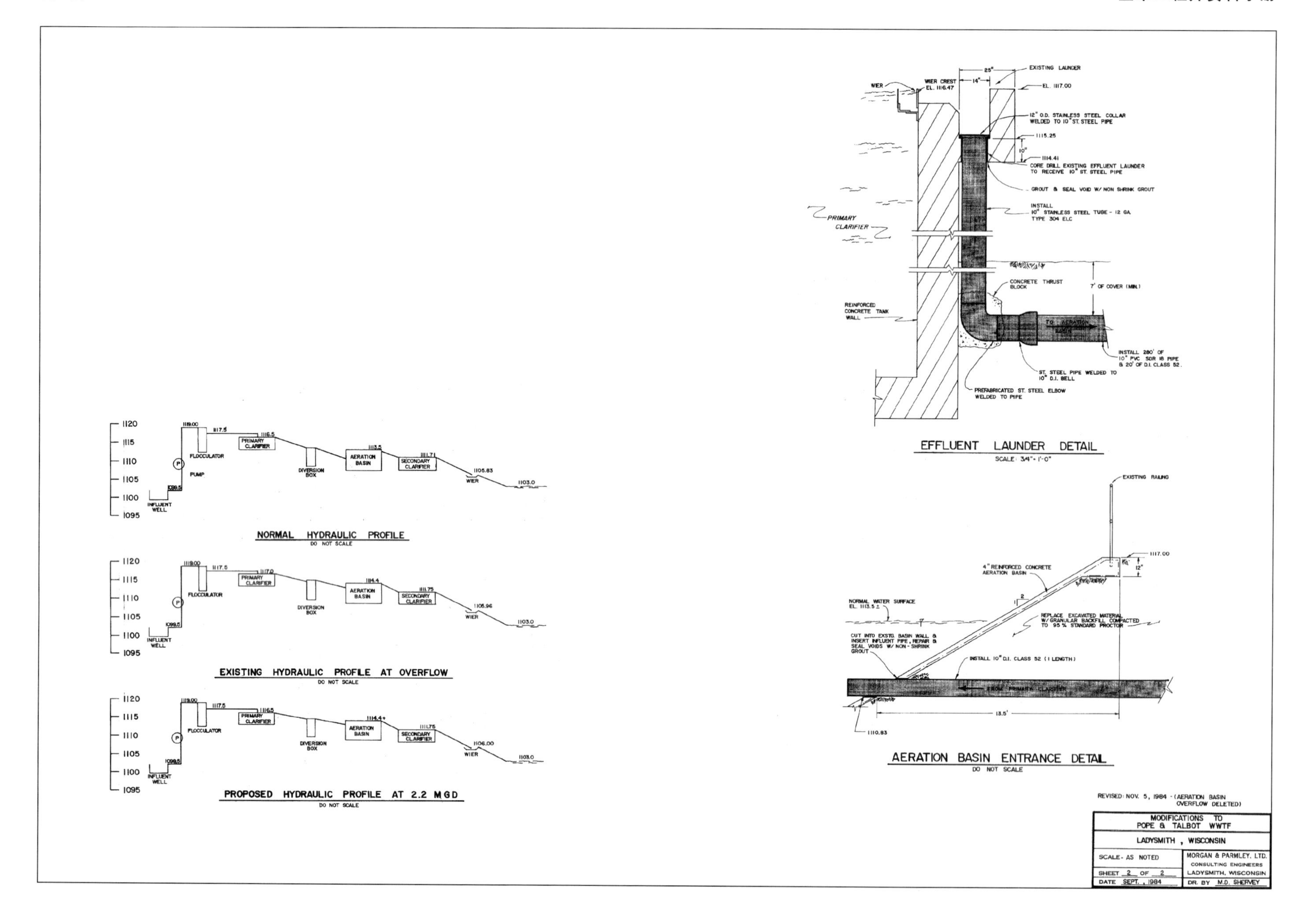

EFFLUENT LAUNDER DETAIL
SCALE: 3/4"=1'-0"
AERATION BASIN ENTRANCE DETAIL
DO NOT SCALE
NORMAL HYDRAULIC PROFILE
DO NOT SCALE
EXISTING HYDRAULIC PROFILE AT OVERFLOW
DO NOT SCALE
PROPOSED HYDRAULIC PROFILE AT 2.2 MGD
DO NOT SCALE
REVISED: NOV. 5, 1984 - (AERATION BASIN OVERFLOW DELETED)
MODIFICATIONS TO POPE & TALBOT WWTF
LADYSMITH, WISCONSIN
SCALE- AS NOTED
SHEET 2 OF 2
DATE SEPT., 1984
MORGAN & PARMLEY, LTD.
CONSULTING ENGINEERS
LADYSMITH, WISCONSIN
DR. BY M.D. SHERVEY

-UTILITY EXTENSIONS-　公用设施扩建工程

SANITARY SEWER, LIFT STATION & TRANSMISSION WATERMAIN

for the

VILLAGE

of

TONY, WISCONSIN

M.& P. PROJECT NO. 97-127

FEBRUARY, 1998

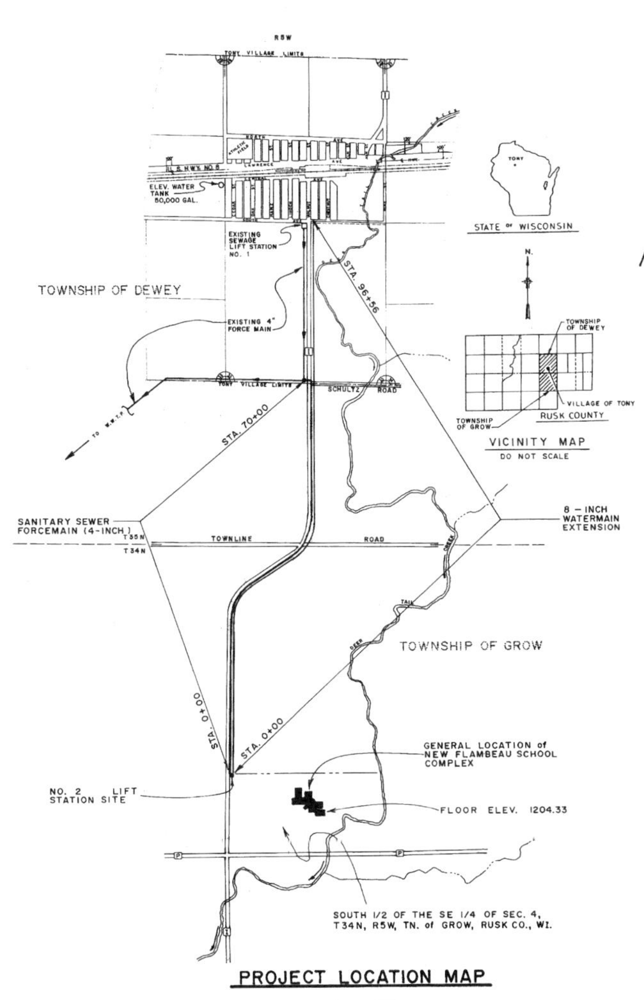

PROJECT LOCATION MAP

REFERENCE SPECIFICATIONS

* WISCONSIN CONSTRUCTION SITE BEST MANAGEMENT PRACTICE HANDBOOK
* STANDARD SPECIFICATIONS FOR ROAD AND BRIDGE CONSTRUCTION 1989 EDITION WIS. D.O.T.
* STANDARD SPECIFICATIONS FOR SEWER & WATER CONSTRUCTION IN WISCONSIN, FIFTH EDITION
* MORGAN & PARMLEY, LTD., ACCOMPANYING SPECIFICATIONS

PROJECT RECORD DRAWINGS: COMPILED FROM CONSTRUCTION DATA FURNISHED TO THE ENGINEER BY CONTRACTOR, OWNER AND/OR ENGINEER'S FIELD REPRESENTATIVE. INFORMATION SHOWN ON THESE DRAWINGS SHOULD BE FIELD VERIFIED BEFORE USING SAME.

DRAWING INDEX

SHEET NO.	DESCRIPTION
1	TITLE SHEET, LOCATION MAPS & NOTES
2	PLAN & PROFILE: STA. 0+00 to 14+00
3	PLAN & PROFILE STA. 14+00 to 28+00
4	PLAN & PROFILE STA. 28+00 to 42+00
5	PLAN & PROFILE STA. 42+00 to 56+00
6	PLAN & PROFILE STA. 56+00 to 70+00
7	PLAN & PROFILE STA. 70+00 to 84+00
8	PLAN & PROFILE STA. 84+00 to 96+37
9	TYPICAL CONSTRUCTION X-SECTIONS, ACCESS DRIVEWAY & DETAILS w/ NOTES & LEGEND
10	SITE PLANS: "A" INTERSECTION SOUTH AVE. & C.T.H. "I"; "B" LIFT STATION
11	LIFT STATION PLAN, X-SECTION & DETAILS
12	EROSION CONTROL DETAILS w/ TYPICAL NOTES This Sheet Not Shown

GENERAL NOTES

1. ALL WATER MAIN SHALL BE PVC DR 18 C900, EXCEPT AS NOTED.
2. ALL WATERMAIN FITTINGS SHALL BE M.J.D.I., WITH TEES AND BENDS BLOCKED AS SHOWN IN THE DETAILS.
3. ALL HYDRANT LEADS SHALL HAVE A 6-INCH VALVE.
4. HYDRANTS SHALL BE WATEROUS PACER, HAVE A 22" BREAKOFF & MEET CITY OF LADYSMITH'S PATTERN FOR THREADS, ROTATION AND NUT.
5. ALL WATERMAIN SHALL HAVE AN 8 FT. COVER (MINIMUM).
6. SANITARY SEWER FORCEMAIN SHALL BE 4" PVC DR 18 C900.
7. SEE SHEET 9 FOR TYPICAL (MINIMUM) HORIZONTAL SEPARATION BETWEEN SANITARY SEWER FORCEMAIN & WATERMAIN.
8. 18" (MINIMUM) VERTICAL DISTANCE (OUT TO OUT) BETWEEN WATERMAIN UNDER SANITARY SEWER FORCEMAIN @ THEIR INTERSECTION.
9. 6" (MINIMUM) VERTICAL DISTANCE (OUT TO OUT) BETWEEN WATERMAIN OVER SANITARY SEWER FORCEMAIN @ THEIR INTERSECTION.
10. ALL ELEVATIONS BASED ON USGS DATUM.
11. EROSION CONTROL MEASURES SHALL BE IN PLACE BEFORE CONSTRUCTION BEGINS. (SEE SHEET 12 & SPECIFICATIONS)
12. THE CONTRACTORS SHALL MINIMIZE EROSION DURING CONSTRUCTION USING GOOD CONSTRUCTION TECHNIQUES AND UTILIZING SILT FENCES AND BALE DITCH CHECKS. THE CONTRACTORS SHALL RELEASE RUNOFF FROM THE SITE IN A NUISANCE FREE MANNER.
13. THE CONTRACTOR IS RESPONSIBLE TO MAINTAIN THE SITE IN A SAFE CONDITION. THE SOLE RESPONSIBILITY FOR WARNING SIGNS, BARRICADES AND ALL ASPECTS OF SAFETY LIE WITH THE CONTRACTOR.
14. EXISTING UTILITIES ARE SHOWN BUT MAY NOT BE ALL INCLUSIVE, CONTRACTOR SHALL HAVE ALL BURIED UTILITIES LOCATED PRIOR TO COMMENCING CONSTRUCTION.
15. EXISTING SEWER FORCEMAIN AND WATERMAIN ARE SHOWN ACCORDING TO EXISTING RECORDS BUT LOCATION SHALL BE VERIFIED IN FIELD.
16. WATERMAIN SHALL BE INSULATED AT ALL CULVERT AND DITCH CROSSINGS WITH 2" THICK EXTRUDED POLYSTYRENE INSULATION BOARD; LAPPED & STAGGERED AT JOINTS.
17. PROPERTY CORNERS KNOWN TO EXIST SHALL NOT BE DISTURBED. DAMAGED CORNERS WILL BE REPLACED AT THE CONTRACTOR'S EXPENSE.
18. ALL WATERMAIN SHALL BE DISINFECTED AND A "SAFE' SAMPLE OBTAINED BEFORE PLACING IN SERVICE.
19. VALVES SHALL BE SET TO GRADE SHOWN ON PLANS, UNLESS DIRECTED OTHERWISE.
20. CONSTRUCTION SHALL COMPLY WITH THE CONDITIONS OF DNR'S APPROVAL LETTER.
21. ALL DISTURBED AREAS SHALL BE SEEDED, FERTILIZED AND MULCHED, OR COVERED W / BASE COURSE AFTER CONSTRUCTION.
22. ALL DISTURBED AREAS TO BE SEEDED SHALL BE RESTORED AS FOLLOWS:
    A) SEED MIXTURE:
       BIRDSFOOT TREFOIL　15#/AC.
       SMOOTH BROMEGRASS　40#/AC.
       TALL FESCUE　25#/AC.
    B) SEED BED - PREPARE A THOROUGHLY TILLED BUT FIRM SEED BED WHEREVER CONDITIONS WILL PERMIT.
    C) FERTILIZER:　500 #/AC.　20-10-10
    D) SEED TO BE SOWN WITH A HYDROSEEDER.
    E) MULCH WITH CLEAN HAY OR STRAW AT THE RATE OF 2 TONS/ACRE. ANCHOR MULCH WITH 75 - 100 GALLONS OF EMUILSIFIED ASPHALT PER TON OF MULCH.
    F) SEED PRIOR TO SEPT. 1ST.
23. CONTRACTOR SHALL DETERMINE CONSTRUCTION SEQUENCE IN ORDER TO MINIMIZE INSTALLATION CONFLICTS.
24. THERE SHALL BE NO DEVIATION FROM THE PLANS WITHOUT THE DESIGN ENGINEER'S APPROVAL.
25. CONSTRUCTION STAKES DESTROYED BY THE CONTRACTOR THAT NEED TO BE REPLACED, WILL BE CHARGED TO THE CONTRACTOR.
26. NO TREES ARE TO BE REMOVED WITHOUT THE APPROVAL OF THE ENGINEER.
27. THE CONTRACTOR MUST REMOVE AND REPLACE ALL EFFECTED STREET SIGNS AFTER PROPERLY SIGNING AND BARRICADING THE CONSTRUCTION AREA. ALL SIGNS MUST BE REPLACED PRIOR TO OPENING THE STREET TO TRAFFIC.
28. CONFLICTING POWER POLES WILL BE MOVED OR STABILIZED BY NSP, IF WITHIN ROAD R/W. HOWEVER, IF POWER POLES ARE BEYOND ROAD R/W, NSP'S CHARGES SHALL BE PAID BY CONTRACTOR.
29. SEE ACCOMPANYING TECHNICAL SPECIFICATIONS AND SECTION 01010 "SPECIAL PROVISIONS".
30. SEE SHEET 9 FOR TYPICAL X-SECTIONS & LEGEND.

-UTILITIES-

SEWER & WATER: VILLAGE OF TONY
BILLY MECHELKE
N5297 LITTLE X RD.
TONY, WI 54563
1-715-532-3183 (W)
1-715-532-7046 (H)

ELECTRIC:
NORTHERN STATES POWER COMPANY
133 NORTH LAKE AVENUE
PHILLIPS, WI 54555
ATTN: JOE PERKINS
1-800-445-1275

TELEPHONE:
AMERITECH
304 S. DEWEY STREET
EAU CLAIRE, WI 54701
ATTN: PAM PARKER
1-800-305-5815

WISCONSIN GAS COMPANY:
104 WEST SOUTH STREET
RICE LAKE, WI 54868
ATTN: DON WEDIN, MGR.
1-715-387-0189

DIGGER'S HOTLINE:
1-800-242-8511

PREPARED BY:
MORGAN & PARMLEY, Ltd.
PROFESSIONAL CONSULTING ENGINEERS
LADYSMITH, WISCONSIN

Courtesy: Morgan & Parmley, Ltd.

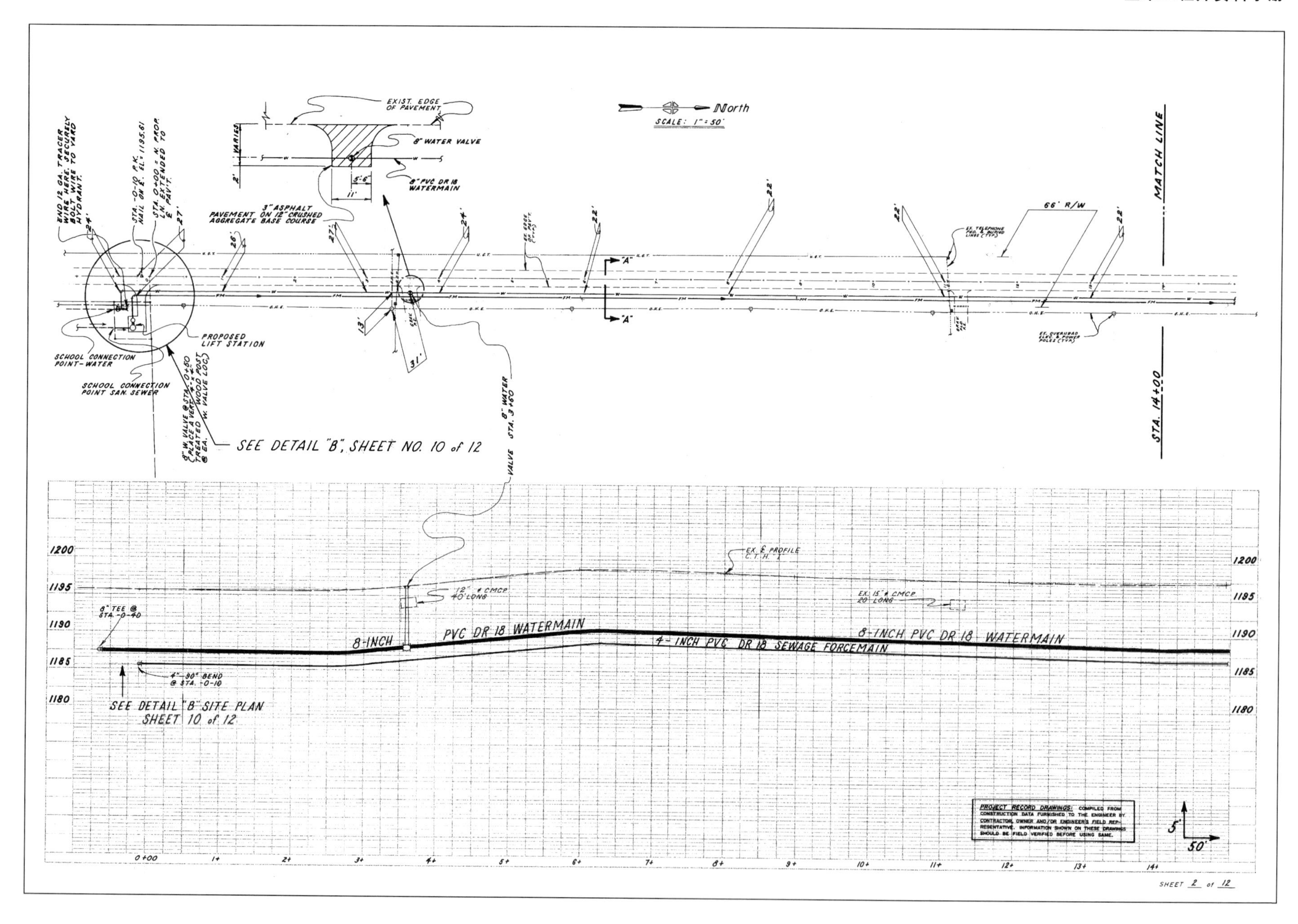

EXIST. EDGE OF PAVEMENT
8" WATER VALVE
8" PVC DR 18 WATERMAIN
3" ASPHALT PAVEMENT ON 12" CRUSHED AGGREGATE BASE COURSE
North
SCALE: 1"=50'
MATCH LINE
66' R/W
STA. 14+00
PROPOSED LIFT STATION
SCHOOL CONNECTION POINT-WATER
SCHOOL CONNECTION POINT SAN. SEWER
SEE DETAIL "B", SHEET NO. 10 of 12
VALVE STA. 3+80 8" WATER
EX. ℄ PROFILE C.T.H. "I"
12" ⌀ CMCP 40' LONG
EX. 15" ⌀ CMCP 20' LONG
8" TEE @ STA. -0-40
8-INCH PVC DR 18 WATERMAIN
4-INCH PVC DR 18 SEWAGE FORCEMAIN
8-INCH PVC DR 18 WATERMAIN
4" 90° BEND @ STA. -0-10
SEE DETAIL "B" SITE PLAN SHEET 10 of 12
PROJECT RECORD DRAWINGS: COMPILED FROM CONSTRUCTION DATA FURNISHED TO THE ENGINEER BY CONTRACTOR, OWNER AND/OR ENGINEER'S FIELD REPRESENTATIVE. INFORMATION SHOWN ON THESE DRAWINGS SHOULD BE FIELD VERIFIED BEFORE USING SAME.
SHEET 2 of 12

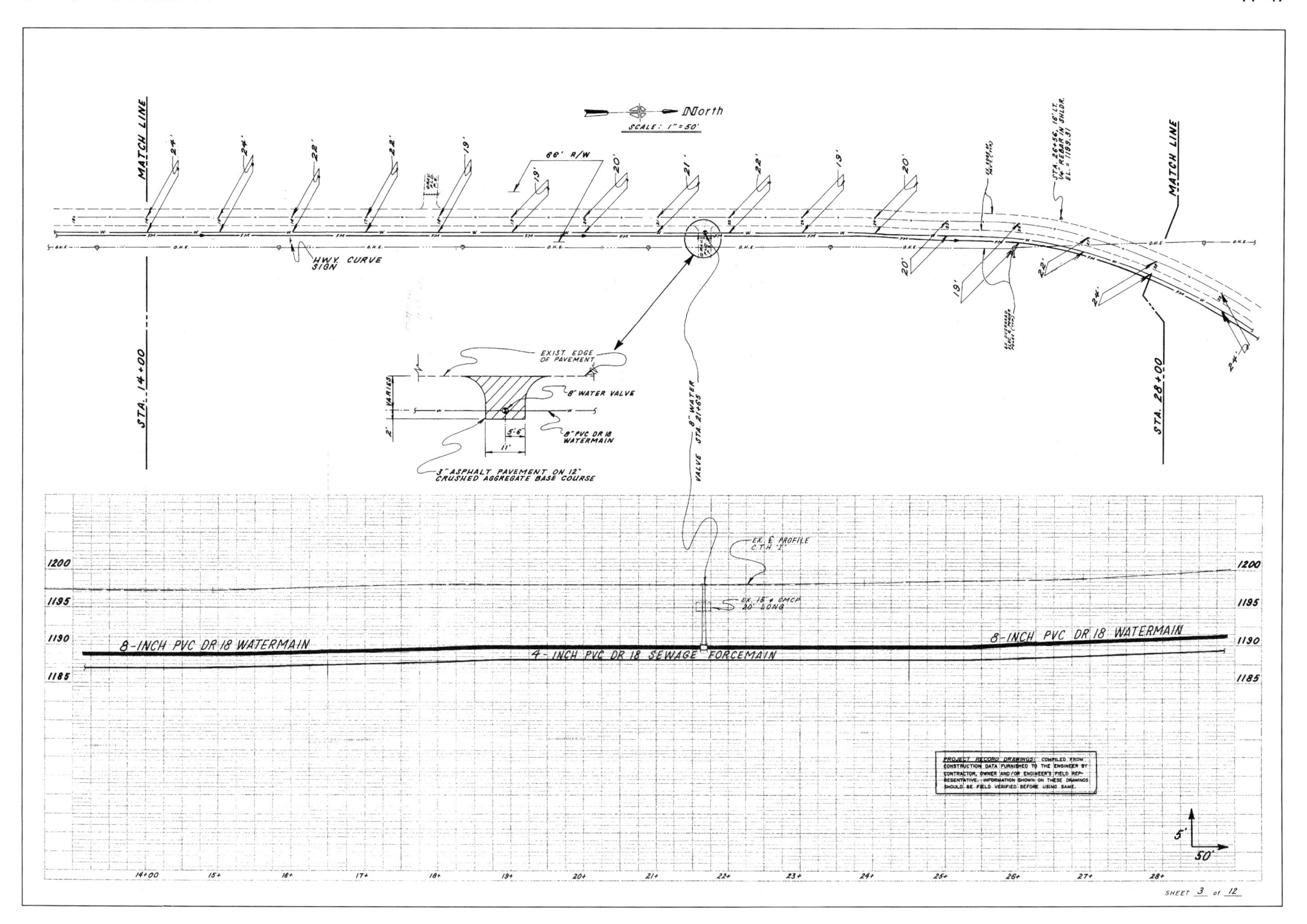
North
SCALE: 1" = 50'
MATCH LINE
66' R/W
HWY. CURVE SIGN
STA. 14+00
STA. 28+00
STA. 26+56, 16' LT. 1/4" REBAR IN SHLDR. EL. = 1199.31
EXIST. EDGE OF PAVEMENT
8" WATER VALVE
8" PVC DR 18 WATERMAIN
VARIES
3" ASPHALT PAVEMENT ON 12" CRUSHED AGGREGATE BASE COURSE
8" WATER VALVE STA. 21+65
EX. ℄ PROFILE C.T.H. "I"
EX. 15" ϕ CMCP 20' LONG
8-INCH PVC DR 18 WATERMAIN
4-INCH PVC DR 18 SEWAGE FORCEMAIN
1200
1195
1190
1185
14+00
15+
16+
17+
18+
19+
20+
21+
22+
23+
24+
25+
26+
27+
28+
5'
50'
PROJECT RECORD DRAWINGS: COMPILED FROM CONSTRUCTION DATA FURNISHED TO THE ENGINEER BY CONTRACTOR, OWNER AND/OR ENGINEER'S FIELD REPRESENTATIVE. INFORMATION SHOWN ON THESE DRAWINGS SHOULD BE FIELD VERIFIED BEFORE USING SAME.
SHEET 3 of 12

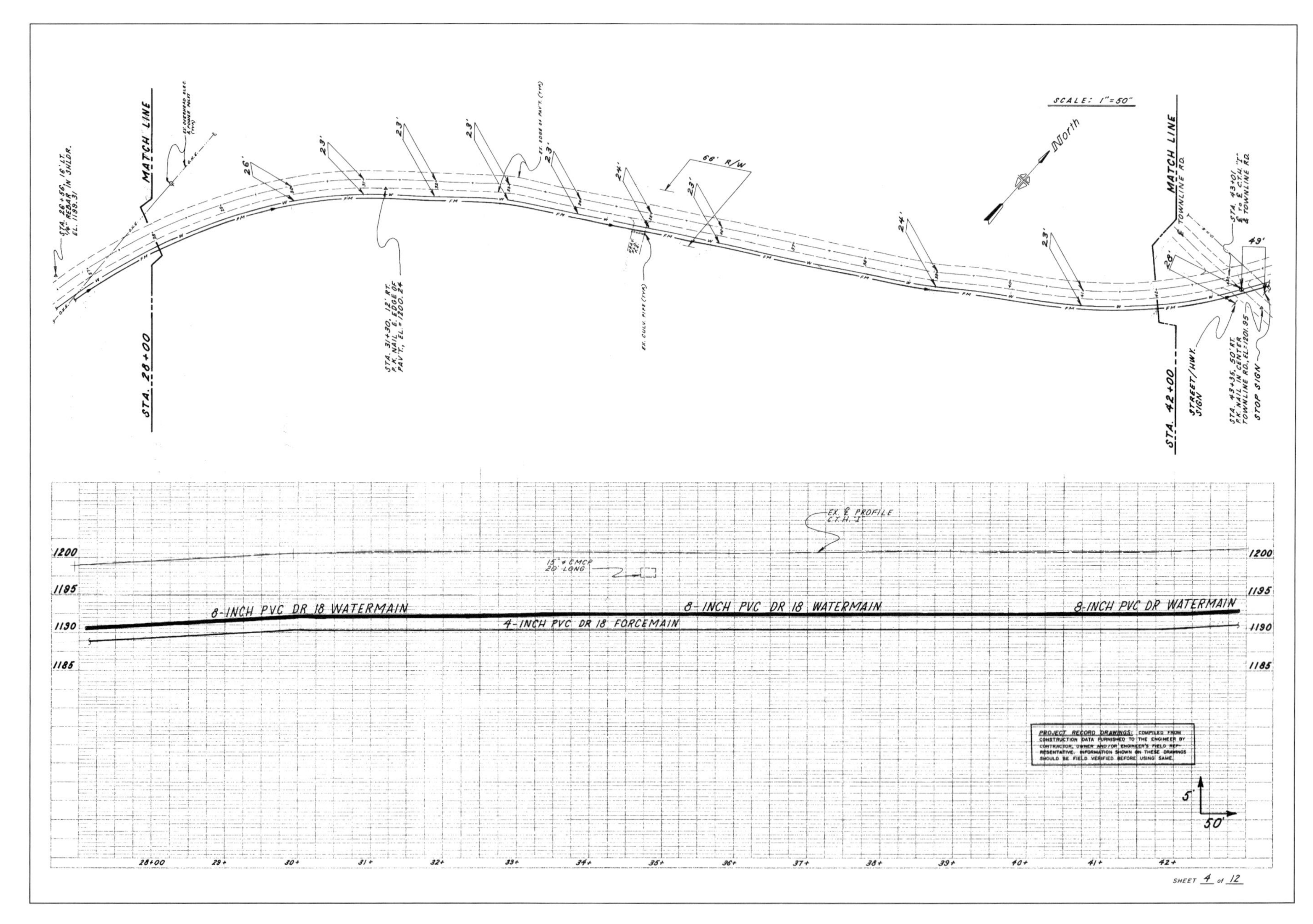

SCALE: 1"=50"
North
MATCH LINE
STA. 28+00
MATCH LINE
STA. 42+00
68' R/W
EX. CULV. PIPE (TYP)
STREET/HWY. SIGN
STOP SIGN
EX. ℄ PROFILE C.T.H. "I"
15" ∅ CMCP 20' LONG
8-INCH PVC DR 18 WATERMAIN
8-INCH PVC DR 18 WATERMAIN
8-INCH PVC DR WATERMAIN
4-INCH PVC DR 18 FORCEMAIN
1200
1195
1190
1185
28+00
29+
30+
31+
32+
33+
34+
35+
36+
37+
38+
39+
40+
41+
42+
PROJECT RECORD DRAWINGS: COMPILED FROM CONSTRUCTION DATA FURNISHED TO THE ENGINEER BY CONTRACTOR, OWNER AND/OR ENGINEER'S FIELD REPRESENTATIVE. INFORMATION SHOWN ON THESE DRAWINGS SHOULD BE FIELD VERIFIED BEFORE USING SAME.
5'
50'
SHEET 4 of 12

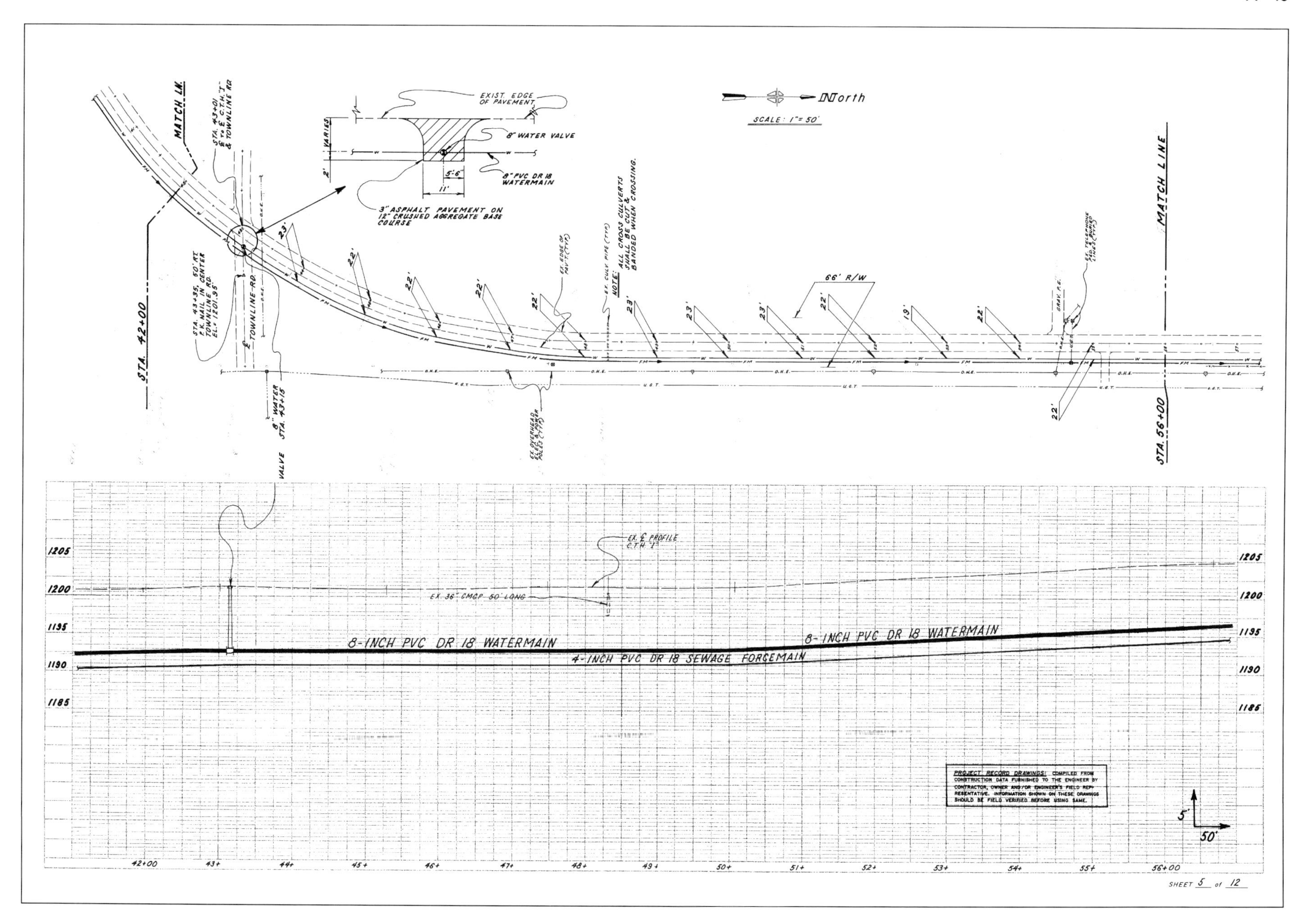

EXIST. EDGE OF PAVEMENT
8" WATER VALVE
8" PVC DR 18 WATERMAIN
3" ASPHALT PAVEMENT ON 12" CRUSHED AGGREGATE BASE COURSE
North
SCALE: 1" = 50'
MATCH LN.
MATCH LINE
STA. 42+00
STA. 56+00
66' R/W
NOTE: ALL CROSS CULVERTS SHALL BE CUT & BANDED WHEN CROSSING.
8-INCH PVC DR 18 WATERMAIN
4-INCH PVC DR 18 SEWAGE FORCEMAIN
EX. 36" CMCP 50' LONG
EX. ℄ PROFILE C.T.H. "I"
PROJECT RECORD DRAWINGS: COMPILED FROM CONSTRUCTION DATA FURNISHED TO THE ENGINEER BY CONTRACTOR, OWNER AND/OR ENGINEER'S FIELD REPRESENTATIVE. INFORMATION SHOWN ON THESE DRAWINGS SHOULD BE FIELD VERIFIED BEFORE USING SAME.
SHEET 5 of 12

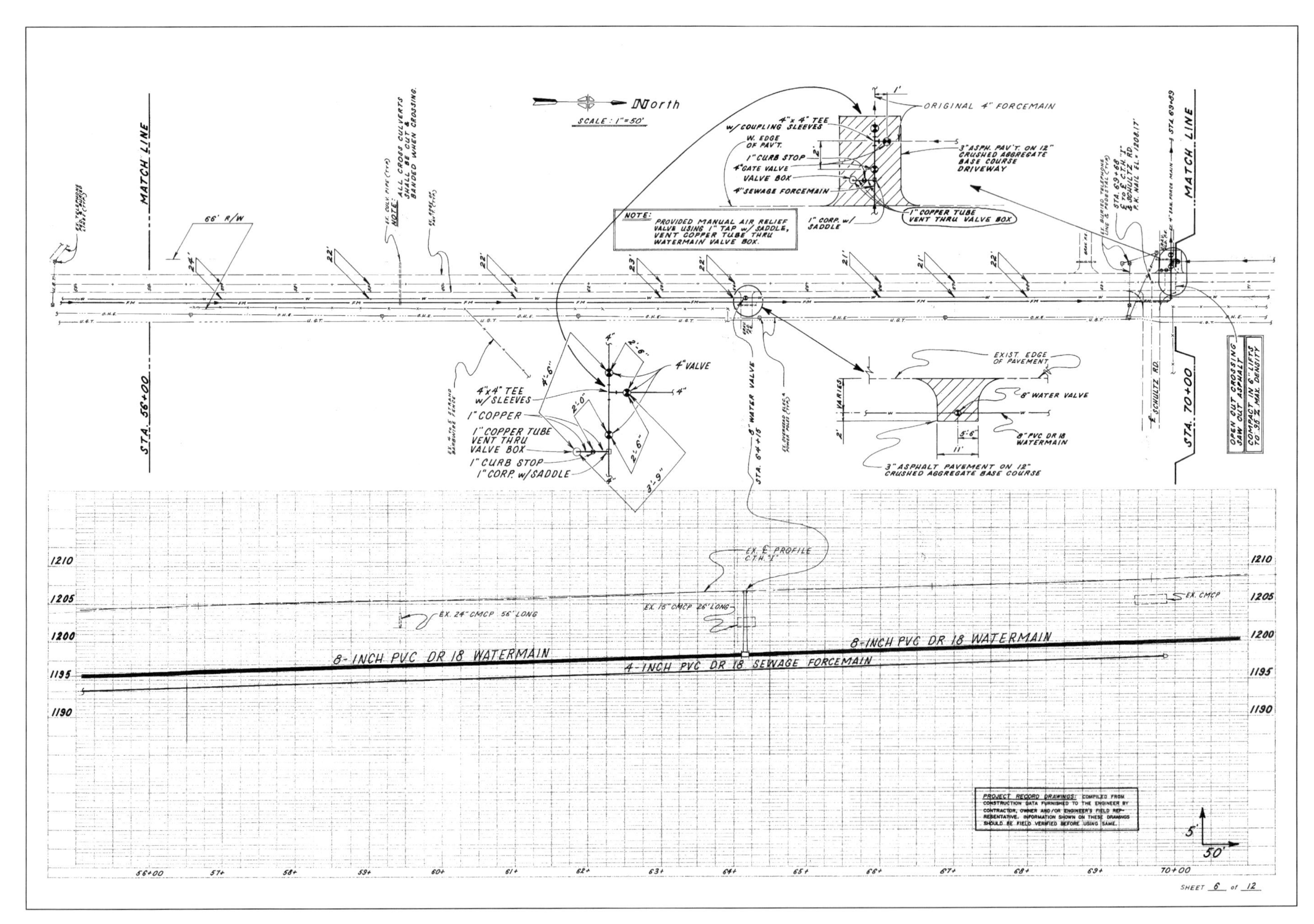
North
SCALE : 1"=50'
MATCH LINE
STA. 56+00
66' R/W
NOTE: ALL CROSS CULVERTS SHALL BE CUT & BANDED WHEN CROSSING.
NOTE: PROVIDED MANUAL AIR RELIEF VALVE USING 1" TAP w/ SADDLE, VENT COPPER TUBE THRU WATERMAIN VALVE BOX.
ORIGINAL 4" FORCEMAIN
4"x 4" TEE w/ COUPLING SLEEVES
W. EDGE OF PAV'T.
1" CURB STOP
4" GATE VALVE
VALVE BOX
4" SEWAGE FORCEMAIN
1" CORP. w/ SADDLE
1" COPPER TUBE VENT THRU VALVE BOX
3" ASPH. PAV'T. ON 12" CRUSHED AGGREGATE BASE COURSE DRIVEWAY
4"x4" TEE w/ SLEEVES
1" COPPER
1" COPPER TUBE VENT THRU VALVE BOX
1" CURB STOP
1" CORP. w/ SADDLE
4" VALVE
8" WATER VALVE STA. 64+15
EXIST EDGE OF PAVEMENT
8" WATER VALVE
8" PVC DR 18 WATERMAIN
VARIES
3" ASPHALT PAVEMENT ON 12" CRUSHED AGGREGATE BASE COURSE
℄ SCHULTZ RD.
STA. 70+00
OPEN CUT CROSSING SAW CUT ASPHALT
COMPACT IN 6" LIFTS TO 95% MAX. DENSITY
EX. ℄ PROFILE C.T.H. "I"
EX. 24" CMCP 56' LONG
EX. 15" CMCP 26' LONG
EX. CMCP
8-INCH PVC DR 18 WATERMAIN
4-INCH PVC DR 18 SEWAGE FORCEMAIN
1210
1205
1200
1195
1190
56+00
57+
58+
59+
60+
61+
62+
63+
64+
65+
66+
67+
68+
69+
70+00
PROJECT RECORD DRAWINGS: COMPILED FROM CONSTRUCTION DATA FURNISHED TO THE ENGINEER BY CONTRACTOR, OWNER AND/OR ENGINEER'S FIELD REPRESENTATIVE. INFORMATION SHOWN ON THESE DRAWINGS SHOULD BE FIELD VERIFIED BEFORE USING SAME.
5'
50'
SHEET 6 of 12

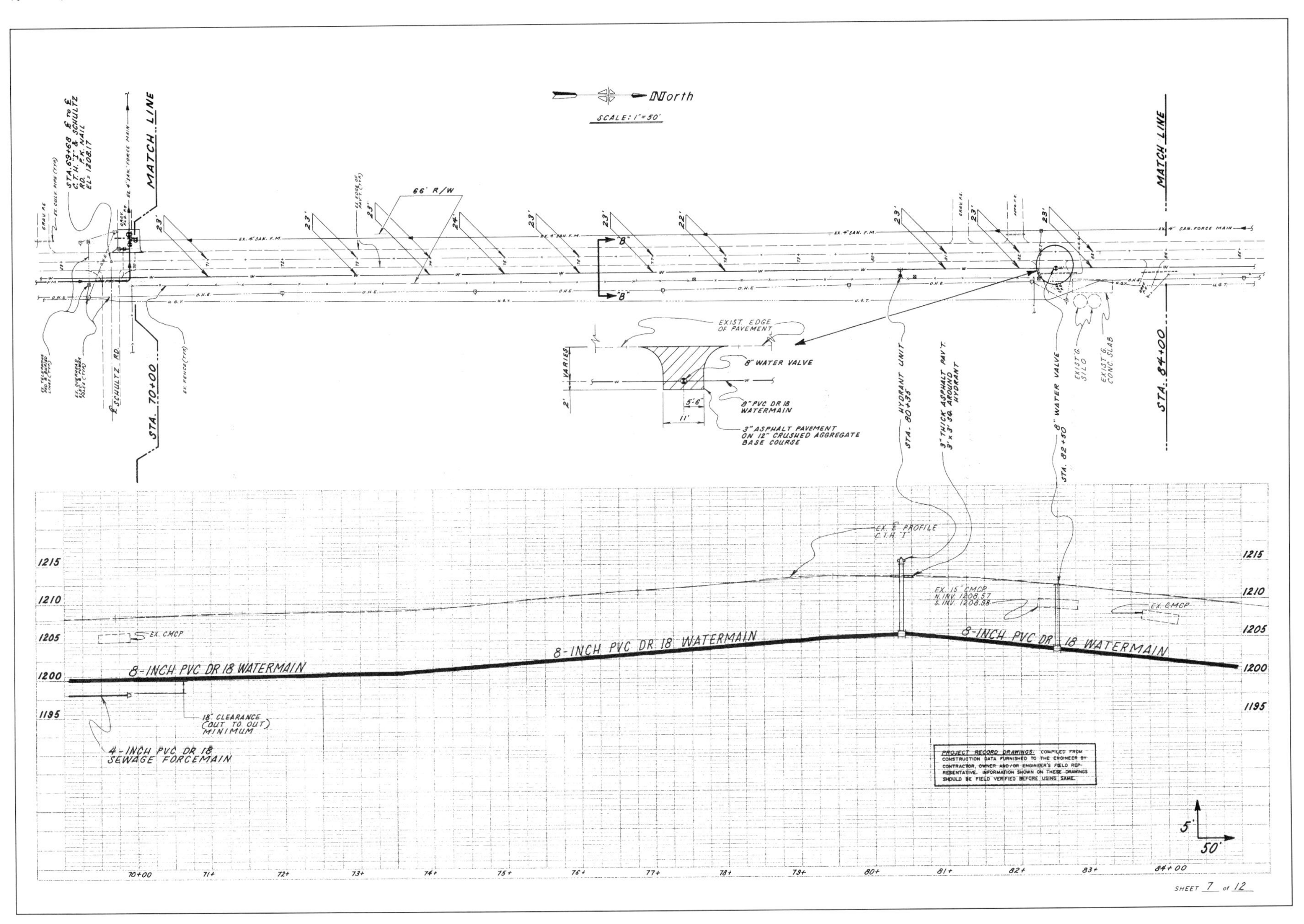
North
SCALE: 1"=50'
MATCH LINE
66' R/W
EXIST EDGE OF PAVEMENT
8" WATER VALVE
8" PVC DR 18 WATERMAIN
3" ASPHALT PAVEMENT ON 12" CRUSHED AGGREGATE BASE COURSE
STA. 70+00
STA. 84+00
HYDRANT UNIT STA. 80+35
3" THICK ASPHALT PAV'T. 3'×3' SQ AROUND HYDRANT
8" WATER VALVE STA. 82+50
EXIST'G. SILO
EXIST'G. CONC. SLAB
EX. ℄ PROFILE C.T.H. "I"
EX. 15" CMCP N. INV. 1208.57 S. INV. 1208.38
EX. CMCP
8-INCH PVC DR 18 WATERMAIN
18" CLEARANCE (OUT TO OUT) MINIMUM
4-INCH PVC DR 18 SEWAGE FORCEMAIN
PROJECT RECORD DRAWINGS: COMPILED FROM CONSTRUCTION DATA FURNISHED TO THE ENGINEER BY CONTRACTOR, OWNER AND/OR ENGINEER'S FIELD REPRESENTATIVE. INFORMATION SHOWN ON THESE DRAWINGS SHOULD BE FIELD VERIFIED BEFORE USING SAME.
SHEET 7 of 12

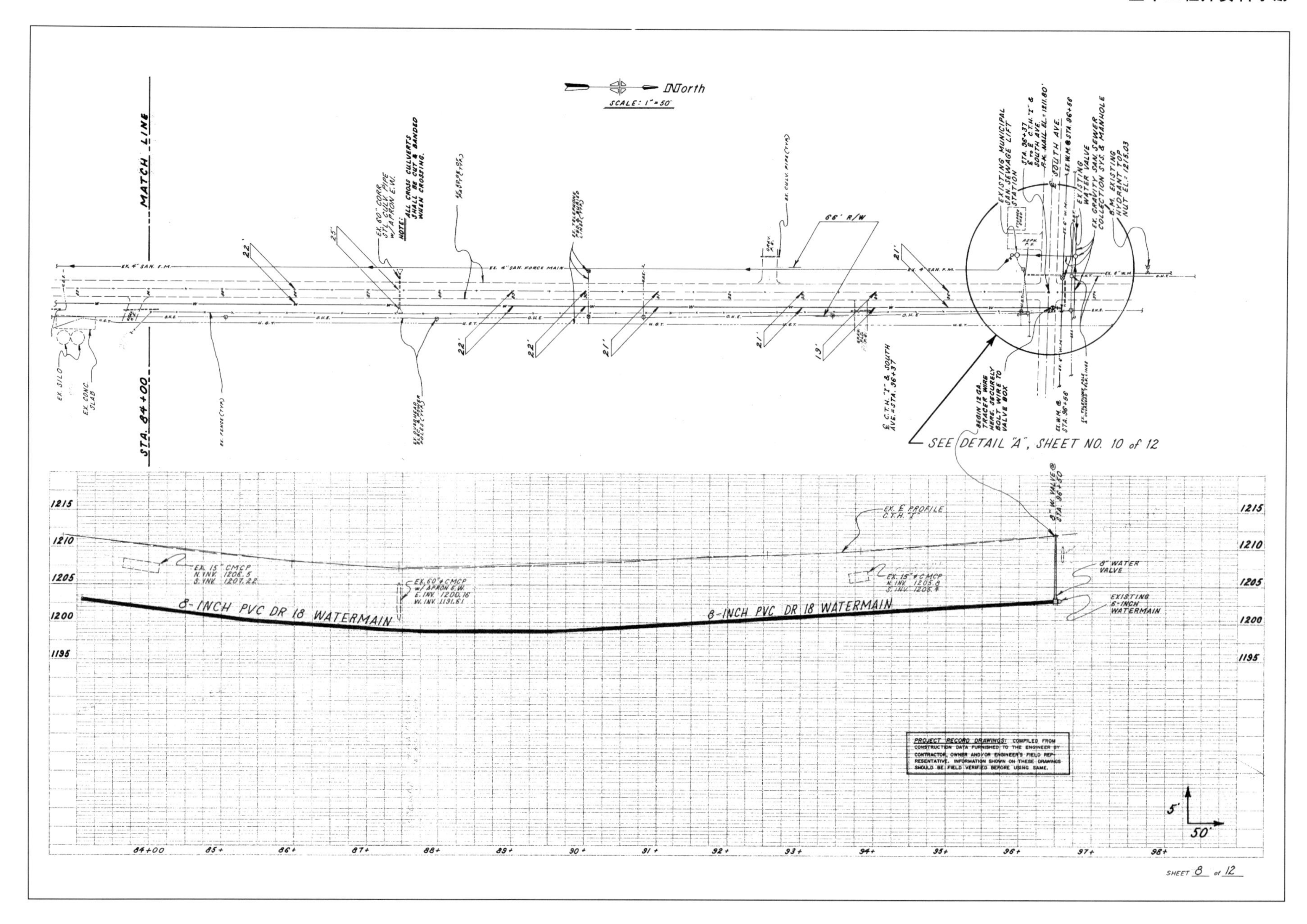
North
SCALE: 1" = 50'
MATCH LINE
STA. 84+00
SEE DETAIL "A", SHEET NO. 10 of 12
EX. 15" CMCP N. INV. 1206.5 S. INV. 1207.22
EX. 60" CMCP w/ APRON E.W. E. INV. 1200.76 W. INV. 1191.61
8-INCH PVC DR 18 WATERMAIN
8-INCH PVC DR 18 WATERMAIN
8" WATER VALVE
EXISTING 6-INCH WATERMAIN
PROJECT RECORD DRAWINGS: COMPILED FROM CONSTRUCTION DATA FURNISHED TO THE ENGINEER BY CONTRACTOR, OWNER AND/OR ENGINEER'S FIELD REPRESENTATIVE. INFORMATION SHOWN ON THESE DRAWINGS SHOULD BE FIELD VERIFIED BEFORE USING SAME.
84+00 85+ 86+ 87+ 88+ 89+ 90+ 91+ 92+ 93+ 94+ 95+ 96+ 97+ 98+
1215 1210 1205 1200 1195
5' 50'
SHEET 8 of 12

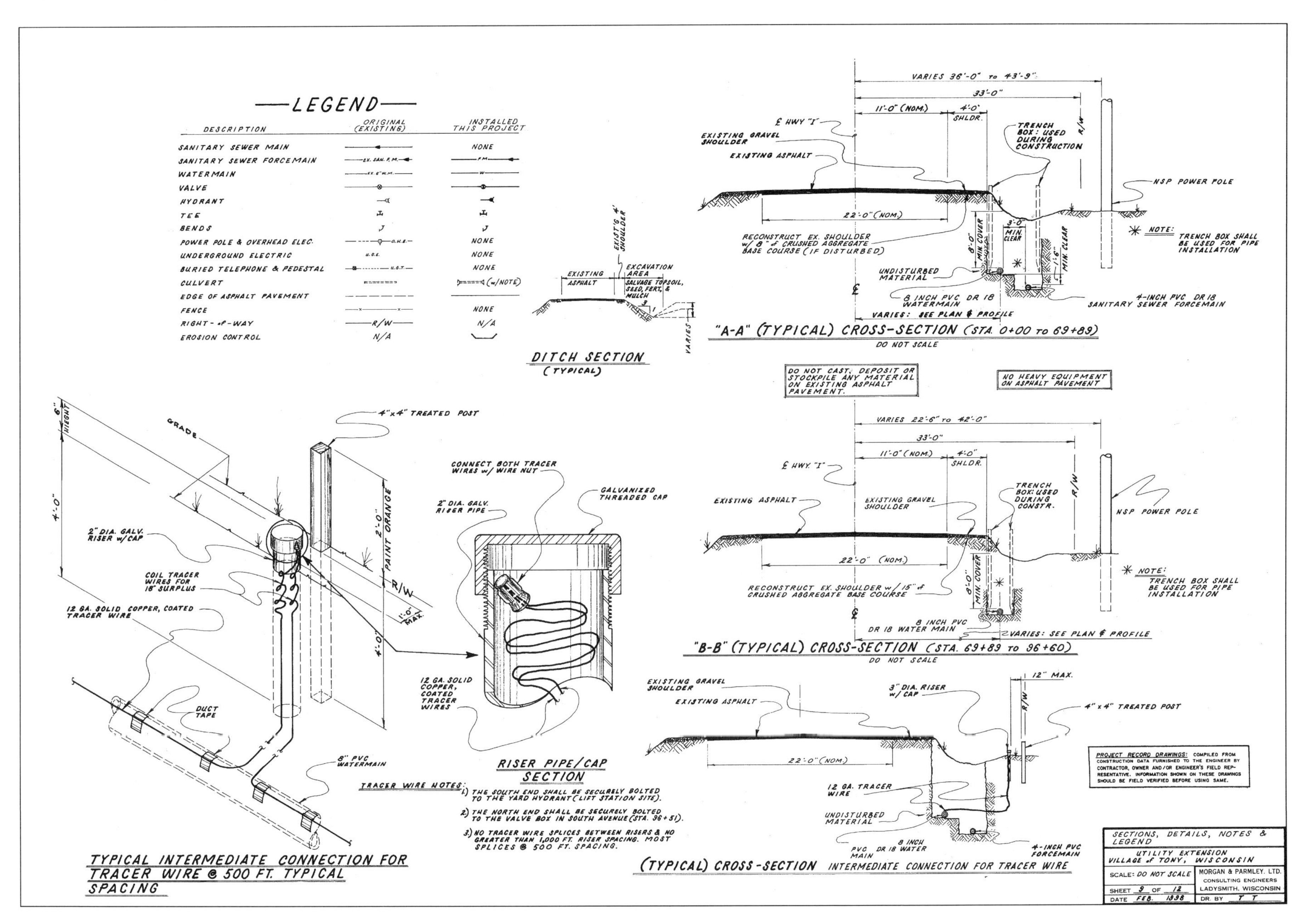
LEGEND
DESCRIPTION
ORIGINAL (EXISTING)
INSTALLED THIS PROJECT
SANITARY SEWER MAIN
SANITARY SEWER FORCEMAIN
WATERMAIN
VALVE
HYDRANT
TEE
BENDS
POWER POLE & OVERHEAD ELEC.
UNDERGROUND ELECTRIC
BURIED TELEPHONE & PEDESTAL
CULVERT
EDGE OF ASPHALT PAVEMENT
FENCE
RIGHT-of-WAY
EROSION CONTROL
DITCH SECTION (TYPICAL)
"A-A" (TYPICAL) CROSS-SECTION (STA. 0+00 TO 69+89)
DO NOT SCALE
"B-B" (TYPICAL) CROSS-SECTION (STA. 69+89 TO 96+60)
DO NOT SCALE
DO NOT CAST, DEPOSIT OR STOCKPILE ANY MATERIAL ON EXISTING ASPHALT PAVEMENT.
NO HEAVY EQUIPMENT ON ASPHALT PAVEMENT
NOTE: TRENCH BOX SHALL BE USED FOR PIPE INSTALLATION
RISER PIPE/CAP SECTION
TRACER WIRE NOTES:
1) THE SOUTH END SHALL BE SECURELY BOLTED TO THE YARD HYDRANT (LIFT STATION SITE).
2) THE NORTH END SHALL BE SECURELY BOLTED TO THE VALVE BOX IN SOUTH AVENUE (STA. 96+51).
3) NO TRACER WIRE SPLICES BETWEEN RISERS & NO GREATER THAN 1,000 FT. RISER SPACING. MOST SPLICES @ 500 FT. SPACING.
TYPICAL INTERMEDIATE CONNECTION FOR TRACER WIRE @ 500 FT. TYPICAL SPACING
(TYPICAL) CROSS-SECTION INTERMEDIATE CONNECTION FOR TRACER WIRE
PROJECT RECORD DRAWINGS: COMPILED FROM CONSTRUCTION DATA FURNISHED TO THE ENGINEER BY CONTRACTOR, OWNER AND/OR ENGINEER'S FIELD REPRESENTATIVE. INFORMATION SHOWN ON THESE DRAWINGS SHOULD BE FIELD VERIFIED BEFORE USING SAME.
SECTIONS, DETAILS, NOTES & LEGEND
UTILITY EXTENSION
VILLAGE of TONY, WISCONSIN
SCALE: DO NOT SCALE
MORGAN & PARMLEY, LTD.
CONSULTING ENGINEERS
LADYSMITH, WISCONSIN
SHEET 9 OF 12
DATE FEB. 1998
DR. BY T T

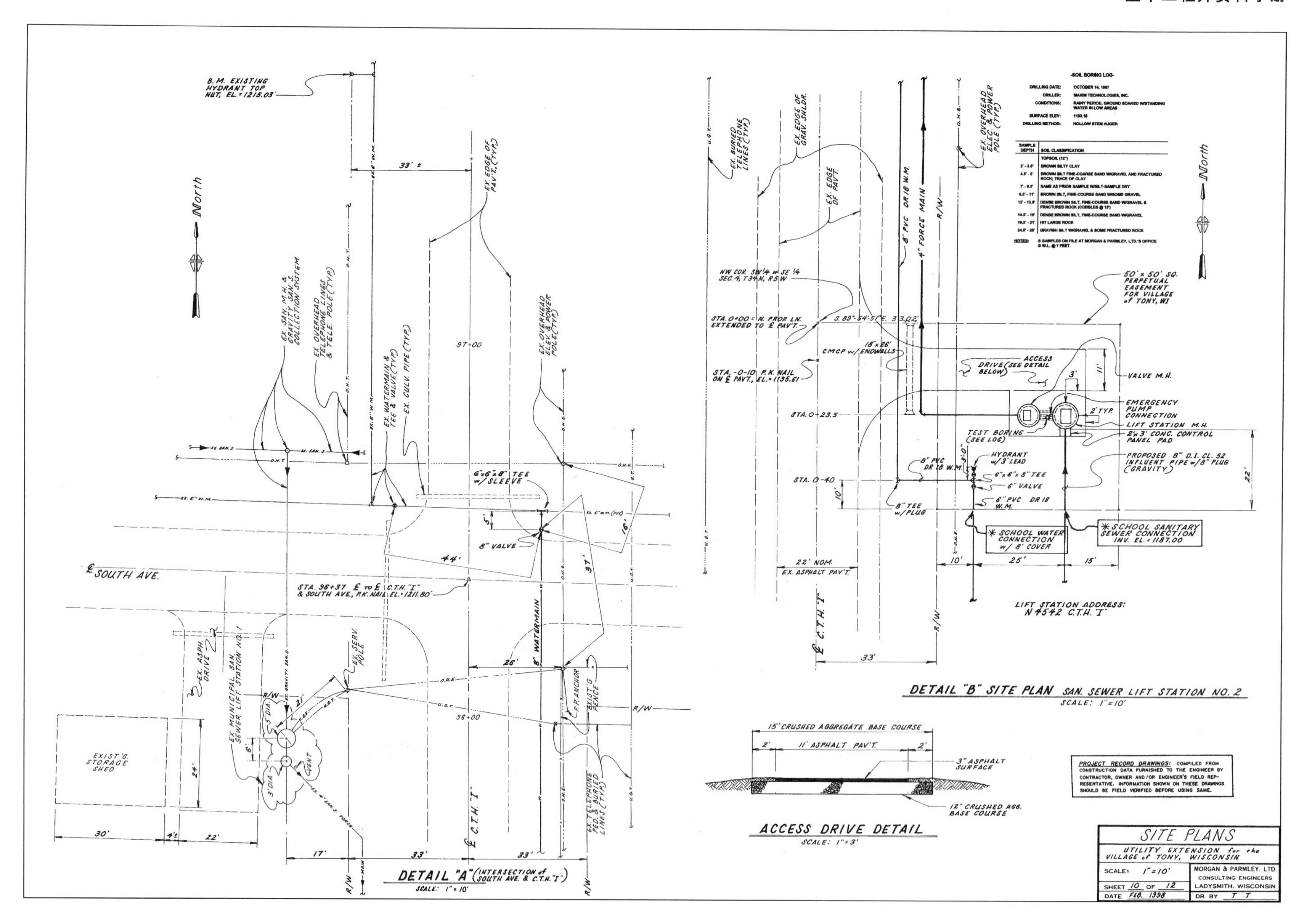
DETAIL "A" (INTERSECTION of SOUTH AVE. & C.T.H. "I")
SCALE: 1" = 10'
DETAIL "B" SITE PLAN SAN. SEWER LIFT STATION NO. 2
SCALE: 1" = 10'
ACCESS DRIVE DETAIL
SCALE: 1" = 3'
-SOIL BORING LOG-
PROJECT RECORD DRAWINGS: COMPILED FROM CONSTRUCTION DATA FURNISHED TO THE ENGINEER BY CONTRACTOR, OWNER AND/OR ENGINEER'S FIELD REPRESENTATIVE. INFORMATION SHOWN ON THESE DRAWINGS SHOULD BE FIELD VERIFIED BEFORE USING SAME.
SITE PLANS
UTILITY EXTENSION for the VILLAGE of TONY, WISCONSIN
SCALE: 1" = 10'
SHEET 10 OF 12
DATE FEB. 1998
MORGAN & PARMLEY, LTD.
CONSULTING ENGINEERS
LADYSMITH, WISCONSIN
DR. BY T T
LIFT STATION ADDRESS: N4542 C.T.H. "I"

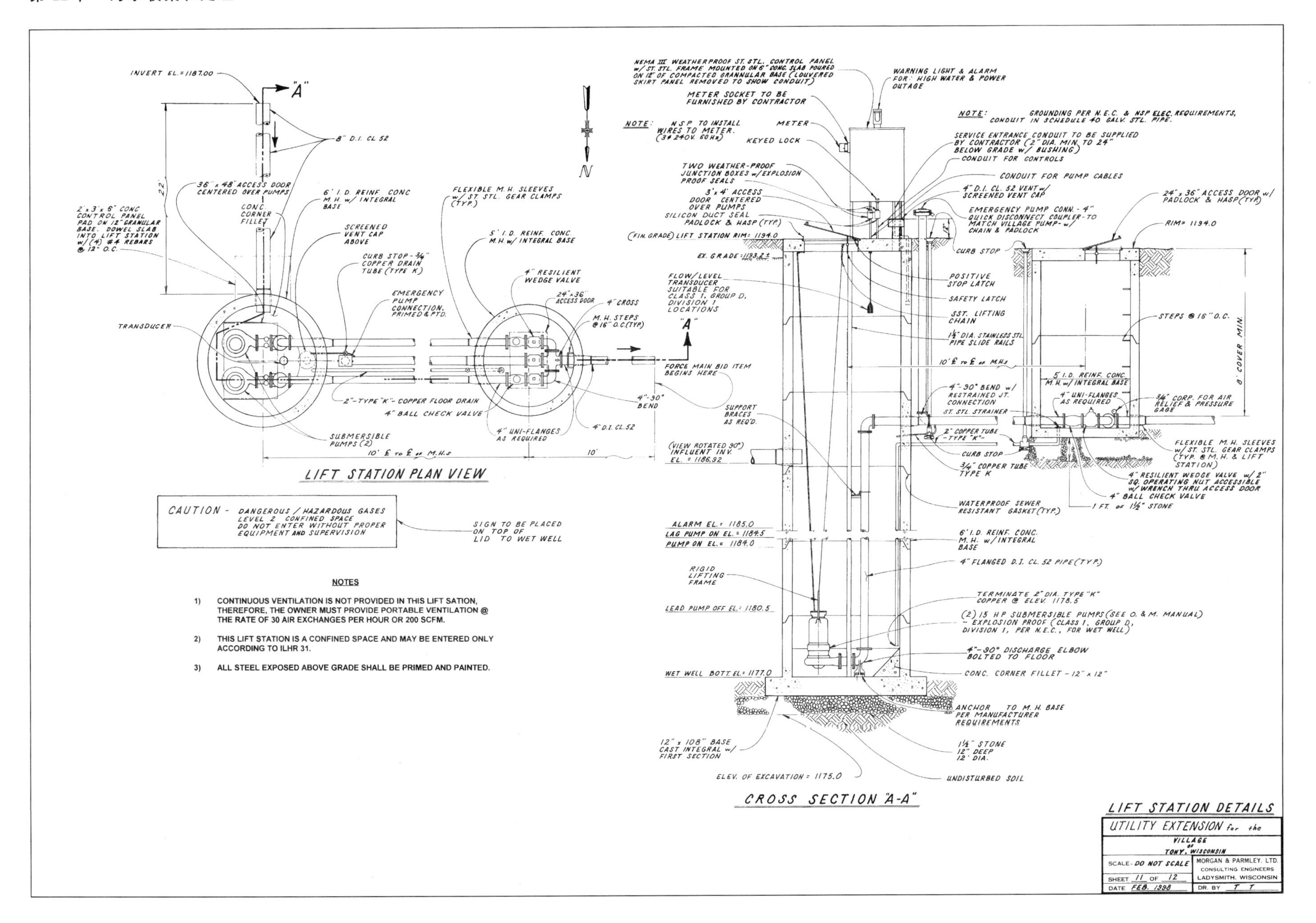

INVERT EL.=1187.00
"A"
8" D.I. CL 52
22'
36" x 48" ACCESS DOOR CENTERED OVER PUMPS
CONC. CORNER FILLET
6' I.D. REINF. CONC M.H. w/ INTEGRAL BASE
2'x3'x6" CONC CONTROL PANEL PAD ON 12" GRANULAR BASE. DOWEL SLAB INTO LIFT STATION w/(4) #4 REBARS @ 12" O.C.
SCREENED VENT CAP ABOVE
FLEXIBLE M.H. SLEEVES w/ ST. STL. GEAR CLAMPS (TYP.)
5' I.D. REINF. CONC. M.H. w/ INTEGRAL BASE
CURB STOP - 3/4" COPPER DRAIN TUBE (TYPE K)
4" RESILIENT WEDGE VALVE
EMERGENCY PUMP CONNECTION, PRIMED & PTD.
24"x36" ACCESS DOOR
4" CROSS
M.H. STEPS @16" O.C.(TYP)
TRANSDUCER
N
2"-TYPE "K"- COPPER FLOOR DRAIN
4" BALL CHECK VALVE
4"-90° BEND
SUBMERSIBLE PUMPS (2)
4" UNI-FLANGES AS REQUIRED
4" D.I. CL.52
10' ℄ TO ℄ OF M.H.S
10'
LIFT STATION PLAN VIEW
CAUTION - DANGEROUS / HAZARDOUS GASES LEVEL 2 CONFINED SPACE DO NOT ENTER WITHOUT PROPER EQUIPMENT AND SUPERVISION
SIGN TO BE PLACED ON TOP OF LID TO WET WELL
NOTES
1) CONTINUOUS VENTILATION IS NOT PROVIDED IN THIS LIFT SATION, THEREFORE, THE OWNER MUST PROVIDE PORTABLE VENTILATION @ THE RATE OF 30 AIR EXCHANGES PER HOUR OR 200 SCFM.
2) THIS LIFT STATION IS A CONFINED SPACE AND MAY BE ENTERED ONLY ACCORDING TO ILHR 31.
3) ALL STEEL EXPOSED ABOVE GRADE SHALL BE PRIMED AND PAINTED.
NEMA III WEATHERPROOF ST. STL. CONTROL PANEL w/ ST. STL. FRAME MOUNTED ON 6" CONC. SLAB POURED ON 12" OF COMPACTED GRANNULAR BASE (LOUVERED SKIRT PANEL REMOVED TO SHOW CONDUIT)
WARNING LIGHT & ALARM FOR: HIGH WATER & POWER OUTAGE
METER SOCKET TO BE FURNISHED BY CONTRACTOR
NOTE: NSP TO INSTALL WIRES TO METER. (3ϕ 240V. 60Hz)
METER
KEYED LOCK
NOTE: GROUNDING PER N.E.C. & NSP ELEC. REQUIREMENTS, CONDUIT IN SCHEDULE 40 GALV. STL. PIPE.
SERVICE ENTRANCE CONDUIT TO BE SUPPLIED BY CONTRACTOR (2" DIA. MIN. TO 24" BELOW GRADE w/ BUSHING)
CONDUIT FOR CONTROLS
CONDUIT FOR PUMP CABLES
TWO WEATHER-PROOF JUNCTION BOXES w/EXPLOSION PROOF SEALS
3'x4' ACCESS DOOR CENTERED OVER PUMPS
SILICON DUCT SEAL
PADLOCK & HASP (TYP.)
4" D.I. CL. 52 VENT w/ SCREENED VENT CAP
EMERGENCY PUMP CONN. - 4" QUICK DISCONNECT COUPLER - TO MATCH VILLAGE PUMP - w/ CHAIN & PADLOCK
24"x36" ACCESS DOOR w/ PADLOCK & HASP (TYP.)
RIM= 1194.0
(FIN. GRADE) LIFT STATION RIM: 1194.0
EX. GRADE = 1193.2±
CURB STOP
POSITIVE STOP LATCH
SAFETY LATCH
SST. LIFTING CHAIN
FLOW/LEVEL TRANSDUCER SUITABLE FOR CLASS 1, GROUP D, DIVISION 1 LOCATIONS
1½" DIA. STAINLESS STL. PIPE SLIDE RAILS
STEPS @ 16" O.C.
8' COVER MIN.
10' ℄ TO ℄ OF M.H.s
5' I.D. REINF. CONC. M.H. w/ INTEGRAL BASE
FORCE MAIN BID ITEM BEGINS HERE
4"-90° BEND w/ RESTRAINED JT. CONNECTION
ST. STL. STRAINER
4" UNI-FLANGES AS REQUIRED
3/4" CORP. FOR AIR RELIEF & PRESSURE GAGE
SUPPORT BRACES AS REQ'D.
2" COPPER TUBE - TYPE "K"
CURB STOP
3/4" COPPER TUBE TYPE K
FLEXIBLE M.H. SLEEVES w/ ST. STL. GEAR CLAMPS (TYP. @ M.H. & LIFT STATION)
(VIEW ROTATED 90°) INFLUENT INV. EL. = 1186.92
4" RESILIENT WEDGE VALVE w/ 2" SQ. OPERATING NUT ACCESSIBLE w/ WRENCH THRU ACCESS DOOR
4" BALL CHECK VALVE
WATERPROOF SEWER RESISTANT GASKET (TYP.)
1 FT. OF 1½" STONE
ALARM EL.= 1185.0
LAG PUMP ON EL.= 1184.5
PUMP ON EL.= 1184.0
6' I.D. REINF. CONC. M.H. w/ INTEGRAL BASE
4" FLANGED D.I. CL. 52 PIPE (TYP.)
RIGID LIFTING FRAME
TERMINATE 2" DIA. TYPE "K" COPPER @ ELEV. 1178.5
LEAD PUMP OFF EL.= 1180.5
(2) 15 HP SUBMERSIBLE PUMPS (SEE O. & M. MANUAL) - EXPLOSION PROOF (CLASS 1, GROUP D, DIVISION 1, PER N.E.C., FOR WET WELL)
4"-90° DISCHARGE ELBOW BOLTED TO FLOOR
WET WELL BOTT. EL.= 1177.0
CONC. CORNER FILLET - 12" x 12"
ANCHOR TO M.H. BASE PER MANUFACTURER REQUIREMENTS
12" x 108" BASE CAST INTEGRAL w/ FIRST SECTION
1½" STONE 12" DEEP 12' DIA.
ELEV. OF EXCAVATION = 1175.0
UNDISTURBED SOIL
CROSS SECTION "A-A"
LIFT STATION DETAILS
UTILITY EXTENSION for the VILLAGE OF TONY, WISCONSIN
SCALE - DO NOT SCALE
MORGAN & PARMLEY, LTD. CONSULTING ENGINEERS LADYSMITH, WISCONSIN
SHEET 11 OF 12
DATE FEB. 1998
DR. BY T T

污水管道和供水干管的扩建工程

SANITARY SEWER & WATER MAIN EXTENSIONS

(W. 11th SREET. N., EASEMENTS, BAKER AVENUE & W. 13th STREET. N.)

CITY
of
LADYSMITH, WISCONSIN
FEBRUARY, 2000

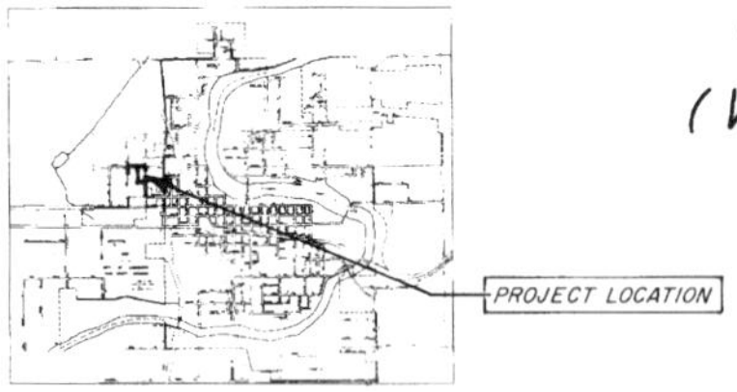

PROJECT LOCATION

GENERAL LOCATION MAP

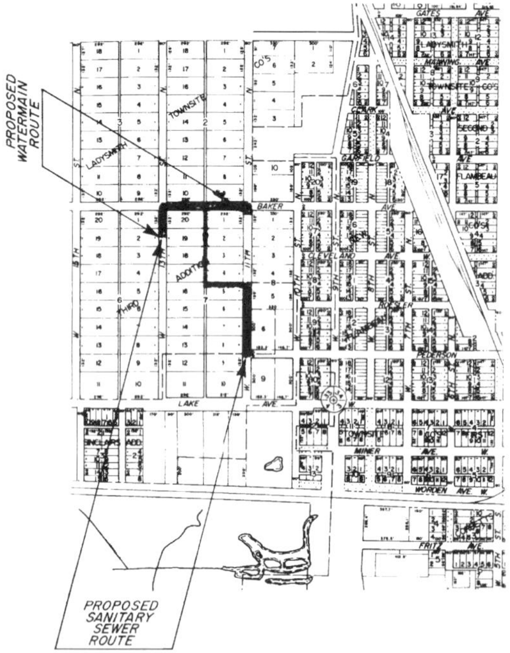

GENERAL NOTES

1. ALL WATER MAIN SHALL BE 8" DI CL 52.
2. ALL WATERMAIN FITTINGS SHALL BE M.J.D.I., W / RESTRAINED JOINT LUGS.
3. ALL HYDRANT LEADS SHALL BE 6-INCH DI CL 52.
4. ALL HYDRANT LEADS SHALL HAVE A 6-INCH VALVE.
5. HYDRANTS SHALL BE WATEROUS WB 67-250 (4" PUMPER NOZZLE), 22" BREAKOFF WITH THREADS MATCHING CITY PATTERN.
6. ALL WATER MAIN SHALL HAVE AN 8 FT. BURY (MIN.).
7. WATER SERVICES SHALL BE 1" TYPE K COPPER W / MUELLER OR FORD BRASS AND LOCATED BY OWNER.
8. SANITARY SERVICES SHALL BE 4" PVC SCH. 40 W / WYES LAID TO PROVIDE FOR MAXIMUM DEPTH OR 10' @ R / W, LOCATION BY OWNER.
9. SEWER MAIN SHALL BE 8" PVC SDR 35.
10. MANHOLES SHALL BE SET WITH A MINIMUM OF 6" OF RINGS W / 9" CASTINGS.
11. MANHOLES SHALL BE 4' DIA. PRECASE CONCRETE W / INTEGRAL BASES AND BOOTS.
12. 8' (MINIMUM) HORIZONTAL DISTANCE (℄ TO ℄) BETWEEN WATERMAIN AND SANITARY SEWER OR STORM SEWER.
13. 18" (MINIMUM) VERTICAL DISTANCE (OUT TO OUT) BETWEEN WATERMAIN UNDER SANITARY SEWER OR STORM SEWER @ THEIR INTERSECTION.
14. 6" (MINIMUM) VERTICAL DISTANCE (OUT TO OUT) BETWEEN WATERMAIN OVER SANITARY SEWER OR STORM SEWER @ THEIR INTERSECTION.
15. ALL ELEVATIONS BASED ON CITY DATUM FROM B.M. SHOWN.
16. EROSION CONTROL MEASURES SHALL BE IN PLACE BEFORE CONSTRUCTION BEGINS.
17. THE CONTRACTORS SHALL MINIMIZE EROSION DURING CONSTRUCTION USING GOOD CONSTRUCTION TECHNIQUES AND UTILIZING SILT FENCES AND BALE DITCH CHECKS. THE CONTRACTOR SHALL RELEASE RUNOFF FROM THE SITE IN A NUISANCE FREE MANNER.
18. THE CONTRACTOR IS RESPONSIBLE TO MAINTAIN THE SITE IN A SAFE CONDITION. THE SOLE RESPONSIBILITY FOR WARNING SIGNS, BARRICADES AND ALL ASPECTS OF SAFETY LIE WITH THE CONTRACTOR.
19. EXISTING UTILITIES ARE SHOWN BUT MAY NOT BE ALL INCLUSIVE, CONTRACTOR SHALL HAVE ALL BURIED UTILITIES LOCATED PRIOR TO COMMENCING CONSTRUCTION.
20. EXISTING SEWER AND WATERMAIN ARE SHOWN ACCORDING TO EXISTING RECORDS BUT LOCATION SHALL BE VERIFIED IN FIELD.
21. WATERMAIN & SERVICES SHALL BE INSULATED AT ALL STORM SEWER, CULVERT AND DITCH CROSSINGS WITH 2" THICK EXTRUDED POLYSTYRENE INSULATION BOARD.
22. PROPERTY CORNERS KNOWN TO EXIST SHALL NOT BE DISTURBED. DAMAGED CORNERS WILL BE REPLACED AT THE CONTRACTOR'S EXPENSE. PRESERVE ALL SURVEY MONUMENTS.
23. ALL WATERMAIN SHALL BE DISINFECTED BEFORE BEING PUT INTO USE.
24. VALVES AND MANHOLES SHALL BE SET TO GRADE SHOWN ON PLANS, UNLESS DIRECTED OTHERWISE.
25. CONSTRUCTION SHALL COMPLY WITH THE CONDITIONS OF DNR APPROVAL LETTER, AND OTHER REGULATORY APPROVALS AS LISTED IN THE SPECIFICATION BOOK.
26. ALL DISTURBED AREAS SHALL BE SEEDED, FERTILIZED AND MULCHED, OR COVERED W / BASE COURSE AFTER CONSTRUCTION, OR RESTORED WITH BITUMINOUS PAVEMENT WHERE APPLICABLE.
27. ALL DISTURBED ARES IN EASEMENT, ALLEY & ROAD DITCH SHALL BE SEEDED, FERTILIZED AND MULCHED.
28. ALL DISTURBED AREAS SHALL BE RESTORED W / SEEDING AS FOLLOWS:

    A) REPLACED ALL SALVAGED TOPSOIL
    B) SEED MIXUTRE: PERMANENT SEEDING – (LAWN TYPE TURF)

    35% KENTUCKY BLUEGRASS
    25% IMPROVED FINE PERENNIAL RYEGRASS
    15% CREEEPING RED FESCUE
    15% IMPROVED HARD FESCUE
    10% WHITE CLOVER

    C) SEED RATE: 2# / 1000 SQ. FT.
    D) MULCH RATE: 3 TON / ACRE
    E) PERMANENT SEEDING: ALLOWWED TO SEPT. 7
    TEMPORARY SEEDING: SEPT. 8 THROUGH Nov. 10
    (4 BU. OATS / ACRE)
    DORMANT SEEDING; SEED PERM. SEEDING INTO TEMP. SEEDING AFTER NOV. 10
    F) SEEDING MAY BE BROADCAST BUT MUST BE COVERED WITH A MAXIMUM OF ½" SOIL. MULCH MUST BE ANCHORED WITH A MULCH TILLER OR BLOWN W / 75-100 GAL. OF ASPHALT / TON OF MULCH.
    G) FERTILIZER: 500# / ACRE 20-10-10

29. CONTRACTOR SHALL DETERMINE CONSTRUCTION SEQUENCE IN ORDER TO MINIMIZE INSTALLATION CONFLICTS WITH PIPE CROSSINGS.
30. THERE SHALL BE NO DEVIATION FROM THE PLANS WITHOUT THE DESIGN ENGINEER'S APPROVAL.
31. CONSTRUCTION STAKES DESTROYED BY THE CONTRACTOR THAT NEED TO BE REPLACED, WILL BE CHARGED TO THE CONTRACTOR.
32. CONTRACTOR SHALL COORDINATOR WORK TO AVOID UNDUE CONFLICT WITH ADJACENT PROPERTY.
33. NO TREES ARE TO BE REMOVED WITHOUT THE APPROVAL OF THE ENGINEER.
34. THE CONTRACTOR MUST REMOVE AND REPLACE ALL EFFECTED STREET SIGNS AFTER PROPERLY SIGNING AND BARRICADING THE CONSTRUCTION AREA. ALL SIGNS MUST BE REPLACED PRIOR TO OPENING THE STREET TO TRAFFIC.
35. CONFLICTING POWER POLES WILL BE MOVED BY NSP.
36. SEE SPECIFICATION BOOK, SECTION 01010 AND FOLLOWING PLAN SHEETS FOR ADDITIONAL DETAILS.

GENERAL SPECIFICATIONS: STANDARD SPECIFICATIONS FOR SEWER AND WATER CONSTRUCTION IN WISCONSIN (5TH EDITION)

WISCONSIN CONSTRUCTION SITE BEST MANAGEMENT PRACTICE HANDBOOK

SPECIFIC SPECIFICATIONS: MORGAN & PARMLEY, LTD. SPEC. BOOK

BIDDING DOCUMENTS: MORGAN & PARMLEY, LTD. SPEC. BOOK

PROPOSED WATERMAIN ROUTE

PROPOSED SANITARY SEWER ROUTE

VICINITY MAP

DO NOT SCALE

UTILITIES

WATER & SEWER:
LADYSMITH DEPARTMENT OF PUBLIC WORKS
ATTN: BILL CHRISTIANSON, DIRECTOR
120 WEST MINER AVENUE
P.O. BOX 431
LADYSMITH, WISCONSIN 54848
TELEPHONE: (715) 532-2601

ELECTRIC:
NORTHERN STATES POWER COMPANY
310 HICKORY HILL LANE
PHILLIPS, WISOCNSIN 54555
ATTN: JOE PERKINS
TELEPHONE: (715) 836-1198

NATURAL GAS:
WISCONSIN GAS COMPANY
104 WEST SOUTH STREET
RICE LAKE, WISCONSIN 54868
ATTN: DON WEDIN
TELEPHONE: (715) 236-2104

CABLE TELEVISION:
CHARTER COMMUNICATIONS
1725 S. MAIN ST.
RICE LAKE, WISCONSIN 54868
TELEPHONE: (800) 262-2578

TELEPHONE:
CENTURYTEL
425 ELLINGSON AVENUE
P.O. BOX 78
HAWKINS, WISCONSIN 54530
TELEPHONE: (715) 585-7707
ATTN: JAMES ARQUETTE

DIGGER'S HOTLINE:
1-800-242-8511

DRAWING INDEX

SHEET NO.	DESCRIPTION
1	TITLE SHEET, LOCATION MAPS, INDEX & NOTES
2	PLAN & PROFILE: SANITARY SEWER
3	PLAN & PROFILE: SANITARY SEWER & WATERMAIN
4	PLAN & PROFILE: SANITARY SEWER & WATERMAIN
5	TYPICAL EROSION CONTROL DETAILS This Sheet Not Shown

PREPARED BY:
MORGAN & PARMLEY, LTD.
PROFESSIONAL CONSULTING ENGINEERS
LADYSMITH, WISCONSIN

Courtesy: Morgan & Parmley, Ltd.

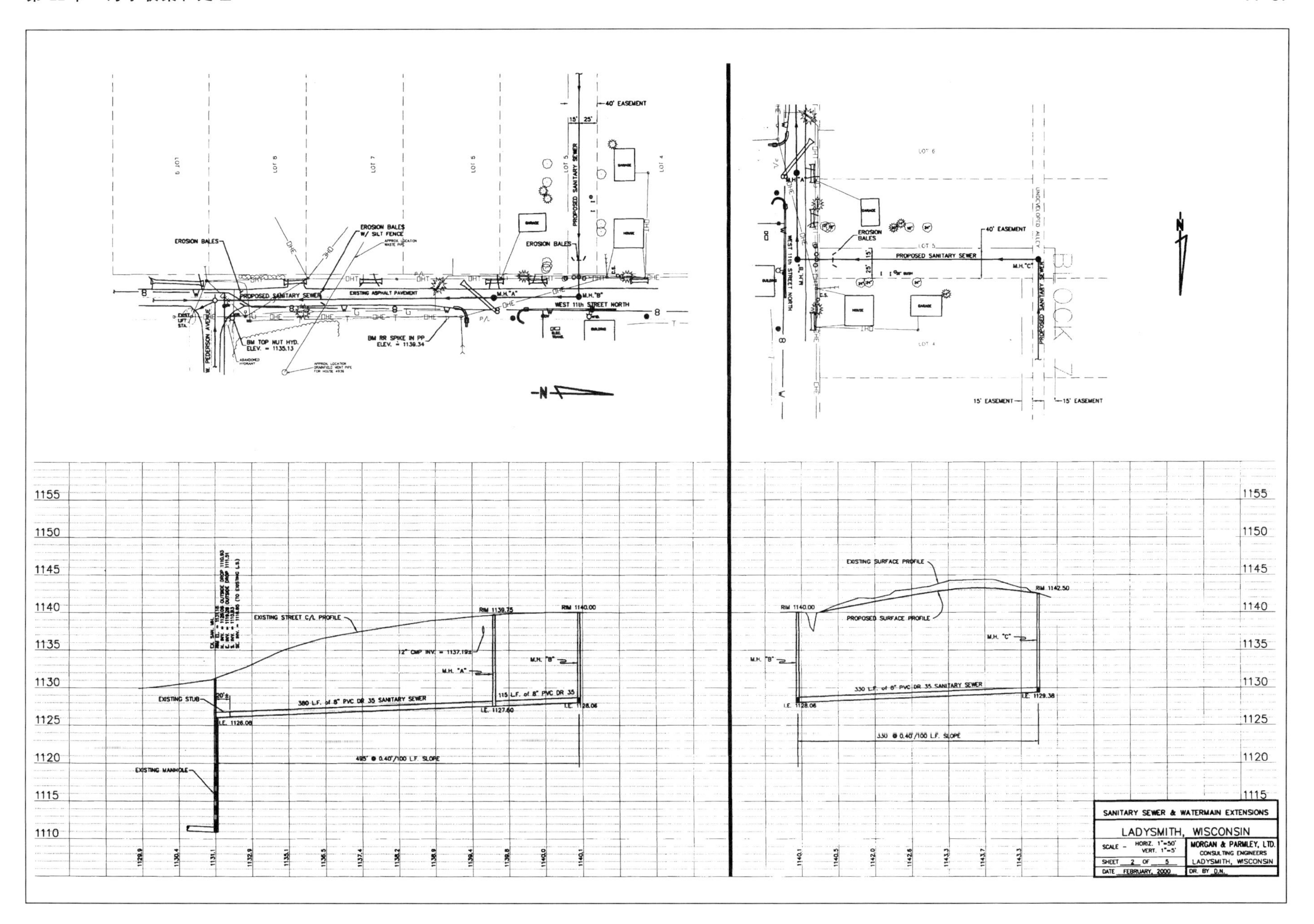
40' EASEMENT
PROPOSED SANITARY SEWER
EROSION BALES
EROSION BALES W/ SILT FENCE
EXISTING ASPHALT PAVEMENT
M.H. "A"
M.H. "B"
M.H. "C"
WEST 11th STREET NORTH
BM TOP NUT HYD. ELEV. = 1135.13
BM RR SPIKE IN PP ELEV. = 1139.34
W. PEDERSON AVENUE
15' EASEMENT
UNDEVELOPED ALLEY
BLOCK 7
1155
1150
1145
1140
1135
1130
1125
1120
1115
1110
EXISTING STREET C/L PROFILE
RIM 1139.75
RIM 1140.00
12" CMP INV. = 1137.19±
EXISTING STUB
380 L.F. of 8" PVC DR 35 SANITARY SEWER
115 L.F. of 8" PVC DR 35
I.E. 1127.60
I.E. 1128.06
I.E. 1126.06
495' @ 0.40'/100 L.F. SLOPE
EXISTING MANHOLE
EXISTING SURFACE PROFILE
PROPOSED SURFACE PROFILE
RIM 1142.50
330 L.F. of 8" PVC DR 35 SANITARY SEWER
I.E. 1129.38
330' @ 0.40'/100 L.F. SLOPE
SANITARY SEWER & WATERMAIN EXTENSIONS
LADYSMITH, WISCONSIN
SCALE – HORIZ. 1"=50' VERT. 1"=5'
SHEET 2 OF 5
DATE FEBRUARY, 2000
MORGAN & PARMLEY, LTD.
CONSULTING ENGINEERS
LADYSMITH, WISCONSIN
DR. BY D.N.

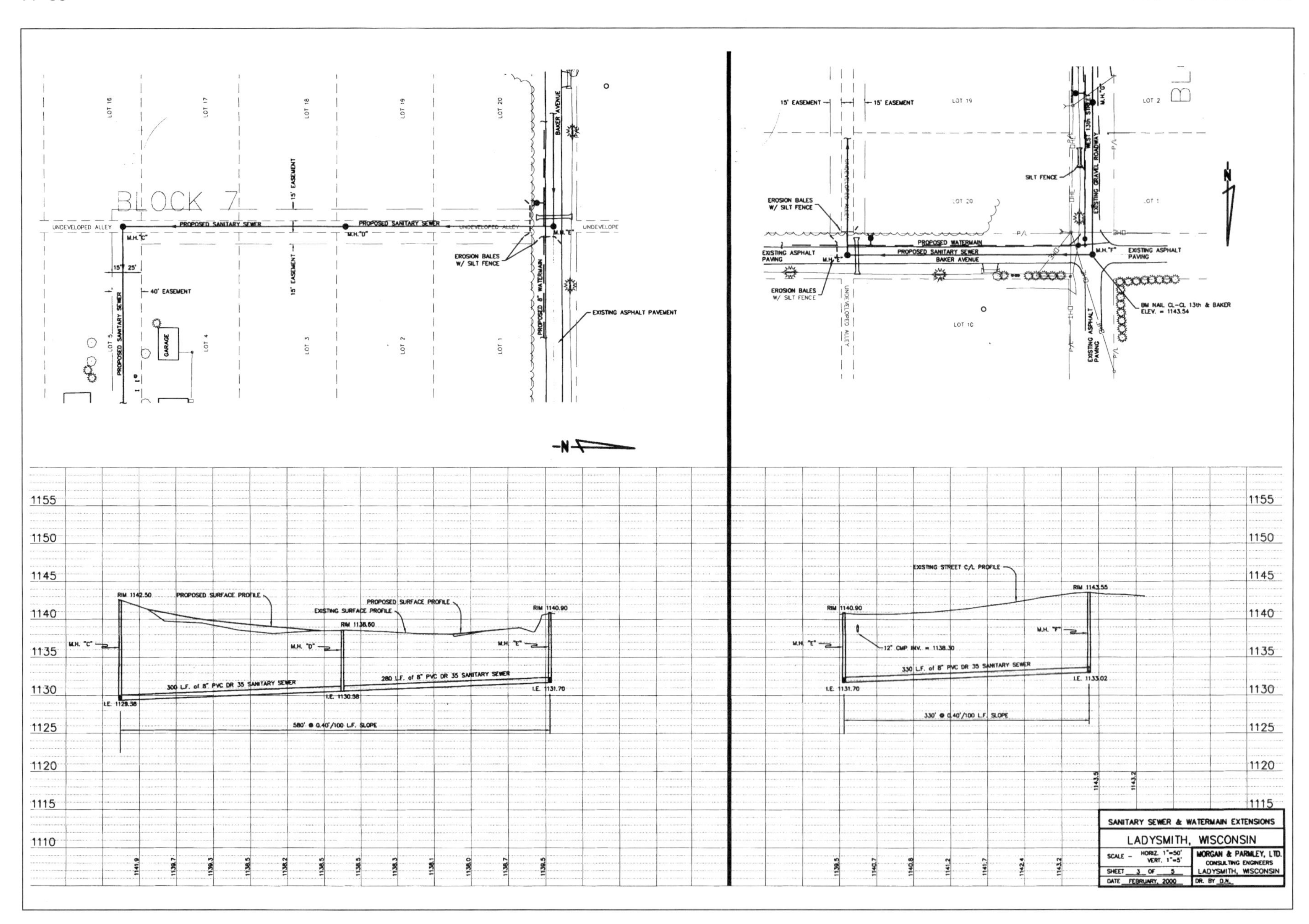

BLOCK 7
UNDEVELOPED ALLEY
PROPOSED SANITARY SEWER
M.H."C"
M.H."D"
M.H."E"
15' EASEMENT
40' EASEMENT
EROSION BALES W/ SILT FENCE
EXISTING ASPHALT PAVEMENT
PROPOSED 8" WATERMAIN
BAKER AVENUE
GARAGE
LOT 16
LOT 17
LOT 18
LOT 19
LOT 20
LOT 5
LOT 4
LOT 3
LOT 2
LOT 1
PROPOSED WATERMAIN
EXISTING ASPHALT PAVING
M.H."F"
SILT FENCE
EXISTING GRAVEL ROADWAY
WEST 13th STREET
BM NAIL CL-CL 13th & BAKER ELEV. = 1143.54
UNDEVELOPED ALLEY
RIM 1142.50
PROPOSED SURFACE PROFILE
EXISTING SURFACE PROFILE
RIM 1138.60
RIM 1140.90
300 L.F. of 8" PVC DR 35 SANITARY SEWER
280 L.F. of 8" PVC DR 35 SANITARY SEWER
I.E. 1129.38
I.E. 1130.58
I.E. 1131.70
580' @ 0.40'/100 L.F. SLOPE
EXISTING STREET C/L PROFILE
RIM 1143.55
12" CMP INV. = 1138.30
330 L.F. of 8" PVC DR 35 SANITARY SEWER
I.E. 1133.02
330' @ 0.40'/100 L.F. SLOPE
SANITARY SEWER & WATERMAIN EXTENSIONS
LADYSMITH, WISCONSIN
SCALE – HORIZ. 1"=50' VERT. 1"=5'
MORGAN & PARMLEY, LTD. CONSULTING ENGINEERS LADYSMITH, WISCONSIN
SHEET 3 OF 5
DATE FEBRUARY, 2000
DR. BY D.N.

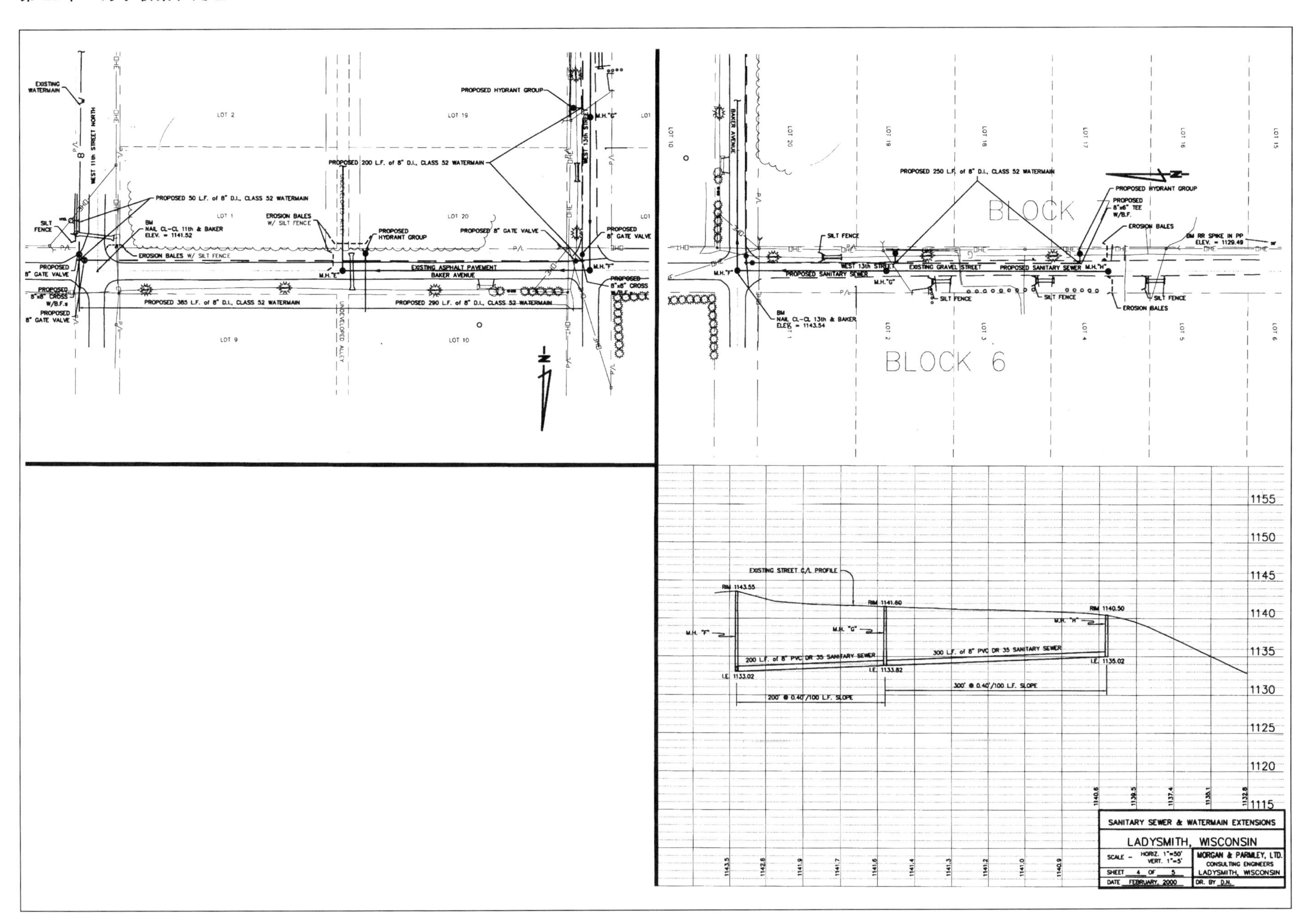
SANITARY SEWER & WATERMAIN EXTENSIONS
LADYSMITH, WISCONSIN
MORGAN & PARMLEY, LTD. CONSULTING ENGINEERS LADYSMITH, WISCONSIN
SCALE – HORIZ. 1"=50' VERT. 1"=5'
SHEET 4 OF 5
DATE FEBRUARY, 2000
DR. BY D.N.
BLOCK 6
BLOCK 7
EXISTING STREET C/L PROFILE
200 L.F. of 8" PVC DR 35 SANITARY SEWER
300 L.F. of 8" PVC DR 35 SANITARY SEWER
200' @ 0.40/100 L.F. SLOPE
300' @ 0.40/100 L.F. SLOPE

设　计

第 12 节　雨水管道系统

例　12

所有降落在市区、道路、停车场和类似的改建场地的雨水、径流、雪融水和相关降水都必须经过适当的处理，否则会导致非常严重的问题。

雨水管道系统的设计是非常复杂的，并在总体设计中集成了附加组件，例如替代物或下水道和供水设施管道系统。

本节包含了数个小型雨水管道系统的设计图。第一套图纸是现有小学场地的初级雨水管道系统图。第二套设计图是一个新建商业综合建筑的停车场。第三套设计图给出了乡村小型装置的设计详图。再后两套设计图是改造市区街道升级它们的雨水排水系统。最后一套设计图给出了一个基本雨水管道扩建工程的详图。

目　录

Preliminary Storm Drainage – Elementary School 初级雨水排水系统：小学

Ladysmith, WI

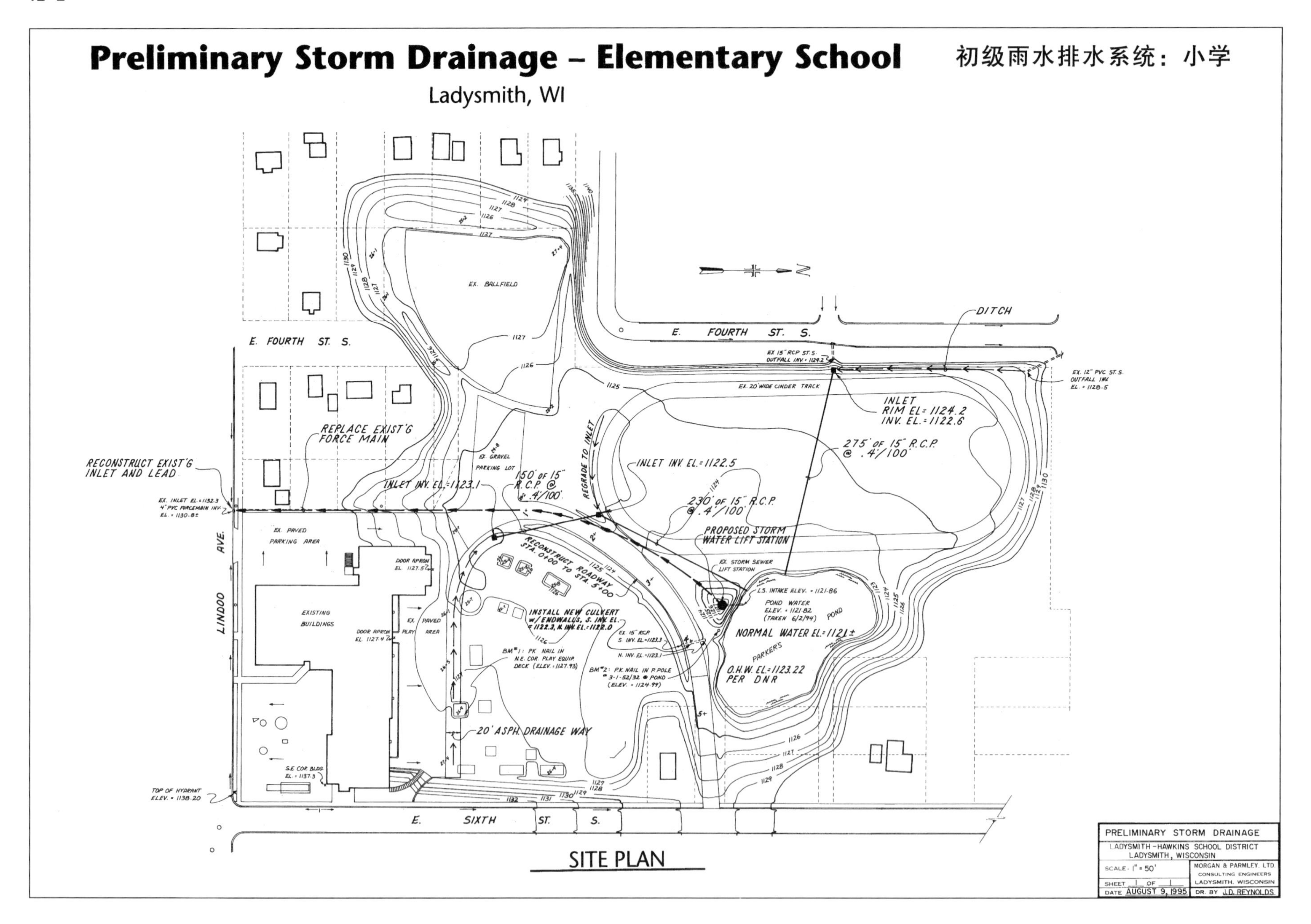

Courtesy: Morgan & Parmley, Ltd.

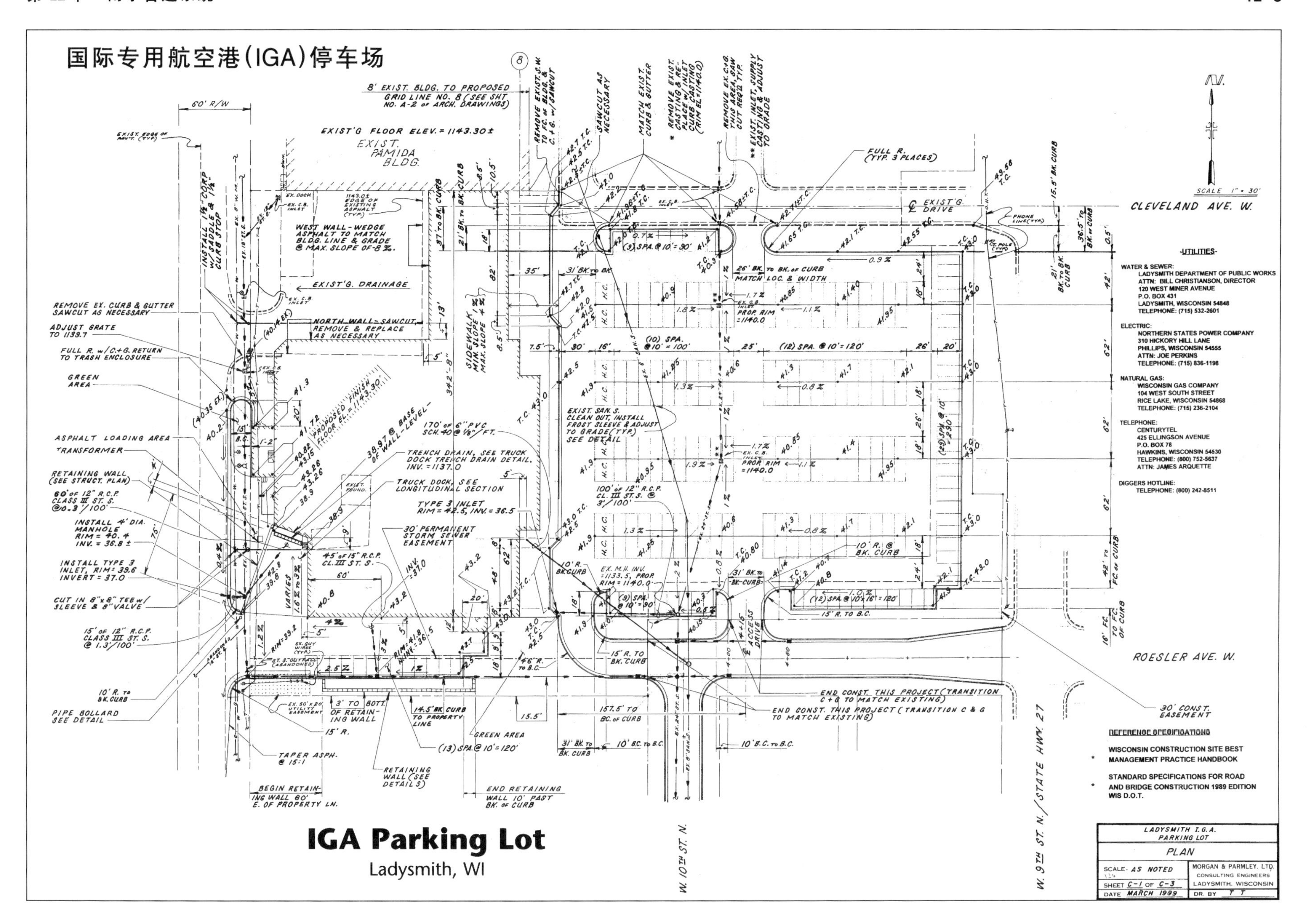

IGA Parking Lot

Ladysmith, WI

Courtesy: Morgan & Parmley, Ltd.

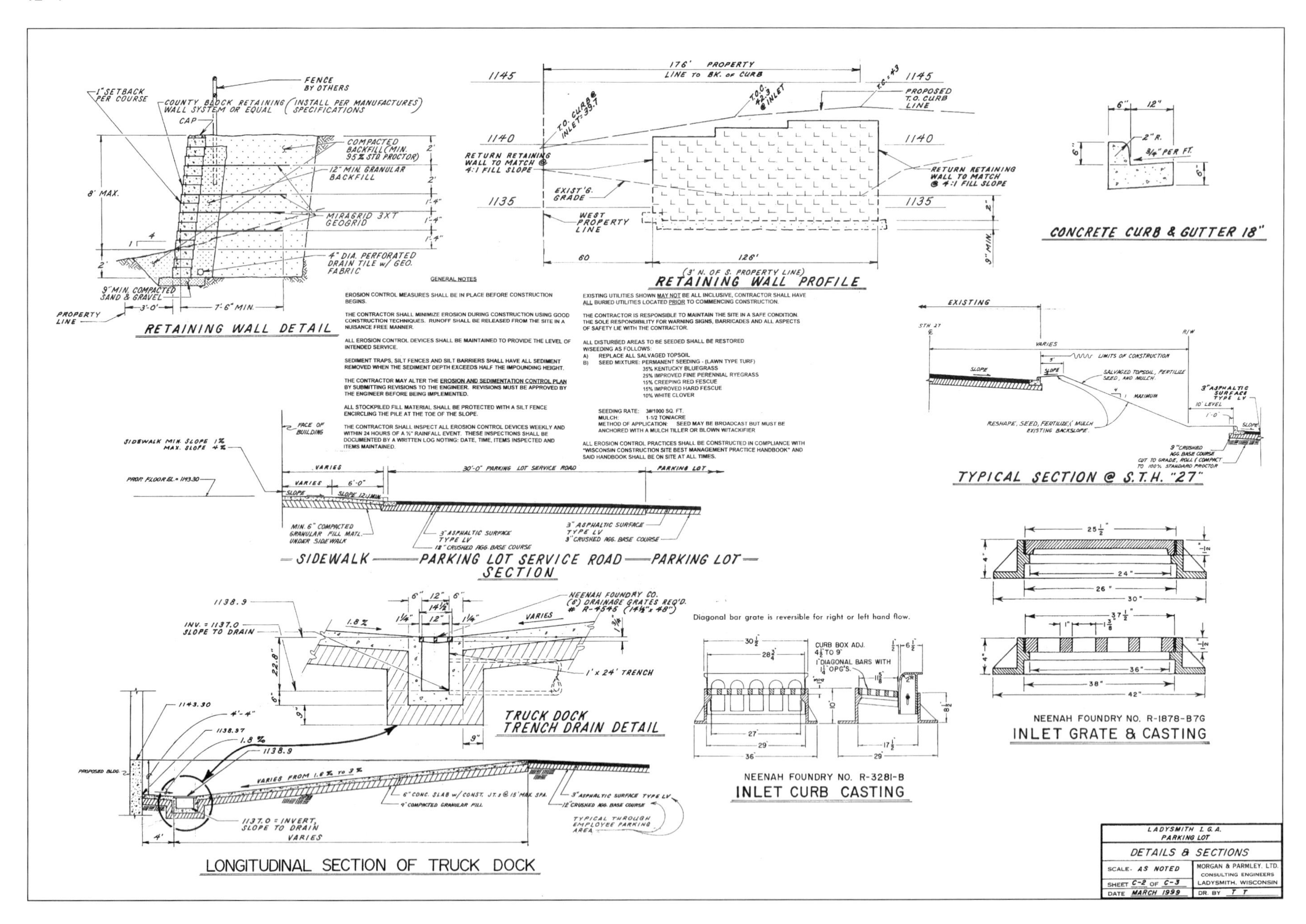
FENCE BY OTHERS
1" SETBACK PER COURSE
COUNTY BLOCK RETAINING WALL SYSTEM OR EQUAL (INSTALL PER MANUFACTURES) SPECIFICATIONS)
CAP
COMPACTED BACKFILL (MIN. 95% STD PROCTOR)
12" MIN. GRANULAR BACKFILL
8' MAX.
MIRAGRID 3XT GEOGRID
4" DIA. PERFORATED DRAIN TILE w/ GEO. FABRIC
9" MIN. COMPACTED SAND & GRAVEL
PROPERTY LINE
3'-0"
7'-6" MIN.
RETAINING WALL DETAIL
176' PROPERTY LINE TO BK. OF CURB
PROPOSED T.O. CURB LINE
RETURN RETAINING WALL TO MATCH @ 4:1 FILL SLOPE
EXIST'G. GRADE
WEST PROPERTY LINE
1145
1140
1135
60
126'
(3' N. OF S. PROPERTY LINE)
RETAINING WALL PROFILE
CONCRETE CURB & GUTTER 18"
GENERAL NOTES
EROSION CONTROL MEASURES SHALL BE IN PLACE BEFORE CONSTRUCTION BEGINS.
THE CONTRACTOR SHALL MINIMIZE EROSION DURING CONSTRUCTION USING GOOD CONSTRUCTION TECHNIQUES. RUNOFF SHALL BE RELEASED FROM THE SITE IN A NUISANCE FREE MANNER.
ALL EROSION CONTROL DEVICES SHALL BE MAINTAINED TO PROVIDE THE LEVEL OF INTENDED SERVICE.
SEDIMENT TRAPS, SILT FENCES AND SILT BARRIERS SHALL HAVE ALL SEDIMENT REMOVED WHEN THE SEDIMENT DEPTH EXCEEDS HALF THE IMPOUNDING HEIGHT.
THE CONTRACTOR MAY ALTER THE EROSION AND SEDIMENTATION CONTROL PLAN BY SUBMITTING REVISIONS TO THE ENGINEER. REVISIONS MUST BE APPROVED BY THE ENGINEER BEFORE BEING IMPLEMENTED.
ALL STOCKPILED FILL MATERIAL SHALL BE PROTECTED WITH A SILT FENCE ENCIRCLING THE PILE AT THE TOE OF THE SLOPE.
THE CONTRACTOR SHALL INSPECT ALL EROSION CONTROL DEVICES WEEKLY AND WITHIN 24 HOURS OF A ½" RAINFALL EVENT. THESE INSPECTIONS SHALL BE DOCUMENTED BY A WRITTEN LOG NOTING: DATE, TIME, ITEMS INSPECTED AND ITEMS MAINTAINED.
EXISTING UTILITIES SHOWN MAY NOT BE ALL INCLUSIVE, CONTRACTOR SHALL HAVE ALL BURIED UTILITIES LOCATED PRIOR TO COMMENCING CONSTRUCTION.
THE CONTRACTOR IS RESPONSIBLE TO MAINTAIN THE SITE IN A SAFE CONDITION. THE SOLE RESPONSIBILITY FOR WARNING SIGNS, BARRICADES AND ALL ASPECTS OF SAFETY LIE WITH THE CONTRACTOR.
ALL DISTURBED AREAS TO BE SEEDED SHALL BE RESTORED W/SEEDING AS FOLLOWS:
A) REPLACE ALL SALVAGED TOPSOIL
B) SEED MIXTURE: PERMANENT SEEDING - (LAWN TYPE TURF)
35% KENTUCKY BLUEGRASS
25% IMPROVED FINE PERENNIAL RYEGRASS
15% CREEPING RED FESCUE
15% IMPROVED HARD FESCUE
10% WHITE CLOVER
SEEDING RATE: 3#/1000 SQ. FT.
MULCH: 1-1/2 TON/ACRE
METHOD OF APPLICATION: SEED MAY BE BROADCAST BUT MUST BE ANCHORED WITH A MULCH TILLER OR BLOWN W/TACKIFIER
ALL EROSION CONTROL PRACTICES SHALL BE CONSTRUCTED IN COMPLIANCE WITH "WISCONSIN CONSTRUCTION SITE BEST MANAGEMENT PRACTICE HANDBOOK" AND SAID HANDBOOK SHALL BE ON SITE AT ALL TIMES.
EXISTING
VARIES
LIMITS OF CONSTRUCTION
SALVAGED TOPSOIL, FERTILIZE SEED, AND MULCH
MAXIMUM
RESHAPE, SEED, FERTILIZE, MULCH EXISTING BACKSLOPE
3" ASPHALTIC SURFACE TYPE LV
9" CRUSHED AGG. BASE COURSE
TYPICAL SECTION @ S.T.H. "27"
SIDEWALK MIN. SLOPE 1% MAX. SLOPE 4%
PROP. FLOOR EL. = 1143.30
FACE OF BUILDING
VARIES
6'-0"
SLOPE 12:1 MIN.
30'-0" PARKING LOT SERVICE ROAD
PARKING LOT
MIN. 6" COMPACTED GRANULAR FILL MATL. UNDER SIDEWALK
3" ASPHALTIC SURFACE TYPE LV
12" CRUSHED AGG. BASE COURSE
3" ASPHALTIC SURFACE TYPE LV
3" CRUSHED AGG. BASE COURSE
SIDEWALK — PARKING LOT SERVICE ROAD — PARKING LOT — SECTION
25½"
24"
26"
30"
37½"
36"
38"
42"
NEENAH FOUNDRY NO. R-1878-B7G
INLET GRATE & CASTING
1138.9
INV. = 1137.0 SLOPE TO DRAIN
NEENAH FOUNDRY CO. (6) DRAINAGE GRATES REQ'D. # R-4545 (14½" x 48")
VARIES
1' x 24' TRENCH
TRUCK DOCK TRENCH DRAIN DETAIL
Diagonal bar grate is reversible for right or left hand flow.
CURB BOX ADJ. 4½ TO 9
1 DIAGONAL BARS WITH 1¼" OPG'S.
NEENAH FOUNDRY NO. R-3281-B
INLET CURB CASTING
1143.30
4'-4"
1138.37
1.8%
1138.9
PROPOSED BLDG.
VARIES FROM 1.8% TO 3%
6" CONC. SLAB w/ CONST. JT.'s @ 15' MAX. SPA.
9" COMPACTED GRANULAR FILL
3" ASPHALTIC SURFACE TYPE LV
12" CRUSHED AGG. BASE COURSE
1137.0 = INVERT, SLOPE TO DRAIN
VARIES
TYPICAL THROUGH EMPLOYEE PARKING AREA
LONGITUDINAL SECTION OF TRUCK DOCK
LADYSMITH I.G.A. PARKING LOT
DETAILS & SECTIONS
SCALE AS NOTED
SHEET C-2 OF C-3
DATE MARCH 1999
MORGAN & PARMLEY, LTD. CONSULTING ENGINEERS LADYSMITH, WISCONSIN
DR. BY T T

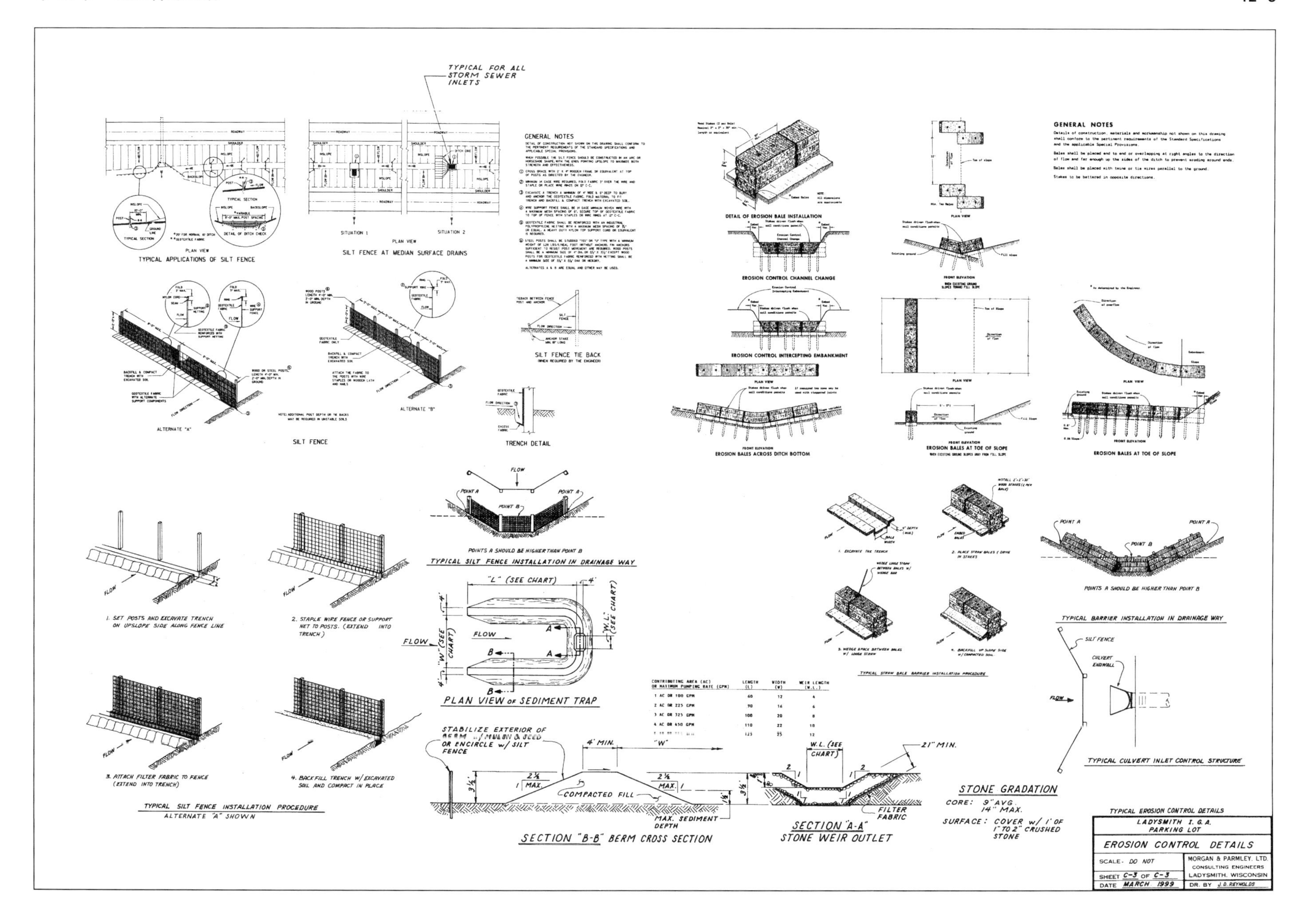
TYPICAL FOR ALL STORM SEWER INLETS
TYPICAL APPLICATIONS OF SILT FENCE
SILT FENCE AT MEDIAN SURFACE DRAINS
GENERAL NOTES
ALTERNATE "A"
ALTERNATE "B"
SILT FENCE
SILT FENCE TIE BACK
TRENCH DETAIL
DETAIL OF EROSION BALE INSTALLATION
EROSION CONTROL CHANNEL CHANGE
EROSION CONTROL INTERCEPTING EMBANKMENT
EROSION BALES ACROSS DITCH BOTTOM
EROSION BALES AT TOE OF SLOPE
GENERAL NOTES
FLOW
POINT A
POINT B
POINTS A SHOULD BE HIGHER THAN POINT B
TYPICAL SILT FENCE INSTALLATION IN DRAINAGE WAY
1. SET POSTS AND EXCAVATE TRENCH ON UPSLOPE SIDE ALONG FENCE LINE
2. STAPLE WIRE FENCE OR SUPPORT NET TO POSTS. (EXTEND INTO TRENCH)
3. ATTACH FILTER FABRIC TO FENCE (EXTEND INTO TRENCH)
4. BACKFILL TRENCH W/ EXCAVATED SOIL AND COMPACT IN PLACE
TYPICAL SILT FENCE INSTALLATION PROCEDURE
ALTERNATE "A" SHOWN
"L" (SEE CHART)
"W" (SEE CHART)
"W.L." (SEE CHART)
PLAN VIEW OF SEDIMENT TRAP
STABILIZE EXTERIOR OF BERM ... OR ENCIRCLE w/ SILT FENCE
4' MIN.
COMPACTED FILL
MAX. SEDIMENT DEPTH
SECTION "B-B" BERM CROSS SECTION
21" MIN.
FILTER FABRIC
SECTION "A-A" STONE WEIR OUTLET
STONE GRADATION
CORE: 9" AVG. 14" MAX.
SURFACE: COVER w/ 1' OF 1" TO 2" CRUSHED STONE
POINTS A SHOULD BE HIGHER THAN POINT B
TYPICAL BARRIER INSTALLATION IN DRAINAGE WAY
SILT FENCE
CULVERT ENDWALL
TYPICAL CULVERT INLET CONTROL STRUCTURE
TYPICAL EROSION CONTROL DETAILS
LADYSMITH I.G.A. PARKING LOT
EROSION CONTROL DETAILS
SCALE- DO NOT
SHEET C-3 OF C-3
DATE MARCH 1999
MORGAN & PARMLEY, LTD.
CONSULTING ENGINEERS
LADYSMITH, WISCONSIN
DR. BY J.D. REYNOLDS

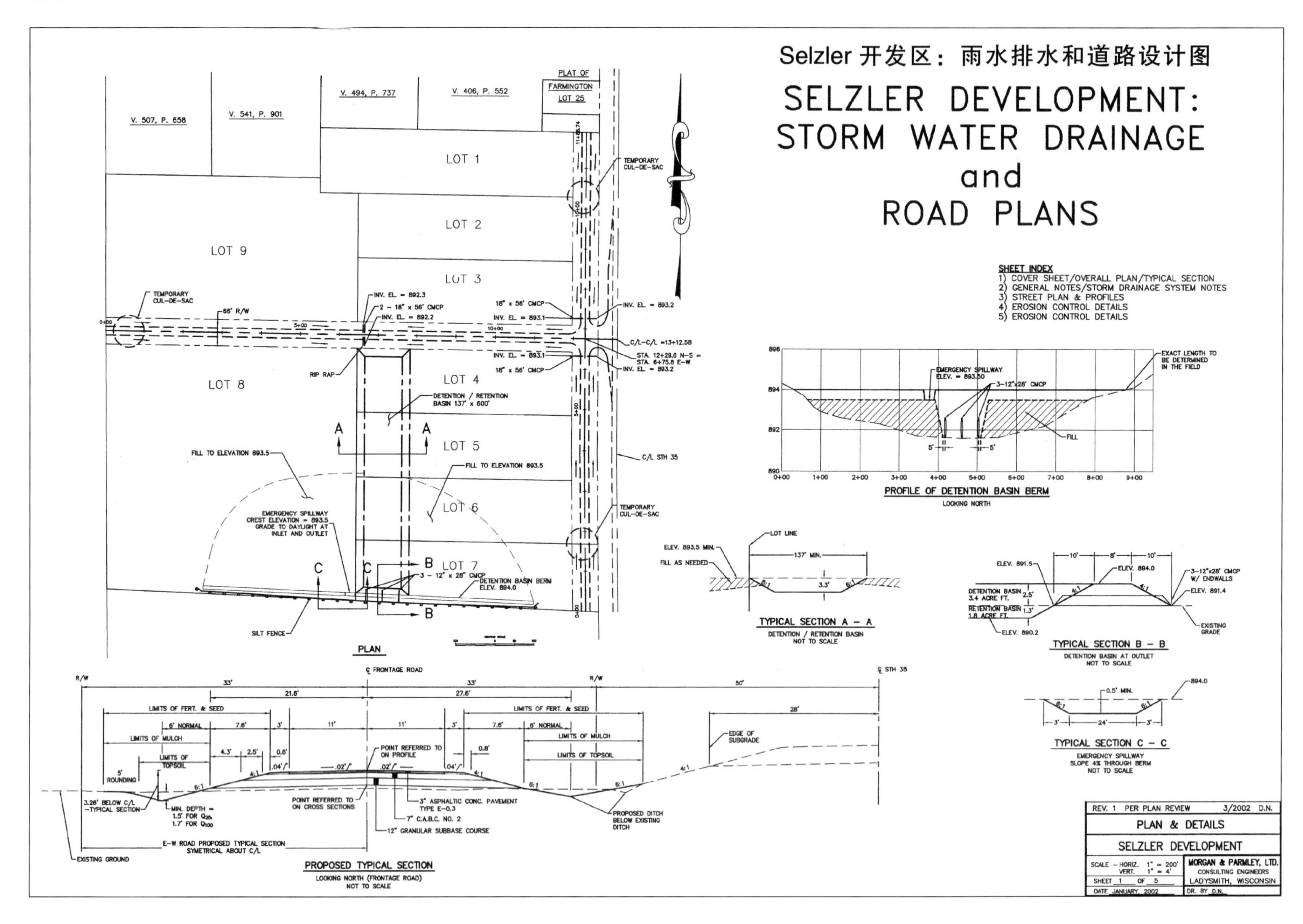

Courtesy: Morgan & Parmley, Ltd.

GENERAL NOTES

1. EROSION CONTROL MEASURES SHALL BE IN PLACE BEFORE CONSTRUCTION BEGINS.
2. THE CONTRACTORS SHALL MINIMIZE EROSION DURING CONSTRUCTION USING GOOD CONSTRUCTION TECHNIQUES AND UTILIZING SILT FENCES AND BALE DITCH CHECKS. THE CONTRACTOR SHALL RELEASE RUNOFF FROM THE SITE IN A NUISANCE FREE MANNER.
3. THE CONTRACTOR IS RESPONSIBLE TO MAINTAIN THE SITE IN A SAFE CONDITION. THE SOLE RESPONSIBILITY FOR WARNING SIGNS, BARRICADES, FLAGGING PERSONNEL AND ALL ASPECTS OF SAFETY LIE WITH THE CONTRACTOR.
4. EXISTING UTILITIES ARE NOT SHOWN. CONTRACTOR SHALL HAVE ALL BURIED UTILITIES LOCATED PRIOR TO COMMENCING CONSTRUCTION.
5. PROPERTY CORNERS KNOWN TO EXIST SHALL NOT BE DISTURBED. DAMAGED CORNERS WILL BE REPLACED AT THE CONTRACTOR'S EXPENSE. PRESERVE ALL SURVEY MONUMENTS.
6. CONSTRUCTION SHALL COMPLY WITH THE CONDITIONS OF DOT, COUNTY AND TOWNSHIP APPROVALS.
7. ALL DISTURBED AREAS SHALL BE SEEDED, FERTILIZED AND MULCHED, OR COVERED W/BASE COURSE AFTER CONSTRUCTION, OR RESTORED WITH BITUMINOUS PAVEMENT WHERE APPLICABLE.
8. THERE SHALL BE NO DEVIATION FROM THE PLANS WITHOUT THE DESIGN ENGINEER'S APPROVAL.
9. CONSTRUCTION STAKES DESTROYED BY THE CONTRACTOR THAT NEED TO BE REPLACED, WILL BE CHARGED TO THE CONTRACTOR.
10. ALL DISTURBED AREAS TO BE SEEDED SHALL BE RESTORED AS FOLLOWS:

    A) REPLACE ALL SALVAGED AND IMPORTED TOPSOIL
    B) SEED MIXTURE: PERMANENT SEEDING-(LAWN TYPE TURF)

        35% KENTUCKY BLUE GRASS
        25% IMPROVED FINE PERENNIAL RYEGRASS
        20% CREEPING RED FESCUE
        20% IMPROVED HARD FESCUE

    C) SEED RATE: 2#/100 SQ. FT.
    D) MULCH RATE: 2 TON/ACRE
    E) PERMANENT SEEDING: ALLOWED TO SEPT. 7
    TEMPORARY SEEDING: SEPT. 8 THROUGH NOV. 10
    (4 BU. OATS/ACRE)
    DORMANT SEEDING: SEED PERM. SEEDING INTO TEMP. SEEDING AFTER NOV. 10
    F) SEEDING MAY BE BROADCAST BUT MUST BE COVERED WITH A MAXIMUM OF ½" SOIL. MULCH MUST BE ANCHORED WITH A MULCH TILLER, TACKIFIER OR NETTING.
    G) FERTILIZER: 500#/ACRE 20-10-10

11. IF MORE THAN 5 ACRES ARE DISTURBED THE CONTRACTOR SHALL OBTAIN A STORM WATER DISCHARGE PERMIT FROM THE DNR.
12. IF IN FIELD CONDITIONS ARE ENCOUNTERED THAT DIFFER FROM THE PLANS, NOTIFY THE DESIGN ENGINEER.

SPECIFICATIONS:

WISCONSIN CONSTRUCTION SITE BEST MANAGEMENT PRACTICE HANDBOOK

STANDARD SPECIFICATIONS FOR ROAD AND BRIDGE CONSTRUCTION 1996 EDITION WIS. D.O.T.

STORM DRAINAGE SYSTEM NOTES.

1. ALL CULVERT PIPES SHALL BE CMCP WITH ENDWALLS.
2. CULVERT PIPE LENGTHS SHALL BE ADJUSTED IN THE FIELD TO MATCH INSLOPES.
3. DITCHES SHALL HAVE A MINIMUM DEPTH OF 1.7'.
4. THE DETENTION BASIN SHALL CONTAIN A MINIMUM OF 3.4 ACRE FEET OF STORAGE ABOVE THE INVERT OF THE OUTLET STRUCTURE AND BELOW THE CREST OF THE EMERGENCY SPILLWAY. THE RETENTION BASIN SHALL CONTAIN A MINIMUM OF 1.8 ACRE FEET BELOW THE INVERT OF THE OUTLET STRUCTURE. THE OWNER MAY SLOPE THE BOTTOM OF THE BASIN TO ACCOMMODATE DEVELOPMENT AND TO FACILITATE DRAINAGE, STORAGE VOLUME MUST BE MAINTAINED AS STATED.
5. THE LENGTH OF THE DETENTION BASIN BERM SHALL BE DETERMINED IN THE FIELD BY INTERSECTING THE EXISTING OR PROPOSED GRADE AT ELEVATION 894.0 ALONG THE SOUTH PROPERTY LINE.
6. STORMWATER FROM THE DETENTION BASIN SHALL LEAVE THE SITE AT A SAFE VELOCITY. SOD, EROSION MAT, TURF REINFORCEMENT OR RIP RAP SHALL PROTECT THE DETENTION BASIN OUTLET AND EMERGENCY SPILLWAY.

DESIGN SUMMARY (TR 55 GR.)

Q_{25} EXISTING	20.6 CFS
Q_{25} PROPOSED	25.5 CFS
MAXIMUM CAPACITY OF OUTLET STRUCTURE	9.5 CFS
DETENTION BASIN STORAGE	3.4 AC. FT.
RETENTION BASIN STORAGE	1.8 AC. FT.
TOTAL STORAGE	5.2 AC. FT.
MAXIMUM CAPACITY OF EMERGENCY SPILLWAY (EXCEEDS Q_{100})	47 CFS
MAXIMUM CAPACITY OF TWIN 18" CMCP Q_{25}	13 CFS

REV. 1 PER PLAN REVIEW	3/2002 D.N.
GENERAL NOTES	
SELZLER DEVELOPMENT	
SCALE - HORIZ. 1" = NONE VERT. 1" = NONE	MORGAN & PARMLEY, LTD. CONSULTING ENGINEERS
SHEET 2 OF 5	LADYSMITH, WISCONSIN
DATE JANUARY, 2002	DR. BY D.N.

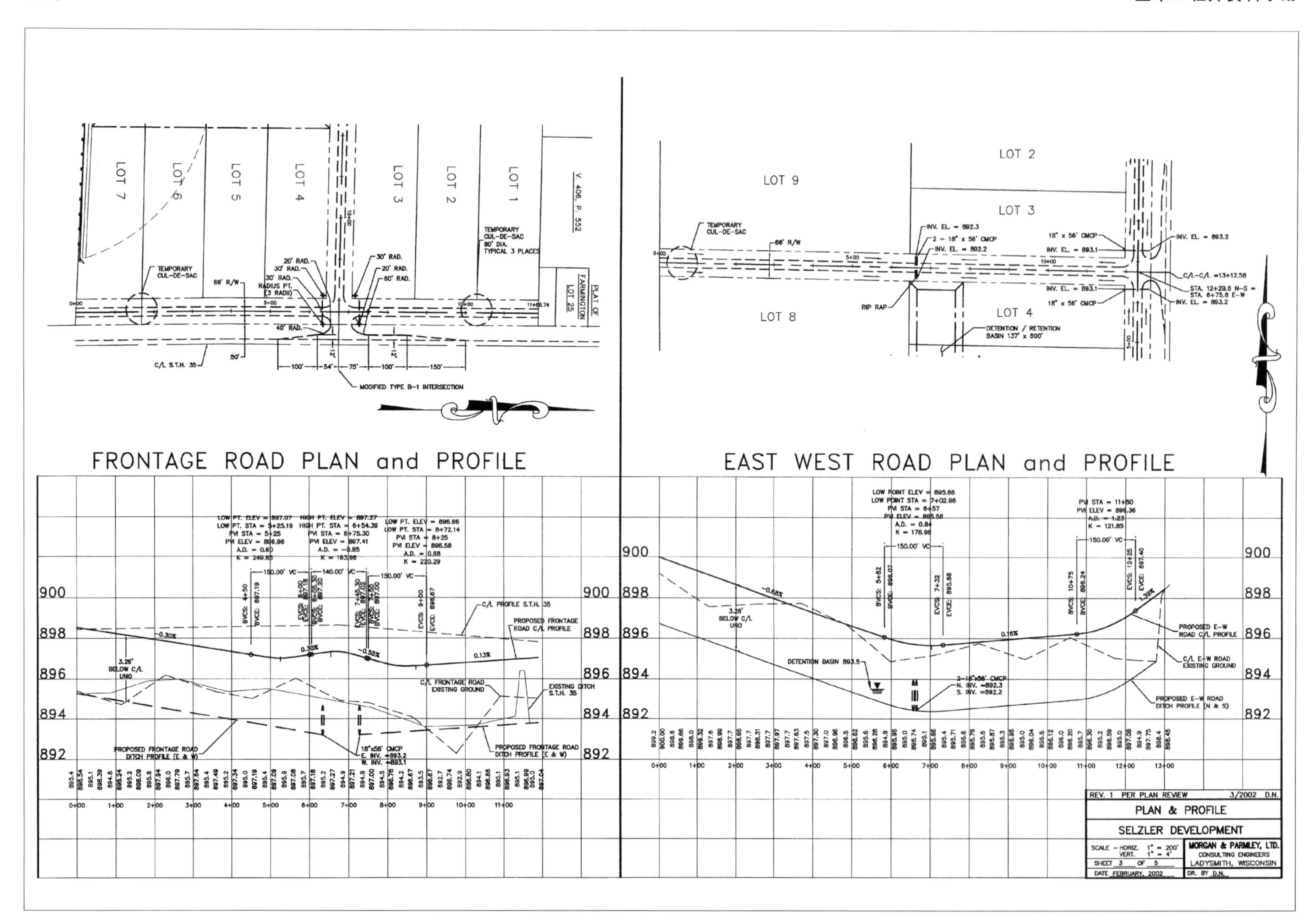

FRONTAGE ROAD PLAN and PROFILE
EAST WEST ROAD PLAN and PROFILE
LOT 7
LOT 6
LOT 5
LOT 4
LOT 3
LOT 2
LOT 1
LOT 9
LOT 8
TEMPORARY CUL-DE-SAC
66' R/W
C/L S.T.H. 35
MODIFIED TYPE B-1 INTERSECTION
DETENTION / RETENTION BASIN 137' x 600'
RIP RAP
PROPOSED FRONTAGE ROAD C/L PROFILE
C/L FRONTAGE ROAD EXISTING GROUND
PROPOSED FRONTAGE ROAD DITCH PROFILE (E & W)
PROPOSED E-W ROAD C/L PROFILE
C/L E-W ROAD EXISTING GROUND
PROPOSED E-W ROAD DITCH PROFILE (N & S)
DETENTION BASIN 893.5
REV. 1 PER PLAN REVIEW 3/2002 D.N.
PLAN & PROFILE
SELZLER DEVELOPMENT
SCALE - HORIZ. 1" = 200'
VERT. 1" = 4'
SHEET 3 OF 5
DATE FEBRUARY, 2002
MORGAN & PARMLEY, LTD.
CONSULTING ENGINEERS
LADYSMITH, WISCONSIN
DR. BY D.N.

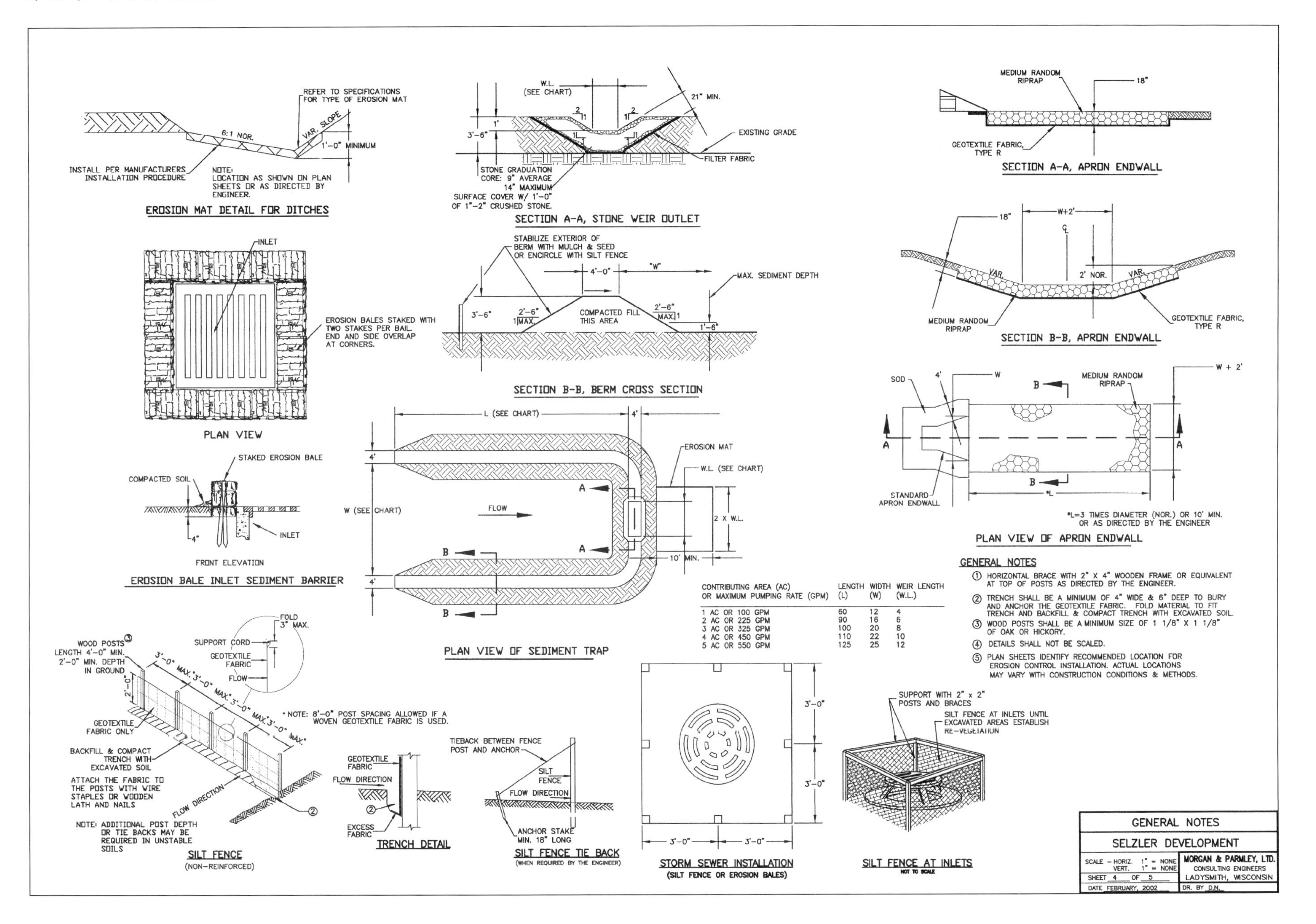
REFER TO SPECIFICATIONS FOR TYPE OF EROSION MAT
6:1 NOR.
VAR. SLOPE
1'-0" MINIMUM
INSTALL PER MANUFACTURERS INSTALLATION PROCEDURE
NOTE: LOCATION AS SHOWN ON PLAN SHEETS OR AS DIRECTED BY ENGINEER.
EROSION MAT DETAIL FOR DITCHES
W.L. (SEE CHART)
21" MIN.
EXISTING GRADE
FILTER FABRIC
STONE GRADUATION CORE: 9" AVERAGE 14" MAXIMUM
SURFACE COVER W/ 1'-0" OF 1"-2" CRUSHED STONE.
SECTION A-A, STONE WEIR OUTLET
MEDIUM RANDOM RIPRAP
18"
GEOTEXTILE FABRIC, TYPE R
SECTION A-A, APRON ENDWALL
INLET
EROSION BALES STAKED WITH TWO STAKES PER BAIL. END AND SIDE OVERLAP AT CORNERS.
PLAN VIEW
STABILIZE EXTERIOR OF BERM WITH MULCH & SEED OR ENCIRCLE WITH SILT FENCE
4'-0"
"W"
MAX. SEDIMENT DEPTH
3'-6"
2'-6" MAX.
COMPACTED FILL THIS AREA
1'-6"
SECTION B-B, BERM CROSS SECTION
W+2'
2' NOR.
VAR.
MEDIUM RANDOM RIPRAP
GEOTEXTILE FABRIC, TYPE R
SECTION B-B, APRON ENDWALL
SOD
W
W + 2'
STANDARD APRON ENDWALL
*L=3 TIMES DIAMETER (NOR.) OR 10' MIN. OR AS DIRECTED BY THE ENGINEER
PLAN VIEW OF APRON ENDWALL
STAKED EROSION BALE
COMPACTED SOIL
INLET
FRONT ELEVATION
EROSION BALE INLET SEDIMENT BARRIER
L (SEE CHART)
EROSION MAT
W.L. (SEE CHART)
W (SEE CHART)
FLOW
2 X W.L.
10' MIN.
PLAN VIEW OF SEDIMENT TRAP
CONTRIBUTING AREA (AC) OR MAXIMUM PUMPING RATE (GPM) | LENGTH (L) | WIDTH (W) | WEIR LENGTH (W.L.)
1 AC OR 100 GPM | 60 | 12 | 4
2 AC OR 225 GPM | 90 | 16 | 6
3 AC OR 325 GPM | 100 | 20 | 8
4 AC OR 450 GPM | 110 | 22 | 10
5 AC OR 550 GPM | 125 | 25 | 12
GENERAL NOTES
① HORIZONTAL BRACE WITH 2" X 4" WOODEN FRAME OR EQUIVALENT AT TOP OF POSTS AS DIRECTED BY THE ENGINEER.
② TRENCH SHALL BE A MINIMUM OF 4" WIDE & 6" DEEP TO BURY AND ANCHOR THE GEOTEXTILE FABRIC. FOLD MATERIAL TO FIT TRENCH AND BACKFILL & COMPACT TRENCH WITH EXCAVATED SOIL.
③ WOOD POSTS SHALL BE A MINIMUM SIZE OF 1 1/8" X 1 1/8" OF OAK OR HICKORY.
④ DETAILS SHALL NOT BE SCALED.
⑤ PLAN SHEETS IDENTIFY RECOMMENDED LOCATION FOR EROSION CONTROL INSTALLATION. ACTUAL LOCATIONS MAY VARY WITH CONSTRUCTION CONDITIONS & METHODS.
FOLD 3" MAX.
SUPPORT CORD
GEOTEXTILE FABRIC
FLOW
WOOD POSTS③ LENGTH 4'-0" MIN. 2'-0" MIN. DEPTH IN GROUND
3'-0" MAX.
* NOTE: 8'-0" POST SPACING ALLOWED IF A WOVEN GEOTEXTILE FABRIC IS USED.
GEOTEXTILE FABRIC ONLY
BACKFILL & COMPACT TRENCH WITH EXCAVATED SOIL
ATTACH THE FABRIC TO THE POSTS WITH WIRE STAPLES OR WOODEN LATH AND NAILS
FLOW DIRECTION
NOTE: ADDITIONAL POST DEPTH OR TIE BACKS MAY BE REQUIRED IN UNSTABLE SOILS
SILT FENCE
(NON-REINFORCED)
GEOTEXTILE FABRIC
FLOW DIRECTION
EXCESS FABRIC
TRENCH DETAIL
TIEBACK BETWEEN FENCE POST AND ANCHOR
SILT FENCE
FLOW DIRECTION
ANCHOR STAKE MIN. 18" LONG
SILT FENCE TIE BACK
(WHEN REQUIRED BY THE ENGINEER)
3'-0"
STORM SEWER INSTALLATION
(SILT FENCE OR EROSION BALES)
SUPPORT WITH 2" x 2" POSTS AND BRACES
SILT FENCE AT INLETS UNTIL EXCAVATED AREAS ESTABLISH RE-VEGETATION
SILT FENCE AT INLETS
NOT TO SCALE
GENERAL NOTES
SELZLER DEVELOPMENT
SCALE - HORIZ. 1" = NONE VERT. 1" = NONE
SHEET 4 OF 5
DATE FEBRUARY, 2002
MORGAN & PARMLEY, LTD.
CONSULTING ENGINEERS
LADYSMITH, WISCONSIN
DR. BY D.N.

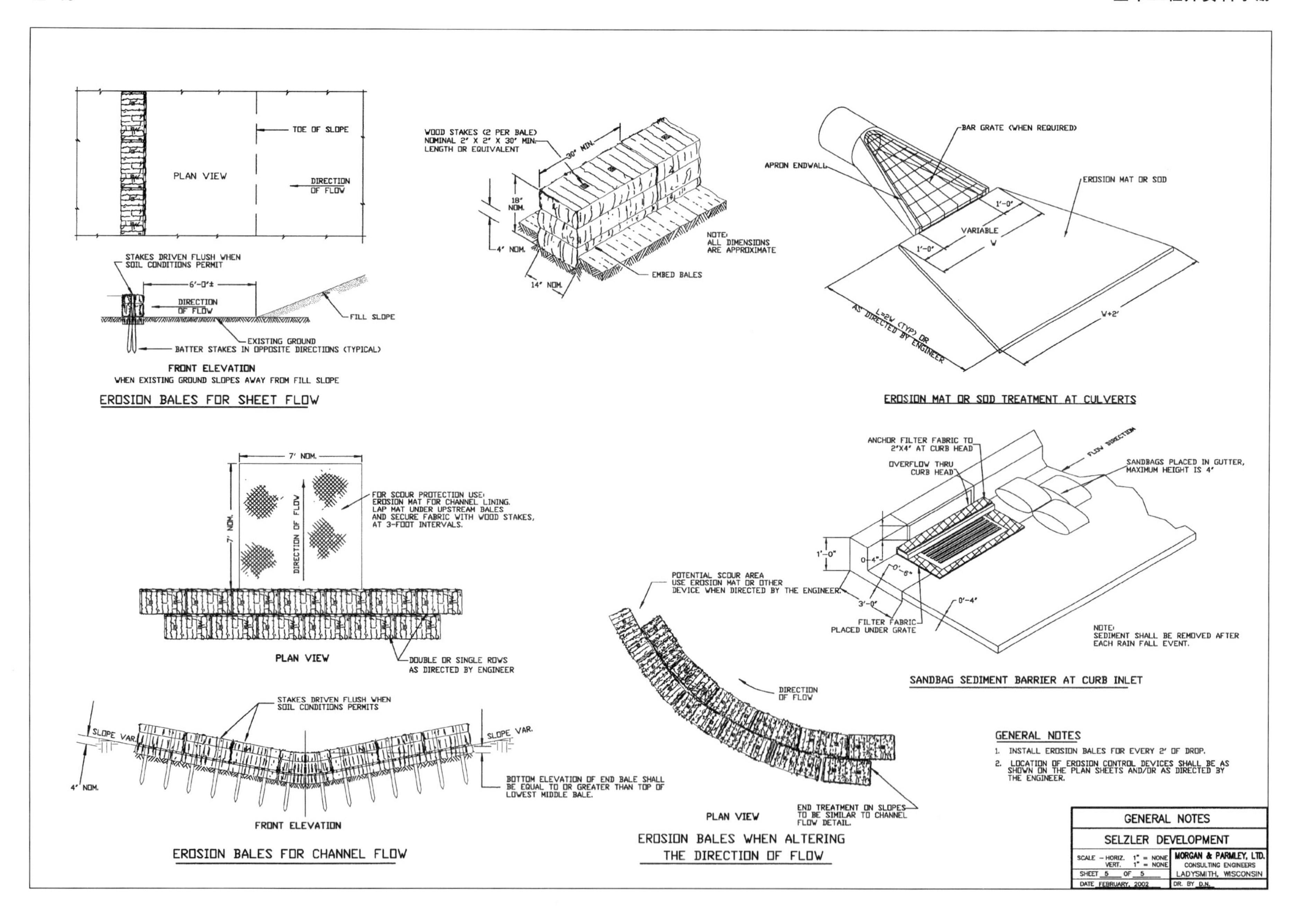

PLAN VIEW
TOE OF SLOPE
DIRECTION OF FLOW
STAKES DRIVEN FLUSH WHEN SOIL CONDITIONS PERMIT
6'-0"±
FILL SLOPE
EXISTING GROUND
BATTER STAKES IN OPPOSITE DIRECTIONS (TYPICAL)
FRONT ELEVATION
WHEN EXISTING GROUND SLOPES AWAY FROM FILL SLOPE
ERISION BALES FOR SHEET FLOW
WOOD STAKES (2 PER BALE) NOMINAL 2" X 2" X 30" MIN. LENGTH OR EQUIVALENT
30" MIN.
18" NOM.
4" NOM.
14" NOM.
NOTE: ALL DIMENSIONS ARE APPROXIMATE
EMBED BALES
BAR GRATE (WHEN REQUIRED)
APRON ENDWALL
EROSION MAT OR SOD
1'-0"
VARIABLE
W
L=2W (TYP) OR AS DIRECTED BY ENGINEER
W+2'
EROSION MAT OR SOD TREATMENT AT CULVERTS
7' NOM.
FOR SCOUR PROTECTION USE: EROSION MAT FOR CHANNEL LINING. LAP MAT UNDER UPSTREAM BALES AND SECURE FABRIC WITH WOOD STAKES, AT 3-FOOT INTERVALS.
DIRECTION OF FLOW
PLAN VIEW
DOUBLE OR SINGLE ROWS AS DIRECTED BY ENGINEER
STAKES DRIVEN FLUSH WHEN SOIL CONDITIONS PERMITS
SLOPE VAR.
4" NOM.
BOTTOM ELEVATION OF END BALE SHALL BE EQUAL TO OR GREATER THAN TOP OF LOWEST MIDDLE BALE.
FRONT ELEVATION
EROSION BALES FOR CHANNEL FLOW
ANCHOR FILTER FABRIC TO 2"X4" AT CURB HEAD
OVERFLOW THRU CURB HEAD
FLOW DIRECTION
SANDBAGS PLACED IN GUTTER, MAXIMUM HEIGHT IS 4"
1'-0"
0'-4"
0'-6"
3'-0"
FILTER FABRIC PLACED UNDER GRATE
NOTE: SEDIMENT SHALL BE REMOVED AFTER EACH RAIN FALL EVENT.
SANDBAG SEDIMENT BARRIER AT CURB INLET
POTENTIAL SCOUR AREA USE EROSION MAT OR OTHER DEVICE WHEN DIRECTED BY THE ENGINEER.
DIRECTION OF FLOW
PLAN VIEW
END TREATMENT ON SLOPES TO BE SIMILAR TO CHANNEL FLOW DETAIL.
EROSION BALES WHEN ALTERING THE DIRECTION OF FLOW
GENERAL NOTES
1. INSTALL EROSION BALES FOR EVERY 2' OF DROP.
2. LOCATION OF EROSION CONTROL DEVICES SHALL BE AS SHOWN ON THE PLAN SHEETS AND/OR AS DIRECTED BY THE ENGINEER.
GENERAL NOTES
SELZLER DEVELOPMENT
SCALE – HORIZ. 1" = NONE VERT. 1" = NONE
MORGAN & PARMLEY, LTD. CONSULTING ENGINEERS LADYSMITH, WISCONSIN
SHEET 5 OF 5
DATE FEBRUARY, 2002
DR. BY D.N.

Mc Sloy 街区改造

CDBG PROJECT NO. PF FY 99-0124

Mc SLOY STREET AREA RECONSTRUCTION

FOR THE VILLAGE OF

GILMAN, WISCONSIN

M. & P. PROJECT NO. 99-120

MARCH 2000

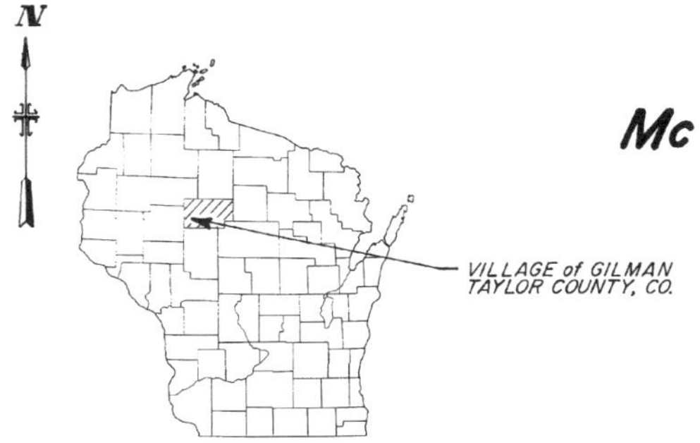

N

VILLAGE of GILMAN
TAYLOR COUNTY, CO.

GENERAL LOCATION MAP
SCALE: DO NOT SCALE

PROJECT SITE

PROJECT LOCATION MAP

SPECIFICATIONS:

WISCONSIN CONSTRUCTION SITE BEST MANAGEMENT PRACTICE HANDBOOK

STANDARD SPECIFICATIONS FOR SEWER AND WATER CONSTRUCTION IN WISCONSIN 5TH EDITION WITH ADDENDUM NO. 1

STANDARD SPECIFICATIONS FOR ROAD AND BRIDGE CONSTRUCTION 1996 EDITION WIS D.O.T.

GENERAL NOTES

1. ALL WATER MAIN SHALL BE 6" PVC DR 18.
2. ALL WATERMAIN FITTINGS SHALL BE M.J.D.I., W / RESTRAINED JOINT LUGS.
3. ALL HYDRANT LEADS SHALL BE 6-INCH VALVE.
4. HYDRANTS SHALL BE WATEROUS WB 67-250 (4" PUMPER NOZZLE), 22" BREAKOFF WITH THREADS MATCHING VILLAGE'S PATTERN.
5. ALL WATER MAIN SHALL HAVE AN 8 FT. BURY (MIN.).
6. WATER SERVICES SHALL BE 1" TYPE K COPPER W / MUELLER OR FORD BRASS.
7. 8' (MINIMUM) HORIZONTAL DISTANCE (℄ TO ℄) BETWEEN WATERMAIN AND SANITARY SEWER OR STORM SEWER.
8. 18" (MINIMUM) VERTICAL DISTANCE (OUT TO OUT) BETWEEN WATERMAIN UNDER SANITARY SEWER OR STORM SEWER @ THEIR INTERSECTION.
9. 6" (MINIMUM) VERTICAL DISTANCE (OUT TO OUT) BETWEEN WATERMAIN OVER SANITARY SEWER OR STORM SEWER @ THEIR INTERSECTION.
10. SANITARY SERVICES SHALL BE 4" SCH. 40 PVC W / WYES LAID TO PROVIDE FOR MAXIMUM DEPTH OR 10' AT THE R / W.
11. SEWERMAIN SHALL BE 8" PVC SDR 35.
12. MANHOLES SHALL BE SET WITH A MINIMUM OF 6" OF RINGS AND 9" CASTINGS.
13. MANHOLES SHALL BE 4' DIAMETER PRECAST CONCRETE W / INTEGRAL BASES AND BOOTS.
14. ALL ELEVATIONS BASED ON USGS DATUM.
15. EROSION CONTROL MEASURES SHALL BE IN PLACE BEFORE CONSTRUCTION BEGINS.
16. THE CONTRACTORS SHALL MINIMIZE EROSION DURING CONSTRUCTION USING GOOD CONSTRUCTION TECHNIQUES AND UTILIZING SILT FENCES AND BALE DITCH CHECKS. THE CONTRACTOR SHALL RELEASE RUNOFF FROM THE SITE IN A NUISANCE FREE MANNER.
17. THE CONTRACTOR IS RESPONSIBLE TO MAINTAIN THE SITE IN A SAFE CONDITION. THE SOLE RESPONSIBILITY FOR WARNING SIGNS, BARRICADES, FLAGGING PERSONNEL AND ALL ASPECTS OF SAFETY LIE WITH THE CONTRACTOR.
18. EXISTING UTILITIES ARE SHOWN BUT MAY NOT BE ALL INCLUSIVE; CONTRACTOR SHALL HAVE ALL BURIED UTILITIES LOCATED PRIOR TO COMMENCING CONSTRUCTION.
19. EXISTING SEWER AND WATERMAIN ARE SHOWN ACCORDING TO EXISTING RECORDS BUT LOCATION SHALL BE VERIFIED IN FIELD.
20. WATERMAIN & SERVICES SHALL BE INSULATED AT ALL STORM SEWER, CULVERT AND DITCH CROSSINGS WITH 2" THICK EXTRUDED POLYSTYRENE INSULATION BOARD.
21. PROPERTY CORNERS KNOWN TO EXIST SHALL NOT BE DISTURBED. DAMAGED CORNERS WILL BE REPLACED AT THE CONTRACTOR'S EXPENSE. PRESERVE ALL SURVEY MONUMENTS.
22. ALL WATERMAIN SHALL BE DISINFECTED BEFORE BEING PUT INTO USE.
23. VALVES AND MANHOLES SHALL BE ADJUSTED TO FINAL GRADE, UNLESS DIRECTED OTHERWISE.
24. CONSTRUCTION SHALL COMPLY WITH THE CONDITIONS OF DNR APPROVAL ELTTER, AND OTHER REGULATORY APPROVALS AS LISTED IN THE SPECIFICATION BOOK.
25. ALL DISTURBED AREAS SHALL BE SEEDED, FERTILIZED AND MULCHED, OR COVERED W / BASE COURSE AFTER CONSTRUCTION, OR RESTORED WITH BITUMINOUS PAVEMENT WHERE APPLICABLE.
26. THERE SHALL BE NO DEVIATION FROM THE PLANS WITHOUT THE DESIGN ENGINEER'S APPROVAL.
27. CONSTRUCTION STAKES DESTROYED BY THE CONTRACTOR THAT NEED TO BE REPLACED, WILL BE CHARGED TO THE CONTRACTOR.
28. ALL DISTURBED AREAS TO BE SEEDED SHALL BE RESTORED AS FOLLOWS:
    A) REPLACED ALL SALVAGED AND IMPORTED TOPSOIL
    B) SEED MIXUTRE: PERMANENT SEEDING – (LAWN TYPE TURF)
        35% KENTUCKY BLUEGRASS
        25% IMPROVED FINE PERENNIAL RYEGRASS
        20% CREEPING RED FESCUE
        20% IMPROVED HARD FESCUE
    C) SEED RATE: 2# / 100 SQ. FT.
    D) MULCH RATE: 2 TON / ACRE
    E) PERMANENT SEEDING: ALLOWED TO SEPT. 7
       TEMPORARY SEEDING: SEPT. 8 THROUGH NOV. 10 (4 BU. OATS / ACRE)
       DORMANT SEEDING; SEED PERM. SEEDING INTO TEMP. SEEDING AFTER NOV. 10
    F) SEEDING MAY BE BRAODCAST BUT MUST BE COVERED WITH A MAXIMUM OF ½" SOIL. MULCH MUST BE ANCHORED WITH A MULCH TILLER, TACKIFIER OR NETTING.
    G) FERTILIZER: 500# / ACRE 20-10-10
29. CONTRACTOR MUST REMOVE AND REPLACE ALL AFFECTED STREET SIGNS AFTER PROPERLY SIGNING AND BARRICADING THE CONSTRUCTION AREA. ALL SIGNS MUST BE REPLACED PRIOR TO OPENING THE STREET TO TRAFFIC.
30. CONTRACTOR SHALL DETERMINE CONSTRUCTION SEQUENCE IN ORDER TO MINIMIZE INSTALLATION CONFLICTS WITH PIPE CROSSINGS.
31. STORM SEWER PIPE SHALL BE CL III RCP OR SMOOTH WALL PE AS NOTED ON THE PLANS.
32. NO TREES ARE TO BE REMOVED WITHOUT THE APPROVAL OF THE ENGINEER.

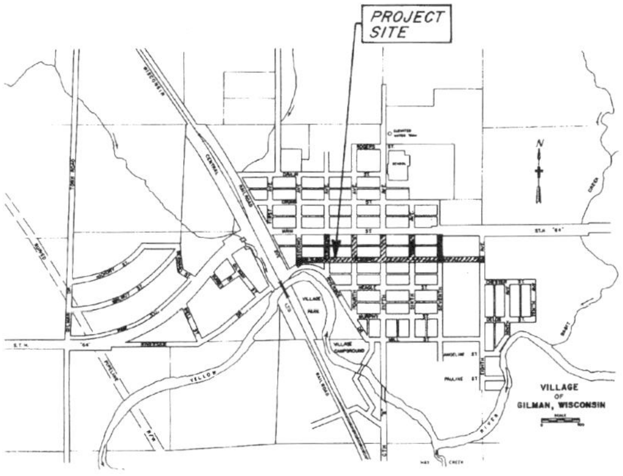

INDEX OF SHEETS

SHEET NO.	DESCRIPTION
1	TITLE SHEET, LOCATION MAP, INDEX, NOTES & TYPICAL CROSS SECTION
2	PLAN & PROFILE – McSLOY STREET STA. 1+00 TO STA. 12+50
3	PLAN & PROFILE – McSLOY STREET STA. 12+50 TO STA. 25+14
4	PLAN & PROFILE – THIRD AVENUE & FOURTH AVENUE
5	PLAN & PROFILE – FIFTH AVENUE & STORM SEWER SCHEDULE & CROSS SECTION
6	PLAN & PROFILE – SIXTH AVENUE & SEVENTH AVENUE
7	TYPICAL EROSION CONTROL DETAILS
8 – 21	CROSS SECTIONS – McSLOY STREET
22 – 23	CROSS SECTIONS – THIRD AVENUE
24 – 25	CROSS SECTIONS – FOURTH AVENUE
26 – 29	CROSS SECTIONS – FIFTH AVENUE
30 – 32	CROSS SECTIONS – SIXTH AVENUE
33 – 35	CROSS SECTIONS – SEVENTH AVENUE

NOTE: Sheets 7 & 9 thru 35 Not Shown

LEGEND

DESCRIPTION	EXISTING	PROPOSED
SANITARY SEWER & MANHOLE		
STORM SEWER & MANHOLE		
CATCH BASIN INLET		
CURB & GUTTER		
SIDEWALK		NONE
WATERMAIN		
HYDRANT		
VALVE		
CURB STOP		
CULVERTS		NONE
POWER POLE		NONE
OVERHEAD ELECTRIC LINE		NONE
BURIED ELECTRIC LINE		NONE
BURIED TELEPHONE CABLE w/ PEDESTAL		NONE
GAS LINE (CAUTION)		NONE
FIBER OPTIC CABLE		NONE
BORING GOOD	△	NONE
BORING BAD	▲	NONE

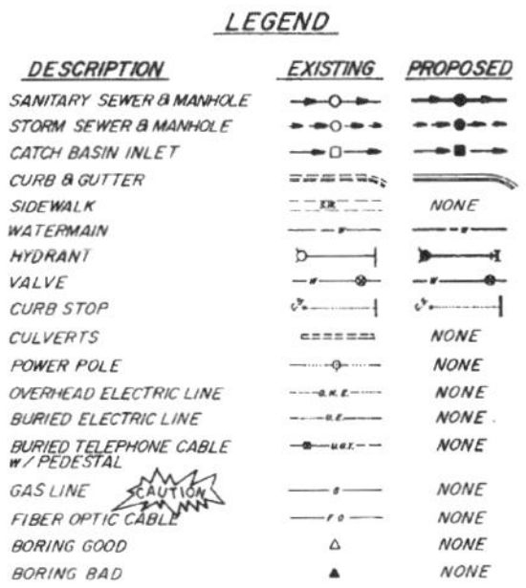

—UTILITIES—

TELEPHONE:
CENTURYTEL OF NORTHERN WISCONSIN, INC.
428 ELLINGSON AVE.
P. O. BOX 78
HAWKINS, WISCONSIN 54530
ATTN: JAMES ARQUETTE
715-585-7707

CABLE:
S & K TV SYSTEMS
508 W. MINER AVE.
LADYSMITH, WISCONSIN 54848
ATTN: RANDY SCOTT
715-532-7321

ELECTRIC:
NORTHERN STATES POWER CO.
500 N. 5th STREET
ABBOTSFORD, WISCONSIN 54405
ATTN: DAVE PEPPER
715-839-2678

DIGGERS HOTLINE:
800-242-8511

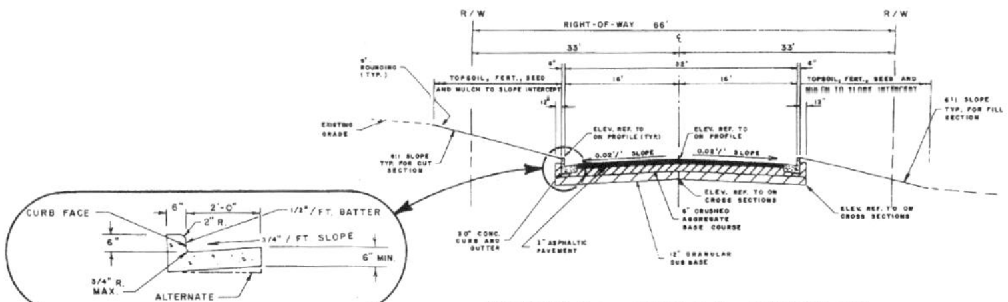

EARTHWORK QUANTITIES

EXCAVATION:	14,550	CU. YD.s
FILL: x 1.3 =	1,300	CU. YD.s
EXCESS WASTE:	13,250	CU. YD.s

TYPICAL CROSS SECTION

PREPARED BY:
MORGAN & PARMLEY, Ltd.
PROFESSIONAL CONSULTING ENGINEERS
LADYSMITH, WISCONSIN

Courtesy: Morgan & Parmley, Ltd.

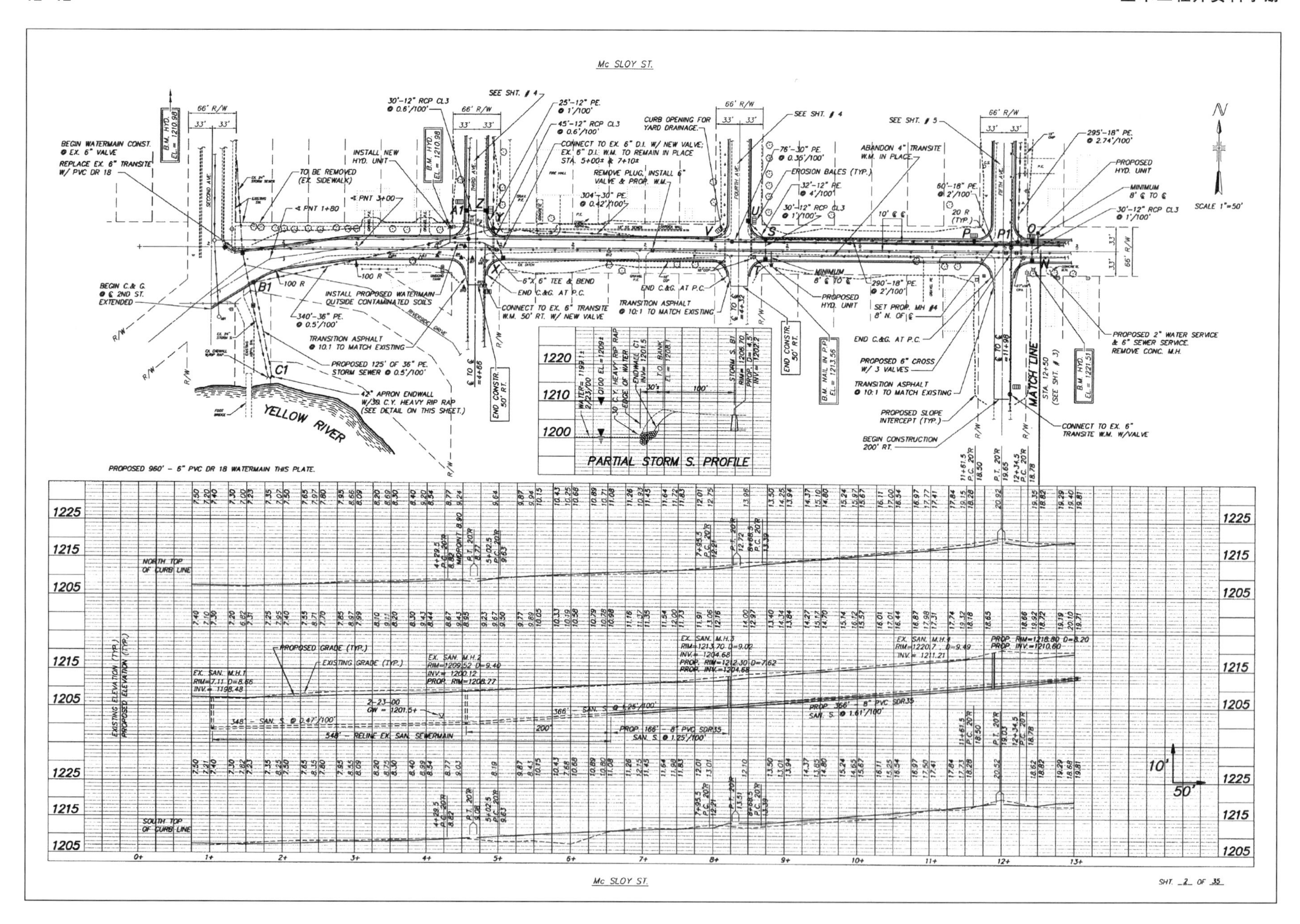
Mc SLOY ST.
PARTIAL STORM S. PROFILE
YELLOW RIVER
MATCH LINE
SCALE 1"=50'
NORTH TOP OF CURB LINE
SOUTH TOP OF CURB LINE
EXISTING ELEVATION (TYP.)
PROPOSED ELEVATION (TYP.)
PROPOSED GRADE (TYP.)
EXISTING GRADE (TYP.)
PROPOSED 960' - 6" PVC DR 18 WATERMAIN THIS PLATE.
548' - RELINE EX. SAN. SEWERMAIN
Mc SLOY ST.
SHT. 2 OF 35

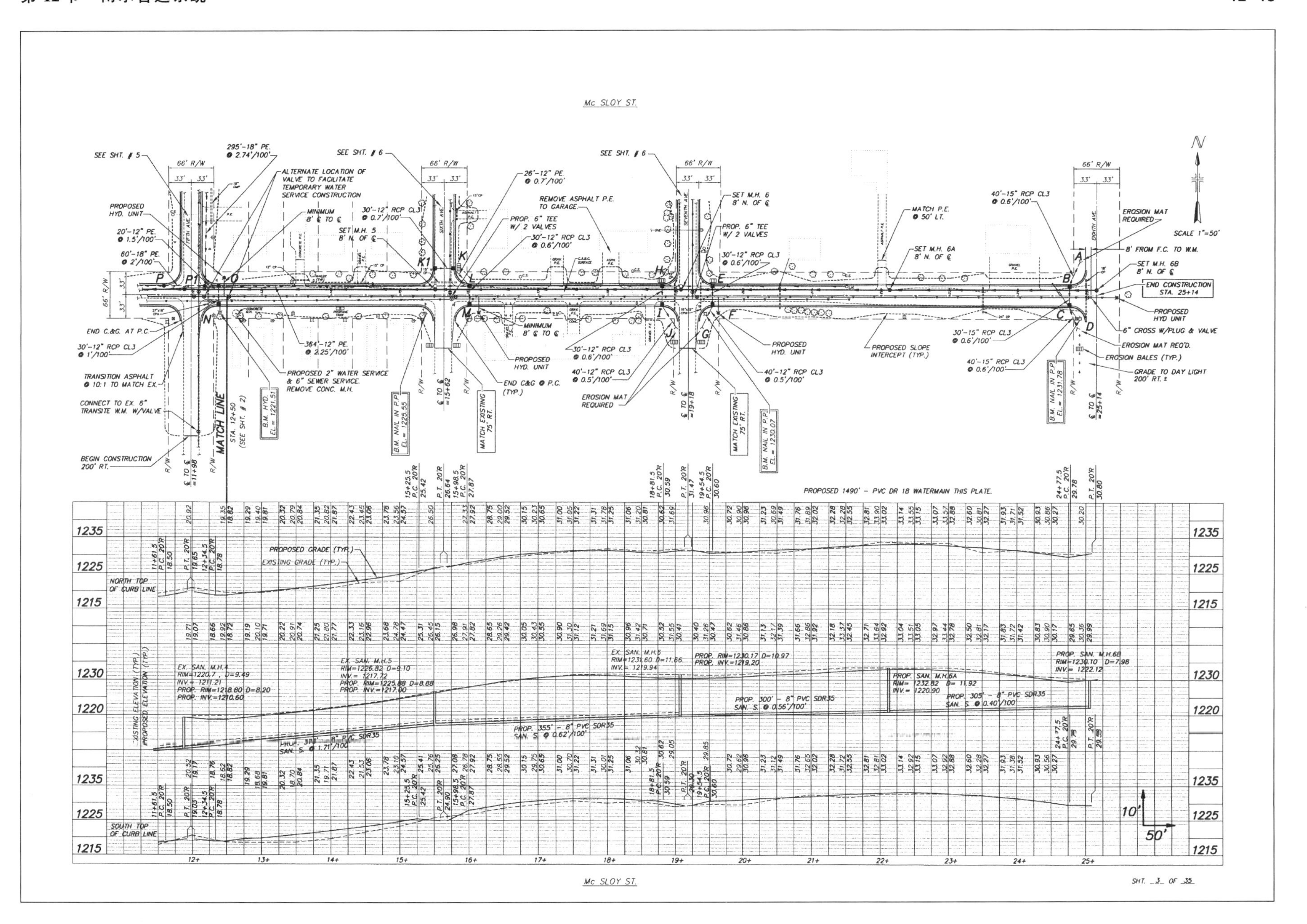

Mc SLOY ST.
SEE SHT. # 5
295'-18" PE. @ 2.74'/100'
66' R/W
33' 33'
ALTERNATE LOCATION OF VALVE TO FACILITATE TEMPORARY WATER SERVICE CONSTRUCTION
SEE SHT. # 6
26'-12" PE. @ 0.7'/100'
REMOVE ASPHALT P.E. TO GARAGE.
SET M.H. 6 8' N. OF ℄
PROP. 6" TEE W/ 2 VALVES
MATCH P.E. @ 50' LT.
40'-15" RCP CL3 @ 0.6'/100'
EROSION MAT REQUIRED
SCALE 1"=50'
PROPOSED HYD. UNIT
MINIMUM 8' ℄ TO ℄
30'-12" RCP CL3 @ 0.7'/100'
SET M.H. 5 8' N. OF ℄
30'-12" RCP CL3 @ 0.6'/100'
20'-12" PE. @ 1.5'/100'
60'-18" PE. @ 2'/100'
SET M.H. 6A 8' N. OF ℄
8' FROM F.C. TO W.M.
SET M.H. 6B 8' N. OF ℄
END CONSTRUCTION STA. 25+14
66' R/W
END C.&G. AT P.C.
30'-12" RCP CL3 @ 1'/100'
364'-12" PE. @ 2.25'/100'
PROPOSED 2" WATER SERVICE & 6" SEWER SERVICE. REMOVE CONC. M.H.
TRANSITION ASPHALT @ 10:1 TO MATCH EX.
CONNECT TO EX. 6" TRANSITE W.M. W/VALVE
BEGIN CONSTRUCTION 200' RT.
MATCH LINE
STA. 12+50 (SEE SHT. # 2)
B.M. HYD. EL.= 1221.51
B.M. NAIL IN P.P. EL.= 1225.55
℄ TO ℄ =15+62
MATCH EXISTING 75' RT.
END C&G @ P.C. (TYP.)
PROPOSED HYD. UNIT
30'-12" RCP CL3 @ 0.6'/100'
40'-12" RCP CL3 @ 0.5'/100'
EROSION MAT REQUIRED
℄ TO ℄ =19+18
MATCH EXISTING 75' RT.
B.M. NAIL IN P.P. EL.= 1230.07
PROPOSED HYD. UNIT
40'-12" RCP CL3 @ 0.5'/100'
PROPOSED SLOPE INTERCEPT (TYP.)
30'-15" RCP CL3 @ 0.6'/100'
40'-15" RCP CL3 @ 0.6'/100'
B.M. NAIL IN P.P. EL.= 1231.78
℄ TO ℄ =25+14
PROPOSED HYD UNIT
6" CROSS W/PLUG & VALVE
EROSION MAT REQ'D.
EROSION BALES (TYP.)
GRADE TO DAY LIGHT 200' RT. ±
℄ TO ℄ =11+98
P P1 O N K1 K L M H E I J G F A B C D
PROPOSED 1490' - PVC DR 18 WATERMAIN THIS PLATE.
PROPOSED GRADE (TYP.)
EXISTING GRADE (TYP.)
NORTH TOP OF CURB LINE
EX. SAN. M.H.4 RIM=1220.7, D=9.49 INV.= 1211.21
PROP. RIM=1218.80 D=8.20 PROP. INV.=1210.60
EX. SAN. M.H.5 RIM=1226.82 D=9.10 INV.= 1217.72
PROP. RIM=1225.88 D=8.88 PROP. INV.=1217.00
EX. SAN. M.H.5 RIM=1231.60 D=11.66 INV.= 1219.94
PROP. RIM=1230.17 D=10.97 PROP. INV.=1219.20
PROP. SAN. M.H.6A RIM= 1232.82 D= 11.92 INV.= 1220.90
PROP. SAN. M.H.6B RIM=1230.10 D=7.98 INV.= 1222.12
PROP. 300' - 8" PVC SDR35 SAN. S. @ 0.56'/100'
PROP. 305' - 8" PVC SDR35 SAN. S. @ 0.40'/100'
PROP. 355' - 8" PVC SDR35 SAN. S. @ 0.62'/100'
PROP. 391' - 8" PVC SDR35 SAN. S. @ 1.71'/100'
EXISTING ELEVATION (TYP.)
PROPOSED ELEVATION (TYP.)
SOUTH TOP OF CURB LINE
1235 1225 1215 1230 1220
12+ 13+ 14+ 15+ 16+ 17+ 18+ 19+ 20+ 21+ 22+ 23+ 24+ 25+
10' 50'
SHT. 3 OF 35

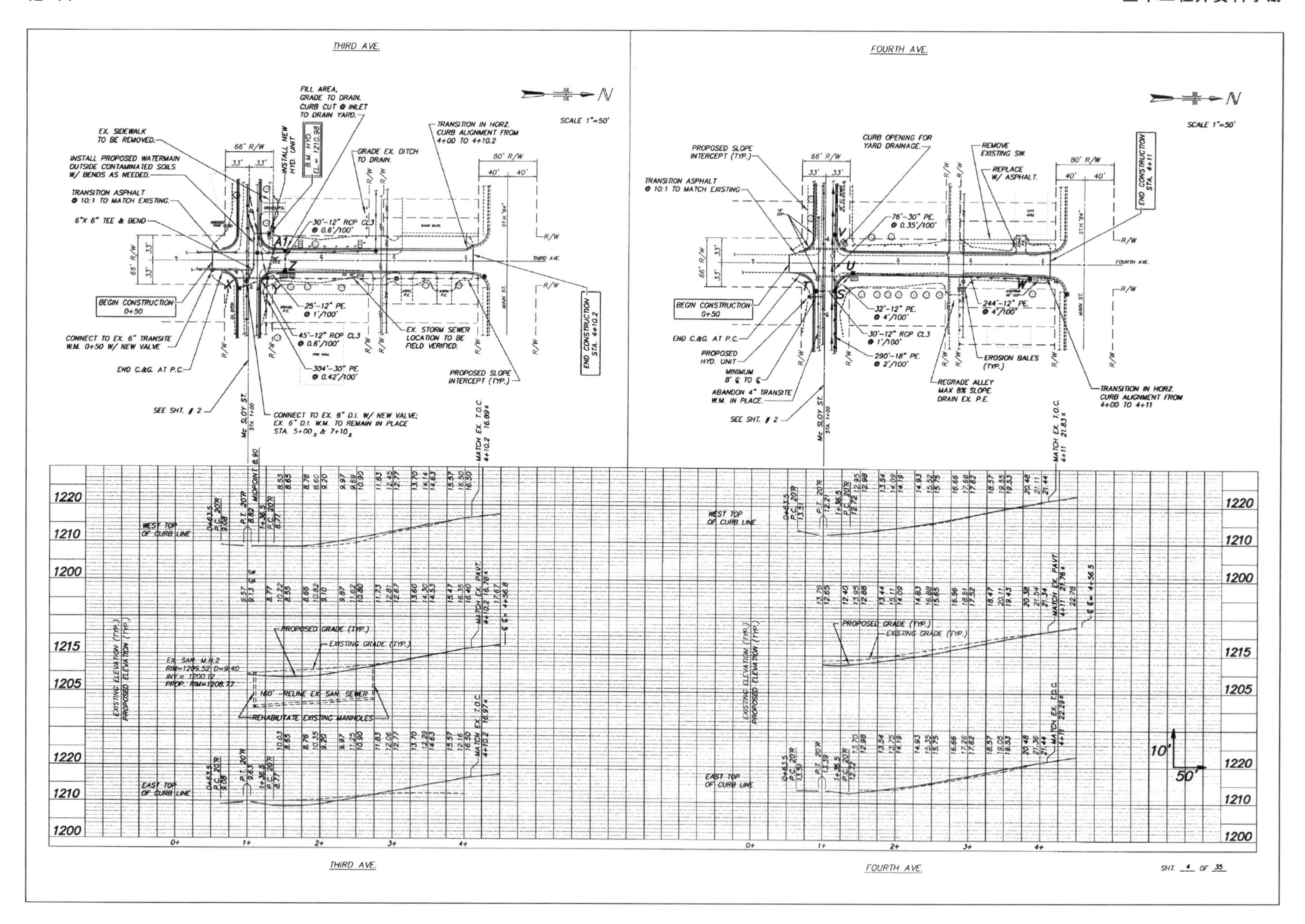
THIRD AVE.
FOURTH AVE.
SCALE 1"=50'
BEGIN CONSTRUCTION 0+50
END CONSTRUCTION STA. 4+10.2
END CONSTRUCTION STA. 4+11
WEST TOP OF CURB LINE
EAST TOP OF CURB LINE
PROPOSED GRADE (TYP.)
EXISTING GRADE (TYP.)
EXISTING ELEVATION (TYP.)
PROPOSED ELEVATION (TYP.)
160' RELINE EX. SAN. SEWER
REHABILITATE EXISTING MANHOLES
SHT. 4 OF 35

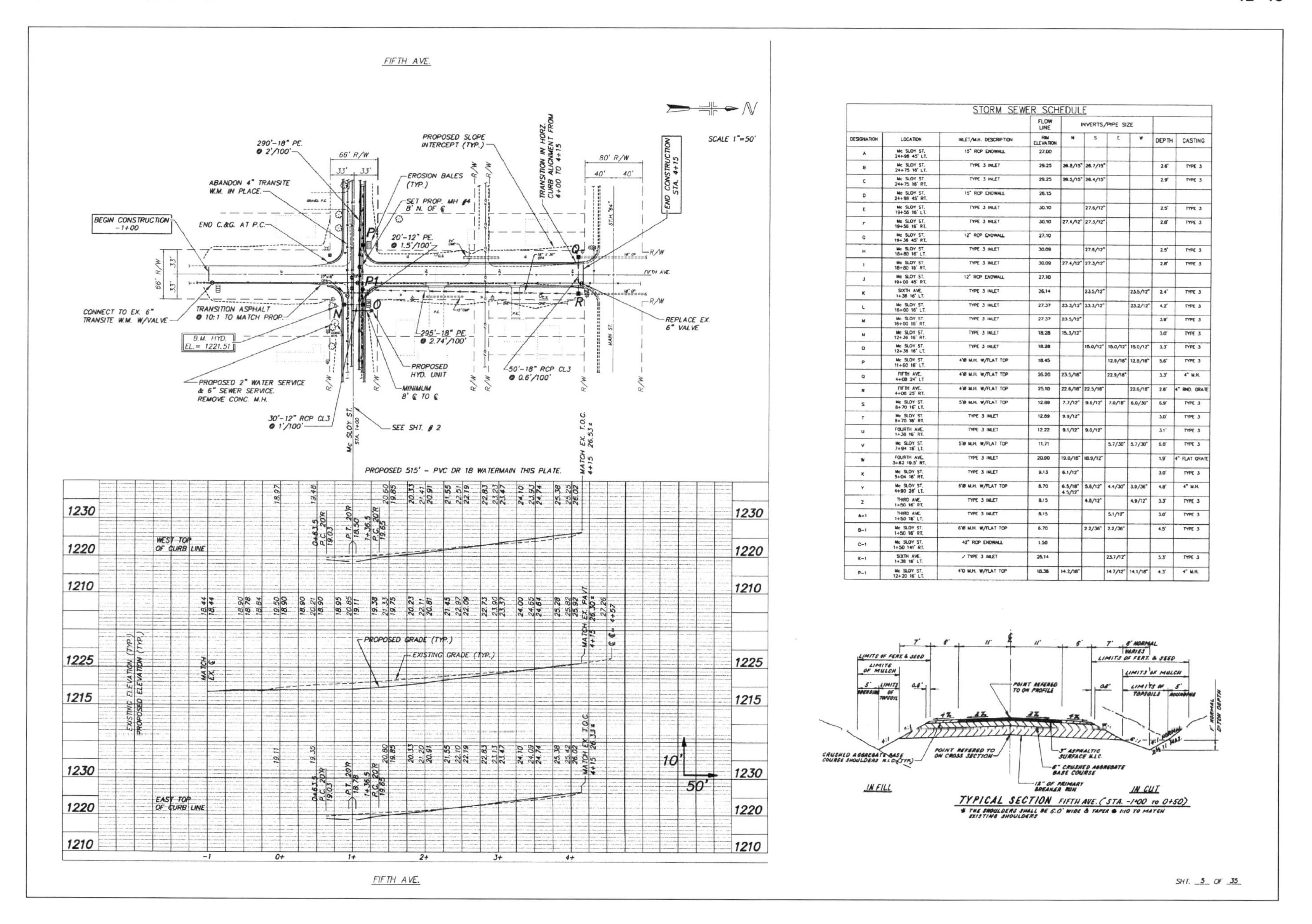

DESIGNATION	LOCATION	INLET/M.H. DESCRIPTION	FLOW LINE RIM ELEVATION	INVERTS/PIPE SIZE N	S	E	W	DEPTH	CASTING
A	Mc SLOY ST. 24+98 45' LT.	15" RCP ENDWALL	27.00						
B	Mc SLOY ST. 24+75 16' LT.	TYPE 3 INLET	29.25	26.8/15"	26.7/15"			2.6'	TYPE 3
C	Mc SLOY ST. 24+75 16' RT.	TYPE 3 INLET	29.25	26.5/15"	26.4/15"			2.9'	TYPE 3
D	Mc SLOY ST. 24+98 45' RT.	15" RCP ENDWALL	26.15						
E	Mc SLOY ST. 19+56 16' LT.	TYPE 3 INLET	30.10		27.6/12"			2.5'	TYPE 3
F	Mc SLOY ST. 19+56 16' RT.	TYPE 3 INLET	30.10	27.4/12"	27.3/12"			2.8'	TYPE 3
G	Mc SLOY ST. 19+38 45' RT.	12" RCP ENDWALL	27.10						
H	Mc SLOY ST. 18+80 16' LT.	TYPE 3 INLET	30.09		27.6/12"			2.5'	TYPE 3
I	Mc SLOY ST. 18+80 16' RT.	TYPE 3 INLET	30.09	27.4/12"	27.3/12"			2.8'	TYPE 3
J	Mc SLOY ST. 19+00 45' RT.	12" RCP ENDWALL	27.10						
K	SIXTH AVE. 1+38 16' LT.	TYPE 3 INLET	26.14		23.5/12"		23.5/12"	2.4'	TYPE 3
L	Mc SLOY ST. 16+00 16' LT.	TYPE 3 INLET	27.37	23.3/12"	23.3/12"		23.2/12"	4.2'	TYPE 3
M	Mc SLOY ST. 16+00 16' RT.	TYPE 3 INLET	27.37	23.5/12"				3.9'	TYPE 3
N	Mc SLOY ST. 12+36 16' RT.	TYPE 3 INLET	18.28	15.3/12"				3.0'	TYPE 3
O	Mc SLOY ST. 12+38 16' LT.	TYPE 3 INLET	18.28		15.0/12"	15.0/12"	15.0/12"	3.3'	TYPE 3
P	Mc SLOY ST. 11+60 16' LT.	4'Ø M.H. W/FLAT TOP	18.45			12.9/18"	12.8/18"	5.6'	TYPE 3
Q	FIFTH AVE. 4+08 24' LT	4'Ø M.H. W/FLAT TOP	26.20	23.5/18"		22.9/18"		3.3'	4" M.H.
R	FIFTH AVE. 4+08 25' RT.	4'Ø M.H. W/FLAT TOP	25.10	22.6/18"	22.5/18"		22.6/18"	2.8'	4" RND. GRATE
S	Mc SLOY ST. 8+70 16' LT.	5'Ø M.H. W/FLAT TOP	12.89	7.7/12"	9.6/12"	7.0/18"	6.0/30"	6.9'	TYPE 3
T	Mc SLOY ST. 8+70 16' RT.	TYPE 3 INLET	12.89	9.9/12"				3.0'	TYPE 3
U	FOURTH AVE. 1+38 16' RT.	TYPE 3 INLET	12.22	9.1/12"	9.0/12"			3.1'	TYPE 3
V	Mc SLOY ST. 7+94 18' LT.	5'Ø M.H. W/FLAT TOP	11.71			5.7/30"	5.7/30"	6.0'	TYPE 3
W	FOURTH AVE. 3+82 19.5' RT.	TYPE 3 INLET	20.90	19.0/18"	18.9/12"			1.9'	4" FLAT GRATE
X	Mc SLOY ST. 5+04 16' RT.	TYPE 3 INLET	9.13	6.1/12"				3.0'	TYPE 3
Y	Mc SLOY ST. 4+90 26' LT.	6'Ø M.H. W/FLAT TOP	8.70	6.5/18" 4.5/12"	5.8/12"	4.4/30"	3.9/36"	4.8'	4" M.H.
Z	THIRD AVE. 1+50 16' RT.	TYPE 3 INLET	8.15		4.8/12"		4.9/12"	3.3'	TYPE 3
A-1	THIRD AVE. 1+50 16' LT.	TYPE 3 INLET	8.15			5.1/12"		3.0'	TYPE 3
B-1	Mc SLOY ST. 1+50 16' RT.	6'Ø M.H. W/FLAT TOP	6.70		2.2/36"	2.2/36"		4.5'	TYPE 3
C-1	Mc SLOY ST. 1+50 141' RT.	42" RCP ENDWALL	1.50						
K-1	SIXTH AVE. 1+38 16' LT.	/ TYPE 3 INLET	26.14			23.7/12"		3.3'	TYPE 3
P-1	Mc SLOY ST. 12+20 16' LT.	4'Ø M.H. W/FLAT TOP	18.38	14.2/18"		14.7/12"	14.1/18"	4.3'	4" M.H.

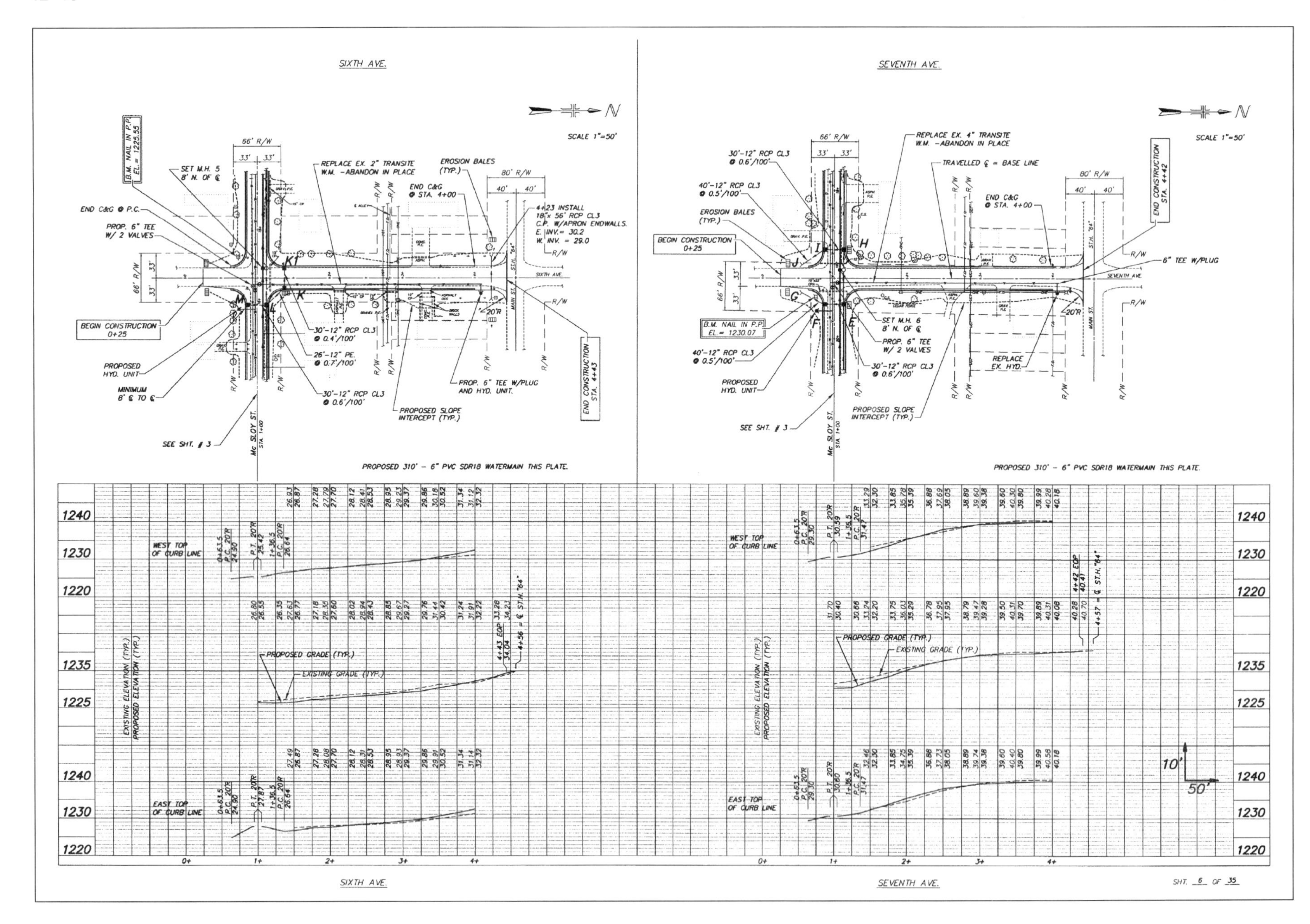
SIXTH AVE.
SEVENTH AVE.
SCALE 1"=50'
REPLACE EX. 2" TRANSITE W.M. -ABANDON IN PLACE
EROSION BALES (TYP.)
END C&G @ STA. 4+00
4+23 INSTALL 18"x 56' RCP CL3 C.P. W/APRON ENDWALLS. E. INV.= 30.2 W. INV. = 29.0
SET M.H. 5 8' N. OF ℄
END C&G @ P.C.
PROP. 6" TEE W/ 2 VALVES
BEGIN CONSTRUCTION 0+25
PROPOSED HYD. UNIT
MINIMUM 8' ℄ TO ℄
30'-12" RCP CL3 @ 0.4'/100'
26'-12" PE. @ 0.7'/100'
30'-12" RCP CL3 @ 0.6'/100'
PROP. 6" TEE W/PLUG AND HYD. UNIT.
PROPOSED SLOPE INTERCEPT (TYP.)
END CONSTRUCTION STA. 4+43
SEE SHT. # 3
Mc SLOY ST.
B.M. NAIL IN P.P. EL.= 1225.55
PROPOSED 310' - 6" PVC SDR18 WATERMAIN THIS PLATE.
REPLACE EX. 4" TRANSITE W.M. -ABANDON IN PLACE
TRAVELLED ℄ = BASE LINE
30'-12" RCP CL3 @ 0.6'/100'
40'-12" RCP CL3 @ 0.5'/100'
END C&G @ STA. 4+00
END CONSTRUCTION STA. 4+42
6" TEE W/PLUG
SET M.H. 6 8' N. OF ℄
B.M. NAIL IN P.P. EL.= 1230.07
REPLACE EX. HYD.
WEST TOP OF CURB LINE
EAST TOP OF CURB LINE
PROPOSED GRADE (TYP.)
EXISTING GRADE (TYP.)
EXISTING ELEVATION (TYP.)
PROPOSED ELEVATION (TYP.)
SHT. 6 OF 35

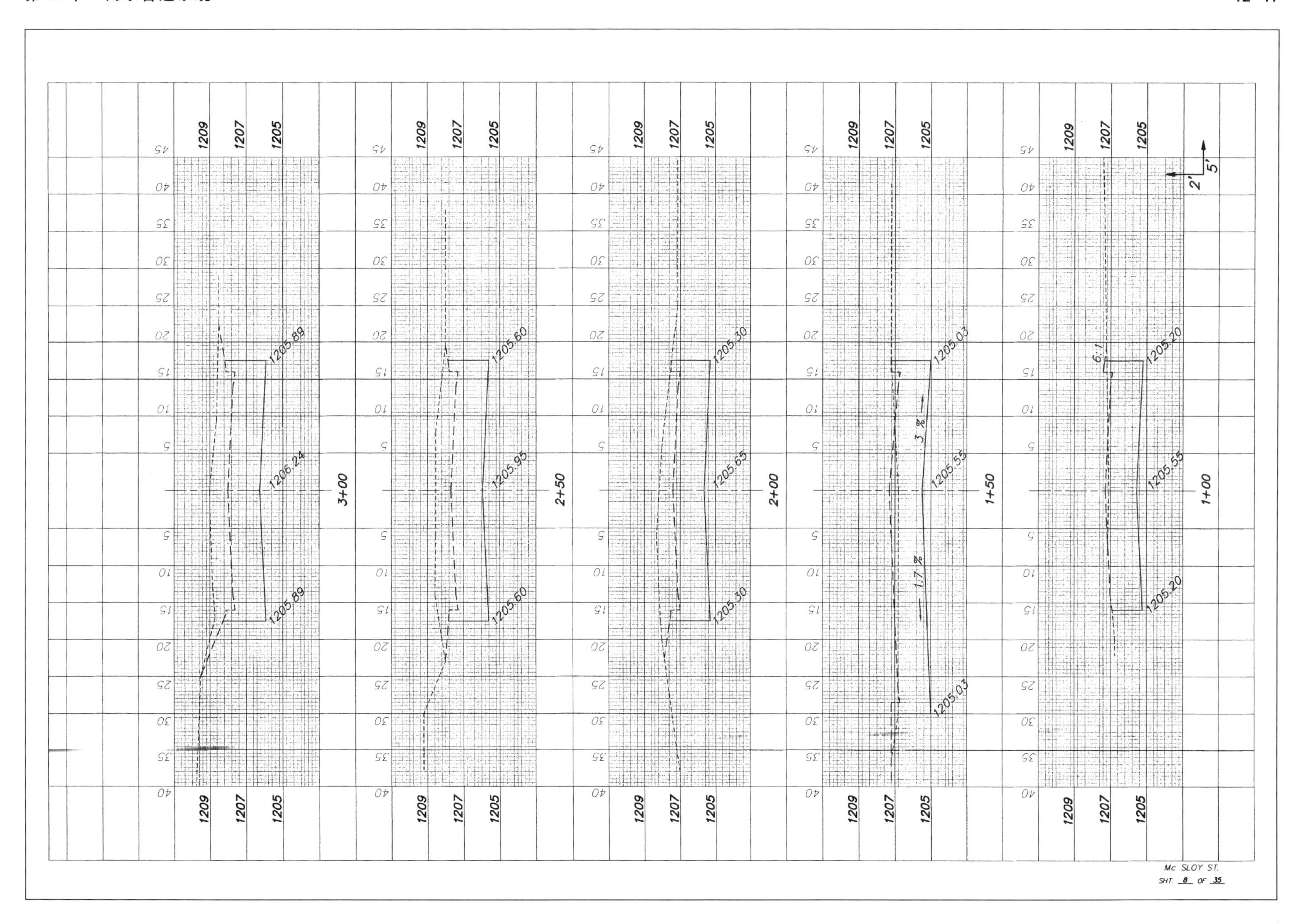
1209
1207
1205
1205.89
1206.24
3+00
1205.60
1205.95
2+50
1205.30
1205.65
2+00
1205.03
3 %
1205.55
1.7 %
1+50
1205.20
6:1
1+00
2'
5'
Mc SLOY ST.
SHT. 8 OF 35

街道改造：霍金斯村

STREET RECONSTRUCTION for the VILLAGE of HAWKINS, WISCONSIN SCANDINAVIA AVENUE

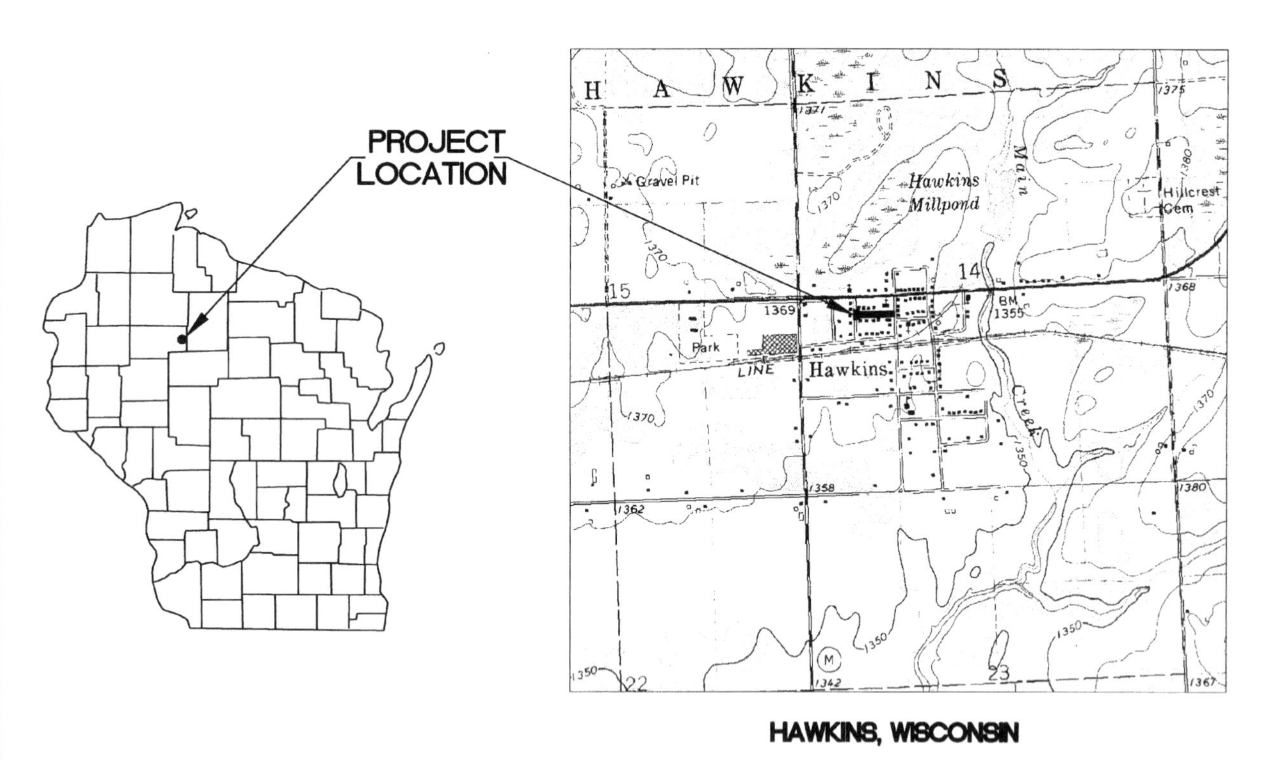

HAWKINS, WISCONSIN

INDEX OF SHEETS

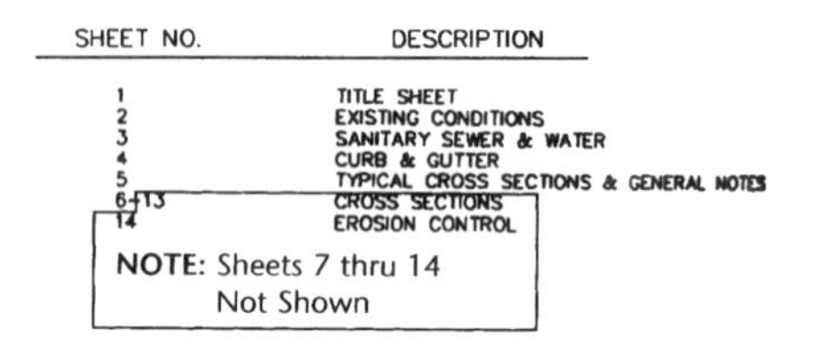

SHEET NO.	DESCRIPTION
1	TITLE SHEET
2	EXISTING CONDITIONS
3	SANITARY SEWER & WATER
4	CURB & GUTTER
5	TYPICAL CROSS SECTIONS & GENERAL NOTES
6-13	CROSS SECTIONS
14	EROSION CONTROL

NOTE: Sheets 7 thru 14 Not Shown

TITLE SHEET	
HAWKINS, WISCONSIN	
SCALE – NONE	MORGAN & PARMLEY, LTD. CONSULTING ENGINEERS LADYSMITH, WISCONSIN
SHEET 1 OF 14	
DATE MARCH, 2000	DR. BY D.A.N.

Courtesy: Morgan & Parmley, Ltd.

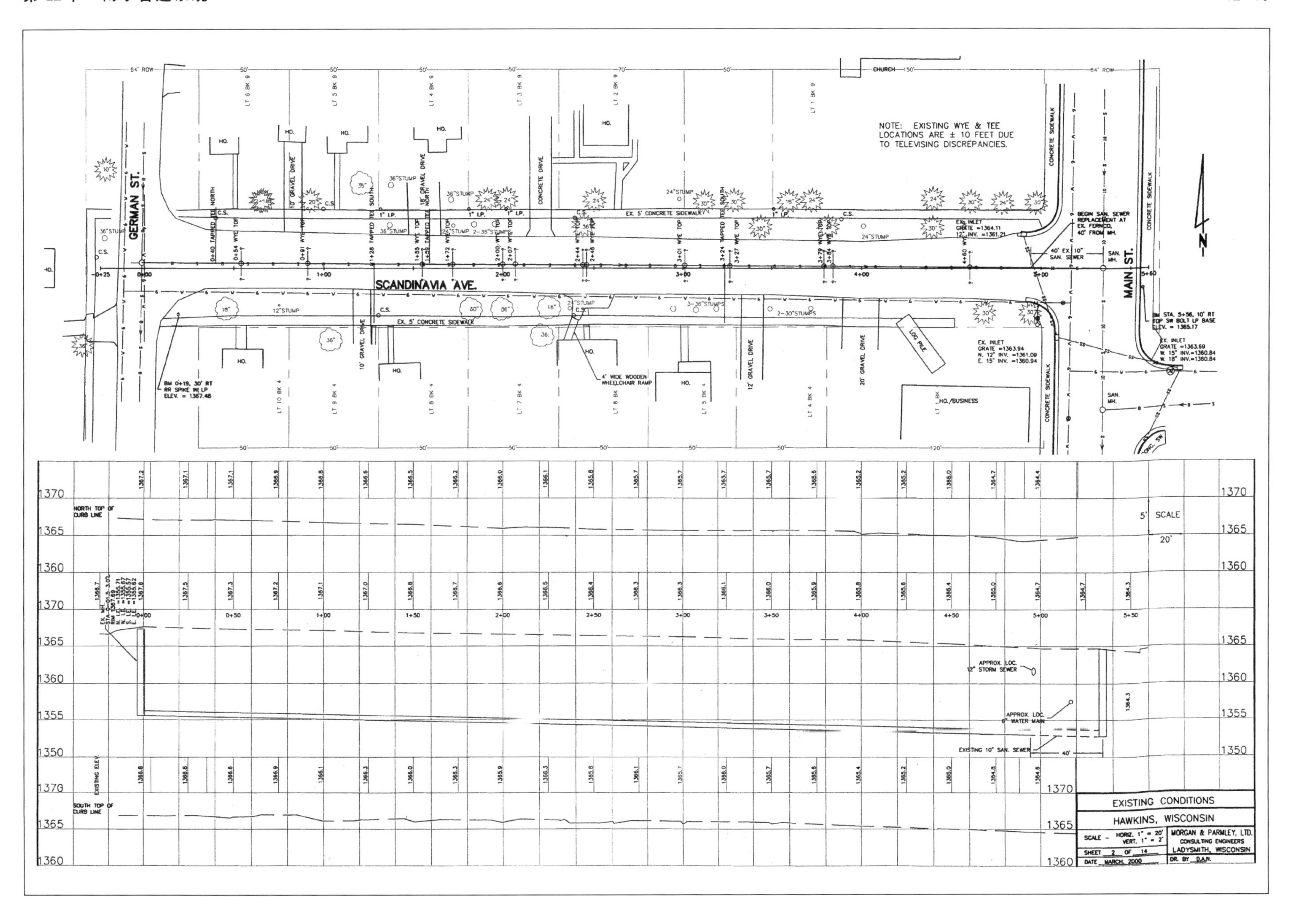
NOTE: EXISTING WYE & TEE LOCATIONS ARE ± 10 FEET DUE TO TELEVISING DISCREPANCIES.
GERMAN ST.
SCANDINAVIA AVE.
MAIN ST.
CHURCH
EX. 5' CONCRETE SIDEWALK
CONCRETE SIDEWALK
CONCRETE DRIVE
LOG PILE
HO./BUSINESS
4' WIDE WOODEN WHEELCHAIR RAMP
BM 0+19, 30' RT RR SPIKE IN LP ELEV. = 1367.48
BEGIN SAN. SEWER REPLACEMENT AT EX. FERNCO, 40' FROM MH.
40' EX. 10" SAN. SEWER
SAN. MH.
EX. INLET GRATE =1364.11 12" INV. =1361.21
EX. INLET GRATE =1363.94 N. 12" INV. =1361.09 E. 15" INV. =1360.94
EX. INLET GRATE =1363.69 W. 15" INV.=1360.84 W. 18" INV.=1360.84
BM STA. 5+56, 10' RT TOP SW BOLT LP BASE ELEV. = 1365.17
NORTH TOP OF CURB LINE
SOUTH TOP OF CURB LINE
EXISTING ELEV.
APPROX. LOC. 12" STORM SEWER
APPROX. LOC. 6" WATER MAIN
EXISTING 10" SAN. SEWER
SCALE 5' 20'
EXISTING CONDITIONS
HAWKINS, WISCONSIN
SCALE - HORIZ. 1" = 20' VERT. 1" = 2'
MORGAN & PARMLEY, LTD. CONSULTING ENGINEERS LADYSMITH, WISCONSIN
SHEET 2 OF 14
DATE MARCH, 2000
DR. BY D.A.N.

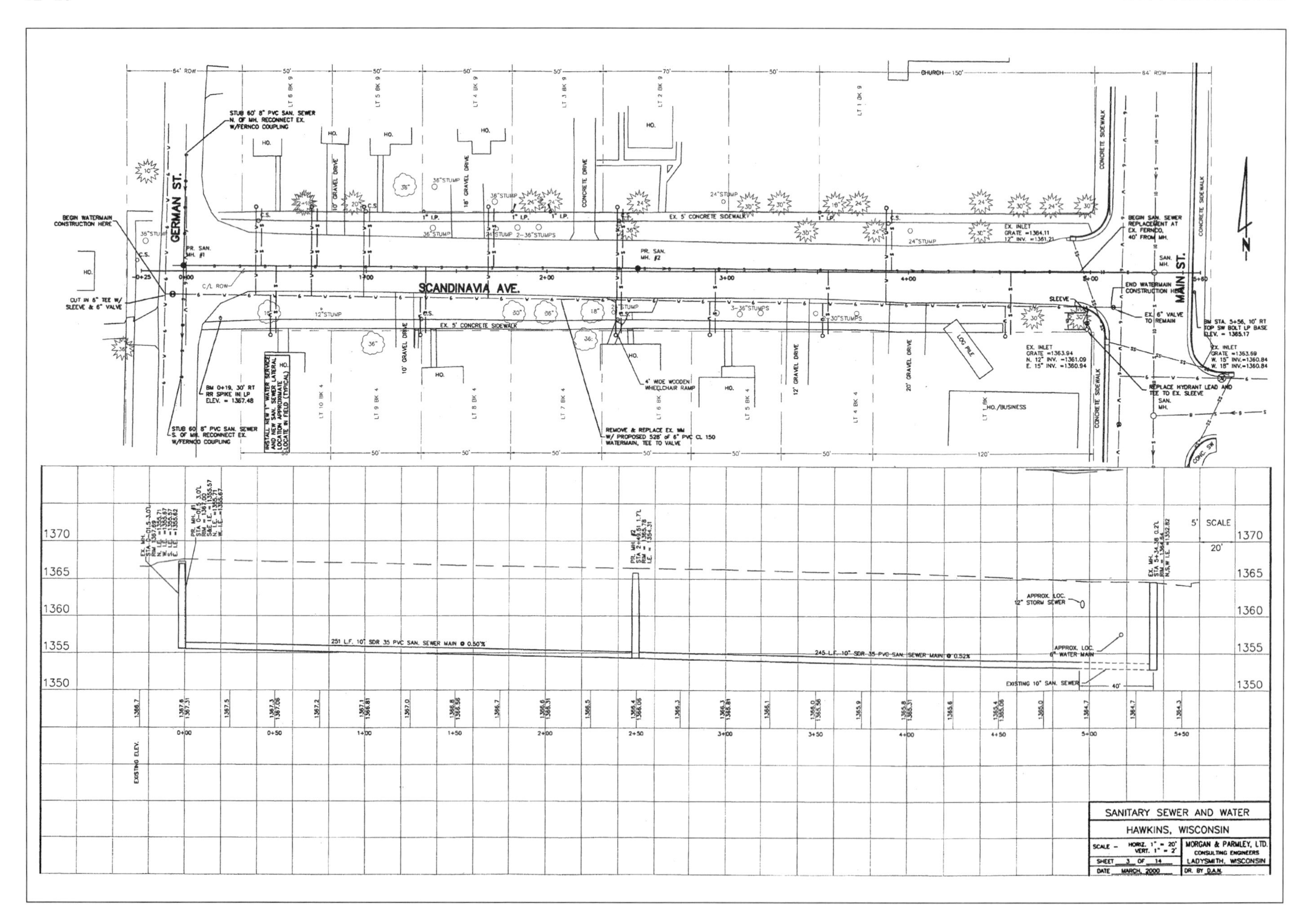

GERMAN ST.
SCANDINAVIA AVE.
MAIN ST.
CHURCH
EX. 5' CONCRETE SIDEWALK
CONCRETE SIDEWALK
CONCRETE DRIVE
10' GRAVEL DRIVE
18' GRAVEL DRIVE
12' GRAVEL DRIVE
20' GRAVEL DRIVE
STUB 60' 8" PVC SAN. SEWER N. OF MH. RECONNECT EX. W/FERNCO COUPLING
STUB 60' 8" PVC SAN. SEWER S. OF MH. RECONNECT EX. W/FERNCO COUPLING
BEGIN WATERMAIN CONSTRUCTION HERE
CUT IN 6" TEE W/ SLEEVE & 6" VALVE
BM 0+19, 30' RT RR SPIKE IN LP ELEV. = 1367.48
INSTALL NEW 1" WATER SERVICE AND NEW SAN. SEWER LATERAL LOCATION APPROXIMATE LOCATE IN FIELD (TYPICAL)
REMOVE & REPLACE EX. MH W/ PROPOSED 528' of 6" PVC CL 150 WATERMAIN, TEE TO VALVE
4" WIDE WOODEN WHEELCHAIR RAMP
PR. SAN. MH. #1
PR. SAN. MH. #2
BEGIN SAN. SEWER REPLACEMENT AT EX. FERNCO, 40' FROM MH.
EX. INLET GRATE =1364.11 12" INV. =1361.21
END WATERMAIN CONSTRUCTION HERE
EX. 6" VALVE TO REMAIN
EX. INLET GRATE =1363.94 N. 12" INV. =1361.09 E. 15" INV. =1360.94
BM STA. 5+56, 10' RT TOP SW BOLT LP BASE ELEV. = 1365.17
EX. INLET GRATE =1363.69 W. 15" INV.=1360.84 W. 18" INV.=1360.84
REPLACE HYDRANT LEAD AND TEE TO EX. SLEEVE
SAN. MH.
SLEEVE
LOG PILE
HO./BUSINESS
251 L.F. 10" SDR 35 PVC SAN. SEWER MAIN @ 0.50%
245 L.F. 10" SDR 35 PVC SAN. SEWER MAIN @ 0.52%
APPROX. LOC. 12" STORM SEWER
APPROX. LOC. 6" WATER MAIN
EXISTING 10" SAN. SEWER
EXISTING ELEV.
SCALE 5' 20'
1370
1365
1360
1355
1350
0+00
0+50
1+00
1+50
2+00
2+50
3+00
3+50
4+00
4+50
5+00
5+50
SANITARY SEWER AND WATER
HAWKINS, WISCONSIN
SCALE – HORIZ. 1" = 20' VERT. 1" = 2'
MORGAN & PARMLEY, LTD. CONSULTING ENGINEERS LADYSMITH, WISCONSIN
SHEET 3 OF 14
DATE MARCH, 2000
DR. BY D.A.N.

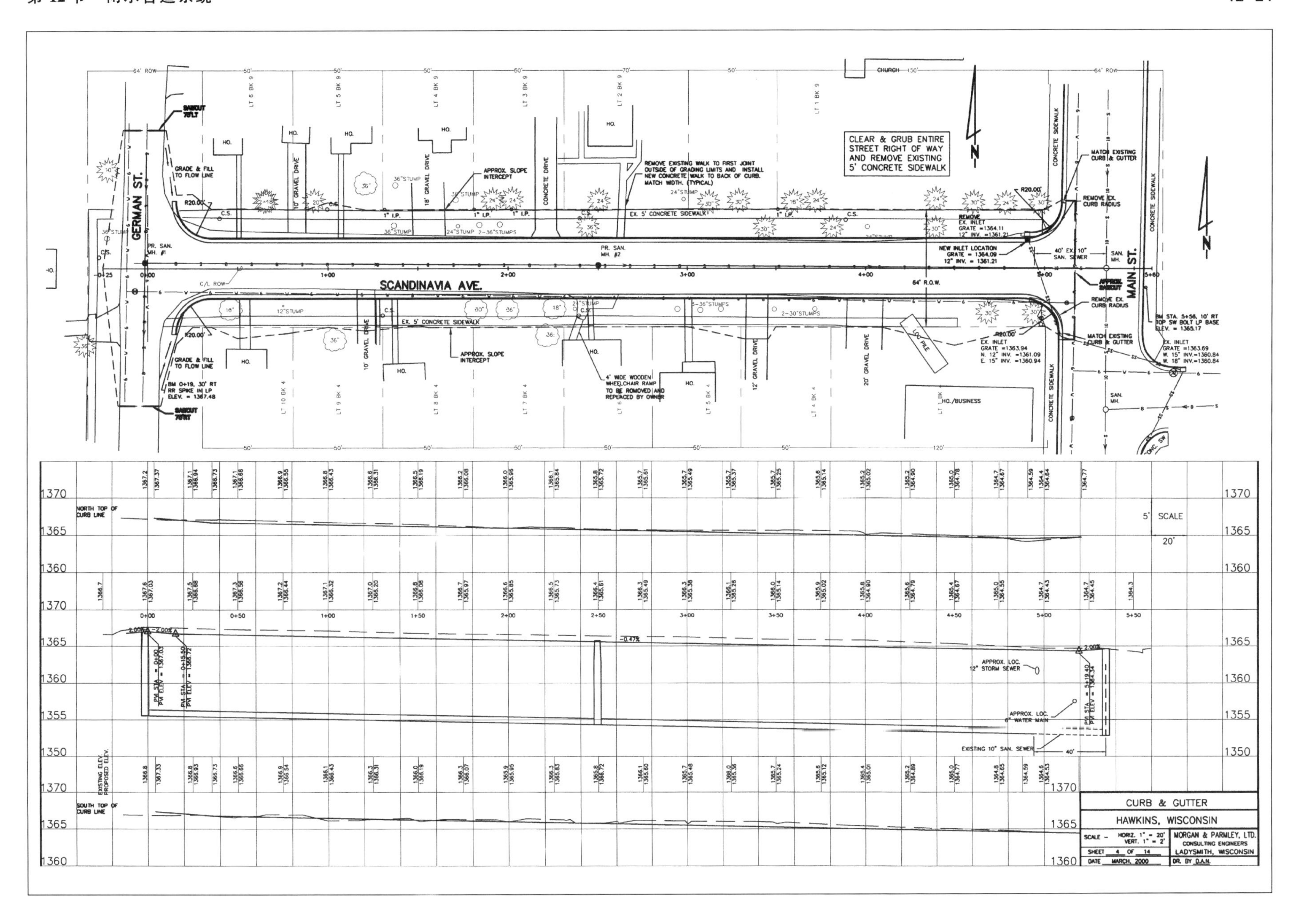
CLEAR & GRUB ENTIRE
STREET RIGHT OF WAY
AND REMOVE EXISTING
5' CONCRETE SIDEWALK
SCANDINAVIA AVE.
GERMAN ST.
MAIN ST.
CHURCH
64' ROW
64' R.O.W.
SAWCUT 75'LT
SAWCUT 75'RT
GRADE & FILL TO FLOW LINE
R20.00'
PR. SAN. MH. #1
PR. SAN. MH. #2
EX. 5' CONCRETE SIDEWALK
APPROX. SLOPE INTERCEPT
REMOVE EXISTING WALK TO FIRST JOINT OUTSIDE OF GRADING LIMITS AND INSTALL NEW CONCRETE WALK TO BACK OF CURB. MATCH WIDTH. (TYPICAL)
NEW INLET LOCATION GRATE = 1364.09 12" INV. = 1361.21
REMOVE EX. INLET GRATE =1364.11 12" INV. =1361.21
40' EX. 10" SAN. SEWER
SAN. MH.
MATCH EXISTING CURB & GUTTER
REMOVE EX. CURB RADIUS
APPROX. SAWCUT
EX. INLET GRATE =1363.94 N. 12" INV. =1361.09 E. 15" INV. =1360.94
EX. INLET GRATE =1363.69 W. 15" INV.=1360.84 W. 18" INV.=1360.84
BM STA. 5+56, 10' RT TOP SW BOLT LP BASE ELEV. = 1365.17
BM 0+19, 30' RT RR SPIKE IN LP ELEV. = 1367.48
4' WIDE WOODEN WHEELCHAIR RAMP TO BE REMOVED AND REPLACED BY OWNER
HO./BUSINESS
CONCRETE SIDEWALK
CONCRETE DRIVE
GRAVEL DRIVE
NORTH TOP OF CURB LINE
SOUTH TOP OF CURB LINE
EXISTING ELEV.
PROPOSED ELEV.
SCALE 5' 20'
-0.47%
APPROX. LOC. 12" STORM SEWER
APPROX. LOC. 6" WATER MAIN
EXISTING 10" SAN. SEWER
PVI STA = 0+00 PVI ELEV = 1367.03
PVI STA = 0+15.50 PVI ELEV = 1366.72
CURB & GUTTER
HAWKINS, WISCONSIN
SCALE – HORIZ. 1" = 20' VERT. 1" = 2'
SHEET 4 OF 14
DATE MARCH, 2000
MORGAN & PARMLEY, LTD. CONSULTING ENGINEERS LADYSMITH, WISCONSIN
DR. BY D.A.N.

GENERAL NOTES

1. ALL WATER MAIN SHALL BE 6" PVC DR 18.
2. ALL WATERMAIN FITTINGS SHALL BE M.J.D.I., W / RESTRAINED JOINT LUGS.
3. ALL HYDRANT LEADS SHALL BE 6-INCH VALVE.
4. HYDRANTS SHALL BE WATEROUS WB 67-250 (4" PUMPER NOZZLE), 22" BREAKOFF WITH THREADS MATCHING VILLAGE'S PATTERN.
5. ALL WATER MAIN SHALL HAVE AN 8 FT. BURY (MIN.).
6. WATER SERVICES SHALL BE 1" TYPE K COPPER W / MUELLER OR FORD BRASS.
7. 8' (MINIMUM) HORIZONTAL DISTANCE (℄ TO ℄) BETWEEN WATERMAIN AND SANITARY SEWER OR STORM SEWER.
8. 18" (MINIMUM) VERTICAL DISTANCE (OUT TO OUT) BETWEEN WATERMAIN UNDER SANITARY SEWER OR STORM SEWER @ THEIR INTERSECTION.
9. 6" (MINIMUM) VERTICAL DISTANCE (OUT TO OUT) BETWEEN WATERMAIN OVER SANITARY SEWER OR STORM SEWER @ THEIR INTERSECTION.
10. SANITARY SERVICES SHALL BE 4" SCH. 40 PVC W / WYES LAID TO PROVIDE FOR MAXIMUM DEPTH OR 10' AT THE R / W.
11. SEWERMAIN SHALL BE PVC SDR 35.
12. MANHOLES SHALL BE SET WITH A MINIMUM OF 6" OF RINGS AND 9" CASTINGS.
13. MANHOLES SHALL BE 4' DIAMETER PRECAST CONCRETE W / INTEGRAL BASES AND BOOTS.
14. ALL ELEVATIONS BASED ON USGS DATUM.
15. EROSION CONTROL MEASURES SHALL BE IN PLACE BEFORE CONSTRUCTION BEGINS.
16. THE CONTRACTORS SHALL MINIMIZE EROSION DURING CONSTRUCTION USING GOOD CONSTRUCTION TECHNIQUES AND UTILIZING SILT FENCES AND BALE DITCH CHECKS. THE CONTRACTOR SHALL RELEASE RUNOFF FROM THE SITE IN A NUISANCE FREE MANNER.
17. THE CONTRACTOR IS RESPONSIBLE TO MAINTAIN THE SITE IN A SAFE CONDITION. THE SOLE RESPONSIBILITY FOR WARNING SIGNS, BARRICADES, FLAGGING PERSONNEL AND ALL ASPECTS OF SAFETY LIE WITH THE CONTRACTOR.
18. EXISTING UTILITIES ARE SHOWN BUT MAY NOT BE ALL INCLUSIVE; CONTRACTOR SHALL HAVE ALL BURIED UTILITIES LOCATED PRIOR TO COMMENCING CONSTRUCTION.
19. EXISTING SEWER AND WATERMAIN ARE SHOWN ACCORDING TO EXISTING RECORDS BUT LOCATION SHALL BE VERIFIED IN FIELD.
20. WATERMAIN & SERVICES SHALL BE INSULATED AT ALL STORM SEWER, CULVERT AND DITCH CROSSINGS WITH 2" THICK EXTRUDED POLYSTYRENE INSULATION BOARD.
21. PROPERTY CORNERS KNOWN TO EXIST SHALL NOT BE DISTURBED. DAMAGED CORNERS WILL BE REPLACED AT THE CONTRACTOR'S EXPENSE. PRESERVE ALL SURVEY MONUMENTS.
22. ALL WATERMAIN SHALL BE DISINFECTED BEFORE BEING PUT INTO USE.
23. VALVES AND MANHOLES SHALL BE ADJUSTED TO FINAL GRADE, UNLESS DIRECTED OTHERWISE.
24. CONSTRUCTION SHALL COMPLY WITH THE CONDITIONS OF DNR APPROVAL ELTTER, AND OTHER REGULATORY APPROVALS AS LISTED IN THE SPECIFICATION BOOK.
25. ALL DISTURBED AREAS SHALL BE SEEDED, FERTILIZED AND MULCHED, OR COVERED W / BASE COURSE AFTER CONSTRUCTION, OR RESTORED WITH BITUMINOUS PAVEMENT WHERE APPLICABLE.
26. THERE SHALL BE NO DEVIATION FROM THE PLANS WITHOUT THE DESIGN ENGINEER'S APPROVAL.
28. ALL DISTURBED AREAS TO BE SEEDED SHALL BE RESTORED AS FOLLOWS:

    A) REPLACED ALL SALVAGED AND IMPORTED TOPSOIL
    B) SEED MIXUTRE: PERMANENT SEEDING – (LAWN TYPE TURF)

        35% KENTUCKY BLUEGRASS
        25% IMPROVED FINE PERENNIAL RYEGRASS
        20% CREEPING RED FESCUE
        20% IMPROVED HARD FESCUE

    C) SEED RATE: 2# / 100 SQ. FT.
    D) MULCH RATE: 2 TON / ACRE
    E) PERMANENT SEEDING: ALLOWED TO SEPT. 7
    TEMPORARY SEEDING: SEPT. 8 THROUGH NOV. 10
    (4 BU. OATS / ACRE)
    DORMANT SEEDING; SEED PERM. SEEDING INTO TEMP. SEEDING AFTER NOV. 10
    F) SEEDING MAY BE BRAODCAST BUT MUST BE COVERED WITH A MAXIMUM OF ½" SOIL. MULCH MUST BE ANCHORED WITH A MULCH TILLER, TACKIFIER OR NETTING.
    G) FERTILIZER: 500# / ACRE 20-10-10
29. CONTRACTOR MUST REMOVE AND REPLACE ALL AFFECTED STREET SIGNS AFTER PROPERLY SIGNING AND BARRICADING THE CONSTRUCTION AREA. ALL SIGNS MUST BE REPLACED PRIOR TO OPENING THE STREET TO TRAFFIC.
30. CONTRACTOR SHALL DETERMINE CONSTRUCTION SEQUENCE IN ORDER TO MINIMIZE INSTALLATION CONFLICTS WITH PIPE CROSSINGS.
31. NO TREES ARE TO BE REMOVED WITHOUT THE APPROVAL OF THE ENGINEER.

SPECIFICATIONS:

WISCONSIN CONSTRUCTION SITE BEST MANAGEMENT PRACTICE HANDBOOK

STANDARD SPECIFICATIONS FOR SEWER AND WATER CONSTRUCTION IN WISCONSIN (5TH EDITION)

STANDARD SPECIFICATIONS FOR ROAD AND BRIDGE CONSTRUCITON 1996 EDITION WIS D.O.T.

-UTILIITES-

WATER & SEWER:
VILLAGE OF HAWKINS
DAVE HETTINGER
HAWKINS VILLAGE HALL
P.O. BOX 108
HAWKINS, WI 54530
1-715-585-6322

ELECTRIC:
NORTHERN STATES
POWER COMPANY
310 HICKORY HILL LANE
PHILLIPS, WI 54555
JOE PERKINS
1-800-445-1275

CABLE TELEVISION:
KRM CABLEVISION
BOX 77
MELLON, WI 54546
1-800-327-4330

TELEPHONE:
CENTURYTEL
425 ELLINGSON AVE.
P.O. BOX 78
HAWKINS, WI 54530
JIM ARQUETTE
1-715-585-7707

DIGGERS HOTLINE:
1-800-242-8511

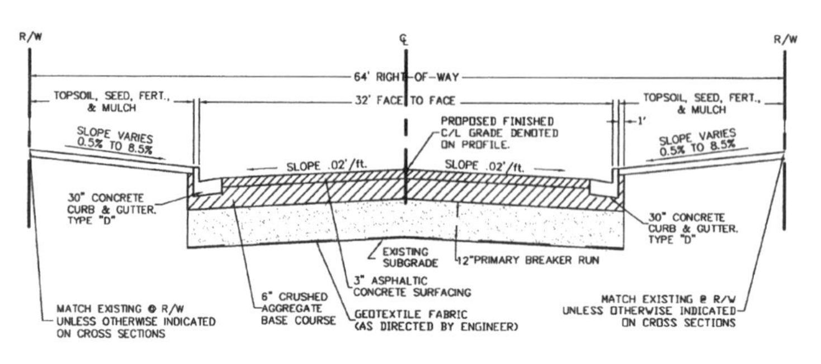

FINISHED STREET TYPICAL SECTION
NOT TO SCALE

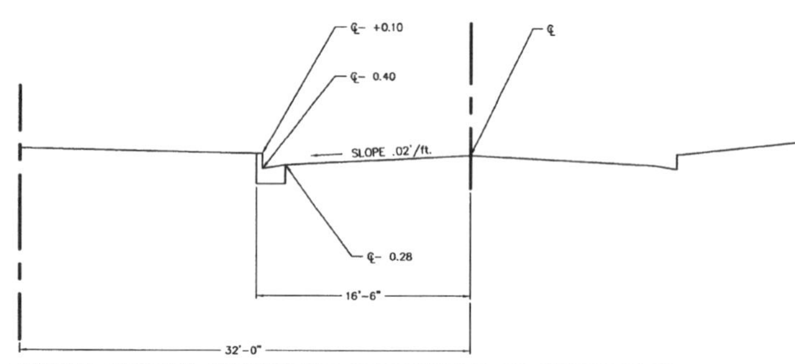

TYPICAL SECTION ELEVATIONS REFERENCED TO CENTERLINE
NOT TO SCALE

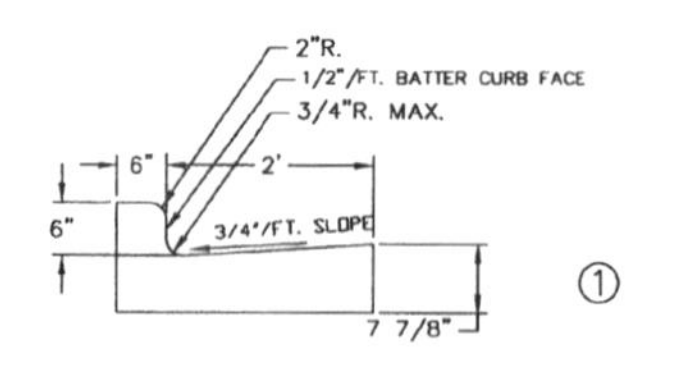

① THE BOTTOM OF CURB AND GUTTER MAY BE CONSTRUCTED EITHER LEVEL OR PARALLEL TO THE SLOPE OF THE SUBGRADE OR BASE COURSE PROVIDED A 6" MIN. GUTTER THICKNESS IS MAINTAINED.

TYPE "D" CURB

TYPICAL CROSS SECTIONS	
HAWKINS, WISCONSIN	
SCALE – NONE	MORGAN & PARMLEY, LTD. CONSULTING ENGINEERS LADYSMITH, WISCONSIN
SHEET 5 OF 14	
DATE MARCH, 2000	DR. BY D.A.N.

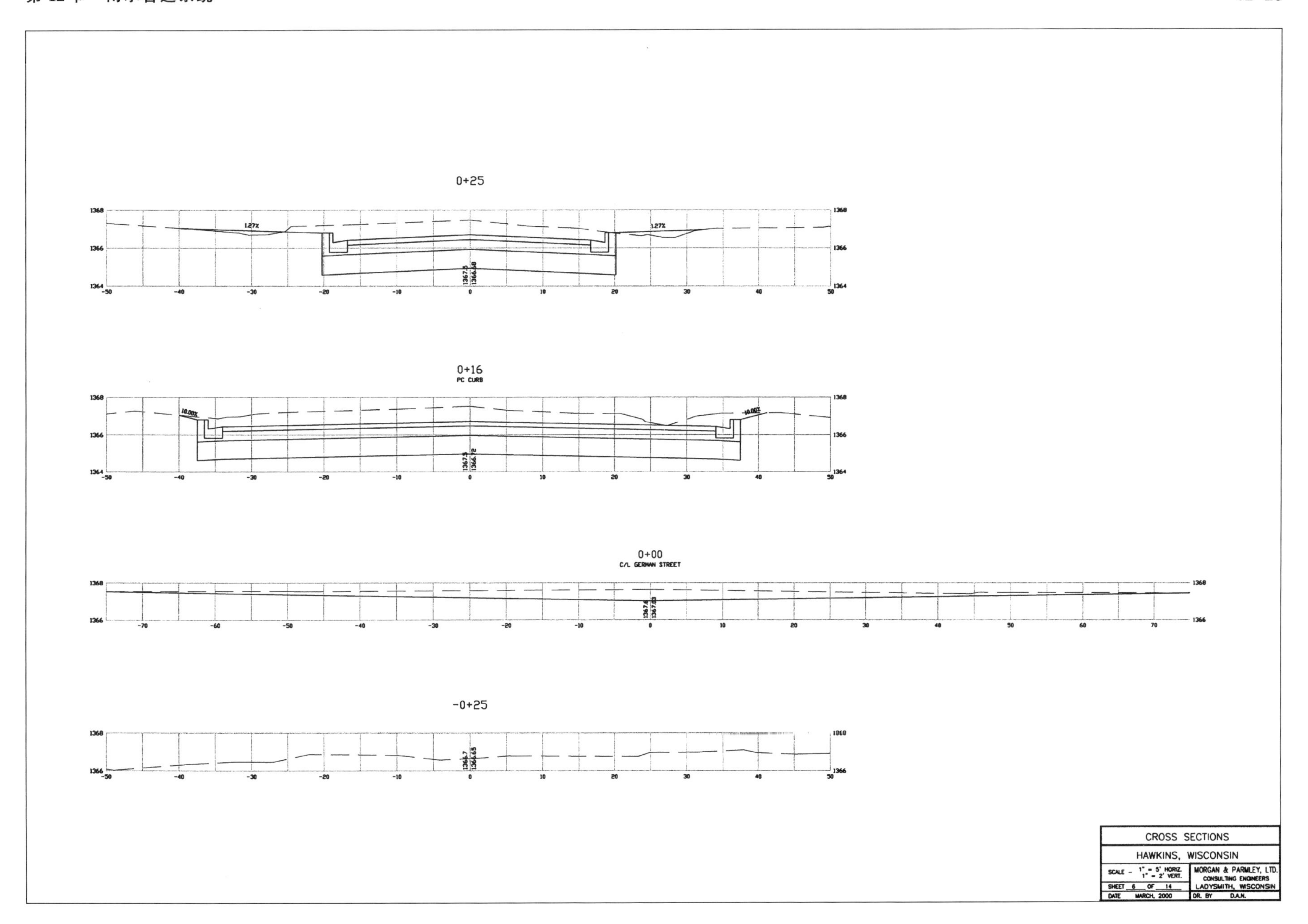

0+25
1.27%
1.27%
1367.5
1366.58
0+16
PC CURB
10.00%
10.00%
1367.5
1366.72
0+00
C/L GERMAN STREET
1367.6
1367.03
-0+25
1366.7
1366.65
CROSS SECTIONS
HAWKINS, WISCONSIN
SCALE - 1" = 5' HORIZ. 1" = 2' VERT.
MORGAN & PARMLEY, LTD.
CONSULTING ENGINEERS
LADYSMITH, WISCONSIN
SHEET 6 OF 14
DATE MARCH, 2000
DR. BY D.A.N.

雨水管道系统扩建工程：海沃德，威斯康星州

STORM SEWER EXTENSION
HAYWARD, WISCONSIN

SECOND ST., WISCONSIN AVE. & THIRD ST.

APRIL, 1980

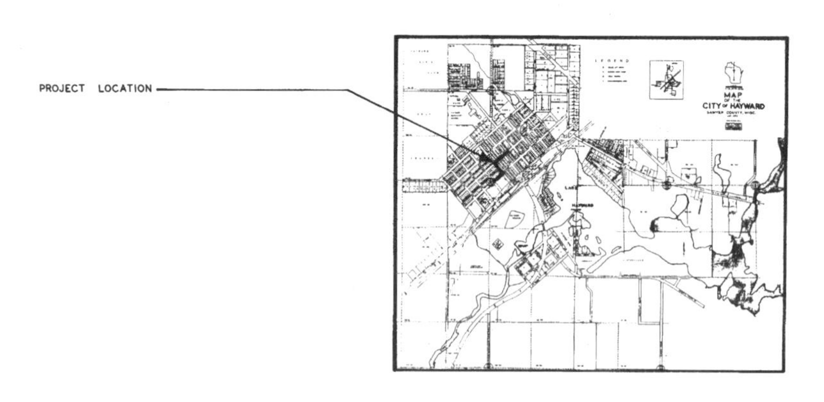

GENERAL MATERIAL LIST

DESCRIPTION	TOTALS
* 24" CONCRETE PIPE, CLASS IV, MORTAR JOINT W/ PERFORATION HOLES	370 L.F.
* 24" CONCRETE PIPE, CLASS III, MORTAR JOINT	575 L.F.
* 18" CONCRETE PIPE, CLASS IV, MORTAR JOINT	370 L.F.
* REINFORCED CONCRETE M.H. - 48" DIA.	52 V.F. (7 UNITS)
* REINFORCED CONCRETE M.H. BASES	7 UNITS
* REINFORCED CONCRETE M.H. COVERS - FLAT 6"	7 UNITS
* CATCH BASINS (24")	100 V.F. (17 UNITS)
* CATCH BASIN BASES	17 UNITS
* CATCH BASIN FRAME & COVER (C.I.)	17 UNITS
* 12" CONCRETE PIPE, CLASS IV (C.B. LEADS)	420 L.F.
* MANHOLE FRAME & COVER (C.I.)	7 UNITS

INDEX OF SHEETS

SHEET	DESCRIPTION
1	TITLE SHEET W/ PROJECT LOCATION
2	PLAN & PROFILE DETAILS
3	TYPICAL STORM SEWER DETAILS

PREPARED BY: MORGAN & PARMLEY, LTD. LADYSMITH, WISCONSIN

Courtesy: Morgan & Parmley, Ltd.

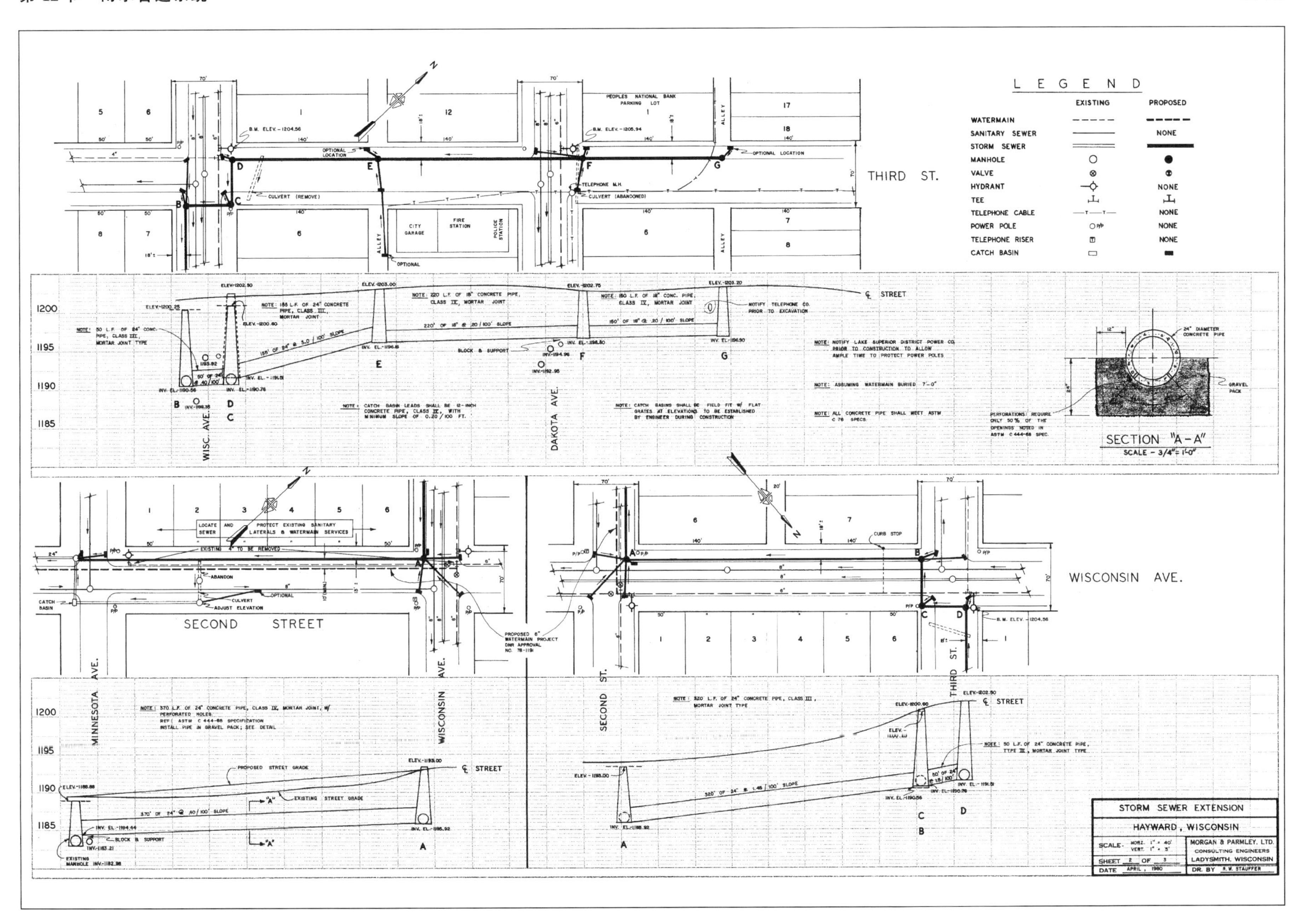
LEGEND
EXISTING
PROPOSED
WATERMAIN
SANITARY SEWER
STORM SEWER
MANHOLE
VALVE
HYDRANT
TEE
TELEPHONE CABLE
POWER POLE
TELEPHONE RISER
CATCH BASIN
NONE
THIRD ST.
WISCONSIN AVE.
SECOND STREET
SECTION "A-A"
SCALE - 3/4"=1'-0"
STORM SEWER EXTENSION
HAYWARD, WISCONSIN
MORGAN & PARMLEY, LTD.
CONSULTING ENGINEERS
LADYSMITH, WISCONSIN
SHEET 2 OF 3
DATE APRIL, 1980

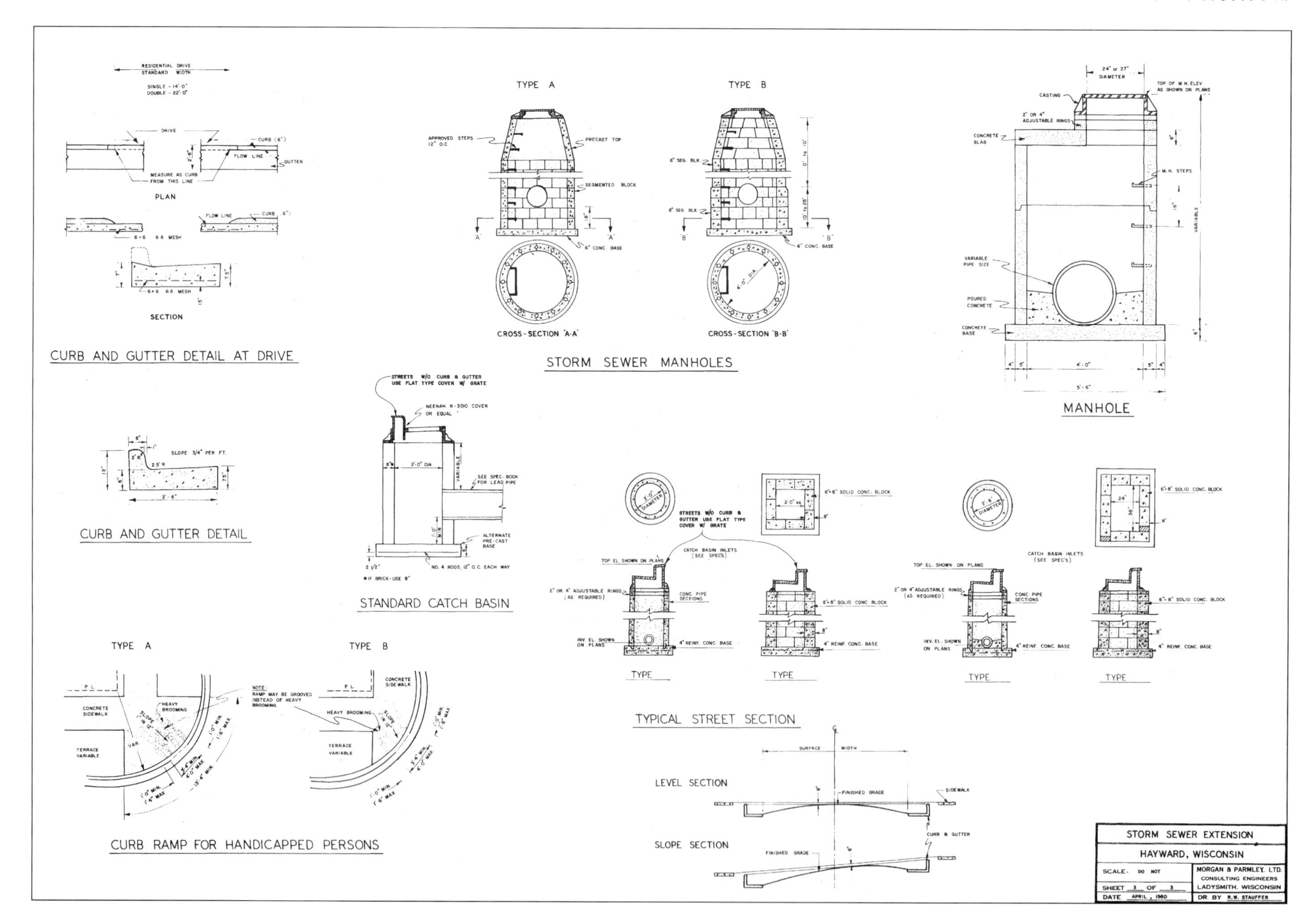
RESIDENTIAL DRIVE STANDARD WIDTH
SINGLE - 14'-0"
DOUBLE - 22'-0"
DRIVE
CURB (6")
FLOW LINE
GUTTER
MEASURE AS CURB FROM THIS LINE
PLAN
6×6 66 MESH
SECTION
CURB AND GUTTER DETAIL AT DRIVE
TYPE A
TYPE B
APPROVED STEPS 12" O.C.
PRECAST TOP
SEGMENTED BLOCK
6" CONC. BASE
6" SEG. BLK
8" SEG. BLK
CROSS-SECTION 'A-A'
CROSS-SECTION 'B-B'
STORM SEWER MANHOLES
24" or 27" DIAMETER
CASTING
TOP OF M.H. ELEV AS SHOWN ON PLANS
2" OR 4" ADJUSTABLE RINGS
CONCRETE SLAB
M.H. STEPS
VARIABLE
VARIABLE PIPE SIZE
POURED CONCRETE
CONCRETE BASE
MANHOLE
SLOPE 3/4" PER FT.
CURB AND GUTTER DETAIL
STREETS W/O CURB & GUTTER USE FLAT TYPE COVER W/ GRATE
NEENAH R-3010 COVER OR EQUAL
2'-0" DIA
SEE SPEC. BOOK FOR LEAD PIPE
ALTERNATE PRE-CAST BASE
NO. 4 RODS, 12" O.C. EACH WAY
STANDARD CATCH BASIN
6"×8" SOLID CONC. BLOCK
CATCH BASIN INLETS (SEE SPECS)
TOP EL. SHOWN ON PLANS
2" OR 4" ADJUSTABLE RINGS (AS REQUIRED)
CONC. PIPE SECTIONS
INV. EL. SHOWN ON PLANS
4" REINF. CONC. BASE
TYPE
TYPICAL STREET SECTION
TYPE A
TYPE B
CONCRETE SIDEWALK
TERRACE VARIABLE
HEAVY BROOMING
NOTE: RAMP MAY BE GROOVED INSTEAD OF HEAVY BROOMING.
CURB RAMP FOR HANDICAPPED PERSONS
SURFACE WIDTH
LEVEL SECTION
FINISHED GRADE
SIDEWALK
SLOPE SECTION
CURB & GUTTER
STORM SEWER EXTENSION
HAYWARD, WISCONSIN
MORGAN & PARMLEY LTD.
CONSULTING ENGINEERS
LADYSMITH, WISCONSIN
SHEET 3 OF 3
DATE APRIL, 1980
DR. BY R.W. STAUFFER

设　计

第 13 节 水坝和水库

例 13

通过建造水坝形成湖泊、池塘和水库来蓄水是经过时间考验的控制调节大自然最常见液体的好办法。建设这些结构和设施用于发电、防洪、灌溉、蓄水、休闲娱乐、航行船闸、土地开发、控制土壤侵蚀和水土保持。

本节给出了三个小水坝的实例。第一例是墨菲(Murphy)水坝建筑，它是一个全新的结构，设计为替代由 WPA 建于 20 世纪 30 年代的旧水坝。这个结构在 20 世纪 70 年代的异常大暴雨中毁坏。20 年后，政府决定修复这一设施。

第二例是私人的低水头水坝的设计图，该水坝为一个疗养胜地创造了一个浅水塘。

最后一例是翻修建于 20 世纪 30 年代早期的一个水坝的一组图纸，新增了光控门和相关部件。

目 录

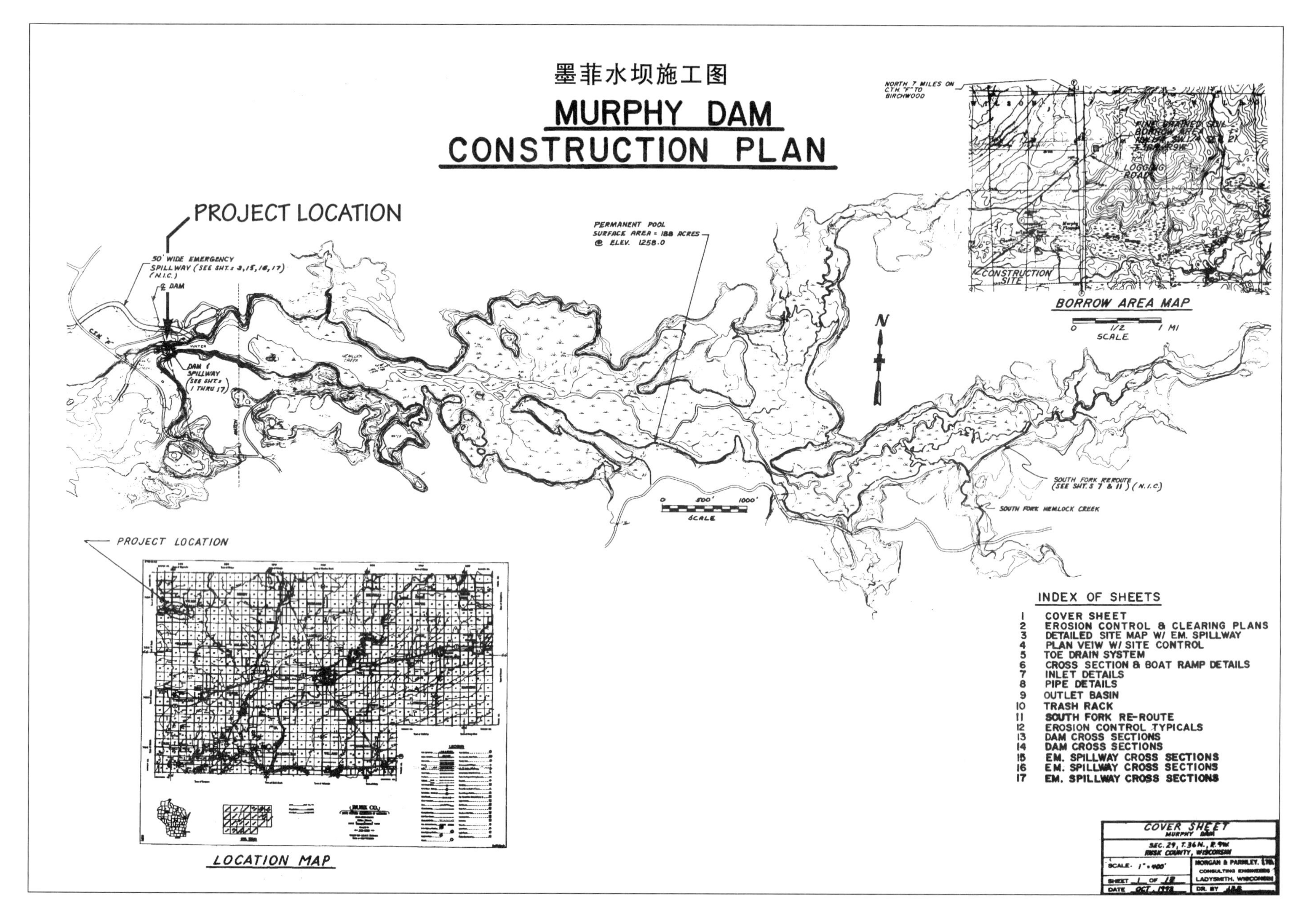

Source: Morgan & Parmley, Ltd.

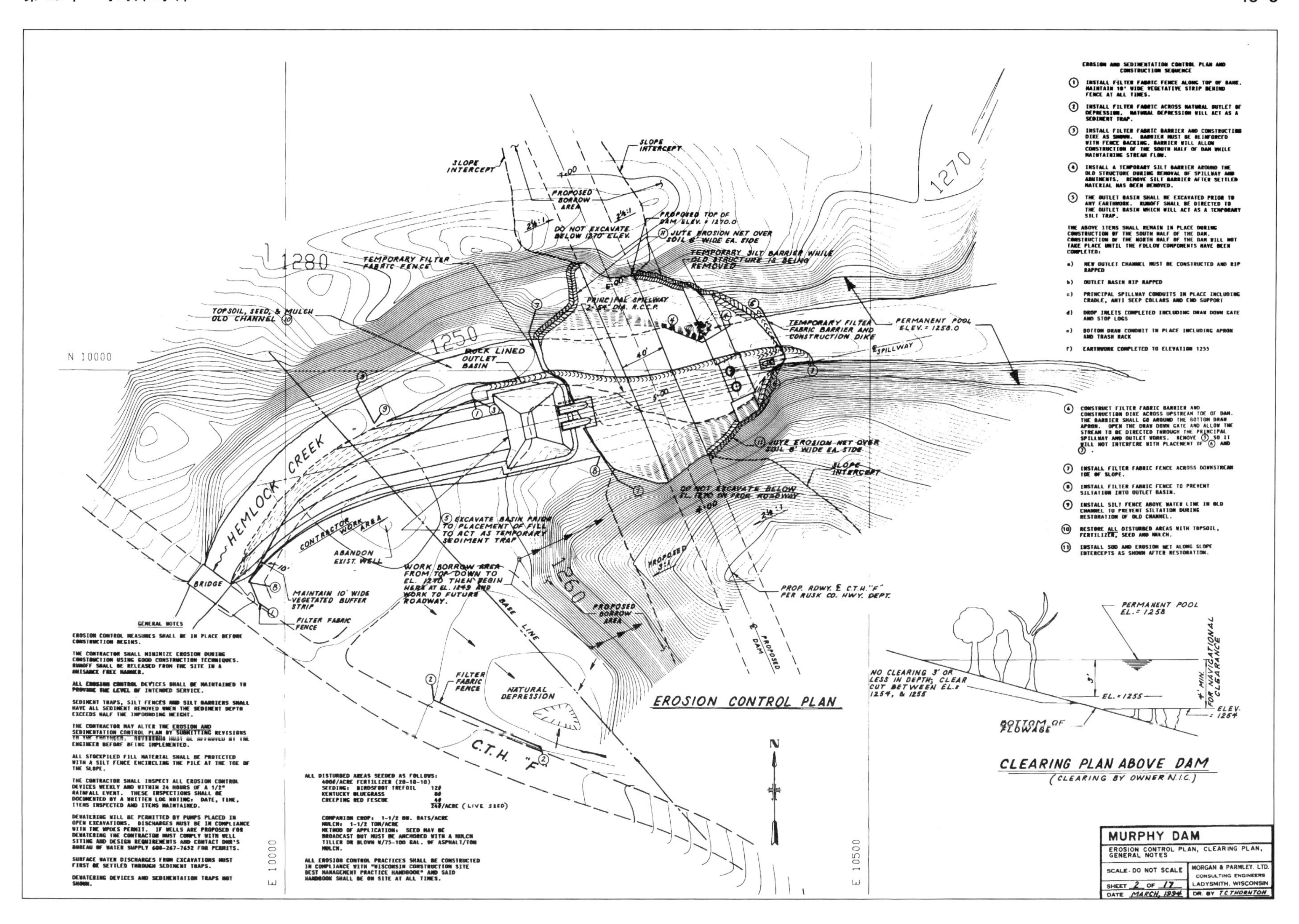
EROSION AND SEDIMENTATION CONTROL PLAN AND CONSTRUCTION SEQUENCE
① INSTALL FILTER FABRIC FENCE ALONG TOP OF BANK. MAINTAIN 10' WIDE VEGETATIVE STRIP BEHIND FENCE AT ALL TIMES.
② INSTALL FILTER FABRIC ACROSS NATURAL OUTLET OF DEPRESSION. NATURAL DEPRESSION WILL ACT AS A SEDIMENT TRAP.
③ INSTALL FILTER FABRIC BARRIER AND CONSTRUCTION DIKE AS SHOWN. BARRIER MUST BE REINFORCED WITH FENCE BACKING. BARRIER WILL ALLOW CONSTRUCTION OF THE SOUTH HALF OF DAM WHILE MAINTAINING STREAM FLOW.
④ INSTALL A TEMPORARY SILT BARRIER AROUND THE OLD STRUCTURE DURING REMOVAL OF SPILLWAY AND ABUTMENTS. REMOVE SILT BARRIER AFTER SETTLED MATERIAL HAS BEEN REMOVED.
⑤ THE OUTLET BASIN SHALL BE EXCAVATED PRIOR TO ANY EARTHWORK. RUNOFF SHALL BE DIRECTED TO THE OUTLET BASIN WHICH WILL ACT AS A TEMPORARY SILT TRAP.
THE ABOVE ITEMS SHALL REMAIN IN PLACE DURING CONSTRUCTION OF THE SOUTH HALF OF THE DAM. CONSTRUCTION OF THE NORTH HALF OF THE DAM WILL NOT TAKE PLACE UNTIL THE FOLLOW COMPONENTS HAVE BEEN COMPLETED:
a) NEW OUTLET CHANNEL MUST BE CONSTRUCTED AND RIP RAPPED
b) OUTLET BASIN RIP RAPPED
c) PRINCIPAL SPILLWAY CONDUITS IN PLACE INCLUDING CRADLE, ANTI SEEP COLLARS AND END SUPPORT
d) DROP INLETS COMPLETED INCLUDING DRAW DOWN GATE AND STOP LOGS
e) BOTTOM DRAW CONDUIT IN PLACE INCLUDING APRON AND TRASH RACK
f) EARTHWORK COMPLETED TO ELEVATION 1255
⑥ CONSTRUCT FILTER FABRIC BARRIER AND CONSTRUCTION DIKE ACROSS UPSTREAM TOE OF DAM. THE BARRIER SHALL GO AROUND THE BOTTOM DRAW APRON. OPEN THE DRAW DOWN GATE AND ALLOW THE STREAM TO BE DIRECTED THROUGH THE PRINCIPAL SPILLWAY AND OUTLET WORKS. REMOVE ③ SO IT WILL NOT INTERFERE WITH PLACEMENT OF ⑥ AND ⑦.
⑦ INSTALL FILTER FABRIC FENCE ACROSS DOWNSTREAM TOE OF SLOPE.
⑧ INSTALL FILTER FABRIC FENCE TO PREVENT SILTATION INTO OUTLET BASIN.
⑨ INSTALL SILT FENCE ABOVE WATER LINE IN OLD CHANNEL TO PREVENT SILTATION DURING RESTORATION OF OLD CHANNEL.
⑩ RESTORE ALL DISTURBED AREAS WITH TOPSOIL, FERTILIZER, SEED AND MULCH.
⑪ INSTALL SOD AND EROSION NET ALONG SLOPE INTERCEPTS AS SHOWN AFTER RESTORATION.
SLOPE INTERCEPT
PROPOSED BORROW AREA
DO NOT EXCAVATE BELOW 1270 ELEV.
PROPOSED TOP OF DAM ELEV. = 1270.0
JUTE EROSION NET OVER SOIL 8' WIDE EA. SIDE
TEMPORARY SILT BARRIER WHILE OLD STRUCTURE IS BEING REMOVED
TEMPORARY FILTER FABRIC FENCE
1280
1270
1250
1260
TOPSOIL, SEED, & MULCH OLD CHANNEL
PRINCIPAL SPILLWAY 2-54" DIA. R.C.C.P.
TEMPORARY FILTER FABRIC BARRIER AND CONSTRUCTION DIKE
PERMANENT POOL ELEV. = 1258.0
SPILLWAY
N 10000
ROCK LINED OUTLET BASIN
HEMLOCK CREEK
DO NOT EXCAVATE BELOW EL. 1270 ON PROP. ROADWAY
EXCAVATE BASIN PRIOR TO PLACEMENT OF FILL TO ACT AS TEMPORARY SEDIMENT TRAP
CONTRACTOR WORK AREA
ABANDON EXIST. WELL
WORK BORROW AREA FROM TOP DOWN TO EL. 1270 THEN BEGIN HERE AT EL. 1243 AND WORK TO FUTURE ROADWAY.
PROPOSED 3:1
PROP. RDWY. ℄ C.T.H. "F" PER RUSK CO. HWY. DEPT.
BRIDGE
MAINTAIN 10' WIDE VEGETATED BUFFER STRIP
FILTER FABRIC FENCE
BASE LINE
PROPOSED BORROW AREA
℄ PROPOSED DAM
NATURAL DEPRESSION
C.T.H. "F"
EROSION CONTROL PLAN
GENERAL NOTES
EROSION CONTROL MEASURES SHALL BE IN PLACE BEFORE CONSTRUCTION BEGINS.
THE CONTRACTOR SHALL MINIMIZE EROSION DURING CONSTRUCTION USING GOOD CONSTRUCTION TECHNIQUES. RUNOFF SHALL BE RELEASED FROM THE SITE IN A NUISANCE FREE MANNER.
ALL EROSION CONTROL DEVICES SHALL BE MAINTAINED TO PROVIDE THE LEVEL OF INTENDED SERVICE.
SEDIMENT TRAPS, SILT FENCES AND SILT BARRIERS SHALL HAVE ALL SEDIMENT REMOVED WHEN THE SEDIMENT DEPTH EXCEEDS HALF THE IMPOUNDING HEIGHT.
THE CONTRACTOR MAY ALTER THE EROSION AND SEDIMENTATION CONTROL PLAN BY SUBMITTING REVISIONS TO THE ENGINEER. REVISIONS MUST BE APPROVED BY THE ENGINEER BEFORE BEING IMPLEMENTED.
ALL STOCKPILED FILL MATERIAL SHALL BE PROTECTED WITH A SILT FENCE ENCIRCLING THE PILE AT THE TOE OF THE SLOPE.
THE CONTRACTOR SHALL INSPECT ALL EROSION CONTROL DEVICES WEEKLY AND WITHIN 24 HOURS OF A 1/2" RAINFALL EVENT. THESE INSPECTIONS SHALL BE DOCUMENTED BY A WRITTEN LOG NOTING: DATE, TIME, ITEMS INSPECTED AND ITEMS MAINTAINED.
DEWATERING WILL BE PERMITTED BY PUMPS PLACED IN OPEN EXCAVATIONS. DISCHARGES MUST BE IN COMPLIANCE WITH THE WPDES PERMIT. IF WELLS ARE PROPOSED FOR DEWATERING THE CONTRACTOR MUST COMPLY WITH WELL SITING AND DESIGN REQUIREMENTS AND CONTACT DNR'S BUREAU OF WATER SUPPLY 608-267-7652 FOR PERMITS.
SURFACE WATER DISCHARGES FROM EXCAVATIONS MUST FIRST BE SETTLED THROUGH SEDIMENT TRAPS.
DEWATERING DEVICES AND SEDIMENTATION TRAPS NOT SHOWN.
ALL DISTURBED AREAS SEEDED AS FOLLOWS:
400#/ACRE FERTILIZER (20-10-10)
SEEDING: BIRDSFOOT TREFOIL 12#
KENTUCKY BLUEGRASS 8#
CREEPING RED FESCUE 4#
24#/ACRE (LIVE SEED)
COMPANION CROP: 1-1/2 BU. OATS/ACRE
MULCH: 1-1/2 TON/ACRE
METHOD OF APPLICATION: SEED MAY BE BROADCAST BUT MUST BE ANCHORED WITH A MULCH TILLER OR BLOWN W/75-100 GAL. OF ASPHALT/TON MULCH.
ALL EROSION CONTROL PRACTICES SHALL BE CONSTRUCTED IN COMPLIANCE WITH "WISCONSIN CONSTRUCTION SITE BEST MANAGEMENT PRACTICE HANDBOOK" AND SAID HANDBOOK SHALL BE ON SITE AT ALL TIMES.
E 10000
E 10500
N
PERMANENT POOL EL. = 1258
NO CLEARING 3' OR LESS IN DEPTH; CLEAR CUT BETWEEN EL.s 1254, & 1255
EL. = 1255
ELEV. = 1254
4' MIN. FOR NAVIGATIONAL CLEARANCE
BOTTOM OF FLOWAGE
CLEARING PLAN ABOVE DAM
(CLEARING BY OWNER N.I.C.)
MURPHY DAM
EROSION CONTROL PLAN, CLEARING PLAN, GENERAL NOTES
SCALE: DO NOT SCALE
MORGAN & PARMLEY, LTD. CONSULTING ENGINEERS LADYSMITH, WISCONSIN
SHEET 2 OF 17
DATE MARCH, 1994
DR. BY T.C. THORNTON

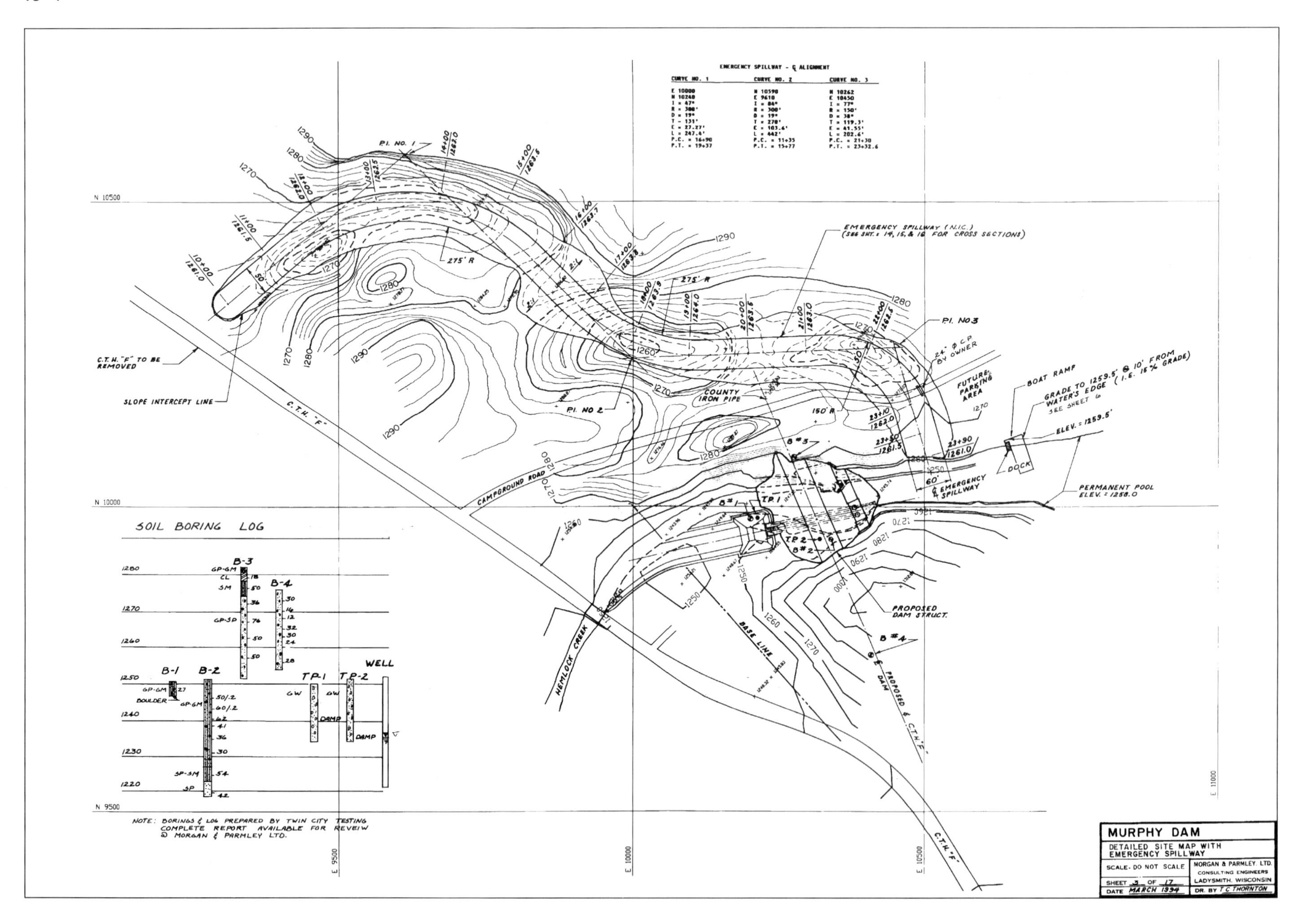
EMERGENCY SPILLWAY - ℄ ALIGNMENT
SOIL BORING LOG
NOTE: BORINGS & LOG PREPARED BY TWIN CITY TESTING COMPLETE REPORT AVAILABLE FOR REVIEW @ MORGAN & PARMLEY LTD.
MURPHY DAM
DETAILED SITE MAP WITH EMERGENCY SPILLWAY
SCALE- DO NOT SCALE
MORGAN & PARMLEY LTD. CONSULTING ENGINEERS LADYSMITH, WISCONSIN
SHEET 3 OF 17
DATE MARCH 1994
DR. BY T.C.THORNTON

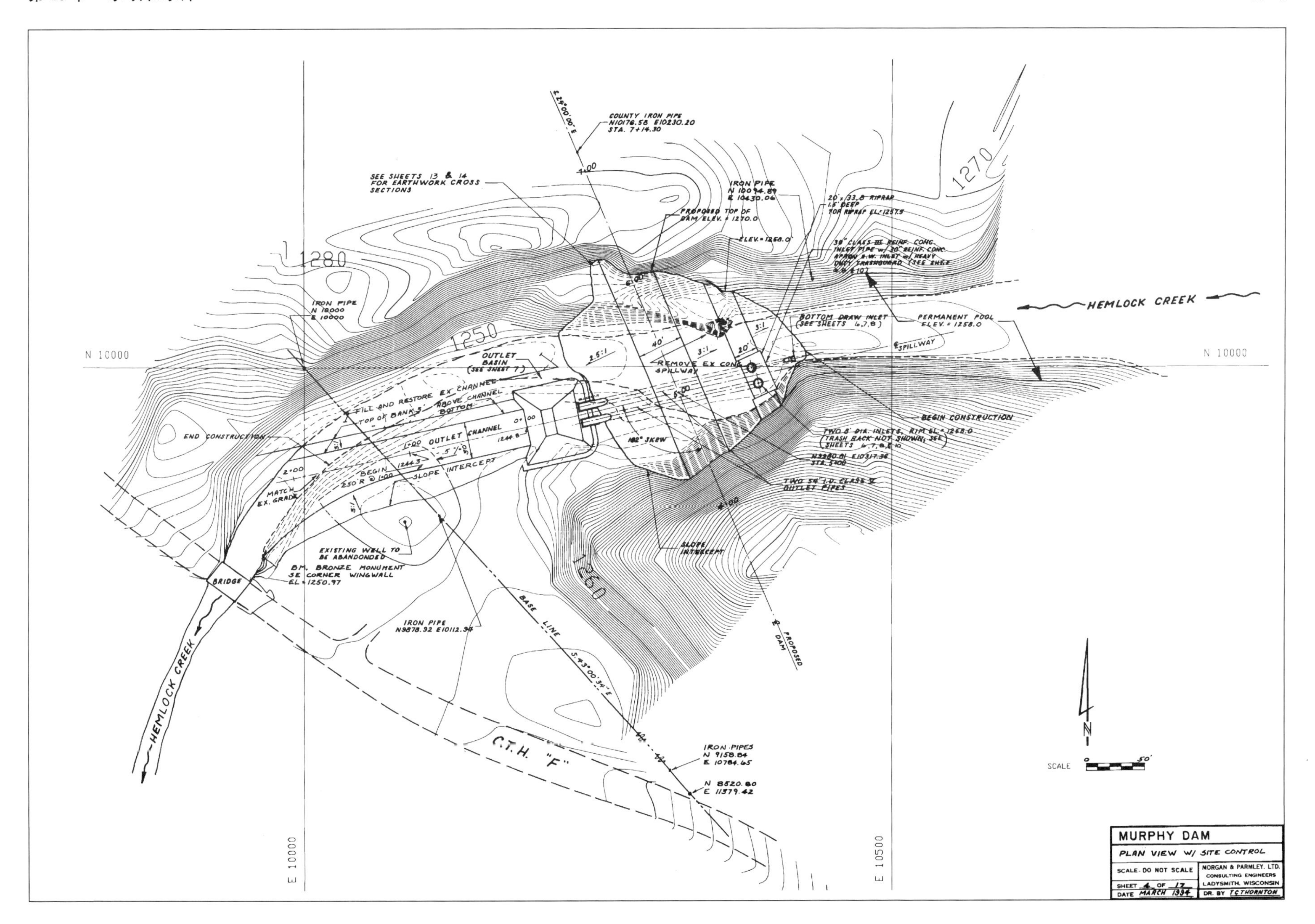

MURPHY DAM
PLAN VIEW W/ SITE CONTROL
SCALE- DO NOT SCALE
MORGAN & PARMLEY, LTD. CONSULTING ENGINEERS LADYSMITH, WISCONSIN
SHEET 4 OF 17
DATE MARCH 1994
DR. BY T.C. THORNTON
HEMLOCK CREEK
PERMANENT POOL ELEV. = 1258.0
BOTTOM DRAW INLET (SEE SHEETS 6,7,8)
BEGIN CONSTRUCTION
END CONSTRUCTION
OUTLET BASIN (SEE SHEET 7)
OUTLET CHANNEL
FILL AND RESTORE EX CHANNEL
REMOVE EX CONC SPILLWAY
SLOPE INTERCEPT
EXISTING WELL TO BE ABANDONED
BRIDGE
C.T.H. "F"
BASE LINE
PROPOSED DAM
N 10000
E 10000
E 10500
SCALE

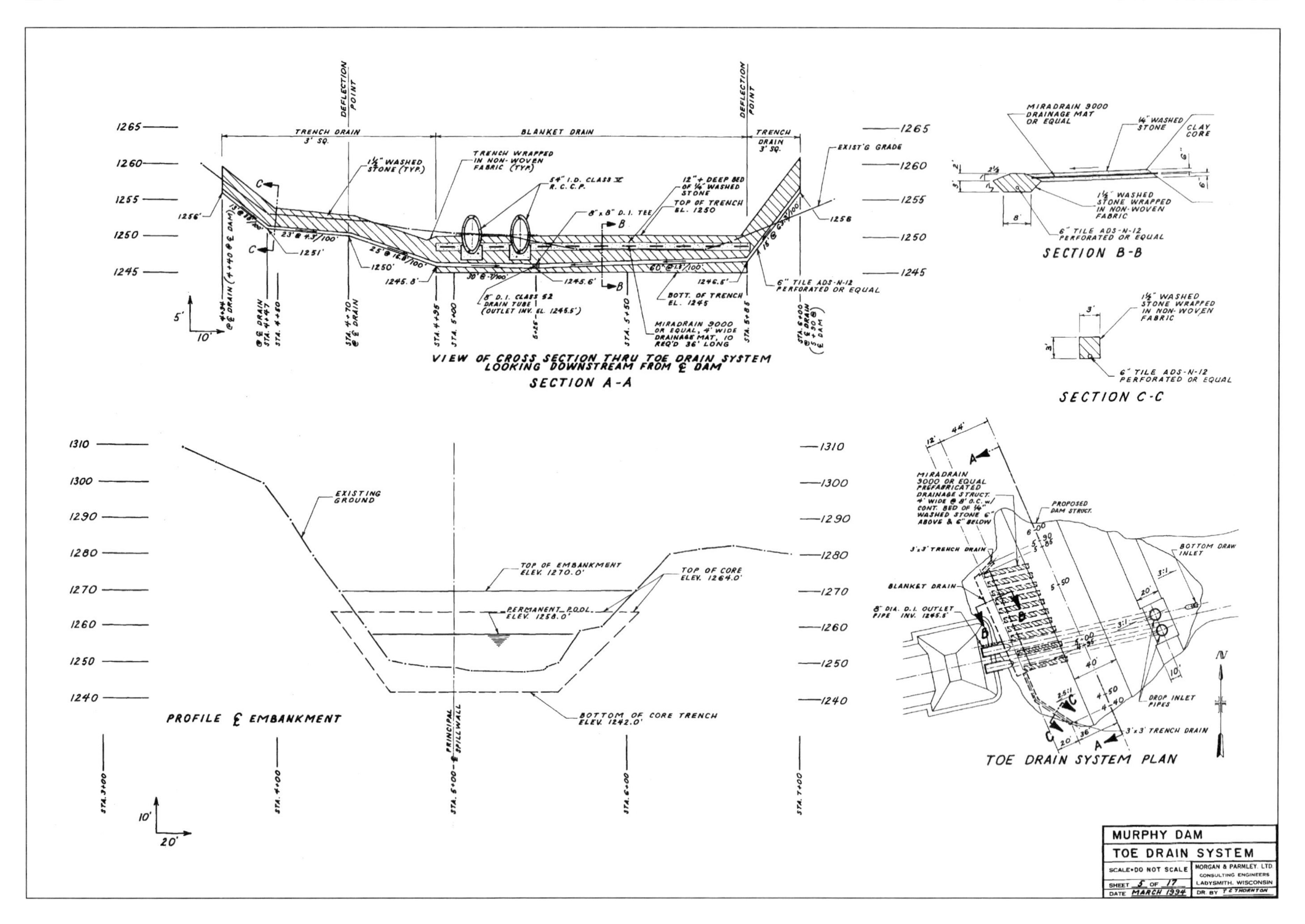
VIEW OF CROSS SECTION THRU TOE DRAIN SYSTEM LOOKING DOWNSTREAM FROM ℄ DAM
SECTION A-A
SECTION B-B
SECTION C-C
PROFILE ℄ EMBANKMENT
TOE DRAIN SYSTEM PLAN
MURPHY DAM
TOE DRAIN SYSTEM
SCALE=DO NOT SCALE
MORGAN & PARMLEY LTD.
CONSULTING ENGINEERS
LADYSMITH, WISCONSIN
SHEET 5 OF 17
DATE MARCH 1994

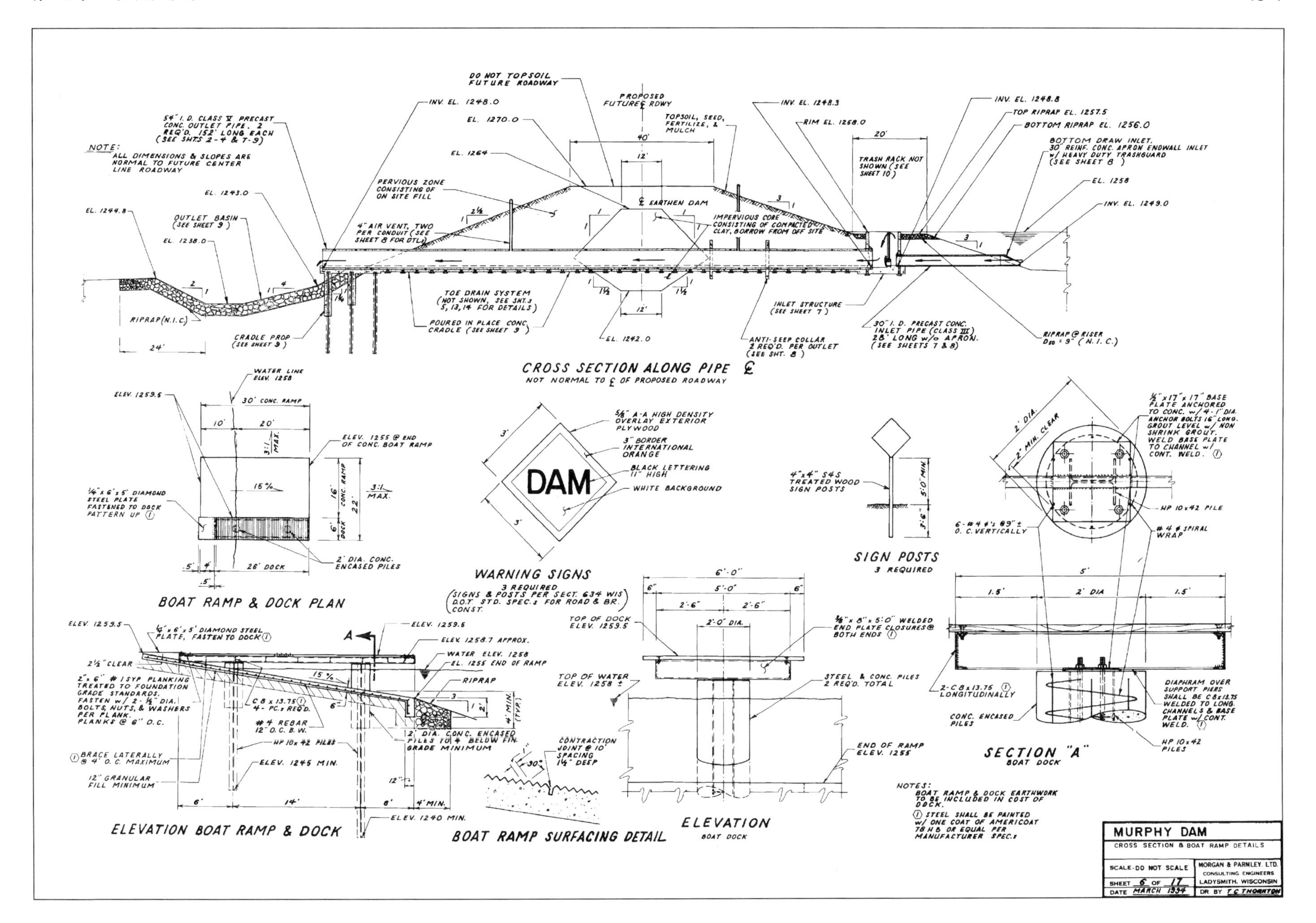
CROSS SECTION ALONG PIPE ℄
NOT NORMAL TO ℄ OF PROPOSED ROADWAY
BOAT RAMP & DOCK PLAN
WARNING SIGNS
3 REQUIRED
SIGN POSTS
3 REQUIRED
SECTION "A"
BOAT DOCK
ELEVATION BOAT RAMP & DOCK
BOAT RAMP SURFACING DETAIL
ELEVATION
BOAT DOCK
DAM
MURPHY DAM
CROSS SECTION & BOAT RAMP DETAILS
MORGAN & PARMLEY LTD.
CONSULTING ENGINEERS
LADYSMITH, WISCONSIN

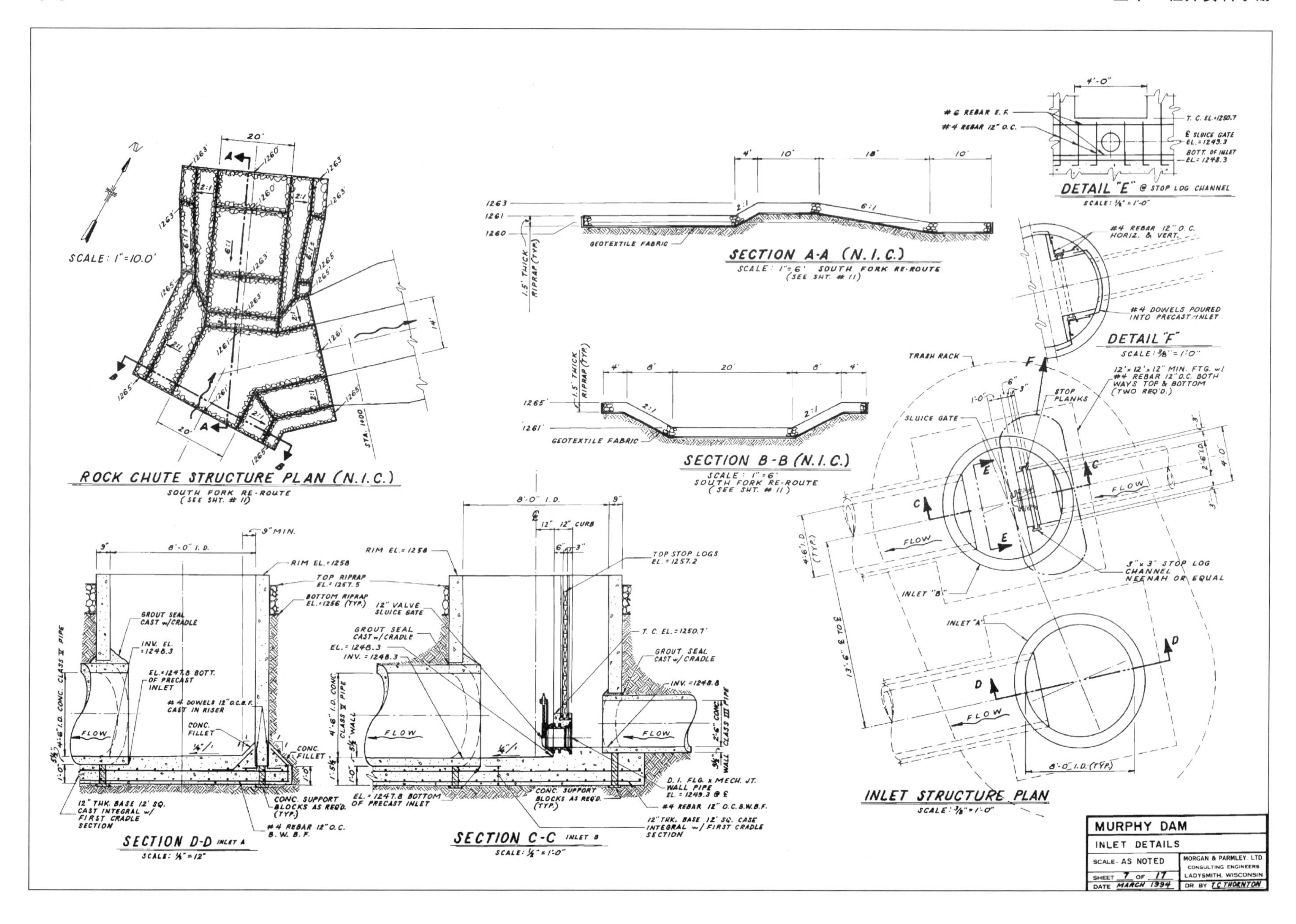
ROCK CHUTE STRUCTURE PLAN (N.I.C.)
SOUTH FORK RE-ROUTE
(SEE SHT. # 11)
SCALE: 1"=10.0'
SECTION A-A (N.I.C.)
SCALE: 1"=6' SOUTH FORK RE-ROUTE
(SEE SHT. # 11)
GEOTEXTILE FABRIC
1.5' THICK RIPRAP (TYP.)
SECTION B-B (N.I.C.)
SCALE: 1"=6'
SOUTH FORK RE-ROUTE
(SEE SHT. # 11)
DETAIL "E" @ STOP LOG CHANNEL
SCALE: 1/2"=1'-0"
6 REBAR E.F.
4 REBAR 12" O.C.
T.C. EL.=1250.7
℄ SLUICE GATE EL.=1249.3
BOTT. OF INLET EL.=1248.3
DETAIL "F"
SCALE: 3/8"=1'-0"
4 REBAR 12" O.C. HORIZ. & VERT.
4 DOWELS POURED INTO PRECAST INLET
TRASH RACK
SLUICE GATE
STOP PLANKS
12"x12"x12" MIN. FTG. w/ # 4 REBAR 12" O.C. BOTH WAYS TOP & BOTTOM (TWO REQ'D.)
3"x3" STOP LOG CHANNEL NEENAH OR EQUAL
INLET "B"
INLET "A"
FLOW
INLET STRUCTURE PLAN
SCALE: 3/8"=1'-0"
SECTION D-D INLET A
SCALE: 1/2"=12"
SECTION C-C INLET B
SCALE: 1/2"=1'-0"
RIM EL.=1258
TOP RIPRAP EL.=1257.5
BOTTOM RIPRAP EL.=1256 (TYP.)
GROUT SEAL CAST w/CRADLE
INV. EL.=1248.3
EL.=1247.8 BOTT. OF PRECAST INLET
# 4 DOWELS 12" O.C.B.F. CAST IN RISER
CONC. FILLET
12" THK. BASE 12' SQ. CAST INTEGRAL w/ FIRST CRADLE SECTION
CONC. SUPPORT BLOCKS AS REQ'D. (TYP.)
4 REBAR 12" O.C. B.W.B.F.
12" VALVE SLUICE GATE
TOP STOP LOGS EL.=1257.2
T.C. EL.=1250.7'
INV.=1248.8
D.I. FLG. x MECH. JT. WALL PIPE EL.=1249.3 @ ℄
MURPHY DAM
INLET DETAILS
SCALE AS NOTED
SHEET 7 OF 17
DATE MARCH 1994
MORGAN & PARMLEY, LTD.
CONSULTING ENGINEERS
LADYSMITH, WISCONSIN
DR. BY T.C. THORNTON

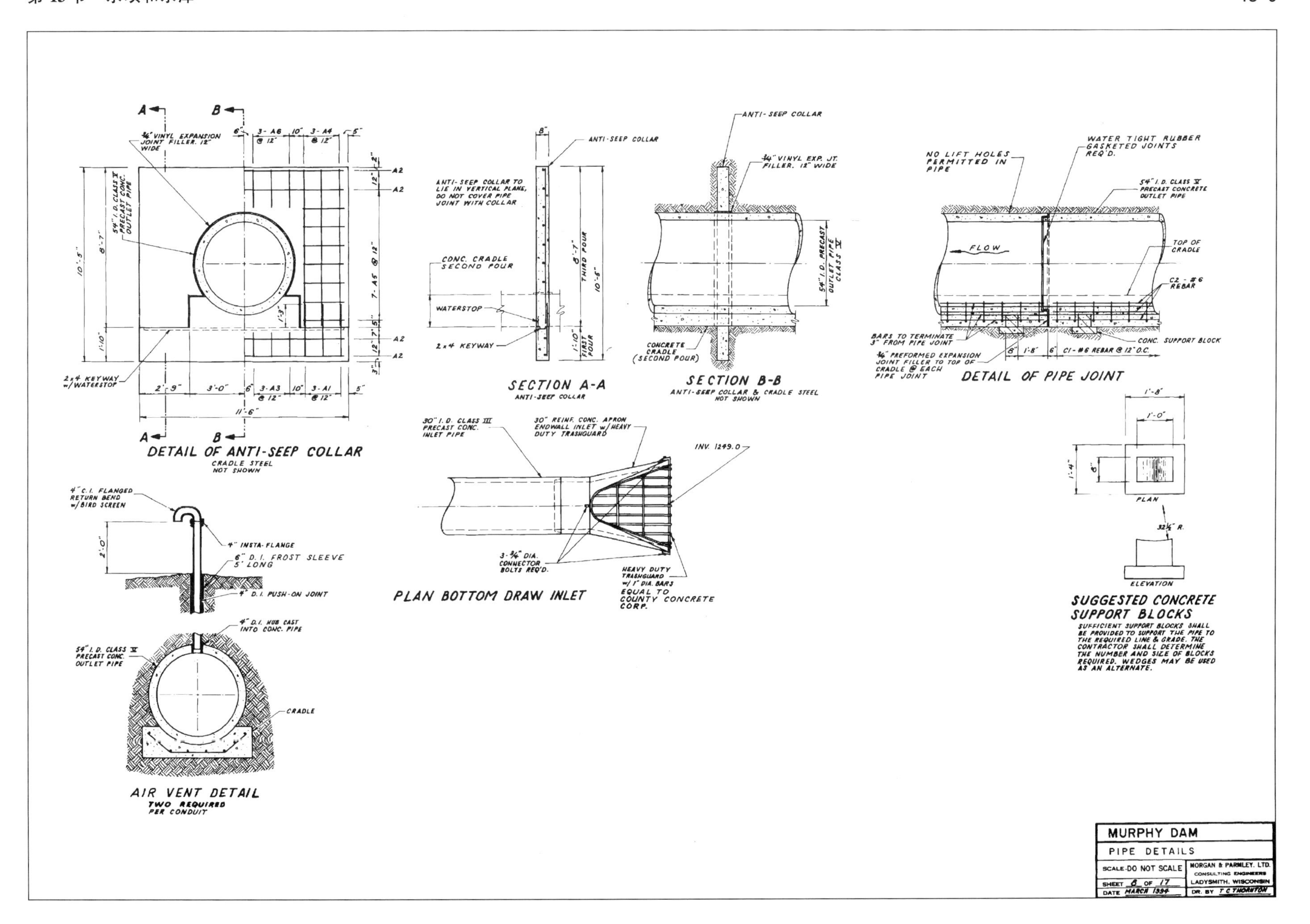
DETAIL OF ANTI-SEEP COLLAR
CRADLE STEEL NOT SHOWN
SECTION A-A
ANTI-SEEP COLLAR
SECTION B-B
ANTI-SEEP COLLAR & CRADLE STEEL NOT SHOWN
DETAIL OF PIPE JOINT
NO LIFT HOLES PERMITTED IN PIPE
WATER TIGHT RUBBER GASKETED JOINTS REQ'D.
54" I.D. CLASS V PRECAST CONCRETE OUTLET PIPE
FLOW
TOP OF CRADLE
CONC. SUPPORT BLOCK
PLAN BOTTOM DRAW INLET
INV. 1249.0
SUGGESTED CONCRETE SUPPORT BLOCKS
SUFFICIENT SUPPORT BLOCKS SHALL BE PROVIDED TO SUPPORT THE PIPE TO THE REQUIRED LINE & GRADE. THE CONTRACTOR SHALL DETERMINE THE NUMBER AND SIZE OF BLOCKS REQUIRED. WEDGES MAY BE USED AS AN ALTERNATE.
PLAN
ELEVATION
AIR VENT DETAIL
TWO REQUIRED PER CONDUIT
CRADLE
MURPHY DAM
PIPE DETAILS
SCALE-DO NOT SCALE
MORGAN & PARMLEY, LTD.
CONSULTING ENGINEERS
LADYSMITH, WISCONSIN
SHEET 8 OF 17
DATE MARCH 1994
DR. BY T C THORNTON

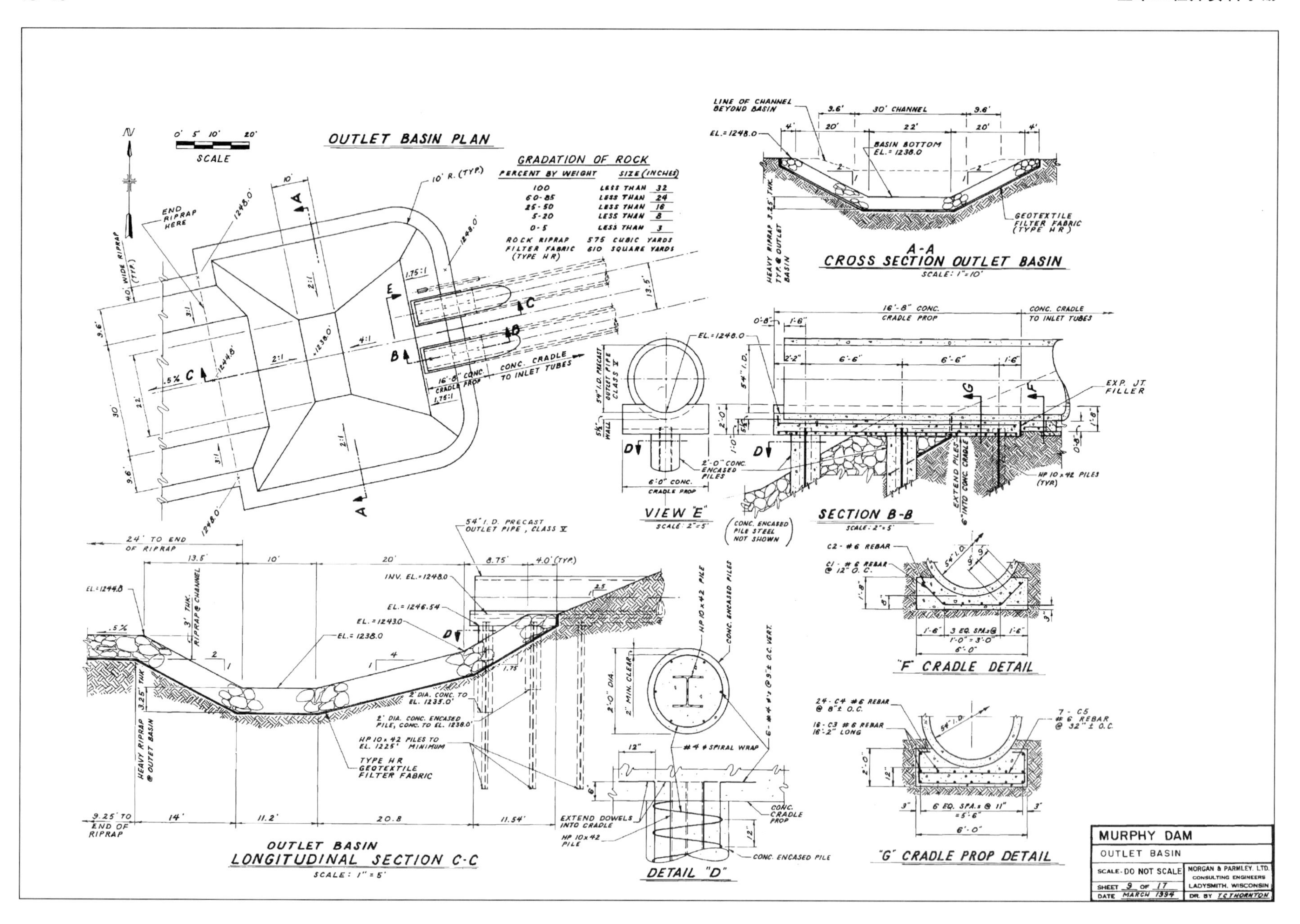
OUTLET BASIN PLAN
SCALE
GRADATION OF ROCK
PERCENT BY WEIGHT SIZE (INCHES)
100 LESS THAN 32
60-85 LESS THAN 24
25-50 LESS THAN 16
5-20 LESS THAN 8
0-5 LESS THAN 3
ROCK RIPRAP 575 CUBIC YARDS
FILTER FABRIC 610 SQUARE YARDS
(TYPE HR)
END RIPRAP HERE
10' R. (TYP.)
CONC. CRADLE TO INLET TUBES
16'-8" CONC. CRADLE PROP
A-A
CROSS SECTION OUTLET BASIN
SCALE: 1"=10'
LINE OF CHANNEL BEYOND BASIN
30' CHANNEL
BASIN BOTTOM EL.=1238.0
GEOTEXTILE FILTER FABRIC (TYPE HR)
HEAVY RIPRAP 3.25' THK. TYP. @ OUTLET BASIN
VIEW "E"
SCALE: 2"=5'
SECTION B-B
SCALE: 2"=5'
EXP. JT. FILLER
HP 10x42 PILES (TYP.)
(CONC. ENCASED PILE STEEL NOT SHOWN)
EXTEND PILES 6" INTO CONC. CRADLE
"F" CRADLE DETAIL
"G" CRADLE PROP DETAIL
OUTLET BASIN
LONGITUDINAL SECTION C-C
SCALE: 1"=5'
54" I.D. PRECAST OUTLET PIPE, CLASS V
TYPE HR GEOTEXTILE FILTER FABRIC
HP 10x42 PILES TO EL. 1225' MINIMUM
2' DIA. CONC. ENCASED PILE, CONC. TO EL. 1238.0'
DETAIL "D"
#4 φ SPIRAL WRAP
EXTEND DOWELS INTO CRADLE
CONC. ENCASED PILE
MURPHY DAM
OUTLET BASIN
SCALE - DO NOT SCALE
MORGAN & PARMLEY, LTD. CONSULTING ENGINEERS LADYSMITH, WISCONSIN
SHEET 9 OF 17
DATE MARCH 1994
DR. BY T.C.THORNTON

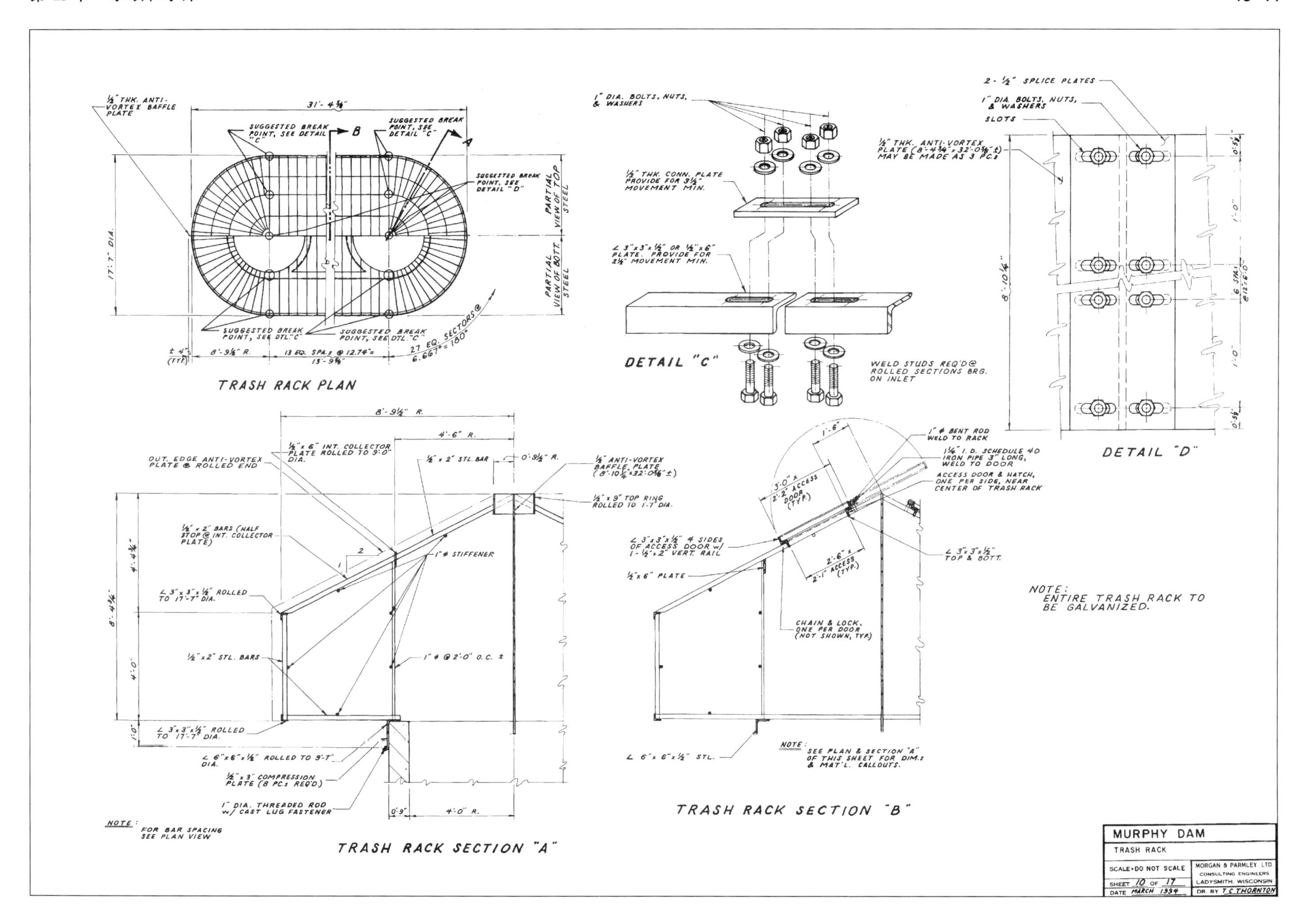
TRASH RACK PLAN
DETAIL "C"
DETAIL "D"
TRASH RACK SECTION "A"
TRASH RACK SECTION "B"
NOTE: ENTIRE TRASH RACK TO BE GALVANIZED.
MURPHY DAM
TRASH RACK
SCALE-DO NOT SCALE
MORGAN & PARMLEY LTD
CONSULTING ENGINEERS
LADYSMITH, WISCONSIN
SHEET 10 OF 17
DATE MARCH 1994
DR. BY T.C.THORNTON

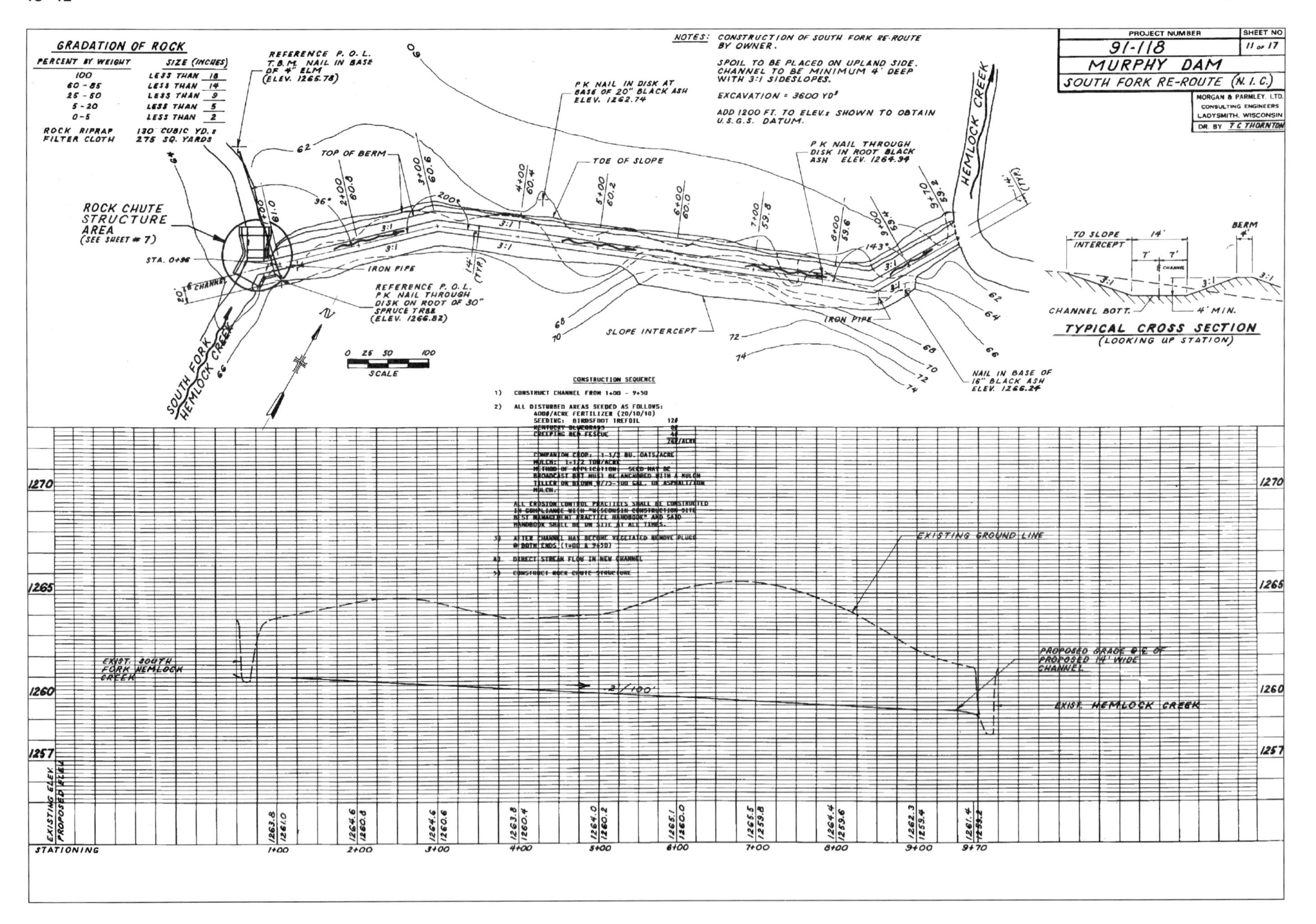
GRADATION OF ROCK
PERCENT BY WEIGHT
SIZE (INCHES)
100 LESS THAN 18
60-85 LESS THAN 14
25-50 LESS THAN 9
5-20 LESS THAN 5
0-5 LESS THAN 2
ROCK RIPRAP 130 CUBIC YD.s
FILTER CLOTH 275 SQ. YARDS
NOTES: CONSTRUCTION OF SOUTH FORK RE-ROUTE BY OWNER.
SPOIL TO BE PLACED ON UPLAND SIDE. CHANNEL TO BE MINIMUM 4' DEEP WITH 3:1 SIDESLOPES.
EXCAVATION = 3600 YD³
ADD 1200 FT. TO ELEV.s SHOWN TO OBTAIN U.S.G.S. DATUM.
PROJECT NUMBER 91-118
SHEET NO 11 of 17
MURPHY DAM
SOUTH FORK RE-ROUTE (N.I.C.)
MORGAN & PARMLEY, LTD. CONSULTING ENGINEERS LADYSMITH, WISCONSIN
DR. BY T C THORNTON
ROCK CHUTE STRUCTURE AREA (SEE SHEET # 7)
REFERENCE P.O.L. T.B.M. NAIL IN BASE OF 4" ELM (ELEV. 1265.78)
P K NAIL IN DISK AT BASE OF 20" BLACK ASH ELEV. 1262.74
P K NAIL THROUGH DISK IN ROOT BLACK ASH ELEV. 1264.94
REFERENCE P.O.L. P K NAIL THROUGH DISK ON ROOT OF 30" SPRUCE TREE (ELEV. 1266.82)
NAIL IN BASE OF 16" BLACK ASH ELEV. 1266.24
TOP OF BERM
TOE OF SLOPE
SLOPE INTERCEPT
IRON PIPE
STA. 0+96
HEMLOCK CREEK
SOUTH FORK HEMLOCK CREEK
SCALE
TYPICAL CROSS SECTION (LOOKING UP STATION)
TO SLOPE INTERCEPT
BERM
CHANNEL BOTT.
4' MIN.
CONSTRUCTION SEQUENCE
1) CONSTRUCT CHANNEL FROM 1+00 - 9+50
2) ALL DISTURBED AREAS SEEDED AS FOLLOWS:
SEEDING: BIRDSFOOT TREFOIL
KENTUCKY BLUEGRASS
CREEPING RED FESCUE
EXISTING GROUND LINE
EXIST. SOUTH FORK HEMLOCK CREEK
PROPOSED GRADE ℄ OF PROPOSED 14' WIDE CHANNEL
EXIST. HEMLOCK CREEK
2/100'
EXISTING ELEV.
PROPOSED ELEV.
STATIONING
1+00 2+00 3+00 4+00 5+00 6+00 7+00 8+00 9+00 9+70
1263.8 1261.0
1264.6 1260.8
1264.6 1260.6
1263.8 1260.4
1264.0 1260.2
1265.1 1260.0
1265.5 1259.8
1264.4 1259.6
1262.3 1259.4
1261.4 1259.2
1270 1265 1260 1257

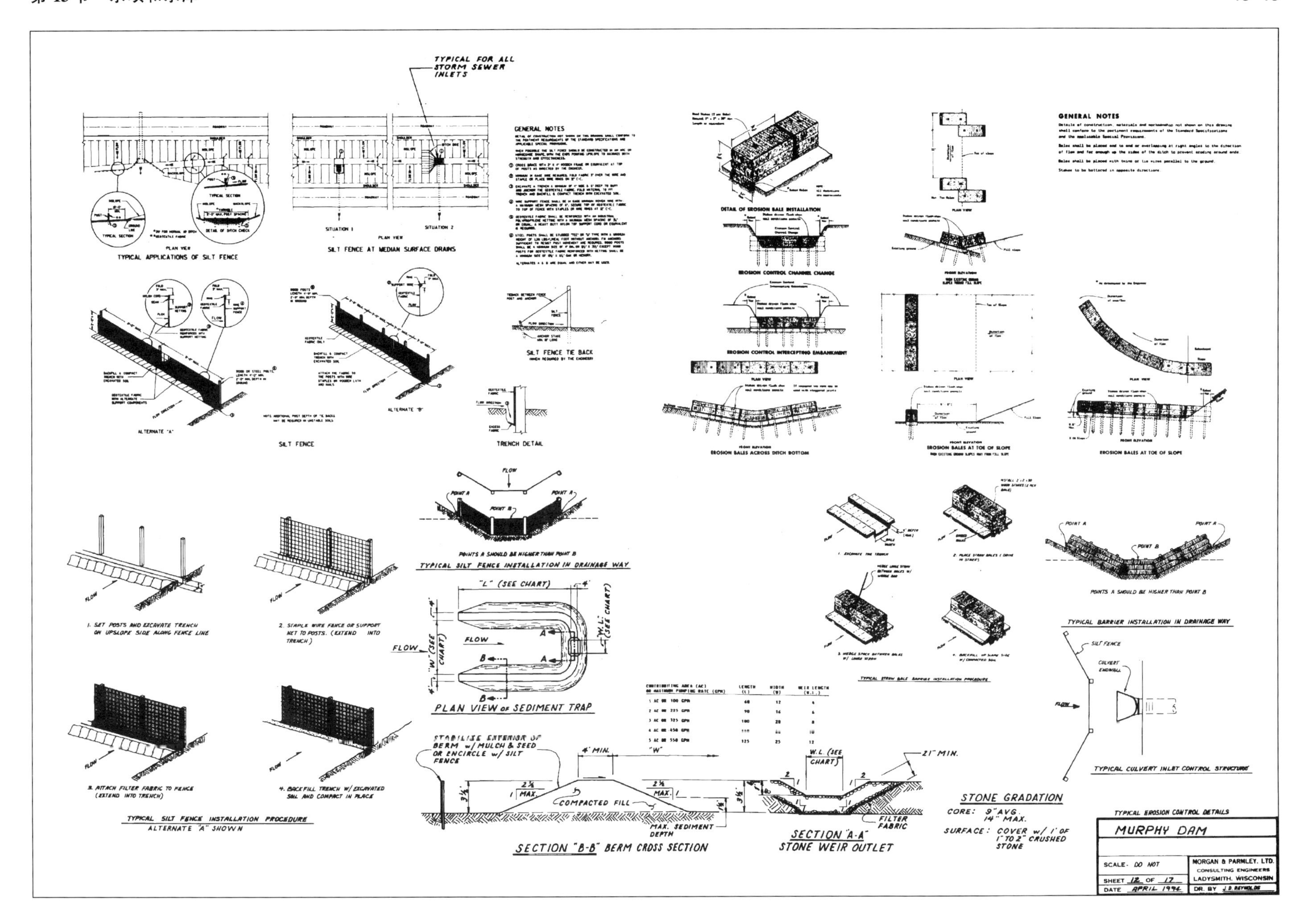

CONTRIBUTING AREA (AC) OR MAXIMUM PUMPING RATE (GPM)	LENGTH (L)	WIDTH (W)	WEIR LENGTH (W.L.)
1 AC OR 100 GPM	60	12	4
2 AC OR 225 GPM	90	16	6
3 AC OR 325 GPM	100	20	8
4 AC OR 450 GPM	[illegible]	[illegible]	10
5 AC OR 550 GPM	125	25	12

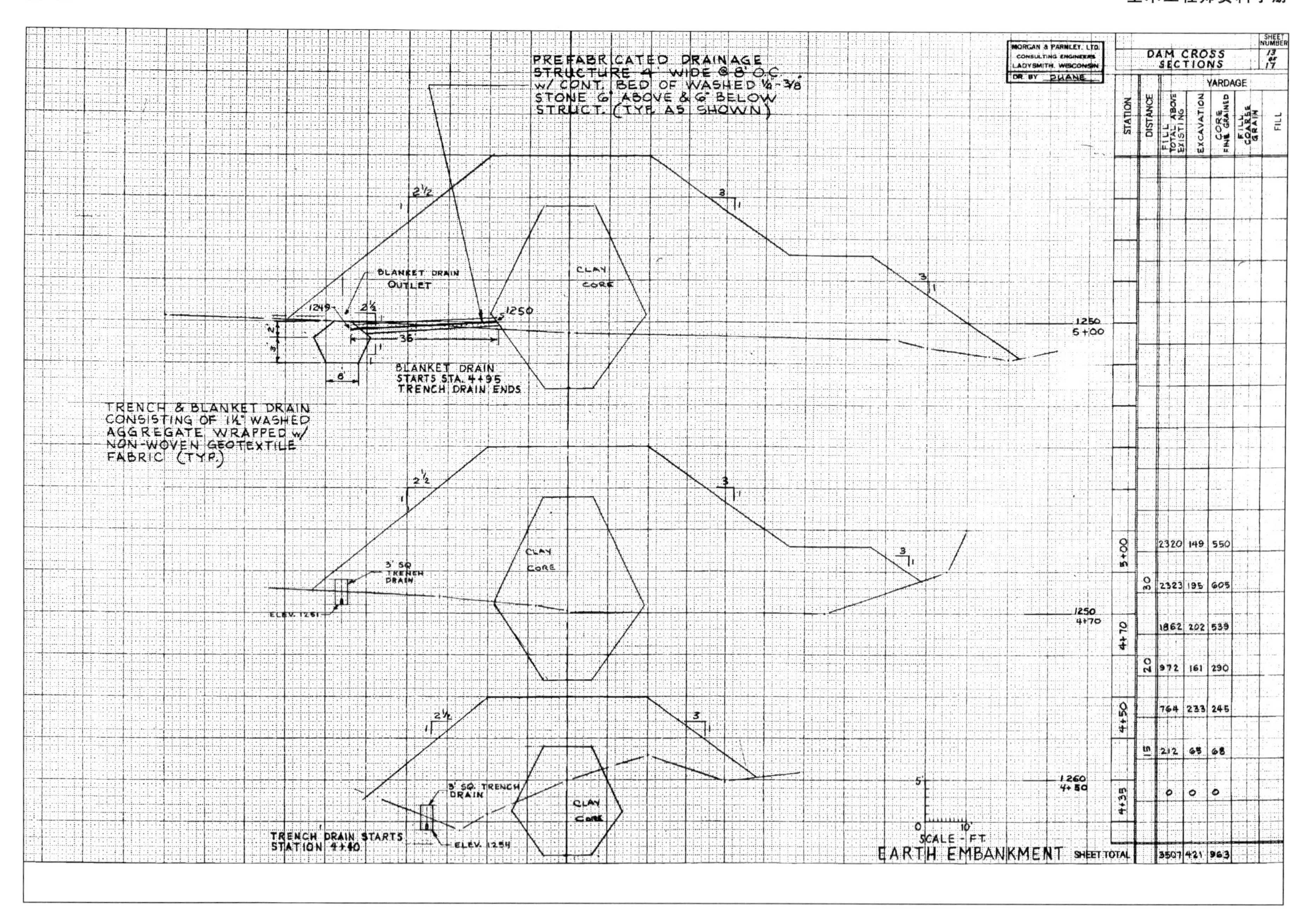
MORGAN & PARMLEY, LTD.
CONSULTING ENGINEERS
LADYSMITH, WISCONSIN
DR. BY DUANE
DAM CROSS SECTIONS
SHEET NUMBER 13 OF 17
YARDAGE
STATION
DISTANCE
FILL TOTAL ABOVE EXISTING
EXCAVATION
CORE FINE GRAINED
FILL COARSE GRAIN
FILL
5+00
30
4+70
20
4+50
15
4+35
2320 149 550
2323 195 605
1862 202 539
972 161 290
764 233 245
212 65 68
0 0 0
SHEET TOTAL
3507 421 963
PREFABRICATED DRAINAGE STRUCTURE 4' WIDE @ 8' O.C. w/ CONT. BED OF WASHED 1/2-3/8 STONE 6" ABOVE & 6" BELOW STRUCT. (TYP. AS SHOWN)
BLANKET DRAIN OUTLET
CLAY CORE
1249
1250
35
BLANKET DRAIN STARTS STA. 4+95
TRENCH DRAIN ENDS
1250
5+00
TRENCH & BLANKET DRAIN CONSISTING OF 1½" WASHED AGGREGATE WRAPPED w/ NON-WOVEN GEOTEXTILE FABRIC (TYP.)
3' SQ. TRENCH DRAIN
ELEV. 1251
1250
4+70
3' SQ. TRENCH DRAIN
ELEV. 1254
TRENCH DRAIN STARTS STATION 4+40
1260
4+50
SCALE - FT.
EARTH EMBANKMENT

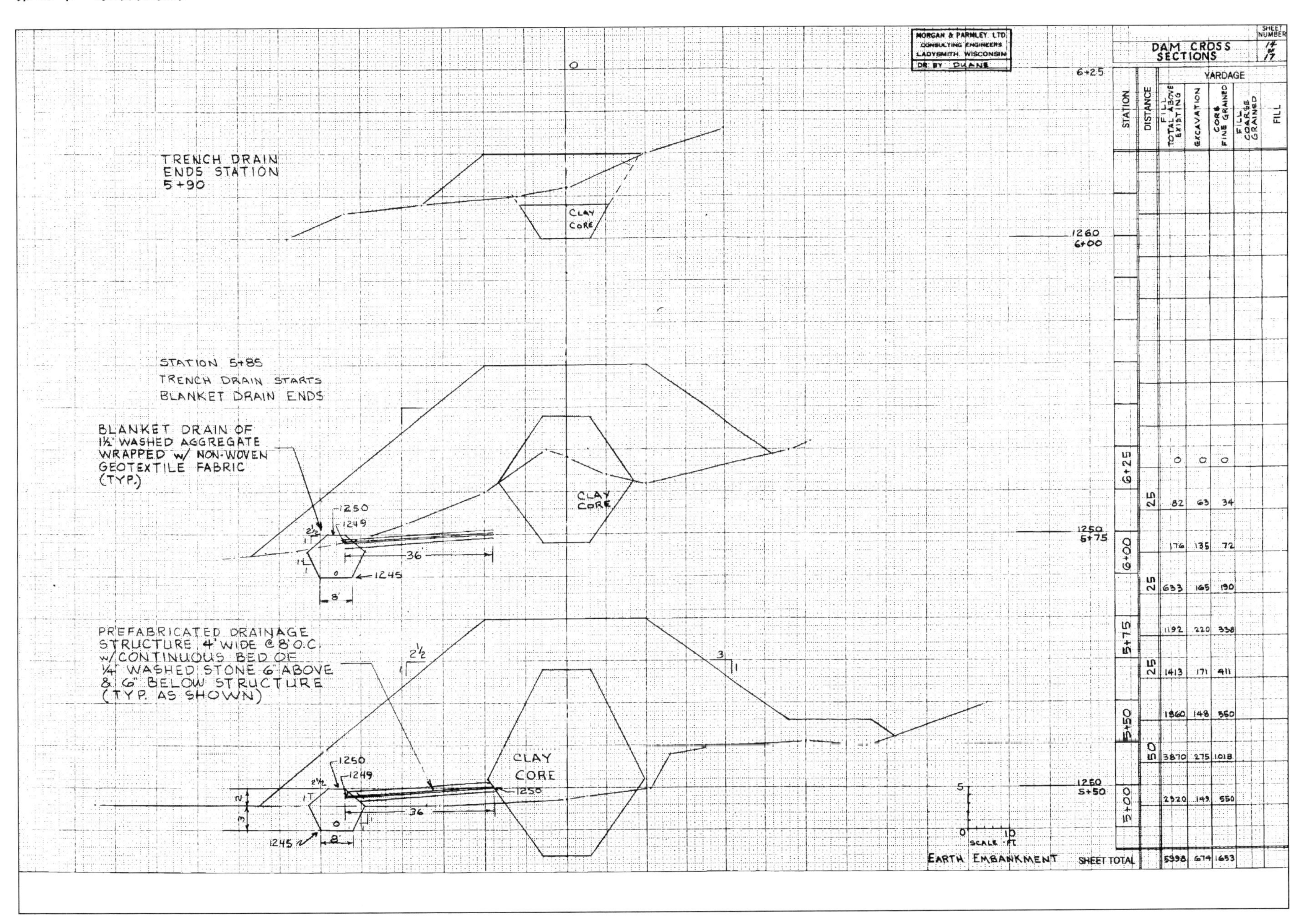

STATION	DISTANCE	FILL TOTAL ABOVE EXISTING	EXCAVATION	CORE FINE GRAINED	FILL COARSE GRAINED	FILL
6+25		0	0	0		
	25	82	63	34		
6+00		176	135	72		
	25	633	165	190		
5+75		1192	220	338		
	25	1413	171	411		
5+50		1860	148	560		
	50	3870	275	1018		
5+00		2920	143	550		
SHEET TOTAL		5998	674	1653		

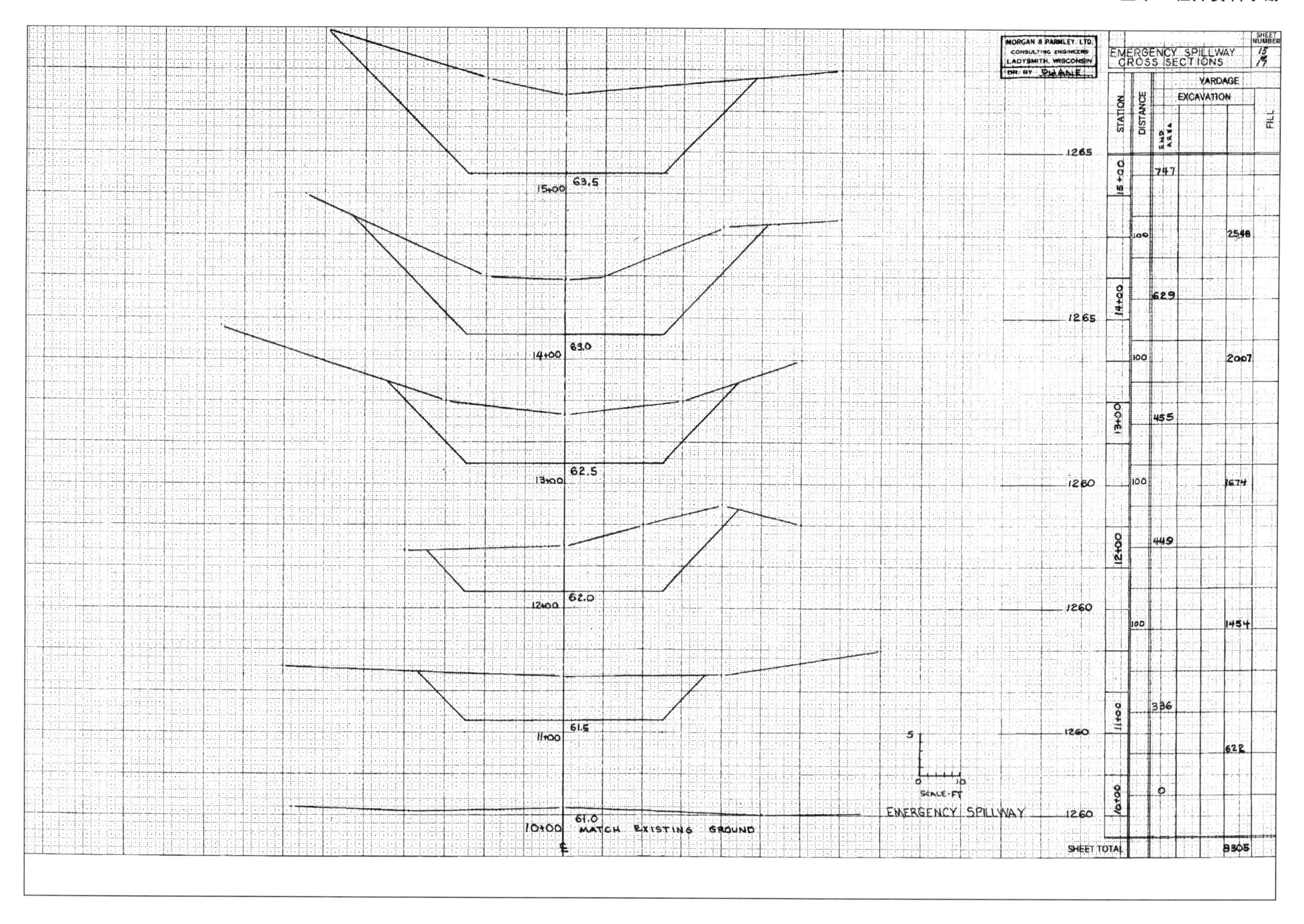
MORGAN & PARMLEY, LTD.
CONSULTING ENGINEERS
LADYSMITH, WISCONSIN
DR. BY DUANE
EMERGENCY SPILLWAY
CROSS SECTIONS
SHEET NUMBER
STATION
DISTANCE
YARDAGE
EXCAVATION
END AREA
FILL
15+00
14+00
13+00
12+00
11+00
10+00
747
629
455
449
336
0
100
2548
2007
1674
1454
622
SHEET TOTAL
8305
1265
1260
63.5
63.0
62.5
62.0
61.5
61.0
10+00 MATCH EXISTING GROUND
EMERGENCY SPILLWAY
SCALE-FT

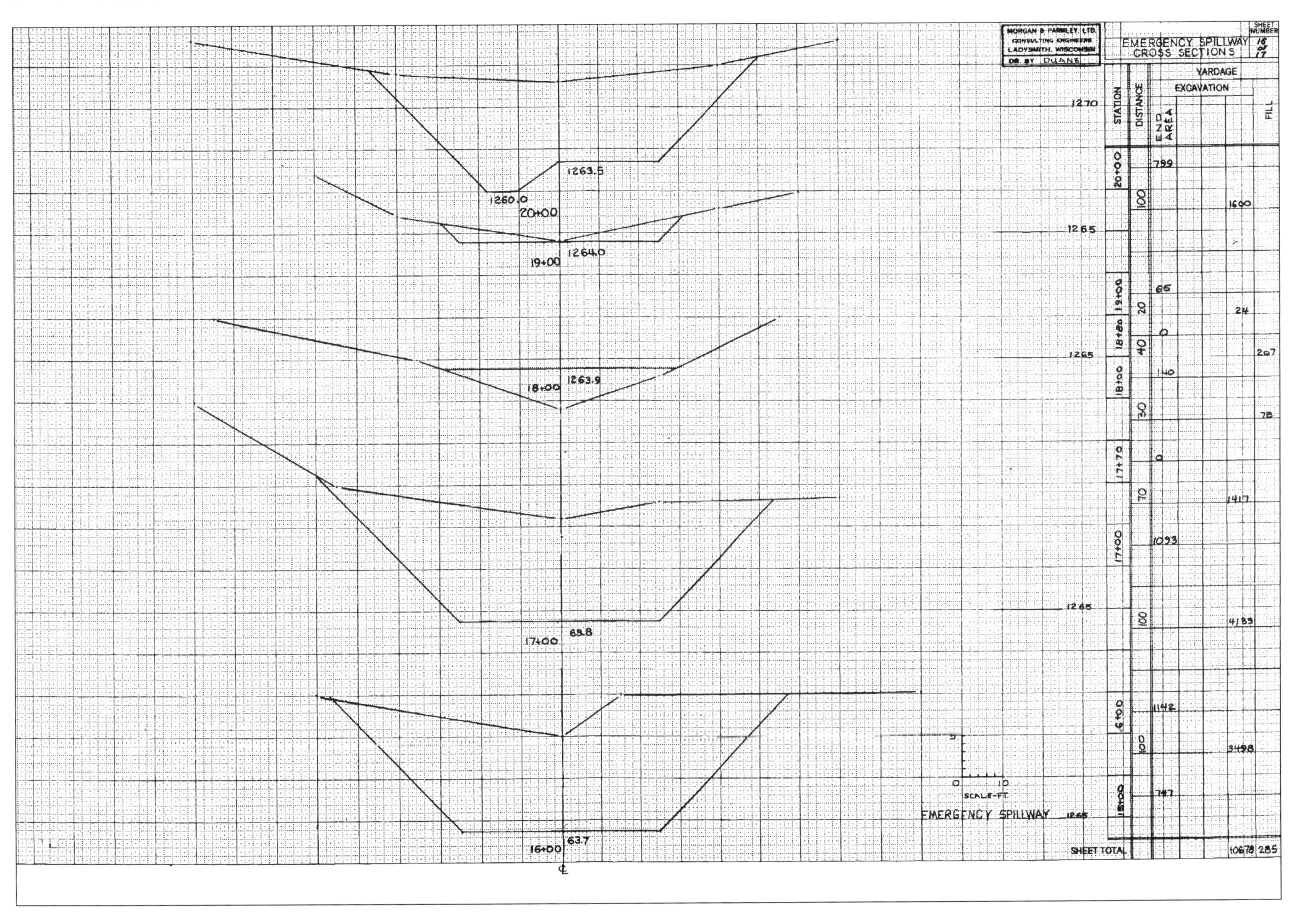
MORGAN & PARMLEY LTD.
CONSULTING ENGINEERS
LADYSMITH, WISCONSIN
DR. BY DUANE
EMERGENCY SPILLWAY
CROSS SECTIONS
SHEET NUMBER
YARDAGE
EXCAVATION
FILL
STATION
DISTANCE
END AREA
1260.0
20+00
1263.5
19+00
1264.0
18+00
1263.9
17+00
63.8
16+00
63.7
1270
1265
SCALE-FT.
EMERGENCY SPILLWAY
SHEET TOTAL

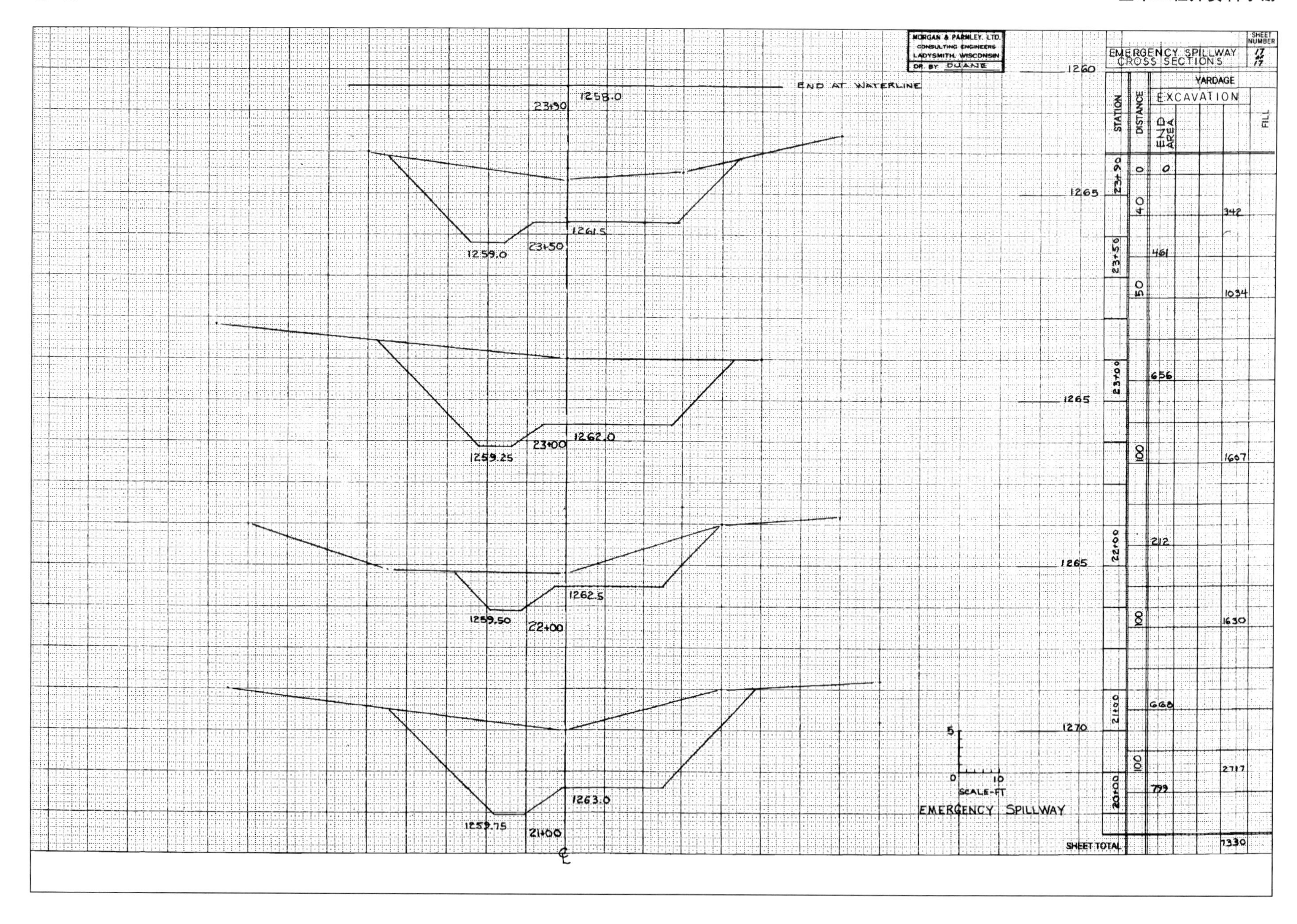
MORGAN & PARMLEY, LTD.
CONSULTING ENGINEERS
LADYSMITH, WISCONSIN
DR. BY DUANE
EMERGENCY SPILLWAY
CROSS SECTIONS
SHEET NUMBER
17
17
YARDAGE
STATION
DISTANCE
EXCAVATION
END AREA
FILL
END AT WATERLINE
1260
1265
1270
23+90
1258.0
1261.5
1259.0
23+50
1262.0
1259.25
23+00
1262.5
1259.50
22+00
1263.0
1259.75
21+00
SCALE-FT
EMERGENCY SPILLWAY
SHEET TOTAL
342
1034
1667
1630
2717
7330
461
656
212
668
799

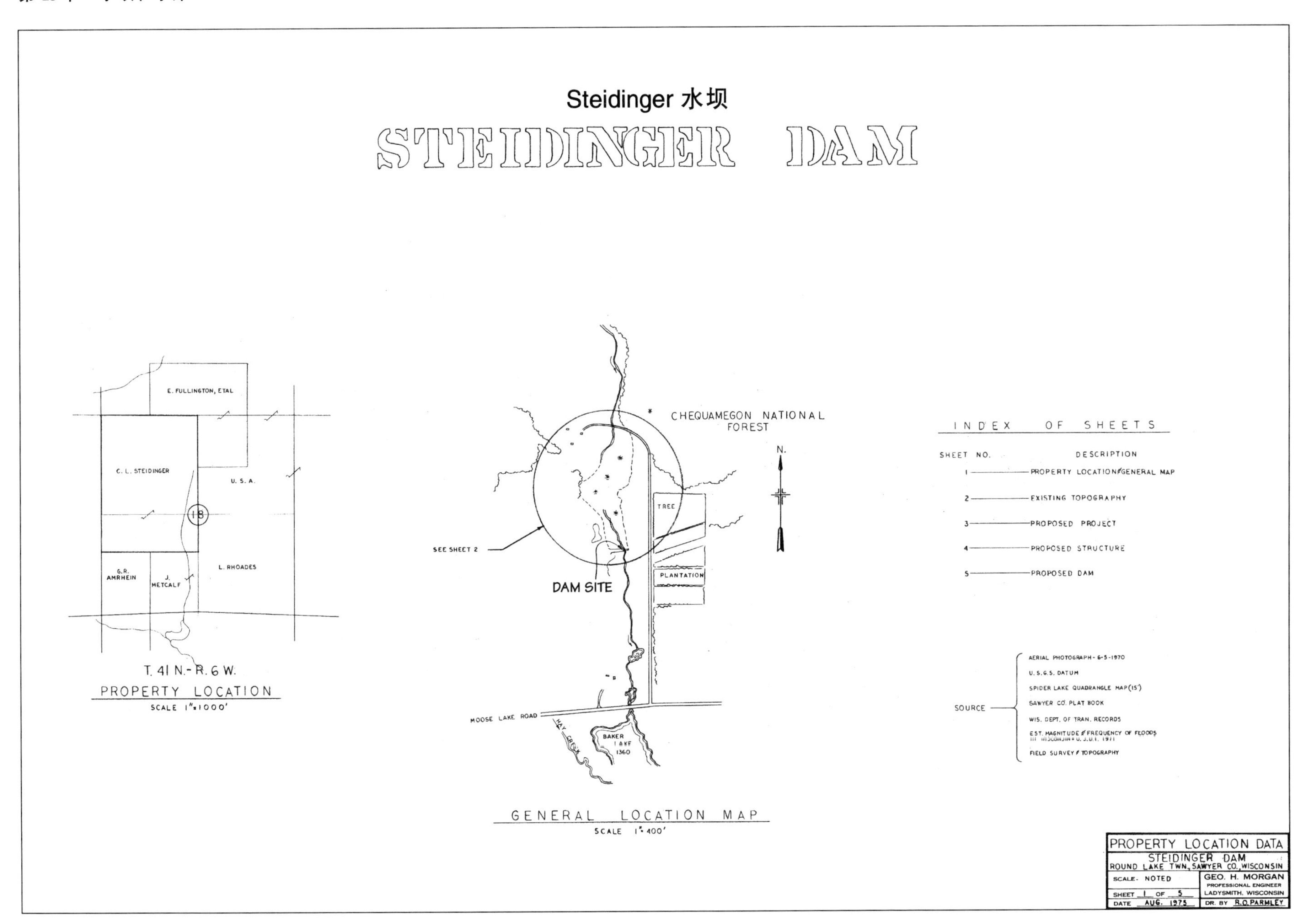

Source: Morgan & Parmley, Ltd.

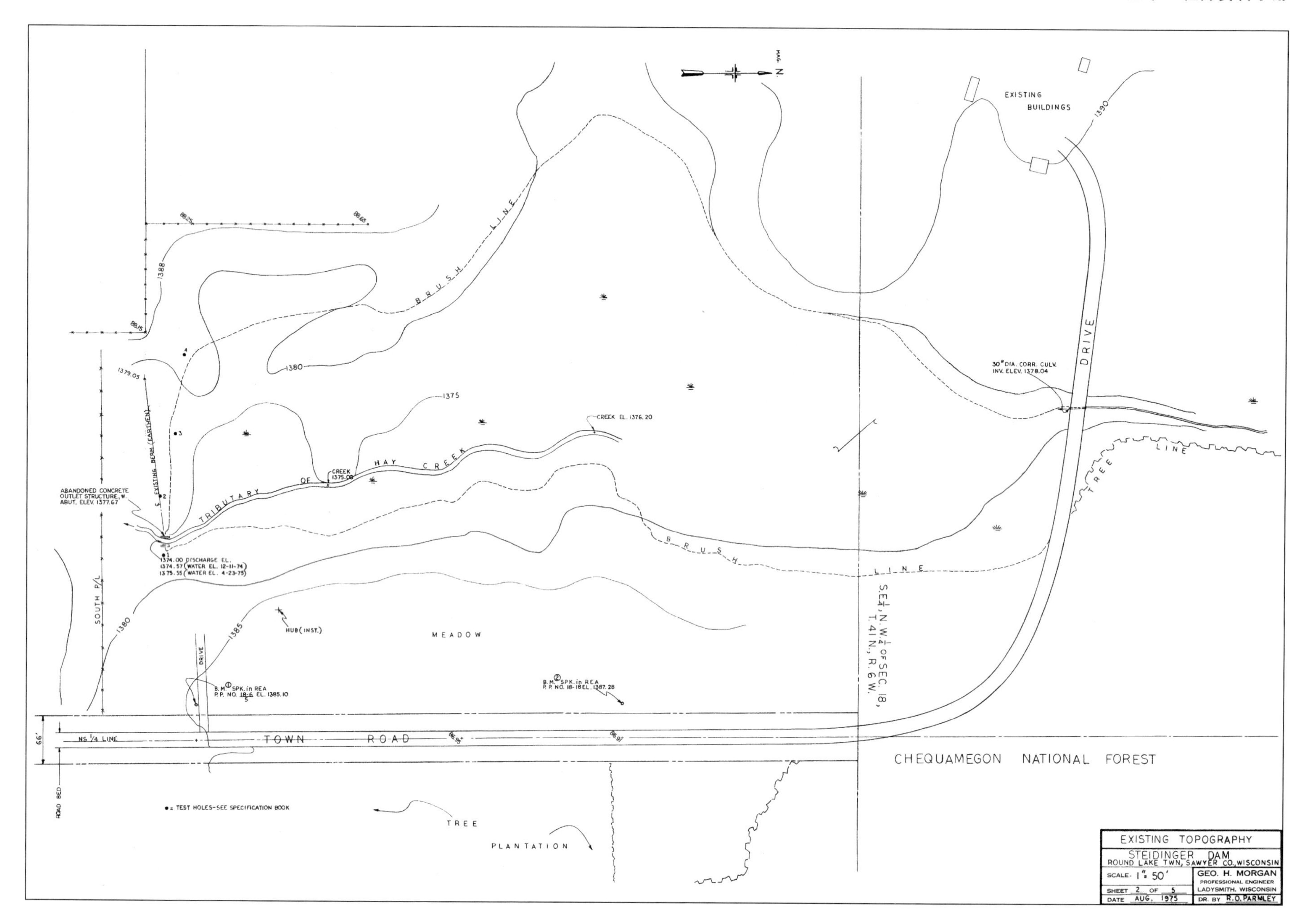
MAG. N.
EXISTING BUILDINGS
1390
1388
1380
1375
BRUSH LINE
CREEK EL. 1376.20
CREEK 1375.00
TRIBUTARY OF HAY CREEK
ABANDONED CONCRETE OUTLET STRUCTURE, W. ABUT. ELEV. 1377.67
EXISTING BERM (EARTHEN)
1374.00 DISCHARGE EL.
1374.57 (WATER EL. 12-11-74)
1375.55 (WATER EL. 4-23-75)
30" DIA. CORR. CULV. INV. ELEV. 1378.04
DRIVE
TREE LINE
SOUTH P/L
1385
HUB (INST.)
MEADOW
B.M. ① SPK. in REA P.P. NO. 18-6 EL. 1385.10
B.M. ② SPK. in REA P.P. NO. 18-18 EL. 1387.28
S.E.¼, N.W.¼ OF SEC. 18, T.41N., R.6W.
66'
NS ¼ LINE
TOWN ROAD
CHEQUAMEGON NATIONAL FOREST
ROAD BED
• = TEST HOLES-SEE SPECIFICATION BOOK
TREE PLANTATION
EXISTING TOPOGRAPHY
STEIDINGER DAM
ROUND LAKE TWN, SAWYER CO. WISCONSIN
SCALE- 1" = 50'
GEO. H. MORGAN
PROFESSIONAL ENGINEER
LADYSMITH, WISCONSIN
SHEET 2 OF 5
DATE AUG. 1975
DR. BY R.O. PARMLEY

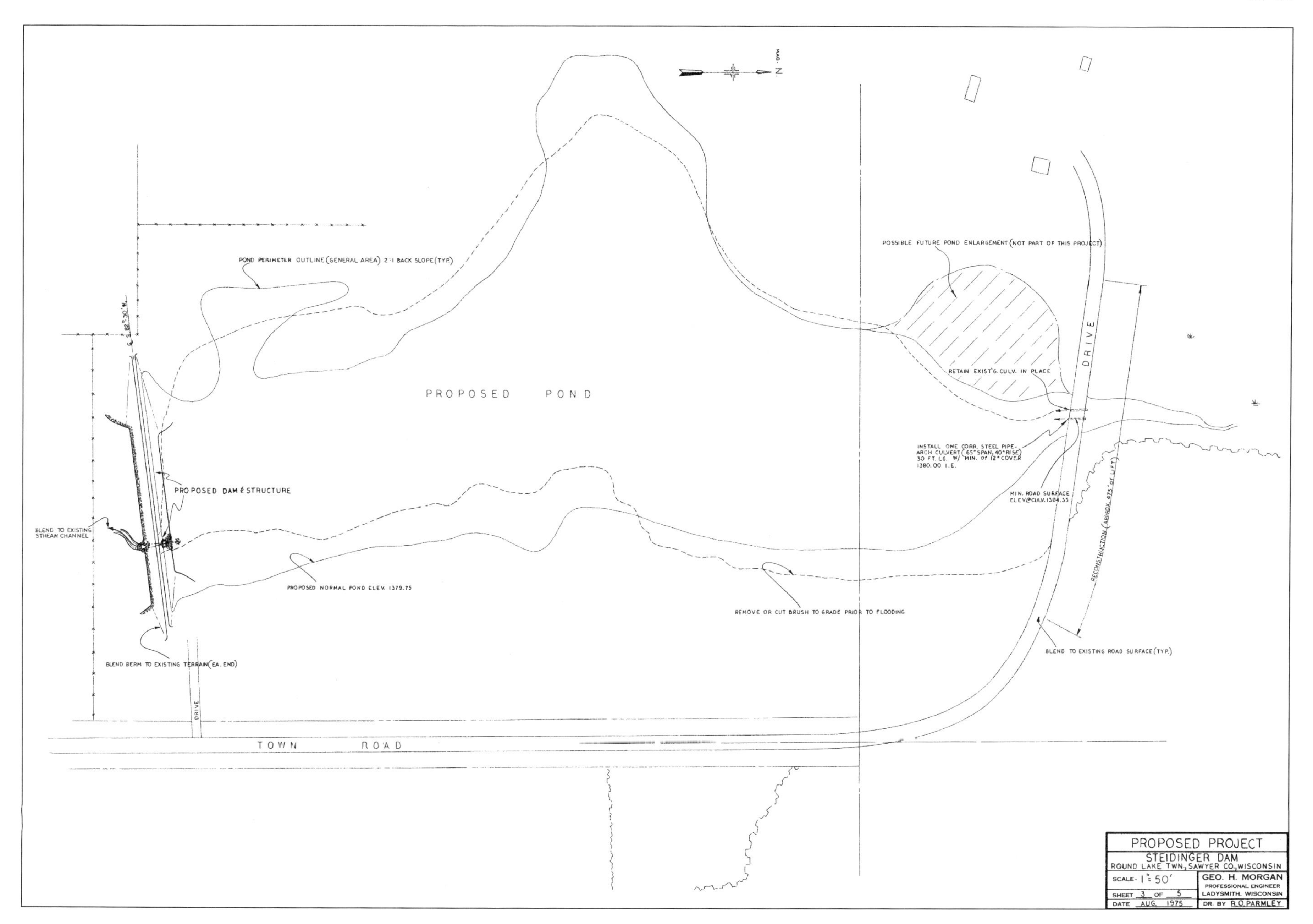
PROPOSED POND
POSSIBLE FUTURE POND ENLARGEMENT (NOT PART OF THIS PROJECT)
POND PERIMETER OUTLINE (GENERAL AREA) 2:1 BACK SLOPE (TYP.)
RETAIN EXIST'G. CULV. IN PLACE
DRIVE
PROPOSED DAM & STRUCTURE
BLEND TO EXISTING STREAM CHANNEL
PROPOSED NORMAL POND ELEV. 1379.75
REMOVE OR CUT BRUSH TO GRADE PRIOR TO FLOODING
BLEND BERM TO EXISTING TERRAIN (EA. END)
BLEND TO EXISTING ROAD SURFACE (TYP.)
MIN. ROAD SURFACE ELEV.@CULV. 1384.35
TOWN ROAD
PROPOSED PROJECT
STEIDINGER DAM
ROUND LAKE TWN., SAWYER CO., WISCONSIN
SCALE - 1" = 50'
SHEET 3 OF 5
DATE AUG. 1975
GEO. H. MORGAN
PROFESSIONAL ENGINEER
LADYSMITH, WISCONSIN
DR. BY R.O. PARMLEY

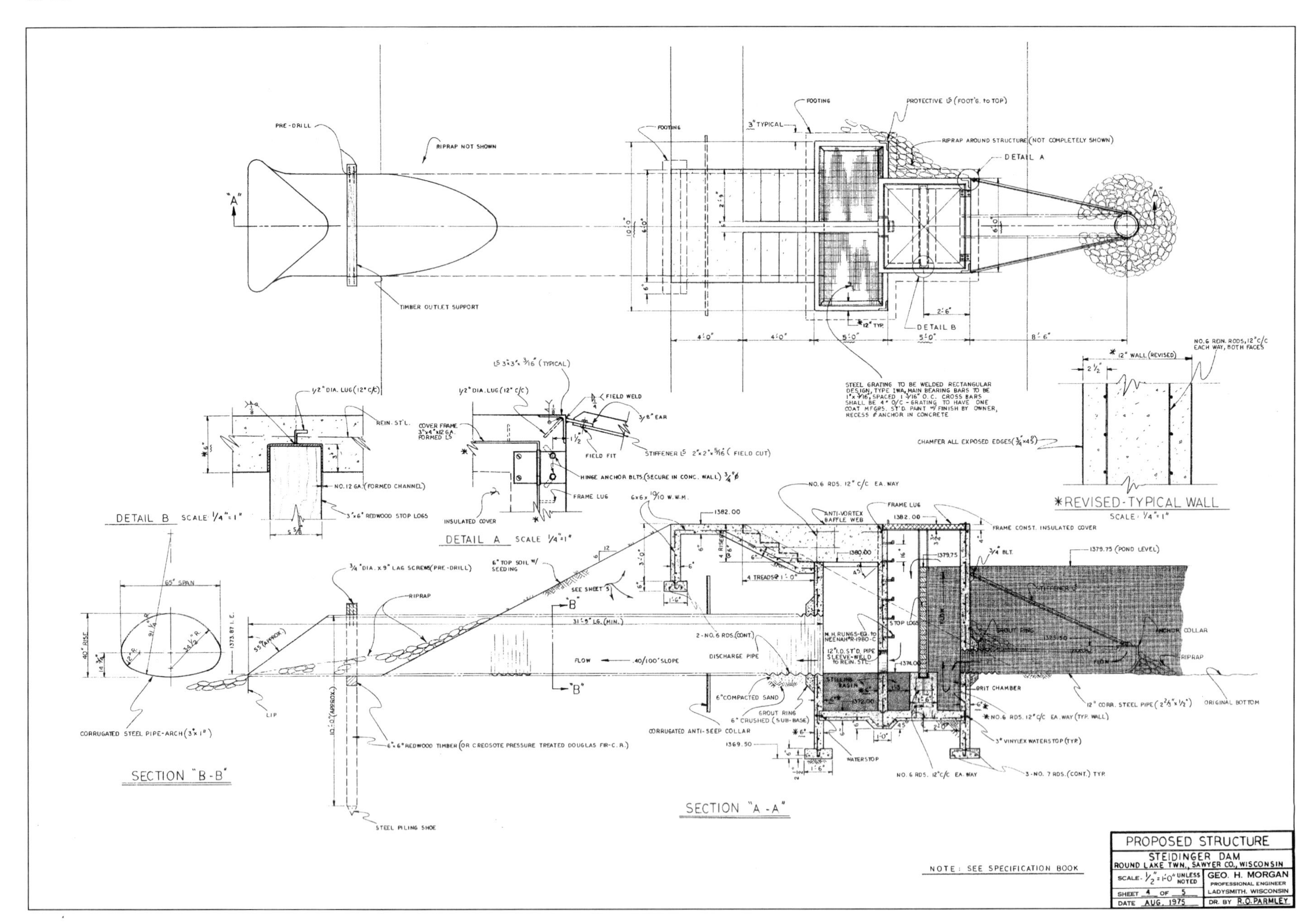
PRE-DRILL
RIPRAP NOT SHOWN
TIMBER OUTLET SUPPORT
FOOTING
3" TYPICAL
PROTECTIVE L5 (FOOT'G. to TOP)
RIPRAP AROUND STRUCTURE (NOT COMPLETELY SHOWN)
DETAIL A
DETAIL B
12" TYP.
4'-0"
4'-0"
5'-0"
5'-0"
8'-6"
2'-6"
STEEL GRATING TO BE WELDED RECTANGULAR DESIGN, TYPE 1WA, MAIN BEARING BARS TO BE 1"x 3/16", SPACED 1 3/16" O.C. CROSS BARS SHALL BE 4" O/C - GRATING TO HAVE ONE COAT MFGRS. ST'D. PAINT W/ FINISH BY OWNER, RECESS & ANCHOR IN CONCRETE
CHAMFER ALL EXPOSED EDGES (3/4"x45°)
12" WALL (REVISED)
NO. 6 REIN. RODS, 12" c/c EACH WAY, BOTH FACES
*REVISED-TYPICAL WALL
SCALE: 1/4"=1"
1/2" DIA. LUG (12" C/C)
REIN. ST'L.
NO. 12 GA. (FORMED CHANNEL)
3"x6" REDWOOD STOP LOGS
DETAIL B SCALE 1/4"=1"
L5 3"x3"x 3/16" (TYPICAL)
1/2" DIA. LUG (12" C/C)
COVER FRAME 3"x4"x12 GA. FORMED L5
FIELD WELD
3/8" EAR
FIELD FIT
STIFFENER L5 2"x2"x 5/16" (FIELD CUT)
HINGE ANCHOR BLTS. (SECURE IN CONC. WALL) 3/4"Ø
FRAME LUG
INSULATED COVER
DETAIL A SCALE 1/4"=1"
6x6x 10/10 W.W.M.
NO. 6 RDS. 12" c/c EA. WAY
FRAME LUG
1382.00
ANTI-VORTEX BAFFLE WEB
FRAME CONST. INSULATED COVER
1379.75 (POND LEVEL)
3/4" BLT.
4 TREADS @ 1'-0"
3/4" DIA. x 9" LAG SCREWS (PRE-DRILL)
6" TOP SOIL W/ SEEDING
SEE SHEET 5
RIPRAP
65" SPAN
40" RISE
1373.87 I.E.
55° (APPROX.)
31'-9" LG. (MIN.)
2-NO. 6 RDS. (CONT.)
DISCHARGE PIPE
FLOW
.40/100' SLOPE
M.H. RUNGS-EQ. TO NEENAH R-1980-C
12" I.D. ST'D. PIPE SLEEVE-WELD TO REIN. STL.
STOP LOGS
GROUT RING
ANCHOR COLLAR
RIPRAP
GRIT CHAMBER
STILLING BASIN
12" CORR. STEEL PIPE (2 2/3"x 1/2")
ORIGINAL BOTTOM
NO. 6 RDS. 12" C/C EA. WAY (TYP. WALL)
3" VINYLEX WATERSTOP (TYP.)
LIP
6" COMPACTED SAND
GROUT RING
6" CRUSHED (SUB-BASE)
CORRUGATED ANTI-SEEP COLLAR
1369.50
WATERSTOP
NO. 6 RDS. 12" C/C EA. WAY
3-NO. 7 RDS. (CONT.) TYP.
CORRUGATED STEEL PIPE-ARCH (3"x 1")
6"x6" REDWOOD TIMBER (OR CREOSOTE PRESSURE TREATED DOUGLAS FIR-C.R.)
STEEL PILING SHOE
SECTION "B-B"
SECTION "A-A"
NOTE: SEE SPECIFICATION BOOK
PROPOSED STRUCTURE
STEIDINGER DAM
ROUND LAKE TWN., SAWYER CO., WISCONSIN
SCALE - 1/2"=1'-0" UNLESS NOTED
GEO. H. MORGAN
PROFESSIONAL ENGINEER
LADYSMITH, WISCONSIN
SHEET 4 OF 5
DATE AUG. 1975
DR. BY R.O. PARMLEY

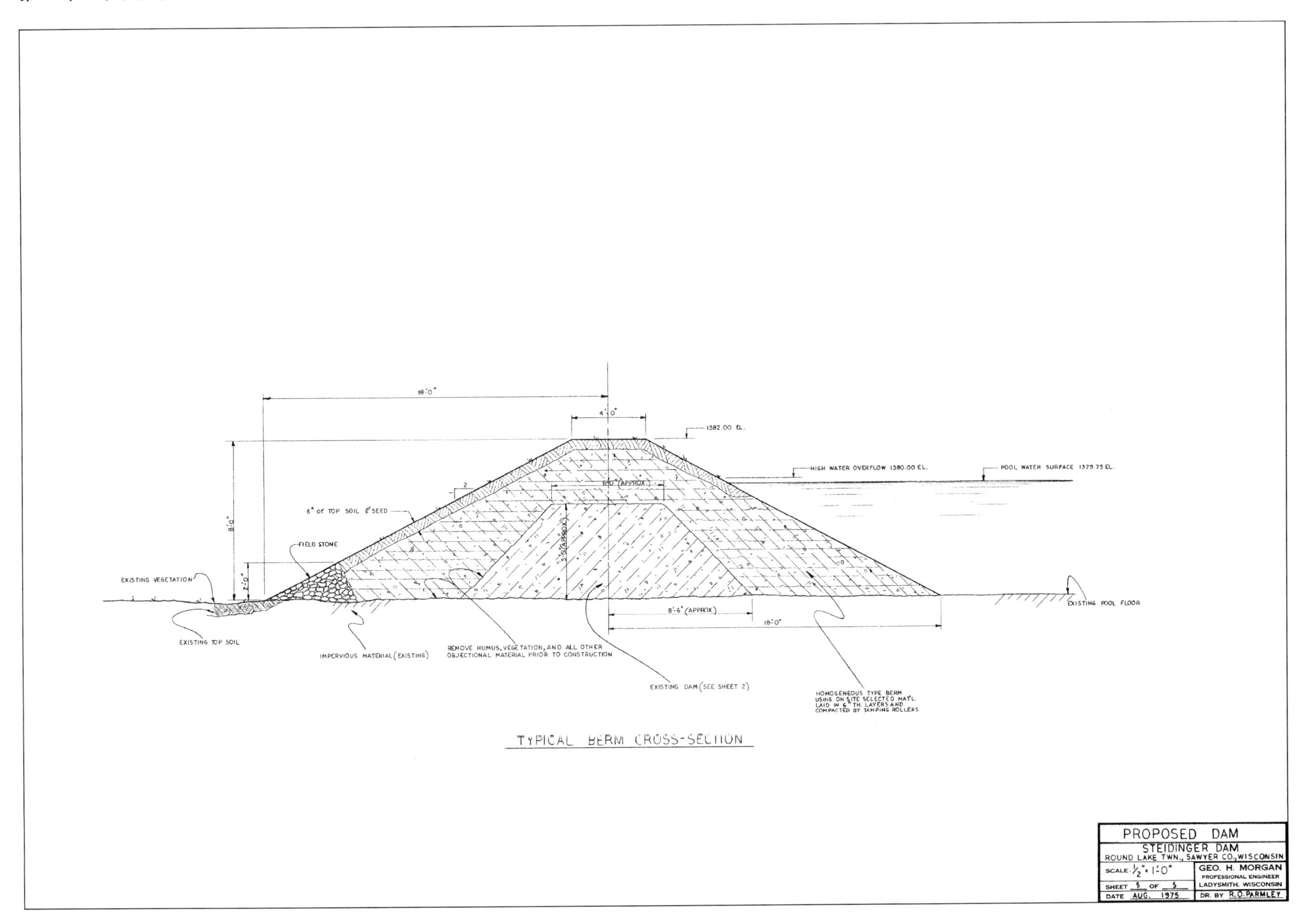
18'-0"
4'-0"
1382.00 EL.
HIGH WATER OVERFLOW 1380.00 EL.
POOL WATER SURFACE 1379.75 EL.
2
6'-0" (APPROX.)
6" of TOP SOIL & SEED
8'-0"
5'-0" (APPROX.)
FIELD STONE
2'-0"
EXISTING VEGETATION
EXISTING POOL FLOOR
8'-6" (APPROX.)
18'-0"
EXISTING TOP SOIL
IMPERVIOUS MATERIAL (EXISTING)
REMOVE HUMUS, VEGETATION, AND ALL OTHER
OBJECTIONAL MATERIAL PRIOR TO CONSTRUCTION
EXISTING DAM (SEE SHEET 2)
HOMOGENEOUS TYPE BERM
USING ON SITE SELECTED MAT'L.
LAID IN 6" TH. LAYERS AND
COMPACTED BY TAMPING ROLLERS
TYPICAL BERM CROSS-SECTION
PROPOSED DAM
STEIDINGER DAM
ROUND LAKE TWN., SAWYER CO., WISCONSIN
SCALE - 1/2" = 1'-0"
GEO. H. MORGAN
PROFESSIONAL ENGINEER
LADYSMITH, WISCONSIN
SHEET 5 OF 5
DATE AUG. 1975
DR. BY R.O. PARMLEY

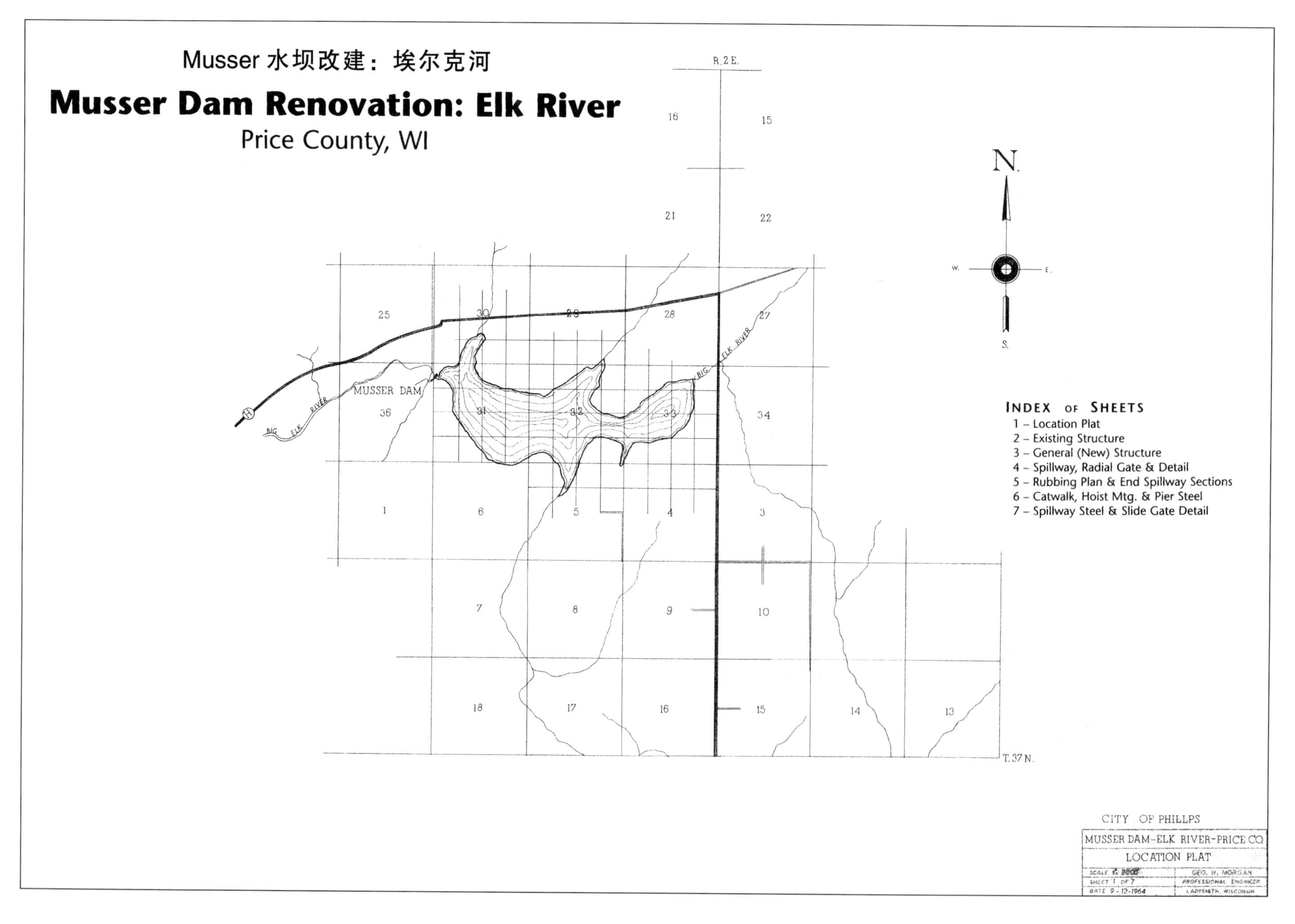

Source: Morgan & Parmley, Ltd.

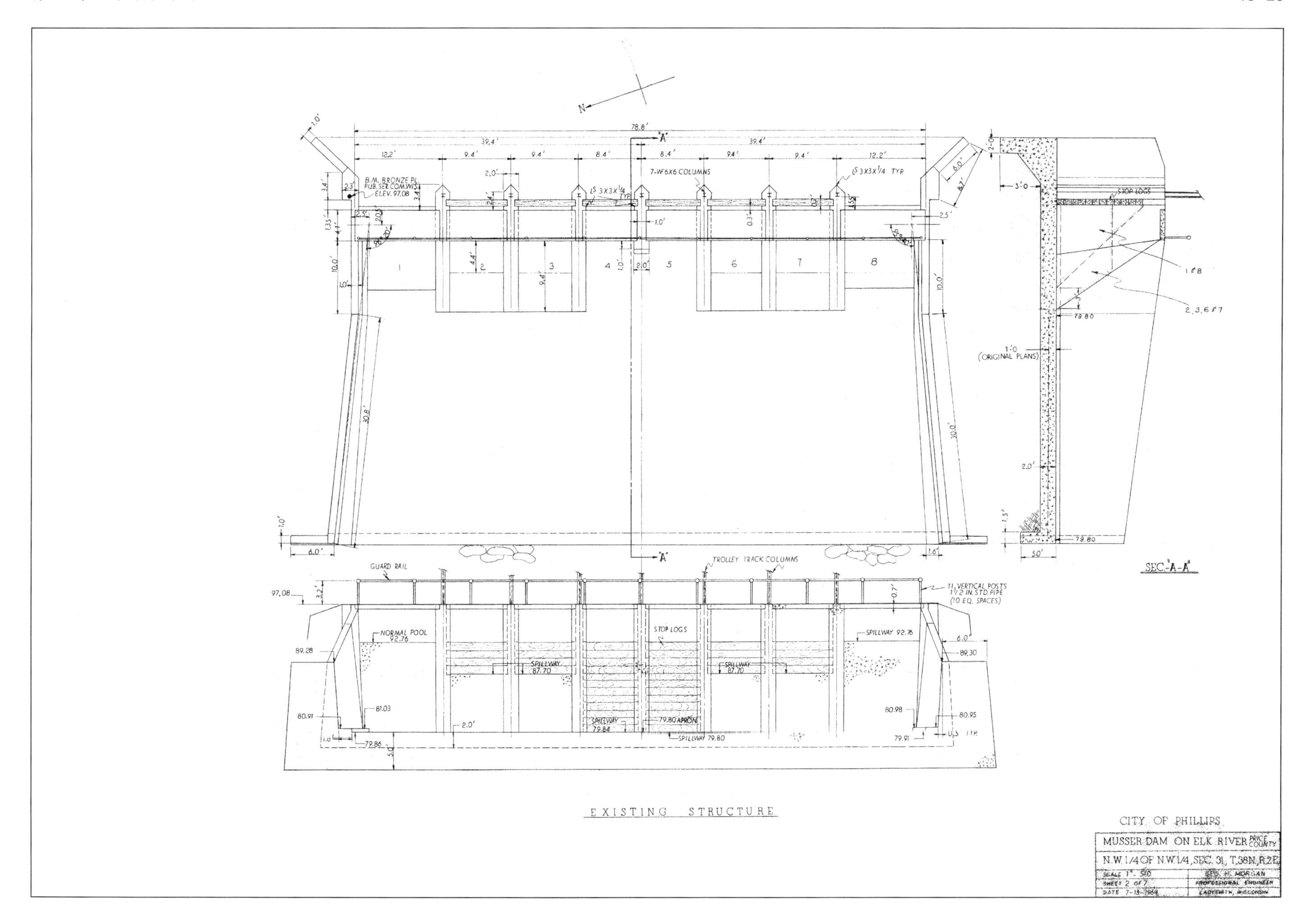

EXISTING STRUCTURE
SEC."A-A"
CITY OF PHILLIPS
MUSSER DAM ON ELK RIVER PRICE COUNTY
N.W.1/4 OF N.W.1/4, SEC. 31, T.38N, R.2E.
SCALE 1" - 5.0
SHEET 2 OF 7
DATE 7-13-1969
GUARD RAIL
TROLLEY TRACK COLUMNS
7-WF6X6 COLUMNS
STOP LOGS
NORMAL POOL 92.76
SPILLWAY 92.76
11, VERTICAL POSTS 1 1/2 IN. STD. PIPE (10 EQ. SPACES)
B.M. BRONZE PL. PUB. SER. COM. WIS. ELEV. 97.08
(ORIGINAL PLANS)

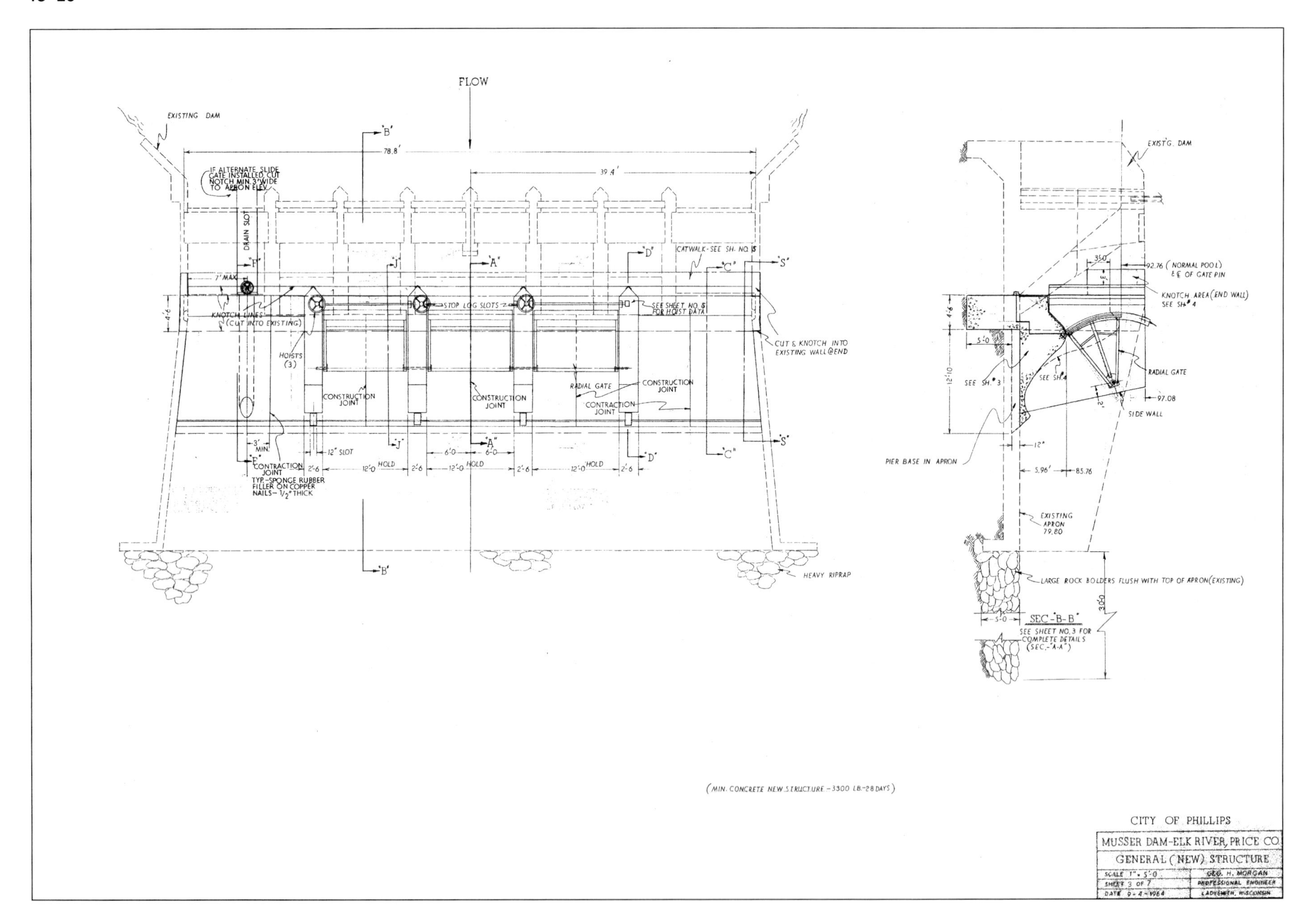
FLOW
EXISTING DAM
78.8'
39.4'
IF ALTERNATE SLIDE GATE INSTALLED CUT NOTCH MIN. 3"WIDE TO APRON ELEV.
DRAIN SLOT
CATWALK-SEE SH. NO. 5
KNOTCH LINES (CUT INTO EXISTING)
STOP LOG SLOTS
SEE SHEET NO. 5 FOR HOIST DATA
CUT & KNOTCH INTO EXISTING WALL@END
HOISTS (3)
CONSTRUCTION JOINT
RADIAL GATE
CONTRACTION JOINT
CONTRACTION JOINT TYP.-SPONGE RUBBER FILLER ON COPPER NAILS-1/2" THICK
12" SLOT
HOLD
HEAVY RIPRAP
EXIST'G. DAM
92.76 (NORMAL POOL) & ℄ OF GATE PIN
KNOTCH AREA (END WALL) SEE SH. 4
SEE SH. 3
SEE SH. 4
RADIAL GATE
97.08
SIDE WALL
PIER BASE IN APRON
EXISTING APRON 79.80
LARGE ROCK BOLDERS FLUSH WITH TOP OF APRON (EXISTING)
SEC-"B-B"
SEE SHEET NO. 3 FOR COMPLETE DETAILS (SEC.-"A-A")
(MIN. CONCRETE NEW STRUCTURE-3300 LB.-28 DAYS)
CITY OF PHILLIPS
MUSSER DAM-ELK RIVER, PRICE CO.
GENERAL (NEW) STRUCTURE
SCALE 1"=5'-0
SHEET 3 OF 7
DATE 9-4-1964
GEO. H. MORGAN
PROFESSIONAL ENGINEER
LADYSMITH, WISCONSIN

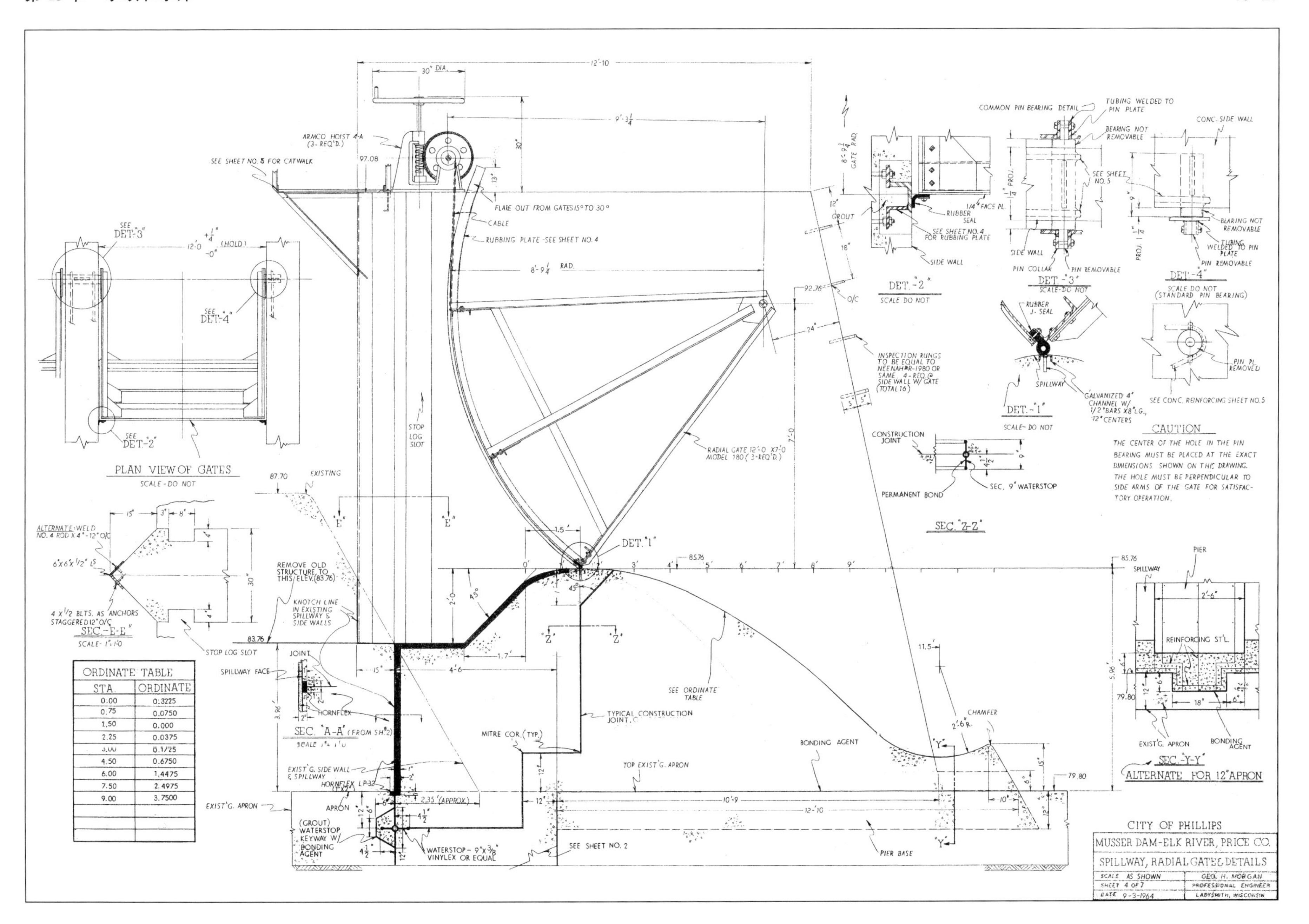

ORDINATE TABLE

STA.	ORDINATE
0.00	0.3225
0.75	0.0750
1.50	0.000
2.25	0.0375
3.00	0.1725
4.50	0.6750
6.00	1.4475
7.50	2.4975
9.00	3.7500

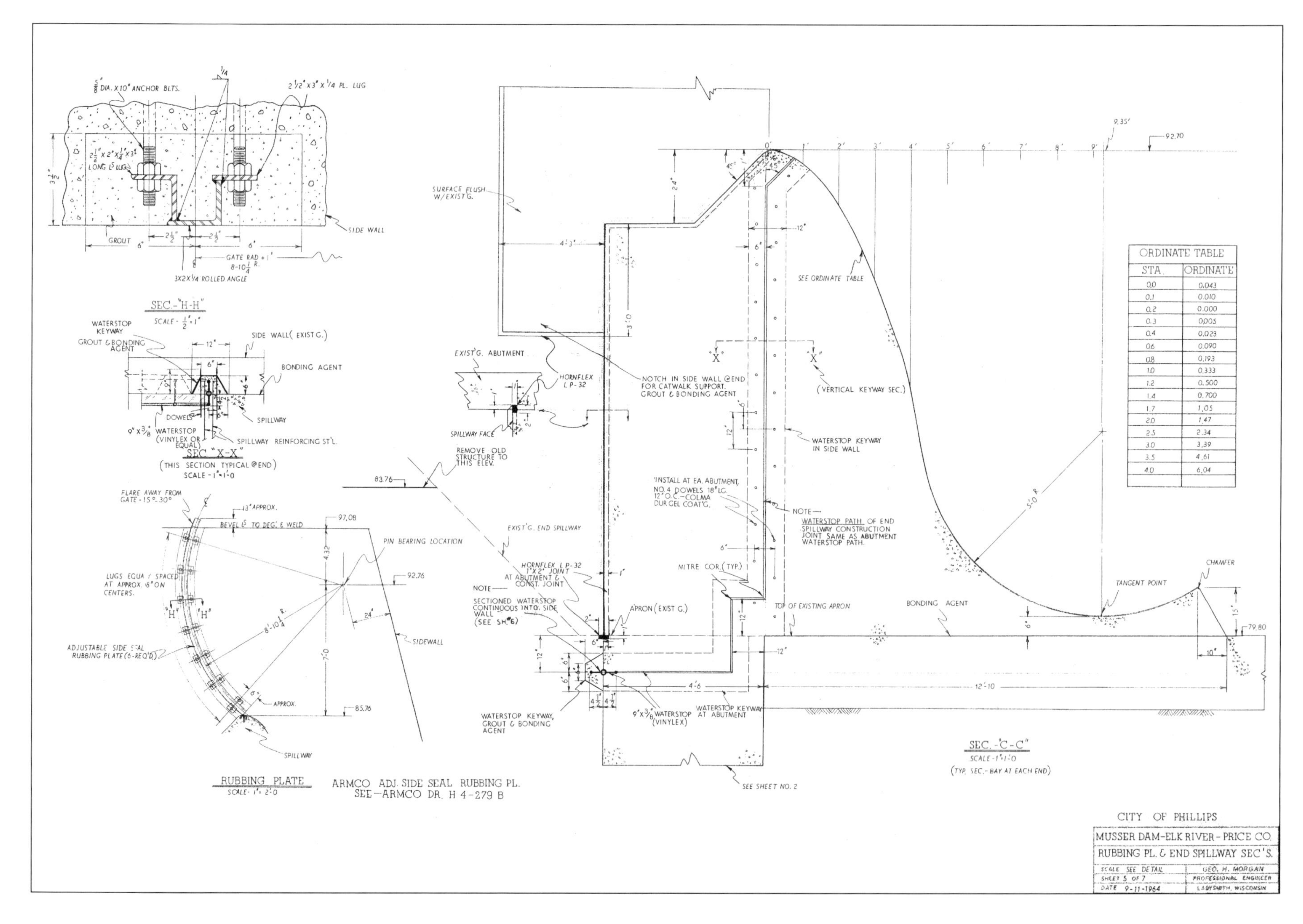

ORDINATE TABLE

STA.	ORDINATE
0.0	0.043
0.1	0.010
0.2	0.000
0.3	0.005
0.4	0.023
0.6	0.090
0.8	0.193
1.0	0.333
1.2	0.500
1.4	0.700
1.7	1.05
2.0	1.47
2.5	2.34
3.0	3.39
3.5	4.61
4.0	6.04

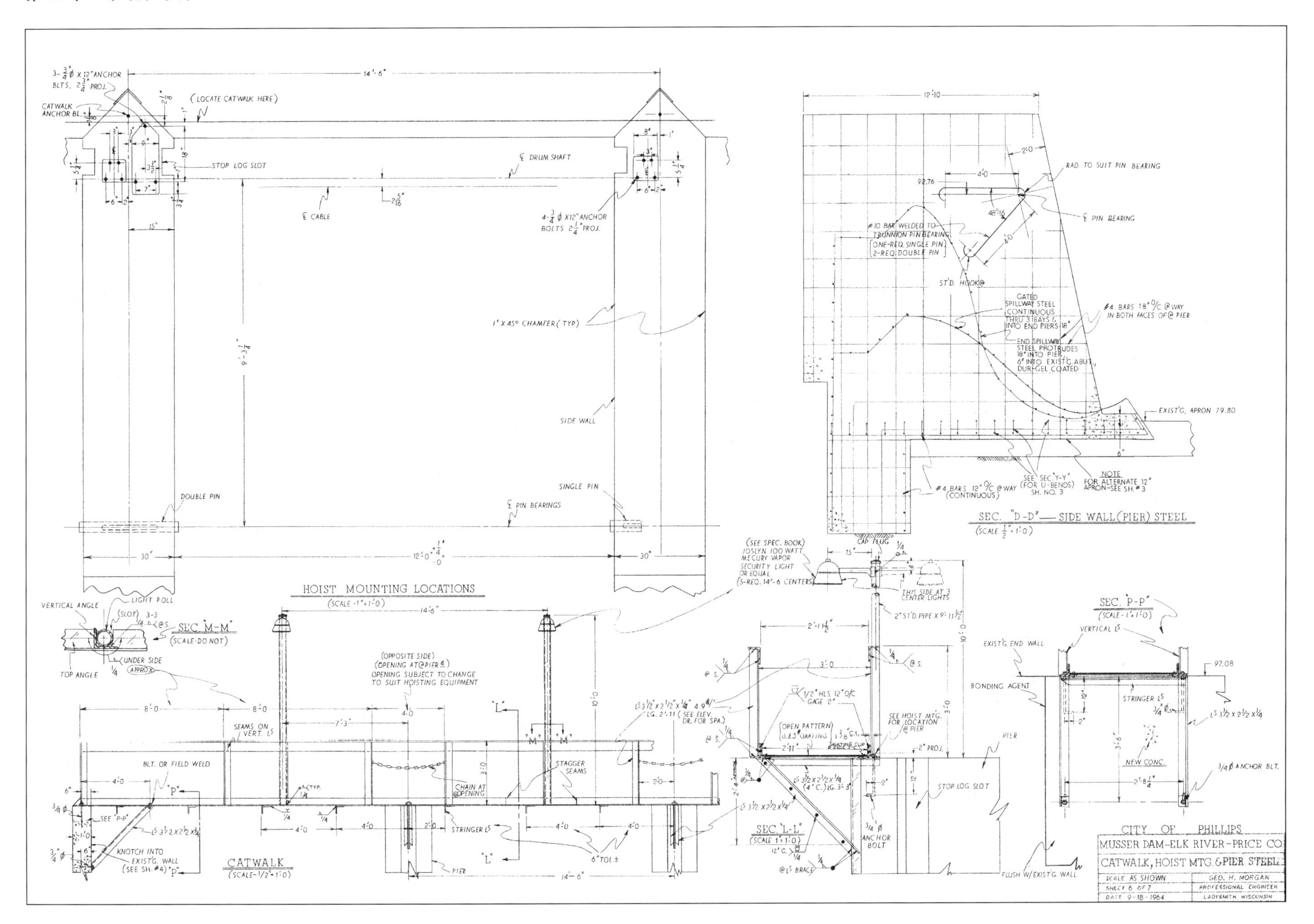
HOIST MOUNTING LOCATIONS
(SCALE -1"=1'-0)
SEC. "D-D" — SIDE WALL(PIER) STEEL
(SCALE 1/2"=1'-0)
CATWALK
(SCALE-1/2"=1'-0)
SEC. "L-L"
(SCALE 1"=1'-0)
SEC. "P-P"
(SCALE-1"=1'-0)
SEC. "M-M"
(SCALE-DO NOT)
CITY OF PHILLIPS
MUSSER DAM-ELK RIVER-PRICE CO.
CATWALK, HOIST MTG.&PIER STEEL
SCALE AS SHOWN
SHEET 6 OF 7
DATE 9-18-1964
GEO. H. MORGAN
PROFESSIONAL ENGINEER
LADYSMITH, WISCONSIN

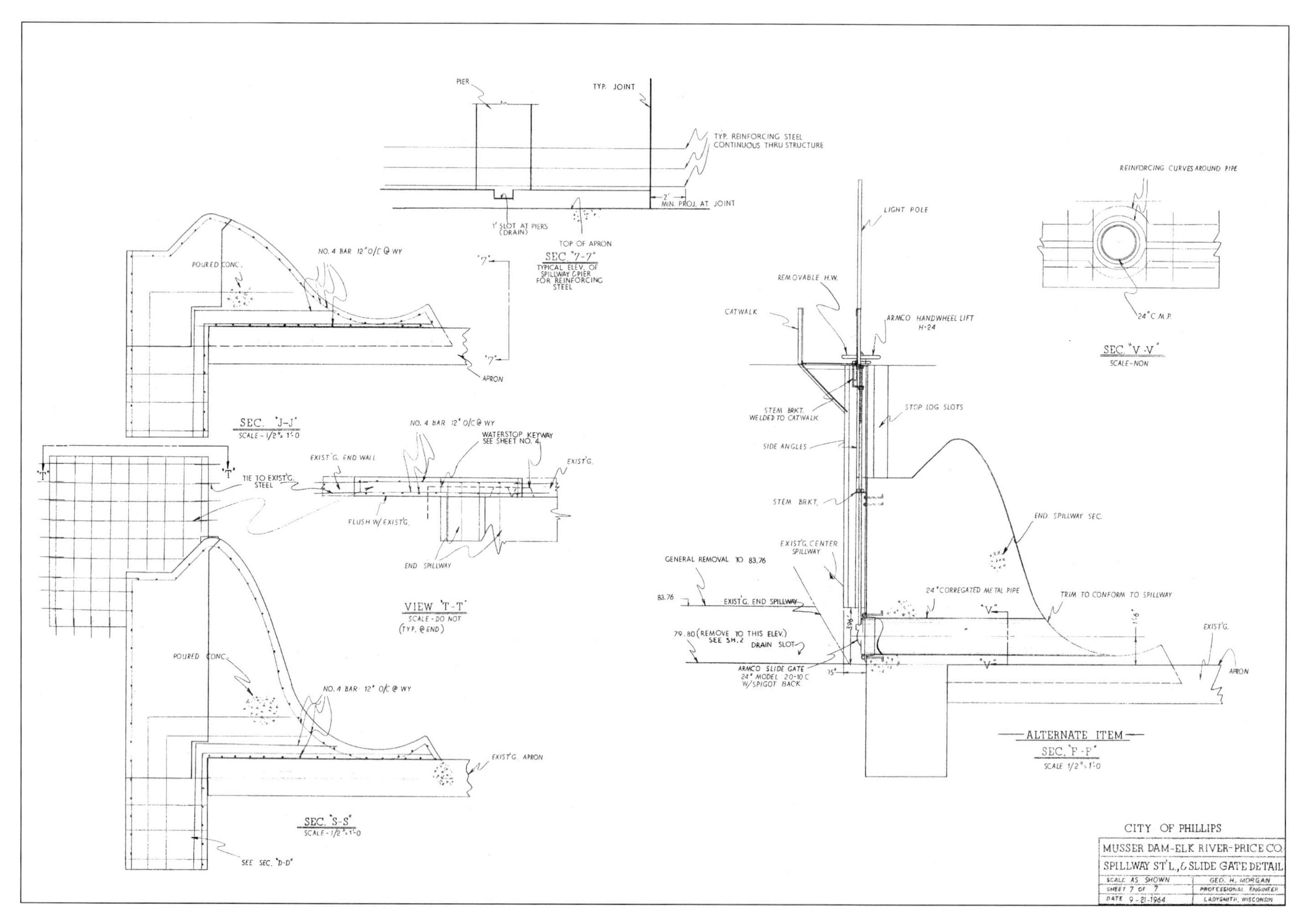
PIER
TYP. JOINT
TYP. REINFORCING STEEL CONTINUOUS THRU STRUCTURE
2' MIN. PROJ. AT JOINT
1' SLOT AT PIERS (DRAIN)
TOP OF APRON
SEC. "7-7"
TYPICAL ELEV. OF SPILLWAY & PIER FOR REINFORCING STEEL
NO. 4 BAR 12" O/C @ WY
POURED CONC.
APRON
SEC. "J-J"
SCALE - 1/2" = 1'-0
NO. 4 BAR 12" O/C @ WY
WATERSTOP KEYWAY SEE SHEET NO. 4
EXIST'G. END WALL
EXIST'G.
TIE TO EXIST'G. STEEL
FLUSH W/ EXIST'G.
END SPILLWAY
VIEW "T-T"
SCALE - DO NOT (TYP. @ END)
POURED CONC.
NO. 4 BAR 12" O/C @ WY
EXIST'G. APRON
SEC. "S-S"
SCALE - 1/2" = 1'-0
SEE SEC. "D-D"
LIGHT POLE
REMOVABLE H.W.
CATWALK
ARMCO HANDWHEEL LIFT H-24
STEM BRKT. WELDED TO CATWALK
STOP LOG SLOTS
SIDE ANGLES
STEM BRKT.
END SPILLWAY SEC.
EXIST'G. CENTER SPILLWAY
GENERAL REMOVAL TO 83.76
83.76
EXIST'G. END SPILLWAY
24" CORREGATED METAL PIPE
TRIM TO CONFORM TO SPILLWAY
79.80 (REMOVE TO THIS ELEV.) SEE SH. 2
DRAIN SLOT
ARMCO SLIDE GATE 24" MODEL 20-10 C W/SPIGOT BACK
EXIST'G.
APRON
ALTERNATE ITEM
SEC. "F-F"
SCALE 1/2" = 1'-0
REINFORCING CURVES AROUND PIPE
24" C.M.P.
SEC. "V-V"
SCALE - NON
CITY OF PHILLIPS
MUSSER DAM-ELK RIVER-PRICE CO.
SPILLWAY ST'L., & SLIDE GATE DETAIL
SCALE AS SHOWN
SHEET 7 OF 7
DATE 9-21-1964
GEO. H. MORGAN
PROFESSIONAL ENGINEER
LADYSMITH, WISCONSIN

设　计

第14节 街道、道路和公路

例 14

公共交通领域，即街道、道路和公路，是土木工程师参与的一个重大领域。

这种工程的大部分是对已经存在的交通道路的重建或改造。后面给出了实际项目设计图的示例。

目 录

工业企业铁路和街道重建

INDUSTRIAL ROADS & STREET RECONSTRUCTION

INDUSTRIAL PARK COMPLEX

FOR

VILLAGE OF GLEN FLORA, WISCONSIN

PROJECT NO.S:
M.& P. NO: 90-126-R
DOT NO: 8441-02-00,71
EDA NO: 06-01-02571
MAY, 1992

UTILITIES

CONTACT - NORTHERN STATES POWER CO.
711 W. 9th Street, North
Ladysmith, WI 54848

Attn: John Rymarkiewicz
715-532-6226

CONTACT - UNIVERSAL TELEPHONE CO.
OF NORTHERN WISCONSIN
Highway "8"
Hawkins, WI 54530

Attn: Jim Arquette
715-585-7707

CONTACT - DIGGER'S HOTLINE

1-800-242-8511

Ticket's Nos.: 1598953
1598962

CONTACT - VILLAGE OF GLEN FLORA WWCTF

Les Evjen, Operator
P.O. Box 253
Glen Flora, WI 54526

715-322-5511

VILLAGE OF GLEN FLORA, RUSK CO, WISCONSIN

GENERAL LOCATION MAP

PROJECT LOCATION

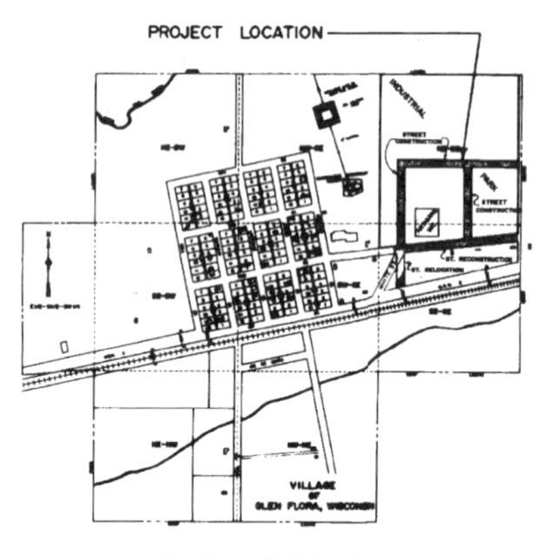

PROJECT LOCATION MAP

INDEX

NO.	DESCRIPTION
1	TITLE SHEET W/ INDEX
2	WHITE AVE. RELOCATION & CONST. 2ND ST. RECONSTRUCTION
3	ORLO AVE. CONSTRUCTION JIPSON ST. CONSTRUCTION
4	TYPICAL DETAILS
5-10	CROSS SECTIONS (SECOND STREET)
11-13	CROSS SECTIONS (ORLO AVENUE)
14-18	CROSS SECTIONS (WHITE AVENUE)
19-23	CROSS SECTIONS (JIPSON STREET)
24	TYPICAL EROSION CONTROL

NOTE: Sheets 6 thru 24 Have Been Omited

PREPARED BY:
MORGAN & PARMLEY LTD.
CONSULTING ENGINEERS
LADYSMITH, WISCONSIN

Courtesy: Morgan & Parmley, Ltd.

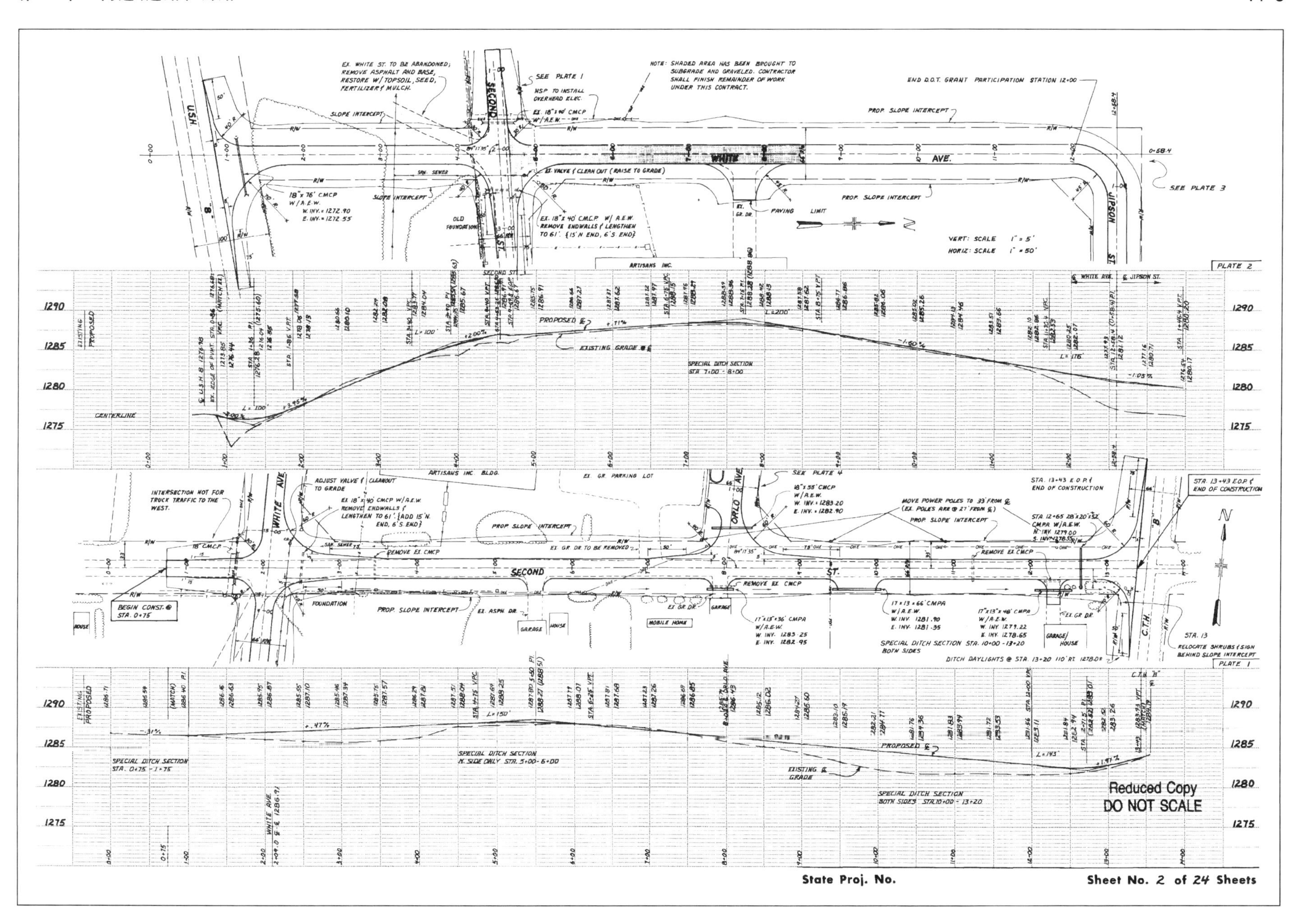

EX. WHITE ST. TO BE ABANDONED; REMOVE ASPHALT AND BASE, RESTORE W/ TOPSOIL, SEED, FERTILIZER & MULCH.
SEE PLATE 1
NOTE: SHADED AREA HAS BEEN BROUGHT TO SUBGRADE AND GRAVELED. CONTRACTOR SHALL FINISH REMAINDER OF WORK UNDER THIS CONTRACT.
END D.O.T. GRANT PARTICIPATION STATION 12+00
PROP. SLOPE INTERCEPT
SLOPE INTERCEPT
WHITE
AVE.
SEE PLATE 3
PROP. SLOPE INTERCEPT
PAVING LIMIT
VERT: SCALE 1" = 5'
HORIZ: SCALE 1" = 50'
ARTISANS INC.
OLD FOUNDATION
PLATE 2
PROPOSED ℄
EXISTING GRADE @ ℄
SPECIAL DITCH SECTION STA. 7+00 - 8+00
CENTERLINE
EXISTING
PROPOSED
1290
1285
1280
1275
ARTISANS INC. BLDG.
EX. GR. PARKING LOT
SEE PLATE 4
INTERSECTION NOT FOR TRUCK TRAFFIC TO THE WEST.
MOVE POWER POLES TO 33' FROM ℄ (EX. POLES ARE @ 27' FROM ℄)
PROP. SLOPE INTERCEPT
STA. 13+43 E.O.P. & END OF CONSTRUCTION
SECOND
ST.
REMOVE EX. CMCP
BEGIN CONST. @ STA. 0+75
FOUNDATION
PROP. SLOPE INTERCEPT
EX. ASPH. DR.
GARAGE
HOUSE
MOBILE HOME
SPECIAL DITCH SECTION STA. 10+00 - 13+20 BOTH SIDES
DITCH DAYLIGHTS @ STA. 13+20 110' RT. 1278.0±
STA. 13 RELOCATE SHRUBS & SIGN BEHIND SLOPE INTERCEPT
C.T.H. "B"
PLATE 1
SPECIAL DITCH SECTION STA. 0+75 - 1+75
SPECIAL DITCH SECTION N. SIDE ONLY STA. 5+00 - 6+00
EXISTING ℄ GRADE
SPECIAL DITCH SECTION BOTH SIDES STA. 10+00 - 13+20
Reduced Copy
DO NOT SCALE
State Proj. No.
Sheet No. 2 of 24 Sheets

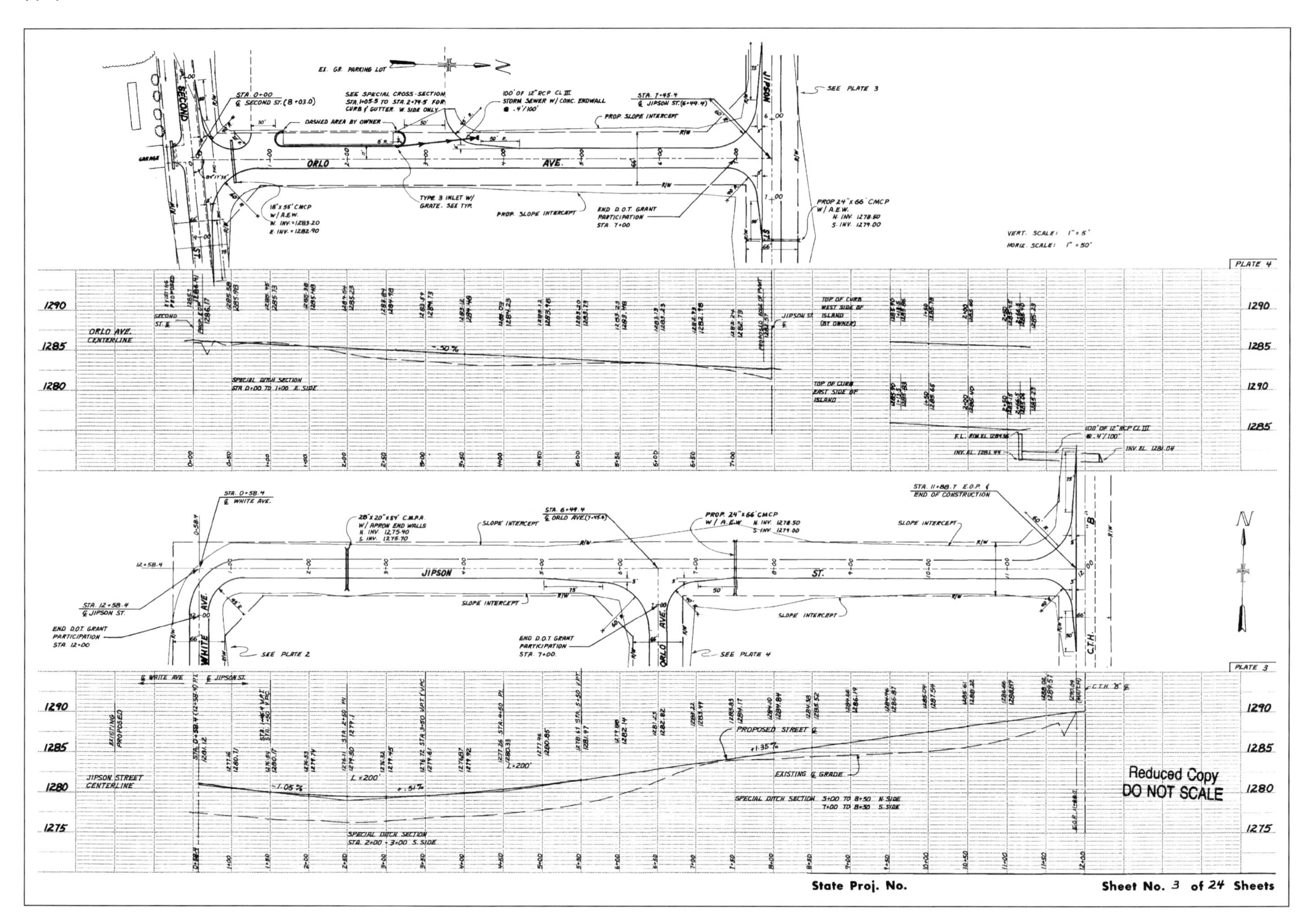
SEE PLATE 3
VERT. SCALE: 1" = 5'
HORIZ. SCALE: 1" = 50'
PLATE 4
ORLO AVE. CENTERLINE
SPECIAL DITCH SECTION STA 0+00 TO 1+00 E. SIDE
TOP OF CURB WEST SIDE OF ISLAND (BY OWNER)
TOP OF CURB EAST SIDE OF ISLAND
JIPSON
ST.
SLOPE INTERCEPT
SEE PLATE 2
SEE PLATE 4
PLATE 3
JIPSON STREET CENTERLINE
PROPOSED STREET ℄
EXISTING ℄ GRADE
SPECIAL DITCH SECTION STA. 2+00 - 3+00 S. SIDE
Reduced Copy
DO NOT SCALE
State Proj. No.
Sheet No. 3 of 24 Sheets

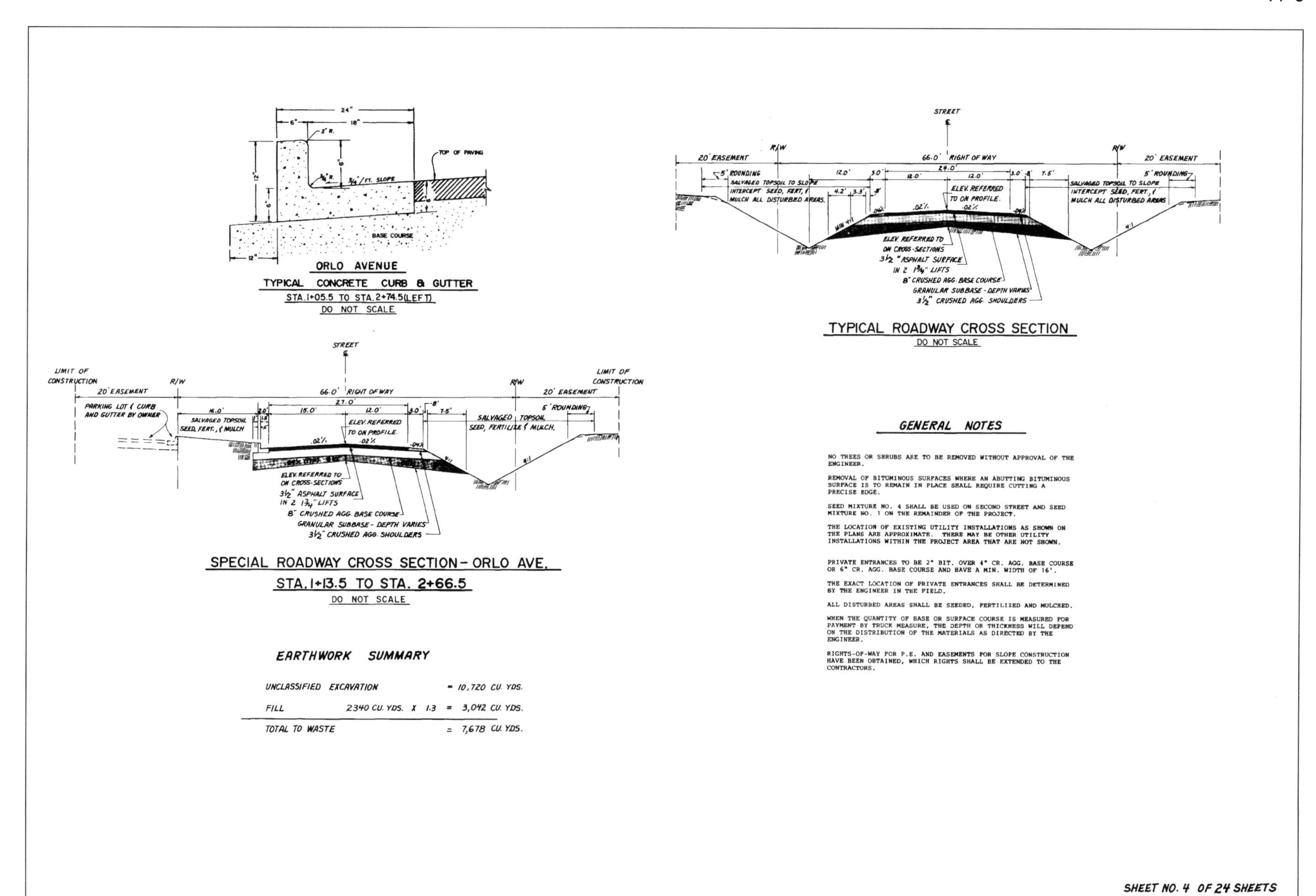
ORLO AVENUE
TYPICAL CONCRETE CURB & GUTTER
STA.1+05.5 TO STA.2+74.5(LEFT)
DO NOT SCALE
TOP OF PAVING
BASE COURSE
3/4"/FT. SLOPE
TYPICAL ROADWAY CROSS SECTION
DO NOT SCALE
STREET
20' EASEMENT
R/W
66.0' RIGHT OF WAY
5' ROUNDING
SALVAGED TOPSOIL TO SLOPE INTERCEPT SEED, FERT, & MULCH ALL DISTURBED AREAS.
ELEV. REFERRED TO ON PROFILE.
ELEV. REFERRED TO ON CROSS-SECTIONS
3½" ASPHALT SURFACE IN 2 1¾" LIFTS
8" CRUSHED AGG. BASE COURSE
GRANULAR SUBBASE - DEPTH VARIES
3½" CRUSHED AGG. SHOULDERS
SPECIAL ROADWAY CROSS SECTION - ORLO AVE.
STA.1+13.5 TO STA. 2+66.5
DO NOT SCALE
LIMIT OF CONSTRUCTION
PARKING LOT & CURB AND GUTTER BY OWNER
SALVAGED TOPSOIL SEED, FERT., & MULCH
SALVAGED TOPSOIL SEED, FERTILIZE & MULCH.
GENERAL NOTES
NO TREES OR SHRUBS ARE TO BE REMOVED WITHOUT APPROVAL OF THE ENGINEER.
REMOVAL OF BITUMINOUS SURFACES WHERE AN ABUTTING BITUMINOUS SURFACE IS TO REMAIN IN PLACE SHALL REQUIRE CUTTING A PRECISE EDGE.
SEED MIXTURE NO. 4 SHALL BE USED ON SECOND STREET AND SEED MIXTURE NO. 1 ON THE REMAINDER OF THE PROJECT.
THE LOCATION OF EXISTING UTILITY INSTALLATIONS AS SHOWN ON THE PLANS ARE APPROXIMATE. THERE MAY BE OTHER UTILITY INSTALLATIONS WITHIN THE PROJECT AREA THAT ARE NOT SHOWN.
PRIVATE ENTRANCES TO BE 2" BIT. OVER 4" CR. AGG. BASE COURSE OR 6" CR. AGG. BASE COURSE AND HAVE A MIN. WIDTH OF 16'.
THE EXACT LOCATION OF PRIVATE ENTRANCES SHALL BE DETERMINED BY THE ENGINEER IN THE FIELD.
ALL DISTURBED AREAS SHALL BE SEEDED, FERTILIZED AND MULCHED.
WHEN THE QUANTITY OF BASE OR SURFACE COURSE IS MEASURED FOR PAYMENT BY TRUCK MEASURE, THE DEPTH OR THICKNESS WILL DEPEND ON THE DISTRIBUTION OF THE MATERIALS AS DIRECTED BY THE ENGINEER.
RIGHTS-OF-WAY FOR P.E. AND EASEMENTS FOR SLOPE CONSTRUCTION HAVE BEEN OBTAINED, WHICH RIGHTS SHALL BE EXTENDED TO THE CONTRACTORS.
EARTHWORK SUMMARY
UNCLASSIFIED EXCAVATION = 10,720 CU. YDS.
FILL 2340 CU. YDS. X 1.3 = 3,042 CU. YDS.
TOTAL TO WASTE = 7,678 CU. YDS.
SHEET NO. 4 OF 24 SHEETS

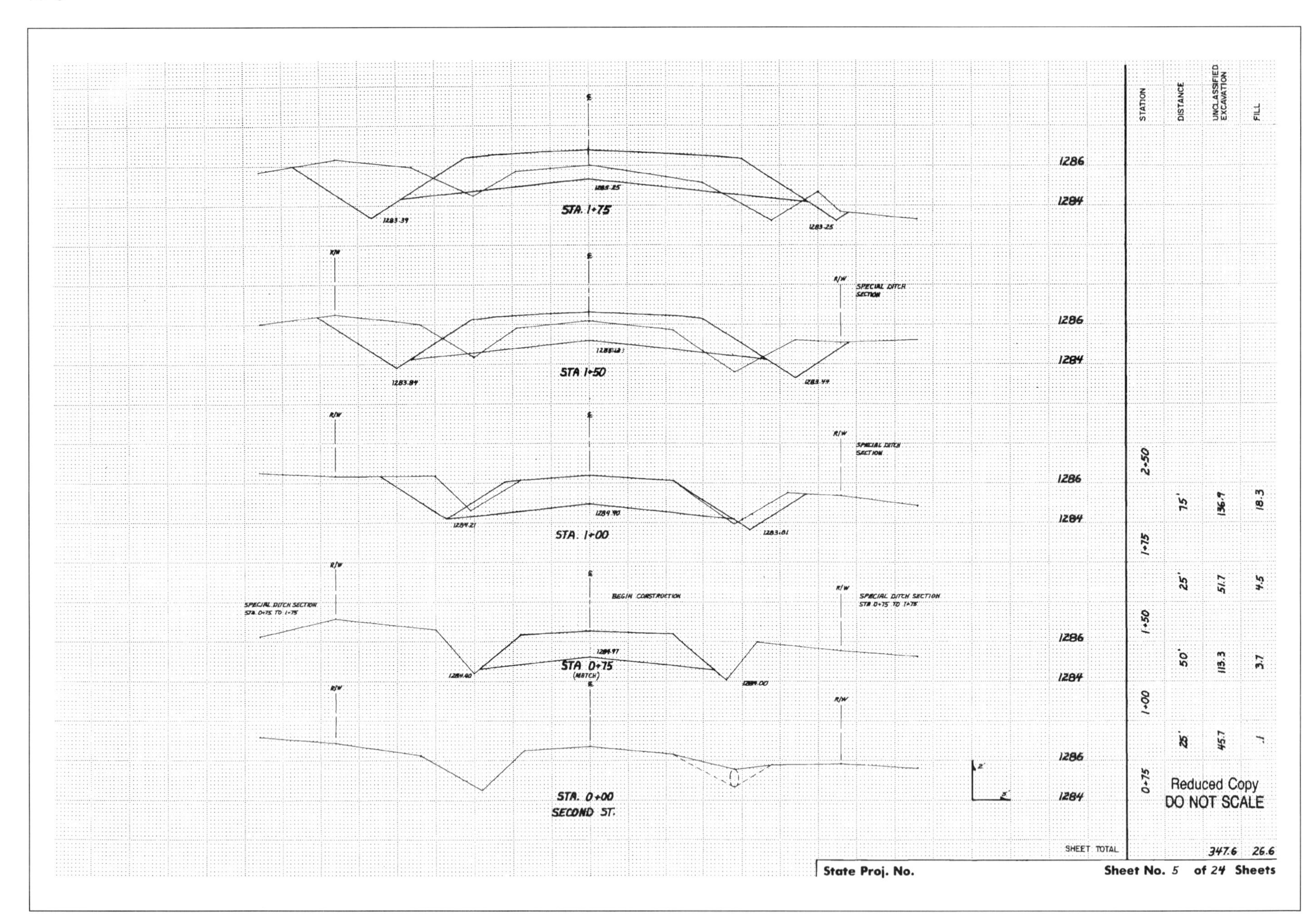
STA. 1+75
1283.39
1285.25
1283.25
STA. 1+50
1283.84
1283.44
STA. 1+00
1284.21
1284.90
1283.81
STA 0+75
(MATCH)
1284.60
1284.97
1284.00
BEGIN CONSTRUCTION
SPECIAL DITCH SECTION
STA. 0+75 TO 1+75
R/W
STA. 0+00
SECOND ST.
1286
1284
STATION
DISTANCE
UNCLASSIFIED EXCAVATION
FILL
2+50
75'
136.9
18.3
1+75
25'
51.7
4.5
1+50
50'
113.3
3.7
1+00
25'
45.7
.1
0+75
Reduced Copy
DO NOT SCALE
SHEET TOTAL
347.6
26.6
State Proj. No.
Sheet No. 5 of 24 Sheets

Doughty 路重建

DOUGHTY ROAD – ROAD RECONSTRUCTION for the CITY OF LADYSMITH

LADYSMITH, WI 54848

T.E.A. GRANT, I.D. 8955-02-71

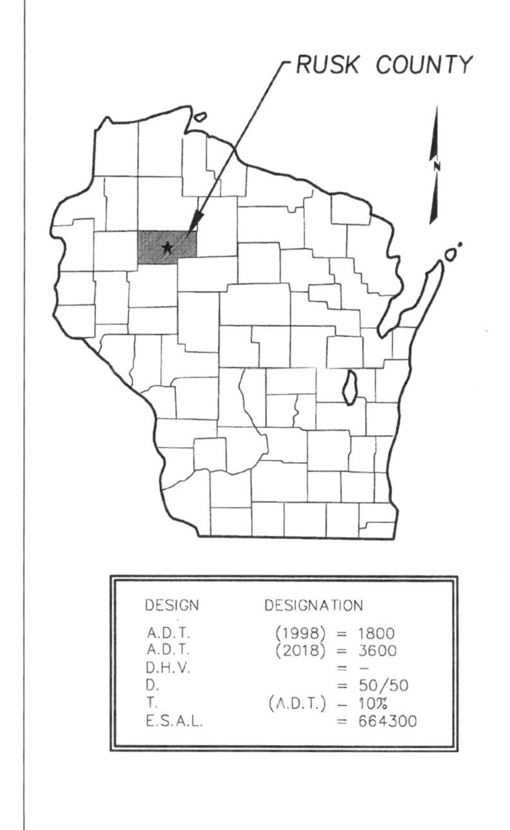

DESIGN	DESIGNATION
A.D.T.	(1998) = 1800
A.D.T.	(2018) = 3600
D.H.V.	= –
D.	= 50/50
T.	(A.D.T.) – 10%
E.S.A.L.	= 664300

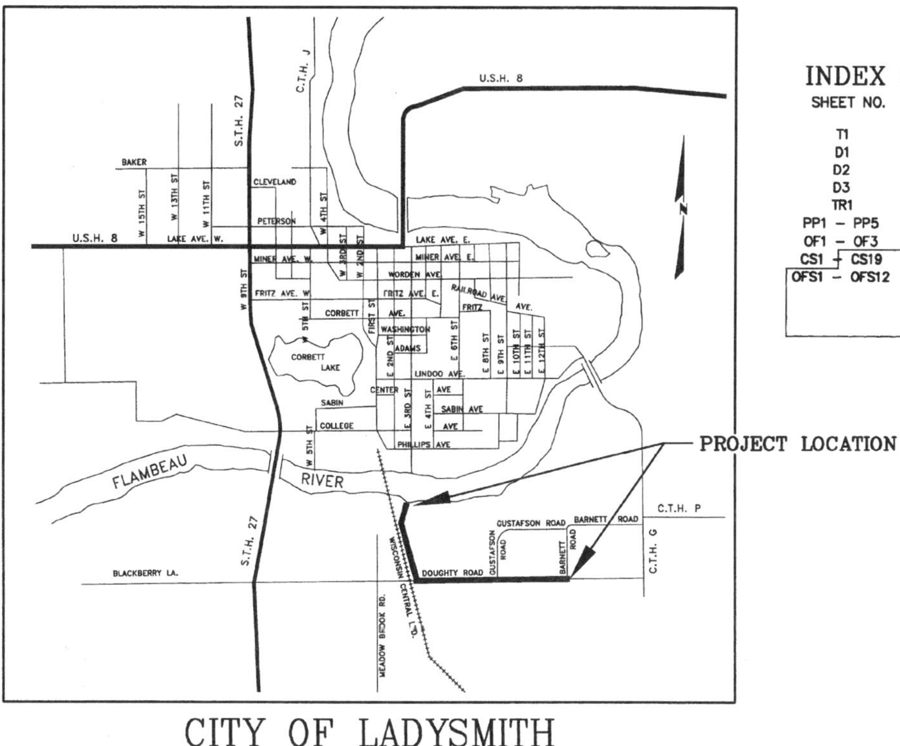

CITY OF LADYSMITH

INDEX OF SHEETS

SHEET NO.	DESCRIPTION
T1	TITLE SHEET
D1	TYPICAL SECTIONS
D2	RAILROAD DETAILS
D3	STORM SEWER DETAILS
TR1	TRAFFIC CONTROL
PP1 – PP5	DOUGHTY RD. PLAN & PROFILES
OF1 – OF3	STORM SEWER OUTFALL PLAN & PROFILES
CS1 – CS19	DOUGHTY ROAD CROSS SECTIONS
OFS1 – OFS12	OUTFALL CROSS SECTIONS

NOTE: Sheets Beyond CS4 Have Been Omitted

REVISION	DATE	DESCRIPTION
2	1/4/99	D1 *
1	7/9/98	MISC. CHANGES

DOUGHTY ROAD ROAD RECONSTRUCTION

SCALE NONE　PROJECT ENGINEER Robert O. Parmley　DRAWN BY M.R.E.　DATE 5-12-98

MORGAN & PARMLEY, LTD.
Professional Consulting Engineers
115 West 2nd Street, Ladysmith, Wisconsin 54848

TELEPHONE 715-532-3721　SHEET DESCRIPTION TITLE SHEET　PROJECT NO. 97-144　SHEET NO. T1

Courtesy: Morgan & Parmley, Ltd.

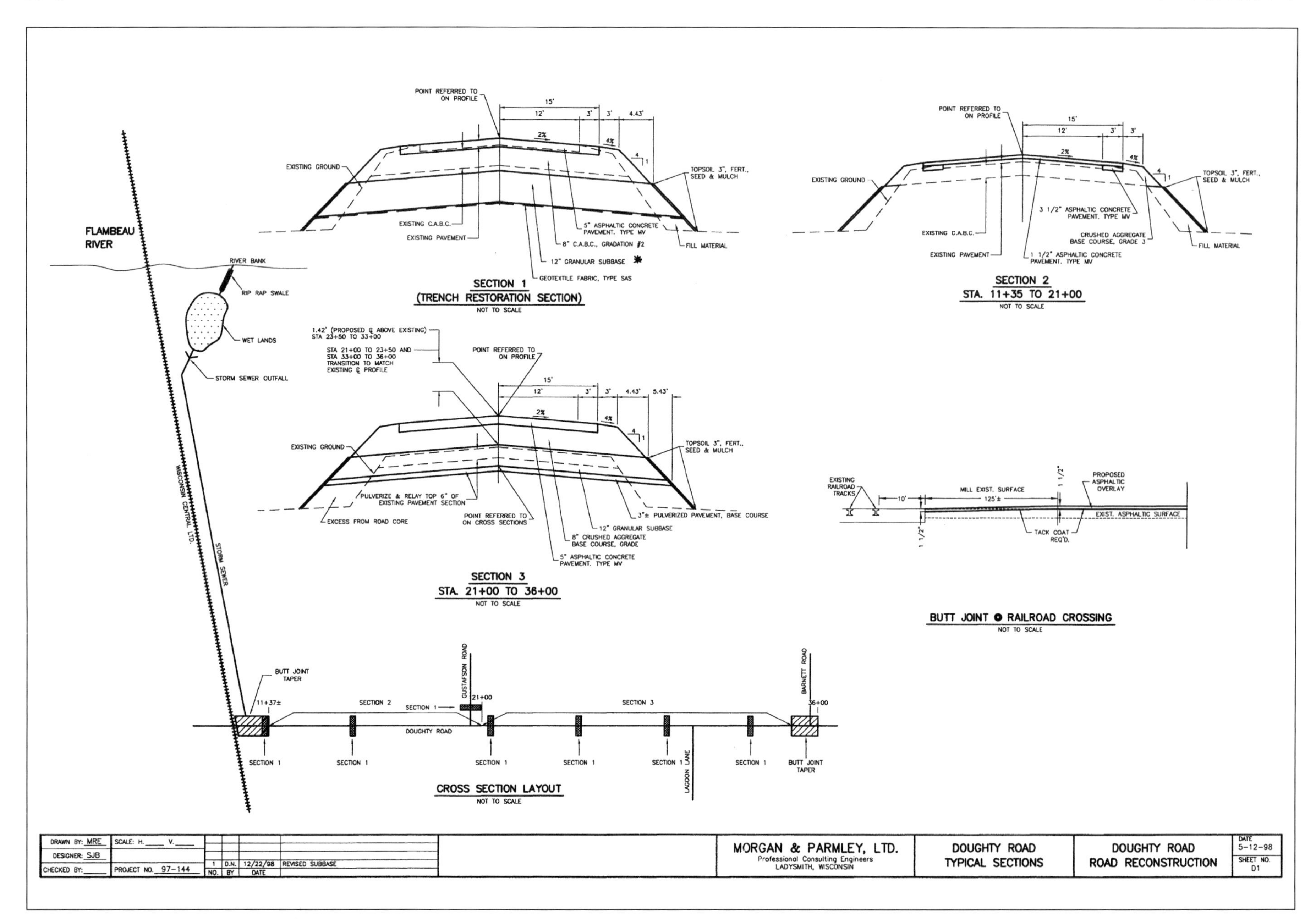

FLAMBEAU RIVER
RIVER BANK
RIP RAP SWALE
WET LANDS
STORM SEWER OUTFALL
WISCONSIN CENTRAL LTD.
STORM SEWER
POINT REFERRED TO ON PROFILE
EXISTING GROUND
TOPSOIL 3", FERT., SEED & MULCH
EXISTING C.A.B.C.
EXISTING PAVEMENT
5" ASPHALTIC CONCRETE PAVEMENT, TYPE MV
8" C.A.B.C., GRADATION #2
12" GRANULAR SUBBASE
GEOTEXTILE FABRIC, TYPE SAS
FILL MATERIAL
SECTION 1
(TRENCH RESTORATION SECTION)
NOT TO SCALE
3 1/2" ASPHALTIC CONCRETE PAVEMENT, TYPE MV
CRUSHED AGGREGATE BASE COURSE, GRADE 3
1 1/2" ASPHALTIC CONCRETE PAVEMENT, TYPE MV
SECTION 2
STA. 11+35 TO 21+00
NOT TO SCALE
1.42' (PROPOSED ℄ ABOVE EXISTING) STA 23+50 TO 33+00
STA 21+00 TO 23+50 AND STA 33+00 TO 36+00 TRANSITION TO MATCH EXISTING ℄ PROFILE
PULVERIZE & RELAY TOP 6" OF EXISTING PAVEMENT SECTION
EXCESS FROM ROAD CORE
POINT REFERRED TO ON CROSS SECTIONS
3"± PULVERIZED PAVEMENT, BASE COURSE
12" GRANULAR SUBBASE
8" CRUSHED AGGREGATE BASE COURSE, GRADE
5" ASPHALTIC CONCRETE PAVEMENT, TYPE MV
SECTION 3
STA. 21+00 TO 36+00
NOT TO SCALE
EXISTING RAILROAD TRACKS
MILL EXIST. SURFACE
125'±
PROPOSED ASPHALTIC OVERLAY
EXIST. ASPHALTIC SURFACE
TACK COAT REQ'D.
BUTT JOINT @ RAILROAD CROSSING
NOT TO SCALE
BUTT JOINT TAPER
GUSTAFSON ROAD
BARNETT ROAD
DOUGHTY ROAD
LAGOON LANE
SECTION 1
SECTION 2
SECTION 3
CROSS SECTION LAYOUT
NOT TO SCALE
DRAWN BY: MRE
DESIGNER: SJB
CHECKED BY:
SCALE: H. V.
PROJECT NO. 97-144
1 D.N. 12/22/98 REVISED SUBBASE
NO. BY DATE
MORGAN & PARMLEY, LTD.
Professional Consulting Engineers
LADYSMITH, WISCONSIN
DOUGHTY ROAD TYPICAL SECTIONS
DOUGHTY ROAD ROAD RECONSTRUCTION
DATE 5-12-98
SHEET NO. D1

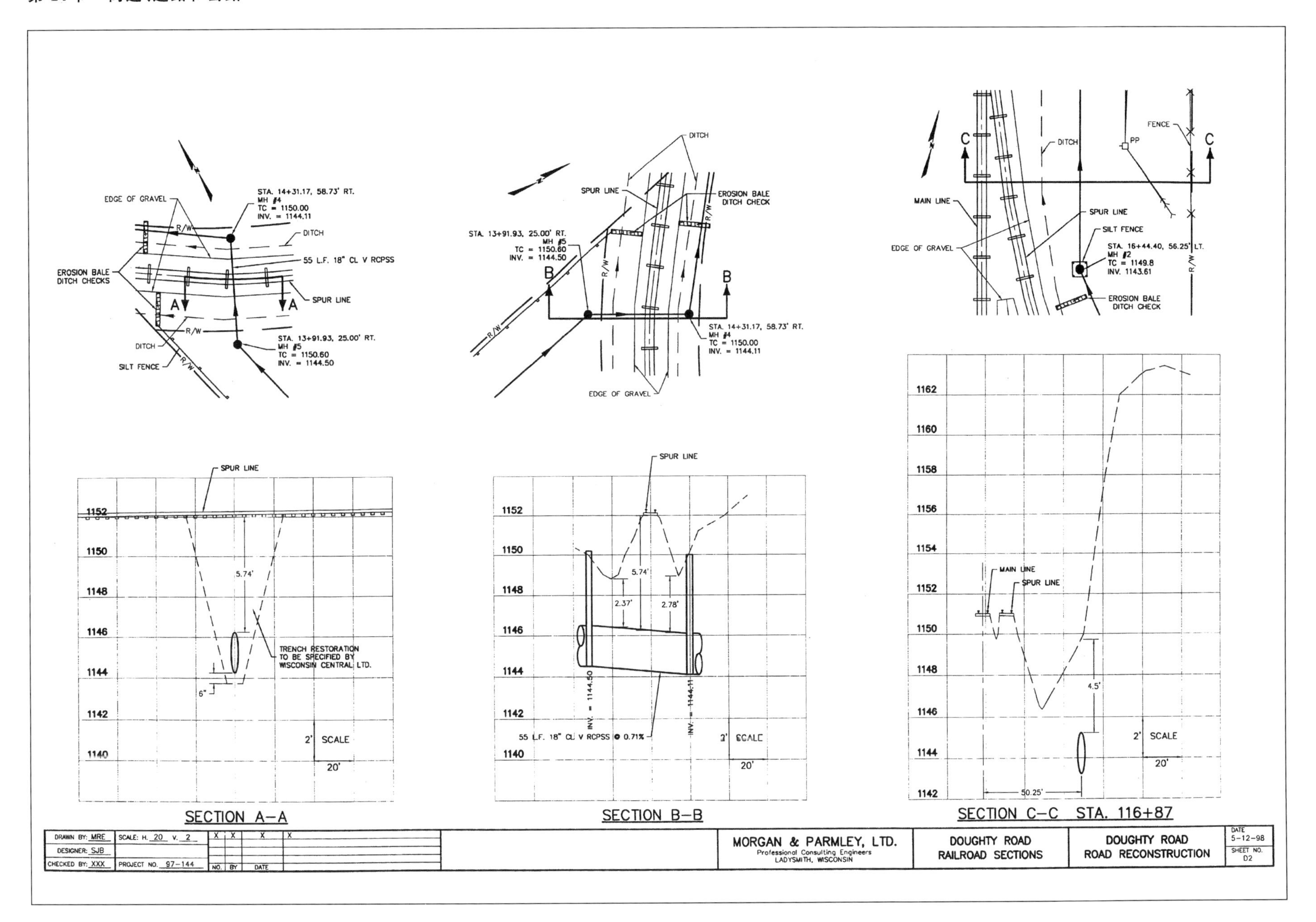
EDGE OF GRAVEL
STA. 14+31.17, 58.73' RT.
MH #4
TC = 1150.00
INV. = 1144.11
DITCH
55 L.F. 18" CL V RCPSS
EROSION BALE
DITCH CHECKS
SPUR LINE
STA. 13+91.93, 25.00' RT.
MH #5
TC = 1150.60
INV. = 1144.50
SILT FENCE
EROSION BALE
DITCH CHECK
EDGE OF GRAVEL
MAIN LINE
FENCE
PP
STA. 16+44.40, 56.25' LT.
MH #2
TC = 1149.8
INV. 1143.61
TRENCH RESTORATION
TO BE SPECIFIED BY
WISCONSIN CENTRAL LTD.
55 L.F. 18" CL V RCPSS @ 0.71%
SECTION A–A
SECTION B–B
SECTION C–C STA. 116+87
DRAWN BY: MRE
DESIGNER: SJB
CHECKED BY: XXX
SCALE: H. 20 V. 2
PROJECT NO. 97-144
MORGAN & PARMLEY, LTD.
Professional Consulting Engineers
LADYSMITH, WISCONSIN
DOUGHTY ROAD
RAILROAD SECTIONS
DOUGHTY ROAD
ROAD RECONSTRUCTION
DATE
5-12-98
SHEET NO.
D2

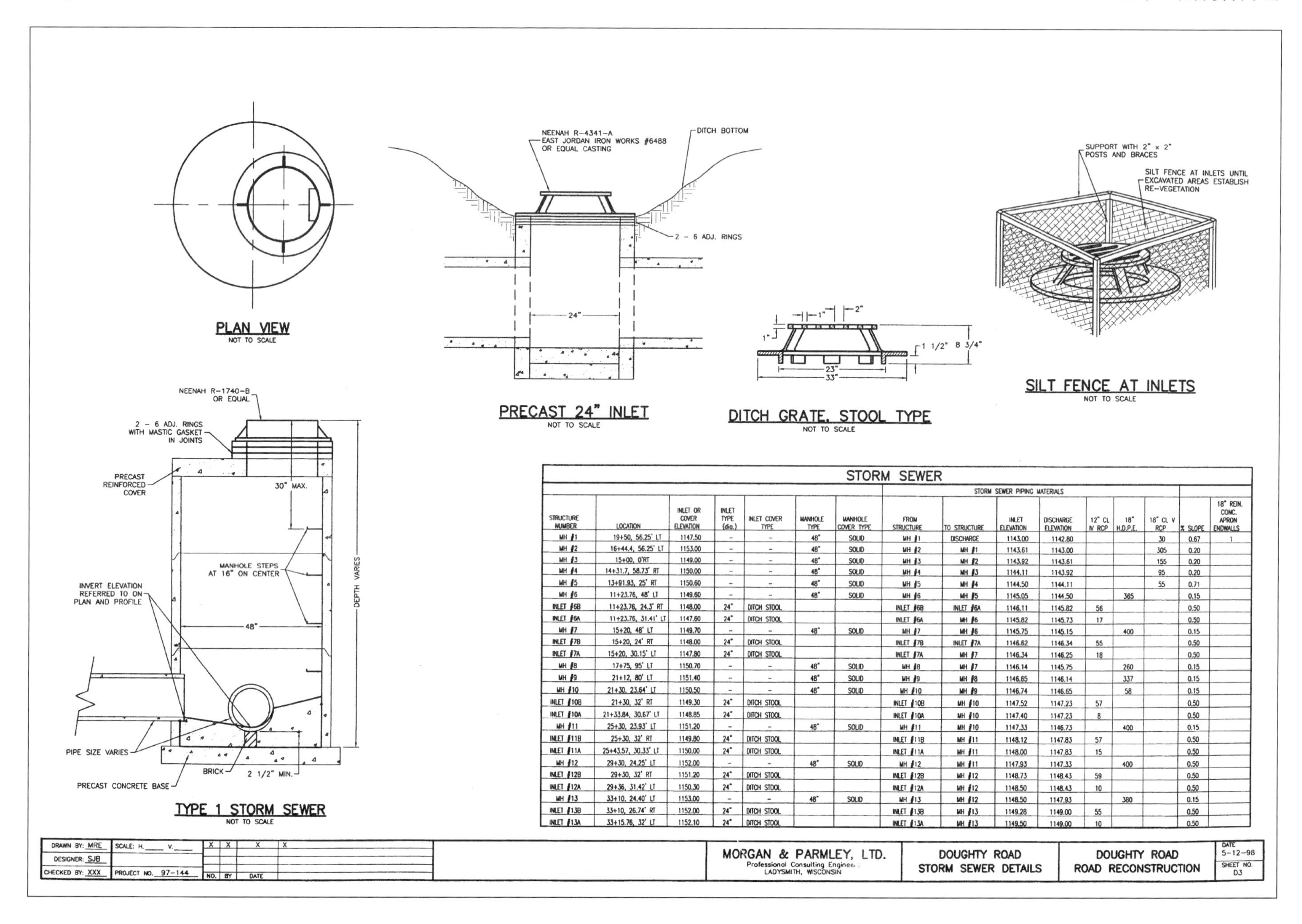

STORM SEWER

STRUCTURE NUMBER	LOCATION	INLET OR COVER ELEVATION	INLET TYPE (dia.)	INLET COVER TYPE	MANHOLE TYPE	MANHOLE COVER TYPE	FROM STRUCTURE	TO STRUCTURE	INLET ELEVATION	DISCHARGE ELEVATION	12" CL IV RCP	18" H.D.P.E.	18" CL V RCP	% SLOPE	18" REIN. CONC. APRON ENDWALLS
							STORM SEWER PIPING MATERIALS								
MH #1	19+50, 56.25' LT	1147.50	–	–	48"	SOLID	MH #1	DISCHARGE	1143.00	1142.80			30	0.67	1
MH #2	16+44.4, 56.25' LT	1153.00	–	–	48"	SOLID	MH #2	MH #1	1143.61	1143.00			305	0.20	
MH #3	15+00, 0'RT	1149.00	–	–	48"	SOLID	MH #3	MH #2	1143.92	1143.61			155	0.20	
MH #4	14+31.7, 58.73' RT	1150.00	–	–	48"	SOLID	MH #4	MH #3	1144.11	1143.92			95	0.20	
MH #5	13+91.93, 25' RT	1150.60	–	–	48"	SOLID	MH #5	MH #4	1144.50	1144.11			55	0.71	
MH #6	11+23.76, 48' LT	1149.60	–	–	48"	SOLID	MH #6	MH #5	1145.05	1144.50		365		0.15	
INLET #6B	11+23.76, 24.3' RT	1148.00	24"	DITCH STOOL			INLET #6B	INLET #6A	1146.11	1145.82	56			0.50	
INLET #6A	11+23.76, 31.41' LT	1147.60	24"	DITCH STOOL			INLET #6A	MH #6	1145.82	1145.73	17			0.50	
MH #7	15+20, 48' LT	1149.70	–	–	48"	SOLID	MH #7	MH #6	1145.75	1145.15		400		0.15	
INLET #7B	15+20, 24' RT	1148.00	24"	DITCH STOOL			INLET #7B	INLET #7A	1146.62	1146.34	55			0.50	
INLET #7A	15+20, 30.15' LT	1147.80	24"	DITCH STOOL			INLET #7A	MH #7	1146.34	1146.25	18			0.50	
MH #8	17+75, 95' LT	1150.70	–	–	48"	SOLID	MH #8	MH #7	1146.14	1145.75		260		0.15	
MH #9	21+12, 80' LT	1151.40	–	–	48"	SOLID	MH #9	MH #8	1146.65	1146.14		337		0.15	
MH #10	21+30, 23.64' LT	1150.50	–	–	48"	SOLID	MH #10	MH #9	1146.74	1146.65		58		0.15	
INLET #10B	21+30, 32' RT	1149.30	24"	DITCH STOOL			INLET #10B	MH #10	1147.52	1147.23	57			0.50	
INLET #10A	21+33.84, 30.67' LT	1148.85	24"	DITCH STOOL			INLET #10A	MH #10	1147.40	1147.23	8			0.50	
MH #11	25+30, 23.93' LT	1151.20	–	–	48"	SOLID	MH #11	MH #10	1147.33	1146.73		400		0.15	
INLET #11B	25+30, 32' RT	1149.80	24"	DITCH STOOL			INLET #11B	MH #11	1148.12	1147.83	57			0.50	
INLET #11A	25+43.57, 30.33' LT	1150.00	24"	DITCH STOOL			INLET #11A	MH #11	1148.00	1147.83	15			0.50	
MH #12	29+30, 24.25' LT	1152.00	–	–	48"	SOLID	MH #12	MH #11	1147.93	1147.33		400		0.50	
INLET #12B	29+30, 32' RT	1151.20	24"	DITCH STOOL			INLET #12B	MH #12	1148.73	1148.43	59			0.50	
INLET #12A	29+36, 31.42' LT	1150.30	24"	DITCH STOOL			INLET #12A	MH #12	1148.50	1148.43	10			0.50	
MH #13	33+10, 24.40' LT	1153.00	–	–	48"	SOLID	MH #13	MH #12	1148.50	1147.93		380		0.15	
INLET #13B	33+10, 26.74' RT	1152.00	24"	DITCH STOOL			INLET #13B	MH #13	1149.28	1149.00	55			0.50	
INLET #13A	33+15.76, 32' LT	1152.10	24"	DITCH STOOL			INLET #13A	MH #13	1149.50	1149.00	10			0.50	

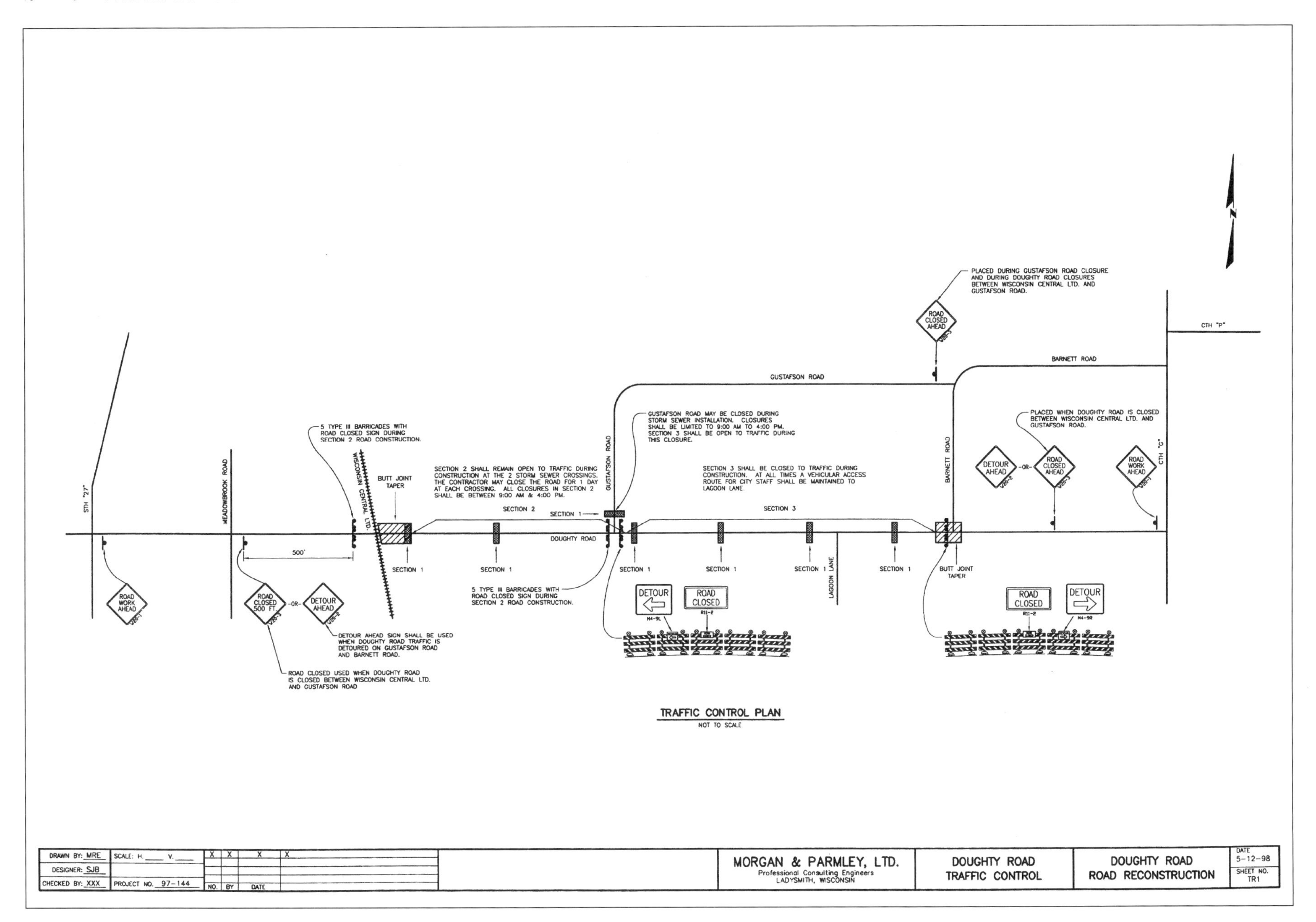

STH "27"
MEADOWBROOK ROAD
ROAD WORK AHEAD
W20-1
ROAD CLOSED 500 FT.
W20-3
-OR-
DETOUR AHEAD
W20-2
500'
5 TYPE III BARRICADES WITH ROAD CLOSED SIGN DURING SECTION 2 ROAD CONSTRUCTION.
WISCONSIN CENTRAL LTD.
BUTT JOINT TAPER
SECTION 2 SHALL REMAIN OPEN TO TRAFFIC DURING CONSTRUCTION AT THE 2 STORM SEWER CROSSINGS. THE CONTRACTOR MAY CLOSE THE ROAD FOR 1 DAY AT EACH CROSSING. ALL CLOSURES IN SECTION 2 SHALL BE BETWEEN 9:00 AM & 4:00 PM.
SECTION 2
SECTION 1
DOUGHTY ROAD
SECTION 1
SECTION 1
5 TYPE III BARRICADES WITH ROAD CLOSED SIGN DURING SECTION 2 ROAD CONSTRUCTION.
DETOUR AHEAD SIGN SHALL BE USED WHEN DOUGHTY ROAD TRAFFIC IS DETOURED ON GUSTAFSON ROAD AND BARNETT ROAD.
ROAD CLOSED USED WHEN DOUGHTY ROAD IS CLOSED BETWEEN WISCONSIN CENTRAL LTD. AND GUSTAFSON ROAD
GUSTAFSON ROAD
GUSTAFSON ROAD MAY BE CLOSED DURING STORM SEWER INSTALLATION. CLOSURES SHALL BE LIMITED TO 9:00 AM TO 4:00 PM. SECTION 3 SHALL BE OPEN TO TRAFFIC DURING THIS CLOSURE.
SECTION 3 SHALL BE CLOSED TO TRAFFIC DURING CONSTRUCTION. AT ALL TIMES A VEHICULAR ACCESS ROUTE FOR CITY STAFF SHALL BE MAINTAINED TO LAGOON LANE.
SECTION 3
SECTION 1
SECTION 1
SECTION 1
SECTION 1
LAGOON LANE
DETOUR
M4-9L
ROAD CLOSED
R11-2
PLACED DURING GUSTAFSON ROAD CLOSURE AND DURING DOUGHTY ROAD CLOSURES BETWEEN WISCONSIN CENTRAL LTD. AND GUSTAFSON ROAD.
ROAD CLOSED AHEAD
W20-3
BARNETT ROAD
CTH "P"
BARNETT ROAD
PLACED WHEN DOUGHTY ROAD IS CLOSED BETWEEN WISCONSIN CENTRAL LTD. AND GUSTAFSON ROAD.
DETOUR AHEAD
W20-2
-OR-
ROAD CLOSED AHEAD
W20-3
ROAD WORK AHEAD
W20-1
CTH "C"
BUTT JOINT TAPER
ROAD CLOSED
R11-2
DETOUR
M4-9R
N
TRAFFIC CONTROL PLAN
NOT TO SCALE
DRAWN BY: MRE
DESIGNER: SJB
CHECKED BY: XXX
SCALE: H. V.
PROJECT NO. 97-144
NO. BY DATE
MORGAN & PARMLEY, LTD.
Professional Consulting Engineers
LADYSMITH, WISCONSIN
DOUGHTY ROAD TRAFFIC CONTROL
DOUGHTY ROAD ROAD RECONSTRUCTION
DATE 5-12-98
SHEET NO. TR1

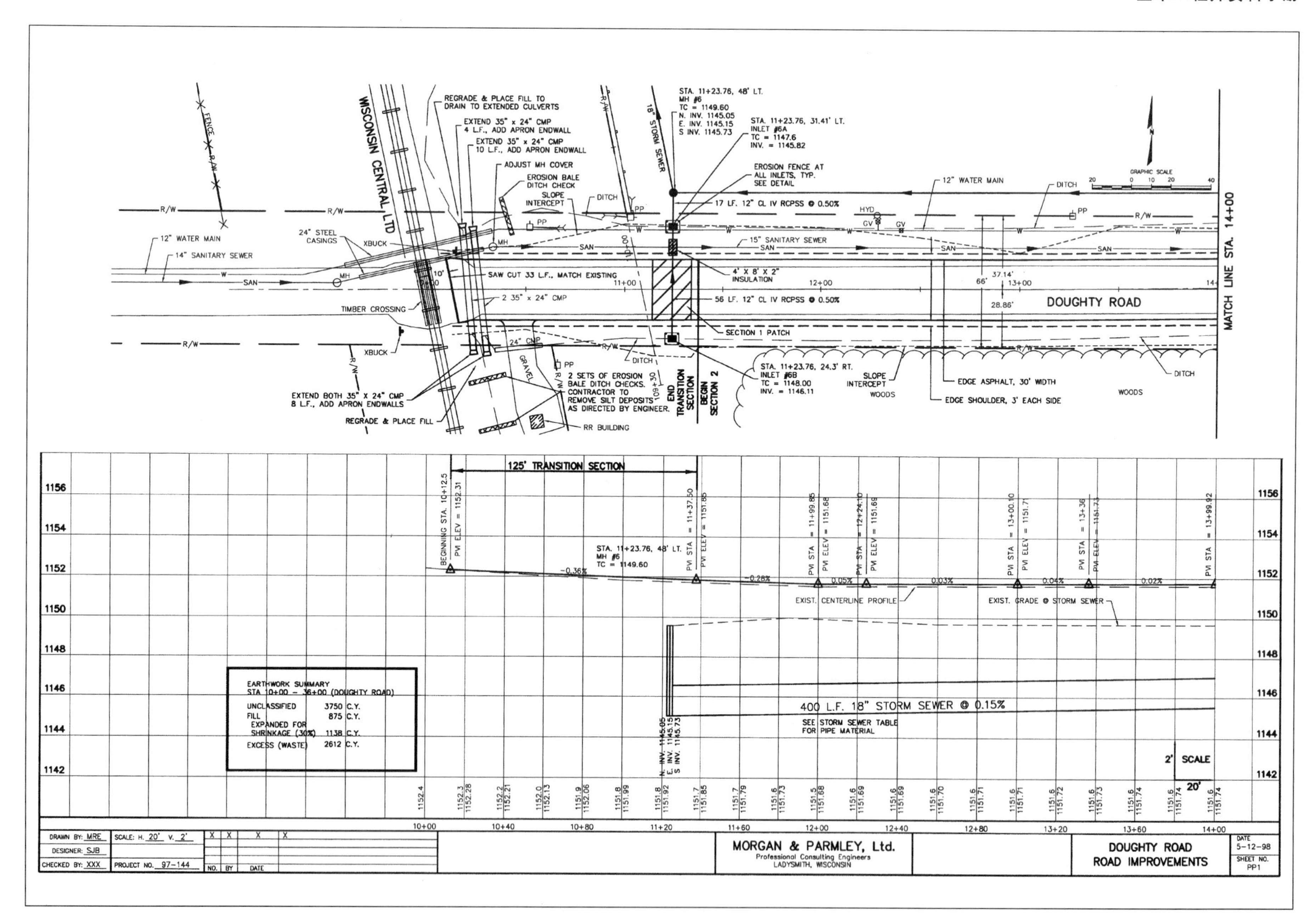
WISCONSIN CENTRAL LTD
DOUGHTY ROAD
MATCH LINE STA. 14+00
125' TRANSITION SECTION
400 L.F. 18" STORM SEWER @ 0.15%
SEE STORM SEWER TABLE FOR PIPE MATERIAL
EXIST. CENTERLINE PROFILE
EXIST. GRADE @ STORM SEWER
EARTHWORK SUMMARY
STA 10+00 - 36+00 (DOUGHTY ROAD)
UNCLASSIFIED 3750 C.Y.
FILL 875 C.Y.
EXPANDED FOR SHRINKAGE (30%) 1138 C.Y.
EXCESS (WASTE) 2612 C.Y.
MORGAN & PARMLEY, Ltd.
Professional Consulting Engineers
LADYSMITH, WISCONSIN
DOUGHTY ROAD
ROAD IMPROVEMENTS
DATE 5-12-98
SHEET NO. PP1
DRAWN BY: MRE
DESIGNER: SJB
CHECKED BY: XXX
SCALE: H. 20' V. 2'
PROJECT NO. 97-144

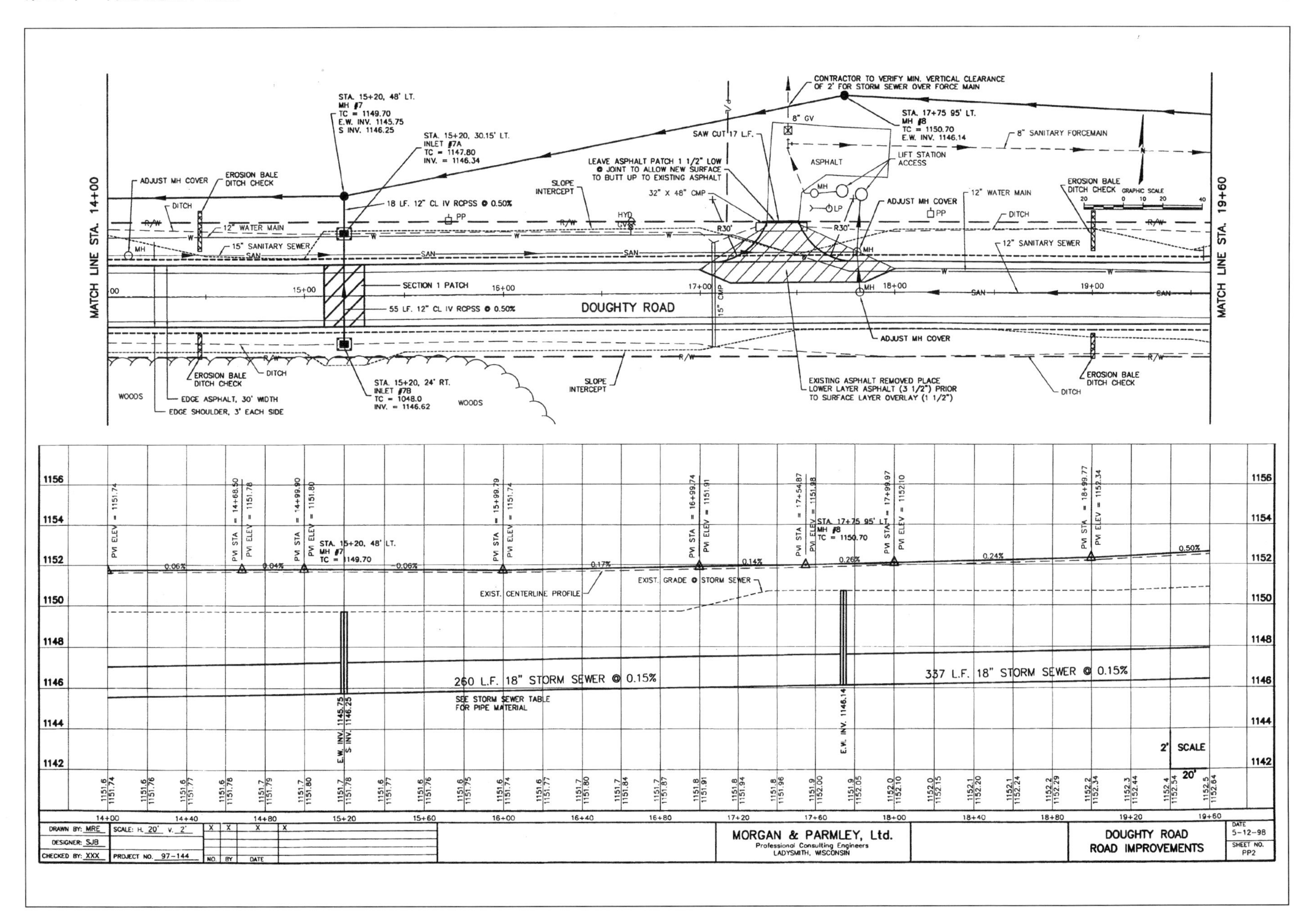

MATCH LINE STA. 14+00
MATCH LINE STA. 19+60
DOUGHTY ROAD
CONTRACTOR TO VERIFY MIN. VERTICAL CLEARANCE OF 2' FOR STORM SEWER OVER FORCE MAIN
STA. 15+20, 48' LT. MH #7 TC = 1149.70 E.W. INV. 1145.75 S INV. 1146.25
STA. 15+20, 30.15' LT. INLET #7A TC = 1147.80 INV. = 1146.34
STA. 17+75 95' LT. MH #8 TC = 1150.70 E.W. INV. 1146.14
STA. 15+20, 24' RT. INLET #7B TC = 1048.0 INV. = 1146.62
18 LF. 12" CL IV RCPSS @ 0.50%
55 LF. 12" CL IV RCPSS @ 0.50%
SECTION 1 PATCH
8" SANITARY FORCEMAIN
12" WATER MAIN
15" SANITARY SEWER
12" SANITARY SEWER
ADJUST MH COVER
EROSION BALE DITCH CHECK
SLOPE INTERCEPT
SAW CUT 17 L.F.
LEAVE ASPHALT PATCH 1 1/2" LOW @ JOINT TO ALLOW NEW SURFACE TO BUTT UP TO EXISTING ASPHALT
32" X 48" CMP
LIFT STATION ACCESS
ASPHALT
EXISTING ASPHALT REMOVED PLACE LOWER LAYER ASPHALT (3 1/2") PRIOR TO SURFACE LAYER OVERLAY (1 1/2")
EDGE ASPHALT, 30' WIDTH
EDGE SHOULDER, 3' EACH SIDE
WOODS
DITCH
260 L.F. 18" STORM SEWER @ 0.15%
337 L.F. 18" STORM SEWER @ 0.15%
SEE STORM SEWER TABLE FOR PIPE MATERIAL
EXIST. CENTERLINE PROFILE
EXIST. GRADE @ STORM SEWER
MORGAN & PARMLEY, Ltd.
Professional Consulting Engineers
LADYSMITH, WISCONSIN
DOUGHTY ROAD
ROAD IMPROVEMENTS
DRAWN BY: MRE
DESIGNER: SJB
CHECKED BY: XXX
SCALE: H. 20' V. 2'
PROJECT NO. 97-144
DATE 5-12-98
SHEET NO. PP2

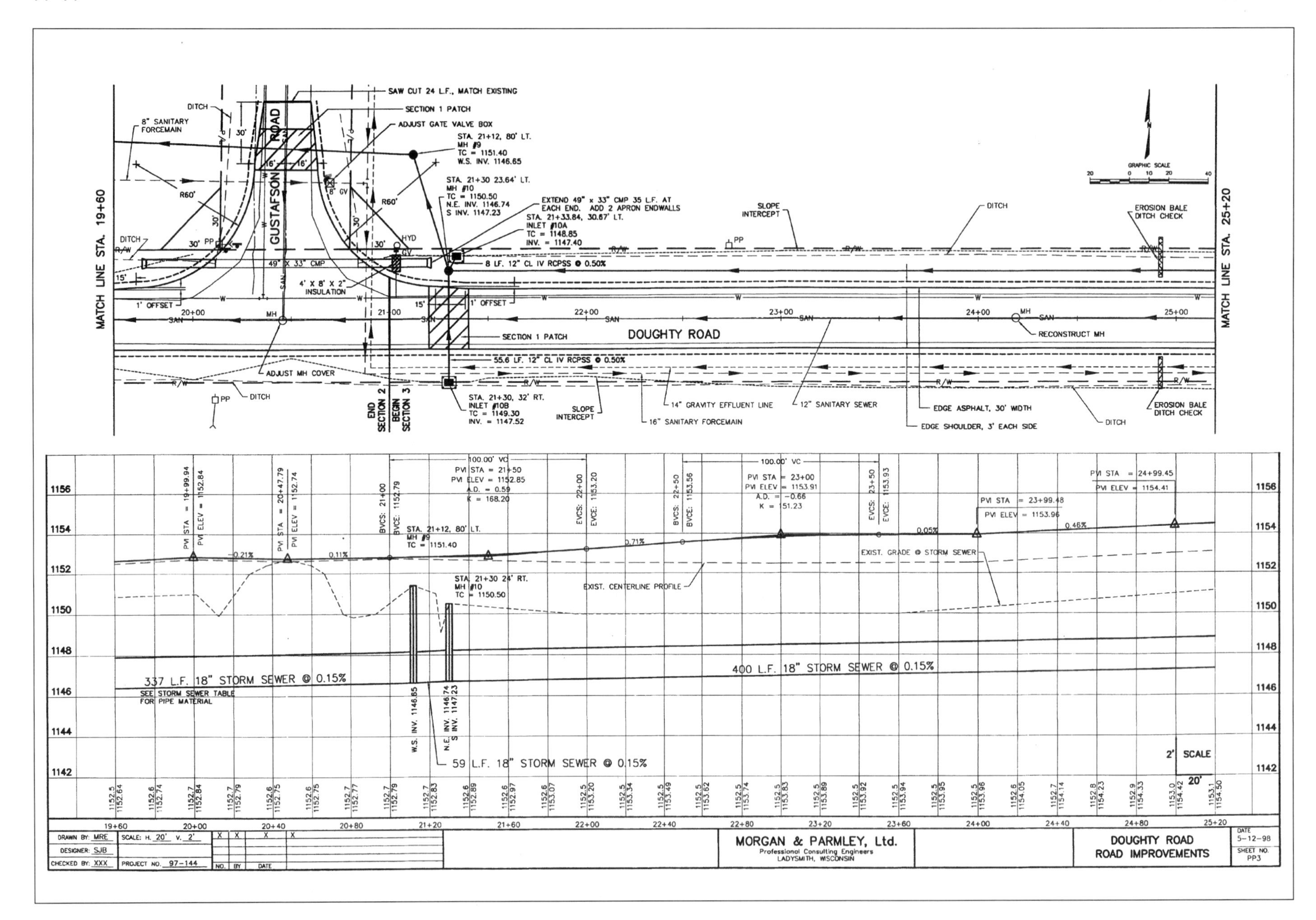
SAW CUT 24 L.F., MATCH EXISTING
SECTION 1 PATCH
ADJUST GATE VALVE BOX
STA. 21+12, 80' LT. MH #9 TC = 1151.40 W.S. INV. 1146.65
STA. 21+30 23.64' LT. MH #10 TC = 1150.50 N.E. INV. 1146.74 S INV. 1147.23
EXTEND 49" x 33" CMP 35 L.F. AT EACH END. ADD 2 APRON ENDWALLS
STA. 21+33.84, 30.67' LT. INLET #10A TC = 1148.85 INV. = 1147.40
8 LF. 12" CL IV RCPSS @ 0.50%
55.6 LF. 12" CL IV RCPSS @ 0.50%
STA. 21+30, 32' RT. INLET #10B TC = 1149.30 INV. = 1147.52
GUSTAFSON ROAD
DOUGHTY ROAD
MATCH LINE STA. 19+60
MATCH LINE STA. 25+20
8" SANITARY FORCEMAIN
DITCH
SLOPE INTERCEPT
EROSION BALE DITCH CHECK
4' X 8' X 2" INSULATION
1' OFFSET
ADJUST MH COVER
RECONSTRUCT MH
END SECTION 2
BEGIN SECTION 3
14" GRAVITY EFFLUENT LINE
12" SANITARY SEWER
16" SANITARY FORCEMAIN
EDGE ASPHALT, 30' WIDTH
EDGE SHOULDER, 3' EACH SIDE
GRAPHIC SCALE
PVI STA = 19+99.94 PVI ELEV = 1152.84
PVI STA = 20+47.79 PVI ELEV = 1152.74
100.00' VC PVI STA = 21+50 PVI ELEV = 1152.85 A.D. = 0.59 K = 168.20
100.00' VC PVI STA = 23+00 PVI ELEV = 1153.91 A.D. = -0.66 K = 151.23
PVI STA = 23+99.48 PVI ELEV = 1153.96
PVI STA = 24+99.45 PVI ELEV = 1154.41
EXIST. GRADE @ STORM SEWER
EXIST. CENTERLINE PROFILE
337 L.F. 18" STORM SEWER @ 0.15%
SEE STORM SEWER TABLE FOR PIPE MATERIAL
400 L.F. 18" STORM SEWER @ 0.15%
59 L.F. 18" STORM SEWER @ 0.15%
W.S. INV. 1146.65
N.E. INV. 1146.74 S INV. 1147.23
MORGAN & PARMLEY, Ltd.
Professional Consulting Engineers
LADYSMITH, WISCONSIN
DOUGHTY ROAD ROAD IMPROVEMENTS
DRAWN BY: MRE
DESIGNER: SJB
CHECKED BY: XXX
SCALE: H. 20' V. 2'
PROJECT NO. 97-144
DATE 5-12-98
SHEET NO. PP3

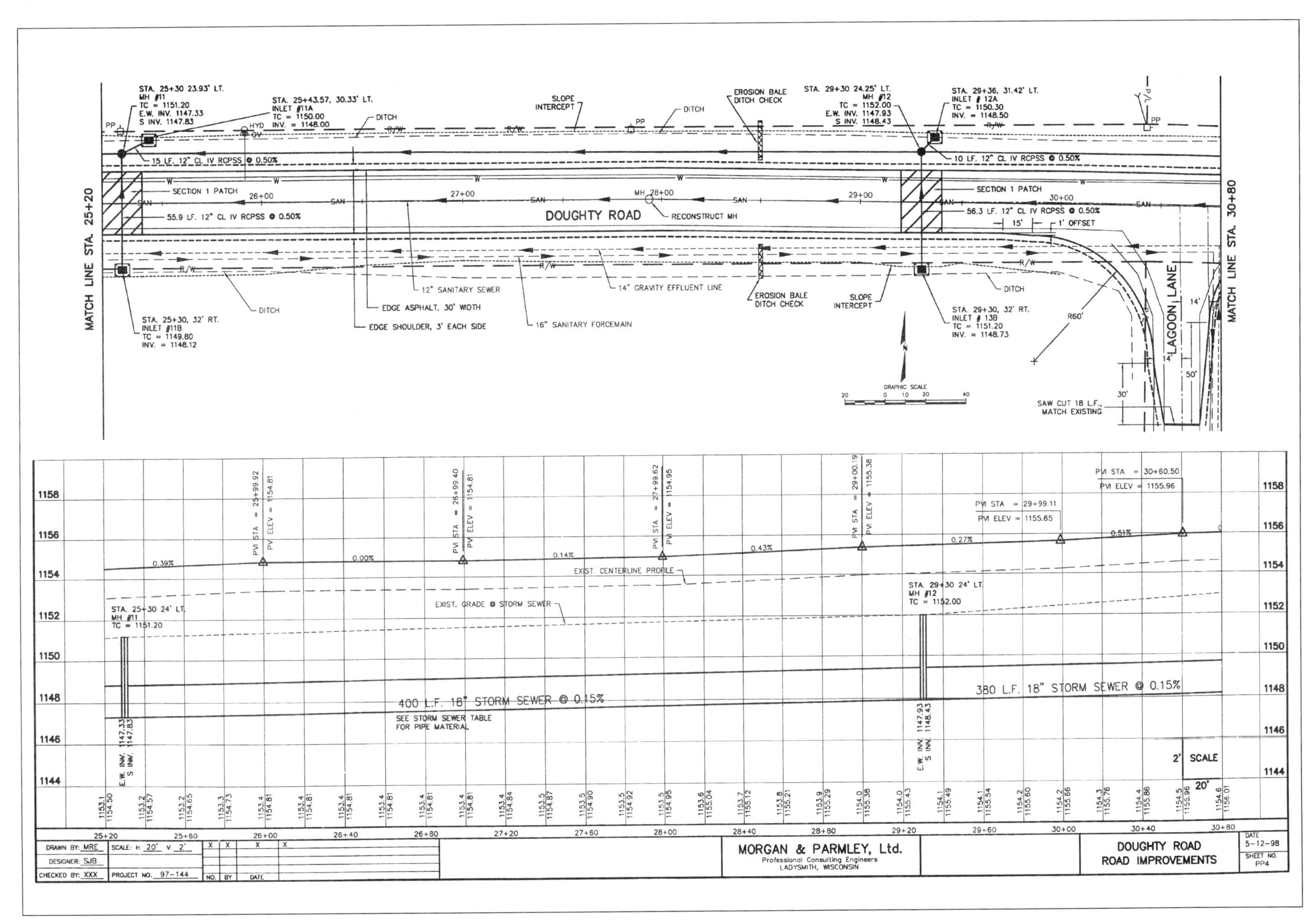
MATCH LINE STA. 25+20
MATCH LINE STA. 30+80
DOUGHTY ROAD
LAGOON LANE
STA. 25+30 23.93' LT. MH #11 TC = 1151.20 E.W. INV. 1147.33 S INV. 1147.83
STA. 25+43.57, 30.33' LT. INLET #11A TC = 1150.00 INV. = 1148.00
STA. 29+30 24.25' LT. MH #12 TC = 1152.00 E.W. INV. 1147.93 S INV. 1148.43
STA. 29+36, 31.42' LT. INLET # 12A TC = 1150.30 INV. = 1148.50
STA. 25+30, 32' RT. INLET #11B TC = 1149.80 INV. = 1148.12
STA. 29+30, 32' RT. INLET # 13B TC = 1151.20 INV. = 1148.73
15 LF. 12" CL IV RCPSS @ 0.50%
10 LF. 12" CL IV RCPSS @ 0.50%
55.9 LF. 12" CL IV RCPSS @ 0.50%
56.3 LF. 12" CL IV RCPSS @ 0.50%
SECTION 1 PATCH
RECONSTRUCT MH
12" SANITARY SEWER
14" GRAVITY EFFLUENT LINE
16" SANITARY FORCEMAIN
EDGE ASPHALT, 30' WIDTH
EDGE SHOULDER, 3' EACH SIDE
EROSION BALE DITCH CHECK
SLOPE INTERCEPT
DITCH
SAW CUT 18 L.F., MATCH EXISTING
1' OFFSET
R60'
GRAPHIC SCALE
400 L.F. 18" STORM SEWER @ 0.15%
380 L.F. 18" STORM SEWER @ 0.15%
SEE STORM SEWER TABLE FOR PIPE MATERIAL
EXIST. CENTERLINE PROFILE
EXIST. GRADE @ STORM SEWER
PVI STA = 30+60.50 PVI ELEV = 1155.96
PVI STA = 29+99.11 PVI ELEV = 1155.65
MORGAN & PARMLEY, Ltd.
Professional Consulting Engineers
LADYSMITH, WISCONSIN
DOUGHTY ROAD ROAD IMPROVEMENTS
DATE 5-12-98
SHEET NO. PP4
DRAWN BY: MRE
DESIGNER: SJB
CHECKED BY: XXX
SCALE: H 20' V 2'
PROJECT NO. 97-144

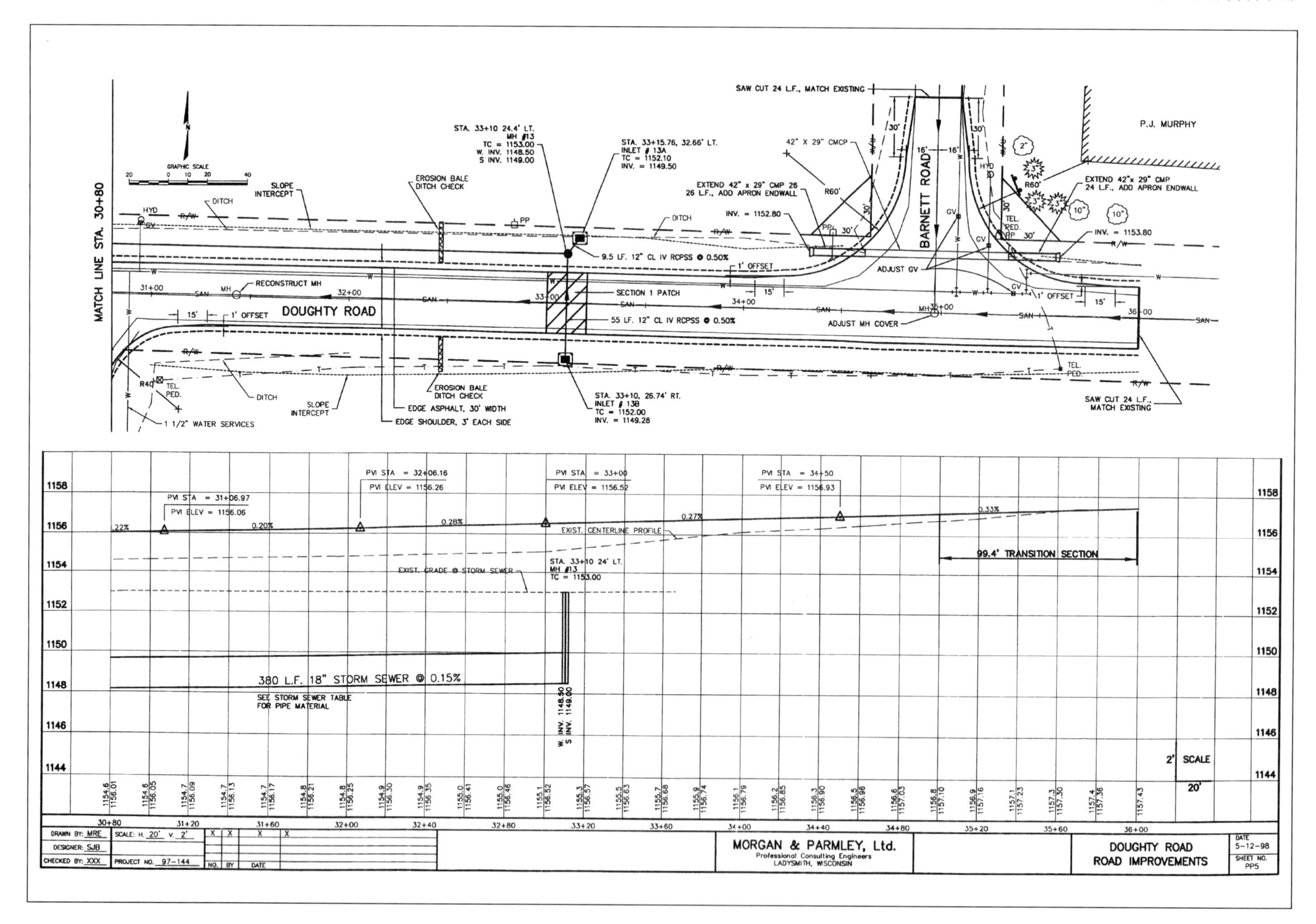

MATCH LINE STA. 30+80
DOUGHTY ROAD
BARNETT ROAD
P.J. MURPHY
STA. 33+10 24.4' LT. MH #13 TC = 1153.00 W. INV. 1148.50 S INV. 1149.00
STA. 33+15.76, 32.66' LT. INLET # 13A TC = 1152.10 INV. = 1149.50
STA. 33+10, 26.74' RT. INLET # 13B TC = 1152.00 INV. = 1149.28
EROSION BALE DITCH CHECK
SLOPE INTERCEPT
SAW CUT 24 L.F., MATCH EXISTING
EXTEND 42" x 29" CMP 26 L.F., ADD APRON ENDWALL
EXTEND 42" x 29" CMP 24 L.F., ADD APRON ENDWALL
9.5 LF. 12" CL IV RCPSS @ 0.50%
55 LF. 12" CL IV RCPSS @ 0.50%
RECONSTRUCT MH
SECTION 1 PATCH
ADJUST GV
ADJUST MH COVER
EDGE ASPHALT, 30' WIDTH
EDGE SHOULDER, 3' EACH SIDE
1 1/2" WATER SERVICES
PVI STA = 31+06.97 PVI ELEV = 1156.06
PVI STA = 32+06.16 PVI ELEV = 1156.26
PVI STA = 33+00 PVI ELEV = 1156.52
PVI STA = 34+50 PVI ELEV = 1156.93
EXIST. CENTERLINE PROFILE
99.4' TRANSITION SECTION
EXIST. GRADE @ STORM SEWER
380 L.F. 18" STORM SEWER @ 0.15%
SEE STORM SEWER TABLE FOR PIPE MATERIAL
MORGAN & PARMLEY, Ltd. Professional Consulting Engineers LADYSMITH, WISCONSIN
DOUGHTY ROAD ROAD IMPROVEMENTS
DATE 5-12-98
SHEET NO. PP5
DRAWN BY: MRE
DESIGNER: SJB
CHECKED BY: XXX
SCALE: H. 20' V. 2'
PROJECT NO. 97-144

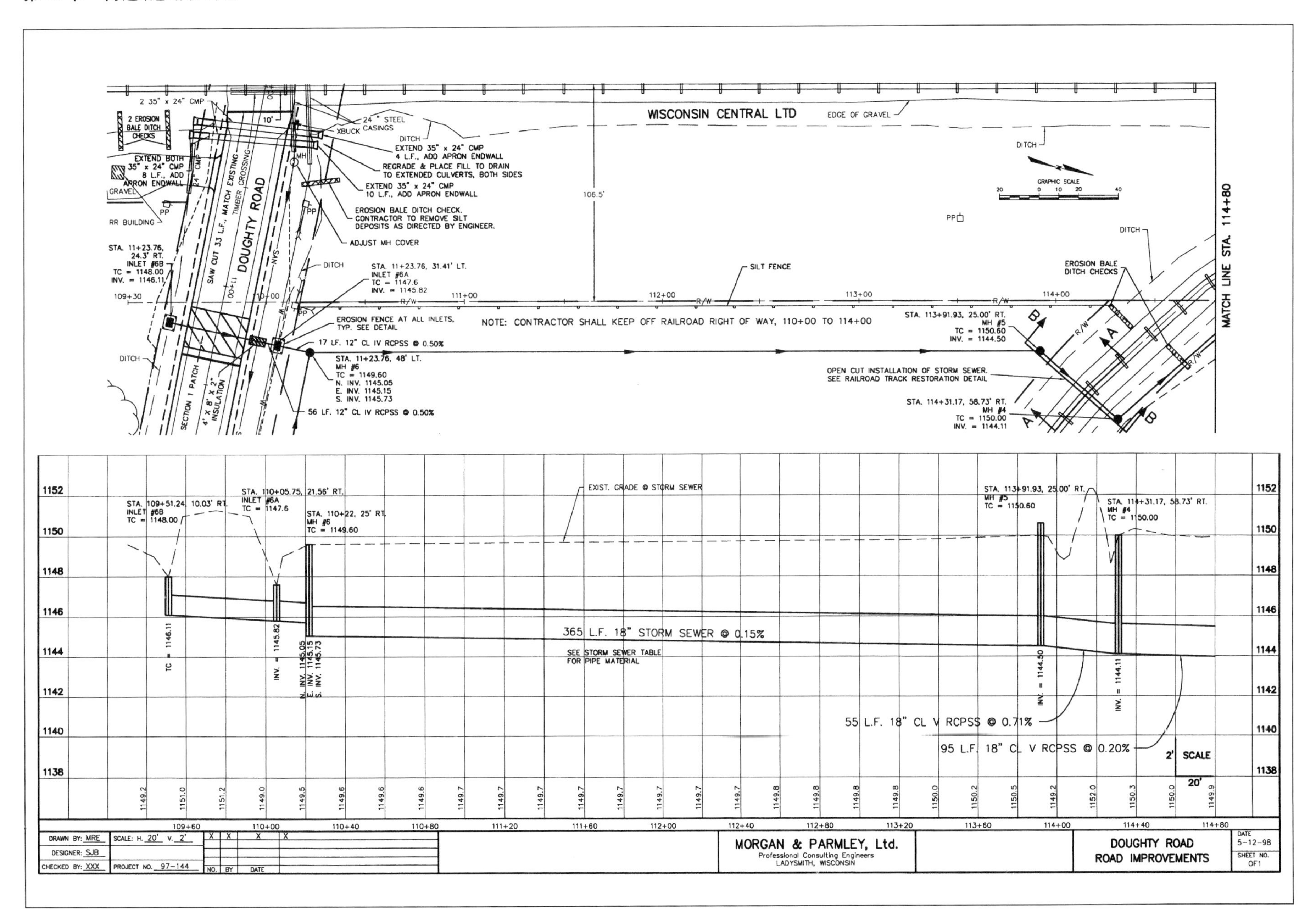

WISCONSIN CENTRAL LTD
EDGE OF GRAVEL
DOUGHTY ROAD
NOTE: CONTRACTOR SHALL KEEP OFF RAILROAD RIGHT OF WAY, 110+00 TO 114+00
MATCH LINE STA. 114+80
365 L.F. 18" STORM SEWER @ 0.15%
SEE STORM SEWER TABLE FOR PIPE MATERIAL
55 L.F. 18" CL V RCPSS @ 0.71%
95 L.F. 18" CL V RCPSS @ 0.20%
MORGAN & PARMLEY, Ltd.
Professional Consulting Engineers
LADYSMITH, WISCONSIN
DOUGHTY ROAD
ROAD IMPROVEMENTS

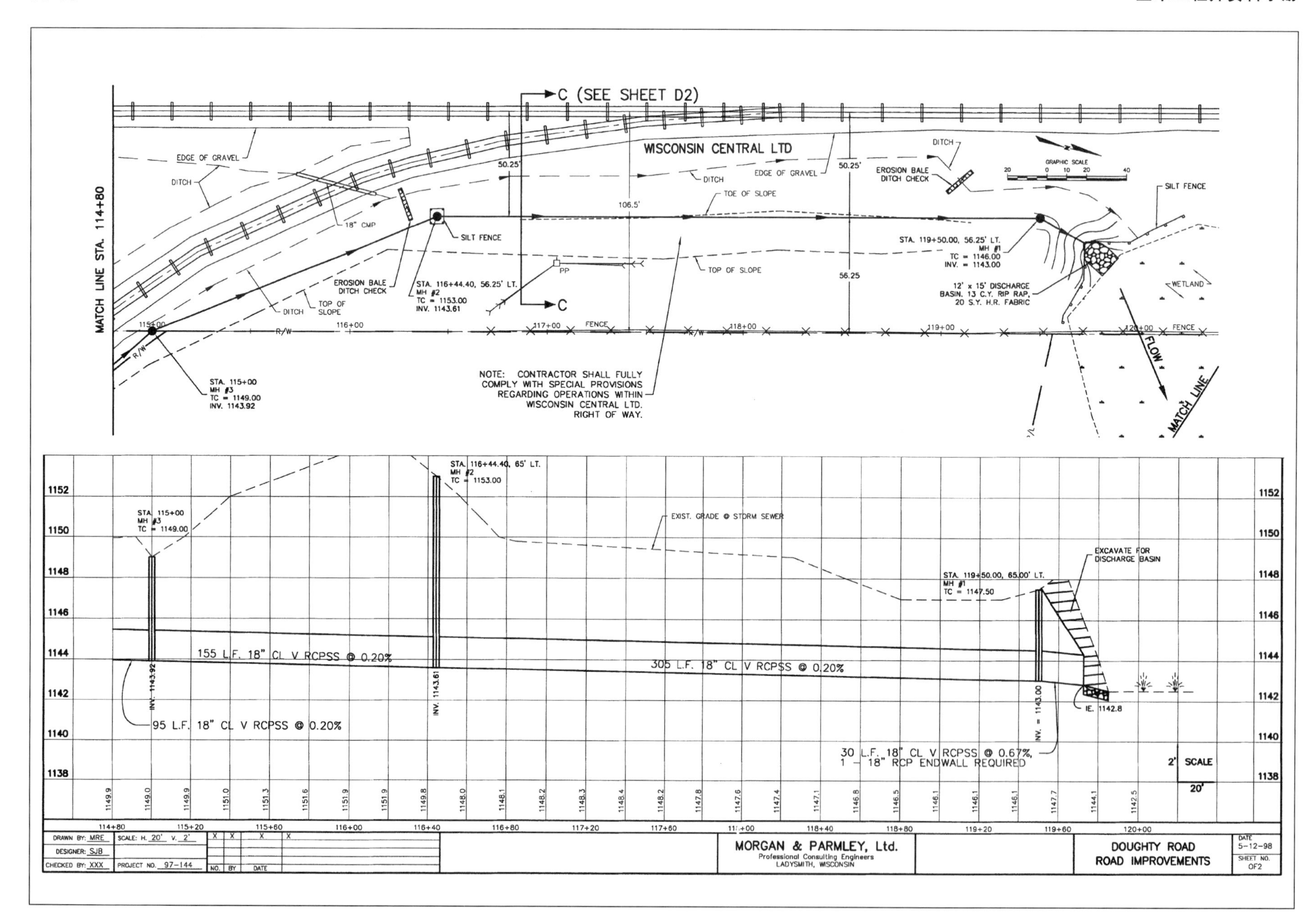

C (SEE SHEET D2)
WISCONSIN CENTRAL LTD
MATCH LINE STA. 114+80
EDGE OF GRAVEL
DITCH
18" CMP
SILT FENCE
EROSION BALE DITCH CHECK
TOE OF SLOPE
TOP OF SLOPE
STA. 116+44.40, 56.25' LT. MH #2 TC = 1153.00 INV. 1143.61
STA. 115+00 MH #3 TC = 1149.00 INV. 1143.92
STA. 119+50.00, 56.25' LT. MH #1 TC = 1146.00 INV. = 1143.00
12' x 15' DISCHARGE BASIN, 13 C.Y. RIP RAP, 20 S.Y. H.R. FABRIC
WETLAND
FLOW
MATCH LINE
NOTE: CONTRACTOR SHALL FULLY COMPLY WITH SPECIAL PROVISIONS REGARDING OPERATIONS WITHIN WISCONSIN CENTRAL LTD. RIGHT OF WAY.
GRAPHIC SCALE
EXIST. GRADE @ STORM SEWER
EXCAVATE FOR DISCHARGE BASIN
STA. 119+50.00, 65.00' LT. MH #1 TC = 1147.50
155 L.F. 18" CL V RCPSS @ 0.20%
305 L.F. 18" CL V RCPSS @ 0.20%
95 L.F. 18" CL V RCPSS @ 0.20%
30 L.F. 18" CL V RCPSS @ 0.67%, 1 - 18" RCP ENDWALL REQUIRED
IE. 1142.8
2' SCALE 20'
DRAWN BY: MRE
DESIGNER: SJB
CHECKED BY: XXX
SCALE: H. 20' V. 2'
PROJECT NO. 97-144
MORGAN & PARMLEY, Ltd.
Professional Consulting Engineers
LADYSMITH, WISCONSIN
DOUGHTY ROAD
ROAD IMPROVEMENTS
DATE 5-12-98
SHEET NO. OF2

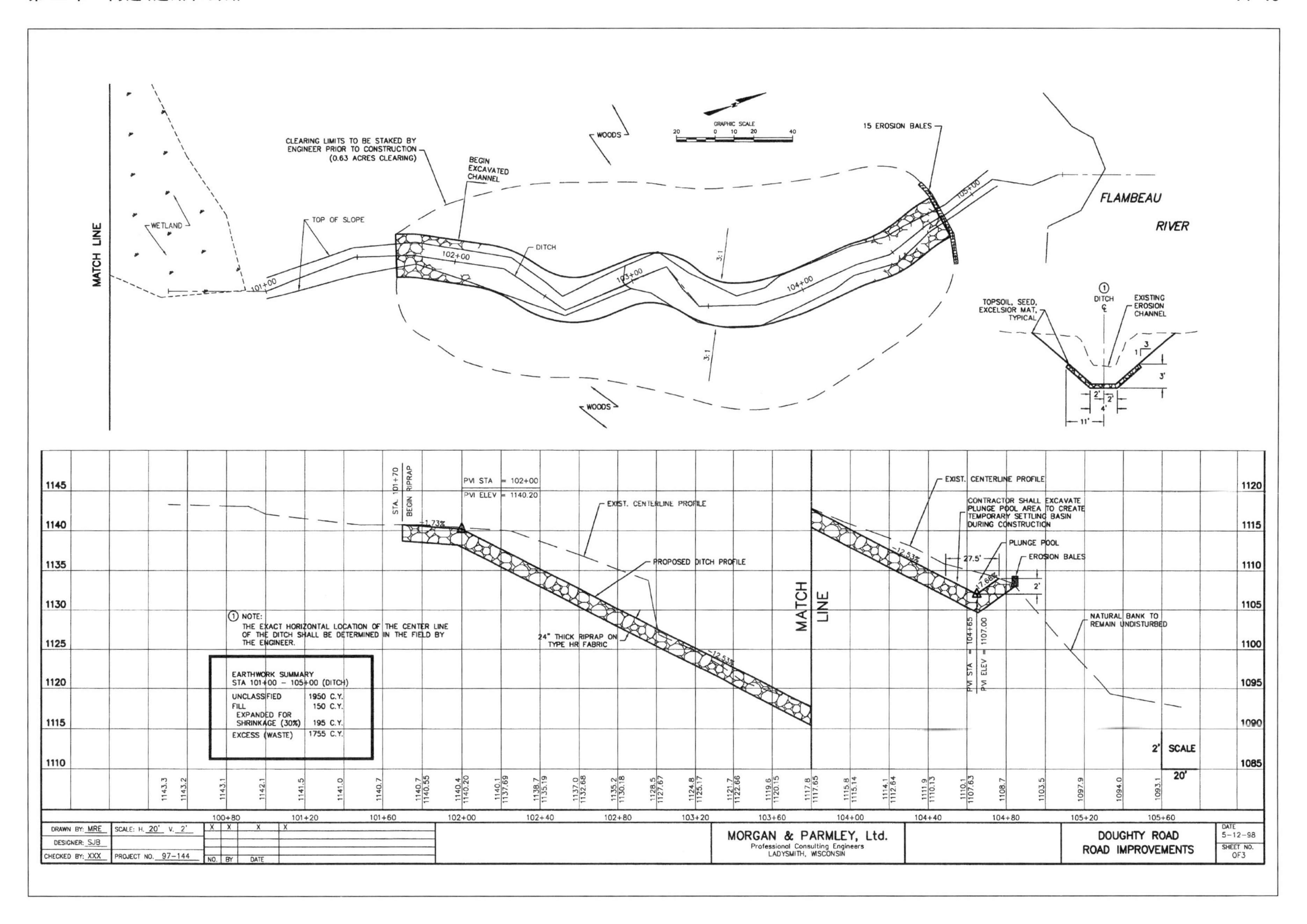
MATCH LINE
WETLAND
CLEARING LIMITS TO BE STAKED BY ENGINEER PRIOR TO CONSTRUCTION (0.63 ACRES CLEARING)
TOP OF SLOPE
BEGIN EXCAVATED CHANNEL
DITCH
WOODS
GRAPHIC SCALE
15 EROSION BALES
FLAMBEAU RIVER
TOPSOIL, SEED, EXCELSIOR MAT, TYPICAL
DITCH
EXISTING EROSION CHANNEL
PVI STA = 102+00
PVI ELEV = 1140.20
STA. 101+70 BEGIN RIPRAP
EXIST. CENTERLINE PROFILE
PROPOSED DITCH PROFILE
24" THICK RIPRAP ON TYPE HR FABRIC
NOTE: THE EXACT HORIZONTAL LOCATION OF THE CENTER LINE OF THE DITCH SHALL BE DETERMINED IN THE FIELD BY THE ENGINEER.
EARTHWORK SUMMARY
STA 101+00 – 105+00 (DITCH)
UNCLASSIFIED 1950 C.Y.
FILL 150 C.Y.
EXPANDED FOR SHRINKAGE (30%) 195 C.Y.
EXCESS (WASTE) 1755 C.Y.
MATCH LINE
CONTRACTOR SHALL EXCAVATE PLUNGE POOL AREA TO CREATE TEMPORARY SETTLING BASIN DURING CONSTRUCTION
PLUNGE POOL
EROSION BALES
NATURAL BANK TO REMAIN UNDISTURBED
PVI STA = 104+65
PVI ELEV = 1107.00
SCALE
DRAWN BY: MRE
DESIGNER: SJB
CHECKED BY: XXX
SCALE: H. 20' V. 2'
PROJECT NO. 97-144
MORGAN & PARMLEY, Ltd.
Professional Consulting Engineers
LADYSMITH, WISCONSIN
DOUGHTY ROAD
ROAD IMPROVEMENTS
DATE 5-12-98
SHEET NO. OF3

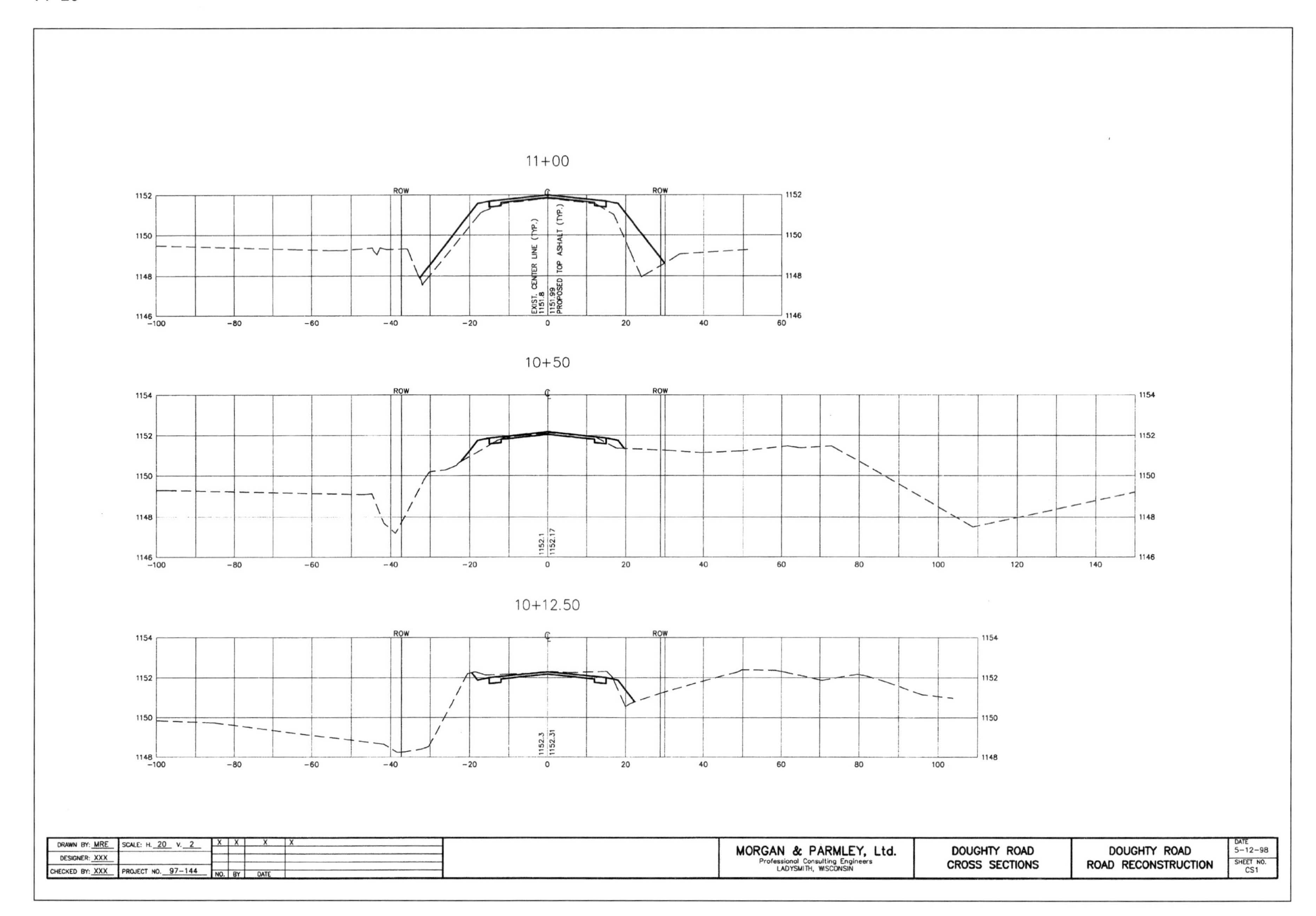
11+00
ROW
ROW
EXIST. CENTER LINE (TYP.)
1151.8
1151.99
PROPOSED TOP ASHALT (TYP.)
10+50
1152.1
1152.17
10+12.50
1152.3
1152.31
DRAWN BY: MRE
DESIGNER: XXX
CHECKED BY: XXX
SCALE: H. 20 V. 2
PROJECT NO. 97-144
NO. BY DATE
MORGAN & PARMLEY, Ltd.
Professional Consulting Engineers
LADYSMITH, WISCONSIN
DOUGHTY ROAD
CROSS SECTIONS
DOUGHTY ROAD
ROAD RECONSTRUCTION
DATE
5-12-98
SHEET NO.
CS1

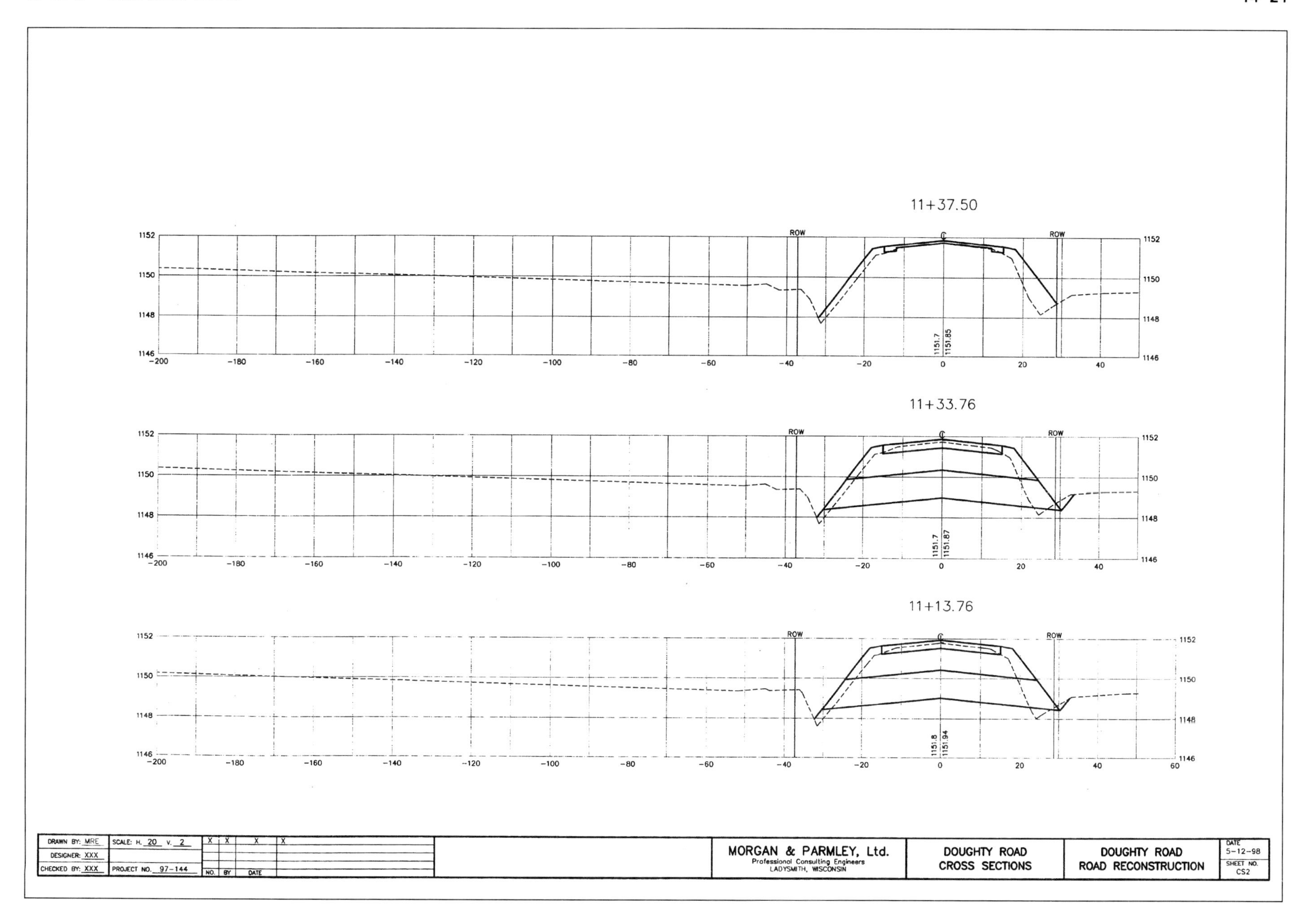

11+37.50
11+33.76
11+13.76
ROW
1152
1150
1148
1146
151.7
1151.85
1151.87
1151.8
1151.94
DRAWN BY: MRE
DESIGNER: XXX
CHECKED BY: XXX
SCALE: H. 20 V. 2
PROJECT NO. 97-144
NO.
BY
DATE
MORGAN & PARMLEY, Ltd.
Professional Consulting Engineers
LADYSMITH, WISCONSIN
DOUGHTY ROAD
CROSS SECTIONS
DOUGHTY ROAD
ROAD RECONSTRUCTION
DATE
5-12-98
SHEET NO.
CS2

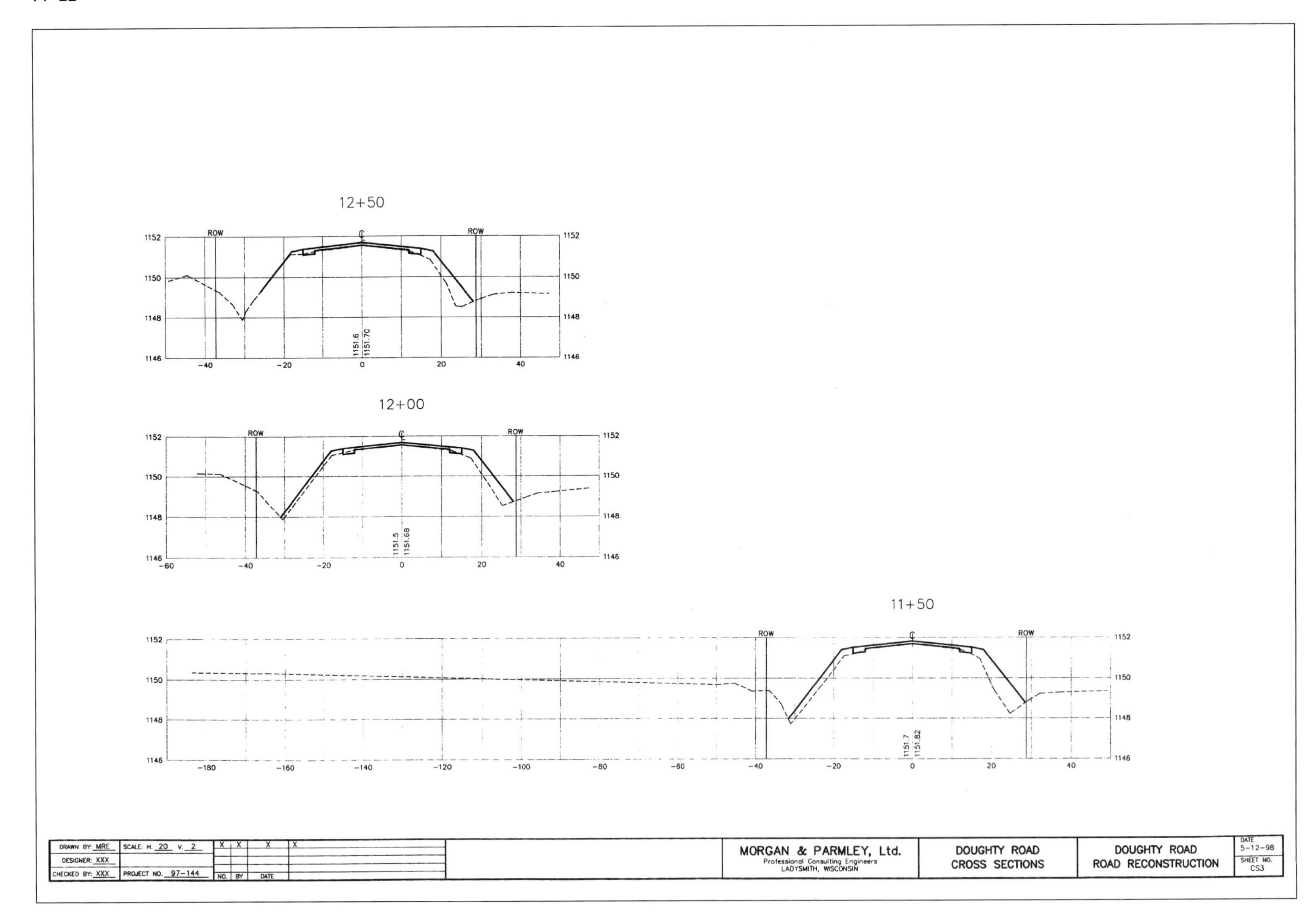
12+50
ROW
ROW
1152
1150
1148
1146
1151.6
1151.70
-40
-20
0
20
40
12+00
1151.5
1151.68
-60
11+50
1151.7
1151.82
-180
-160
-140
-120
-100
-80
-60
DRAWN BY: MRE
DESIGNER: XXX
CHECKED BY: XXX
SCALE: H. 20 V. 2
PROJECT NO. 97-144
NO. BY DATE
MORGAN & PARMLEY, Ltd.
Professional Consulting Engineers
LADYSMITH, WISCONSIN
DOUGHTY ROAD
CROSS SECTIONS
DOUGHTY ROAD
ROAD RECONSTRUCTION
DATE
5-12-98
SHEET NO.
CS3

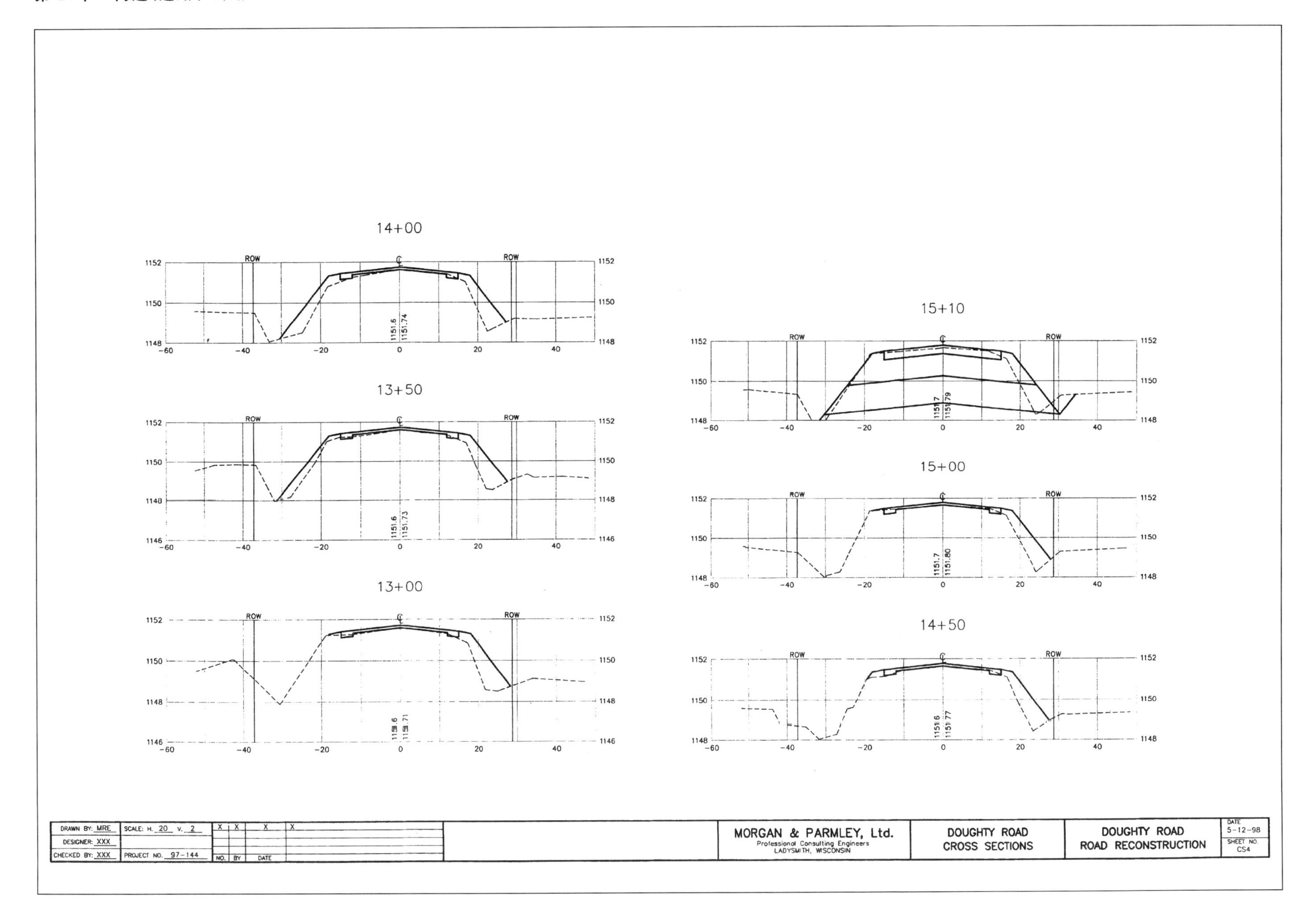
14+00
13+50
13+00
15+10
15+00
14+50
ROW
DRAWN BY: MRE
DESIGNER: XXX
CHECKED BY: XXX
SCALE: H. 20 V. 2
PROJECT NO. 97-144
NO. BY DATE
MORGAN & PARMLEY, Ltd.
Professional Consulting Engineers
LADYSMITH, WISCONSIN
DOUGHTY ROAD
CROSS SECTIONS
DOUGHTY ROAD
ROAD RECONSTRUCTION
DATE
5-12-98
SHEET NO.
CS4

Rames 路改造

INDEX OF SHEETS

STATE OF WISCONSIN
DEPARTMENT OF TRANSPORTATION
PLAN OF PROPOSED IMPROVEMENT

RAMES ROAD – NORTH 1.163 MILES

C.T.H. A
RUSK COUNTY

STATE PROJECT	FEDERAL PROJECT	
	PROJECT	CONTRACT
8787-05-71		

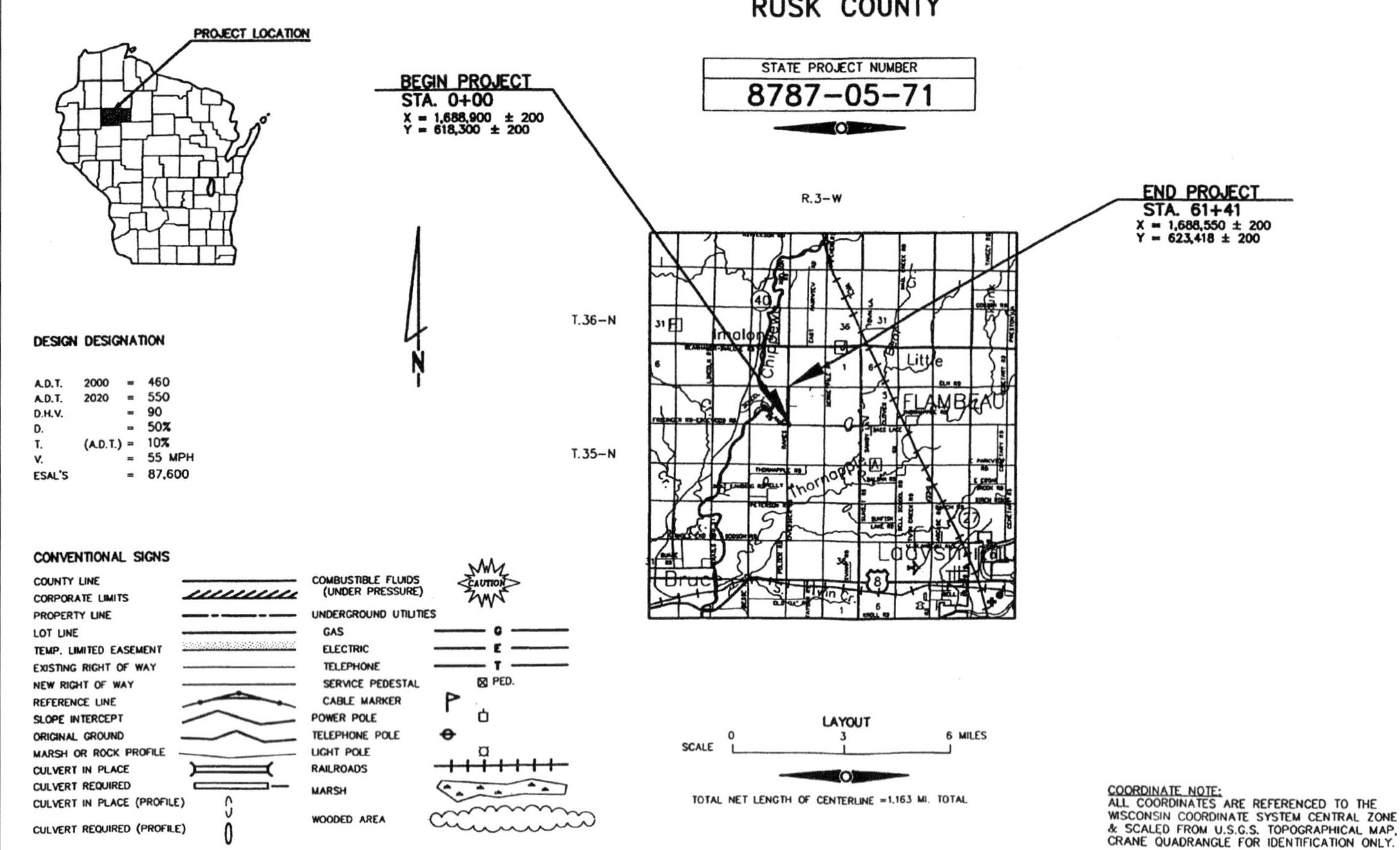

ACCEPTED FOR
COUNTY of RUSK
1-20-2000 Highway Commissioner
(Date) (Signature & Title of Official)

ORIGINAL PLAN PREPARED BY:
MORGAN & PARMLEY, LTD.
CONSULTING ENGINEERS, LADYSMITH, WISCONSIN

WISCONSIN
LARRY F. GOTHAM
E-19463
NEW AUBURN, WI
PROFESSIONAL ENGINEER

1/20/00
(Date) (Signature)

STATE OF WISCONSIN
DEPARTMENT OF TRANSPORTATION

PREPARED BY
Surveyor MORGAN & PARMLEY, LTD.
Designer MORGAN & PARMLEY, LTD.
District Examiner CHRISTINE KOSKI
District Supervisor RICK WASHKUHN
Project Dev. Engineer
C.O. Examiner

Courtesy: Morgan & Parmley, Ltd.

STATE PROJECT NUMBER	SHEET NO.
8787-05-71	2

TYPICAL SECTIONS-DETAILED SUMMARY OF MISCELLANEOUS QUANTITIES

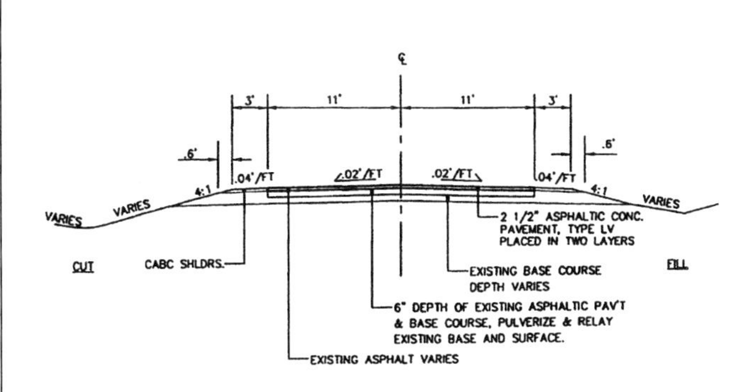

TYPICAL SECTION

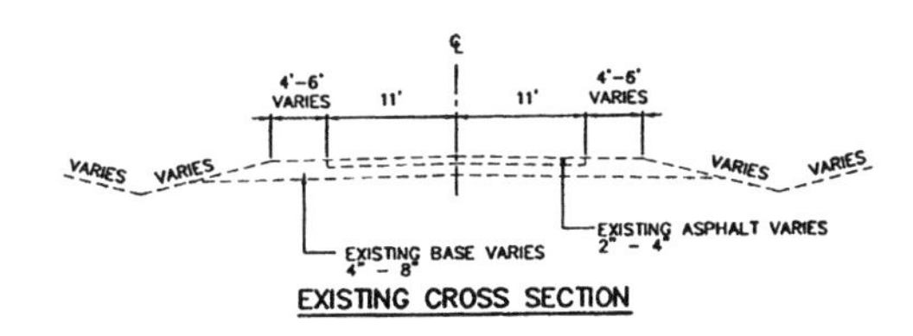

EXISTING CROSS SECTION

CALL DIGGERS' HOTLINE

1-800-242-8511

TOLL FREE

2040 W. WISCONSIN AVE.
SUITE 10
MILWAUKEE, WI 53233

RUSK COUNTY SURVEYOR
DAVE KAISER, RLS
(715) 532-2165

OTHER CONTACTS

DNR LIASON

DAN MICHELS
D.N.R. NORTHWESTERN DISTRICT H.Q.
P.O. BOX 309
SPOONER, WI. 54801 (715) 635-4228

DESIGN CONSULTANT

LARRY GOTHAM
MORGAN & PARMLEY LTD.
115 W. 2ND STREET SOUTH
LADYSMITH WI. 54848 (715) 532-3721

UTILITIES

JUMP RIVER ELECTRIC CO-OP

ATTN: HANK LEW
1102 W. 9TH STREET NORTH
LADYSMITH, WI. 54848
(715) 532-5524

UTC OF NORTHERN WISCONSIN CENTURY TELEPHONE

ATTN: JAMES ARQUETTE
425 ELLINGSON AVE
P.O. BOX 78
HAWKINS, WI. 54530
(800) 752-5637

DETAILED SUMMARY OF MISCELLANEOUS QUANTITIES

ASPHALTIC MATERIAL FOR PLANT MIXES

STATION	TO	STATION	LOCATION	TON
0+00		61+41	MAINLINE	129
P.E.'S & SIDEROADS		------	UNDISTRIBUTED	3
TOTAL				132

TEMPORARY PAVEMENT MARKING, 4"

STATION	TO	STATION	LOCATION	L.F.
0+00		61+41	CENTERLINE	9,800
TOTAL				9,800

PAVEMENT MARKING, 4", PAINT

STATION	TO	STATION	LOCATION	L.F.
0+00		61+41	CENTERLINE	9,800
0+00		61+41	RT. & LT. EDGE LINES	12,300
TOTAL				22,100

PULVERIZE AND RELAY EXISTING BASE & SURFACE

STATION	TO	STATION	LOCATION	S.Y.
0+00		61+41	MAINLINE	15,080
		------	UNDISTRIBUTED	20
TOTAL				15,100

CRUSHED AGGREGATE BASE COURSE

STATION	TO	STATION	LOCATION	C.Y.
0+00		61+41	MAINLINE	750
0+00		61+41	RT. & LT. SHLDR.	450
P.E.'S & SIDEROADS			UNDISTRIBUTED	100
TOTAL				1,300

ASPHALTIC MATERIAL FOR TACK COAT

STATION	TO	STATION	APPLICATIONS	RATE	GAL
0+00		61+41	1	0.025 GAL/SQ. YD.	375
P.E.'S & SIDEROADS		------	1	0.025 GAL/SQ. YD.	10
TOTAL					385

ASPHALTIC CONCRETE PAVEMENT, TYPE LV

STATION	TO	STATION	LOCATION	TON
0+00		61+41	MAINLINE	2,150
P.E.'S & SIDEROADS		------	UNDISTRIBUTED	50
TOTAL				2,200

GENERAL NOTES:

PRIOR TO PULVERIZING AND RELAYING, CRUSHED AGGREGATE BASE COURSE SHALL BE EVENLY PLACED ACROSS THE ROADWAY IN A VOLUME SUFFICIENT TO FILL EXISTING RUTS AND CREATE A 0.02'/FT CROWN, OR TO CONFORM TO SUPERELEVATION RATE.

THRU ALL PRIVATE ENTRANCES AND ON THE INSIDE OF ALL HORIZONTAL CURVES, THE EDGE OF ASPHALT SHALL BE EXTENDED 18".

THERE ARE UTILITY FACILITIES IN THE PROJECT AREA WHICH ARE NOT SHOWN ON THE PLANS. THERE ARE NO KNOWN CONFLICTS.

THE 2 1/2" ASPHALTIC CONCRETE PAVEMENT SHALL BE PLACED IN ONE 1 1/4" LOWER LAYER AND ONE 1 1/4" UPPER LAYER.

TACK COAT SHALL BE APPLIED BETWEEN THE ASPHALT LAYERS AT A RATE OF 0.025 GAL/SQ. YD.

STANDARD DETAIL DRAWINGS

15C8-8a	PAVEMENT MARKINGS (MAINLINE)
15C8-8b	PAVEMENT MARKINGS (INTERSECTIONS)
15C12-2	TRAFFIC CONTROL FOR LANE CLOSURES (SUITABLE FOR MOVING OPERATIONS)

LIST OF STANDARD ABBREVIATIONS

ABUT	ABUTMENT
AC	ACRES
AGG	AGGREGATE
AH	AHEAD
ADT	AVERAGE DAILY TRAFFIC
AVE	AVENUE
AVG	AVERAGE
ASPH	ASPHALTIC
BK	BACK
BM	BENCHMARK
Δ	CENTRAL ANGLE
℄, C/L	CENTERLINE
C & G	CURB AND GUTTER
CABC	CRUSHED AGGREGATE BASE COURSE
CONC	CONCRETE
CONST	CONSTRUCTION
COR	CORNER
CORR	CORRUGATED
CSCP	CORRUGATED STEEL CULVERT PIPE
CSPA	CORRUGATED STEEL PIPE ARCH
CTH	COUNTY TRUNK HIGHWAY
CP	CULVERT PIPE
CY	CUBIC YARD
CWT	HUNDRED WEIGHT
DIA	DIAMETER
D	DEGREE OF CURVE
DHV	DESIGN HOURLY VOLUME
DWY	DRIVEWAY
EBS	EXC. BELOW SUBGRADE
ELEV, EL	ELEVATION
ELEC	ELECTRICAL
EXC	EXCAVATION
EXIST	EXISTING
E	EAST
FE	FIELD ENTRANCE

FF	FACE TO FACE
FL, F/L	FLOW LINE
G	GARAGE
GN	GRID NORTH
H	HOUSE
HOR	HORIZONTAL
HWY	HIGHWAY
HYD	HYDRANDT
I	INTERSECTION ANGLE
INTERS	INTERSECTION
INV	INVERT
IP	IRON PIN OR PIPE
LC	LONG CHORD OF CURVE
LEN	LENGTH
LF	LINEAR FOOT
LHF	LEFT HAND FORWARD
L	LENGTH OF CURVE
LHE	LIMITED HIGHWAY EASEMENT
LT	LEFT
LS	LUMP SUM
m	METER
mm	MILLIMETERS
MH	MANHOLE
N	NORTH
PAVT	PAVEMENT
PC	POINT OF CURVATURE
PE	PRIVATE ENTRANCE
PI	POINT OF INTERSECTION
PL	PROPERTY LINE
PP	POWER POLE
PROP	PROPOSED
PT	POINT OF TANGENCY
R	RANGE, RADIUS
RCCP	REINFORCED CONCRETE CULVERT PIPE
RD	ROAD
REBAR	REINFORCEMENT BAR

REM	REMOVE
REQD	REQUIRED
RDWY	ROADWAY
RHF	RIGHT HAND FORWARD
RL, R/L	REFERENCE LINE
RR	RAILROAD
RT	RIGHT
R/W	RIGHT-OF-WAY
S	SOUTH
SAN S	SANITARY SEWER
SDD	STANDARD DETAIL DRAWING
SE	SOUTHEAST, SUPERELEVATION
SHLDR	SHOULDER
SPECS	SPECIFICATIONS
SQ	SQUARE
SS	STORM SEWER
STH	STATE TRUNK HIGHWAY
ST	STREET
STA	STATION
SW	SIDEWALK
TAN	TANGENT
T	TANGENT LENGTH OF CURVE, TRUCKS
TC	TOP OF CURB
T℄, T/L	TRANSIT LINE
TEL	TELEPHONE
TEMP	TEMPORARY
TLE	TEMPORARY LIMITED EASEMENT
TYP	TYPICAL
UNCL	UNCLASSIFIED
USH	UNITED STATES HIGHWAY
UG	UNDERGROUND
V	DESIGN SPEED
VAR	VARIABLE
VERT	VERTICAL
W	WEST
YD	YARD

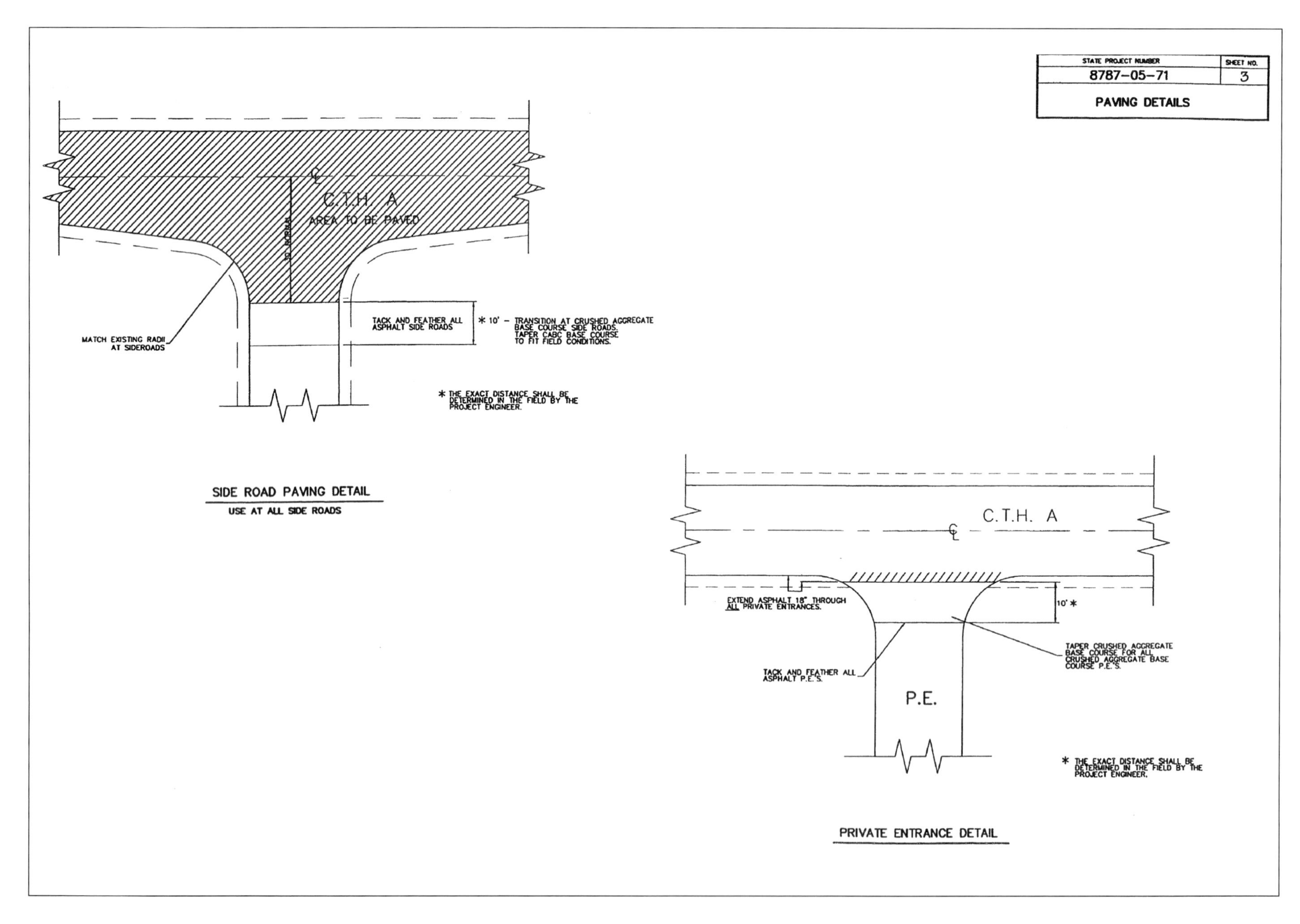
STATE PROJECT NUMBER
SHEET NO.
8787-05-71
3
PAVING DETAILS
C.T.H A
AREA TO BE PAVED
TO NORMAL
TACK AND FEATHER ALL ASPHALT SIDE ROADS
* 10' - TRANSITION AT CRUSHED AGGREGATE BASE COURSE SIDE ROADS. TAPER CABC BASE COURSE TO FIT FIELD CONDITIONS.
MATCH EXISTING RADII AT SIDEROADS
* THE EXACT DISTANCE SHALL BE DETERMINED IN THE FIELD BY THE PROJECT ENGINEER.
SIDE ROAD PAVING DETAIL
USE AT ALL SIDE ROADS
C.T.H. A
EXTEND ASPHALT 18" THROUGH ALL PRIVATE ENTRANCES.
10' *
TAPER CRUSHED AGGREGATE BASE COURSE FOR ALL CRUSHED AGGREGATE BASE COURSE P.E.'S.
TACK AND FEATHER ALL ASPHALT P.E.'S.
P.E.
* THE EXACT DISTANCE SHALL BE DETERMINED IN THE FIELD BY THE PROJECT ENGINEER.
PRIVATE ENTRANCE DETAIL

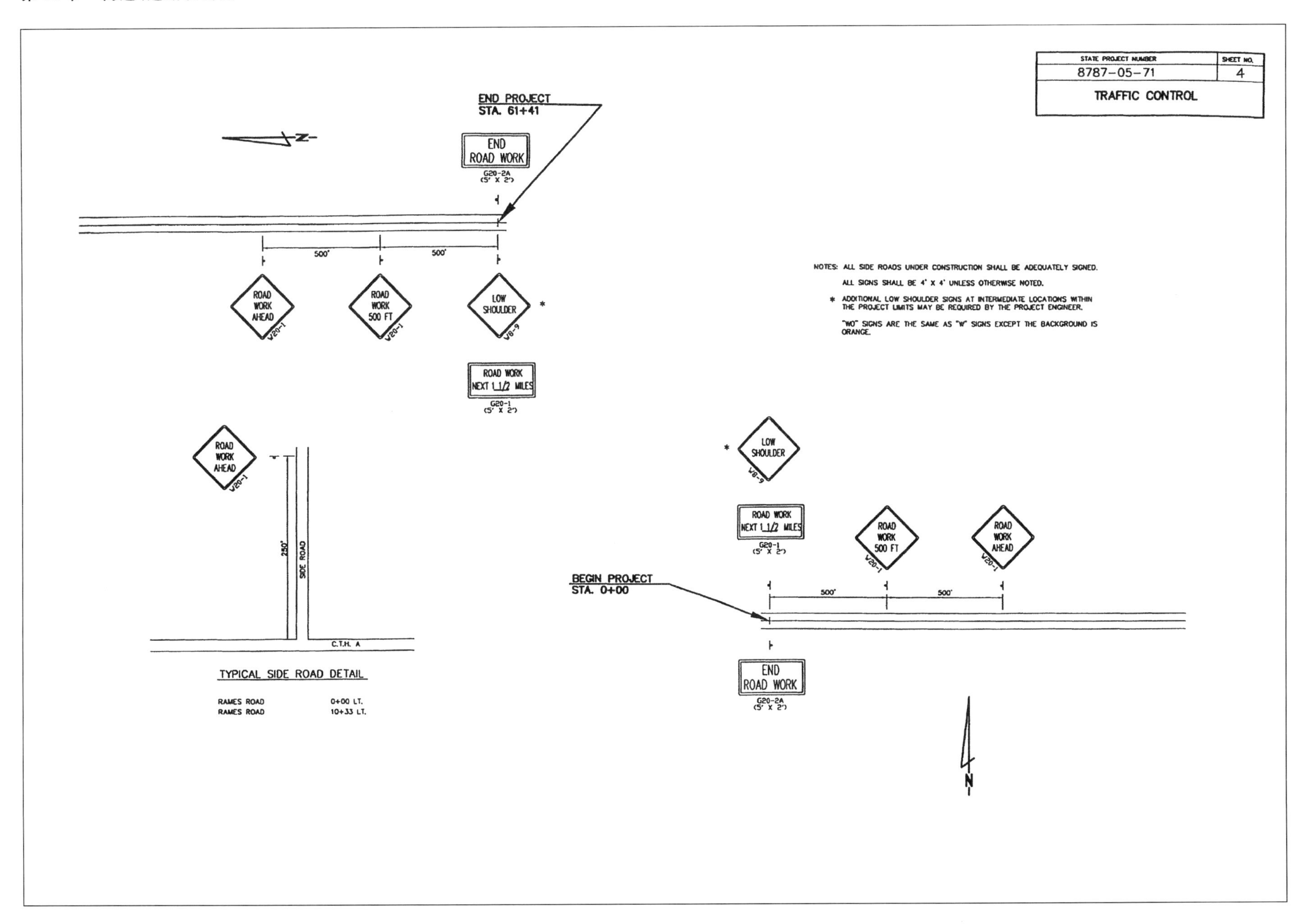
STATE PROJECT NUMBER
8787-05-71
SHEET NO.
4
TRAFFIC CONTROL
END PROJECT
STA. 61+41
END
ROAD WORK
G20-2A
(5' X 2')
500'
500'
ROAD
WORK
AHEAD
W20-1
ROAD
WORK
500 FT
W20-1
LOW
SHOULDER
W8-9
*
ROAD WORK
NEXT 1 1/2 MILES
G20-1
(5' X 2')
NOTES: ALL SIDE ROADS UNDER CONSTRUCTION SHALL BE ADEQUATELY SIGNED.
ALL SIGNS SHALL BE 4' X 4' UNLESS OTHERWISE NOTED.
* ADDITIONAL LOW SHOULDER SIGNS AT INTERMEDIATE LOCATIONS WITHIN THE PROJECT LIMITS MAY BE REQUIRED BY THE PROJECT ENGINEER.
"WO" SIGNS ARE THE SAME AS "W" SIGNS EXCEPT THE BACKGROUND IS ORANGE.
250'
SIDE ROAD
C.T.H. A
TYPICAL SIDE ROAD DETAIL
RAMES ROAD 0+00 LT.
RAMES ROAD 10+33 LT.
BEGIN PROJECT
STA. 0+00
N

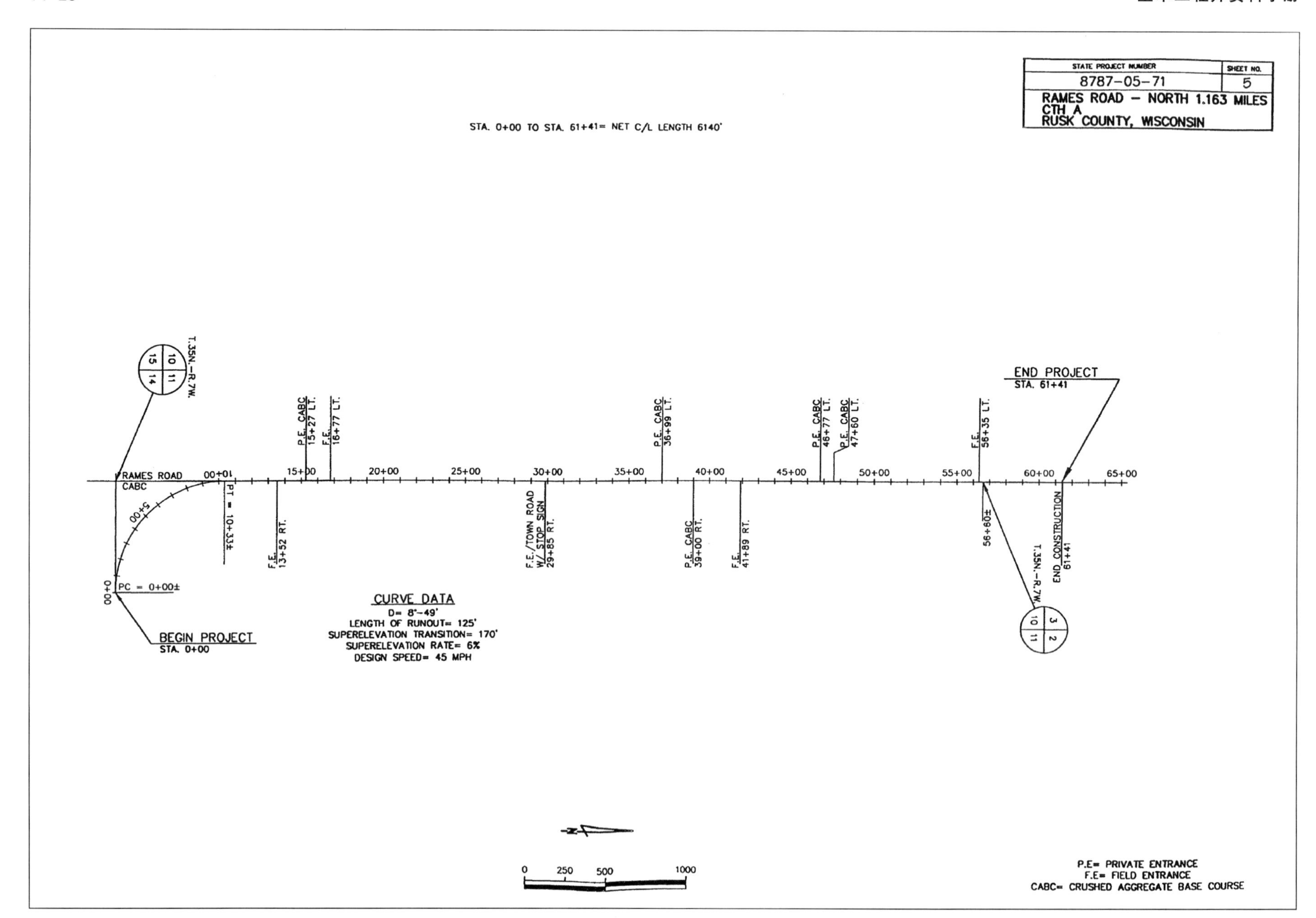
STATE PROJECT NUMBER
8787-05-71
SHEET NO.
5
RAMES ROAD – NORTH 1.163 MILES
CTH A
RUSK COUNTY, WISCONSIN
STA. 0+00 TO STA. 61+41= NET C/L LENGTH 6140'
T.35N.-R.7W.
RAMES ROAD
CABC
PT = 10+33±
PC = 0+00±
BEGIN PROJECT
STA. 0+00
END PROJECT
STA. 61+41
P.E. CABC 15+27 LT.
F.E. 16+77 LT.
F.E. 13+52 RT.
F.E./TOWN ROAD W/ STOP SIGN 29+85 RT.
P.E. CABC 36+99 LT.
P.E. CABC 39+00 RT.
F.E. 41+89 RT.
P.E. CABC 46+77 LT.
P.E. CABC 47+60 LT.
F.E. 56+35 LT.
56+60±
END CONSTRUCTION 61+41
CURVE DATA
D= 8°-49'
LENGTH OF RUNOUT= 125'
SUPERELEVATION TRANSITION= 170'
SUPERELEVATION RATE= 6%
DESIGN SPEED= 45 MPH
P.E= PRIVATE ENTRANCE
F.E= FIELD ENTRANCE
CABC= CRUSHED AGGREGATE BASE COURSE

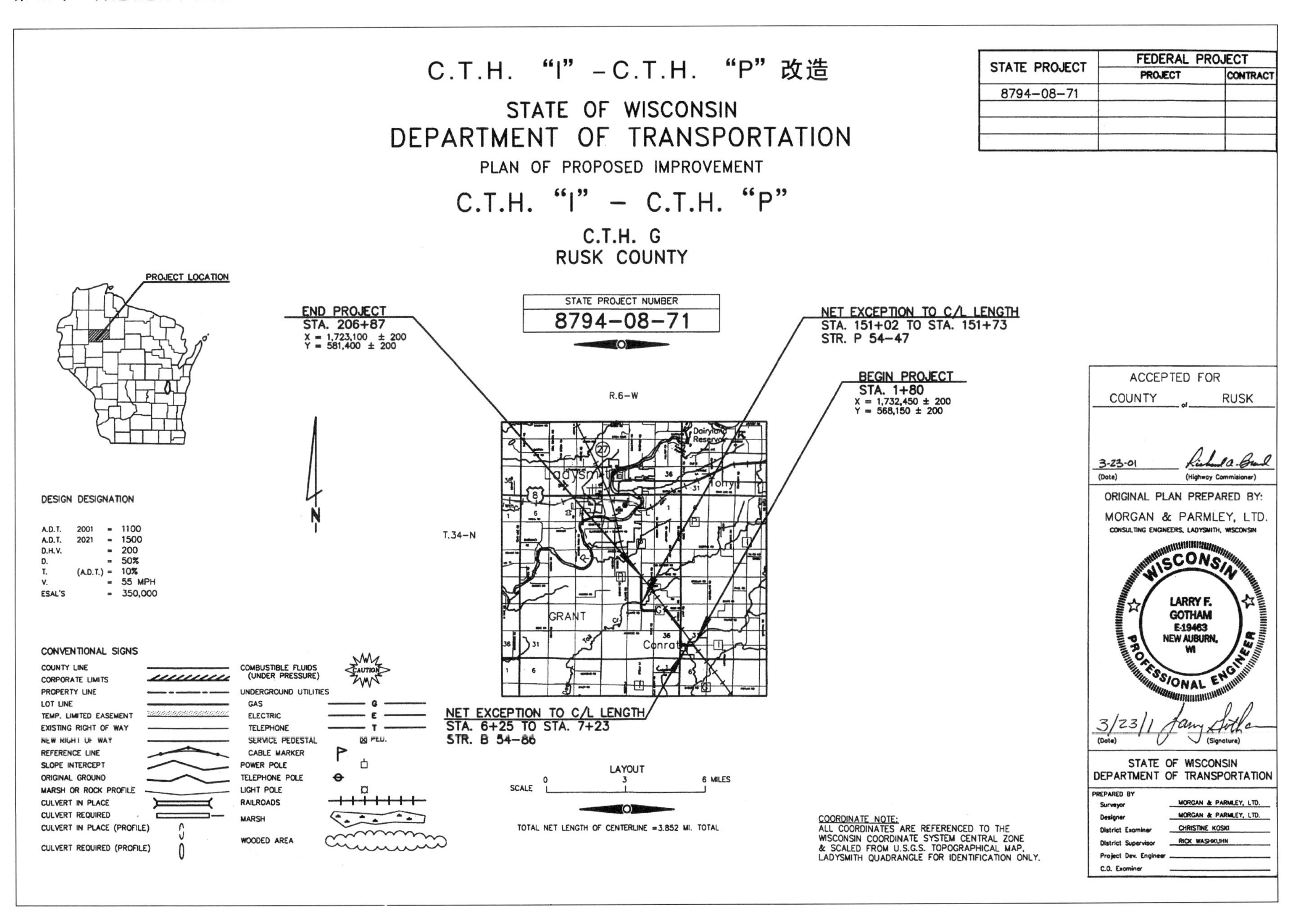

Courtesy: Morgan & Parmley, Ltd.

STATE PROJECT NUMBER	SHEET NO.
8794-08-71	2

TYPICAL SECTIONS—DETAILED SUMMARY OF MISCELLANEOUS QUANTITIES

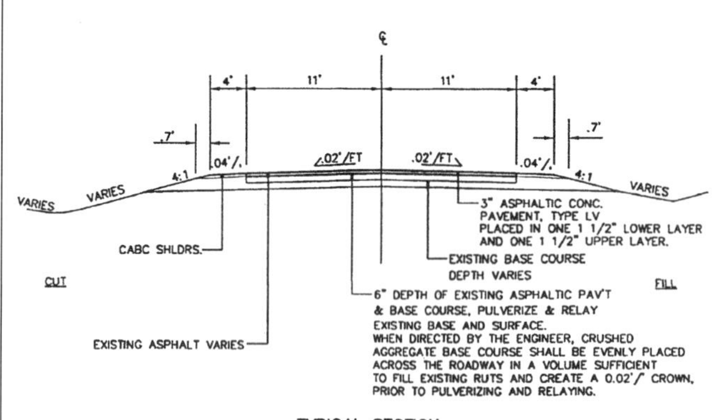

TYPICAL SECTION

STA. 1+80 TO STA. 206+87
(NOT TO SCALE)

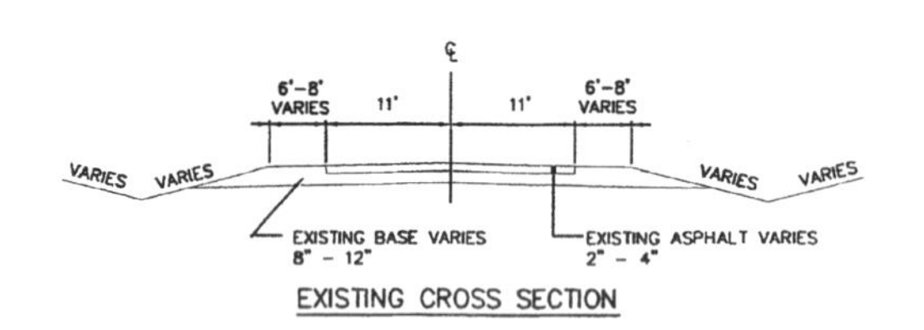

EXISTING CROSS SECTION

CALL DIGGERS' HOTLINE

1-800-242-8511

TOLL FREE

2040 W. WISCONSIN AVE.
SUITE 10
MILWAUKEE, WI 53233

RUSK COUNTY SURVEYOR
DAVE KAISER, RLS
(715) 532-2165

DETAILED SUMMARY OF MISCELLANEOUS QUANTITIES

ASPHALTIC MATERIAL FOR PLANT MIXES

STATION	TO	STATION	LOCATION	TON
1+80		206+87	MAINLINE	516
P.E.'S & SIDEROADS		——	UNDISTRIBUTED	12
TOTAL				528

ASPHALTIC MATERIAL FOR TACK COAT

STATION	TO	STATION	APPLICATIONS	RATE	GAL
1+80		206+87	1	0.025 GAL/SQ. YD.	1,300
P.E.'S & SIDEROADS		——	1	0.025 GAL/SQ. YD.	20
TOTAL					1,320

ASPHALTIC CONCRETE PAVEMENT, TYPE LV

STATION	TO	STATION	LOCATION	TON
1+80		206+87	MAINLINE	8,600
P.E.'S & SIDEROADS		——	UNDISTRIBUTED	200
TOTAL				8,800

CRUSHED AGGREGATE BASE COURSE

STATION	TO	STATION	LOCATION	TON
1+80		206+87	MAINLINE	3,500
1+80		206+87	RT. & LT. SHLDR.	3,000
P.E.'S & SIDEROADS			UNDISTRIBUTED	350
TOTAL				6,850

PULVERIZE AND RELAY EXISTING BASE & SURFACE

STATION	TO	STATION	LOCATION	S.Y.
1+80		206+87	MAINLINE	50,950
		——	UNDISTRIBUTED	50
TOTAL				51,000

PAVEMENT MARKING, 4-INCH, PAINT

STATION	TO	STATION	LOCATION	L.F.
1+80		206+87	RT. & LT. EDGE LINES	41,000
TOTAL				41,000

PAVEMENT MARKING, SAME DAY, 4-INCH, PAINT

STATION	TO	STATION	LOCATION	L.F.
1+80		206+87	CENTERLINE	30,000
TOTAL				30,000

SAWING EXISTING PAVEMENT

STATION	LOCATION	L.F.
1+80	RT. & LT.	22
206+87	RT. & LT.	22
	UNDISTRIBUTED	56
TOTAL		100

TEMPORARY PAVEMENT MARKING, 4-INCH

STATION	TO	STATION	LOCATION	L.F.
1+80		206+87	CENTERLINE	20,000
TOTAL				20,000

GENERAL NOTES:

THROUGH ALL PRIVATE ENTRANCES AND ON THE INSIDE OF ALL HORIZONTAL CURVES, THE EDGE OF ASPHALT SHALL BE EXTENDED 18".

THERE ARE UTILITY FACILITIES IN THE PROJECT AREA WHICH ARE NOT SHOWN ON THE PLANS. THERE ARE NO KNOWN CONFLICTS.

TACK COAT SHALL BE APPLIED BETWEEN THE ASPHALT LAYERS AT A RATE OF 0.025 GAL/SQ. YD.

OTHER CONTACTS

DNR LIASON

DAN MICHELS
D.N.R. NORTHWESTERN DISTRICT H.Q.
P.O. BOX 309
SPOONER, WI. 54801 (715) 635-4228

DESIGN CONSULTANT

LARRY GOTHAM
MORGAN & PARMLEY LTD.
115 W. 2ND STREET SOUTH
LADYSMITH WI. 54848 (715) 532-3721

UTILITIES

EXCEL

ATTN: JOE PERKINS
310 HICKORY HILL LANE
PHILLIPS, WI. 54555
(715) 836-1198

JUMP RIVER ELECTRIC CO-OP

ATTN: HANK LEW
1102 W. 9TH STREET
LADYSMITH, WI. 54848
(715) 532-5524

CENTURYTEL

ATTN: JAMES ARQUETTE
425 ELLINGTON ROAD
P.O. BOX 78
HAWKINS, WI. 54530
(715) 752-5637

STANDARD DETAIL DRAWINGS

15C4-1	TRAFFIC CONTROL ADVANCE WARNING
15C6-4	SIGNING AND MARKING AT 2-LANE BRIDGE
15C8-9a	PAVEMENT MARKINGS (MAINLINE)
15C8-9b	PAVEMENT MARKINGS (INTERSECTIONS)
15C12-2	TRAFFIC CONTROL FOR LANE CLOSURES (SUITABLE FOR MOVING OPERATIONS)

LIST OF STANDARD ABBREVIATIONS

ABUT	ABUTMENT	FF	FACE TO FACE	REM	REMOVE
AC	ACRES	FL, F/L	FLOW LINE	REQD	REQUIRED
AGG	AGGREGATE	G	GARAGE	RDWY	ROADWAY
AH	AHEAD	GN	GRID NORTH	RHF	RIGHT HAND FORWARD
ADT	AVERAGE DAILY TRAFFIC	H	HOUSE	RL, R/L	REFERENCE LINE
AVE	AVENUE	HOR	HORIZONTAL	RR	RAILROAD
AVG	AVERAGE	HWY	HIGHWAY	RT	RIGHT
ASPH	ASPHALTIC	HYD	HYDRANDT	R/W	RIGHT-OF-WAY
BK	BACK	I	INTERSECTION ANGLE	S	SOUTH
BM	BENCHMARK	INTERS	INTERSECTION	SAN S	SANITARY SEWER
△	CENTRAL ANGLE	INV	INVERT	SDD	STANDARD DETAIL DRAWING
℄, C/L	CENTERLINE	IP	IRON PIN OR PIPE	SE	SOUTHEAST, SUPERELEVATION
C & G	CURB AND GUTTER	LC	LONG CHORD OF CURVE	SHLDR	SHOULDER
CABC	CRUSHED AGGREGATE BASE COURSE	LEN	LENGTH	SPECS	SPECIFICATIONS
CONC	CONCRETE	LF	LINEAR FOOT	SQ	SQUARE
CONST	CONSTRUCTION	LHF	LEFT HAND FORWARD	SS	STORM SEWER
COR	CORNER	L	LENGTH OF CURVE	STH	STATE TRUNK HIGHWAY
CORR	CORRUGATED	LHE	LIMITED HIGHWAY EASEMENT	ST	STREET
CSCP	CORRUGATED STEEL CULVERT PIPE	LT	LEFT	STA	STATION
		LS	LUMP SUM	SW	SIDEWALK
CSPA	CORRUGATED STEEL PIPE ARCH	m	METER	TAN	TANGENT
		mm	MILLIMETERS	T	TANGENT LENGTH OF CURVE, TRUCKS
CTH	COUNTY TRUNK HIGHWAY	MH	MANHOLE	TC	TOP OF CURB
CP	CULVERT PIPE	N	NORTH	TL, T/L	TRANSIT LINE
CY	CUBIC YARD	PAVT	PAVEMENT	TEL	TELEPHONE
CWT	HUNDRED WEIGHT	PC	POINT OF CURVATURE	TEMP	TEMPORARY
DIA	DIAMETER	PE	PRIVATE ENTRANCE	TLE	TEMPORARY LIMITED EASEMENT
D	DEGREE OF CURVE	PI	POINT OF INTERSECTION	TYP	TYPICAL
DHV	DESIGN HOURLY VOLUME	PL	PROPERTY LINE	UNCL	UNCLASSIFIED
DWY	DRIVEWAY	PP	POWER POLE	USH	UNITED STATES HIGHWAY
EBS	EXC. BELOW SUBGRADE	PROP	PROPOSED	UG	UNDERGROUND
ELEV, EL	ELEVATION	PT	POINT OF TANGENCY	V	DESIGN SPEED
ELEC	ELECTRICAL	R	RANGE, RADIUS	VAR	VARIABLE
EXC	EXCAVATION	RCCP	REINFORCED CONCRETE CULVERT PIPE	VERT	VERTICAL
EXIST	EXISTING			W	WEST
E	EAST	RD	ROAD	YD	YARD
FE	FIELD ENTRANCE	REBAR	REINFORCEMENT BAR		

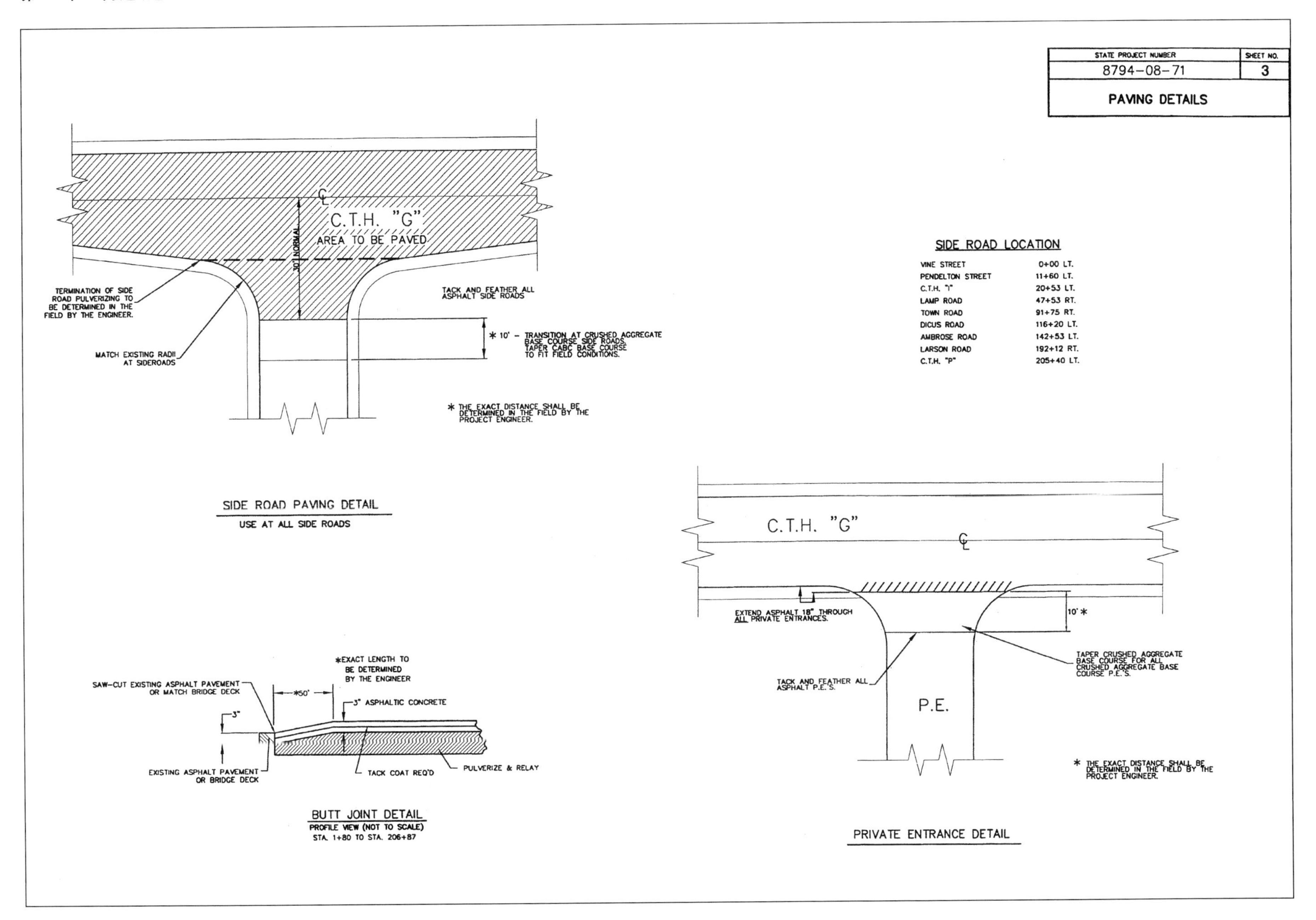

STATE PROJECT NUMBER
SHEET NO.
8794-08-71
3
PAVING DETAILS
C.T.H. "G"
AREA TO BE PAVED
30' NORMAL
TERMINATION OF SIDE ROAD PULVERIZING TO BE DETERMINED IN THE FIELD BY THE ENGINEER.
TACK AND FEATHER ALL ASPHALT SIDE ROADS
* 10' - TRANSITION AT CRUSHED AGGREGATE BASE COURSE SIDE ROADS. TAPER CABC BASE COURSE TO FIT FIELD CONDITIONS.
MATCH EXISTING RADII AT SIDEROADS
* THE EXACT DISTANCE SHALL BE DETERMINED IN THE FIELD BY THE PROJECT ENGINEER.
SIDE ROAD PAVING DETAIL
USE AT ALL SIDE ROADS
SIDE ROAD LOCATION
VINE STREET 0+00 LT.
PENDELTON STREET 11+60 LT.
C.T.H. "I" 20+53 LT.
LAMP ROAD 47+53 RT.
TOWN ROAD 91+75 RT.
DICUS ROAD 116+20 LT.
AMBROSE ROAD 142+53 LT.
LARSON ROAD 192+12 RT.
C.T.H. "P" 205+40 LT.
C.T.H. "G"
EXTEND ASPHALT 18" THROUGH ALL PRIVATE ENTRANCES.
10' *
TAPER CRUSHED AGGREGATE BASE COURSE FOR ALL CRUSHED AGGREGATE BASE COURSE P.E.'S.
TACK AND FEATHER ALL ASPHALT P.E.'S.
P.E.
* THE EXACT DISTANCE SHALL BE DETERMINED IN THE FIELD BY THE PROJECT ENGINEER.
PRIVATE ENTRANCE DETAIL
*EXACT LENGTH TO BE DETERMINED BY THE ENGINEER
SAW-CUT EXISTING ASPHALT PAVEMENT OR MATCH BRIDGE DECK
*50'
3" ASPHALTIC CONCRETE
3"
EXISTING ASPHALT PAVEMENT OR BRIDGE DECK
TACK COAT REQ'D
PULVERIZE & RELAY
BUTT JOINT DETAIL
PROFILE VIEW (NOT TO SCALE)
STA. 1+80 TO STA. 206+87

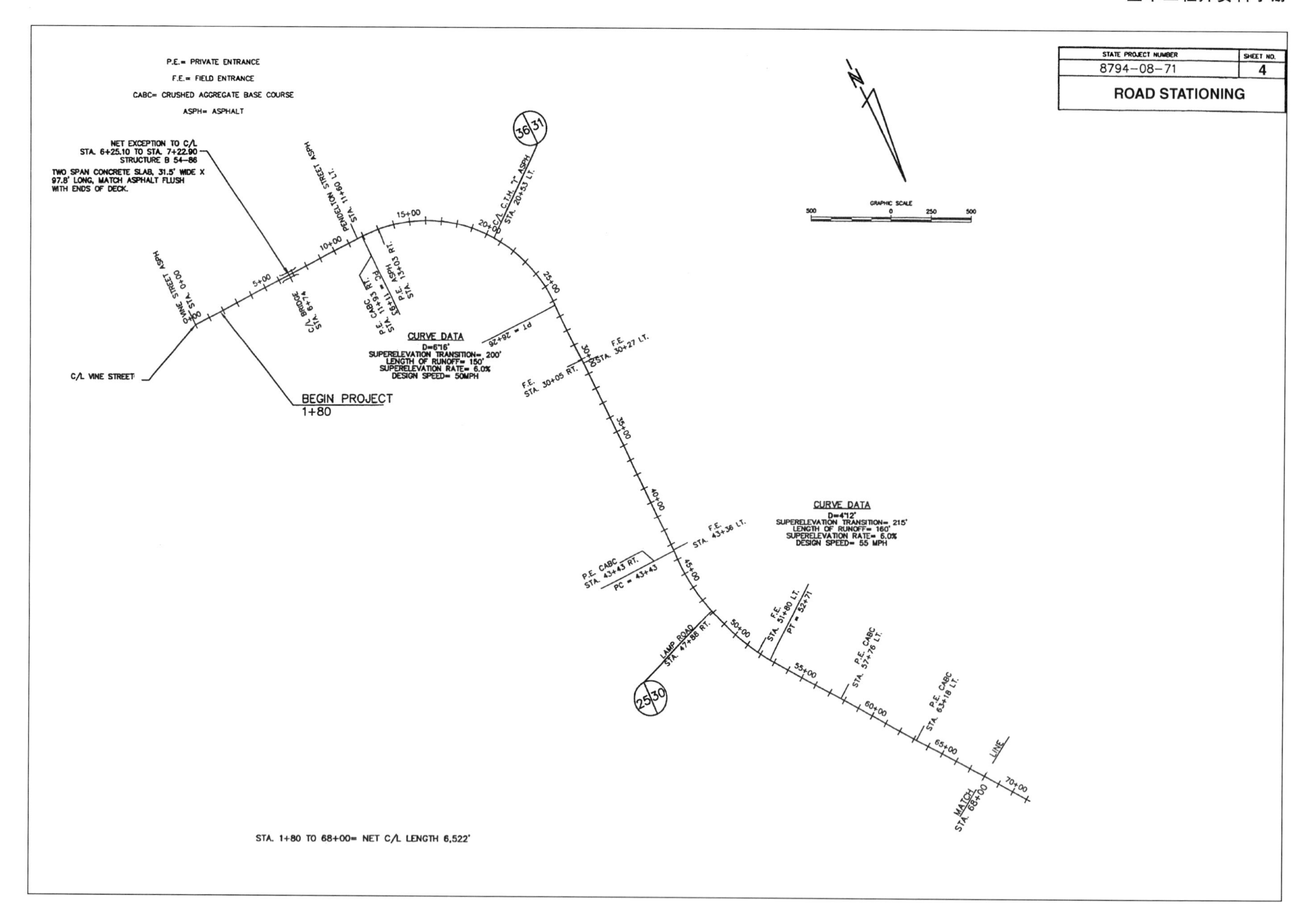
STATE PROJECT NUMBER
8794-08-71
SHEET NO.
4
ROAD STATIONING
P.E.= PRIVATE ENTRANCE
F.E.= FIELD ENTRANCE
CABC= CRUSHED AGGREGATE BASE COURSE
ASPH= ASPHALT
NET EXCEPTION TO C/L
STA. 6+25.10 TO STA. 7+22.90
STRUCTURE B 54-86
TWO SPAN CONCRETE SLAB, 31.5' WIDE X
97.8' LONG, MATCH ASPHALT FLUSH
WITH ENDS OF DECK.
GRAPHIC SCALE
CURVE DATA
D=6°16'
SUPERELEVATION TRANSITION= 200'
LENGTH OF RUNOFF= 150'
SUPERELEVATION RATE= 6.0%
DESIGN SPEED= 50MPH
C/L VINE STREET
BEGIN PROJECT
1+80
CURVE DATA
D=4°12'
SUPERELEVATION TRANSITION= 215'
LENGTH OF RUNOFF= 160'
SUPERELEVATION RATE= 6.0%
DESIGN SPEED= 55 MPH
MATCH LINE
STA. 68+00
STA. 1+80 TO 68+00= NET C/L LENGTH 6,522'

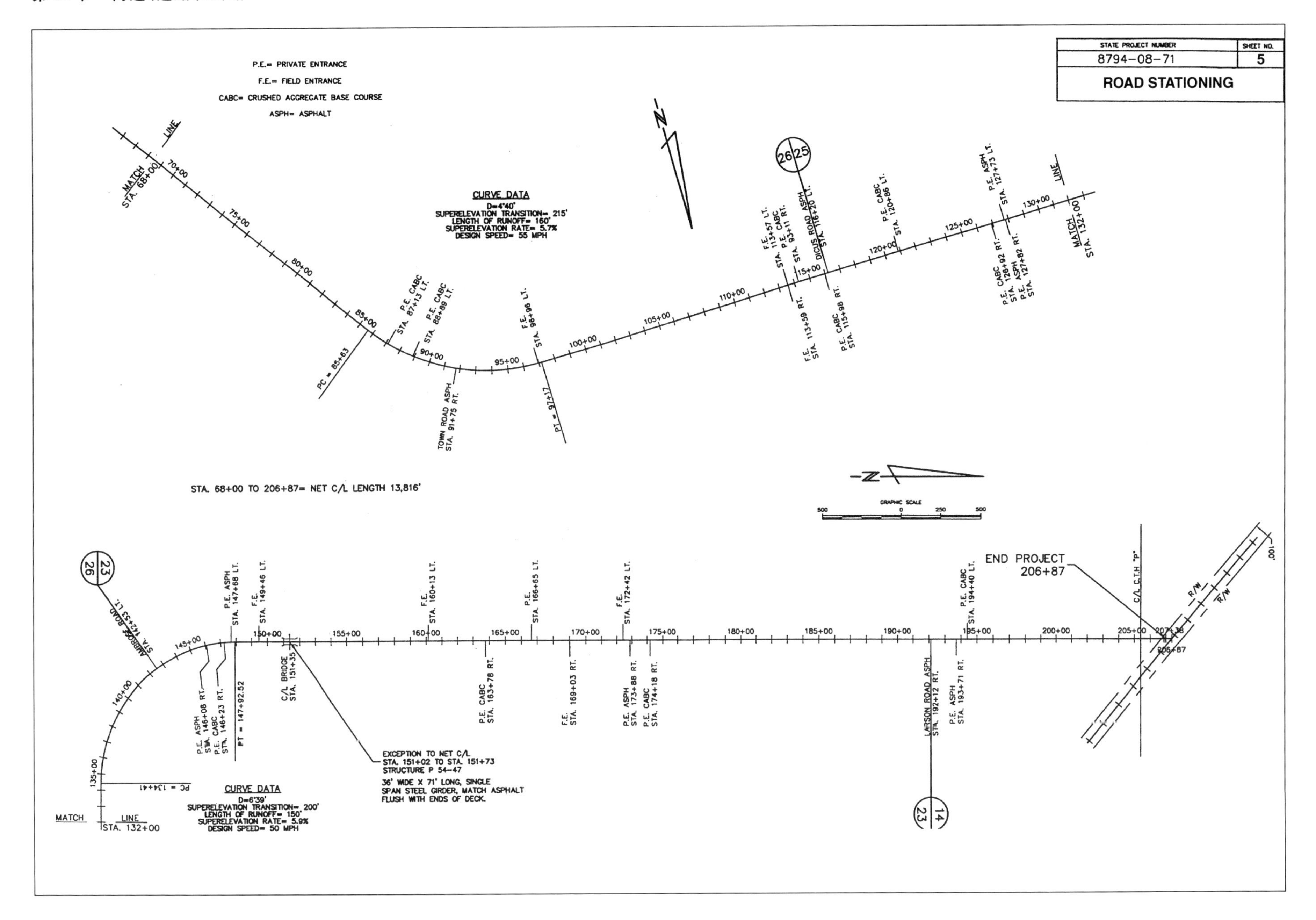
STATE PROJECT NUMBER
8794-08-71
SHEET NO.
5
ROAD STATIONING
P.E.= PRIVATE ENTRANCE
F.E.= FIELD ENTRANCE
CABC= CRUSHED AGGREGATE BASE COURSE
ASPH= ASPHALT
CURVE DATA
D=4°40'
SUPERELEVATION TRANSITION= 215'
LENGTH OF RUNOFF= 160'
SUPERELEVATION RATE= 5.7%
DESIGN SPEED= 55 MPH
STA. 68+00 TO 206+87= NET C/L LENGTH 13,816'
GRAPHIC SCALE
END PROJECT
206+87
EXCEPTION TO NET C/L
STA. 151+02 TO STA. 151+73
STRUCTURE P 54-47
36' WIDE X 71' LONG, SINGLE
SPAN STEEL GIRDER, MATCH ASPHALT
FLUSH WITH ENDS OF DECK.
CURVE DATA
D=6°39'
SUPERELEVATION TRANSITION= 200'
LENGTH OF RUNOFF= 150'
SUPERELEVATION RATE= 5.9%
DESIGN SPEED= 50 MPH
MATCH LINE STA. 132+00
MATCH LINE STA. 68+00
PC = 85+63
PT = 97+17
TOWN ROAD ASPH STA. 91+75 RT.
DICUS ROAD ASPH STA. 116+20 LT.
LARSON ROAD ASPH STA. 192+12 RT.
C/L BRIDGE STA. 151+35
PT = 147+92.52
PC = 134+41
C/L C.T.H "P"

设　计

第 15 节　桥梁

例 15

大多数的桥梁设计必须满足各州和联邦的标准。施工之前，设计图必须经过交通管理部门的审查。第一套图纸反映的是当前的格式，而第二套图纸代表的是 25 年前的典型样式。然而，设计和大样仍然保持相对一致。第三套图纸给出了木桥墩设计的详图。最后一套桥梁设计图纸是跨越通行水道的机动雪车桥。这类结构往往适用地方规范，而不是公路标准。

目 录

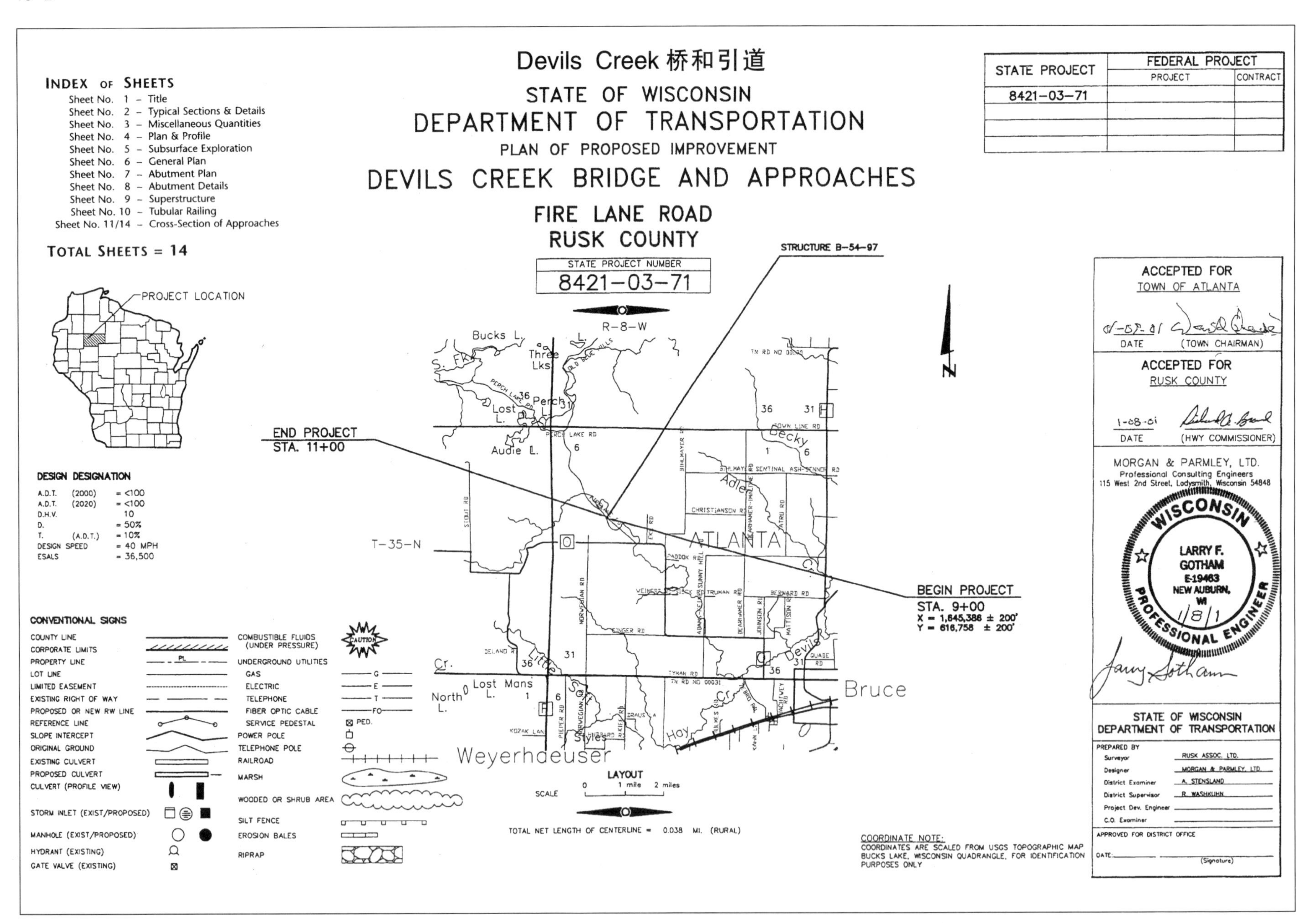

Courtesy: Morgan & Parmley, Ltd.

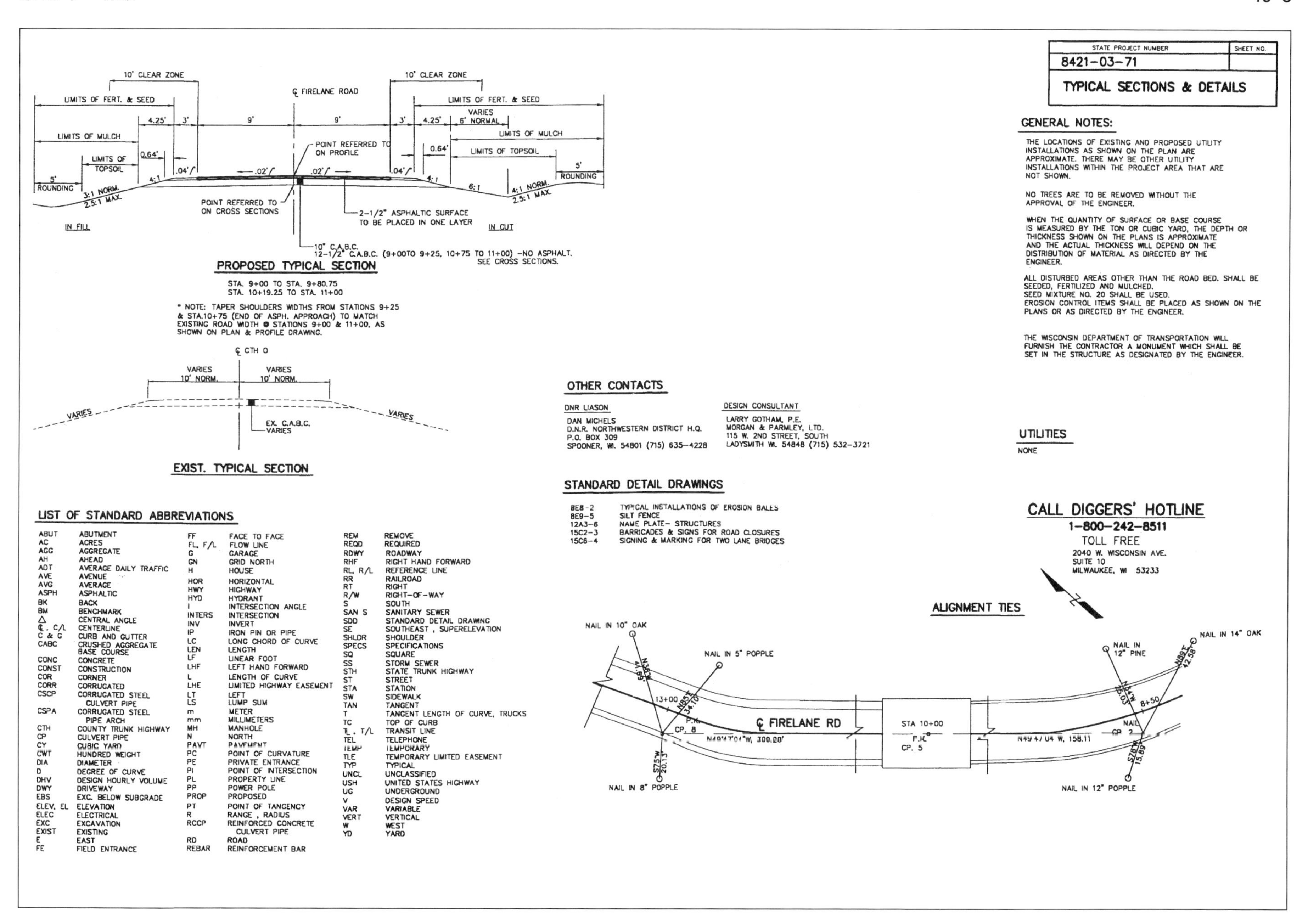
STATE PROJECT NUMBER
SHEET NO.
8421-03-71
TYPICAL SECTIONS & DETAILS
10' CLEAR ZONE
℄ FIRELANE ROAD
LIMITS OF FERT. & SEED
4.25'
3'
9'
VARIES
6' NORMAL
LIMITS OF MULCH
POINT REFERRED TO ON PROFILE
0.64'
LIMITS OF TOPSOIL
5'
ROUNDING
3:1 NORM.
2.5:1 MAX.
4:1
6:1
4:1 NORM.
POINT REFERRED TO ON CROSS SECTIONS
2-1/2" ASPHALTIC SURFACE TO BE PLACED IN ONE LAYER
IN FILL
IN CUT
10" C.A.B.C.
12-1/2" C.A.B.C. (9+00TO 9+25, 10+75 TO 11+00) -NO ASPHALT.
SEE CROSS SECTIONS.
PROPOSED TYPICAL SECTION
STA. 9+00 TO STA. 9+80.75
STA. 10+19.25 TO STA. 11+00
* NOTE: TAPER SHOULDERS WIDTHS FROM STATIONS 9+25 & STA.10+75 (END OF ASPH. APPROACH) TO MATCH EXISTING ROAD WIDTH @ STATIONS 9+00 & 11+00, AS SHOWN ON PLAN & PROFILE DRAWING.
℄ CTH O
VARIES
10' NORM.
EX. C.A.B.C.
VARIES
EXIST. TYPICAL SECTION
GENERAL NOTES:
THE LOCATIONS OF EXISTING AND PROPOSED UTILITY INSTALLATIONS AS SHOWN ON THE PLAN ARE APPROXIMATE. THERE MAY BE OTHER UTILITY INSTALLATIONS WITHIN THE PROJECT AREA THAT ARE NOT SHOWN.
NO TREES ARE TO BE REMOVED WITHOUT THE APPROVAL OF THE ENGINEER.
WHEN THE QUANTITY OF SURFACE OR BASE COURSE IS MEASURED BY THE TON OR CUBIC YARD, THE DEPTH OR THICKNESS SHOWN ON THE PLANS IS APPROXIMATE AND THE ACTUAL THICKNESS WILL DEPEND ON THE DISTRIBUTION OF MATERIAL AS DIRECTED BY THE ENGINEER.
ALL DISTURBED AREAS OTHER THAN THE ROAD BED. SHALL BE SEEDED, FERTILIZED AND MULCHED.
SEED MIXTURE NO. 20 SHALL BE USED.
EROSION CONTROL ITEMS SHALL BE PLACED AS SHOWN ON THE PLANS OR AS DIRECTED BY THE ENGINEER.
THE WISCONSIN DEPARTMENT OF TRANSPORTATION WILL FURNISH THE CONTRACTOR A MONUMENT WHICH SHALL BE SET IN THE STRUCTURE AS DESIGNATED BY THE ENGINEER.
OTHER CONTACTS
DNR LIASON
DAN MICHELS
D.N.R. NORTHWESTERN DISTRICT H.Q.
P.O. BOX 309
SPOONER, WI. 54801 (715) 635-4228
DESIGN CONSULTANT
LARRY GOTHAM, P.E.
MORGAN & PARMLEY, LTD.
115 W. 2ND STREET, SOUTH
LADYSMITH WI. 54848 (715) 532-3721
UTILITIES
NONE
STANDARD DETAIL DRAWINGS
8E8-2 TYPICAL INSTALLATIONS OF EROSION BALES
8E9-5 SILT FENCE
12A3-6 NAME PLATE- STRUCTURES
15C2-3 BARRICADES & SIGNS FOR ROAD CLOSURES
15C6-4 SIGNING & MARKING FOR TWO LANE BRIDGES
CALL DIGGERS' HOTLINE
1-800-242-8511
TOLL FREE
2040 W. WISCONSIN AVE.
SUITE 10
MILWAUKEE, WI 53233
ALIGNMENT TIES
NAIL IN 10" OAK
NAIL IN 5" POPPLE
NAIL IN 12" PINE
NAIL IN 14" OAK
NAIL IN 8" POPPLE
NAIL IN 12" POPPLE
13+00
8+50
P.K.
CP. 8
CP. 2
℄ FIRELANE RD
STA 10+00
P.K.
CP. 5
N49°47'04"W, 300.00'
N49°47'04"W, 158.11'
LIST OF STANDARD ABBREVIATIONS
ABUT ABUTMENT
AC ACRES
AGG AGGREGATE
AH AHEAD
ADT AVERAGE DAILY TRAFFIC
AVE AVENUE
AVG AVERAGE
ASPH ASPHALTIC
BK BACK
BM BENCHMARK
Δ CENTRAL ANGLE
℄, C/L CENTERLINE
C & G CURB AND GUTTER
CABC CRUSHED AGGREGATE BASE COURSE
CONC CONCRETE
CONST CONSTRUCTION
COR CORNER
CORR CORRUGATED
CSCP CORRUGATED STEEL CULVERT PIPE
CSPA CORRUGATED STEEL PIPE ARCH
CTH COUNTY TRUNK HIGHWAY
CP CULVERT PIPE
CY CUBIC YARD
CWT HUNDRED WEIGHT
DIA DIAMETER
D DEGREE OF CURVE
DHV DESIGN HOURLY VOLUME
DWY DRIVEWAY
EBS EXC. BELOW SUBGRADE
ELEV, EL ELEVATION
ELEC ELECTRICAL
EXC EXCAVATION
EXIST EXISTING
E EAST
FE FIELD ENTRANCE
FF FACE TO FACE
FL, F/L FLOW LINE
G GARAGE
GN GRID NORTH
H HOUSE
HOR HORIZONTAL
HWY HIGHWAY
HYD HYDRANT
I INTERSECTION ANGLE
INTERS INTERSECTION
INV INVERT
IP IRON PIN OR PIPE
LC LONG CHORD OF CURVE
LEN LENGTH
LF LINEAR FOOT
LHF LEFT HAND FORWARD
L LENGTH OF CURVE
LHE LIMITED HIGHWAY EASEMENT
LT LEFT
LS LUMP SUM
m METER
mm MILLIMETERS
MH MANHOLE
N NORTH
PAVT PAVEMENT
PC POINT OF CURVATURE
PE PRIVATE ENTRANCE
PI POINT OF INTERSECTION
PL PROPERTY LINE
PP POWER POLE
PROP PROPOSED
PT POINT OF TANGENCY
R RANGE , RADIUS
RCCP REINFORCED CONCRETE CULVERT PIPE
RD ROAD
REBAR REINFORCEMENT BAR
REM REMOVE
REQD REQUIRED
RDWY ROADWAY
RHF RIGHT HAND FORWARD
RL, R/L REFERENCE LINE
RR RAILROAD
RT RIGHT
R/W RIGHT-OF-WAY
S SOUTH
SAN S SANITARY SEWER
SDD STANDARD DETAIL DRAWING
SE SOUTHEAST , SUPERELEVATION
SHLDR SHOULDER
SPECS SPECIFICATIONS
SQ SQUARE
SS STORM SEWER
STH STATE TRUNK HIGHWAY
ST STREET
STA STATION
SW SIDEWALK
TAN TANGENT
T TANGENT LENGTH OF CURVE, TRUCKS
TC TOP OF CURB
TL, T/L TRANSIT LINE
TEL TELEPHONE
TEMP TEMPORARY
TLE TEMPORARY LIMITED EASEMENT
TYP TYPICAL
UNCL UNCLASSIFIED
USH UNITED STATES HIGHWAY
UG UNDERGROUND
V DESIGN SPEED
VAR VARIABLE
VERT VERTICAL
W WEST
YD YARD

STATE PROJECT NUMBER	SHEET NO.
8421-03-71	
FIRELANE ROAD MISCELLANEOUS QUANTITIES	

CLEARING

STA.	TO	STA.	LOC.	STA.
9+00		9+75	LT. & RT.	.75
10+50		11+00	LT. & RT.	.5
			TOTAL =	1.25

GRUBBING

STA.	TO	STA.	LOC.	STA.
9+00		9+75	LT. & RT.	.75
10+50		11+00	LT. & RT.	.5
			TOTAL =	1.25

EARTHWORK SUMMARY

STA.	TO	STA.	COMMON EXCAV. C.Y.	FILL EXP. 25% C.Y.	SELECT BORROW C.Y.	WASTE C.Y.
9+00		9+61	30	248	218	0
10+39		11+00	42	102	60	0
		TOTALS	72	350	278	0

ASPHALTIC SURFACE

STA.	TO	STA.	TONS
9+25		9+80.75	16
10+19.25		10+75	16
		TOTALS =	32

CRUSHED AGGREGATE BASE COURSE

STA.	TO	STA.	CU. YDS.
9+00		9+80.75	100
10+19.25		11+00	100
PARKING LOT ENTRANCE			20
		TOTAL =	220

CRUSHED AGGREGATE BASE COURSE FOR SHOULDERS

STA.	TO	STA.	CU. YDS.
9+00		9+80.75	6
10+19.25		11+00	6
		TOTAL =	12

TOPSOIL

STA.	TO	STA.	LOC.	TOPSOIL S.Y.
9+00		9+80.75	LT.	156
9+00		9+80.75	RT.	138
10+19.25		11+00	LT.	91
10+19.25		11+00	RT.	142
			TOTAL =	527

MULCH

STA.	TO	STA.	LOC.	MULCH S.Y.
9+00		9+80.75	LT.	211
9+00		9+80.75	RT.	188
10+19.25		11+00	LT.	139
10+19.25		11+00	RT.	193
			TOTAL =	731

FERTILIZER

STA.	TO	STA.	LOC.	FERTILIZER TYPE B CWT
9+00		9+80.75	LT./RT.	20
10+19.25		11+00	LT./RT.	15
			TOTAL =	35

SEEDING, MIXTURE NO. 20

STA.	TO	STA.	LOC.	SEED #20 LB.
9+00		9+80.75	LT./RT.	9
10+19.25		11+00	LT./RT.	7
			TOTAL =	16

EROSION MAT CLASS II, TYPE B

STA.	TO	STA.	LOC.	DELIVERED S.Y.	INSTALLED S.Y.
9+60		9+80	LT.	26	26
9+60		9+80	RT.	29	29
10+19		10+39	LT.	19	19
10+19		10+39	RT.	17	17
UNDISTRIBUTED				9	9
			TOTALS	100	100

SILT FENCE

STA.	TO	STA.	LOC.	DELIVERED L.F.	INSTALLED L.F.	MAINTENANCE L.F.
9+00		9+80.75	LT./RT.	209	209	209
10+19.25		11+00	LT./RT.	183	183	183
UNDISTRIBUTED				58	58	58
			TOTALS	450	450	450

EROSION BALES

LOCATION	DELIVERED EACH	INSTALLED EACH
UNDISTRIBUTED	10	10

WOOD POSTS, 4 x 4 INCH x 10 FEET

LOCATION	EACH
END OF DECK (OBJECT MARKER)	4

SIGNS, TYPE II, REFLECTIVE (W5-52 LT./RT.)

LOCATION	S.F.
END OF DECK (OBJECT MARKER)	12

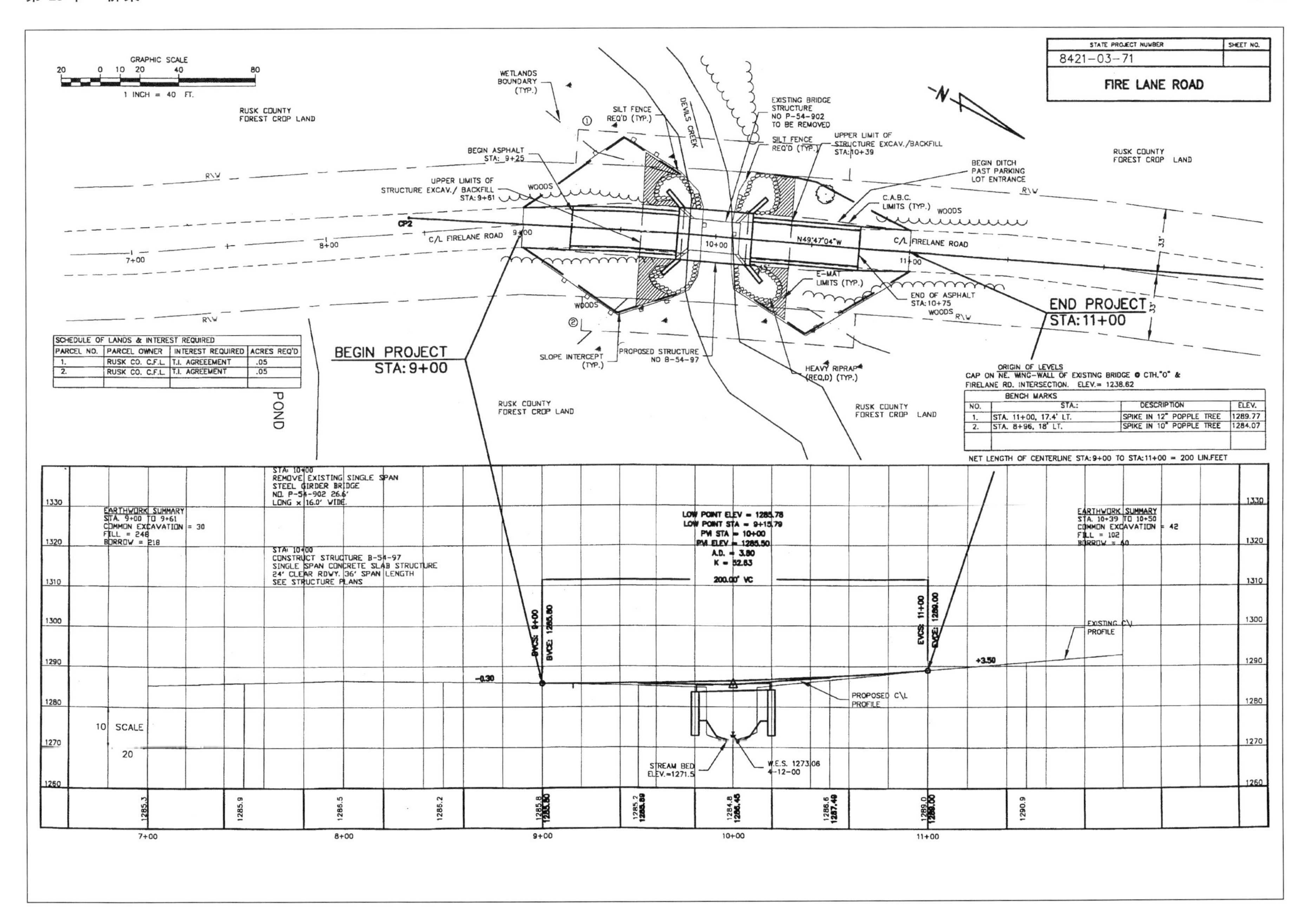

SCHEDULE OF LANDS & INTEREST REQUIRED			
PARCEL NO.	PARCEL OWNER	INTEREST REQUIRED	ACRES REQ'D
1.	RUSK CO. C.F.L.	T.I. AGREEEMENT	.05
2.	RUSK CO. C.F.L.	T.I. AGREEMENT	.05

BENCH MARKS

NO.	STA.:	DESCRIPTION	ELEV.
1.	STA. 11+00, 17.4' LT.	SPIKE IN 12" POPPLE TREE	1289.77
2.	STA. 8+96, 18' LT.	SPIKE IN 10" POPPLE TREE	1284.07

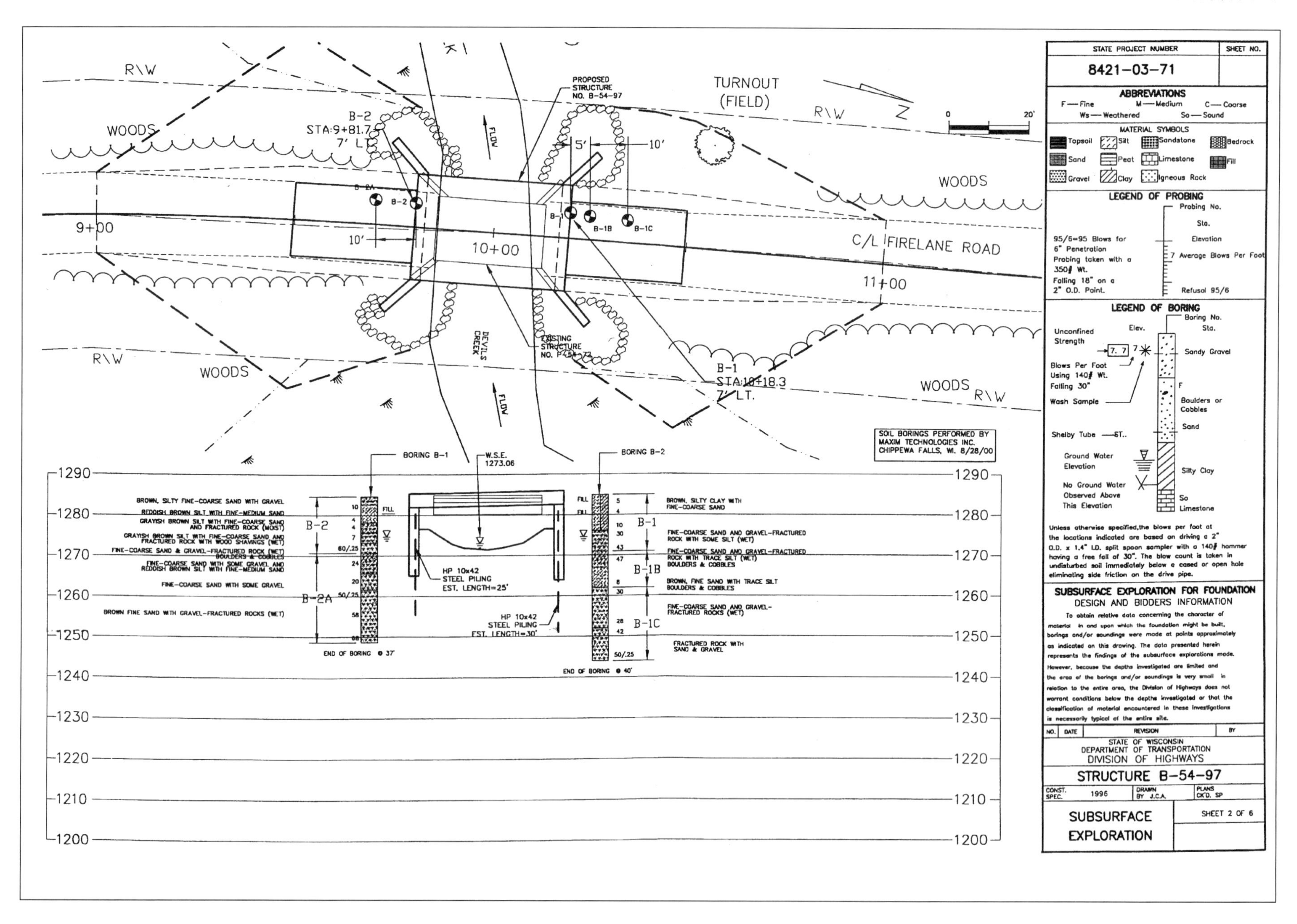

STATE PROJECT NUMBER
SHEET NO.
8421-03-71
ABBREVIATIONS
MATERIAL SYMBOLS
LEGEND OF PROBING
LEGEND OF BORING
SUBSURFACE EXPLORATION FOR FOUNDATION
DESIGN AND BIDDERS INFORMATION
STATE OF WISCONSIN
DEPARTMENT OF TRANSPORTATION
DIVISION OF HIGHWAYS
STRUCTURE B-54-97
SUBSURFACE EXPLORATION
SHEET 2 OF 6
TURNOUT (FIELD)
C/L FIRELANE ROAD
WOODS
BORING B-1
BORING B-2
SOIL BORINGS PERFORMED BY MAXIM TECHNOLOGIES INC. CHIPPEWA FALLS, WI. 8/28/00

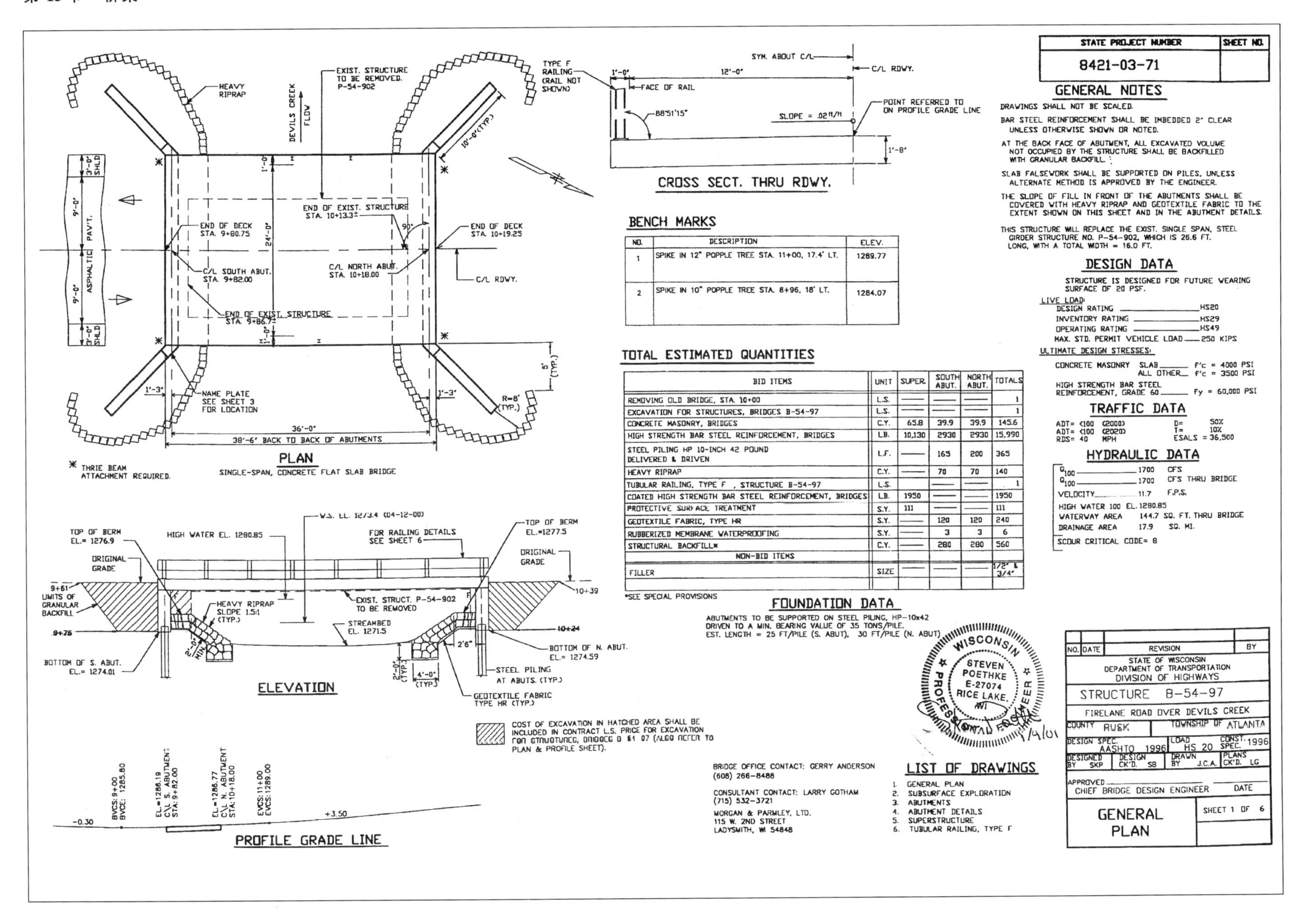

STATE PROJECT NUMBER
SHEET NO.
8421-03-71
GENERAL NOTES
DRAWINGS SHALL NOT BE SCALED.
BAR STEEL REINFORCEMENT SHALL BE IMBEDDED 2" CLEAR UNLESS OTHERWISE SHOWN OR NOTED.
AT THE BACK FACE OF ABUTMENT, ALL EXCAVATED VOLUME NOT OCCUPIED BY THE STRUCTURE SHALL BE BACKFILLED WITH GRANULAR BACKFILL.
SLAB FALSEWORK SHALL BE SUPPORTED ON PILES, UNLESS ALTERNATE METHOD IS APPROVED BY THE ENGINEER.
THE SLOPE OF FILL IN FRONT OF THE ABUTMENTS SHALL BE COVERED WITH HEAVY RIPRAP AND GEOTEXTILE FABRIC TO THE EXTENT SHOWN ON THIS SHEET AND IN THE ABUTMENT DETAILS.
THIS STRUCTURE WILL REPLACE THE EXIST. SINGLE SPAN, STEEL GIRDER STRUCTURE NO. P-54-902, WHICH IS 26.6 FT. LONG, WITH A TOTAL WIDTH = 16.0 FT.
DESIGN DATA
STRUCTURE IS DESIGNED FOR FUTURE WEARING SURFACE OF 20 PSF.
LIVE LOAD:
DESIGN RATING HS20
INVENTORY RATING HS29
OPERATING RATING HS49
MAX. STD. PERMIT VEHICLE LOAD 250 KIPS
ULTIMATE DESIGN STRESSES:
CONCRETE MASONRY SLAB f'c = 4000 PSI
ALL OTHER f'c = 3500 PSI
HIGH STRENGTH BAR STEEL REINFORCEMENT, GRADE 60 Fy = 60,000 PSI
TRAFFIC DATA
ADT= <100 (2000) D= 50%
ADT= <100 (2020) T= 10%
RDS= 40 MPH ESALS = 36,500
HYDRAULIC DATA
Q100 1700 CFS
Q100 1700 CFS THRU BRIDGE
VELOCITY 11.7 F.P.S.
HIGH WATER 100 EL. 1280.85
WATERWAY AREA 144.7 SQ. FT. THRU BRIDGE
DRAINAGE AREA 17.9 SQ. MI.
SCOUR CRITICAL CODE= 8
CROSS SECT. THRU RDWY.
SYM. ABOUT C/L
C/L RDWY.
TYPE F RAILING (RAIL NOT SHOWN)
1'-0"
12'-0"
FACE OF RAIL
88°51'15"
SLOPE = .02 ft/ft
POINT REFERRED TO ON PROFILE GRADE LINE
1'-8"
BENCH MARKS
NO. DESCRIPTION ELEV.
1 SPIKE IN 12" POPPLE TREE STA. 11+00, 17.4' LT. 1289.77
2 SPIKE IN 10" POPPLE TREE STA. 8+96, 18' LT. 1284.07
TOTAL ESTIMATED QUANTITIES
BID ITEMS | UNIT | SUPER. | SOUTH ABUT. | NORTH ABUT. | TOTALS
REMOVING OLD BRIDGE, STA. 10+00 | L.S. | — | — | — | 1
EXCAVATION FOR STRUCTURES, BRIDGES B-54-97 | L.S. | — | — | — | 1
CONCRETE MASONRY, BRIDGES | C.Y. | 65.8 | 39.9 | 39.9 | 145.6
HIGH STRENGTH BAR STEEL REINFORCEMENT, BRIDGES | LB. | 10,130 | 2930 | 2930 | 15,990
STEEL PILING HP 10-INCH 42 POUND DELIVERED & DRIVEN | L.F. | — | 165 | 200 | 365
HEAVY RIPRAP | C.Y. | — | 70 | 70 | 140
TUBULAR RAILING, TYPE F, STRUCTURE B-54-97 | L.S. | | — | — | 1
COATED HIGH STRENGTH BAR STEEL REINFORCEMENT, BRIDGES | LB. | 1950 | — | — | 1950
PROTECTIVE SURFACE TREATMENT | S.Y. | 111 | — | — | 111
GEOTEXTILE FABRIC, TYPE HR | S.Y. | — | 120 | 120 | 240
RUBBERIZED MEMBRANE WATERPROOFING | S.Y. | — | 3 | 3 | 6
STRUCTURAL BACKFILL* | C.Y. | — | 280 | 280 | 560
NON-BID ITEMS
FILLER | SIZE | — | — | — | 1/2" & 3/4"
*SEE SPECIAL PROVISIONS
FOUNDATION DATA
ABUTMENTS TO BE SUPPORTED ON STEEL PILING, HP-10x42 DRIVEN TO A MIN. BEARING VALUE OF 35 TONS/PILE.
EST. LENGTH = 25 FT/PILE (S. ABUT), 30 FT/PILE (N. ABUT)
PLAN
SINGLE-SPAN, CONCRETE FLAT SLAB BRIDGE
HEAVY RIPRAP
EXIST. STRUCTURE TO BE REMOVED. P-54-902
DEVILS CREEK FLOW
10'-0" (TYP.)
END OF DECK STA. 9+80.75
END OF EXIST. STRUCTURE STA. 10+13.3±
END OF DECK STA. 10+19.25
C/L SOUTH ABUT. STA. 9+82.00
C/L NORTH ABUT. STA. 10+18.00
C/L RDWY.
END OF EXIST. STRUCTURE STA. 9+86.7±
ASPHALTIC PAV'T.
3'-0" SHLD
9'-0"
24'-0"
1'-0"
5' (TYP.)
R=8' (TYP.)
1'-3"
NAME PLATE SEE SHEET 3 FOR LOCATION
36'-0"
38'-6" BACK TO BACK OF ABUTMENTS
* THRIE BEAM ATTACHMENT REQUIRED.
ELEVATION
TOP OF BERM EL.= 1276.9
HIGH WATER EL. 1280.85
W.S. EL. 1273.4 (04-12-00)
FOR RAILING DETAILS SEE SHEET 6
TOP OF BERM EL.=1277.5
ORIGINAL GRADE
9+61 LIMITS OF GRANULAR BACKFILL
10+39
HEAVY RIPRAP SLOPE 1.5:1 (TYP.)
EXIST. STRUCT. P-54-902 TO BE REMOVED
STREAMBED EL. 1271.5
9+76
10+24
BOTTOM OF N. ABUT. EL.= 1274.59
BOTTOM OF S. ABUT. EL.= 1274.01
STEEL PILING AT ABUTS. (TYP.)
GEOTEXTILE FABRIC TYPE HR (TYP.)
2'-6"
4'-0" (TYP.)
COST OF EXCAVATION IN HATCHED AREA SHALL BE INCLUDED IN CONTRACT L.S. PRICE FOR EXCAVATION FOR STRUCTURES, BRIDGES B-54-97 (ALSO REFER TO PLAN & PROFILE SHEET).
PROFILE GRADE LINE
BVCS: 9+00 BVCE: 1285.80
EL.=1286.19 C/L S. ABUTMENT STA: 9+82.00
EL.=1286.77 C/L N. ABUTMENT STA: 10+18.00
EVCS: 11+00 EVCS: 1289.00
-0.30
+3.50
BRIDGE OFFICE CONTACT: GERRY ANDERSON (608) 266-8488
CONSULTANT CONTACT: LARRY GOTHAM (715) 532-3721
MORGAN & PARMLEY, LTD.
115 W. 2ND STREET
LADYSMITH, WI 54848
WISCONSIN STEVEN POETHKE E-27074 RICE LAKE, WI PROFESSIONAL ENGINEER
LIST OF DRAWINGS
1. GENERAL PLAN
2. SUBSURFACE EXPLORATION
3. ABUTMENTS
4. ABUTMENT DETAILS
5. SUPERSTRUCTURE
6. TUBULAR RAILING, TYPE F
NO. DATE REVISION BY
STATE OF WISCONSIN DEPARTMENT OF TRANSPORTATION DIVISION OF HIGHWAYS
STRUCTURE B-54-97
FIRELANE ROAD OVER DEVILS CREEK
COUNTY RUSK TOWNSHIP OF ATLANTA
DESIGN SPEC. AASHTO 1996 LOAD HS 20 CONST. SPEC. 1996
DESIGNED BY SKP DESIGN CK'D. SB DRAWN BY J.C.A. PLANS CK'D. LG
APPROVED CHIEF BRIDGE DESIGN ENGINEER DATE
GENERAL PLAN
SHEET 1 OF 6

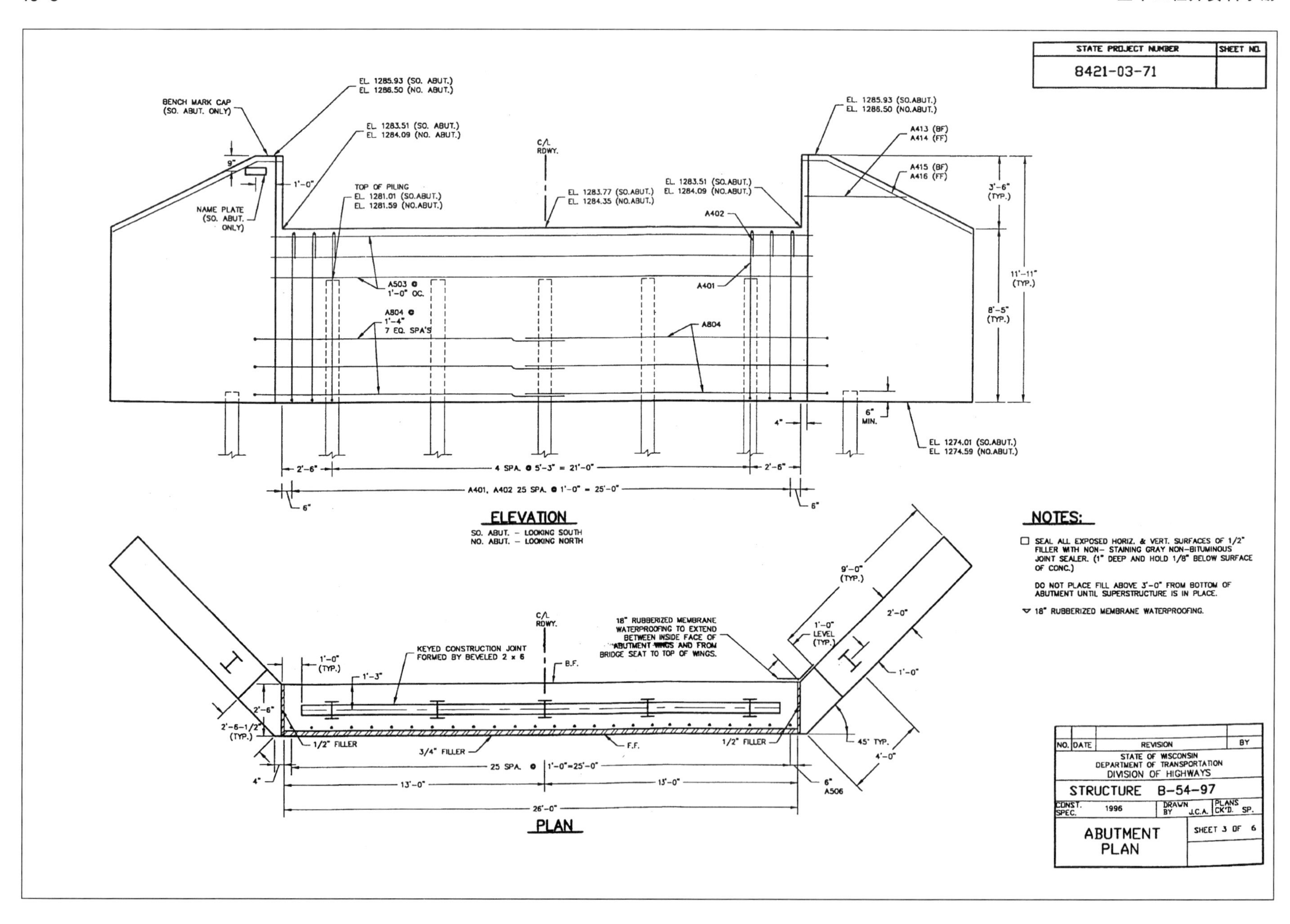

STATE PROJECT NUMBER
SHEET NO.
8421-03-71
BENCH MARK CAP (SO. ABUT. ONLY)
EL. 1285.93 (SO. ABUT.) EL. 1286.50 (NO. ABUT.)
EL. 1283.51 (SO. ABUT.) EL. 1284.09 (NO. ABUT.)
NAME PLATE (SO. ABUT. ONLY)
TOP OF PILING EL. 1281.01 (SO.ABUT.) EL. 1281.59 (NO.ABUT.)
C/L RDWY.
EL. 1283.77 (SO.ABUT.) EL. 1284.35 (NO.ABUT.)
A503 @ 1'-0" OC.
A804 @ 1'-4" 7 EQ. SPA'S
A401
A402
A804
A413 (BF) A414 (FF)
A415 (BF) A416 (FF)
3'-6" (TYP.)
11'-11" (TYP.)
8'-5" (TYP.)
6" MIN.
EL. 1274.01 (SO.ABUT.) EL. 1274.59 (NO.ABUT.)
4 SPA. @ 5'-3" = 21'-0"
A401, A402 25 SPA. @ 1'-0" = 25'-0"
ELEVATION
SO. ABUT. – LOOKING SOUTH
NO. ABUT. – LOOKING NORTH
NOTES:
SEAL ALL EXPOSED HORIZ. & VERT. SURFACES OF 1/2" FILLER WITH NON- STAINING GRAY NON-BITUMINOUS JOINT SEALER. (1" DEEP AND HOLD 1/8" BELOW SURFACE OF CONC.)
DO NOT PLACE FILL ABOVE 3'-0" FROM BOTTOM OF ABUTMENT UNTIL SUPERSTRUCTURE IS IN PLACE.
18" RUBBERIZED MEMBRANE WATERPROOFING.
KEYED CONSTRUCTION JOINT FORMED BY BEVELED 2 x 6
18" RUBBERIZED MEMBRANE WATERPROOFING TO EXTEND BETWEEN INSIDE FACE OF ABUTMENT WINGS AND FROM BRIDGE SEAT TO TOP OF WINGS.
B.F.
F.F.
1/2" FILLER
3/4" FILLER
9'-0" (TYP.)
1'-0" LEVEL (TYP.)
45° TYP.
25 SPA. @ 1'-0"=25'-0"
13'-0"
26'-0"
A506
PLAN
NO. DATE REVISION BY
STATE OF WISCONSIN DEPARTMENT OF TRANSPORTATION DIVISION OF HIGHWAYS
STRUCTURE B-54-97
CONST. SPEC. 1996
DRAWN BY J.C.A.
PLANS CK'D. SP.
ABUTMENT PLAN
SHEET 3 OF 6

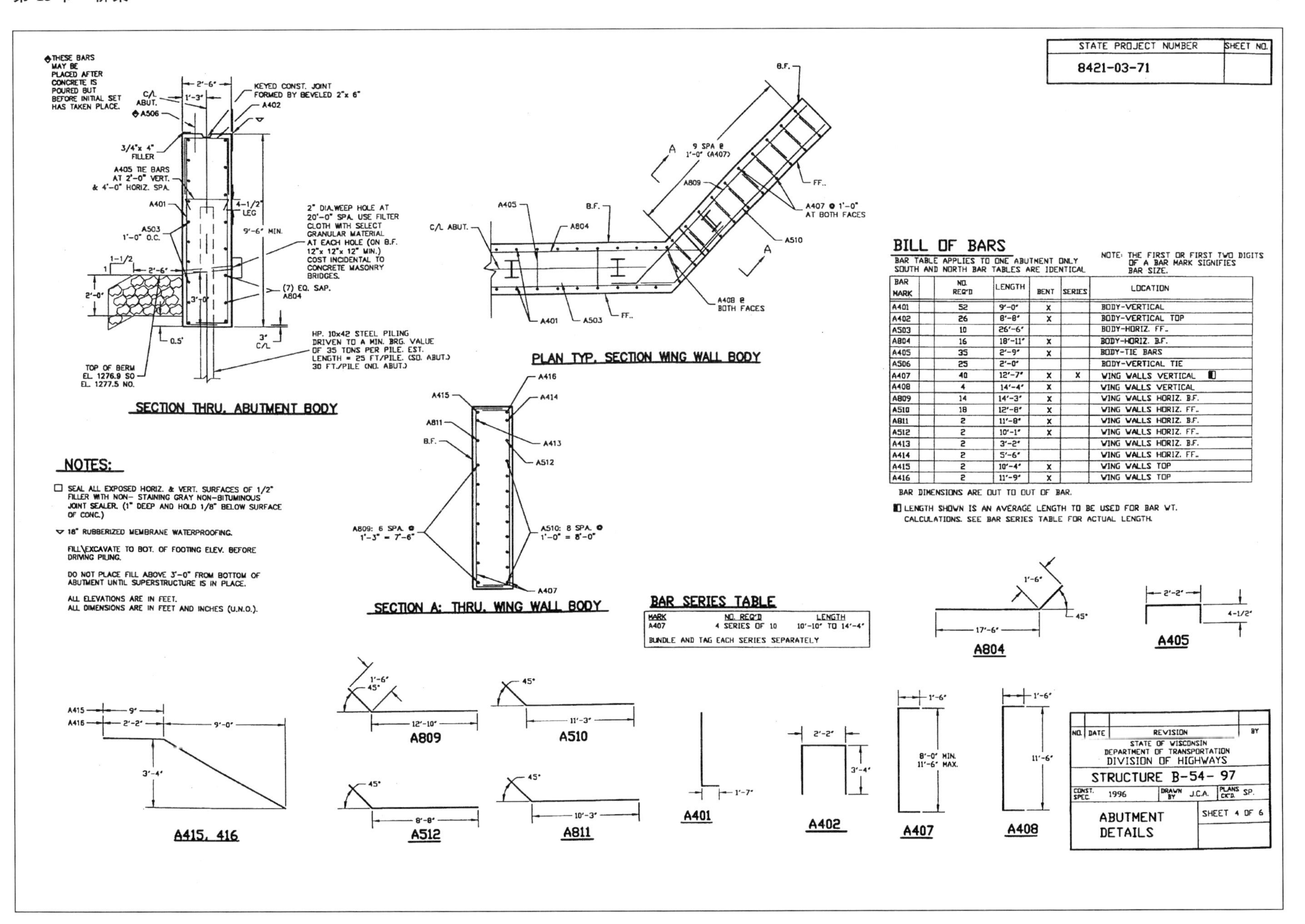

BILL OF BARS

BAR MARK		NO. REQ'D	LENGTH	BENT	SERIES	LOCATION
A401		52	9'-0"	X		BODY-VERTICAL
A402		26	8'-8"	X		BODY-VERTICAL TOP
A503		10	26'-6"			BODY-HORIZ. FF..
A804		16	18'-11"	X		BODY-HORIZ. B.F.
A405		35	2'-9"	X		BODY-TIE BARS
A506		25	2'-0"			BODY-VERTICAL TIE
A407		40	12'-7"	X	X	WING WALLS VERTICAL ▮
A408		4	14'-4"	X		WING WALLS VERTICAL
A809		14	14'-3"	X		WING WALLS HORIZ. B.F.
A510		18	12'-8"	X		WING WALLS HORIZ. FF..
A811		2	11'-8"	X		WING WALLS HORIZ. B.F.
A512		2	10'-1"	X		WING WALLS HORIZ. FF..
A413		2	3'-2"			WING WALLS HORIZ. B.F.
A414		2	5'-6"			WING WALLS HORIZ. FF..
A415		2	10'-4"	X		WING WALLS TOP
A416		2	11'-9"	X		WING WALLS TOP

BAR SERIES TABLE

MARK	NO. REQ'D	LENGTH
A407	4 SERIES OF 10	10'-10" TO 14'-4"

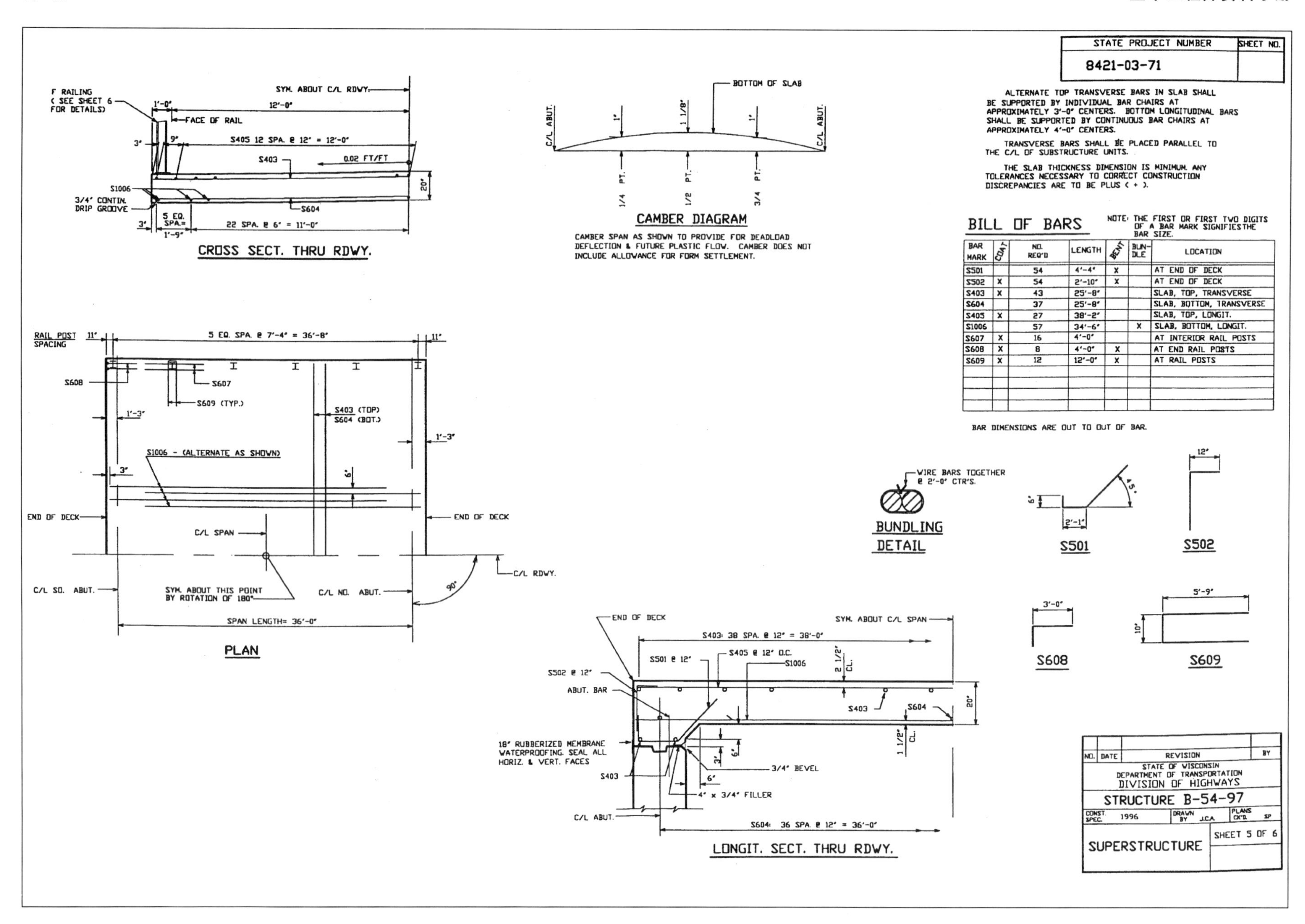

BILL OF BARS

NOTE: THE FIRST OR FIRST TWO DIGITS OF A BAR MARK SIGNIFIES THE BAR SIZE.

BAR MARK	COAT	NO. REQ'D	LENGTH	BENT	BUNDLE	LOCATION
S501		54	4'-4"	X		AT END OF DECK
S502	X	54	2'-10"	X		AT END OF DECK
S403	X	43	25'-8"			SLAB, TOP, TRANSVERSE
S604		37	25'-8"			SLAB, BOTTOM, TRANSVERSE
S405	X	27	38'-2"			SLAB, TOP, LONGIT.
S1006		57	34'-6"		X	SLAB, BOTTOM, LONGIT.
S607	X	16	4'-0"			AT INTERIOR RAIL POSTS
S608	X	8	4'-0"	X		AT END RAIL POSTS
S609	X	12	12'-0"	X		AT RAIL POSTS

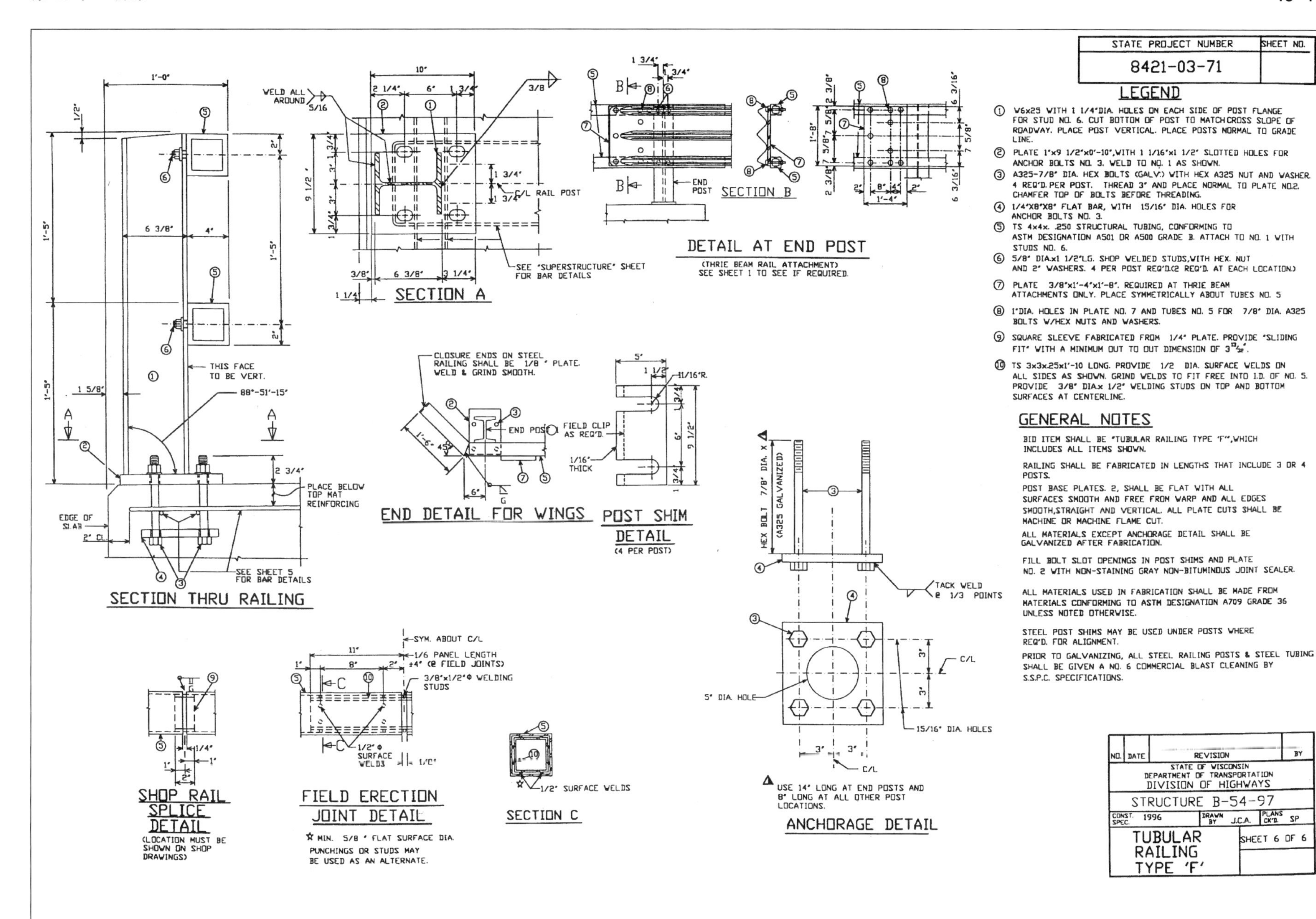

STATE PROJECT NUMBER
SHEET NO.
8421-03-71
LEGEND
① W6x25 WITH 1 1/4"DIA. HOLES ON EACH SIDE OF POST FLANGE FOR STUD NO. 6. CUT BOTTOM OF POST TO MATCH CROSS SLOPE OF ROADWAY. PLACE POST VERTICAL. PLACE POSTS NORMAL TO GRADE LINE.
② PLATE 1"x9 1/2"x0'-10",WITH 1 1/16"x1 1/2" SLOTTED HOLES FOR ANCHOR BOLTS NO. 3. WELD TO NO. 1 AS SHOWN.
③ A325-7/8" DIA. HEX BOLTS (GALV.) WITH HEX A325 NUT AND WASHER. 4 REQ'D. PER POST. THREAD 3" AND PLACE NORMAL TO PLATE NO.2. CHAMFER TOP OF BOLTS BEFORE THREADING.
④ 1/4"X8"X8" FLAT BAR, WITH 15/16" DIA. HOLES FOR ANCHOR BOLTS NO. 3.
⑤ TS 4x4x. .250 STRUCTURAL TUBING, CONFORMING TO ASTM DESIGNATION A501 OR A500 GRADE B. ATTACH TO NO. 1 WITH STUDS NO. 6.
⑥ 5/8" DIA.x1 1/2"LG. SHOP WELDED STUDS,WITH HEX. NUT AND 2" WASHERS. 4 PER POST REQ'D.(2 REQ'D. AT EACH LOCATION.)
⑦ PLATE 3/8"x1'-4"x1'-8". REQUIRED AT THRIE BEAM ATTACHMENTS ONLY. PLACE SYMMETRICALLY ABOUT TUBES NO. 5
⑧ 1"DIA. HOLES IN PLATE NO. 7 AND TUBES NO. 5 FOR 7/8" DIA. A325 BOLTS W/HEX NUTS AND WASHERS.
⑨ SQUARE SLEEVE FABRICATED FROM 1/4" PLATE. PROVIDE "SLIDING FIT" WITH A MINIMUM OUT TO OUT DIMENSION OF 3 13/32".
⑩ TS 3x3x.25x1'-10 LONG. PROVIDE 1/2 DIA. SURFACE WELDS ON ALL SIDES AS SHOWN. GRIND WELDS TO FIT FREE INTO I.D. OF NO. 5. PROVIDE 3/8" DIA.x 1/2" WELDING STUDS ON TOP AND BOTTOM SURFACES AT CENTERLINE.
GENERAL NOTES
BID ITEM SHALL BE "TUBULAR RAILING TYPE 'F'",WHICH INCLUDES ALL ITEMS SHOWN.
RAILING SHALL BE FABRICATED IN LENGTHS THAT INCLUDE 3 OR 4 POSTS.
POST BASE PLATES. 2, SHALL BE FLAT WITH ALL SURFACES SMOOTH AND FREE FROM WARP AND ALL EDGES SMOOTH,STRAIGHT AND VERTICAL. ALL PLATE CUTS SHALL BE MACHINE OR MACHINE FLAME CUT.
ALL MATERIALS EXCEPT ANCHORAGE DETAIL SHALL BE GALVANIZED AFTER FABRICATION.
FILL BOLT SLOT OPENINGS IN POST SHIMS AND PLATE NO. 2 WITH NON-STAINING GRAY NON-BITUMINOUS JOINT SEALER.
ALL MATERIALS USED IN FABRICATION SHALL BE MADE FROM MATERIALS CONFORMING TO ASTM DESIGNATION A709 GRADE 36 UNLESS NOTED OTHERWISE.
STEEL POST SHIMS MAY BE USED UNDER POSTS WHERE REQ'D. FOR ALIGNMENT.
PRIOR TO GALVANIZING, ALL STEEL RAILING POSTS & STEEL TUBING SHALL BE GIVEN A NO. 6 COMMERCIAL BLAST CLEANING BY S.S.P.C. SPECIFICATIONS.
SECTION THRU RAILING
THIS FACE TO BE VERT.
88°-51'-15"
EDGE OF SLAB
PLACE BELOW TOP MAT REINFORCING
SEE SHEET 5 FOR BAR DETAILS
SECTION A
WELD ALL AROUND
C/L RAIL POST
SEE "SUPERSTRUCTURE" SHEET FOR BAR DETAILS
DETAIL AT END POST
(THRIE BEAM RAIL ATTACHMENT)
SEE SHEET 1 TO SEE IF REQUIRED.
SECTION B
END POST
CLOSURE ENDS ON STEEL RAILING SHALL BE 1/8" PLATE. WELD & GRIND SMOOTH.
END POST
FIELD CLIP AS REQ'D.
END DETAIL FOR WINGS
POST SHIM DETAIL
(4 PER POST)
1/16" THICK
HEX BOLT 7/8" DIA. X (A325 GALVANIZED)
TACK WELD @ 1/3 POINTS
5" DIA. HOLE
15/16" DIA. HOLES
USE 14" LONG AT END POSTS AND 8" LONG AT ALL OTHER POST LOCATIONS.
ANCHORAGE DETAIL
SYM. ABOUT C/L
1/6 PANEL LENGTH ±4" (@ FIELD JOINTS)
3/8"x1/2"Φ WELDING STUDS
1/2"Φ SURFACE WELDS
SHOP RAIL SPLICE DETAIL
(LOCATION MUST BE SHOWN ON SHOP DRAWINGS)
FIELD ERECTION JOINT DETAIL
☆ MIN. 5/8" FLAT SURFACE DIA. PUNCHINGS OR STUDS MAY BE USED AS AN ALTERNATE.
SECTION C
1/2" SURFACE WELDS
NO. DATE REVISION BY
STATE OF WISCONSIN
DEPARTMENT OF TRANSPORTATION
DIVISION OF HIGHWAYS
STRUCTURE B-54-97
CONST. SPEC. 1996
DRAWN BY J.C.A.
PLANS CK'D. SP
TUBULAR RAILING TYPE 'F'
SHEET 6 OF 6

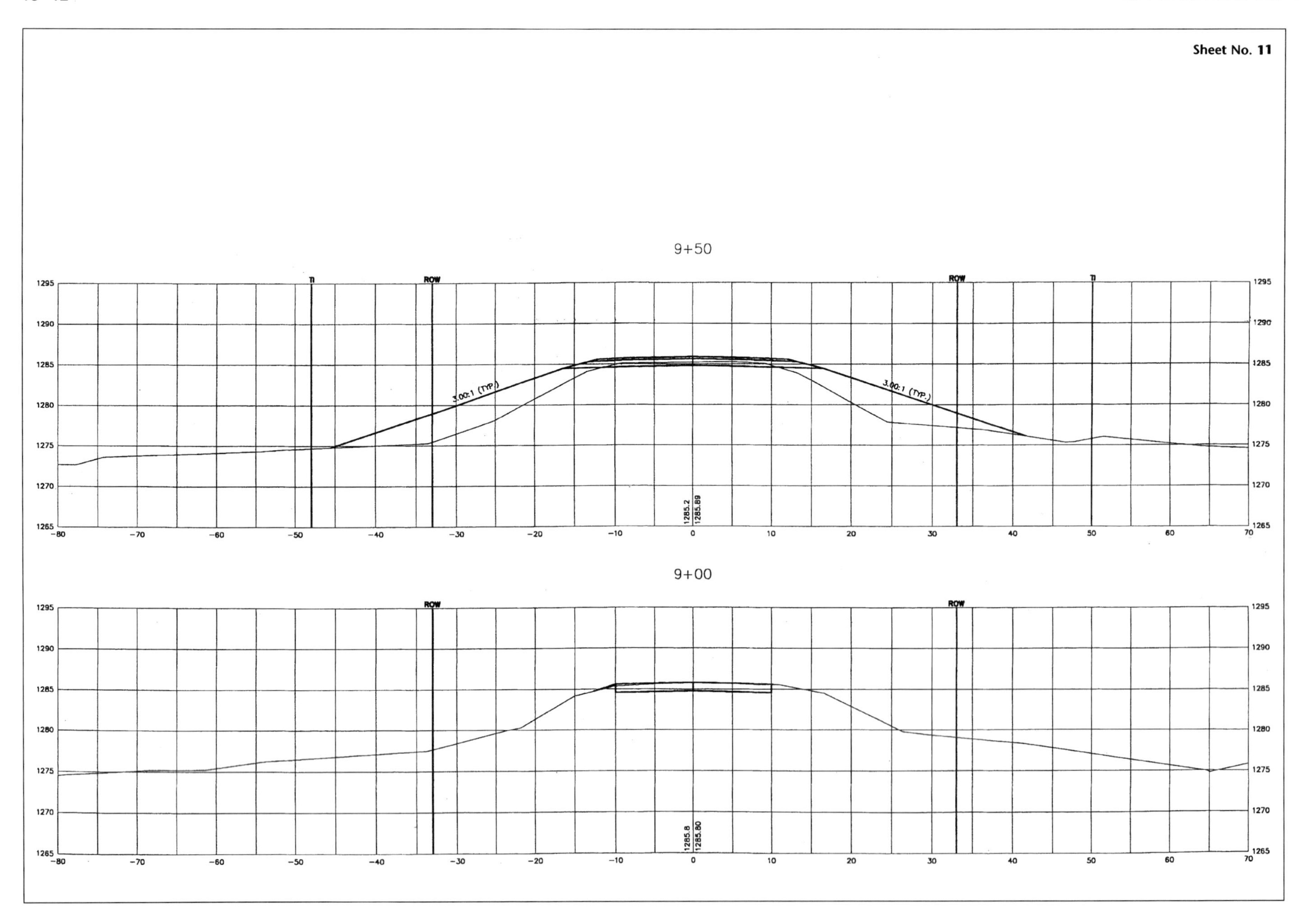
Sheet No. 11
9+50
TI
ROW
ROW
TI
3.00:1 (TYP.)
3.00:1 (TYP.)
1285.2
1285.89
9+00
ROW
ROW
1285.8
1285.80

STA:10+00 CENTERLINE OF STRUCTURE

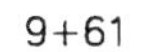

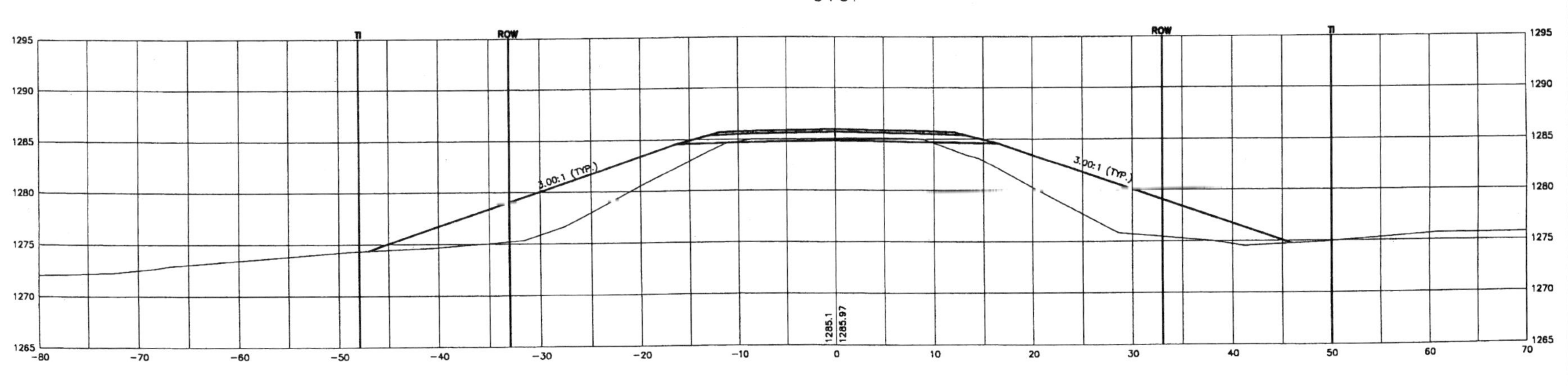

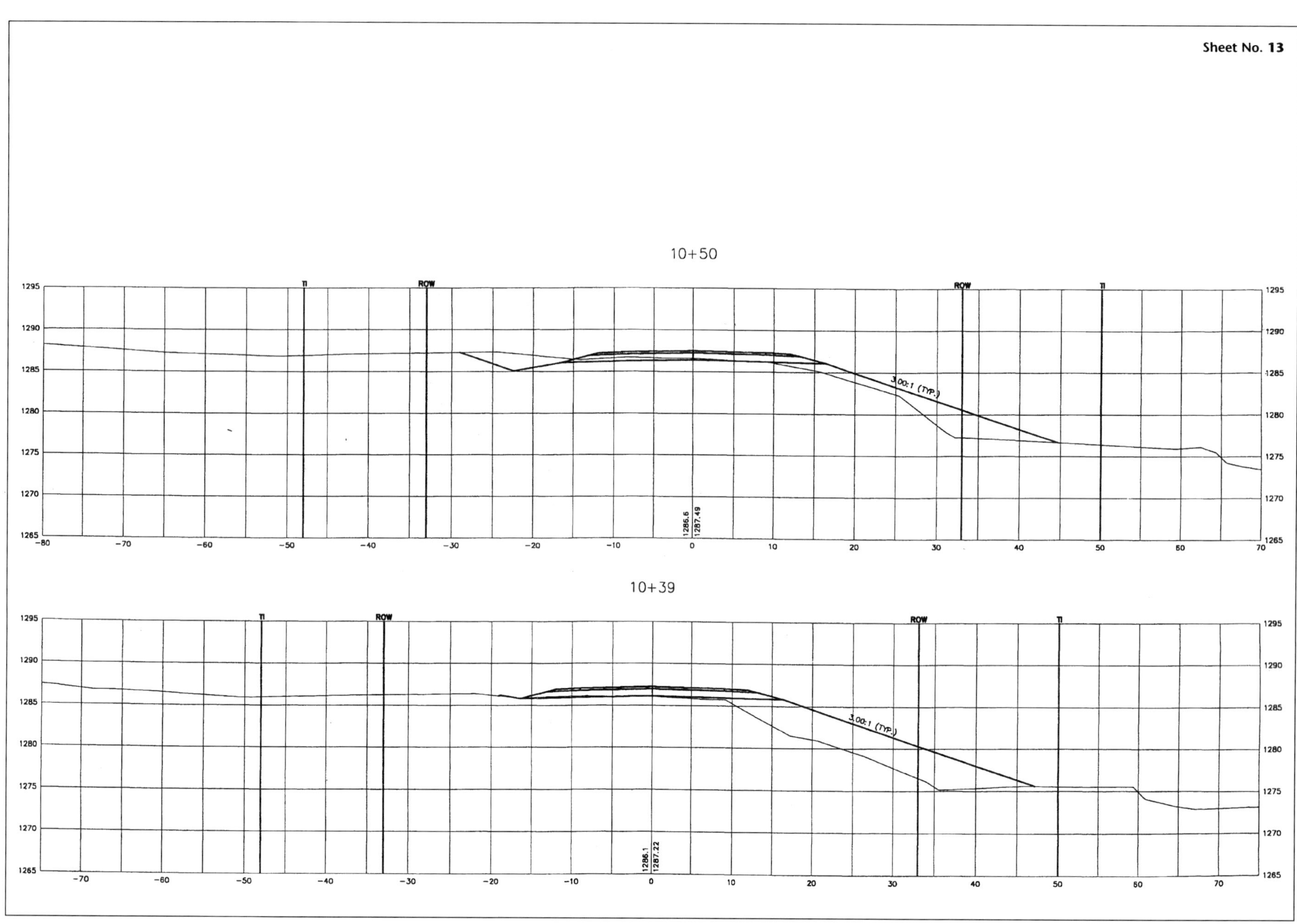
Sheet No. 13
10+50
TI
ROW
ROW
TI
3.00:1 (TYP.)
1286.6
1287.49
10+39
TI
ROW
ROW
TI
3.00:1 (TYP.)
1286.1
1287.22

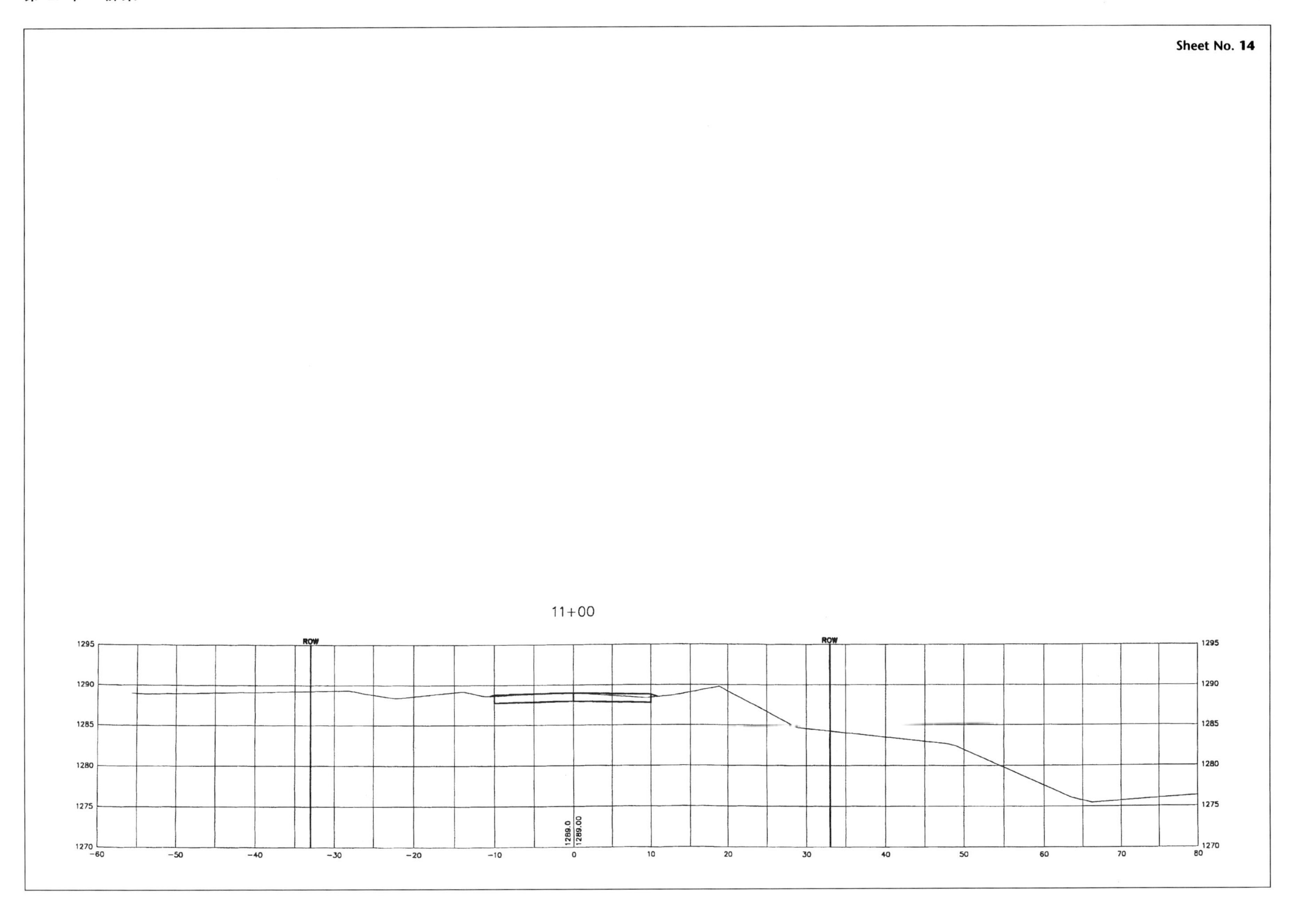
Sheet No. 14
11+00
ROW
ROW
1295
1290
1285
1280
1275
1270
-60
-50
-40
-30
-20
-10
0
10
20
30
40
50
60
70
80
1289.0
1289.00

Cloverland 公路桥

CLOVERLAND ROAD BRIDGE

— MAIN CREEK —

TOWN OF GROW
RUSK COUNTY, WI.

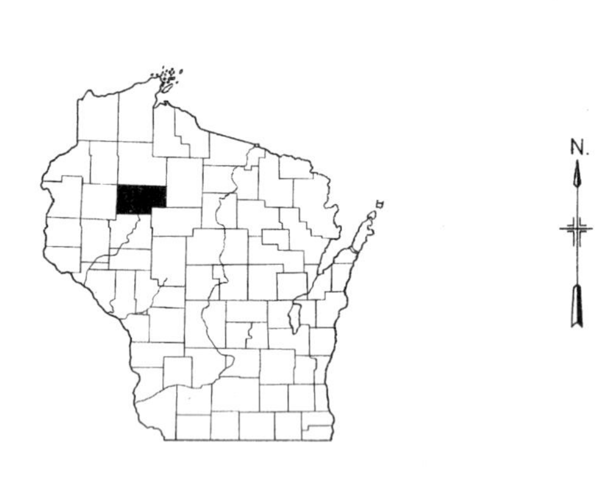

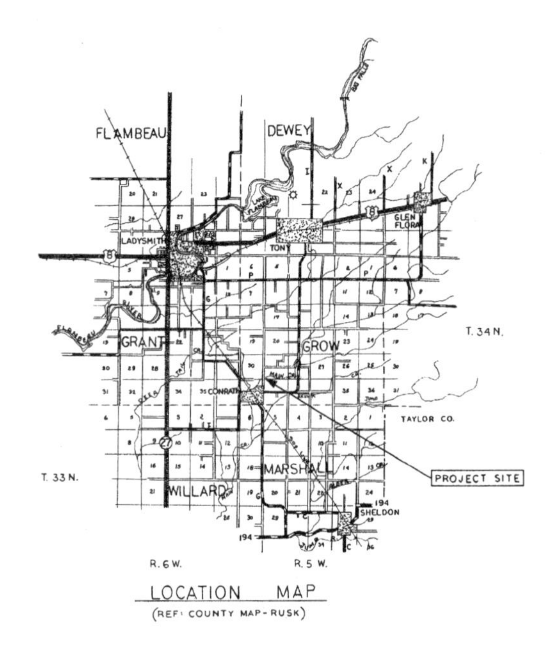

LOCATION MAP
(REF: COUNTY MAP-RUSK)

INDEX OF SHEETS

NO.	DESCRIPTION
1	TITLE SHEET & INDEX W/ LOCATION MAP
2	TOPOGRAPHY & GENERAL MAP
3	PLAN & ROAD PROFILE
4	GENERAL PLAN W/ DESIGN & HYDRAULIC DATA
5	SUBSURFACE EXPLORATION
6	FOUNDATIONS
7	ABUTMENTS
8	SUPERSTRUCTURE W/ DETAILS
9	PRESTRESSED GIRDER DETAILS
10	TUBULAR STEEL RAILING

MAY 12, 1976

CLOVERLAND ROAD BRIDGE	
TITLE SHEET AND INDEX	
SCALE- DO NOT	MORGAN & PARMLEY, LTD. CONSULTING ENGINEERS LADYSMITH, WISCONSIN
SHEET 1 OF 10	
DATE MAY 1976	DR. BY R.O.PARMLEY

Courtesy: Morgan & Parmley, Ltd.

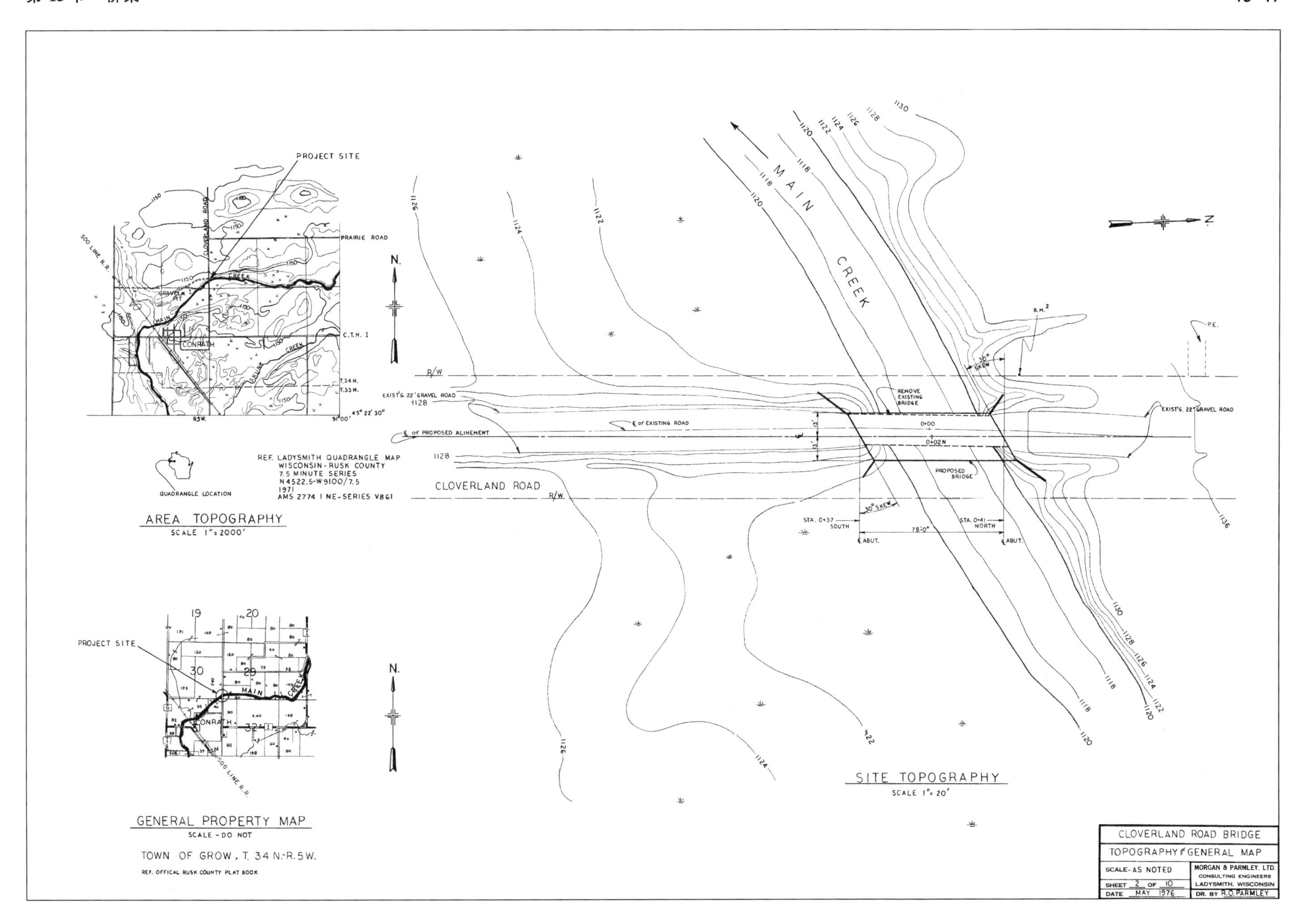
PROJECT SITE
PRAIRIE ROAD
CLOVERLAND ROAD
CREEK
MAIN
CONRATH
C.T.H. I
T.34 N.
T.33 N.
R5W.
91°00' 45° 22' 30"
500 LINE R.R.
QUADRANGLE LOCATION
REF. LADYSMITH QUADRANGLE MAP
WISCONSIN - RUSK COUNTY
7.5 MINUTE SERIES
N 4522.5-W 9100/7.5
1971
AMS 2774 I NE-SERIES V861
AREA TOPOGRAPHY
SCALE 1" = 2000'
PROJECT SITE
GENERAL PROPERTY MAP
SCALE - DO NOT
TOWN OF GROW, T. 34 N.-R.5 W.
REF. OFFICAL RUSK COUNTY PLAT BOOK
MAIN CREEK
R/W
EXIST'G 22' GRAVEL ROAD
℄ of EXISTING ROAD
℄ of PROPOSED ALINEMENT
REMOVE EXISTING BRIDGE
PROPOSED BRIDGE
CLOVERLAND ROAD
30° SKEW
STA. 0+37 SOUTH
STA. 0+41 NORTH
78'-0"
℄ ABUT.
B.M.
P.E.
SITE TOPOGRAPHY
SCALE 1" = 20'
CLOVERLAND ROAD BRIDGE
TOPOGRAPHY & GENERAL MAP
SCALE - AS NOTED
SHEET 2 OF 10
DATE MAY 1976
MORGAN & PARMLEY, LTD.
CONSULTING ENGINEERS
LADYSMITH, WISCONSIN
DR. BY R.O. PARMLEY

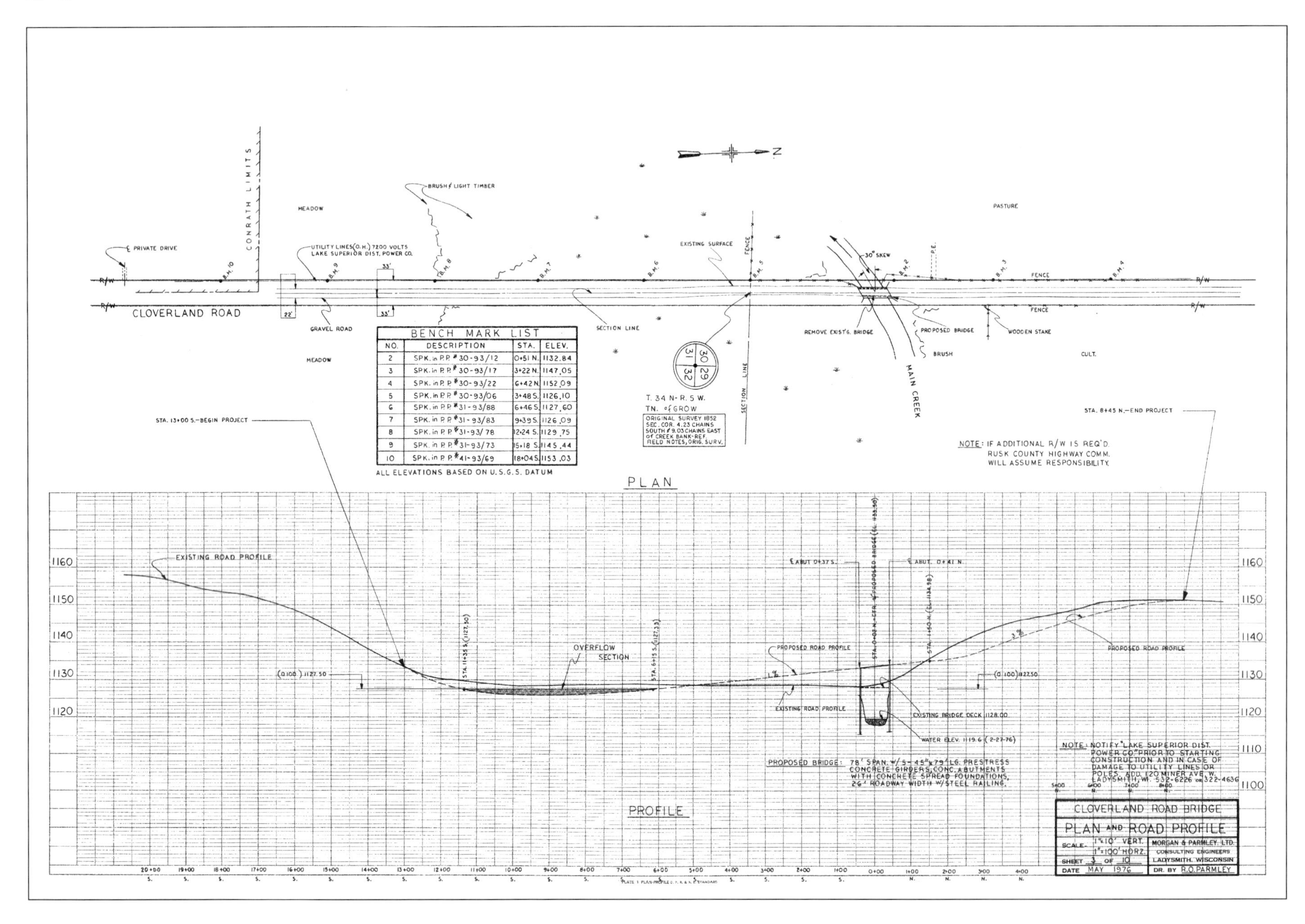

NO.	DESCRIPTION	STA.	ELEV.
2	SPK. in P.P. #30-93/12	0+51 N.	1132.84
3	SPK. in P.P. #30-93/17	3+22 N.	1147.05
4	SPK. in P.P. #30-93/22	6+42 N.	1152.09
5	SPK. in P.P. #30-93/06	3+48 S.	1126.10
6	SPK. in P.P. #31-93/88	6+46 S.	1127.60
7	SPK. in P.P. #31-93/83	9+39 S.	1126.09
8	SPK. in P.P. #31-93/78	12+24 S.	1129.75
9	SPK. in P.P. #31-93/73	15+18 S.	1145.44
10	SPK. in P.P. #41-93/69	18+04 S.	1153.03

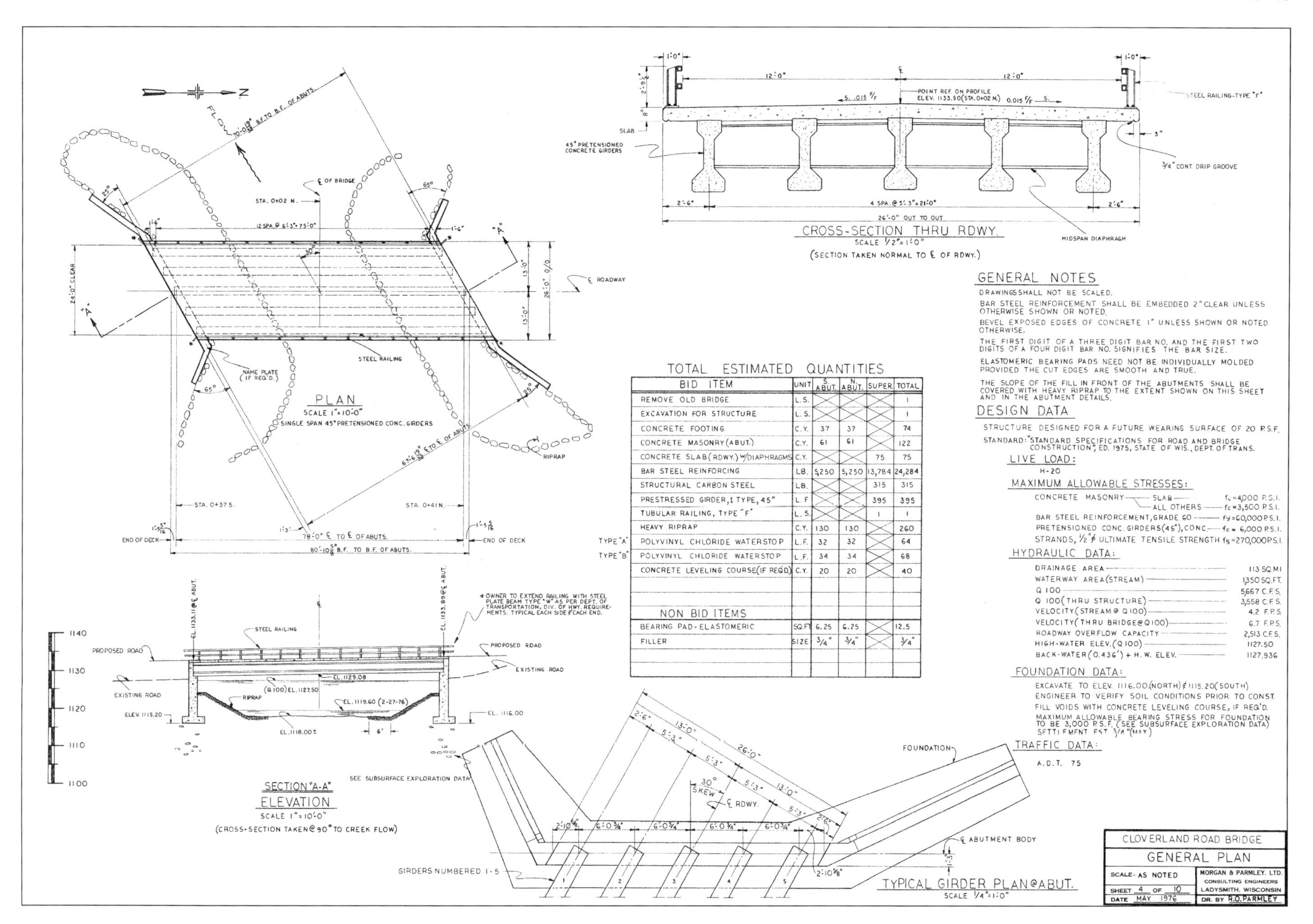

TOTAL ESTIMATED QUANTITIES

	BID ITEM	UNIT	S. ABUT.	N. ABUT.	SUPER.	TOTAL
	REMOVE OLD BRIDGE	L.S.				1
	EXCAVATION FOR STRUCTURE	L.S.				1
	CONCRETE FOOTING	C.Y.	37	37		74
	CONCRETE MASONRY (ABUT.)	C.Y.	61	61		122
	CONCRETE SLAB (RDWY.) W/DIAPHRAGMS	C.Y.			75	75
	BAR STEEL REINFORCING	LB.	5,250	5,250	13,784	24,284
	STRUCTURAL CARBON STEEL	LB.			315	315
	PRESTRESSED GIRDER, I TYPE, 45"	L.F.			395	395
	TUBULAR RAILING, TYPE "F"	L.S.			1	1
	HEAVY RIPRAP	C.Y.	130	130		260
TYPE "A"	POLYVINYL CHLORIDE WATERSTOP	L.F.	32	32		64
TYPE "B"	POLYVINYL CHLORIDE WATERSTOP	L.F.	34	34		68
	CONCRETE LEVELING COURSE (IF REQ'D.)	C.Y.	20	20		40
	NON BID ITEMS					
	BEARING PAD-ELASTOMERIC	SQ.FT.	6.25	6.25		12.5
	FILLER	SIZE	3/4"	3/4"		3/4"

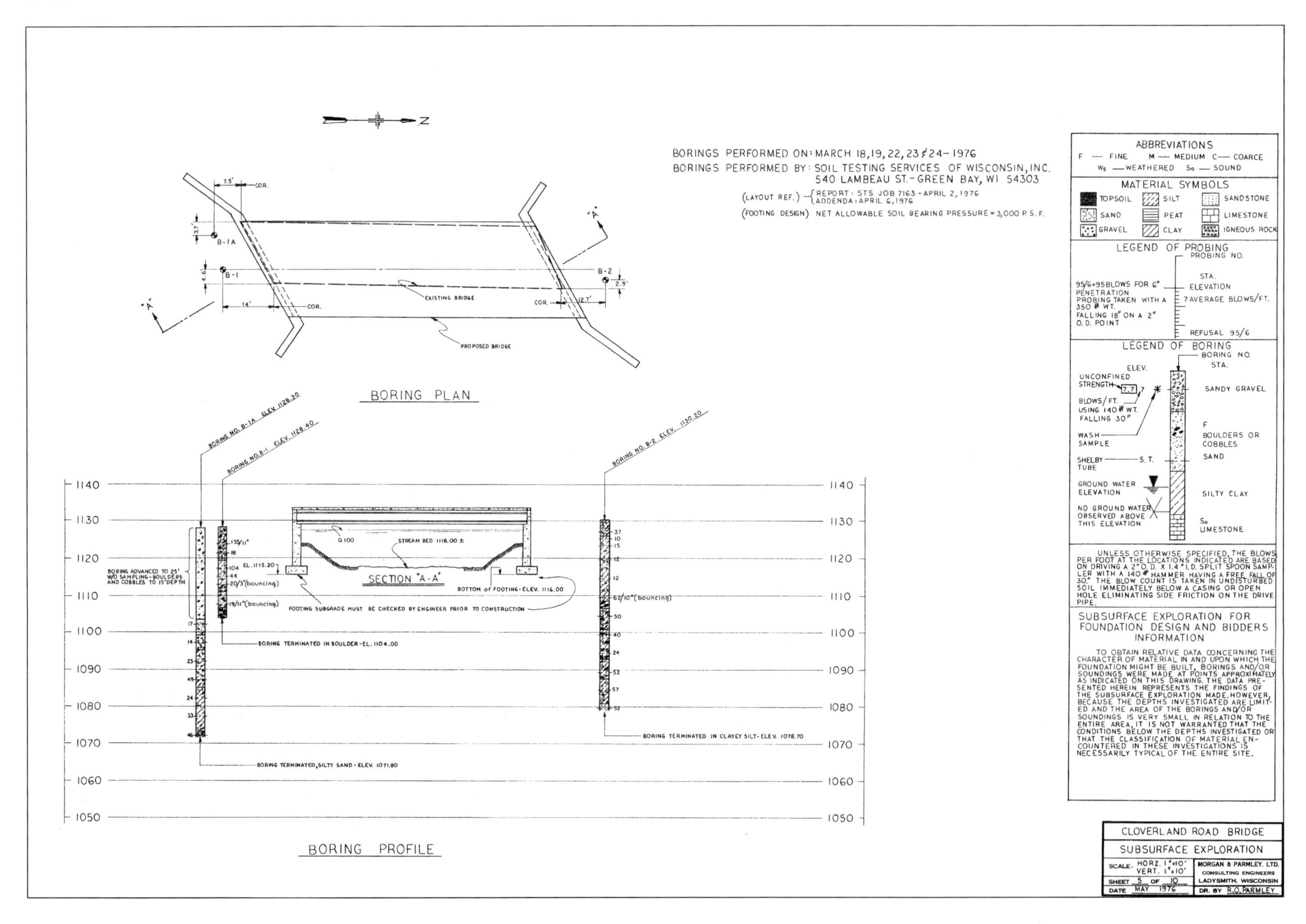
BORINGS PERFORMED ON: MARCH 18,19,22,23 & 24-1976
BORINGS PERFORMED BY: SOIL TESTING SERVICES OF WISCONSIN, INC. 540 LAMBEAU ST.-GREEN BAY, WI 54303
(LAYOUT REF.) REPORT: STS JOB 7163-APRIL 2, 1976 ADDENDA: APRIL 6, 1976
(FOOTING DESIGN) NET ALLOWABLE SOIL BEARING PRESSURE = 3,000 P.S.F.
BORING PLAN
EXISTING BRIDGE
PROPOSED BRIDGE
BORING PROFILE
SECTION "A-A"
STREAM BED 1118.00 ±
BOTTOM of FOOTING-ELEV. 1116.00
FOOTING SUBGRADE MUST BE CHECKED BY ENGINEER PRIOR TO CONSTRUCTION
BORING NO. B-1A ELEV. 1128.30
BORING NO. B-1 ELEV. 1128.40
BORING NO. B-2 ELEV. 1130.20
BORING ADVANCED TO 25' W/O SAMPLING-BOULDERS AND COBBLES TO 15' DEPTH
BORING TERMINATED IN BOULDER-EL. 1104.00
BORING TERMINATED, SILTY SAND-ELEV. 1071.80
BORING TERMINATED IN CLAYEY SILT-ELEV. 1078.70
ABBREVIATIONS
F — FINE M — MEDIUM C — COARCE
Ws — WEATHERED So — SOUND
MATERIAL SYMBOLS
TOPSOIL SILT SANDSTONE
SAND PEAT LIMESTONE
GRAVEL CLAY IGNEOUS ROCK
LEGEND OF PROBING
LEGEND OF BORING
UNLESS OTHERWISE SPECIFIED, THE BLOWS PER FOOT AT THE LOCATIONS INDICATED ARE BASED ON DRIVING A 2" O.D. X 1.4" I.D. SPLIT SPOON SAMPLER WITH A 140# HAMMER HAVING A FREE FALL OF 30." THE BLOW COUNT IS TAKEN IN UNDISTURBED SOIL IMMEDIATELY BELOW A CASING OR OPEN HOLE ELIMINATING SIDE FRICTION ON THE DRIVE PIPE.
SUBSURFACE EXPLORATION FOR FOUNDATION DESIGN AND BIDDERS INFORMATION
TO OBTAIN RELATIVE DATA CONCERNING THE CHARACTER OF MATERIAL IN AND UPON WHICH THE FOUNDATION MIGHT BE BUILT, BORINGS AND/OR SOUNDINGS WERE MADE AT POINTS APPROXIMATELY AS INDICATED ON THIS DRAWING. THE DATA PRESENTED HEREIN REPRESENTS THE FINDINGS OF THE SUBSURFACE EXPLORATION MADE. HOWEVER, BECAUSE THE DEPTHS INVESTIGATED ARE LIMITED AND THE AREA OF THE BORINGS AND/OR SOUNDINGS IS VERY SMALL IN RELATION TO THE ENTIRE AREA, IT IS NOT WARRANTED THAT THE CONDITIONS BELOW THE DEPTHS INVESTIGATED OR THAT THE CLASSIFICATION OF MATERIAL ENCOUNTERED IN THESE INVESTIGATIONS IS NECESSARILY TYPICAL OF THE ENTIRE SITE.
CLOVERLAND ROAD BRIDGE
SUBSURFACE EXPLORATION
SCALE: HORZ. 1"=10' VERT. 1"=10'
SHEET 5 OF 10
DATE MAY 1976
MORGAN & PARMLEY, LTD. CONSULTING ENGINEERS LADYSMITH, WISCONSIN
DR. BY R.O. PARMLEY

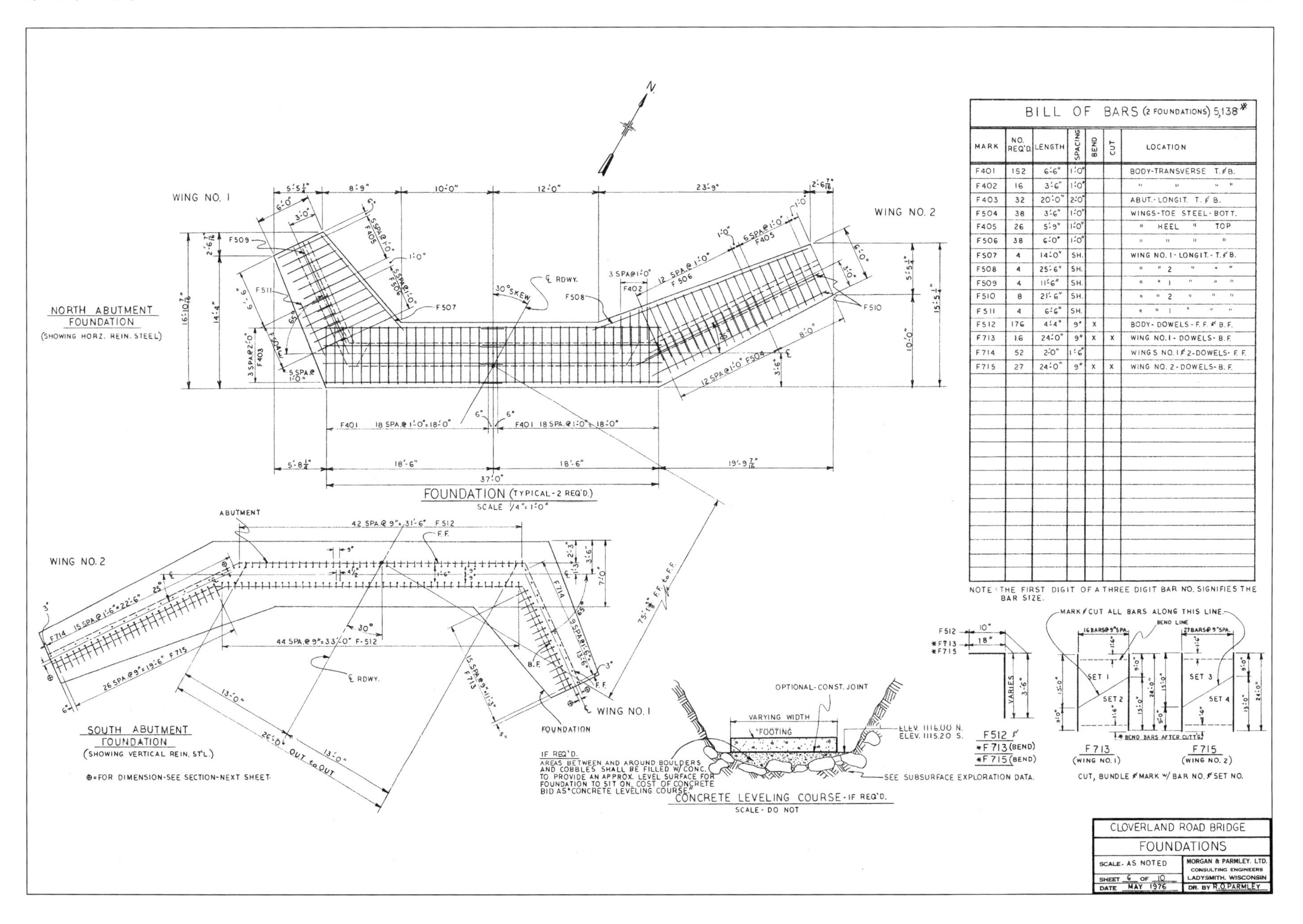

BILL OF BARS (2 FOUNDATIONS) 5,138#

MARK	NO. REQ'D.	LENGTH	SPACING	BEND	CUT	LOCATION
F401	152	6'-6"	1'-0"			BODY-TRANSVERSE T.&B.
F402	16	3'-6"	1'-0"			" " " "
F403	32	20'-0"	2'-0"			ABUT.-LONGIT. T.&B.
F504	38	3'-6"	1'-0"			WINGS-TOE STEEL-BOTT.
F405	26	5'-9"	1'-0"			" HEEL " TOP
F506	38	6'-0"	1'-0"			" " " "
F507	4	14'-0"	SH.			WING NO. 1-LONGIT.-T.&B.
F508	4	25'-6"	SH.			" " 2 " " "
F509	4	11'-6"	SH.			" " 1 " " "
F510	8	21'-6"	SH.			" " 2 " " "
F511	4	6'-6"	SH.			" " 1 " " "
F512	176	4'-4"	9"	X		BODY-DOWELS-F.F.&B.F.
F713	16	24'-0"	9"	X	X	WING NO. 1-DOWELS-B.F.
F714	52	2'-0"	1'-6"			WINGS NO. 1&2-DOWELS-F.F.
F715	27	24'-0"	9"	X	X	WING NO. 2-DOWELS-B.F.

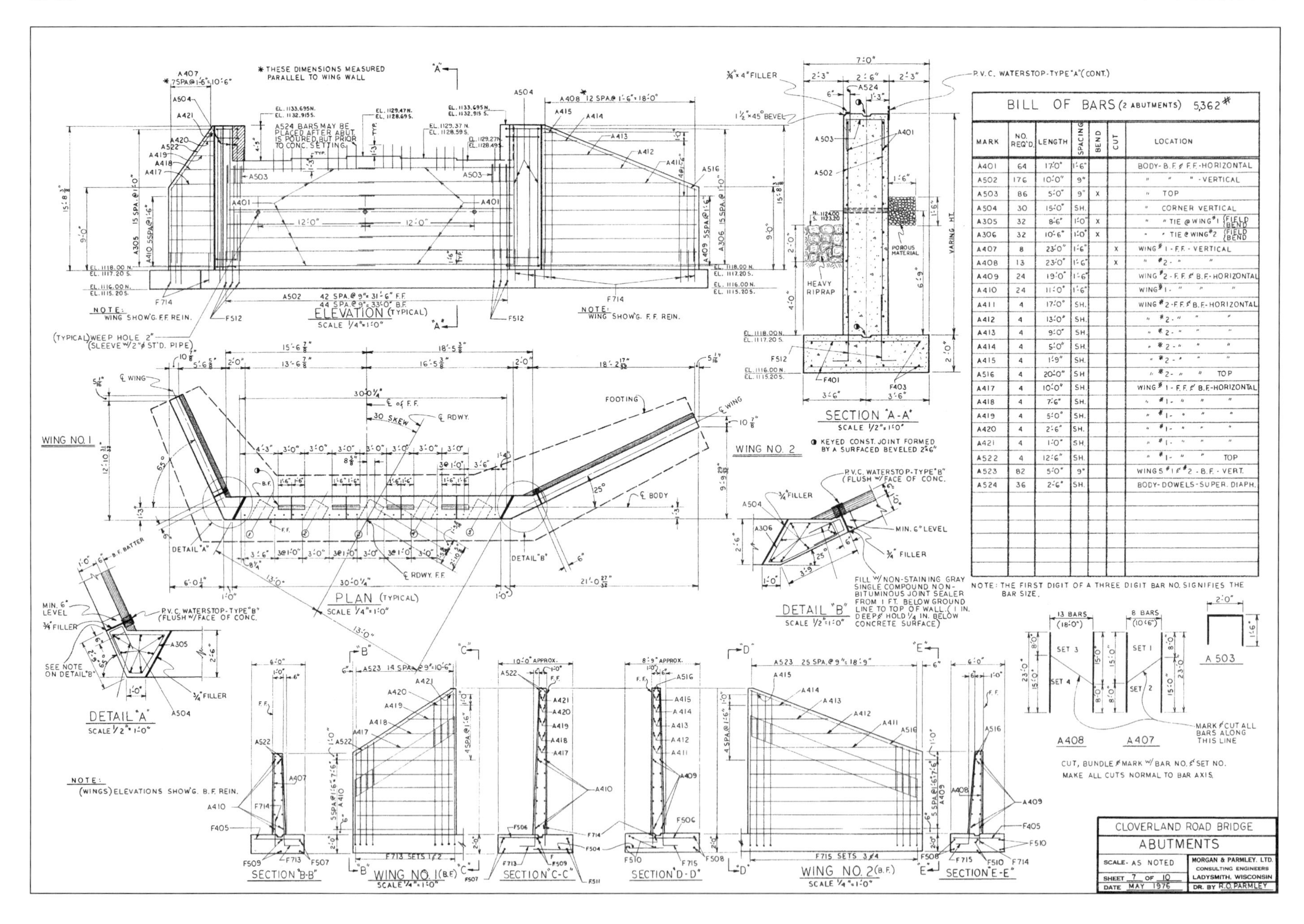

BILL OF BARS (2 ABUTMENTS) 5,362#

MARK	NO. REQ'D.	LENGTH	SPACING	BEND	CUT	LOCATION
A401	64	17'-0"	1'-6"			BODY-B.F. & F.F.-HORIZONTAL
A502	176	10'-0"	9"			" " " -VERTICAL
A503	86	5'-0"	9"	X		" TOP
A504	30	15'-0"	SH.			" CORNER VERTICAL
A305	32	8'-6"	1'-0"	X		" " TIE @ WING #1 (FIELD BEND)
A306	32	10'-6"	1'-0"	X		" " TIE @ WING #2 (FIELD BEND)
A407	8	23'-0"	1'-6"		X	WING #1-F.F.-VERTICAL
A408	13	23'-0"	1'-6"		X	" #2- " "
A409	24	19'-0"	1'-6"			WING #2-F.F. & B.F.-HORIZONTAL
A410	24	11'-0"	1'-6"			WING #1- " " "
A411	4	17'-0"	SH.			WING #2-F.F. & B.F.-HORIZONTAL
A412	4	13'-0"	SH.			" #2- " " "
A413	4	9'-0"	SH.			" #2- " " "
A414	4	5'-0"	SH.			" #2- " " "
A415	4	1'-9"	SH.			" #2- " " "
A516	4	20'-0"	SH			" #2- " " TOP
A417	4	10'-0"	SH.			WING #1-F.F. & B.F.-HORIZONTAL
A418	4	7'-6"	SH.			" #1- " " "
A419	4	5'-0"	SH.			" #1- " " "
A420	4	2'-6"	SH.			" #1- " " "
A421	4	1'-0"	SH.			" #1- " " "
A522	4	12'-6"	SH.			" #1- " " TOP
A523	82	5'-0"	9"			WINGS #1 & #2-B.F.-VERT.
A524	36	2'-6"	SH.			BODY-DOWELS-SUPER. DIAPH.

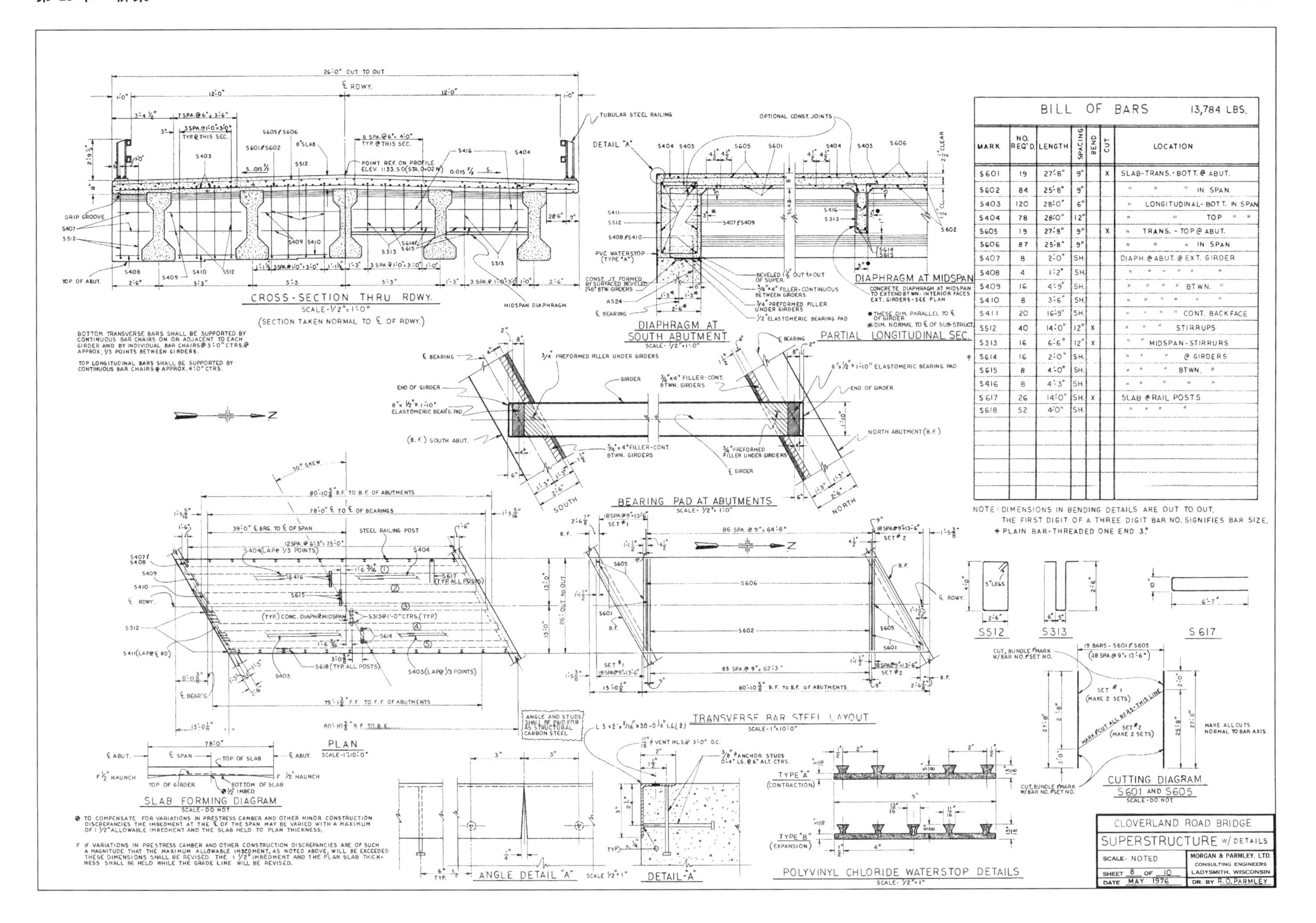

BILL OF BARS　13,784 LBS.

MARK	NO. REQ'D.	LENGTH	SPACING	BEND	CUT	LOCATION
S601	19	27'-8"	9"		X	SLAB-TRANS.-BOTT. @ ABUT.
S602	84	25'-8"	9"			" " " IN SPAN
S403	120	28'-0"	6"			" LONGITUDINAL-BOTT. IN SPAN
S404	78	28'-0"	12"			" " TOP " "
S605	19	27'-8"	9"		X	" TRANS.-TOP @ ABUT.
S606	87	25'-8"	9"			" " " IN SPAN
S407	8	2'-0"	SH.			DIAPH. @ ABUT. @ EXT. GIRDER
S408	4	1'-2"	SH.			" " " " " "
S409	16	4'-9"	SH.			" " " " BTWN. "
S410	8	3'-6"	SH.			" " " " " "
S411	20	16'-9"	SH.			" " " " CONT. BACKFACE
S512	40	14'-0"	12"	X		" " " STIRRUPS
S313	16	6'-6"	12"	X		" " MIDSPAN-STIRRURS
S614	16	2'-0"	SH.			" " " @ GIRDERS
S615	8	4'-0"	SH.			" " " BTWN. "
S416	8	4'-3"	SH.			" " " " "
S617	26	14'-0"	SH.	X		SLAB @ RAIL POSTS
S618	52	4'-0"	SH.			" " " "

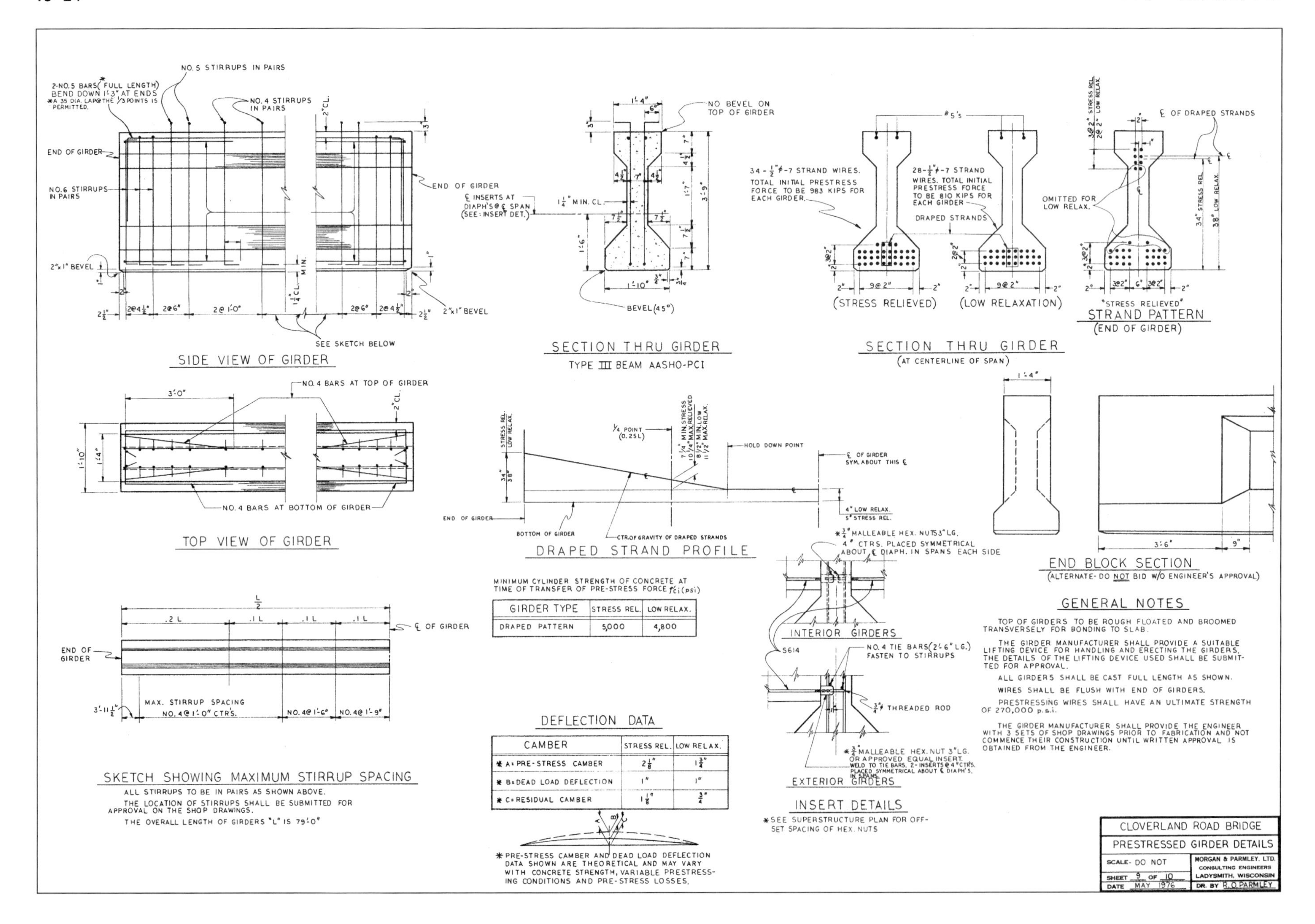
NO. 5 STIRRUPS IN PAIRS
2-NO. 5 BARS* (FULL LENGTH) BEND DOWN 1'-3" AT ENDS
* A 35 DIA. LAP @ THE 1/3 POINTS IS PERMITTED.
NO. 4 STIRRUPS IN PAIRS
END OF GIRDER
NO. 6 STIRRUPS IN PAIRS
2"x1" BEVEL
SEE SKETCH BELOW
SIDE VIEW OF GIRDER
NO BEVEL ON TOP OF GIRDER
℄ INSERTS AT DIAPH'S @ ℄ SPAN (SEE: INSERT DET.)
1¼" MIN. CL.
BEVEL (45°)
SECTION THRU GIRDER
TYPE III BEAM AASHO-PCI
#5's
34-½"φ-7 STRAND WIRES. TOTAL INITIAL PRESTRESS FORCE TO BE 983 KIPS FOR EACH GIRDER.
28-½"φ-7 STRAND WIRES. TOTAL INITIAL PRESTRESS FORCE TO BE 810 KIPS FOR EACH GIRDER
DRAPED STRANDS
(STRESS RELIEVED)
(LOW RELAXATION)
SECTION THRU GIRDER
(AT CENTERLINE OF SPAN)
℄ OF DRAPED STRANDS
OMITTED FOR LOW RELAX.
"STRESS RELIEVED"
STRAND PATTERN
(END OF GIRDER)
NO. 4 BARS AT TOP OF GIRDER
NO. 4 BARS AT BOTTOM OF GIRDER
TOP VIEW OF GIRDER
¼ POINT (0.25L)
HOLD DOWN POINT
℄ OF GIRDER SYM. ABOUT THIS ℄
END OF GIRDER
BOTTOM OF GIRDER
CTR. OF GRAVITY OF DRAPED STRANDS
4" LOW RELAX.
5" STRESS REL.
DRAPED STRAND PROFILE
MINIMUM CYLINDER STRENGTH OF CONCRETE AT TIME OF TRANSFER OF PRE-STRESS FORCE f'ci (psi)
GIRDER TYPE | STRESS REL. | LOW RELAX.
DRAPED PATTERN | 5,000 | 4,800
END BLOCK SECTION
(ALTERNATE - DO NOT BID W/O ENGINEER'S APPROVAL)
L/2
℄ OF GIRDER
END OF GIRDER
MAX. STIRRUP SPACING
NO. 4 @ 1'-0" CTR'S.
NO. 4 @ 1'-6"
NO. 4 @ 1'-9"
SKETCH SHOWING MAXIMUM STIRRUP SPACING
ALL STIRRUPS TO BE IN PAIRS AS SHOWN ABOVE.
THE LOCATION OF STIRRUPS SHALL BE SUBMITTED FOR APPROVAL ON THE SHOP DRAWINGS.
THE OVERALL LENGTH OF GIRDERS "L" IS 79'-0"
DEFLECTION DATA
CAMBER | STRESS REL. | LOW RELAX.
* A = PRE-STRESS CAMBER | 2⅛" | 1¾"
* B = DEAD LOAD DEFLECTION | 1" | 1"
* C = RESIDUAL CAMBER | 1⅛" | ¾"
* PRE-STRESS CAMBER AND DEAD LOAD DEFLECTION DATA SHOWN ARE THEORETICAL AND MAY VARY WITH CONCRETE STRENGTH, VARIABLE PRESTRESSING CONDITIONS AND PRE-STRESS LOSSES.
* ¾" MALLEABLE HEX. NUTS 3" LG. 4" CTRS. PLACED SYMMETRICAL ABOUT ℄ DIAPH. IN SPANS EACH SIDE
INTERIOR GIRDERS
S614
NO. 4 TIE BARS (2'-6" LG.) FASTEN TO STIRRUPS
¾"φ THREADED ROD
* ¾" MALLEABLE HEX. NUT 3" LG. OR APPROVED EQUAL INSERT. WELD TO TIE BARS. 2-INSERTS @ 4" CTRS. PLACED SYMMETRICAL ABOUT ℄ DIAPH'S. IN SPANS.
EXTERIOR GIRDERS
INSERT DETAILS
* SEE SUPERSTRUCTURE PLAN FOR OFF-SET SPACING OF HEX. NUTS
GENERAL NOTES
TOP OF GIRDERS TO BE ROUGH FLOATED AND BROOMED TRANSVERSELY FOR BONDING TO SLAB.
THE GIRDER MANUFACTURER SHALL PROVIDE A SUITABLE LIFTING DEVICE FOR HANDLING AND ERECTING THE GIRDERS. THE DETAILS OF THE LIFTING DEVICE USED SHALL BE SUBMITTED FOR APPROVAL.
ALL GIRDERS SHALL BE CAST FULL LENGTH AS SHOWN.
WIRES SHALL BE FLUSH WITH END OF GIRDERS.
PRESTRESSING WIRES SHALL HAVE AN ULTIMATE STRENGTH OF 270,000 p.s.i.
THE GIRDER MANUFACTURER SHALL PROVIDE THE ENGINEER WITH 3 SETS OF SHOP DRAWINGS PRIOR TO FABRICATION AND NOT COMMENCE THEIR CONSTRUCTION UNTIL WRITTEN APPROVAL IS OBTAINED FROM THE ENGINEER.
CLOVERLAND ROAD BRIDGE
PRESTRESSED GIRDER DETAILS
SCALE - DO NOT
SHEET 9 OF 10
DATE MAY 1976
MORGAN & PARMLEY, LTD. CONSULTING ENGINEERS LADYSMITH, WISCONSIN
DR. BY R.O. PARMLEY

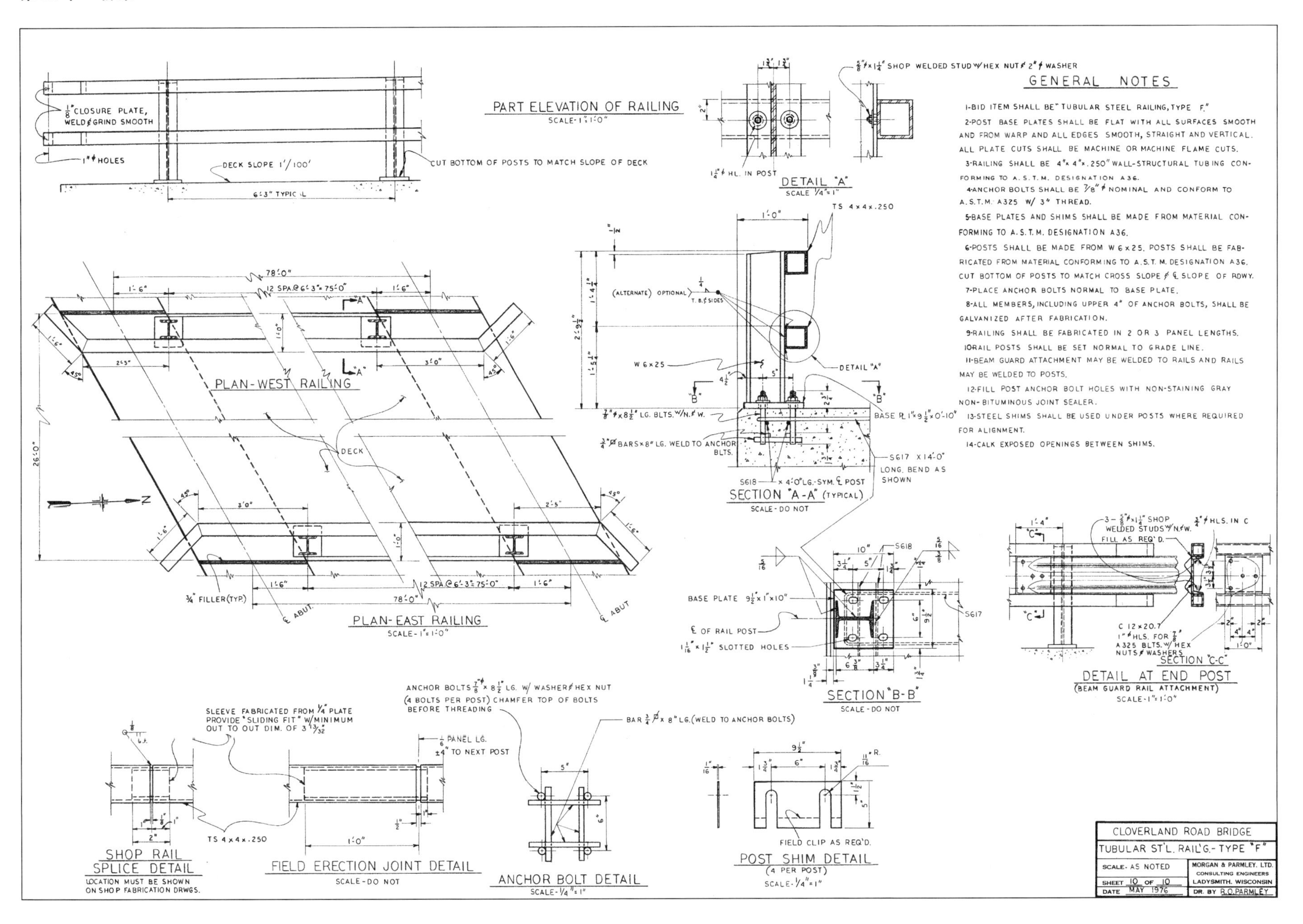
PART ELEVATION OF RAILING
SCALE-1"=1'-0"
1/8" CLOSURE PLATE, WELD & GRIND SMOOTH
1"ϕ HOLES
DECK SLOPE 1'/100'
CUT BOTTOM OF POSTS TO MATCH SLOPE OF DECK
6'-3" TYPICAL
DETAIL "A"
SCALE 1/4"=1"
5/8"ϕ×1 1/4" SHOP WELDED STUD W/HEX NUT & 2"ϕ WASHER
1 1/4"ϕ HL. IN POST
GENERAL NOTES
1-BID ITEM SHALL BE "TUBULAR STEEL RAILING, TYPE F."
2-POST BASE PLATES SHALL BE FLAT WITH ALL SURFACES SMOOTH AND FROM WARP AND ALL EDGES SMOOTH, STRAIGHT AND VERTICAL. ALL PLATE CUTS SHALL BE MACHINE OR MACHINE FLAME CUTS.
3-RAILING SHALL BE 4"×4"×.250" WALL-STRUCTURAL TUBING CONFORMING TO A.S.T.M. DESIGNATION A36.
4-ANCHOR BOLTS SHALL BE 7/8"ϕ NOMINAL AND CONFORM TO A.S.T.M. A325 W/ 3" THREAD.
5-BASE PLATES AND SHIMS SHALL BE MADE FROM MATERIAL CONFORMING TO A.S.T.M. DESIGNATION A36.
6-POSTS SHALL BE MADE FROM W 6×25. POSTS SHALL BE FABRICATED FROM MATERIAL CONFORMING TO A.S.T.M. DESIGNATION A36. CUT BOTTOM OF POSTS TO MATCH CROSS SLOPE & ℄ SLOPE OF RDWY.
7-PLACE ANCHOR BOLTS NORMAL TO BASE PLATE.
8-ALL MEMBERS, INCLUDING UPPER 4" OF ANCHOR BOLTS, SHALL BE GALVANIZED AFTER FABRICATION.
9-RAILING SHALL BE FABRICATED IN 2 OR 3 PANEL LENGTHS.
10-RAIL POSTS SHALL BE SET NORMAL TO GRADE LINE.
11-BEAM GUARD ATTACHMENT MAY BE WELDED TO RAILS AND RAILS MAY BE WELDED TO POSTS.
12-FILL POST ANCHOR BOLT HOLES WITH NON-STAINING GRAY NON-BITUMINOUS JOINT SEALER.
13-STEEL SHIMS SHALL BE USED UNDER POSTS WHERE REQUIRED FOR ALIGNMENT.
14-CALK EXPOSED OPENINGS BETWEEN SHIMS.
TS 4×4×.250
(ALTERNATE) OPTIONAL
T.B.& SIDES
W 6×25
DETAIL "A"
7/8"ϕ×8 1/2" LG. BLTS. W/N.& W.
3/4"ϕ BARS×8" LG. WELD TO ANCHOR BLTS.
BASE PL 1"×9 1/2"×0'-10"
S617 ×14'-0" LONG, BEND AS SHOWN
S618 ×4'-0" LG.-SYM. ℄ POST
SECTION "A-A" (TYPICAL)
SCALE-DO NOT
78'-0"
12 SPA.@6'-3"=75'-0"
PLAN-WEST RAILING
DECK
PLAN-EAST RAILING
SCALE-1"=1'-0"
3/4" FILLER (TYP.)
℄ ABUT.
℄ ABUT
26'-0"
N
BASE PLATE 9 1/2"×1"×10"
℄ OF RAIL POST
1 1/16"×1 1/2" SLOTTED HOLES
S618
S617
SECTION "B-B"
SCALE-DO NOT
3-5/8"ϕ×1 1/2" SHOP WELDED STUDS W/N.& W.
3/4"ϕ HLS. IN C
FILL AS REQ'D.
C 12×20.7
1"ϕ HLS. FOR 7/8" A325 BLTS. W/HEX NUTS & WASHERS
SECTION "C-C"
DETAIL AT END POST
(BEAM GUARD RAIL ATTACHMENT)
SCALE-1"=1'-0"
ANCHOR BOLTS 7/8"ϕ×8 1/2" LG. W/ WASHER & HEX NUT (4 BOLTS PER POST) CHAMFER TOP OF BOLTS BEFORE THREADING
SLEEVE FABRICATED FROM 1/4" PLATE PROVIDE "SLIDING FIT" W/MINIMUM OUT TO OUT DIM. OF 3 13/32"
1/6 PANEL LG. ±4" TO NEXT POST
BAR 3/4"ϕ×8" LG. (WELD TO ANCHOR BOLTS)
TS 4×4×.250
SHOP RAIL SPLICE DETAIL
LOCATION MUST BE SHOWN ON SHOP FABRICATION DRWGS.
FIELD ERECTION JOINT DETAIL
SCALE-DO NOT
ANCHOR BOLT DETAIL
SCALE-1/4"=1"
FIELD CLIP AS REQ'D.
POST SHIM DETAIL
(4 PER POST)
SCALE-1/4"=1"
CLOVERLAND ROAD BRIDGE
TUBULAR ST'L. RAIL'G.-TYPE "F"
SCALE-AS NOTED
SHEET 10 OF 10
DATE MAY 1976
MORGAN & PARMLEY, LTD.
CONSULTING ENGINEERS
LADYSMITH, WISCONSIN
DR. BY R.O. PARMLEY

Cummings 桥
Cummings Bridge

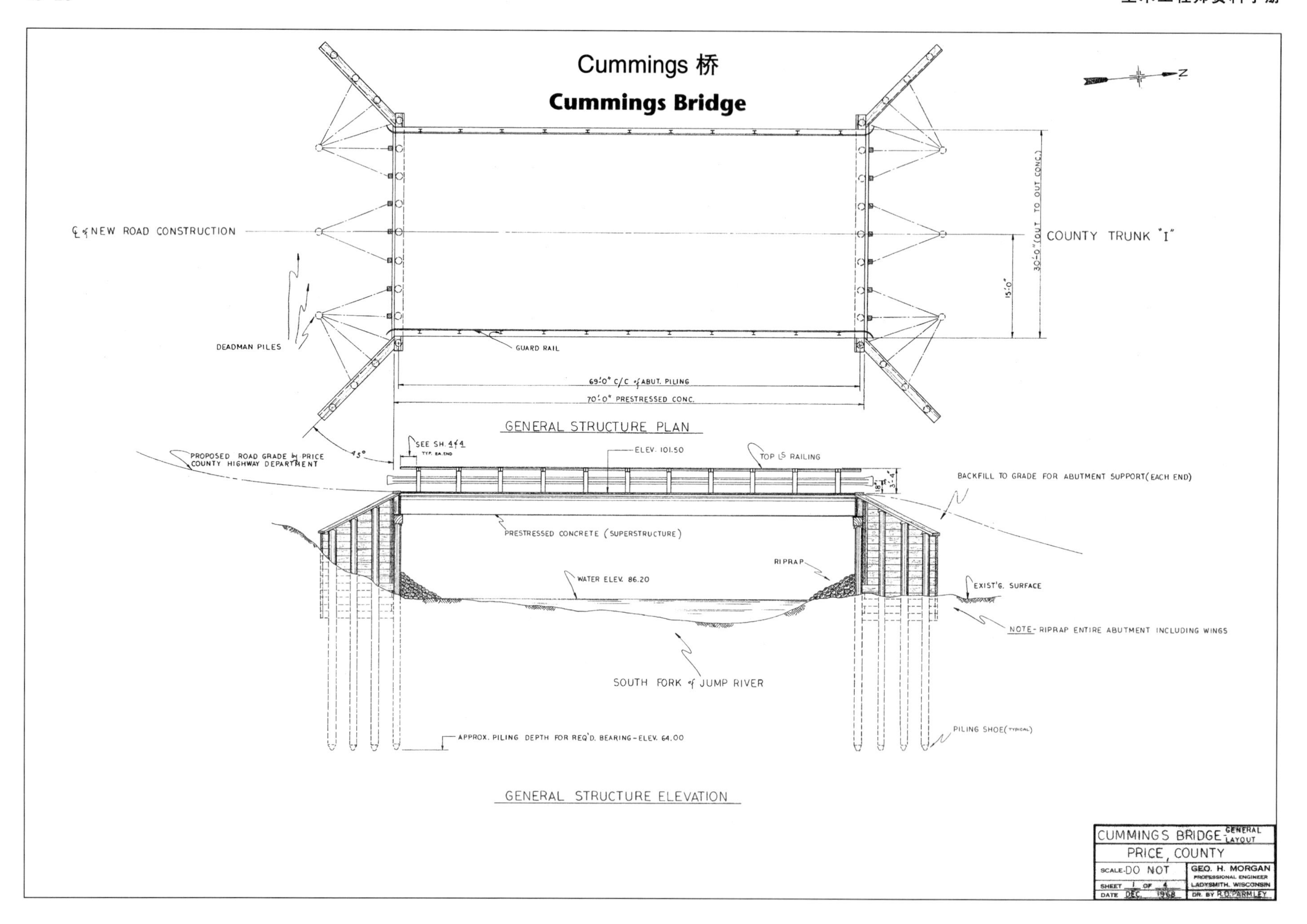

Courtesy: Morgan & Parmley, Ltd.

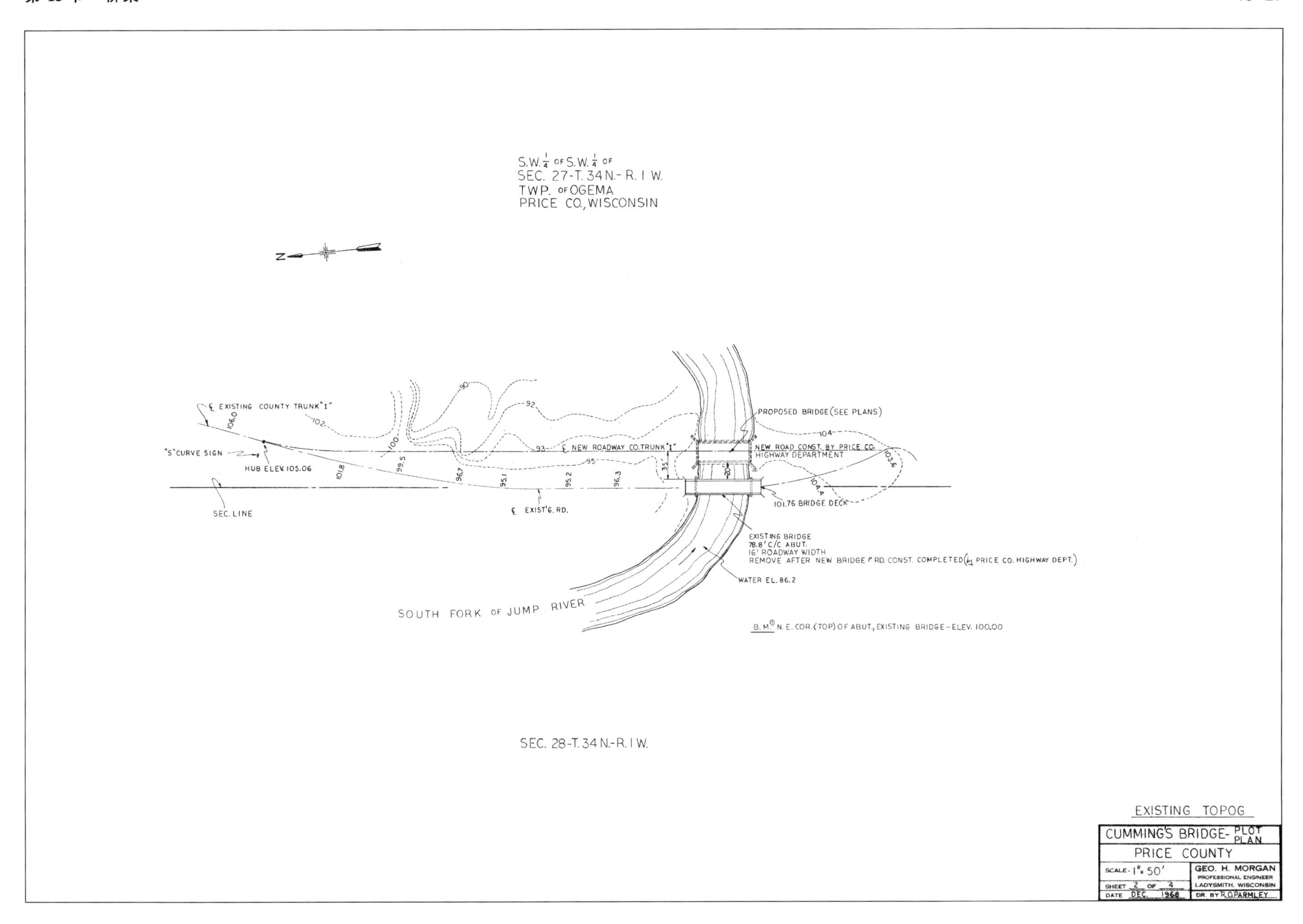
S.W.¼ OF S.W.¼ OF
SEC. 27-T. 34 N.- R. I W.
TWP. OF OGEMA
PRICE CO., WISCONSIN
N
℄ EXISTING COUNTY TRUNK "I"
"S" CURVE SIGN
HUB ELEV. 105.06
SEC. LINE
℄ NEW ROADWAY CO. TRUNK "I"
℄ EXIST'G. RD.
PROPOSED BRIDGE (SEE PLANS)
NEW ROAD CONST. BY PRICE CO. HIGHWAY DEPARTMENT
101.76 BRIDGE DECK
EXISTING BRIDGE
78.8' C/C ABUT.
16' ROADWAY WIDTH
REMOVE AFTER NEW BRIDGE & RD. CONST. COMPLETED (by PRICE CO. HIGHWAY DEPT.)
WATER EL. 86.2
SOUTH FORK OF JUMP RIVER
B.M. N. E. COR. (TOP) OF ABUT., EXISTING BRIDGE - ELEV. 100.00
SEC. 28-T. 34 N.-R. I W.
EXISTING TOPOG
CUMMING'S BRIDGE- PLOT PLAN
PRICE COUNTY
SCALE- 1" = 50'
SHEET 2 OF 4
DATE DEC. 1968
GEO. H. MORGAN
PROFESSIONAL ENGINEER
LADYSMITH, WISCONSIN
DR. BY R.O. PARMLEY

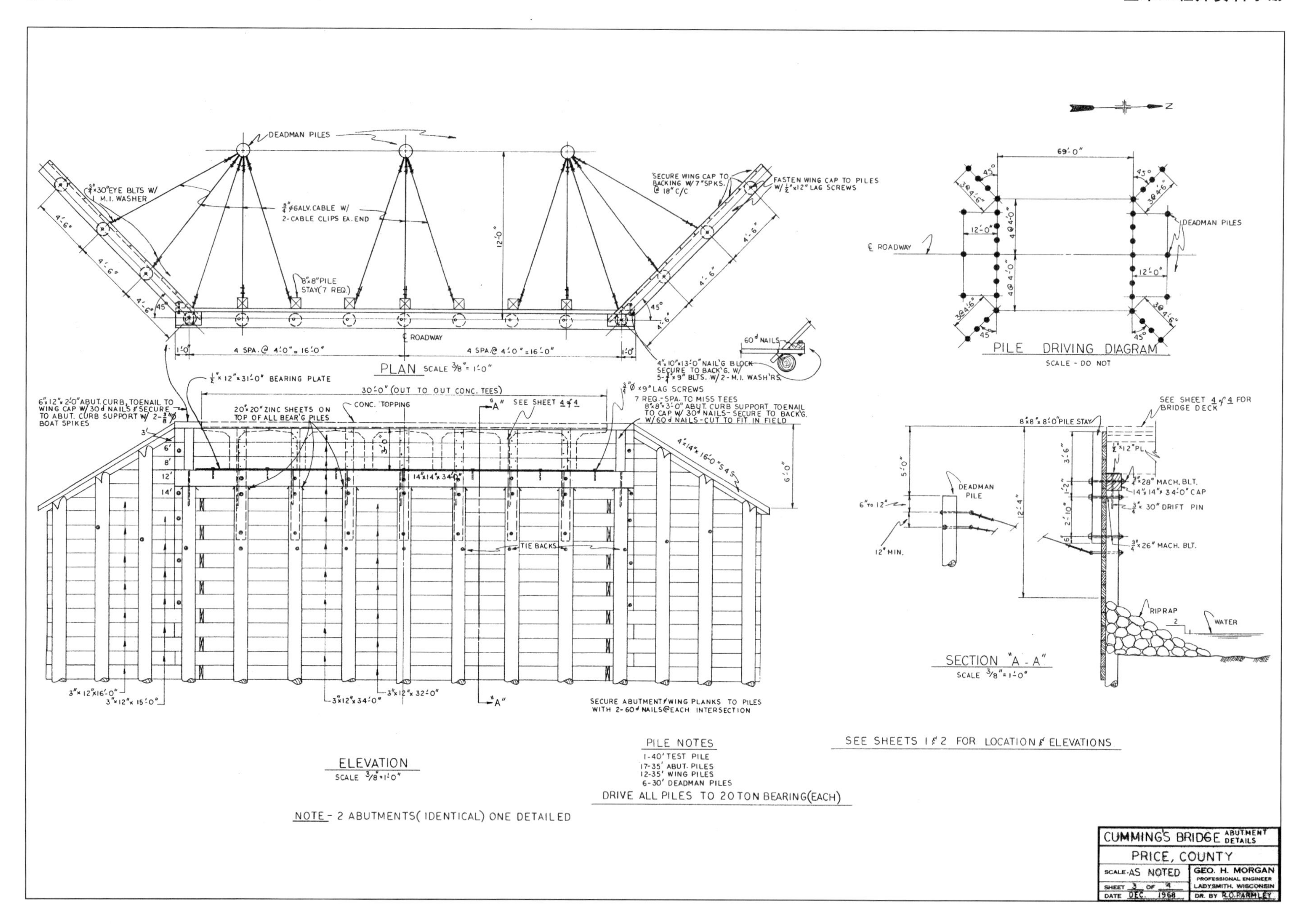
DEADMAN PILES
PLAN SCALE 3/8" = 1'-0"
ELEVATION SCALE 3/8" = 1'-0"
PILE DRIVING DIAGRAM
SCALE - DO NOT
SECTION "A - A"
SCALE 3/8" = 1'-0"
PILE NOTES
1-40' TEST PILE
17-35' ABUT. PILES
12-35' WING PILES
6-30' DEADMAN PILES
DRIVE ALL PILES TO 20 TON BEARING(EACH)
NOTE - 2 ABUTMENTS(IDENTICAL) ONE DETAILED
SEE SHEETS 1 & 2 FOR LOCATION & ELEVATIONS
SECURE ABUTMENT & WING PLANKS TO PILES WITH 2-60d NAILS@EACH INTERSECTION
TIE BACKS
RIPRAP
WATER
CUMMINGS BRIDGE ABUTMENT DETAILS
PRICE, COUNTY
SCALE-AS NOTED
GEO. H. MORGAN
PROFESSIONAL ENGINEER
LADYSMITH, WISCONSIN
SHEET 3 OF 4
DATE DEC. 1968
DR. BY R.O.PARMLEY

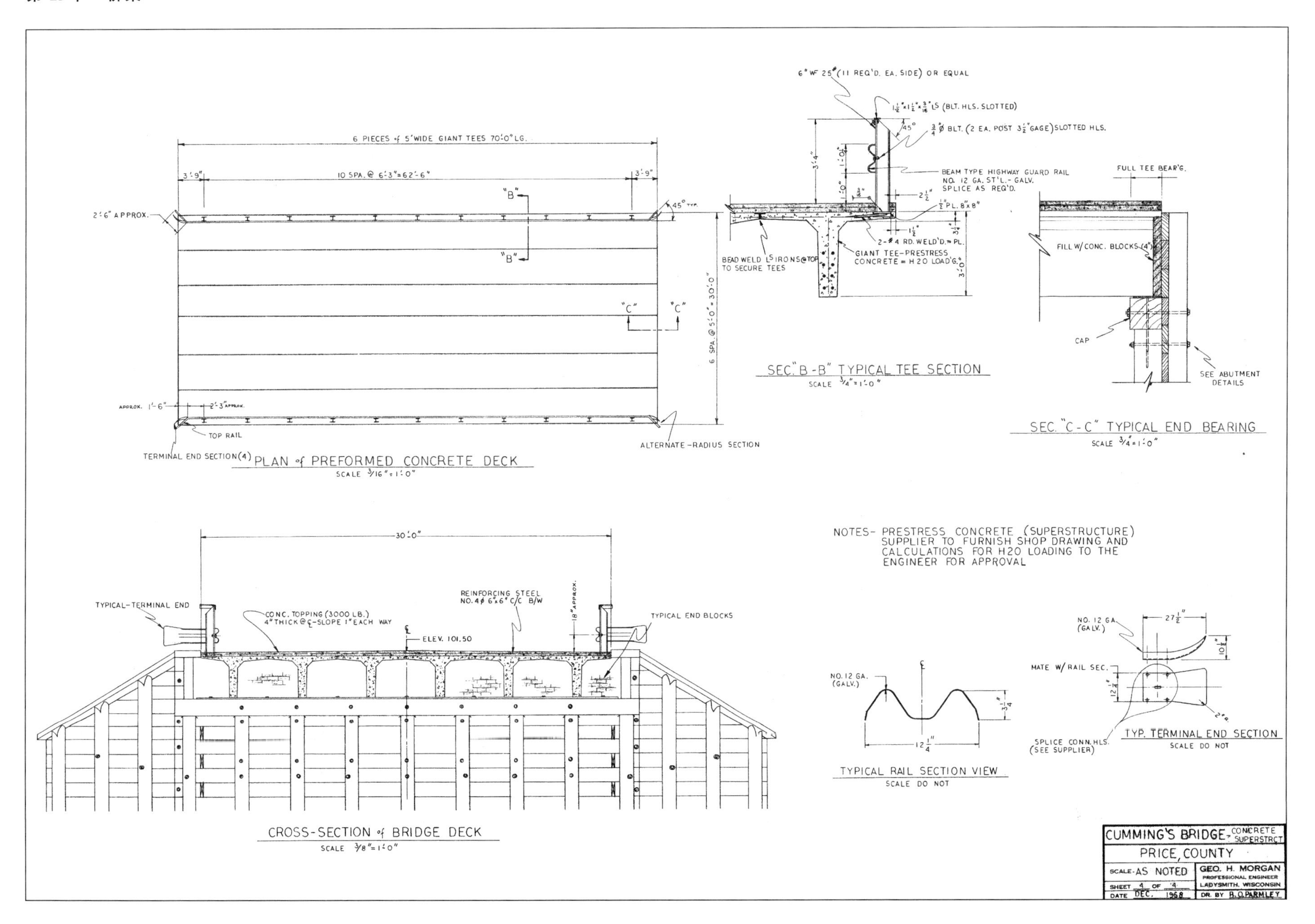
6 PIECES of 5' WIDE GIANT TEES 70'-0" LG.
10 SPA. @ 6'-3" = 62'-6"
PLAN of PREFORMED CONCRETE DECK
SCALE 3/16" = 1'-0"
TERMINAL END SECTION (4)
TOP RAIL
ALTERNATE-RADIUS SECTION
SEC. "B-B" TYPICAL TEE SECTION
SCALE 3/4" = 1'-0"
BEAM TYPE HIGHWAY GUARD RAIL NO. 12 GA. ST'L. - GALV. SPLICE AS REQ'D.
GIANT TEE-PRESTRESS CONCRETE = H20 LOAD'G.
BEAD WELD L^S IRONS @ TOP TO SECURE TEES
SEC. "C-C" TYPICAL END BEARING
SCALE 3/4" = 1'-0"
FULL TEE BEAR'G.
FILL W/ CONC. BLOCKS (4")
CAP
SEE ABUTMENT DETAILS
NOTES- PRESTRESS CONCRETE (SUPERSTRUCTURE) SUPPLIER TO FURNISH SHOP DRAWING AND CALCULATIONS FOR H20 LOADING TO THE ENGINEER FOR APPROVAL
TYPICAL-TERMINAL END
CONC. TOPPING (3000 LB.) 4" THICK @ ℄-SLOPE 1" EACH WAY
REINFORCING STEEL NO. 4 ø 6"x6" C/C B/W
ELEV. 101.50
TYPICAL END BLOCKS
CROSS-SECTION of BRIDGE DECK
SCALE 3/8" = 1'-0"
TYPICAL RAIL SECTION VIEW
SCALE DO NOT
TYP. TERMINAL END SECTION
SCALE DO NOT
MATE W/ RAIL SEC.
SPLICE CONN. HLS. (SEE SUPPLIER)
CUMMING'S BRIDGE - CONCRETE SUPERSTRCT
PRICE, COUNTY
SCALE-AS NOTED
GEO. H. MORGAN
PROFESSIONAL ENGINEER
LADYSMITH, WISCONSIN
SHEET 4 OF 4
DATE DEC. 1968
DR. BY B. O. PARMLEY

Devils Creek 机动雪车桥

DEVILS CREEK SNOWMOBILE BRIDGE

RUSK COUNTY FORESTRY DEPARTMENT

RUSK COUNTY, WISCONSIN

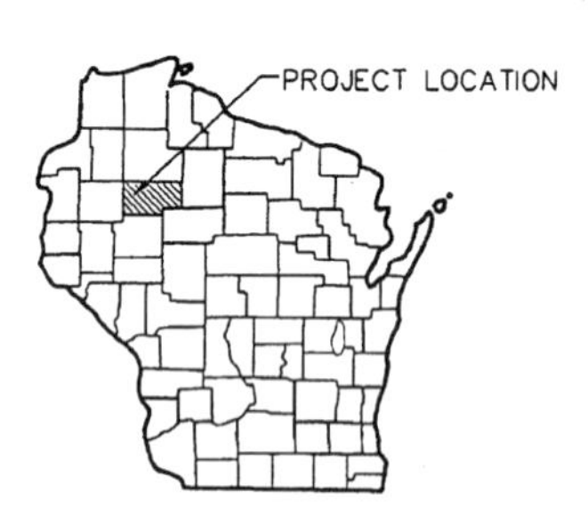

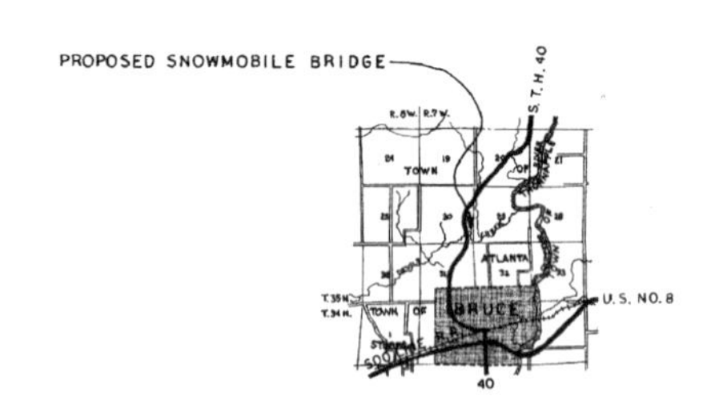

PROJECT LOCATION

SCALE 1 MILE=

I N D E X

1 —— PROJECT LOCATION & INDEX

2 —— EXISTING SITE PLAN, TOPOGRAPHY, CROSS-SECTIONS & GENERAL MAP

3 —— HYDRAULIC CROSS-SECTION, PLAN VIEW & RELATED DATA

4 —— BRIDGE PLAN, ELEVATION & SECTION

5 —— BRIDGE ABUTMENT, DETAILS, PILING NOTES & RAIL POSTS

NOTE- SEE SPECIFICATION BOOK

SNOWMOBILE BRIDGE	
DEVILS CREEK CROSSING RUSK COUNTY, WISCONSIN	
SCALE- NOTED	GEO. H. MORGAN PROFESSIONAL ENGINEER LADYSMITH, WISCONSIN
SHEET 1 OF 5	
DATE JAN. 1976	DR. BY R.O. PARMLEY

Courtesy: Morgan & Parmley, Ltd.

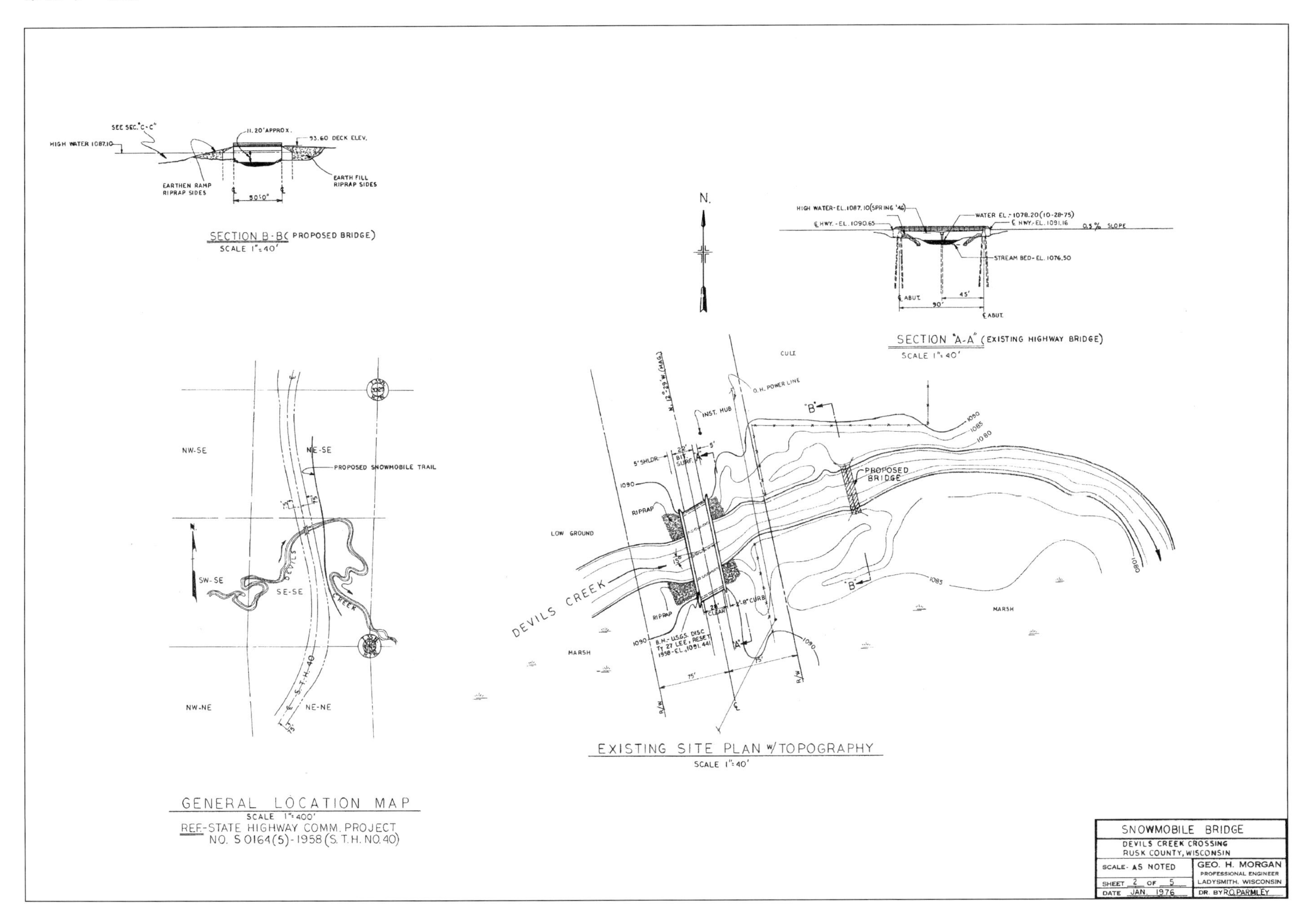
SECTION B-B (PROPOSED BRIDGE)
SCALE 1"=40'
SECTION "A-A" (EXISTING HIGHWAY BRIDGE)
SCALE 1"=40'
EXISTING SITE PLAN W/TOPOGRAPHY
SCALE 1"=40'
DEVILS CREEK
GENERAL LOCATION MAP
SCALE 1"=400'
REF-STATE HIGHWAY COMM. PROJECT
NO. S 0164(5)-1958 (S.T.H. NO.40)
SNOWMOBILE BRIDGE
DEVILS CREEK CROSSING
RUSK COUNTY, WISCONSIN
SCALE- AS NOTED
SHEET 2 OF 5
DATE JAN. 1976
GEO. H. MORGAN
PROFESSIONAL ENGINEER
LADYSMITH, WISCONSIN
DR. BY R.O. PARMLEY

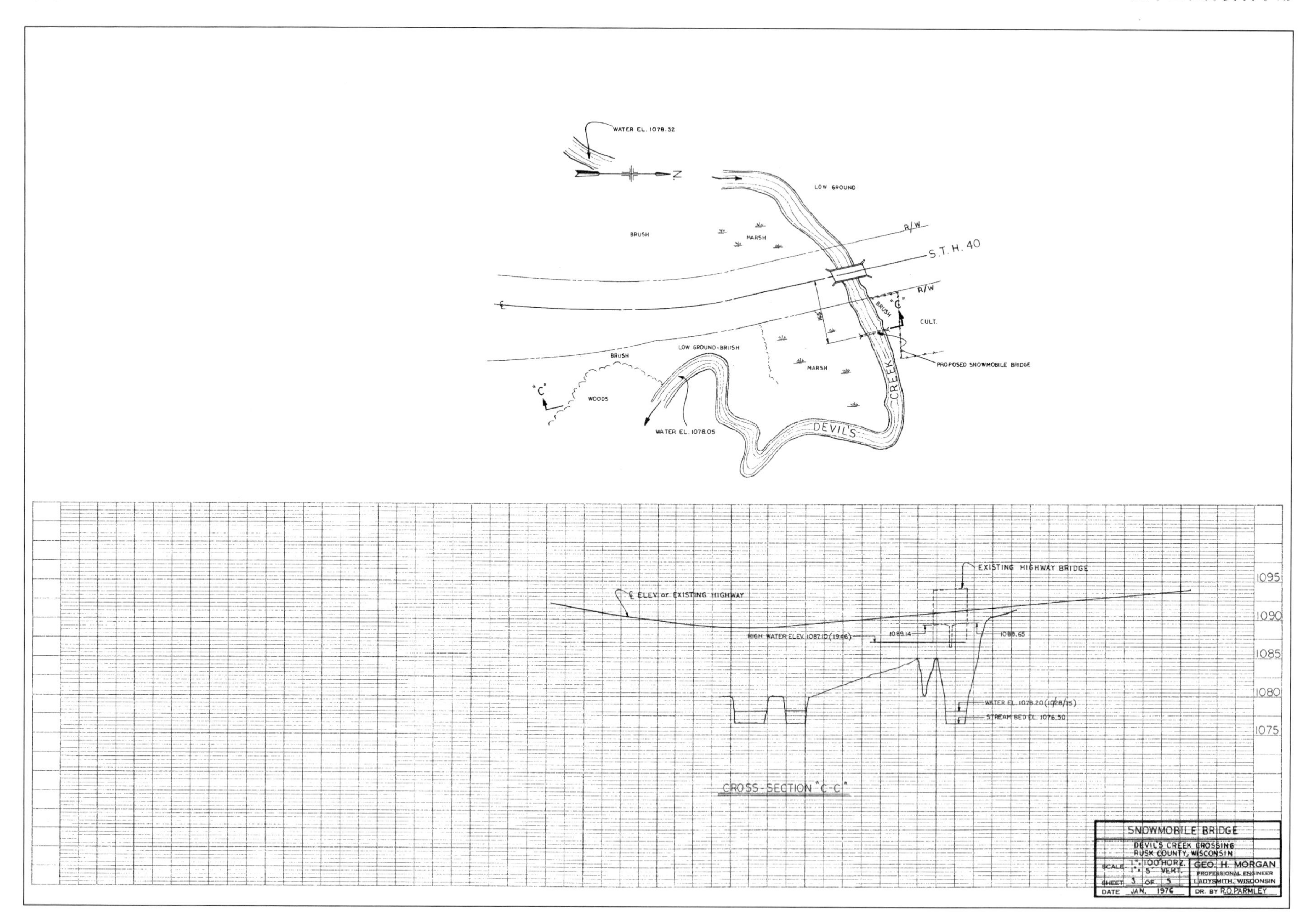
WATER EL. 1078.32
N
LOW GROUND
BRUSH
MARSH
R/W
S.T.H. 40
R/W
BRUSH
"C"
CULT.
155
LOW GROUND-BRUSH
BRUSH
MARSH
PROPOSED SNOWMOBILE BRIDGE
"C"
WOODS
CREEK
DEVIL'S
WATER EL. 1078.05
EXISTING HIGHWAY BRIDGE
℄ ELEV. OF EXISTING HIGHWAY
HIGH WATER ELEV. 1087.10 (1966)
1089.14
1088.65
WATER EL. 1078.20 (10/28/75)
STREAM BED EL. 1076.50
1095
1090
1085
1080
1075
CROSS-SECTION "C-C"
SNOWMOBILE BRIDGE
DEVIL'S CREEK CROSSING
RUSK COUNTY, WISCONSIN
SCALE 1"=100' HORZ. 1"=5' VERT.
GEO. H. MORGAN
PROFESSIONAL ENGINEER
LADYSMITH, WISCONSIN
SHEET 3 OF 5
DATE JAN. 1976
DR. BY R.O. PARMLEY

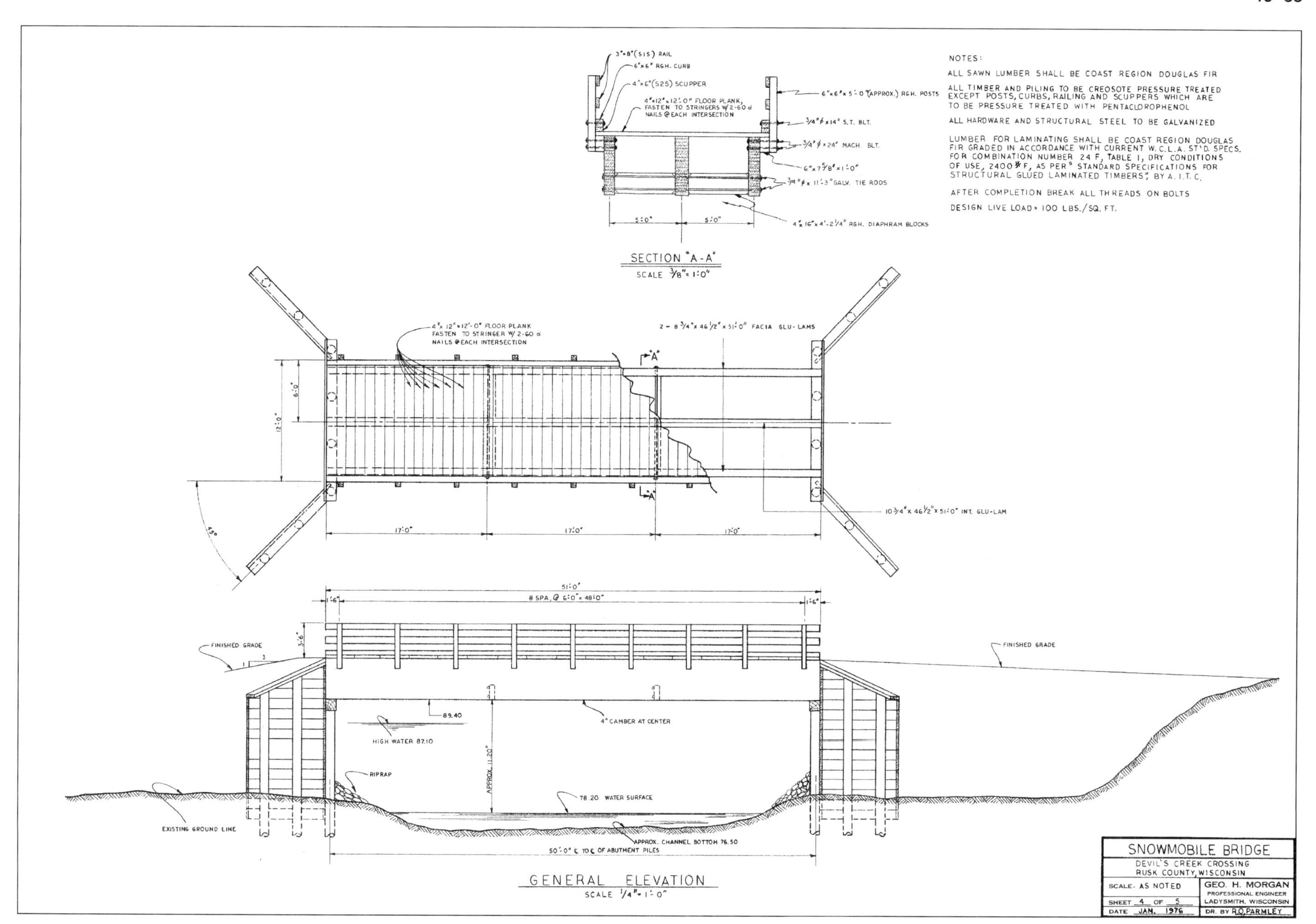
3"×8"(S1S) RAIL
6"×6" RGH. CURB
4"×6"(S2S) SCUPPER
4"×12"×12'-0" FLOOR PLANK, FASTEN TO STRINGERS W/2-60 d NAILS @ EACH INTERSECTION
6"×6"×5'-0"(APPROX.) RGH. POSTS
3/4"φ×14" S.T. BLT.
3/4"φ×24" MACH. BLT.
6"×7 5/8"×1'-0"
3/4"φ×11'-3" GALV. TIE RODS
5'-0"
5'-0"
4"×16"×4'-2 1/4" RGH. DIAPHRAM BLOCKS
SECTION "A-A"
SCALE 3/8"=1'-0"
NOTES:
ALL SAWN LUMBER SHALL BE COAST REGION DOUGLAS FIR
ALL TIMBER AND PILING TO BE CREOSOTE PRESSURE TREATED EXCEPT POSTS, CURBS, RAILING AND SCUPPERS WHICH ARE TO BE PRESSURE TREATED WITH PENTACLOROPHENOL
ALL HARDWARE AND STRUCTURAL STEEL TO BE GALVANIZED
LUMBER FOR LAMINATING SHALL BE COAST REGION DOUGLAS FIR GRADED IN ACCORDANCE WITH CURRENT W.C.L.A. ST'D. SPECS. FOR COMBINATION NUMBER 24 F, TABLE 1, DRY CONDITIONS OF USE, 2400#F, AS PER "STANDARD SPECIFICATIONS FOR STRUCTURAL GLUED LAMINATED TIMBERS", BY A.I.T.C.
AFTER COMPLETION BREAK ALL THREADS ON BOLTS
DESIGN LIVE LOAD=100 LBS./SQ. FT.
4"×12"×12'-0" FLOOR PLANK FASTEN TO STRINGER W/2-60 d NAILS @ EACH INTERSECTION
2-8 3/4"×46 1/2"×51'-0" FACIA GLU-LAMS
"A"
6'-0"
12'-0"
45°
17'-0"
17'-0"
17'-0"
10 3/4"×46 1/2"×51'-0" INT. GLU-LAM
51'-0"
8 SPA. @ 6'-0"=48'-0"
1'-6"
1'-6"
3'-6"
FINISHED GRADE
FINISHED GRADE
89.40
4" CAMBER AT CENTER
HIGH WATER 87.10
APPROX. 11.20'
RIPRAP
78.20 WATER SURFACE
EXISTING GROUND LINE
APPROX. CHANNEL BOTTOM 76.50
50'-0" ℄ TO ℄ OF ABUTMENT PILES
GENERAL ELEVATION
SCALE 1/4"=1'-0"
SNOWMOBILE BRIDGE
DEVIL'S CREEK CROSSING
RUSK COUNTY, WISCONSIN
SCALE- AS NOTED
GEO. H. MORGAN
PROFESSIONAL ENGINEER
LADYSMITH, WISCONSIN
SHEET 4 OF 5
DATE JAN. 1976
DR. BY R.O. PARMLEY

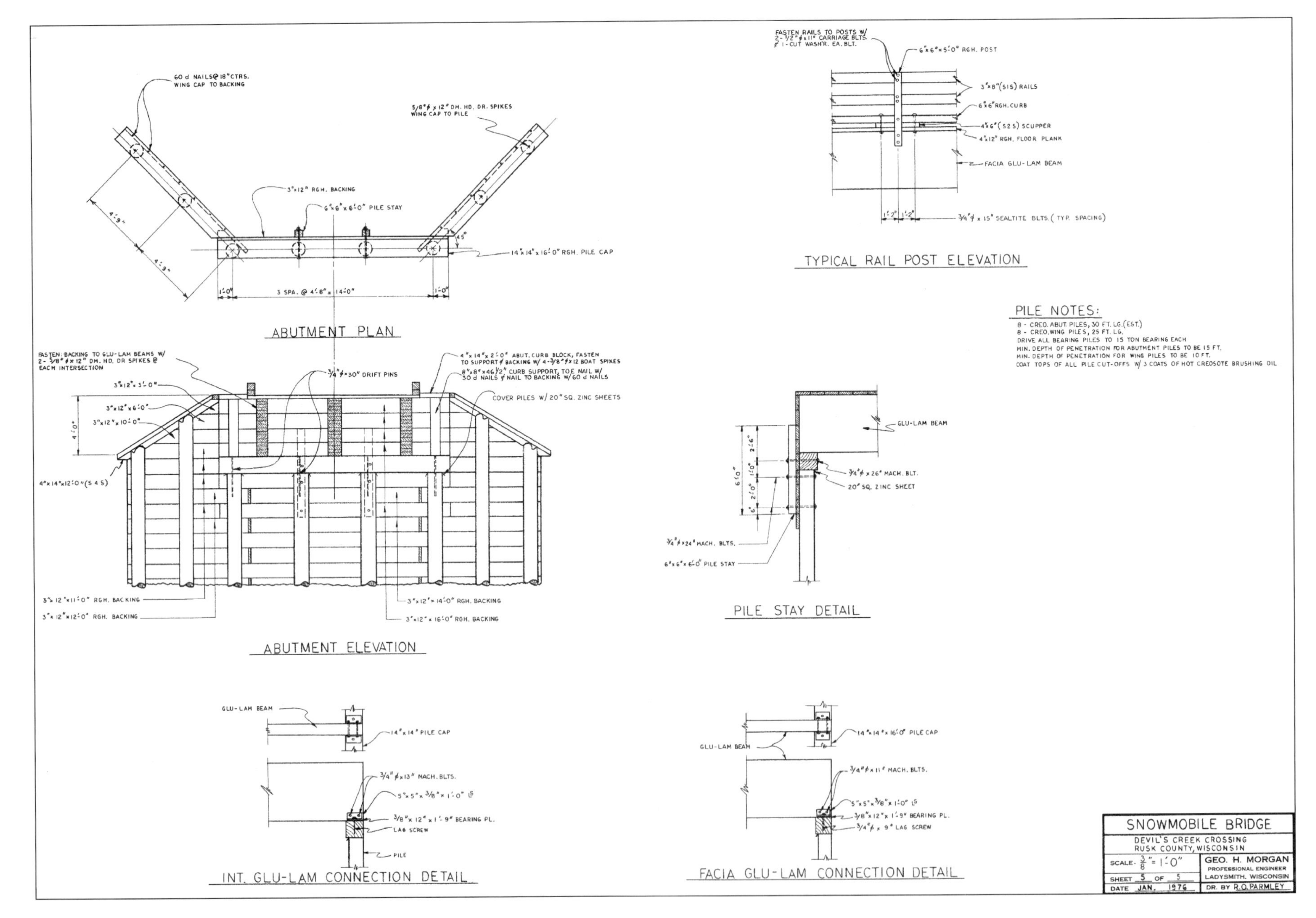

ABUTMENT PLAN
ABUTMENT ELEVATION
TYPICAL RAIL POST ELEVATION
PILE NOTES:
8 - CREO. ABUT. PILES, 30 FT. LG. (EST.)
8 - CREO. WING PILES, 25 FT. LG.
DRIVE ALL BEARING PILES TO 15 TON BEARING EACH
MIN. DEPTH OF PENETRATION FOR ABUTMENT PILES TO BE 15 FT.
MIN. DEPTH OF PENETRATION FOR WING PILES TO BE 10 FT.
COAT TOPS OF ALL PILE CUT-OFFS W/ 3 COATS OF HOT CREOSOTE BRUSHING OIL
PILE STAY DETAIL
INT. GLU-LAM CONNECTION DETAIL
FACIA GLU-LAM CONNECTION DETAIL
SNOWMOBILE BRIDGE
DEVIL'S CREEK CROSSING
RUSK COUNTY, WISCONSIN
GEO. H. MORGAN
PROFESSIONAL ENGINEER
LADYSMITH, WISCONSIN
DATE JAN. 1976
DR. BY R.O. PARMLEY

设 计

第 16 节 机场

例 16

下面是一个机场改建项目的例图，它包括 67 张图纸。因为本手册的篇幅有限，书中仅收录了基本的图纸和大量包含许多典型标准大样的剖面图。然而，读者可以参考标题页查看其中的“图纸目录”来对项目设计图纸的所有范畴有一个大概的估计。

目 录

Black River Falls 区域机场

BLACK RIVER FALLS AREA AIRPORT
JACKSON COUNTY, WISCONSIN
A. I. P. 3-55-0008-01

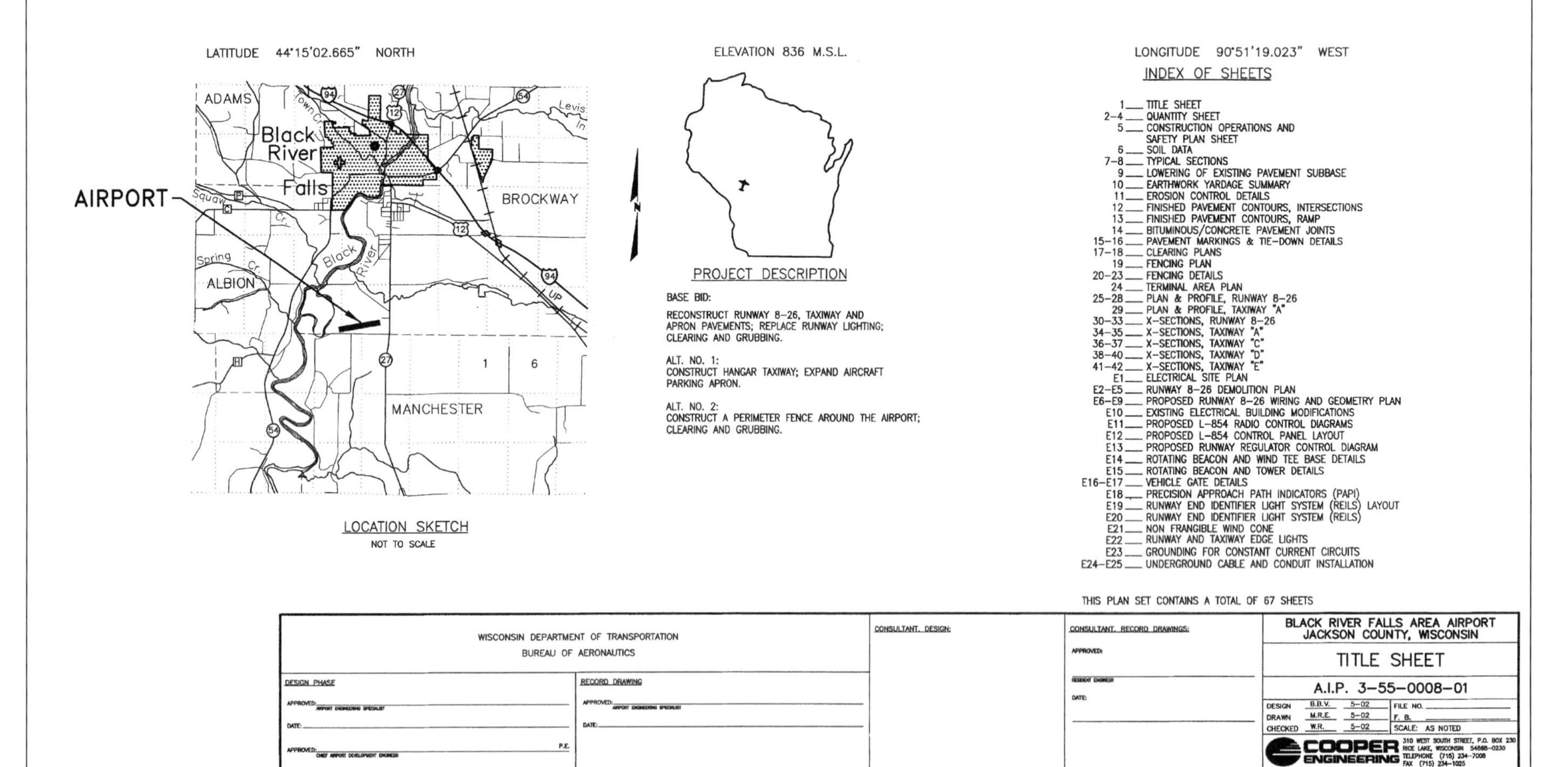

Source: Cooper Engineering, Co.

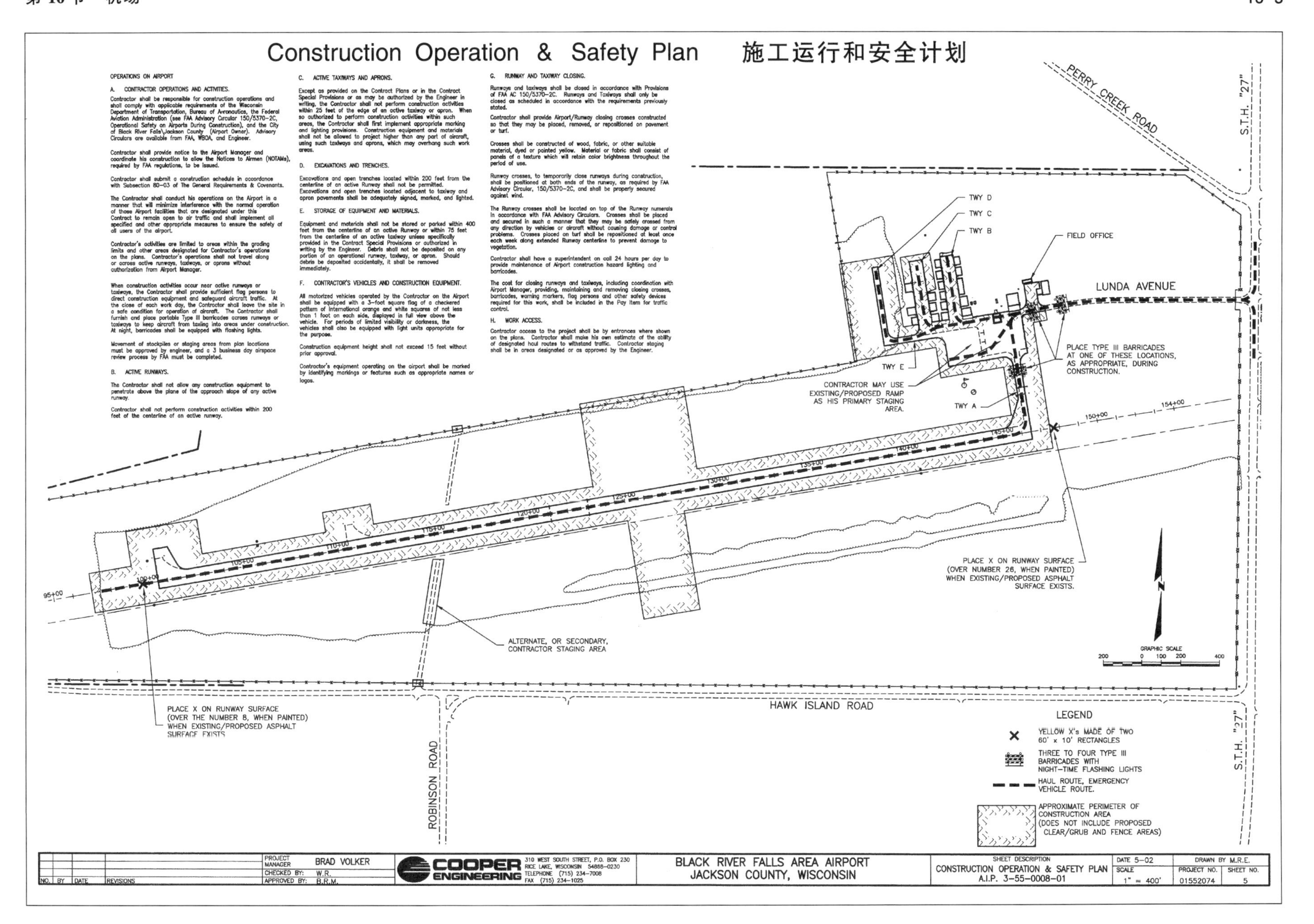
Construction Operation & Safety Plan　施工运行和安全计划
OPERATIONS ON AIRPORT
A. CONTRACTOR OPERATIONS AND ACTIVITIES.
Contractor shall be responsible for construction operations and shall comply with applicable requirements of the Wisconsin Department of Transportation, Bureau of Aeronautics, the Federal Aviation Administration (see FAA Advisory Circular 150/5370-2C, Operational Safety on Airports During Construction), and the City of Black River Falls\Jackson County (Airport Owner). Advisory Circulars are available from FAA, WBOA, and Engineer.
Contractor shall provide notice to the Airport Manager and coordinate his construction to allow the Notices to Airmen (NOTAMs), required by FAA regulations, to be issued.
Contractor shall submit a construction schedule in accordance with Subsection 80-03 of The General Requirements & Covenants.
The Contractor shall conduct his operations on the Airport in a manner that will minimize interference with the normal operation of those Airport facilities that are designated under this Contract to remain open to air traffic and shall implement all specified and other appropriate measures to ensure the safety of all users of the airport.
Contractor's activities are limited to areas within the grading limits and other areas designated for Contractor's operations on the plans. Contractor's operations shall not travel along or across active runways, taxiways, or aprons without authorization from Airport Manager.
When construction activities occur near active runways or taxiways, the Contractor shall provide sufficient flag persons to direct construction equipment and safeguard aircraft traffic. At the close of each work day, the Contractor shall leave the site in a safe condition for operation of aircraft. The Contractor shall furnish and place portable Type III barricades across runways or taxiways to keep aircraft from taxiing into areas under construction. At night, barricades shall be equipped with flashing lights.
Movement of stockpiles or staging areas from plan locations must be approved by engineer, and a 3 business day airspace review process by FAA must be completed.
B. ACTIVE RUNWAYS.
The Contractor shall not allow any construction equipment to penetrate above the plane of the approach slope of any active runway.
Contractor shall not perform construction activities within 200 feet of the centerline of an active runway.
C. ACTIVE TAXIWAYS AND APRONS.
Except as provided on the Contract Plans or in the Contract Special Provisions or as may be authorized by the Engineer in writing, the Contractor shall not perform construction activities within 25 feet of the edge of an active taxiway or apron. When so authorized to perform construction activities within such areas, the Contractor shall first implement appropriate marking and lighting provisions. Construction equipment and materials shall not be allowed to project higher than any part of aircraft, using such taxiways and aprons, which may overhang such work areas.
D. EXCAVATIONS AND TRENCHES.
Excavations and open trenches located within 200 feet from the centerline of an active Runway shall not be permitted. Excavations and open trenches located adjacent to taxiway and apron pavements shall be adequately signed, marked, and lighted.
E. STORAGE OF EQUIPMENT AND MATERIALS.
Equipment and materials shall not be stored or parked within 400 feet from the centerline of an active Runway or within 75 feet from the centerline of an active taxiway unless specifically provided in the Contract Special Provisions or authorized in writing by the Engineer. Debris shall not be deposited on any portion of an operational runway, taxiway, or apron. Should debris be deposited accidentally, it shall be removed immediately.
F. CONTRACTOR'S VEHICLES AND CONSTRUCTION EQUIPMENT.
All motorized vehicles operated by the Contractor on the Airport shall be equipped with a 3-foot square flag of a checkered pattern of international orange and white squares of not less than 1 foot on each side, displayed in full view above the vehicle. For periods of limited visibility or darkness, the vehicles shall also be equipped with light units appropriate for the purpose.
Construction equipment height shall not exceed 15 feet without prior approval.
Contractor's equipment operating on the airport shall be marked by identifying markings or features such as appropriate names or logos.
G. RUNWAY AND TAXIWAY CLOSING.
Runways and taxiways shall be closed in accordance with Provisions of FAA AC 150/5370-2C. Runways and Taxiways shall only be closed as scheduled in accordance with the requirements previously stated.
Contractor shall provide Airport/Runway closing crosses constructed so that they may be placed, removed, or repositioned on pavement or turf.
Crosses shall be constructed of wood, fabric, or other suitable material, dyed or painted yellow. Material or fabric shall consist of panels of a texture which will retain color brightness throughout the period of use.
Runway crosses, to temporarily close runways during construction, shall be positioned at both ends of the runway, as required by FAA Advisory Circular, 150/5370-2C, and shall be properly secured against wind.
The Runway crosses shall be located on top of the Runway numerals in accordance with FAA Advisory Circulars. Crosses shall be placed and secured in such a manner that they may be safely crossed from any direction by vehicles or aircraft without causing damage or control problems. Crosses placed on turf shall be repositioned at least once each week along extended Runway centerline to prevent damage to vegetation.
Contractor shall have a superintendent on call 24 hours per day to provide maintenance of Airport construction hazard lighting and barricades.
The cost for closing runways and taxiways, including coordination with Airport Manager, providing, maintaining and removing closing crosses, barricades, warning markers, flag persons and other safety devices required for this work, shall be included in the Pay Item for traffic control.
H. WORK ACCESS.
Contractor access to the project shall be by entrances where shown on the plans. Contractor shall make his own estimate of the ability of designated haul routes to withstand traffic. Contractor staging shall be in areas designated or as approved by the Engineer.
PERRY CREEK ROAD
S.T.H. "27"
TWY D
TWY C
TWY B
FIELD OFFICE
LUNDA AVENUE
PLACE TYPE III BARRICADES AT ONE OF THESE LOCATIONS, AS APPROPRIATE, DURING CONSTRUCTION.
TWY E
CONTRACTOR MAY USE EXISTING/PROPOSED RAMP AS HIS PRIMARY STAGING AREA.
TWY A
95+00
100+00
105+00
110+00
115+00
120+00
125+00
130+00
135+00
140+00
145+00
150+00
154+00
PLACE X ON RUNWAY SURFACE (OVER NUMBER 26, WHEN PAINTED) WHEN EXISTING/PROPOSED ASPHALT SURFACE EXISTS.
ALTERNATE, OR SECONDARY, CONTRACTOR STAGING AREA
N
GRAPHIC SCALE
200 0 100 200 400
HAWK ISLAND ROAD
PLACE X ON RUNWAY SURFACE (OVER THE NUMBER 8, WHEN PAINTED) WHEN EXISTING/PROPOSED ASPHALT SURFACE EXISTS
ROBINSON ROAD
LEGEND
YELLOW X's MADE OF TWO 60' x 10' RECTANGLES
THREE TO FOUR TYPE III BARRICADES WITH NIGHT-TIME FLASHING LIGHTS
HAUL ROUTE, EMERGENCY VEHICLE ROUTE.
APPROXIMATE PERIMETER OF CONSTRUCTION AREA (DOES NOT INCLUDE PROPOSED CLEAR/GRUB AND FENCE AREAS)
NO. BY DATE REVISIONS
PROJECT MANAGER BRAD VOLKER
CHECKED BY: W.R.
APPROVED BY: B.R.M.
COOPER ENGINEERING
310 WEST SOUTH STREET, P.O. BOX 230
RICE LAKE, WISCONSIN 54868-0230
TELEPHONE (715) 234-7008
FAX (715) 234-1025
BLACK RIVER FALLS AREA AIRPORT
JACKSON COUNTY, WISCONSIN
SHEET DESCRIPTION
CONSTRUCTION OPERATION & SAFETY PLAN
A.I.P. 3-55-0008-01
DATE 5-02
DRAWN BY M.R.E.
SCALE 1" = 400'
PROJECT NO. 01552074
SHEET NO. 5

Typical Section 典型剖面

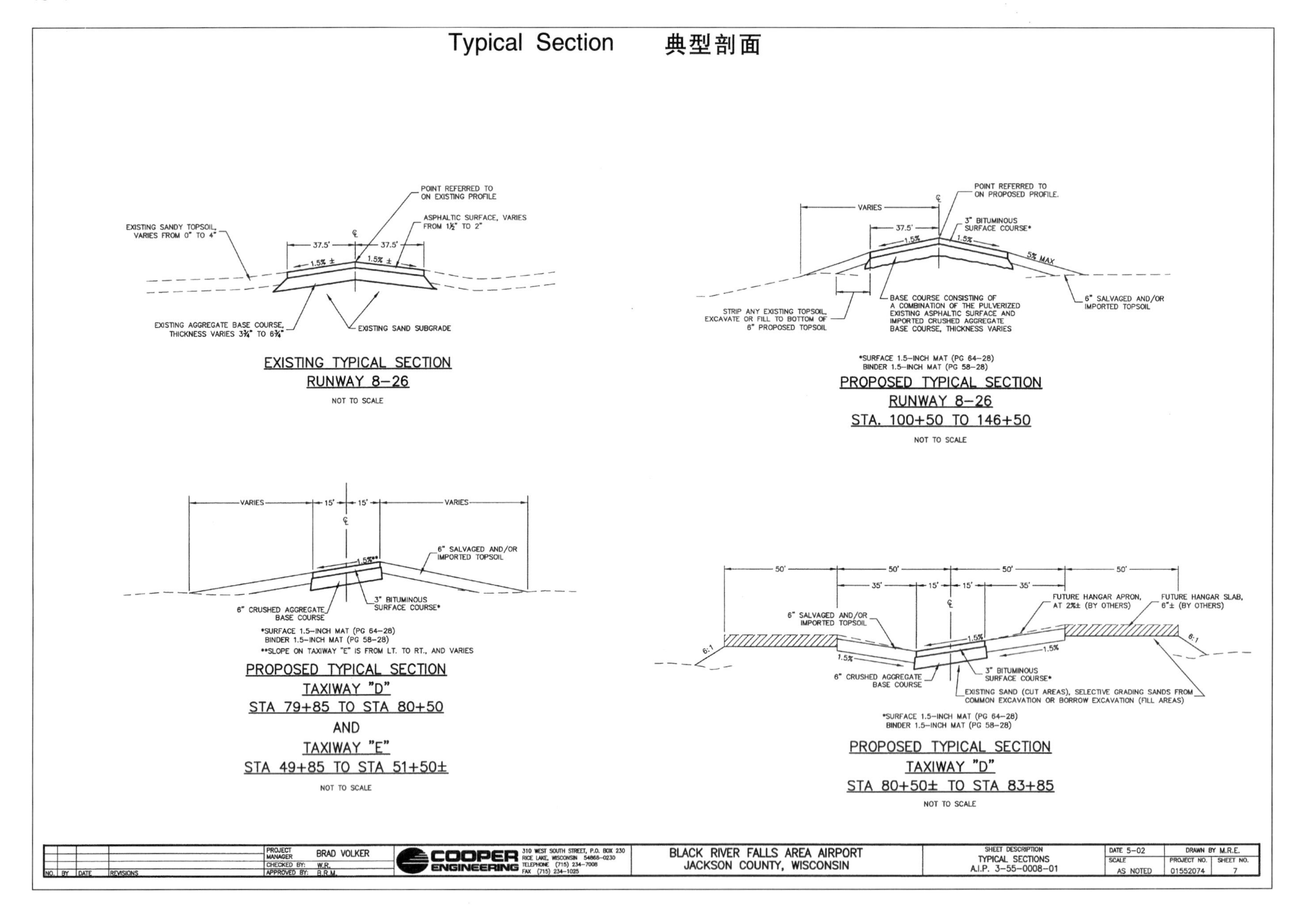

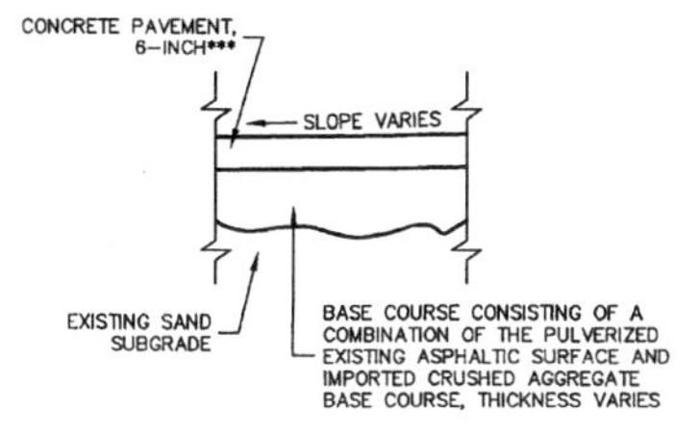

***SAW-CUT JOINTS AS SHOWN ON SHEET NO. 14. SAW-CUT DEPTH TO BE ¼ OF THE CONCRETE THICKNESS. (INCIDENTAL)

PROPOSED TYPICAL SECTION
CONCRETE FUELING PAD

NOT TO SCALE

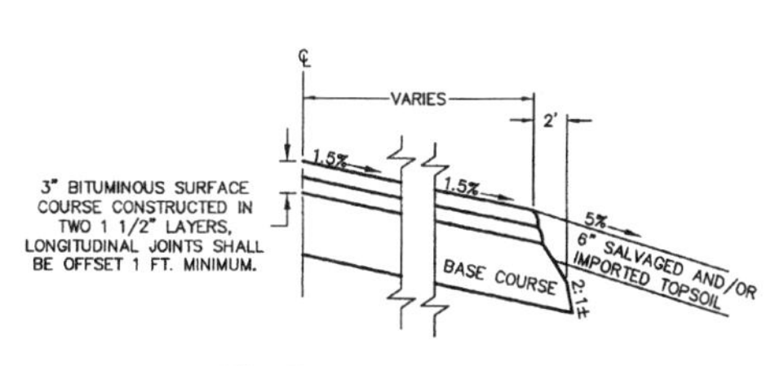

EDGE OF PAVEMENT DETAIL

NOT TO SCALE

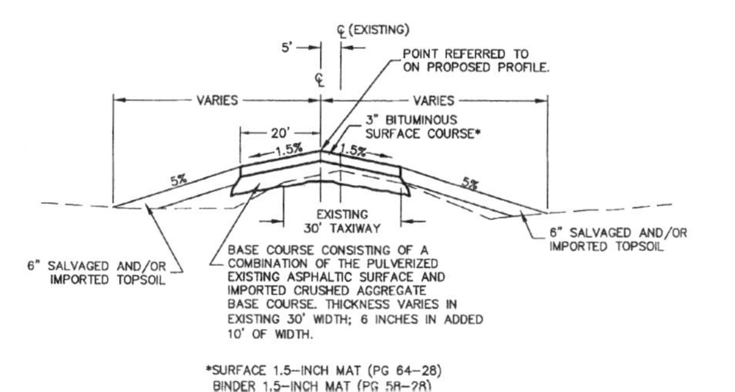

*SURFACE 1.5-INCH MAT (PG 64-28)
BINDER 1.5-INCH MAT (PG 58-28)

PROPOSED TYPICAL SECTION
TAXIWAY "A"

NOT TO SCALE

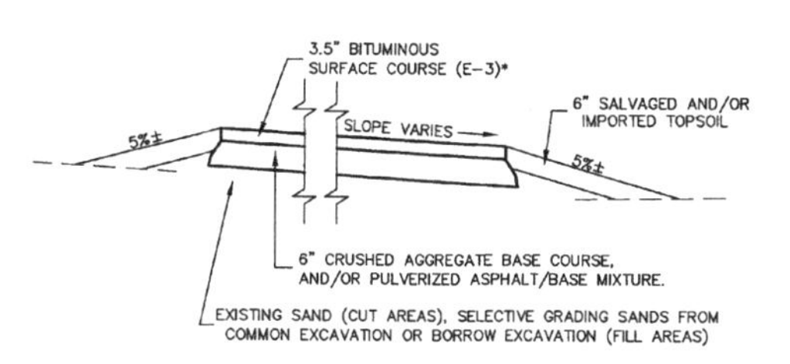

*SURFACE 1.5-INCH MAT (PG 64-28)
BINDER 1.5-INCH MAT (PG 58-28)

PROPOSED TYPICAL SECTION
AIRCRAFT PARKING APRON AND
HANGAR TAXIWAY EXPANSIONS

NOT TO SCALE

NO.	BY	DATE	REVISIONS					
PROJECT MANAGER	BRAD VOLKER	COOPER ENGINEERING 310 WEST SOUTH STREET, P.O. BOX 230 RICE LAKE, WISCONSIN 54868-0230 TELEPHONE (715) 234-7008 FAX (715) 234-1025	BLACK RIVER FALLS AREA AIRPORT JACKSON COUNTY, WISCONSIN	SHEET DESCRIPTION TYPICAL SECTIONS A.I.P. 3-55-0008-01	DATE 5-02	DRAWN BY M.R.E.		
CHECKED BY:	W.R.				SCALE AS NOTED	PROJECT NO. 01552074	SHEET NO. 8	
APPROVED BY:	B.R.M.							

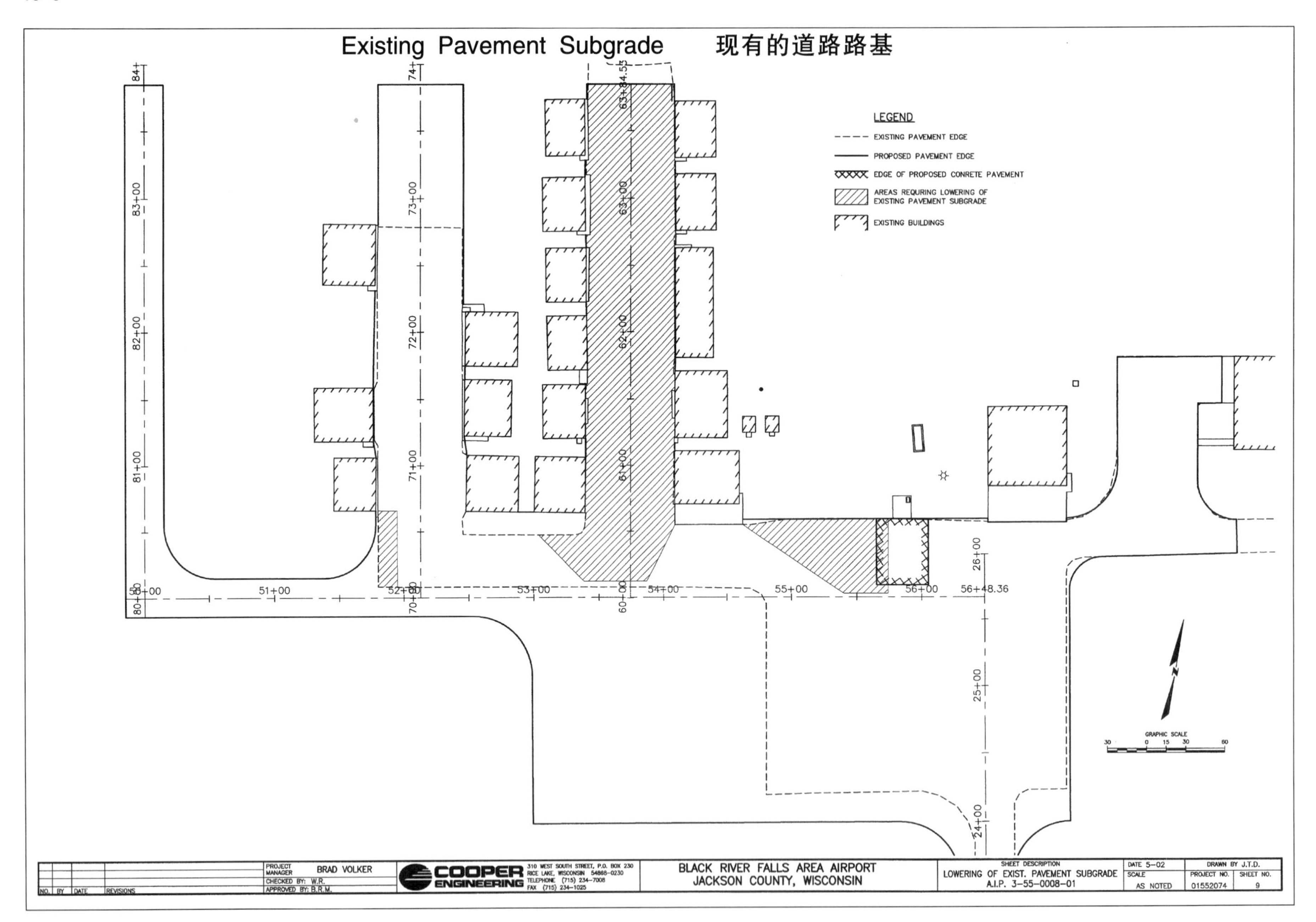
Existing Pavement Subgrade 现有的道路路基
LEGEND
EXISTING PAVEMENT EDGE
PROPOSED PAVEMENT EDGE
EDGE OF PROPOSED CONRETE PAVEMENT
AREAS REQURING LOWERING OF EXISTING PAVEMENT SUBGRADE
EXISTING BUILDINGS
GRAPHIC SCALE
PROJECT MANAGER BRAD VOLKER
CHECKED BY: W.R.
APPROVED BY: B.R.M.
COOPER ENGINEERING
310 WEST SOUTH STREET, P.O. BOX 230
RICE LAKE, WISCONSIN 54868-0230
TELEPHONE (715) 234-7008
FAX (715) 234-1025
BLACK RIVER FALLS AREA AIRPORT
JACKSON COUNTY, WISCONSIN
SHEET DESCRIPTION
LOWERING OF EXIST. PAVEMENT SUBGRADE
A.I.P. 3-55-0008-01
DATE 5-02
SCALE AS NOTED
DRAWN BY J.T.D.
PROJECT NO. 01552074
SHEET NO. 9
NO. BY DATE REVISIONS

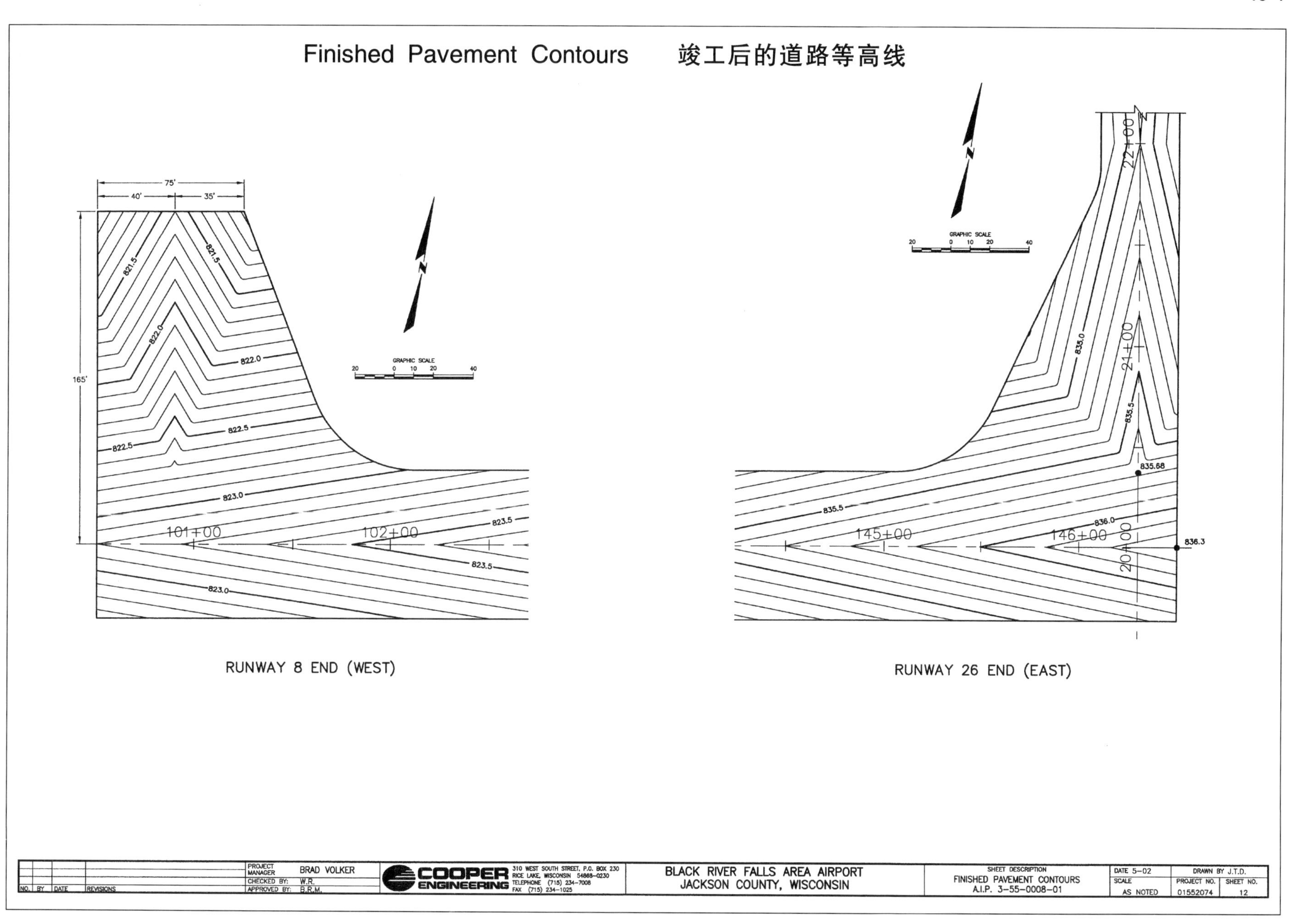
Finished Pavement Contours　竣工后的道路等高线
75'
40'
35'
165'
821.5
822.0
822.5
823.0
823.5
101+00
102+00
GRAPHIC SCALE
RUNWAY 8 END (WEST)
835.0
835.5
835.68
836.0
836.3
145+00
146+00
20+00
21+00
22+00
RUNWAY 26 END (EAST)
PROJECT MANAGER BRAD VOLKER
CHECKED BY: W.R.
APPROVED BY: B.R.M.
COOPER ENGINEERING
310 WEST SOUTH STREET, P.O. BOX 230
RICE LAKE, WISCONSIN 54868-0230
TELEPHONE (715) 234-7008
FAX (715) 234-1025
BLACK RIVER FALLS AREA AIRPORT
JACKSON COUNTY, WISCONSIN
SHEET DESCRIPTION
FINISHED PAVEMENT CONTOURS
A.I.P. 3-55-0008-01
DATE 5-02
DRAWN BY J.T.D.
SCALE AS NOTED
PROJECT NO. 01552074
SHEET NO. 12
NO. BY DATE REVISIONS

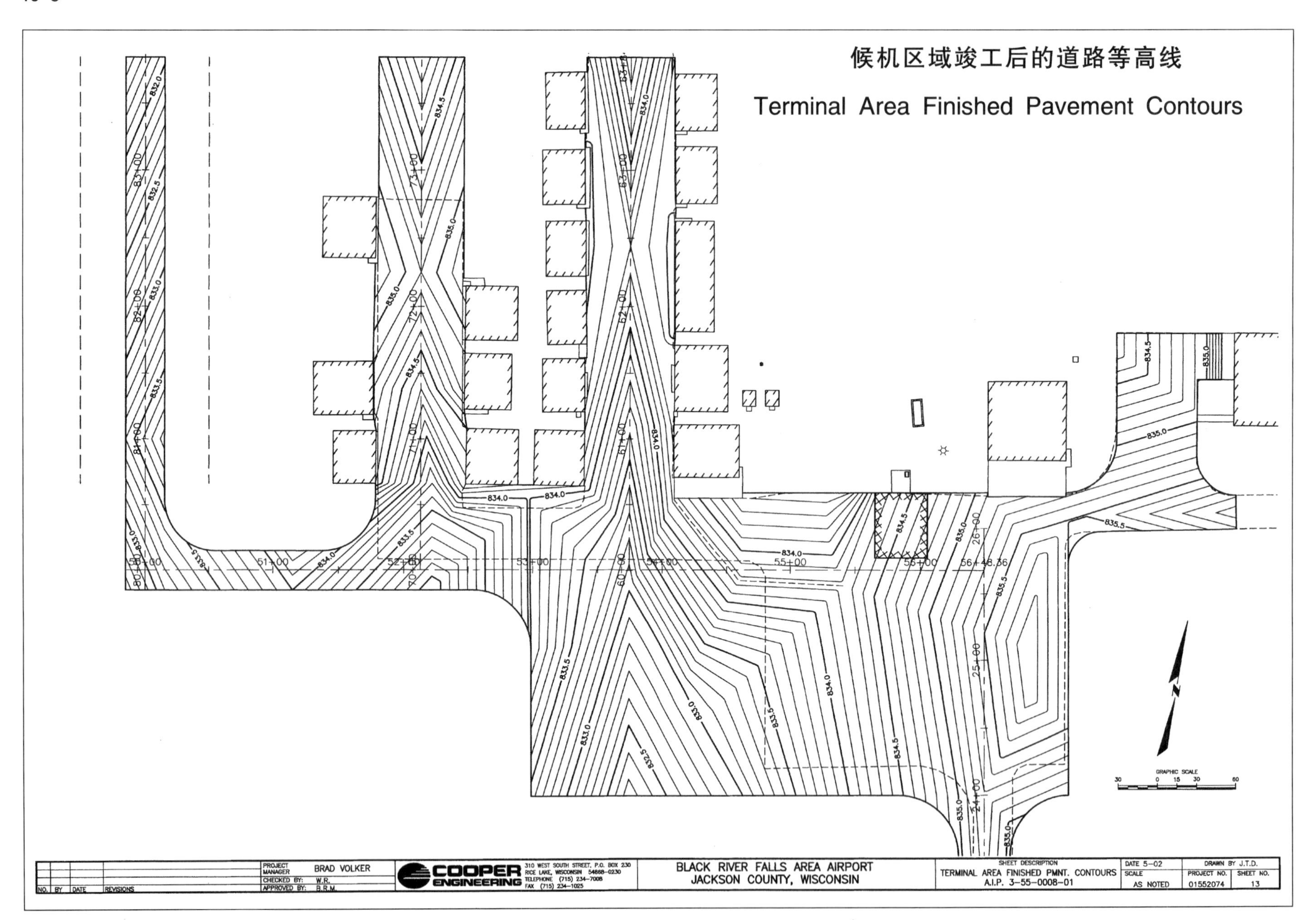
候机区域竣工后的道路等高线
Terminal Area Finished Pavement Contours
BLACK RIVER FALLS AREA AIRPORT
JACKSON COUNTY, WISCONSIN
TERMINAL AREA FINISHED PMNT. CONTOURS
A.I.P. 3-55-0008-01
PROJECT MANAGER BRAD VOLKER
CHECKED BY: W.R.
APPROVED BY: B.R.M.
COOPER ENGINEERING
DATE 5-02
DRAWN BY J.T.D.
SCALE AS NOTED
PROJECT NO. 01552074
SHEET NO. 13

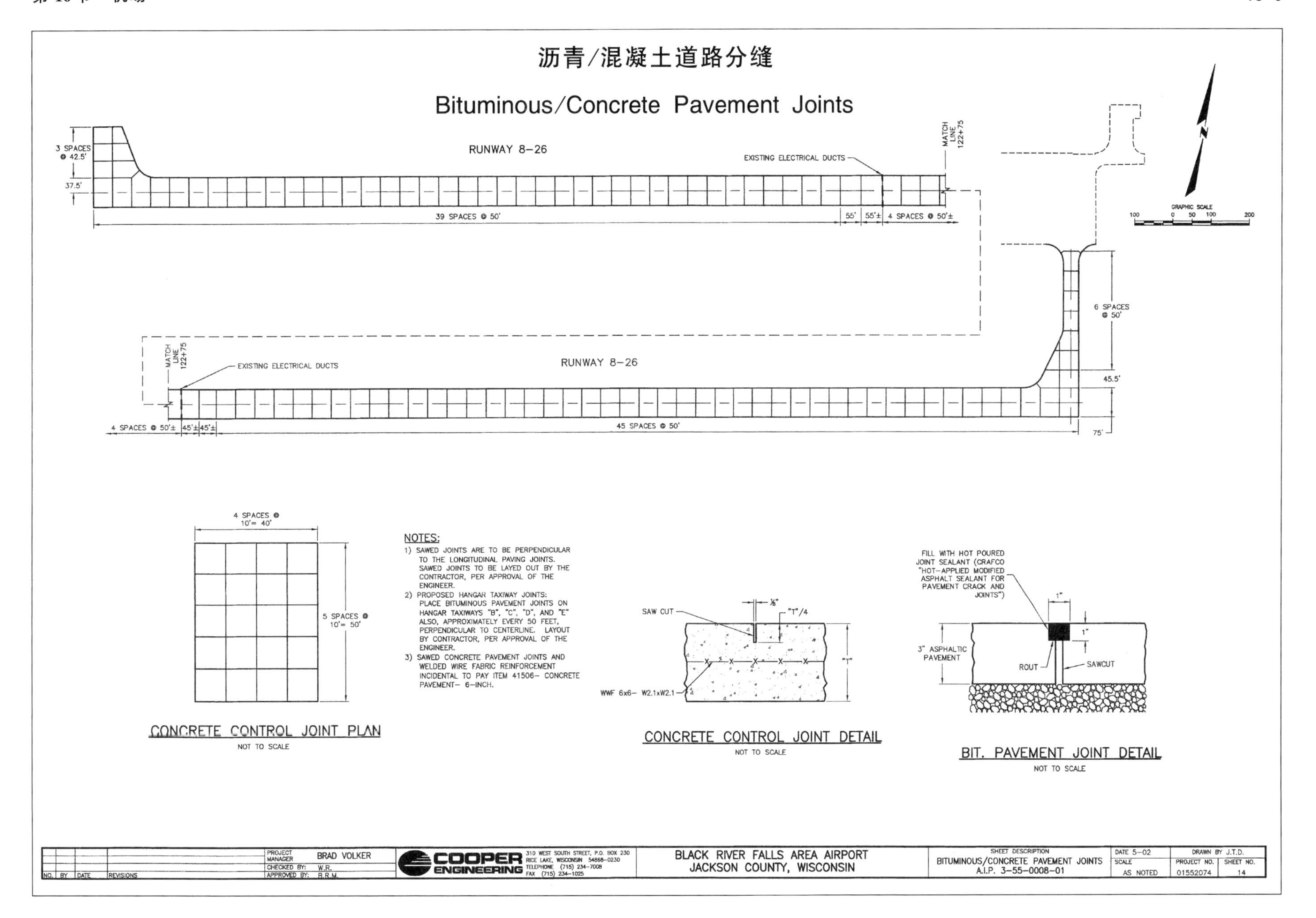

沥青/混凝土道路分缝
Bituminous/Concrete Pavement Joints
RUNWAY 8-26
3 SPACES @ 42.5'
37.5'
EXISTING ELECTRICAL DUCTS
MATCH LINE 122+75
39 SPACES @ 50'
55'
55'±
4 SPACES @ 50'±
GRAPHIC SCALE
100 0 50 100 200
6 SPACES @ 50'
45.5'
75'
4 SPACES @ 50'±
45'±
45'±
45 SPACES @ 50'
4 SPACES @ 10'= 40'
5 SPACES @ 10'= 50'
CONCRETE CONTROL JOINT PLAN
NOT TO SCALE
NOTES:
1) SAWED JOINTS ARE TO BE PERPENDICULAR TO THE LONGITUDINAL PAVING JOINTS. SAWED JOINTS TO BE LAYED OUT BY THE CONTRACTOR, PER APPROVAL OF THE ENGINEER.
2) PROPOSED HANGAR TAXIWAY JOINTS: PLACE BITUMINOUS PAVEMENT JOINTS ON HANGAR TAXIWAYS "B", "C", "D", AND "E" ALSO, APPROXIMATELY EVERY 50 FEET, PERPENDICULAR TO CENTERLINE. LAYOUT BY CONTRACTOR, PER APPROVAL OF THE ENGINEER.
3) SAWED CONCRETE PAVEMENT JOINTS AND WELDED WIRE FABRIC REINFORCEMENT INCIDENTAL TO PAY ITEM 41506- CONCRETE PAVEMENT- 6-INCH.
SAW CUT
⅛"
"T"/4
"T"
WWF 6x6- W2.1xW2.1
CONCRETE CONTROL JOINT DETAIL
NOT TO SCALE
FILL WITH HOT POURED JOINT SEALANT (CRAFCO "HOT-APPLIED MODIFIED ASPHALT SEALANT FOR PAVEMENT CRACK AND JOINTS")
1"
1"
3" ASPHALTIC PAVEMENT
ROUT
SAWCUT
BIT. PAVEMENT JOINT DETAIL
NOT TO SCALE
NO. BY DATE REVISIONS
PROJECT MANAGER BRAD VOLKER
CHECKED BY: W.R.
APPROVED BY: B.R.M.
COOPER ENGINEERING
310 WEST SOUTH STREET, P.O. BOX 230 RICE LAKE, WISCONSIN 54868-0230 TELEPHONE (715) 234-7008 FAX (715) 234-1025
BLACK RIVER FALLS AREA AIRPORT
JACKSON COUNTY, WISCONSIN
SHEET DESCRIPTION
BITUMINOUS/CONCRETE PAVEMENT JOINTS
A.I.P. 3-55-0008-01
DATE 5-02
DRAWN BY J.T.D.
SCALE AS NOTED
PROJECT NO. 01552074
SHEET NO. 14

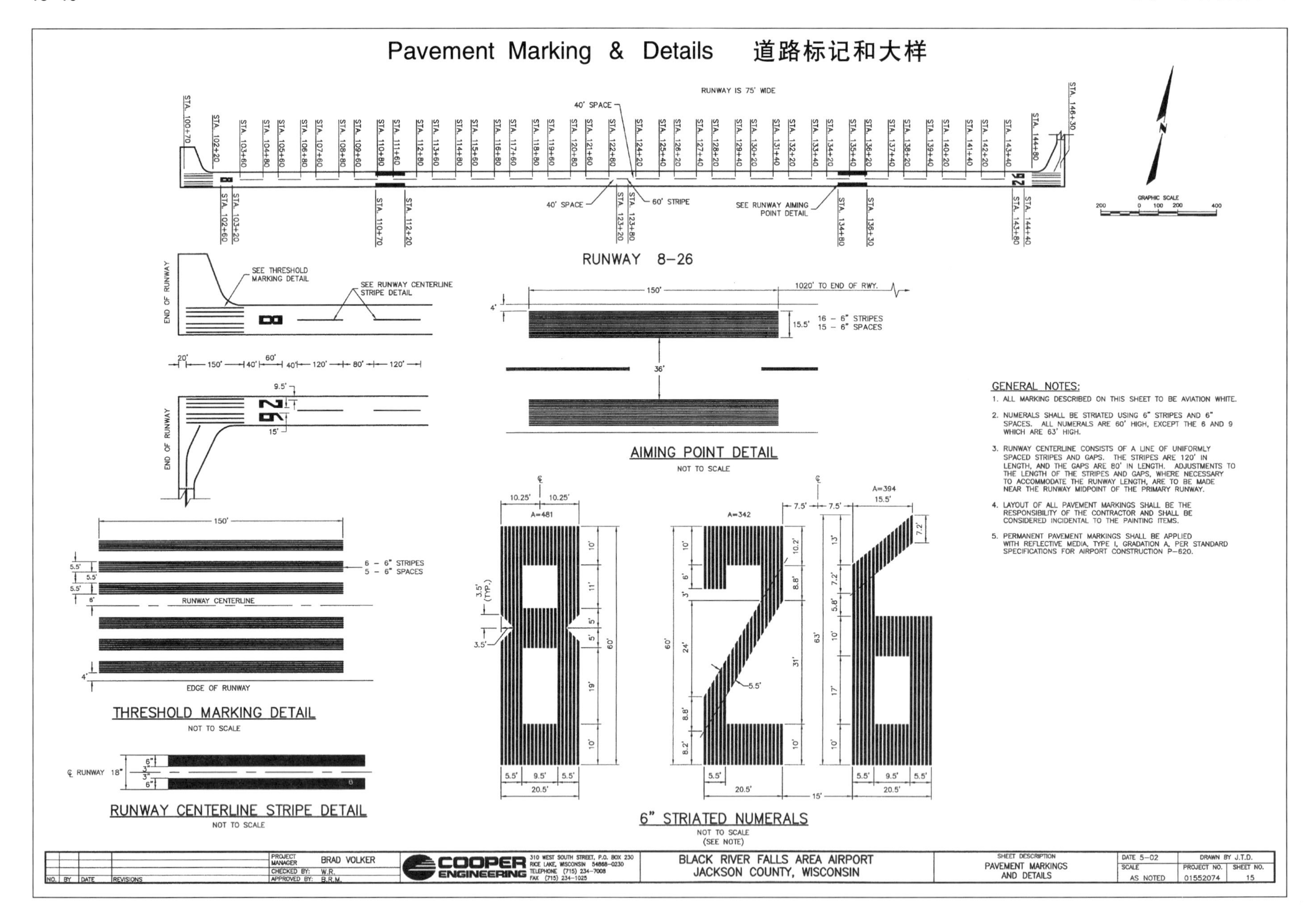
Pavement Marking & Details 道路标记和大样
RUNWAY IS 75' WIDE
RUNWAY 8-26
SEE THRESHOLD MARKING DETAIL
SEE RUNWAY CENTERLINE STRIPE DETAIL
SEE RUNWAY AIMING POINT DETAIL
END OF RUNWAY
AIMING POINT DETAIL
NOT TO SCALE
16 - 6" STRIPES
15 - 6" SPACES
1020' TO END OF RWY.
GENERAL NOTES:
1. ALL MARKING DESCRIBED ON THIS SHEET TO BE AVIATION WHITE.
2. NUMERALS SHALL BE STRIATED USING 6" STRIPES AND 6" SPACES. ALL NUMERALS ARE 60' HIGH, EXCEPT THE 6 AND 9 WHICH ARE 63' HIGH.
3. RUNWAY CENTERLINE CONSISTS OF A LINE OF UNIFORMLY SPACED STRIPES AND GAPS. THE STRIPES ARE 120' IN LENGTH, AND THE GAPS ARE 80' IN LENGTH. ADJUSTMENTS TO THE LENGTH OF THE STRIPES AND GAPS, WHERE NECESSARY TO ACCOMMODATE THE RUNWAY LENGTH, ARE TO BE MADE NEAR THE RUNWAY MIDPOINT OF THE PRIMARY RUNWAY.
4. LAYOUT OF ALL PAVEMENT MARKINGS SHALL BE THE RESPONSIBILITY OF THE CONTRACTOR AND SHALL BE CONSIDERED INCIDENTAL TO THE PAINTING ITEMS.
5. PERMANENT PAVEMENT MARKINGS SHALL BE APPLIED WITH REFLECTIVE MEDIA, TYPE I, GRADATION A, PER STANDARD SPECIFICATIONS FOR AIRPORT CONSTRUCTION P-620.
THRESHOLD MARKING DETAIL
NOT TO SCALE
6 - 6" STRIPES
5 - 6" SPACES
RUNWAY CENTERLINE
EDGE OF RUNWAY
RUNWAY CENTERLINE STRIPE DETAIL
NOT TO SCALE
6" STRIATED NUMERALS
NOT TO SCALE
(SEE NOTE)
PROJECT MANAGER BRAD VOLKER
CHECKED BY: W.R.
APPROVED BY: B.R.M.
COOPER ENGINEERING
310 WEST SOUTH STREET, P.O. BOX 230
RICE LAKE, WISCONSIN 54868-0230
TELEPHONE (715) 234-7008
FAX (715) 234-1025
BLACK RIVER FALLS AREA AIRPORT
JACKSON COUNTY, WISCONSIN
SHEET DESCRIPTION
PAVEMENT MARKINGS AND DETAILS
DATE 5-02
DRAWN BY J.T.D.
SCALE AS NOTED
PROJECT NO. 01552074
SHEET NO. 15
NO. BY DATE REVISIONS

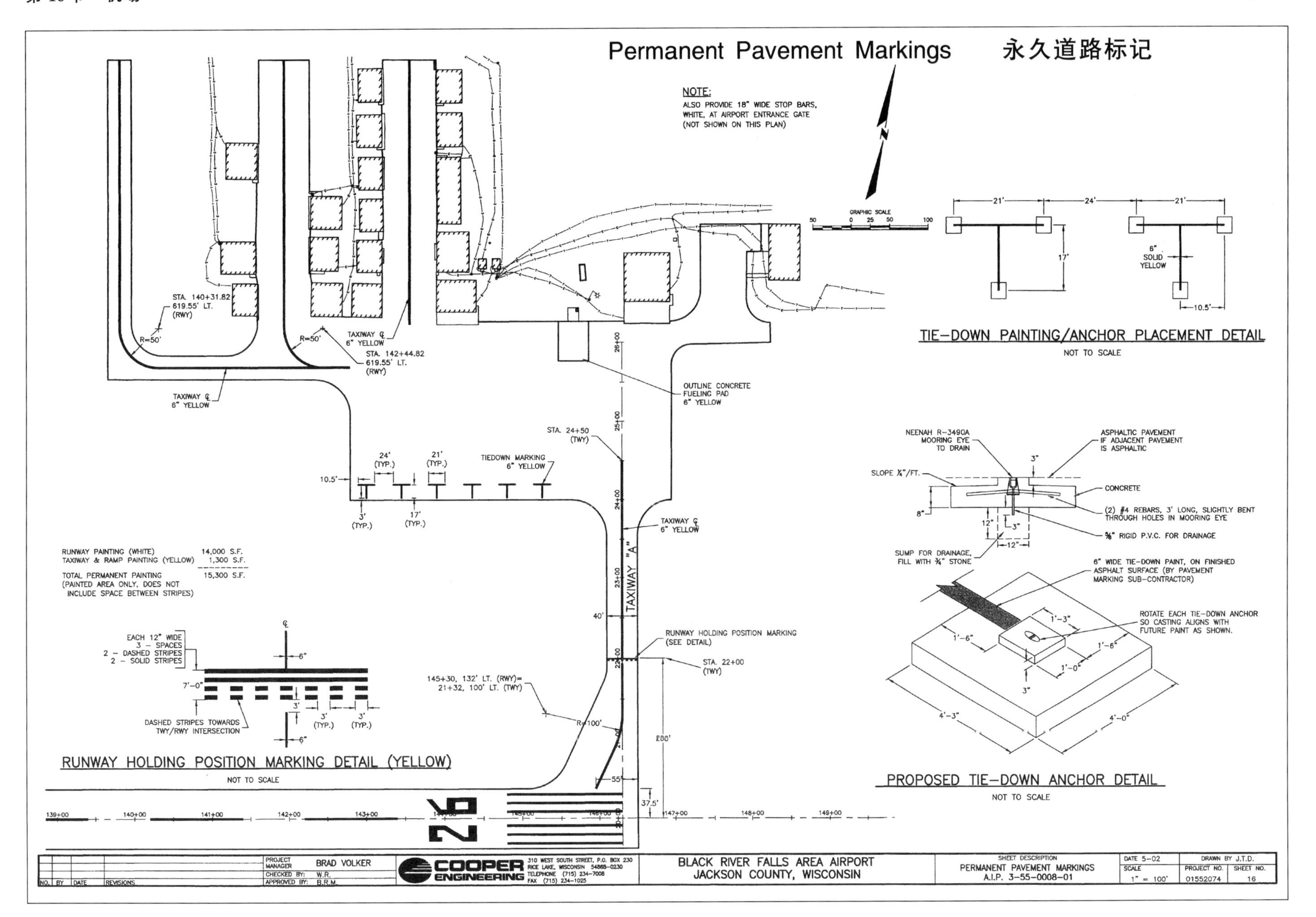
Permanent Pavement Markings　永久道路标记
NOTE:
ALSO PROVIDE 18" WIDE STOP BARS, WHITE, AT AIRPORT ENTRANCE GATE (NOT SHOWN ON THIS PLAN)
TIE-DOWN PAINTING/ANCHOR PLACEMENT DETAIL
NOT TO SCALE
PROPOSED TIE-DOWN ANCHOR DETAIL
NOT TO SCALE
RUNWAY HOLDING POSITION MARKING DETAIL (YELLOW)
NOT TO SCALE
RUNWAY PAINTING (WHITE) 14,000 S.F.
TAXIWAY & RAMP PAINTING (YELLOW) 1,300 S.F.
TOTAL PERMANENT PAINTING 15,300 S.F.
(PAINTED AREA ONLY, DOES NOT INCLUDE SPACE BETWEEN STRIPES)
TAXIWAY "A"
BLACK RIVER FALLS AREA AIRPORT
JACKSON COUNTY, WISCONSIN
PERMANENT PAVEMENT MARKINGS
A.I.P. 3-55-0008-01
COOPER ENGINEERING
BRAD VOLKER

Clearing Plan Sheet 净场平面图

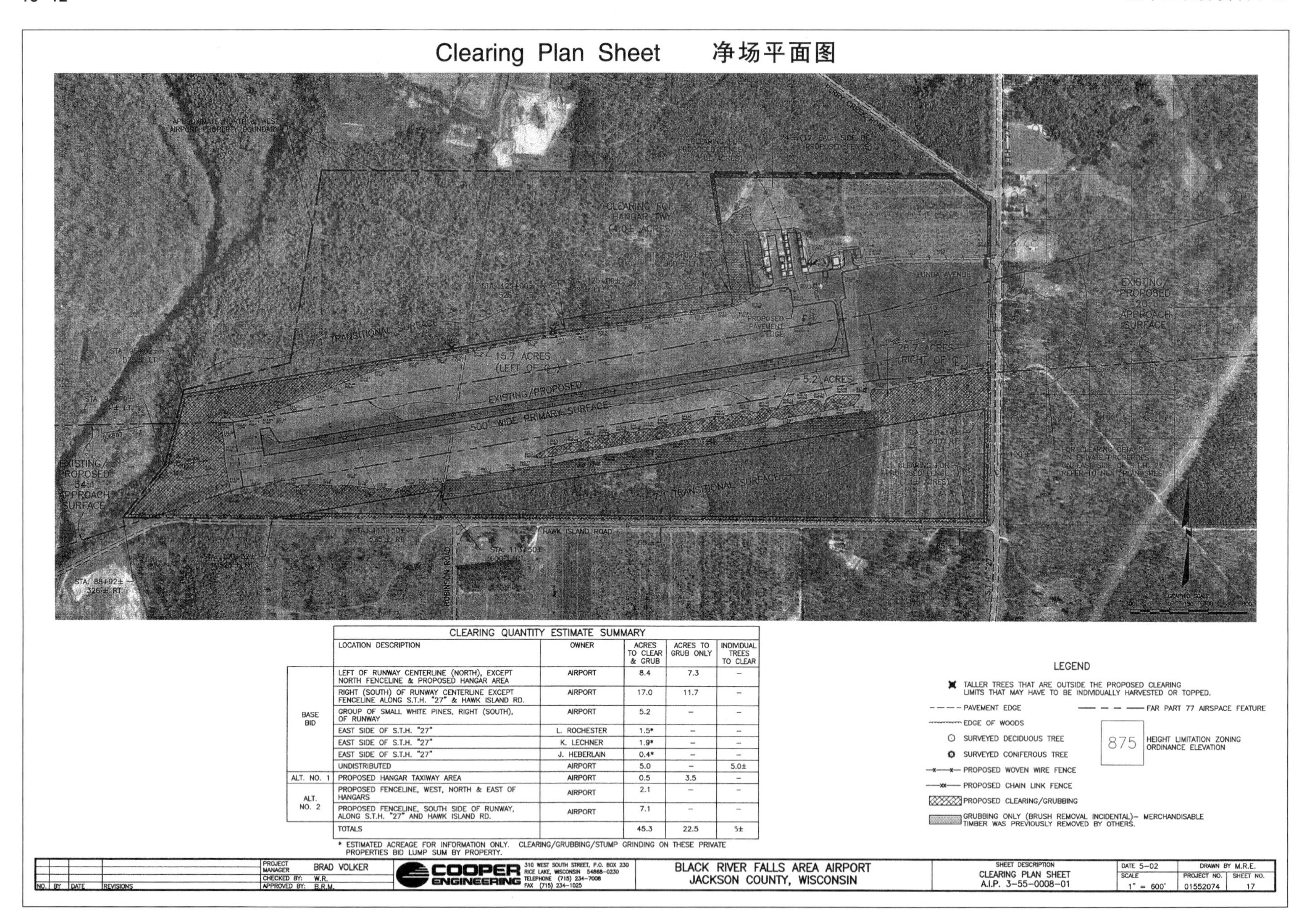

CLEARING QUANTITY ESTIMATE SUMMARY

	LOCATION DESCRIPTION	OWNER	ACRES TO CLEAR & GRUB	ACRES TO GRUB ONLY	INDIVIDUAL TREES TO CLEAR
BASE BID	LEFT OF RUNWAY CENTERLINE (NORTH), EXCEPT NORTH FENCELINE & PROPOSED HANGAR AREA	AIRPORT	8.4	7.3	–
	RIGHT (SOUTH) OF RUNWAY CENTERLINE EXCEPT FENCELINE ALONG S.T.H. "27" & HAWK ISLAND RD.	AIRPORT	17.0	11.7	–
	GROUP OF SMALL WHITE PINES, RIGHT (SOUTH), OF RUNWAY	AIRPORT	5.2	–	–
	EAST SIDE OF S.T.H. "27"	L. ROCHESTER	1.5*	–	–
	EAST SIDE OF S.T.H. "27"	K. LECHNER	1.9*	–	–
	EAST SIDE OF S.T.H. "27"	J. HEBERLAIN	0.4*	–	–
	UNDISTRIBUTED	AIRPORT	5.0	–	5.0±
ALT. NO. 1	PROPOSED HANGAR TAXIWAY AREA	AIRPORT	0.5	3.5	–
ALT. NO. 2	PROPOSED FENCELINE, WEST, NORTH & EAST OF HANGARS	AIRPORT	2.1	–	–
	PROPOSED FENCELINE, SOUTH SIDE OF RUNWAY, ALONG S.T.H. "27" AND HAWK ISLAND RD.	AIRPORT	7.1	–	–
	TOTALS		45.3	22.5	5±

* ESTIMATED ACREAGE FOR INFORMATION ONLY. CLEARING/GRUBBING/STUMP GRINDING ON THESE PRIVATE PROPERTIES BID LUMP SUM BY PROPERTY.

Obstruction Clearing Plan　净场障碍平面图
LUNDA AVENUE
POWERLINE
L. ROCHESTER
K. LECHNER
J. HEERLAIN
STATE OF WISCONSIN
PROPERTY BOUNDARIES SHOWN ARE APPROXIMATE
154+00
150+00
PROJECT MANAGER BRAD VOLKER
CHECKED BY: W.R.
APPROVED BY: B.R.M.
NO. BY DATE REVISIONS
COOPER ENGINEERING
310 WEST SOUTH STREET, P.O. BOX 230
RICE LAKE, WISCONSIN 54868-0230
TELEPHONE (715) 234-7008
FAX (715) 234-1025
BLACK RIVER FALLS AREA AIRPORT
JACKSON COUNTY, WISCONSIN
SHEET DESCRIPTION
OBSTRUCTION CLEARING PLAN-
EAST OF S.T.H. 27
DATE 5-02
DRAWN BY J.T.D.
SCALE AS NOTED
PROJECT NO. 01552074
SHEET NO. 18

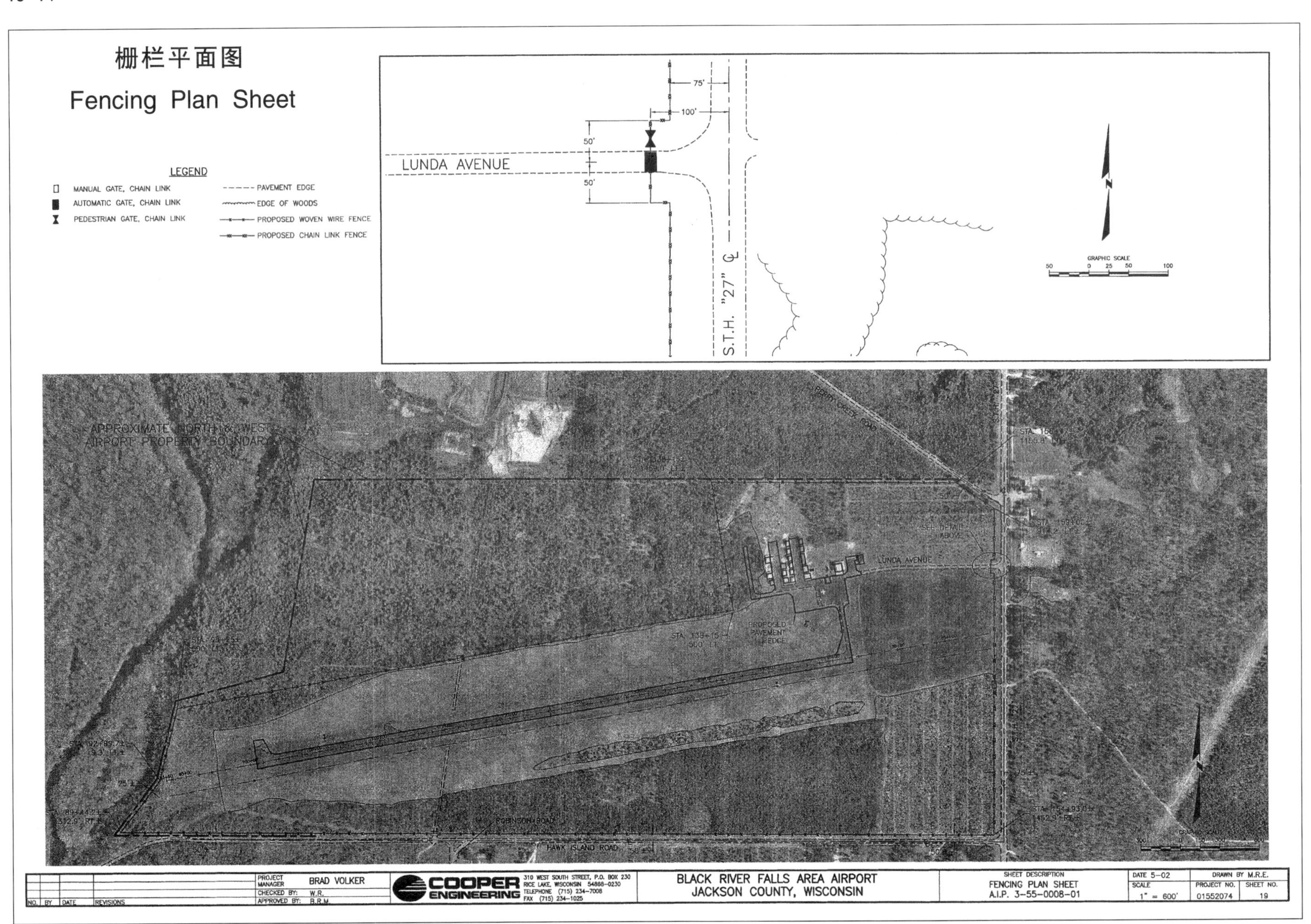

栅栏平面图
Fencing Plan Sheet
LEGEND
MANUAL GATE, CHAIN LINK
AUTOMATIC GATE, CHAIN LINK
PEDESTRIAN GATE, CHAIN LINK
PAVEMENT EDGE
EDGE OF WOODS
PROPOSED WOVEN WIRE FENCE
PROPOSED CHAIN LINK FENCE
LUNDA AVENUE
75'
100'
50'
50'
S.T.H. "27" ℄
GRAPHIC SCALE
50 0 25 50 100
APPROXIMATE NORTH & WEST AIRPORT PROPERTY BOUNDARY
SEE DETAIL ABOVE
LUNDA AVENUE
PROPOSED PAVEMENT EDGE
ROBINSON ROAD
HAWK ISLAND ROAD
PROJECT MANAGER BRAD VOLKER
CHECKED BY: W.R.
APPROVED BY: B.R.M.
NO. BY DATE REVISIONS
COOPER ENGINEERING
310 WEST SOUTH STREET, P.O. BOX 230
RICE LAKE, WISCONSIN 54868-0230
TELEPHONE (715) 234-7008
FAX (715) 234-1025
BLACK RIVER FALLS AREA AIRPORT
JACKSON COUNTY, WISCONSIN
SHEET DESCRIPTION
FENCING PLAN SHEET
A.I.P. 3-55-0008-01
DATE 5-02
DRAWN BY M.R.E.
SCALE 1" = 600'
PROJECT NO. 01552074
SHEET NO. 19

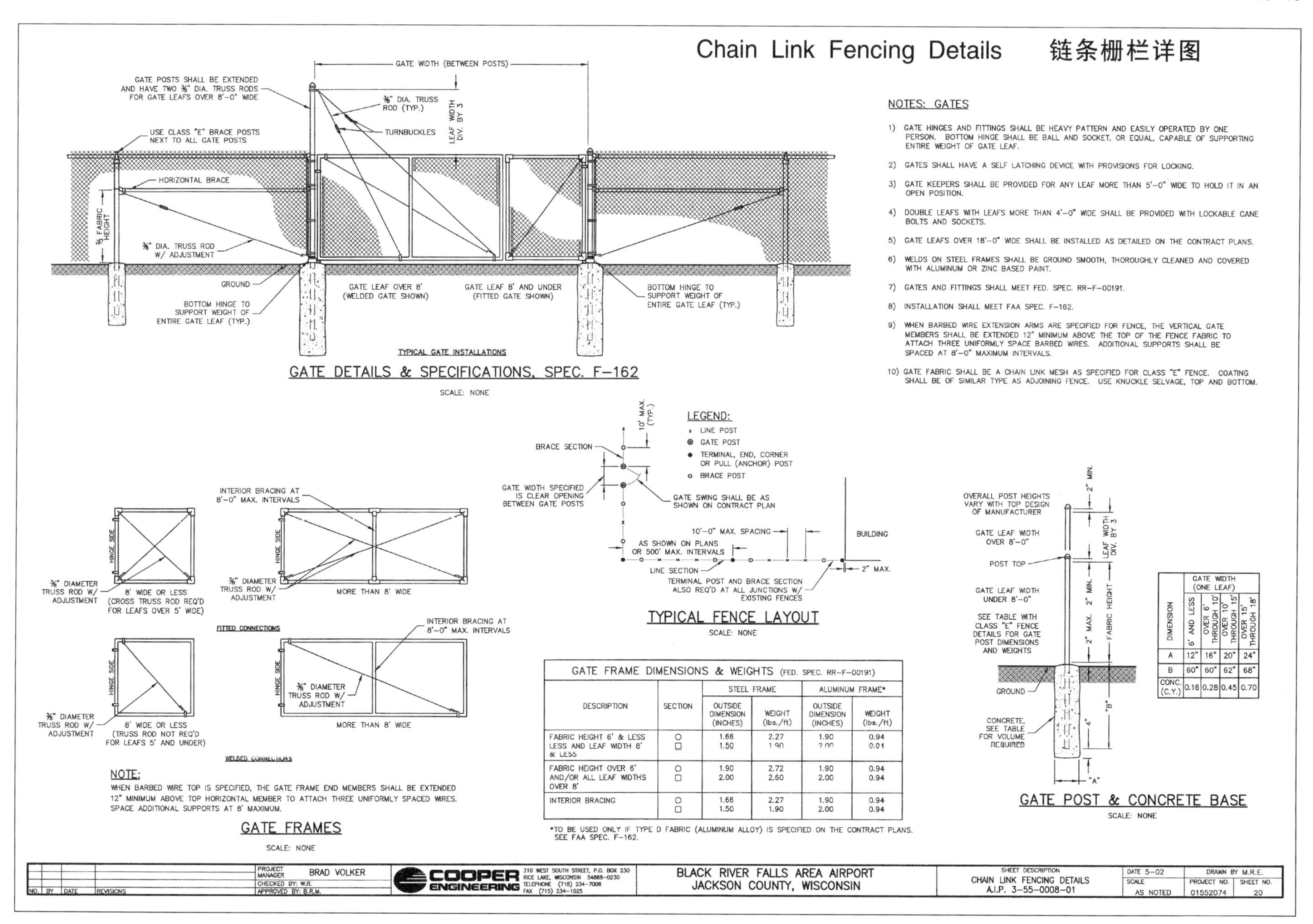

GATE FRAME DIMENSIONS & WEIGHTS (FED. SPEC. RR-F-00191)

DESCRIPTION	SECTION	STEEL FRAME		ALUMINUM FRAME*	
		OUTSIDE DIMENSION (INCHES)	WEIGHT (lbs./ft)	OUTSIDE DIMENSION (INCHES)	WEIGHT (lbs./ft)
FABRIC HEIGHT 6' & LESS LESS AND LEAF WIDTH 8' & LESS	○ □	1.66 1.50	2.27 1.90	1.90 2.00	0.94 0.94
FABRIC HEIGHT OVER 6' AND/OR ALL LEAF WIDTHS OVER 8'	○ □	1.90 2.00	2.72 2.60	1.90 2.00	0.94 0.94
INTERIOR BRACING	○ □	1.66 1.50	2.27 1.90	1.90 2.00	0.94 0.94

DIMENSION	GATE WIDTH (ONE LEAF)			
	6' AND LESS	OVER 6' THROUGH 10'	OVER 10' THROUGH 15'	OVER 15' THROUGH 18'
A	12"	16"	20"	24"
B	60"	60"	62"	68"
CONC. (C.Y.)	0.16	0.28	0.45	0.70

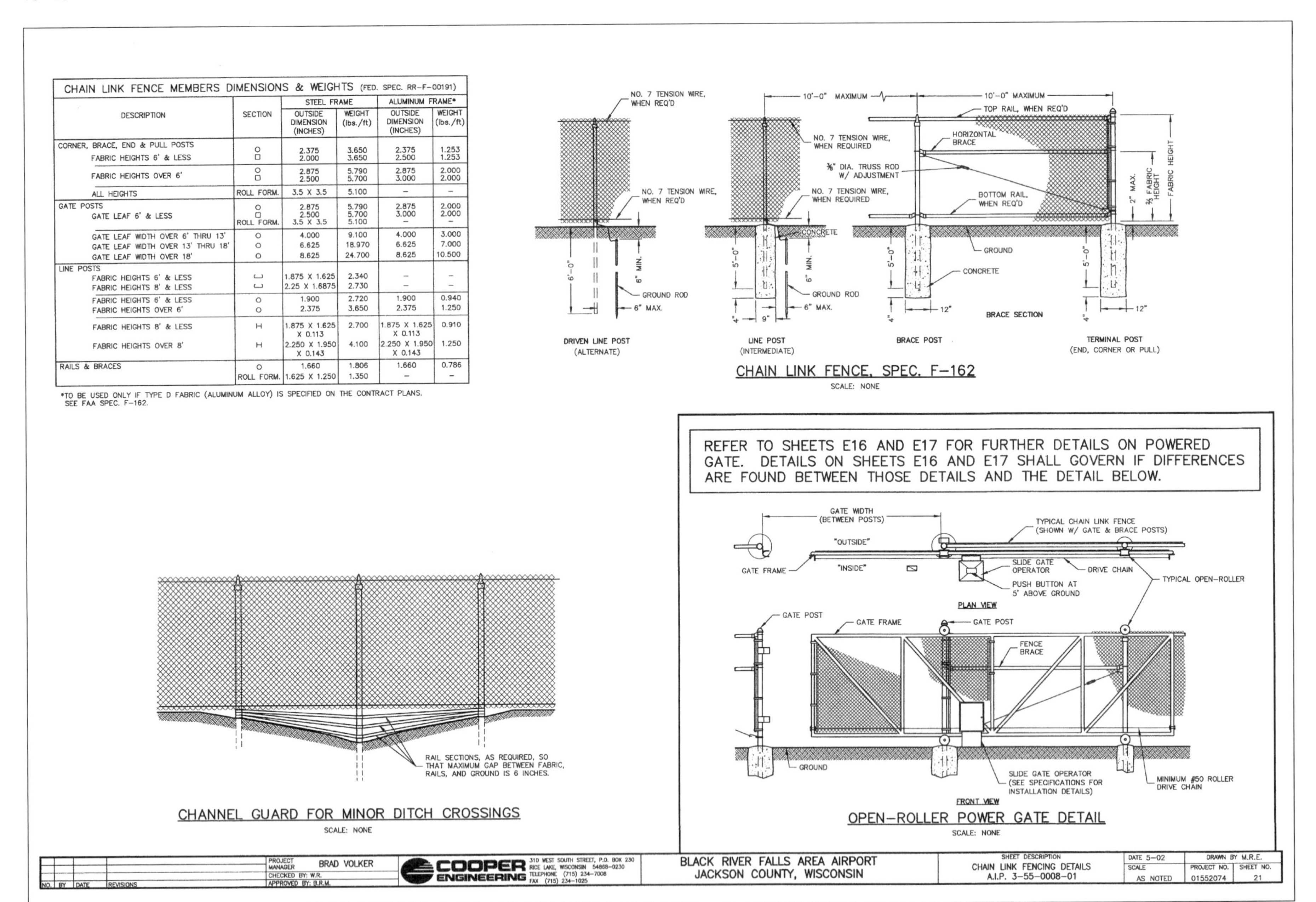

CHAIN LINK FENCE MEMBERS DIMENSIONS & WEIGHTS (FED. SPEC. RR-F-00191)

DESCRIPTION	SECTION	STEEL FRAME		ALUMINUM FRAME*	
		OUTSIDE DIMENSION (INCHES)	WEIGHT (lbs./ft)	OUTSIDE DIMENSION (INCHES)	WEIGHT (lbs./ft)
CORNER, BRACE, END & PULL POSTS FABRIC HEIGHTS 6' & LESS	○ □	2.375 2.000	3.650 3.650	2.375 2.500	1.253 1.253
FABRIC HEIGHTS OVER 6'	○ □	2.875 2.500	5.790 5.700	2.875 3.000	2.000 2.000
ALL HEIGHTS	ROLL FORM.	3.5 X 3.5	5.100	–	–
GATE POSTS GATE LEAF 6' & LESS	○ □ ROLL FORM.	2.875 2.500 3.5 X 3.5	5.790 5.700 5.100	2.875 3.000 –	2.000 2.000 –
GATE LEAF WIDTH OVER 6' THRU 13'	○	4.000	9.100	4.000	3.000
GATE LEAF WIDTH OVER 13' THRU 18'	○	6.625	18.970	6.625	7.000
GATE LEAF WIDTH OVER 18'	○	8.625	24.700	8.625	10.500
LINE POSTS FABRIC HEIGHTS 6' & LESS	⊔	1.875 X 1.625	2.340	–	–
FABRIC HEIGHTS 8' & LESS	⊔	2.25 X 1.6875	2.730	–	–
FABRIC HEIGHTS 6' & LESS	○	1.900	2.720	1.900	0.940
FABRIC HEIGHTS OVER 6'	○	2.375	3.650	2.375	1.250
FABRIC HEIGHTS 8' & LESS	H	1.875 X 1.625 X 0.113	2.700	1.875 X 1.625 X 0.113	0.910
FABRIC HEIGHTS OVER 8'	H	2.250 X 1.950 X 0.143	4.100	2.250 X 1.950 X 0.143	1.250
RAILS & BRACES	○ ROLL FORM.	1.660 1.625 X 1.250	1.806 1.350	1.660 –	0.786 –

*TO BE USED ONLY IF TYPE D FABRIC (ALUMINUM ALLOY) IS SPECIFIED ON THE CONTRACT PLANS. SEE FAA SPEC. F-162.

CHAIN LINK FENCE, SPEC. F-162

SCALE: NONE

OPEN-ROLLER POWER GATE DETAIL

SCALE: NONE

CHANNEL GUARD FOR MINOR DITCH CROSSINGS

SCALE: NONE

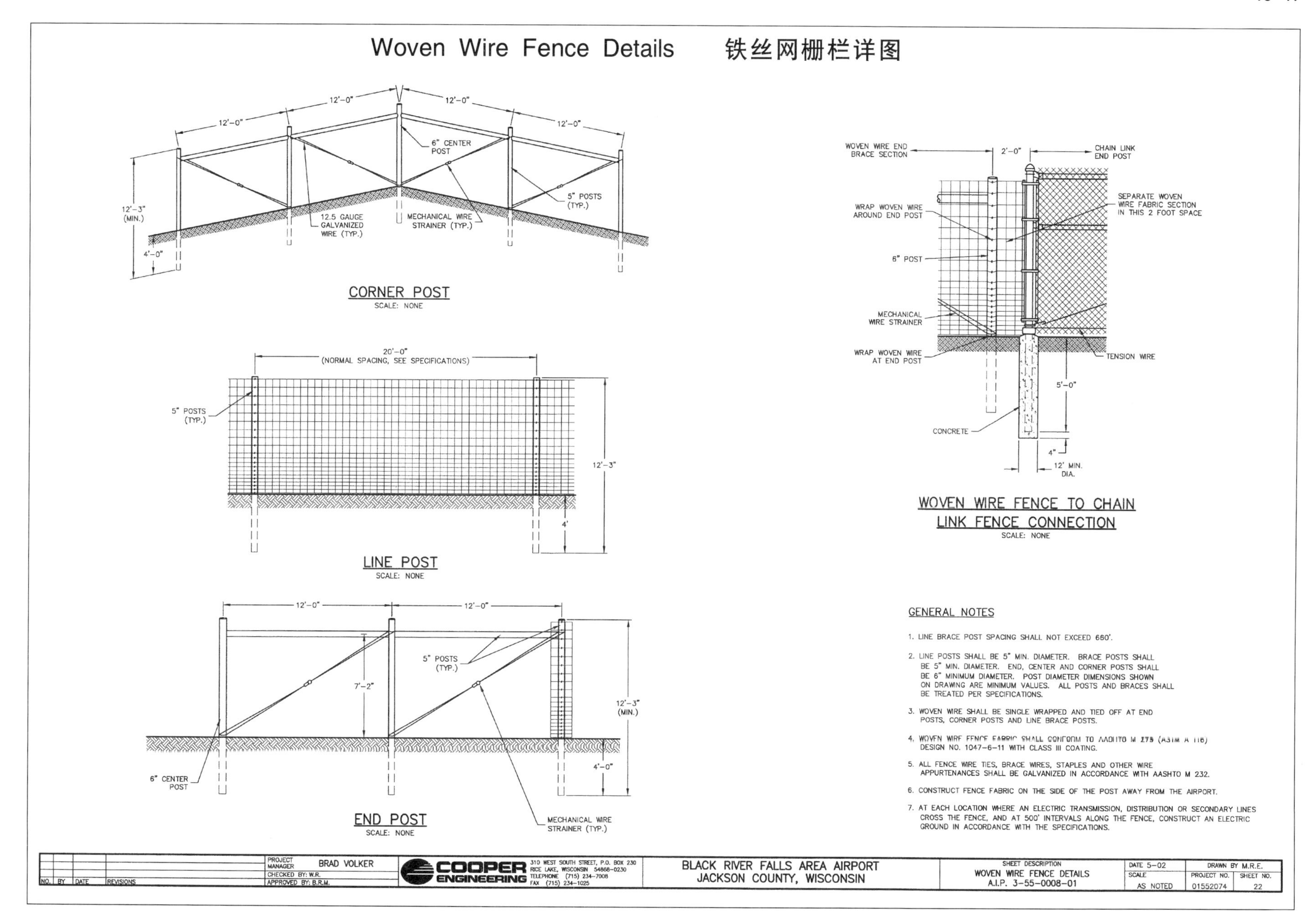
Woven Wire Fence Details　铁丝网栅栏详图
12'-0"
12'-0"
12'-0"
12'-0"
6" CENTER POST
5" POSTS (TYP.)
12'-3" (MIN.)
4'-0"
12.5 GAUGE GALVANIZED WIRE (TYP.)
MECHANICAL WIRE STRAINER (TYP.)
CORNER POST
SCALE: NONE
20'-0" (NORMAL SPACING, SEE SPECIFICATIONS)
5" POSTS (TYP.)
12'-3"
4'
LINE POST
SCALE: NONE
12'-0"
12'-0"
5" POSTS (TYP.)
7'-2"
12'-3" (MIN.)
4'-0"
6" CENTER POST
MECHANICAL WIRE STRAINER (TYP.)
END POST
SCALE: NONE
WOVEN WIRE END BRACE SECTION
2'-0"
CHAIN LINK END POST
SEPARATE WOVEN WIRE FABRIC SECTION IN THIS 2 FOOT SPACE
WRAP WOVEN WIRE AROUND END POST
6" POST
MECHANICAL WIRE STRAINER
WRAP WOVEN WIRE AT END POST
TENSION WIRE
5'-0"
CONCRETE
4"
12' MIN. DIA.
WOVEN WIRE FENCE TO CHAIN LINK FENCE CONNECTION
SCALE: NONE
GENERAL NOTES
1. LINE BRACE POST SPACING SHALL NOT EXCEED 660'.
2. LINE POSTS SHALL BE 5" MIN. DIAMETER. BRACE POSTS SHALL BE 5" MIN. DIAMETER. END, CENTER AND CORNER POSTS SHALL BE 6" MINIMUM DIAMETER. POST DIAMETER DIMENSIONS SHOWN ON DRAWING ARE MINIMUM VALUES. ALL POSTS AND BRACES SHALL BE TREATED PER SPECIFICATIONS.
3. WOVEN WIRE SHALL BE SINGLE WRAPPED AND TIED OFF AT END POSTS, CORNER POSTS AND LINE BRACE POSTS.
4. WOVEN WIRE FENCE FABRIC SHALL CONFORM TO AASHTO M 279 (ASTM A 116) DESIGN NO. 1047-6-11 WITH CLASS III COATING.
5. ALL FENCE WIRE TIES, BRACE WIRES, STAPLES AND OTHER WIRE APPURTENANCES SHALL BE GALVANIZED IN ACCORDANCE WITH AASHTO M 232.
6. CONSTRUCT FENCE FABRIC ON THE SIDE OF THE POST AWAY FROM THE AIRPORT.
7. AT EACH LOCATION WHERE AN ELECTRIC TRANSMISSION, DISTRIBUTION OR SECONDARY LINES CROSS THE FENCE, AND AT 500' INTERVALS ALONG THE FENCE, CONSTRUCT AN ELECTRIC GROUND IN ACCORDANCE WITH THE SPECIFICATIONS.
NO. BY DATE REVISIONS
PROJECT MANAGER BRAD VOLKER
CHECKED BY: W.R.
APPROVED BY: B.R.M.
COOPER ENGINEERING
310 WEST SOUTH STREET, P.O. BOX 230
RICE LAKE, WISCONSIN 54868-0230
TELEPHONE (715) 234-7008
FAX (715) 234-1025
BLACK RIVER FALLS AREA AIRPORT
JACKSON COUNTY, WISCONSIN
SHEET DESCRIPTION
WOVEN WIRE FENCE DETAILS
A.I.P. 3-55-0008-01
DATE 5-02
SCALE AS NOTED
DRAWN BY M.R.E.
PROJECT NO. 01552074
SHEET NO. 22

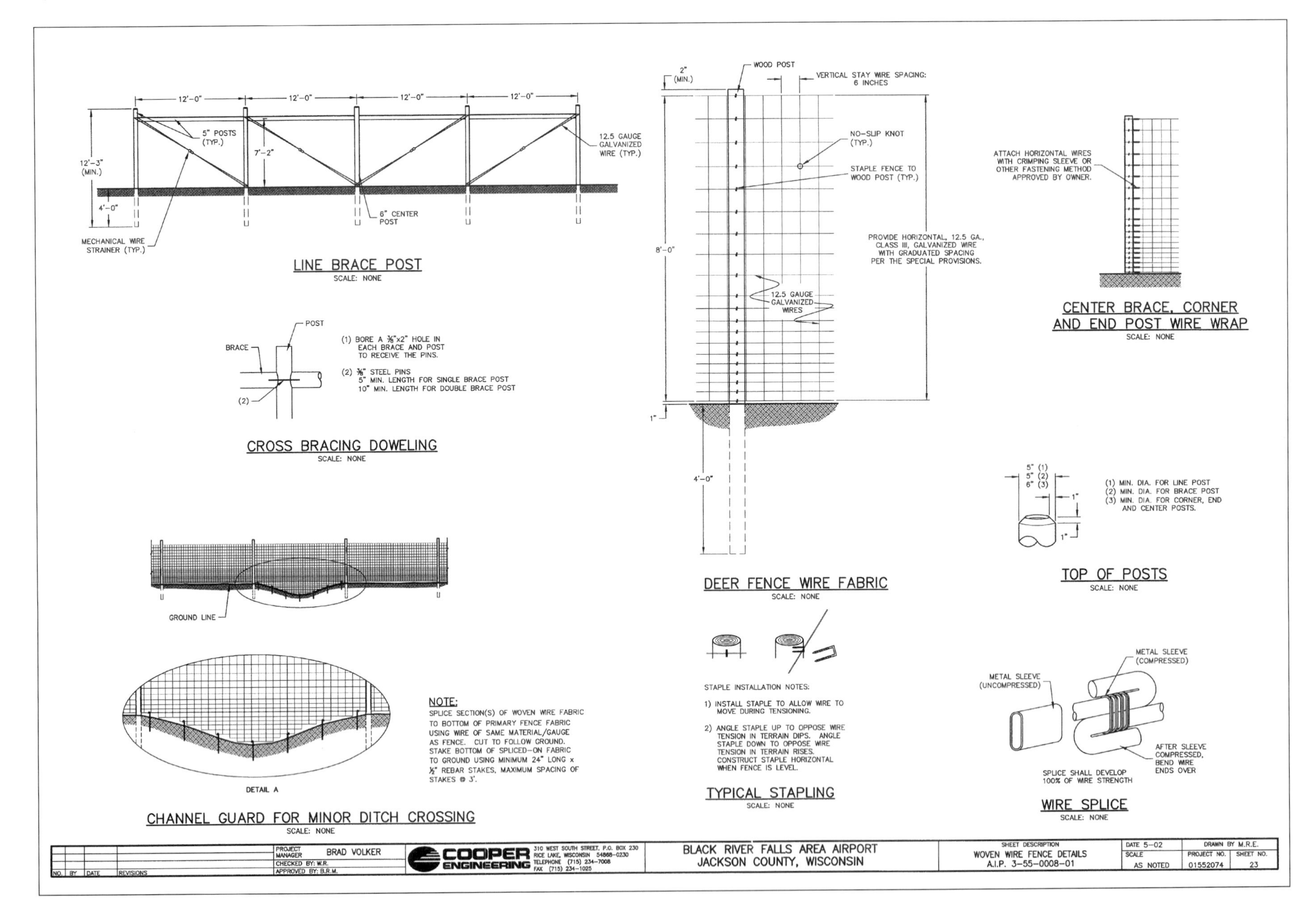
LINE BRACE POST
SCALE: NONE
12'-0"
5" POSTS (TYP.)
12.5 GAUGE GALVANIZED WIRE (TYP.)
12'-3" (MIN.)
7'-2"
4'-0"
6" CENTER POST
MECHANICAL WIRE STRAINER (TYP.)
CROSS BRACING DOWELING
SCALE: NONE
POST
BRACE
(2)
(1) BORE A ⅜"x2" HOLE IN EACH BRACE AND POST TO RECEIVE THE PINS.
(2) ⅜" STEEL PINS 5" MIN. LENGTH FOR SINGLE BRACE POST 10" MIN. LENGTH FOR DOUBLE BRACE POST
GROUND LINE
DETAIL A
CHANNEL GUARD FOR MINOR DITCH CROSSING
SCALE: NONE
NOTE:
SPLICE SECTION(S) OF WOVEN WIRE FABRIC TO BOTTOM OF PRIMARY FENCE FABRIC USING WIRE OF SAME MATERIAL/GAUGE AS FENCE. CUT TO FOLLOW GROUND. STAKE BOTTOM OF SPLICED-ON FABRIC TO GROUND USING MINIMUM 24" LONG x ½" REBAR STAKES, MAXIMUM SPACING OF STAKES @ 3'.
WOOD POST
2" (MIN.)
VERTICAL STAY WIRE SPACING: 6 INCHES
NO-SLIP KNOT (TYP.)
STAPLE FENCE TO WOOD POST (TYP.)
PROVIDE HORIZONTAL 12.5 GA., CLASS III, GALVANIZED WIRE WITH GRADUATED SPACING PER THE SPECIAL PROVISIONS.
8'-0"
12.5 GAUGE GALVANIZED WIRES
1"
4'-0"
DEER FENCE WIRE FABRIC
SCALE: NONE
STAPLE INSTALLATION NOTES:
1) INSTALL STAPLE TO ALLOW WIRE TO MOVE DURING TENSIONING.
2) ANGLE STAPLE UP TO OPPOSE WIRE TENSION IN TERRAIN DIPS. ANGLE STAPLE DOWN TO OPPOSE WIRE TENSION IN TERRAIN RISES. CONSTRUCT STAPLE HORIZONTAL WHEN FENCE IS LEVEL.
TYPICAL STAPLING
SCALE: NONE
ATTACH HORIZONTAL WIRES WITH CRIMPING SLEEVE OR OTHER FASTENING METHOD APPROVED BY OWNER.
CENTER BRACE, CORNER AND END POST WIRE WRAP
SCALE: NONE
5" (1)
5" (2)
6" (3)
1"
(1) MIN. DIA. FOR LINE POST
(2) MIN. DIA. FOR BRACE POST
(3) MIN. DIA. FOR CORNER, END AND CENTER POSTS.
TOP OF POSTS
SCALE: NONE
METAL SLEEVE (UNCOMPRESSED)
METAL SLEEVE (COMPRESSED)
AFTER SLEEVE COMPRESSED, BEND WIRE ENDS OVER
SPLICE SHALL DEVELOP 100% OF WIRE STRENGTH
WIRE SPLICE
SCALE: NONE
NO. BY DATE REVISIONS
PROJECT MANAGER BRAD VOLKER
CHECKED BY: W.R.
APPROVED BY: B.R.M.
COOPER ENGINEERING
310 WEST SOUTH STREET, P.O. BOX 230 RICE LAKE, WISCONSIN 54868-0230
TELEPHONE (715) 234-7008
FAX (715) 234-1025
BLACK RIVER FALLS AREA AIRPORT
JACKSON COUNTY, WISCONSIN
SHEET DESCRIPTION
WOVEN WIRE FENCE DETAILS
A.I.P. 3-55-0008-01
DATE 5-02
DRAWN BY M.R.E.
SCALE AS NOTED
PROJECT NO. 01552074
SHEET NO. 23

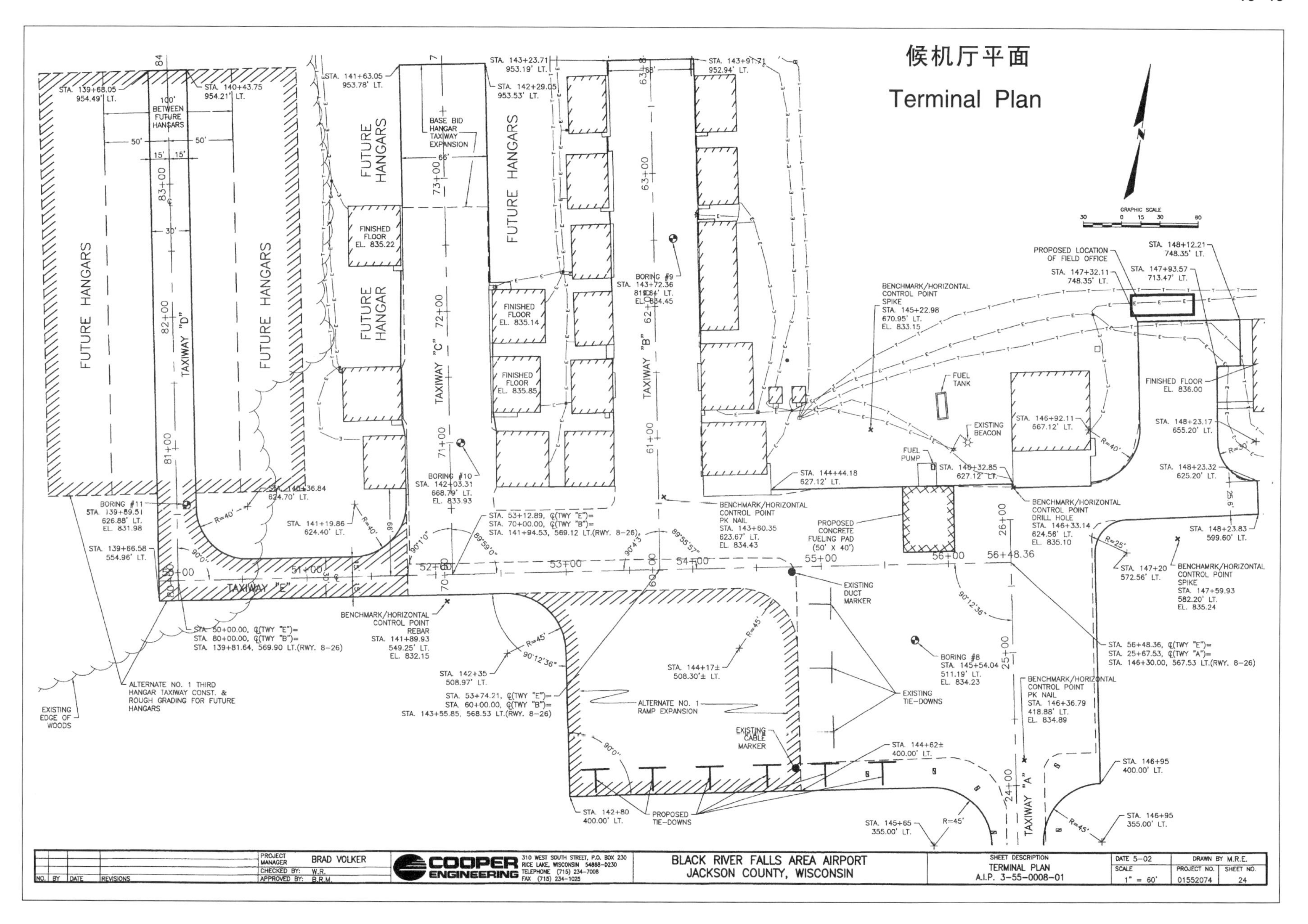

候机厅平面
Terminal Plan
FUTURE HANGARS
FUTURE HANGAR
TAXIWAY "D"
TAXIWAY "C"
TAXIWAY "B"
TAXIWAY "E"
TAXIWAY "A"
FINISHED FLOOR
BASE BID HANGAR TAXIWAY EXPANSION
PROPOSED LOCATION OF FIELD OFFICE
FUEL TANK
FUEL PUMP
EXISTING BEACON
PROPOSED CONCRETE FUELING PAD (50' X 40')
EXISTING DUCT MARKER
EXISTING TIE-DOWNS
EXISTING CABLE MARKER
ALTERNATE NO. 1 RAMP EXPANSION
PROPOSED TIE-DOWNS
ALTERNATE NO. 1 THIRD HANGAR TAXIWAY CONST. & ROUGH GRADING FOR FUTURE HANGARS
EXISTING EDGE OF WOODS
BLACK RIVER FALLS AREA AIRPORT
JACKSON COUNTY, WISCONSIN
COOPER ENGINEERING
TERMINAL PLAN
A.I.P. 3-55-0008-01

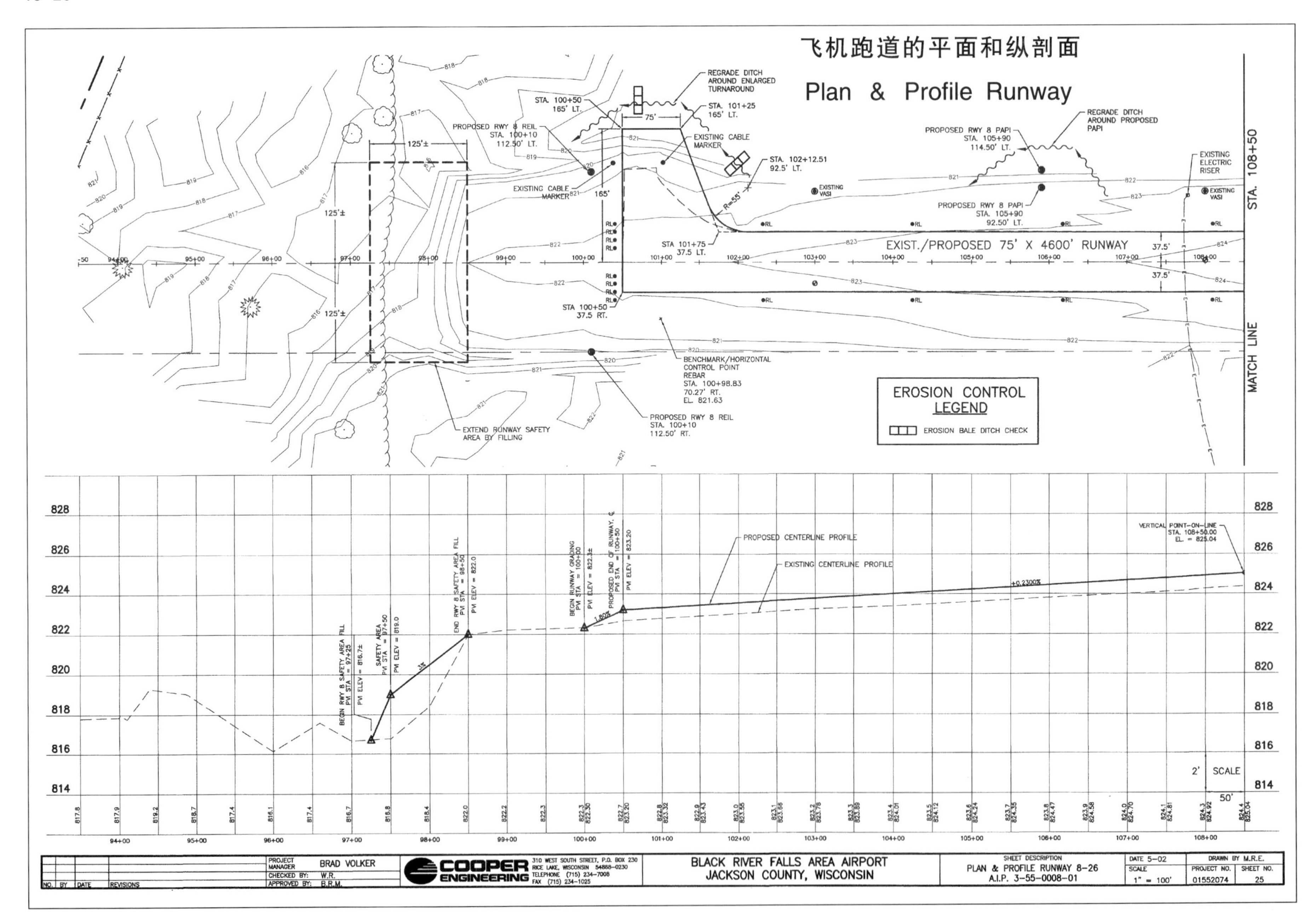
飞机跑道的平面和纵剖面
Plan & Profile Runway
REGRADE DITCH AROUND ENLARGED TURNAROUND
STA. 100+50 165' LT.
STA. 101+25 165' LT.
PROPOSED RWY 8 REIL STA. 100+10 112.50' LT.
EXISTING CABLE MARKER
STA. 102+12.51 92.5' LT.
PROPOSED RWY 8 PAPI STA. 105+90 114.50' LT.
REGRADE DITCH AROUND PROPOSED PAPI
PROPOSED RWY 8 PAPI STA. 105+90 92.50' LT.
EXISTING ELECTRIC RISER
EXISTING VASI
STA 101+75 37.5 LT.
EXIST./PROPOSED 75' X 4600' RUNWAY
STA 100+50 37.5 RT.
BENCHMARK/HORIZONTAL CONTROL POINT REBAR STA. 100+98.83 70.27' RT. EL. 821.63
PROPOSED RWY 8 REIL STA. 100+10 112.50' RT.
EXTEND RUNWAY SAFETY AREA BY FILLING
EROSION CONTROL LEGEND
EROSION BALE DITCH CHECK
STA. 108+50
MATCH LINE
PROPOSED CENTERLINE PROFILE
EXISTING CENTERLINE PROFILE
VERTICAL POINT-ON-LINE STA. 108+50.00 EL. = 825.04
BEGIN RWY 8 SAFETY AREA FILL PVI STA = 97+25 PVI ELEV = 816.7±
SAFETY AREA PVI STA = 97+50 PVI ELEV = 819.0
END RWY 8 SAFETY AREA FILL PVI STA = 98+50 PVI ELEV = 822.0
BEGIN RUNWAY GRADING PVI STA = 100+00 PVI ELEV = 822.3±
PROPOSED END OF RUNWAY ℄ PVI STA = 100+50 PVI ELEV = 823.20
+0.2300%
2' SCALE 50'
PROJECT MANAGER BRAD VOLKER
CHECKED BY: W.R.
APPROVED BY: B.R.M.
COOPER ENGINEERING
310 WEST SOUTH STREET, P.O. BOX 230 RICE LAKE, WISCONSIN 54868-0230 TELEPHONE (715) 234-7008 FAX (715) 234-1025
BLACK RIVER FALLS AREA AIRPORT JACKSON COUNTY, WISCONSIN
SHEET DESCRIPTION PLAN & PROFILE RUNWAY 8-26 A.I.P. 3-55-0008-01
DATE 5-02
DRAWN BY M.R.E.
SCALE 1" = 100'
PROJECT NO. 01552074
SHEET NO. 25
NO. BY DATE REVISIONS

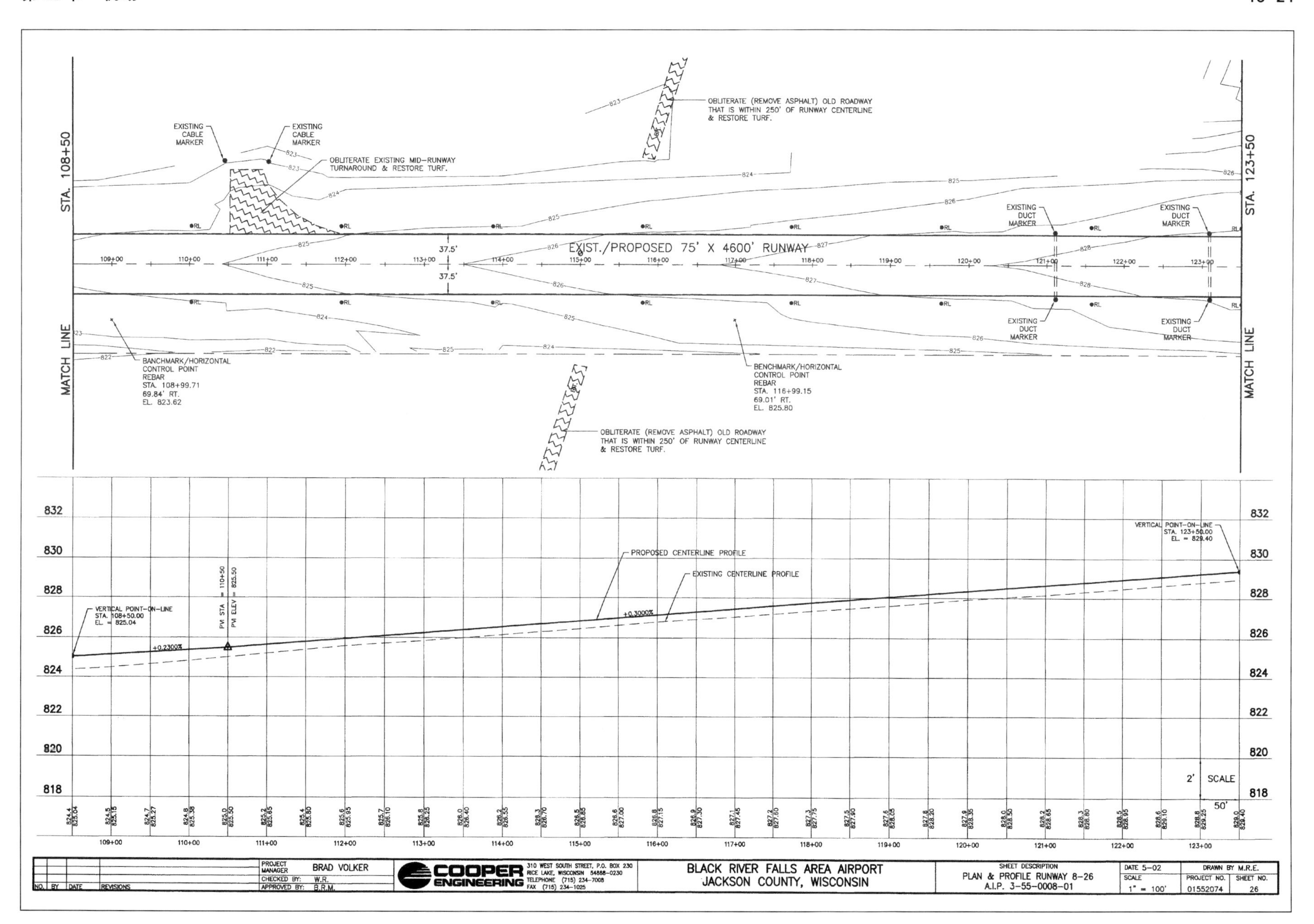
EXIST./PROPOSED 75' X 4600' RUNWAY
STA. 108+50
MATCH LINE
STA. 123+50
MATCH LINE
EXISTING CABLE MARKER
EXISTING CABLE MARKER
OBLITERATE EXISTING MID-RUNWAY TURNAROUND & RESTORE TURF.
OBLITERATE (REMOVE ASPHALT) OLD ROADWAY THAT IS WITHIN 250' OF RUNWAY CENTERLINE & RESTORE TURF.
EXISTING DUCT MARKER
BENCHMARK/HORIZONTAL CONTROL POINT REBAR STA. 108+99.71 69.84' RT. EL. 823.62
BENCHMARK/HORIZONTAL CONTROL POINT REBAR STA. 116+99.15 69.01' RT. EL. 825.80
37.5'
VERTICAL POINT-ON-LINE STA. 108+50.00 EL. = 825.04
PVI STA = 110+50
PVI ELEV = 825.50
+0.2300%
PROPOSED CENTERLINE PROFILE
EXISTING CENTERLINE PROFILE
+0.3000%
VERTICAL POINT-ON-LINE STA. 123+50.00 EL. = 829.40
2' SCALE 50'
PROJECT MANAGER BRAD VOLKER
CHECKED BY: W.R.
APPROVED BY: B.R.M.
NO. BY DATE REVISIONS
COOPER ENGINEERING
310 WEST SOUTH STREET, P.O. BOX 230
RICE LAKE, WISCONSIN 54868-0230
TELEPHONE (715) 234-7008
FAX (715) 234-1025
BLACK RIVER FALLS AREA AIRPORT
JACKSON COUNTY, WISCONSIN
SHEET DESCRIPTION
PLAN & PROFILE RUNWAY 8-26
A.I.P. 3-55-0008-01
DATE 5-02
SCALE 1" = 100'
DRAWN BY M.R.E.
PROJECT NO. 01552074
SHEET NO. 26

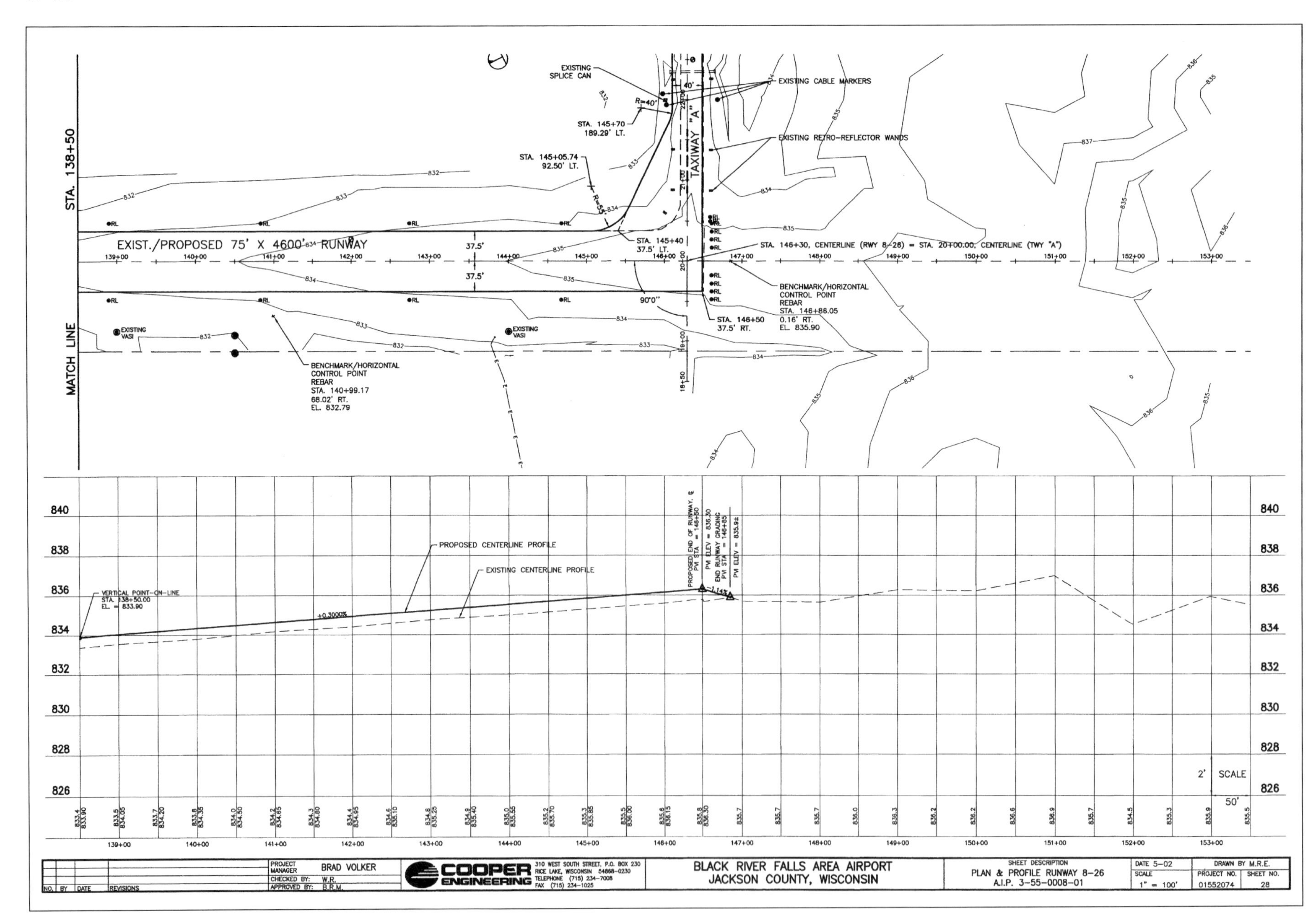
EXISTING SPLICE CAN
EXISTING CABLE MARKERS
EXISTING RETRO-REFLECTOR WANDS
STA. 145+70 189.29' LT.
STA. 145+05.74 92.50' LT.
TAXIWAY "A"
STA. 138+50
EXIST./PROPOSED 75' X 4600' RUNWAY
STA. 145+40 37.5' LT.
STA. 146+30, CENTERLINE (RWY 8-26) = STA. 20+00.00, CENTERLINE (TWY "A")
BENCHMARK/HORIZONTAL CONTROL POINT REBAR STA. 146+86.05 0.16' RT. EL. 835.90
STA. 146+50 37.5' RT.
MATCH LINE
EXISTING VASI
BENCHMARK/HORIZONTAL CONTROL POINT REBAR STA. 140+99.17 68.02' RT. EL. 832.79
PROPOSED CENTERLINE PROFILE
EXISTING CENTERLINE PROFILE
VERTICAL POINT-ON-LINE STA. 138+50.00 EL. = 833.90
+0.3000%
PROPOSED END OF RUNWAY PVI STA = 146+50 PVI ELEV = 836.30
END RUNWAY GRADING PVI STA = 146+85 PVI ELEV = 835.9±
SCALE
PROJECT MANAGER BRAD VOLKER
CHECKED BY: W.R.
APPROVED BY: B.R.M.
COOPER ENGINEERING
310 WEST SOUTH STREET, P.O. BOX 230 RICE LAKE, WISCONSIN 54868-0230 TELEPHONE (715) 234-7008 FAX (715) 234-1025
BLACK RIVER FALLS AREA AIRPORT
JACKSON COUNTY, WISCONSIN
SHEET DESCRIPTION PLAN & PROFILE RUNWAY 8-26 A.I.P. 3-55-0008-01
DATE 5-02
SCALE 1" = 100'
DRAWN BY M.R.E.
PROJECT NO. 01552074
SHEET NO. 28
NO. BY DATE REVISIONS

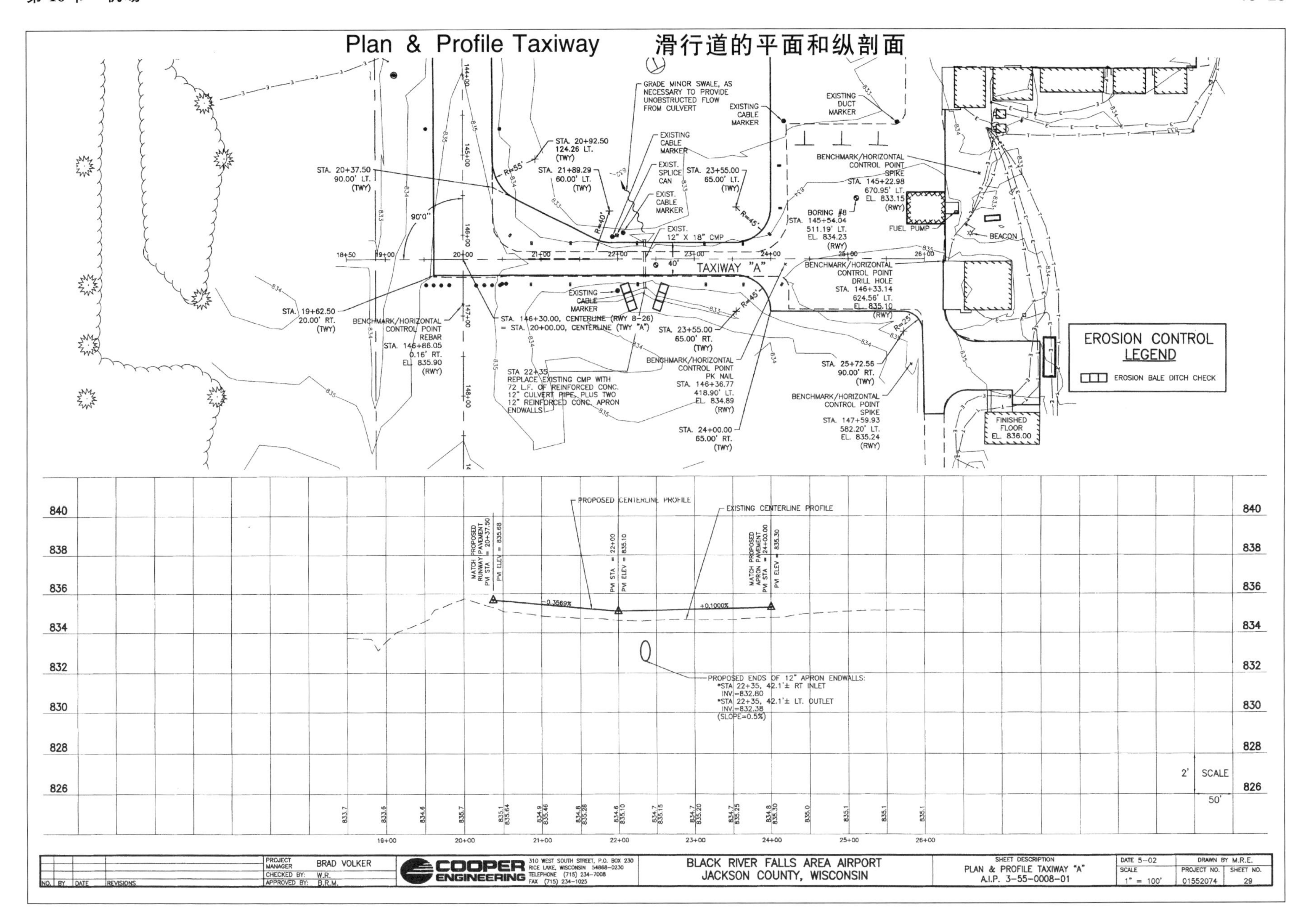
Plan & Profile Taxiway　滑行道的平面和纵剖面
TAXIWAY "A"
EROSION CONTROL LEGEND
EROSION BALE DITCH CHECK
PROPOSED CENTERLINE PROFILE
EXISTING CENTERLINE PROFILE
PROPOSED ENDS OF 12" APRON ENDWALLS:
*STA 22+35, 42.1'± RT INLET INV=832.80
*STA 22+35, 42.1'± LT. OUTLET INV=832.38
(SLOPE=0.5%)
BLACK RIVER FALLS AREA AIRPORT
JACKSON COUNTY, WISCONSIN
PLAN & PROFILE TAXIWAY "A"
A.I.P. 3-55-0008-01
COOPER ENGINEERING

飞机跑道剖面图

Runway Cross Sections

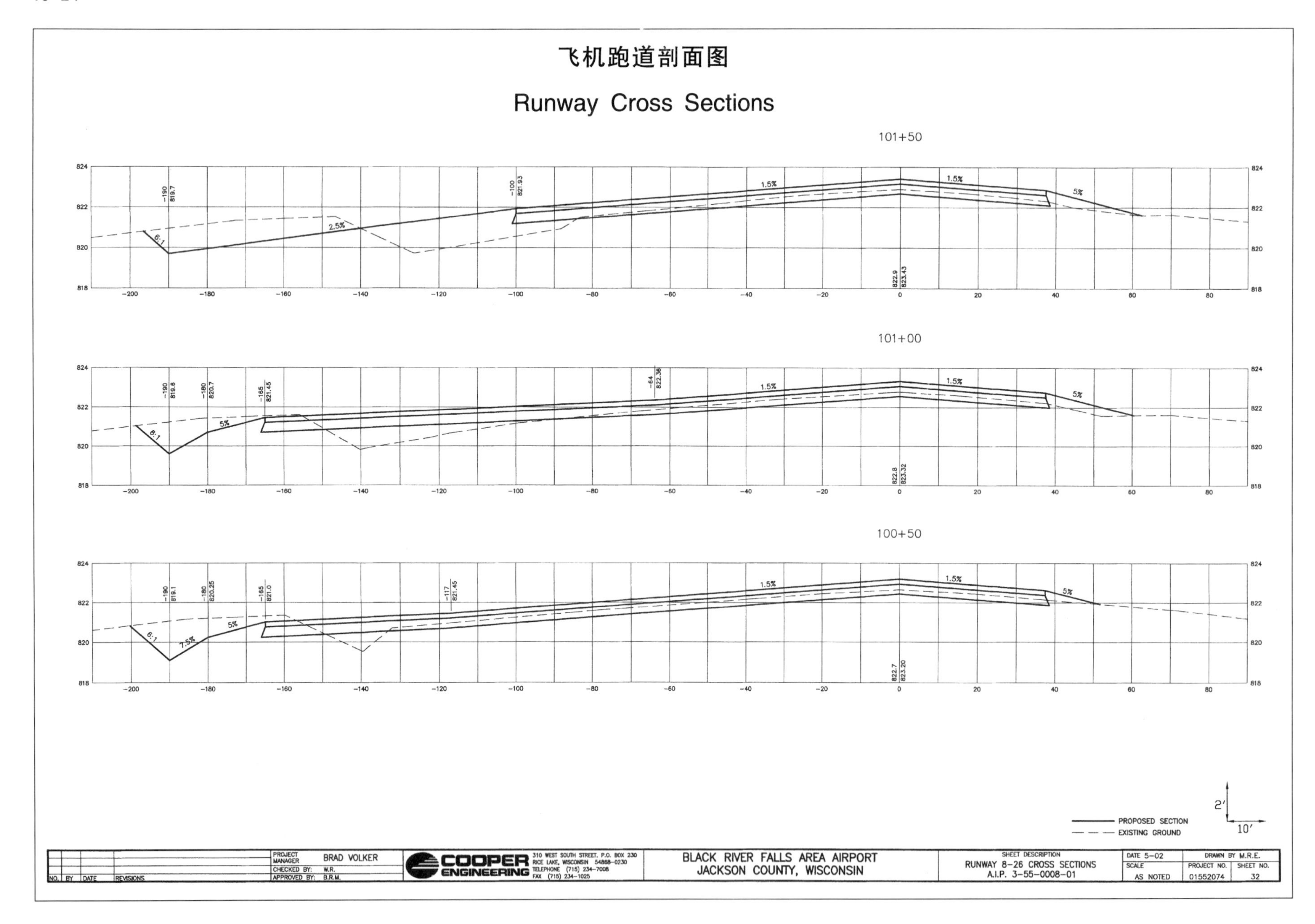

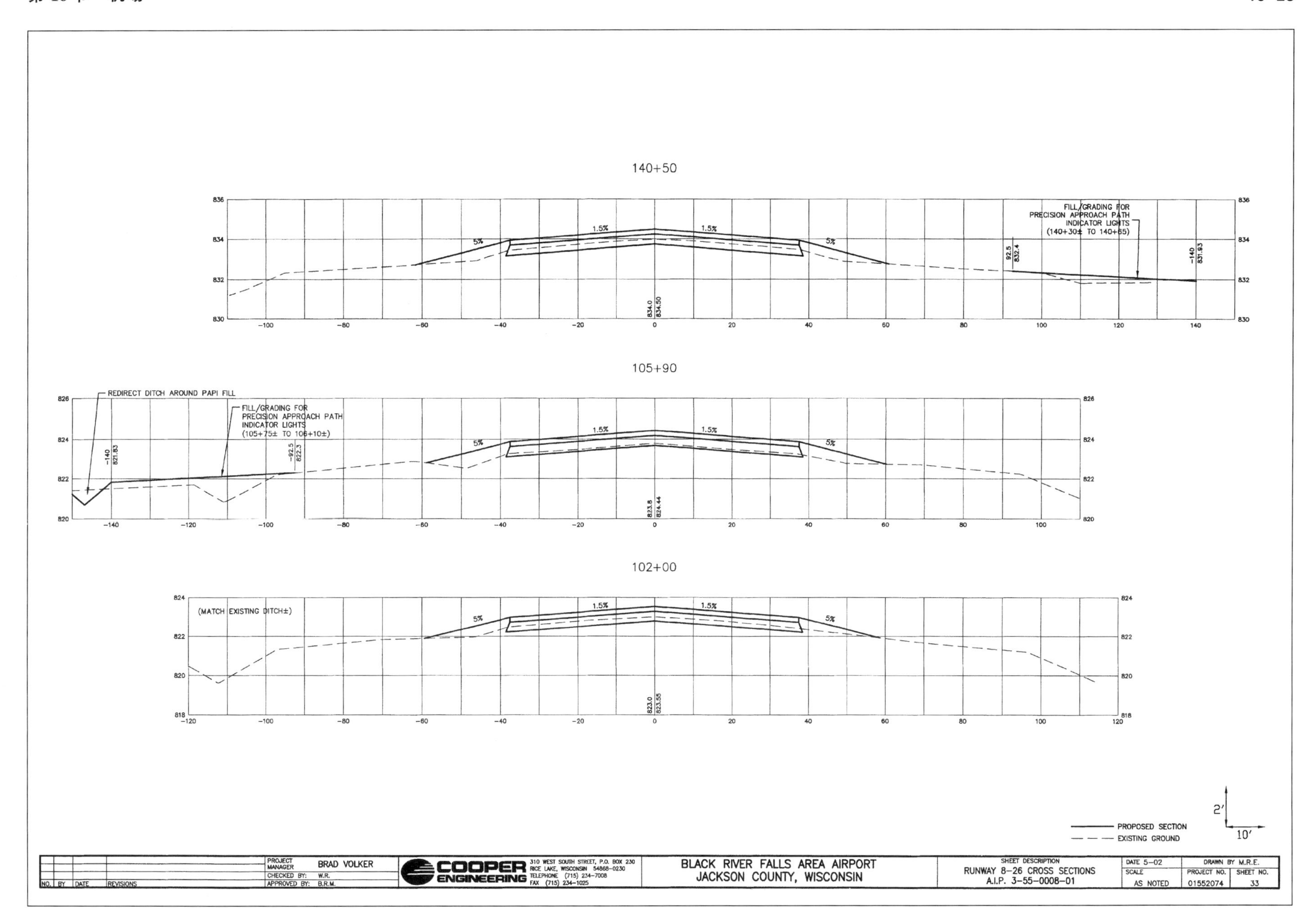

140+50
FILL/GRADING FOR PRECISION APPROACH PATH INDICATOR LIGHTS (140+30± TO 140+65)
1.5%
1.5%
5%
5%
105+90
REDIRECT DITCH AROUND PAPI FILL
FILL/GRADING FOR PRECISION APPROACH PATH INDICATOR LIGHTS (105+75± TO 106+10±)
102+00
(MATCH EXISTING DITCH±)
PROPOSED SECTION
EXISTING GROUND
2′
10′
PROJECT MANAGER BRAD VOLKER
CHECKED BY: W.R.
APPROVED BY: B.R.M.
NO. BY DATE REVISIONS
COOPER ENGINEERING
310 WEST SOUTH STREET, P.O. BOX 230
RICE LAKE, WISCONSIN 54868-0230
TELEPHONE (715) 234-7008
FAX (715) 234-1025
BLACK RIVER FALLS AREA AIRPORT
JACKSON COUNTY, WISCONSIN
SHEET DESCRIPTION
RUNWAY 8-26 CROSS SECTIONS
A.I.P. 3-55-0008-01
DATE 5-02
DRAWN BY M.R.E.
SCALE AS NOTED
PROJECT NO. 01552074
SHEET NO. 33

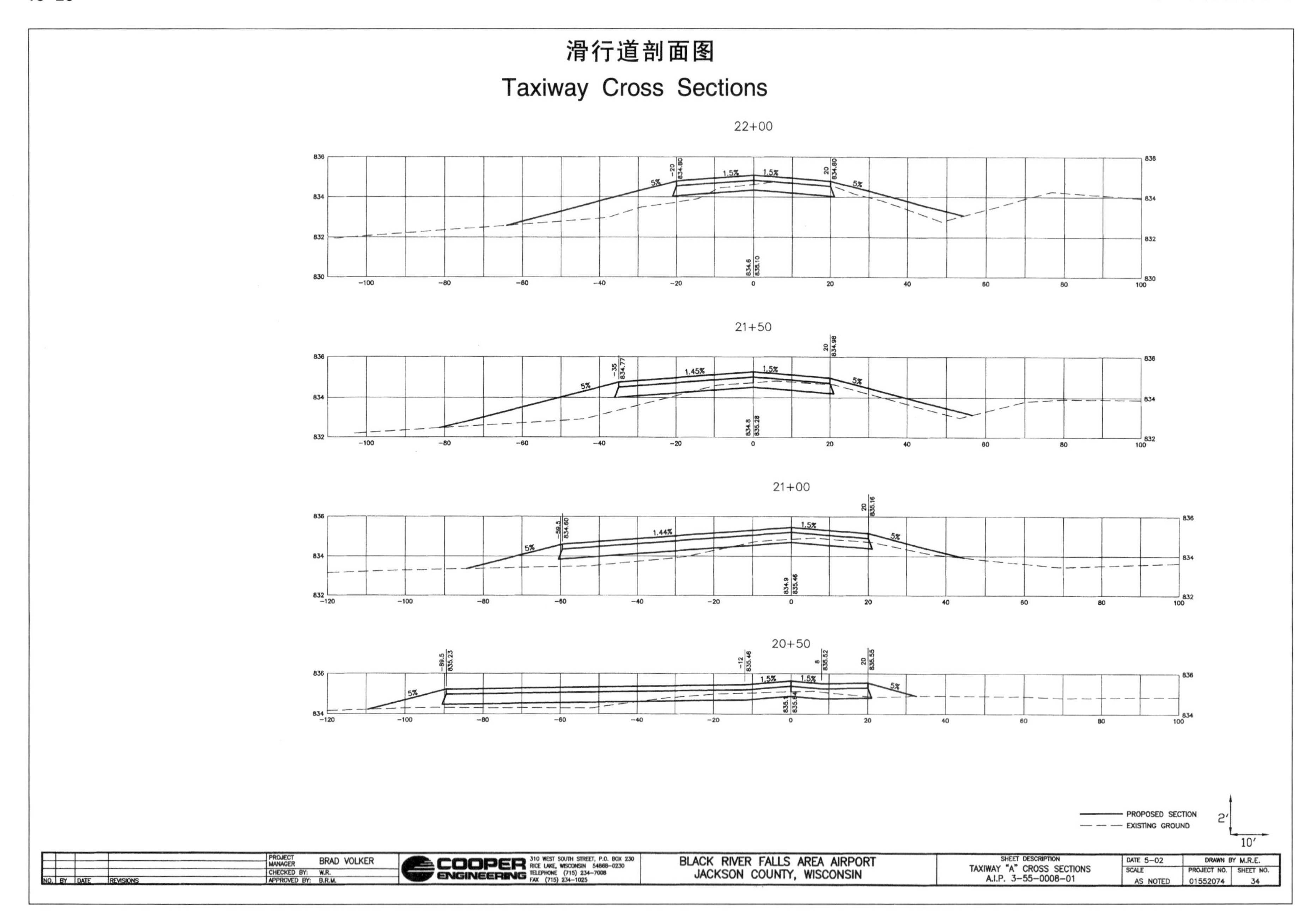

滑行道剖面图
Taxiway Cross Sections
22+00
21+50
21+00
20+50
PROPOSED SECTION
EXISTING GROUND
PROJECT MANAGER BRAD VOLKER
CHECKED BY: W.R.
APPROVED BY: B.R.M.
COOPER ENGINEERING
310 WEST SOUTH STREET, P.O. BOX 230
RICE LAKE, WISCONSIN 54868-0230
TELEPHONE (715) 234-7008
FAX (715) 234-1025
BLACK RIVER FALLS AREA AIRPORT
JACKSON COUNTY, WISCONSIN
SHEET DESCRIPTION
TAXIWAY "A" CROSS SECTIONS
A.I.P. 3-55-0008-01
DATE 5-02
DRAWN BY M.R.E.
SCALE AS NOTED
PROJECT NO. 01552074
SHEET NO. 34

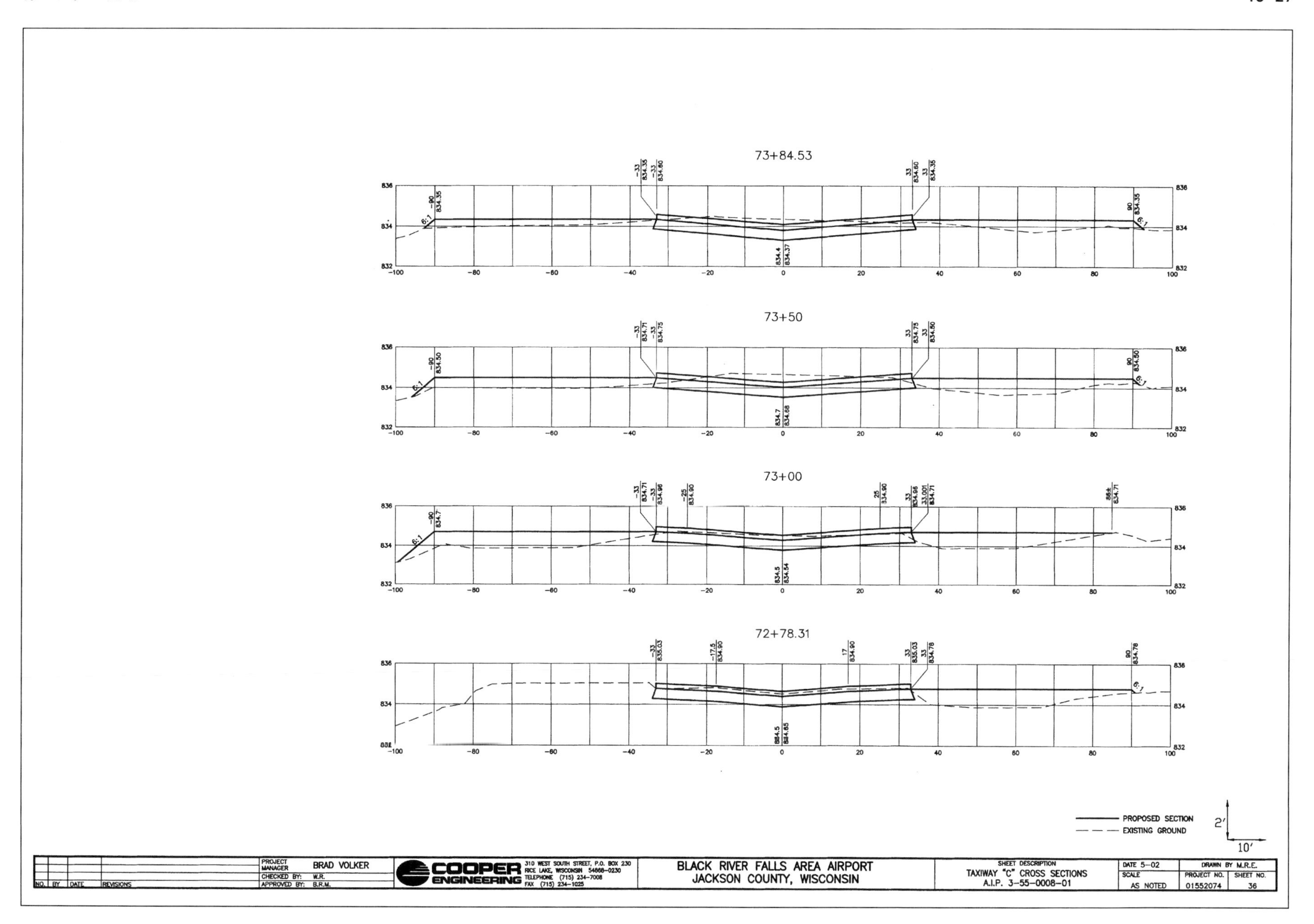
73+84.53
73+50
73+00
72+78.31
PROPOSED SECTION
EXISTING GROUND
2′
10′
PROJECT MANAGER BRAD VOLKER
CHECKED BY: W.R.
APPROVED BY: B.R.M.
NO. BY DATE REVISIONS
COOPER ENGINEERING
310 WEST SOUTH STREET, P.O. BOX 230
RICE LAKE, WISCONSIN 54868-0230
TELEPHONE (715) 234-7008
FAX (715) 234-1025
BLACK RIVER FALLS AREA AIRPORT
JACKSON COUNTY, WISCONSIN
SHEET DESCRIPTION
TAXIWAY "C" CROSS SECTIONS
A.I.P. 3-55-0008-01
DATE 5-02
DRAWN BY M.R.E.
SCALE AS NOTED
PROJECT NO. 01552074
SHEET NO. 36

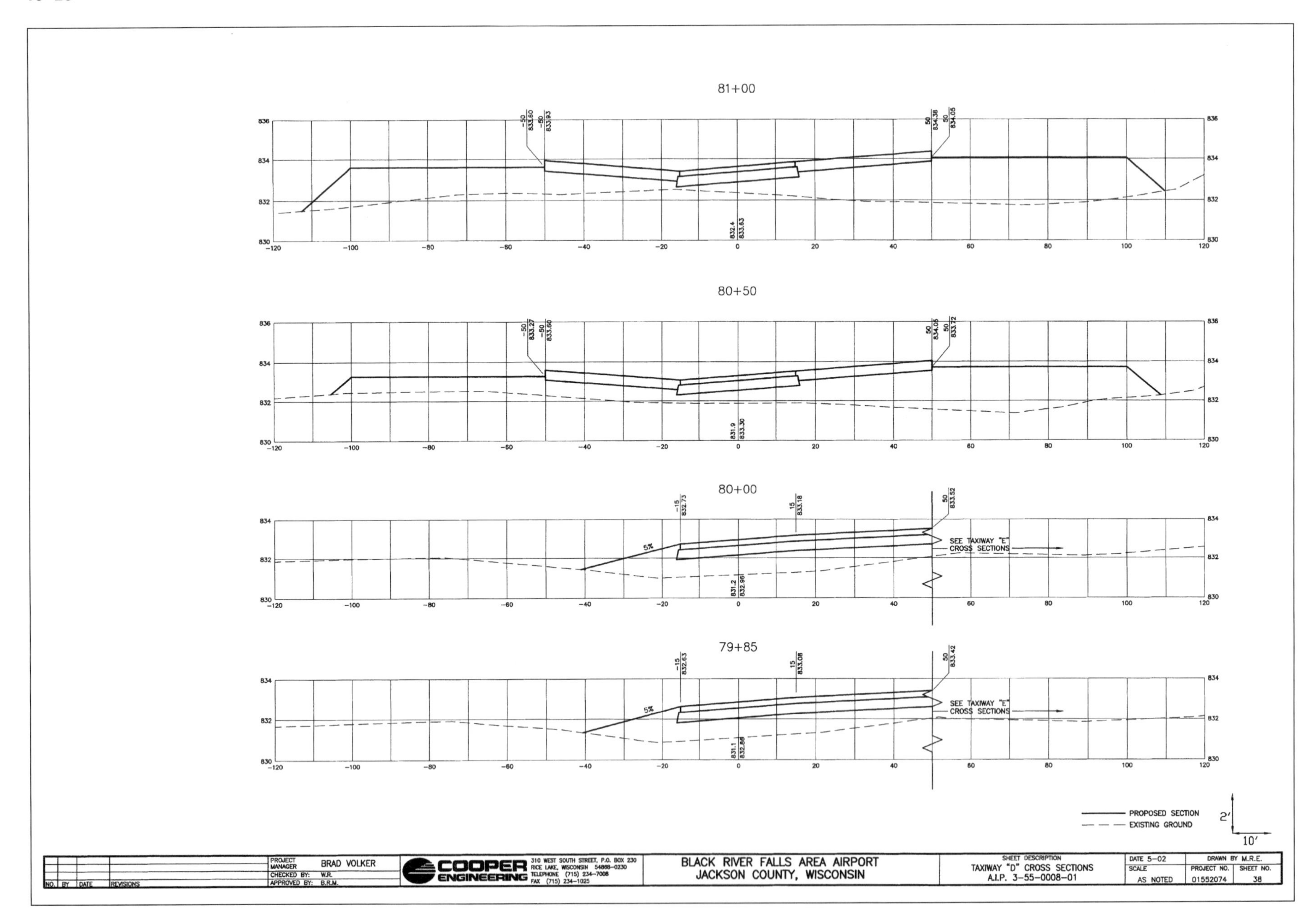
81+00
80+50
80+00
79+85
5%
SEE TAXIWAY "E" CROSS SECTIONS
PROPOSED SECTION
EXISTING GROUND
2′
10′
PROJECT MANAGER BRAD VOLKER
CHECKED BY: W.R.
APPROVED BY: B.R.M.
NO. BY DATE REVISIONS
COOPER ENGINEERING
310 WEST SOUTH STREET, P.O. BOX 230
RICE LAKE, WISCONSIN 54868-0230
TELEPHONE (715) 234-7008
FAX (715) 234-1025
BLACK RIVER FALLS AREA AIRPORT
JACKSON COUNTY, WISCONSIN
SHEET DESCRIPTION
TAXIWAY "D" CROSS SECTIONS
A.I.P. 3-55-0008-01
DATE 5-02
DRAWN BY M.R.E.
SCALE AS NOTED
PROJECT NO. 01552074
SHEET NO. 38

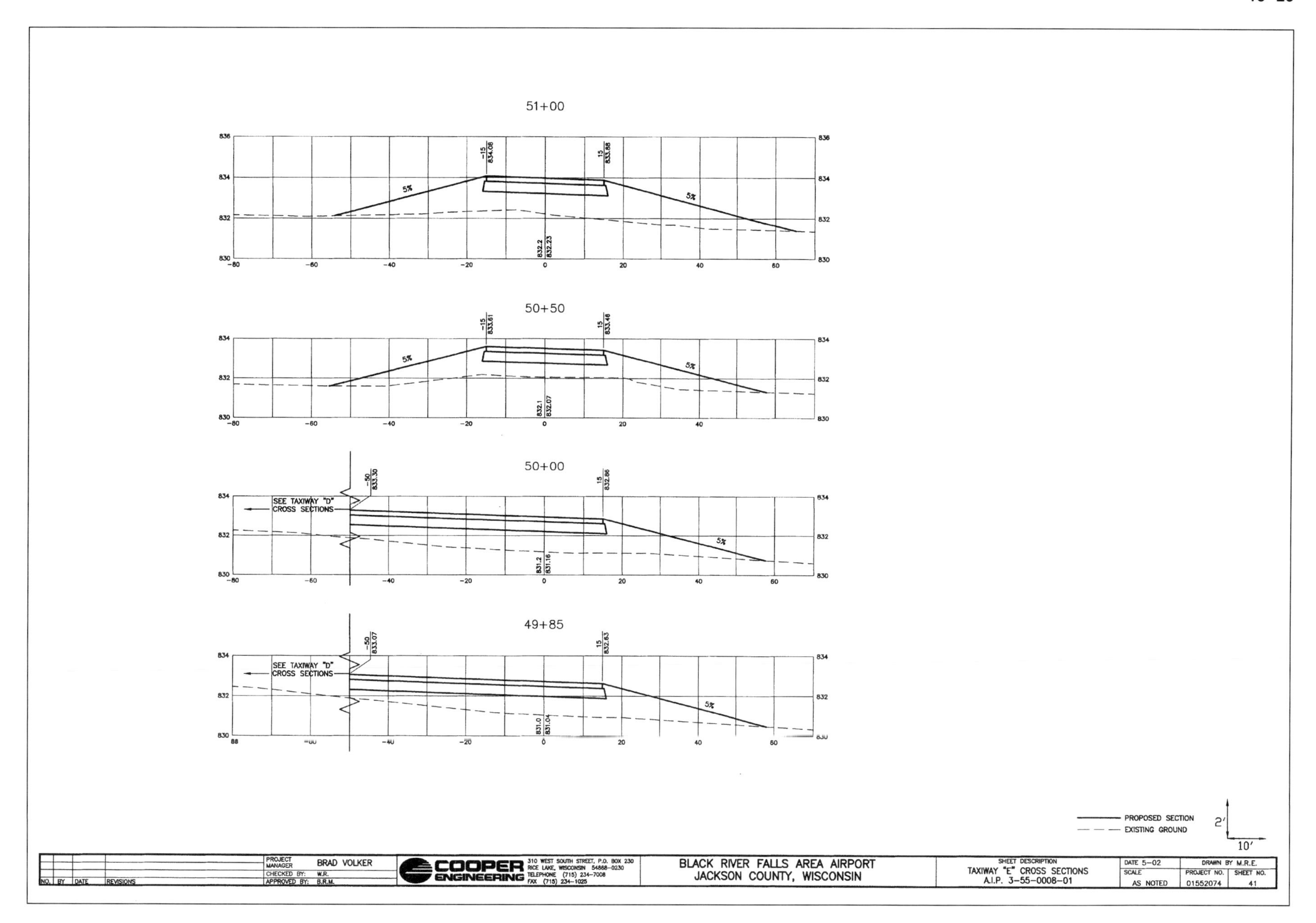
51+00
50+50
50+00
49+85
SEE TAXIWAY "D" CROSS SECTIONS
5%
PROPOSED SECTION
EXISTING GROUND
BRAD VOLKER
COOPER ENGINEERING
BLACK RIVER FALLS AREA AIRPORT
JACKSON COUNTY, WISCONSIN
TAXIWAY "E" CROSS SECTIONS
A.I.P. 3-55-0008-01

设　计

第17节　体育设施

例　17

体育设施可能是非常特殊的实践领域，但是有时一般的从业者被要求参与这些项目。所以，本节包括了大量的案例用于读者参考。

第一套设计图是翻新现有跑道并将其从1/4英里改造为400米。接下来的三个设计图位于同一场地，但却是扩建的设施。最后一个示例是包含有短网拍墙球场地的综合设施，这表明了某项体育设计可能仅是整个项目的一部分。

目　录

体育场跑道——米制转换和翻新

LADYSMITH HIGH SCHOOL ATHLETIC TRACK
-METRIC CONVERSION & REHABILITATION-

LADYSMITH / HAWKINS SCHOOL DISTRICT

PROJECT NO. 93-131

JUNE, 1994

(REVISED: MAR. '95)

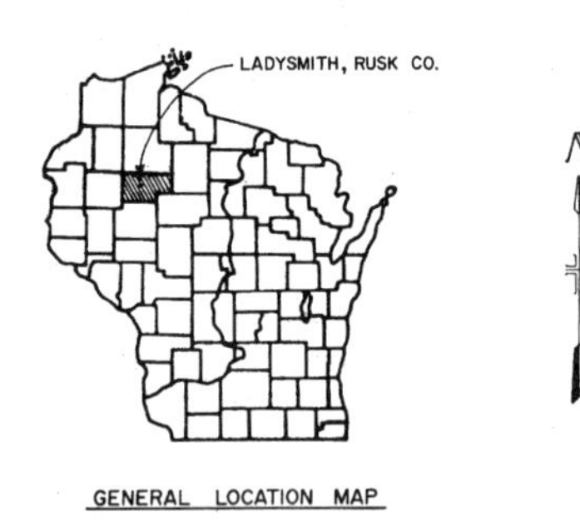

GENERAL LOCATION MAP

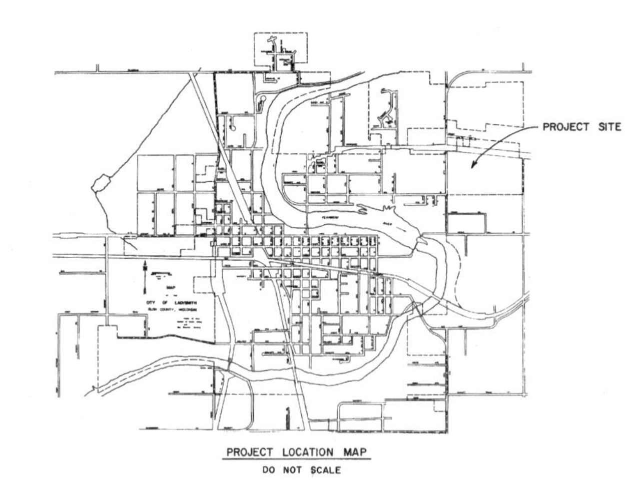

PROJECT LOCATION MAP

DO NOT SCALE

INDEX OF SHEETS

NO.	DESCRIPTION
1	TITLE SHEET W/ LOCATION MAPS
2	EXISTING ATHLETIC FACILITY LAYOUT
3	METRIC CONVERSION & REHABILITATION OF TRACK WITH CONSTRUCTION DETAILS
4	POLE VAULT & LONG/TRIPLE JUMP CONSTRUCTION DETAILS
5	TYPICAL EROSION CONTROL DETAILS NOTE: THIS SHEET NOT INCLUDED

NOTES

REFER TO ACCOMPANYING SPECIFICATIONS AND BIDDING DOCUMENTS FOR SPECIFIC WRITTEN DETAILS.

ENGINEER SHALL FIELD STAKE ALL BASIC HORIZONTAL AND VERTICAL CONTROLS.

-UTILITIES-

CONTACT: Ladysmith Director of Public Works (Water/Sewer)
Bill Christianson, - Director
120 W. Miner Avenue
Ladysmith, Wisconsin 54848
Telephone: (715) 532-2601

CONTACT: Northern States Power Company (Electrical)
711 W. 9th Street, N.
Ladysmith, Wisconsin 54848
Telephone: (715) 532-6226

CONTACT: Wisconsin Gas Company
Telephone: 1-800-242-8511 (Diggers Hotline)

CONTACT: Marcus Communications (TV Cable)
Brian Gustafson
219 W. Miner Avenue
Ladysmith, Wisconsin 54848
Telephone: (715) 532-6040

CONTACT: Ameritech (Telephone Company)
Telephone: 1-800-242-8511 (Diggers Hotline)

CONTACT: Bob Jenness (School Maintenance Supervisor)
Telephone: (715) 532-5277

PREPARED BY:
MORGAN & PARMLEY, LTD
CONSULTING ENGINEERS
LADYSMITH, WISCONSIN

Courtesy: Morgan & Parmley, Ltd.

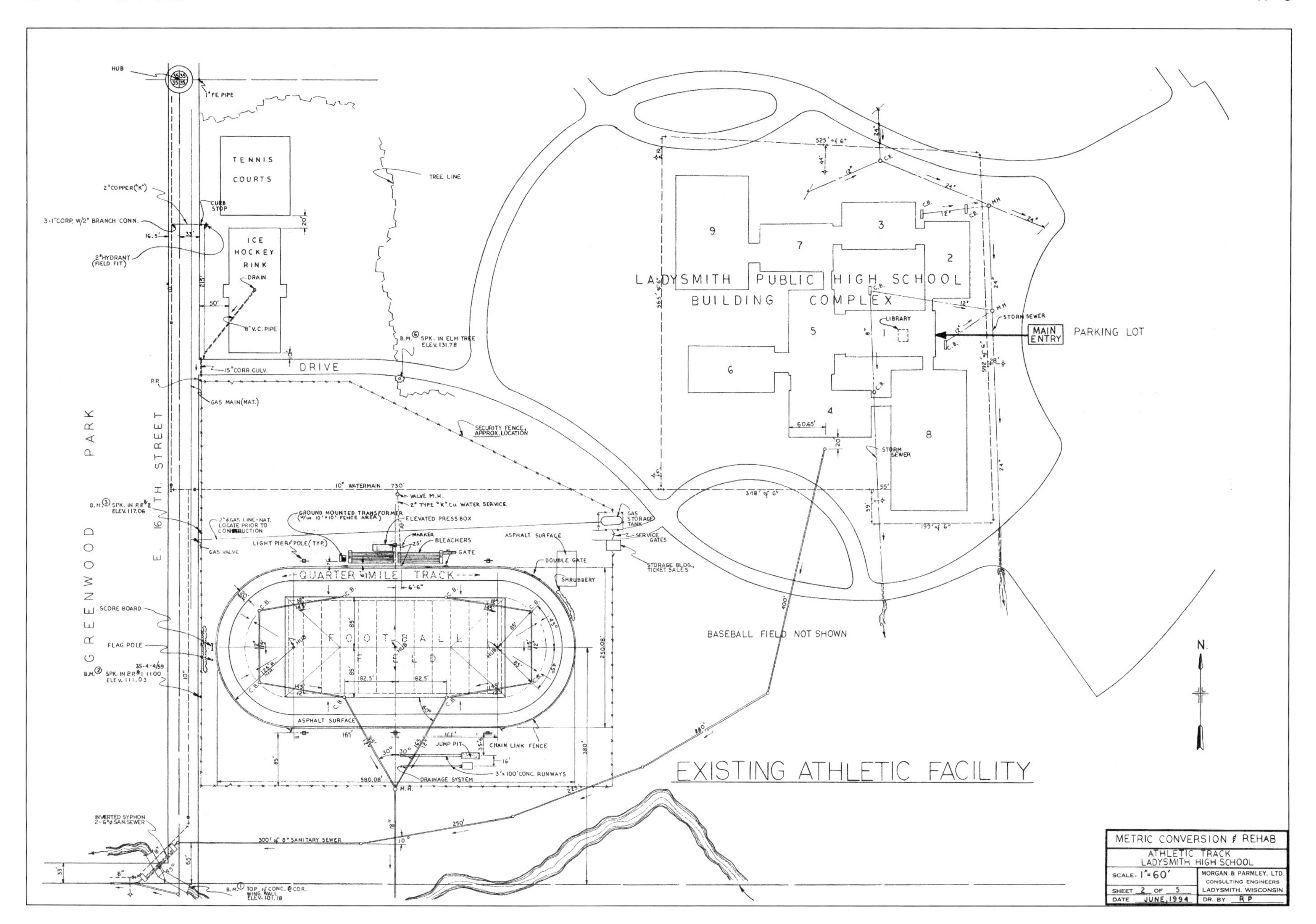
EXISTING ATHLETIC FACILITY
LADYSMITH PUBLIC HIGH SCHOOL
BUILDING COMPLEX
MAIN ENTRY
PARKING LOT
TENNIS COURTS
ICE HOCKEY RINK
DRIVE
QUARTER MILE TRACK
FOOTBALL
BASEBALL FIELD NOT SHOWN
GREENWOOD PARK
E. 16 TH. STREET
TREE LINE
SECURITY FENCE APPROX. LOCATION
10" WATERMAIN 730'
300' of 8" SANITARY SEWER
N.
METRIC CONVERSION & REHAB
ATHLETIC TRACK
LADYSMITH HIGH SCHOOL
SCALE- 1"=60'
SHEET 2 OF 5
DATE JUNE, 1994
MORGAN & PARMLEY, LTD.
CONSULTING ENGINEERS
LADYSMITH, WISCONSIN
DR. BY R.P.

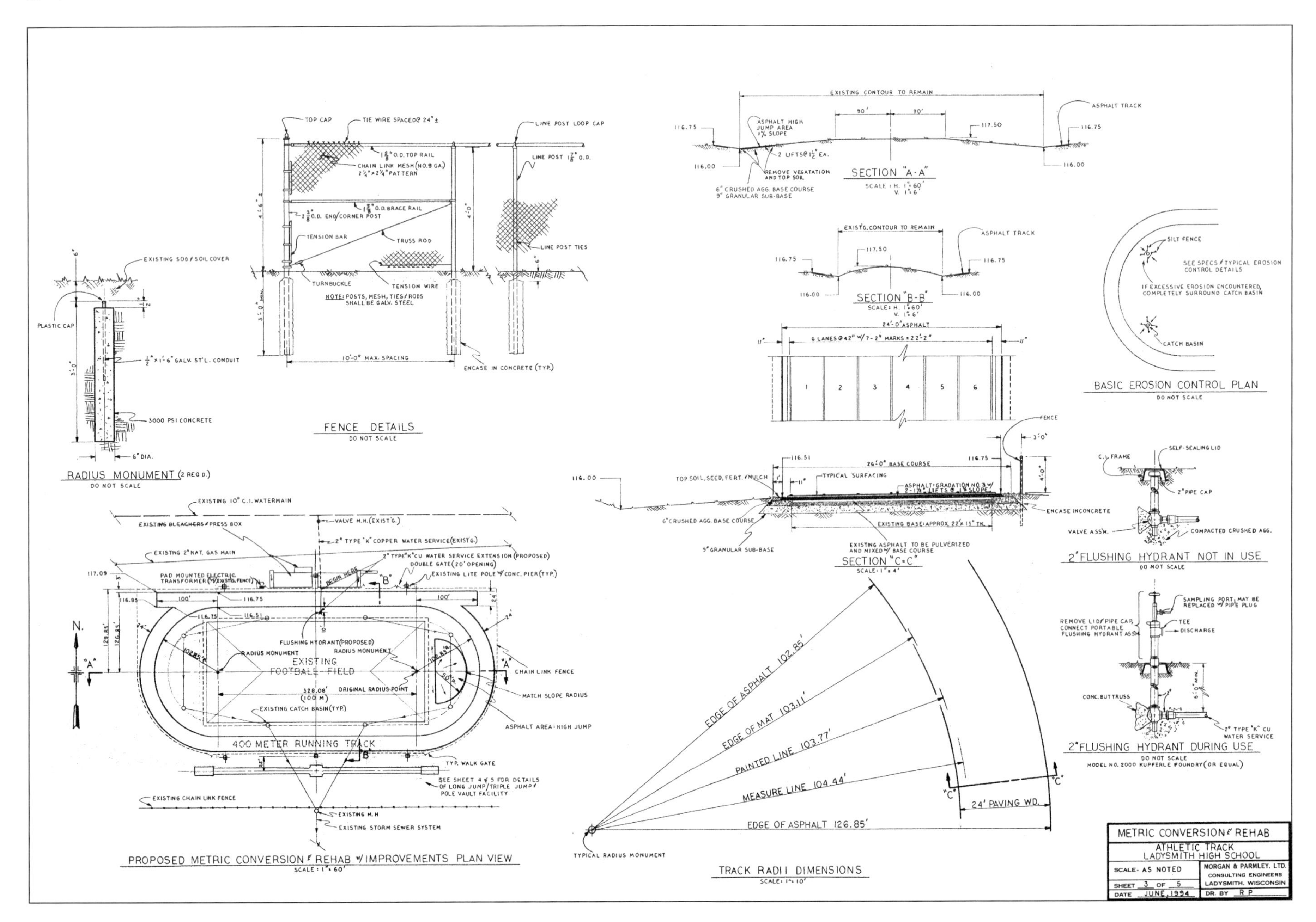

FENCE DETAILS
DO NOT SCALE
RADIUS MONUMENT (2 REQ'D.)
DO NOT SCALE
SECTION "A-A"
SECTION "B-B"
SECTION "C-C"
BASIC EROSION CONTROL PLAN
DO NOT SCALE
2" FLUSHING HYDRANT NOT IN USE
DO NOT SCALE
2" FLUSHING HYDRANT DURING USE
DO NOT SCALE
MODEL NO. 2000 KUPFERLE FOUNDRY (OR EQUAL)
PROPOSED METRIC CONVERSION & REHAB W/ IMPROVEMENTS PLAN VIEW
SCALE: 1"=60'
TRACK RADII DIMENSIONS
SCALE: 1"=10'
EXISTING FOOTBALL FIELD
400 METER RUNNING TRACK
EDGE OF ASPHALT 102.85'
EDGE OF MAT 103.11'
PAINTED LINE 103.77'
MEASURE LINE 104.44'
EDGE OF ASPHALT 126.85'
24' PAVING WD.
TYPICAL RADIUS MONUMENT
METRIC CONVERSION & REHAB
ATHLETIC TRACK
LADYSMITH HIGH SCHOOL
SCALE: AS NOTED
MORGAN & PARMLEY, LTD.
CONSULTING ENGINEERS
LADYSMITH, WISCONSIN
SHEET 3 OF 5
DATE JUNE, 1994
DR. BY R P

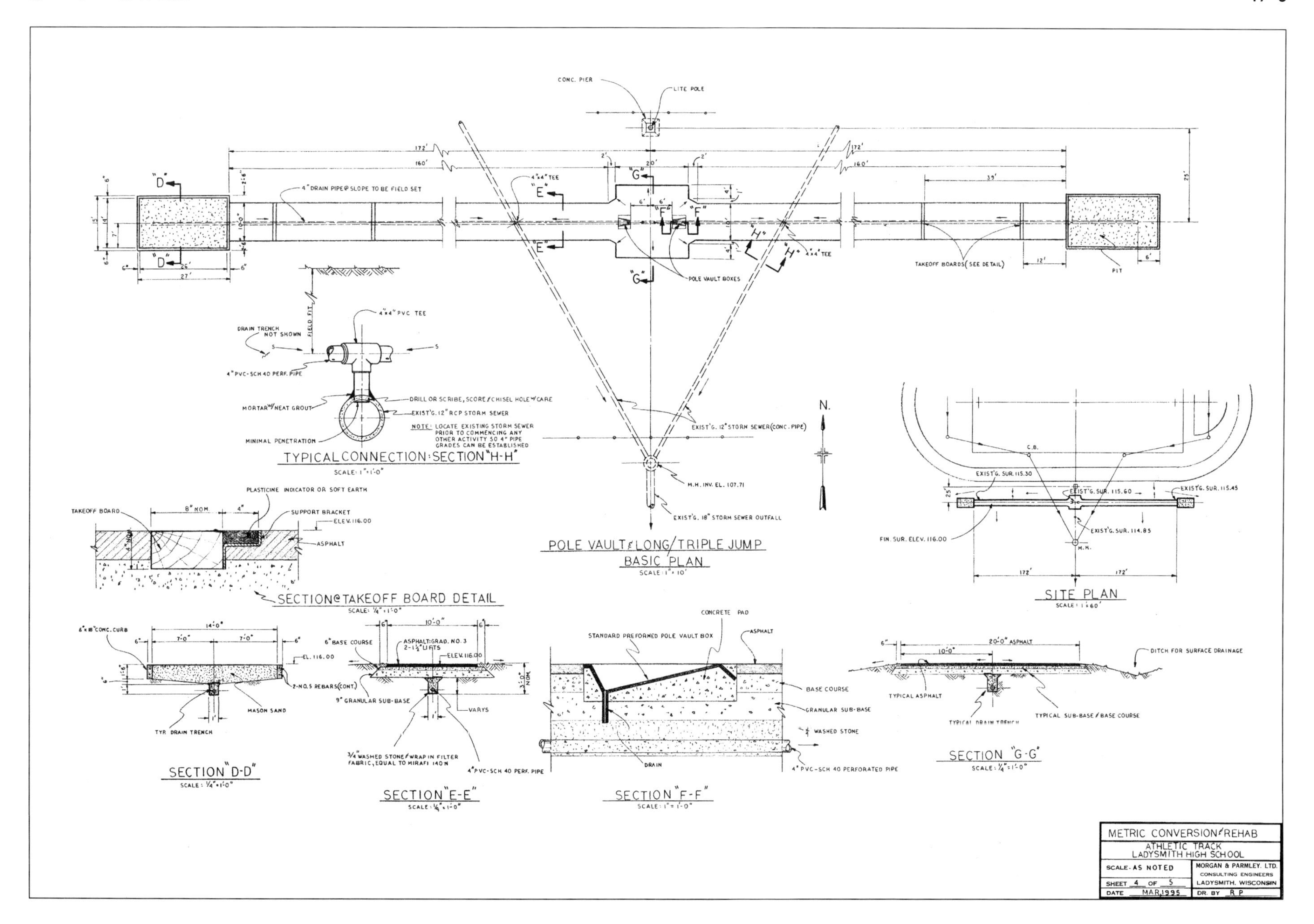
POLE VAULT/LONG/TRIPLE JUMP
BASIC PLAN
SCALE: 1" = 10'
TYPICAL CONNECTION: SECTION "H-H"
SCALE: 1" = 1'-0"
SECTION @ TAKEOFF BOARD DETAIL
SCALE: 1/4" = 1'-0"
SITE PLAN
SCALE: 1" = 60'
SECTION "D-D"
SCALE: 1/4" = 1'-0"
SECTION "E-E"
SCALE: 1/4" = 1'-0"
SECTION "F-F"
SCALE: 1" = 1'-0"
SECTION "G-G"
SCALE: 1/4" = 1'-0"
METRIC CONVERSION/REHAB
ATHLETIC TRACK
LADYSMITH HIGH SCHOOL
SCALE: AS NOTED
SHEET 4 OF 5
DATE MAR 1995
MORGAN & PARMLEY, LTD.
CONSULTING ENGINEERS
LADYSMITH, WISCONSIN
DR. BY R P

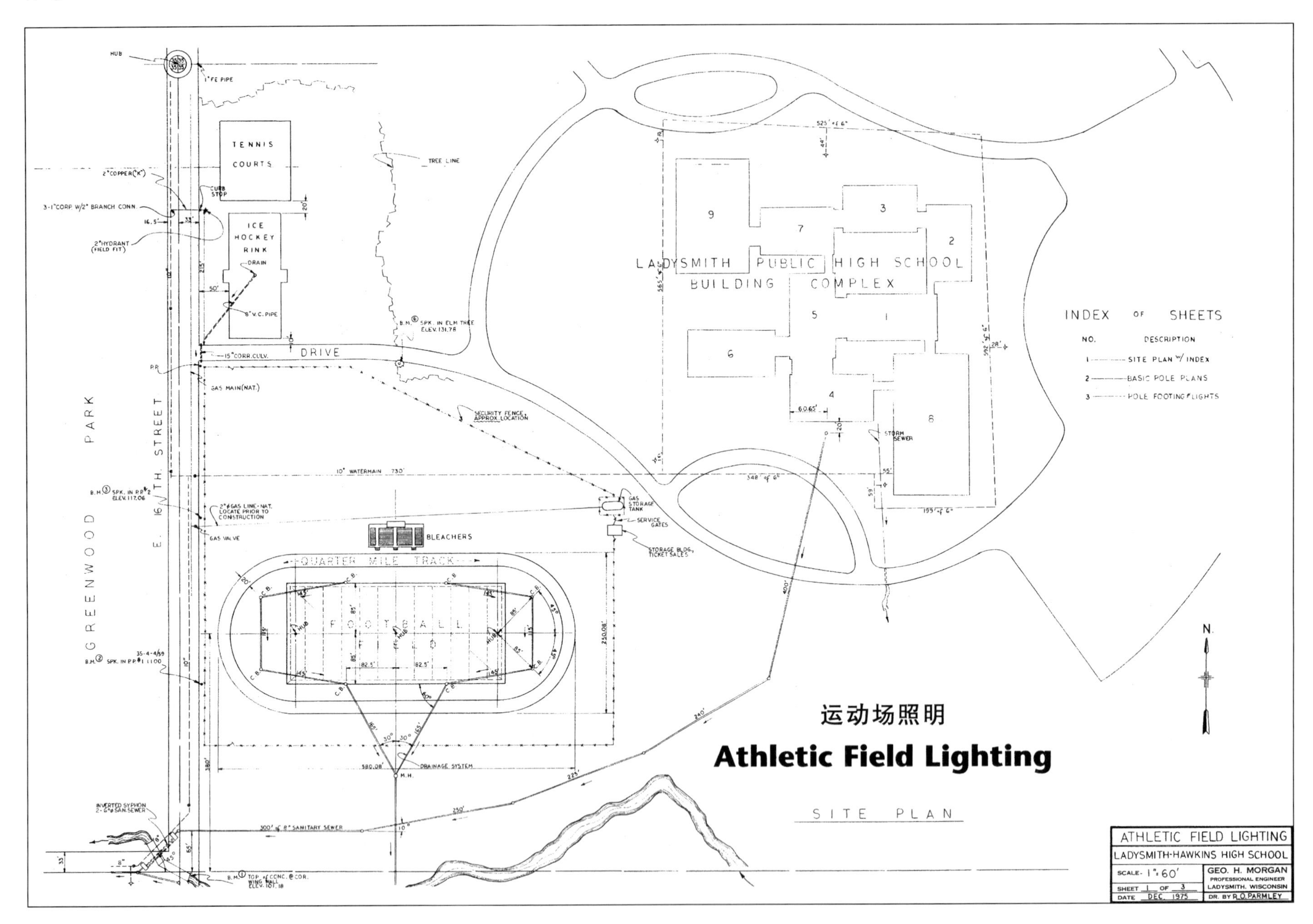

Courtesy: Morgan & Parmley, Ltd.

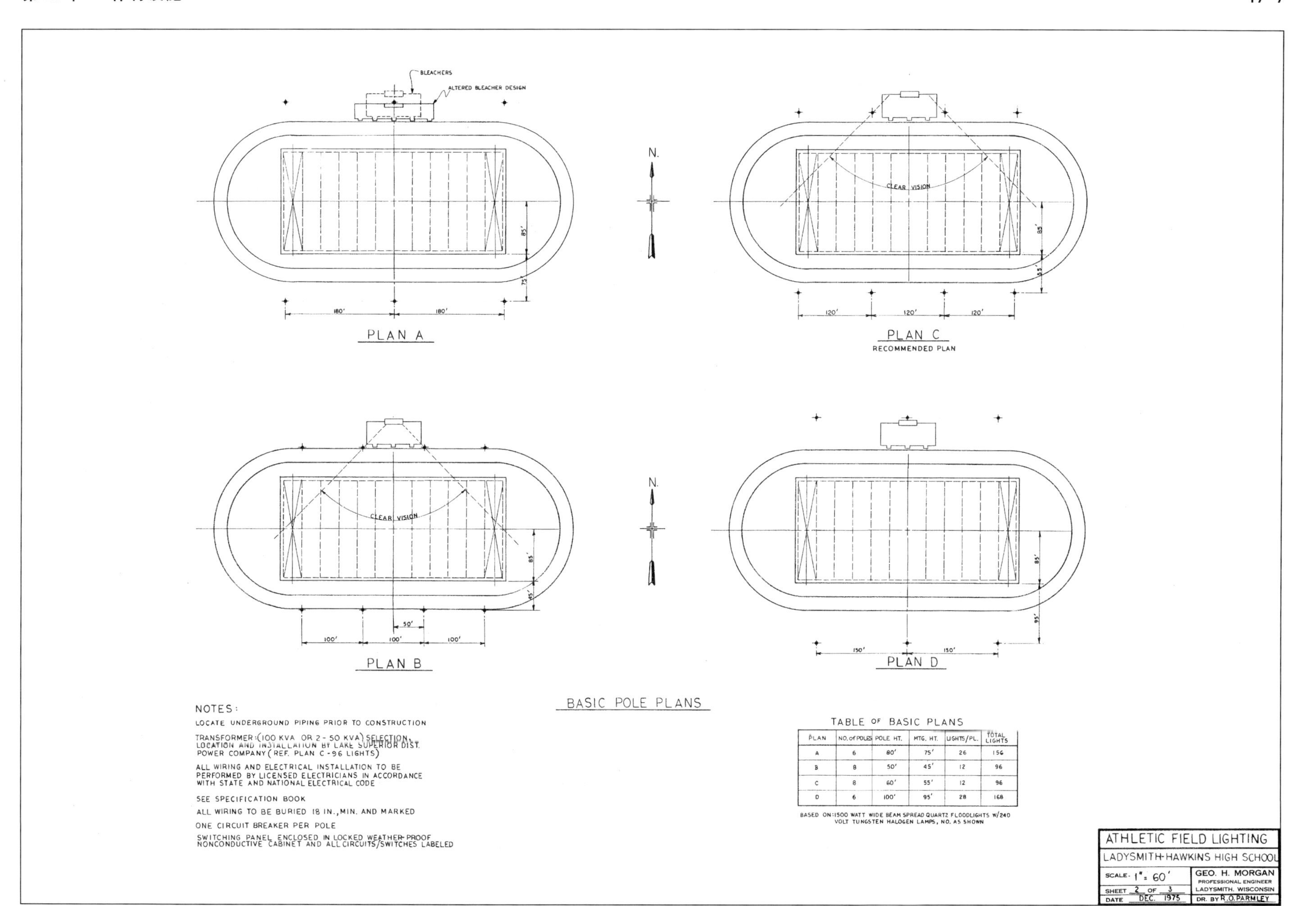

TABLE OF BASIC PLANS

PLAN	NO. of POLES	POLE HT.	MTG. HT.	LIGHTS/PL.	TOTAL LIGHTS
A	6	80'	75'	26	156
B	8	50'	45'	12	96
C	8	60'	55'	12	96
D	6	100'	95'	28	168

BASED ON: 1500 WATT WIDE BEAM SPREAD QUARTZ FLOODLIGHTS W/240 VOLT TUNGSTEN HALOGEN LAMPS, NO. AS SHOWN

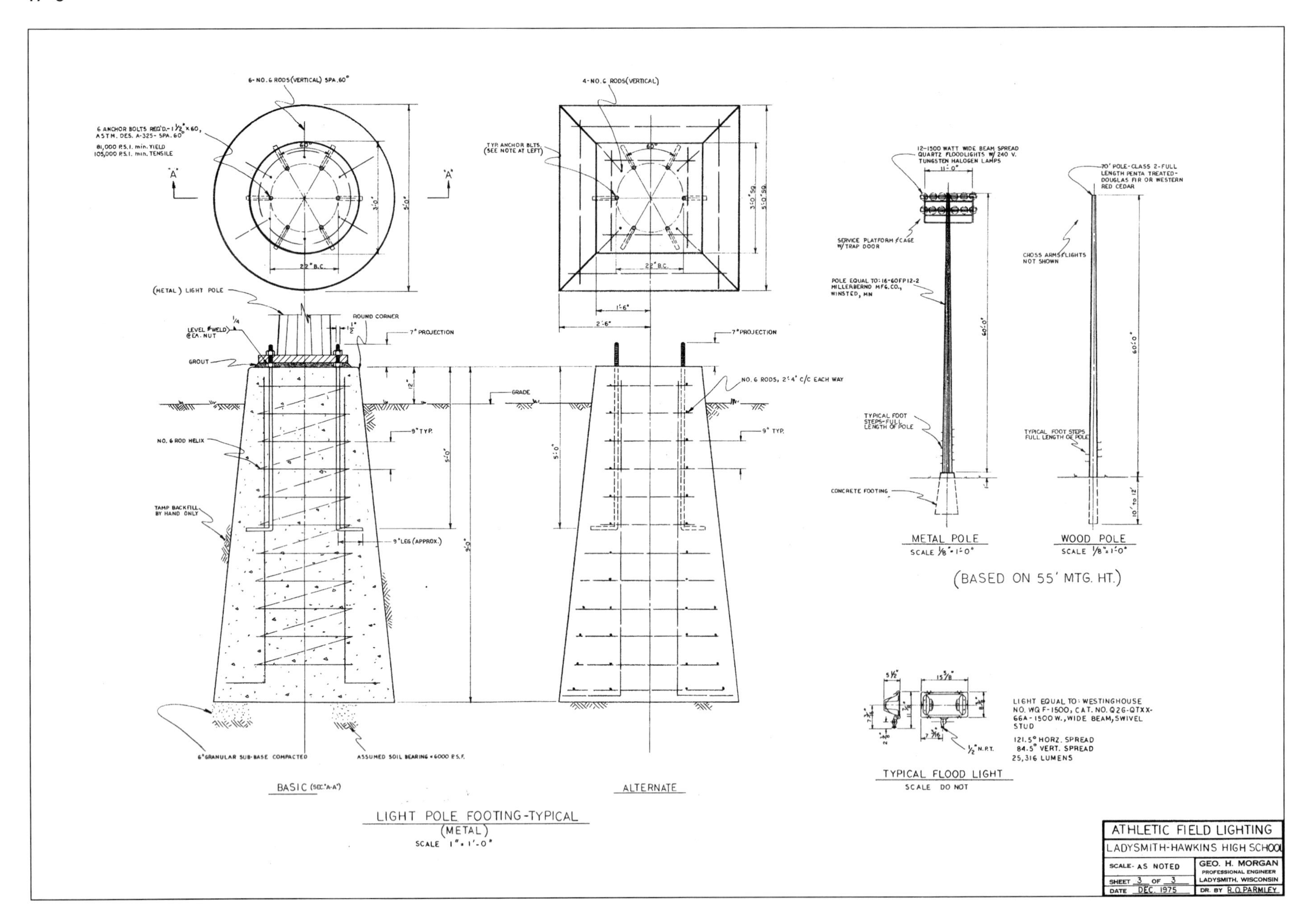

6-NO.6 RODS(VERTICAL) SPA. 60°
4-NO.6 RODS(VERTICAL)
6 ANCHOR BOLTS REQ'D.-1 1/2"×60, A.S.T.M. DES. A-325- SPA. 60°
81,000 P.S.I. min. YIELD
105,000 P.S.I. min. TENSILE
TYP. ANCHOR BLTS. (SEE NOTE AT LEFT)
22" B.C.
(METAL) LIGHT POLE
ROUND CORNER
7" PROJECTION
LEVEL WELD @EA. NUT
GROUT
GRADE
NO. 6 RODS, 2'-4" C/C EACH WAY
9" TYP.
NO. 6 ROD HELIX
TAMP BACKFILL BY HAND ONLY
9" LEG (APPROX.)
6" GRANULAR SUB-BASE COMPACTED
ASSUMED SOIL BEARING = 6000 P.S.F.
BASIC (SEC. "A-A")
ALTERNATE
LIGHT POLE FOOTING-TYPICAL (METAL)
SCALE 1" = 1'-0"
12-1500 WATT WIDE BEAM SPREAD QUARTZ FLOODLIGHTS W/ 240 V. TUNGSTEN HALOGEN LAMPS
SERVICE PLATFORM & CAGE W/ TRAP DOOR
POLE EQUAL TO: 16-60FP12-2 MILLERBERND MFG. CO., WINSTED, MN
TYPICAL FOOT STEPS-FULL LENGTH OF POLE
CONCRETE FOOTING
70' POLE-CLASS 2-FULL LENGTH PENTA TREATED-DOUGLAS FIR OR WESTERN RED CEDAR
CROSS ARMS & LIGHTS NOT SHOWN
TYPICAL FOOT STEPS FULL LENGTH OF POLE
METAL POLE
SCALE 1/8" = 1'-0"
WOOD POLE
SCALE 1/8" = 1'-0"
(BASED ON 55' MTG. HT.)
LIGHT EQUAL TO: WESTINGHOUSE NO. WG F-1500, CAT. NO. Q26-QTXX-66A-1500 W., WIDE BEAM, SWIVEL STUD
121.5° HORZ. SPREAD
84.5° VERT. SPREAD
25,316 LUMENS
TYPICAL FLOOD LIGHT
SCALE DO NOT
ATHLETIC FIELD LIGHTING
LADYSMITH-HAWKINS HIGH SCHOOL
SCALE- AS NOTED
SHEET 3 OF 3
DATE DEC. 1975
GEO. H. MORGAN
PROFESSIONAL ENGINEER
LADYSMITH, WISCONSIN
DR. BY R.O. PARMLEY

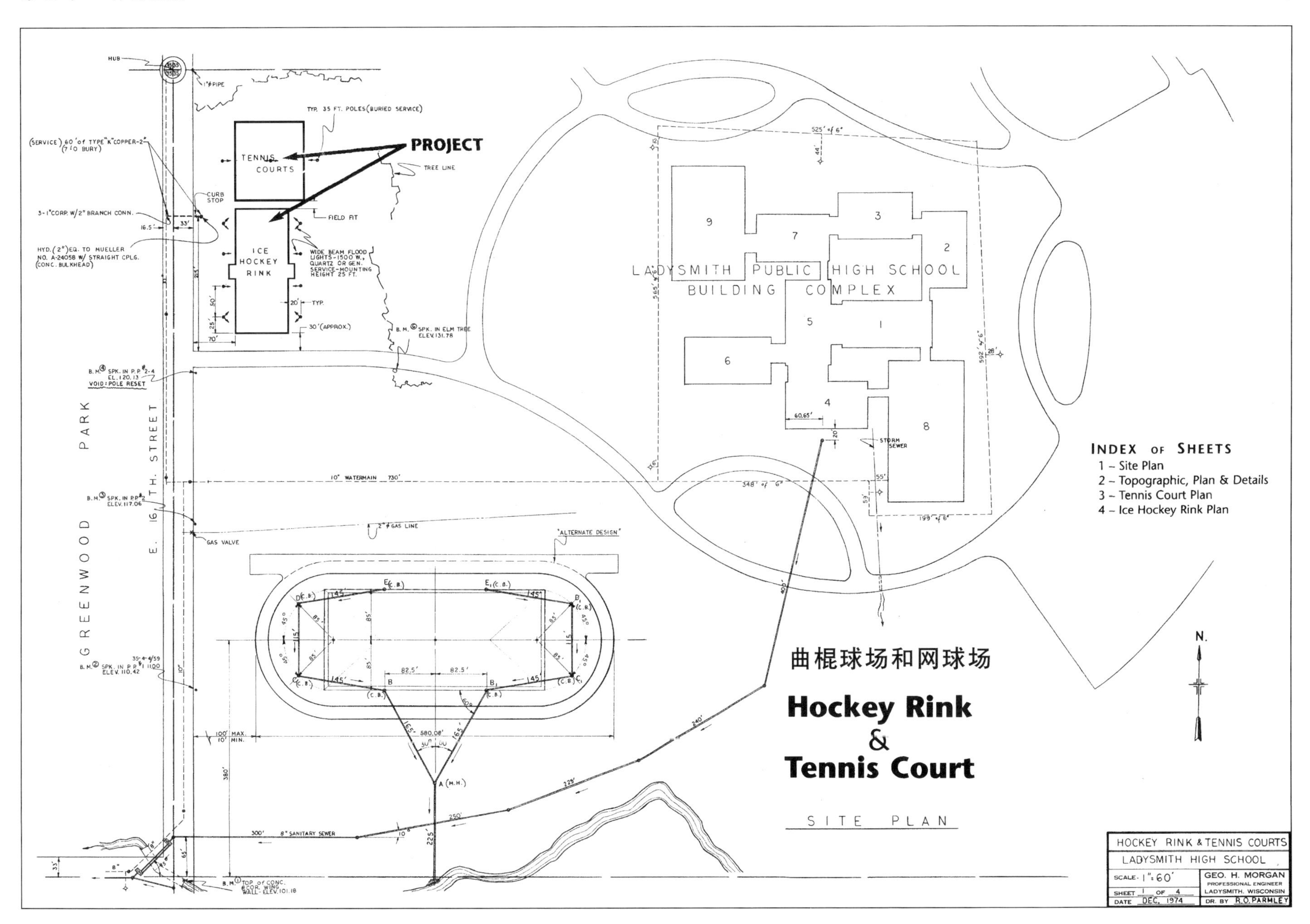

Courtesy: Morgan & Parmley, Ltd.

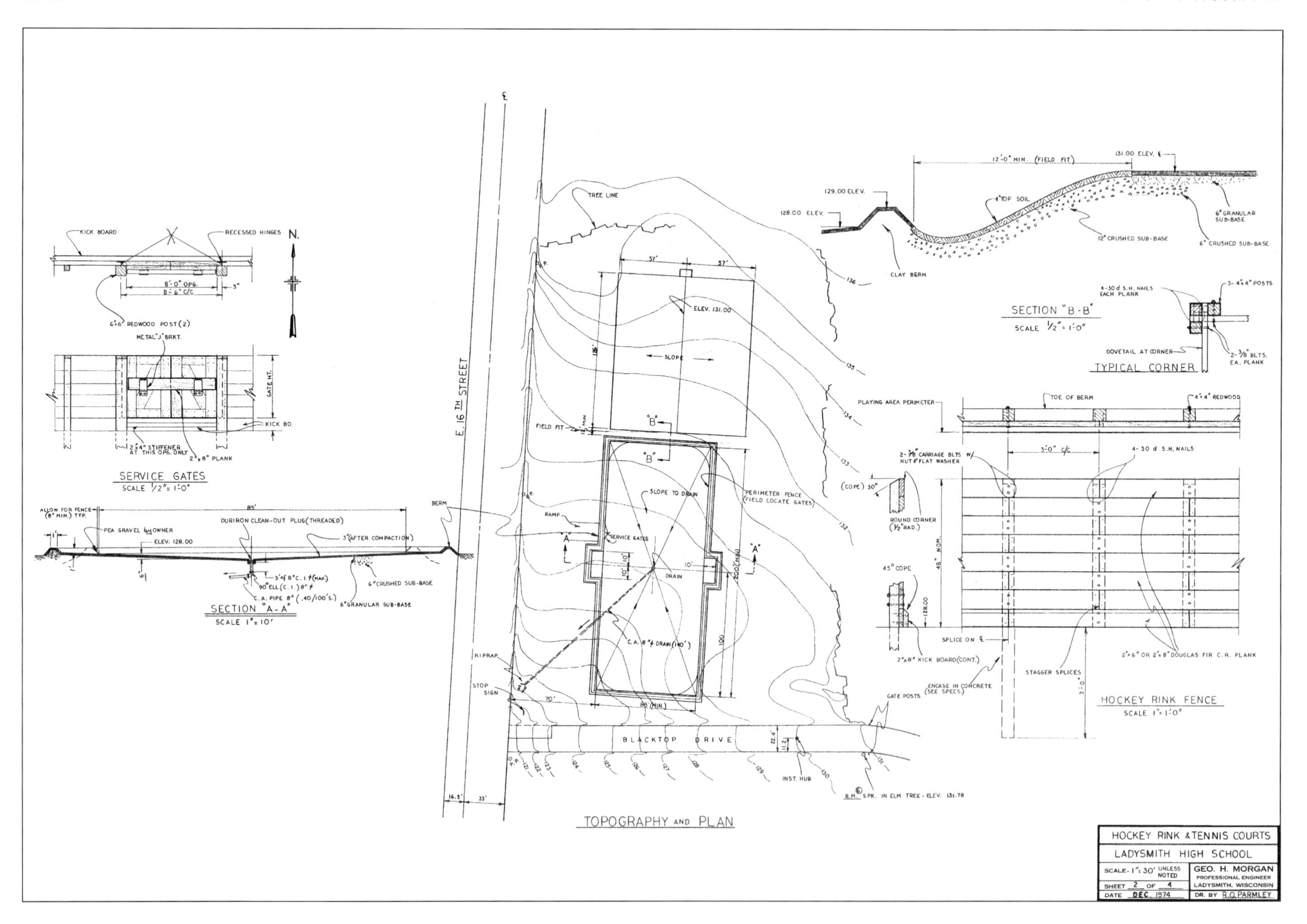
KICK BOARD
RECESSED HINGES
N.
8'-0" OPG.
8'-6" C/C
3"
6"x6" REDWOOD POST (2)
METAL "J" BRKT.
GATE HT.
KICK BD.
2"x4" STIFFENER AT THIS OPG. ONLY
2"x8" PLANK
SERVICE GATES
SCALE 1/2"=1'-0"
ALLOW FOR FENCE (8" MIN.) TYP.
85'
PEA GRAVEL by OWNER
DURIRON CLEAN-OUT PLUG (THREADED)
ELEV. 128.00
3"(AFTER COMPACTION)
BERM
90° ELL (C.I.) 8"φ
C.A. PIPE 8" (.40/100'S.)
6" CRUSHED SUB-BASE
6" GRANULAR SUB-BASE
SECTION "A-A"
SCALE 1"=10'
E. 16TH STREET
TREE LINE
57'
57'
ELEV. 131.00
SLOPE
FIELD FIT
SLOPE TO DRAIN
PERIMETER FENCE (FIELD LOCATE GATES)
RAMP
SERVICE GATES
DRAIN
200'(MIN)
C.A. 8"φ DRAIN (140')
RIPRAP
STOP SIGN
70'
85'(MIN)
BLACKTOP DRIVE
INST. HUB
B.M. SPK. IN ELM TREE - ELEV. 131.78
16.5'
33'
TOPOGRAPHY AND PLAN
131.00 ELEV.
12'-0" MIN. (FIELD FIT)
129.00 ELEV.
128.00 ELEV.
4" TOP SOIL
6" GRANULAR SUB-BASE
12" CRUSHED SUB-BASE
6" CRUSHED SUB-BASE
CLAY BERM
SECTION "B-B"
SCALE 1/2"=1'-0"
4-30d S.H. NAILS EACH PLANK
3-4"x4" POSTS
DOVETAIL AT CORNER
2-3/8" BLTS. EA. PLANK
TYPICAL CORNER
PLAYING AREA PERIMETER
TOE OF BERM
4"x4" REDWOOD
3'-0" C/C
4-30 d S.H. NAILS
2-3/8" CARRIAGE BLTS W/ NUT & FLAT WASHER
(COPE) 30°
ROUND CORNER (1/2" RAD.)
48" NOM.
45° COPE
128.00
SPLICE ON ℄
2"x8" KICK BOARD (CONT.)
ENCASE IN CONCRETE (SEE SPECS.)
GATE POSTS
STAGGER SPLICES
2"x6" OR 2"x8" DOUGLAS FIR C.R. PLANK
HOCKEY RINK FENCE
SCALE 1"=1'-0"
HOCKEY RINK & TENNIS COURTS
LADYSMITH HIGH SCHOOL
SCALE- 1"=30' UNLESS NOTED
SHEET 2 OF 4
DATE DEC. 1974
GEO. H. MORGAN
PROFESSIONAL ENGINEER
LADYSMITH, WISCONSIN
DR. BY R.O. PARMLEY

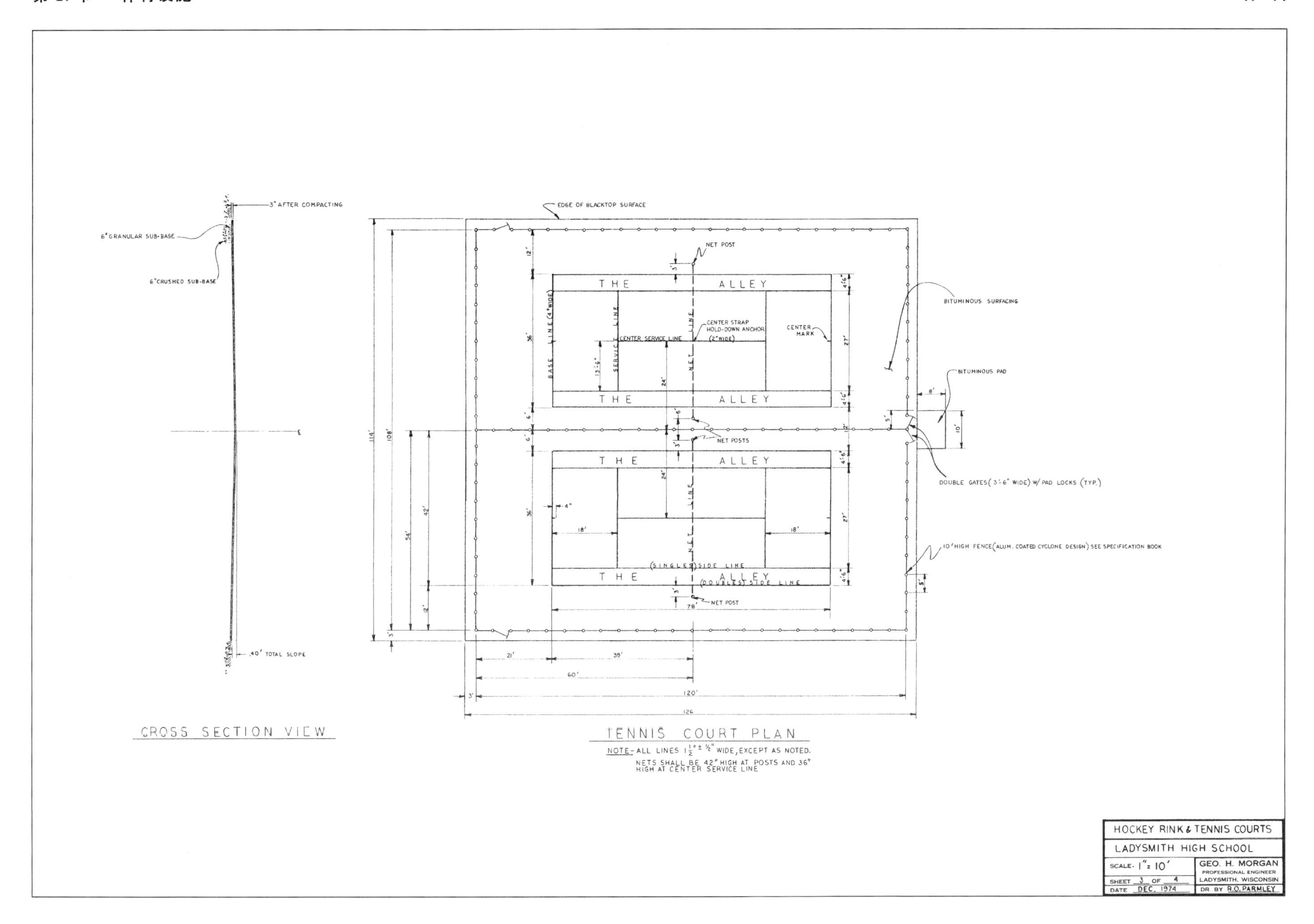
3" AFTER COMPACTING
6" GRANULAR SUB-BASE
6" CRUSHED SUB-BASE
.40' TOTAL SLOPE
CROSS SECTION VIEW
EDGE OF BLACKTOP SURFACE
NET POST
THE ALLEY
CENTER STRAP HOLD-DOWN ANCHOR (2" WIDE)
CENTER SERVICE LINE
CENTER MARK
BASE LINE (4" WIDE)
SERVICE LINE
NET LINE
BITUMINOUS SURFACING
BITUMINOUS PAD
NET POSTS
DOUBLE GATES (3'-6" WIDE) W/ PAD LOCKS (TYP.)
10' HIGH FENCE (ALUM. COATED CYCLONE DESIGN) SEE SPECIFICATION BOOK
(SINGLES) SIDE LINE
(DOUBLES) SIDE LINE
TENNIS COURT PLAN
NOTE- ALL LINES 1½" ± ½" WIDE, EXCEPT AS NOTED.
NETS SHALL BE 42" HIGH AT POSTS AND 36" HIGH AT CENTER SERVICE LINE
HOCKEY RINK & TENNIS COURTS
LADYSMITH HIGH SCHOOL
SCALE- 1" = 10'
SHEET 3 OF 4
DATE DEC. 1974
GEO. H. MORGAN
PROFESSIONAL ENGINEER
LADYSMITH, WISCONSIN
DR BY R.O. PARMLEY

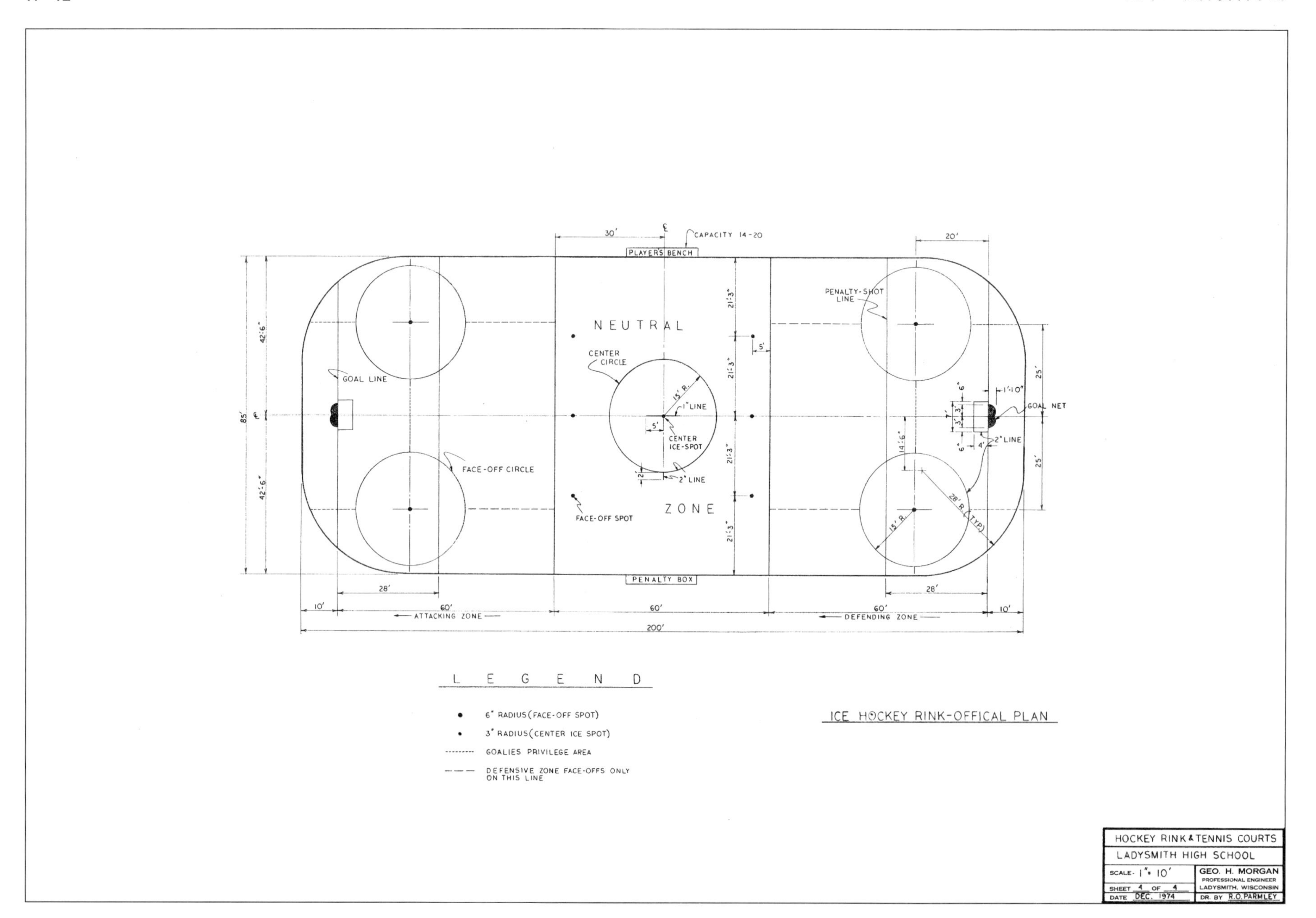
CAPACITY 14-20
PLAYER'S BENCH
PENALTY BOX
NEUTRAL
ZONE
CENTER CIRCLE
15' R.
1" LINE
CENTER ICE-SPOT
2" LINE
FACE-OFF SPOT
FACE-OFF CIRCLE
GOAL LINE
PENALTY-SHOT LINE
GOAL NET
28' R. (TYP.)
15' R.
ATTACKING ZONE
DEFENDING ZONE
200'
L E G E N D
6" RADIUS (FACE-OFF SPOT)
3" RADIUS (CENTER ICE SPOT)
GOALIES PRIVILEGE AREA
DEFENSIVE ZONE FACE-OFFS ONLY ON THIS LINE
ICE HOCKEY RINK-OFFICAL PLAN
HOCKEY RINK & TENNIS COURTS
LADYSMITH HIGH SCHOOL
SCALE- 1" = 10'
SHEET 4 OF 4
DATE DEC. 1974
GEO. H. MORGAN
PROFESSIONAL ENGINEER
LADYSMITH, WISCONSIN
DR. BY R.O. PARMLEY

高级中学球场：地面层平面

High School Ball Fields
(Grading Plan)

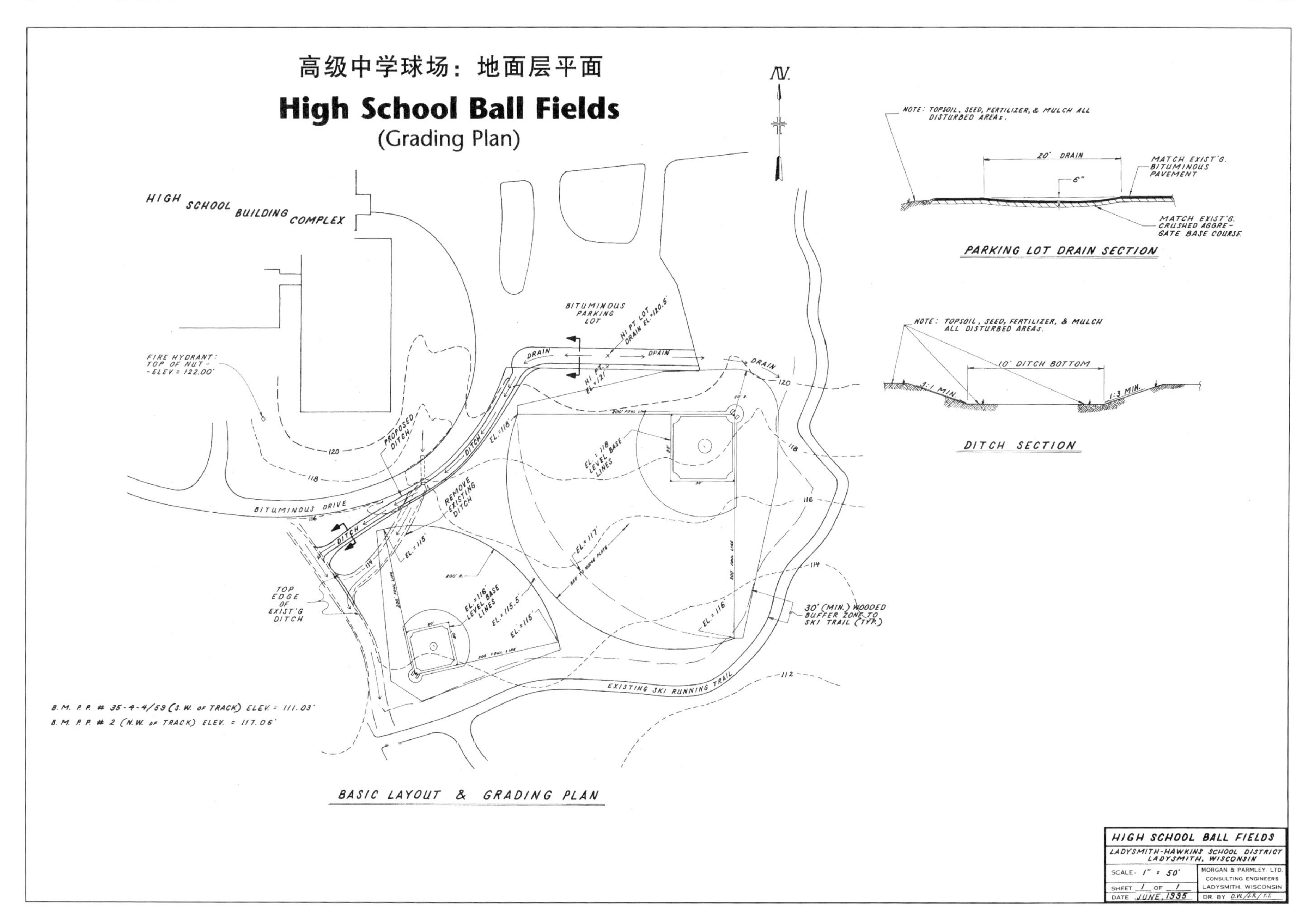

Courtesy: Morgan & Parmley, Ltd.

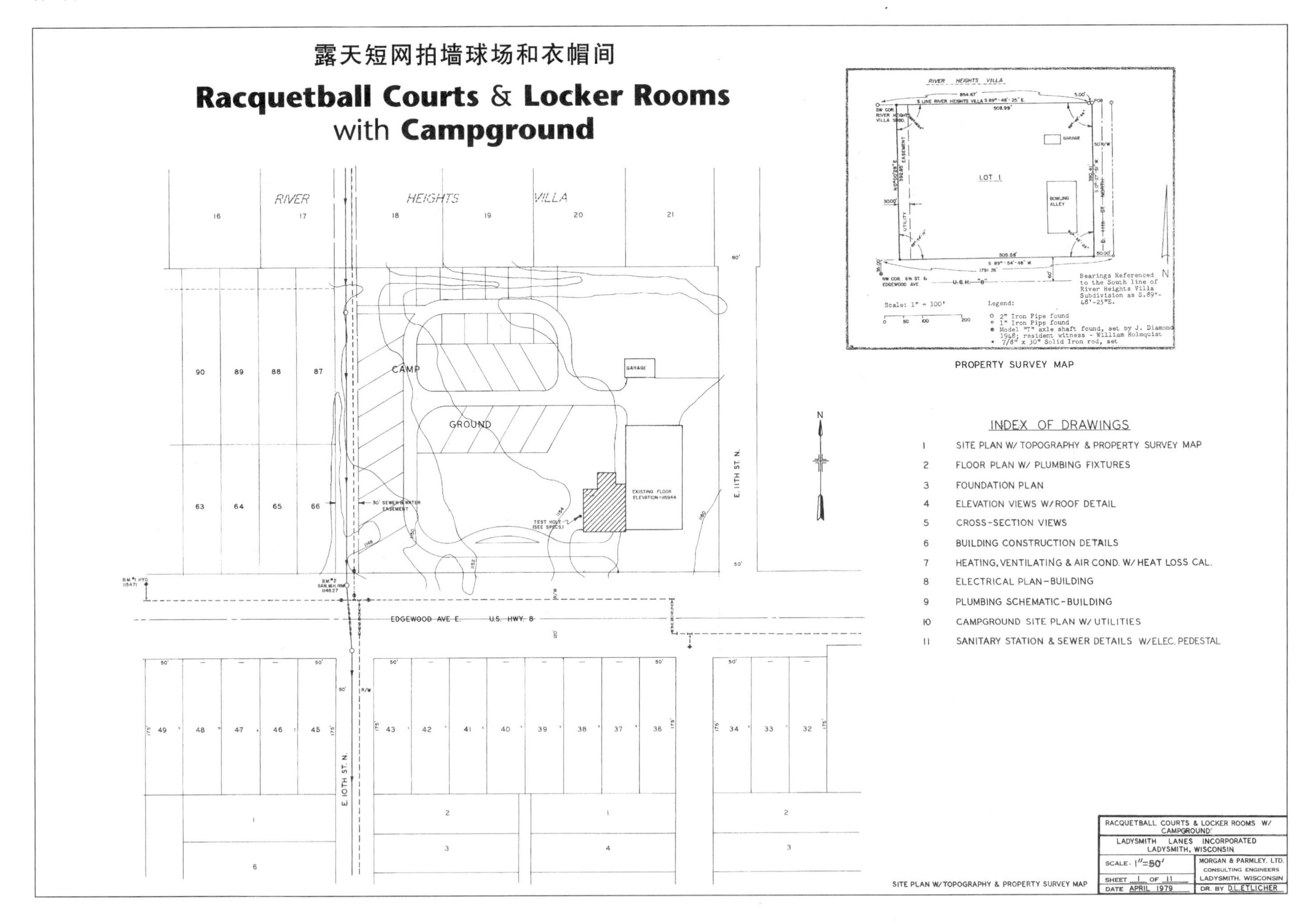

Courtesy: Morgan & Parmley, Ltd.

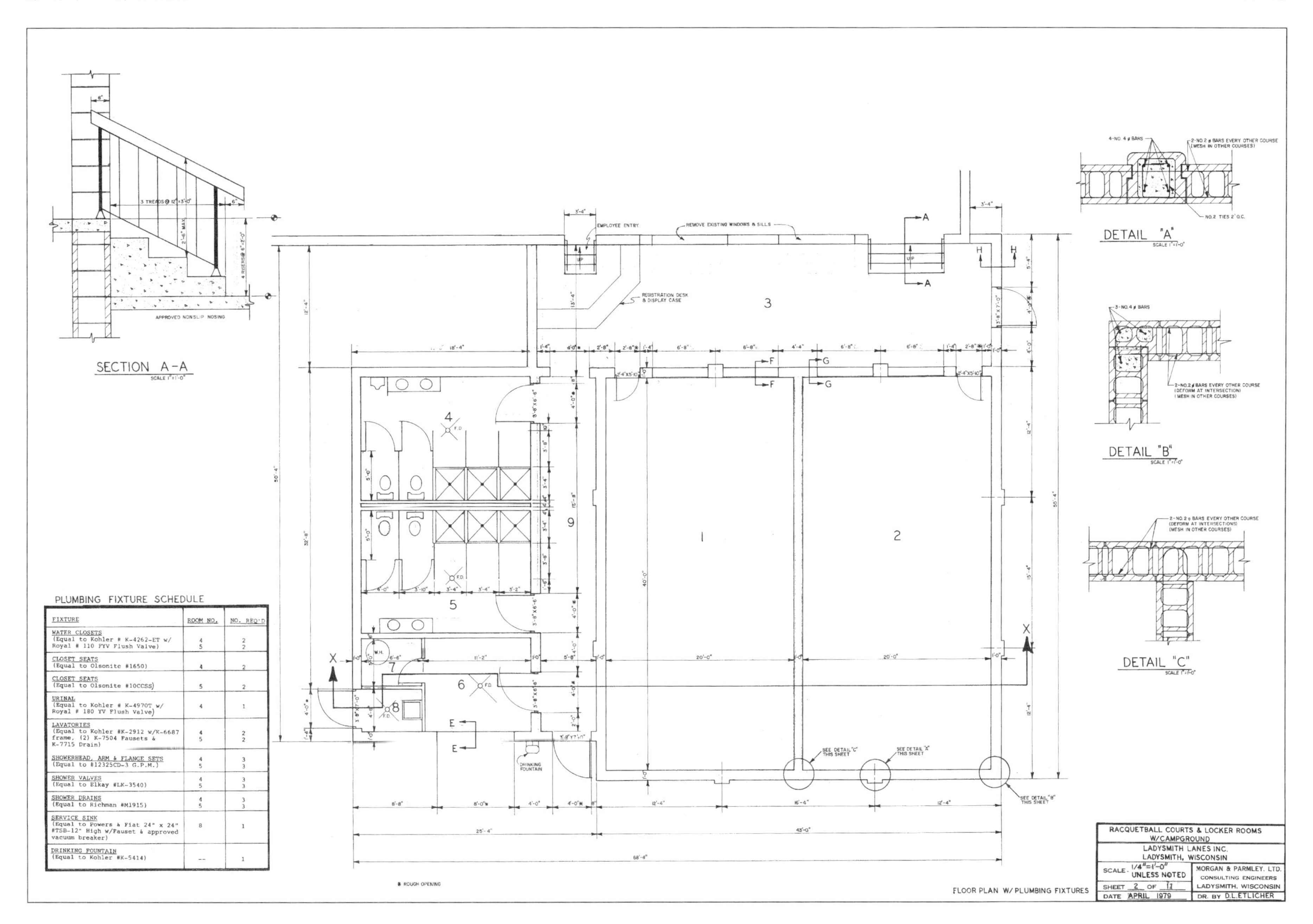

PLUMBING FIXTURE SCHEDULE

FIXTURE	ROOM NO.	NO. REQ'D
WATER CLOSETS (Equal to Kohler # K-4262-ET w/ Royal # 110 FYV Flush Valve)	4 5	2 2
CLOSET SEATS (Equal to Olsonite #1650)	4	2
CLOSET SEATS (Equal to Olsonite #10CCSS)	5	2
URINAL (Equal to Kohler # K-4970T w/ Royal # 180 YV Flush Valve)	4	1
LAVATORIES (Equal to Kohler #K-2912 w/K-6687 frame, (2) K-7504 Fausets & K-7715 Drain)	4 5	2 2
SHOWERHEAD, ARM & FLANGE SETS (Equal to #12325CD-3 G.P.M.)	4 5	3 3
SHOWER VALVES (Equal to Elkay #LK-3540)	4 5	3 3
SHOWER DRAINS (Equal to Richman #M1915)	4 5	3 3
SERVICE SINK (Equal to Powers & Fiat 24" x 24" #TSB-12" High w/Fauset & approved vacuum breaker)	8	1
DRINKING FOUNTAIN (Equal to Kohler #K-5414)	--	1

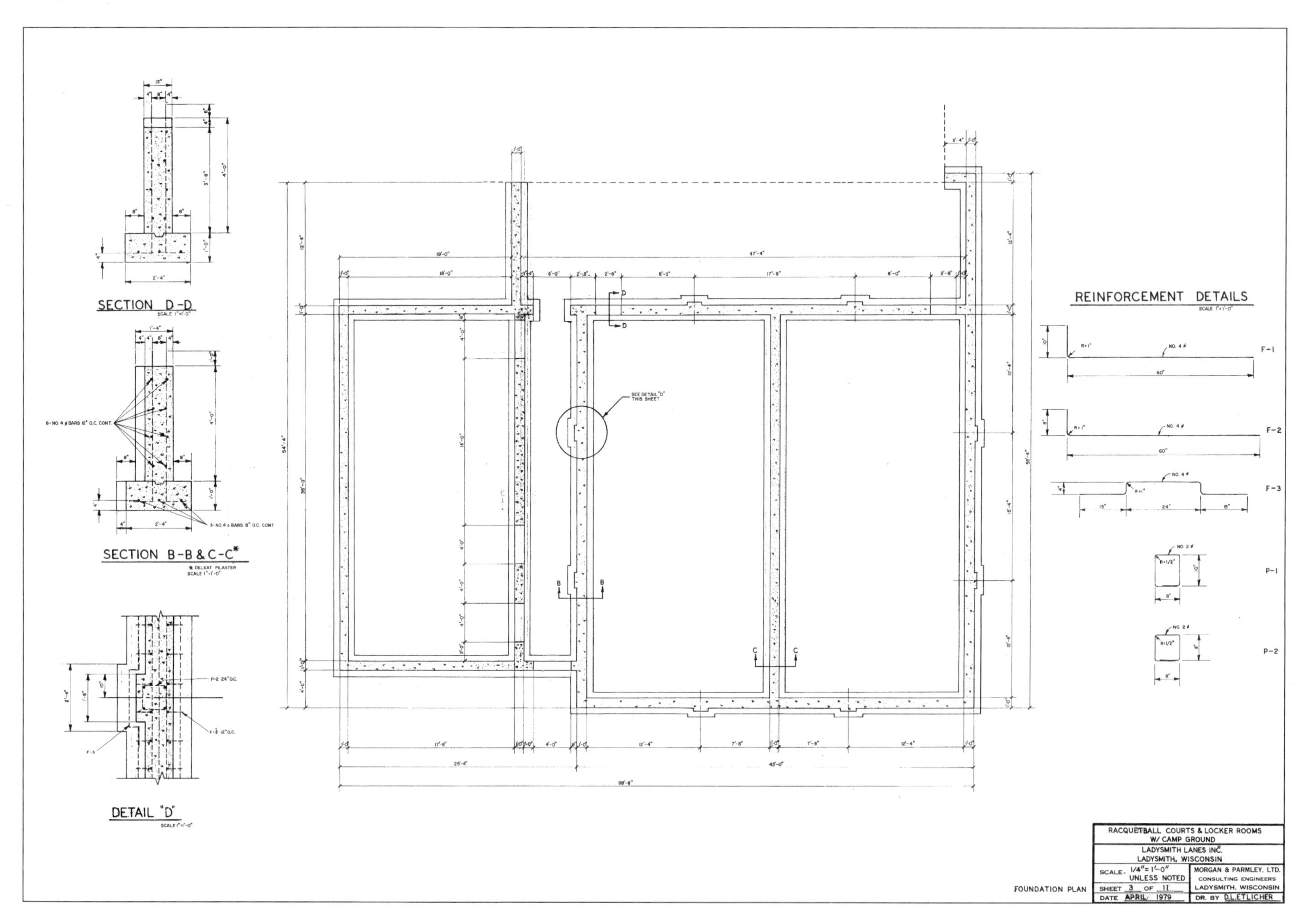
SECTION D-D
SECTION B-B & C-C*
DETAIL "D"
REINFORCEMENT DETAILS
RACQUETBALL COURTS & LOCKER ROOMS W/ CAMP GROUND
LADYSMITH LANES INC.
LADYSMITH, WISCONSIN
SCALE 1/4"=1'-0" UNLESS NOTED
MORGAN & PARMLEY, LTD. CONSULTING ENGINEERS LADYSMITH, WISCONSIN
SHEET 3 OF 11
DATE APRIL 1979
DR. BY D.L.ETLICHER
FOUNDATION PLAN

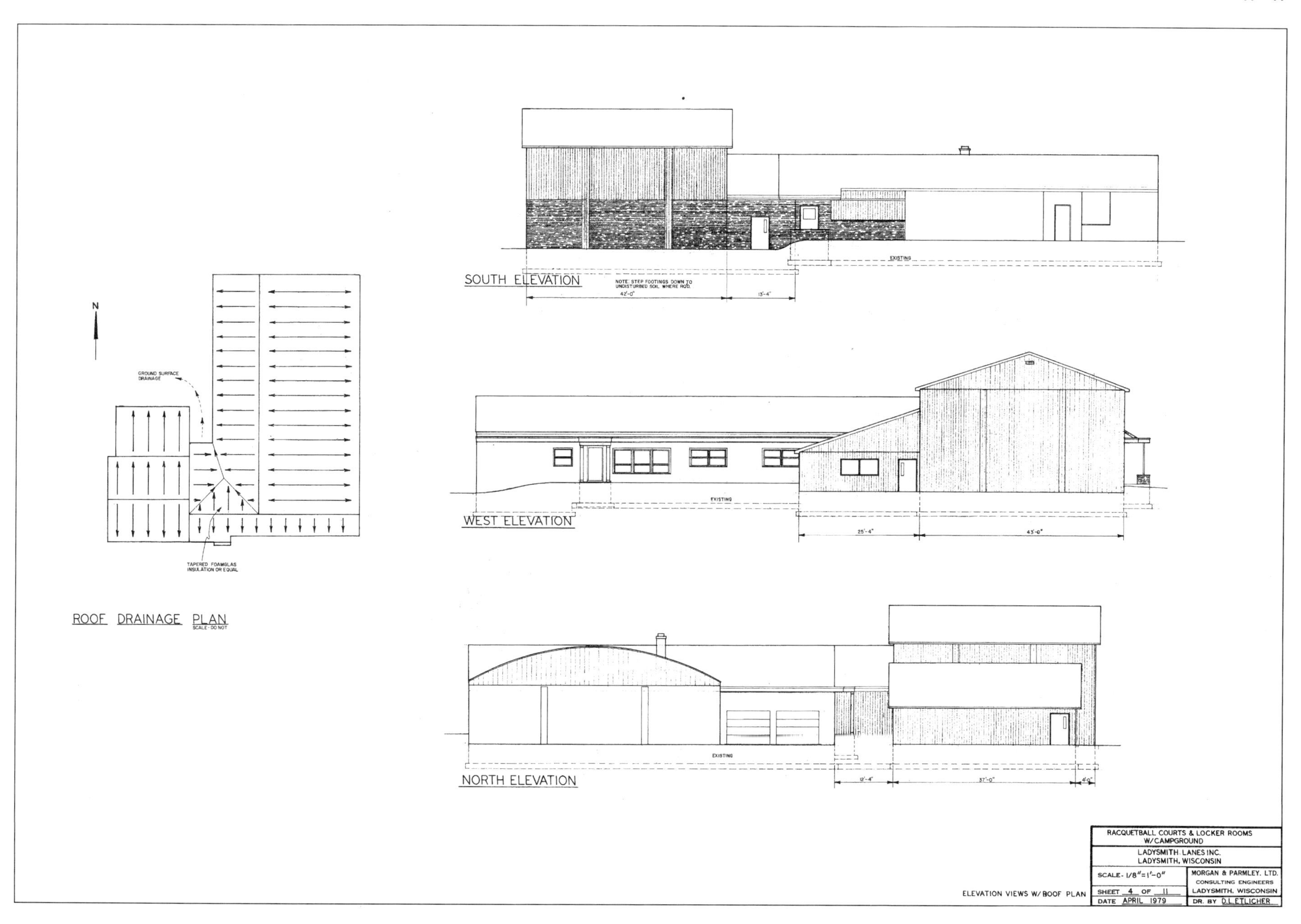
EXISTING
SOUTH ELEVATION
NOTE: STEP FOOTINGS DOWN TO
UNDISTURBED SOIL WHERE RQ'D.
42'-0"
13'-4"
N
GROUND SURFACE
DRAINAGE
TAPERED FOAMGLAS
INSULATION OR EQUAL
ROOF DRAINAGE PLAN
SCALE - DO NOT
EXISTING
WEST ELEVATION
25'-4"
43'-0"
EXISTING
NORTH ELEVATION
12'-4"
37'-0"
4'-0"
RACQUETBALL COURTS & LOCKER ROOMS
W/CAMPGROUND
LADYSMITH LANES INC.
LADYSMITH, WISCONSIN
SCALE - 1/8"=1'-0"
MORGAN & PARMLEY, LTD.
CONSULTING ENGINEERS
LADYSMITH, WISCONSIN
ELEVATION VIEWS W/ BOOF PLAN
SHEET 4 OF 11
DATE APRIL 1979
DR. BY D.L.ETLICHER

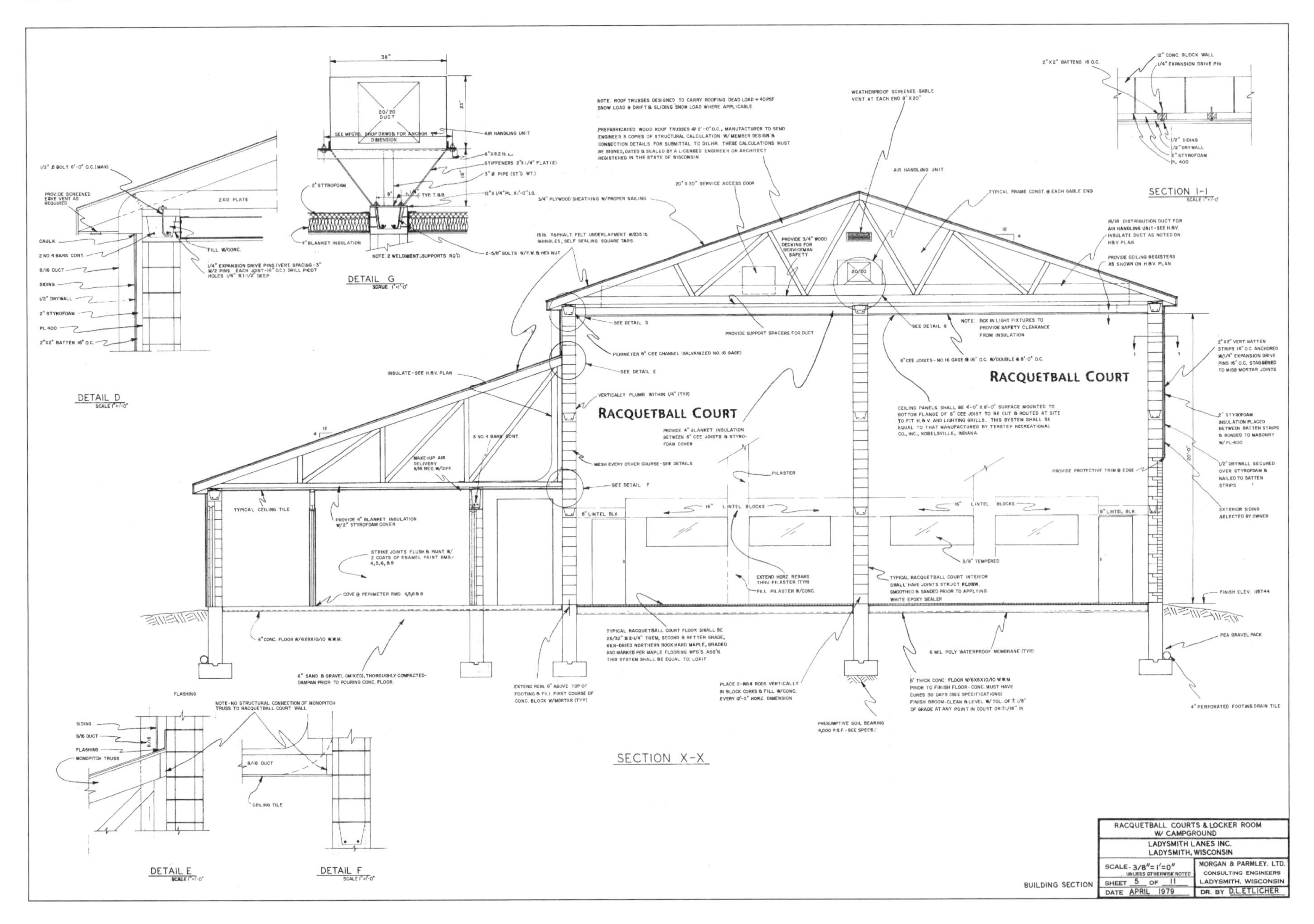
DETAIL D
SCALE 1"=1'-0"
1/2" Ø BOLT 6'-0" O.C. (MAX)
PROVIDE SCREENED EAVE VENT AS REQUIRED
2X12 PLATE
CAULK
2 NO. 4 BARS CONT.
8/16 DUCT
SIDING
1/2" DRYWALL
2" STYROFOAM
PL 400
2"X2" BATTEN 16" O.C.
FILL W/CONC.
DETAIL G
SCALE 1"=1'-0"
38"
20/20 DUCT
AIR HANDLING UNIT
2" STYROFOAM
4" BLANKET INSULATION
NOTE: 2 WELDMENT SUPPORTS RQ'D.
2-5/8" BOLTS W/F.W. & HEX NUT
NOTE: ROOF TRUSSES DESIGNED TO CARRY ROOFING DEAD LOAD + 40 PSF SNOW LOAD & DRIFT & SLIDING SNOW LOAD WHERE APPLICABLE.
WEATHERPROOF SCREENED GABLE VENT AT EACH END 8"X20"
AIR HANDLING UNIT
20"X30" SERVICE ACCESS DOOR
3/4" PLYWOOD SHEATHING W/PROPER NAILING
TYPICAL FRAME CONST. @ EACH GABLE END
SECTION 1-1
2"X2" BATTENS 16 O.C.
12" CONC. BLOCK WALL
1/4" EXPANSION DRIVE PIN
1/2" SIDING
1/2" DRYWALL
2" STYROFOAM
PL 400
SEE DETAIL D
SEE DETAIL E
SEE DETAIL F
SEE DETAIL G
PROVIDE SUPPORT SPACERS FOR DUCT
INSULATE-SEE H.&V. PLAN
RACQUETBALL COURT
RACQUETBALL COURT
VERTICALLY PLUMB WITHIN 1/4" (TYP)
3 NO. 4 BARS CONT.
MESH EVERY OTHER COURSE-SEE DETAILS
PILASTER
TYPICAL CEILING TILE
PROVIDE 4" BLANKET INSULATION W/2" STYROFOAM COVER
8" LINTEL BLK
16" LINTEL BLOCKS
3/8" TEMPERED
PROVIDE PROTECTIVE TRIM @ EDGE
FINISH ELEV. 187.44
EXTERIOR SIDING SELECTED BY OWNER
PEA GRAVEL PACK
6 MIL POLY WATERPROOF MEMBRANE (TYP)
4" PERFORATED FOOTING DRAIN TILE
4" CONC. FLOOR W/6X6X10/10 W.W.M.
PRESUMPTIVE SOIL BEARING 4,000 P.S.F.-SEE SPECS.
SECTION X-X
FLASHING
NOTE-NO STRUCTURAL CONNECTION OF MONOPITCH TRUSS TO RACQUETBALL COURT WALL
SIDING
8/16 DUCT
FLASHING
MONOPITCH TRUSS
CEILING TILE
DETAIL E
SCALE 1"=1'-0"
DETAIL F
SCALE 1"=1'-0"
RACQUETBALL COURTS & LOCKER ROOM W/ CAMPGROUND
LADYSMITH LANES INC.
LADYSMITH, WISCONSIN
SCALE- 3/8"=1'=0" UNLESS OTHERWISE NOTED
SHEET 5 OF 11
DATE APRIL 1979
MORGAN & PARMLEY, LTD.
CONSULTING ENGINEERS
LADYSMITH, WISCONSIN
DR. BY D.L.ETLICHER
BUILDING SECTION

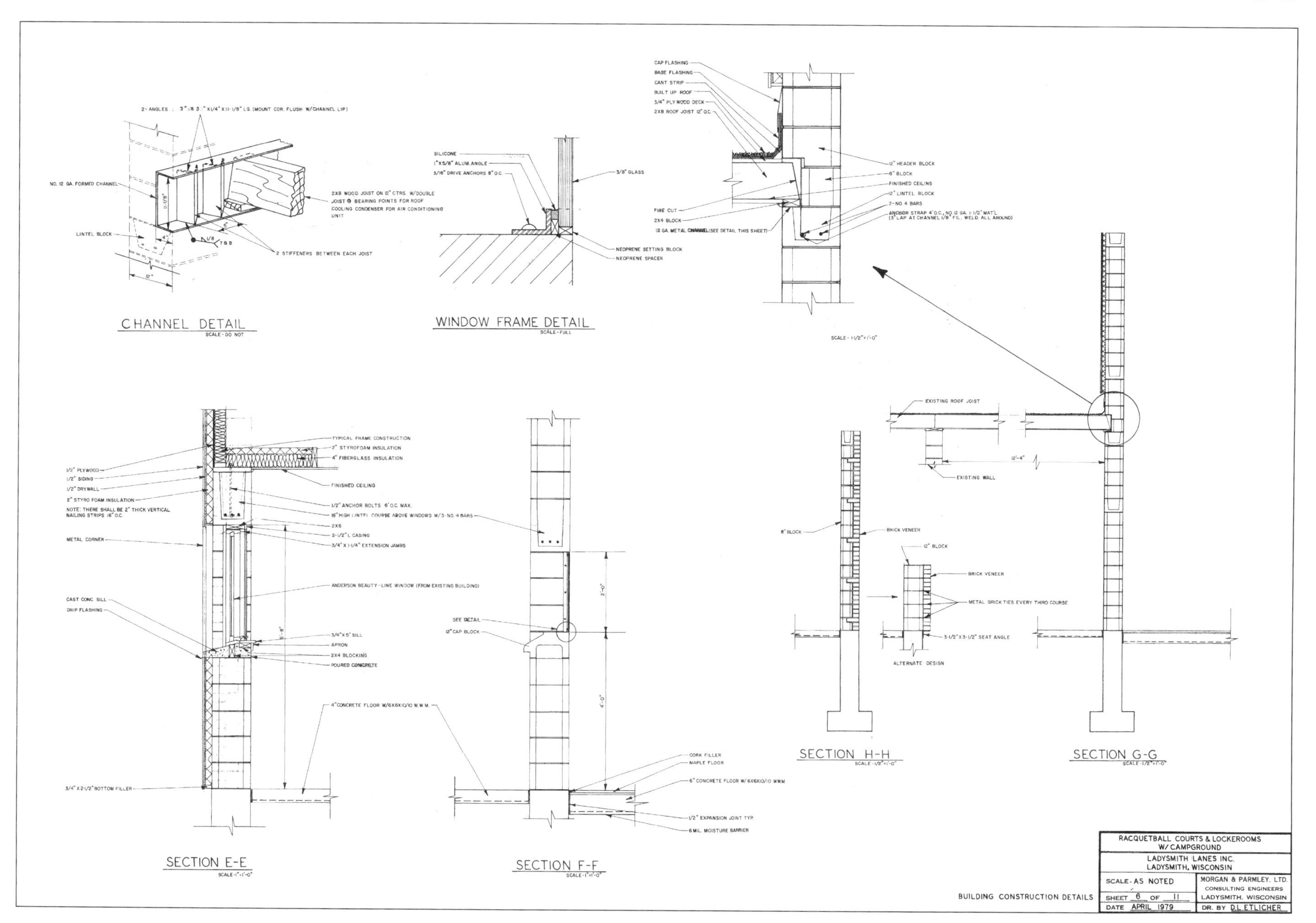
CHANNEL DETAIL
SCALE - DO NOT
WINDOW FRAME DETAIL
SCALE - FULL
SECTION E-E
SCALE - 1"=1'-0"
SECTION F-F
SCALE - 1"=1'-0"
SECTION H-H
SECTION G-G
RACQUETBALL COURTS & LOCKEROOMS W/ CAMPGROUND
LADYSMITH LANES INC.
LADYSMITH, WISCONSIN
SCALE - AS NOTED
SHEET 6 OF 11
DATE APRIL 1979
MORGAN & PARMLEY, LTD.
CONSULTING ENGINEERS
LADYSMITH, WISCONSIN
DR. BY D.L. ETLICHER
BUILDING CONSTRUCTION DETAILS

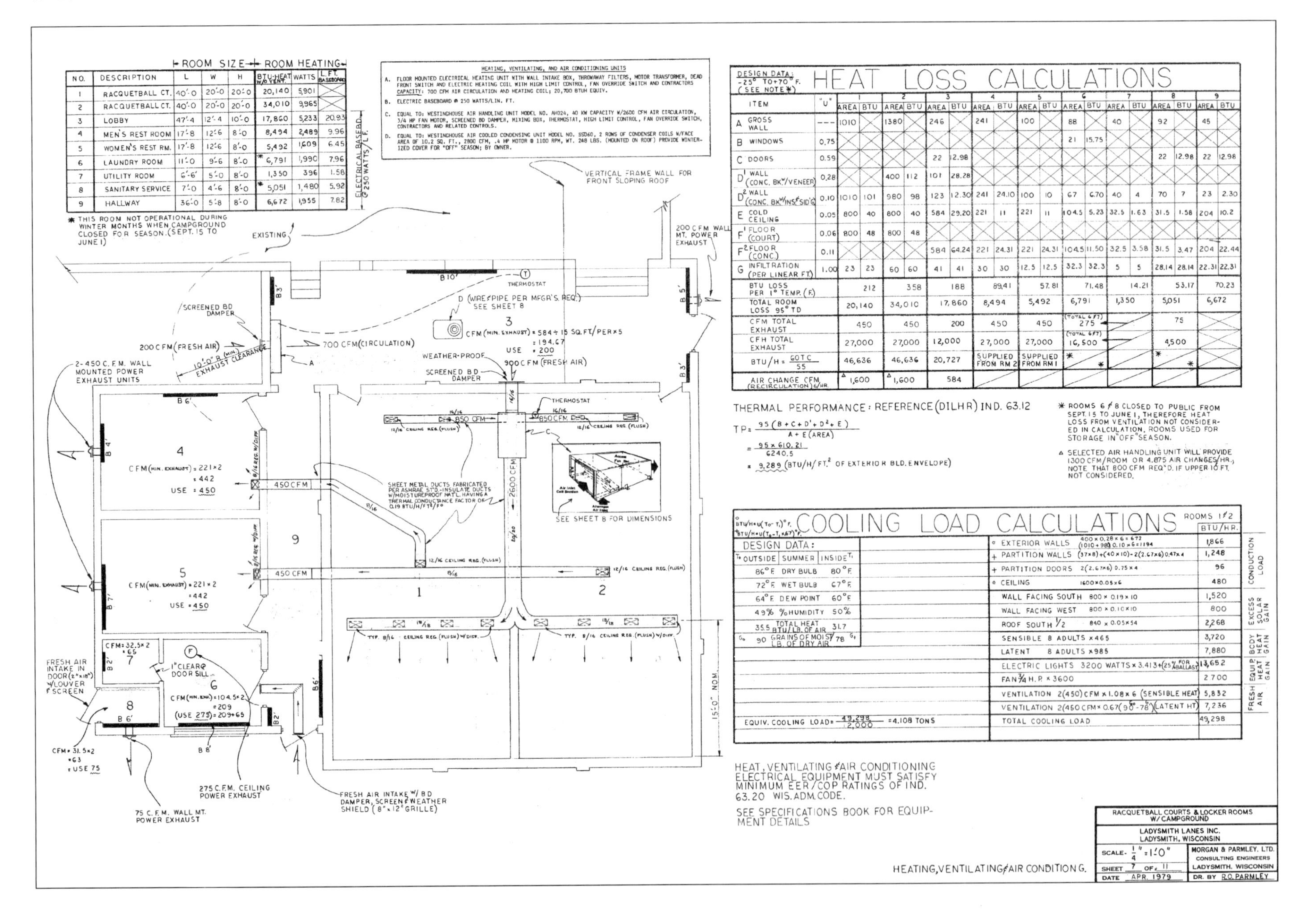

ROOM SIZE
ROOM HEATING
HEATING, VENTILATING, AND AIR CONDITIONING UNITS
HEAT LOSS CALCULATIONS
DESIGN DATA: -25° TO +70° F. (SEE NOTE *)
THERMAL PERFORMANCE: REFERENCE (DILHR) IND. 63.12
COOLING LOAD CALCULATIONS
ROOMS 1 & 2
DESIGN DATA:
TOTAL COOLING LOAD
HEAT, VENTILATING & AIR CONDITIONING ELECTRICAL EQUIPMENT MUST SATISFY MINIMUM EER/COP RATINGS OF IND. 63.20 WIS. ADM. CODE.
SEE SPECIFICATIONS BOOK FOR EQUIPMENT DETAILS
SEE SHEET 8 FOR DIMENSIONS
VERTICAL FRAME WALL FOR FRONT SLOPING ROOF
RACQUETBALL COURTS & LOCKER ROOMS W/ CAMPGROUND
LADYSMITH LANES INC.
LADYSMITH, WISCONSIN
MORGAN & PARMLEY, LTD.
CONSULTING ENGINEERS
LADYSMITH, WISCONSIN
SCALE- 1/4" = 1'-0"
SHEET 7 OF 11
DATE APR. 1979
DR. BY R.O. PARMLEY
HEATING, VENTILATING & AIR CONDITIONG.

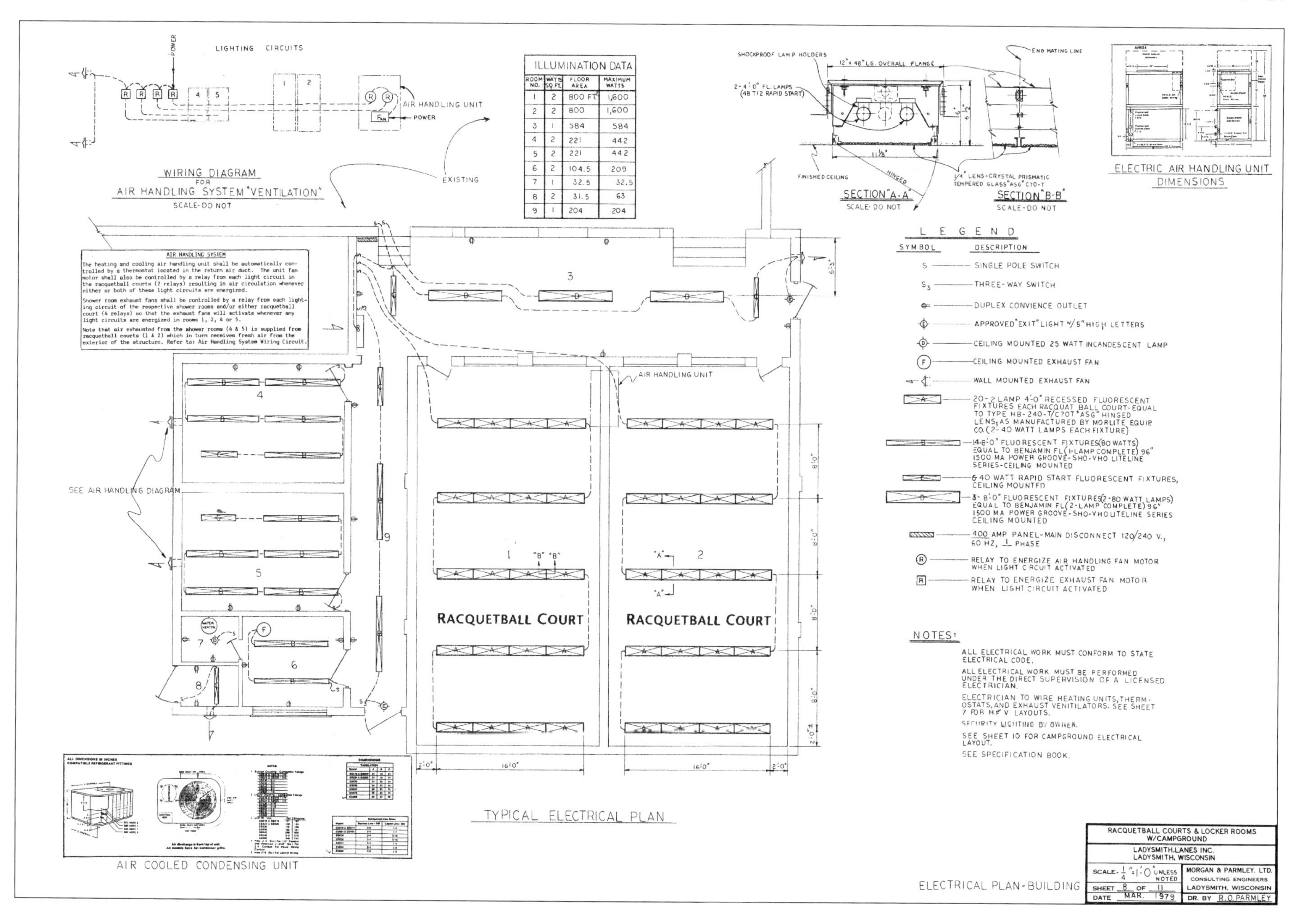
LIGHTING CIRCUITS
POWER
AIR HANDLING UNIT
POWER
EXISTING
WIRING DIAGRAM
FOR
AIR HANDLING SYSTEM "VENTILATION"
SCALE-DO NOT
ILLUMINATION DATA
ROOM NO. | WATTS SQ.FT. | FLOOR AREA | MAXIMUM WATTS
1 | 2 | 800 FT² | 1,600
2 | 2 | 800 | 1,600
3 | 1 | 584 | 584
4 | 2 | 221 | 442
5 | 2 | 221 | 442
6 | 2 | 104.5 | 209
7 | 1 | 32.5 | 32.5
8 | 2 | 31.5 | 63
9 | 1 | 204 | 204
SHOCKPROOF LAMP HOLDERS
12" x 48" LG. OVERALL FLANGE
2-4'-0" FL. LAMPS (48 T12 RAPID START)
END MATING LINE
FINISHED CEILING
HINGED
1/4" LENS-CRYSTAL PRISMATIC TEMPERED GLASS "ASG" C70-T
SECTION "A-A"
SCALE-DO NOT
SECTION "B-B"
SCALE-DO NOT
ELECTRIC AIR HANDLING UNIT DIMENSIONS
AIR HANDLING SYSTEM
The heating and cooling air handling unit shall be automatically controlled by a thermostat located in the return air duct. The unit fan motor shall also be controlled by a relay from each light circuit in the racquetball courts (2 relays) resulting in air circulation whenever either or both of these light circuits are energized.
Shower room exhaust fans shall be controlled by a relay from each lighting circuit of the respective shower rooms and/or either racquetball court (4 relays) so that the exhaust fans will activate whenever any light circuits are energized in rooms 1, 2, 4 or 5.
Note that air exhausted from the shower rooms (4 & 5) is supplied from racquetball courts (1 & 2) which in turn receives fresh air from the exterior of the structure. Refer to: Air Handling System Wiring Circuit.
SEE AIR HANDLING DIAGRAM
AIR HANDLING UNIT
RACQUETBALL COURT
RACQUETBALL COURT
L E G E N D
SYMBOL
DESCRIPTION
S — SINGLE POLE SWITCH
S3 — THREE-WAY SWITCH
DUPLEX CONVIENCE OUTLET
APPROVED "EXIT" LIGHT W/5" HIGH LETTERS
CEILING MOUNTED 25 WATT INCANDESCENT LAMP
CEILING MOUNTED EXHAUST FAN
WALL MOUNTED EXHAUST FAN
20-2 LAMP 4'-0" RECESSED FLUORESCENT FIXTURES EACH RACQUAT BALL COURT-EQUAL TO TYPE HB-240-T/C70T "ASG" HINGED LENS, AS MANUFACTURED BY MORLITE EQUIP CO. (2-40 WATT LAMPS EACH FIXTURE)
14-8'-0" FLUORESCENT FIXTURES (80 WATTS) EQUAL TO BENJAMIN FL (1-LAMP COMPLETE) 96" 1500 MA POWER GROOVE-SHO-VHO LITELINE SERIES-CEILING MOUNTED
6 40 WATT RAPID START FLUORESCENT FIXTURES, CEILING MOUNTED
3-8'-0" FLUORESCENT FIXTURES (2-80 WATT LAMPS) EQUAL TO BENJAMIN FL (2-LAMP COMPLETE) 96" 1500 MA POWER GROOVE-SHO-VHO LITELINE SERIES CEILING MOUNTED
400 AMP PANEL-MAIN DISCONNECT 120/240 V., 60 HZ, 1 PHASE
RELAY TO ENERGIZE AIR HANDLING FAN MOTOR WHEN LIGHT CIRCUIT ACTIVATED
RELAY TO ENERGIZE EXHAUST FAN MOTOR WHEN LIGHT CIRCUIT ACTIVATED
NOTES:
ALL ELECTRICAL WORK MUST CONFORM TO STATE ELECTRICAL CODE.
ALL ELECTRICAL WORK MUST BE PERFORMED UNDER THE DIRECT SUPERVISION OF A LICENSED ELECTRICIAN.
ELECTRICIAN TO WIRE HEATING UNITS, THERMOSTATS, AND EXHAUST VENITILATORS. SEE SHEET 7 FOR H & V LAYOUTS.
SECURITY LIGHTING BY OWNER.
SEE SHEET 10 FOR CAMPGROUND ELECTRICAL LAYOUT.
SEE SPECIFICATION BOOK.
2'-0" 16'-0" 16'-0" 2'-0"
TYPICAL ELECTRICAL PLAN
AIR COOLED CONDENSING UNIT
ELECTRICAL PLAN-BUILDING
RACQUETBALL COURTS & LOCKER ROOMS W/CAMPGROUND
LADYSMITH LANES INC.
LADYSMITH, WISCONSIN
SCALE- 1/4"=1'-0" UNLESS NOTED
SHEET 8 OF 11
DATE MAR. 1979
MORGAN & PARMLEY, LTD.
CONSULTING ENGINEERS
LADYSMITH, WISCONSIN
DR. BY R.O. PARMLEY

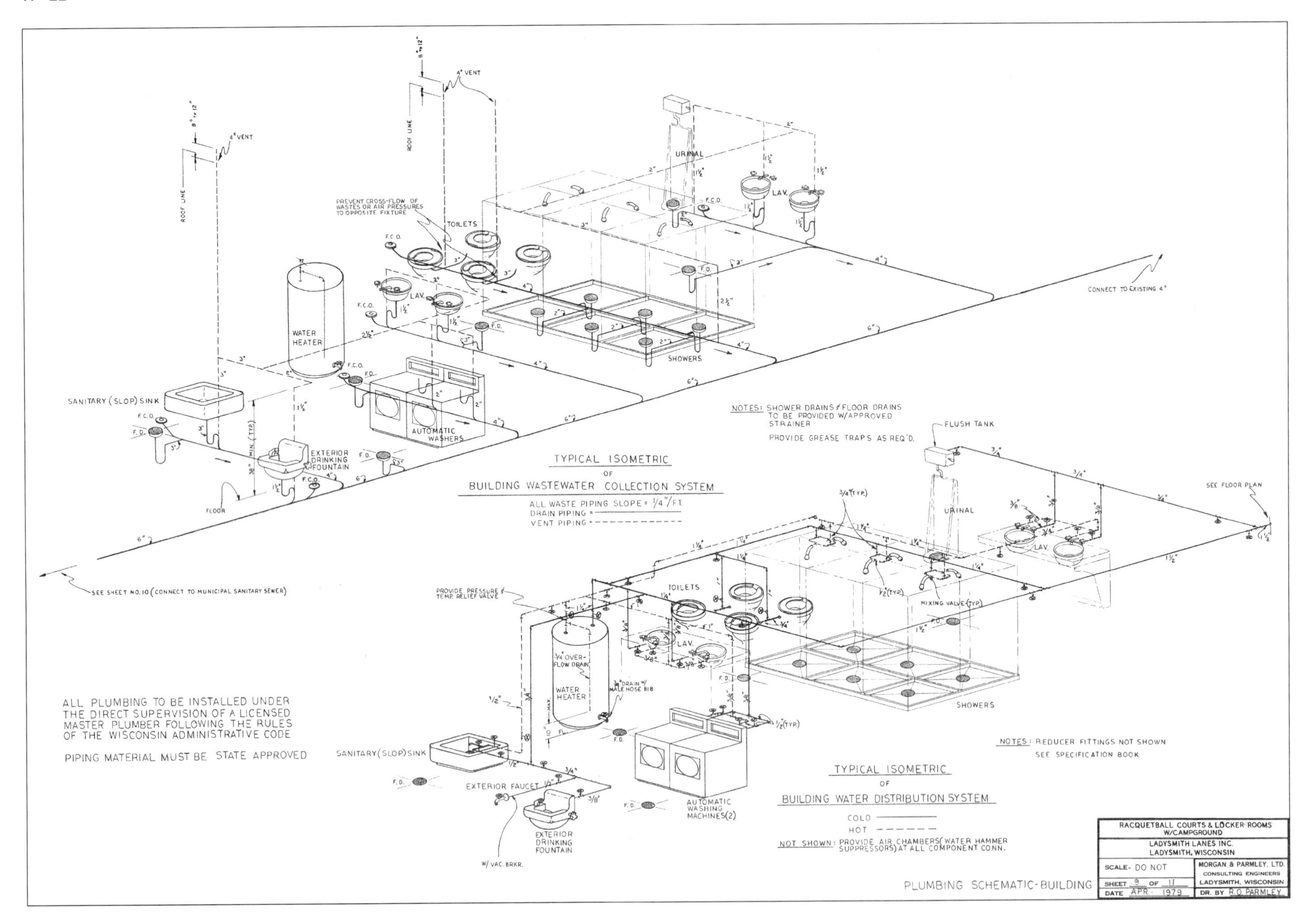
TYPICAL ISOMETRIC OF BUILDING WASTEWATER COLLECTION SYSTEM
ALL WASTE PIPING SLOPE = 1/4"/FT.
DRAIN PIPING =
VENT PIPING =
NOTES: SHOWER DRAINS & FLOOR DRAINS TO BE PROVIDED W/APPROVED STRAINER
PROVIDE GREASE TRAPS AS REQ'D.
PREVENT CROSS-FLOW OF WASTES OR AIR PRESSURES TO OPPOSITE FIXTURE
TOILETS
URINAL
LAV.
WATER HEATER
SANITARY (SLOP) SINK
EXTERIOR DRINKING FOUNTAIN
AUTOMATIC WASHERS
SHOWERS
ROOF LINE
4" VENT
FLOOR
CONNECT TO EXISTING 4"
SEE SHEET NO.10 (CONNECT TO MUNICIPAL SANITARY SEWER)
TYPICAL ISOMETRIC OF BUILDING WATER DISTRIBUTION SYSTEM
COLD
HOT
NOT SHOWN: PROVIDE AIR CHAMBERS (WATER HAMMER SUPPRESSORS) AT ALL COMPONENT CONN.
NOTES: REDUCER FITTINGS NOT SHOWN SEE SPECIFICATION BOOK
FLUSH TANK
SEE FLOOR PLAN
MIXING VALVE (TYP.)
PROVIDE PRESSURE & TEMP. RELIEF VALVE
3/4" OVERFLOW DRAIN
3/4" DRAIN W/ MALE HOSE BIB
EXTERIOR FAUCET
W/ VAC. BRKR.
AUTOMATIC WASHING MACHINES(2)
ALL PLUMBING TO BE INSTALLED UNDER THE DIRECT SUPERVISION OF A LICENSED MASTER PLUMBER FOLLOWING THE RULES OF THE WISCONSIN ADMINISTRATIVE CODE
PIPING MATERIAL MUST BE STATE APPROVED
PLUMBING SCHEMATIC-BUILDING
RACQUETBALL COURTS & LOCKER ROOMS W/CAMPGROUND
LADYSMITH LANES INC. LADYSMITH, WISCONSIN
SCALE- DO NOT
SHEET 9 OF 11
DATE APR. 1979
MORGAN & PARMLEY, LTD. CONSULTING ENGINEERS LADYSMITH, WISCONSIN
DR. BY R.O. PARMLEY

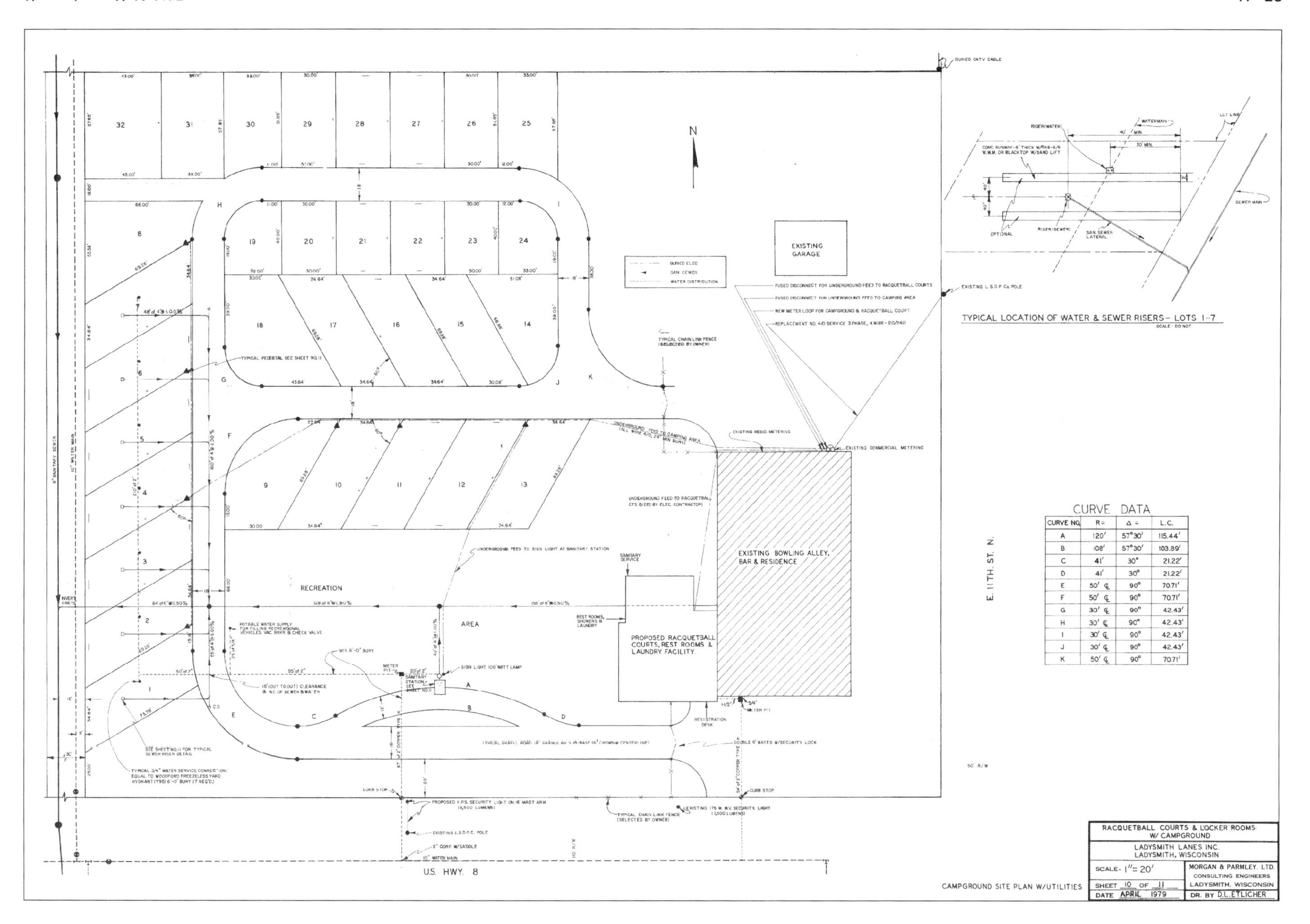

RACQUETBALL COURTS & LOCKER ROOMS W/ CAMPGROUND
LADYSMITH LANES INC. LADYSMITH, WISCONSIN
SCALE- 1"= 20'
SHEET 10 OF 11
DATE APRIL 1979
MORGAN & PARMLEY, LTD. CONSULTING ENGINEERS LADYSMITH, WISCONSIN
DR. BY D.L. ETLICHER
CAMPGROUND SITE PLAN W/UTILITIES
TYPICAL LOCATION OF WATER & SEWER RISERS – LOTS 1-7
SCALE - DO NOT
CURVE DATA
CURVE NO. | R= | Δ = | L.C.
A | 120' | 57°30' | 115.44'
B | 108' | 57°30' | 103.89'
C | 41' | 30° | 21.22'
D | 41' | 30° | 21.22'
E | 50' ℄ | 90° | 70.71'
F | 50' ℄ | 90° | 70.71'
G | 30' ℄ | 90° | 42.43'
H | 30' ℄ | 90° | 42.43'
I | 30' ℄ | 90° | 42.43'
J | 30' ℄ | 90° | 42.43'
K | 50' ℄ | 90° | 70.71'
EXISTING GARAGE
EXISTING BOWLING ALLEY, BAR & RESIDENCE
PROPOSED RACQUETBALL COURTS, REST ROOMS & LAUNDRY FACILITY
RECREATION
AREA
E. 11TH. ST. N.
U.S. HWY. 8
N
BURIED CATV CABLE
EXISTING L.S.O.P. Co. POLE
FUSED DISCONNECT FOR UNDERGROUND FEED TO RACQUETBALL COURTS
FUSED DISCONNECT FOR UNDERGROUND FEED TO CAMPING AREA
NEW METER LOOP FOR CAMPGROUND & RACQUETBALL COURT
REPLACEMENT NO. 4/0 SERVICE 3 PHASE, 4 WIRE - 120/240
TYPICAL CHAIN LINK FENCE (SELECTED BY OWNER)
TYPICAL PEDESTAL SEE SHEET NO. 11
EXISTING RESID. METERING
EXISTING COMMERCIAL METERING
UNDERGROUND FEED TO RACQUETBALL CTS. SIZED BY ELEC. CONTRACTOR
UNDERGROUND FEED TO SIGN LIGHT AT SANITARY STATION
SANITARY SERVICE
REST ROOMS, SHOWERS & LAUNDRY
POTABLE WATER SUPPLY FOR FILLING RECREATIONAL VEHICLES. VAC. BRKR. & CHECK VALVE
MIN. 6'-0" BURY
METER PIT
SANITARY STATION
SIGN LIGHT 100 WATT LAMP
18' (OUT TO OUT) CLEARANCE
SEE SHEET NO. 11 FOR TYPICAL SEWER RISER DETAIL
TYPICAL 3/4" WATER SERVICE CONNECTION EQUAL TO WOODFORD FREEZELESS YARD HYDRANT (Y95) 6'-0" BURY (7 REQ'D)
REGISTRATION DESK
METER PIT
DOUBLE 6' GATES W/SECURITY LOCK
CURB STOP
PROPOSED H.P.S. SECURITY LIGHT ON 16' MAST ARM (8,500 LUMENS)
EXISTING L.S.O.P.C. POLE
2" CORP. W/SADDLE
10" WATER MAIN
EXISTING 175 W. M.V. SECURITY LIGHT (7,500 LUMENS)
TYPICAL CHAIN LINK FENCE (SELECTED BY OWNER)
8" SANITARY SEWER
10" WATER MAIN
RISER(WATER)
WATERMAIN
LOT LINE
SEWER MAIN
OPTIONAL
RISER(SEWER)
SAN. SEWER LATERAL

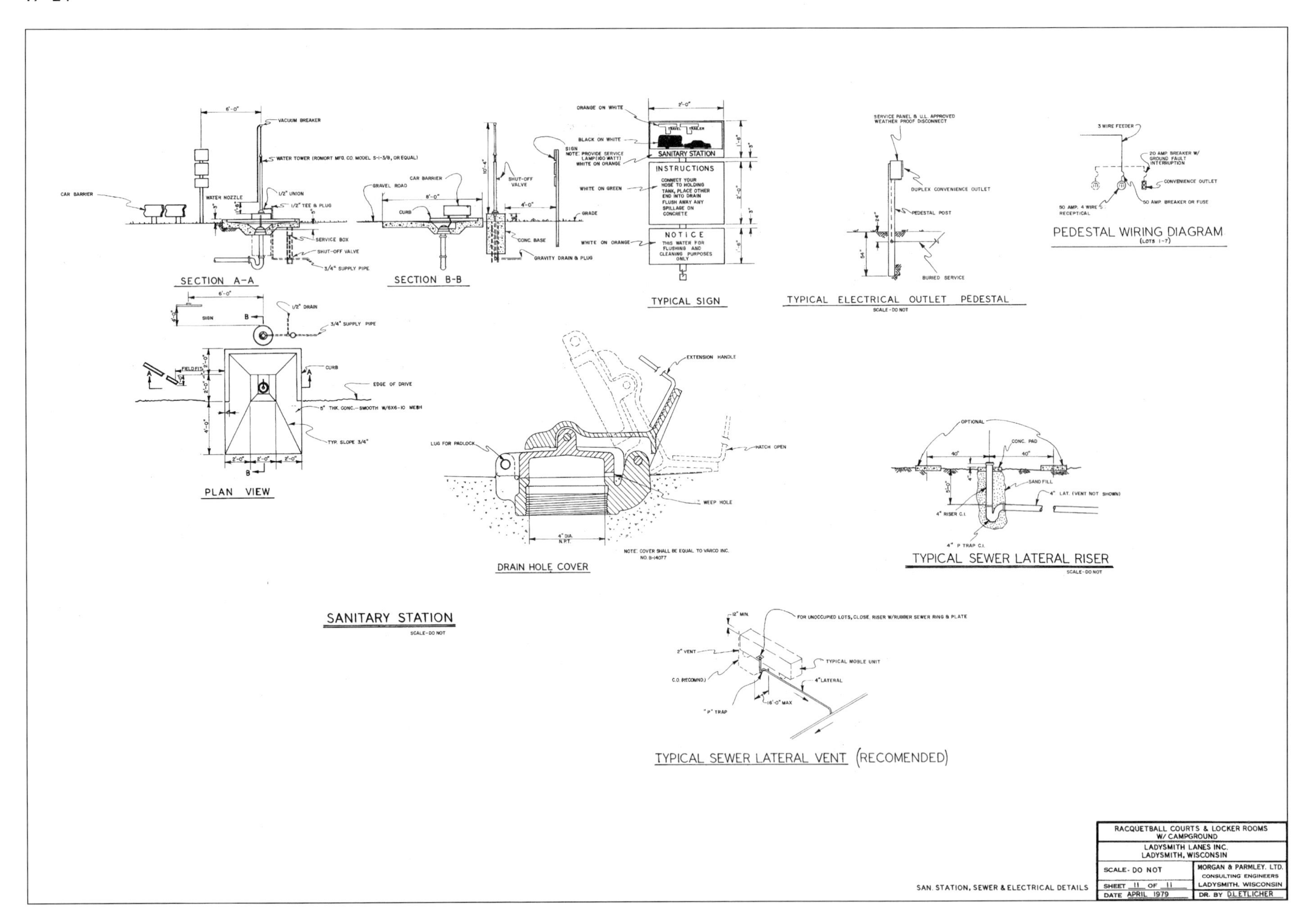
SECTION A-A
SECTION B-B
PLAN VIEW
TYPICAL SIGN
SANITARY STATION
INSTRUCTIONS
CONNECT YOUR HOSE TO HOLDING TANK, PLACE OTHER END INTO DRAIN FLUSH AWAY ANY SPILLAGE ON CONCRETE
NOTICE
THIS WATER FOR FLUSHING AND CLEANING PURPOSES ONLY
TYPICAL ELECTRICAL OUTLET PEDESTAL
SCALE-DO NOT
PEDESTAL WIRING DIAGRAM
(LOTS 1-7)
DRAIN HOLE COVER
TYPICAL SEWER LATERAL RISER
SCALE-DO NOT
SANITARY STATION
SCALE-DO NOT
TYPICAL SEWER LATERAL VENT (RECOMENDED)
RACQUETBALL COURTS & LOCKER ROOMS W/ CAMPGROUND
LADYSMITH LANES INC.
LADYSMITH, WISCONSIN
SCALE- DO NOT
SHEET 11 OF 11
DATE APRIL 1979
MORGAN & PARMLEY, LTD.
CONSULTING ENGINEERS
LADYSMITH, WISCONSIN
DR. BY D.L.ETLICHER
SAN. STATION, SEWER & ELECTRICAL DETAILS

设　计

第 18 节　拖车式活动住屋和露营场所

例　18

最近 30 年，拖车式活动住屋和露营场所在全国各地快速发展。随着人口的增长以及对田园式消遣的需求，这种类型的设计将进一步增加。

第一套设计图给出了待建拖车式活动住屋，它与市政污水管道和与这些设施相连的干管系统相邻。第二套图纸只包括了市政污水管道，这样还需要私用的供水和配水网络，如设计图所示。第三套设计图给出了除了电力服务以外就没有任何市政服务的乡间露营场所的详图。最后一套图纸示例显示了对本节第一套图纸所示的已有设施的加建。值得注意的是，名称从 Parkview 移动住宅公园转变为 Leatherman 拖车式活动住屋。

目　录

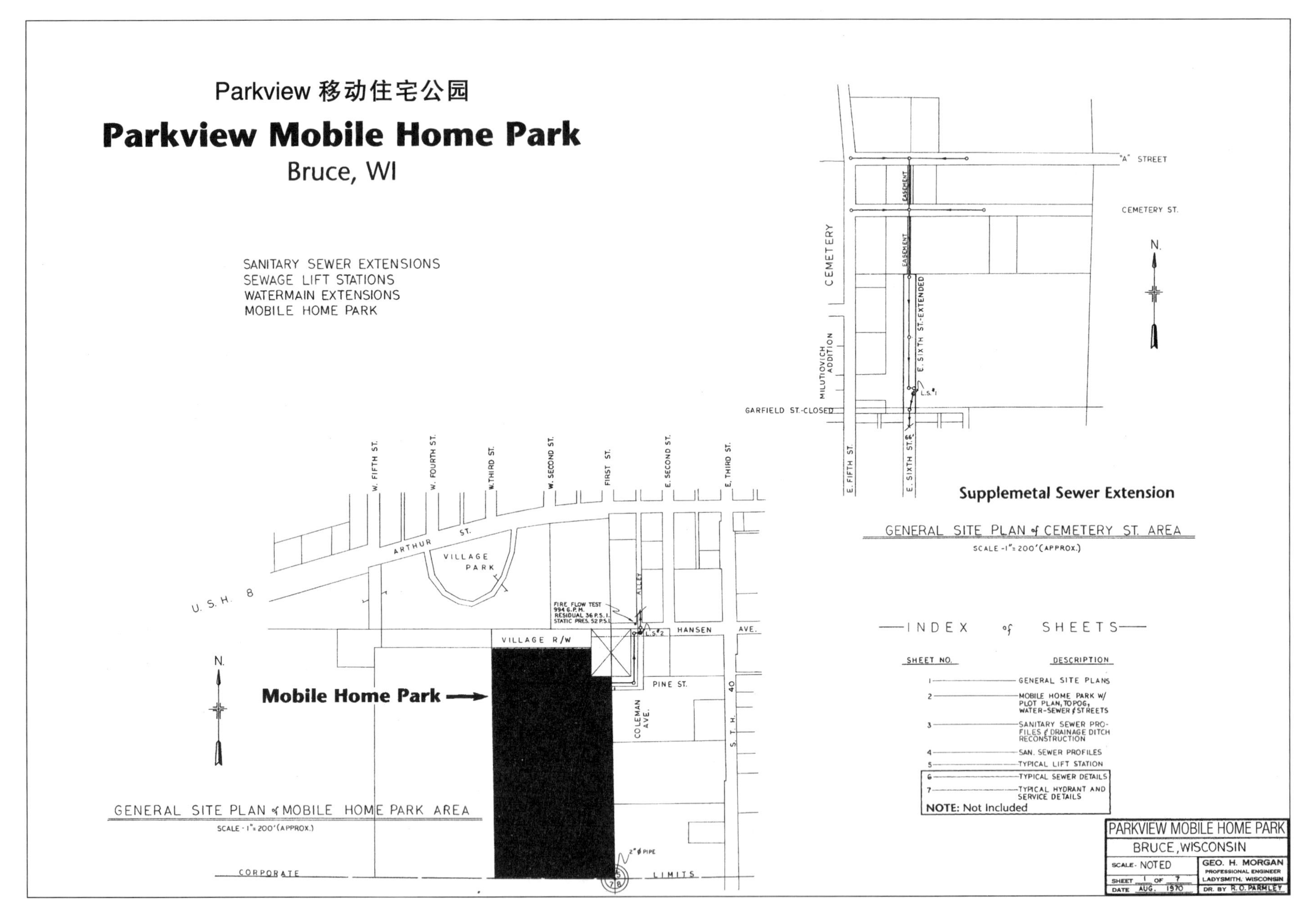

Courtesy: Morgan & Parmley, Ltd.

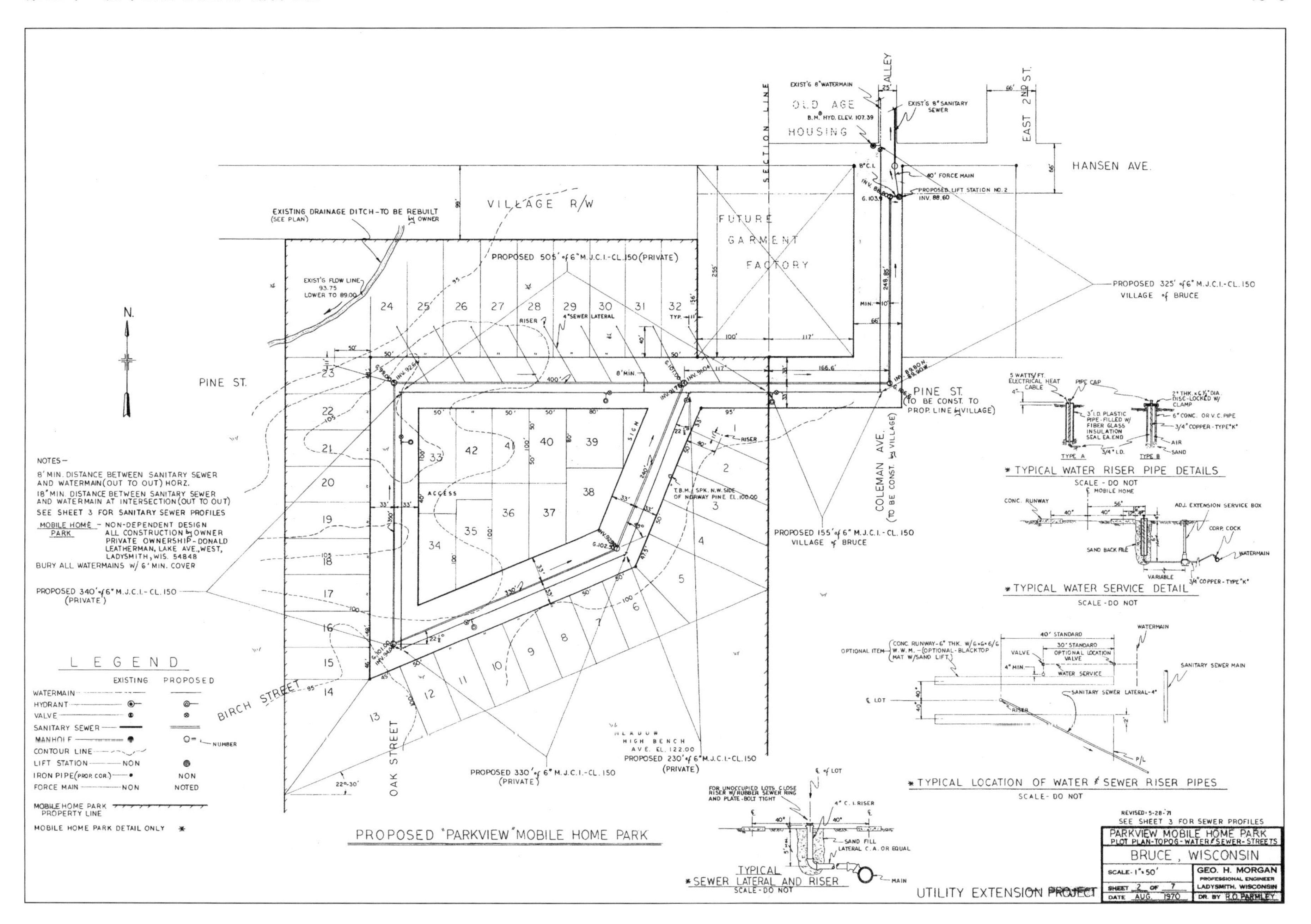

PROPOSED "PARKVIEW" MOBILE HOME PARK
PARKVIEW MOBILE HOME PARK
PLOT PLAN-TOPOG-WATER & SEWER-STREETS
BRUCE, WISCONSIN
UTILITY EXTENSION PROJECT
GEO. H. MORGAN
PROFESSIONAL ENGINEER
LADYSMITH, WISCONSIN
SCALE - 1" = 50'
SHEET 2 OF 7
DATE AUG. 1970
DR. BY R.O. PARMLEY
REVISED 5-28-'71
SEE SHEET 3 FOR SEWER PROFILES
NOTES—
8' MIN. DISTANCE BETWEEN SANITARY SEWER AND WATERMAIN (OUT TO OUT) HORZ.
18" MIN. DISTANCE BETWEEN SANITARY SEWER AND WATERMAIN AT INTERSECTION (OUT TO OUT)
SEE SHEET 3 FOR SANITARY SEWER PROFILES
MOBILE HOME PARK — NON-DEPENDENT DESIGN ALL CONSTRUCTION by OWNER PRIVATE OWNERSHIP-DONALD LEATHERMAN, LAKE AVE., WEST, LADYSMITH, WIS. 54848
BURY ALL WATERMAINS W/ 6' MIN. COVER
LEGEND
EXISTING PROPOSED
WATERMAIN
HYDRANT
VALVE
SANITARY SEWER
MANHOLE
NUMBER
CONTOUR LINE
LIFT STATION NON
IRON PIPE (PROP. COR.) NON
FORCE MAIN NON NOTED
MOBILE HOME PARK PROPERTY LINE
MOBILE HOME PARK DETAIL ONLY *
*TYPICAL WATER RISER PIPE DETAILS
SCALE - DO NOT
*TYPICAL WATER SERVICE DETAIL
SCALE - DO NOT
*TYPICAL LOCATION OF WATER & SEWER RISER PIPES
SCALE - DO NOT
TYPICAL *SEWER LATERAL AND RISER
SCALE - DO NOT
PINE ST.
HANSEN AVE.
EAST 2ND ST.
ALLEY
BIRCH STREET
OAK STREET
COLEMAN AVE. (TO BE CONST. by VILLAGE)
PINE ST. (TO BE CONST. TO PROP. LINE by VILLAGE)
VILLAGE R/W
FUTURE GARMENT FACTORY
OLD AGE HOUSING
SECTION LINE
N.
EXISTING DRAINAGE DITCH—TO BE REBUILT (SEE PLAN) by OWNER
EXIST'G FLOW LINE 93.75 LOWER TO 89.00
PROPOSED 505' of 6" M.J.C.I.-CL.150 (PRIVATE)
PROPOSED 325' of 6" M.J.C.I.-CL.150 VILLAGE of BRUCE
PROPOSED 155' of 6" M.J.C.I.-CL.150 VILLAGE of BRUCE
PROPOSED 340' of 6" M.J.C.I.-CL.150 (PRIVATE)
PROPOSED 330' of 6" M.J.C.I.-CL.150 (PRIVATE)
PROPOSED 230' of 6" M.J.C.I.-CL.150 (PRIVATE)
EXIST'G 8" WATERMAIN
EXIST'G 8" SANITARY SEWER
B.M. HYD. ELEV. 107.39
40' FORCE MAIN
PROPOSED LIFT STATION NO. 2 INV. 88.60
T.B.M. SPK. N.W. SIDE OF NORWAY PINE EL. 100.00
HIGH BENCH AVE. EL. 122.00
ACCESS
SIGN
RISER
4" SEWER LATERAL

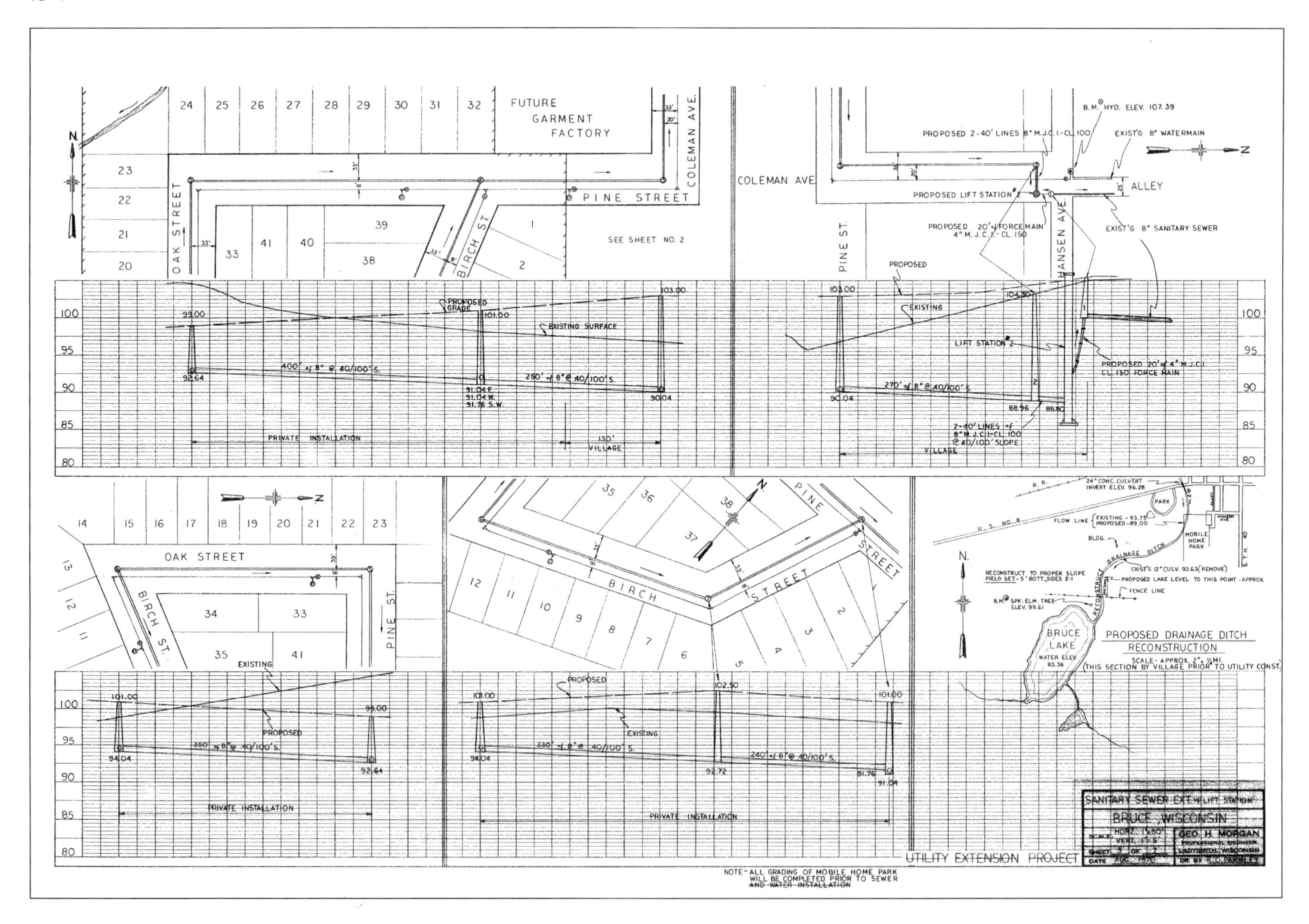
FUTURE GARMENT FACTORY
COLEMAN AVE.
PINE STREET
OAK STREET
BIRCH ST.
SEE SHEET NO. 2
PROPOSED GRADE
EXISTING SURFACE
400' of 8" @ .40/100'S.
250' of 8" @ .40/100'S.
PRIVATE INSTALLATION
VILLAGE
B.M. HYD. ELEV. 107.39
PROPOSED 2-40' LINES 8" M.J.C.I.-CL.100
EXIST'G 8" WATERMAIN
ALLEY
PROPOSED LIFT STATION 2
PROPOSED 20' FORCE MAIN 4" M.J.C.I.-CL.150
EXIST'G 8" SANITARY SEWER
HANSEN AVE.
PINE ST.
PROPOSED
EXISTING
LIFT STATION 2
PROPOSED 20' of 4" M.J.C.I. CL.150 FORCE MAIN
270' of 8" @ .40/100'S.
2-40' LINES of 8" M.J.C.I.-CL.100 @ .40/100' SLOPE
PINE ST.
350' of 8" @ .40/100'S.
BIRCH STREET
PINE STREET
330' of 8" @ .40/100'S.
240' of 8" @ .40/100'S.
PRIVATE INSTALLATION
24" CONC. CULVERT INVERT ELEV. 96.28
PARK
U.S. NO. 8
FLOW LINE EXISTING-93.75 PROPOSED-89.00
MOBILE HOME PARK
BLDG.
DRAINAGE DITCH
EXIST'G 12" CULV. 93.63 (REMOVE)
RECONSTRUCT TO PROPER SLOPE FIELD SET-5' BOTT. SIDES 2:1
PROPOSED LAKE LEVEL TO THIS POINT-APPROX.
FENCE LINE
B.M. SPK. ELM. TREE ELEV. 99.61
BRUCE LAKE
WATER ELEV. 83.36
PROPOSED DRAINAGE DITCH RECONSTRUCTION
SCALE-APPROX. 2"=1/4 MI.
(THIS SECTION BY VILLAGE PRIOR TO UTILITY CONST.)
S.T.H. 40
SANITARY SEWER EXT. W/LIFT STATION
BRUCE, WISCONSIN
GEO. H. MORGAN
PROFESSIONAL ENGINEER
LADYSMITH, WISCONSIN
DR. BY R.O. PARMLEY
UTILITY EXTENSION PROJECT
NOTE-ALL GRADING OF MOBILE HOME PARK WILL BE COMPLETED PRIOR TO SEWER AND WATER INSTALLATION

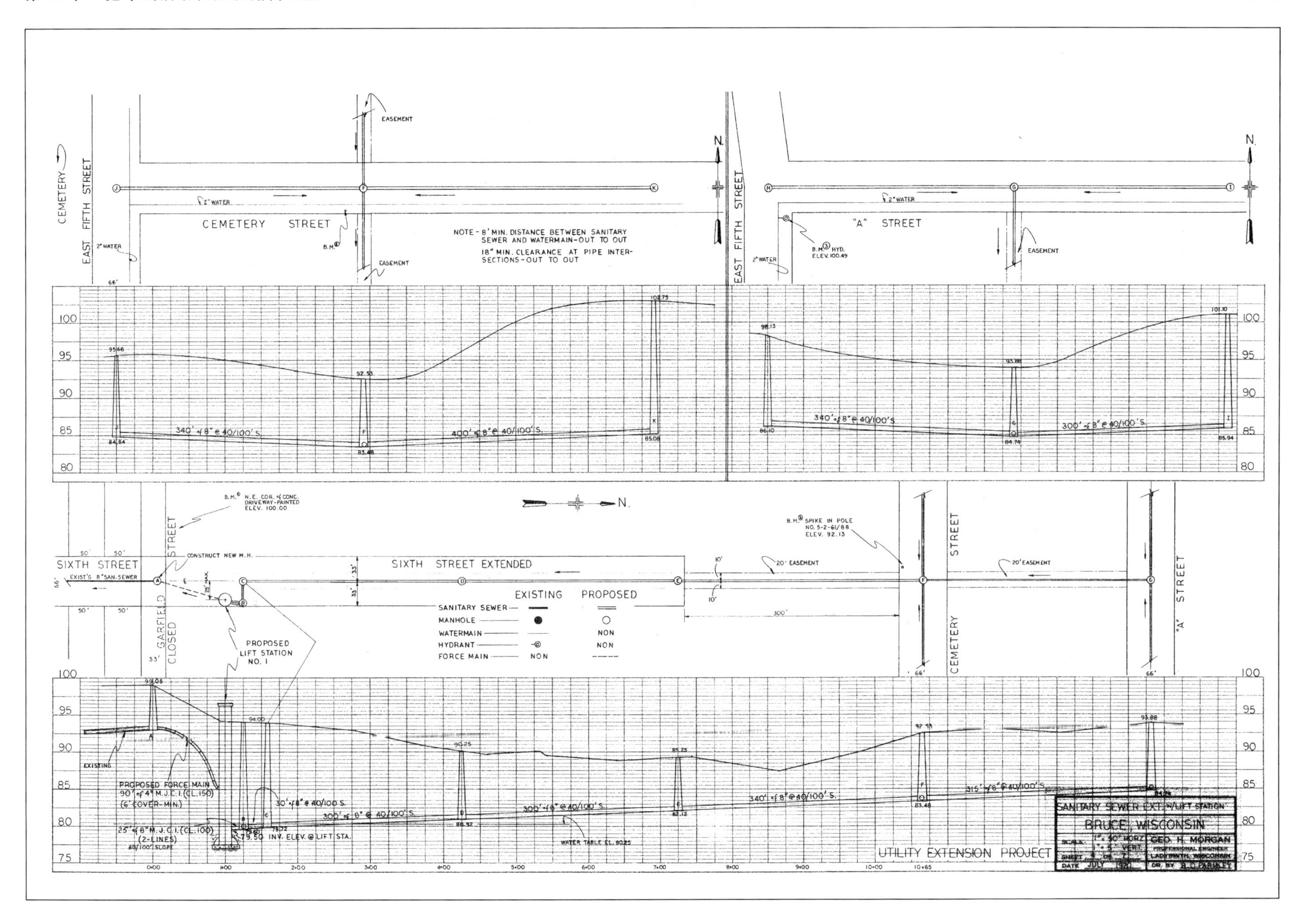
CEMETERY
EAST FIFTH STREET
CEMETERY STREET
2" WATER
EASEMENT
B.M.②
NOTE - 8' MIN. DISTANCE BETWEEN SANITARY SEWER AND WATERMAIN-OUT TO OUT
18" MIN. CLEARANCE AT PIPE INTERSECTIONS-OUT TO OUT
"A" STREET
B.M.③ HYD. ELEV. 100.49
N.
340' of 8" @ 40/100'S.
400' of 8" @ 40/100'S.
300' of 8" @ 40/100'S.
B.M.① N.E. COR. of CONC. DRIVEWAY-PAINTED ELEV. 100.00
B.M.⑤ SPIKE IN POLE NO. 5-2-61/88 ELEV. 92.13
SIXTH STREET
EXIST'G 8" SAN. SEWER
CONSTRUCT NEW M.H.
SIXTH STREET EXTENDED
20' EASEMENT
GARFIELD STREET CLOSED
CEMETERY STREET
"A" STREET
EXISTING
PROPOSED
SANITARY SEWER
MANHOLE
WATERMAIN
HYDRANT
FORCE MAIN
NON
PROPOSED LIFT STATION NO. 1
EXISTING
PROPOSED FORCE MAIN 90' of 4" M.J.C.I. (CL.150) (6' COVER-MIN.)
25' of 8" M.J.C.I. (CL.100) (2-LINES) 40/100' SLOPE
79.50 INV. ELEV. @ LIFT STA.
30' of 8" @ 40/100 S.
300' of 8" @ 40/100'S.
340' of 8" @ 40/100'S.
315' of 8" @ 40/100'S.
WATER TABLE EL. 80.25
UTILITY EXTENSION PROJECT
SANITARY SEWER EXT. W/LIFT STATION
BRUCE, WISCONSIN
GEO. H. MORGAN
PROFESSIONAL ENGINEER
DATE JULY 1970

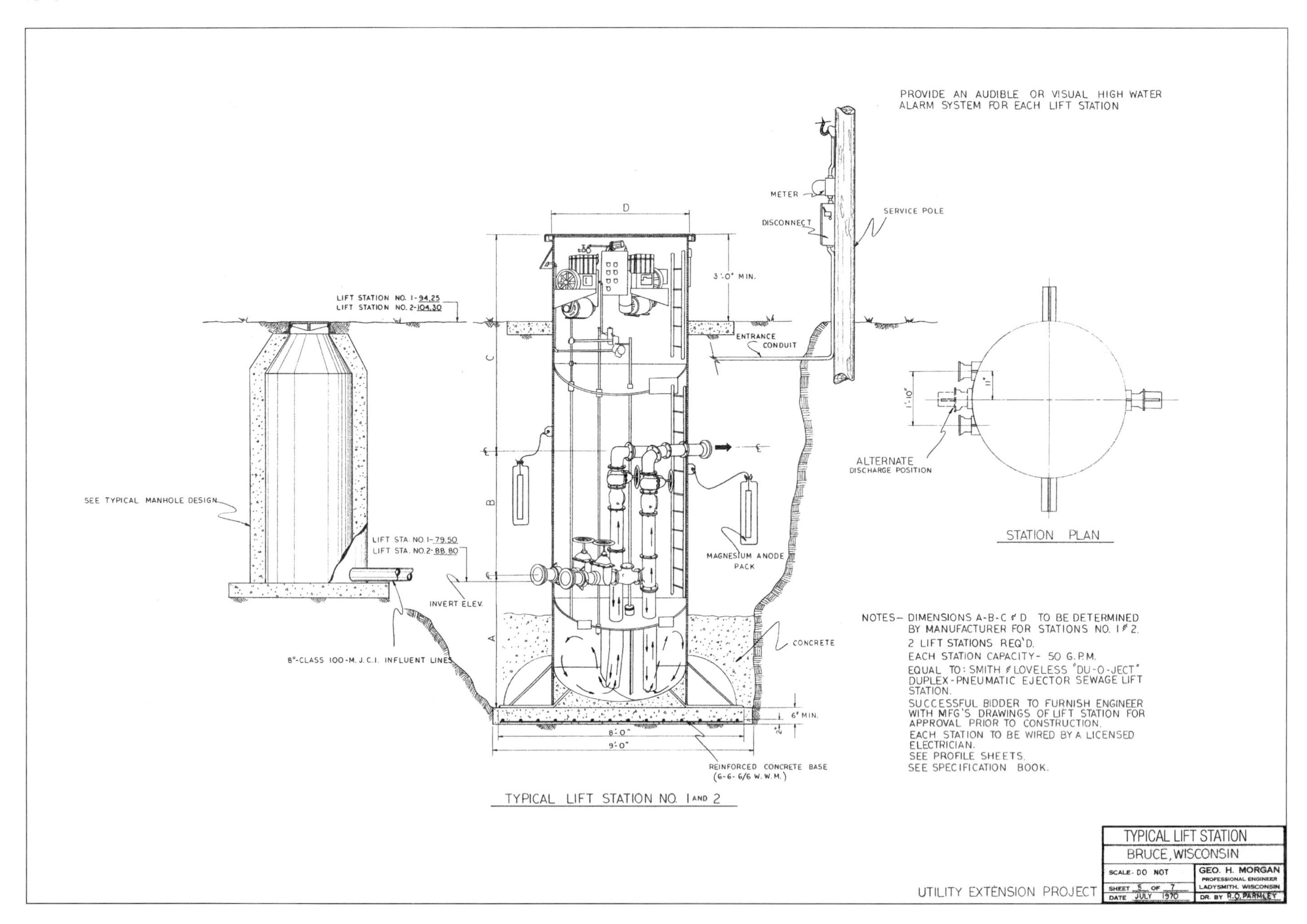

PROVIDE AN AUDIBLE OR VISUAL HIGH WATER ALARM SYSTEM FOR EACH LIFT STATION
METER
DISCONNECT
SERVICE POLE
D
3'-0" MIN.
LIFT STATION NO. 1-94.25
LIFT STATION NO. 2-104.30
ENTRANCE CONDUIT
C
B
A
SEE TYPICAL MANHOLE DESIGN
LIFT STA. NO. 1-79.50
LIFT STA. NO. 2-88.80
INVERT ELEV.
8"-CLASS 100-M.J.C.I. INFLUENT LINES
MAGNESIUM ANODE PACK
CONCRETE
6" MIN.
8'-0"
9'-0"
REINFORCED CONCRETE BASE (6-6-6/6 W.W.M.)
TYPICAL LIFT STATION NO. 1 AND 2
1'-10"
ALTERNATE DISCHARGE POSITION
STATION PLAN
NOTES- DIMENSIONS A-B-C & D TO BE DETERMINED BY MANUFACTURER FOR STATIONS NO. 1 & 2.
2 LIFT STATIONS REQ'D.
EACH STATION CAPACITY- 50 G.P.M.
EQUAL TO: SMITH & LOVELESS "DU-O-JECT" DUPLEX-PNEUMATIC EJECTOR SEWAGE LIFT STATION.
SUCCESSFUL BIDDER TO FURNISH ENGINEER WITH MFG'S DRAWINGS OF LIFT STATION FOR APPROVAL PRIOR TO CONSTRUCTION.
EACH STATION TO BE WIRED BY A LICENSED ELECTRICIAN.
SEE PROFILE SHEETS.
SEE SPECIFICATION BOOK.
TYPICAL LIFT STATION
BRUCE, WISCONSIN
UTILITY EXTENSION PROJECT
SCALE- DO NOT
SHEET 5 OF 7
DATE JULY 1970
GEO. H. MORGAN
PROFESSIONAL ENGINEER
LADYSMITH, WISCONSIN
DR. BY R.O. PARMLEY

Sheldon 拖车式活动住屋公园

SHELDON TRAILER PARK

NONDEPENDENT MOBILE HOME PARK - OWER: J. HAMMEN

SANITARY SEWER SYSTEM, WELL, WATER
DISTRIBUTION SYSTEM, ELECTRICAL PLAN,
SERVICE CONNECTIONS & PARK DESIGN

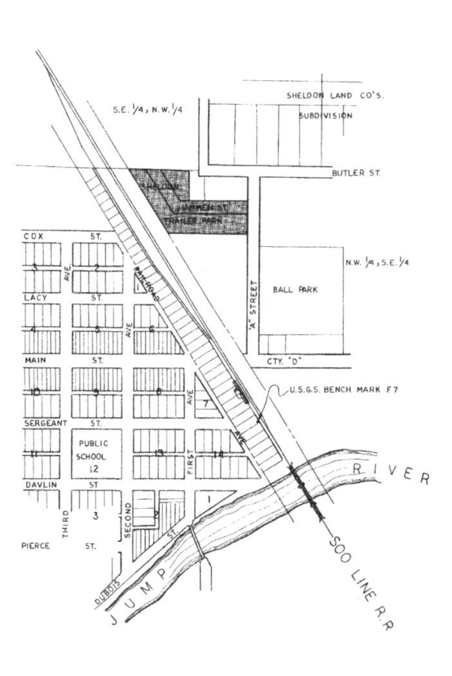

MAP of VILLAGE of SHELDON, WISCONSIN

INDEX of SHEETS

SHEET NO.	DESCRIPTION
1	TITLE SHEET, LOCATION MAP AND INDEX
2	SANITARY SEWER W/ PROFILES AND ISOMETRIC
3	WELL & WATER SYSTEM W/ TYP. RISERS
4	ELECTRICAL PLAN
5	TYPICAL SEWER DETAILS

PROPOSED UTILITIES	
SHELDON TRAILER PARK SHELDON, WI	
SCALE- 1" = 300'	GEO. H. MORGAN PROFESSIONAL ENGINEER
SHEET 1 OF 5	LADYSMITH, WISCONSIN
DATE OCT. 1975	DR. BY R.O. PARMLEY

Courtesy: Morgan & Parmley, Ltd.

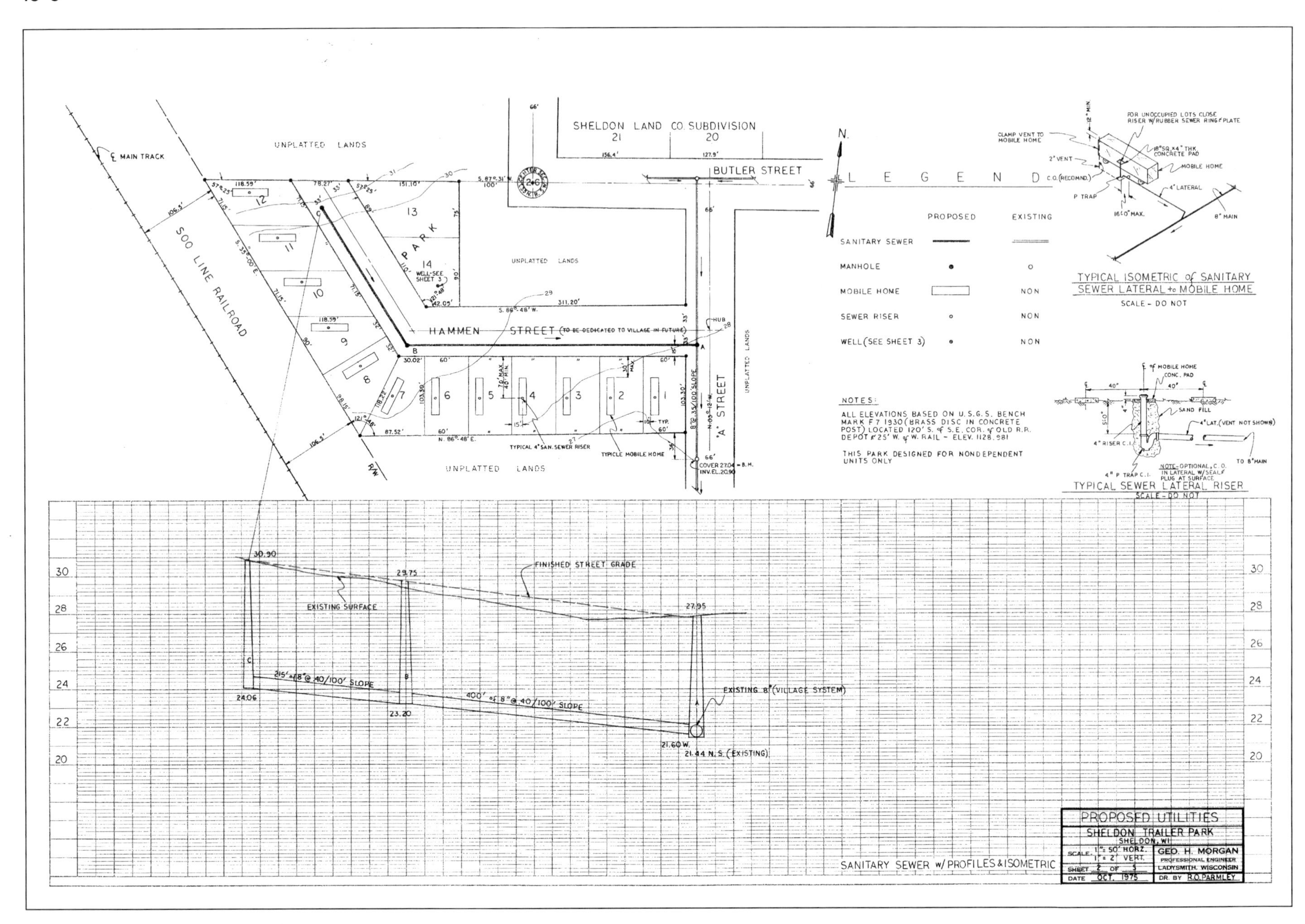

SHELDON LAND CO. SUBDIVISION
BUTLER STREET
HAMMEN STREET (TO BE DEDICATED TO VILLAGE IN FUTURE)
'A' STREET
SOO LINE RAILROAD
UNPLATTED LANDS
PARK
LEGEND
PROPOSED
EXISTING
SANITARY SEWER
MANHOLE
MOBILE HOME
SEWER RISER
WELL (SEE SHEET 3)
NON
NOTES:
ALL ELEVATIONS BASED ON U.S.G.S. BENCH MARK F 7 1930 (BRASS DISC IN CONCRETE POST) LOCATED 120' S. of S.E. COR. of OLD R.R. DEPOT & 25' W. of W. RAIL - ELEV. 1128.981
THIS PARK DESIGNED FOR NONDEPENDENT UNITS ONLY
TYPICAL ISOMETRIC of SANITARY SEWER LATERAL to MOBILE HOME
SCALE - DO NOT
TYPICAL SEWER LATERAL RISER
SCALE - DO NOT
FINISHED STREET GRADE
EXISTING SURFACE
215' of 8" @ .40/100' SLOPE
400' of 8" @ .40/100' SLOPE
EXISTING 8" (VILLAGE SYSTEM)
21.60 W.
21.44 N.S. (EXISTING)
PROPOSED UTILITIES
SHELDON TRAILER PARK
SHELDON, WI
GEO. H. MORGAN
PROFESSIONAL ENGINEER
LADYSMITH, WISCONSIN
SANITARY SEWER w/ PROFILES & ISOMETRIC
DATE OCT. 1975
DR. BY R.O. PARMLEY

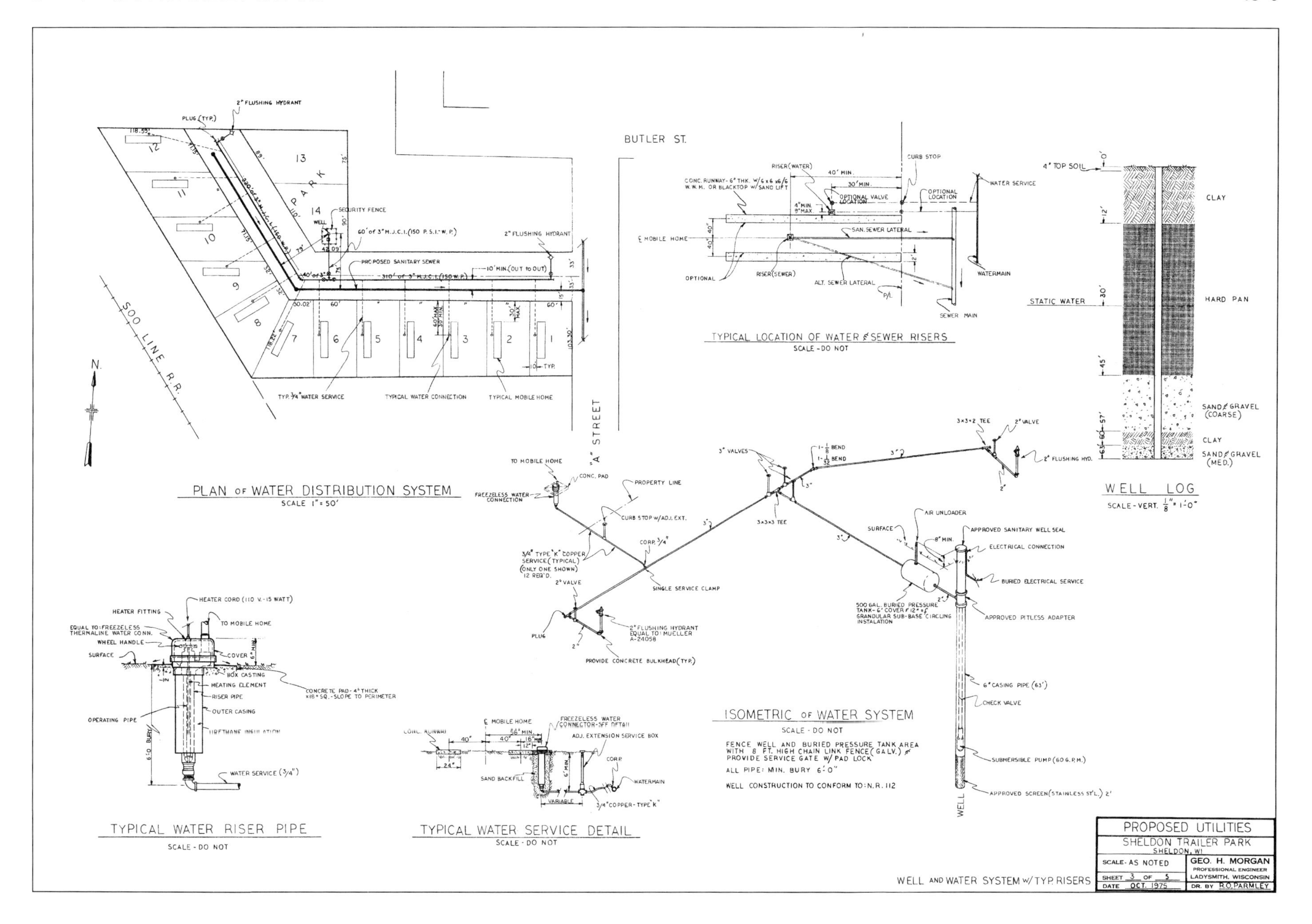
PLAN OF WATER DISTRIBUTION SYSTEM
SCALE 1" = 50'
TYPICAL LOCATION OF WATER & SEWER RISERS
SCALE - DO NOT
WELL LOG
ISOMETRIC OF WATER SYSTEM
SCALE - DO NOT
FENCE WELL AND BURIED PRESSURE TANK AREA WITH 8 FT. HIGH CHAIN LINK FENCE (GALV.) & PROVIDE SERVICE GATE W/ PAD LOCK
ALL PIPE: MIN. BURY 6'-0"
WELL CONSTRUCTION TO CONFORM TO: N.R. 112
TYPICAL WATER RISER PIPE
SCALE - DO NOT
TYPICAL WATER SERVICE DETAIL
SCALE - DO NOT
PROPOSED UTILITIES
SHELDON TRAILER PARK
SHELDON, WI
SCALE - AS NOTED
SHEET 3 OF 5
DATE OCT. 1975
GEO. H. MORGAN
PROFESSIONAL ENGINEER
LADYSMITH, WISCONSIN
DR. BY R.O. PARMLEY
WELL AND WATER SYSTEM W/ TYP. RISERS

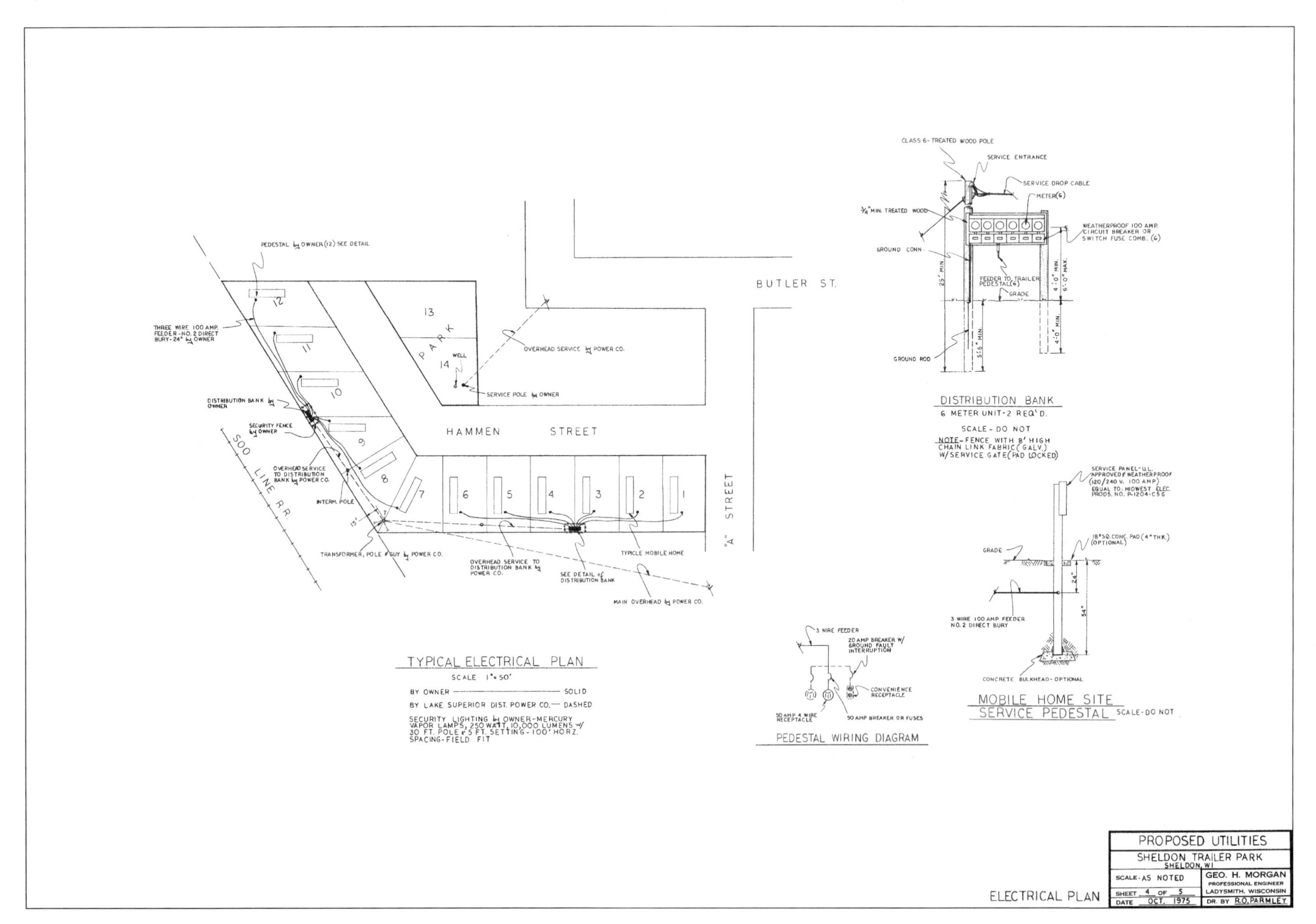

TYPICAL ELECTRICAL PLAN
SCALE 1"=50'
BY OWNER — SOLID
BY LAKE SUPERIOR DIST. POWER CO. — DASHED
SECURITY LIGHTING by OWNER-MERCURY VAPOR LAMPS, 250 WATT, 10,000 LUMENS w/ 30 FT. POLE & 5 FT. SETTING-100' HORZ. SPACING-FIELD FIT
BUTLER ST.
HAMMEN STREET
"A" STREET
SOO LINE R.R.
PARK
DISTRIBUTION BANK
6 METER UNIT-2 REQ'D.
SCALE-DO NOT
NOTE-FENCE WITH 8' HIGH CHAIN LINK FABRIC (GALV.) W/SERVICE GATE (PAD LOCKED)
MOBILE HOME SITE SERVICE PEDESTAL
SCALE-DO NOT
PEDESTAL WIRING DIAGRAM
PROPOSED UTILITIES
SHELDON TRAILER PARK
SHELDON, WI
SCALE-AS NOTED
SHEET 4 OF 5
DATE OCT. 1975
GEO. H. MORGAN
PROFESSIONAL ENGINEER
LADYSMITH, WISCONSIN
DR. BY R.O. PARMLEY
ELECTRICAL PLAN

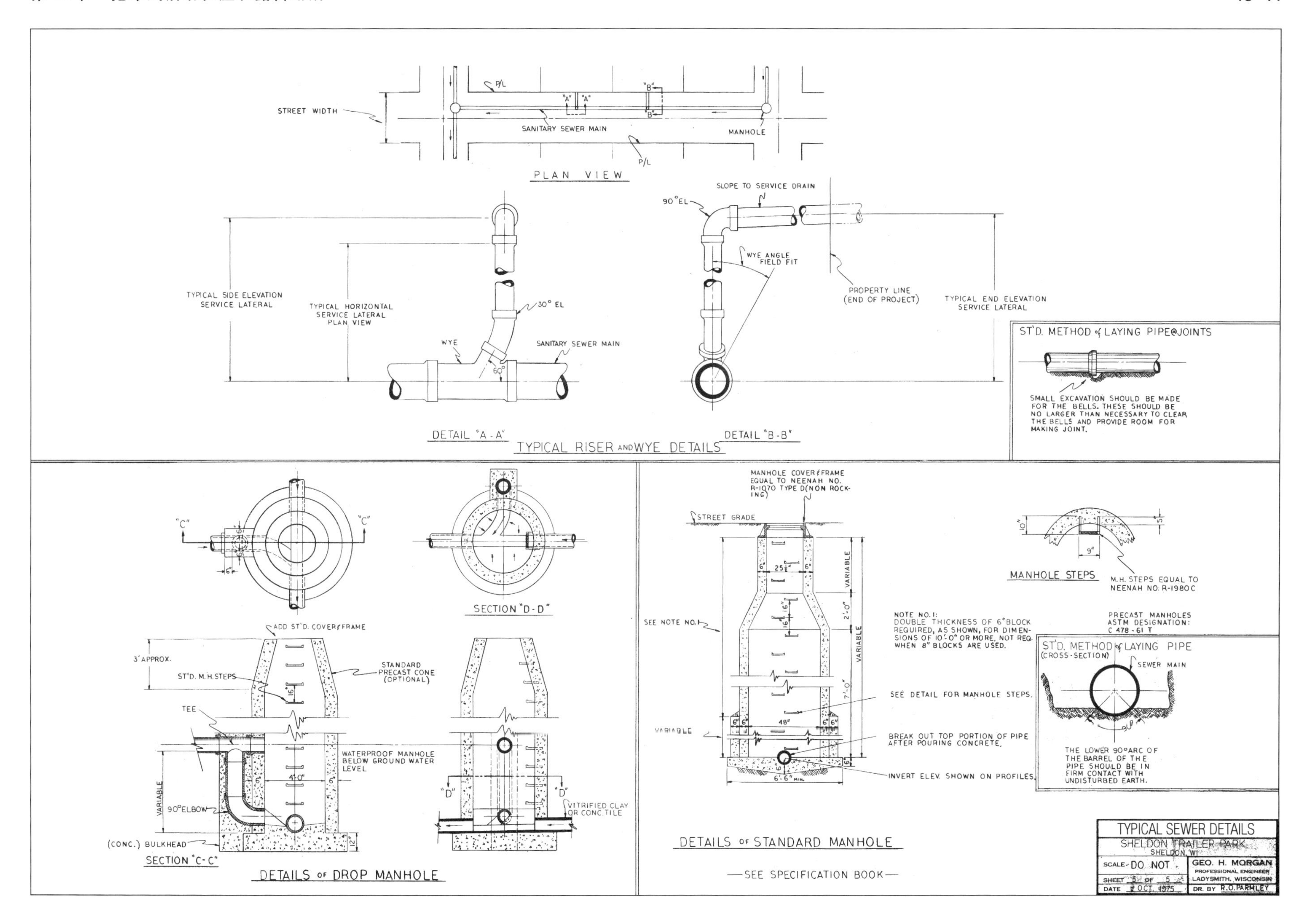
PLAN VIEW
STREET WIDTH
SANITARY SEWER MAIN
MANHOLE
TYPICAL RISER AND WYE DETAILS
DETAIL "A-A"
DETAIL "B-B"
ST'D. METHOD of LAYING PIPE@JOINTS
SMALL EXCAVATION SHOULD BE MADE FOR THE BELLS. THESE SHOULD BE NO LARGER THAN NECESSARY TO CLEAR THE BELLS AND PROVIDE ROOM FOR MAKING JOINT.
DETAILS of DROP MANHOLE
SECTION "C-C"
SECTION "D-D"
DETAILS of STANDARD MANHOLE
MANHOLE STEPS
M.H. STEPS EQUAL TO NEENAH NO. R-1980C
PRECAST MANHOLES ASTM DESIGNATION: C 478-61 T
ST'D. METHOD of LAYING PIPE (CROSS-SECTION)
THE LOWER 90° ARC OF THE BARREL OF THE PIPE SHOULD BE IN FIRM CONTACT WITH UNDISTURBED EARTH.
—SEE SPECIFICATION BOOK—
TYPICAL SEWER DETAILS
SHELDON TRAILER PARK
SHELDON, WI
GEO. H. MORGAN
PROFESSIONAL ENGINEER
LADYSMITH, WISCONSIN
DR. BY R.O.PARMLEY

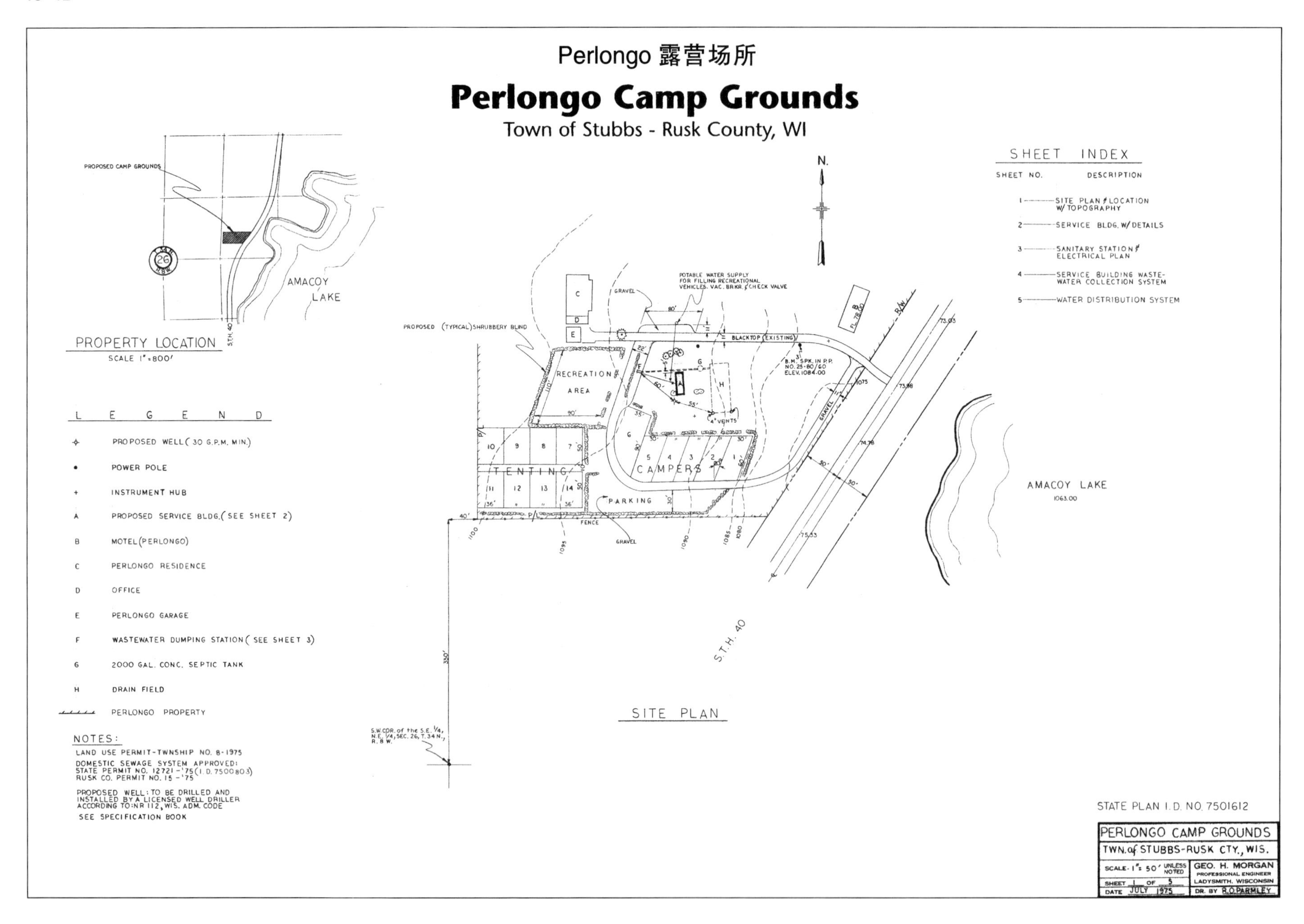

Courtesy: Morgan & Parmley, Ltd.

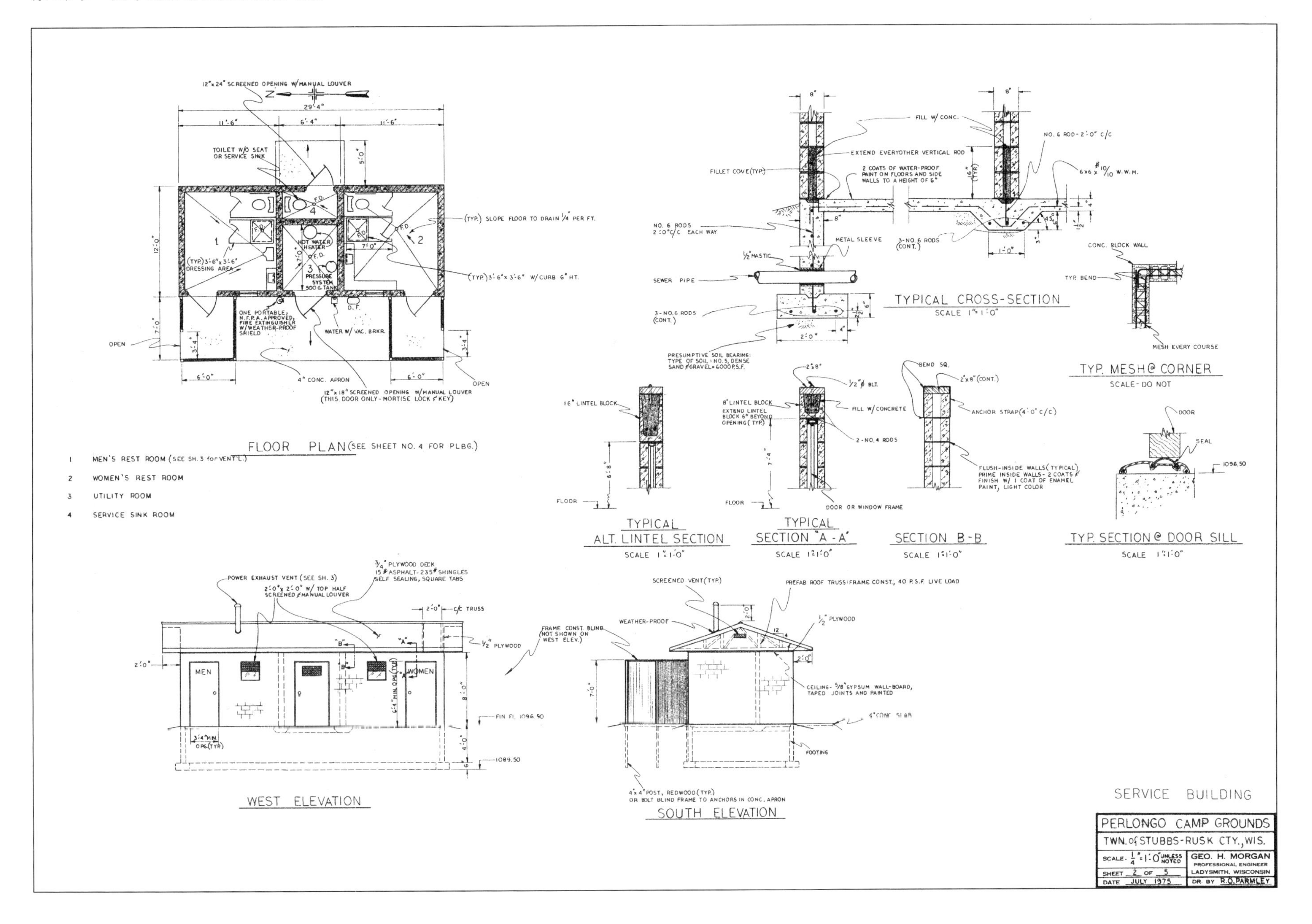

FLOOR PLAN (SEE SHEET NO. 4 FOR PLBG.)
1 MEN'S REST ROOM (SEE SH. 3 for VENT'L.)
2 WOMEN'S REST ROOM
3 UTILITY ROOM
4 SERVICE SINK ROOM
12"x24" SCREENED OPENING W/MANUAL LOUVER
TOILET W/O SEAT OR SERVICE SINK
(TYP.) SLOPE FLOOR TO DRAIN 1/4" PER FT.
(TYP.) 3'-6"x3'-6" DRESSING AREA
(TYP.) 3'-6"x3'-6" W/CURB 6" HT.
HOT WATER HEATER
PRESSURE SYSTEM 500 G. TANK
ONE PORTABLE, N.F.P.A. APPROVED, FIRE EXTINGUISHER W/WEATHER-PROOF SHIELD
WATER W/ VAC. BRKR.
4" CONC. APRON
12"x18" SCREENED OPENING W/MANUAL LOUVER (THIS DOOR ONLY-MORTISE LOCK & KEY)
OPEN
TYPICAL CROSS-SECTION
SCALE 1"=1'-0"
FILL W/ CONC.
EXTEND EVERYOTHER VERTICAL ROD
FILLET COVE (TYP.)
2 COATS OF WATER-PROOF PAINT ON FLOORS AND SIDE WALLS TO A HEIGHT OF 6"
NO. 6 ROD-2'-0" c/c
6x6x 10/10 W.W.M.
NO. 6 RODS 2'-0" c/c EACH WAY
METAL SLEEVE
3-NO. 6 RODS (CONT.)
1/2" MASTIC
SEWER PIPE
PRESUMPTIVE SOIL BEARING: TYPE OF SOIL: NO. 5, DENSE SAND & GRAVEL=6000 P.S.F.
CONC. BLOCK WALL
TYP. BEND
MESH EVERY COURSE
TYP. MESH @ CORNER
SCALE-DO NOT
16" LINTEL BLOCK
TYPICAL ALT. LINTEL SECTION
SCALE 1"=1'-0"
8" LINTEL BLOCK
EXTEND LINTEL BLOCK 6" BEYOND OPENING (TYP.)
1/2" φ BLT.
FILL W/ CONCRETE
2-NO. 4 RODS
DOOR OR WINDOW FRAME
TYPICAL SECTION "A-A"
SCALE 1"=1'-0"
BEND SQ.
2"x8" (CONT.)
ANCHOR STRAP (4'-0" c/c)
FLUSH-INSIDE WALLS (TYPICAL) PRIME INSIDE WALLS-2 COATS & FINISH W/ 1 COAT OF ENAMEL PAINT, LIGHT COLOR
SECTION B-B
SCALE 1"=1'-0"
DOOR
SEAL
1096.50
TYP. SECTION @ DOOR SILL
SCALE 1"=1'-0"
FLOOR
POWER EXHAUST VENT (SEE SH. 3)
2'-0"x2'-0" W/ TOP HALF SCREENED & MANUAL LOUVER
3/4" PLYWOOD DECK 15# ASPHALT-235# SHINGLES SELF SEALING, SQUARE TABS
c/c TRUSS
1/2" PLYWOOD
MEN
WOMEN
6'-4" MIN. OPG. (TYP.)
3'-4" MIN. OPG. (TYP.)
FIN. FL. 1096.50
1089.50
WEST ELEVATION
SCREENED VENT (TYP.)
PREFAB ROOF TRUSS: FRAME CONST., 40 P.S.F. LIVE LOAD
WEATHER-PROOF
FRAME CONST. BLIND (NOT SHOWN ON WEST ELEV.)
CEILING-5/8" GYPSUM WALL-BOARD, TAPED JOINTS AND PAINTED
4" CONC. SLAB
FOOTING
4"x4" POST, REDWOOD (TYP.) OR BOLT BLIND FRAME TO ANCHORS IN CONC. APRON
SOUTH ELEVATION
SERVICE BUILDING
PERLONGO CAMP GROUNDS
TWN. of STUBBS-RUSK CTY., WIS.
SCALE-1/4"=1'-0" UNLESS NOTED
SHEET 2 OF 5
DATE JULY 1975
GEO. H. MORGAN
PROFESSIONAL ENGINEER
LADYSMITH, WISCONSIN
DR. BY R.O. PARMLEY

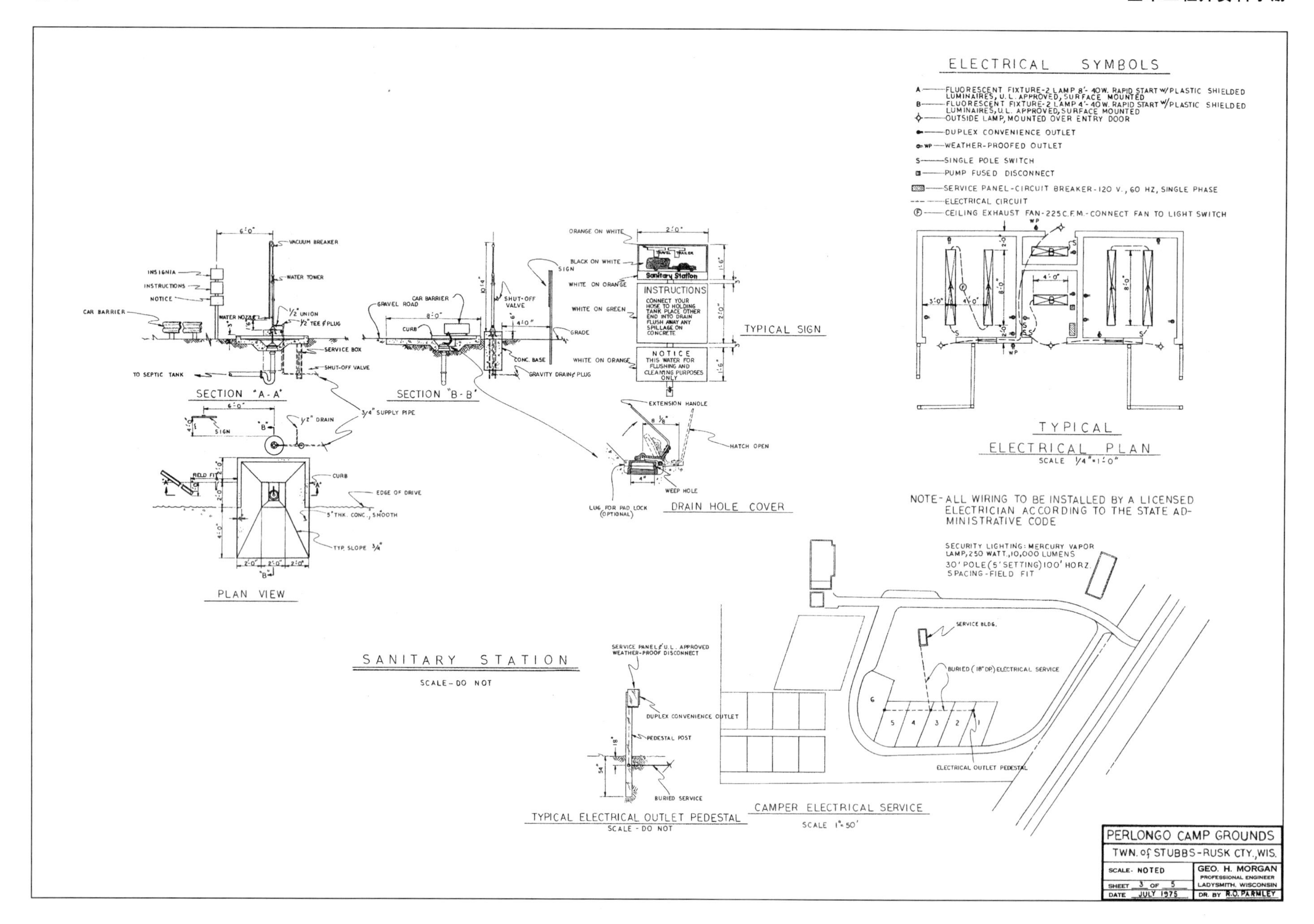
ELECTRICAL SYMBOLS
A — FLUORESCENT FIXTURE-2 LAMP 8'-40W. RAPID START w/PLASTIC SHIELDED LUMINAIRES, U.L. APPROVED, SURFACE MOUNTED
B — FLUORESCENT FIXTURE-2 LAMP 4'-40W. RAPID START w/PLASTIC SHIELDED LUMINAIRES, U.L. APPROVED, SURFACE MOUNTED
OUTSIDE LAMP, MOUNTED OVER ENTRY DOOR
DUPLEX CONVENIENCE OUTLET
WP — WEATHER-PROOFED OUTLET
S — SINGLE POLE SWITCH
PUMP FUSED DISCONNECT
SERVICE PANEL-CIRCUIT BREAKER-120 V., 60 HZ, SINGLE PHASE
ELECTRICAL CIRCUIT
F — CEILING EXHAUST FAN-225 C.F.M.-CONNECT FAN TO LIGHT SWITCH
TYPICAL ELECTRICAL PLAN
SCALE 1/4"=1'-0"
NOTE-ALL WIRING TO BE INSTALLED BY A LICENSED ELECTRICIAN ACCORDING TO THE STATE ADMINISTRATIVE CODE
SECURITY LIGHTING: MERCURY VAPOR LAMP, 250 WATT, 10,000 LUMENS
30' POLE (5' SETTING) 100' HORZ. SPACING-FIELD FIT
SERVICE BLDG.
BURIED (18" DP.) ELECTRICAL SERVICE
ELECTRICAL OUTLET PEDESTAL
CAMPER ELECTRICAL SERVICE
SCALE 1"=50'
VACUUM BREAKER
INSIGNIA
INSTRUCTIONS
NOTICE
WATER TOWER
CAR BARRIER
WATER NOZZLE
1/2" UNION
1/2" TEE & PLUG
SERVICE BOX
SHUT-OFF VALVE
TO SEPTIC TANK
SECTION "A-A"
GRAVEL ROAD
CURB
SHUT-OFF VALVE
GRADE
CONC. BASE
GRAVITY DRAIN & PLUG
SECTION "B-B"
3/4" SUPPLY PIPE
1/2" DRAIN
SIGN
FIELD FIT
EDGE OF DRIVE
5" THK. CONC., SMOOTH
TYP. SLOPE 3/4"
PLAN VIEW
ORANGE ON WHITE
BLACK ON WHITE
WHITE ON ORANGE
WHITE ON GREEN
WHITE ON ORANGE
Sanitary Station
INSTRUCTIONS
CONNECT YOUR HOSE TO HOLDING TANK PLACE OTHER END INTO DRAIN FLUSH AWAY ANY SPILLAGE ON CONCRETE
NOTICE
THIS WATER FOR FLUSHING AND CLEANING PURPOSES ONLY
TYPICAL SIGN
EXTENSION HANDLE
HATCH OPEN
WEEP HOLE
LUG FOR PAD LOCK (OPTIONAL)
DRAIN HOLE COVER
SANITARY STATION
SCALE-DO NOT
SERVICE PANEL & U.L. APPROVED WEATHER-PROOF DISCONNECT
DUPLEX CONVENIENCE OUTLET
PEDESTAL POST
BURIED SERVICE
TYPICAL ELECTRICAL OUTLET PEDESTAL
SCALE-DO NOT
PERLONGO CAMP GROUNDS
TWN. of STUBBS-RUSK CTY., WIS.
SCALE- NOTED
GEO. H. MORGAN
PROFESSIONAL ENGINEER
LADYSMITH, WISCONSIN
SHEET 3 OF 5
DATE JULY 1975
DR. BY R.O. PARMLEY

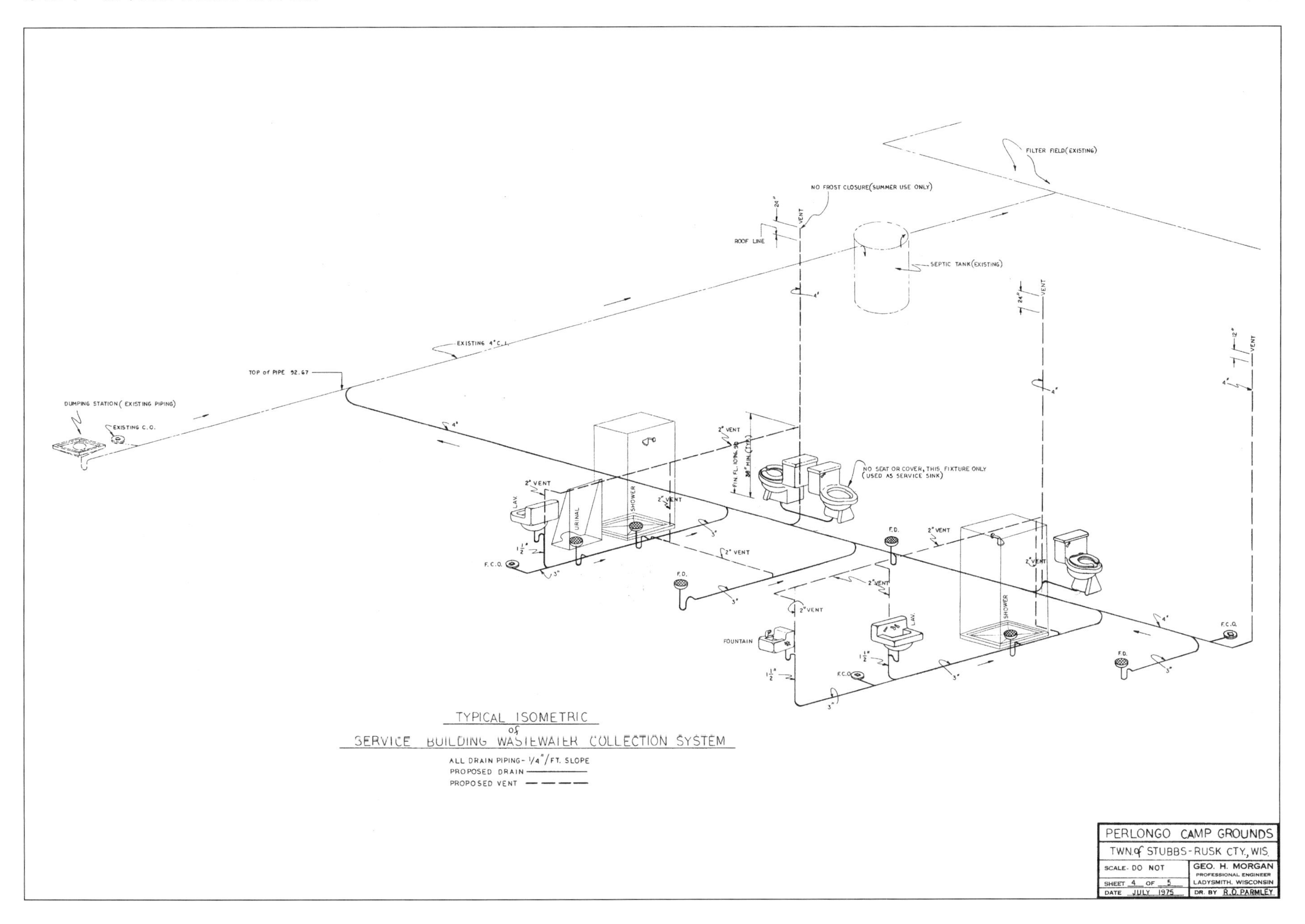
FILTER FIELD(EXISTING)
NO FROST CLOSURE(SUMMER USE ONLY)
ROOF LINE
VENT
SEPTIC TANK(EXISTING)
EXISTING 4" C.I.
TOP of PIPE 52.67
DUMPING STATION(EXISTING PIPING)
EXISTING C.O.
2" VENT
LAV.
URINAL
SHOWER
F.C.O.
F.D.
FOUNTAIN
NO SEAT OR COVER, THIS FIXTURE ONLY (USED AS SERVICE SINK)
TYPICAL ISOMETRIC of SERVICE BUILDING WASTEWATER COLLECTION SYSTEM
ALL DRAIN PIPING- 1/4"/FT. SLOPE
PROPOSED DRAIN
PROPOSED VENT
PERLONGO CAMP GROUNDS
TWN.of STUBBS-RUSK CTY, WIS.
SCALE- DO NOT
SHEET 4 OF 5
DATE JULY 1975
GEO. H. MORGAN
PROFESSIONAL ENGINEER
LADYSMITH, WISCONSIN
DR. BY R.O. PARMLEY

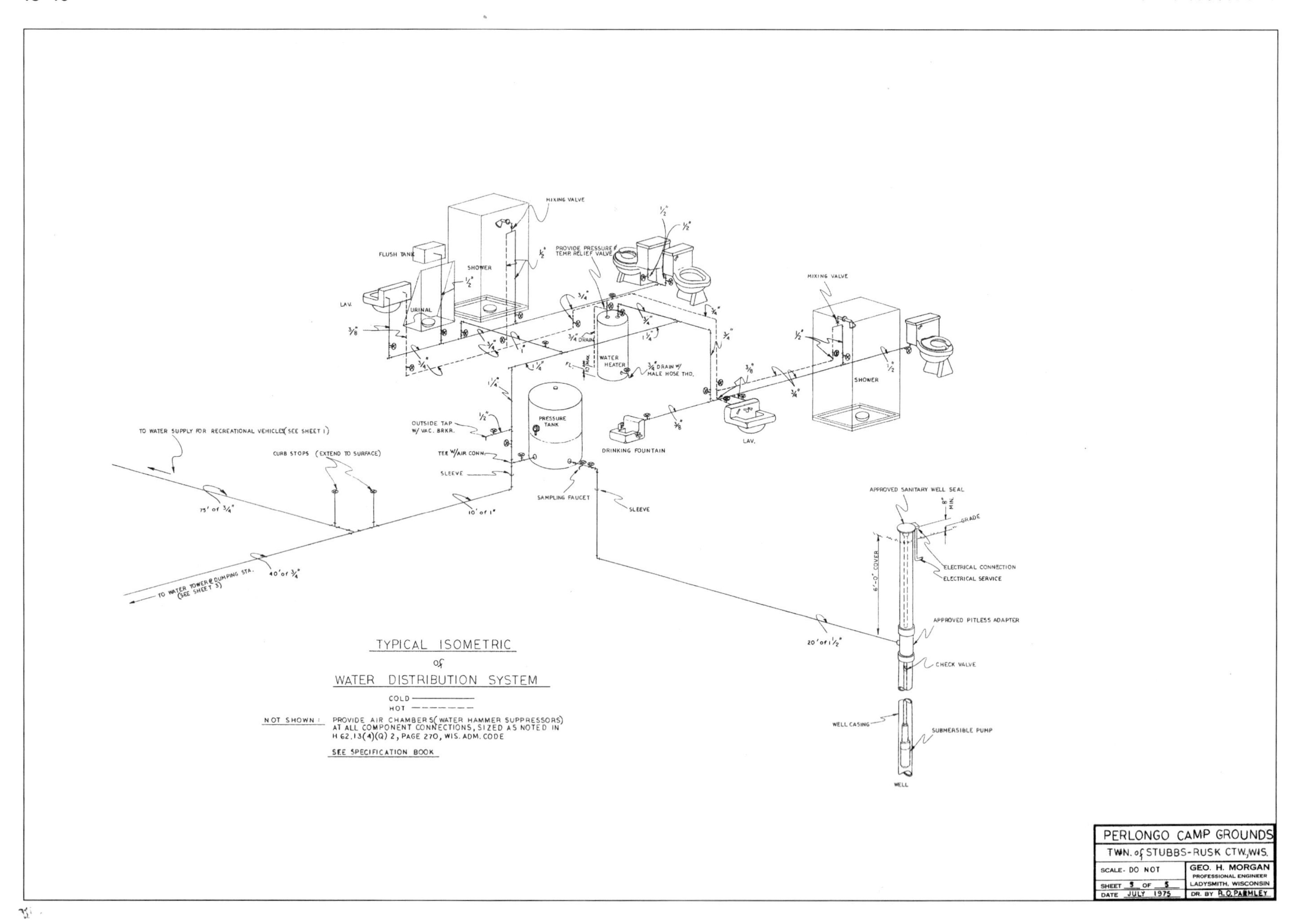

TYPICAL ISOMETRIC of WATER DISTRIBUTION SYSTEM
COLD
HOT
NOT SHOWN: PROVIDE AIR CHAMBERS (WATER HAMMER SUPPRESSORS) AT ALL COMPONENT CONNECTIONS, SIZED AS NOTED IN H 62.13(4)(Q) 2, PAGE 270, WIS. ADM. CODE
SEE SPECIFICATION BOOK
MIXING VALVE
FLUSH TANK
SHOWER
LAV.
URINAL
PROVIDE PRESSURE & TEMP. RELIEF VALVE
WATER HEATER
3/4" DRAIN
3/4" DRAIN W/ MALE HOSE THD.
PRESSURE TANK
OUTSIDE TAP W/ VAC. BRKR.
TEE W/AIR CONN.
SLEEVE
SAMPLING FAUCET
DRINKING FOUNTAIN
TO WATER SUPPLY FOR RECREATIONAL VEHICLES (SEE SHEET 1)
CURB STOPS (EXTEND TO SURFACE)
75' of 3/4"
10' of 1"
40' of 3/4"
TO WATER TOWER & DUMPING STA. (SEE SHEET 3)
20' of 1 1/2"
APPROVED SANITARY WELL SEAL
GRADE
6'-0" COVER
ELECTRICAL CONNECTION
ELECTRICAL SERVICE
APPROVED PITLESS ADAPTER
CHECK VALVE
WELL CASING
SUBMERSIBLE PUMP
WELL
PERLONGO CAMP GROUNDS
TWN. of STUBBS-RUSK CTW, WIS.
SCALE- DO NOT
SHEET 5 OF 5
DATE JULY 1975
GEO. H. MORGAN
PROFESSIONAL ENGINEER
LADYSMITH, WISCONSIN
DR. BY R.O. PARMLEY

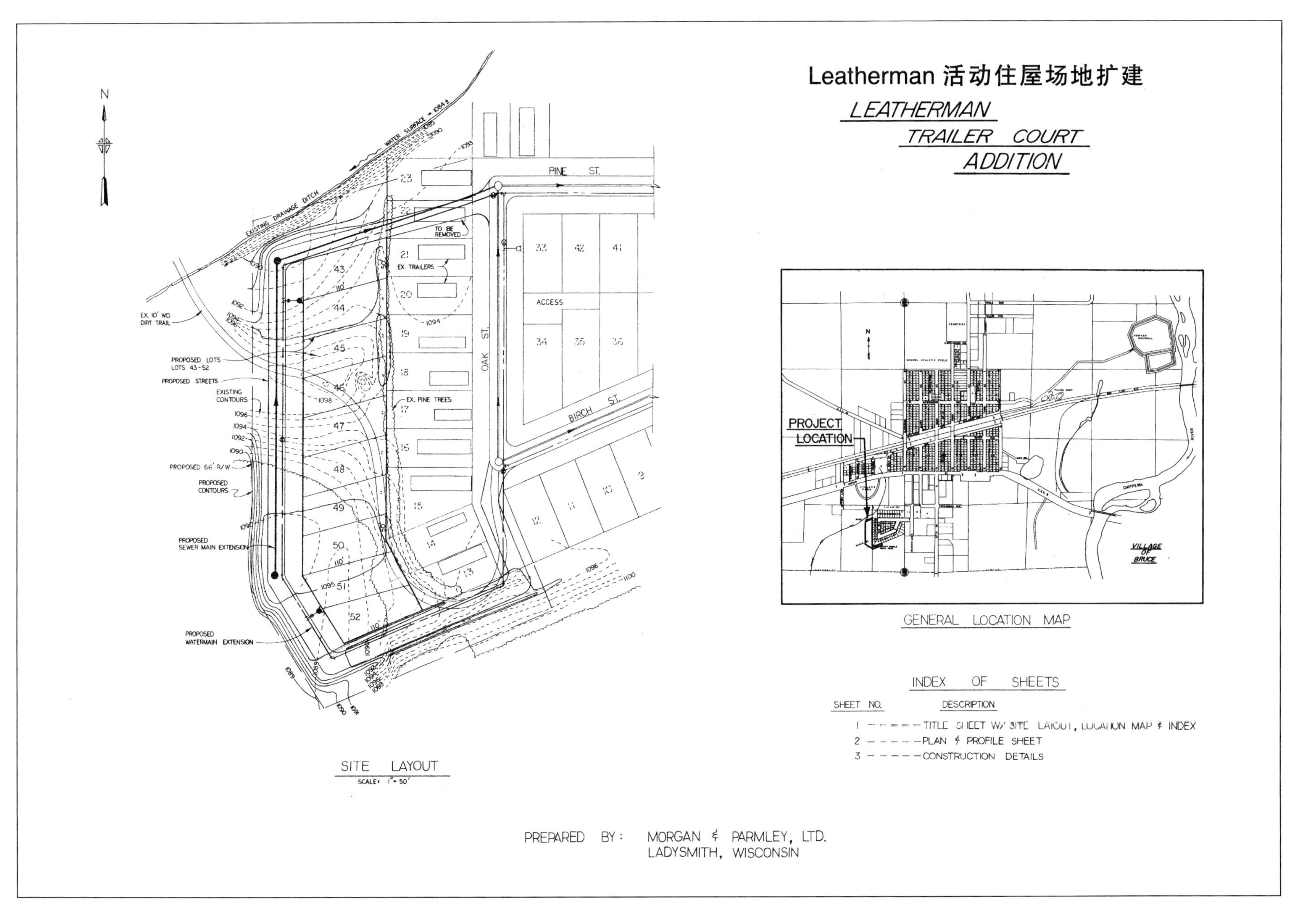

Courtesy: Morgan & Parmley, Ltd.

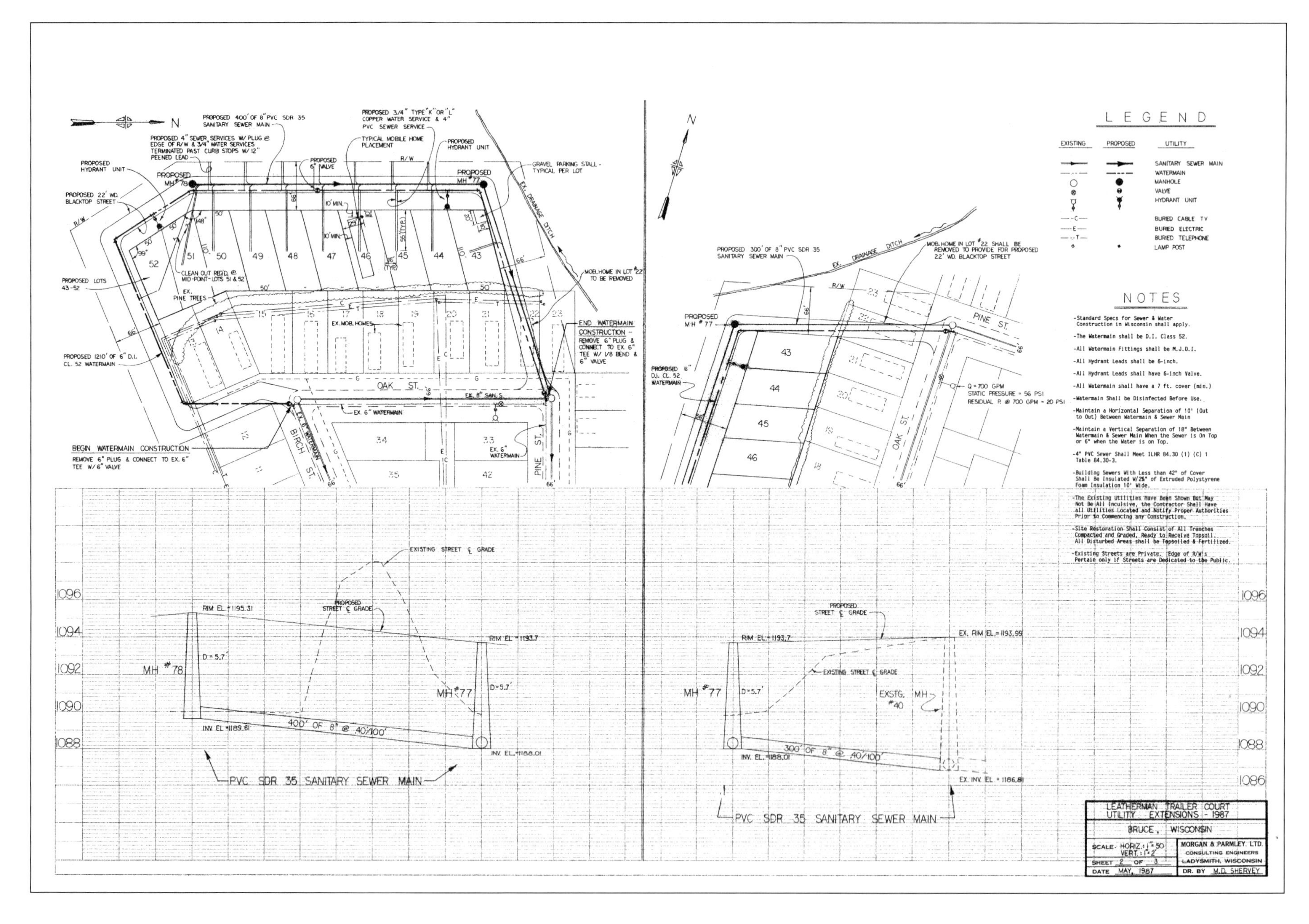
PROPOSED 400' OF 8" PVC SDR 35 SANITARY SEWER MAIN
PROPOSED 3/4" TYPE "K" OR "L" COPPER WATER SERVICE & 4" PVC SEWER SERVICE
PROPOSED 4" SEWER SERVICES W/ PLUG @ EDGE OF R/W & 3/4" WATER SERVICES TERMINATED PAST CURB STOPS W/ 12" PEENED LEAD
TYPICAL MOBILE HOME PLACEMENT
PROPOSED HYDRANT UNIT
PROPOSED MH #78
PROPOSED MH #77
PROPOSED 6" VALVE
GRAVEL PARKING STALL - TYPICAL PER LOT
EX. DRAINAGE DITCH
PROPOSED 22' WD. BLACKTOP STREET
PROPOSED LOTS 43-52
CLEAN OUT REQ'D. @ MID-POINT-LOTS 51 & 52
EX. PINE TREES
EX. MOB. HOMES
MOBL. HOME IN LOT #22 TO BE REMOVED
END WATERMAIN CONSTRUCTION - REMOVE 6" PLUG & CONNECT TO EX. 6" TEE W/ 1/8 BEND & 6" VALVE
PROPOSED 1210' OF 6" D.I. CL. 52 WATERMAIN
OAK ST.
EX. 8" SAN. S.
EX. 6" WATERMAIN
BIRCH ST.
PINE ST.
BEGIN WATERMAIN CONSTRUCTION
REMOVE 6" PLUG & CONNECT TO EX. 6" TEE W/ 6" VALVE
PROPOSED 300' OF 8" PVC SDR 35 SANITARY SEWER MAIN
MOBL. HOME IN LOT #22 SHALL BE REMOVED TO PROVIDE FOR PROPOSED 22' WD. BLACKTOP STREET
PROPOSED 6" D.I. CL. 52 WATERMAIN
Q = 700 GPM
STATIC PRESSURE = 56 PSI
RESIDUAL P. @ 700 GPM = 20 PSI
LEGEND
EXISTING
PROPOSED
UTILITY
SANITARY SEWER MAIN
WATERMAIN
MANHOLE
VALVE
HYDRANT UNIT
BURIED CABLE TV
BURIED ELECTRIC
BURIED TELEPHONE
LAMP POST
NOTES
-Standard Specs for Sewer & Water Construction in Wisconsin shall apply.
-The Watermain shall be D.I. Class 52.
-All Watermain Fittings shall be M.J.D.I.
-All Hydrant Leads shall be 6-inch.
-All Hydrant Leads shall have 6-inch Valve.
-All Watermain shall have a 7 ft. cover (min.)
-Watermain Shall be Disinfected Before Use.
-Maintain a Horizontal Separation of 10' (Out to Out) Between Watermain & Sewer Main
-Maintain a Vertical Separation of 18" Between Watermain & Sewer Main When the Sewer is On Top or 6" when the Water is on Top.
-4" PVC Sewer Shall Meet ILHR 84.30 (1) (C) 1 Table 84.30-3.
-Building Sewers With Less than 42" of Cover Shall Be Insulated W/2½" of Extruded Polystyrene Foam Insulation 10' Wide.
-The Existing Utilities Have Been Shown But May Not Be All Inculsive, the Contractor Shall Have all Utilities Located and Notify Proper Authorities Prior to Commencing any Construction.
-Site Restoration Shall Consist of All Trenches Compacted and Graded, Ready to Receive Topsoil. All Disturbed Areas shall be Topsoiled & Fertilized.
-Existing Streets are Private. Edge of R/W's Pertain only if Streets are Dedicated to the Public.
EXISTING STREET ℄ GRADE
PROPOSED STREET ℄ GRADE
RIM EL. = 1195.31
RIM EL. = 1193.7
D = 5.7'
MH #78
MH #77
INV. EL. = 1189.61
INV. EL. = 1188.01
400' OF 8" @ .40/100'
PVC SDR 35 SANITARY SEWER MAIN
EX. RIM EL. = 1193.99
EXSTG. MH #40
300' OF 8" @ .40/100'
EX. INV. EL. = 1186.81
1096
1094
1092
1090
1088
1086
LEATHERMAN TRAILER COURT UTILITY EXTENSIONS - 1987
BRUCE, WISCONSIN
SCALE- HORIZ.: 1" = 50 VERT.: 1" = 2'
SHEET 2 OF 3
DATE MAY, 1987
MORGAN & PARMLEY, LTD. CONSULTING ENGINEERS LADYSMITH, WISCONSIN
DR. BY M.D. SHERVEY

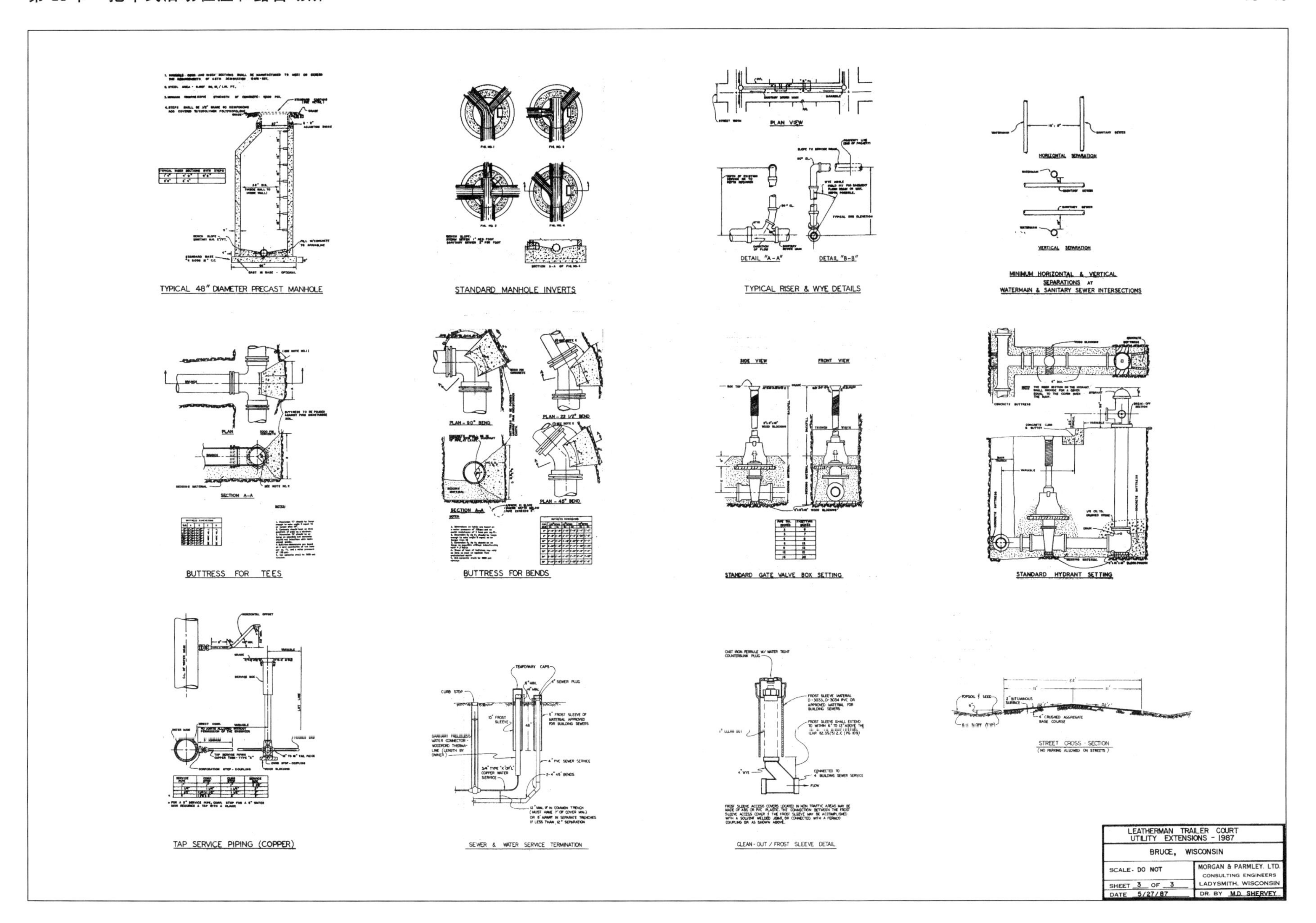
TYPICAL 48" DIAMETER PRECAST MANHOLE
STANDARD MANHOLE INVERTS
PLAN VIEW
DETAIL "A-A"
DETAIL "B-B"
TYPICAL RISER & WYE DETAILS
HORIZONTAL SEPARATION
VERTICAL SEPARATION
MINIMUM HORIZONTAL & VERTICAL SEPARATIONS AT WATERMAIN & SANITARY SEWER INTERSECTIONS
PLAN
SECTION A-A
BUTTRESS FOR TEES
PLAN - 90° BEND
PLAN - 22 1/2° BEND
PLAN - 45° BEND
BUTTRESS FOR BENDS
SIDE VIEW
FRONT VIEW
STANDARD GATE VALVE BOX SETTING
STANDARD HYDRANT SETTING
TAP SERVICE PIPING (COPPER)
SEWER & WATER SERVICE TERMINATION
CLEAN-OUT / FROST SLEEVE DETAIL
STREET CROSS - SECTION
(NO PARKING ALLOWED ON STREETS)
LEATHERMAN TRAILER COURT
UTILITY EXTENSIONS - 1987
BRUCE, WISCONSIN
SCALE- DO NOT
SHEET 3 OF 3
DATE 5/27/87
MORGAN & PARMLEY LTD.
CONSULTING ENGINEERS
LADYSMITH, WISCONSIN
DR. BY M.D. SHERVEY

设 计

第 19 节　改造和修复

例 19

改造是对已有结构或者系统做修改以提升它的功能或者扩展它的使用功能。

修复工作需要将结构或者系统恢复到它以前的情况，而且通常在修复工程中做基本的改进。

后面几页给出了结构和系统改造和修复的设计图示例。

目 录

Storage Structure for Fly Ash Collector 粉煤灰收集器的储藏结构

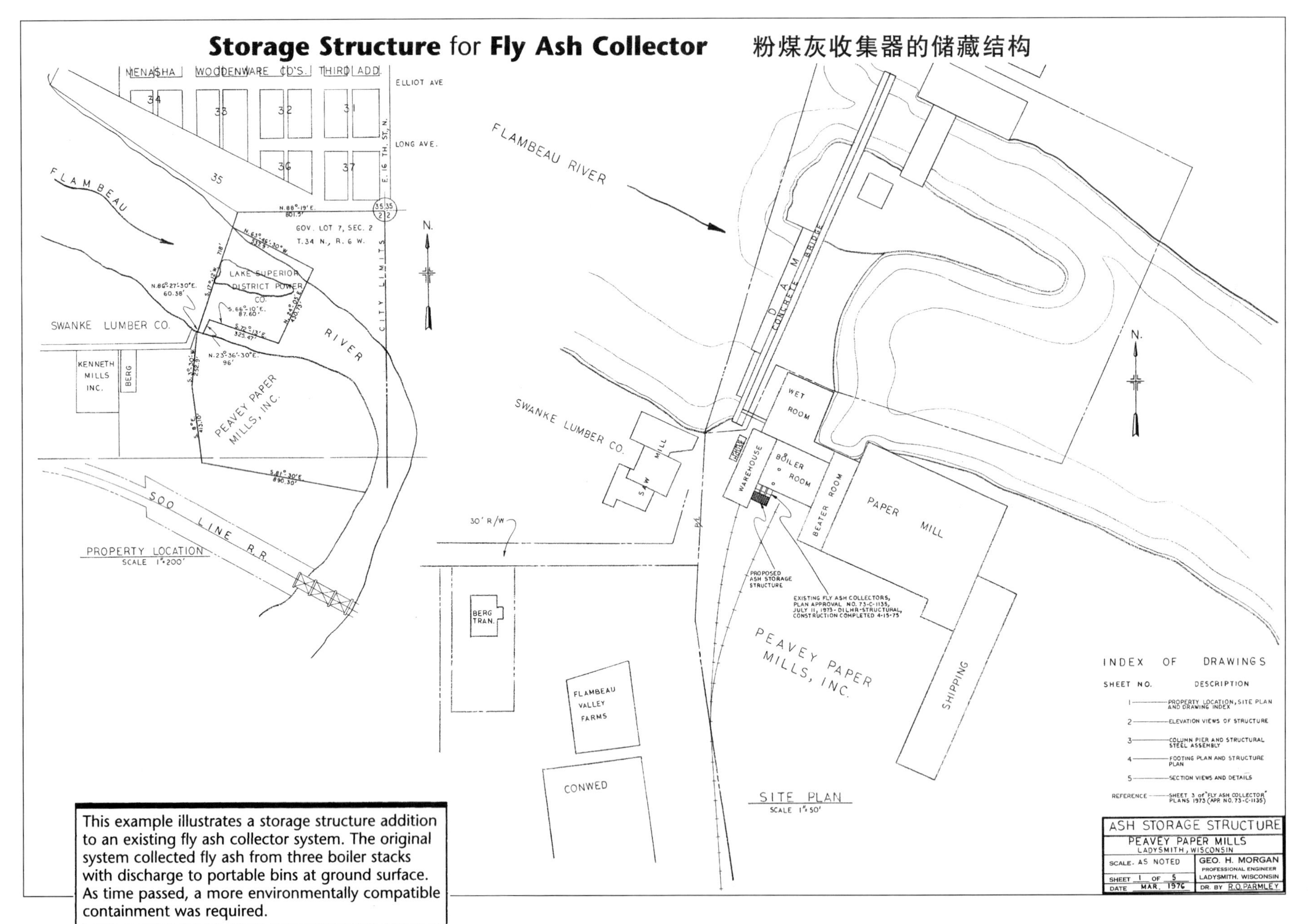

This example illustrates a storage structure addition to an existing fly ash collector system. The original system collected fly ash from three boiler stacks with discharge to portable bins at ground surface. As time passed, a more environmentally compatible containment was required.

Source: Morgan & Parmley, Ltd.

Storage Structure for Fly Ash Collector 粉煤灰收集器的储藏结构

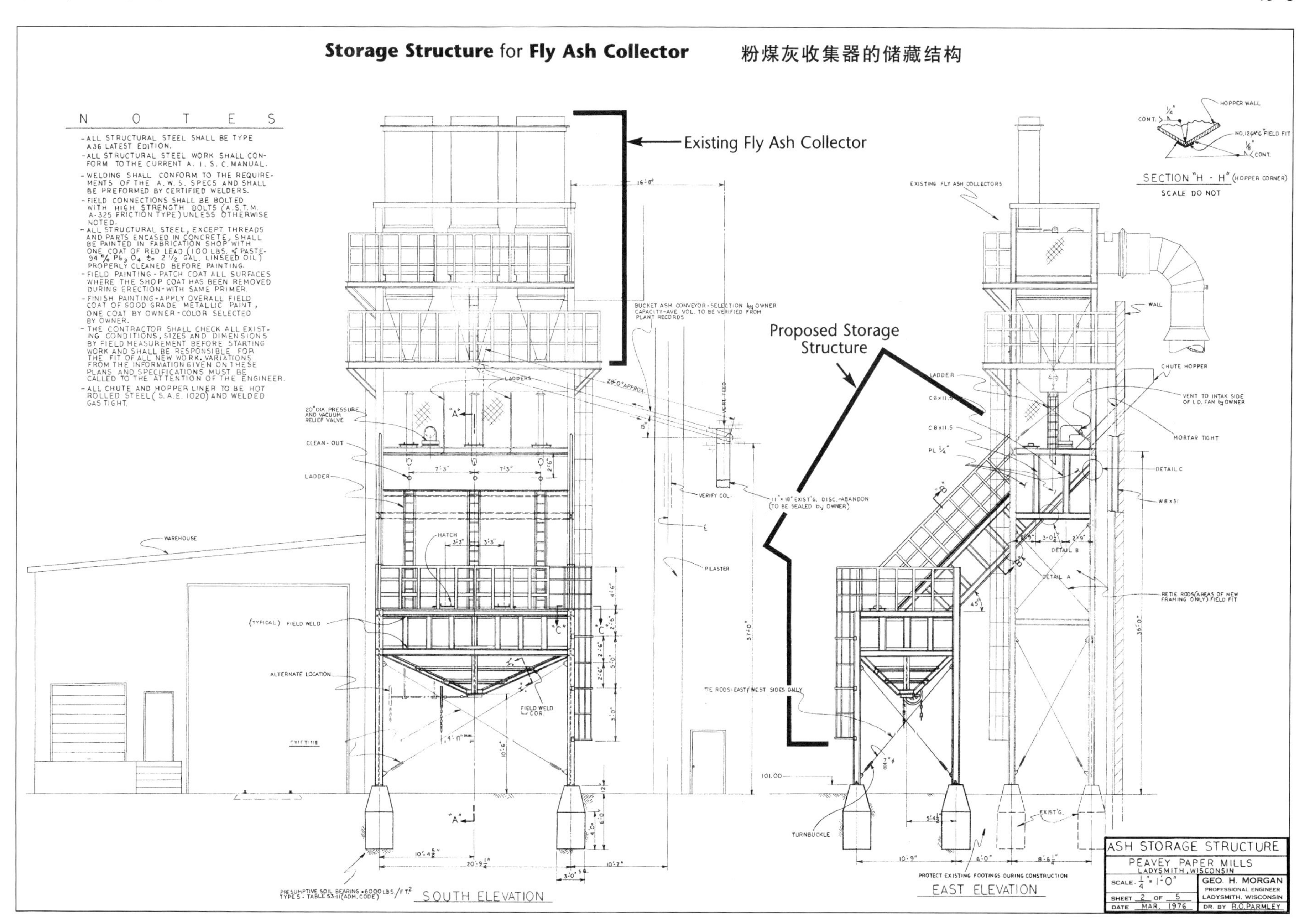

Storage Structure for Fly Ash Collector　粉煤灰收集器的储藏结构

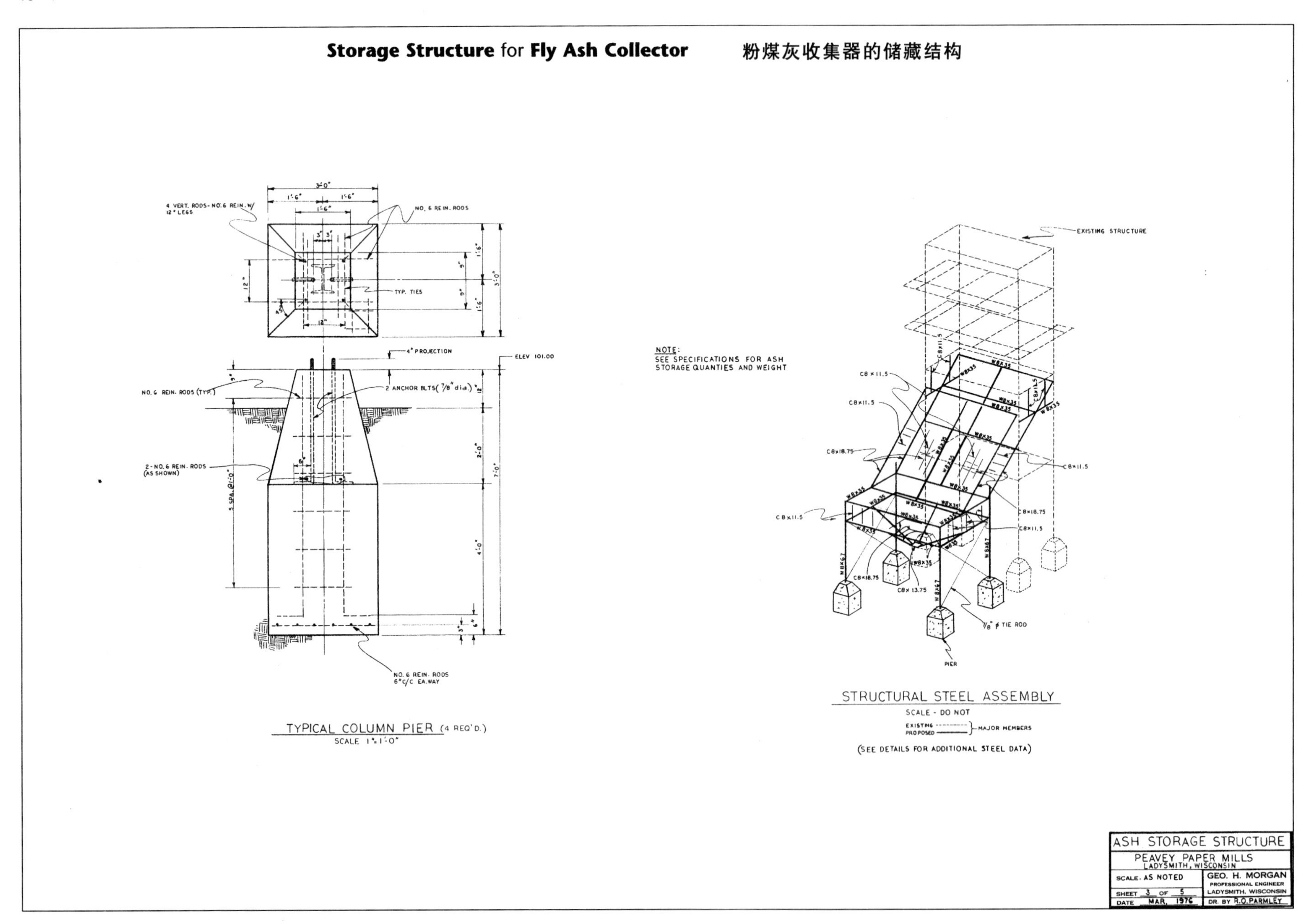

Storage Structure for Fly Ash Collector　粉煤灰收集器的储藏结构

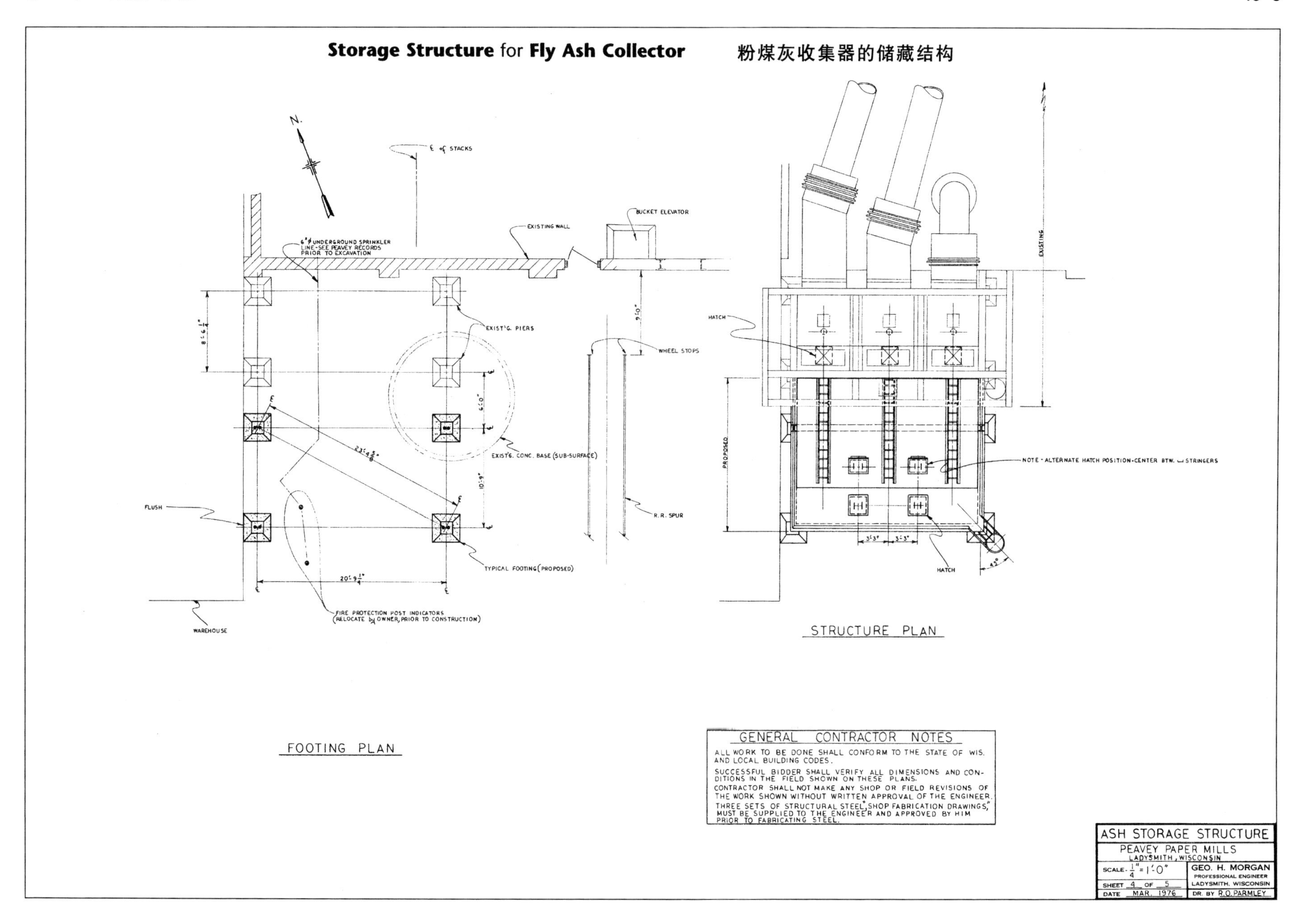

Storage Structure for Fly Ash Collector 粉煤灰收集器的储藏结构

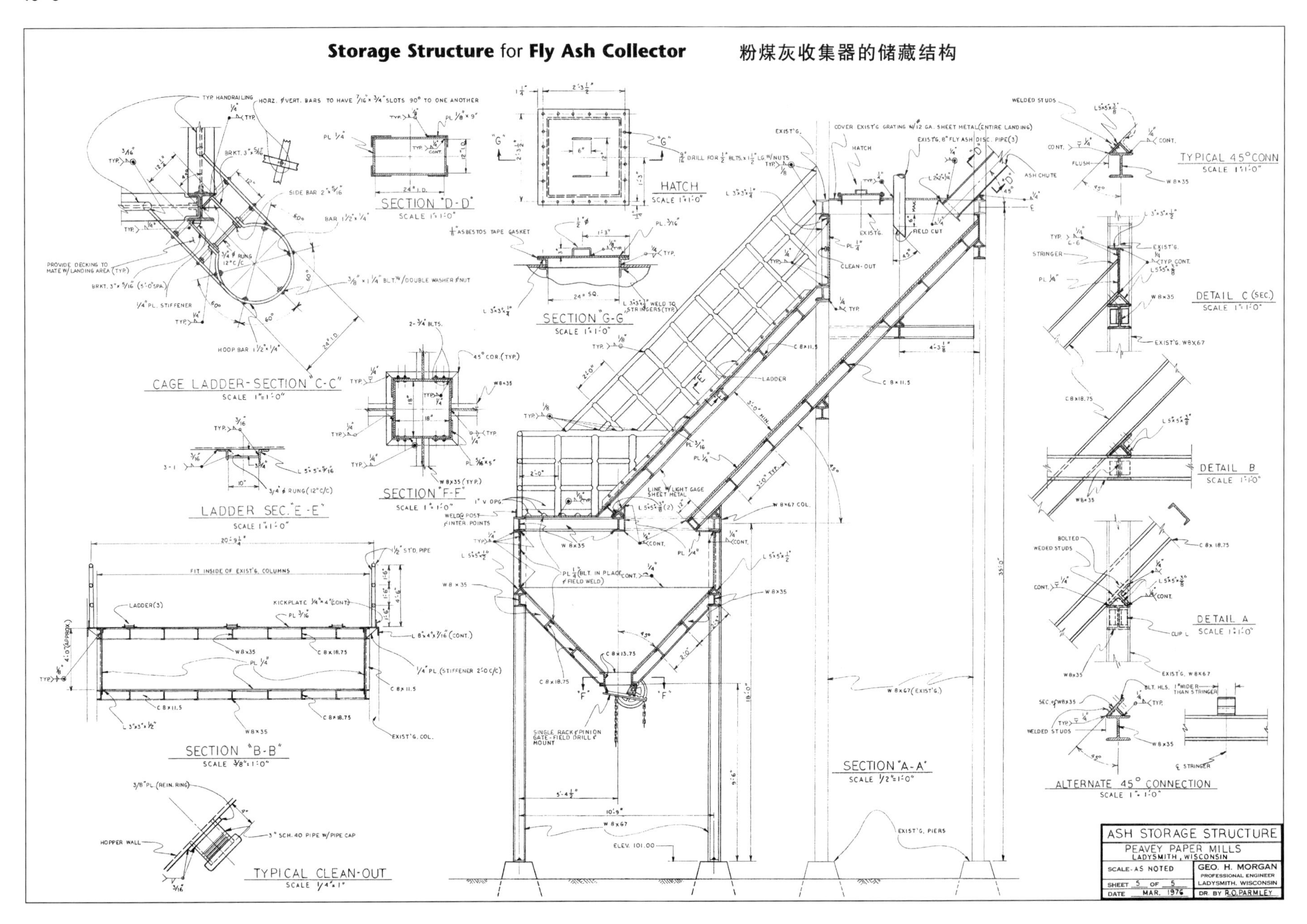

Salt Storage Shed Truss Retrofit　储盐货棚桁架改造

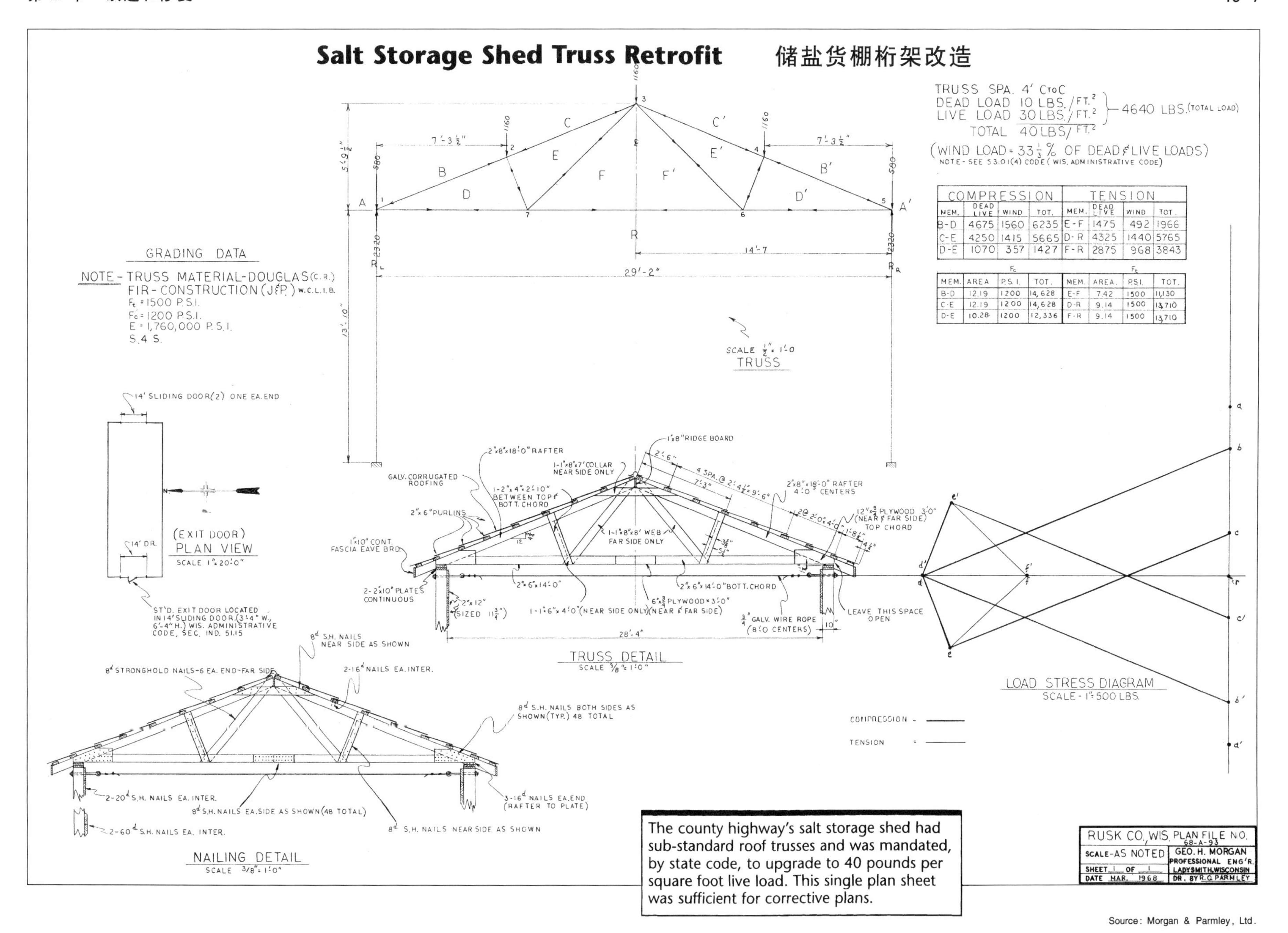

COMPRESSION				TENSION			
MEM.	DEAD LIVE	WIND	TOT.	MEM.	DEAD LIVE	WIND	TOT.
B-D	4675	1560	6235	E-F	1475	492	1966
C-E	4250	1415	5665	D-R	4325	1440	5765
D-E	1070	357	1427	F-R	2875	968	3843

F_c				F_t			
MEM.	AREA	P.S.I.	TOT.	MEM.	AREA.	P.S.I.	TOT.
B-D	12.19	1200	14,628	E-F	7.42	1500	11,130
C-E	12.19	1200	14,628	D-R	9.14	1500	13,710
D-E	10.28	1200	12,336	F-R	9.14	1500	13,710

The county highway's salt storage shed had sub-standard roof trusses and was mandated, by state code, to upgrade to 40 pounds per square foot live load. This single plan sheet was sufficient for corrective plans.

Source: Morgan & Parmley, Ltd.

废水处理设施(WWTF)扩建和紫外线消毒设施

WWTF ADDITION & U.V. DISINFECTION FACILITY

E.D.A. PROJECT NO. 06-01-02571

VILLAGE
OF
GLEN FLORA, WISCONSIN
DEC., 1992

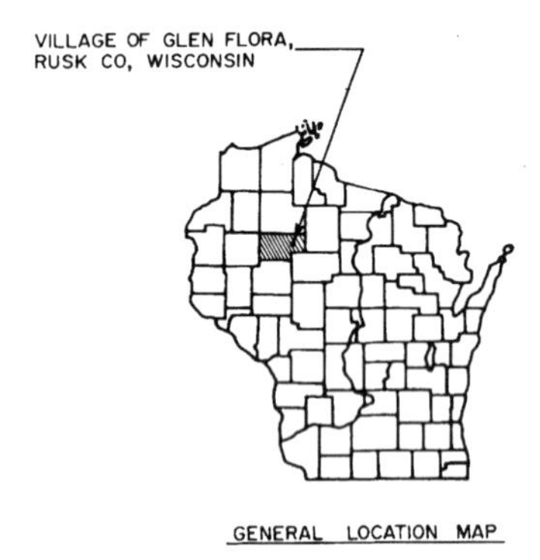

GENERAL LOCATION MAP

THESE PLANS REPRESENT ONLY 1 SECTION OF THE TOTAL PROJECT

THE TOTAL PROJECT EDA NO. 06-01-02571 IS PARTIALLY FUNDED BY A 60% GRANT FROM THE UNITED STATES ECONOMIC DEVELOPMENT ADMINISTRATION IN THE TOTAL GRANT AMOUNT OF $590,520

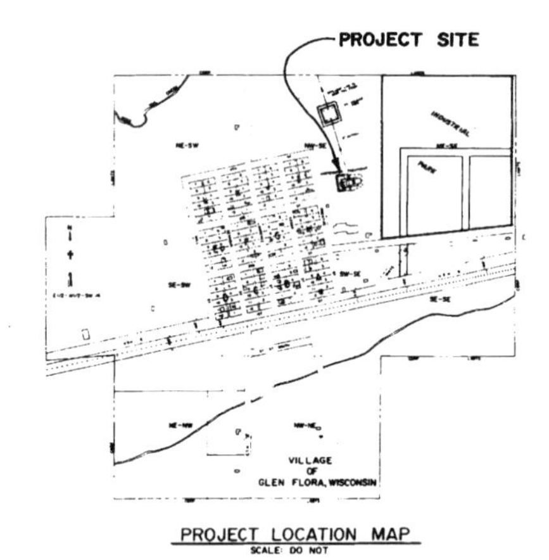

PROJECT LOCATION MAP
SCALE: DO NOT

UTILITIES

CONTACT - NORTHERN STATES POWER CO.
711 W. 9th Street, North
Ladysmith, WI. 54848

Attn: John Rymarkiewicz
715-532-6226

CONTACT - UNIVERSAL TELEPHONE CO.
OF NORTHERN WISCONSIN
Highway "8"
Hawkins, WI 54530

Attn: Jim Arquette
715-585-7707

CONTACT - DIGGERS HOTLINE

1-800 242-8511

CONTACT - VILLAGE OF GLEN FLORA WWCTF

Les Evjen, Operator
P.O. Box 253
Glen Flora, WI. 54526

715-322-5511

CONTACT - GLEN FLORA ELEMENTARY SCHOOL
SCHOOL WELL & PIPELINE
Larry Johnson, Custodian

715-322-5271

INDEX OF SHEETS

NO.	DESCRIPTION
1	TITLE SHEET, GENERAL LOCATION & INDEX
2	TREATMENT SYSTEM LAYOUT & HYDRAULIC PROFILE
3	FILTER BED NO. 5 & PIPING DETAILS
4	DOSING TANK & METERING MANHOLE MODIFICATIONS
5	U.V. DISINFECTION SYSTEM & BUILDING
6	TYPICAL EROSION CONTROL DETAILS

PREPARED BY:
MORGAN & PARMLEY LTD.
CONSULTING ENGINEERS
LADYSMITH, WISCONSIN

The wastewater treatment facility (WWTF) had to expand its hydraulic capacity and change its method of effluent disinfection from hazardous chlorination to the safe ultraviolet (U.V.) process.

Source: Morgan & Parmley, Ltd.

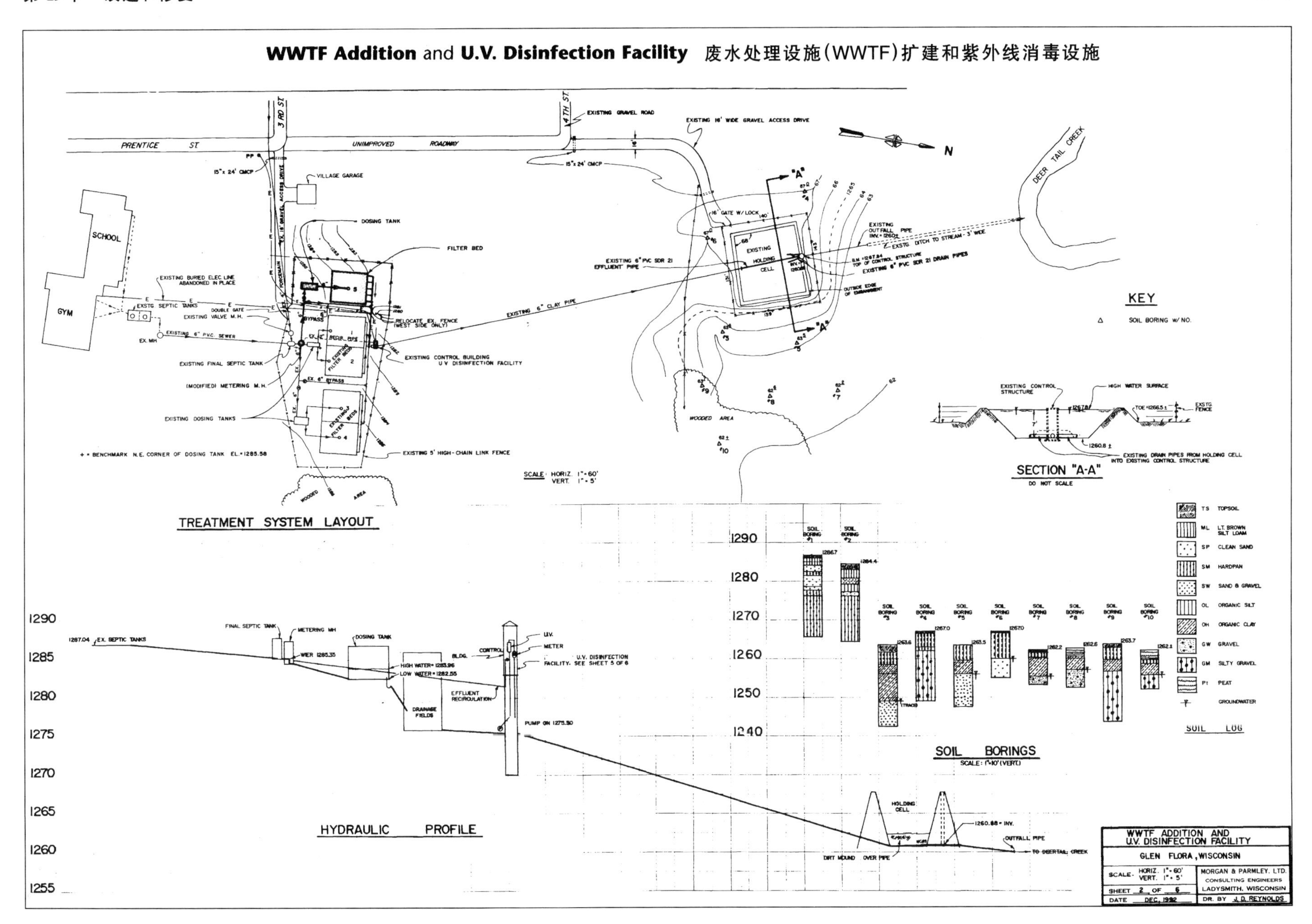
WWTF Addition and U.V. Disinfection Facility 废水处理设施(WWTF)扩建和紫外线消毒设施
TREATMENT SYSTEM LAYOUT
HYDRAULIC PROFILE
SOIL BORINGS
SCALE: 1"=10' (VERT.)
SECTION "A-A"
DO NOT SCALE
KEY
SOIL BORING W/ NO.
SOIL LOG
TS TOPSOIL
ML LT. BROWN SILT LOAM
SP CLEAN SAND
SM HARDPAN
SW SAND & GRAVEL
OL ORGANIC SILT
OH ORGANIC CLAY
GW GRAVEL
GM SILTY GRAVEL
PT PEAT
GROUNDWATER
PRENTICE ST.
UNIMPROVED ROADWAY
3 RD ST.
4 TH ST.
DEER TAIL CREEK
SCHOOL
GYM
VILLAGE GARAGE
DOSING TANK
FILTER BED
EXISTING GRAVEL ROAD
EXISTING 16' WIDE GRAVEL ACCESS DRIVE
EXISTING 6" CLAY PIPE
EXISTING CONTROL BUILDING U.V. DISINFECTION FACILITY
EXISTING FINAL SEPTIC TANK
(MODIFIED) METERING M.H.
EXISTING DOSING TANKS
EXISTING 5' HIGH-CHAIN LINK FENCE
+ = BENCHMARK N.E. CORNER OF DOSING TANK EL.=1285.58
SCALE: HORIZ. 1"=60' VERT. 1"=5'
EXISTING HOLDING CELL
WOODED AREA
FINAL SEPTIC TANK
METERING MH
DOSING TANK
U.V.
METER
U.V. DISINFECTION FACILITY; SEE SHEET 5 OF 6
DRAINAGE FIELDS
EFFLUENT RECIRCULATION
PUMP ON 1275.50
HOLDING CELL
OUTFALL PIPE
DIRT MOUND OVER PIPE
TO DEERTAIL CREEK
WWTF ADDITION AND U.V. DISINFECTION FACILITY
GLEN FLORA, WISCONSIN
SCALE: HORIZ. 1"=60' VERT. 1"=5'
MORGAN & PARMLEY, LTD. CONSULTING ENGINEERS LADYSMITH, WISCONSIN
SHEET 2 OF 6
DATE DEC. 1992
DR. BY J.D. REYNOLDS

WWTF Addition and U.V. Disinfection Facility 废水处理设施(WWTF)扩建和紫外线消毒设施

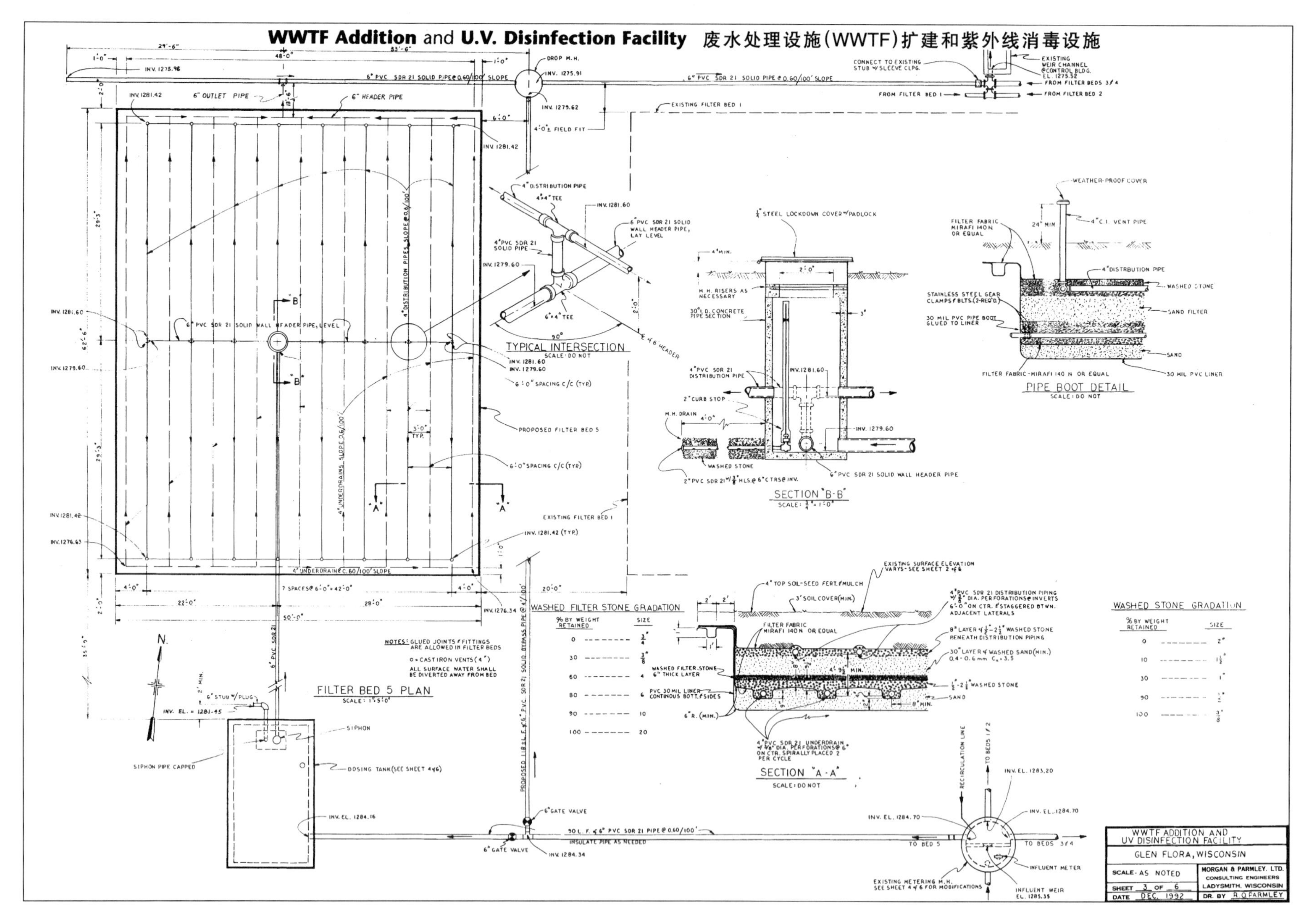

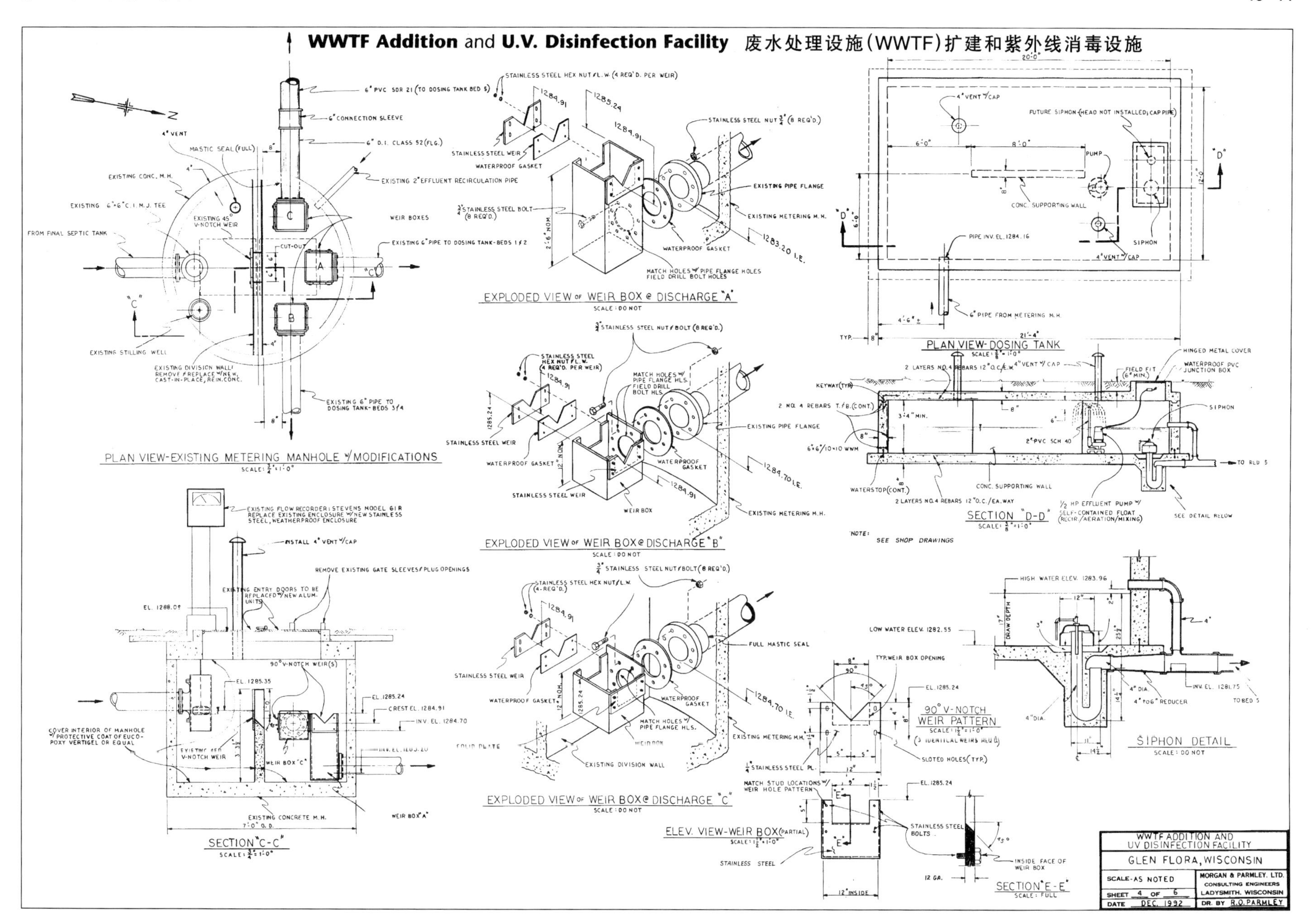

WWTF Addition and U.V. Disinfection Facility 废水处理设施(WWTF)扩建和紫外线消毒设施

WWTF Addition and U.V. Disinfection Facility 废水处理设施(WWTF)扩建和紫外线消毒设施

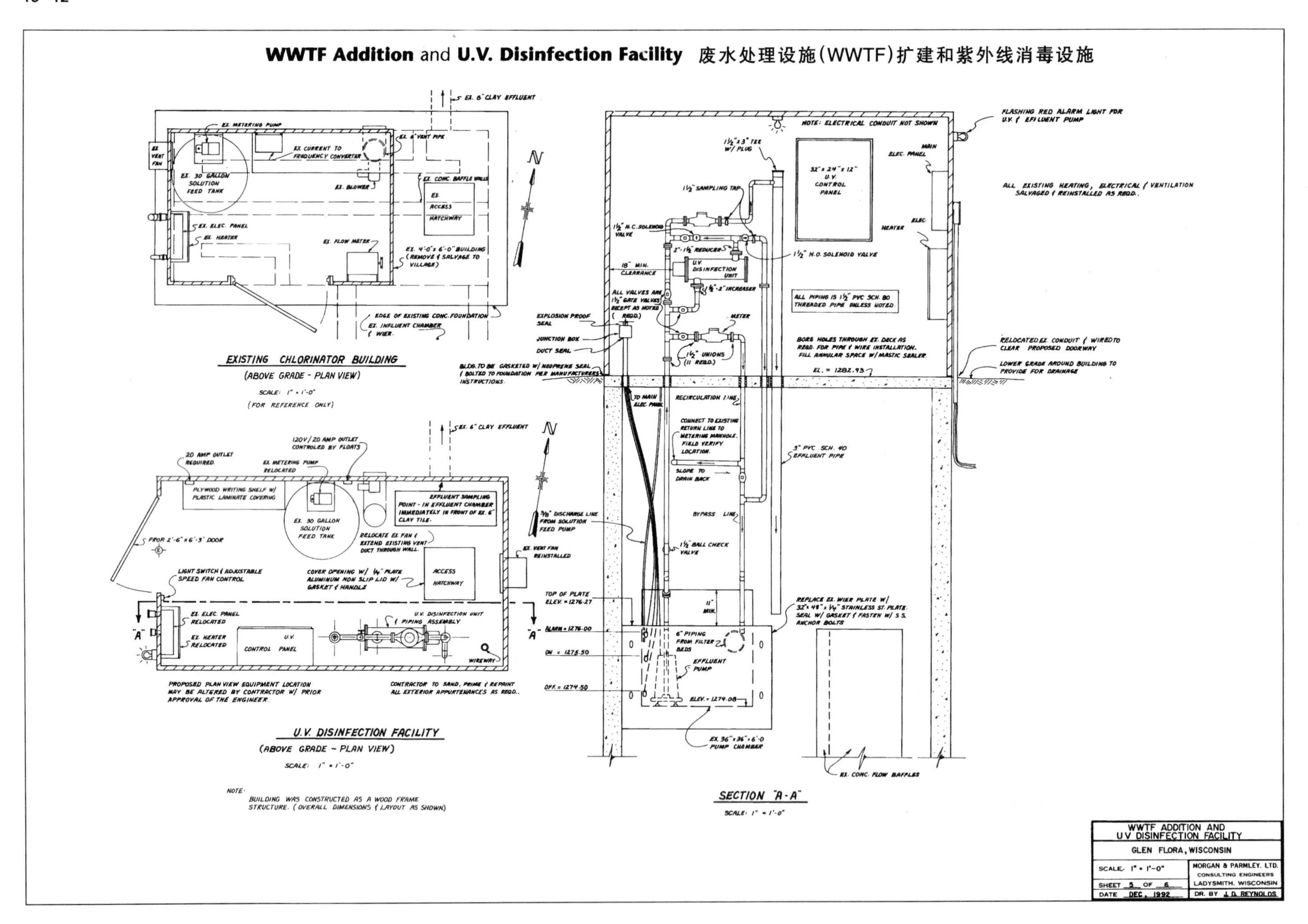

GENERAL LOCATION MAP

传输干管改线

TRANSMISSION MAIN RE-ROUTE
RUSK COUNTY AIRPORT

VILLAGE

OF

TONY, WISCONSIN

MUNIPAL WATER SYSTEM

M. & P. PROJECT NO. 99-110

JUNE, 1999

—DRAWING INDEX—

SHEET NO.	DESCRIPTION
1	TITLE SHEET, LOCATION MAPS & GENERAL NOTES
2	PROJECT SITE PLAN & CONSTRUCTION DETAILS
3	EROSION CONTROL DETAILS w/ TYPICAL NOTES

REFERENCE SPECIFICATIONS*

*STANDARD SPECIFICATIONS FOR SEWER AND WATER CONSTRUCTION IN WISCONSIN (5TH EDITION) W / ADDENDUMS 1 & 2

*WISCONSIN CONSTRUCTION SITE BEST MANAGEMENT PRACTICE HANDBOOK

*MORGAN & PARMLEY, LTD. SUPPLEMENTAL SPECIFICATIONS

GENERAL NOTES

1. ALL WATERMAIN SHALL BE DUCTILE IRON CL 52.
2. ALL WATERMAIN FITTINGS SHALL BE M.J.D.I. W / RESTRAINED JOINT LUGS AND MEET AWWA C153.
3. ALL HYDRANT LEADS SHALL BE 6-INCH DUCTILE IRON CL 52.
4. ALL HYDRANT LEADS SHALL HAVE A 6-INCH GATE VALVE.
5. HYDRANTS SHALL BE WATEROUS WB 67, 8.5 FT. BURY (4" PUMPER NOZZLE & 2-2½" NOZZLES), 22" BREAKOFF WITH THREADS MATCHING VILLAGE OF TONY PATTERN; I.E. CITY OF LADYSMITH FIRE DEPARTMENT.
6. ALL WATERMAIN SHALL HAVE AN 8 FT. BURY (MIN.) AND BE CABLE BONDED FOR CONTINUITY.
7. WATER SERVICES SHALL BE 1" TYPE K COPPER W / MUELLER OR FORD BRASS AND LOCATED BY OWNER.
8. ALL SERVICE CORPORATIONS SHALL HAVE A SADDLE CLAMP.
9. 8 FT. (MINIMUM) HORIZONTAL DISTANCE (℄ TO ℄) BETWEEN WATERMAIN AND SANITARY SEWER OR STORM SEWER.
10. 18-INCH (MINIMUM) VERTICAL DISTANCE (OUT TO OUT) BETWEEN WATERMAIN UNDER SANITARY SEWER OR STORM SEWER @ THEIR INTERSECTION.
11. 6-INCH (MINIMUM) VERTICAL DISTANCE (OUT TO OUT) BETWEEN WATERMAIN OVER SANITARY SEWER OR STORM SEWER @ THEIR INTERSECTION.
12. ALL ELEVATIONS BASED ON USGS DATUM.
13. ALL GATE VALVES IN TRANSMISSION MAIN SHALL HAVE A MAXIMUM SPACING OF 800 FEET.
14. EROSION CONTROL MEASURES SHALL BE IN PLACE BEFORE CONSTRUCTION BEINGS.
15. THE CONTRACTOR SHALL MINIMIZE EROSION DURING CONSTRUCTION USING GOOD CONSTRUCTION TECHNIQUES AND UTILIZING SILT FENCES AND BALE DITCH CHECKS. THE CONTRACTOR SHALL RELEASE RUNOFF FROM THE SITE IN A NUISANCE FREE MANNER.
16. THE CONTRACTOR IS RESPONSIBLE TO MAINTAIN THE SITE IN A SAFE CONDITION. THE SOLE RESPONSIBILITY FOR WARNING SIGNS, BARRICADES AND ALL ASPECTS OF SAFETY LIE WITH THE CONTRACTOR.
17. EXISTING UTILITIES SHOWN MAY NOT BE ALL INCLUSIVE, CONTRACTOR SHALL HAVE ALL BURIED UTILITIES LOCATED PRIOR TO COMMENCING CONSTRUCTION.
18. EXISTING WATER TRANSMISSION MAIN SHOWN ACCORDING TO EXISTING RECORDS, BUT LOCATION SHALL BE VERIFIED IN FIELD.
19. WATERMAIN & SERVICES SHALL BE INSULATED AT ALL STORM SEWER, CULVERT AND DITCH CROSSINGS WITH 2" THICK EXTRUDED POLYSTYRENE INSULATION BOARD.
20. PROPERTY CORNERS KNOWN TO EXIST SHALL NOT BE DISTURBED. DAMAGED CORNERS WILL BE REPLACED AT THE CONTRACTOR'S EXPENSE.
21. ALL WATERMAIN SHALL BE DISINFECTED BEFORE BEING PUT INTO USE.
22. ALL VALVE RISER CAPS SHALL BE SET TO GRADE SHOWN ON THE PLANS, UNLESS DIRECTED OTHERWISE.
23. CONSTRUCTION SHALL COMPLY WITH THE CONDITIONS OF DNR APPROVAL LETTER.
24. ALL DISTURBED AREAS SHALL BE SEEDED, FERTILIZED AND MULCHED, OR COVERED W / BASE COURSE AFTER CONSTRUCTION.
25. ALL DISTURBED AREAS SHALL BE RESTORED W / SEEDING AS FOLLOWS:
    A) REPLACE ALL SALVAGED TOPSOIL
    B) SEED MIXTURE: PERMANENT SEEDING - (LAWN TYPE TURF)
       35% KENTUCKY BLUEGRASS
       25% IMPROVED FINE PERENNIAL RYEGRASS
       15% CREEPING RED FESCUE
       15% IMPROVED HARD FESCUE
       10% WHITE CLOVER
    C) SEED RATE: 2# / 1000 SQ. FT.
    D) MULCH RATE: 3 TON / ACRE
    E) PERMANENT SEEDING: ALLOWED TO SEPT. 7
       TEMPORARY SEEDING: SEPT. 8 THROUGH NOV. 10 - (4 BU. OATS / ACRE)
       DORMANT SEEDING: SEED PERM. SEEDING INTO TEMP. SEEDING AFTER NOV. 10
    F) SEEDING MAY BE BROADCAST BUT MUST BE COVERED WITH A MAXIMUM OF ½" SOIL. MULCH MUST BE ANCHORED WITH A MULCH TILLER, TACKIFIER OR NETTING.
    G) FERTILIZER: 500# / ACRE 20-10-10
26. CONTRACTOR SHALL DETERMINE CONSTRUCTION SEQUENCE IN ORDER TO MINIMIZE INSTALLATION CONFLICTS WITH PIPE CROSSINGS.
27. THERE SHALL BE NO DEVIATION FROM THE PLANS WITHOUT THE DESIGN ENGINEER'S APPROVAL.
28. CONSTRUCTION STAKES DESTROYED BY THE CONTRACTOR THAT NEED TO BE REPLACED, WILL BE CHARGED TO THE CONTRACTOR.
29. NO TREES ARE TO BE REMOVED WITHOUT THE APPROVAL OF THE ENGINEER.
30. CONTRACTOR SHALL COORDINATE AND SCHEDULE HIS WORK WITH VILLAGE & ENGINEER TO INSURE MINIMAL CONFLICT WITH WATER UTILITY DURING CONNECTION TO EXISTING TRANSMISSION MAIN. THEREFORE, THE CONTRACTOR SHALL SUPPLY A PLAN & SEQUENCE METHOD FOR TIE-IN, TESTING AND DISINFECTION FOR ENGINEER'S APPROVAL, PRIOR TO COMMENCING CONSTRUCTION.

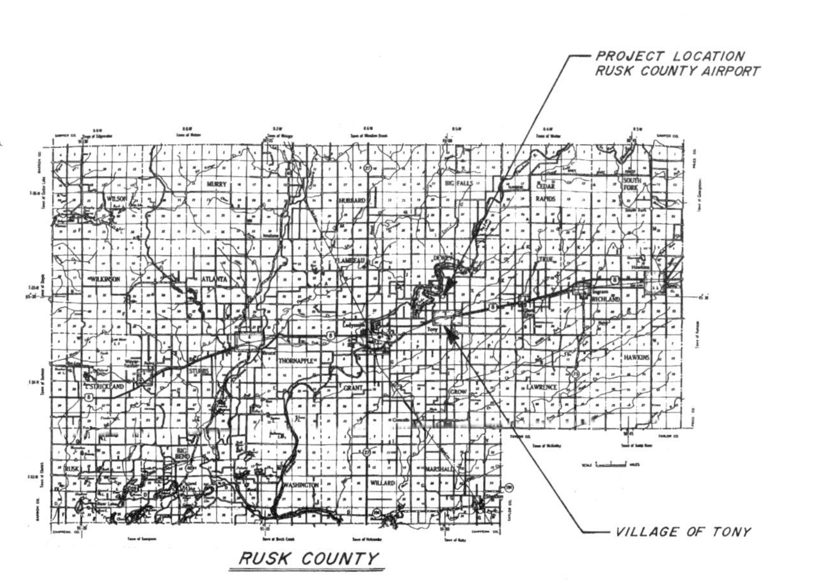

RUSK COUNTY

-UTILITIES-

SEWER & WATER: VILLAGE OF TONY
BILLY MECHELKE
N5297 LITTLE X RD.
TONY, WI 54563
715/532-3183 (W)
715/532-7046 (H)

ELECTRIC:
NORTHERN STATES POWER COMPANY
310 HICKORY HILL LANE
PHILLIPS, WISCONSIN 54555
ATTN: JOE PERKINS
715/836-1198

TELEPHONE:
CENTURYTEL
425 ELLINGSON AVENUE
P.O. BOX 78
HAWKINS, WI 54530
800/752-5637
ATTN: JAMES ARQUETTE

DIGGER'S HOTLINE:
800/242-8511

PREPARED BY:
MORGAN & PARMLEY, LTD.
PROFESSIONAL CONSULTING ENGINEERS
LADYSMITH, WISCONSIN

The airport initiated plans to construct a north-south cross runway which would intersect the existing potable water transmission main. This conflict needed to be resolved prior to constructing the new cross runway. The only logical solution to permanently protect the water supply was to reroute the transmission main, leaving the abandoned section in-place.

Source: Morgan & Parmley, Ltd.

Transmission Main Reroute 传输干管改线

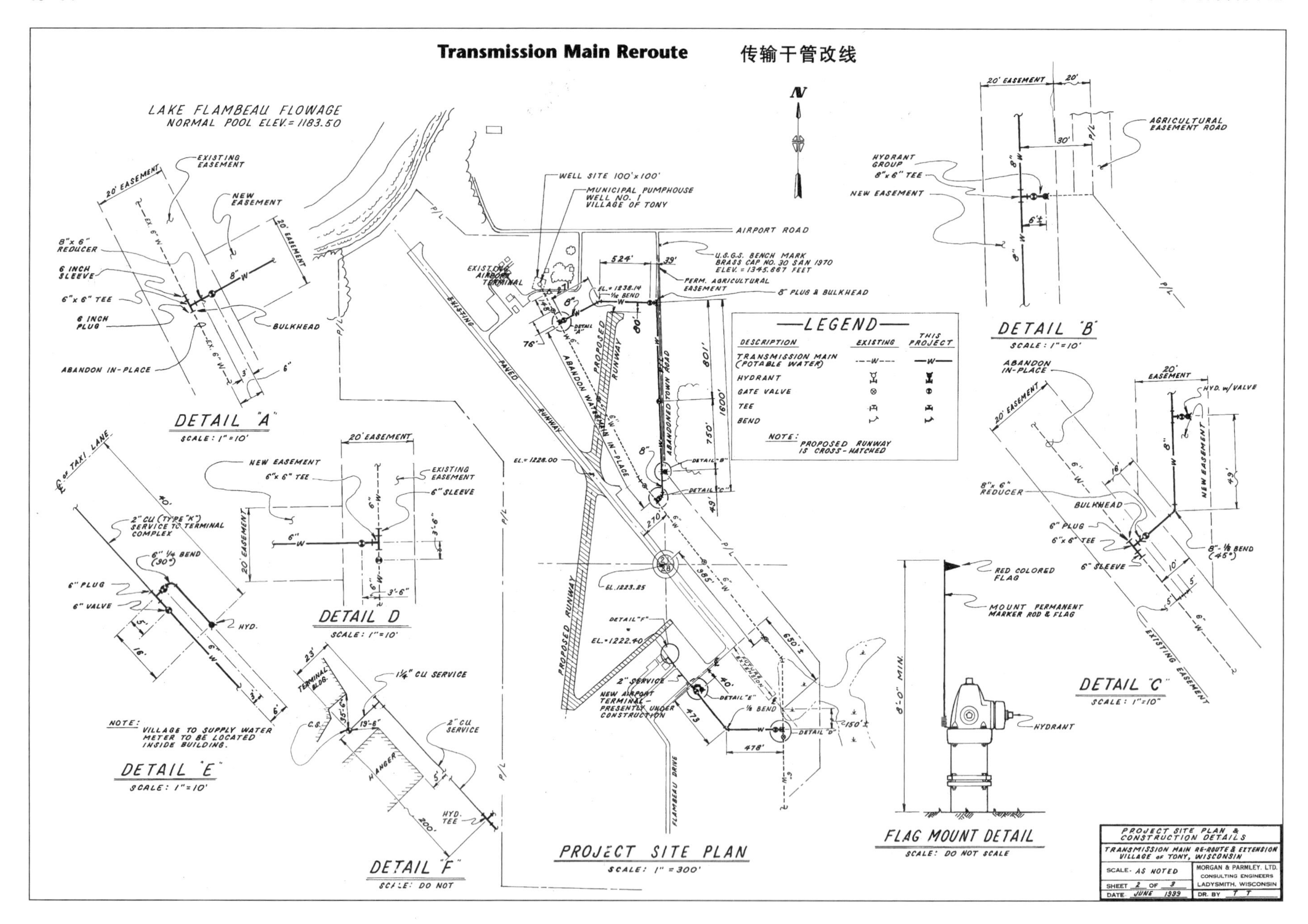

Transmission Main Reroute　传输干管改线

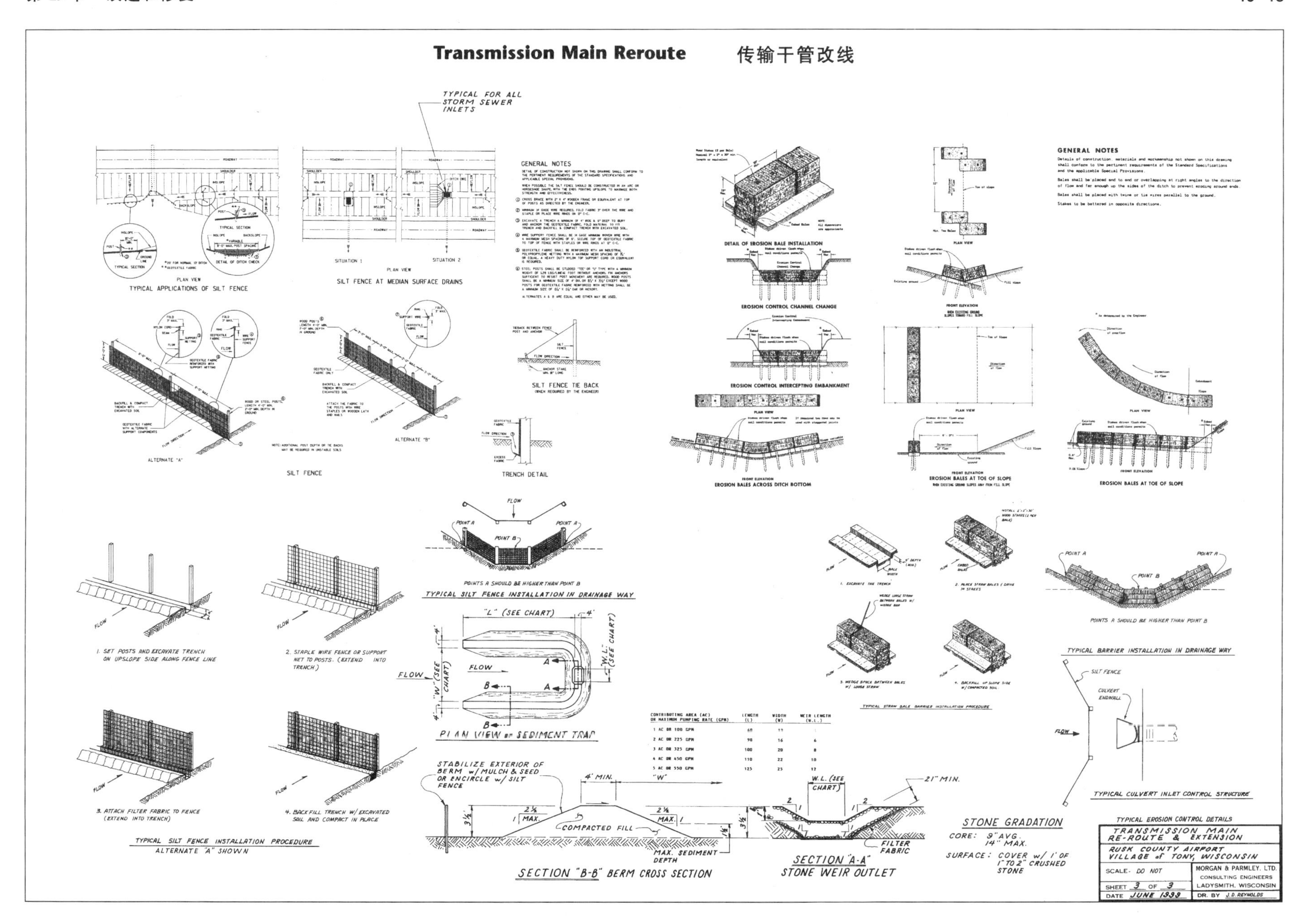

Proposed Addition with Boiler Installation 建议的扩建，包括锅炉安装

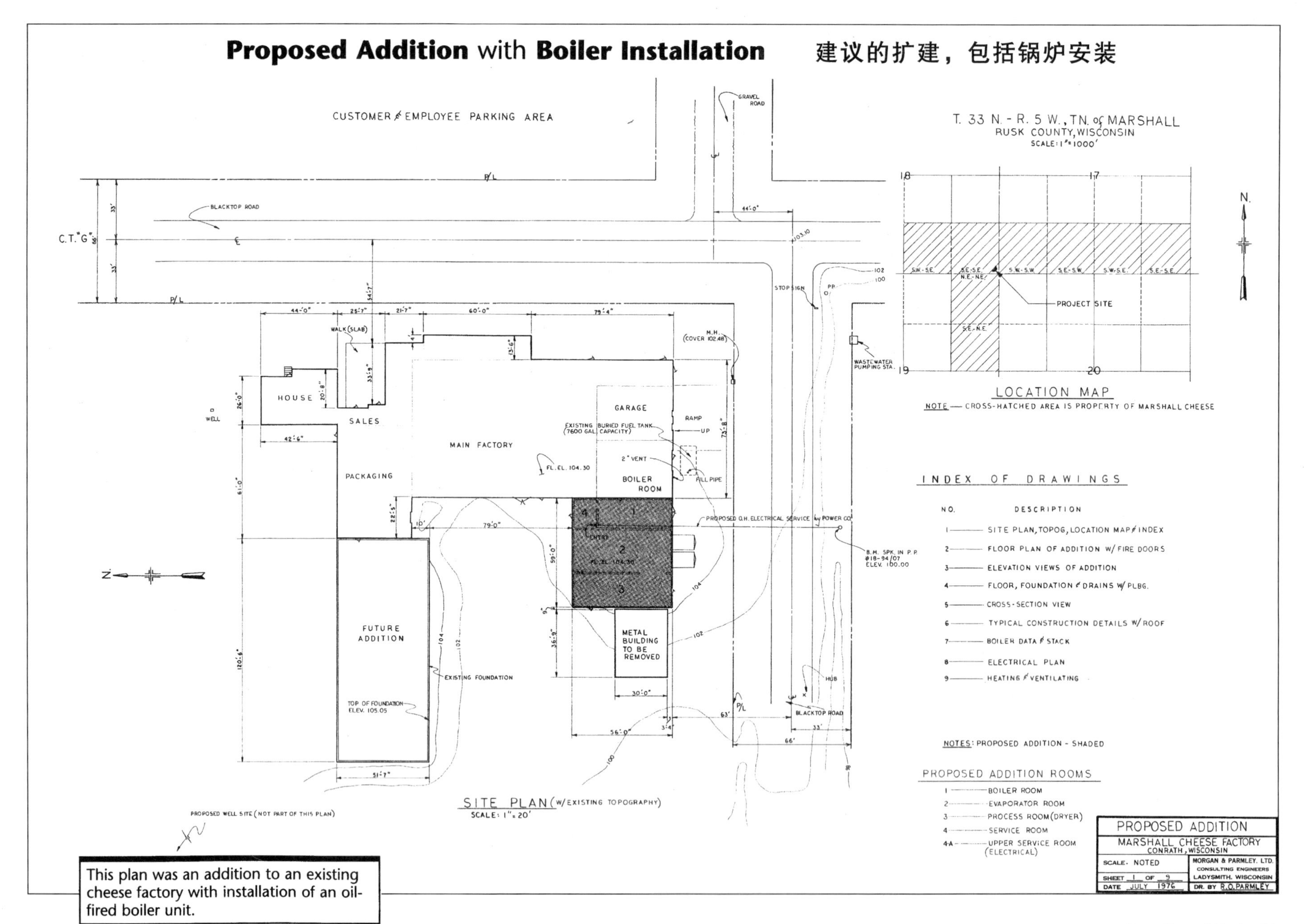

This plan was an addition to an existing cheese factory with installation of an oil-fired boiler unit.

Source: Morgan & Parmley, Ltd.

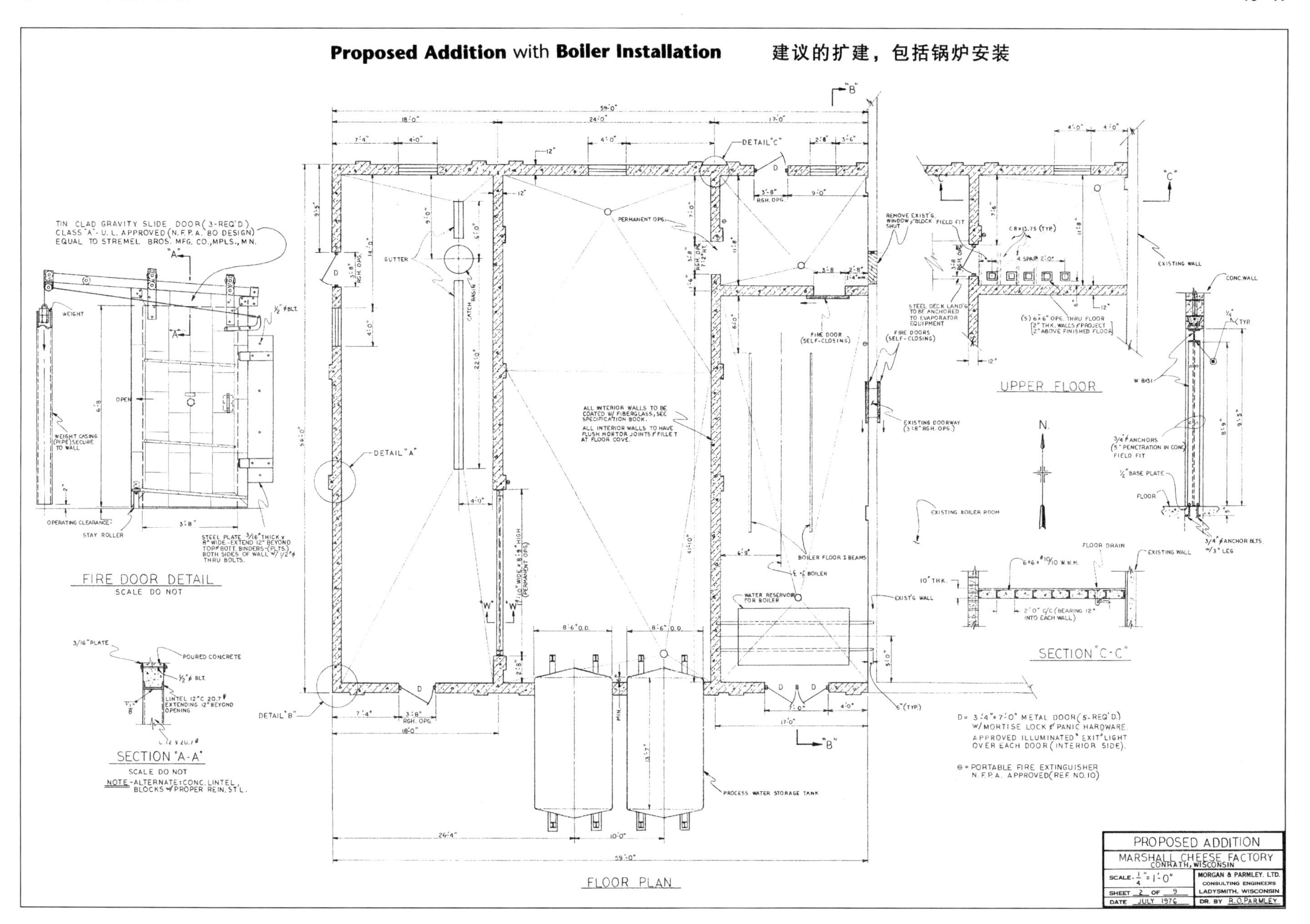

Proposed Addition with Boiler Installation 建议的扩建，包括锅炉安装
FIRE DOOR DETAIL
SCALE DO NOT
SECTION "A-A"
SCALE DO NOT
NOTE-ALTERNATE: CONC. LINTEL, BLOCKS W/PROPER REIN. ST'L.
FLOOR PLAN
UPPER FLOOR
SECTION "C-C"
D= 3'-4"×7'-0" METAL DOOR (5-REQ'D.) W/MORTISE LOCK & PANIC HARDWARE. APPROVED ILLUMINATED "EXIT" LIGHT OVER EACH DOOR (INTERIOR SIDE).
⊖ = PORTABLE FIRE EXTINGUISHER N.F.P.A. APPROVED (REF. NO.10)
ALL INTERIOR WALLS TO BE COATED W/ FIBERGLASS, SEE SPECIFICATION BOOK.
ALL INTERIOR WALLS TO HAVE FLUSH MORTOR JOINTS & FILLET AT FLOOR COVE.
PROCESS WATER STORAGE TANK
PROPOSED ADDITION
MARSHALL CHEESE FACTORY
CONRATH, WISCONSIN
MORGAN & PARMLEY, LTD.
CONSULTING ENGINEERS
LADYSMITH, WISCONSIN
DR. BY R.O.PARMLEY

Proposed Addition with Boiler Installation 建议的扩建，包括锅炉安装

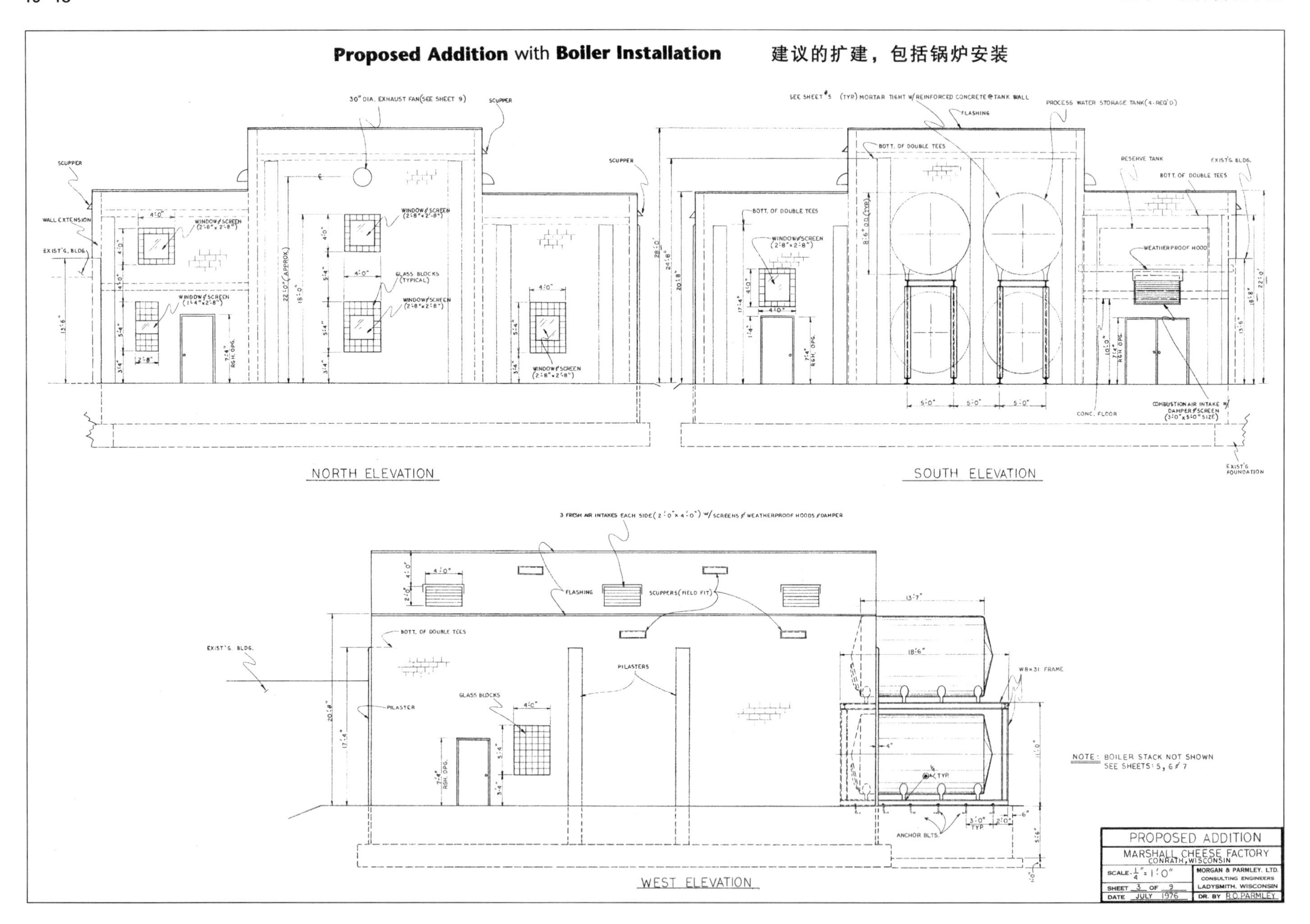

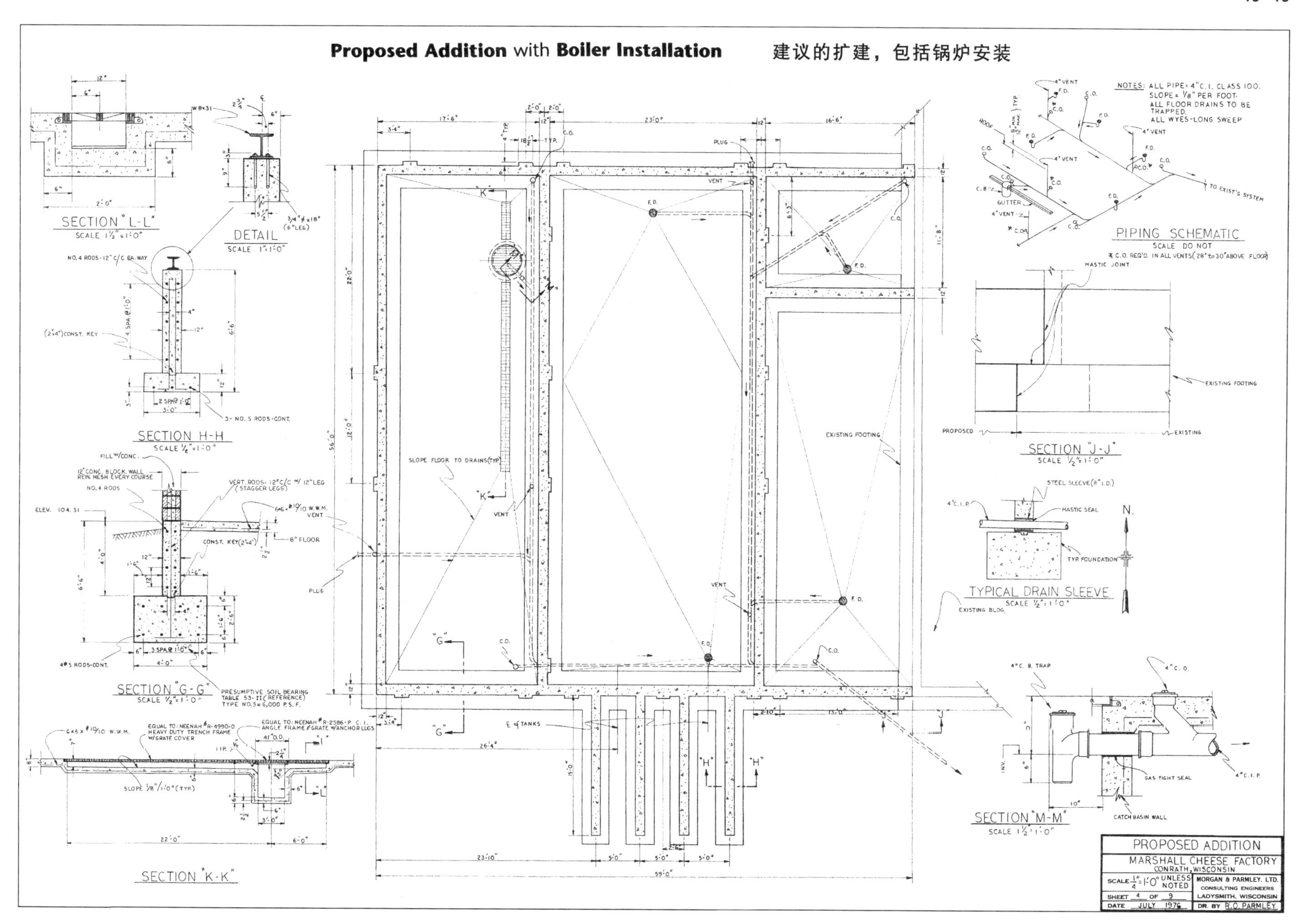

Proposed Addition with Boiler Installation　建议的扩建，包括锅炉安装
SECTION "L-L"
SCALE 1½"=1'-0"
DETAIL
SCALE 1"=1'-0"
SECTION H-H
SCALE ½"=1'-0"
SECTION "G-G"
SCALE ½"=1'-0"
PRESUMPTIVE SOIL BEARING TABLE 53-II (REFERENCE) TYPE NO.5=6,000 P.S.F.
SECTION "K-K"
SLOPE FLOOR TO DRAINS (TYP)
EXISTING FOOTING
PIPING SCHEMATIC
SCALE DO NOT
NOTES: ALL PIPE=4"C.I. CLASS 100. SLOPE=⅛" PER FOOT. ALL FLOOR DRAINS TO BE TRAPPED. ALL WYES-LONG SWEEP
TO EXIST'G SYSTEM
SECTION "J-J"
SCALE ½"=1'-0"
TYPICAL DRAIN SLEEVE
SCALE ½"=1'-0"
SECTION "M-M"
SCALE 1½"=1'-0"
PROPOSED ADDITION
MARSHALL CHEESE FACTORY
CONRATH, WISCONSIN
MORGAN & PARMLEY, LTD.
CONSULTING ENGINEERS
LADYSMITH, WISCONSIN
SHEET 4 OF 9
DATE JULY 1976
DR. BY R.O. PARMLEY

Proposed Addition with Boiler Installation 建议的扩建，包括锅炉安装

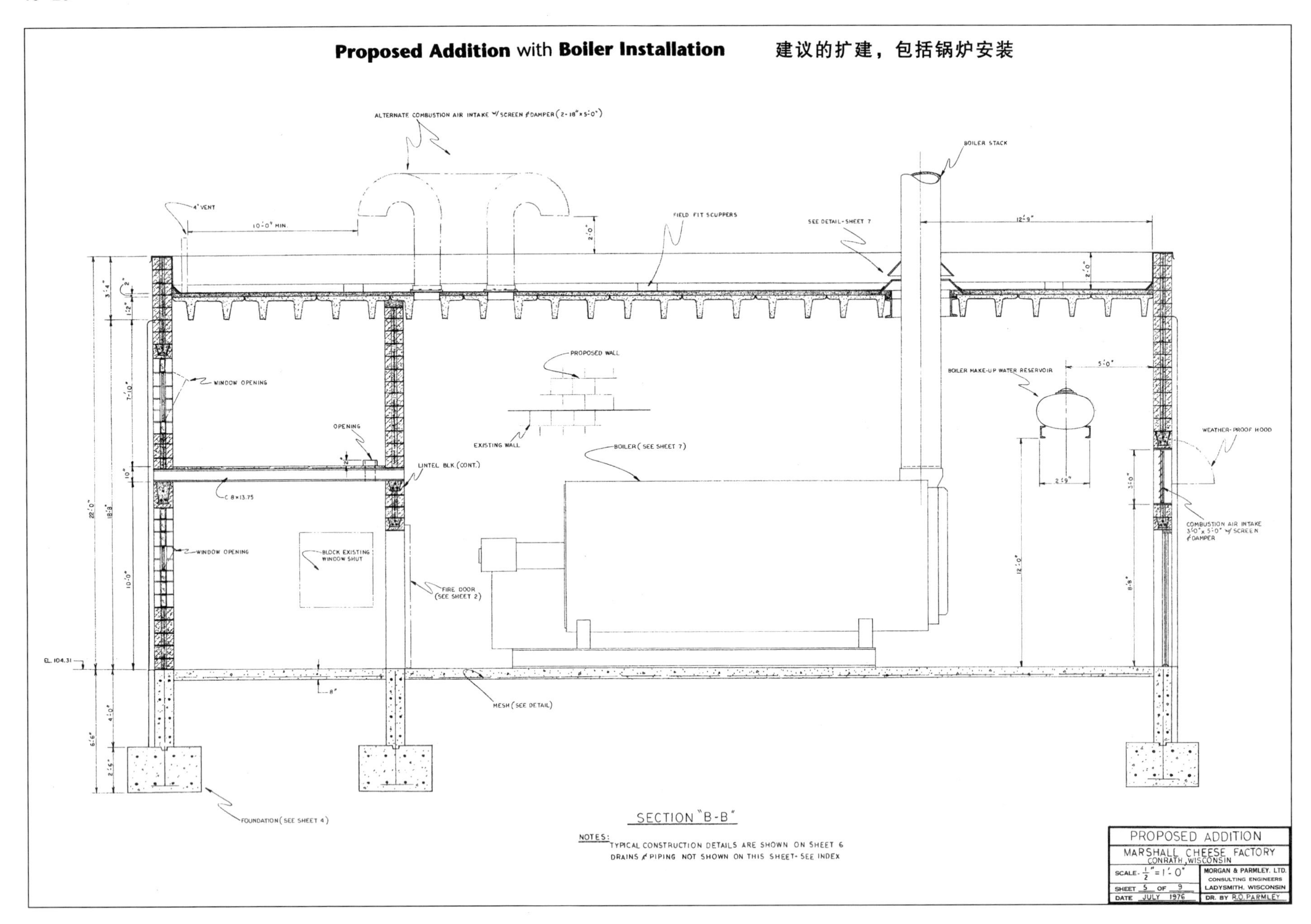

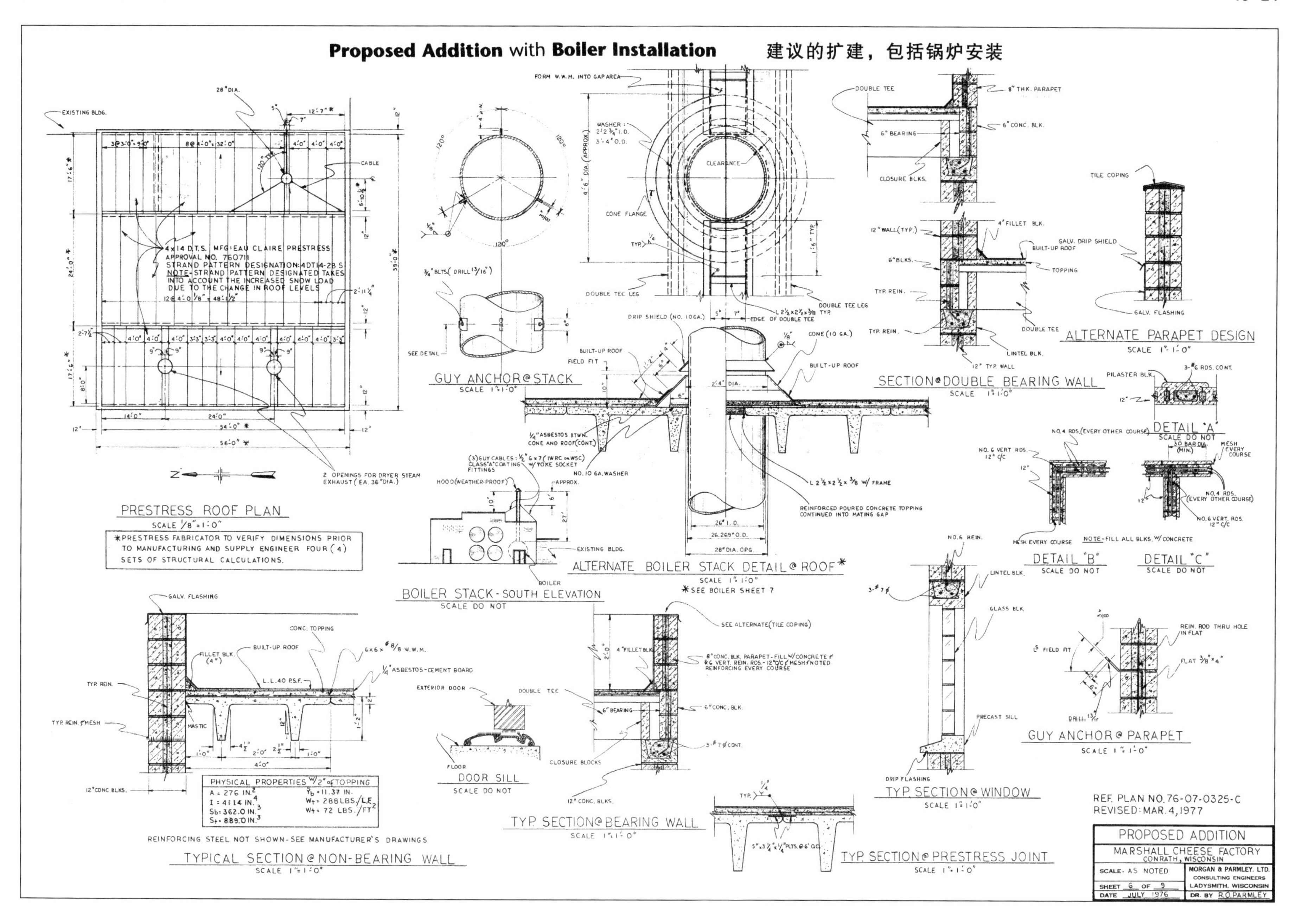
Proposed Addition with Boiler Installation 建议的扩建，包括锅炉安装
PRESTRESS ROOF PLAN
SCALE 1/8"=1'-0"
*PRESTRESS FABRICATOR TO VERIFY DIMENSIONS PRIOR TO MANUFACTURING AND SUPPLY ENGINEER FOUR (4) SETS OF STRUCTURAL CALCULATIONS.
GUY ANCHOR@STACK
SCALE 1"=1'-0"
BOILER STACK-SOUTH ELEVATION
SCALE DO NOT
ALTERNATE BOILER STACK DETAIL@ROOF*
SCALE 1"=1'-0"
*SEE BOILER SHEET 7
SECTION@DOUBLE BEARING WALL
SCALE 1"=1'-0"
ALTERNATE PARAPET DESIGN
SCALE 1"=1'-0"
DETAIL "A"
DETAIL "B"
DETAIL "C"
TYPICAL SECTION@NON-BEARING WALL
SCALE 1"=1'-0"
DOOR SILL
SCALE DO NOT
TYP. SECTION@BEARING WALL
SCALE 1"=1'-0"
TYP. SECTION@WINDOW
SCALE 1"=1'-0"
TYP. SECTION@PRESTRESS JOINT
SCALE 1"=1'-0"
GUY ANCHOR@PARAPET
SCALE 1"=1'-0"
REINFORCING STEEL NOT SHOWN-SEE MANUFACTURER'S DRAWINGS
REF. PLAN NO. 76-07-0325-C
REVISED: MAR. 4, 1977
PROPOSED ADDITION
MARSHALL CHEESE FACTORY
CONRATH, WISCONSIN
MORGAN & PARMLEY, LTD.
CONSULTING ENGINEERS
LADYSMITH, WISCONSIN
SCALE-AS NOTED
SHEET 6 OF 9
DATE JULY 1976
DR. BY R.O. PARMLEY

Proposed Addition with Boiler Installation 建议的扩建，包括锅炉安装

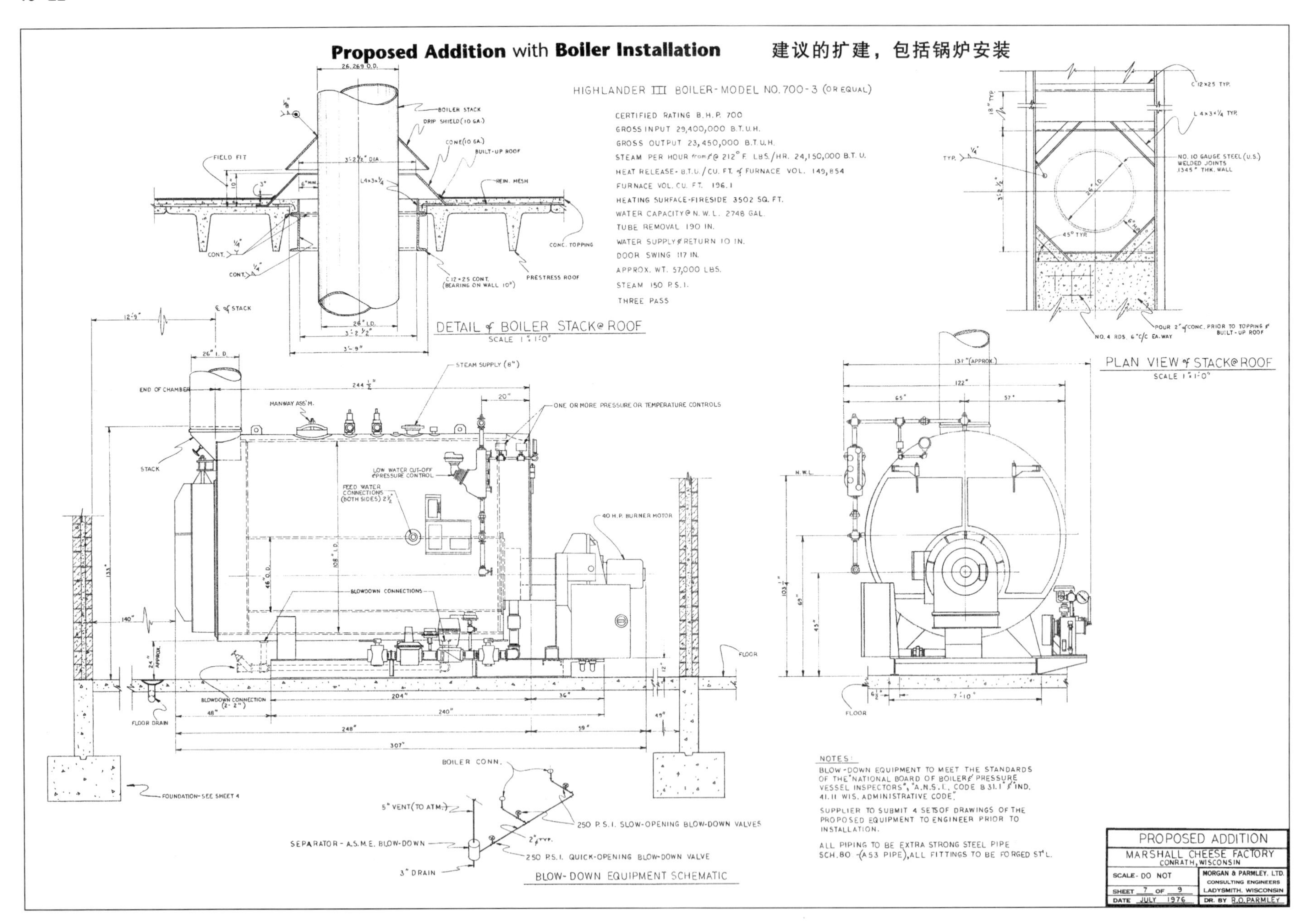

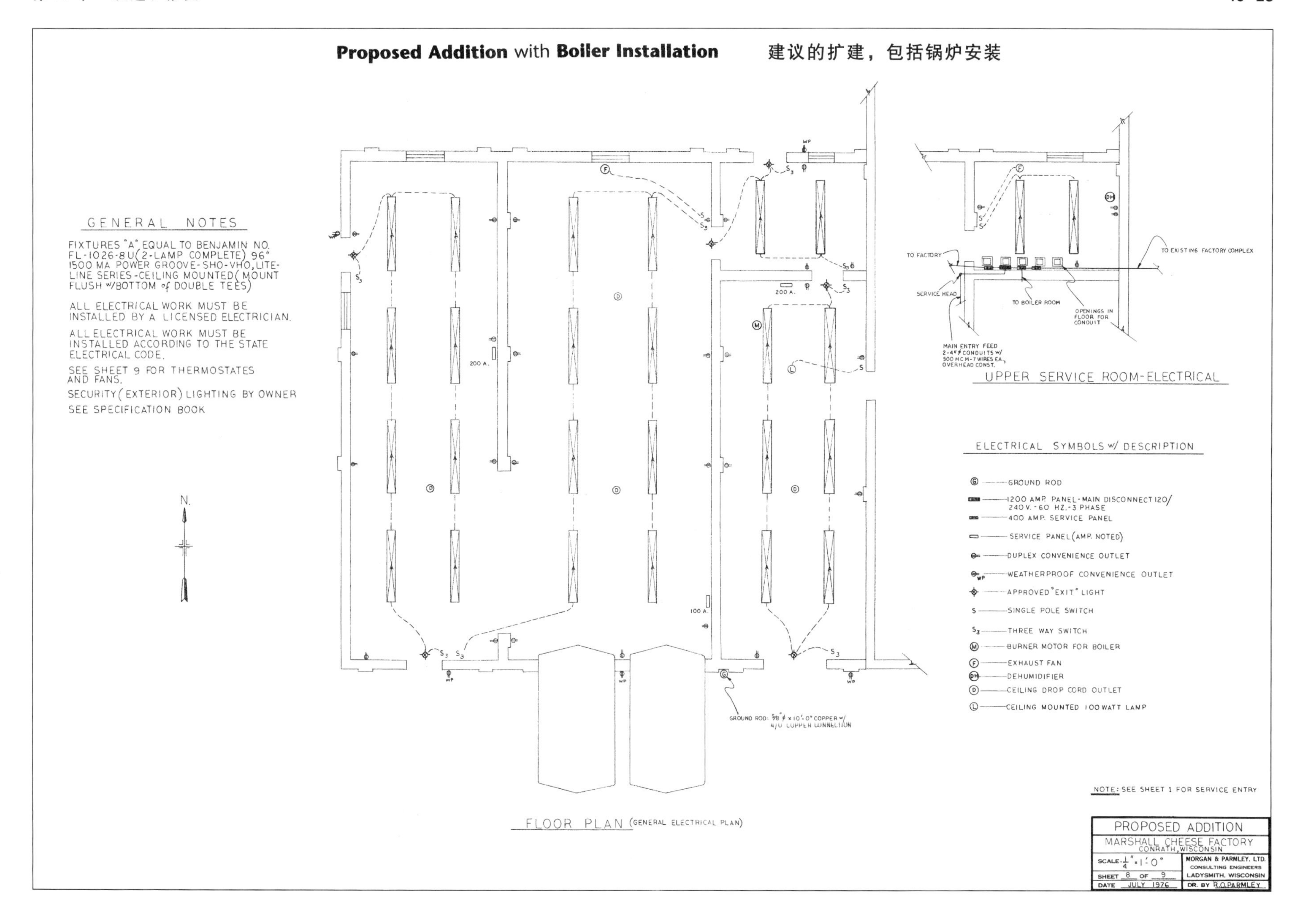
Proposed Addition with **Boiler Installation**　　建议的扩建，包括锅炉安装
GENERAL NOTES
FIXTURES "A" EQUAL TO BENJAMIN NO. FL-1026-8U(2-LAMP COMPLETE) 96" 1500 MA POWER GROOVE-SHO-VHO, LITE-LINE SERIES-CEILING MOUNTED(MOUNT FLUSH w/BOTTOM of DOUBLE TEES)
ALL ELECTRICAL WORK MUST BE INSTALLED BY A LICENSED ELECTRICIAN.
ALL ELECTRICAL WORK MUST BE INSTALLED ACCORDING TO THE STATE ELECTRICAL CODE.
SEE SHEET 9 FOR THERMOSTATES AND FANS.
SECURITY(EXTERIOR) LIGHTING BY OWNER
SEE SPECIFICATION BOOK
N.
200 A.
100 A.
GROUND ROD: 5/8"ϕ×10'-0" COPPER w/ 4/0 COPPER CONNECTION
FLOOR PLAN (GENERAL ELECTRICAL PLAN)
TO FACTORY
TO EXISTING FACTORY COMPLEX
SERVICE HEAD
TO BOILER ROOM
OPENINGS IN FLOOR FOR CONDUIT
MAIN ENTRY FEED 2-4"ϕ CONDUITS w/ 500 MCM-7 WIRES EA., OVERHEAD CONST.
UPPER SERVICE ROOM-ELECTRICAL
ELECTRICAL SYMBOLS w/ DESCRIPTION
GROUND ROD
1200 AMP. PANEL-MAIN DISCONNECT 120/240V.-60 HZ.-3 PHASE
400 AMP. SERVICE PANEL
SERVICE PANEL(AMP. NOTED)
DUPLEX CONVENIENCE OUTLET
WEATHERPROOF CONVENIENCE OUTLET
APPROVED "EXIT" LIGHT
SINGLE POLE SWITCH
THREE WAY SWITCH
BURNER MOTOR FOR BOILER
EXHAUST FAN
DEHUMIDIFIER
CEILING DROP CORD OUTLET
CEILING MOUNTED 100 WATT LAMP
NOTE: SEE SHEET 1 FOR SERVICE ENTRY
PROPOSED ADDITION
MARSHALL CHEESE FACTORY
CONRATH, WISCONSIN
SCALE-1/4"=1'-0"
SHEET 8 OF 9
DATE JULY 1976
MORGAN & PARMLEY, LTD.
CONSULTING ENGINEERS
LADYSMITH, WISCONSIN
DR. BY R.O.PARMLEY

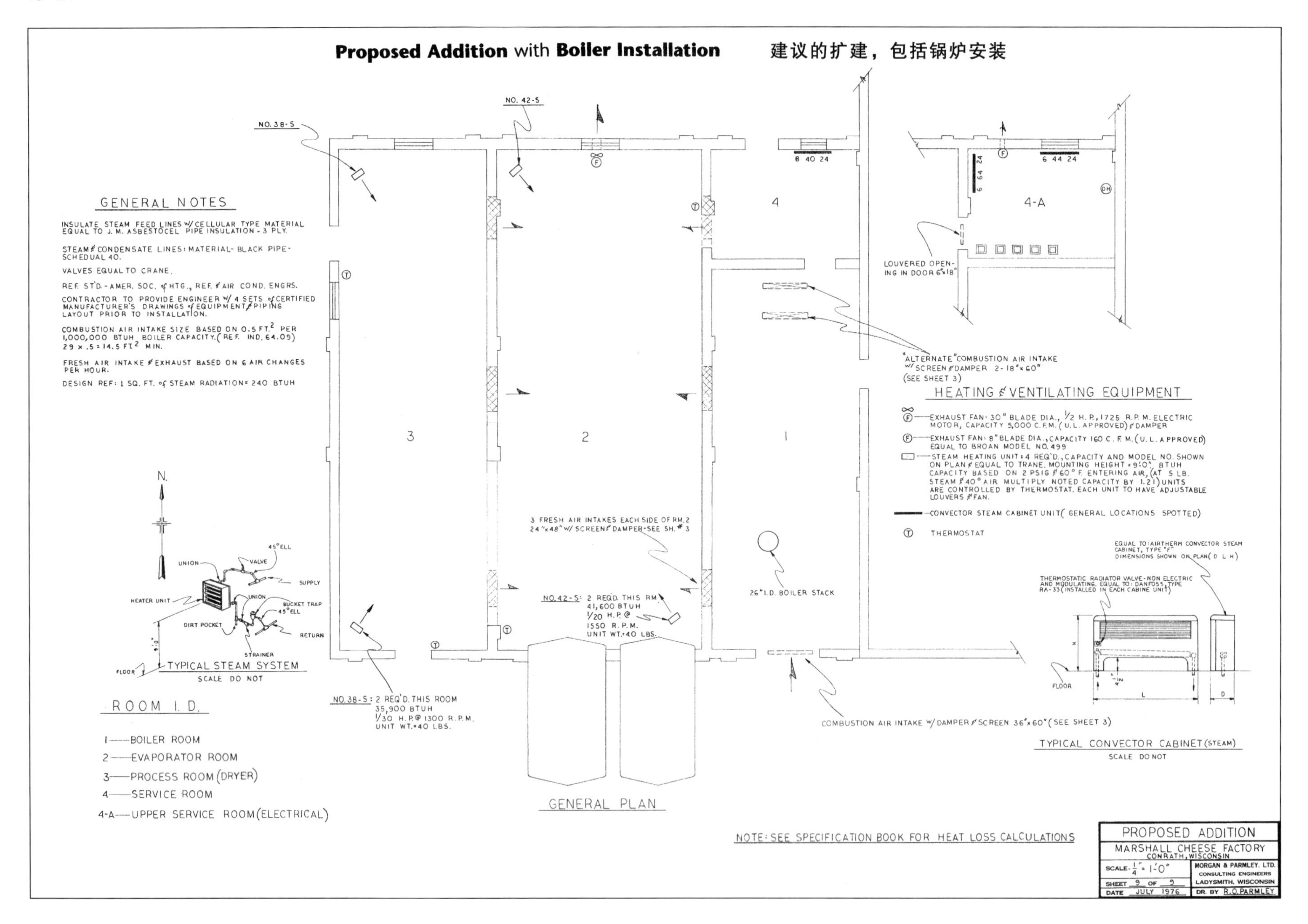
Proposed Addition with Boiler Installation
建议的扩建，包括锅炉安装
GENERAL NOTES
INSULATE STEAM FEED LINES W/ CELLULAR TYPE MATERIAL EQUAL TO J. M. ASBESTOCEL PIPE INSULATION - 3 PLY.
STEAM & CONDENSATE LINES: MATERIAL- BLACK PIPE- SCHEDUAL 40.
VALVES EQUAL TO CRANE.
REF. ST'D.- AMER. SOC. of HTG., REF. & AIR COND. ENGRS.
CONTRACTOR TO PROVIDE ENGINEER W/ 4 SETS of CERTIFIED MANUFACTURER'S DRAWINGS of EQUIPMENT & PIPING LAYOUT PRIOR TO INSTALLATION.
COMBUSTION AIR INTAKE SIZE BASED ON 0.5 FT.2 PER 1,000,000 BTUH BOILER CAPACITY.(REF. IND. 64.05) 29 x .5 = 14.5 FT.2 MIN.
FRESH AIR INTAKE & EXHAUST BASED ON 6 AIR CHANGES PER HOUR.
DESIGN REF: 1 SQ. FT. of STEAM RADIATION = 240 BTUH
NO. 38-S
NO. 42-S
N.
UNION
VALVE
45°ELL
SUPPLY
HEATER UNIT
UNION
BUCKET TRAP
45°ELL
DIRT POCKET
RETURN
STRAINER
FLOOR
TYPICAL STEAM SYSTEM
SCALE DO NOT
ROOM I. D.
1——BOILER ROOM
2——EVAPORATOR ROOM
3——PROCESS ROOM (DRYER)
4——SERVICE ROOM
4-A——UPPER SERVICE ROOM (ELECTRICAL)
NO. 38-S: 2 REQ'D. THIS ROOM 35,900 BTUH 1/30 H.P.@ 1300 R.P.M. UNIT WT.=40 LBS.
3 FRESH AIR INTAKES EACH SIDE OF RM. 2 24"x48" W/ SCREEN & DAMPER-SEE SH. # 3
NO. 42-S: 2 REQ'D. THIS RM. 41,600 BTUH 1/20 H.P.@ 1550 R.P.M. UNIT WT.=40 LBS.
3
2
1
4
4-A
8 40 24
6 44 24
6 64 24
LOUVERED OPENING IN DOOR 6"x18"
"ALTERNATE" COMBUSTION AIR INTAKE W/ SCREEN & DAMPER 2-18"x60" (SEE SHEET 3)
26" I.D. BOILER STACK
COMBUSTION AIR INTAKE W/ DAMPER & SCREEN 36"x60" (SEE SHEET 3)
GENERAL PLAN
HEATING & VENTILATING EQUIPMENT
EXHAUST FAN: 30" BLADE DIA., 1/2 H.P., 1725 R.P.M. ELECTRIC MOTOR, CAPACITY 5,000 C.F.M. (U.L. APPROVED) & DAMPER
EXHAUST FAN: 8" BLADE DIA., CAPACITY 160 C.F.M. (U.L. APPROVED) EQUAL TO BROAN MODEL NO. 499
STEAM HEATING UNIT: 4 REQ'D., CAPACITY AND MODEL NO. SHOWN ON PLAN & EQUAL TO TRANE. MOUNTING HEIGHT = 9'-0". BTUH CAPACITY BASED ON 2 PSIG & 60° F. ENTERING AIR, (AT 5 LB. STEAM & 40° AIR MULTIPLY NOTED CAPACITY BY 1.21) UNITS ARE CONTROLLED BY THERMOSTAT. EACH UNIT TO HAVE ADJUSTABLE LOUVERS & FAN.
CONVECTOR STEAM CABINET UNIT (GENERAL LOCATIONS SPOTTED)
THERMOSTAT
EQUAL TO AIRTHERM CONVECTOR STEAM CABINET, TYPE "F" DIMENSIONS SHOWN ON PLAN (D L H)
THERMOSTATIC RADIATOR VALVE-NON ELECTRIC AND MODULATING. EQUAL TO DANFOSS, TYPE RA-33 (INSTALLED IN EACH CABINET UNIT)
FLOOR
L
D
TYPICAL CONVECTOR CABINET (STEAM)
SCALE DO NOT
NOTE: SEE SPECIFICATION BOOK FOR HEAT LOSS CALCULATIONS
PROPOSED ADDITION
MARSHALL CHEESE FACTORY
CONRATH, WISCONSIN
SCALE- 1/4" = 1'-0"
SHEET 9 OF 9
DATE JULY 1976
MORGAN & PARMLEY, LTD.
CONSULTING ENGINEERS
LADYSMITH, WISCONSIN
DR. BY R.O.PARMLEY

DRUMMOND

BAYFIELD COUNTY, WISCONSIN

GENERAL LOCATION MAP

水塔修复

ELEVATED WATER TANK REHABILITATION

DRUMMOND SANITARY DISTRICT

DRUMMOND, WISCONSIN

M. & P. PROJECT NO. 99-157

APRIL 2000

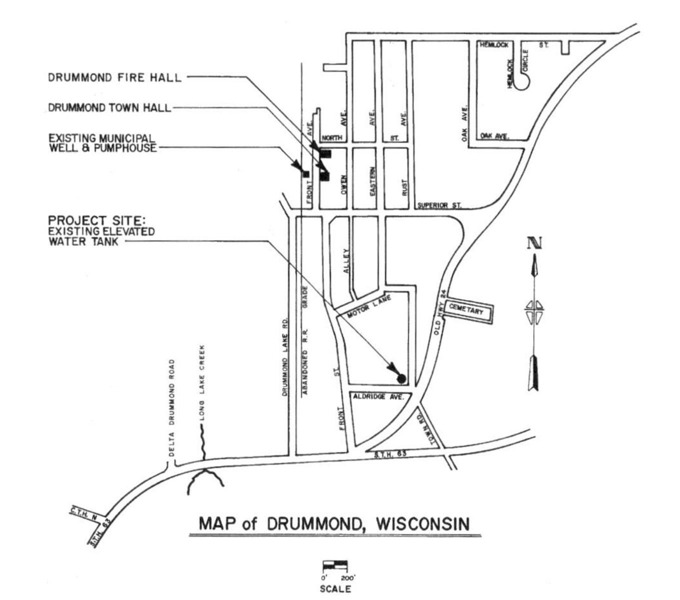

MAP of DRUMMOND, WISCONSIN

——NOTES——

REFER TO ACCOMPANYING SPECIFICATIONS AND REGULATORY APPROVALS FOR SPECIFIC DETAILS AND PROCEDURES.

REFER TO SECTION 01010-SUMMARY OF WORK-FOR PROPOSED WORK AND SPECIAL PROVISIONS.

PAINTING SYSTEMS AND REMOVAL / APPLICATION PROCEDURES SHALL BE CONSISTANT WITH CURRENT AWWA STANDARDS AND APPLICABLE REGULATORY RULES & GUIDLINES.

PRIOR TO PLACING FACILITY BACK INTO SERVICE, THE ENTIRE RESERVOIR FACILITY SHALL BE COMPLETELY DISINFECTED, AS OUTLINED IN THE CURRENT AWWA STANDARD C652. SAFE WATER SAMPLES SHALL BE OBTAINED FROM A CERTIFIED LAB BEFORE PLACING INTO SERVICE.

BEFORE TAKING ELEVATED TANK OFF THE DISTRIBUTION SYSTEM, CONTRACTOR SHALL GIVE THE WATERWORKS OPERATOR AND THE FIRE DEPARTMENT ADEQUATE ADVANCE NOTICE.

——UTILITIES——

MUNICIPAL WATER SYSTEM:

DRUMMOND SANITARY DISTRICT
P.O. BOX 43 — FRONT AVE.
DRUMMOND, WISCONSIN 54832
MARK JEROME: WATERWORKS OPERATOR
TELE. NO. 715 / 739-6741

TELEPHONE COMPANY:

CHEQUAMEGON TELEPHONE CO.
1st AVENUE — P. O. BOX 67
CABLE, WISCONSIN 54821
TELE. NO. 800 / 250-8927

ELECTRICAL SERVICE:

(A.K.A.) NORTHERN STATES POWER CO. (XCEL ENERGY)
16048 ELECTRIC AVENUE
HAYWARD, WISCONSIN 54843
KEN DISHER: FIELD ENGINEER
TELE. NO. 800 / 895-4999

—FIRE DEPARTMENT—

DRUMMOND FIRE & RESCUE
WADE SPEARS, CHIEF
TELE. NO. 715 / 739-6696

——DIGGERS HOTLINE——

TELE. NO. 800/242-8511

——INDEX OF SHEETS——

SHEET NO.	DESCRIPTION
1	TITLE SHEET, GENERAL LOCATION MAP, PROJECT LOCATION MAP, UTILITIES LIST, SHEET INDEX & NOTES
2	EXISTING TANK GEOMETRY AND CONDITIONS w/ REHABILITATION DETAILS
3	PROPOSED CIRCULATION PUMP SYSTEM AND RETROFITTING DETAILS

PREPARED BY:
MORGAN & PARMLEY, Ltd.
PROFESSIONAL CONSULTING ENGINEERS
115 W. 2nd STREET SOUTH
LADYSMITH, WISCONSIN 54848

A 20 year old elevated potable water storage tank required a major rehabilitation. The tank insulation had seriously deteriorated from ultraviolet rays, the paint contained lead, cadmium, and chromium, and there was need for some structural upgrades.

Source: Morgan & Parmley, Ltd.

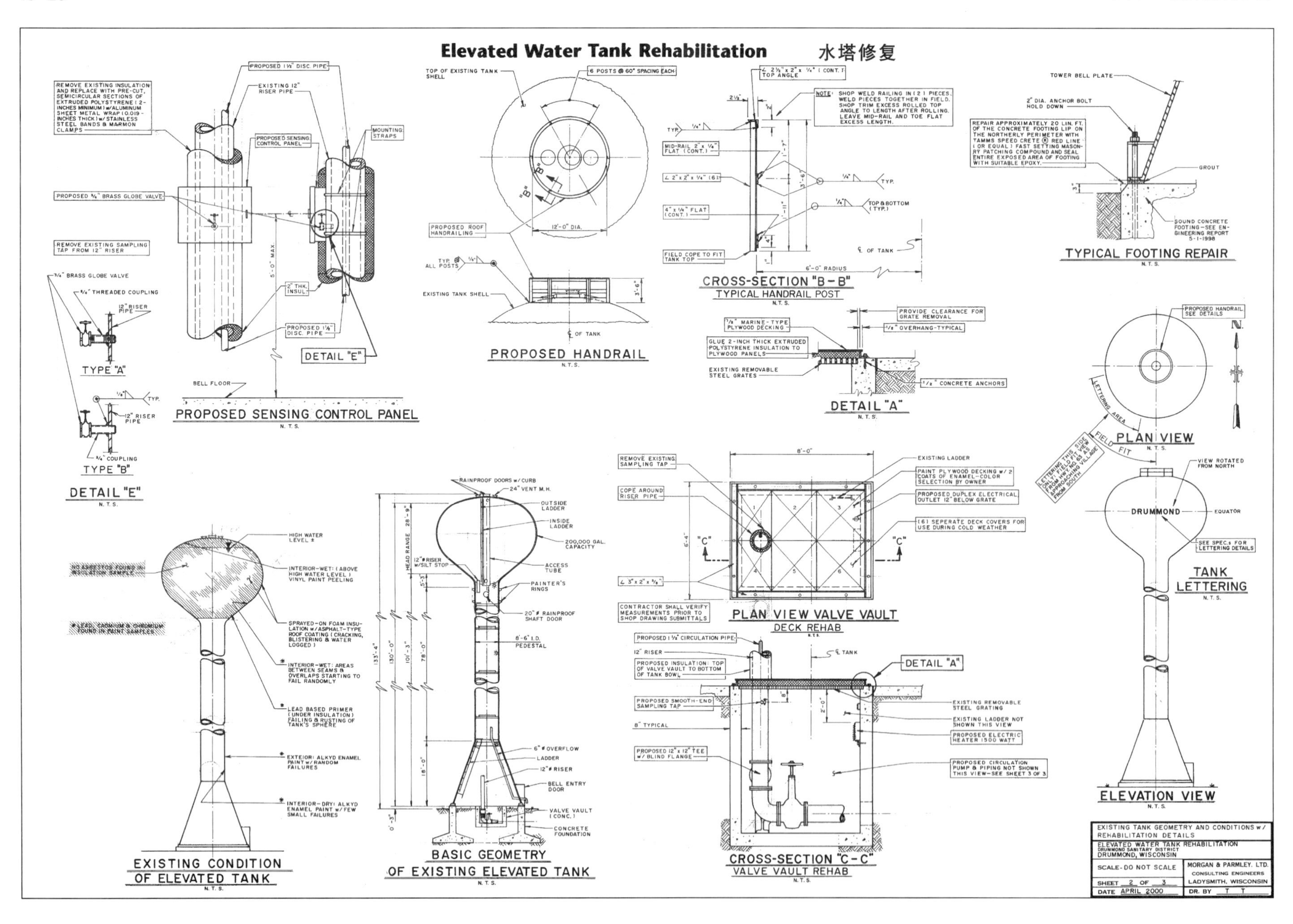

Elevated Water Tank Rehabilitation 水塔修复
PROPOSED SENSING CONTROL PANEL
DETAIL "E"
TYPE "A"
TYPE "B"
PROPOSED HANDRAIL
CROSS-SECTION "B-B"
TYPICAL HANDRAIL POST
TYPICAL FOOTING REPAIR
DETAIL "A"
PLAN VIEW
PLAN VIEW VALVE VAULT
DECK REHAB
CROSS-SECTION "C-C"
VALVE VAULT REHAB
TANK LETTERING
ELEVATION VIEW
EXISTING CONDITION OF ELEVATED TANK
BASIC GEOMETRY OF EXISTING ELEVATED TANK
DRUMMOND
EXISTING TANK GEOMETRY AND CONDITIONS w/ REHABILITATION DETAILS
ELEVATED WATER TANK REHABILITATION
DRUMMOND, WISCONSIN
SCALE-DO NOT SCALE
MORGAN & PARMLEY, LTD.
CONSULTING ENGINEERS
LADYSMITH, WISCONSIN
SHEET 2 OF 3
DATE APRIL 2000

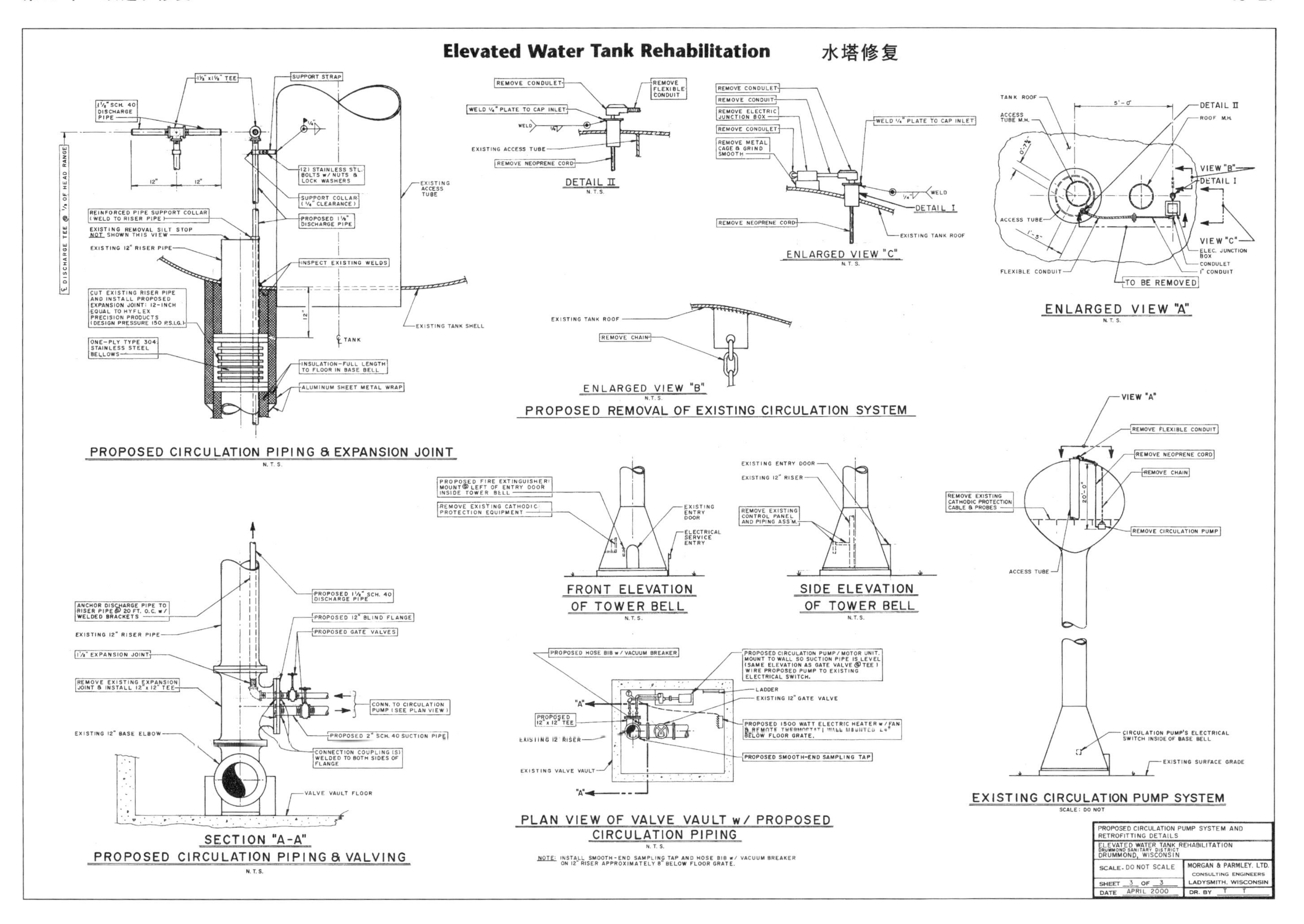
Elevated Water Tank Rehabilitation　水塔修复
1½" x 1½" TEE
SUPPORT STRAP
1½" SCH. 40 DISCHARGE PIPE
(2) STAINLESS STL. BOLTS w/ NUTS & LOCK WASHERS
SUPPORT COLLAR (¼" CLEARANCE)
PROPOSED 1½" DISCHARGE PIPE
EXISTING ACCESS TUBE
REINFORCED PIPE SUPPORT COLLAR (WELD TO RISER PIPE)
EXISTING REMOVAL SILT STOP NOT SHOWN THIS VIEW
EXISTING 12" RISER PIPE
INSPECT EXISTING WELDS
CUT EXISTING RISER PIPE AND INSTALL PROPOSED EXPANSION JOINT: 12-INCH EQUAL TO HYFLEX PRECISION PRODUCTS (DESIGN PRESSURE 150 P.S.I.G.)
EXISTING TANK SHELL
ONE-PLY TYPE 304 STAINLESS STEEL BELLOWS
INSULATION-FULL LENGTH TO FLOOR IN BASE BELL
ALUMINUM SHEET METAL WRAP
℄ DISCHARGE TEE @ ⅓ OF HEAD RANGE
℄ TANK
PROPOSED CIRCULATION PIPING & EXPANSION JOINT
N.T.S.
REMOVE CONDULET
REMOVE FLEXIBLE CONDUIT
WELD ¼" PLATE TO CAP INLET
WELD
EXISTING ACCESS TUBE
REMOVE NEOPRENE CORD
DETAIL II
N.T.S.
REMOVE CONDULET
REMOVE CONDUIT
REMOVE ELECTRIC JUNCTION BOX
REMOVE CONDULET
REMOVE METAL CAGE & GRIND SMOOTH
WELD ¼" PLATE TO CAP INLET
DETAIL I
REMOVE NEOPRENE CORD
EXISTING TANK ROOF
ENLARGED VIEW "C"
N.T.S.
TANK ROOF
ACCESS TUBE M.H.
5'-0"
DETAIL II
ROOF M.H.
VIEW "B"
DETAIL I
ACCESS TUBE
VIEW "C"
ELEC. JUNCTION BOX
CONDULET
1" CONDUIT
FLEXIBLE CONDUIT
TO BE REMOVED
ENLARGED VIEW "A"
N.T.S.
EXISTING TANK ROOF
REMOVE CHAIN
ENLARGED VIEW "B"
N.T.S.
PROPOSED REMOVAL OF EXISTING CIRCULATION SYSTEM
VIEW "A"
REMOVE FLEXIBLE CONDUIT
REMOVE NEOPRENE CORD
REMOVE CHAIN
REMOVE EXISTING CATHODIC PROTECTION CABLE & PROBES
REMOVE CIRCULATION PUMP
ACCESS TUBE
PROPOSED FIRE EXTINGUISHER: MOUNT @ LEFT OF ENTRY DOOR INSIDE TOWER BELL
REMOVE EXISTING CATHODIC PROTECTION EQUIPMENT
EXISTING ENTRY DOOR
ELECTRICAL SERVICE ENTRY
FRONT ELEVATION OF TOWER BELL
N.T.S.
EXISTING ENTRY DOOR
EXISTING 12" RISER
REMOVE EXISTING CONTROL PANEL AND PIPING ASS'M.
SIDE ELEVATION OF TOWER BELL
N.T.S.
PROPOSED 1½" SCH. 40 DISCHARGE PIPE
ANCHOR DISCHARGE PIPE TO RISER PIPE @ 20 FT. O.C. w/ WELDED BRACKETS
PROPOSED 12" BLIND FLANGE
PROPOSED GATE VALVES
EXISTING 12" RISER PIPE
1½" EXPANSION JOINT
REMOVE EXISTING EXPANSION JOINT & INSTALL 12" x 12" TEE
CONN. TO CIRCULATION PUMP (SEE PLAN VIEW)
EXISTING 12" BASE ELBOW
PROPOSED 2" SCH. 40 SUCTION PIPE
CONNECTION COUPLING(S) WELDED TO BOTH SIDES OF FLANGE
VALVE VAULT FLOOR
SECTION "A-A"
PROPOSED CIRCULATION PIPING & VALVING
N.T.S.
PROPOSED HOSE BIB w/ VACUUM BREAKER
PROPOSED CIRCULATION PUMP/MOTOR UNIT. MOUNT TO WALL SO SUCTION PIPE IS LEVEL (SAME ELEVATION AS GATE VALVE @ TEE) WIRE PROPOSED PUMP TO EXISTING ELECTRICAL SWITCH.
LADDER
EXISTING 12" GATE VALVE
"A"
PROPOSED 12" x 12" TEE
EXISTING 12" RISER
PROPOSED 1500 WATT ELECTRIC HEATER w/ FAN & REMOTE THERMOSTAT
EXISTING VALVE VAULT
PROPOSED SMOOTH-END SAMPLING TAP
PLAN VIEW OF VALVE VAULT w/ PROPOSED CIRCULATION PIPING
N.T.S.
NOTE: INSTALL SMOOTH-END SAMPLING TAP AND HOSE BIB w/ VACUUM BREAKER ON 12" RISER APPROXIMATELY 8" BELOW FLOOR GRATE.
CIRCULATION PUMP'S ELECTRICAL SWITCH INSIDE OF BASE BELL
EXISTING SURFACE GRADE
EXISTING CIRCULATION PUMP SYSTEM
SCALE: DO NOT
PROPOSED CIRCULATION PUMP SYSTEM AND RETROFITTING DETAILS
ELEVATED WATER TANK REHABILITATION
DRUMMOND SANITARY DISTRICT
DRUMMOND, WISCONSIN
SCALE: DO NOT SCALE
MORGAN & PARMLEY, LTD.
CONSULTING ENGINEERS
LADYSMITH, WISCONSIN
SHEET 3 OF 3
DATE APRIL 2000
DR. BY T T

设　计

第 20 节 特殊项目

例 20

特殊项目就是那些不能适合已确定的常规实践领域而且通常是仅此一次的项目。它们通常是结合了多种基本学科的混合组合体来达到特殊项目的目标。结构、供暖通风和空调、电力、铅管和管道系统，仅是完成此专业所需要的组成部分中的几个。较大范围的设计技巧需要下级咨询工程师在总咨询工程师的指导下开展工作，完全领会设计图和技术说明。这些条件展示在本节后面两个设计图中，标题为铁路场地改良项目和鱼苗孵化场。

目 录

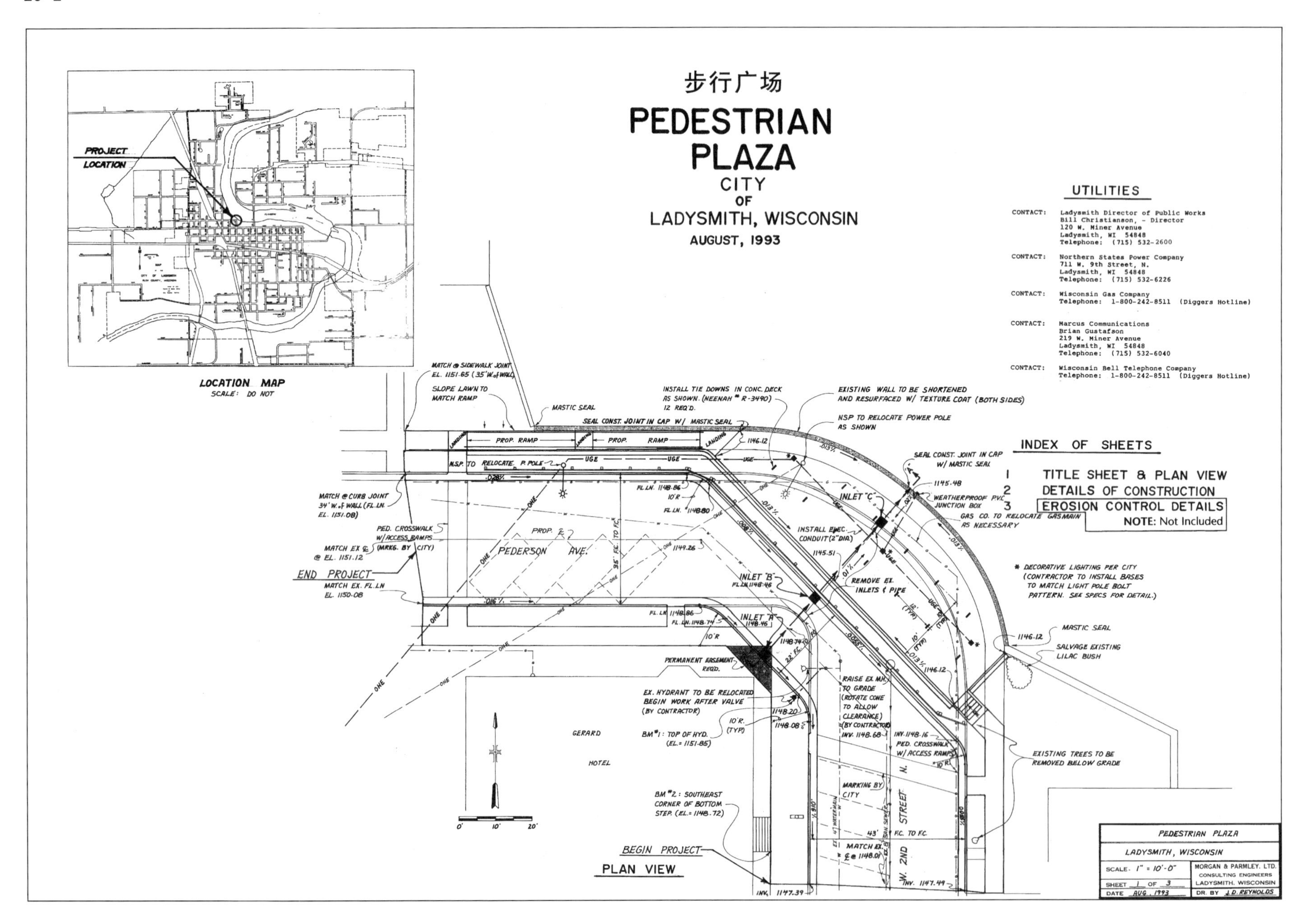

Courtesy: Morgan & Parmley, Ltd.

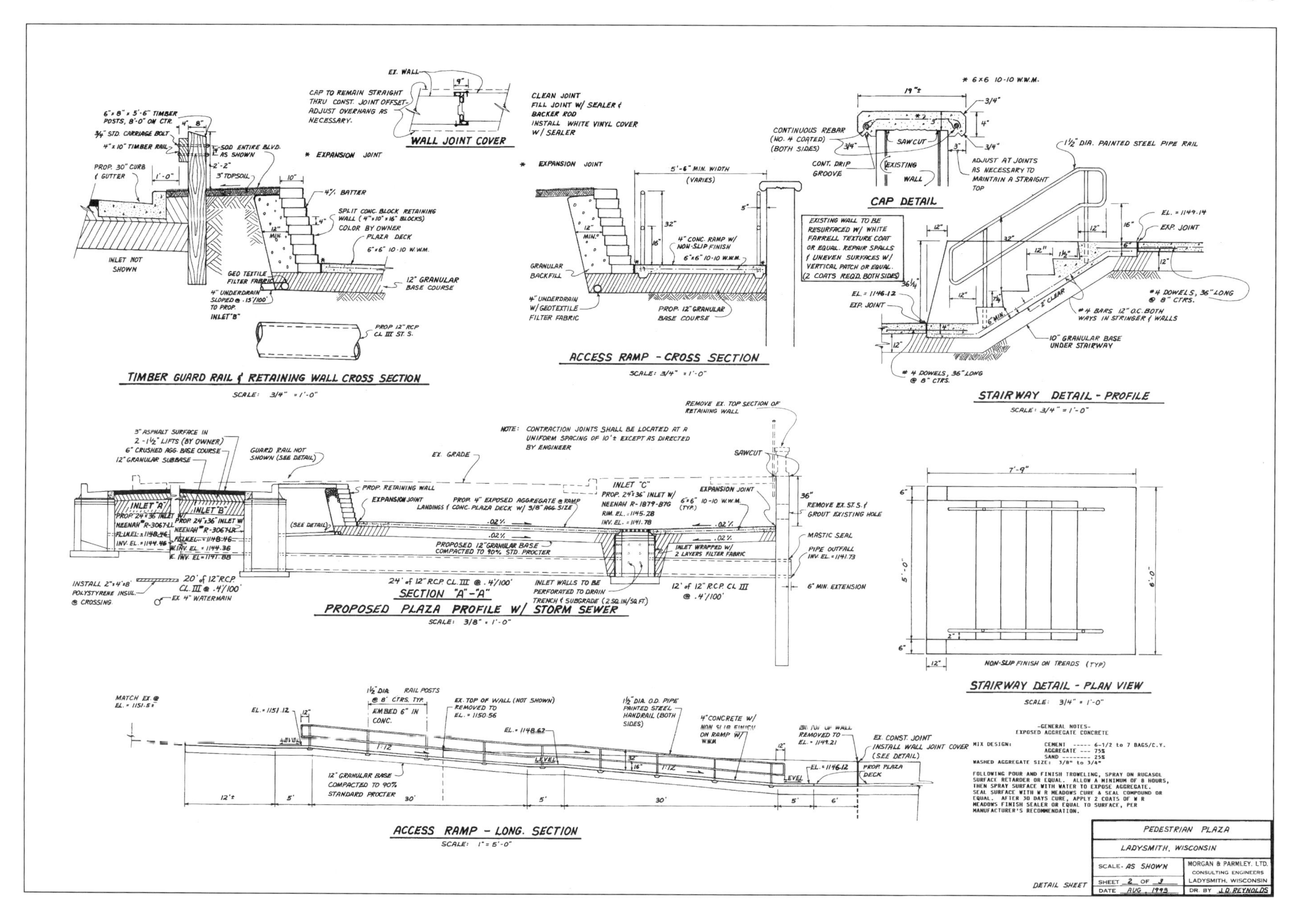
TIMBER GUARD RAIL & RETAINING WALL CROSS SECTION
SCALE: 3/4" = 1'-0"
WALL JOINT COVER
ACCESS RAMP - CROSS SECTION
SCALE: 3/4" = 1'-0"
CAP DETAIL
STAIRWAY DETAIL - PROFILE
SCALE: 3/4" = 1'-0"
SECTION "A"-"A"
PROPOSED PLAZA PROFILE W/ STORM SEWER
SCALE: 3/8" = 1'-0"
STAIRWAY DETAIL - PLAN VIEW
SCALE: 3/4" = 1'-0"
ACCESS RAMP - LONG. SECTION
SCALE: 1" = 5'-0"
PEDESTRIAN PLAZA
LADYSMITH, WISCONSIN
MORGAN & PARMLEY, LTD.
CONSULTING ENGINEERS
LADYSMITH, WISCONSIN
DETAIL SHEET

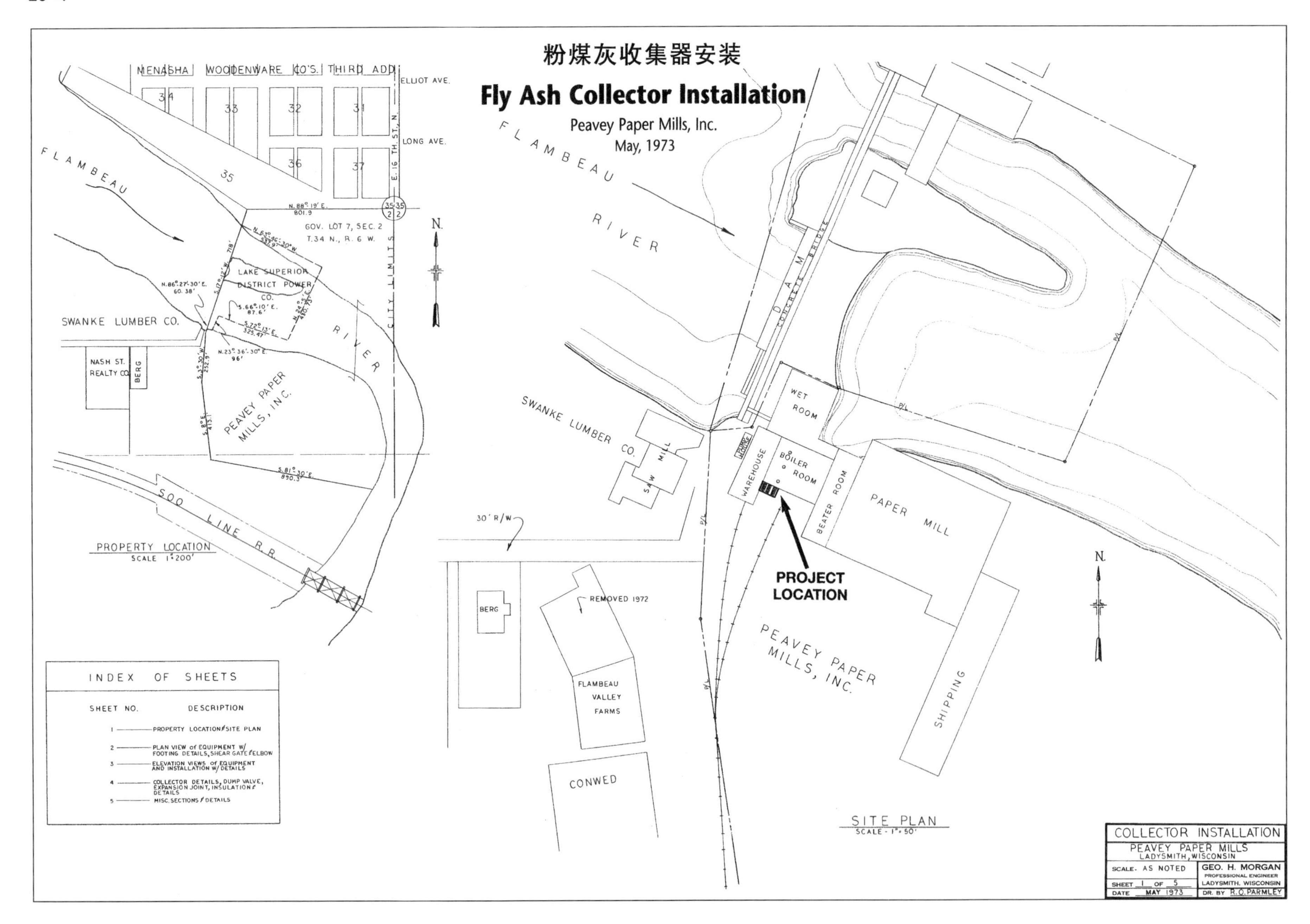

Courtesy: Morgan & Parmley, Ltd.

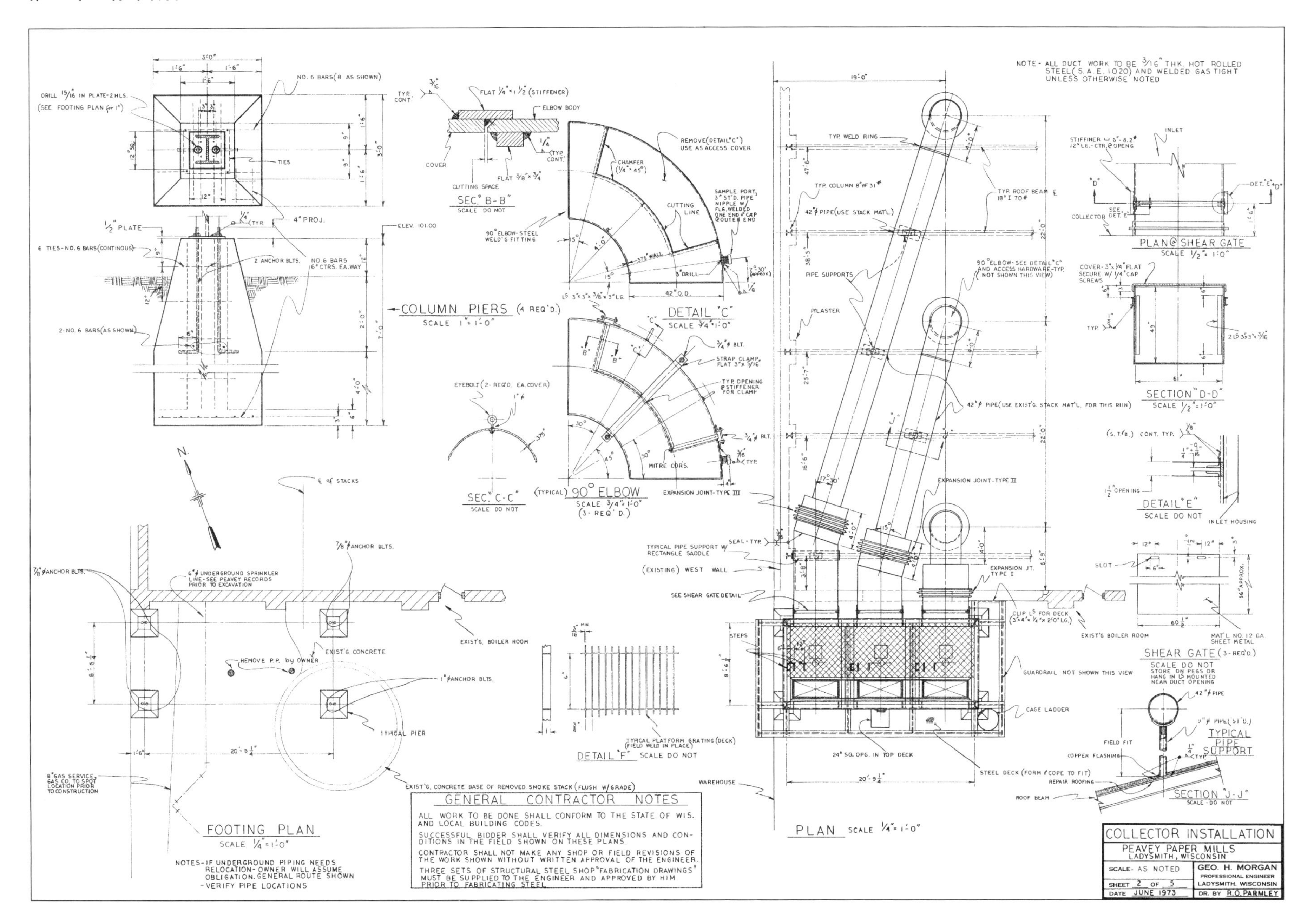
NOTE- ALL DUCT WORK TO BE 3/16" THK. HOT ROLLED STEEL(S.A.E. 1020) AND WELDED GAS TIGHT UNLESS OTHERWISE NOTED
COLUMN PIERS (4 REQ'D.)
SCALE 1"=1'-0"
DRILL 15/16 IN PLATE-2 HLS.
(SEE FOOTING PLAN for 1")
NO. 6 BARS(8 AS SHOWN)
TIES
1/2" PLATE
4" PROJ.
ELEV. 101.00
6 TIES-NO. 6 BARS(CONTINOUS)
2 ANCHOR BLTS.
NO. 6 BARS 6" CTRS. EA. WAY
2-NO. 6 BARS(AS SHOWN)
SEC. B-B
SCALE DO NOT
FLAT 1/4"x1 1/2"(STIFFENER)
ELBOW BODY
COVER
CUTTING SPACE
FLAT 3/8"x3/4
DETAIL "C"
SCALE 3/4"=1'-0"
REMOVE(DETAIL "C") USE AS ACCESS COVER
CHAMFER
CUTTING LINE
SAMPLE PORT, 3" STD. PIPE NIPPLE W/ FLG. WELDED ONE END & CAP @ OUTER END
90° ELBOW-STEEL WELD'G FITTING
3" DRILL
42" O.D.
STRAP CLAMP, FLAT 3"x5/16
TYP. OPENING & STIFFENER FOR CLAMP
EYEBOLT (2-REQ'D. EA. COVER)
SEC. C-C
SCALE DO NOT
(TYPICAL) 90° ELBOW
SCALE 3/4"=1'-0"
(3-REQ'D.)
MITRE CORS.
EXPANSION JOINT-TYPE III
E of STACKS
7/8" ANCHOR BLTS.
6" UNDERGROUND SPRINKLER LINE-SEE PEAVEY RECORDS PRIOR TO EXCAVATION
EXIST'G. BOILER ROOM
EXIST'G. CONCRETE
REMOVE P.P. by OWNER
1" ANCHOR BLTS.
TYPICAL PIER
8" GAS SERVICE-GAS CO. TO SPOT LOCATION PRIOR TO CONSTRUCTION
EXIST'G. CONCRETE BASE OF REMOVED SMOKE STACK (FLUSH W/GRADE)
FOOTING PLAN
SCALE 1/4"=1'-0"
NOTES-IF UNDERGROUND PIPING NEEDS RELOCATION-OWNER WILL ASSUME OBLIGATION. GENERAL ROUTE SHOWN -VERIFY PIPE LOCATIONS
DETAIL "F" SCALE DO NOT
TYPICAL PLATFORM GRATING (DECK) (FIELD WELD IN PLACE)
GENERAL CONTRACTOR NOTES
ALL WORK TO BE DONE SHALL CONFORM TO THE STATE OF WIS. AND LOCAL BUILDING CODES.
SUCCESSFUL BIDDER SHALL VERIFY ALL DIMENSIONS AND CONDITIONS IN THE FIELD SHOWN ON THESE PLANS.
CONTRACTOR SHALL NOT MAKE ANY SHOP OR FIELD REVISIONS OF THE WORK SHOWN WITHOUT WRITTEN APPROVAL OF THE ENGINEER.
THREE SETS OF STRUCTURAL STEEL SHOP "FABRICATION DRAWINGS" MUST BE SUPPLIED TO THE ENGINEER AND APPROVED BY HIM PRIOR TO FABRICATING STEEL.
TYP. WELD RING
TYP. COLUMN 8" WF 31#
42" PIPE (USE STACK MAT'L)
TYP. ROOF BEAM 18" I 70#
PIPE SUPPORTS
PILASTER
90° ELBOW-SEE DETAIL "C" AND ACCESS HARDWARE-TYP. (NOT SHOWN THIS VIEW)
42" PIPE (USE EXIST'G. STACK MAT'L. FOR THIS RUN)
EXPANSION JOINT-TYPE II
TYPICAL PIPE SUPPORT W/ RECTANGLE SADDLE
(EXISTING) WEST WALL
SEE SHEAR GATE DETAIL
EXPANSION JT. TYPE I
CLIP L5 FOR DECK
STEPS
GUARDRAIL NOT SHOWN THIS VIEW
CAGE LADDER
24" SQ. OPG. IN TOP DECK
STEEL DECK (FORM & COPE TO FIT)
WAREHOUSE
PLAN SCALE 1/4"=1'-0"
INLET
PLAN @ SHEAR GATE
SCALE 1/2"=1'-0"
COLLECTOR
SECTION "D-D"
SCALE 1/2"=1'-0"
DETAIL "E"
SCALE DO NOT
INLET HOUSING
SLOT
EXIST'G BOILER ROOM
MAT'L NO. 12 GA SHEET METAL
SHEAR GATE (3-REQ'D.)
SCALE DO NOT
STORE ON PEGS OR HANG IN L5 MOUNTED NEAR DUCT OPENING
42" PIPE
FIELD FIT
COPPER FLASHING
REPAIR ROOFING
ROOF BEAM
TYPICAL PIPE SUPPORT
SECTION "J-J"
SCALE-DO NOT
COLLECTOR INSTALLATION
PEAVEY PAPER MILLS
LADYSMITH, WISCONSIN
SCALE- AS NOTED
GEO. H. MORGAN
PROFESSIONAL ENGINEER
LADYSMITH, WISCONSIN
SHEET 2 OF 5
DATE JUNE 1973
DR. BY R.O. PARMLEY

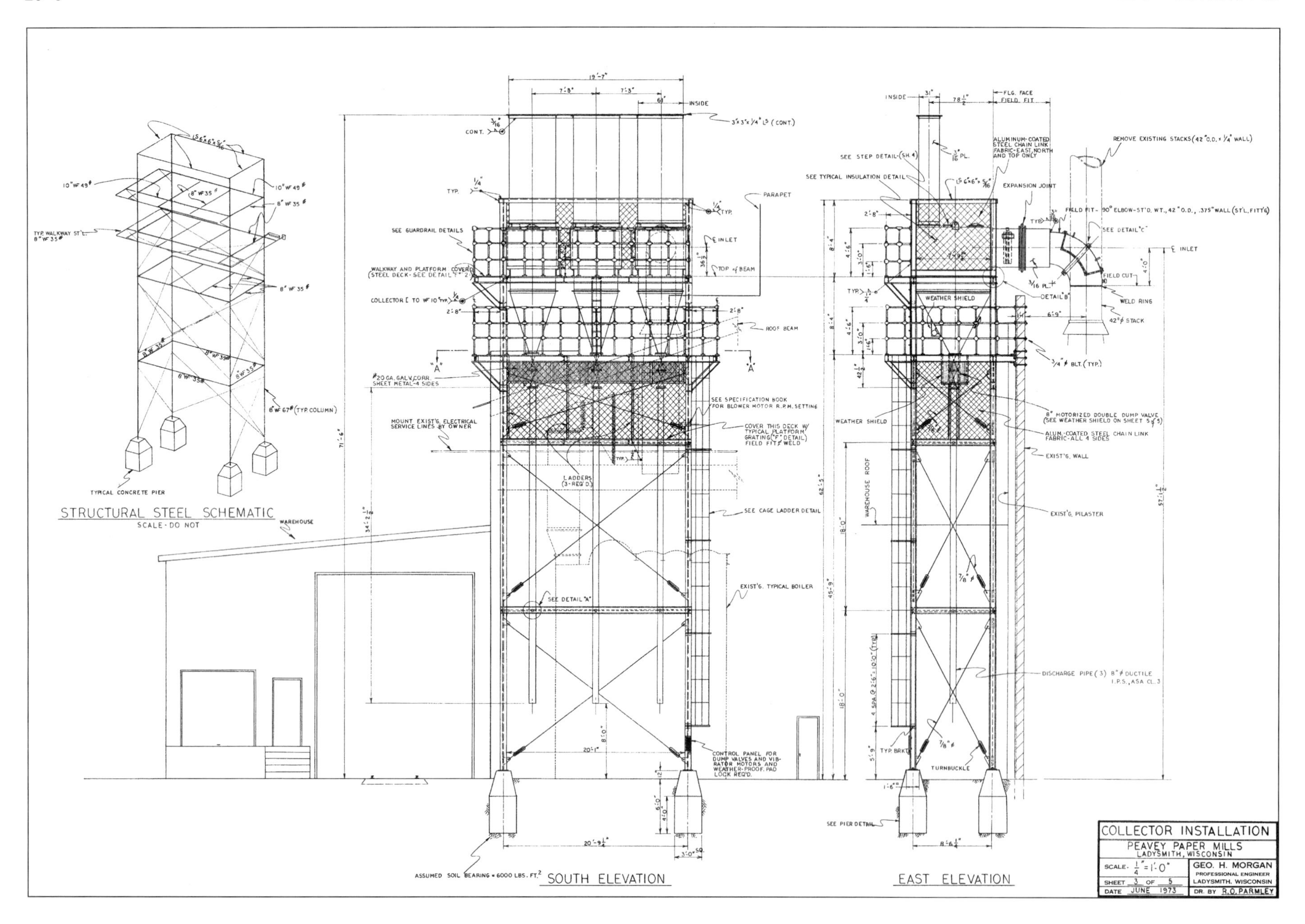
STRUCTURAL STEEL SCHEMATIC
SCALE - DO NOT
SOUTH ELEVATION
EAST ELEVATION
COLLECTOR INSTALLATION
PEAVEY PAPER MILLS
LADYSMITH, WISCONSIN
GEO. H. MORGAN
PROFESSIONAL ENGINEER
LADYSMITH, WISCONSIN
SHEET 3 OF 5
DATE JUNE 1973
DR. BY R.O. PARMLEY
ASSUMED SOIL BEARING = 6000 LBS. FT.²
TYPICAL CONCRETE PIER
WAREHOUSE
REMOVE EXISTING STACKS (42" O.D. x 1/4" WALL)
EXPANSION JOINT
WEATHER SHIELD
EXIST'G. WALL
EXIST'G. PILASTER
EXIST'G. TYPICAL BOILER
SEE CAGE LADDER DETAIL
ROOF BEAM
PARAPET
TURNBUCKLE
SEE PIER DETAIL

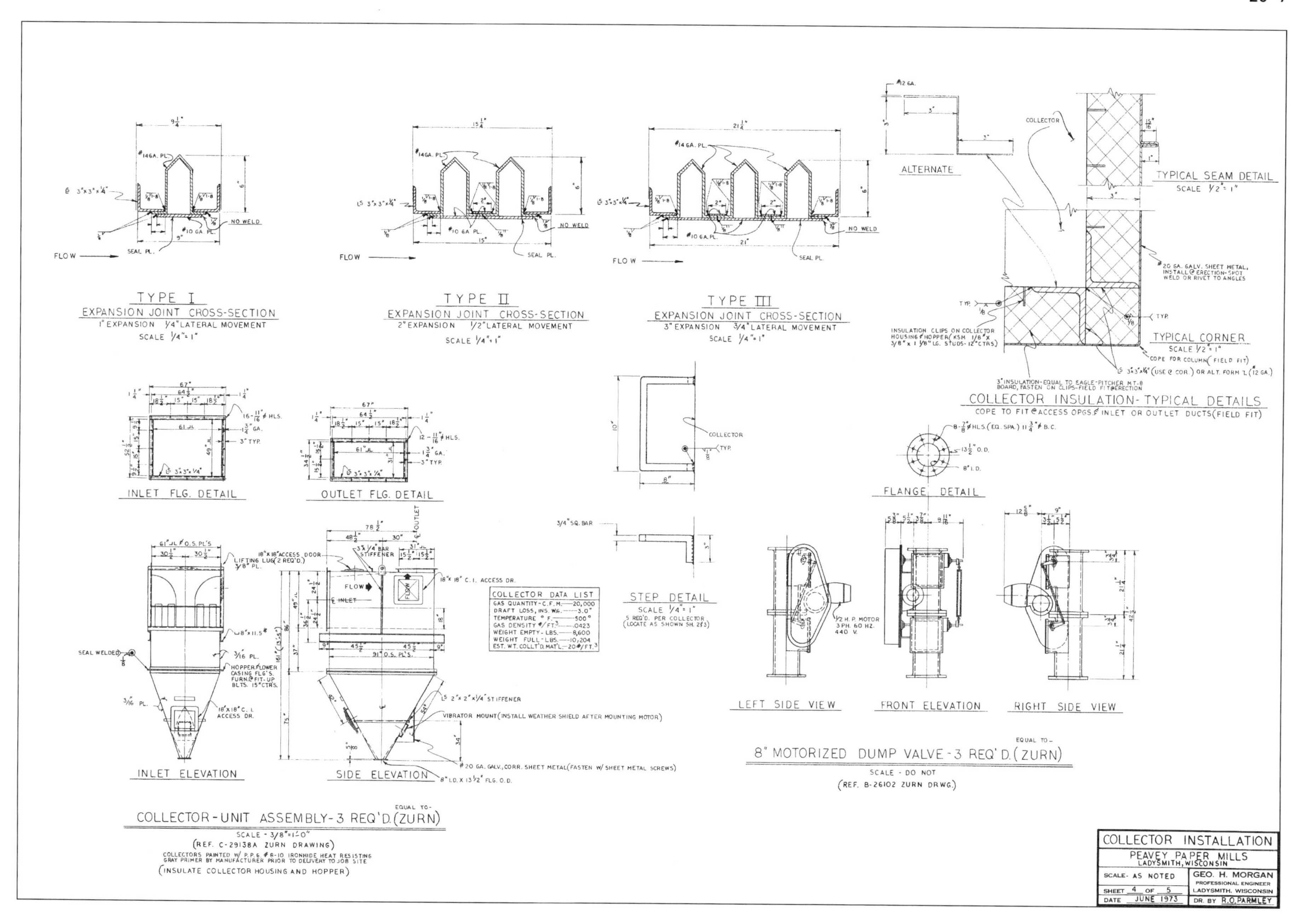
TYPE I
EXPANSION JOINT CROSS-SECTION
1" EXPANSION 1/4" LATERAL MOVEMENT
SCALE 1/4"=1"
TYPE II
EXPANSION JOINT CROSS-SECTION
2" EXPANSION 1/2" LATERAL MOVEMENT
SCALE 1/4"=1"
TYPE III
EXPANSION JOINT CROSS-SECTION
3" EXPANSION 3/4" LATERAL MOVEMENT
SCALE 1/4"=1"
FLOW
SEAL PL.
NO WELD
ALTERNATE
TYPICAL SEAM DETAIL
SCALE 1/2"=1"
TYPICAL CORNER
SCALE 1/2"=1"
COLLECTOR INSULATION-TYPICAL DETAILS
COPE TO FIT @ ACCESS OPGS & INLET OR OUTLET DUCTS (FIELD FIT)
INLET FLG. DETAIL
OUTLET FLG. DETAIL
FLANGE DETAIL
STEP DETAIL
SCALE 1/4"=1"
COLLECTOR DATA LIST
GAS QUANTITY-C.F.M. 20,000
DRAFT LOSS, INS. WG. 3.0"
TEMPERATURE °F. 500°
GAS DENSITY #/FT.3 .0423
WEIGHT EMPTY-LBS. 8,600
WEIGHT FULL-LBS. 10,204
EST. WT. COLLT'D. MAT'L. 20#/FT.3
LEFT SIDE VIEW
FRONT ELEVATION
RIGHT SIDE VIEW
8" MOTORIZED DUMP VALVE-3 REQ'D. (ZURN)
SCALE - DO NOT
(REF. B-26102 ZURN DRWG.)
INLET ELEVATION
SIDE ELEVATION
COLLECTOR-UNIT ASSEMBLY-3 REQ'D. (ZURN)
SCALE - 3/8"=1'-0"
(REF. C-29138A ZURN DRAWING)
COLLECTORS PAINTED W/ P.P.G. #6-10 IRONHIDE HEAT RESISTING GRAY PRIMER BY MANUFACTURER PRIOR TO DELIVERY TO JOB SITE
(INSULATE COLLECTOR HOUSING AND HOPPER)
COLLECTOR INSTALLATION
PEAVEY PAPER MILLS
LADYSMITH, WISCONSIN
SCALE- AS NOTED
SHEET 4 OF 5
DATE JUNE 1973
GEO. H. MORGAN
PROFESSIONAL ENGINEER
LADYSMITH, WISCONSIN
DR. BY R.O. PARMLEY

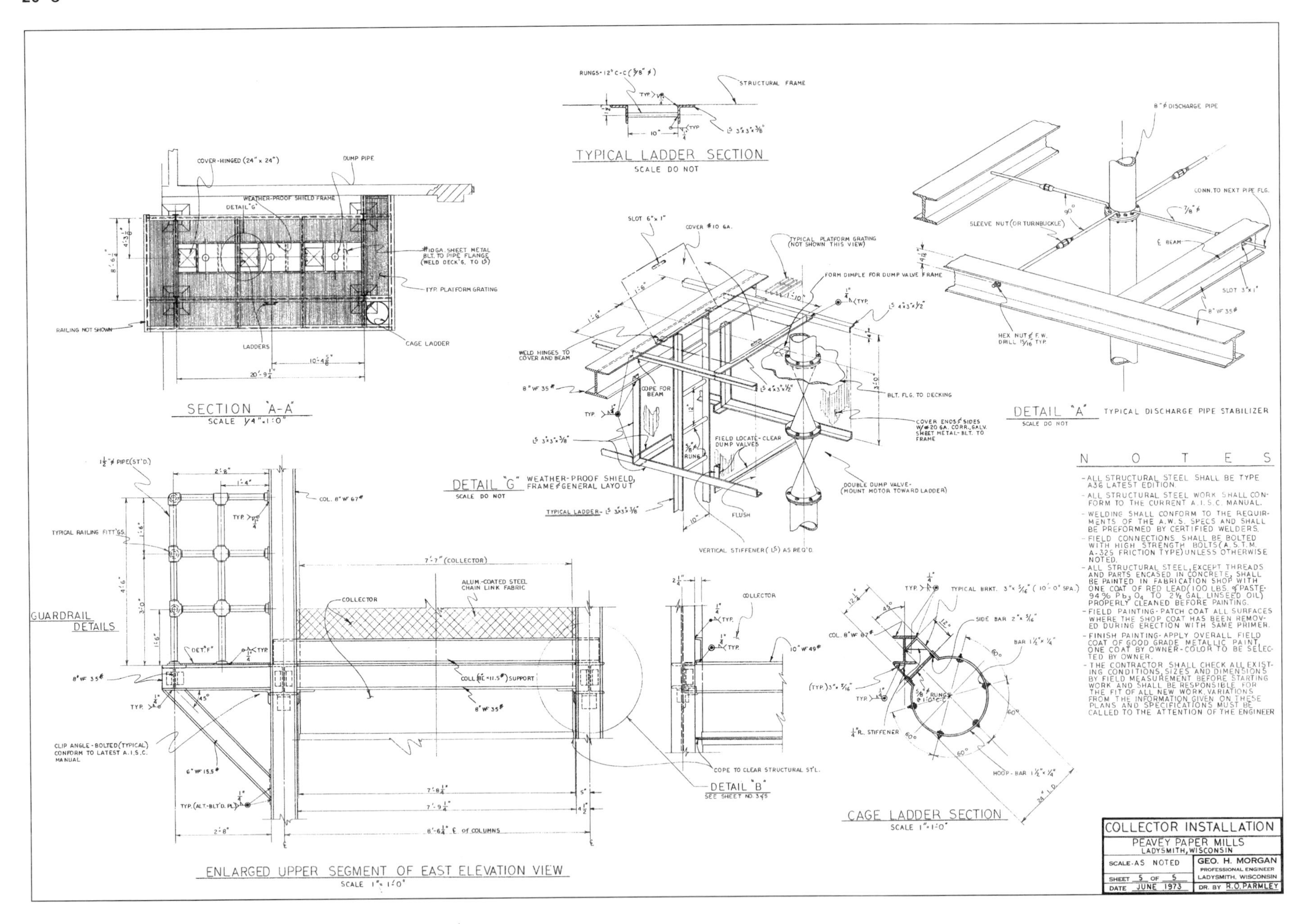
RUNGS-12" C-C (3/8" ϕ)
STRUCTURAL FRAME
TYPICAL LADDER SECTION
SCALE DO NOT
8" ϕ DISCHARGE PIPE
COVER-HINGED (24" x 24")
DUMP PIPE
WEATHER-PROOF SHIELD FRAME
DETAIL "G"
#10GA. SHEET METAL BLT. TO PIPE FLANGE (WELD DECK'G. TO ∟S)
TYP. PLATFORM GRATING
RAILING NOT SHOWN
LADDERS
CAGE LADDER
SECTION "A-A"
SCALE 1/4"=1'-0"
SLOT 6"x1"
COVER #10 GA.
TYPICAL PLATFORM GRATING (NOT SHOWN THIS VIEW)
FORM DIMPLE FOR DUMP VALVE FRAME
CONN. TO NEXT PIPE FLG.
SLEEVE NUT (OR TURNBUCKLE)
℄ BEAM
SLOT 3"x1"
8" WF 35#
HEX NUT & F.W. DRILL 13/16 TYP.
WELD HINGES TO COVER AND BEAM
COPE FOR BEAM
BLT. FLG. TO DECKING
COVER ENDS & SIDES W/#20 GA. CORR. GALV. SHEET METAL-BLT. TO FRAME
FIELD LOCATE-CLEAR DUMP VALVES
DETAIL "A" TYPICAL DISCHARGE PIPE STABILIZER
SCALE DO NOT
DETAIL "G" WEATHER-PROOF SHIELD, FRAME & GENERAL LAYOUT
SCALE DO NOT
TYPICAL LADDER-∟S 3x3x3/8"
FLUSH
DOUBLE DUMP VALVE- (MOUNT MOTOR TOWARD LADDER)
VERTICAL STIFFENER (∟S) AS REQ'D.
NOTES
-ALL STRUCTURAL STEEL SHALL BE TYPE A36 LATEST EDITION.
-ALL STRUCTURAL STEEL WORK SHALL CONFORM TO THE CURRENT A.I.S.C. MANUAL.
-WELDING SHALL CONFORM TO THE REQUIREMENTS OF THE A.W.S. SPECS AND SHALL BE PREFORMED BY CERTIFIED WELDERS.
-FIELD CONNECTIONS SHALL BE BOLTED WITH HIGH STRENGTH BOLTS (A.S.T.M. A-325 FRICTION TYPE) UNLESS OTHERWISE NOTED.
-ALL STRUCTURAL STEEL, EXCEPT THREADS AND PARTS ENCASED IN CONCRETE, SHALL BE PAINTED IN FABRICATION SHOP WITH ONE COAT OF RED LEAD (100 LBS. of PASTE-94% Pb3O4 TO 2½ GAL LINSEED OIL) PROPERLY CLEANED BEFORE PAINTING.
-FIELD PAINTING-PATCH COAT ALL SURFACES WHERE THE SHOP COAT HAS BEEN REMOVED DURING ERECTION WITH SAME PRIMER.
-FINISH PAINTING-APPLY OVERALL FIELD COAT OF GOOD GRADE METALLIC PAINT, ONE COAT BY OWNER-COLOR TO BE SELECTED BY OWNER.
-THE CONTRACTOR SHALL CHECK ALL EXISTING CONDITIONS, SIZES AND DIMENSIONS BY FIELD MEASUREMENT BEFORE STARTING WORK AND SHALL BE RESPONSIBLE FOR THE FIT OF ALL NEW WORK. VARIATIONS FROM THE INFORMATION GIVEN ON THESE PLANS AND SPECIFICATIONS MUST BE CALLED TO THE ATTENTION OF THE ENGINEER
1½" ϕ PIPE (ST'D.)
COL. 8" WF 67#
TYPICAL RAILING FITT'GS.
GUARDRAIL DETAILS
7'-7" (COLLECTOR)
ALUM.-COATED STEEL CHAIN LINK FABRIC
COLLECTOR
COLL. (℄=11.5#) SUPPORT
8" WF 35#
CLIP ANGLE-BOLTED (TYPICAL) CONFORM TO LATEST A.I.S.C. MANUAL
6" WF 15.5#
8'-6¼" ℄ of COLUMNS
ENLARGED UPPER SEGMENT OF EAST ELEVATION VIEW
SCALE 1"=1'-0"
10" WF 49#
COPE TO CLEAR STRUCTURAL ST'L.
DETAIL "B"
SEE SHEET NO. 3 of 5
TYPICAL BRKT. 3"x 5/16" (10'-0" SPA.)
SIDE BAR 2"x 3/16"
BAR 1½"x ¼"
1/4" PL. STIFFENER
HOOP-BAR 1½"x ¼"
24" I.D.
CAGE LADDER SECTION
SCALE 1"=1'-0"
COLLECTOR INSTALLATION
PEAVEY PAPER MILLS
LADYSMITH, WISCONSIN
SCALE-AS NOTED
GEO. H. MORGAN
PROFESSIONAL ENGINEER
LADYSMITH, WISCONSIN
SHEET 5 OF 5
DATE JUNE 1973
DR. BY R.O. PARMLEY

铁路场地改良项目

LADYSMITH RAIL SITE IMPROVEMENT PROJECT
CITY OF LADYSMITH INDUSTRIAL PARK

M & P PROJECT NO. 99-118

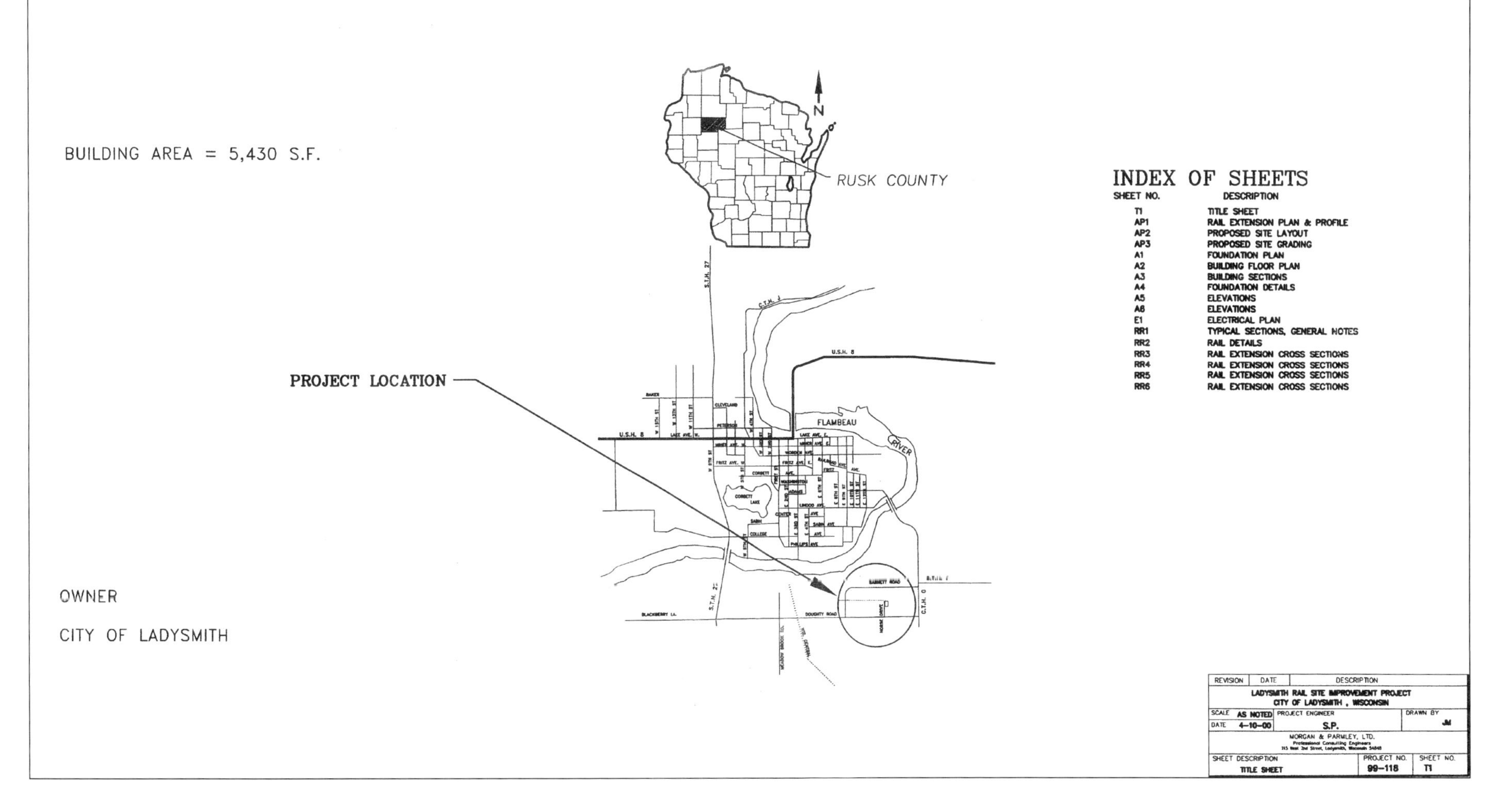

Courtesy: Morgan & Parmley, Ltd.

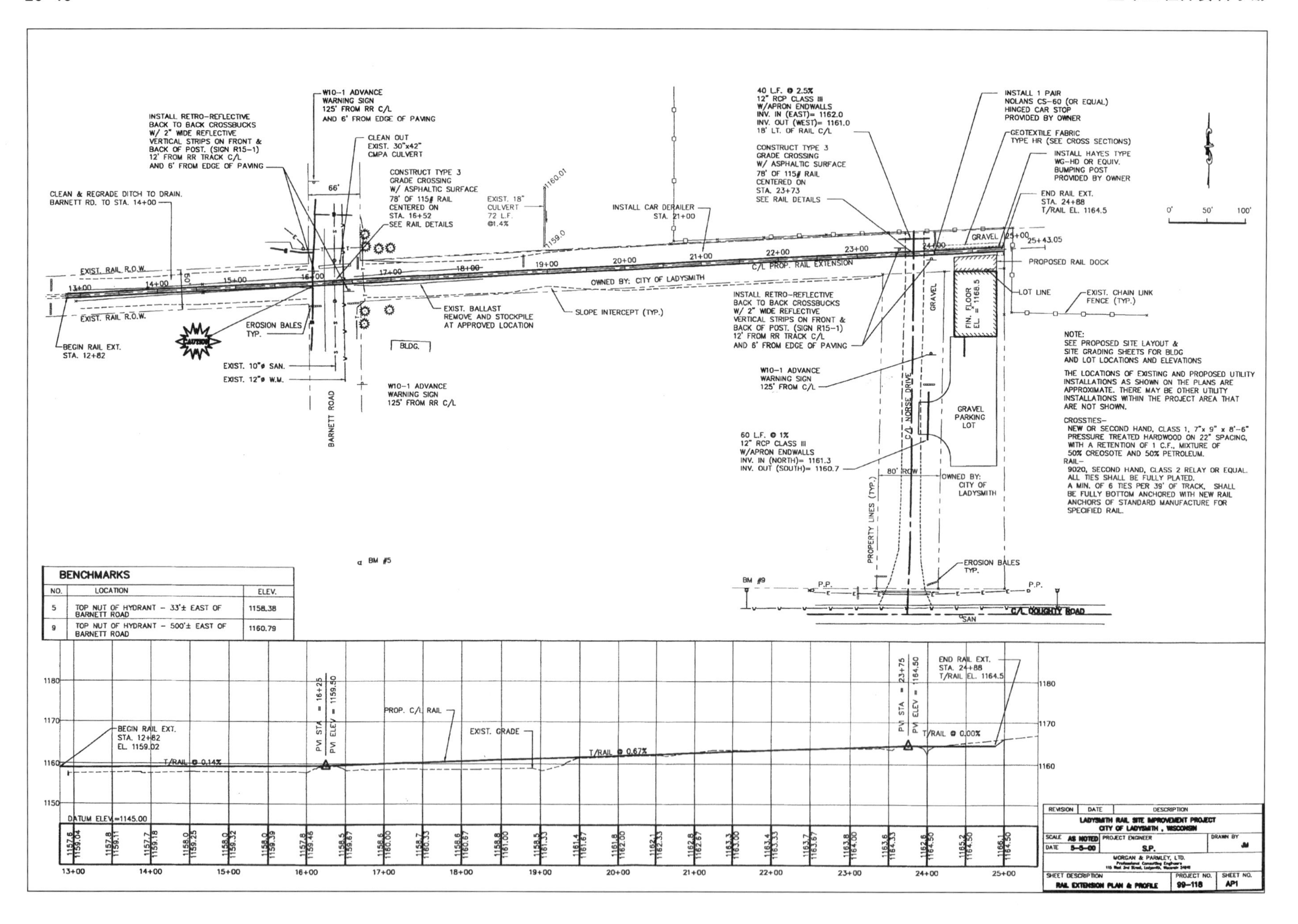
W10-1 ADVANCE WARNING SIGN 125' FROM RR C/L AND 6' FROM EDGE OF PAVING
INSTALL RETRO-REFLECTIVE BACK TO BACK CROSSBUCKS W/ 2" WIDE REFLECTIVE VERTICAL STRIPS ON FRONT & BACK OF POST. (SIGN R15-1) 12' FROM RR TRACK C/L AND 6' FROM EDGE OF PAVING
CLEAN OUT EXIST. 30"x42" CMPA CULVERT
CONSTRUCT TYPE 3 GRADE CROSSING W/ ASPHALTIC SURFACE 78' OF 115# RAIL CENTERED ON STA. 16+52 SEE RAIL DETAILS
EXIST. 18" CULVERT 72 L.F. @1.4%
CLEAN & REGRADE DITCH TO DRAIN. BARNETT RD. TO STA. 14+00
INSTALL CAR DERAILER STA. 21+00
40 L.F. @ 2.5% 12" RCP CLASS III W/APRON ENDWALLS INV. IN (EAST)= 1162.0 INV. OUT (WEST)= 1161.0 18' LT. OF RAIL C/L
CONSTRUCT TYPE 3 GRADE CROSSING W/ ASPHALTIC SURFACE 78' OF 115# RAIL CENTERED ON STA. 23+73 SEE RAIL DETAILS
INSTALL 1 PAIR NOLANS CS-60 (OR EQUAL) HINGED CAR STOP PROVIDED BY OWNER
GEOTEXTILE FABRIC TYPE HR (SEE CROSS SECTIONS)
INSTALL HAYES TYPE WG-HD OR EQUIV. BUMPING POST PROVIDED BY OWNER
END RAIL EXT. STA. 24+88 T/RAIL EL. 1164.5
0' 50' 100'
EXIST. RAIL R.O.W.
C/L PROP. RAIL EXTENSION
OWNED BY: CITY OF LADYSMITH
PROPOSED RAIL DOCK
GRAVEL
25+43.05
LOT LINE
EXIST. CHAIN LINK FENCE (TYP.)
FIN. FLOOR EL. = 1168.5
BEGIN RAIL EXT. STA. 12+82
EROSION BALES TYP.
EXIST. BALLAST REMOVE AND STOCKPILE AT APPROVED LOCATION
SLOPE INTERCEPT (TYP.)
BLDG.
EXIST. 10"φ SAN.
EXIST. 12"φ W.M.
W10-1 ADVANCE WARNING SIGN 125' FROM RR C/L
BARNETT ROAD
INSTALL RETRO-REFLECTIVE BACK TO BACK CROSSBUCKS W/ 2" WIDE REFLECTIVE VERTICAL STRIPS ON FRONT & BACK OF POST. (SIGN R15-1) 12' FROM RR TRACK C/L AND 6' FROM EDGE OF PAVING
W10-1 ADVANCE WARNING SIGN 125' FROM C/L
C/L NORSE DRIVE
GRAVEL PARKING LOT
60 L.F. @ 1% 12" RCP CLASS III W/APRON ENDWALLS INV. IN (NORTH)= 1161.3 INV. OUT (SOUTH)= 1160.7
80' ROW
OWNED BY: CITY OF LADYSMITH
PROPERTY LINES (TYP.)
NOTE:
SEE PROPOSED SITE LAYOUT & SITE GRADING SHEETS FOR BLDG AND LOT LOCATIONS AND ELEVATIONS
THE LOCATIONS OF EXISTING AND PROPOSED UTILITY INSTALLATIONS AS SHOWN ON THE PLANS ARE APPROXIMATE. THERE MAY BE OTHER UTILITY INSTALLATIONS WITHIN THE PROJECT AREA THAT ARE NOT SHOWN.
CROSSTIES–
NEW OR SECOND HAND, CLASS 1, 7"x 9" x 8'-6" PRESSURE TREATED HARDWOOD ON 22" SPACING, WITH A RETENTION OF 1 C.F., MIXTURE OF 50% CREOSOTE AND 50% PETROLEUM.
RAIL–
9020, SECOND HAND, CLASS 2 RELAY OR EQUAL. ALL TIES SHALL BE FULLY PLATED. A MIN. OF 6 TIES PER 39' OF TRACK. SHALL BE FULLY BOTTOM ANCHORED WITH NEW RAIL ANCHORS OF STANDARD MANUFACTURE FOR SPECIFIED RAIL.
BM #5
BM #9
EROSION BALES TYP.
P.P.
C/L DOUGHTY ROAD
SAN
BENCHMARKS
NO. | LOCATION | ELEV.
5 | TOP NUT OF HYDRANT – 33'± EAST OF BARNETT ROAD | 1158.38
9 | TOP NUT OF HYDRANT – 500'± EAST OF BARNETT ROAD | 1160.79
BEGIN RAIL EXT. STA. 12+82 EL. 1159.02
T/RAIL @ 0.14%
PVI STA = 16+25
PVI ELEV = 1159.50
PROP. C/L RAIL
EXIST. GRADE
T/RAIL @ 0.67%
PVI STA = 23+75
PVI ELEV = 1164.50
T/RAIL @ 0.00%
END RAIL EXT. STA. 24+88 T/RAIL EL. 1164.5
DATUM ELEV.=1145.00
13+00 14+00 15+00 16+00 17+00 18+00 19+00 20+00 21+00 22+00 23+00 24+00 25+00
REVISION DATE DESCRIPTION
LADYSMITH RAIL SITE IMPROVEMENT PROJECT
CITY OF LADYSMITH, WISCONSIN
SCALE AS NOTED
DATE 5-5-00
PROJECT ENGINEER S.P.
DRAWN BY JM
MORGAN & PARMLEY, LTD.
SHEET DESCRIPTION RAIL EXTENSION PLAN & PROFILE
PROJECT NO. 99-118
SHEET NO. AP1

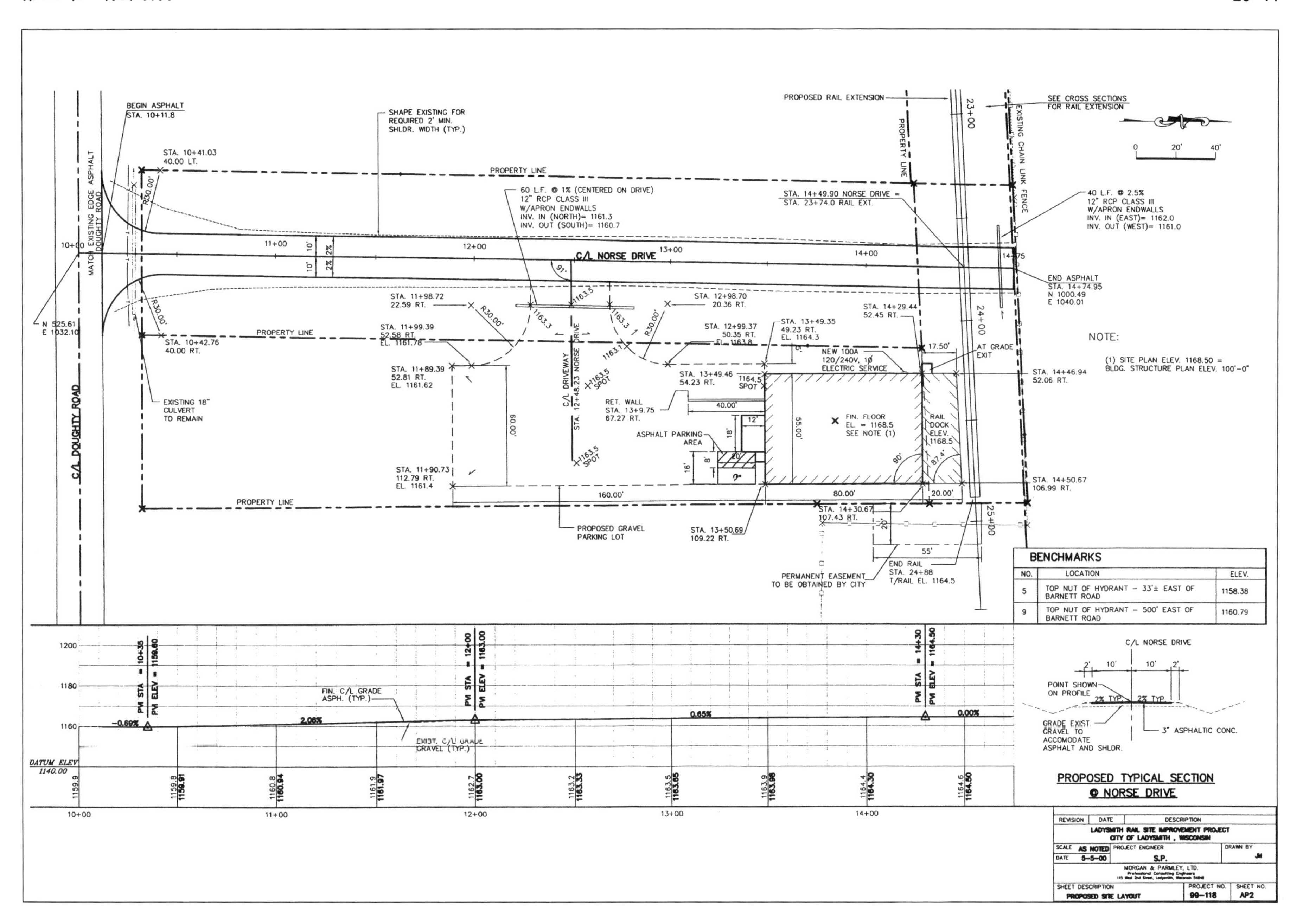

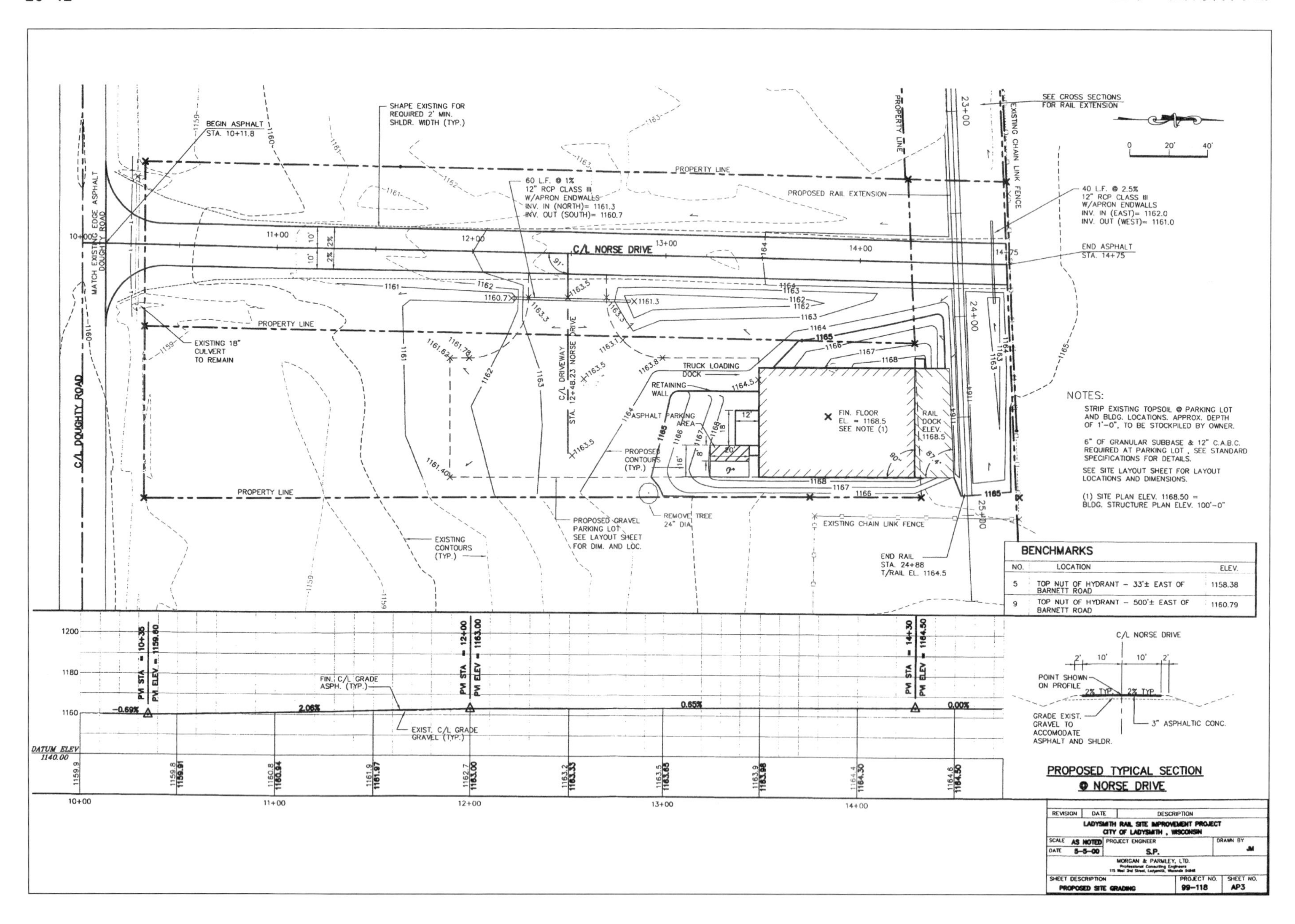

SEE CROSS SECTIONS FOR RAIL EXTENSION
BEGIN ASPHALT STA. 10+11.8
SHAPE EXISTING FOR REQUIRED 2' MIN. SHLDR. WIDTH (TYP.)
PROPERTY LINE
PROPOSED RAIL EXTENSION
60 L.F. @ 1% 12" RCP CLASS III W/APRON ENDWALLS INV. IN (NORTH)= 1161.3 INV. OUT (SOUTH)= 1160.7
40 L.F. @ 2.5% 12" RCP CLASS III W/APRON ENDWALLS INV. IN (EAST)= 1162.0 INV. OUT (WEST)= 1161.0
C/L NORSE DRIVE
END ASPHALT STA. 14+75
MATCH EXISTING EDGE ASPHALT DOUGHTY ROAD
C/L DOUGHTY ROAD
EXISTING CHAIN LINK FENCE
EXISTING 18" CULVERT TO REMAIN
TRUCK LOADING DOCK
RETAINING WALL
ASPHALT PARKING AREA
FIN. FLOOR EL. = 1168.5 SEE NOTE (1)
RAIL DOCK ELEV. 1168.5
PROPOSED CONTOURS (TYP.)
C/L DRIVEWAY STA. 12+48.23 NORSE DRIVE
REMOVE TREE 24" DIA.
PROPOSED GRAVEL PARKING LOT SEE LAYOUT SHEET FOR DIM. AND LOC.
EXISTING CONTOURS (TYP.)
END RAIL STA. 24+88 T/RAIL EL. 1164.5
NOTES:
STRIP EXISTING TOPSOIL @ PARKING LOT AND BLDG. LOCATIONS. APPROX. DEPTH OF 1'-0". TO BE STOCKPILED BY OWNER.
6" OF GRANULAR SUBBASE & 12" C.A.B.C. REQUIRED AT PARKING LOT, SEE STANDARD SPECIFICATIONS FOR DETAILS.
SEE SITE LAYOUT SHEET FOR LOCATIONS AND DIMENSIONS.
(1) SITE PLAN ELEV. 1168.50 = BLDG. STRUCTURE PLAN ELEV. 100'-0"
BENCHMARKS
NO. LOCATION ELEV.
5 TOP NUT OF HYDRANT - 33'± EAST OF BARNETT ROAD 1158.38
9 TOP NUT OF HYDRANT - 500'± EAST OF BARNETT ROAD 1160.79
PVI STA = 10+35 PVI ELEV = 1159.60
PVI STA = 12+00 PVI ELEV = 1163.00
PVI STA = 14+30 PVI ELEV = 1164.50
FIN. C/L GRADE ASPH. (TYP.)
EXIST. C/L GRADE GRAVEL (TYP.)
DATUM ELEV 1140.00
C/L NORSE DRIVE
POINT SHOWN ON PROFILE
GRADE EXIST. GRAVEL TO ACCOMODATE ASPHALT AND SHLDR.
3" ASPHALTIC CONC.
PROPOSED TYPICAL SECTION @ NORSE DRIVE
LADYSMITH RAIL SITE IMPROVEMENT PROJECT
CITY OF LADYSMITH, WISCONSIN
SCALE AS NOTED
DATE 5-5-00
PROJECT ENGINEER S.P.
MORGAN & PARMLEY, LTD.
SHEET DESCRIPTION PROPOSED SITE GRADING
PROJECT NO. 99-118
SHEET NO. AP3

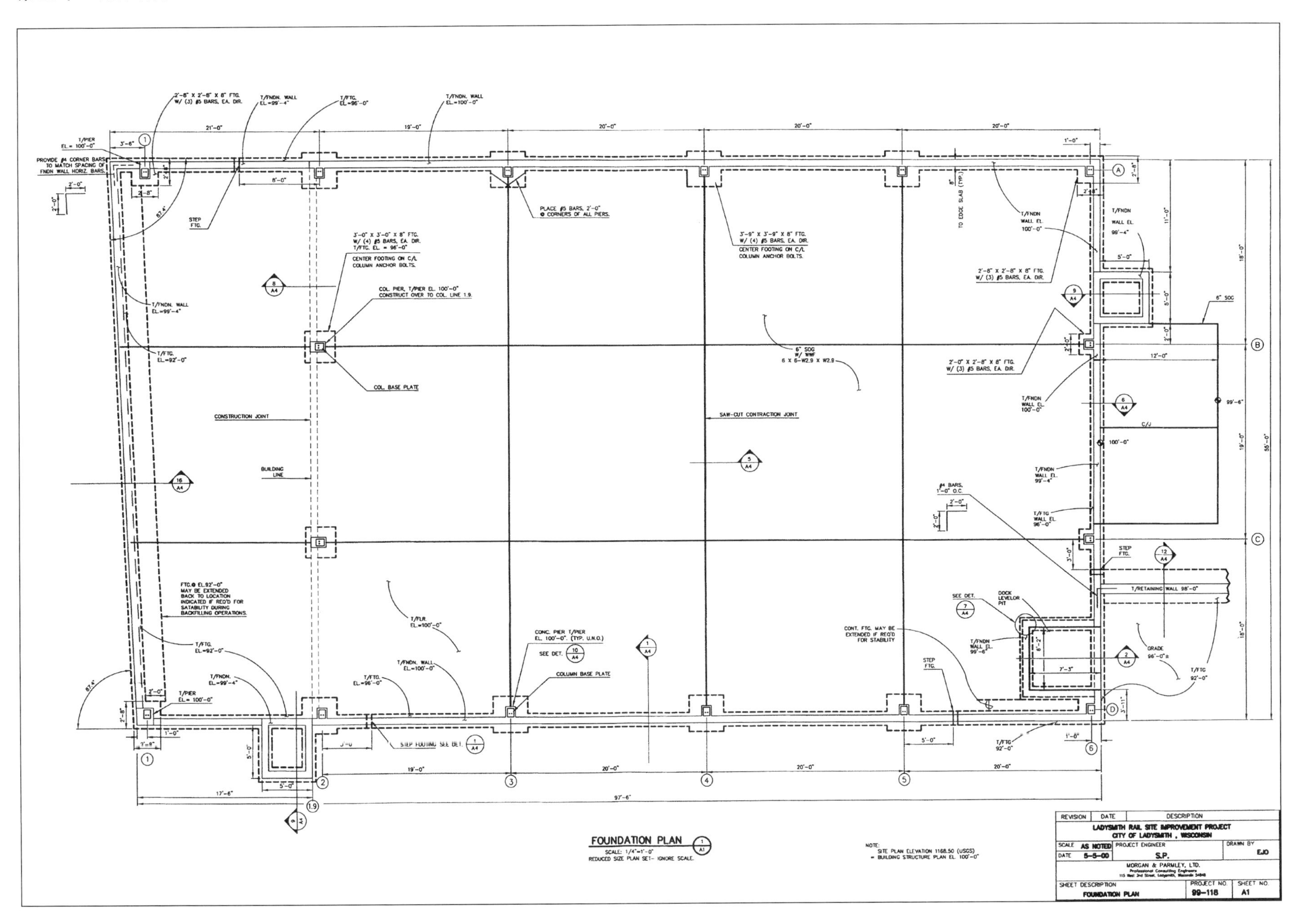
FOUNDATION PLAN
SCALE: 1/4"=1'-0"
REDUCED SIZE PLAN SET- IGNORE SCALE.
NOTE:
SITE PLAN ELEVATION 1168.50 (USGS)
= BUILDING STRUCTURE PLAN EL. 100'-0"
LADYSMITH RAIL SITE IMPROVEMENT PROJECT
CITY OF LADYSMITH , WISCONSIN
SCALE AS NOTED
DATE 5-5-00
PROJECT ENGINEER S.P.
DRAWN BY EJO
MORGAN & PARMLEY, LTD.
SHEET DESCRIPTION FOUNDATION PLAN
PROJECT NO. 99-118
SHEET NO. A1
CONSTRUCTION JOINT
SAW-CUT CONTRACTION JOINT
BUILDING LINE
COL. BASE PLATE
COLUMN BASE PLATE
STEP FTG.
97'-6"
55'-0"

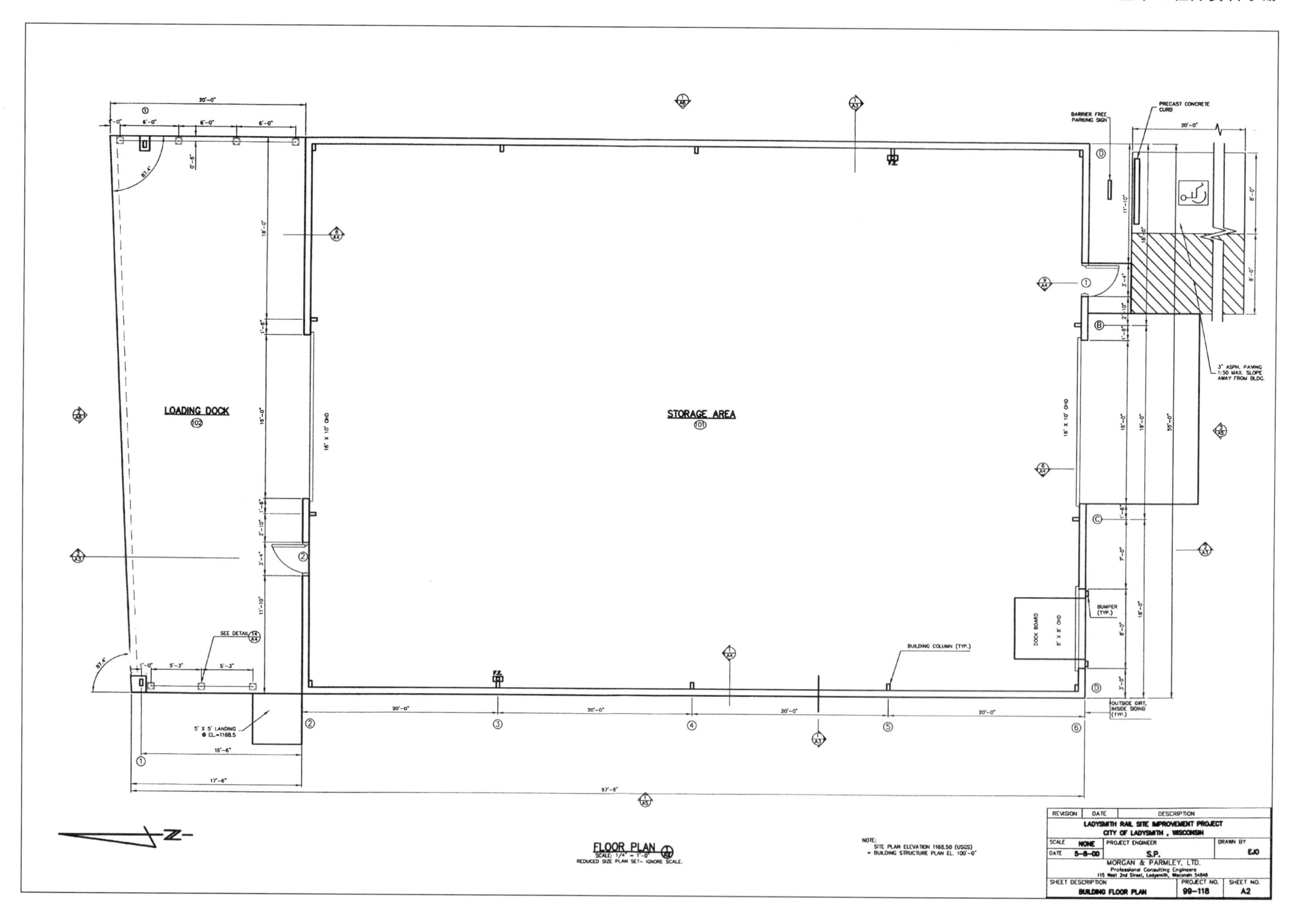
LOADING DOCK
102
STORAGE AREA
101
16' X 10' OHD
8' X 8' OHD
DOCK BOARD
BARRIER FREE PARKING SIGN
PRECAST CONCRETE CURB
3" ASPH. PAVING 1:50 MAX. SLOPE AWAY FROM BLDG.
BUMPER (TYP.)
BUILDING COLUMN (TYP.)
SEE DETAIL
5' X 5' LANDING @ EL.=1168.5
OUTSIDE GIRT, INSIDE SIDING (TYP.)
FLOOR PLAN
SCALE: 1/4" = 1'-0"
REDUCED SIZE PLAN SET- IGNORE SCALE.
NOTE:
SITE PLAN ELEVATION 1168.50 (USGS)
= BUILDING STRUCTURE PLAN EL. 100'-0"
REVISION
DATE
DESCRIPTION
LADYSMITH RAIL SITE IMPROVEMENT PROJECT
CITY OF LADYSMITH , WISCONSIN
SCALE NONE
DATE 5-8-00
PROJECT ENGINEER S.P.
DRAWN BY EJO
MORGAN & PARMLEY, LTD.
Professional Consulting Engineers
115 West 2nd Street, Ladysmith, Wisconsin 54848
SHEET DESCRIPTION
BUILDING FLOOR PLAN
PROJECT NO. 99-118
SHEET NO. A2
97'-6"
17'-6"
16'-6"
20'-0"
55'-0"

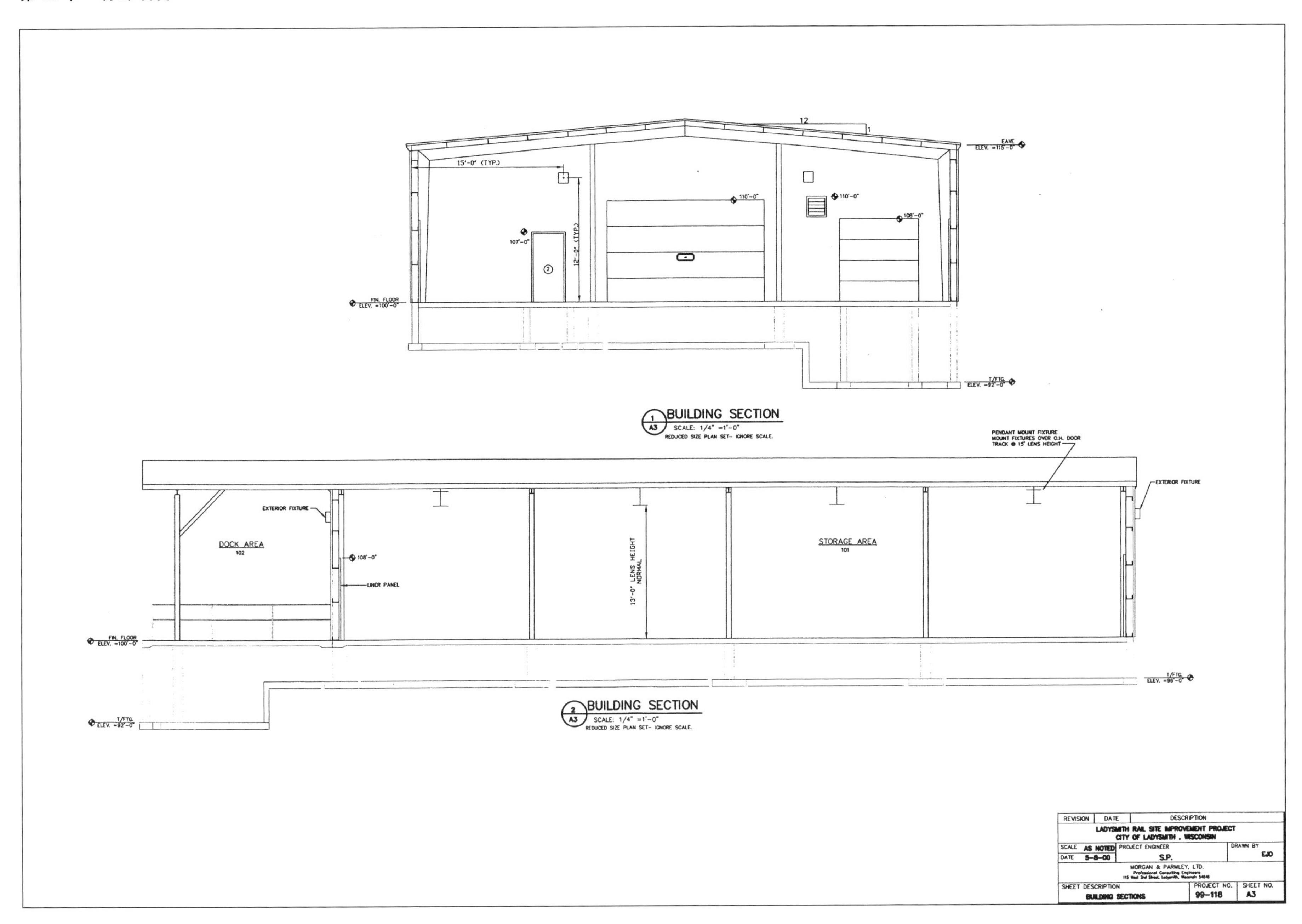
BUILDING SECTION
SCALE: 1/4" =1'-0"
REDUCED SIZE PLAN SET- IGNORE SCALE.
15'-0" (TYP.)
12'-0" (TYP.)
EAVE ELEV. =115'-0"
FIN. FLOOR ELEV. =100'-0"
T/FTG. ELEV. =92'-0"
110'-0"
108'-0"
107'-0"
BUILDING SECTION
SCALE: 1/4" =1'-0"
REDUCED SIZE PLAN SET- IGNORE SCALE.
PENDANT MOUNT FIXTURE MOUNT FIXTURES OVER O.H. DOOR TRACK @ 15' LENS HEIGHT
EXTERIOR FIXTURE
DOCK AREA
102
STORAGE AREA
101
LINER PANEL
13'-0" LENS HEIGHT NORMAL
T/FTG. ELEV. =98'-0"
LADYSMITH RAIL SITE IMPROVEMENT PROJECT
CITY OF LADYSMITH, WISCONSIN
SCALE AS NOTED
DATE 5-8-00
PROJECT ENGINEER S.P.
DRAWN BY EJO
MORGAN & PARMLEY, LTD.
SHEET DESCRIPTION BUILDING SECTIONS
PROJECT NO. 99-118
SHEET NO. A3

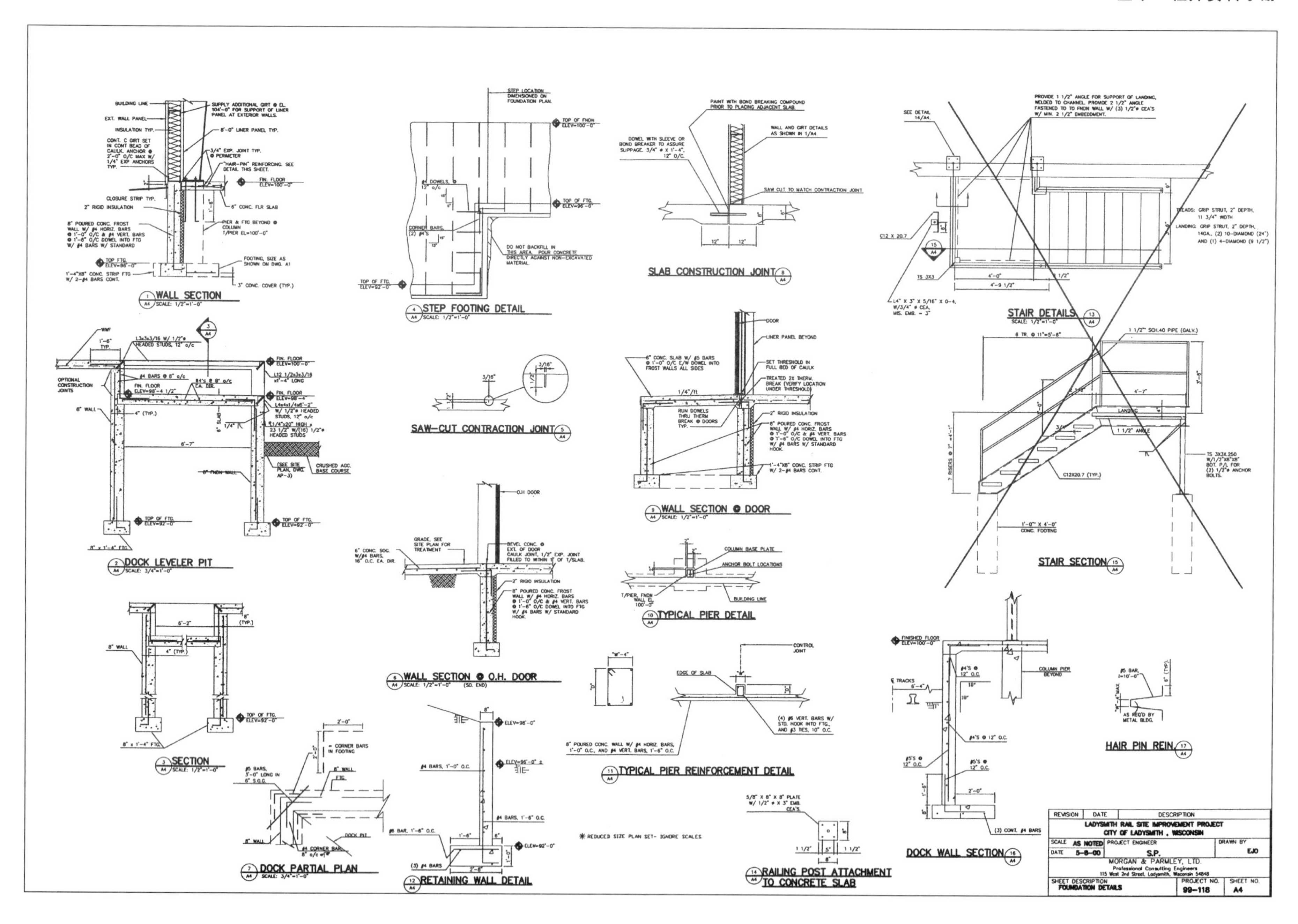
1 A4 WALL SECTION
SCALE: 1/2"=1'-0"
4 A4 STEP FOOTING DETAIL
SCALE: 1/2"=1'-0"
SLAB CONSTRUCTION JOINT 8 A4
STAIR DETAILS 13 A4
SCALE: 1/2"=1'-0"
2 A4 DOCK LEVELER PIT
SCALE: 3/4"=1'-0"
SAW-CUT CONTRACTION JOINT 5 A4
9 A4 WALL SECTION @ DOOR
SCALE: 1/2"=1'-0"
STAIR SECTION 15 A4
10 A4 TYPICAL PIER DETAIL
3 A4 SECTION
SCALE: 1/2"=1'-0"
6 A4 WALL SECTION @ O.H. DOOR
SCALE: 1/2"=1'-0" (SO. END)
11 A4 TYPICAL PIER REINFORCEMENT DETAIL
HAIR PIN REIN. 17 A4
7 A4 DOCK PARTIAL PLAN
SCALE: 3/4"=1'-0"
12 A4 RETAINING WALL DETAIL
14 A4 RAILING POST ATTACHMENT TO CONCRETE SLAB
DOCK WALL SECTION 16 A4
REDUCED SIZE PLAN SET- IGNORE SCALES.
REVISION
DATE
DESCRIPTION
LADYSMITH RAIL SITE IMPROVEMENT PROJECT
CITY OF LADYSMITH , WISCONSIN
SCALE AS NOTED
PROJECT ENGINEER
S.P.
DRAWN BY
EJO
DATE 5-8-00
MORGAN & PARMLEY, LTD.
Professional Consulting Engineers
SHEET DESCRIPTION
FOUNDATION DETAILS
PROJECT NO. 99-118
SHEET NO. A4

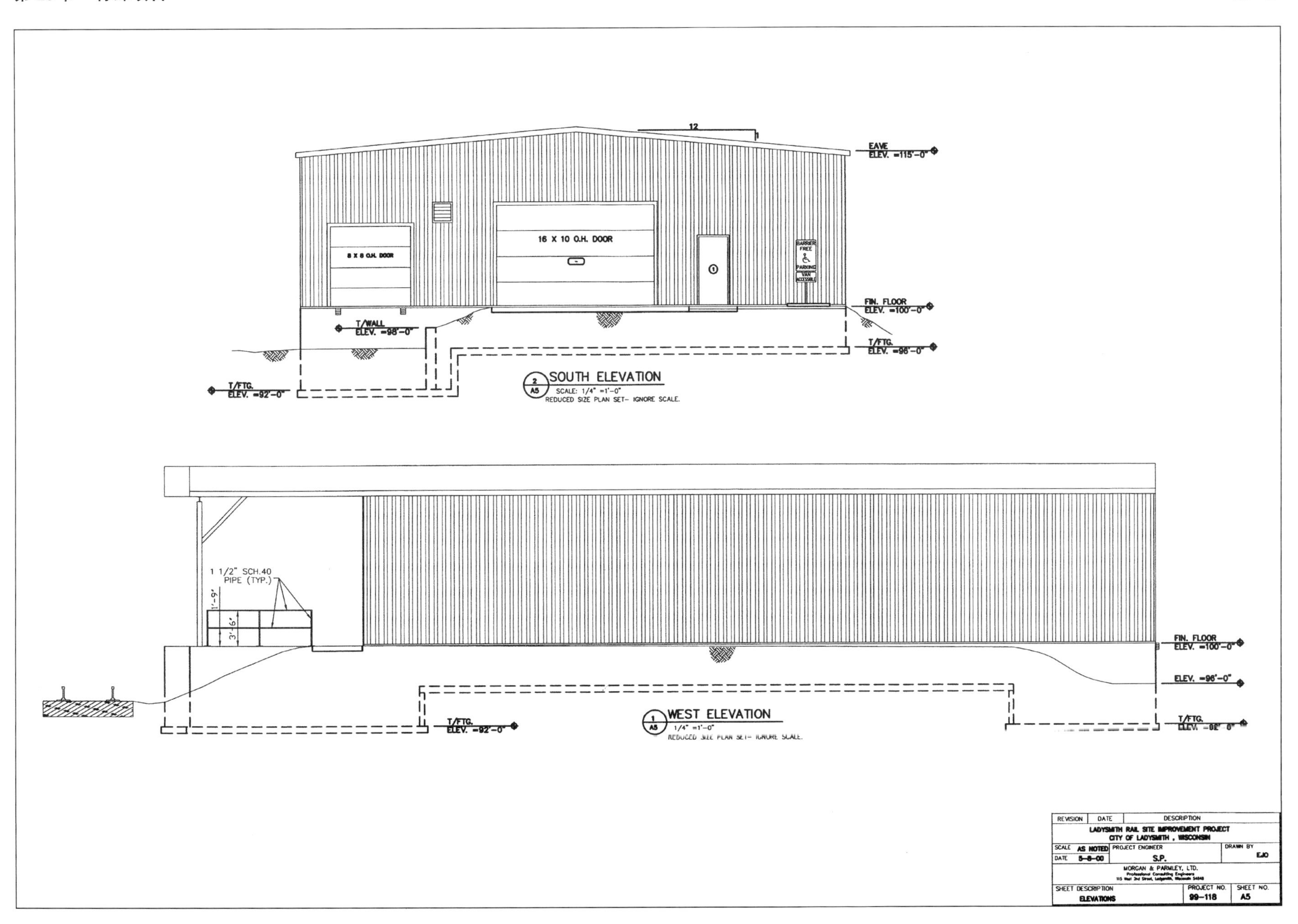

EAVE ELEV. =115'-0"
16 X 10 O.H. DOOR
8 X 8 O.H. DOOR
BARRIER FREE PARKING VAN ACCESSIBLE
FIN. FLOOR ELEV. =100'-0"
T/WALL ELEV. =98'-0"
T/FTG. ELEV. =96'-0"
T/FTG. ELEV. =92'-0"
SOUTH ELEVATION
SCALE: 1/4" =1'-0"
REDUCED SIZE PLAN SET- IGNORE SCALE.
1 1/2" SCH.40 PIPE (TYP.)
FIN. FLOOR ELEV. =100'-0"
ELEV. =96'-0"
T/FTG. ELEV. =92'-0"
WEST ELEVATION
1/4" =1'-0"
REVISION
DATE
DESCRIPTION
LADYSMITH RAIL SITE IMPROVEMENT PROJECT
CITY OF LADYSMITH , WISCONSIN
SCALE AS NOTED
DATE 5-8-00
PROJECT ENGINEER S.P.
DRAWN BY EJO
MORGAN & PARMLEY, LTD.
Professional Consulting Engineers
SHEET DESCRIPTION ELEVATIONS
PROJECT NO. 99-118
SHEET NO. A5

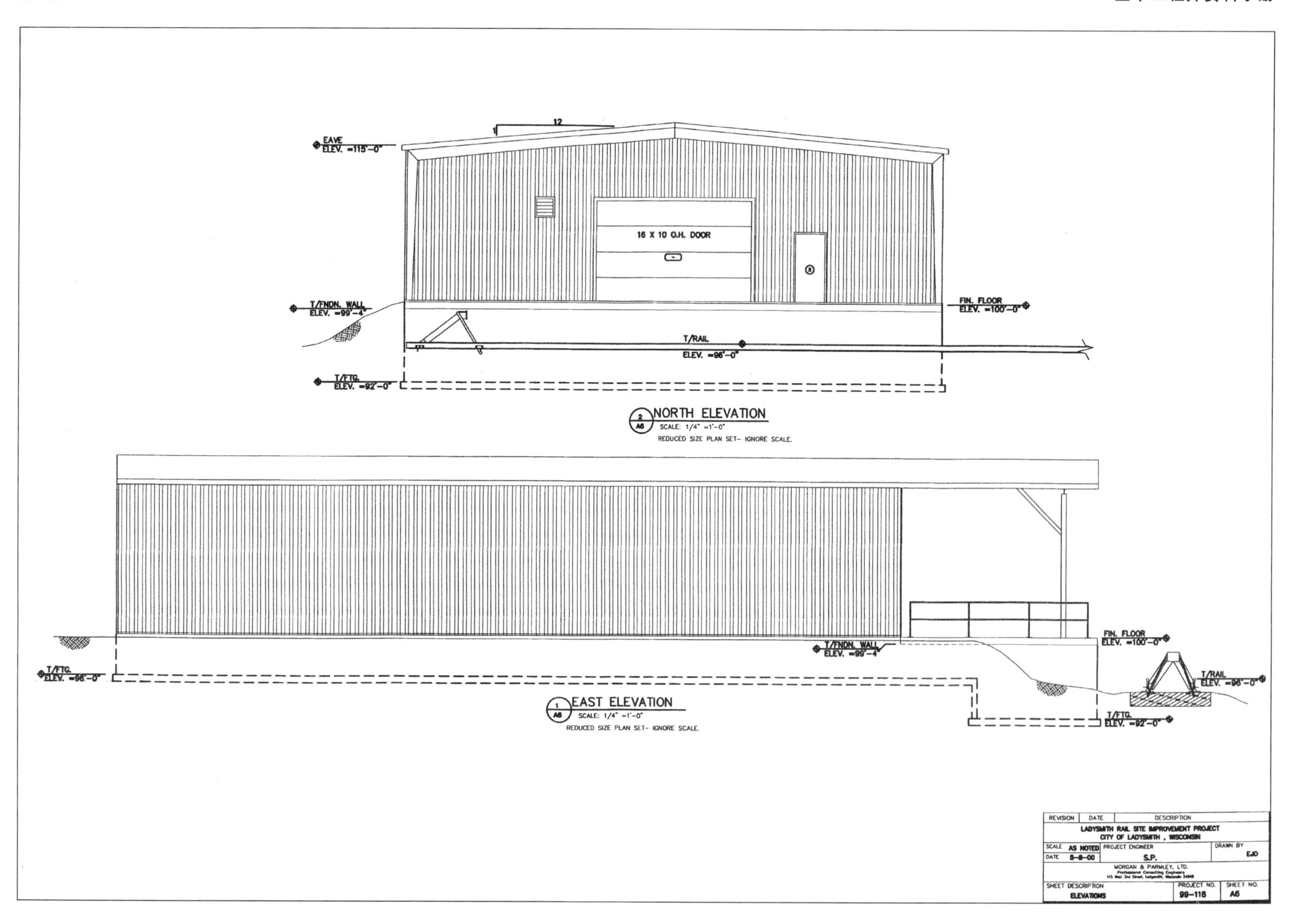

12
1
EAVE
ELEV. =115'-0"
16 X 10 O.H. DOOR
T/FNDN. WALL
ELEV. =99'-4"
FIN. FLOOR
ELEV. =100'-0"
T/RAIL
ELEV. =96'-0"
T/FTG.
ELEV. =92'-0"
2 A6 NORTH ELEVATION
SCALE: 1/4" =1'-0"
REDUCED SIZE PLAN SET- IGNORE SCALE.
T/FTG.
ELEV. =96'-0"
T/FNDN. WALL
ELEV. =99'-4"
FIN. FLOOR
ELEV. =100'-0"
T/RAIL
ELEV. =96'-0"
T/FTG.
ELEV. =92'-0"
1 A6 EAST ELEVATION
SCALE: 1/4" =1'-0"
REDUCED SIZE PLAN SET- IGNORE SCALE.
REVISION
DATE
DESCRIPTION
LADYSMITH RAIL SITE IMPROVEMENT PROJECT
CITY OF LADYSMITH , WISCONSIN
SCALE AS NOTED
DATE 5-8-00
PROJECT ENGINEER
S.P.
DRAWN BY
EJO
MORGAN & PARMLEY, LTD.
SHEET DESCRIPTION
ELEVATIONS
PROJECT NO.
99-118
SHEET NO.
A6

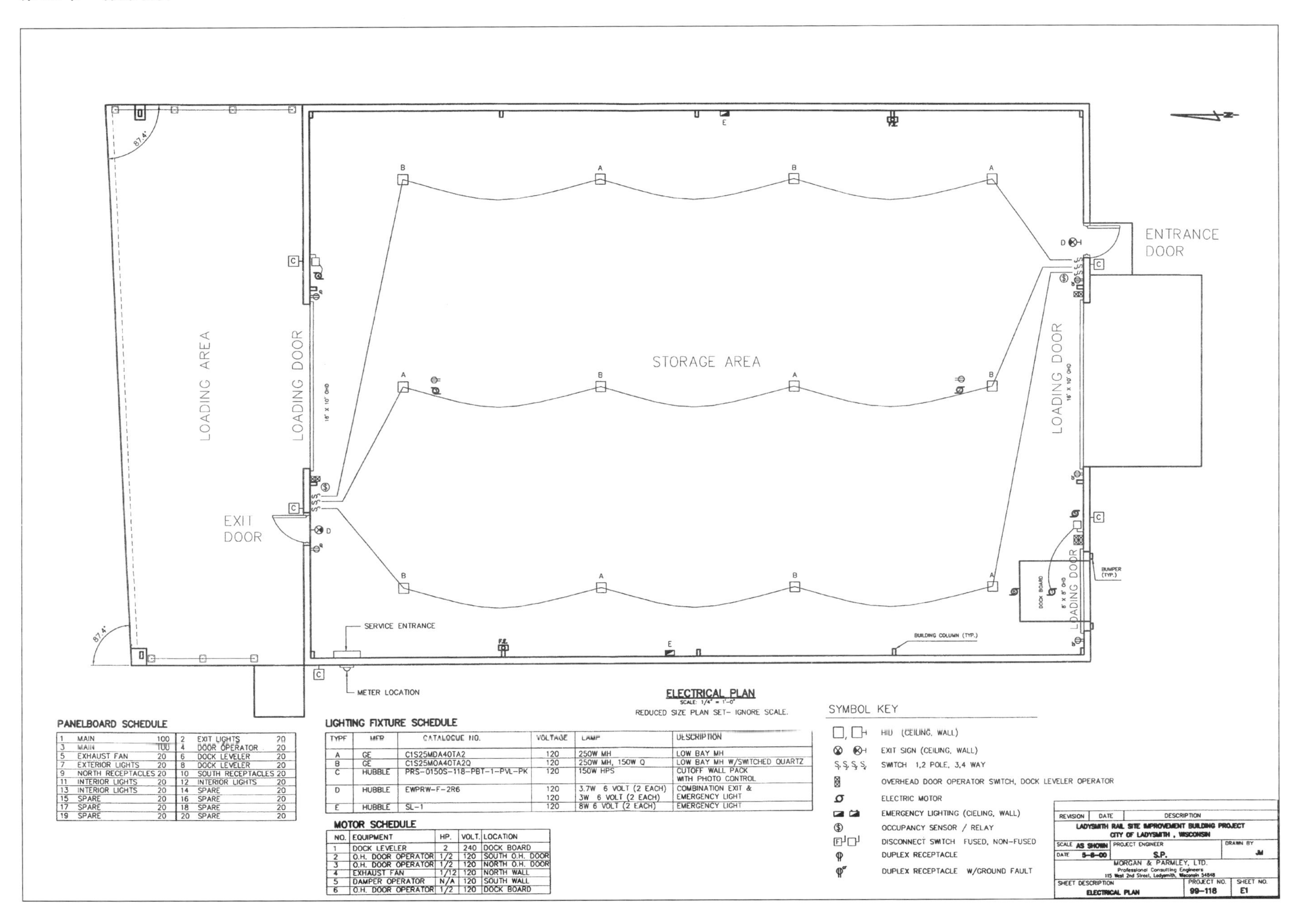

PANELBOARD SCHEDULE

1	MAIN	100	2	EXIT LIGHTS	20
3	MAIN	100	4	DOOR OPERATOR	20
5	EXHAUST FAN	20	6	DOCK LEVELER	20
7	EXTERIOR LIGHTS	20	8	DOCK LEVELER	20
9	NORTH RECEPTACLES	20	10	SOUTH RECEPTACLES	20
11	INTERIOR LIGHTS	20	12	INTERIOR LIGHTS	20
13	INTERIOR LIGHTS	20	14	SPARE	20
15	SPARE	20	16	SPARE	20
17	SPARE	20	18	SPARE	20
19	SPARE	20	20	SPARE	20

LIGHTING FIXTURE SCHEDULE

TYPE	MFR	CATALOGUE NO.	VOLTAGE	LAMP	DESCRIPTION
A	GE	C1S25MDA40TA2	120	250W MH	LOW BAY MH
B	GE	C1S25MOA40TA2Q	120	250W MH, 150W Q	LOW BAY MH W/SWITCHED QUARTZ
C	HUBBLE	PRS-0150S-118-PBT-1-PVL-PK	120	150W HPS	CUTOFF WALL PACK WITH PHOTO CONTROL
D	HUBBLE	EWPRW-F-2R6	120 120	3.7W 6 VOLT (2 EACH) 3W 6 VOLT (2 EACH)	COMBINATION EXIT & EMERGENCY LIGHT
E	HUBBLE	SL-1	120	8W 6 VOLT (2 EACH)	EMERGENCY LIGHT

MOTOR SCHEDULE

NO.	EQUIPMENT	HP.	VOLT.	LOCATION
1	DOCK LEVELER	2	240	DOCK BOARD
2	O.H. DOOR OPERATOR	1/2	120	SOUTH O.H. DOOR
3	O.H. DOOR OPERATOR	1/2	120	NORTH O.H. DOOR
4	EXHAUST FAN	1/12	120	NORTH WALL
5	DAMPER OPERATOR	N/A	120	SOUTH WALL
6	O.H. DOOR OPERATOR	1/2	120	DOCK BOARD

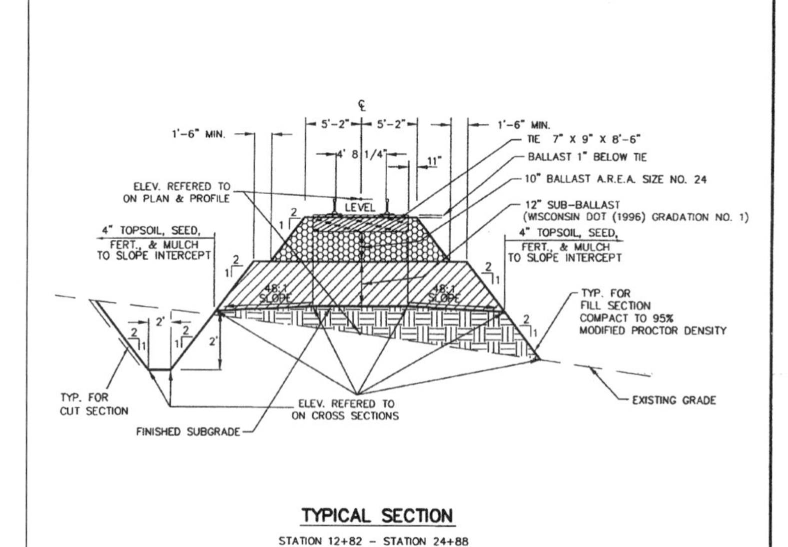

TYPICAL SECTION

STATION 12+82 – STATION 24+88

BULIDING DESIGN INFORMATION

Safe Soil Bearing Capacity = 2000 psf

Pre-engineered Metal Building Design Loading
(In accordance with Wisconsin Administrative Code, Chapter 53)

Roof Live Loading (Snow): 40 psf
Roof Collateral Live Load: 3 psf
Wind Loading:
- Structure, wall components: 20 psf
- Uplift pressure, normal to surface: 20 psf
- Roof overhangs, eaves: 30 psf uplift

Building Design :
Metal building supplier to provide State of Wisconsin Department of Commerce, Safety and Buildings with building submittal for approval in accordance with requirements of the Administrative Code and Specification Section 13200. Plans and calculations shall be prepared by a professional engineer registered to practice in the State of Wisconsin.

Concrete & Concrete Reinforcing Design Stresses (See Specifications)

Thermal Performance ? Energy Calculations
Building is without space heating and therefor exempt from provisions of Administrative Code, Chapter 63.

GENERAL NOTES

ADDITIONAL NOTES ON FOLLOWING SHEETS

1. THE LOCATIONS OF EXISTING AND PROPOSED UTILITY INSTALLATIONS AS SHOWN ON THE PLANS ARE APPROXIMATE. THERE MAY BE OTHER UTILITY INSTALLATIONS WITHIN THE PROJECT AREA THAT ARE NOT SHOWN.
2. THE SEED MIXTURE TO BE USED ON DESIGNATED AREAS SHALL CONFORM TO THE PERTINENT REQUIREMENTS OF THE SPECIFICATIONS.
3. ALL DISTURBED AREAS OTHER THAN THE RAILROAD TRACK BED SHALL BE SEEDED, FERTILIZED AND MULCHED.
4. PRIOR TO THE PLACEMENT OF FILL MATERIAL; THE BOTTOM OF THE EXCAVATION SHALL BE EVALUATED BY A QUALIFIED SOILS ENGINEER TO DETERMINE IF THE EXPOSED SOILS OFFER SUFFICIENT BEARING CAPACITY FOR THE ANTICIPATED LOADS.
5. ALL COMPACTION SHALL BE 95 % OF MAXIMUM MODIFIED PROCTOR DENSITY (ASTM D-1557)
6. THE RAILROAD EXCAVATION AND EMBANKMENT WILL BE BROUGHT TO FINISHED GRADES BY THE EARTHWORK CONTRACTOR.
7. THE EARTHWORK CONTRACTOR WILL BE RESPONSIBLE FOR THE CONSTRUCTION OF ALL EMBANKMENT TO FINISHED SUBGRADE , INSTALLATION OF CULVERTS AND DITCHES AND RESTORATION OF ALL DISTURBED AREAS WITH TOPSOIL , SEED AND FERTILIZER , OR CRUSHED BASE COURSE AND PAVEMENT.
8. THE RAILROAD TRACK CONTRACTOR WILL BE RESPONSIBLE FOR THE PLACEMENT OF SUB-BALLAST , BALLAST , TIES , RAILS AND APPURTENANCES ON BASE PREPARED BY THE EARTHWORK CONTRACTOR.
9. WORK ZONE LIMITS SHALL BE 25' EACH SIDE OF THE RAIL CENTERLINE.
10. THE CONTRACTOR SHALL REMOVE AND STOCKPILE ALL TOPSOIL AND ORGANIC MATERIAL AFTER CLEARING & GRUBBING THE SITE. PRIOR TO PLACING ANY EMBANKMENT MATERIAL, THE CONTRACTOR SHALL SCARIFY THE TOP 6" OF MATERIAL AND COMPACT TO 95% MODIFIED DENSITY.
11. IN A CUT SECTION THE CONTRACTOR SHALL SCARIFY THE 6" OF MATERIAL BELOW THE SUBGRADE AND RECOMPACT TO 95% MODIFIED DENSITY.

MISC. QUANTITIES

4850 C.Y. .... COMMON EXCAVATION
465 C.Y. .... FILL MATERIAL
3830 S.Y. .... SALVAGED TOPSOIL, FERTILIZER, SEED, AND MULCH
90 C.Y. .... C.A.B.C. (SHOULDER MATERIAL – NORSE DRIVE) EXP. 30%
560 C.Y. .... C.A.B.C. (PARKING LOT) EXP. 30%
280 C.Y. .... GRANULAR FILL (PARKING LOT) EXP. 30%
875 C.Y. .... GRANULAR FILL (BLDG & DOCK) EXP. 30%
180 TONS. .... 3" SINGLE AGGREGATE ASPHALTIC PAVEMENT
100 L.F. .... 12" REINFORCED CONCRETE PIPE CLASS III
4 Ea. .... 12" REINFORCED CONCRETE PIPE CLASS III APRON ENDWALL
70 Ea. .... EROSION BALES

RAIL CONSTRUCTION

STATION 12+82 TO 24+88

2,100 L.F. .... RE9020 90 # SECOND HAND CLASS 2, RELAY RAIL (1,050 LF. OF TRACK)
312 L.F. .... RE11525 115 # SECOND HAND CLASS 2, RELAY RAIL (156' OF RAILROAD TRACK) (ROAD CROSSING)

675 Ea. .... NEW OR SECOND HAND RAILROAD TIES 7" X 9" X 8'-6" CLASS 1
850 C.Y. .... 2-1/2"-3/4" CLEAN STONE (A.R.E.A. SIZE NO. 24)(BALLAST)
50 S.Y. .... GEOTEXTILE FABRIC , TYPE HR
1302 C.Y. .... SUBBALLAST (WDOT GRADATION NO. 1)

1 EA. .... HAYS WG – HD (OR EQUAL) BUMPING POST (PROVIDED BY OWNER)
1 PAIR .... NOLAN'S MODEL CS-60 HINGED CAR STOPS (PROVIDED BY OWNER)

EARTHWORK SUMMARY

	CUT C.Y.	FILL C.Y.	NET WASTE C.Y.
BUILDING	950	215	109
RAIL	3900	250	3650
TOTALS	4850	465	3759

UTILITIES

UTILITIES WILL BE MOVED OR ADJUSTED BY THEIR OWNERS UNLESS OTHERWISE CALLED FOR ON THE PLANS.

*** CALL DIGGERS HOTLINE FOR LOCATION 1-800-242-8511**

THE UTILITY OWNERS ARE:

CONTACT: ELECTRIC
XCEL ENERGY
310 HICKORY HILL LANE
PHILLIPS , WISCONSIN 54555
ATTEN: JOE PERKINS
TELEPHONE: (715) 836-1198

CONTACT: NATURAL GAS
WISCONSIN GAS COMPANY
104 WEST SOUTH STREET
RICE LAKE , WISCONSIN 54868
ATTEN: DON WEDIN
TELEPHONE: (800) 925-2104

CONTACT: TELEPHONE
CENTURYTEL
425 ELLINGSON AVENUE
P.O. BOX 78
HAWKINS, WI 54530
ATTEN: JAMES ARQUETTE
TELEPHONE: (715) 585-7707

CONTACT: WATER & SEWER
LADYSMITH DEPT. OF PUBLIC WORKS
120 WEST MINOR AVENUE
P.O. BOX 431
LADYSMITH, WI 54848
ATTEN: BILL CHRISTIANSON, DIRECTOR
TELEPHONE: (715) 532-2601

CONTACT: CABLE TELEVISION
CHARTER COMMUNICATIONS
P.O. BOX 539
RICE LAKE, WISCONSIN
TELEPHONE: (800) 262-2578

REVISION	DATE	DESCRIPTION
LADYSMITH RAIL SITE IMPROVEMENT PROJECT CITY OF LADYSMITH , WISCONSIN		
SCALE AS NOTED DATE 4-24-00	PROJECT ENGINEER S.P.	DRAWN BY JM
MORGAN & PARMLEY, LTD. Professional Consulting Engineers 115 West 2nd Street, Ladysmith, Wisconsin 54848		
SHEET DESCRIPTION TYPICAL SECTIONS, GENERAL NOTES	PROJECT NO. 99-118	SHEET NO. RR1

3°52'49" AT NORSE DRIVE
2°07'29" AT BARNETT ROAD
SKEW
STATION 16+25 OR 23+75
ALL TIES BETWEEN STATION 15+85 TO 16+65
AND STATION 16+35 TO 24+15
19 1/4" ON CENTER
COUNTERSINK 5/8"X12" CAMRAIL TORX TRUSS WASHER HEAD TIMBER SCREW EVERY OTHER TIE
CENTERLINE OF ROAD
EDGE OF GRAVEL
15' ALL ROADWAY CROWN REMOVED
ASPH. SURFACE

TYPE 3 GRADE CROSSING
BARNETT ROAD AND NORSE DRIVE CROSSINGS
CENTERLINE STATION 16+25 AND 23+75

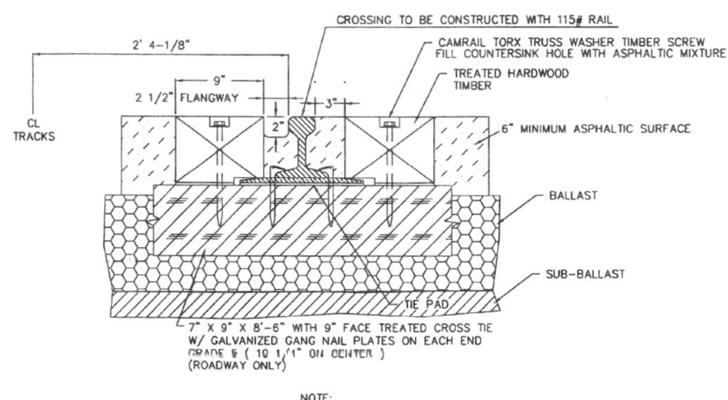

NOTE:
CROSSING BID TO INCLUDE RAIL, TIE PADS, TREATED HARDWOOD TIMBERS AND ALL OTHER NECESSARY ITEMS INCLUDING CROSSBUCKS AND ADVANCED WARNING SIGNS
CONTRACTOR TO INSTALL 6" ASPHALTIC SURFACE THROUGH CROSSING AND BETWEEN TREATED HARDWOOD TIMBERS AS NOTED ELSEWHERE

SECTION C – C

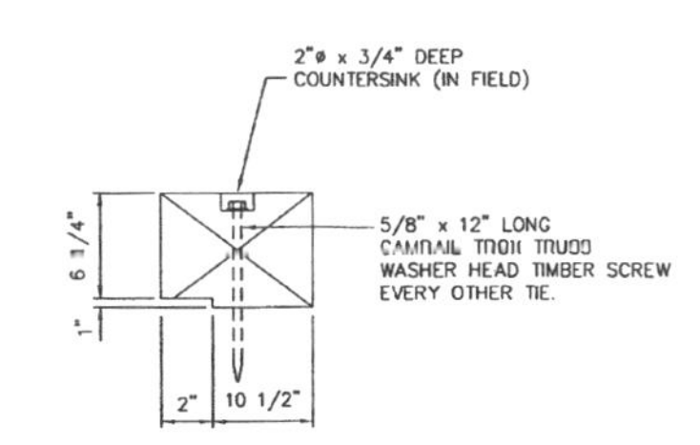

CROSSING PLANK DETAIL
FOR 9020 RAIL

GENERAL NOTES

DETAILS OF CONSTRUCTION , MATERIALS AND WORKMANSHIP NOT SHOWN ON THIS DRAWING SHALL CONFORM TO THE PERTINENT REQUIREMENTS OF THE STANDARD SPECIFICATIONS , THE APPLICABLE SPECIAL PROVISIONS, WISCONSIN DOT STANDARD SPECIFICATIONS AND FACILITIES DEVELOPEMENT MANUAL FOR TYPE 3 GRADE CROSSINGS. THE A.R.E.A DESIGN MANUAL AND WCL SPECIFICATIONS FOR GRADING AND CONSTRUCTION OF INDUSTRIAL TRACKS.

5/8"X12" LONG CAMRAIL TROX TRUSS WASHER HEAD TIMBER SCREWS EVERY OTHER TIE AND AT THE END OF EACH TIMBER.

CROSSING TO BE CONSTRUCTED OF 115 # NUMBER 2 RELAY RAIL AND SHALL BE FULLY TIE PLATED WITH FOUR SPIKES PER TIE PLATE AND FULLY BOX ANCHORED

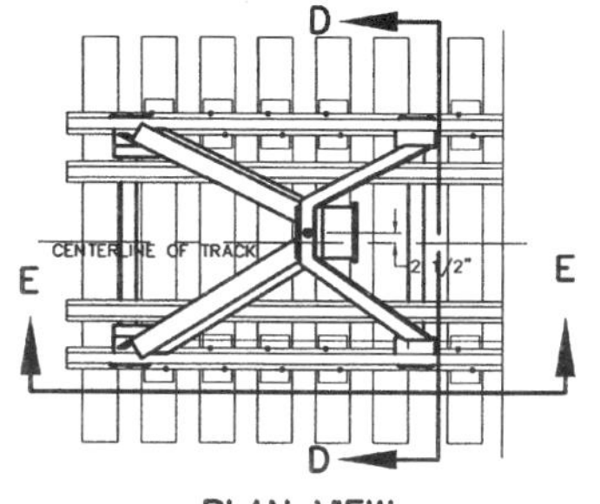

PLAN VIEW
BUMPING POST DETAIL
HAYES TYPE WG-HD OR EQUAL

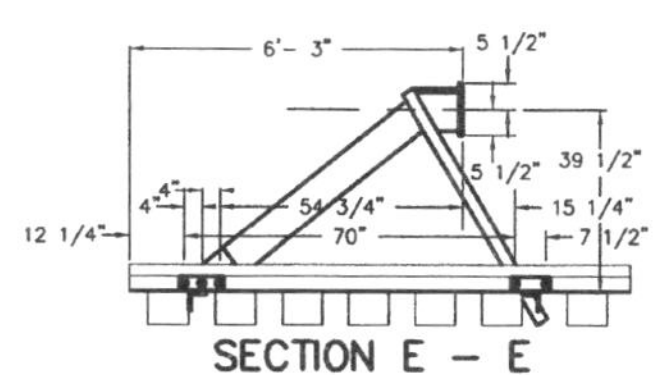

SECTION E – E

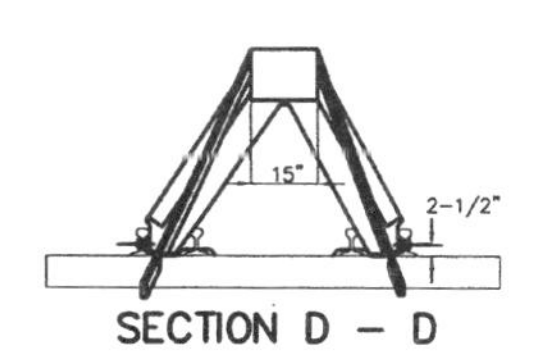

SECTION D – D

REVISION	DATE	DESCRIPTION
LADYSMITH RAIL SITE IMPROVEMENT PROJECT CITY OF LADYSMITH , WISCONSIN		
SCALE AS NOTED DATE 4-24-00	PROJECT ENGINEER S.P.	DRAWN BY JM
MORGAN & PARMLEY, LTD. Professional Consulting Engineers 115 West 2nd Street, Ladysmith, Wisconsin 54848		
SHEET DESCRIPTION RAIL DETAILS	PROJECT NO. 99-118	SHEET NO RR2

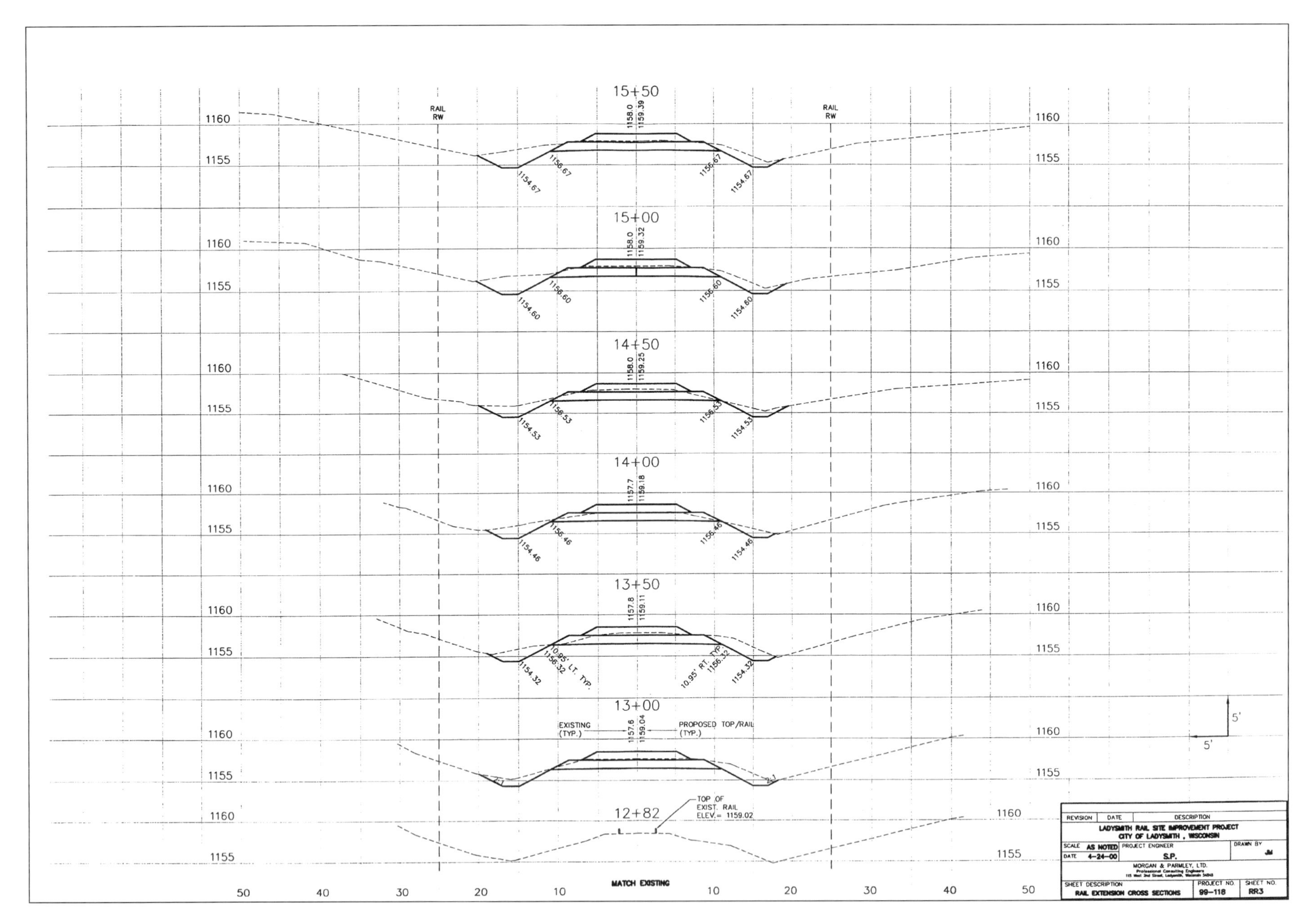

15+50
15+00
14+50
14+00
13+50
13+00
12+82
RAIL RW
1160
1155
EXISTING (TYP.)
PROPOSED TOP/RAIL (TYP.)
TOP OF EXIST. RAIL ELEV.= 1159.02
10.95' LT. TYP.
10.95' RT. TYP.
MATCH EXISTING
50 40 30 20 10 10 20 30 40 50
5'
REVISION DATE DESCRIPTION
LADYSMITH RAIL SITE IMPROVEMENT PROJECT
CITY OF LADYSMITH, WISCONSIN
SCALE AS NOTED
DATE 4-24-00
PROJECT ENGINEER S.P.
DRAWN BY
MORGAN & PARMLEY, LTD.
Professional Consulting Engineers
SHEET DESCRIPTION RAIL EXTENSION CROSS SECTIONS
PROJECT NO. 99-118
SHEET NO. RR3

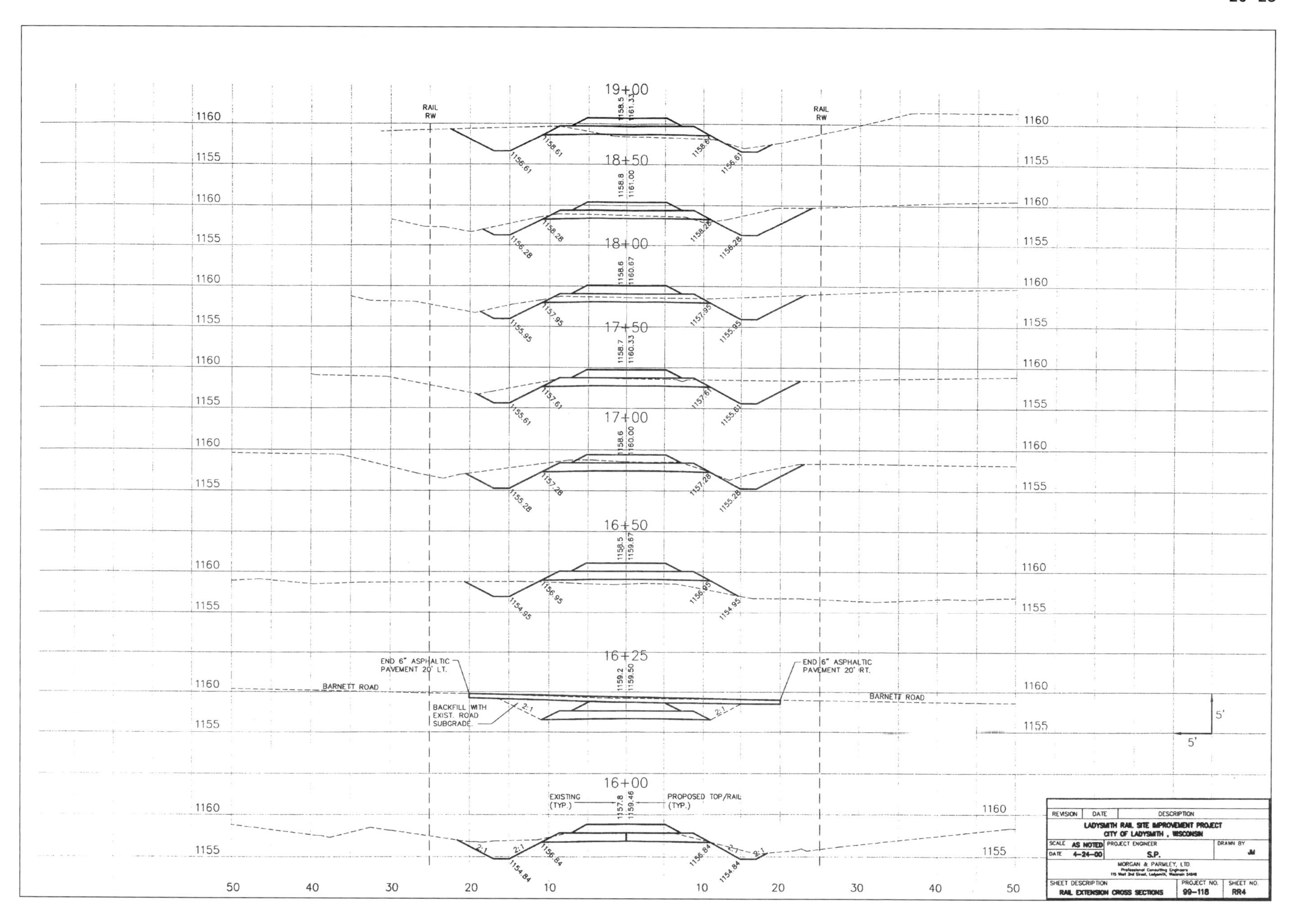
19+00
18+50
18+00
17+50
17+00
16+50
16+25
16+00
RAIL RW
END 6" ASPHALTIC PAVEMENT 20' LT.
END 6" ASPHALTIC PAVEMENT 20' RT.
BARNETT ROAD
BACKFILL WITH EXIST. ROAD SUBGRADE
EXISTING (TYP.)
PROPOSED TOP/RAIL (TYP.)
REVISION DATE DESCRIPTION
LADYSMITH RAIL SITE IMPROVEMENT PROJECT
CITY OF LADYSMITH, WISCONSIN
SCALE AS NOTED
DATE 4-24-00
PROJECT ENGINEER S.P.
DRAWN BY JM
MORGAN & PARMLEY, LTD.
SHEET DESCRIPTION RAIL EXTENSION CROSS SECTIONS
PROJECT NO. 99-118
SHEET NO. RR4

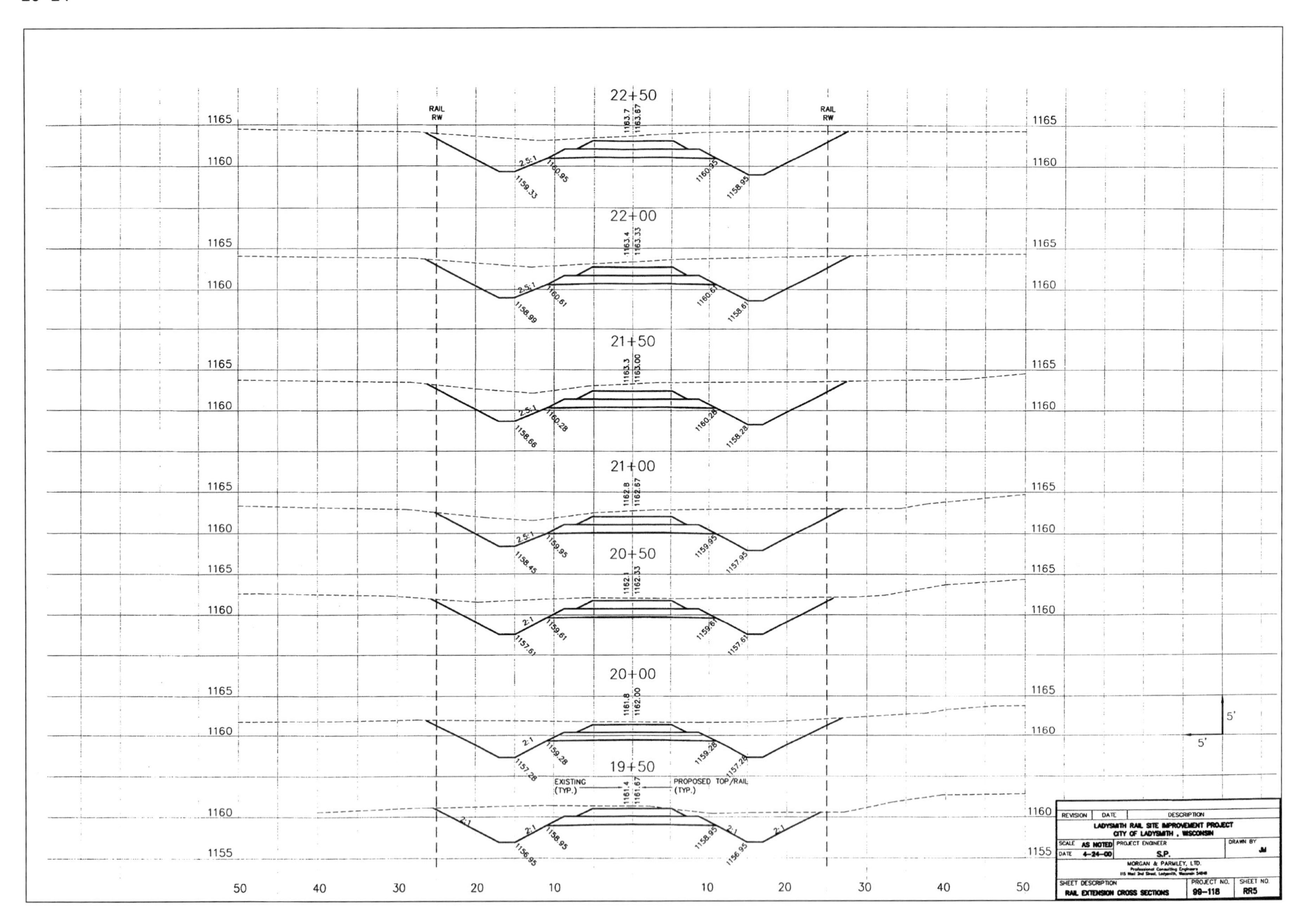
22+50
22+00
21+50
21+00
20+50
20+00
19+50
RAIL RW
EXISTING (TYP.)
PROPOSED TOP/RAIL (TYP.)
LADYSMITH RAIL SITE IMPROVEMENT PROJECT
CITY OF LADYSMITH, WISCONSIN
SCALE AS NOTED
DATE 4-24-00
PROJECT ENGINEER S.P.
DRAWN BY JM
MORGAN & PARMLEY, LTD.
Professional Consulting Engineers
SHEET DESCRIPTION RAIL EXTENSION CROSS SECTIONS
PROJECT NO. 99-118
SHEET NO. RR5

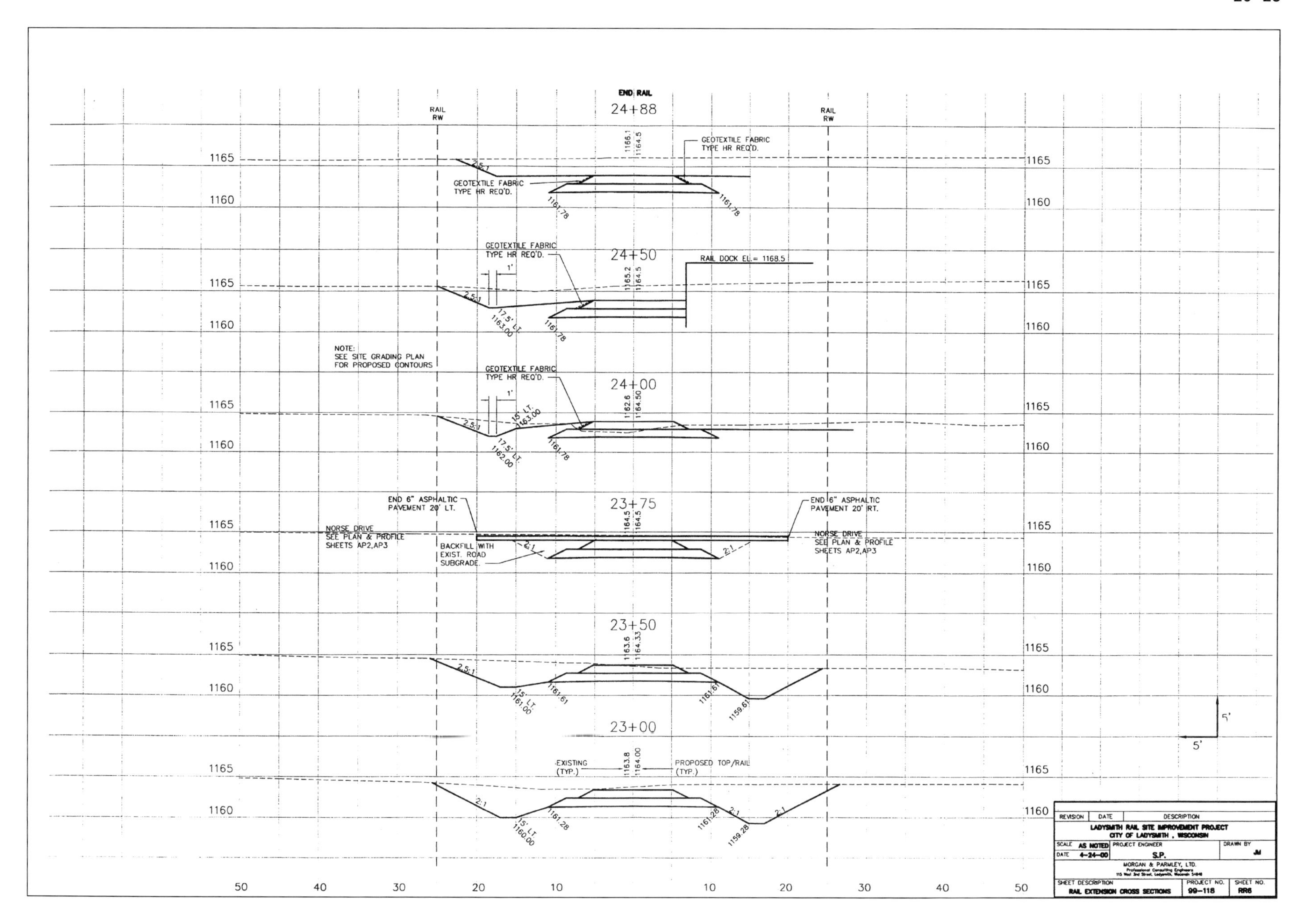
END RAIL
24+88
RAIL RW
GEOTEXTILE FABRIC TYPE HR REQ'D.
24+50
RAIL DOCK EL.= 1168.5
NOTE: SEE SITE GRADING PLAN FOR PROPOSED CONTOURS
24+00
23+75
END 6" ASPHALTIC PAVEMENT 20' LT.
END 6" ASPHALTIC PAVEMENT 20' RT.
NORSE DRIVE SEE PLAN & PROFILE SHEETS AP2,AP3
BACKFILL WITH EXIST. ROAD SUBGRADE.
23+50
23+00
EXISTING (TYP.)
PROPOSED TOP/RAIL (TYP.)
LADYSMITH RAIL SITE IMPROVEMENT PROJECT
CITY OF LADYSMITH, WISCONSIN
SCALE AS NOTED
DATE 4-24-00
PROJECT ENGINEER S.P.
DRAWN BY JM
MORGAN & PARMLEY, LTD.
SHEET DESCRIPTION RAIL EXTENSION CROSS SECTIONS
PROJECT NO. 99-118
SHEET NO. RR6

部落鱼苗孵化场

EDA Project No. 06-01-02846

WILLIAM J. POUPART TRIBAL FISH HATCHERY

FOR THE

LAC DU FLAMBEAU BAND OF LAKE SUPERIOR CHIPPEWA INDIANS OF WISCONSIN

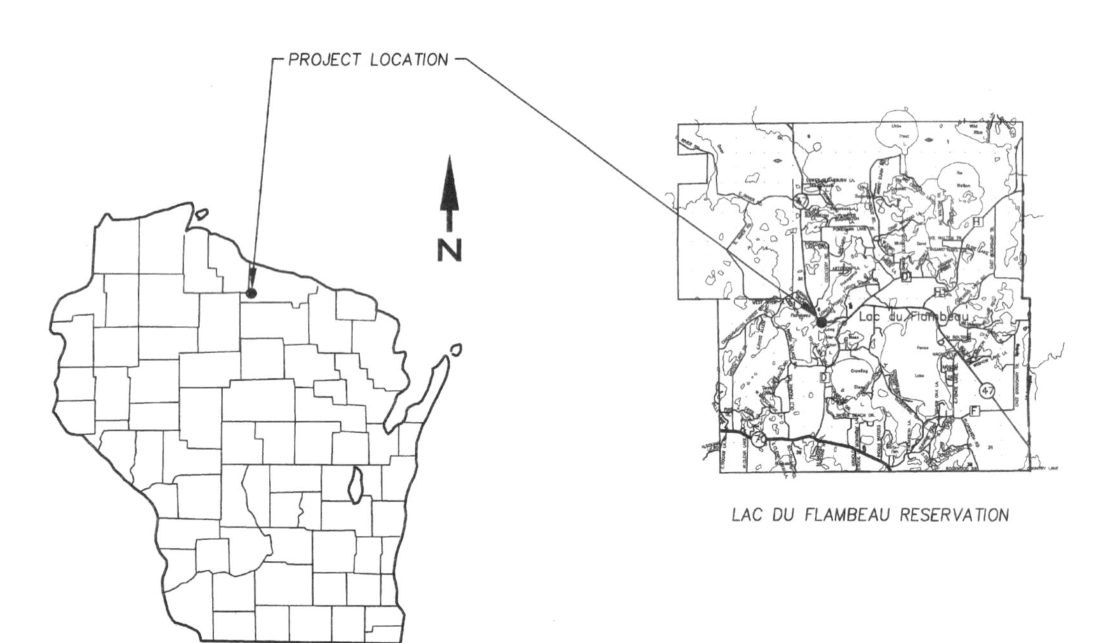

LAC DU FLAMBEAU RESERVATION

TABLE OF CONTENTS

REVISION	DATE	DESCRIPTION

WILLIAM J. POUPART TRIBAL FISH HATCHERY
LAC DU FLAMBEAU, WISCONSIN

SCALE NONE　DATE 7-30-98　PROJECT ENGINEER BRUCE MARKGREN　DRAWN BY D.A.N.

COOPER ENGINEERING COMPANY
310 WEST SOUTH STREET RICE LAKE, WISCONSIN

TELEPHONE 715-234-7008　SHEET DESCRIPTION TITLE SHEET　PROJECT NO. CE97008　SHEET NO. G-1

Courtesy: Cooper Engineering Co., Ine.

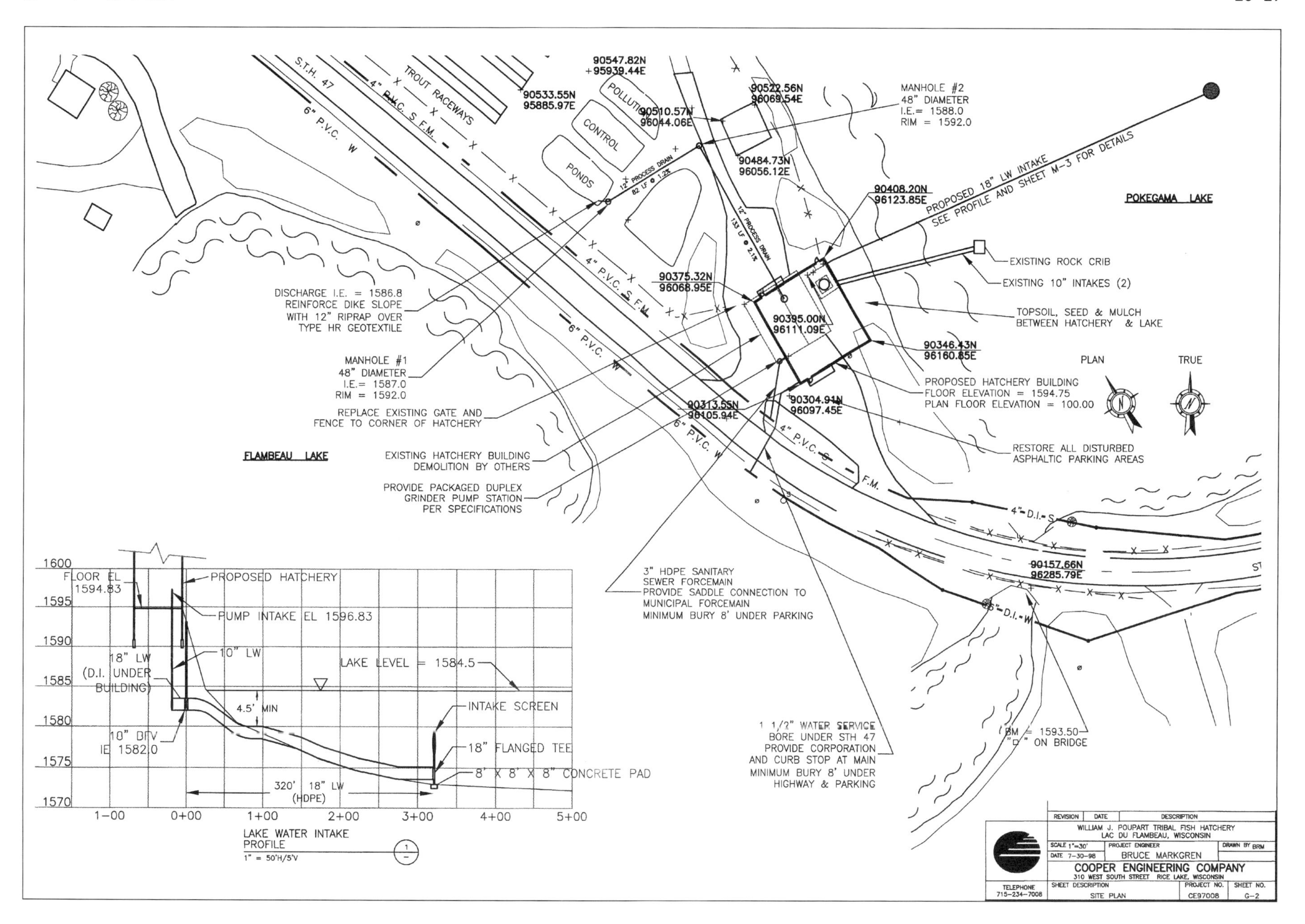
S.T.H. 47
4" P.V.C. S.F.M.
6" P.V.C. W
TROUT RACEWAYS
POLLUTION CONTROL PONDS
90547.82N
+95939.44E
90533.55N
95885.97E
90510.57N
96044.06E
90522.56N
96069.54E
90484.73N
96056.12E
MANHOLE #2
48" DIAMETER
I.E.= 1588.0
RIM = 1592.0
12" PROCESS DRAIN
PROPOSED 18" LW INTAKE
SEE PROFILE AND SHEET M-3 FOR DETAILS
POKEGAMA LAKE
90408.20N
96123.85E
EXISTING ROCK CRIB
EXISTING 10" INTAKES (2)
TOPSOIL, SEED & MULCH
BETWEEN HATCHERY & LAKE
90375.32N
96068.95E
90395.00N
96111.09E
90346.43N
96160.85E
DISCHARGE I.E. = 1586.8
REINFORCE DIKE SLOPE
WITH 12" RIPRAP OVER
TYPE HR GEOTEXTILE
MANHOLE #1
48" DIAMETER
I.E.= 1587.0
RIM = 1592.0
REPLACE EXISTING GATE AND
FENCE TO CORNER OF HATCHERY
90313.55N
96105.94E
90304.91N
96097.45E
PROPOSED HATCHERY BUILDING
FLOOR ELEVATION = 1594.75
PLAN FLOOR ELEVATION = 100.00
PLAN
TRUE
RESTORE ALL DISTURBED
ASPHALTIC PARKING AREAS
FLAMBEAU LAKE
EXISTING HATCHERY BUILDING
DEMOLITION BY OTHERS
PROVIDE PACKAGED DUPLEX
GRINDER PUMP STATION
PER SPECIFICATIONS
4" D.I.-S
6" D.I.-W
90157.66N
96285.79E
3" HDPE SANITARY
SEWER FORCEMAIN
PROVIDE SADDLE CONNECTION TO
MUNICIPAL FORCEMAIN
MINIMUM BURY 8' UNDER PARKING
1 1/2" WATER SERVICE
BORE UNDER STH 47
PROVIDE CORPORATION
AND CURB STOP AT MAIN
MINIMUM BURY 8' UNDER
HIGHWAY & PARKING
BM = 1593.50
"□" ON BRIDGE
FLOOR EL
1594.83
PROPOSED HATCHERY
PUMP INTAKE EL 1596.83
10" LW
18" LW
(D.I. UNDER
BUILDING)
LAKE LEVEL = 1584.5
4.5' MIN
INTAKE SCREEN
10" BFV
IE 1582.0
18" FLANGED TEE
8' X 8' X 8" CONCRETE PAD
320' 18" LW
(HDPE)
1600
1595
1590
1585
1580
1575
1570
1+00
0+00
1+00
2+00
3+00
4+00
5+00
LAKE WATER INTAKE
PROFILE
1" = 50'H/5'V
REVISION
DATE
DESCRIPTION
WILLIAM J. POUPART TRIBAL FISH HATCHERY
LAC DU FLAMBEAU, WISCONSIN
SCALE 1"=30'
PROJECT ENGINEER
DRAWN BY BRM
DATE 7-30-98
BRUCE MARKGREN
COOPER ENGINEERING COMPANY
310 WEST SOUTH STREET RICE LAKE, WISCONSIN
TELEPHONE
715-234-7008
SHEET DESCRIPTION
SITE PLAN
PROJECT NO.
CE97008
SHEET NO.
G-2

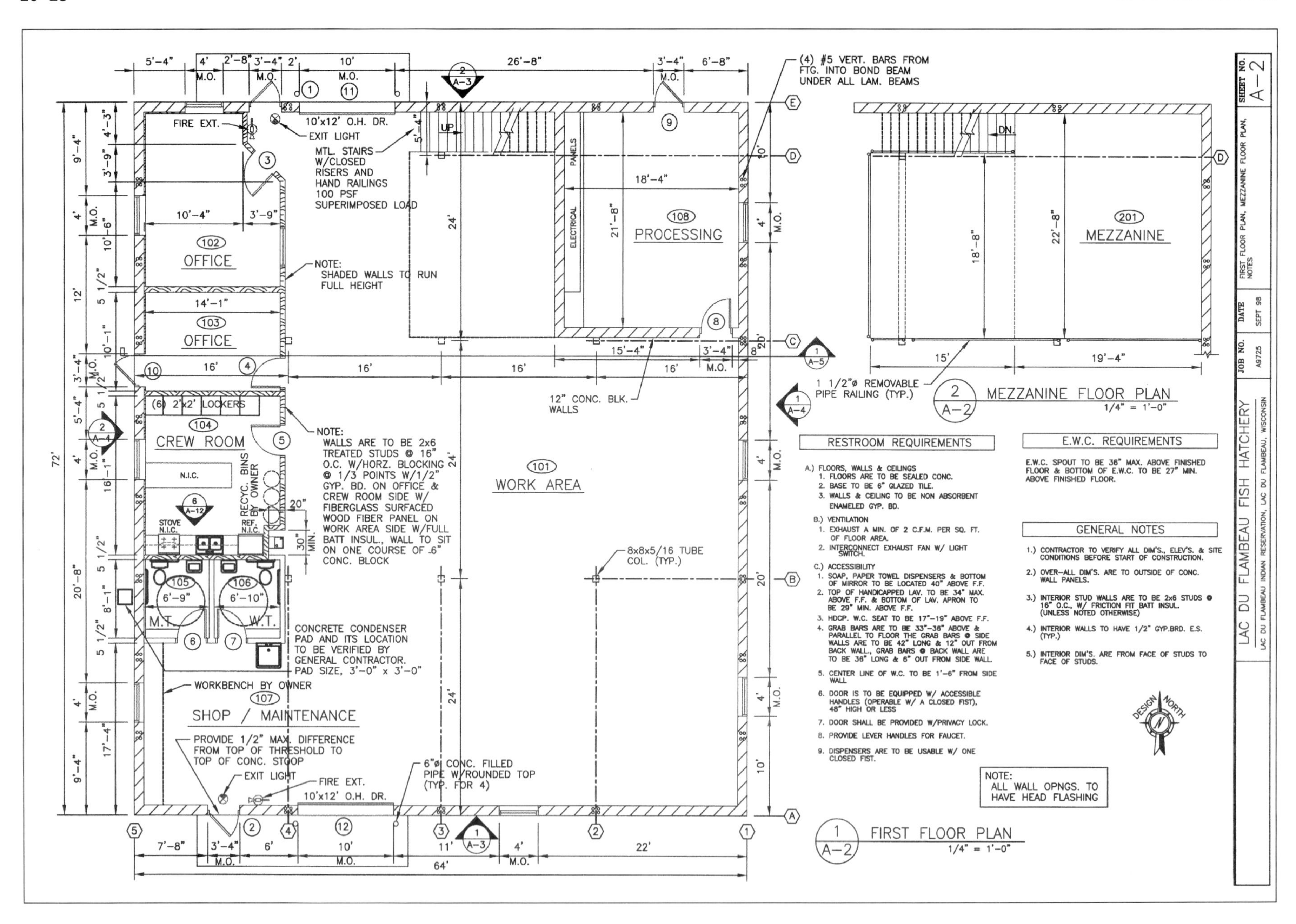

(4) #5 VERT. BARS FROM FTG. INTO BOND BEAM UNDER ALL LAM. BEAMS
FIRE EXT.
10'x12' O.H. DR.
EXIT LIGHT
MTL. STAIRS W/CLOSED RISERS AND HAND RAILINGS 100 PSF SUPERIMPOSED LOAD
UP
DN.
PANELS
ELECTRICAL
108 PROCESSING
201 MEZZANINE
102 OFFICE
103 OFFICE
NOTE: SHADED WALLS TO RUN FULL HEIGHT
(6) 2'x2' LOCKERS
104 CREW ROOM
N.I.C.
RECYC. BINS BY OWNER
STOVE N.I.C.
REF. N.I.C.
NOTE: WALLS ARE TO BE 2x6 TREATED STUDS @ 16" O.C. W/HORZ. BLOCKING @ 1/3 POINTS W/1/2" GYP. BD. ON OFFICE & CREW ROOM SIDE W/ FIBERGLASS SURFACED WOOD FIBER PANEL ON WORK AREA SIDE W/FULL BATT INSUL., WALL TO SIT ON ONE COURSE OF 6" CONC. BLOCK
12" CONC. BLK. WALLS
101 WORK AREA
8x8x5/16 TUBE COL. (TYP.)
105 M.T.
106 W.T.
CONCRETE CONDENSER PAD AND ITS LOCATION TO BE VERIFIED BY GENERAL CONTRACTOR. PAD SIZE, 3'-0" x 3'-0"
WORKBENCH BY OWNER
107 SHOP / MAINTENANCE
PROVIDE 1/2" MAX. DIFFERENCE FROM TOP OF THRESHOLD TO TOP OF CONC. STOOP
EXIT LIGHT
FIRE EXT.
6"ø CONC. FILLED PIPE W/ROUNDED TOP (TYP. FOR 4)
M.O.
1 1/2"ø REMOVABLE PIPE RAILING (TYP.)
2 A-2 MEZZANINE FLOOR PLAN 1/4" = 1'-0"
1 A-2 FIRST FLOOR PLAN 1/4" = 1'-0"
RESTROOM REQUIREMENTS
A.) FLOORS, WALLS & CEILINGS
1. FLOORS ARE TO BE SEALED CONC.
2. BASE TO BE 6" GLAZED TILE.
3. WALLS & CEILING TO BE NON ABSORBENT ENAMELED GYP. BD.
B.) VENTILATION
1. EXHAUST A MIN. OF 2 C.F.M. PER SQ. FT. OF FLOOR AREA.
2. INTERCONNECT EXHAUST FAN W/ LIGHT SWITCH.
C.) ACCESSIBILITY
1. SOAP, PAPER TOWEL DISPENSERS & BOTTOM OF MIRROR TO BE LOCATED 40" ABOVE F.F.
2. TOP OF HANDICAPPED LAV. TO BE 34" MAX. ABOVE F.F. & BOTTOM OF LAV. APRON TO BE 29" MIN. ABOVE F.F.
3. HDCP. W.C. SEAT TO BE 17"-19" ABOVE F.F.
4. GRAB BARS ARE TO BE 33"-36" ABOVE & PARALLEL TO FLOOR THE GRAB BARS @ SIDE WALLS ARE TO BE 42" LONG & 12" OUT FROM BACK WALL., GRAB BARS @ BACK WALL ARE TO BE 36" LONG & 6" OUT FROM SIDE WALL.
5. CENTER LINE OF W.C. TO BE 1'-6" FROM SIDE WALL
6. DOOR IS TO BE EQUIPPED W/ ACCESSIBLE HANDLES (OPERABLE W/ A CLOSED FIST), 48" HIGH OR LESS
7. DOOR SHALL BE PROVIDED W/PRIVACY LOCK.
8. PROVIDE LEVER HANDLES FOR FAUCET.
9. DISPENSERS ARE TO BE USABLE W/ ONE CLOSED FIST.
E.W.C. REQUIREMENTS
E.W.C. SPOUT TO BE 36" MAX. ABOVE FINISHED FLOOR & BOTTOM OF E.W.C. TO BE 27" MIN. ABOVE FINISHED FLOOR.
GENERAL NOTES
1.) CONTRACTOR TO VERIFY ALL DIM'S., ELEV'S. & SITE CONDITIONS BEFORE START OF CONSTRUCTION.
2.) OVER-ALL DIM'S. ARE TO OUTSIDE OF CONC. WALL PANELS.
3.) INTERIOR STUD WALLS ARE TO BE 2x6 STUDS @ 16" O.C., W/ FRICTION FIT BATT INSUL. (UNLESS NOTED OTHERWISE)
4.) INTERIOR WALLS TO HAVE 1/2" GYP.BRD. E.S. (TYP.)
5.) INTERIOR DIM'S. ARE FROM FACE OF STUDS TO FACE OF STUDS.
DESIGN NORTH
NOTE: ALL WALL OPNGS. TO HAVE HEAD FLASHING
LAC DU FLAMBEAU FISH HATCHERY
LAC DU FLAMBEAU INDIAN RESERVATION, LAC DU FLAMBEAU, WISCONSIN
JOB NO. A9725
DATE SEPT 98
FIRST FLOOR PLAN, MEZZANINE FLOOR PLAN, NOTES
SHEET NO. A-2

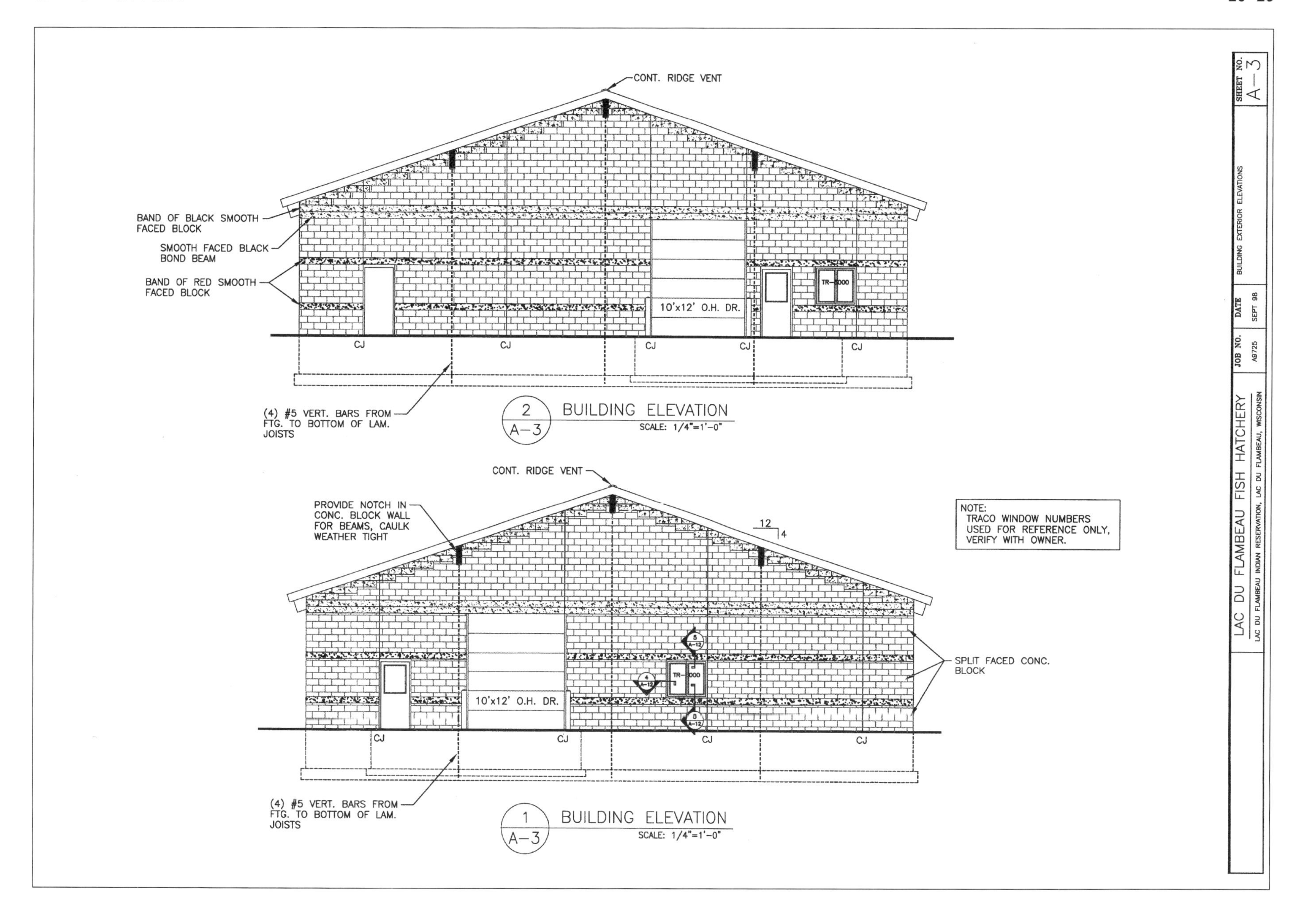
CONT. RIDGE VENT
BAND OF BLACK SMOOTH FACED BLOCK
SMOOTH FACED BLACK BOND BEAM
BAND OF RED SMOOTH FACED BLOCK
TR-3000
10'x12' O.H. DR.
CJ
(4) #5 VERT. BARS FROM FTG. TO BOTTOM OF LAM. JOISTS
2
A-3
BUILDING ELEVATION
SCALE: 1/4"=1'-0"
CONT. RIDGE VENT
PROVIDE NOTCH IN CONC. BLOCK WALL FOR BEAMS, CAULK WEATHER TIGHT
12
4
NOTE:
TRACO WINDOW NUMBERS USED FOR REFERENCE ONLY, VERIFY WITH OWNER.
SPLIT FACED CONC. BLOCK
10'x12' O.H. DR.
(4) #5 VERT. BARS FROM FTG. TO BOTTOM OF LAM. JOISTS
1
A-3
BUILDING ELEVATION
SCALE: 1/4"=1'-0"
SHEET NO.
A-3
BUILDING EXTERIOR ELEVATIONS
DATE
SEPT 98
JOB NO.
A9725
LAC DU FLAMBEAU FISH HATCHERY
LAC DU FLAMBEAU INDIAN RESERVATION, LAC DU FLAMBEAU, WISCONSIN

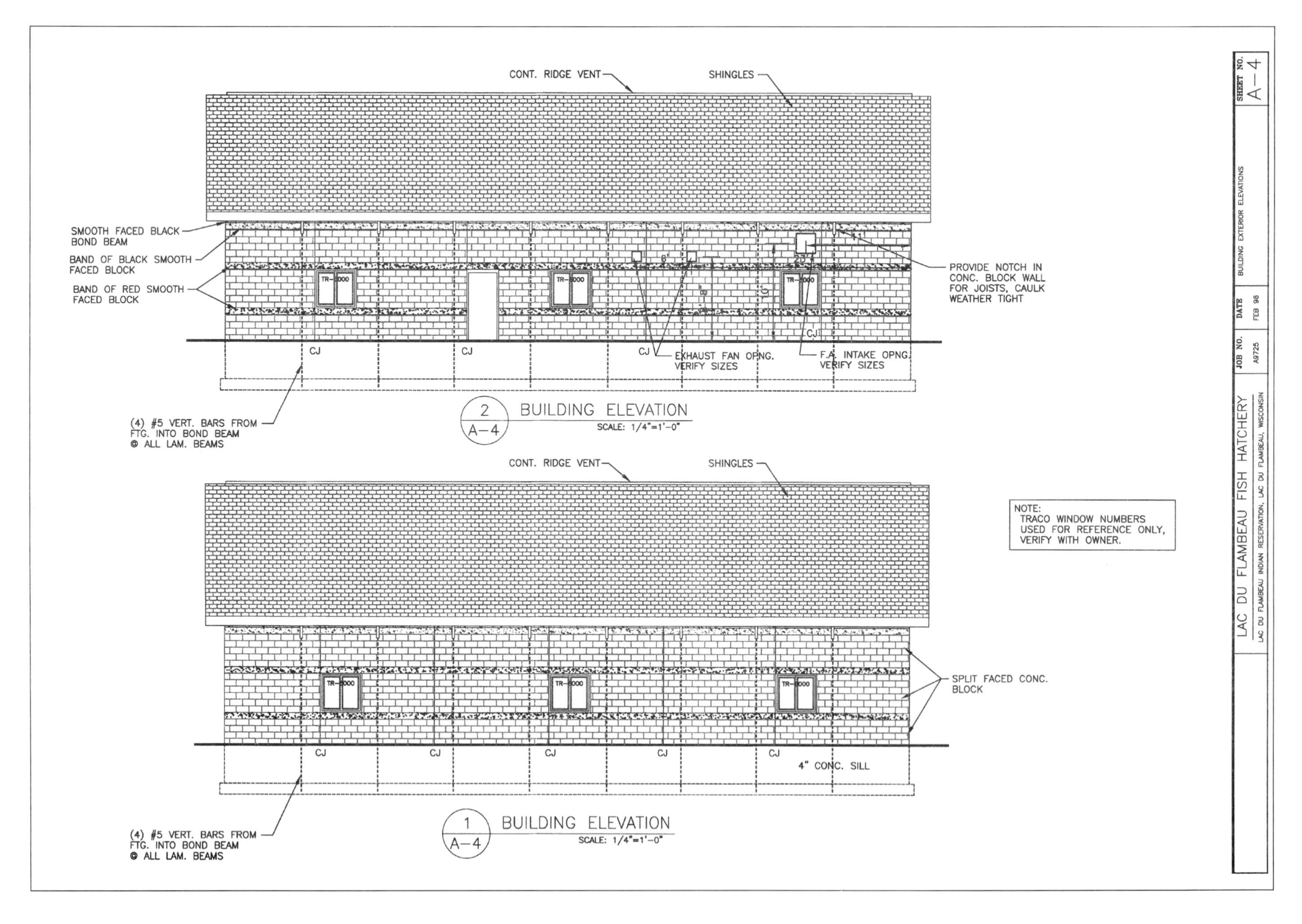

CONT. RIDGE VENT
SHINGLES
SMOOTH FACED BLACK BOND BEAM
BAND OF BLACK SMOOTH FACED BLOCK
BAND OF RED SMOOTH FACED BLOCK
TR-5000
PROVIDE NOTCH IN CONC. BLOCK WALL FOR JOISTS, CAULK WEATHER TIGHT
CJ
EXHAUST FAN OPNG. VERIFY SIZES
F.A. INTAKE OPNG. VERIFY SIZES
(4) #5 VERT. BARS FROM FTG. INTO BOND BEAM @ ALL LAM. BEAMS
2 A-4 BUILDING ELEVATION
SCALE: 1/4"=1'-0"
NOTE: TRACO WINDOW NUMBERS USED FOR REFERENCE ONLY, VERIFY WITH OWNER.
SPLIT FACED CONC. BLOCK
4" CONC. SILL
1 A-4 BUILDING ELEVATION
SCALE: 1/4"=1'-0"
SHEET NO. A-4
BUILDING EXTERIOR ELEVATIONS
DATE FEB 98
JOB NO. A9725
LAC DU FLAMBEAU FISH HATCHERY
LAC DU FLAMBEAU INDIAN RESERVATION, LAC DU FLAMBEAU, WISCONSIN

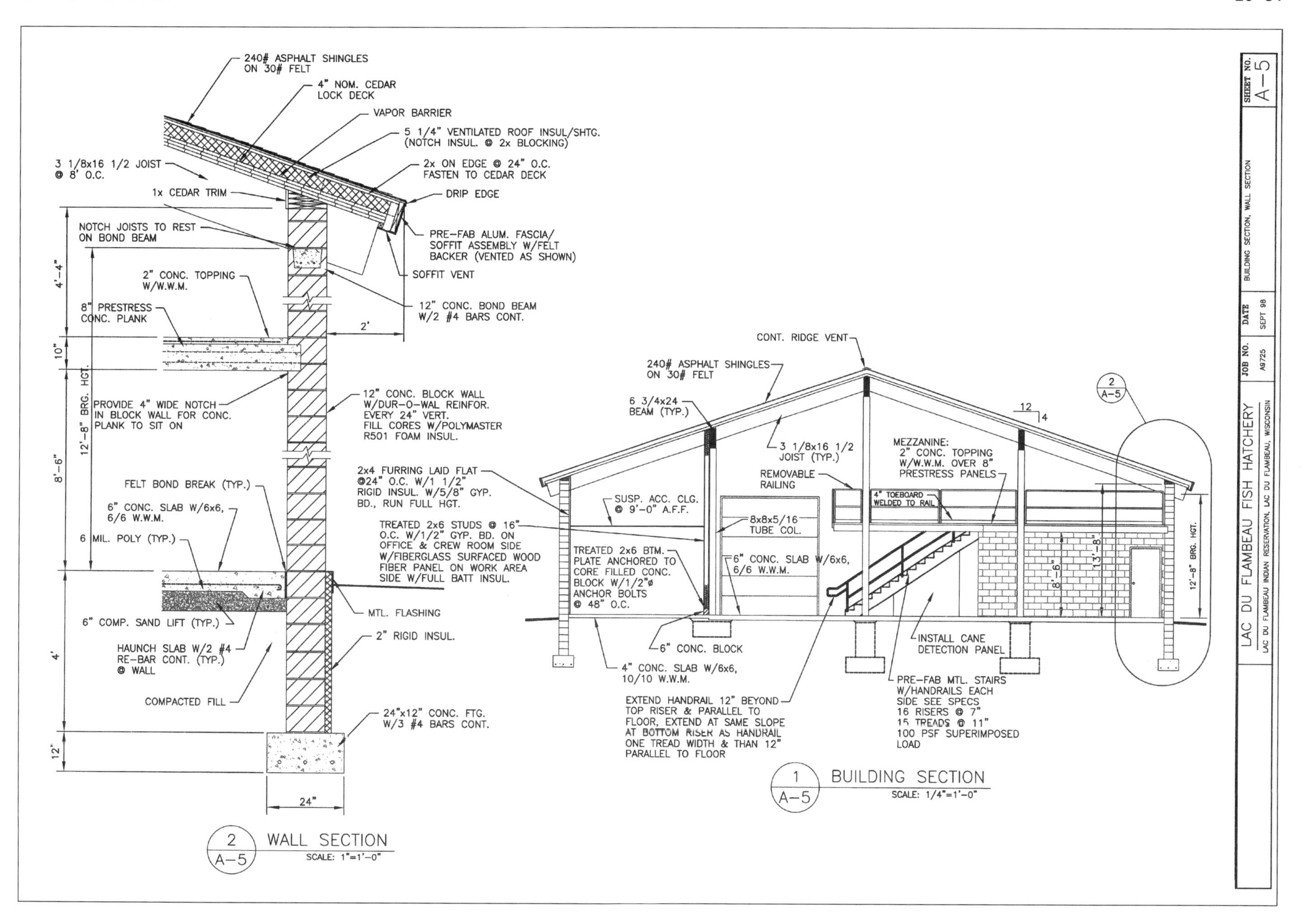
240# ASPHALT SHINGLES ON 30# FELT
4" NOM. CEDAR LOCK DECK
VAPOR BARRIER
5 1/4" VENTILATED ROOF INSUL/SHTG. (NOTCH INSUL. @ 2x BLOCKING)
3 1/8x16 1/2 JOIST @ 8' O.C.
2x ON EDGE @ 24" O.C. FASTEN TO CEDAR DECK
1x CEDAR TRIM
DRIP EDGE
NOTCH JOISTS TO REST ON BOND BEAM
PRE-FAB ALUM. FASCIA/ SOFFIT ASSEMBLY W/FELT BACKER (VENTED AS SHOWN)
2" CONC. TOPPING W/W.W.M.
SOFFIT VENT
8" PRESTRESS CONC. PLANK
12" CONC. BOND BEAM W/2 #4 BARS CONT.
2'
4'-4"
10"
12'-8" BRG. HGT.
8'-6"
PROVIDE 4" WIDE NOTCH IN BLOCK WALL FOR CONC. PLANK TO SIT ON
12" CONC. BLOCK WALL W/DUR-O-WAL REINFOR. EVERY 24" VERT. FILL CORES W/POLYMASTER R501 FOAM INSUL.
2x4 FURRING LAID FLAT @24" O.C. W/1 1/2" RIGID INSUL. W/5/8" GYP. BD., RUN FULL HGT.
FELT BOND BREAK (TYP.)
6" CONC. SLAB W/6x6, 6/6 W.W.M.
TREATED 2x6 STUDS @ 16" O.C. W/1/2" GYP. BD. ON OFFICE & CREW ROOM SIDE W/FIBERGLASS SURFACED WOOD FIBER PANEL ON WORK AREA SIDE W/FULL BATT INSUL.
6 MIL. POLY (TYP.)
MTL. FLASHING
6" COMP. SAND LIFT (TYP.)
2" RIGID INSUL.
HAUNCH SLAB W/2 #4 RE-BAR CONT. (TYP.) @ WALL
4'
COMPACTED FILL
24"x12" CONC. FTG. W/3 #4 BARS CONT.
12"
24"
2 A-5 WALL SECTION
SCALE: 1"=1'-0"
CONT. RIDGE VENT
240# ASPHALT SHINGLES ON 30# FELT
6 3/4x24 BEAM (TYP.)
3 1/8x16 1/2 JOIST (TYP.)
12
4
2 A-5
MEZZANINE: 2" CONC. TOPPING W/W.W.M. OVER 8" PRESTRESS PANELS
REMOVABLE RAILING
4" TOEBOARD WELDED TO RAIL
SUSP. ACC. CLG. @ 9'-0" A.F.F.
8x8x5/16 TUBE COL.
TREATED 2x6 BTM. PLATE ANCHORED TO CORE FILLED CONC. BLOCK W/1/2"ø ANCHOR BOLTS @ 48" O.C.
6" CONC. SLAB W/6x6, 6/6 W.W.M.
8'-6"
13'-8"
12'-8" BRG. HGT.
6" CONC. BLOCK
INSTALL CANE DETECTION PANEL
4" CONC. SLAB W/6x6, 10/10 W.W.M.
EXTEND HANDRAIL 12" BEYOND TOP RISER & PARALLEL TO FLOOR, EXTEND AT SAME SLOPE AT BOTTOM RISER AS HANDRAIL ONE TREAD WIDTH & THAN 12" PARALLEL TO FLOOR
PRE-FAB MTL. STAIRS W/HANDRAILS EACH SIDE SEE SPECS 16 RISERS @ 7" 15 TREADS @ 11" 100 PSF SUPERIMPOSED LOAD
1 A-5 BUILDING SECTION
SCALE: 1/4"=1'-0"
SHEET NO. A-5
BUILDING SECTION, WALL SECTION
DATE SEPT 98
JOB NO. A9725
LAC DU FLAMBEAU FISH HATCHERY
LAC DU FLAMBEAU INDIAN RESERVATION, LAC DU FLAMBEAU, WISCONSIN

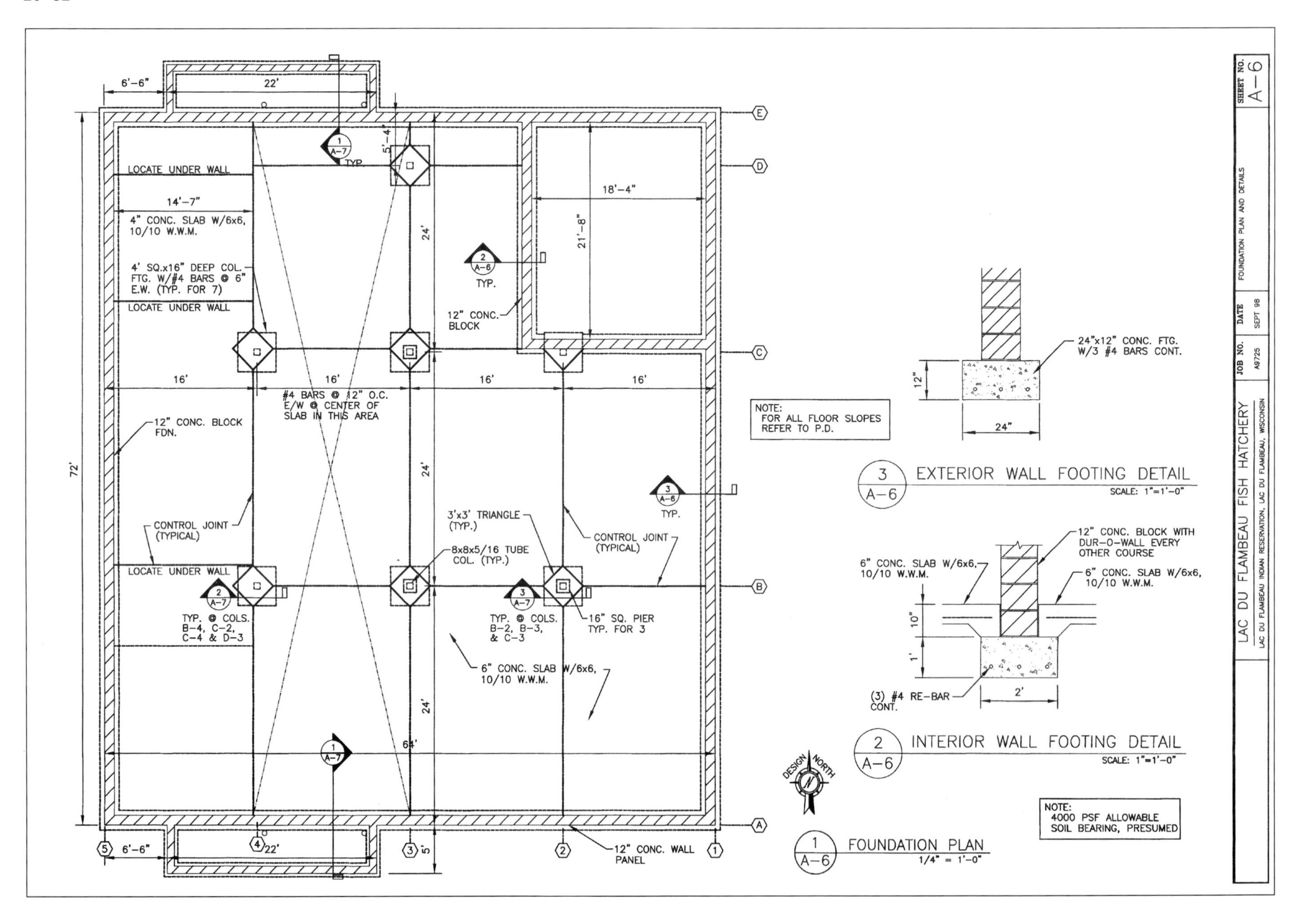

6'-6"
22'
LOCATE UNDER WALL
14'-7"
4" CONC. SLAB W/6x6, 10/10 W.W.M.
4' SQ.x16" DEEP COL. FTG. W/#4 BARS @ 6" E.W. (TYP. FOR 7)
LOCATE UNDER WALL
18'-4"
21'-8"
24'
12" CONC. BLOCK
16'
#4 BARS @ 12" O.C. E/W @ CENTER OF SLAB IN THIS AREA
12" CONC. BLOCK FDN.
72'
CONTROL JOINT (TYPICAL)
3'x3' TRIANGLE (TYP.)
CONTROL JOINT (TYPICAL)
8x8x5/16 TUBE COL. (TYP.)
LOCATE UNDER WALL
TYP. @ COLS. B-4, C-2, C-4 & D-3
TYP. @ COLS. B-2, B-3, & C-3
16" SQ. PIER TYP. FOR 3
6" CONC. SLAB W/6x6, 10/10 W.W.M.
64'
12" CONC. WALL PANEL
NOTE: FOR ALL FLOOR SLOPES REFER TO P.D.
24"x12" CONC. FTG. W/3 #4 BARS CONT.
3 A-6 EXTERIOR WALL FOOTING DETAIL
SCALE: 1"=1'-0"
12" CONC. BLOCK WITH DUR-O-WALL EVERY OTHER COURSE
6" CONC. SLAB W/6x6, 10/10 W.W.M.
6" CONC. SLAB W/6x6, 10/10 W.W.M.
(3) #4 RE-BAR CONT.
2 A-6 INTERIOR WALL FOOTING DETAIL
SCALE: 1"=1'-0"
DESIGN NORTH
NOTE: 4000 PSF ALLOWABLE SOIL BEARING, PRESUMED
1 A-6 FOUNDATION PLAN
1/4" = 1'-0"
SHEET NO. A-6
FOUNDATION PLAN AND DETAILS
DATE SEPT 98
JOB NO. A9725
LAC DU FLAMBEAU FISH HATCHERY
LAC DU FLAMBEAU INDIAN RESERVATION, LAC DU FLAMBEAU, WISCONSIN

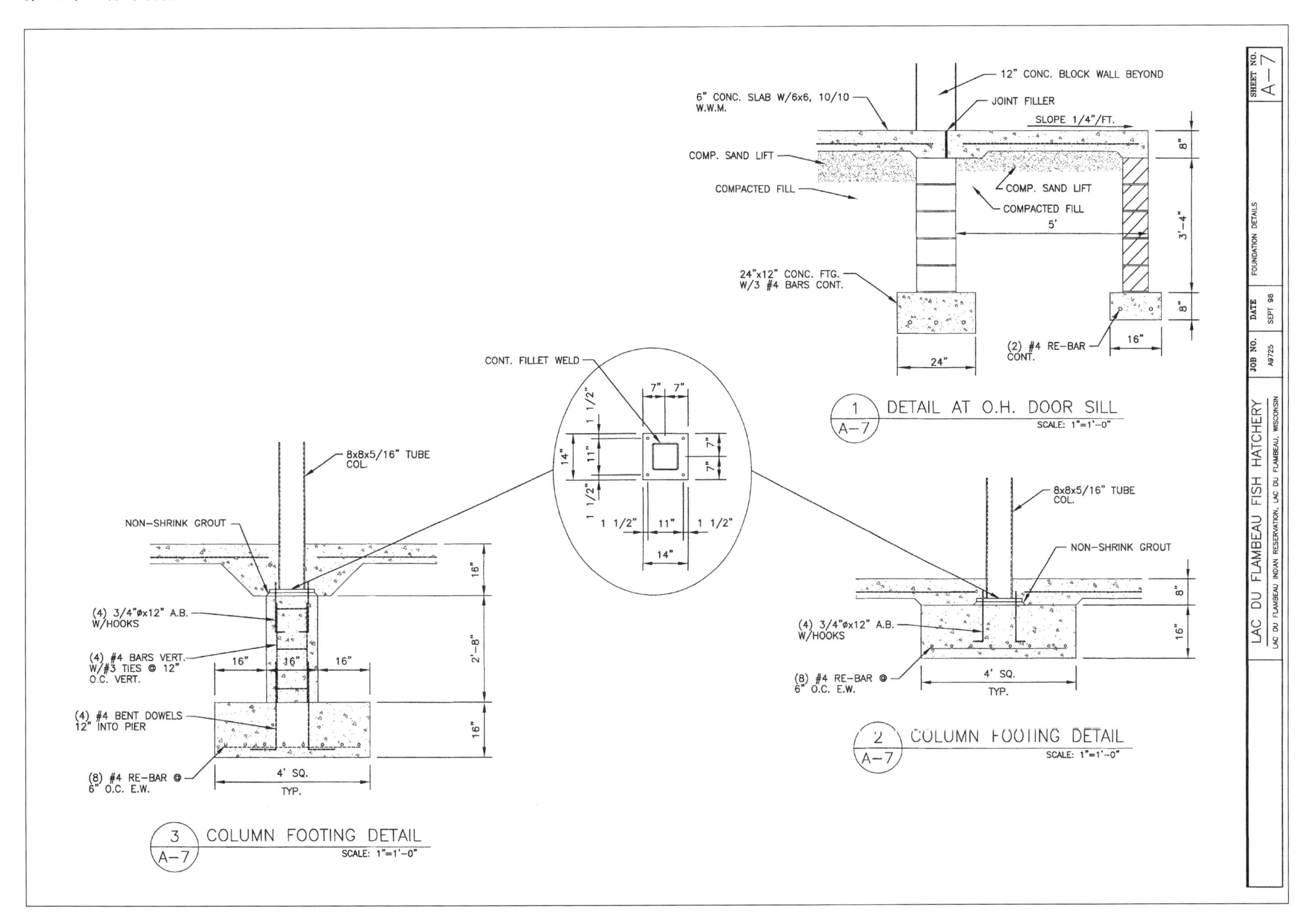
12" CONC. BLOCK WALL BEYOND
6" CONC. SLAB W/6x6, 10/10 W.W.M.
JOINT FILLER
SLOPE 1/4"/FT.
COMP. SAND LIFT
COMPACTED FILL
COMP. SAND LIFT
COMPACTED FILL
5'
8"
3'-4"
8"
24"x12" CONC. FTG. W/3 #4 BARS CONT.
(2) #4 RE-BAR CONT.
16"
24"
1 A-7 DETAIL AT O.H. DOOR SILL
SCALE: 1"=1'-0"
CONT. FILLET WELD
7" 7"
1 1/2"
11"
14"
1 1/2"
7"
7"
1 1/2" 11" 1 1/2"
14"
8x8x5/16" TUBE COL.
NON-SHRINK GROUT
(4) 3/4"øx12" A.B. W/HOOKS
(4) #4 BARS VERT. W/#3 TIES @ 12" O.C. VERT.
16" 16" 16"
16"
2'-8"
16"
(4) #4 BENT DOWELS 12" INTO PIER
(8) #4 RE-BAR @ 6" O.C. E.W.
4' SQ. TYP.
3 A-7 COLUMN FOOTING DETAIL
SCALE: 1"=1'-0"
8x8x5/16" TUBE COL.
NON-SHRINK GROUT
(4) 3/4"øx12" A.B. W/HOOKS
(8) #4 RE-BAR @ 6" O.C. E.W.
8"
16"
4' SQ. TYP.
2 A-7 COLUMN FOOTING DETAIL
SCALE: 1"=1'-0"
SHEET NO. A-7
FOUNDATION DETAILS
DATE SEPT 98
JOB NO. A9725
LAC DU FLAMBEAU FISH HATCHERY
LAC DU FLAMBEAU INDIAN RESERVATION, LAC DU FLAMBEAU, WISCONSIN

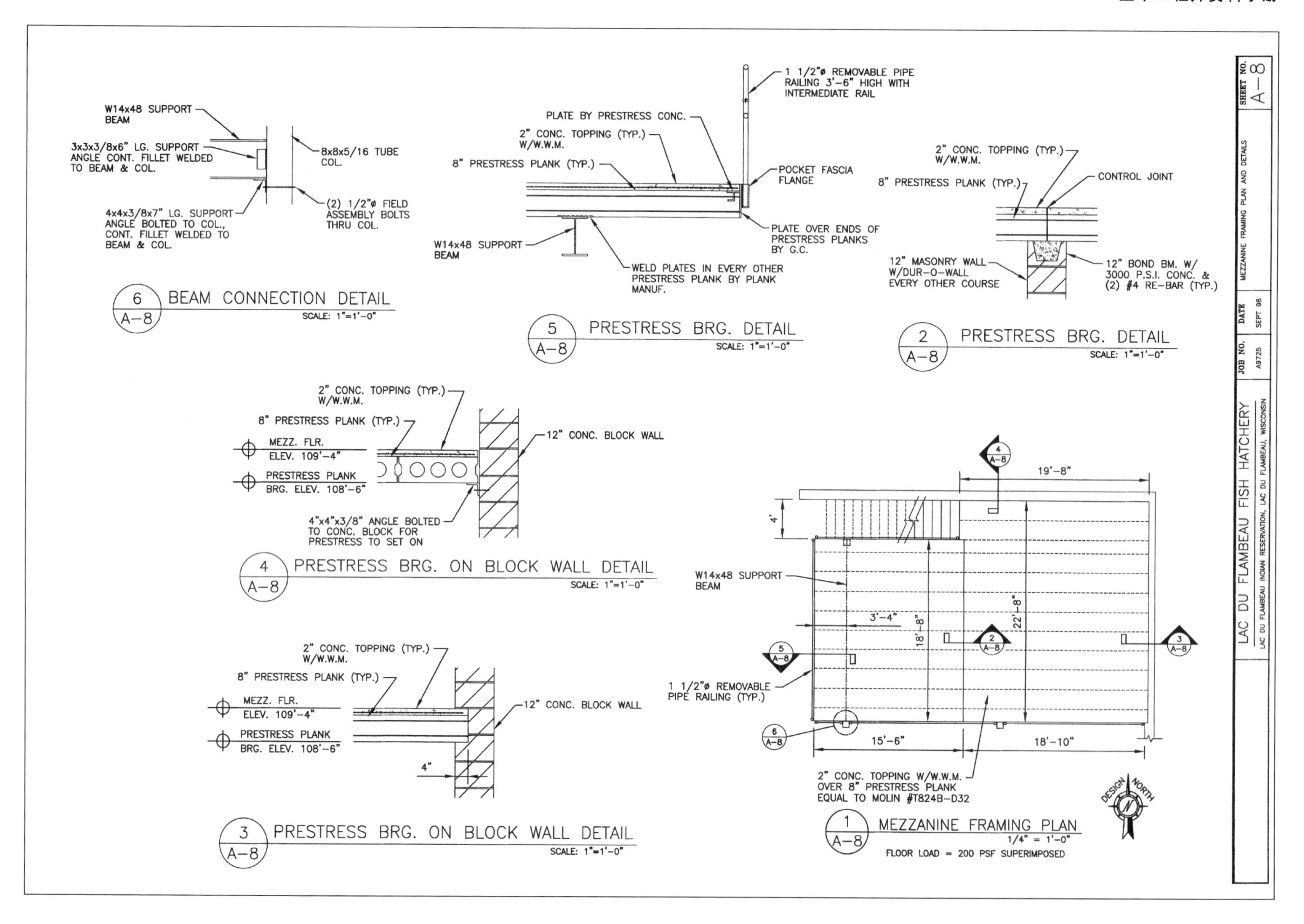

W14x48 SUPPORT BEAM
3x3x3/8x6" LG. SUPPORT ANGLE CONT. FILLET WELDED TO BEAM & COL.
8x8x5/16 TUBE COL.
4x4x3/8x7" LG. SUPPORT ANGLE BOLTED TO COL., CONT. FILLET WELDED TO BEAM & COL.
(2) 1/2"ø FIELD ASSEMBLY BOLTS THRU COL.
6 A-8 BEAM CONNECTION DETAIL
SCALE: 1"=1'-0"
1 1/2"ø REMOVABLE PIPE RAILING 3'-6" HIGH WITH INTERMEDIATE RAIL
PLATE BY PRESTRESS CONC.
2" CONC. TOPPING (TYP.) W/W.W.M.
8" PRESTRESS PLANK (TYP.)
POCKET FASCIA FLANGE
PLATE OVER ENDS OF PRESTRESS PLANKS BY G.C.
W14x48 SUPPORT BEAM
WELD PLATES IN EVERY OTHER PRESTRESS PLANK BY PLANK MANUF.
5 A-8 PRESTRESS BRG. DETAIL
SCALE: 1"=1'-0"
2" CONC. TOPPING (TYP.) W/W.W.M.
8" PRESTRESS PLANK (TYP.)
CONTROL JOINT
12" MASONRY WALL W/DUR-O-WALL EVERY OTHER COURSE
12" BOND BM. W/ 3000 P.S.I. CONC. & (2) #4 RE-BAR (TYP.)
2 A-8 PRESTRESS BRG. DETAIL
SCALE: 1"=1'-0"
2" CONC. TOPPING (TYP.) W/W.W.M.
8" PRESTRESS PLANK (TYP.)
MEZZ. FLR. ELEV. 109'-4"
PRESTRESS PLANK BRG. ELEV. 108'-6"
12" CONC. BLOCK WALL
4"x4"x3/8" ANGLE BOLTED TO CONC. BLOCK FOR PRESTRESS TO SET ON
4 A-8 PRESTRESS BRG. ON BLOCK WALL DETAIL
SCALE: 1"=1'-0"
2" CONC. TOPPING (TYP.) W/W.W.M.
8" PRESTRESS PLANK (TYP.)
MEZZ. FLR. ELEV. 109'-4"
PRESTRESS PLANK BRG. ELEV. 108'-6"
12" CONC. BLOCK WALL
4"
3 A-8 PRESTRESS BRG. ON BLOCK WALL DETAIL
SCALE: 1"=1'-0"
19'-8"
4'
W14x48 SUPPORT BEAM
3'-4"
18'-8"
22'-8"
1 1/2"ø REMOVABLE PIPE RAILING (TYP.)
15'-6"
18'-10"
2" CONC. TOPPING W/W.W.M. OVER 8" PRESTRESS PLANK EQUAL TO MOLIN #T824B-D32
DESIGN NORTH
1 A-8 MEZZANINE FRAMING PLAN
1/4" = 1'-0"
FLOOR LOAD = 200 PSF SUPERIMPOSED
SHEET NO. A-8
MEZZANINE FRAMING PLAN AND DETAILS
DATE SEPT 98
JOB NO. A9725
LAC DU FLAMBEAU FISH HATCHERY
LAC DU FLAMBEAU INDIAN RESERVATION, LAC DU FLAMBEAU, WISCONSIN

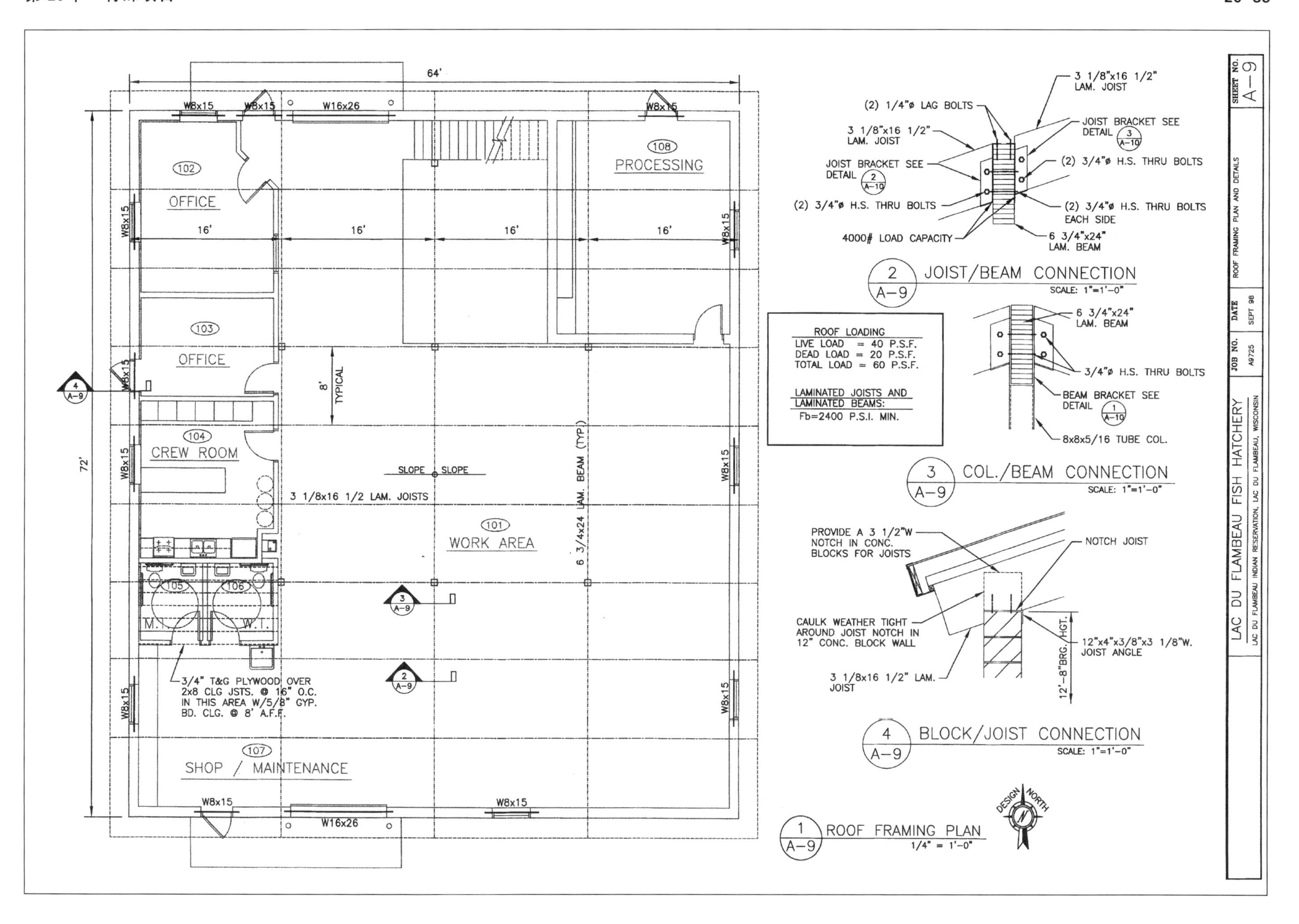

64'
72'
W8x15
W16x26
102
OFFICE
103
OFFICE
104
CREW ROOM
105
106
M.T.
W.T.
101
WORK AREA
107
SHOP / MAINTENANCE
108
PROCESSING
16'
8' TYPICAL
SLOPE SLOPE
3 1/8x16 1/2 LAM. JOISTS
6 3/4x24 LAM. BEAM (TYP)
3/4" T&G PLYWOOD OVER 2x8 CLG JSTS. @ 16" O.C. IN THIS AREA W/5/8" GYP. BD. CLG. @ 8' A.F.F.
ROOF LOADING
LIVE LOAD = 40 P.S.F.
DEAD LOAD = 20 P.S.F.
TOTAL LOAD = 60 P.S.F.
LAMINATED JOISTS AND LAMINATED BEAMS:
Fb=2400 P.S.I. MIN.
3 1/8"x16 1/2" LAM. JOIST
(2) 1/4"ø LAG BOLTS
JOIST BRACKET SEE DETAIL 3 A-10
JOIST BRACKET SEE DETAIL 2 A-10
(2) 3/4"ø H.S. THRU BOLTS
(2) 3/4"ø H.S. THRU BOLTS EACH SIDE
4000# LOAD CAPACITY
6 3/4"x24" LAM. BEAM
2 A-9 JOIST/BEAM CONNECTION
SCALE: 1"=1'-0"
3/4"ø H.S. THRU BOLTS
BEAM BRACKET SEE DETAIL 1 A-10
8x8x5/16 TUBE COL.
3 A-9 COL./BEAM CONNECTION
SCALE: 1"=1'-0"
PROVIDE A 3 1/2"W NOTCH IN CONC. BLOCKS FOR JOISTS
NOTCH JOIST
CAULK WEATHER TIGHT AROUND JOIST NOTCH IN 12" CONC. BLOCK WALL
3 1/8x16 1/2" LAM. JOIST
12"x4"x3/8"x3 1/8"W. JOIST ANGLE
12'-8"BRG. HGT.
4 A-9 BLOCK/JOIST CONNECTION
SCALE: 1"=1'-0"
1 A-9 ROOF FRAMING PLAN
1/4" = 1'-0"
DESIGN NORTH
SHEET NO. A-9
ROOF FRAMING PLAN AND DETAILS
DATE SEPT 98
JOB NO. A9725
LAC DU FLAMBEAU FISH HATCHERY
LAC DU FLAMBEAU INDIAN RESERVATION, LAC DU FLAMBEAU, WISCONSIN

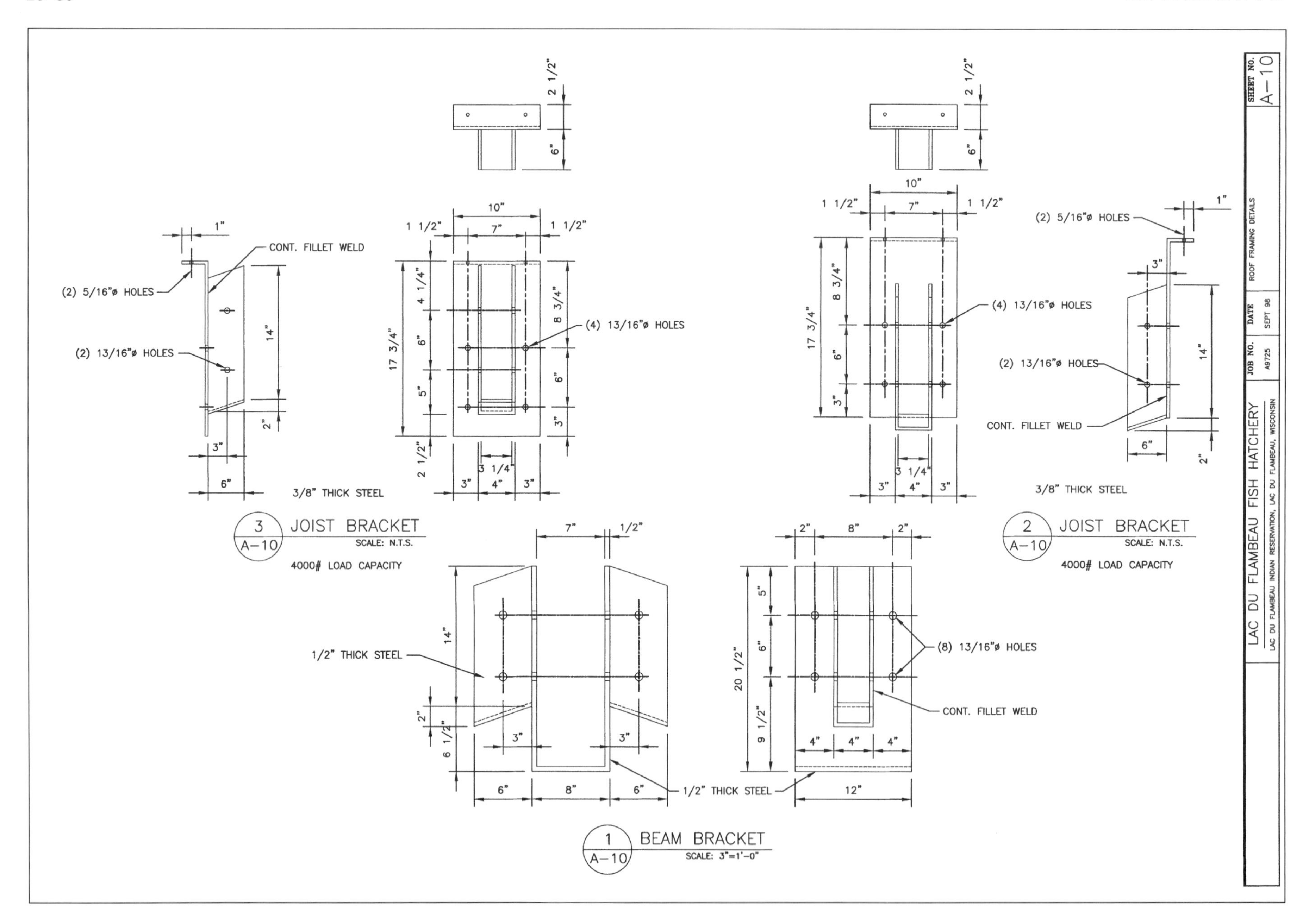

3 JOIST BRACKET
A-10
SCALE: N.T.S.
4000# LOAD CAPACITY
3/8" THICK STEEL
CONT. FILLET WELD
(2) 5/16"ø HOLES
(2) 13/16"ø HOLES
(4) 13/16"ø HOLES
2 JOIST BRACKET
A-10
SCALE: N.T.S.
4000# LOAD CAPACITY
1 BEAM BRACKET
A-10
SCALE: 3"=1'-0"
1/2" THICK STEEL
(8) 13/16"ø HOLES
LAC DU FLAMBEAU FISH HATCHERY
LAC DU FLAMBEAU INDIAN RESERVATION, LAC DU FLAMBEAU, WISCONSIN
JOB NO. A9725
DATE SEPT 98
ROOF FRAMING DETAILS
SHEET NO. A-10

ROOM FINISH SCHEDULE										
ROOM NO.	ROOM NAME	FLOOR	BASE	MATERIALS					CEILING HEIGHT	NOTES
				WALLS				CEILING		
				NORTH	SOUTH	EAST	WEST			
101	WORK AREA	CONC.	–	PCB	PCB	PCB	F.P.	WD. DECK	VARIES	
102	OFFICE	CONC.	–	P.G.B.	P.G.B.	P.G.B.	P.G.B.	SUSP.	9'–0"	
103	OFFICE	CONC.	–	P.G.B.	P.G.B.	P.G.B.	P.G.B.	SUSP.	9'–0"	
104	CREW ROOM	CONC.	–	P.G.B.	P.G.B.	P.G.B.	P.G.B.	SUSP.	9'–0"	
105	MENS TOILET	CONC.	–	P.G.B.	P.G.B.	P.G.B.	P.G.B.	P.G.B.	8'–0"	
106	WOMENS TOILET	CONC.	–	P.G.B.	P.G.B.	P.G.B.	P.G.B.	P.G.B.	8'–0"	
107	SHOP/MAINTENANCE	CONC.	–	F.P.	PCB	–	PCB	WD. DECK	VARIES	
108	PROCESSING	CONC.	–	PCB	PCB	PCB	PCB	CONC. DECK	8'–6"	
201	MEZZANINE	CONC.	–	PCB	–	PCB	–	WD. DECK	VARIES	

ABBREVIATIONS

CONC.– SEALED CONCRETE
PCB– PAINTED CONC. BLOCK
P.G.B.– PAINTED GYPSUM BOARD
F.P.– FIBERGLASS SURFACED WOOD FIBER PANEL

DOOR & FRAME SCHEDULE																
DOOR NO.	DOOR							FRAME						FIRE LABEL	HDWR. GROUP	NOTES
	SIZE			MATL.	TYPE	GLASS	U-CUT OR LOUVER	MATL.	ELEV.	DEPTH	DETAILS					
	WIDTH	HEIGHT	THICK								HEAD	JAMB	SILL			
1	3'–0"	7'–0"	1–3/4"	H.M.	C	A	–	H.M.	2	5 3/4"	4"	2"	–	–	–	2
2	3'–0"	7'–0"	1–3/4"	H.M.	C	A	–	H.M.	2	5 3/4"	4"	2"	–	–	–	2
3	3'–0"	7'–0"	1–3/4"	H.M.	D	B	16"x8"	H.M.	1	5 3/4"	2"	2"	–	–	–	–
4	3'–0"	7'–0"	1–3/4"	H.M.	D	B	16"x8"	H.M.	1	5 3/4"	2"	2"	–	–	–	–
5	3'–0"	7'–0"	1–3/4"	H.M.	D	B	16"x8"	H.M.	1	5 3/4"	2"	2"	–	–	–	–
6	3'–0"	7'–0"	1–3/4"	H.M.	B	–	16"x8"	H.M.	1	5 3/4"	2"	2"	–	–	–	–
7	3'–0"	7'–0"	1–3/4"	H.M.	B	–	16"x8"	H.M.	1	5 3/4"	2"	2"	–	–	–	–
8	3'–0"	7'–0"	1–3/4"	H.M.	A	–	–	H.M.	1	5 3/4"	2"	2"	–	B	–	–
0	3'–0"	7'–0"	1–3/4"	H.M.	A	–	–	H.M.	2	5 3/4"	4"	2"	–	–	–	2
10	3'–0"	7'–0"	1–3/4"	H.M.	A	–	–	H.M.	2	5 3/4"	4"	2"	–	–	–	–
11	10'–0"	12'–0"	–	MTL.	E	–	–	–	–	–	–	–	–	–	–	1
11	10'–0"	12'–0"	–	MTL.	E	–	–	–	–	–	–	–	–	–	–	1

NOTES
1.) PROVIDE AUTOMATIC DOOR OPENER
2.) SEE DETAILS 1 & 2/A-12

GLASS TYPES FOR DOORS
A=1" INSULATED TEMPERED GLASS, CLEAR
B=1/4" SAFETY GLASS, CLEAR

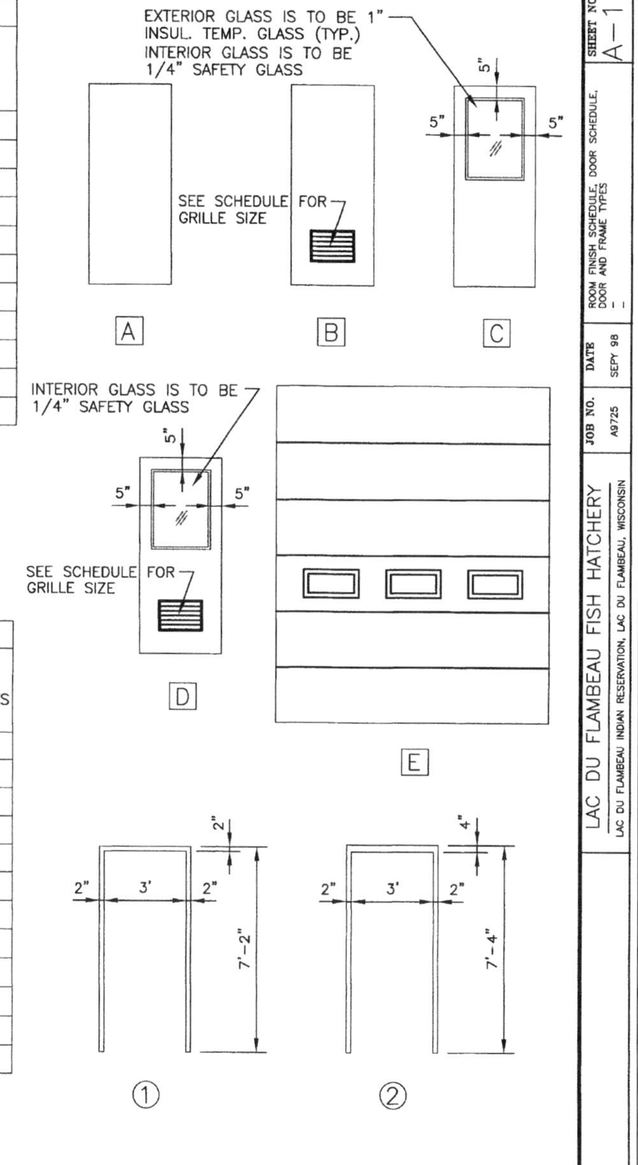

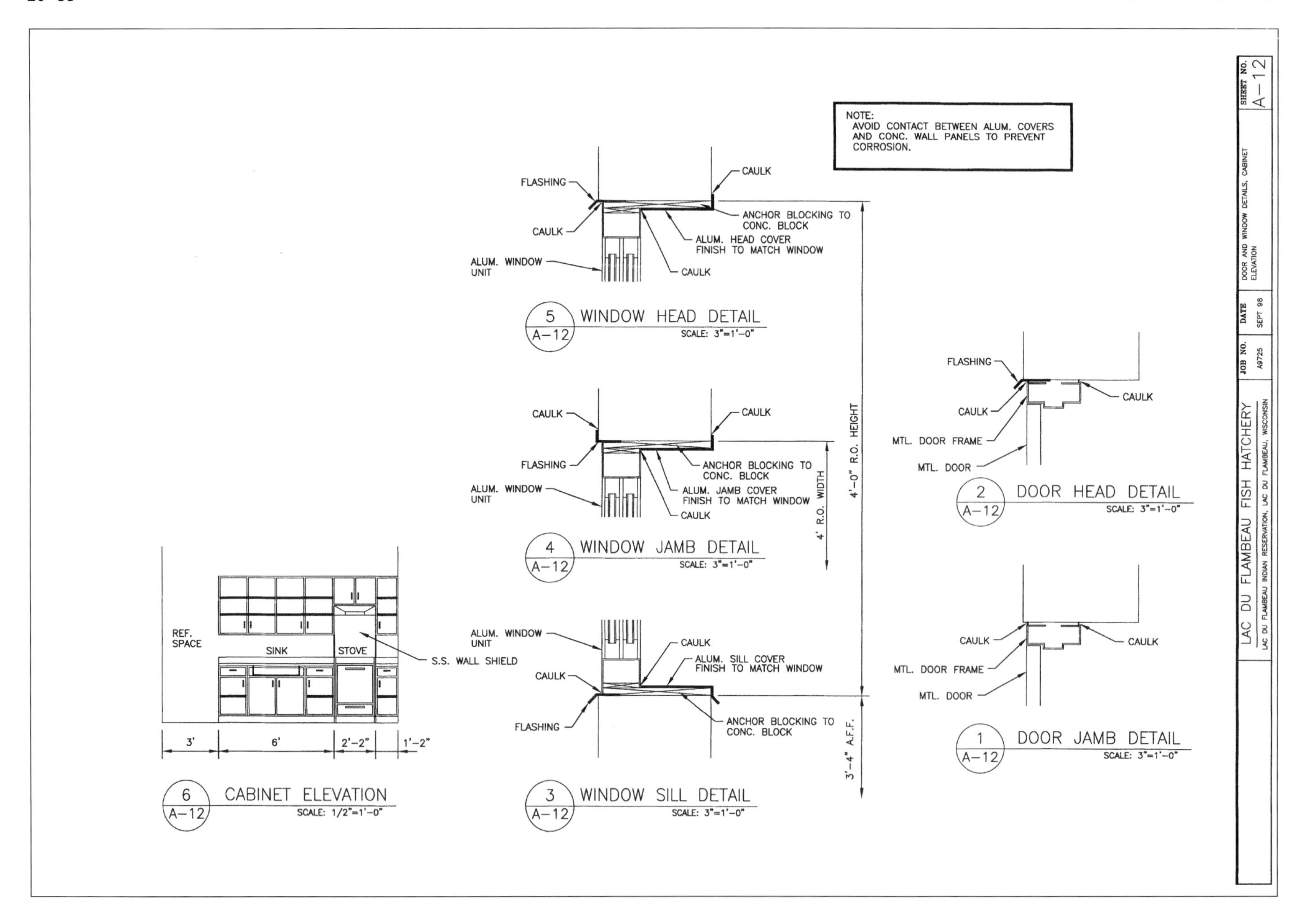
NOTE:
AVOID CONTACT BETWEEN ALUM. COVERS AND CONC. WALL PANELS TO PREVENT CORROSION.
FLASHING
CAULK
ANCHOR BLOCKING TO CONC. BLOCK
ALUM. HEAD COVER FINISH TO MATCH WINDOW
ALUM. WINDOW UNIT
5 A-12 WINDOW HEAD DETAIL
SCALE: 3"=1'-0"
ALUM. JAMB COVER FINISH TO MATCH WINDOW
4 A-12 WINDOW JAMB DETAIL
SCALE: 3"=1'-0"
4' R.O. WIDTH
4'-0" R.O. HEIGHT
3'-4" A.F.F.
ALUM. SILL COVER FINISH TO MATCH WINDOW
3 A-12 WINDOW SILL DETAIL
SCALE: 3"=1'-0"
REF. SPACE
SINK
STOVE
S.S. WALL SHIELD
3'
6'
2'-2"
1'-2"
6 A-12 CABINET ELEVATION
SCALE: 1/2"=1'-0"
MTL. DOOR FRAME
MTL. DOOR
2 A-12 DOOR HEAD DETAIL
SCALE: 3"=1'-0"
1 A-12 DOOR JAMB DETAIL
SCALE: 3"=1'-0"
LAC DU FLAMBEAU FISH HATCHERY
LAC DU FLAMBEAU INDIAN RESERVATION, LAC DU FLAMBEAU, WISCONSIN
JOB NO. A9725
DATE SEPT 98
DOOR AND WINDOW DETAILS, CABINET ELEVATION
SHEET NO. A-12

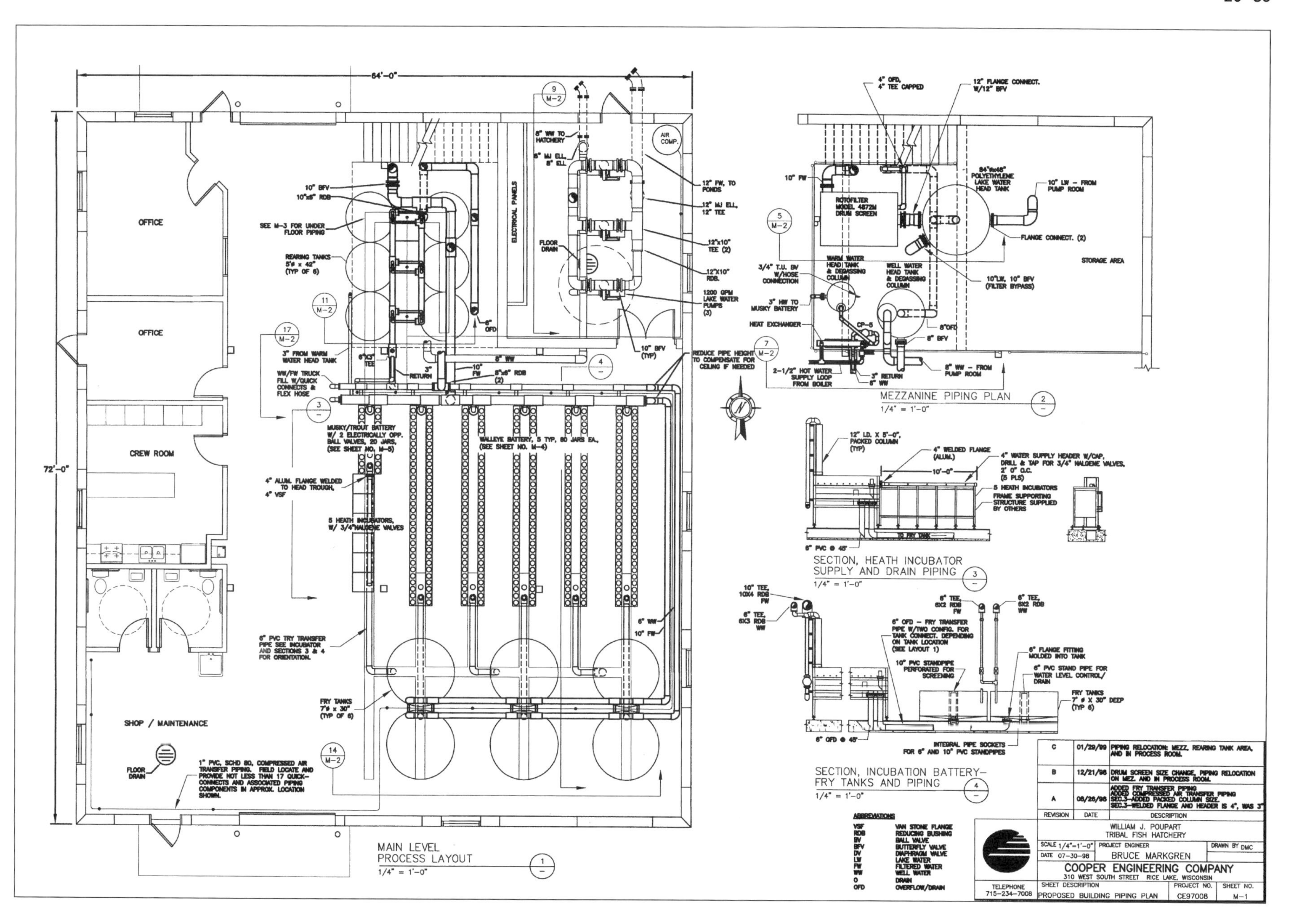
MAIN LEVEL
PROCESS LAYOUT
1/4" = 1'-0"
MEZZANINE PIPING PLAN
1/4" = 1'-0"
SECTION, HEATH INCUBATOR
SUPPLY AND DRAIN PIPING
1/4" = 1'-0"
SECTION, INCUBATION BATTERY-
FRY TANKS AND PIPING
1/4" = 1'-0"
OFFICE
OFFICE
CREW ROOM
SHOP / MAINTENANCE
STORAGE AREA
ELECTRICAL PANELS
64'-0"
72'-0"
WILLIAM J. POUPART
TRIBAL FISH HATCHERY
BRUCE MARKGREN
COOPER ENGINEERING COMPANY
310 WEST SOUTH STREET RICE LAKE, WISCONSIN
TELEPHONE 715-234-7008
PROPOSED BUILDING PIPING PLAN
CE97008
M-1

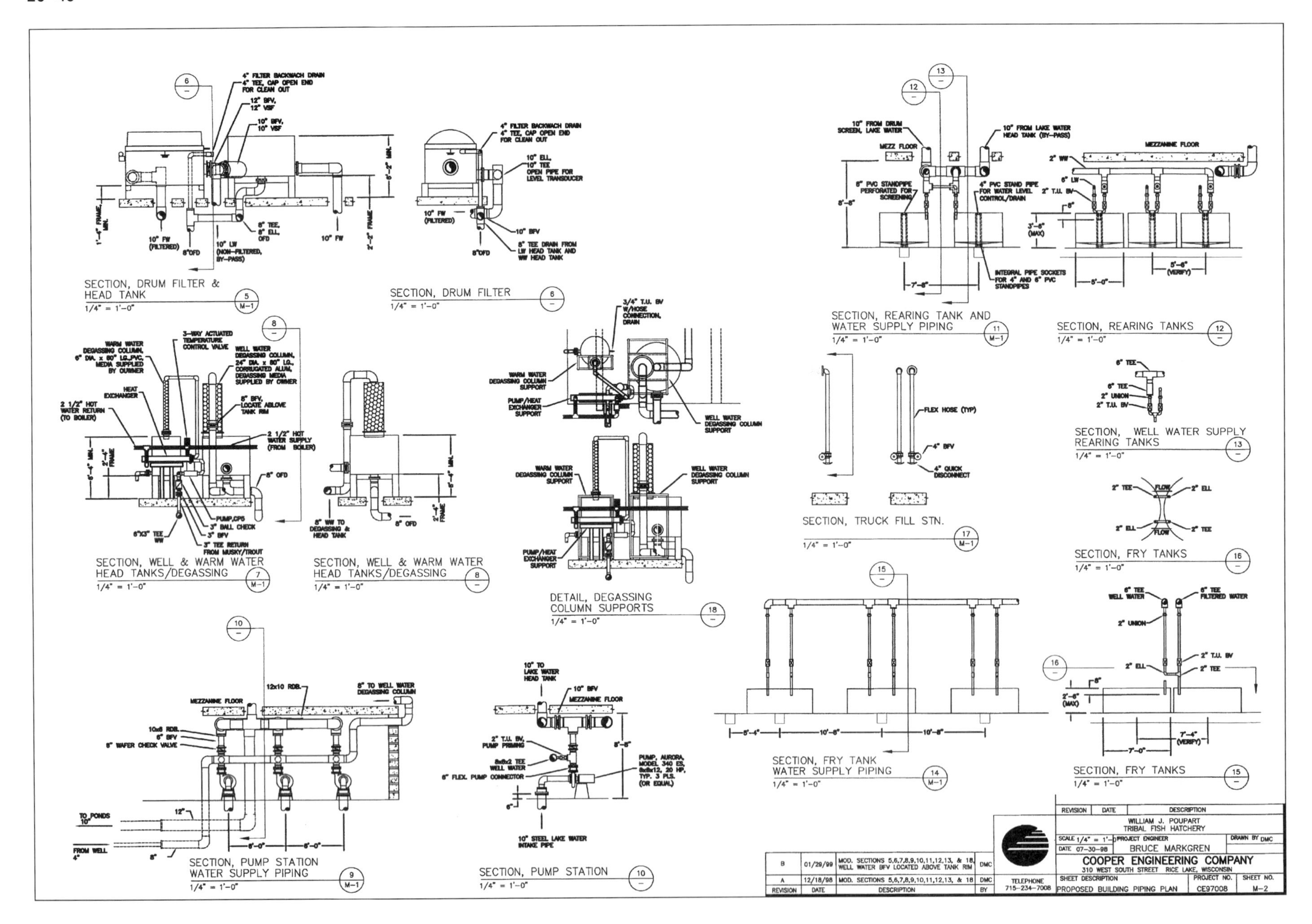
SECTION, DRUM FILTER & HEAD TANK
1/4" = 1'-0"
SECTION, DRUM FILTER
1/4" = 1'-0"
SECTION, REARING TANK AND WATER SUPPLY PIPING
1/4" = 1'-0"
SECTION, REARING TANKS
1/4" = 1'-0"
SECTION, WELL & WARM WATER HEAD TANKS/DEGASSING
1/4" = 1'-0"
SECTION, WELL & WARM WATER HEAD TANKS/DEGASSING
1/4" = 1'-0"
DETAIL, DEGASSING COLUMN SUPPORTS
1/4" = 1'-0"
SECTION, WELL WATER SUPPLY REARING TANKS
1/4" = 1'-0"
SECTION, TRUCK FILL STN.
1/4" = 1'-0"
SECTION, FRY TANKS
1/4" = 1'-0"
SECTION, PUMP STATION WATER SUPPLY PIPING
1/4" = 1'-0"
SECTION, PUMP STATION
1/4" = 1'-0"
SECTION, FRY TANK WATER SUPPLY PIPING
1/4" = 1'-0"
SECTION, FRY TANKS
1/4" = 1'-0"
WILLIAM J. POUPART TRIBAL FISH HATCHERY
BRUCE MARKGREN
COOPER ENGINEERING COMPANY
310 WEST SOUTH STREET RICE LAKE, WISCONSIN
TELEPHONE 715-234-7008
PROPOSED BUILDING PIPING PLAN
CE97008
M-2

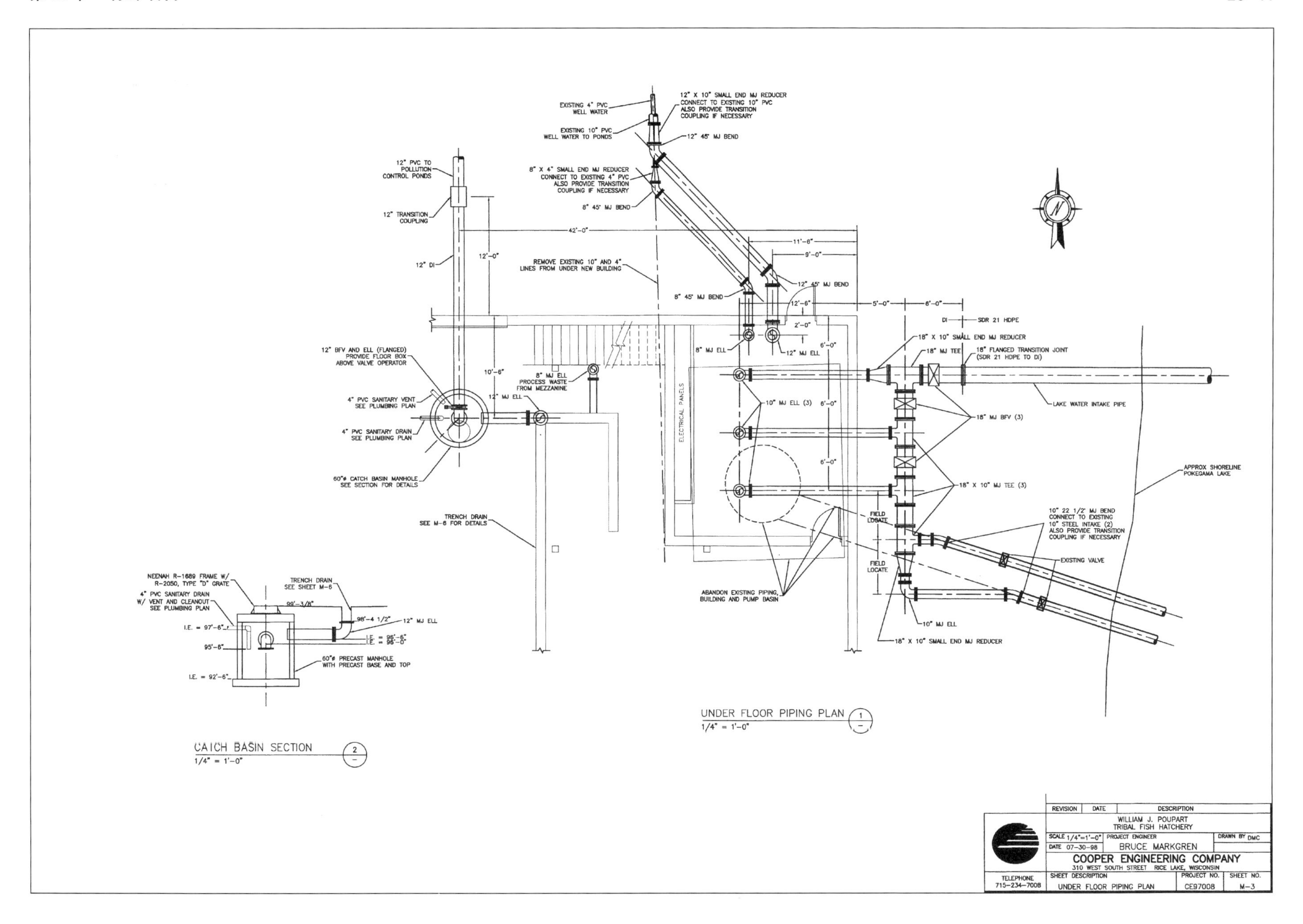
UNDER FLOOR PIPING PLAN
1/4" = 1'-0"
CATCH BASIN SECTION
1/4" = 1'-0"
LAKE WATER INTAKE PIPE
APPROX SHORELINE POKEGAMA LAKE
ABANDON EXISTING PIPING, BUILDING AND PUMP BASIN
ELECTRICAL PANELS
WILLIAM J. POUPART TRIBAL FISH HATCHERY
BRUCE MARKGREN
COOPER ENGINEERING COMPANY
310 WEST SOUTH STREET RICE LAKE, WISCONSIN
TELEPHONE 715-234-7008
UNDER FLOOR PIPING PLAN
CE97008
M-3

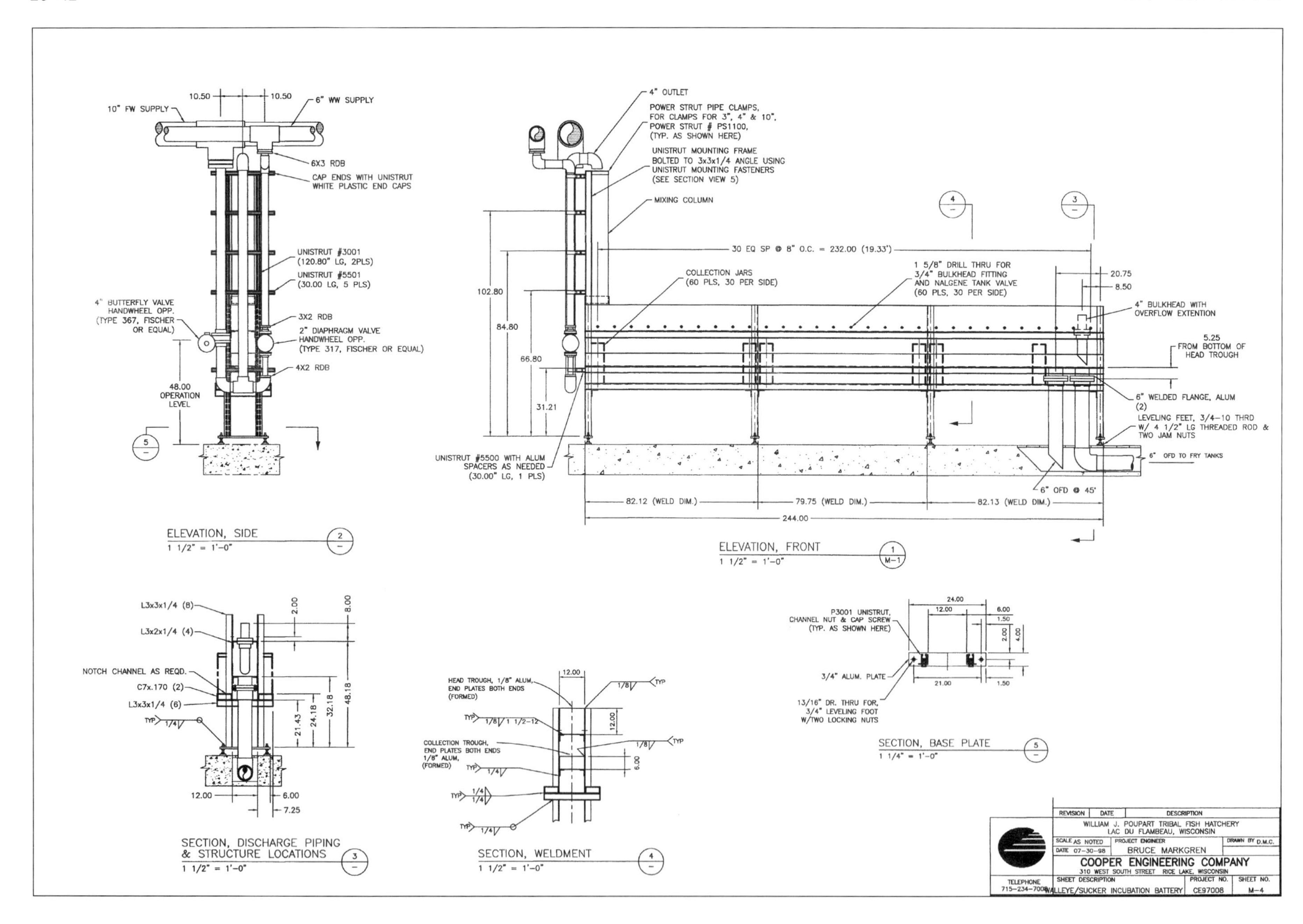

ELEVATION, SIDE
1 1/2" = 1'-0"
ELEVATION, FRONT
1 1/2" = 1'-0"
SECTION, DISCHARGE PIPING & STRUCTURE LOCATIONS
1 1/2" = 1'-0"
SECTION, WELDMENT
1 1/2" = 1'-0"
SECTION, BASE PLATE
1 1/4" = 1'-0"
10" FW SUPPLY
6" WW SUPPLY
6X3 RDB
CAP ENDS WITH UNISTRUT WHITE PLASTIC END CAPS
UNISTRUT #3001 (120.80" LG, 2PLS)
UNISTRUT #5501 (30.00 LG, 5 PLS)
4" BUTTERFLY VALVE HANDWHEEL OPP. (TYPE 367, FISCHER OR EQUAL)
3X2 RDB
2" DIAPHRAGM VALVE HANDWHEEL OPP. (TYPE 317, FISCHER OR EQUAL)
4X2 RDB
48.00 OPERATION LEVEL
4" OUTLET
POWER STRUT PIPE CLAMPS, FOR CLAMPS FOR 3", 4" & 10", POWER STRUT # PS1100, (TYP. AS SHOWN HERE)
UNISTRUT MOUNTING FRAME BOLTED TO 3x3x1/4 ANGLE USING UNISTRUT MOUNTING FASTENERS (SEE SECTION VIEW 5)
MIXING COLUMN
30 EQ SP @ 8" O.C. = 232.00 (19.33')
COLLECTION JARS (60 PLS, 30 PER SIDE)
1 5/8" DRILL THRU FOR 3/4" BULKHEAD FITTING AND NALGENE TANK VALVE (60 PLS, 30 PER SIDE)
4" BULKHEAD WITH OVERFLOW EXTENTION
5.25 FROM BOTTOM OF HEAD TROUGH
6" WELDED FLANGE, ALUM (2)
LEVELING FEET, 3/4-10 THRD W/ 4 1/2" LG THREADED ROD & TWO JAM NUTS
6" OFD TO FRY TANKS
6" OFD @ 45°
UNISTRUT #5500 WITH ALUM SPACERS AS NEEDED (30.00" LG, 1 PLS)
82.12 (WELD DIM.)
79.75 (WELD DIM.)
82.13 (WELD DIM.)
244.00
L3x3x1/4 (8)
L3x2x1/4 (4)
NOTCH CHANNEL AS REQD.
C7x.170 (2)
L3x3x1/4 (6)
HEAD TROUGH, 1/8" ALUM, END PLATES BOTH ENDS (FORMED)
COLLECTION TROUGH, END PLATES BOTH ENDS 1/8" ALUM, (FORMED)
P3001 UNISTRUT, CHANNEL NUT & CAP SCREW (TYP. AS SHOWN HERE)
3/4" ALUM. PLATE
13/16" DR. THRU FOR, 3/4" LEVELING FOOT W/TWO LOCKING NUTS
REVISION DATE DESCRIPTION
WILLIAM J. POUPART TRIBAL FISH HATCHERY
LAC DU FLAMBEAU, WISCONSIN
SCALE AS NOTED
PROJECT ENGINEER BRUCE MARKGREN
DRAWN BY D.M.C.
DATE 07-30-98
COOPER ENGINEERING COMPANY
310 WEST SOUTH STREET RICE LAKE, WISCONSIN
TELEPHONE 715-234-7008
SHEET DESCRIPTION WALLEYE/SUCKER INCUBATION BATTERY
PROJECT NO. CE97008
SHEET NO. M-4

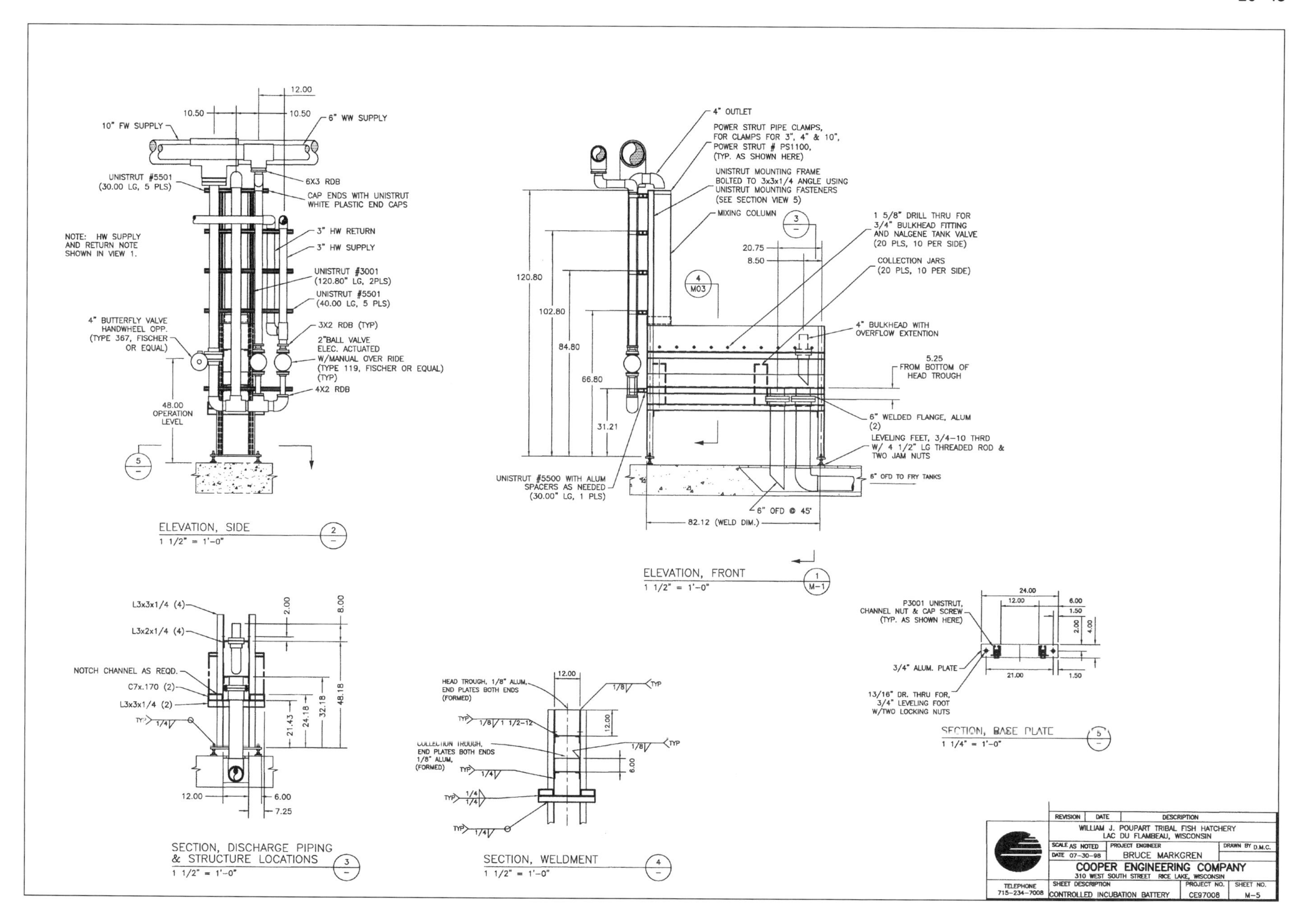

ELEVATION, SIDE
1 1/2" = 1'-0"
10" FW SUPPLY
6" WW SUPPLY
UNISTRUT #5501 (30.00 LG, 5 PLS)
6X3 RDB
CAP ENDS WITH UNISTRUT WHITE PLASTIC END CAPS
NOTE: HW SUPPLY AND RETURN NOTE SHOWN IN VIEW 1.
3" HW RETURN
3" HW SUPPLY
UNISTRUT #3001 (120.80" LG, 2PLS)
UNISTRUT #5501 (40.00 LG, 5 PLS)
4" BUTTERFLY VALVE HANDWHEEL OPP. (TYPE 367, FISCHER OR EQUAL)
3X2 RDB (TYP)
2"BALL VALVE ELEC. ACTUATED W/MANUAL OVER RIDE (TYPE 119, FISCHER OR EQUAL) (TYP)
4X2 RDB
48.00 OPERATION LEVEL
ELEVATION, FRONT
1 1/2" = 1'-0"
4" OUTLET
POWER STRUT PIPE CLAMPS, FOR CLAMPS FOR 3", 4" & 10", POWER STRUT # PS1100, (TYP. AS SHOWN HERE)
UNISTRUT MOUNTING FRAME BOLTED TO 3x3x1/4 ANGLE USING UNISTRUT MOUNTING FASTENERS (SEE SECTION VIEW 5)
MIXING COLUMN
1 5/8" DRILL THRU FOR 3/4" BULKHEAD FITTING AND NALGENE TANK VALVE (20 PLS, 10 PER SIDE)
COLLECTION JARS (20 PLS, 10 PER SIDE)
4" BULKHEAD WITH OVERFLOW EXTENTION
5.25 FROM BOTTOM OF HEAD TROUGH
6" WELDED FLANGE, ALUM (2)
LEVELING FEET, 3/4-10 THRD W/ 4 1/2" LG THREADED ROD & TWO JAM NUTS
6" OFD TO FRY TANKS
6" OFD @ 45°
82.12 (WELD DIM.)
UNISTRUT #5500 WITH ALUM SPACERS AS NEEDED (30.00" LG, 1 PLS)
SECTION, DISCHARGE PIPING & STRUCTURE LOCATIONS
1 1/2" = 1'-0"
L3x3x1/4 (4)
L3x2x1/4 (4)
NOTCH CHANNEL AS REQD.
C7x.170 (2)
L3x3x1/4 (2)
SECTION, WELDMENT
1 1/2" = 1'-0"
HEAD TROUGH, 1/8" ALUM, END PLATES BOTH ENDS (FORMED)
COLLECTION TROUGH, END PLATES BOTH ENDS 1/8" ALUM, (FORMED)
SECTION, BASE PLATE
1 1/4" = 1'-0"
P3001 UNISTRUT, CHANNEL NUT & CAP SCREW (TYP. AS SHOWN HERE)
3/4" ALUM. PLATE
13/16" DR. THRU FOR, 3/4" LEVELING FOOT W/TWO LOCKING NUTS
WILLIAM J. POUPART TRIBAL FISH HATCHERY
LAC DU FLAMBEAU, WISCONSIN
SCALE AS NOTED
DATE 07-30-98
PROJECT ENGINEER BRUCE MARKGREN
DRAWN BY D.M.C.
COOPER ENGINEERING COMPANY
310 WEST SOUTH STREET RICE LAKE, WISCONSIN
TELEPHONE 715-234-7008
SHEET DESCRIPTION CONTROLLED INCUBATION BATTERY
PROJECT NO. CE97008
SHEET NO. M-5

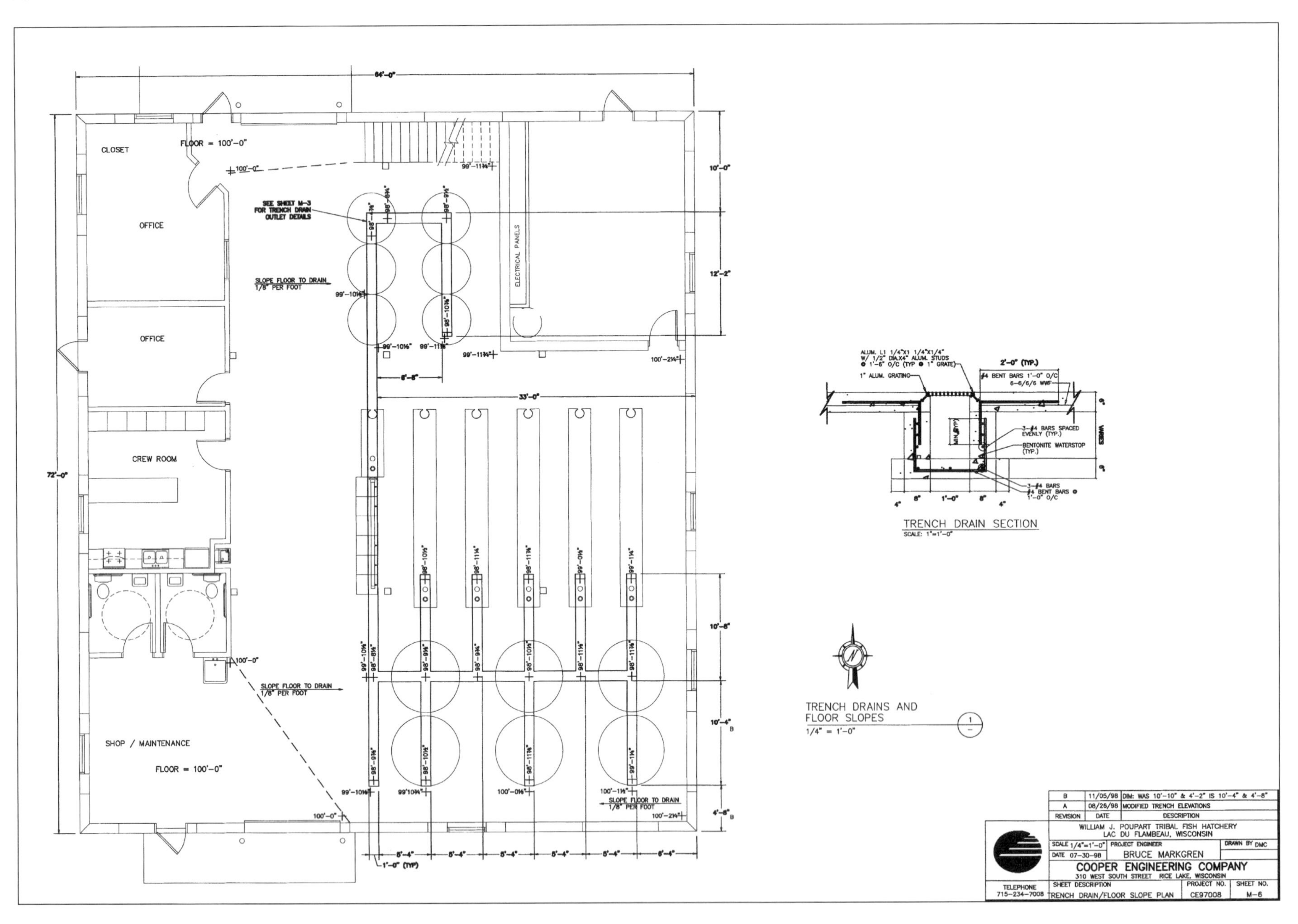
CLOSET
FLOOR = 100'-0"
OFFICE
OFFICE
CREW ROOM
SHOP / MAINTENANCE
FLOOR = 100'-0"
ELECTRICAL PANELS
SEE SHEET M-3 FOR TRENCH DRAIN OUTLET DETAILS
SLOPE FLOOR TO DRAIN 1/8" PER FOOT
TRENCH DRAIN SECTION
SCALE: 1"=1'-0"
TRENCH DRAINS AND FLOOR SLOPES
1/4" = 1'-0"
WILLIAM J. POUPART TRIBAL FISH HATCHERY
LAC DU FLAMBEAU, WISCONSIN
BRUCE MARKGREN
COOPER ENGINEERING COMPANY
310 WEST SOUTH STREET RICE LAKE, WISCONSIN
TELEPHONE 715-234-7008
TRENCH DRAIN/FLOOR SLOPE PLAN
CE97008
M-6

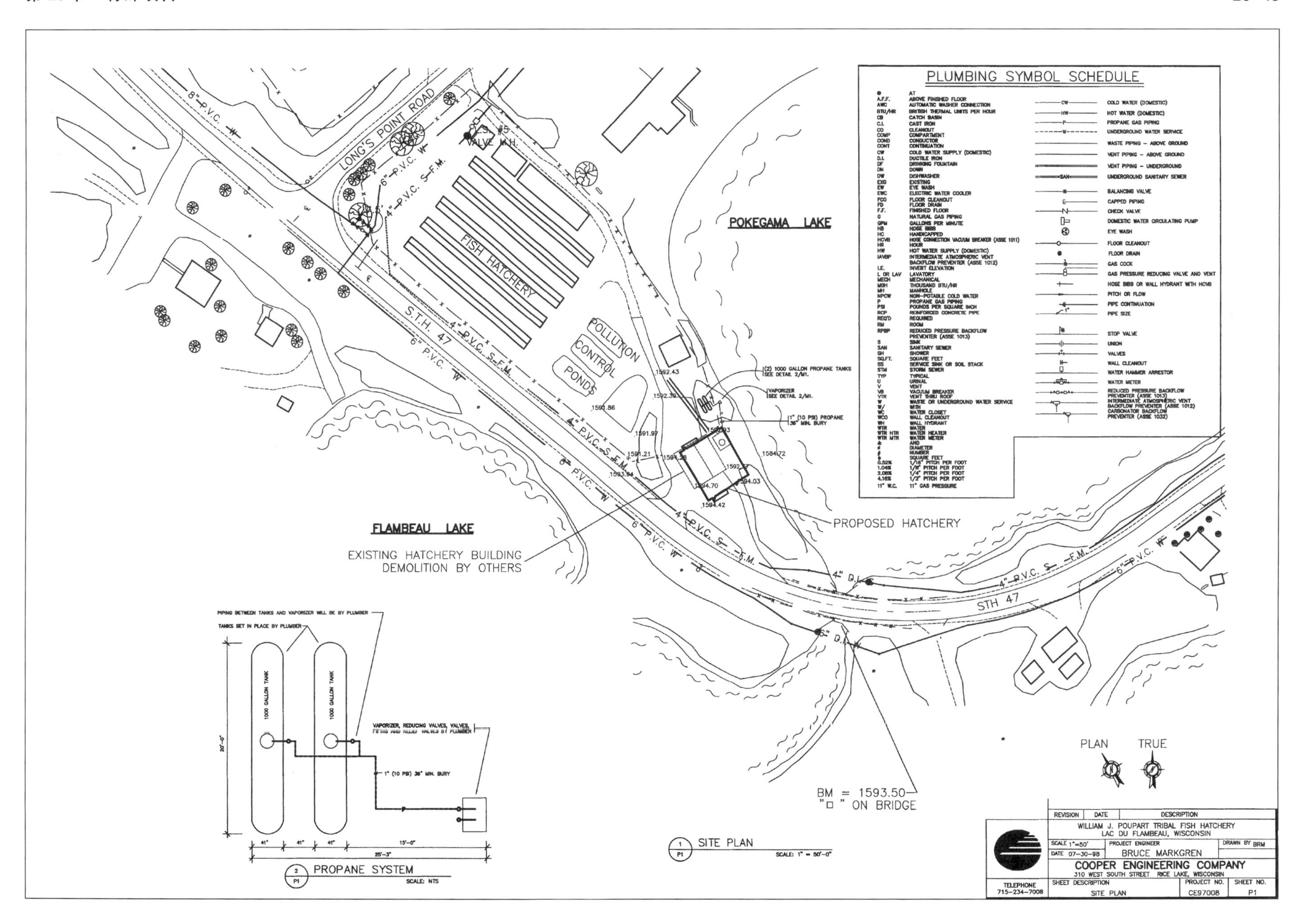
PLUMBING SYMBOL SCHEDULE
@ AT
A.F.F. ABOVE FINISHED FLOOR
AWC AUTOMATIC WASHER CONNECTION
BTU/HR BRITISH THERMAL UNITS PER HOUR
CB CATCH BASIN
C.I. CAST IRON
CO CLEANOUT
COMP COMPARTMENT
COND CONDUCTOR
CONT CONTINUATION
CW COLD WATER SUPPLY (DOMESTIC)
D.I. DUCTILE IRON
DF DRINKING FOUNTAIN
DN DOWN
DW DISHWASHER
EXG EXISTING
EW EYE WASH
EWC ELECTRIC WATER COOLER
FCO FLOOR CLEANOUT
FD FLOOR DRAIN
F.F. FINISHED FLOOR
G NATURAL GAS PIPING
GPM GALLONS PER MINUTE
HB HOSE BIBS
HC HANDICAPPED
HCVB HOSE CONNECTION VACUUM BREAKER (ASSE 1011)
HR HOUR
HW HOT WATER SUPPLY (DOMESTIC)
IAVBP INTERMEDIATE ATMOSPHERIC VENT BACKFLOW PREVENTER (ASSE 1012)
I.E. INVERT ELEVATION
L OR LAV LAVATORY
MECH MECHANICAL
MBH THOUSAND BTU/HR
MH MANHOLE
NPCW NON-POTABLE COLD WATER
P PROPANE GAS PIPING
PSI POUNDS PER SQUARE INCH
RCP REINFORCED CONCRETE PIPE
REQ'D REQUIRED
RM ROOM
RPBP REDUCED PRESSURE BACKFLOW PREVENTER (ASSE 1013)
S SINK
SAN SANITARY SEWER
SH SHOWER
SQ.FT. SQUARE FEET
SS SERVICE SINK OR SOIL STACK
STM STORM SEWER
TYP TYPICAL
U URINAL
V VENT
VB VACUUM BREAKER
VTR VENT THRU ROOF
W WASTE OR UNDERGROUND WATER SERVICE
W/ WITH
WC WATER CLOSET
WCO WALL CLEANOUT
WH WALL HYDRANT
WTR WATER
WTR HTR WATER HEATER
WTR MTR WATER METER
& AND
DIAMETER
NUMBER
SQUARE FEET
0.52% 1/16" PITCH PER FOOT
1.04% 1/8" PITCH PER FOOT
2.08% 1/4" PITCH PER FOOT
4.16% 1/2" PITCH PER FOOT
11" W.C. 11" GAS PRESSURE
CW COLD WATER (DOMESTIC)
HW HOT WATER (DOMESTIC)
P PROPANE GAS PIPING
W UNDERGROUND WATER SERVICE
WASTE PIPING – ABOVE GROUND
VENT PIPING – ABOVE GROUND
VENT PIPING – UNDERGROUND
SAN UNDERGROUND SANITARY SEWER
BALANCING VALVE
CAPPED PIPING
CHECK VALVE
DOMESTIC WATER CIRCULATING PUMP
EYE WASH
FLOOR CLEANOUT
FLOOR DRAIN
GAS COCK
GAS PRESSURE REDUCING VALVE AND VENT
HOSE BIBB OR WALL HYDRANT WITH HCVB
PITCH OR FLOW
PIPE CONTINUATION
1" PIPE SIZE
STOP VALVE
UNION
VALVES
WALL CLEANOUT
WATER HAMMER ARRESTOR
WATER METER
REDUCED PRESSURE BACKFLOW PREVENTER (ASSE 1013)
INTERMEDIATE ATMOSPHERIC VENT BACKFLOW PREVENTER (ASSE 1012)
CARBONATOR BACKFLOW PREVENTER (ASSE 1032)
8" P.V.C. W
LONG'S POINT ROAD
L.S. #5 VALVE M.H.
6" P.V.C. W
6" P.V.C. S.F.M.
4" P.V.C. S.F.M.
FISH HATCHERY
POKEGAMA LAKE
POLLUTION CONTROL PONDS
S.T.H. 47
4" P.V.C. S.F.M.
6" P.V.C. W
(2) 1000 GALLON PROPANE TANKS SEE DETAIL 2/M1.
VAPORIZER SEE DETAIL 2/M1.
1" (10 PSI) PROPANE 36" MIN. BURY
1592.43
1592.39
1591.86
1591.97
1591.21
1593.84
1584.72
1594.70
1594.03
1594.42
PROPOSED HATCHERY
FLAMBEAU LAKE
EXISTING HATCHERY BUILDING DEMOLITION BY OTHERS
4" D.I.
4" P.V.C. S.F.M.
6" P.V.C. W
STH 47
6" D.I. W
PLAN
TRUE
BM = 1593.50
"□" ON BRIDGE
PIPING BETWEEN TANKS AND VAPORIZER WILL BE BY PLUMBER
TANKS SET IN PLACE BY PLUMBER
1000 GALLON TANK
1000 GALLON TANK
VAPORIZER, REDUCING VALVES, VALVES, PIPING AND RELIEF VALVES BY PLUMBER
1" (10 PSI) 36" MIN. BURY
20'-0"
41" 41" 41" 15'-0"
25'-3"
2 P1 PROPANE SYSTEM SCALE: NTS
1 P1 SITE PLAN SCALE: 1" = 50'-0"
REVISION DATE DESCRIPTION
WILLIAM J. POUPART TRIBAL FISH HATCHERY
LAC DU FLAMBEAU, WISCONSIN
SCALE 1"=50'
PROJECT ENGINEER
DRAWN BY BRM
DATE 07-30-98
BRUCE MARKGREN
COOPER ENGINEERING COMPANY
310 WEST SOUTH STREET RICE LAKE, WISCONSIN
TELEPHONE 715-234-7008
SHEET DESCRIPTION SITE PLAN
PROJECT NO. CE97008
SHEET NO. P1

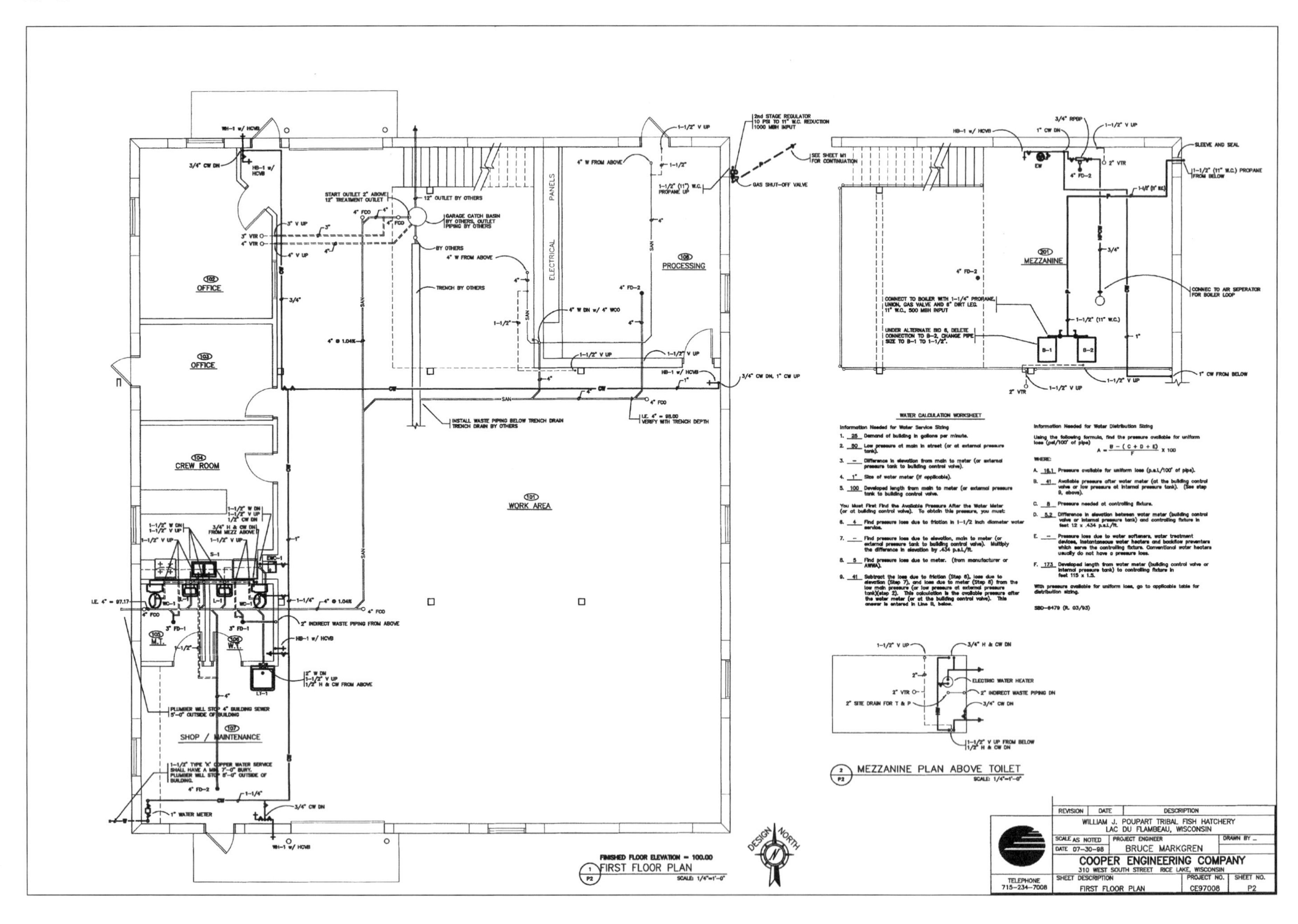
OFFICE
OFFICE
CREW ROOM
WORK AREA
PROCESSING
ELECTRICAL
PANELS
SHOP / MAINTENANCE
MEZZANINE
FINISHED FLOOR ELEVATION = 100.00
FIRST FLOOR PLAN
SCALE: 1/4"=1'-0"
DESIGN NORTH
2nd STAGE REGULATOR
10 PSI TO 11" W.C. REDUCTION
1000 MBH INPUT
SEE SHEET M1 FOR CONTINUATION
GAS SHUT-OFF VALVE
CONNECT TO BOILER WITH 1-1/4" PROPANE, UNION, GAS VALVE AND 6" DIRT LEG. 11" W.C., 500 MBH INPUT
UNDER ALTERNATE BID 8, DELETE CONNECTION TO B-2, CHANGE PIPE SIZE TO B-1 TO 1-1/2".
CONNECT TO AIR SEPERATOR FOR BOILER LOOP
SLEEVE AND SEAL
1" CW FROM BELOW
INSTALL WASTE PIPING BELOW TRENCH DRAIN TRENCH DRAIN BY OTHERS
TRENCH BY OTHERS
GARAGE CATCH BASIN BY OTHERS, OUTLET PIPING BY OTHERS
PLUMBER WILL STOP 4" BUILDING SEWER 5'-0" OUTSIDE OF BUILDING
1-1/2" TYPE "K" COPPER WATER SERVICE SHALL HAVE A MIN. 7'-0" BURY. PLUMBER WILL STOP 5'-0" OUTSIDE OF BUILDING.
1" WATER METER
WATER CALCULATION WORKSHEET
Information Needed for Water Service Sizing
1. 25 Demand of building in gallons per minute.
2. 50 Low pressure at main in street (or at external pressure tank).
3. – Difference in elevation from main to meter (or external pressure tank to building control valve).
4. 1" Size of water meter (if applicable).
5. 100 Developed length from main to meter (or external pressure tank to building control valve.
You Must First Find the Available Pressure After the Water Meter (or at building control valve). To obtain this pressure, you must:
6. 4 Find pressure loss due to friction in 1-1/2 inch diameter water service.
7. – Find pressure loss due to elevation, main to meter (or external pressure tank to building control valve). Multiply the difference in elevation by .434 p.s.i./ft.
8. 5 Find pressure loss due to meter. (from manufacturer or AWWA).
9. 41 Subtract the loss due to friction (Step 6), loss due to elevation (Step 7), and loss due to meter (Step 8) from the low main pressure (or low pressure at external pressure tank)(step 2). This calculation is the available pressure after the water meter (or at the building control valve). This answer is entered in Line B, below.
Information Needed for Water Distribution Sizing
Using the following formula, find the pressure available for uniform loss (psi/100' of pipe)
A = (B − (C + D + E)) / F X 100
WHERE:
A. 16.1 Pressure available for uniform loss (p.s.i./100' of pipe).
B. 41 Available pressure after water meter (at the building control valve or low pressure at internal pressure tank). (See step 9, above).
C. 8 Pressure needed at controlling fixture.
D. 5.2 Difference in elevation between water meter (building control valve or internal pressure tank) and controlling fixture in feet 12 x .434 p.s.i./ft.
E. – Pressure loss due to water softeners, water treatment devices, instantaneous water heaters and backflow preventers which serve the controlling fixture. Conventional water heaters usually do not have a pressure loss.
F. 173 Developed length from water meter (building control valve or internal pressure tank) to controlling fixture in feet 115 x 1.5.
With pressure available for uniform loss, go to applicable table for distribution sizing.
SBD-6479 (R. 03/93)
ELECTRIC WATER HEATER
2" INDIRECT WASTE PIPING DN
2" SITE DRAIN FOR T & P
MEZZANINE PLAN ABOVE TOILET
SCALE: 1/4"=1'-0"
REVISION
DATE
DESCRIPTION
WILLIAM J. POUPART TRIBAL FISH HATCHERY
LAC DU FLAMBEAU, WISCONSIN
SCALE AS NOTED
DATE 07-30-98
PROJECT ENGINEER
BRUCE MARKGREN
DRAWN BY –
COOPER ENGINEERING COMPANY
310 WEST SOUTH STREET RICE LAKE, WISCONSIN
TELEPHONE 715-234-7008
SHEET DESCRIPTION
FIRST FLOOR PLAN
PROJECT NO. CE97008
SHEET NO. P2

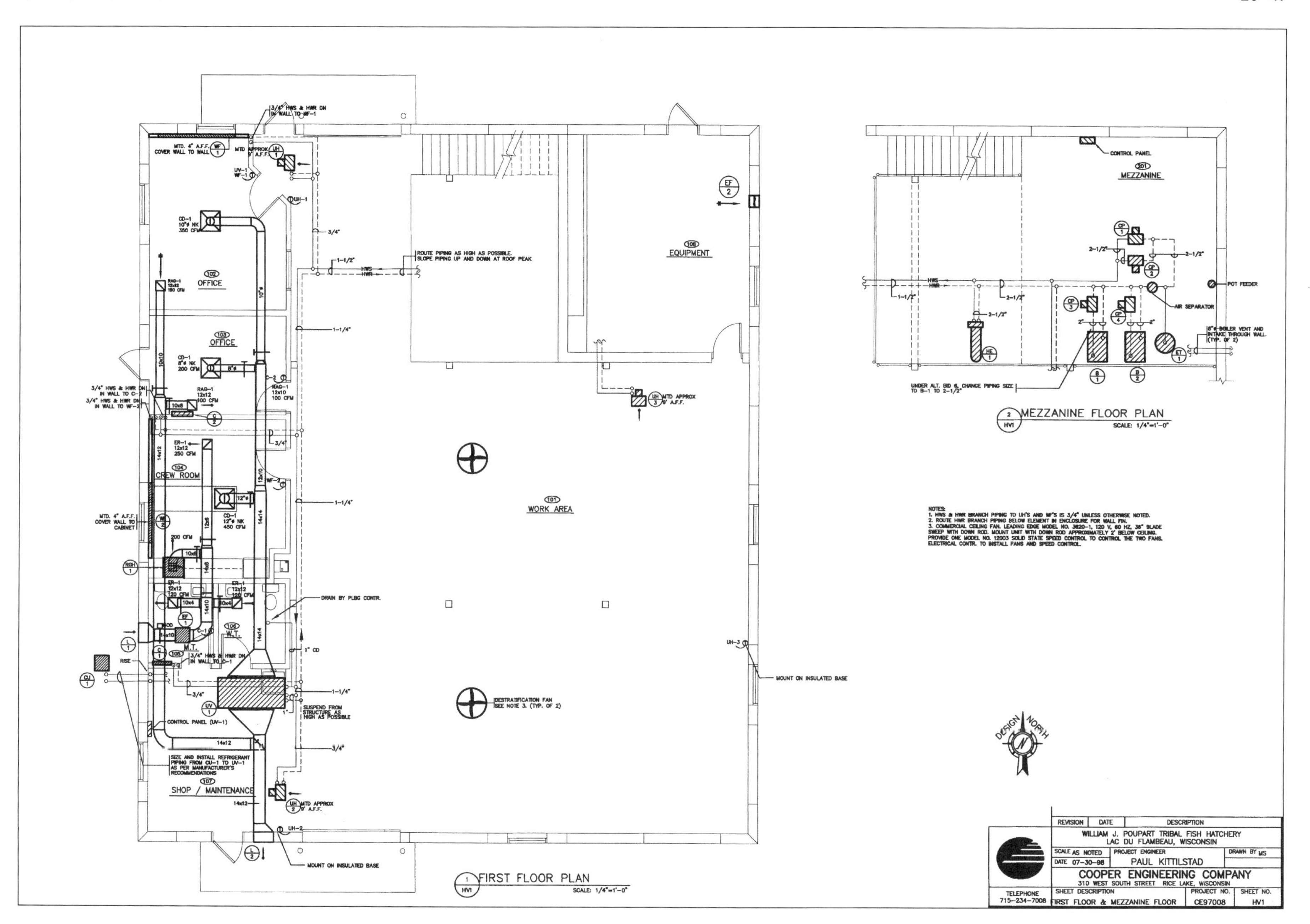

102 OFFICE
103 OFFICE
104 CREW ROOM
105 M.T.
106 W.T.
107 SHOP / MAINTENANCE
101 WORK AREA
108 EQUIPMENT
201 MEZZANINE
CONTROL PANEL
POT FEEDER
AIR SEPARATOR
ROUTE PIPING AS HIGH AS POSSIBLE. SLOPE PIPING UP AND DOWN AT ROOF PEAK
DRAIN BY PLBG CONTR.
MOUNT ON INSULATED BASE
CONTROL PANEL (UV-1)
DESTRATIFICATION FAN SEE NOTE 3. (TYP. OF 2)
1 FIRST FLOOR PLAN HV1 SCALE: 1/4"=1'-0"
2 MEZZANINE FLOOR PLAN HV1 SCALE: 1/4"=1'-0"
NOTES:
1. HWS & HWR BRANCH PIPING TO UH'S AND WF'S IS 3/4" UNLESS OTHERWISE NOTED.
2. ROUTE HWR BRANCH PIPING BELOW ELEMENT IN ENCLOSURE FOR WALL FIN.
3. COMMERCIAL CEILING FAN. LEADING EDGE MODEL NO. 3620-1, 120 V, 60 HZ, 36" BLADE SWEEP WITH DOWN ROD. MOUNT UNIT WITH DOWN ROD APPROXIMATELY 2' BELOW CEILING. PROVIDE ONE MODEL NO. 12003 SOLID STATE SPEED CONTROL TO CONTROL THE TWO FANS. ELECTRICAL CONTR. TO INSTALL FANS AND SPEED CONTROL.
DESIGN NORTH
REVISION DATE DESCRIPTION
WILLIAM J. POUPART TRIBAL FISH HATCHERY
LAC DU FLAMBEAU, WISCONSIN
SCALE AS NOTED
DATE 07-30-98
PROJECT ENGINEER PAUL KITTILSTAD
DRAWN BY MS
COOPER ENGINEERING COMPANY
310 WEST SOUTH STREET RICE LAKE, WISCONSIN
TELEPHONE 715-234-7008
SHEET DESCRIPTION FIRST FLOOR & MEZZANINE FLOOR
PROJECT NO. CE97008
SHEET NO. HV1

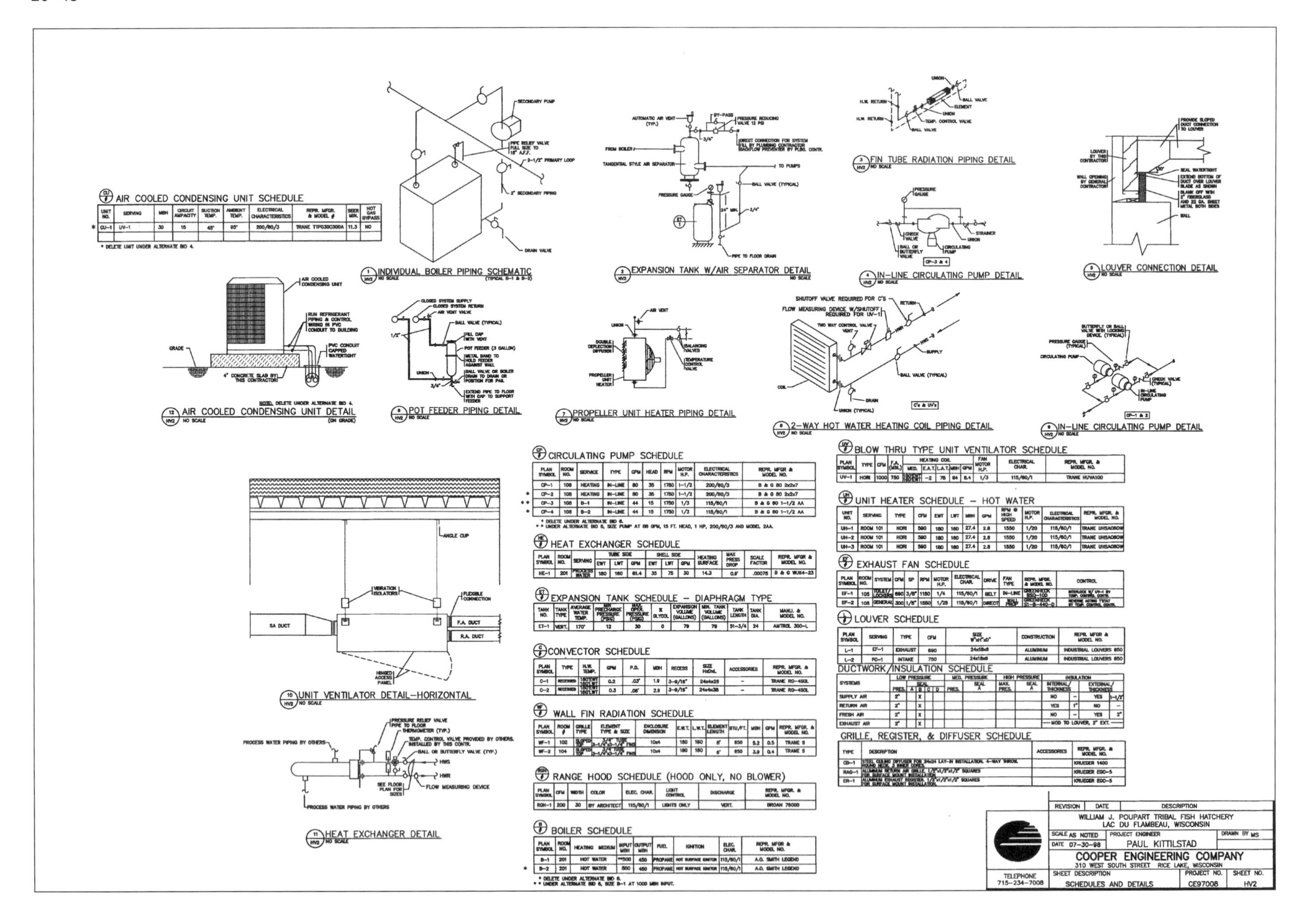

AIR COOLED CONDENSING UNIT SCHEDULE

UNIT NO.	SERVING	MBH	CIRCUIT AMPACITY	SUCTION TEMP.	AMBIENT TEMP.	ELECTRICAL CHARACTERISTICS	REPR. MFGR. & MODEL #	SEER MIN.	HOT GAS BYPASS
* CU-1	UV-1	30	15	45°	95°	200/60/3	TRANE TTP030C300A	11.3	NO

* DELETE UNIT UNDER ALTERNATE BID 4.

CIRCULATING PUMP SCHEDULE

PLAN SYMBOL	ROOM NO.	SERVICE	TYPE	GPM	HEAD	RPM	MOTOR H.P.	ELECTRICAL CHARACTERISTICS	REPR. MFGR & MODEL NO.
CP-1	108	HEATING	IN-LINE	80	35	1750	1-1/2	200/60/3	B & G 80 2x2x7
* CP-2	108	HEATING	IN-LINE	80	35	1750	1-1/2	200/60/3	B & G 80 2x2x7
** CP-3	108	B-1	IN-LINE	44	15	1750	1/3	115/60/1	B & G 60 1-1/2 AA
* CP-4	108	B-2	IN-LINE	44	15	1750	1/3	115/60/1	B & G 60 1-1/2 AA

* DELETE UNDER ALTERNATE BID 6.
** UNDER ALTERNATE BID 6, SIZE PUMP AT 88 GPM, 15 FT. HEAD, 1 HP, 200/60/3 AND MODEL 2AA.

HEAT EXCHANGER SCHEDULE

PLAN SYMBOL	ROOM NO.	SERVING	TUBE SIDE EWT	TUBE SIDE LWT	TUBE SIDE GPM	SHELL SIDE EWT	SHELL SIDE LWT	SHELL SIDE GPM	HEATING SURFACE	MAX PRESS DROP	SCALE FACTOR	REPR. MFGR & MODEL NO.
HE-1	201	PROCESS WATER	180	160	61.4	35	75	30	14.3	0.9'	.00075	B & G WU64-23

EXPANSION TANK SCHEDULE – DIAPHRAGM TYPE

TANK NO.	TANK TYPE	AVERAGE WATER TEMP.	MIN. PRECHARGE PRESSURE (PSIG)	MAX. OPER. PRESSURE (PSIG)	% GLYCOL	EXPANSION VOLUME (GALLONS)	MIN. TANK VOLUME (GALLONS)	TANK LENGTH	TANK DIA.	MANU. & MODEL NO.
ET-1	VERT.	170°	12	30	0	79	79	51-3/4	24	AMTROL 300-L

CONVECTOR SCHEDULE

PLAN SYMBOL	TYPE	H.W. TEMP.	GPM	P.D.	MBH	RECESS	SIZE HxDxL	ACCESSORIES	REPR. MFGR. & MODEL NO.
C-1	RECESSED	180°EWT 160°LWT	0.2	.03'	1.9	3-9/16"	24x4x28	–	TRANE RG-4SOL
C-2	RECESSED	180°EWT 160°LWT	0.3	.06'	2.9	3-9/16"	24x4x38	–	TRANE RG-4SOL

WALL FIN RADIATION SCHEDULE

PLAN SYMBOL	ROOM #	GRILLE TYPE	ELEMENT TYPE & SIZE	ENCLOSURE DIMENSION	E.W.T.	L.W.T.	ELEMENT LENGTH	BTU/FT.	MBH	GPM	REPR. MFGR. & MODEL NO.
WF-1	102	SLOPED TOP	3/4" TUBE 3-1/4"x3-1/4" FINS	10x4	180	160	8'	650	5.2	0.5	TRANE S
WF-2	104	SLOPED TOP	3/4" TUBE 3-1/4"x3-1/4" FINS	10x4	180	160	6'	650	3.9	0.4	TRANE S

RANGE HOOD SCHEDULE (HOOD ONLY, NO BLOWER)

PLAN SYMBOL	CFM	WIDTH	COLOR	ELEC. CHAR.	LIGHT CONTROL	DISCHARGE	REPR. MFGR. & MODEL NO.
RGH-1	200	30	BY ARCHITECT	115/60/1	LIGHTS ONLY	VERT.	BROAN 76000

BOILER SCHEDULE

PLAN SYMBOL	ROOM NO.	HEATING MEDIUM	INPUT MBH	OUTPUT MBH	FUEL	IGNITION	ELEC. CHAR.	REPR. MFGR & MODEL NO.
B-1	201	HOT WATER	**500	450	PROPANE	HOT SURFACE IGNITOR	115/60/1	A.O. SMITH LEGEND
* B-2	201	HOT WATER	500	450	PROPANE	HOT SURFACE IGNITOR	115/60/1	A.O. SMITH LEGEND

* DELETE UNDER ALTERNATE BID 6.
** UNDER ALTERNATE BID 6, SIZE B-1 AT 1000 MBH INPUT.

BLOW THRU TYPE UNIT VENTILATOR SCHEDULE

PLAN SYMBOL	TYPE	CFM	F.A. (MIN.)	HEATING COIL MED.	HEATING COIL E.A.T.	HEATING COIL L.A.T.	HEATING COIL MBH	HEATING COIL GPM	FAN MOTOR H.P.	ELECTRICAL CHAR.	REPR. MFGR. & MODEL NO.
UV-1	HORI	1000	750	180°EWT 160°LWT	-2	75	84	8.4	1/3	115/60/1	TRANE HUVA100

UNIT HEATER SCHEDULE – HOT WATER

UNIT NO.	SERVING	TYPE	CFM	EWT	LWT	MBH	GPM	RPM @ HIGH SPEED	MOTOR H.P.	ELECTRICAL CHARACTERISTICS	REPR. MFGR. & MODEL NO.
UH-1	ROOM 101	HORI	590	180	160	27.4	2.8	1550	1/20	115/60/1	TRANE UHSA060W
UH-2	ROOM 101	HORI	590	180	160	27.4	2.8	1550	1/20	115/60/1	TRANE UHSA060W
UH-3	ROOM 101	HORI	590	180	160	27.4	2.8	1550	1/20	115/60/1	TRANE UHSA060W

EXHAUST FAN SCHEDULE

PLAN SYMBOL	ROOM NO.	SYSTEM	CFM	SP	RPM	MOTOR H.P.	ELECTRICAL CHAR.	DRIVE	FAN TYPE	REPR. MFGR. & MODEL NO.	CONTROL
EF-1	105	TOILET/LOCKERS	690	3/8"	1180	1/4	115/60/1	BELT	IN-LINE	GREENHECK BSQ-100	INTERLOCK W/ UV-1 BY TEMP. CONTROL CONTR.
EF-2	108	GENERAL	300	1/8"	1550	1/25	115/60/1	DIRECT	WALL PROP	GREENHECK S1-8-440-D	REVERSE ACTING T'STAT BY TEMP. CONTROL CONTR.

LOUVER SCHEDULE

PLAN SYMBOL	SERVING	TYPE	CFM	SIZE W"xH"xD"	CONSTRUCTION	REPR. MFGR & MODEL NO.
L-1	EF-1	EXHAUST	690	24x18x6	ALUMINUM	INDUSTRIAL LOUVERS 650
L-2	FC-1	INTAKE	750	24x18x6	ALUMINUM	INDUSTRIAL LOUVERS 650

DUCTWORK/INSULATION SCHEDULE

SYSTEMS	LOW PRESSURE PRES.	LOW PRESSURE SEAL A	SEAL B	SEAL C	SEAL D	MED. PRESSURE PRES.	MED. PRESSURE SEAL A	HIGH PRESSURE MAX. PRES.	HIGH PRESSURE SEAL A	INSULATION INTERNAL	INTERNAL THICKNESS	INSULATION EXTERNAL	EXTERNAL THICKNESS
SUPPLY AIR	2"		X							NO	–	YES	1-1/2"
RETURN AIR	2"		X							YES	1"	NO	–
FRESH AIR	2"		X							NO	–	YES	2"
EXHAUST AIR	2"		X							MOD TO LOUVER, 2" EXT.			

GRILLE, REGISTER, & DIFFUSER SCHEDULE

TYPE	DESCRIPTION	ACCESSORIES	REPR. MFGR. & MODEL NO.
CD-1	STEEL CEILING DIFFUSER FOR 24x24 LAY-IN INSTALLATION. 4-WAY THROW. ROUND NECK, 3 INNER CORES.		KRUEGER 1400
RAG-1	ALUMINUM RETURN AIR GRILLE, 1/2"x1/2"x1/2" SQUARES FOR SURFACE MOUNT INSTALLATION.		KRUEGER EGC-5
ER-1	ALUMINUM EXHAUST REGISTER, 1/2"x1/2"x1/2" SQUARES FOR SURFACE MOUNT INSTALLATION.		KRUEGER EGC-5

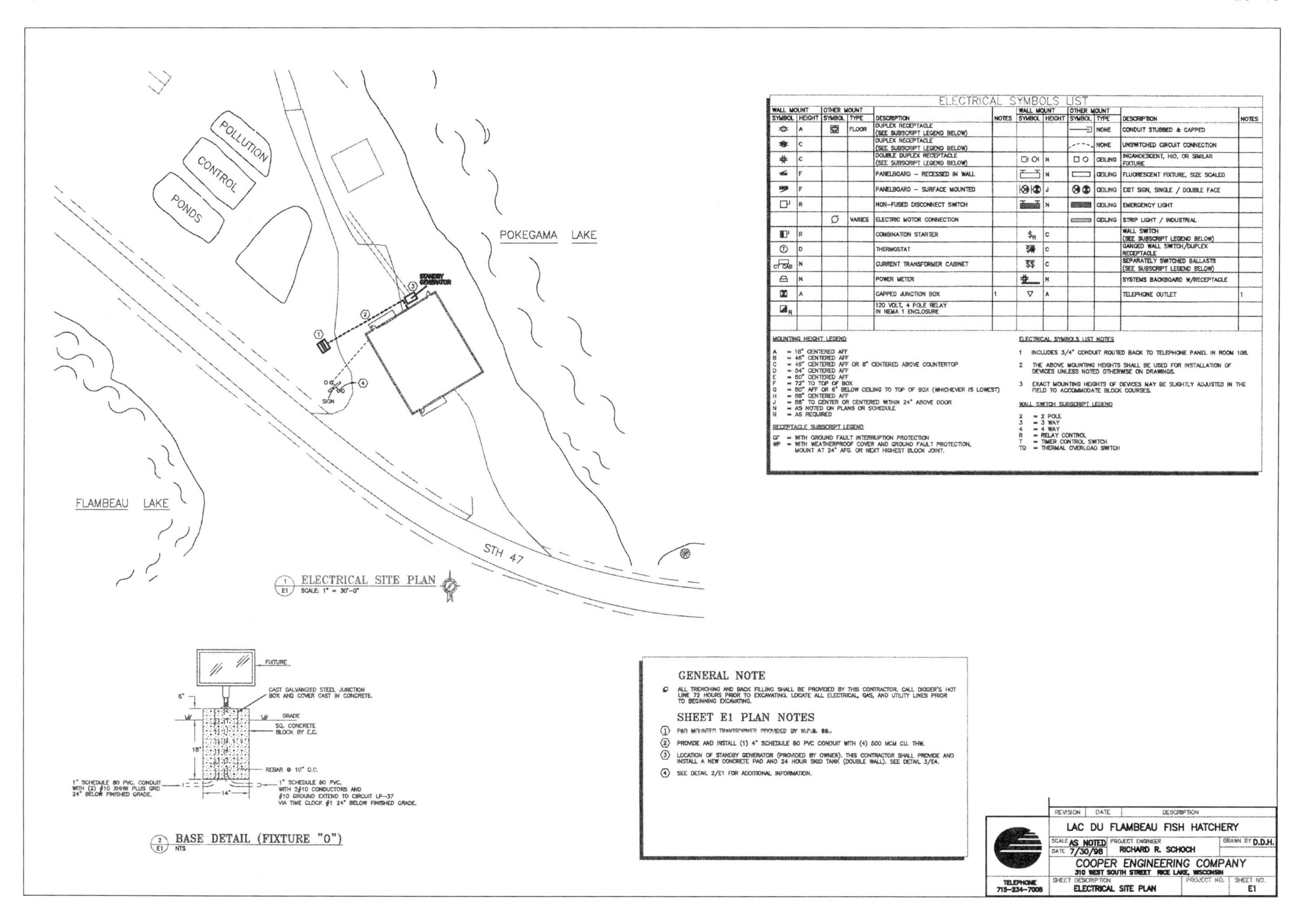

POLLUTION
CONTROL
PONDS
POKEGAMA LAKE
STANDBY GENERATOR
SIGN
FLAMBEAU LAKE
STH 47
1 E1 ELECTRICAL SITE PLAN
SCALE: 1" = 30'-0"
ELECTRICAL SYMBOLS LIST
WALL MOUNT SYMBOL HEIGHT | OTHER MOUNT SYMBOL TYPE | DESCRIPTION | NOTES
A | FLOOR | DUPLEX RECEPTACLE (SEE SUBSCRIPT LEGEND BELOW)
C | DUPLEX RECEPTACLE (SEE SUBSCRIPT LEGEND BELOW)
C | DOUBLE DUPLEX RECEPTACLE (SEE SUBSCRIPT LEGEND BELOW)
F | PANELBOARD – RECESSED IN WALL
F | PANELBOARD – SURFACE MOUNTED
R | NON-FUSED DISCONNECT SWITCH
VARIES | ELECTRIC MOTOR CONNECTION
R | COMBINATION STARTER
D | THERMOSTAT
N | CURRENT TRANSFORMER CABINET
N | POWER METER
A | CAPPED JUNCTION BOX | 1
120 VOLT, 4 POLE RELAY IN NEMA 1 ENCLOSURE
NONE | CONDUIT STUBBED & CAPPED
NONE | UNSWITCHED CIRCUIT CONNECTION
N | CEILING | INCANDESCENT, HID, OR SIMILAR FIXTURE
N | CEILING | FLUORESCENT FIXTURE, SIZE SCALED
J | CEILING | EXIT SIGN, SINGLE / DOUBLE FACE
N | CEILING | EMERGENCY LIGHT
CEILING | STRIP LIGHT / INDUSTRIAL
C | WALL SWITCH (SEE SUBSCRIPT LEGEND BELOW)
C | GANGED WALL SWITCH/DUPLEX RECEPTACLE
C | SEPARATELY SWITCHED BALLASTS (SEE SUBSCRIPT LEGEND BELOW)
N | SYSTEMS BACKBOARD W/RECEPTACLE
A | TELEPHONE OUTLET | 1
MOUNTING HEIGHT LEGEND
A = 18" CENTERED AFF
B = 46" CENTERED AFF
C = 48" CENTERED AFF OR 8" CENTERED ABOVE COUNTERTOP
D = 54" CENTERED AFF
E = 60" CENTERED AFF
F = 72" TO TOP OF BOX
G = 80" AFF OR 6" BELOW CEILING TO TOP OF BOX (WHICHEVER IS LOWEST)
H = 88" CENTERED AFF
J = 88" TO CENTER OR CENTERED WITHIN 24" ABOVE DOOR
N = AS NOTED ON PLANS OR SCHEDULE
R = AS REQUIRED
RECEPTACLE SUBSCRIPT LEGEND
GF = WITH GROUND FAULT INTERRUPTION PROTECTION
WP = WITH WEATHERPROOF COVER AND GROUND FAULT PROTECTION. MOUNT AT 24" AFG. OR NEXT HIGHEST BLOCK JOINT.
ELECTRICAL SYMBOLS LIST NOTES
1 INCLUDES 3/4" CONDUIT ROUTED BACK TO TELEPHONE PANEL IN ROOM 108.
2 THE ABOVE MOUNTING HEIGHTS SHALL BE USED FOR INSTALLATION OF DEVICES UNLESS NOTED OTHERWISE ON DRAWINGS.
3 EXACT MOUNTING HEIGHTS OF DEVICES MAY BE SLIGHTLY ADJUSTED IN THE FIELD TO ACCOMMODATE BLOCK COURSES.
WALL SWITCH SUBSCRIPT LEGEND
2 = 2 POLE
3 = 3 WAY
4 = 4 WAY
R = RELAY CONTROL
T = TIMER CONTROL SWITCH
TO = THERMAL OVERLOAD SWITCH
FIXTURE
CAST GALVANIZED STEEL JUNCTION BOX AND COVER CAST IN CONCRETE.
GRADE
SQ. CONCRETE BLOCK BY E.C.
REBAR @ 10" O.C.
6"
18"
14"
1" SCHEDULE 80 PVC. CONDUIT WITH (2) #10 XHHW PLUS GRD 24" BELOW FINISHED GRADE.
1" SCHEDULE 80 PVC. WITH 2#10 CONDUCTORS AND #10 GROUND EXTEND TO CIRCUIT LP-37 VIA TIME CLOCK #1 24" BELOW FINISHED GRADE.
2 E1 BASE DETAIL (FIXTURE "O")
NTS
GENERAL NOTE
ALL TRENCHING AND BACK FILLING SHALL BE PROVIDED BY THIS CONTRACTOR. CALL DIGGER'S HOT LINE 72 HOURS PRIOR TO EXCAVATING. LOCATE ALL ELECTRICAL, GAS, AND UTILITY LINES PRIOR TO BEGINNING EXCAVATING.
SHEET E1 PLAN NOTES
1 PAD MOUNTED TRANSFORMER PROVIDED BY W.P.S.
2 PROVIDE AND INSTALL (1) 4" SCHEDULE 80 PVC CONDUIT WITH (4) 500 MCM CU. THW.
3 LOCATION OF STANDBY GENERATOR (PROVIDED BY OWNER). THIS CONTRACTOR SHALL PROVIDE AND INSTALL A NEW CONCRETE PAD AND 24 HOUR SKID TANK (DOUBLE WALL). SEE DETAIL 3/E4.
4 SEE DETAIL 2/E1 FOR ADDITIONAL INFORMATION.
REVISION | DATE | DESCRIPTION
LAC DU FLAMBEAU FISH HATCHERY
SCALE AS NOTED
DATE 7/30/98
PROJECT ENGINEER RICHARD R. SCHOCH
DRAWN BY D.D.H.
COOPER ENGINEERING COMPANY
310 WEST SOUTH STREET RICE LAKE, WISCONSIN
TELEPHONE 715-234-7008
SHEET DESCRIPTION ELECTRICAL SITE PLAN
PROJECT NO.
SHEET NO. E1

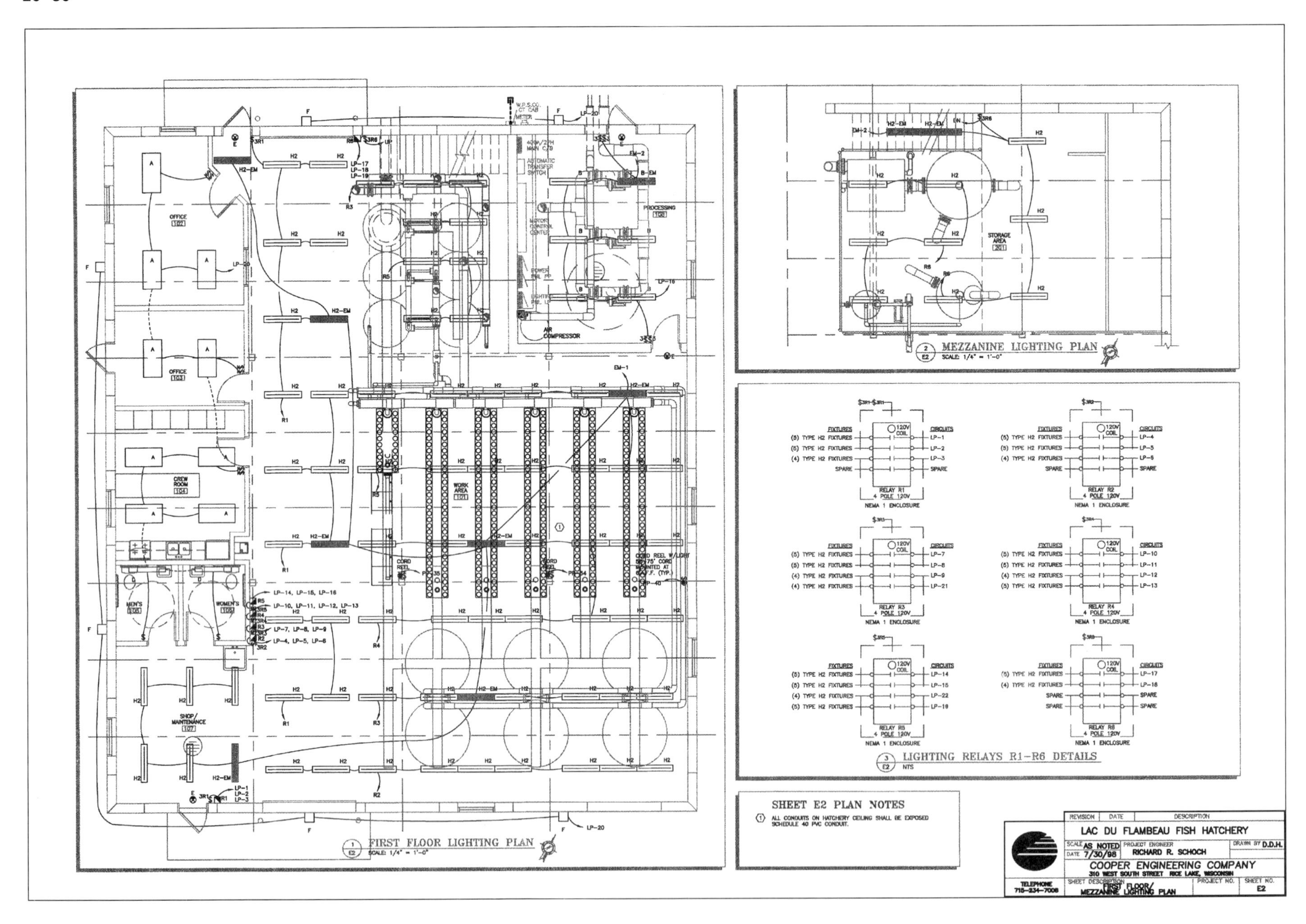
FIRST FLOOR LIGHTING PLAN
SCALE: 1/4" = 1'-0"
MEZZANINE LIGHTING PLAN
SCALE: 1/4" = 1'-0"
LIGHTING RELAYS R1-R6 DETAILS
NTS
SHEET E2 PLAN NOTES
ALL CONDUITS ON HATCHERY CEILING SHALL BE EXPOSED SCHEDULE 40 PVC CONDUIT.
LAC DU FLAMBEAU FISH HATCHERY
SCALE AS NOTED
DATE 7/30/98
PROJECT ENGINEER RICHARD R. SCHOCH
DRAWN BY D.D.H.
COOPER ENGINEERING COMPANY
310 WEST SOUTH STREET RICE LAKE, WISCONSIN
TELEPHONE 715-234-7008
SHEET DESCRIPTION FIRST FLOOR/ MEZZANINE LIGHTING PLAN
SHEET NO. E2

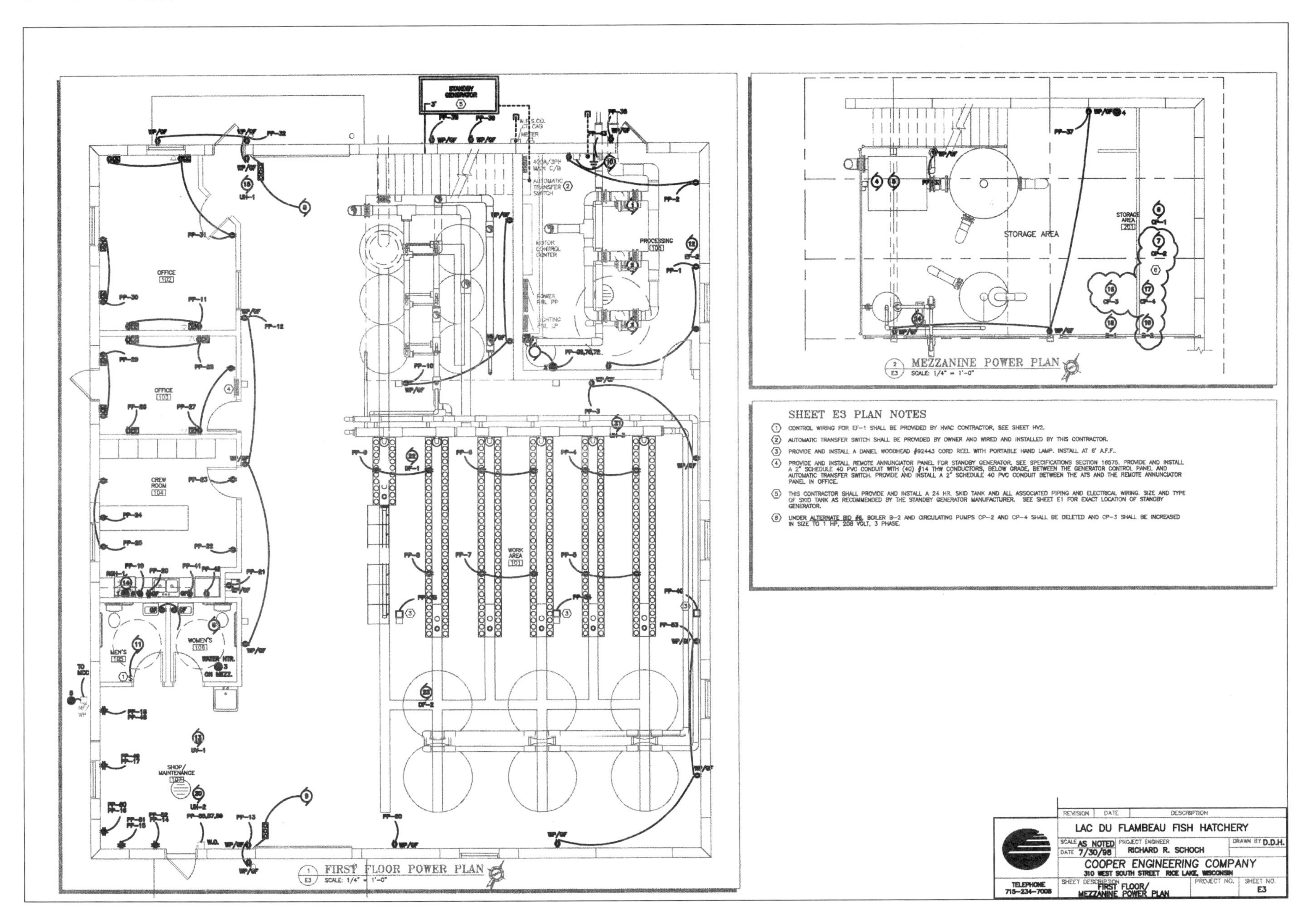
STANDBY GENERATOR
OFFICE 102
OFFICE 103
CREW ROOM 104
MEN'S
WOMEN'S
WATER HTR. ON MEZZ.
SHOP/ MAINTENANCE
WORK AREA 101
PROCESSING
MOTOR CONTROL CENTER
AUTOMATIC TRANSFER SWITCH
FIRST FLOOR POWER PLAN
SCALE: 1/4" = 1'-0"
STORAGE AREA
MEZZANINE POWER PLAN
SCALE: 1/4" = 1'-0"
SHEET E3 PLAN NOTES
1 CONTROL WIRING FOR EF-1 SHALL BE PROVIDED BY HVAC CONTRACTOR, SEE SHEET HV2.
2 AUTOMATIC TRANSFER SWITCH SHALL BE PROVIDED BY OWNER AND WIRED AND INSTALLED BY THIS CONTRACTOR.
3 PROVIDE AND INSTALL A DANIEL WOODHEAD #92443 CORD REEL WITH PORTABLE HAND LAMP. INSTALL AT 6' A.F.F..
4 PROVIDE AND INSTALL REMOTE ANNUNCIATOR PANEL FOR STANDBY GENERATOR. SEE SPECIFICATIONS SECTION 16575. PROVIDE AND INSTALL A 2" SCHEDULE 40 PVC CONDUIT WITH (40) #14 THW CONDUCTORS, BELOW GRADE, BETWEEN THE GENERATOR CONTROL PANEL AND AUTOMATIC TRANSFER SWITCH. PROVIDE AND INSTALL A 2" SCHEDULE 40 PVC CONDUIT BETWEEN THE ATS AND THE REMOTE ANNUNCIATOR PANEL IN OFFICE.
5 THIS CONTRACTOR SHALL PROVIDE AND INSTALL A 24 HR. SKID TANK AND ALL ASSOCIATED PIPING AND ELECTRICAL WIRING. SIZE AND TYPE OF SKID TANK AS RECOMMENDED BY THE STANDBY GENERATOR MANUFACTURER. SEE SHEET E1 FOR EXACT LOCATION OF STANDBY GENERATOR.
6 UNDER ALTERNATE BID #6, BOILER B-2 AND CIRCULATING PUMPS CP-2 AND CP-4 SHALL BE DELETED AND CP-3 SHALL BE INCREASED IN SIZE TO 1 HP, 208 VOLT, 3 PHASE.
REVISION
DATE
DESCRIPTION
LAC DU FLAMBEAU FISH HATCHERY
SCALE AS NOTED
DATE 7/30/98
PROJECT ENGINEER RICHARD R. SCHOCH
DRAWN BY D.D.H.
COOPER ENGINEERING COMPANY
310 WEST SOUTH STREET RICE LAKE, WISCONSIN
TELEPHONE 715-234-7008
SHEET DESCRIPTION FIRST FLOOR/ MEZZANINE POWER PLAN
PROJECT NO.
SHEET NO. E3

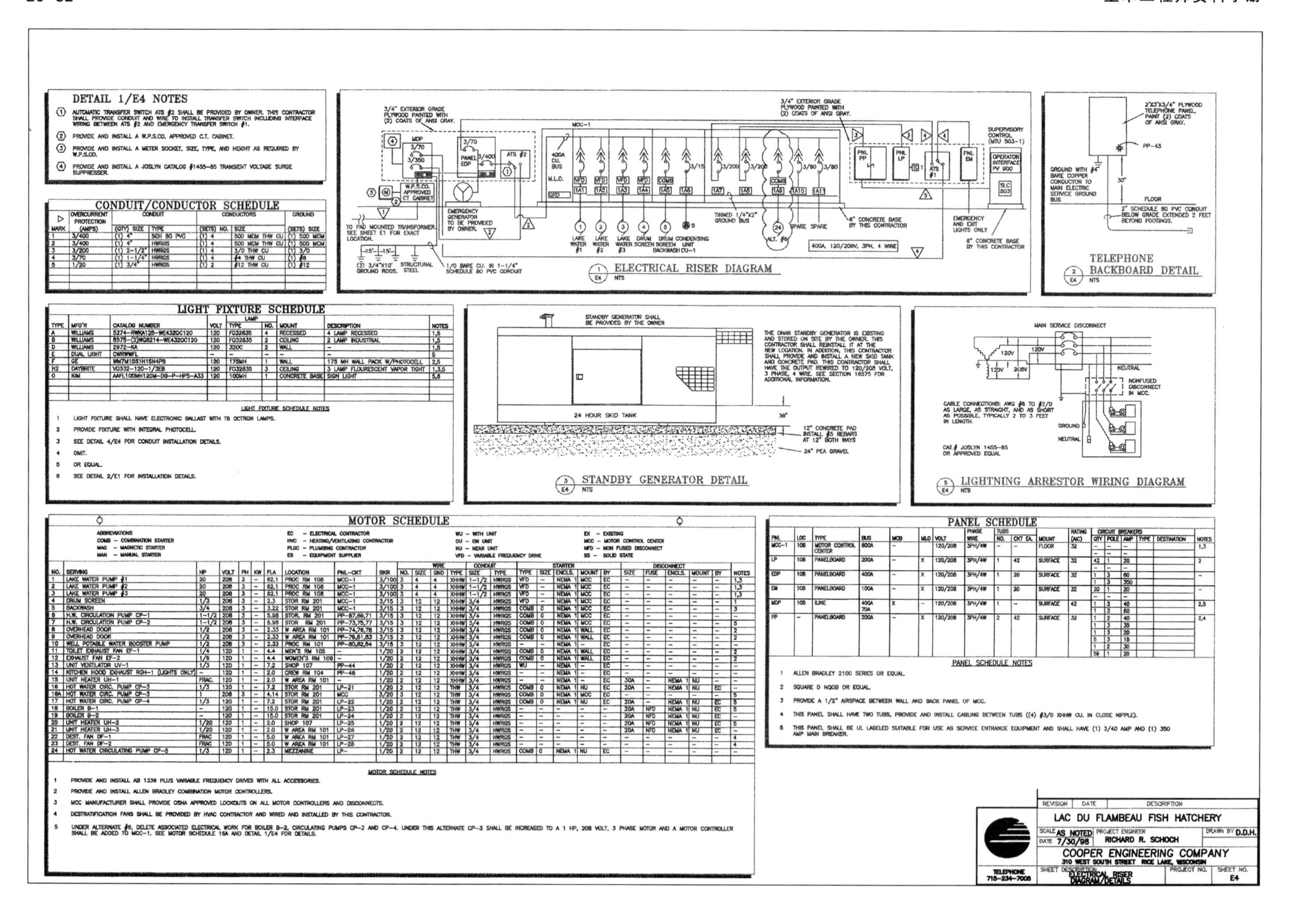
DETAIL 1/E4 NOTES
CONDUIT/CONDUCTOR SCHEDULE
ELECTRICAL RISER DIAGRAM
TELEPHONE BACKBOARD DETAIL
LIGHT FIXTURE SCHEDULE
STANDBY GENERATOR DETAIL
LIGHTNING ARRESTOR WIRING DIAGRAM
MOTOR SCHEDULE
PANEL SCHEDULE
LAC DU FLAMBEAU FISH HATCHERY
RICHARD R. SCHOCH
COOPER ENGINEERING COMPANY
310 WEST SOUTH STREET RICE LAKE, WISCONSIN
TELEPHONE 715-234-7008
E4

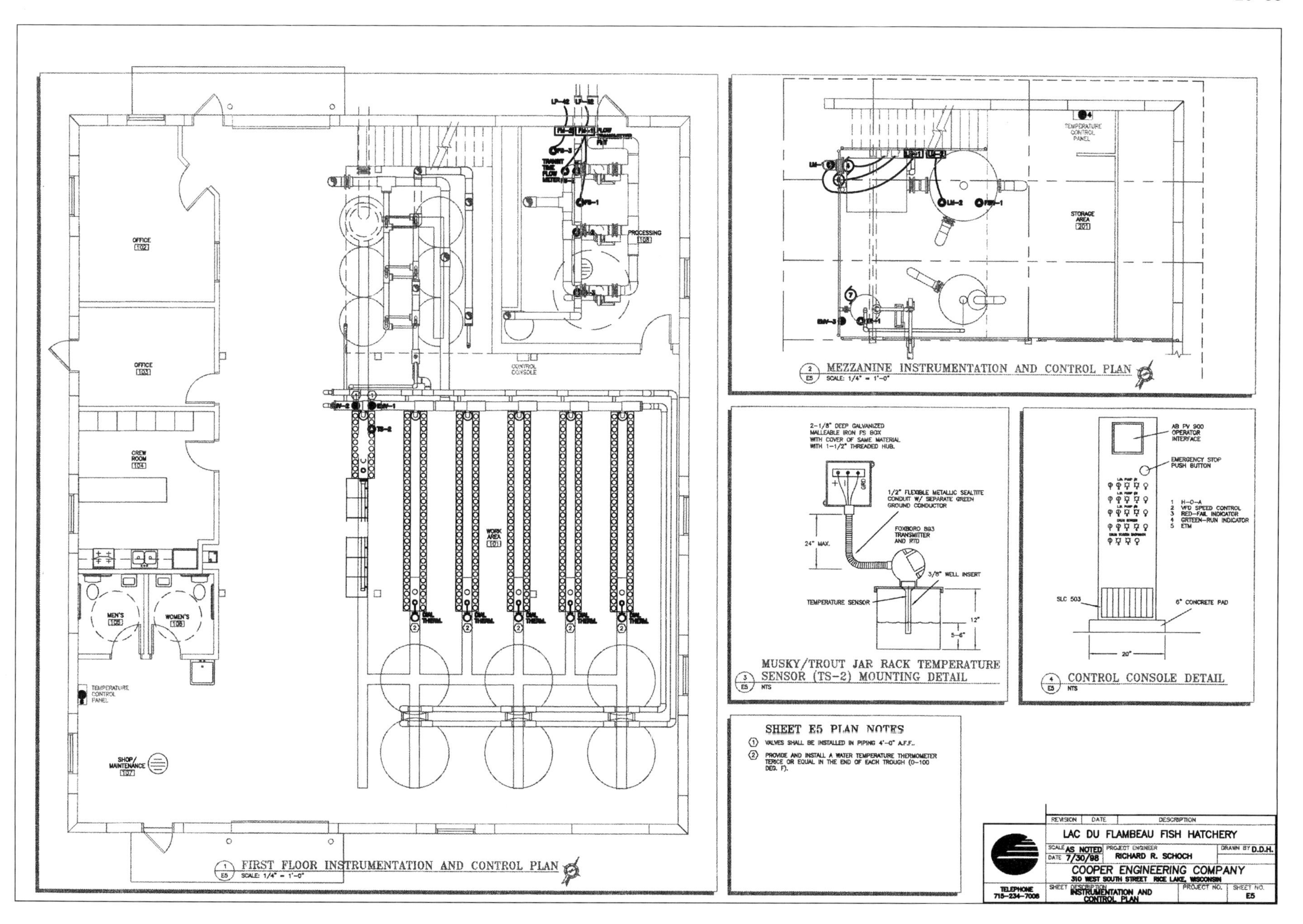
OFFICE
102
OFFICE
103
CREW ROOM
104
MEN'S
105
WOMEN'S
106
TEMPERATURE CONTROL PANEL
SHOP/ MAINTENANCE
107
WORK AREA
101
PROCESSING
108
CONTROL CONSOLE
DIAL THERM.
FIRST FLOOR INSTRUMENTATION AND CONTROL PLAN
SCALE: 1/4" = 1'-0"
TEMPERATURE CONTROL PANEL
STORAGE AREA
201
MEZZANINE INSTRUMENTATION AND CONTROL PLAN
SCALE: 1/4" = 1'-0"
2-1/8" DEEP GALVANIZED MALLEABLE IRON FS BOX WITH COVER OF SAME MATERIAL WITH 1-1/2" THREADED HUB.
1/2" FLEXIBLE METALLIC SEALTITE CONDUIT W/ SEPARATE GREEN GROUND CONDUCTOR
FOXBORO 893 TRANSMITTER AND RTD
24" MAX.
3/8" WELL INSERT
TEMPERATURE SENSOR
12"
5-6"
MUSKY/TROUT JAR RACK TEMPERATURE SENSOR (TS-2) MOUNTING DETAIL
NTS
AB PV 900 OPERATOR INTERFACE
EMERGENCY STOP PUSH BUTTON
1 H-O-A
2 VFD SPEED CONTROL
3 RED-FAIL INDICATOR
4 GRTEEN-RUN INDICATOR
5 ETM
SLC 503
6" CONCRETE PAD
20"
CONTROL CONSOLE DETAIL
NTS
SHEET E5 PLAN NOTES
1 VALVES SHALL BE INSTALLED IN PIPING 4'-0" A.F.F..
2 PROVIDE AND INSTALL A WATER TEMPERATURE THERMOMETER TERICE OR EQUAL IN THE END OF EACH TROUGH (0-100 DEG. F).
REVISION
DATE
DESCRIPTION
LAC DU FLAMBEAU FISH HATCHERY
SCALE AS NOTED
DATE 7/30/98
PROJECT ENGINEER RICHARD R. SCHOCH
DRAWN BY D.D.H.
COOPER ENGINEERING COMPANY
310 WEST SOUTH STREET RICE LAKE, WISCONSIN
TELEPHONE 715-234-7008
SHEET DESCRIPTION INSTRUMENTATION AND CONTROL PLAN
PROJECT NO.
SHEET NO. E5

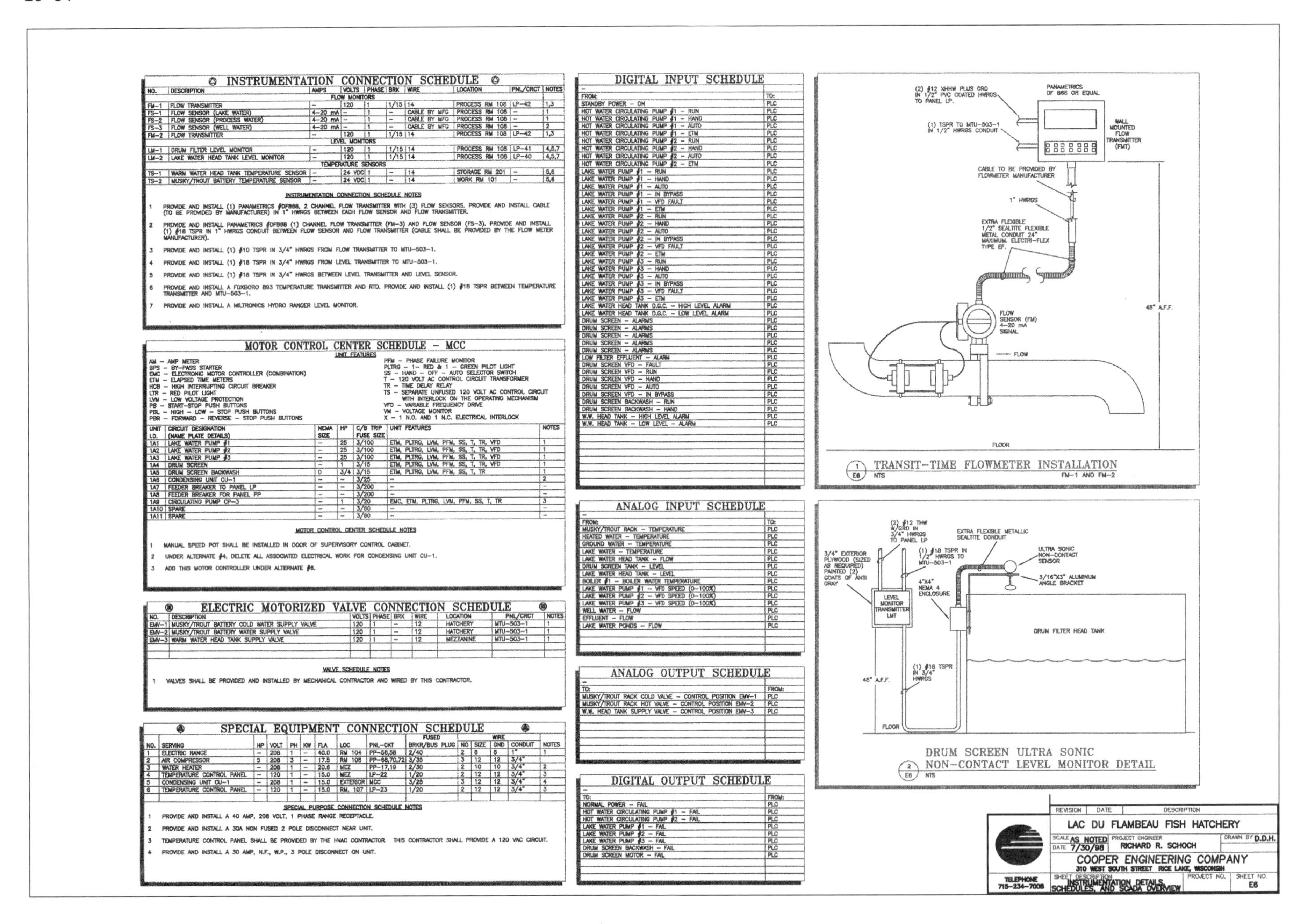

INSTRUMENTATION CONNECTION SCHEDULE

NO.	DESCRIPTION	AMPS	VOLTS	PHASE	BRK	WIRE	LOCATION	PNL/CRCT	NOTES
	FLOW MONITORS								
FM-1	FLOW TRANSMITTER	–	120	1	1/15	14	PROCESS RM 108	LP-42	1,3
FS-1	FLOW SENSOR (LAKE WATER)	4-20 mA	–	1	–	CABLE BY MFG	PROCESS RM 108	–	1
FS-2	FLOW SENSOR (PROCESS WATER)	4-20 mA	–	1	–	CABLE BY MFG	PROCESS RM 108	–	1
FS-3	FLOW SENSOR (WELL WATER)	4-20 mA	–	1	–	CABLE BY MFG	PROCESS RM 108	–	2
FM-2	FLOW TRANSMITTER	–	120	1	1/15	14	PROCESS RM 108	LP-42	1,3
	LEVEL MONITORS								
LM-1	DRUM FILTER LEVEL MONITOR	–	120	1	1/15	14	PROCESS RM 108	LP-41	4,5,7
LM-2	LAKE WATER HEAD TANK LEVEL MONITOR	–	120	1	1/15	14	PROCESS RM 108	LP-40	4,5,7
	TEMPERATURE SENSORS								
TS-1	WARM WATER HEAD TANK TEMPERATURE SENSOR	–	24 VDC	1	–	14	STORAGE RM 201	–	5,6
TS-2	MUSKY/TROUT BATTERY TEMPERATURE SENSOR	–	24 VDC	1	–	14	WORK RM 101	–	5,6

INSTRUMENTATION CONNECTION SCHEDULE NOTES

1. PROVIDE AND INSTALL (1) PANAMETRICS #DF868, 2 CHANNEL FLOW TRANSMITTER WITH (3) FLOW SENSORS. PROVIDE AND INSTALL CABLE (TO BE PROVIDED BY MANUFACTURER) IN 1" HWRGS BETWEEN EACH FLOW SENSOR AND FLOW TRANSMITTER.
2. PROVIDE AND INSTALL PANAMETRICS #DF868 (1) CHANNEL FLOW TRANSMITTER (FM-3) AND FLOW SENSOR (FS-3). PROVIDE AND INSTALL (1) #18 TSPR IN 1" HWRGS CONDUIT BETWEEN FLOW SENSOR AND FLOW TRANSMITTER (CABLE SHALL BE PROVIDED BY THE FLOW METER MANUFACTURER).
3. PROVIDE AND INSTALL (1) #10 TSPR IN 3/4" HWRGS FROM FLOW TRANSMITTER TO MTU-503-1.
4. PROVIDE AND INSTALL (1) #18 TSPR IN 3/4" HWRGS FROM LEVEL TRANSMITTER TO MTU-503-1.
5. PROVIDE AND INSTALL (1) #18 TSPR IN 3/4" HWRGS BETWEEN LEVEL TRANSMITTER AND LEVEL SENSOR.
6. PROVIDE AND INSTALL A FOXBORO 893 TEMPERATURE TRANSMITTER AND RTD. PROVIDE AND INSTALL (1) #18 TSPR BETWEEN TEMPERATURE TRANSMITTER AND MTU-503-1.
7. PROVIDE AND INSTALL A MILTRONICS HYDRO RANGER LEVEL MONITOR.

MOTOR CONTROL CENTER SCHEDULE – MCC

UNIT FEATURES

AM – AMP METER
BPS – BY-PASS STARTER
EMC – ELECTRONIC MOTOR CONTROLLER (COMBINATION)
ETM – ELAPSED TIME METERS
HCB – HIGH INTERRUPTING CIRCUIT BREAKER
LTR – RED PILOT LIGHT
LVM – LOW VOLTAGE PROTECTION
PB – START-STOP PUSH BUTTONS
PBL – HIGH – LOW – STOP PUSH BUTTONS
PBR – FORWARD – REVERSE – STOP PUSH BUTTONS
PFM – PHASE FAILURE MONITOR
PLTRG – 1 – RED & 1 – GREEN PILOT LIGHT
SS – HAND – OFF – AUTO SELECTOR SWITCH
T – 120 VOLT AC CONTROL CIRCUIT TRANSFORMER
TR – TIME DELAY RELAY
TS – SEPARATE UNFUSED 120 VOLT AC CONTROL CIRCUIT WITH INTERLOCK ON THE OPERATING MECHANISM
VFD – VARIABLE FREQUENCY DRIVE
VM – VOLTAGE MONITOR
X – 1 N.O. AND 1 N.C. ELECTRICAL INTERLOCK

UNIT I.D.	CIRCUIT DESIGNATION (NAME PLATE DETAILS)	NEMA SIZE	HP	C/B TRIP FUSE SIZE	UNIT FEATURES	NOTES
1A1	LAKE WATER PUMP #1	–	25	3/100	ETM, PLTRG, LVM, PFM, SS, T, TR, VFD	1
1A2	LAKE WATER PUMP #2	–	25	3/100	ETM, PLTRG, LVM, PFM, SS, T, TR, VFD	1
1A3	LAKE WATER PUMP #3	–	25	3/100	ETM, PLTRG, LVM, PFM, SS, T, TR, VFD	1
1A4	DRUM SCREEN	–	1	3/15	ETM, PLTRG, LVM, PFM, SS, T, TR, VFD	1
1A5	DRUM SCREEN BACKWASH	0	3/4	3/15	ETM, PLTRG, LVM, PFM, SS, T, TR	1
1A6	CONDENSING UNIT CU-1	–	–	3/25	–	2
1A7	FEEDER BREAKER TO PANEL LP	–	–	3/200	–	–
1A8	FEEDER BREAKER FOR PANEL PP	–	–	3/200	–	–
1A9	CIRCULATING PUMP CP-3	–	1	3/20	EMC, ETM, PLTRG, LVM, PFM, SS, T, TR	3
1A10	SPARE	–	–	3/80	–	–
1A11	SPARE	–	–	3/80	–	–

MOTOR CONTROL CENTER SCHEDULE NOTES

1. MANUAL SPEED POT SHALL BE INSTALLED IN DOOR OF SUPERVISORY CONTROL CABINET.
2. UNDER ALTERNATE #4, DELETE ALL ASSOCIATED ELECTRICAL WORK FOR CONDENSING UNIT CU-1.
3. ADD THIS MOTOR CONTROLLER UNDER ALTERNATE #6.

ELECTRIC MOTORIZED VALVE CONNECTION SCHEDULE

NO.	DESCRIPTION	VOLTS	PHASE	BRK	WIRE	LOCATION	PNL/CRCT	NOTES
EMV-1	MUSKY/TROUT BATTERY COLD WATER SUPPLY VALVE	120	1	–	12	HATCHERY	MTU-503-1	1
EMV-2	MUSKY/TROUT BATTERY WATER SUPPLY VALVE	120	1	–	12	HATCHERY	MTU-503-1	1
EMV-3	WARM WATER HEAD TANK SUPPLY VALVE	120	1	–	12	MEZZANINE	MTU-503-1	1

VALVE SCHEDULE NOTES

1. VALVES SHALL BE PROVIDED AND INSTALLED BY MECHANICAL CONTRACTOR AND WIRED BY THIS CONTRACTOR.

SPECIAL EQUIPMENT CONNECTION SCHEDULE

NO.	SERVING	HP	VOLT	PH	KW	FLA	LOC	PNL-CKT	FUSED BRKR/BUS PLUG	WIRE NO	WIRE SIZE	WIRE GND	CONDUIT	NOTES
1	ELECTRIC RANGE	–	208	1	–	40.0	RM 104	PP-56,58	2/40	2	8	8	1"	1
2	AIR COMPRESSOR	5	208	3	–	17.5	RM 108	PP-68,70,72	3/35	3	12	12	3/4"	
3	WATER HEATER	–	208	1	–	20.6	MEZ	PP-17,19	2/30	2	10	10	3/4"	2
4	TEMPERATURE CONTROL PANEL	–	120	1	–	15.0	MEZ	LP-22	1/20	2	12	12	3/4"	3
5	CONDENSING UNIT CU-1	–	208	1	–	15.0	EXTERIOR	MCC	3/25	3	12	12	3/4"	4
6	TEMPERATURE CONTROL PANEL	–	120	1	–	15.0	RM. 107	LP-23	1/20	2	12	12	3/4"	3

SPECIAL PURPOSE CONNECTION SCHEDULE NOTES

1. PROVIDE AND INSTALL A 40 AMP, 208 VOLT, 1 PHASE RANGE RECEPTACLE.
2. PROVIDE AND INSTALL A 30A NON FUSED 2 POLE DISCONNECT NEAR UNIT.
3. TEMPERATURE CONTROL PANEL SHALL BE PROVIDED BY THE HVAC CONTRACTOR. THIS CONTRACTOR SHALL PROVIDE A 120 VAC CIRCUIT.
4. PROVIDE AND INSTALL A 30 AMP, N.F., W.P., 3 POLE DISCONNECT ON UNIT.

DIGITAL INPUT SCHEDULE

FROM:	TO:
STANDBY POWER – ON	PLC
HOT WATER CIRCULATING PUMP #1 – RUN	PLC
HOT WATER CIRCULATING PUMP #1 – HAND	PLC
HOT WATER CIRCULATING PUMP #1 – AUTO	PLC
HOT WATER CIRCULATING PUMP #1 – ETM	PLC
HOT WATER CIRCULATING PUMP #2 – RUN	PLC
HOT WATER CIRCULATING PUMP #2 – HAND	PLC
HOT WATER CIRCULATING PUMP #2 – AUTO	PLC
HOT WATER CIRCULATING PUMP #2 – ETM	PLC
LAKE WATER PUMP #1 – RUN	PLC
LAKE WATER PUMP #1 – HAND	PLC
LAKE WATER PUMP #1 – AUTO	PLC
LAKE WATER PUMP #1 – IN BYPASS	PLC
LAKE WATER PUMP #1 – VFD FAULT	PLC
LAKE WATER PUMP #1 – ETM	PLC
LAKE WATER PUMP #2 – RUN	PLC
LAKE WATER PUMP #2 – HAND	PLC
LAKE WATER PUMP #2 – AUTO	PLC
LAKE WATER PUMP #2 – IN BYPASS	PLC
LAKE WATER PUMP #2 – VFD FAULT	PLC
LAKE WATER PUMP #2 – ETM	PLC
LAKE WATER PUMP #3 – RUN	PLC
LAKE WATER PUMP #3 – HAND	PLC
LAKE WATER PUMP #3 – AUTO	PLC
LAKE WATER PUMP #3 – IN BYPASS	PLC
LAKE WATER PUMP #3 – VFD FAULT	PLC
LAKE WATER PUMP #3 – ETM	PLC
LAKE WATER HEAD TANK D.G.C. – HIGH LEVEL ALARM	PLC
LAKE WATER HEAD TANK D.G.C. – LOW LEVEL ALARM	PLC
DRUM SCREEN – ALARMS	PLC
DRUM SCREEN – ALARMS	PLC
DRUM SCREEN – ALARMS	PLC
DRUM SCREEN – ALARMS	PLC
DRUM SCREEN – ALARMS	PLC
LOW FILTER EFFLUENT – ALARM	PLC
DRUM SCREEN VFD – FAULT	PLC
DRUM SCREEN VFD – RUN	PLC
DRUM SCREEN VFD – HAND	PLC
DRUM SCREEN VFD – AUTO	PLC
DRUM SCREEN VFD – IN BYPASS	PLC
DRUM SCREEN BACKWASH – RUN	PLC
DRUM SCREEN BACKWASH – HAND	PLC
W.W. HEAD TANK – HIGH LEVEL ALARM	PLC
W.W. HEAD TANK – LOW LEVEL – ALARM	PLC

ANALOG INPUT SCHEDULE

FROM:	TO:
MUSKY/TROUT RACK – TEMPERATURE	PLC
HEATED WATER – TEMPERATURE	PLC
GROUND WATER – TEMPERATURE	PLC
LAKE WATER – TEMPERATURE	PLC
LAKE WATER HEAD TANK – FLOW	PLC
DRUM SCREEN TANK – LEVEL	PLC
LAKE WATER HEAD TANK – LEVEL	PLC
BOILER #1 – BOILER WATER TEMPERATURE	PLC
LAKE WATER PUMP #1 – VFD SPEED (0–100%)	PLC
LAKE WATER PUMP #2 – VFD SPEED (0–100%)	PLC
LAKE WATER PUMP #3 – VFD SPEED (0–100%)	PLC
WELL WATER – FLOW	PLC
EFFLUENT – FLOW	PLC
LAKE WATER PONDS – FLOW	PLC

ANALOG OUTPUT SCHEDULE

TO:	FROM:
MUSKY/TROUT RACK COLD VALVE – CONTROL POSITION EMV-1	PLC
MUSKY/TROUT RACK HOT VALVE – CONTROL POSITION EMV-2	PLC
W.W. HEAD TANK SUPPLY VALVE – CONTROL POSITION EMV-3	PLC

DIGITAL OUTPUT SCHEDULE

TO:	FROM:
NORMAL POWER – FAIL	PLC
HOT WATER CIRCULATING PUMP #1 – FAIL	PLC
HOT WATER CIRCULATING PUMP #2 – FAIL	PLC
LAKE WATER PUMP #1 – FAIL	PLC
LAKE WATER PUMP #2 – FAIL	PLC
LAKE WATER PUMP #3 – FAIL	PLC
DRUM SCREEN BACKWASH – FAIL	PLC
DRUM SCREEN MOTOR – FAIL	PLC

1 E6 TRANSIT-TIME FLOWMETER INSTALLATION
NTS FM-1 AND FM-2

2 E6 DRUM SCREEN ULTRA SONIC NON-CONTACT LEVEL MONITOR DETAIL
NTS

REVISION	DATE	DESCRIPTION

LAC DU FLAMBEAU FISH HATCHERY

SCALE AS NOTED | DATE 7/30/98 | PROJECT ENGINEER RICHARD R. SCHOCH | DRAWN BY D.D.H.

COOPER ENGINEERING COMPANY
310 WEST SOUTH STREET RICE LAKE, WISCONSIN
TELEPHONE 715-234-7008

SHEET DESCRIPTION: INSTRUMENTATION DETAILS, SCHEDULES, AND SCADA OVERVIEW | PROJECT NO. | SHEET NO. E6

设　计

第21节 标准详图

例 21

每个专业设计师应当拥有广泛的典型和标准详图，可被重复用于合适的项目。这些插图通常经过了时间考验，并已经成为项目设计图的关键组成部分。一般来说，它们被打印在8 1/2″×11″页面上，并被插入技术说明书中，而不是被复制在设计图纸中。

下面给出了一些标准详图的示例。它们放在这里仅供参考，而不能直接用作最终的权威设计，除非它正确地满足了所有的现行规范、规程和技术要求。

目 录

Erosion Control 腐蚀控制

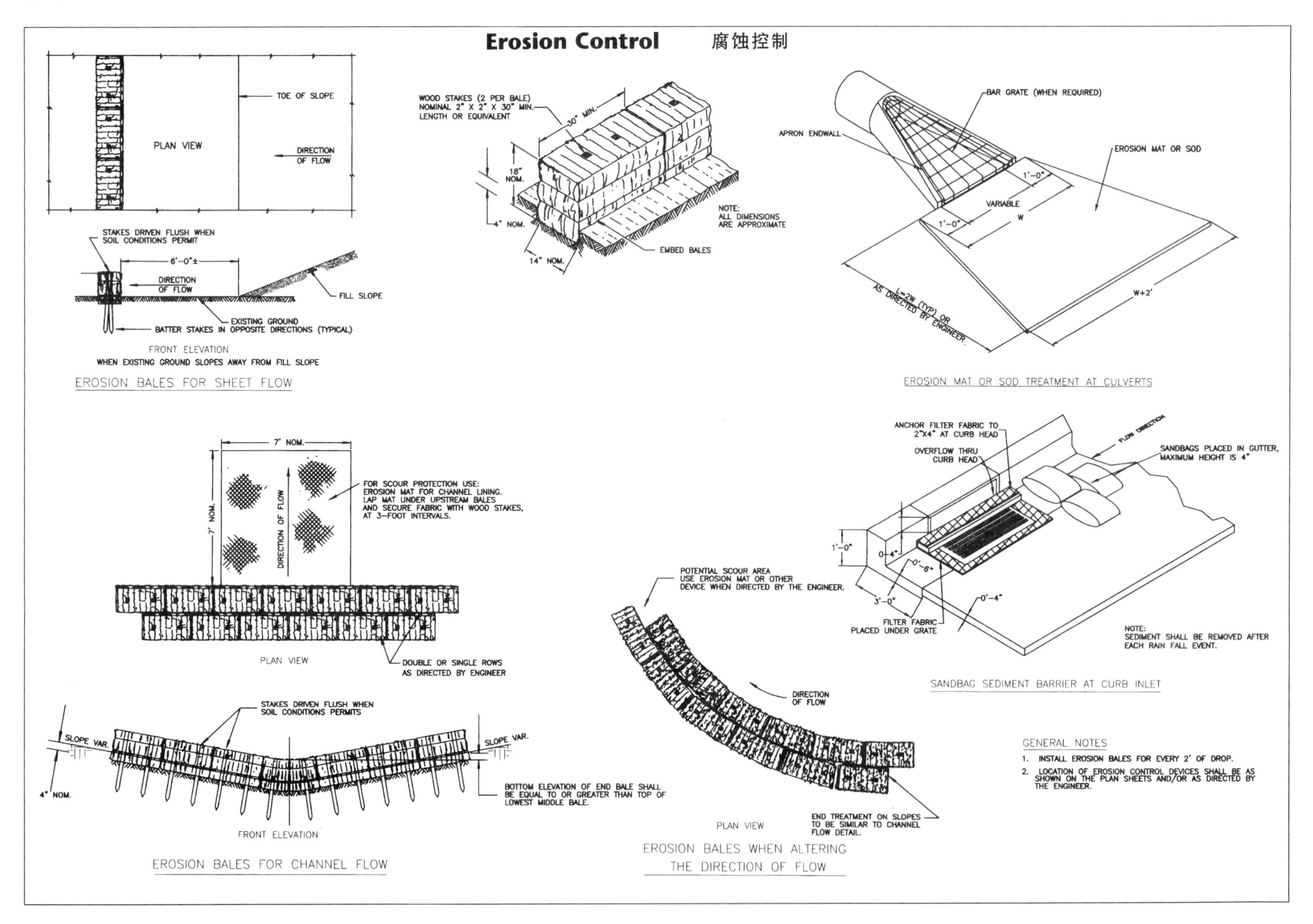

Source: Morgan & Parmley, Ltd.

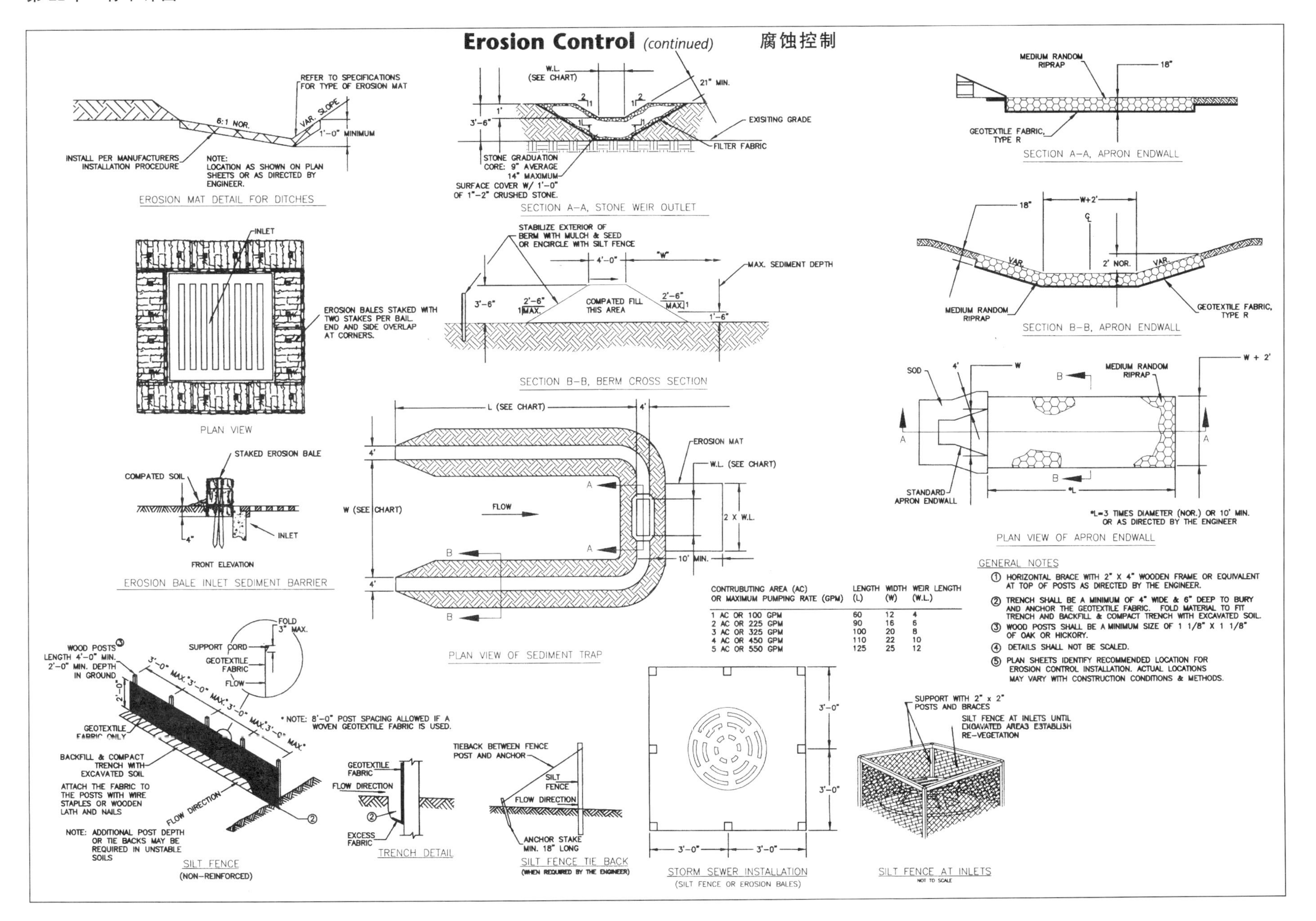

Erosion Control (continued) 腐蚀控制
REFER TO SPECIFICATIONS FOR TYPE OF EROSION MAT
6:1 NOR.
VAR. SLOPE
1'-0" MINIMUM
INSTALL PER MANUFACTURERS INSTALLATION PROCEDURE
NOTE: LOCATION AS SHOWN ON PLAN SHEETS OR AS DIRECTED BY ENGINEER.
EROSION MAT DETAIL FOR DITCHES
W.L. (SEE CHART)
21" MIN.
EXISITING GRADE
FILTER FABRIC
STONE GRADUATION CORE: 9" AVERAGE 14" MAXIMUM
SURFACE COVER W/ 1'-0" OF 1"-2" CRUSHED STONE.
SECTION A-A, STONE WEIR OUTLET
MEDIUM RANDOM RIPRAP
18"
GEOTEXTILE FABRIC, TYPE R
SECTION A-A, APRON ENDWALL
INLET
EROSION BALES STAKED WITH TWO STAKES PER BAIL. END AND SIDE OVERLAP AT CORNERS.
PLAN VIEW
STABILIZE EXTERIOR OF BERM WITH MULCH & SEED OR ENCIRCLE WITH SILT FENCE
4'-0"
"W"
MAX. SEDIMENT DEPTH
3'-6"
2'-6" MAX.
COMPATED FILL THIS AREA
1'-6"
SECTION B-B, BERM CROSS SECTION
W+2'
VAR.
2' NOR.
MEDIUM RANDOM RIPRAP
GEOTEXTILE FABRIC, TYPE R
SECTION B-B, APRON ENDWALL
STAKED EROSION BALE
COMPATED SOIL
INLET
FRONT ELEVATION
EROSION BALE INLET SEDIMENT BARRIER
L (SEE CHART)
EROSION MAT
W.L. (SEE CHART)
W (SEE CHART)
FLOW
2 X W.L.
10' MIN.
PLAN VIEW OF SEDIMENT TRAP
SOD
W
W + 2'
MEDIUM RANDOM RIPRAP
STANDARD APRON ENDWALL
*L=3 TIMES DIAMETER (NOR.) OR 10' MIN. OR AS DIRECTED BY THE ENGINEER
PLAN VIEW OF APRON ENDWALL
CONTRUBUTING AREA (AC) OR MAXIMUM PUMPING RATE (GPM) | LENGTH (L) | WIDTH (W) | WEIR LENGTH (W.L.)
1 AC OR 100 GPM | 60 | 12 | 4
2 AC OR 225 GPM | 90 | 16 | 6
3 AC OR 325 GPM | 100 | 20 | 8
4 AC OR 450 GPM | 110 | 22 | 10
5 AC OR 550 GPM | 125 | 25 | 12
GENERAL NOTES
① HORIZONTAL BRACE WITH 2" X 4" WOODEN FRAME OR EQUIVALENT AT TOP OF POSTS AS DIRECTED BY THE ENGINEER.
② TRENCH SHALL BE A MINIMUM OF 4" WIDE & 6" DEEP TO BURY AND ANCHOR THE GEOTEXTILE FABRIC. FOLD MATERIAL TO FIT TRENCH AND BACKFILL & COMPACT TRENCH WITH EXCAVATED SOIL.
③ WOOD POSTS SHALL BE A MINIMUM SIZE OF 1 1/8" X 1 1/8" OF OAK OR HICKORY.
④ DETAILS SHALL NOT BE SCALED.
⑤ PLAN SHEETS IDENTIFY RECOMMENDED LOCATION FOR EROSION CONTROL INSTALLATION. ACTUAL LOCATIONS MAY VARY WITH CONSTRUCTION CONDITIONS & METHODS.
WOOD POSTS③ LENGTH 4'-0" MIN. 2'-0" MIN. DEPTH IN GROUND
SUPPORT CORD
FOLD 3" MAX.
GEOTEXTILE FABRIC
FLOW
3'-0" MAX.
*NOTE: 8'-0" POST SPACING ALLOWED IF A WOVEN GEOTEXTILE FABRIC IS USED.
GEOTEXTILE FABRIC ONLY
BACKFILL & COMPACT TRENCH WITH EXCAVATED SOIL
ATTACH THE FABRIC TO THE POSTS WITH WIRE STAPLES OR WOODEN LATH AND NAILS
FLOW DIRECTION
NOTE: ADDITIONAL POST DEPTH OR TIE BACKS MAY BE REQUIRED IN UNSTABLE SOILS
SILT FENCE (NON-REINFORCED)
GEOTEXTILE FABRIC
FLOW DIRECTION
EXCESS FABRIC
TRENCH DETAIL
TIEBACK BETWEEN FENCE POST AND ANCHOR
SILT FENCE
FLOW DIRECTION
ANCHOR STAKE MIN. 18" LONG
SILT FENCE TIE BACK (WHEN REQUIRED BY THE ENGINEER)
3'-0"
STORM SEWER INSTALLATION (SILT FENCE OR EROSION BALES)
SUPPORT WITH 2" x 2" POSTS AND BRACES
SILT FENCE AT INLETS UNTIL EXCAVATED AREAS ESTABLISH RE-VEGETATION
SILT FENCE AT INLETS
NOT TO SCALE

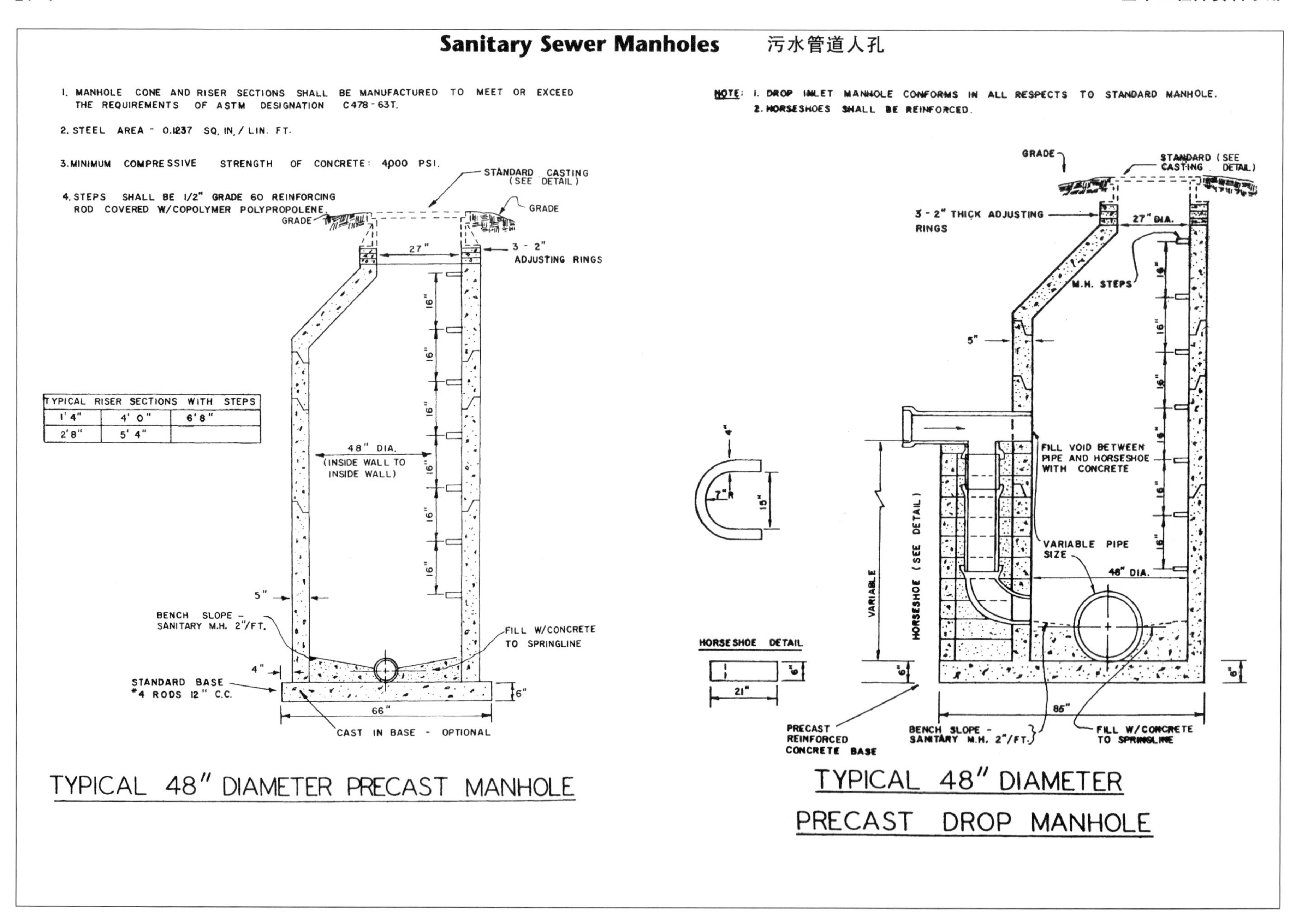

Source: Morgan & Parmley, Ltd.

典型柔性人孔套管

Typical Flexible Manhole Sleeve

SLEEVE DIMENSIONS			
PIPE DIA.	"A"	"B"	"C"
4"	6"	14"	6"
6"	8 1/8"	16 1/8"	6 1/2"
8"	10 3/8"	18 3/8"	7 1/2"
10"	12 5/8"	20 5/8"	7 1/2"
12"	14 7/8"	22 7/8"	7 1/2"
15"	18 7/8"	26 7/8"	9"

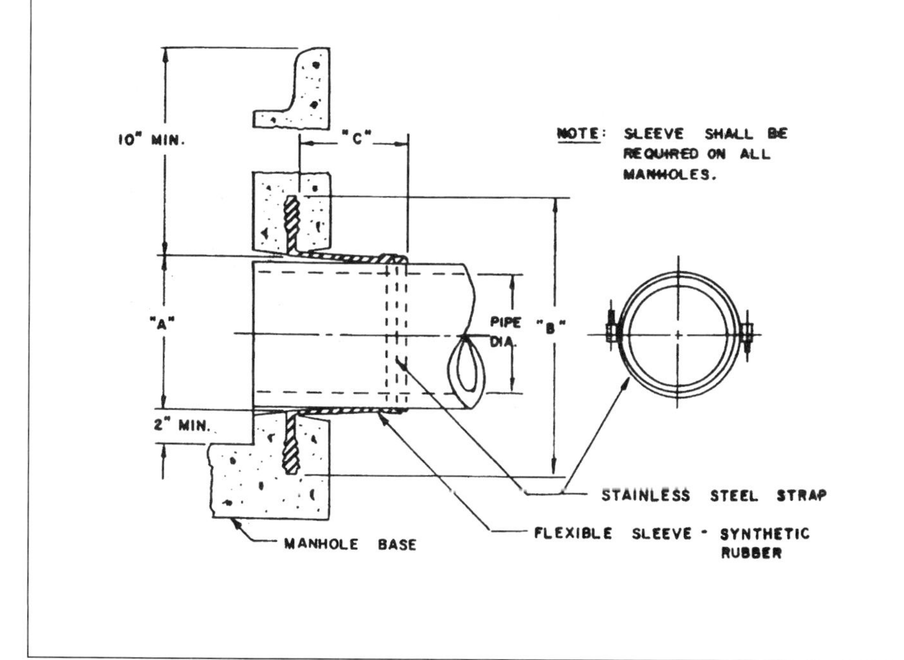

Source: Morgan & Parmley, Ltd.

管道安置标准截面

Standard Section for Laying Pipe

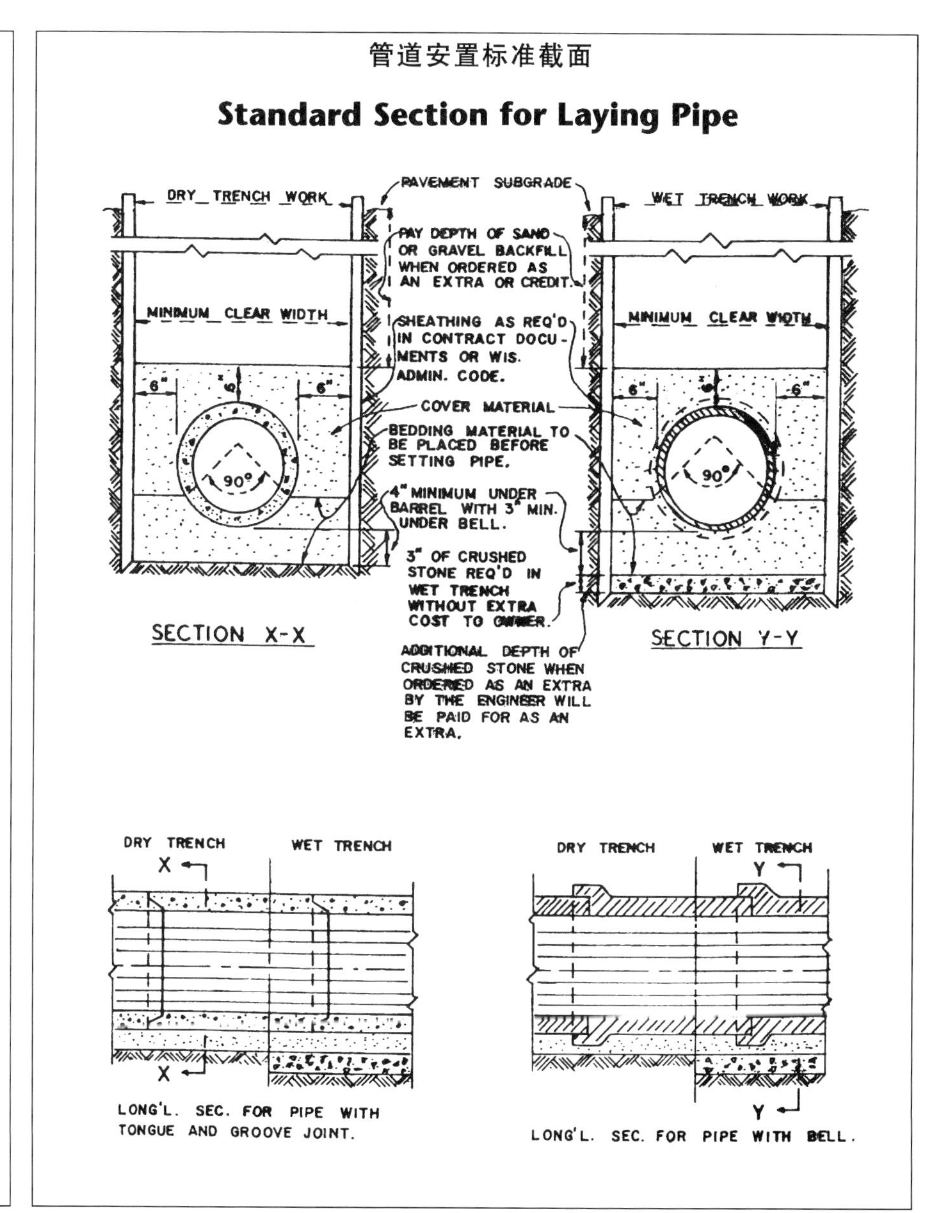

Source: Morgan & Parmley, Ltd.

交叉点的水平和竖向间隔

Horizontal & Vertical Separations at Intersection

WATERMAIN

10'- 0"

SANITARY SEWER

HORIZONTAL SEPARATION

WATERMAIN

6"

SANITARY SEWER

SANITARY SEWER

18"

WATERMAIN

VERTICAL SEPARATION

MINIMUM HORIZONTAL & VERTICAL SEPARATIONS AT WATERMAIN & SANITARY SEWER INTERSECTIONS

Source: Morgan & Parmley, Ltd.

常见管沟详图

Common Trench Detail

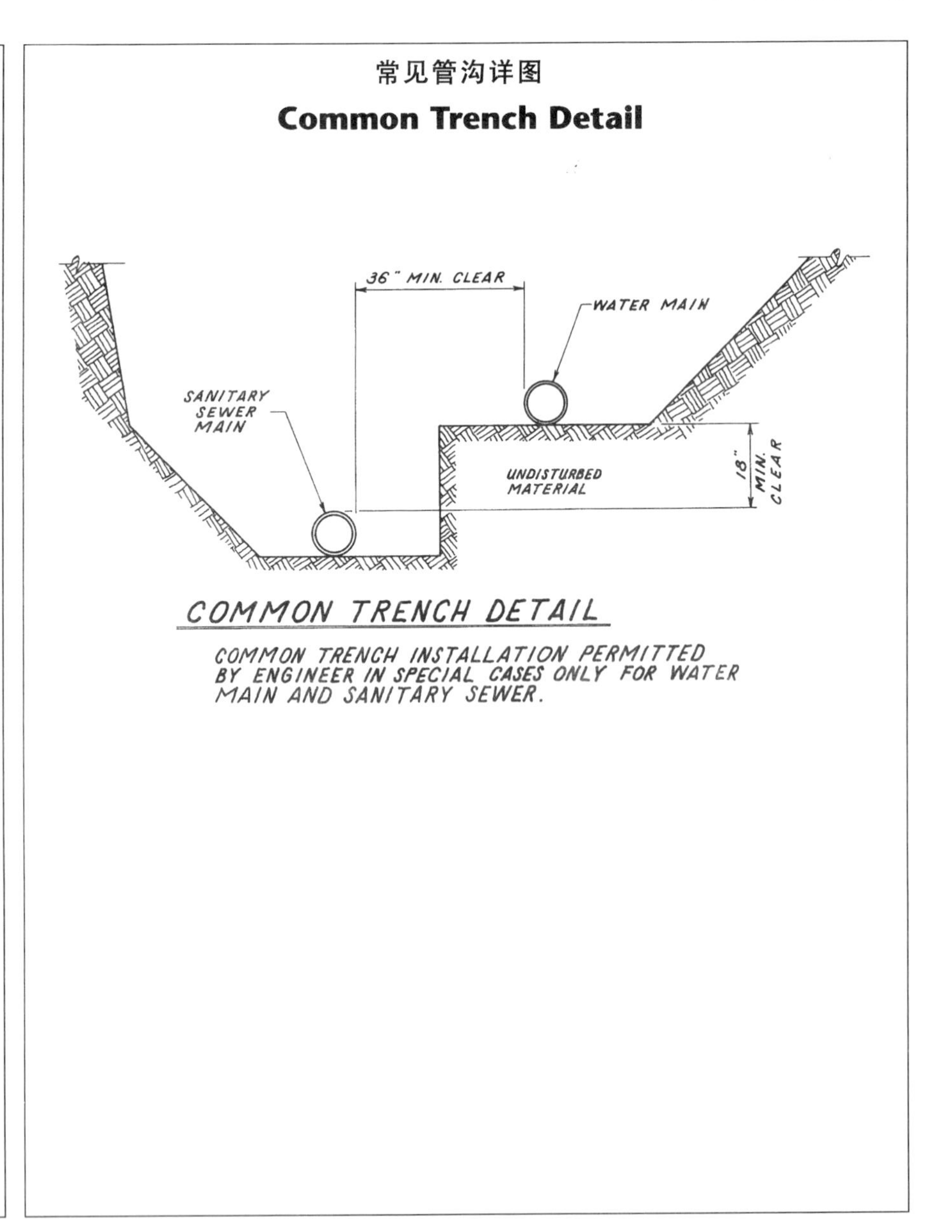

Source: Morgan & Parmley, Ltd.

安置污水管道的压缩截面

Compacted Section for Laying Sewer Pipe

PLACE COMPACTED BACKFILL IN 6"-12" LIFTS FROM TOP OF PIPE BEDDING TO SUBGRADE.

PAVEMENT SUBGRADE

PAY DEPTH OF SAND OR GRAVEL BACKFILL WHEN ORDERED AS AN EXTRA OR CREDIT.

TRENCH WIDTH & SIDE SLOPES PER CODE -OR- SHEATHING AS REQ'D IN CONTRACT DOCUMENTS OR WIS. ADMIN. CODE. SHEATHING SHALL BE LEFT IN PLACE IN COMPACTED AREA.

MINIMUM CLEAR WIDTH

12"

COMPACTED SAND COVER MATERIAL FROM TRENCHWALL OR IMPORTED.

6"

6"

SPRING LINE

BEDDING MATERIAL TO BE PLACED BEFORE SETTING PIPE. 4" MIN. UNDER BARREL, 3" MIN. UNDER BELL.

CAREFULLY COMPACTED BEDDING MATERIAL.

3" OF CRUSHED STONE REQUIRED IN WET TRENCH WITHOUT EXTRA COST TO OWNER.

Source: Morgan & Parmley, Ltd.

典型的立管和三通

Typical Riser & Wye Detail

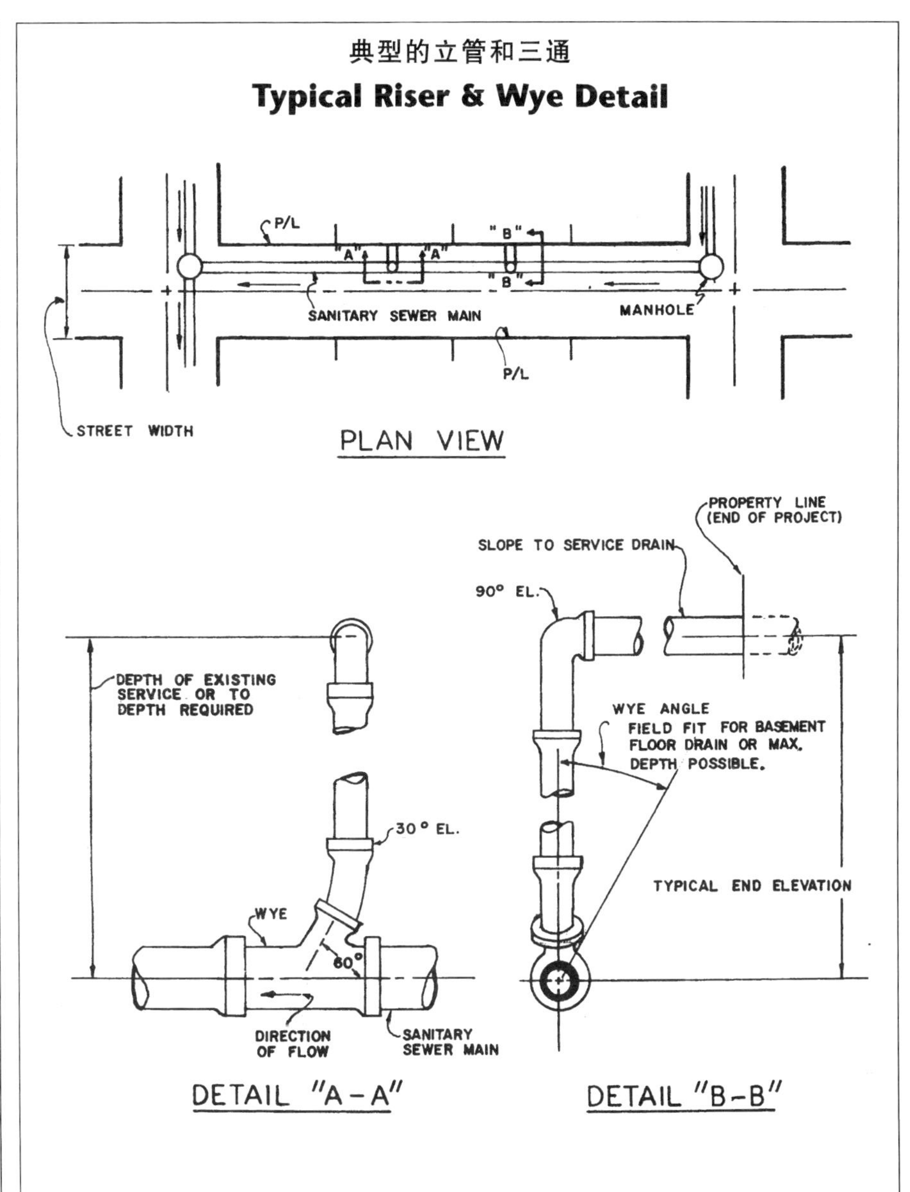

Source: Morgan & Parmley, Ltd.

Typical Hydrant Installation 典型的消火栓安装

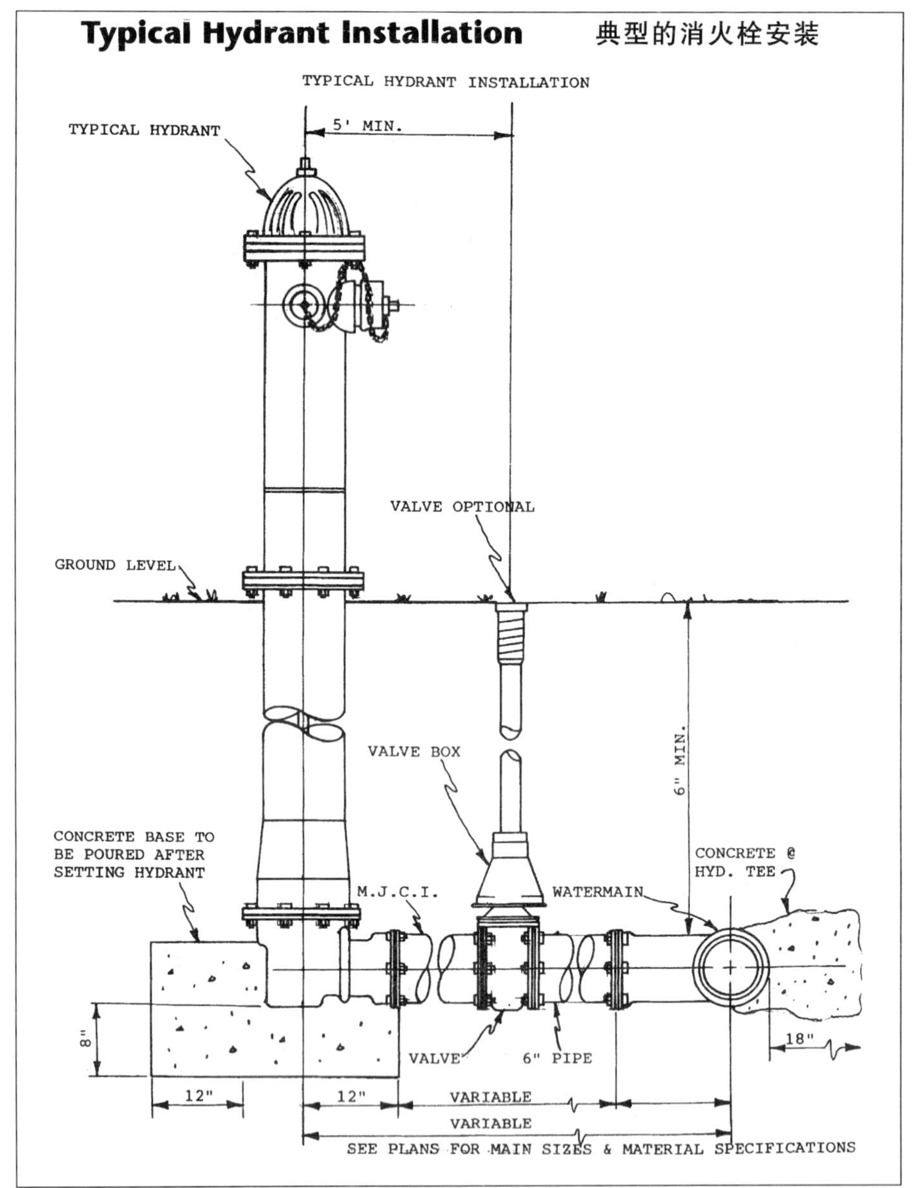

Source: Morgan & Parmley, Ltd.

典型的干式消火栓安装 Typical Dry Hydrant Installation

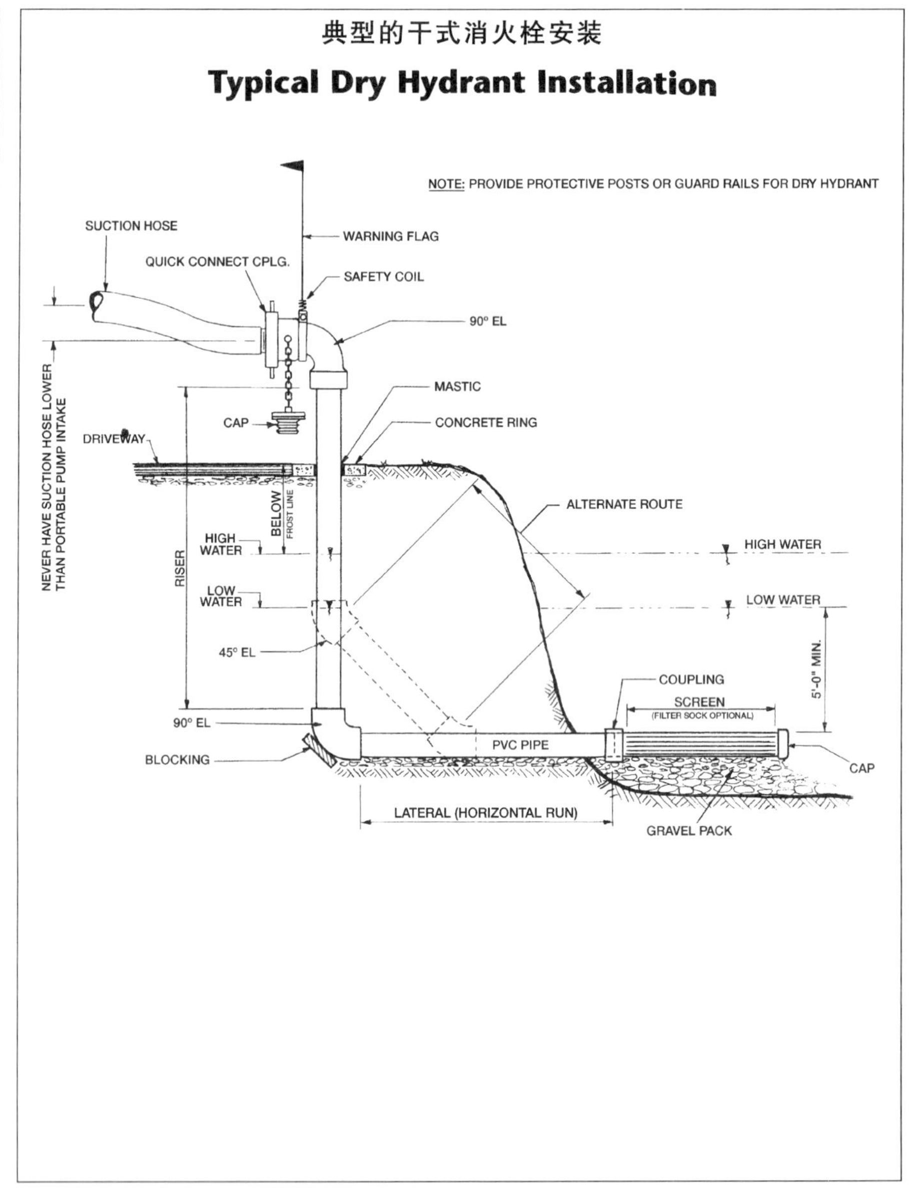

Source: *Hydraulics Field Manual*, 2/e
Robert O. Parmley, P.E.
McGraw-Hill © 2001

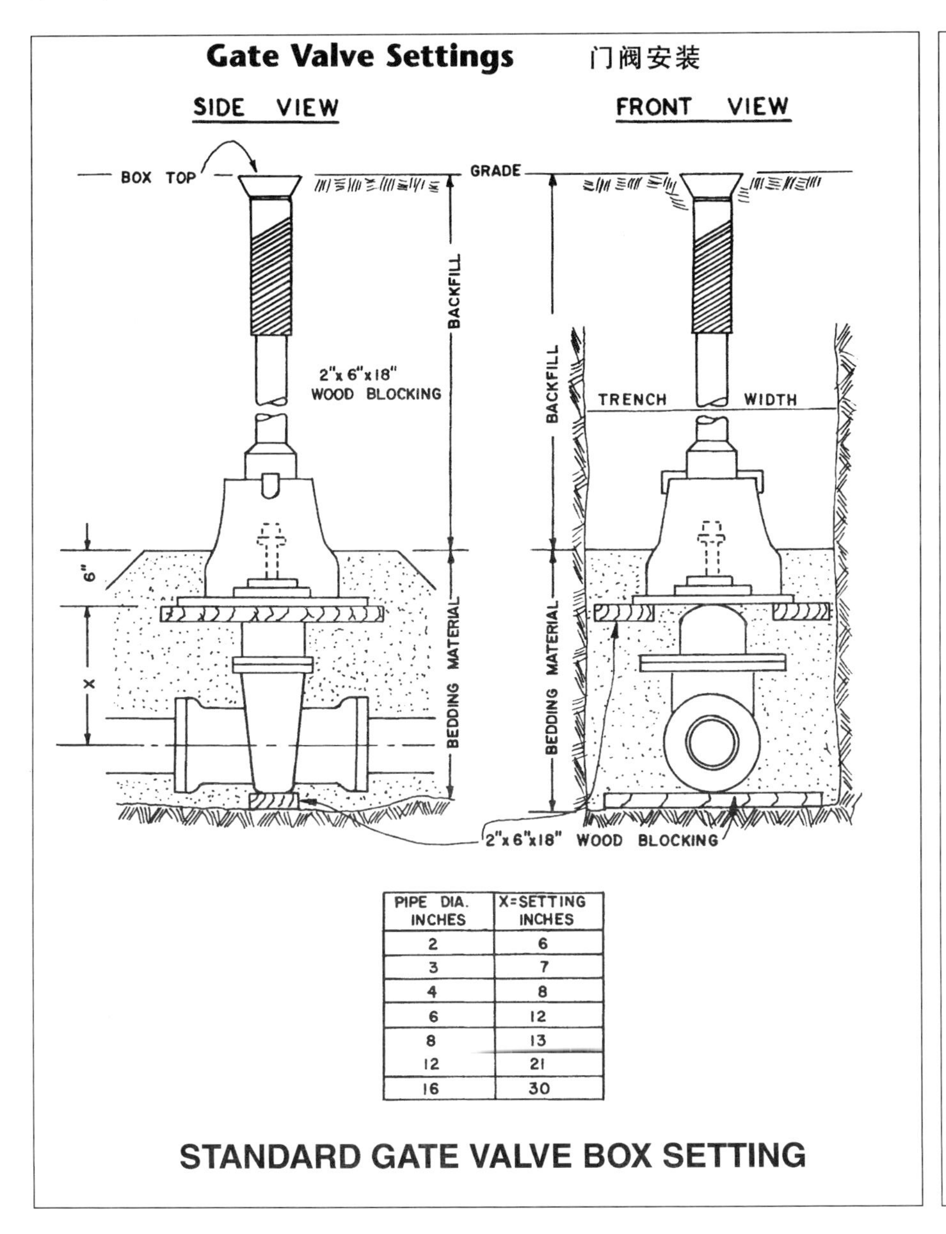

PIPE DIA. INCHES	X=SETTING INCHES
2	6
3	7
4	8
6	12
8	13
12	21
16	30

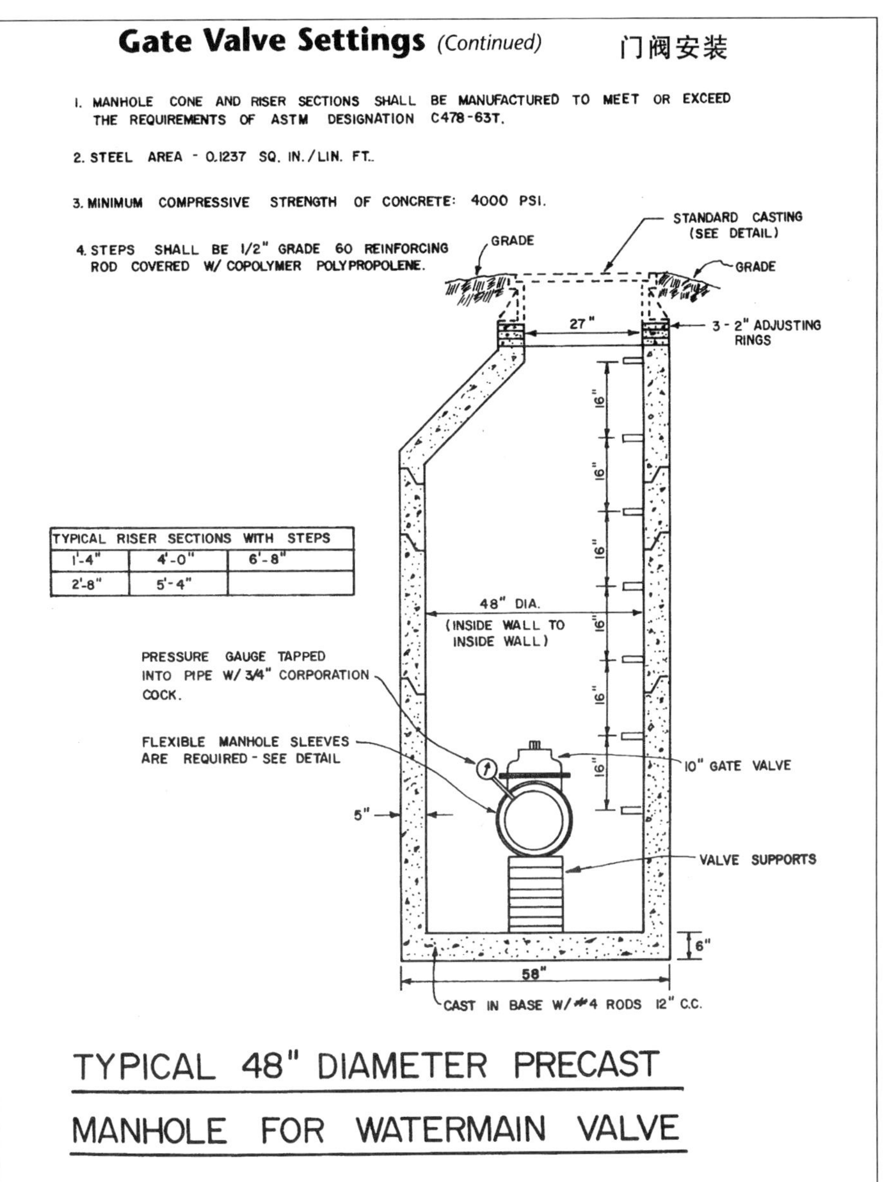

TYPICAL RISER SECTIONS WITH STEPS		
1'-4"	4'-0"	6'-8"
2'-8"	5'-4"	

Source: Morgan & Parmley, Ltd.

Buttresses for Tees 丁字接头和弯管的支持物

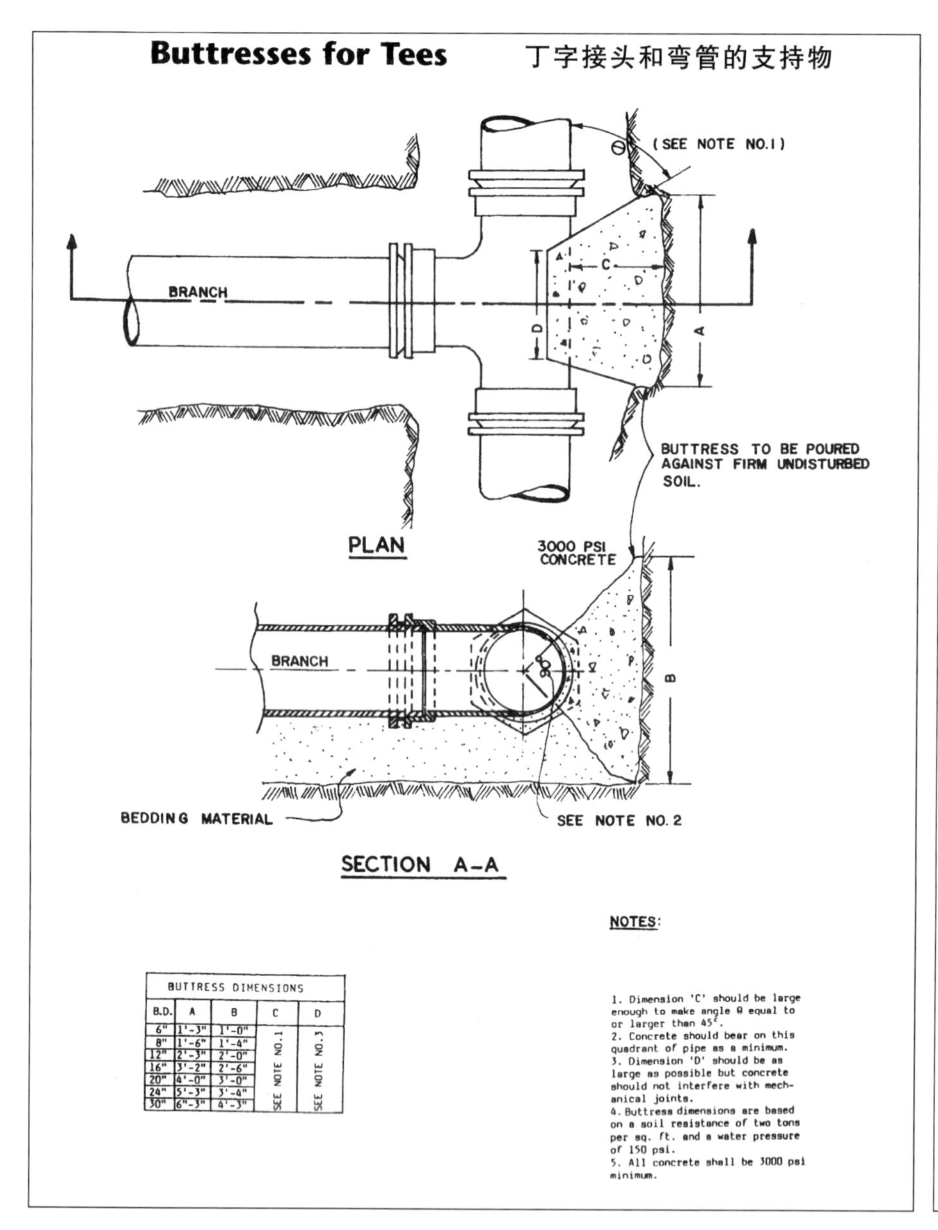

BUTTRESS DIMENSIONS				
B.D.	A	B	C	D
6"	1'-3"	1'-0"	SEE NOTE NO.1	SEE NOTE NO.3
8"	1'-6"	1'-4"		
12"	2'-3"	2'-0"		
16"	3'-2"	2'-6"		
20"	4'-0"	3'-0"		
24"	5'-3"	3'-4"		
30"	6"-3"	4'-3"		

NOTES:

1. Dimension 'C' should be large enough to make angle θ equal to or larger than 45°.
2. Concrete should bear on this quadrant of pipe as a minimum.
3. Dimension 'D' should be as large as possible but concrete should not interfere with mechanical joints.
4. Buttress dimensions are based on a soil resistance of two tons per sq. ft. and a water pressure of 150 psi.
5. All concrete shall be 3000 psi minimum.

Source: Morgan & Parmley, Ltd.

Buttresses for Bends 丁字接头和弯管的支持物

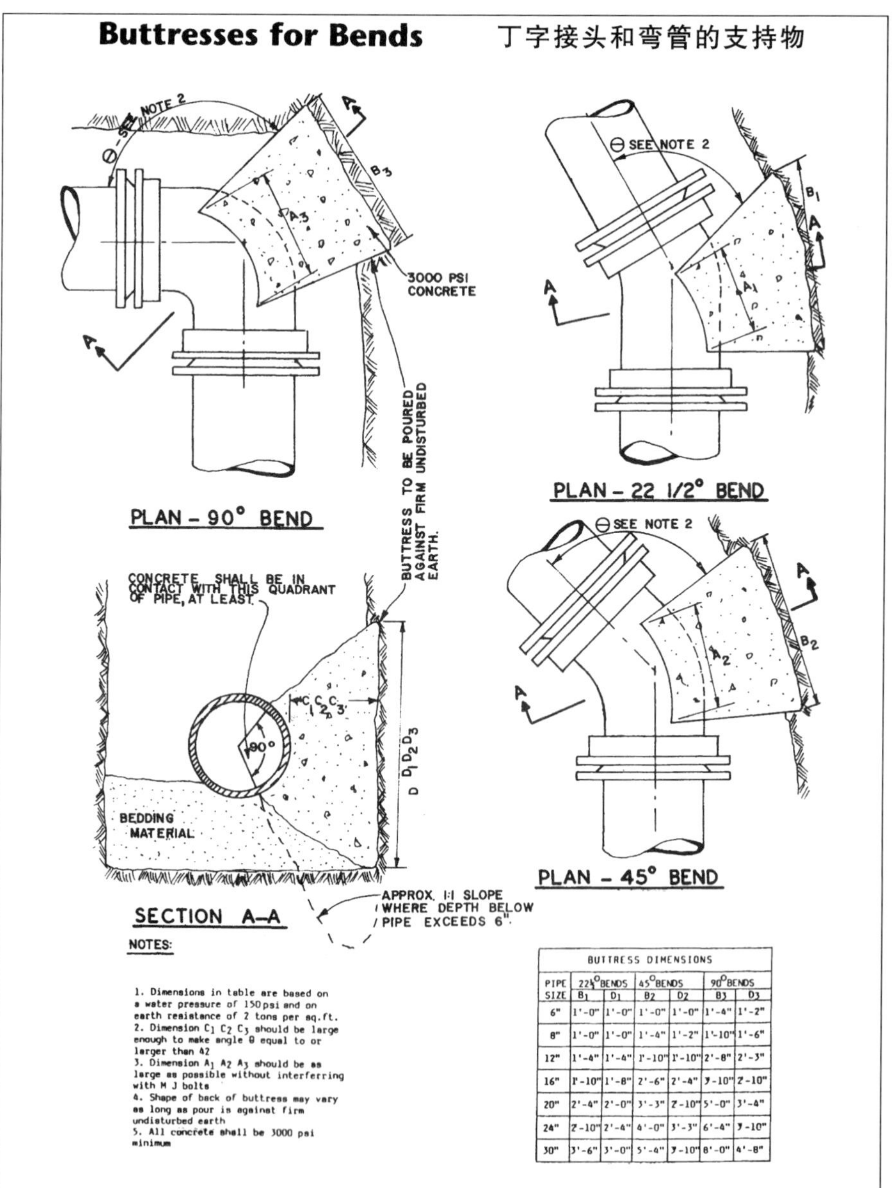

NOTES:

1. Dimensions in table are based on a water pressure of 150 psi and on earth resistance of 2 tons per sq.ft.
2. Dimension C_1 C_2 C_3 should be large enough to make angle θ equal to or larger than 42
3. Dimension A_1 A_2 A_3 should be as large as possible without interferring with M J bolts
4. Shape of back of buttress may vary as long as pour is against firm undisturbed earth
5. All concrete shall be 3000 psi minimum

BUTTRESS DIMENSIONS						
PIPE SIZE	22½° BENDS		45° BENDS		90° BENDS	
	B_1	D_1	B_2	D_2	B_3	D_3
6"	1'-0"	1'-0"	1'-0"	1'-0"	1'-4"	1'-2"
8"	1'-0"	1'-0"	1'-4"	1'-2"	1'-10"	1'-6"
12"	1'-4"	1'-4"	1'-10"	1'-10"	2'-8"	2'-3"
16"	1'-10"	1'-8"	2'-6"	2'-4"	3'-10"	2'-10"
20"	2'-4"	2'-0"	3'-3"	2'-10"	5'-0"	3'-4"
24"	2'-10"	2'-4"	4'-0"	3'-3"	6'-4"	3'-10"
30"	3'-6"	3'-0"	5'-4"	3'-10"	8'-0"	4'-8"

Source: Morgan & Parmley, Ltd.

Water Services　供水设施

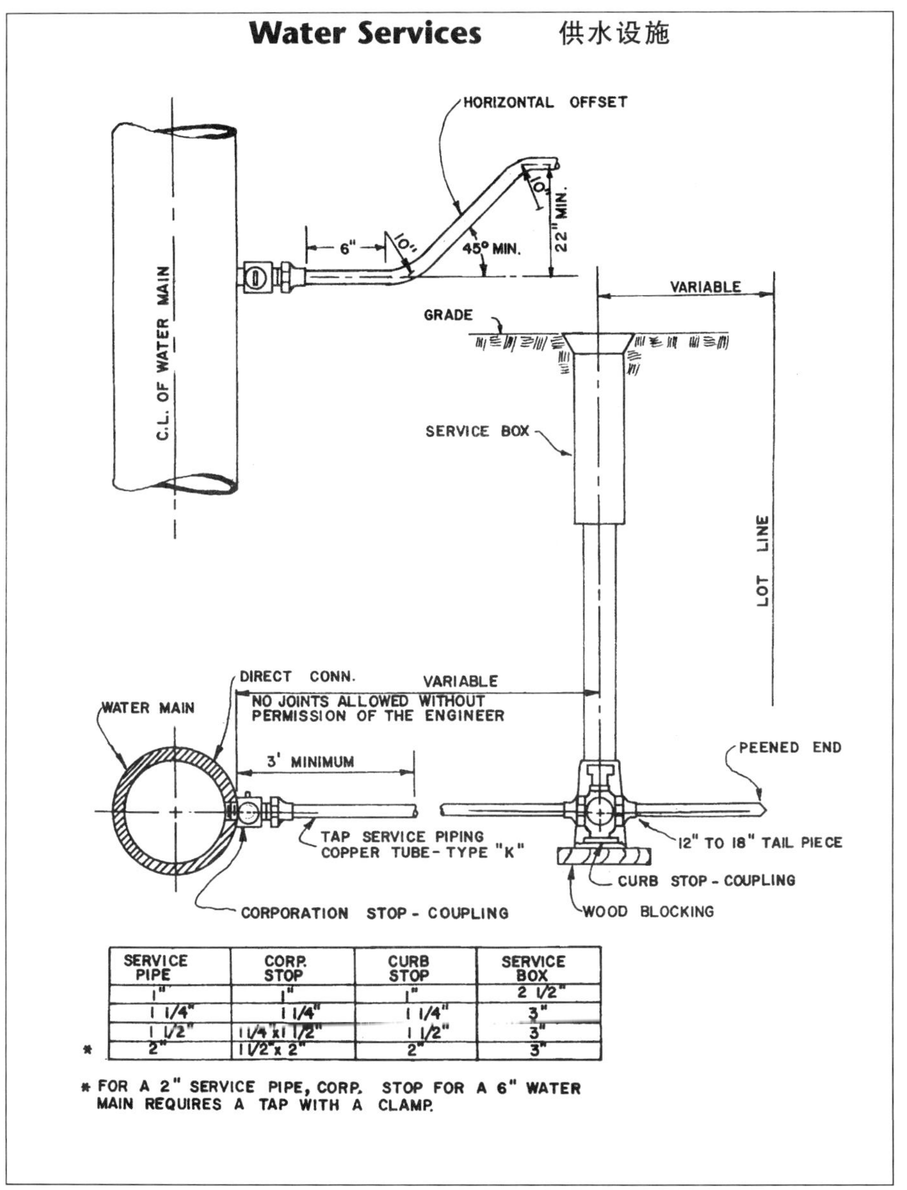

	SERVICE PIPE	CORP. STOP	CURB STOP	SERVICE BOX
	1"	1"	1"	2 1/2"
	1 1/4"	1 1/4"	1 1/4"	3"
	1 1/2"	1 1/4" x 1 1/2"	1 1/2"	3"
*	2"	1 1/2" x 2"	2"	3"

* FOR A 2" SERVICE PIPE, CORP. STOP FOR A 6" WATER MAIN REQUIRES A TAP WITH A CLAMP.

Water Services *(Continued)*　供水设施

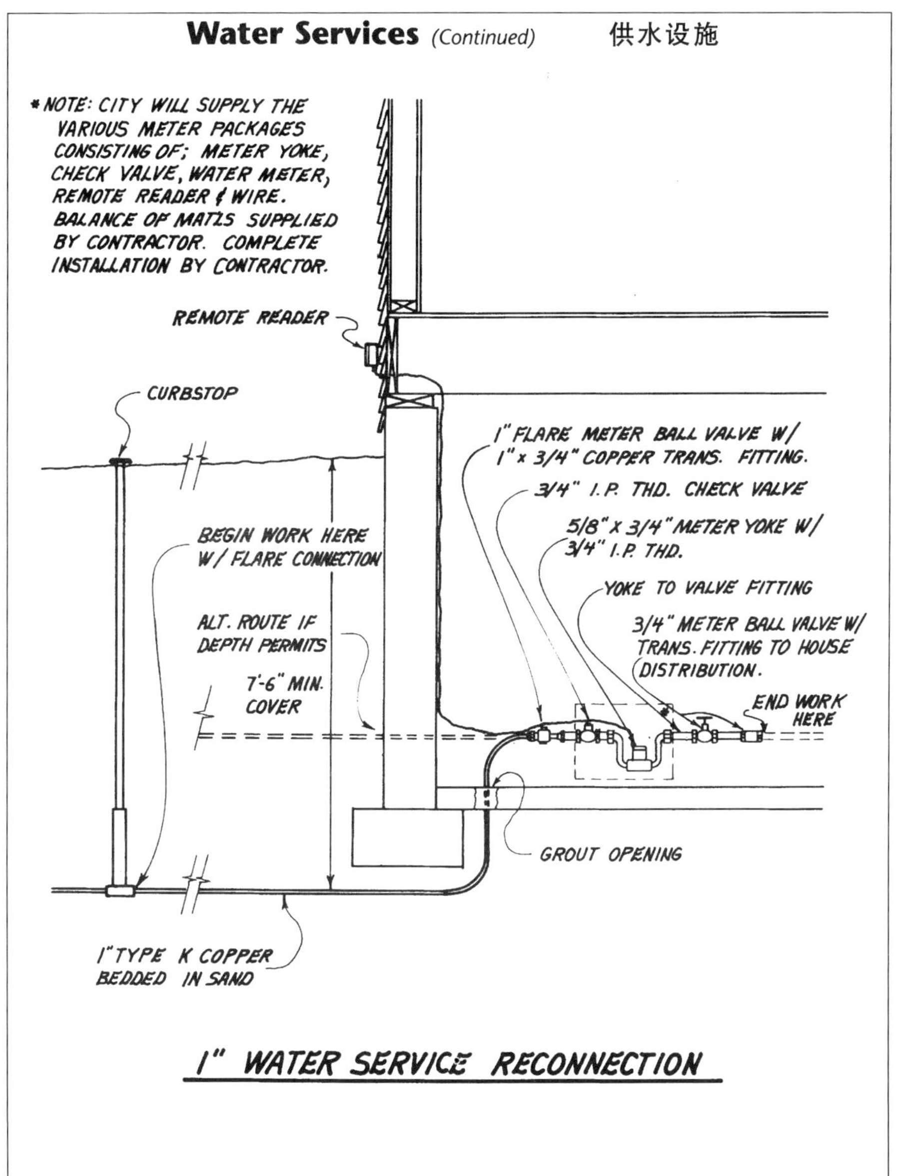

1" WATER SERVICE RECONNECTION

Source: Morgan & Parmley, Ltd.

典型的干管沟截面

Standard Watermain Trench Section

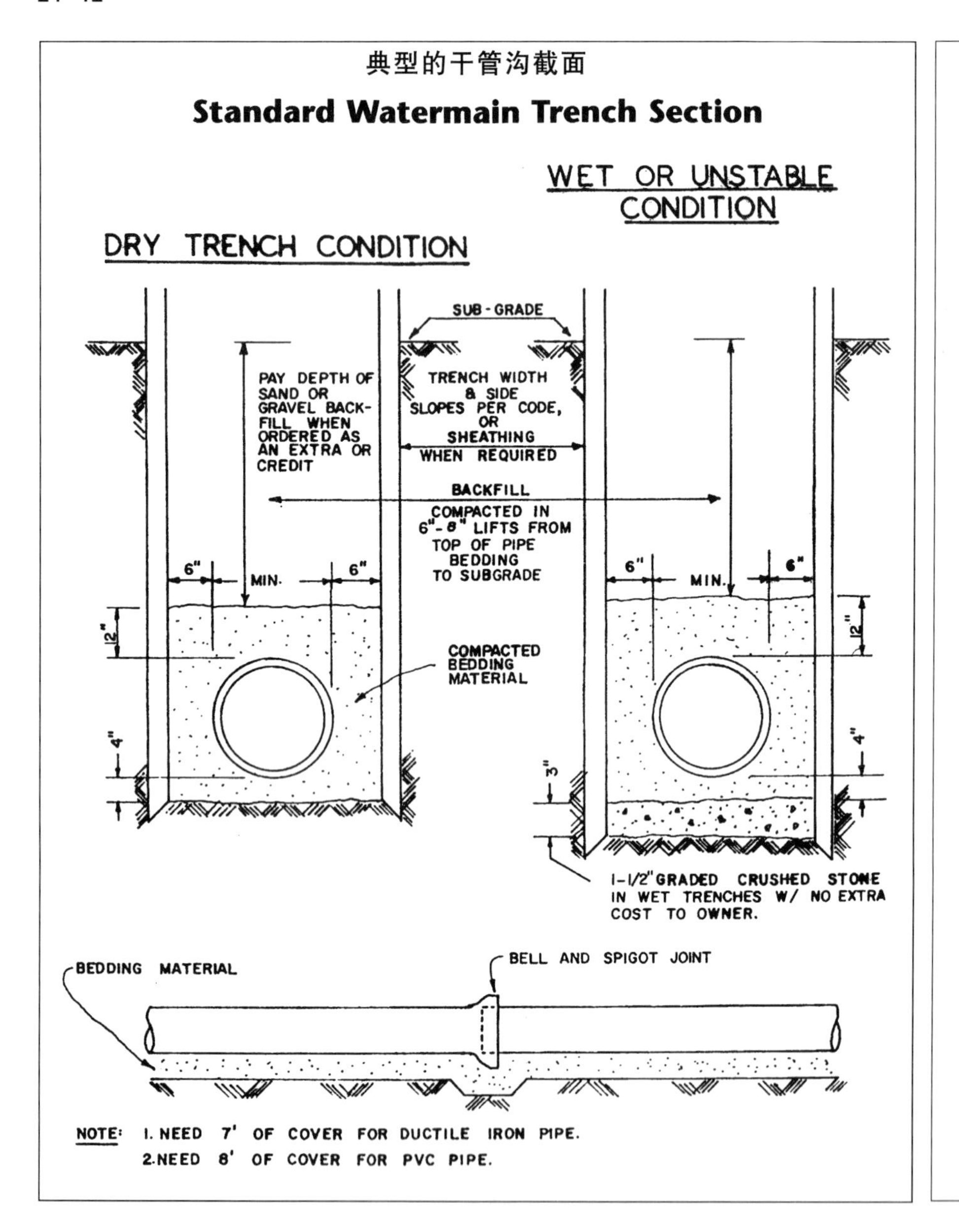

Source: Morgan & Parmley, Ltd.

水管的典型安装详图

Typical Installation Details for Water Pipe

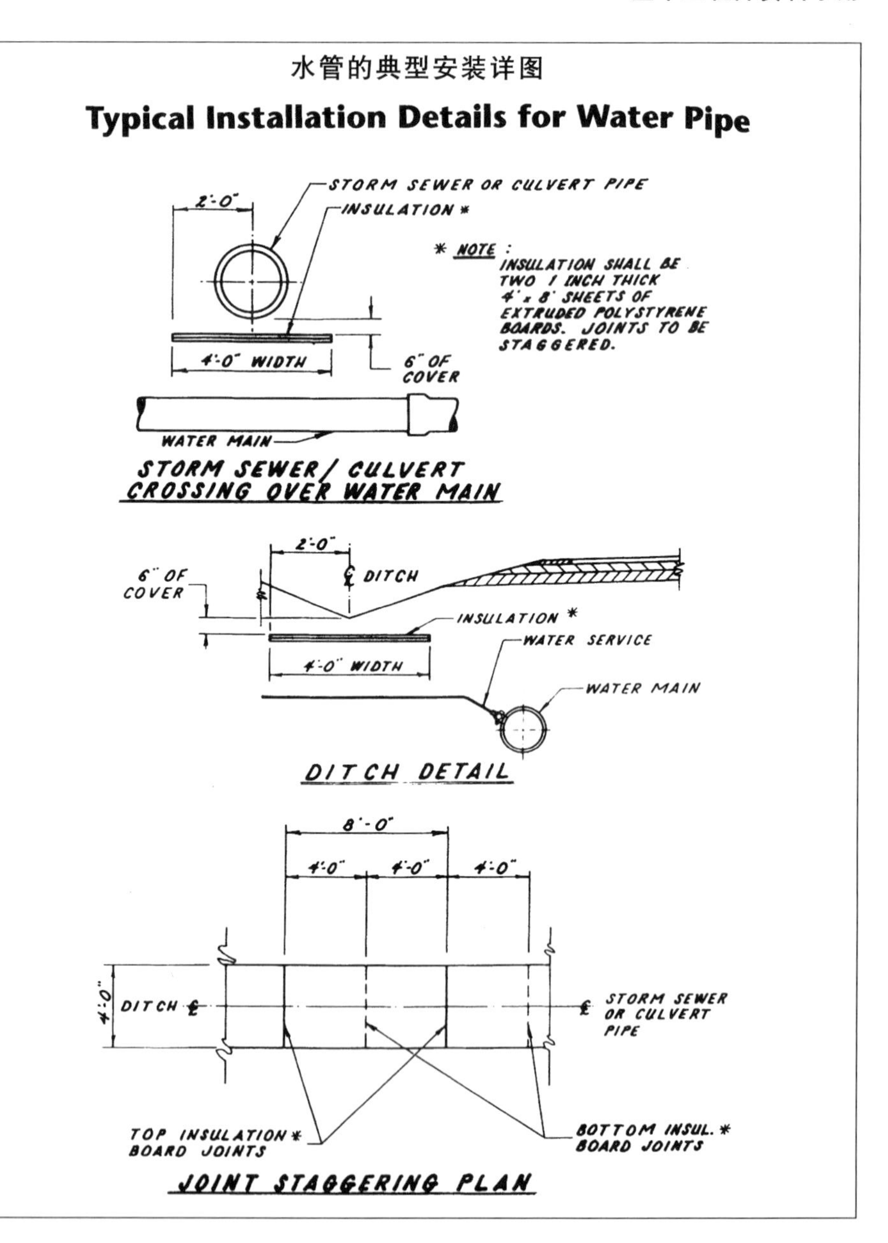

Source: Morgan & Parmley, Ltd.

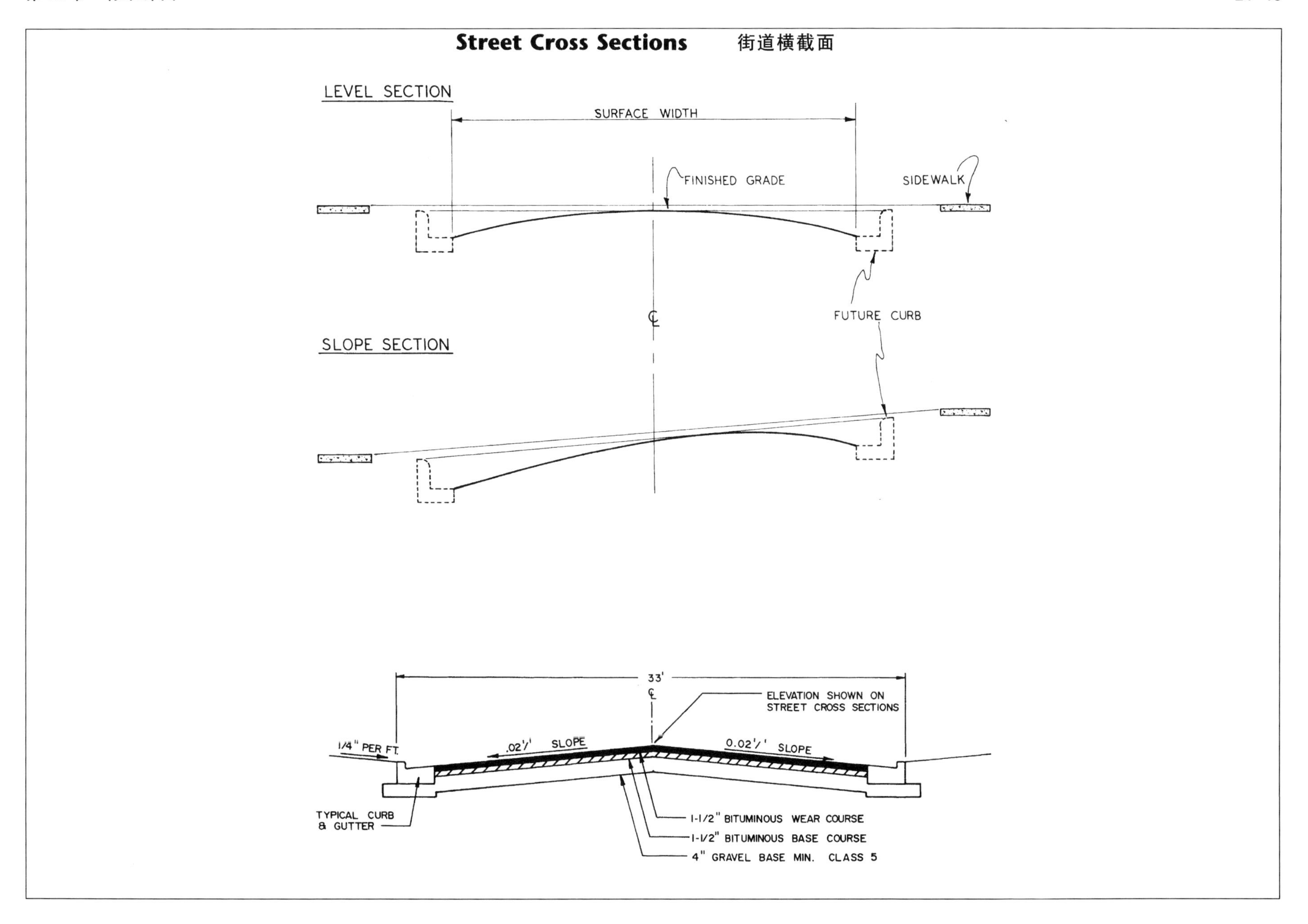

Source: Morgan & Parmley, Ltd.

典型的混凝土路缘石和水槽

Typical Concrete Curb & Gutter

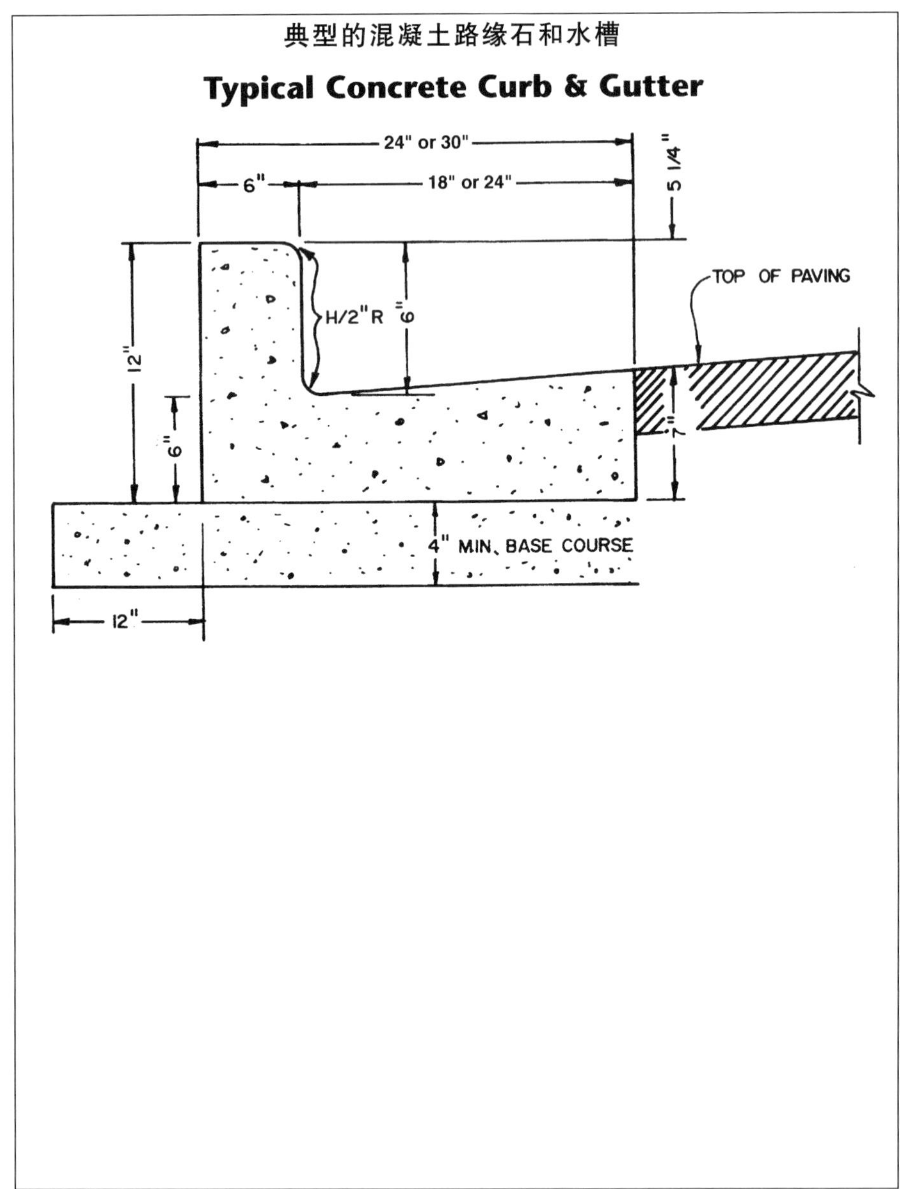

Source: Morgan & Parmley, Ltd.

下水道铁篦子典型详图

Standard Catch Basin Detail

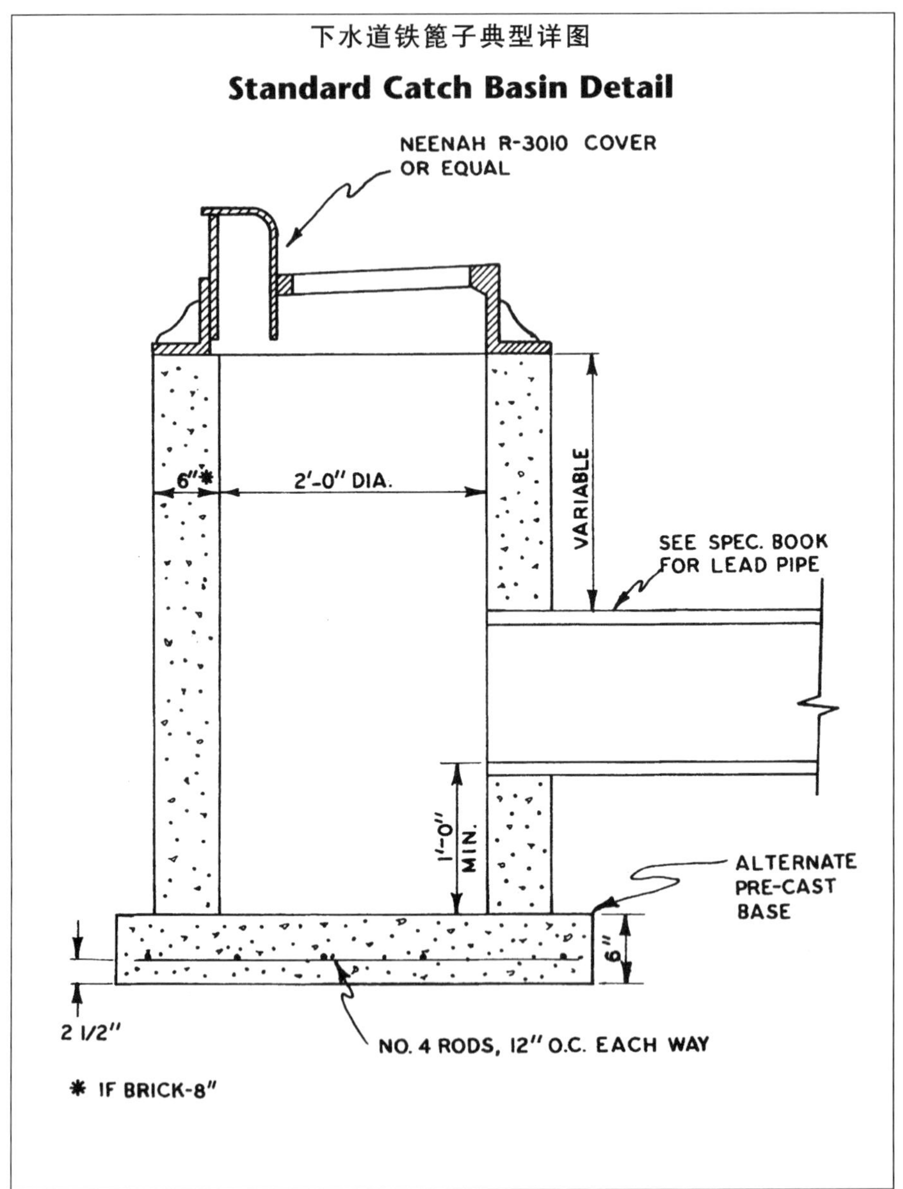

Source: Morgan & Parmley, Ltd.

Typical Inlets with Details　典型进气口详图

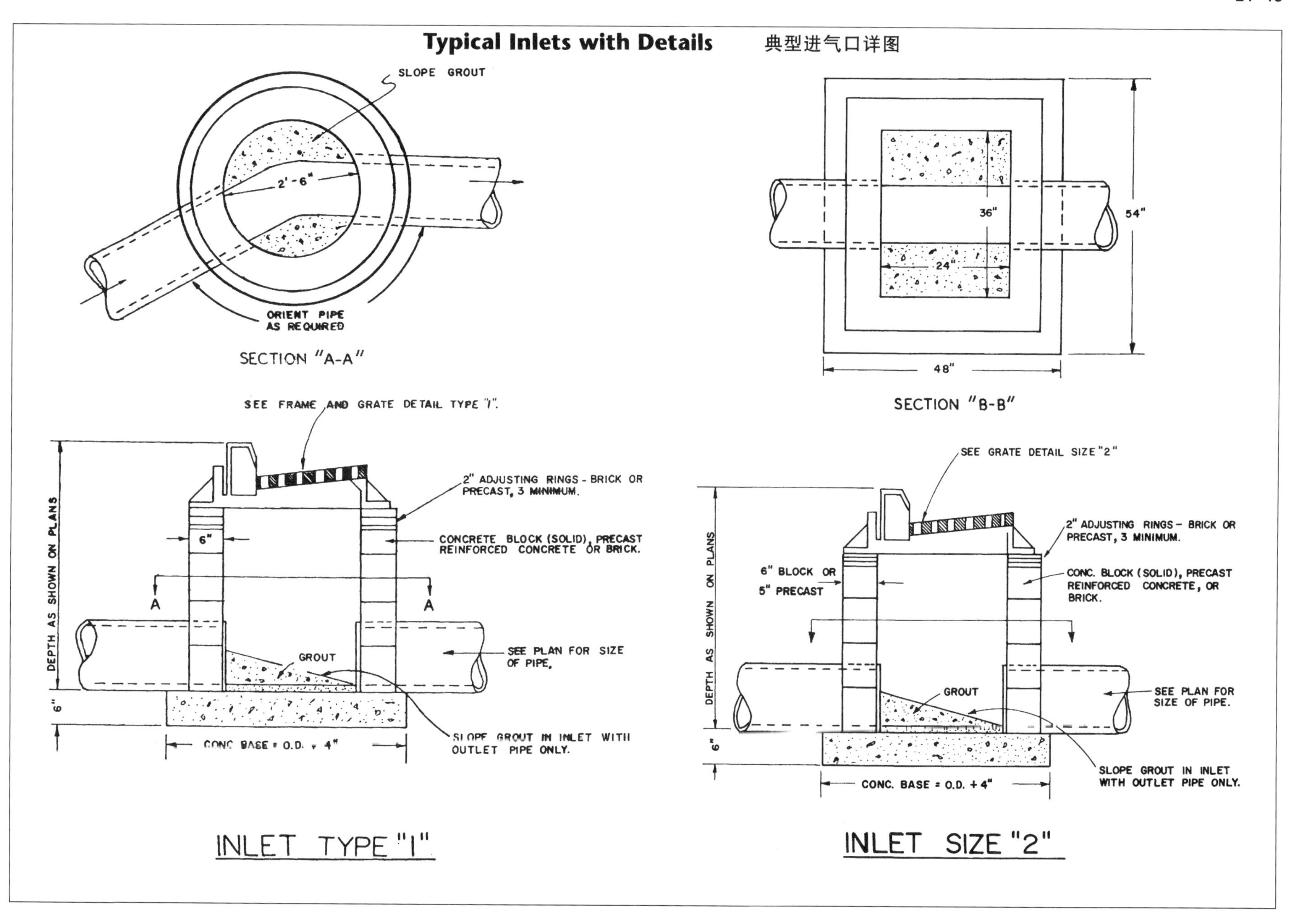

Source: Morgan & Parmley, Ltd.

Typical Inlets with Details *(Continued)* 典型进气口详图

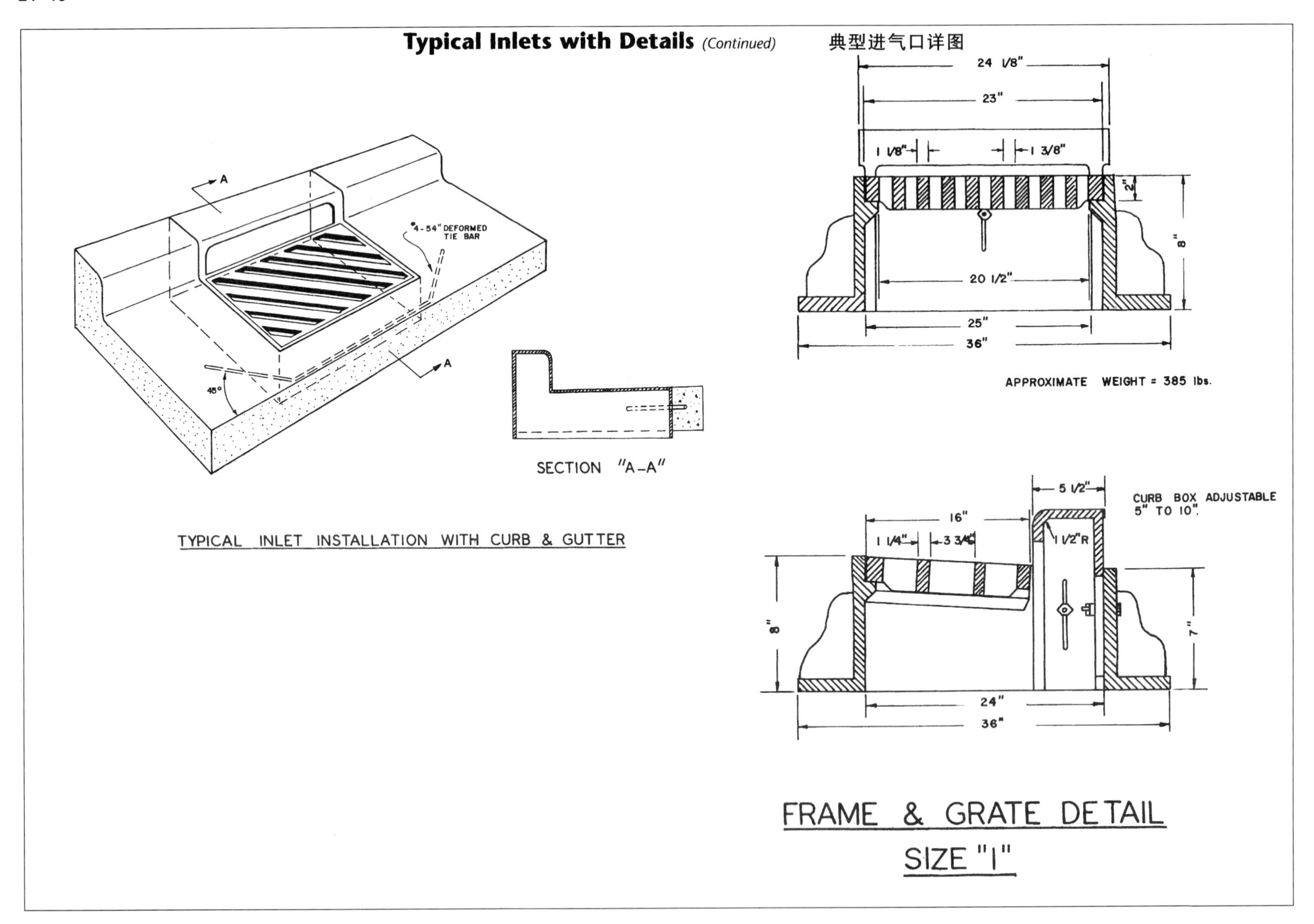

Typical Inlets with Details *(Continued)*　典型进气口详图

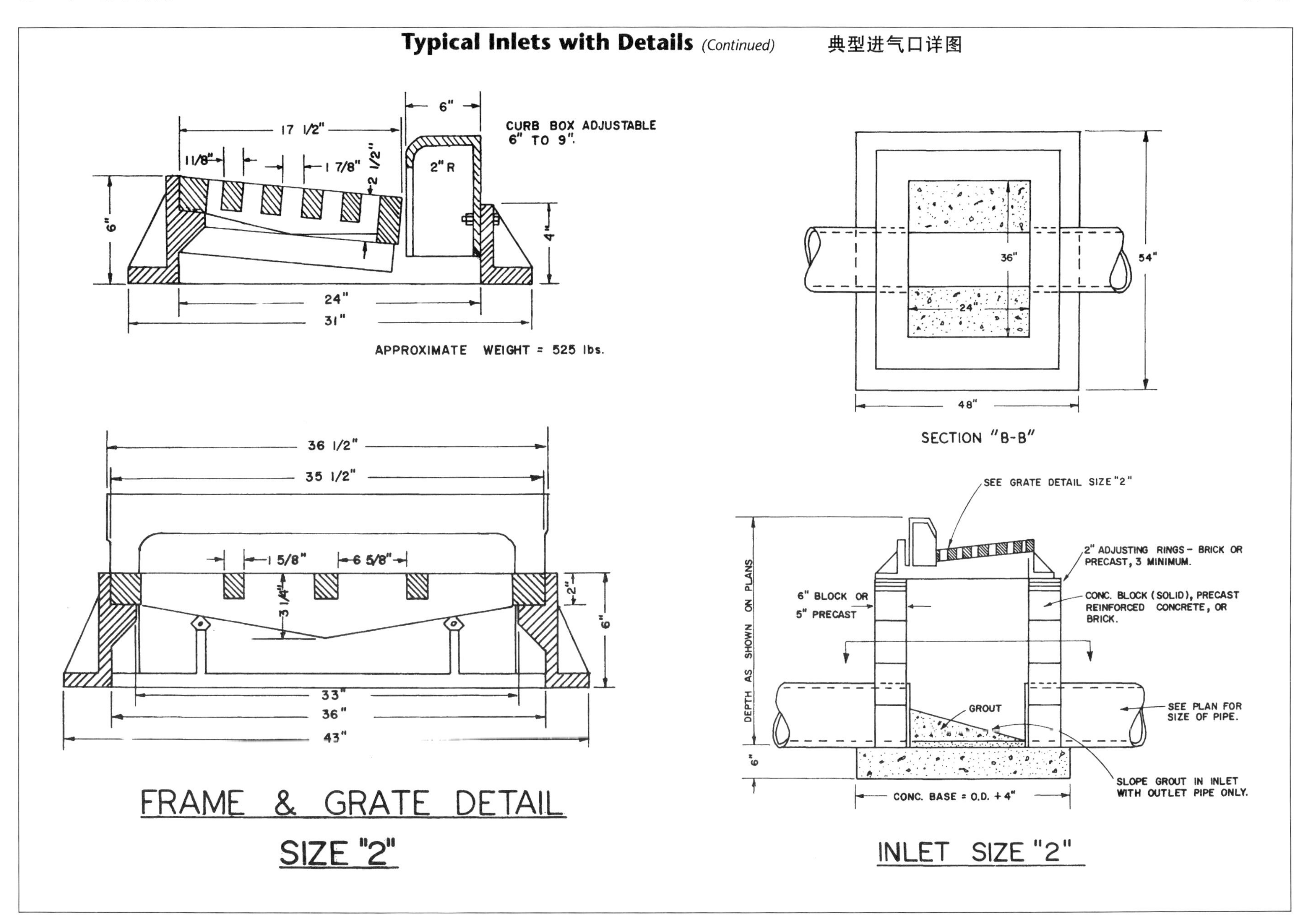

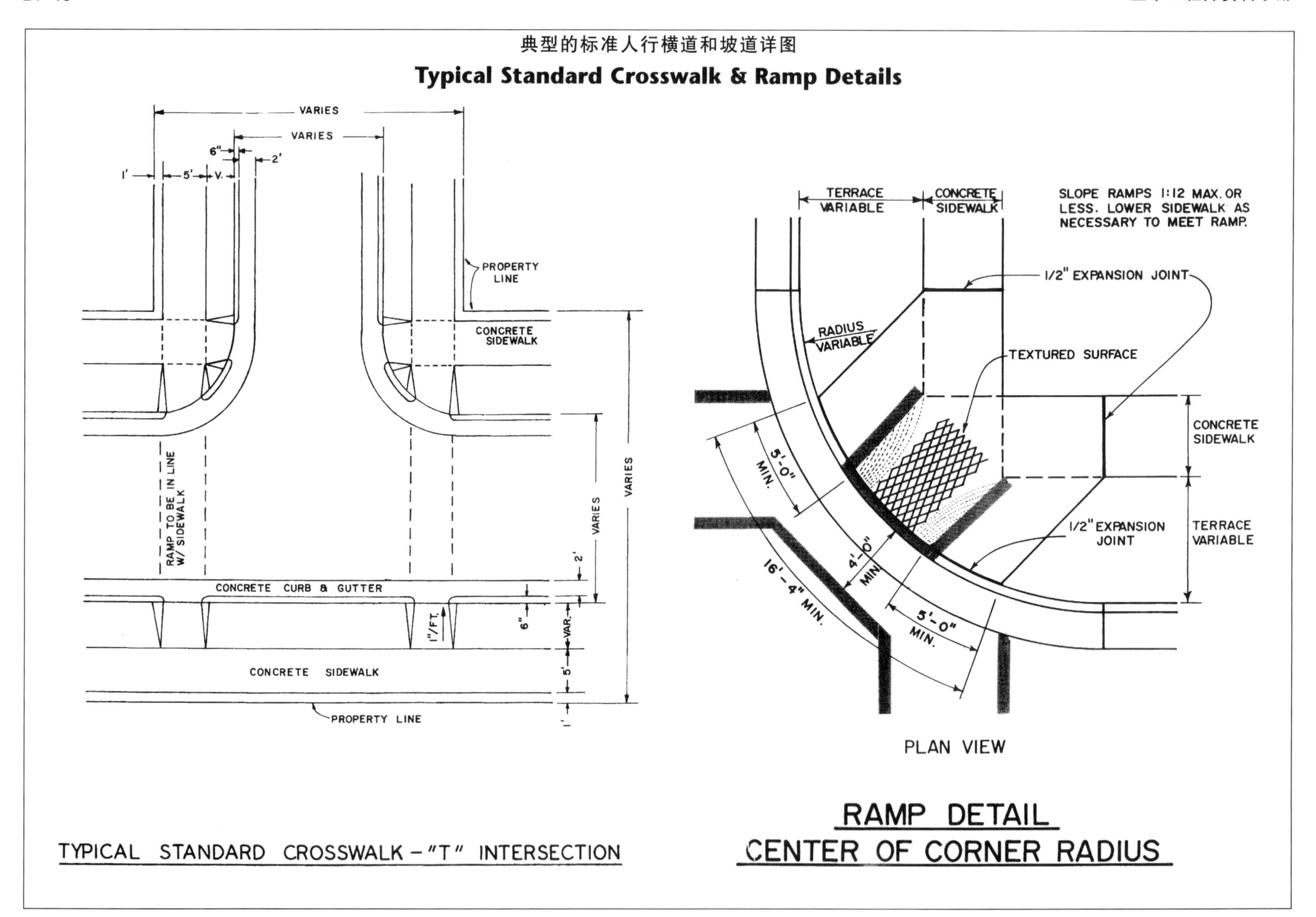

Source: Morgan & Parmley, Ltd.

典型的车道连接段

Typical Driveway Approach

NOTE: 1) INSTALL EXPANSION JOINT @ THE START & FINISH OF A RADIUS.
2) INSTALL EXPANSION JOINT IN CURB LINE EVERY 300 FEET.
3) PLACE CONTRACTION JOINTS IN CURB LINE EVERY 15 FEET.
4) INSTALL EXPANSION JOINT 3 FEET EACH SIDE OF STORM INLETS.

EXISTING DRIVE

6" THICK W/
6×6×10/10 W.W.M.

SLOPE TO MATCH
EXISTING DRIVE

CONCRETE SIDEWALK

6"

8"

SLOPE TO MATCH SIDEWALK

2-#4 REBARS REQ'D. THRU APPROACH

STANDARD WIDTH
SINGLE = 14'-0"
DOUBLE = 22'-0"

CONCRETE CURB & GUTTER

3'-0"
(TYP.)

NOTE: REINFORCEMENT IS REQUIRED FOR DRIVEWAY APPROACHES, REFER TO PROPOSAL FOR SIDEWALK REINFORCING.

Source: Morgan & Parmley, Ltd.

为树木预留的典型的人行道开口

Typical Sidewalk Opening for Trees

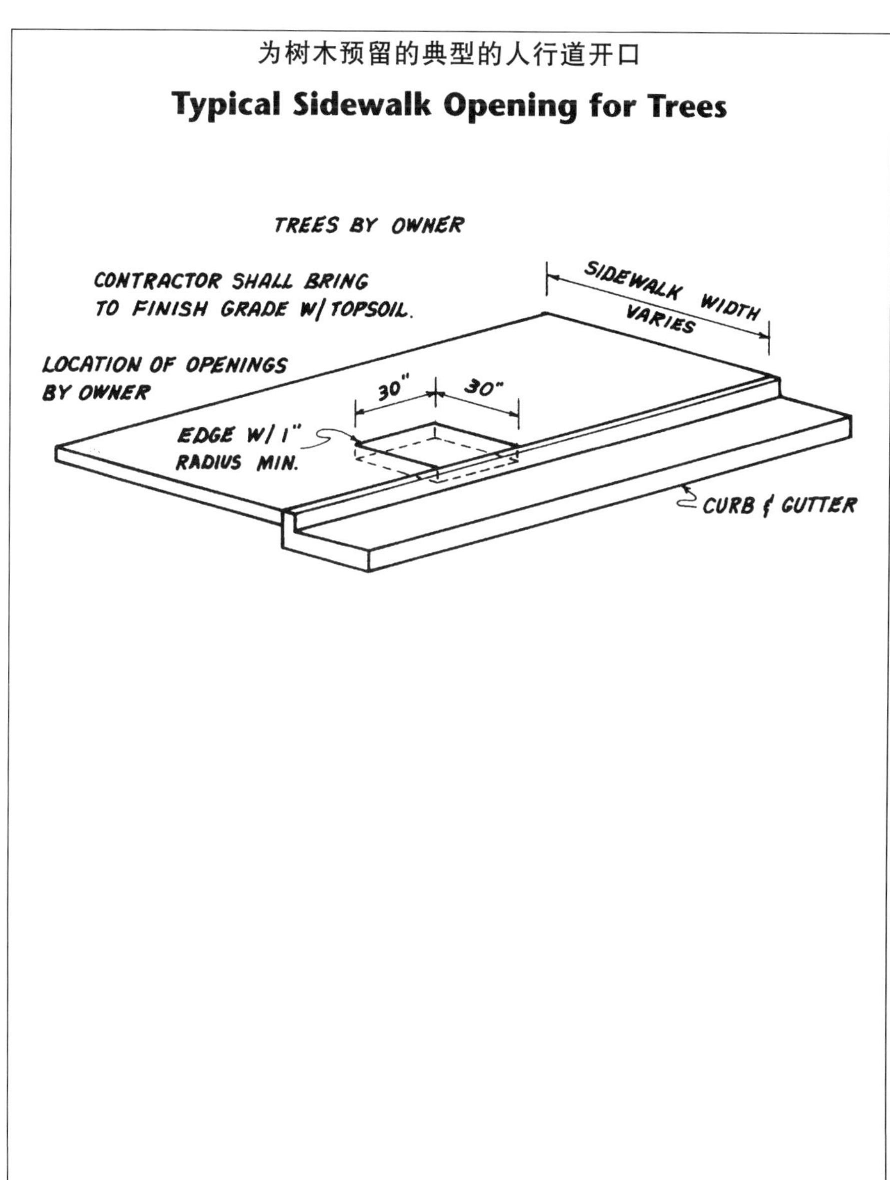

Source: Morgan & Parmley, Ltd.

人行道详图

Sidewalk Details

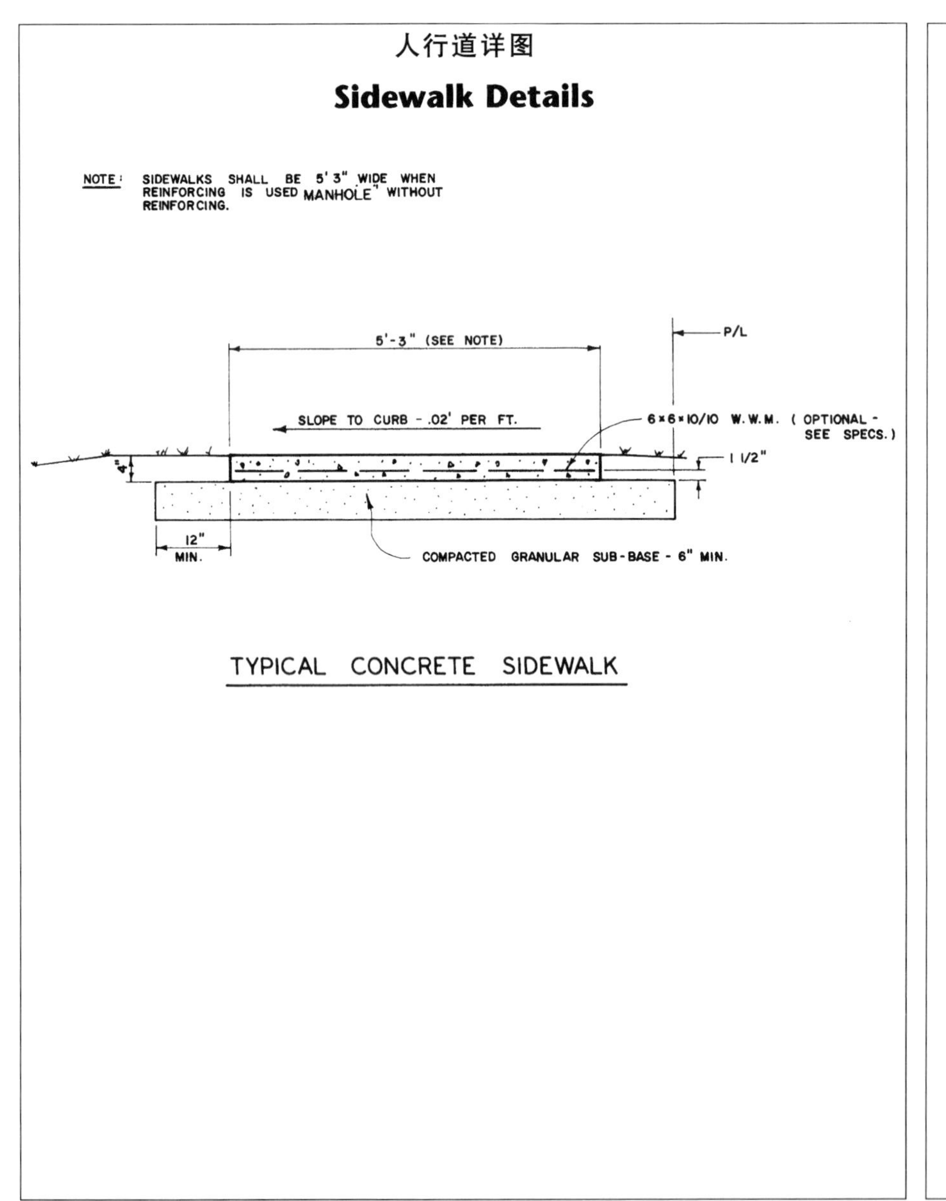

人行道详图

Sidewalk Details *(Continued)*

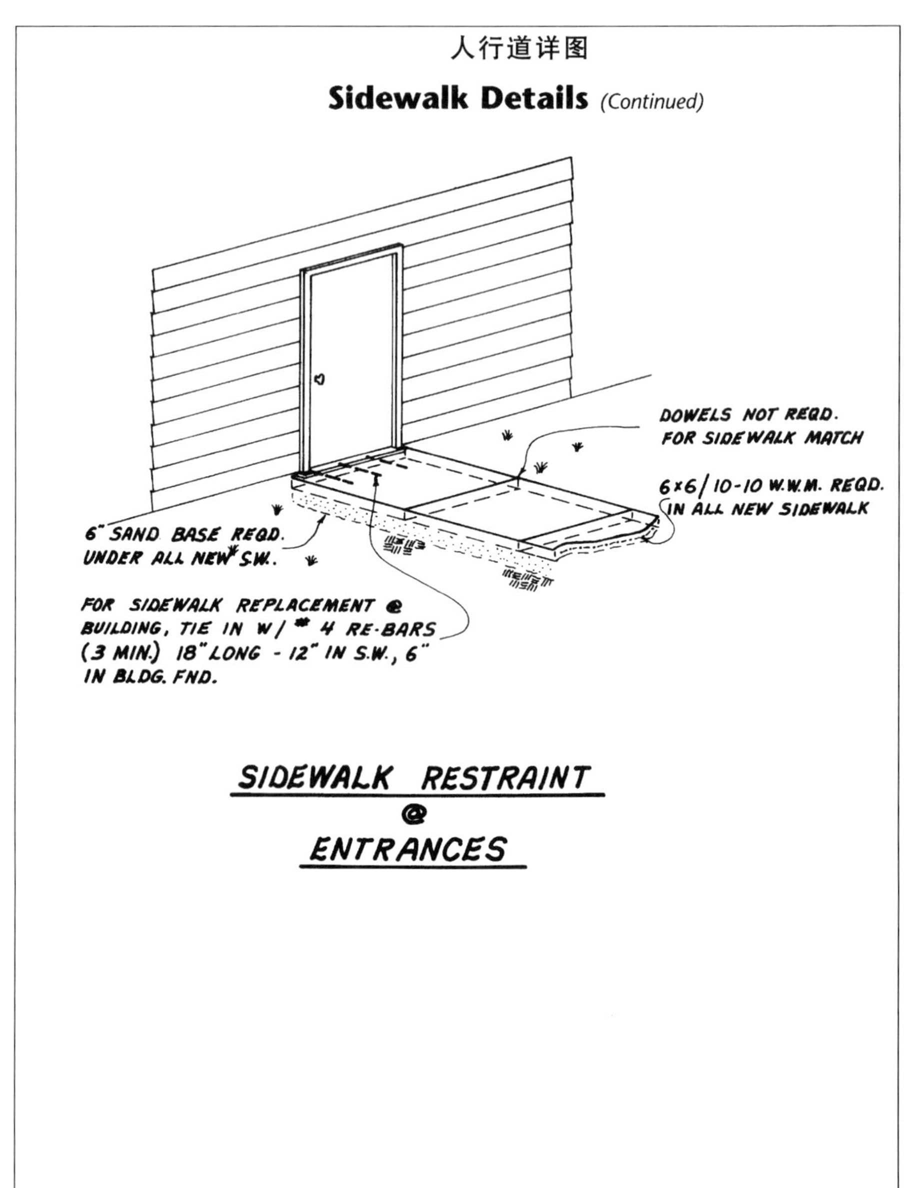

Source: Morgan & Parmley, Ltd.

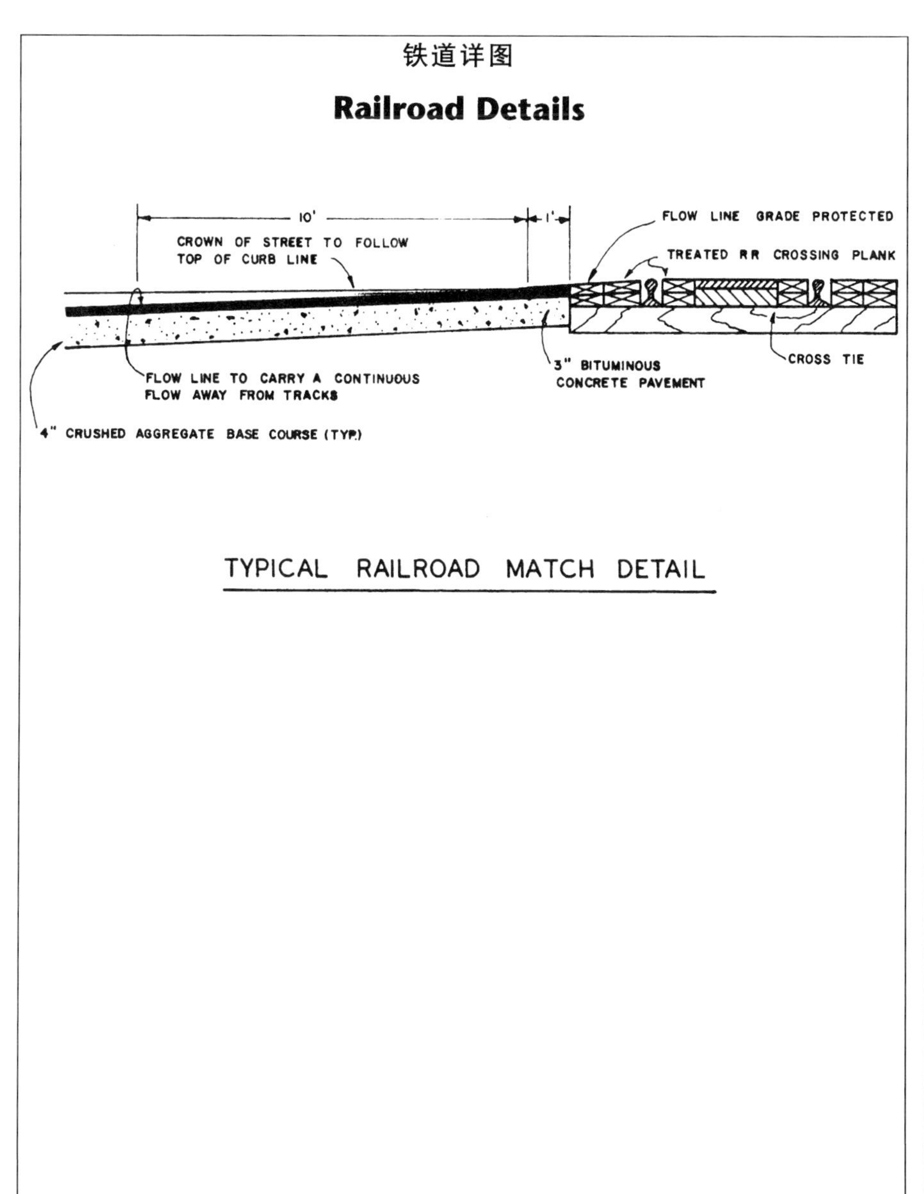

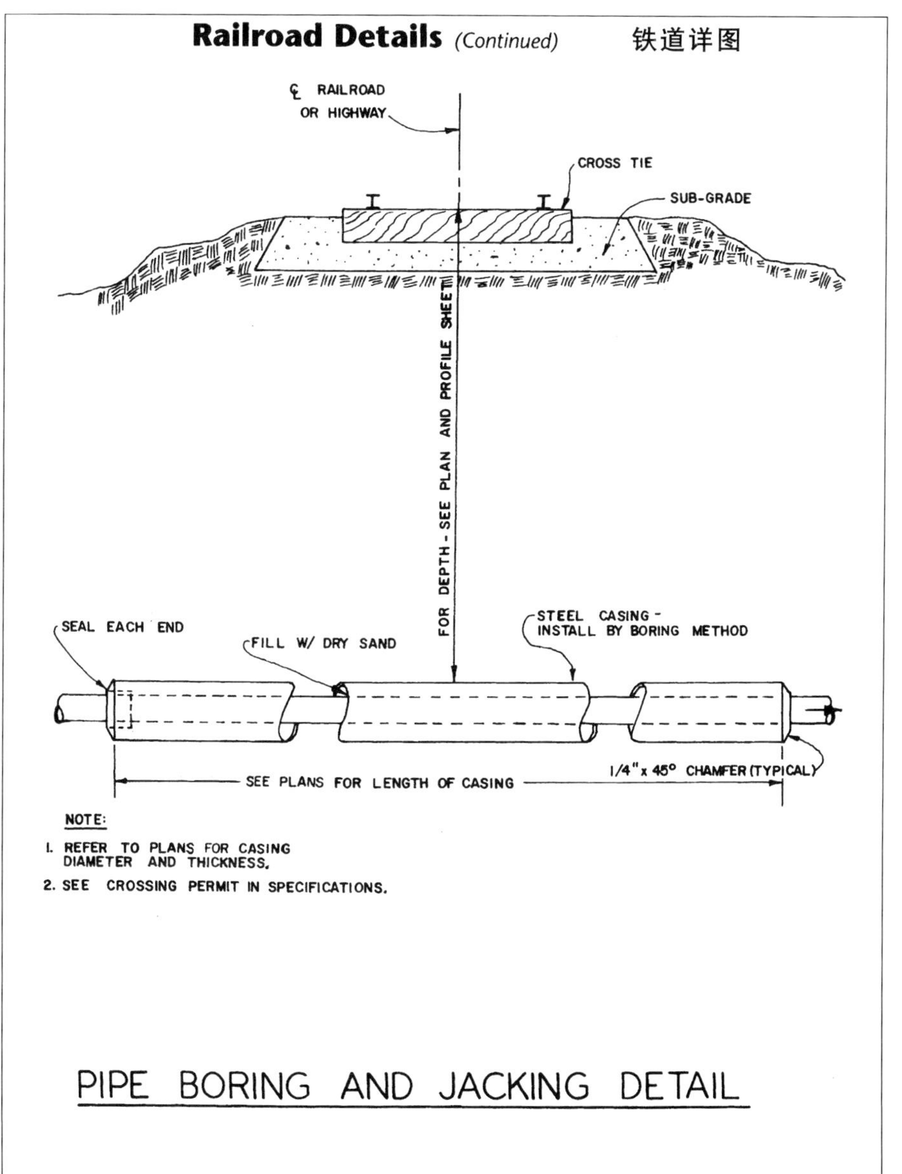

Source: Morgan & Parmley, Ltd.

Common Casing Detail 普通套管详图

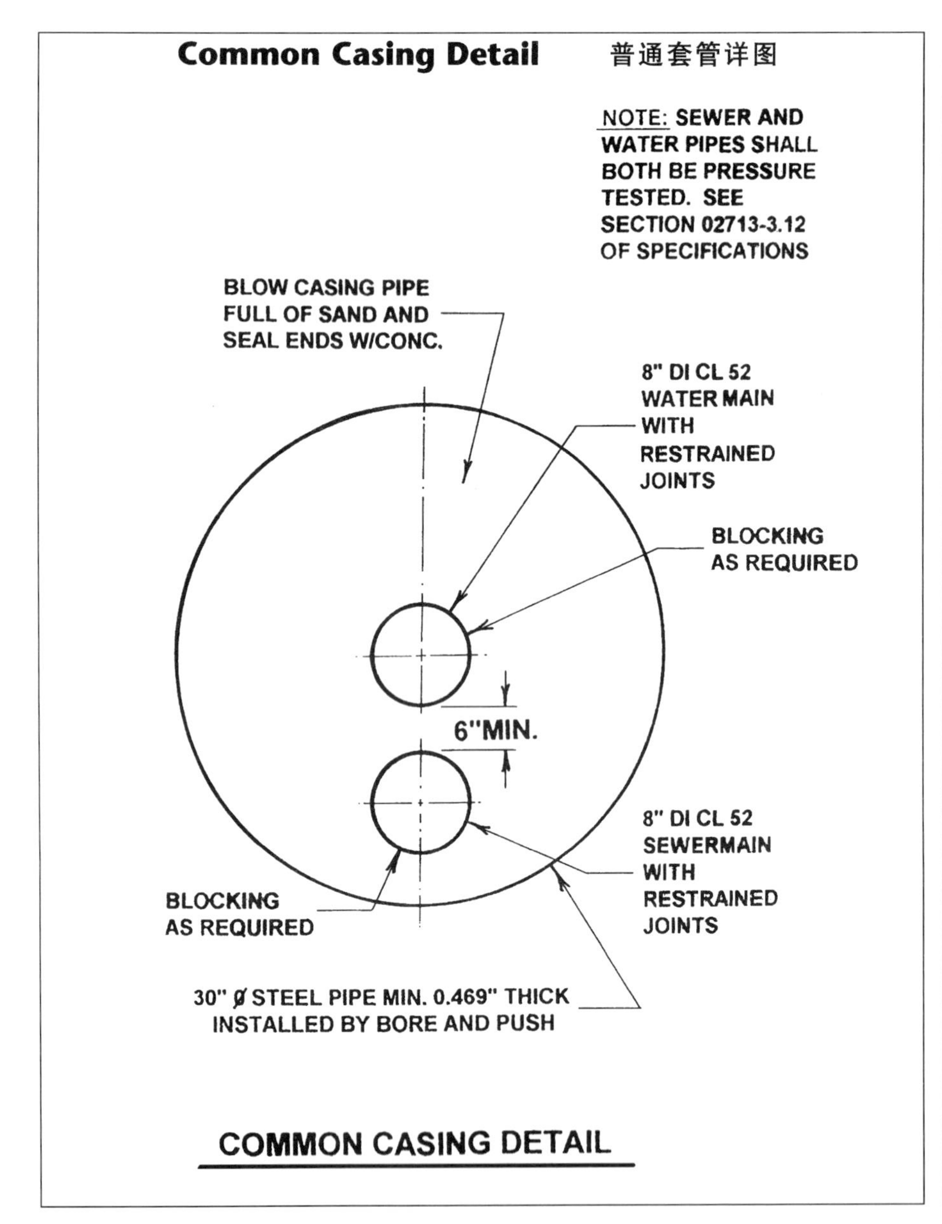

Source: Morgan & Parmley, Ltd.

Pipe Bollard Detail 管状护柱详图

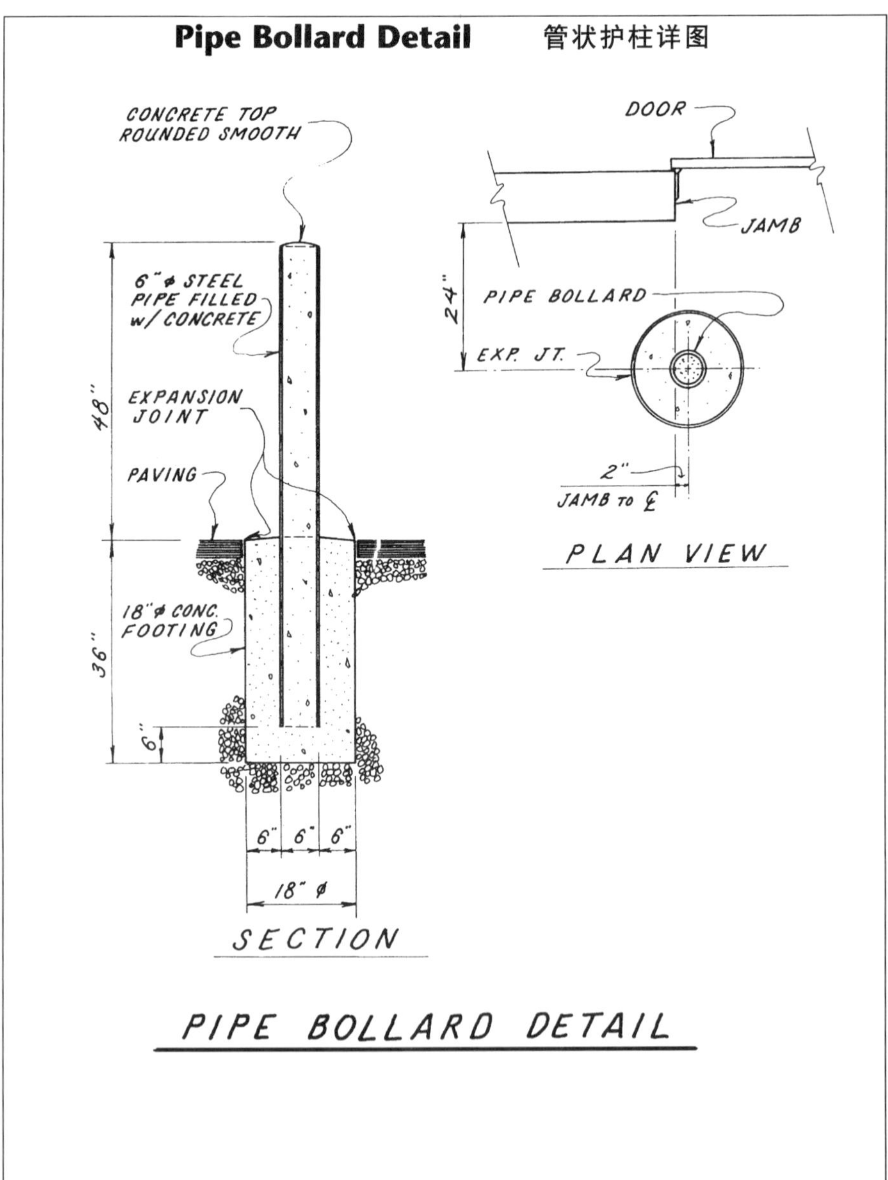

Source: Morgan & Parmley, Ltd.

无障碍设施详图

Handicapped Details

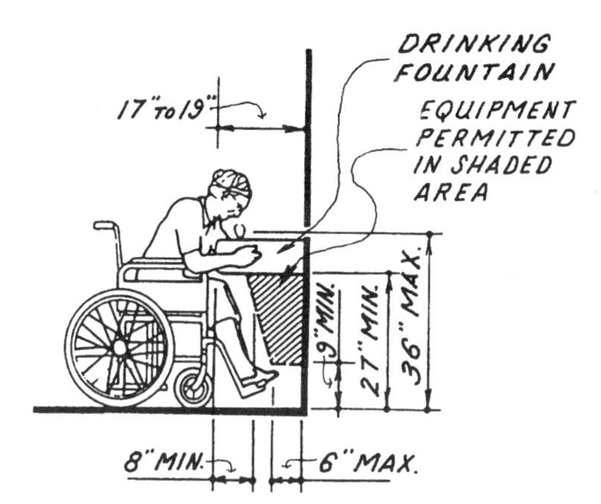

SPOUT HEIGHT & KNEE CLEARANCE

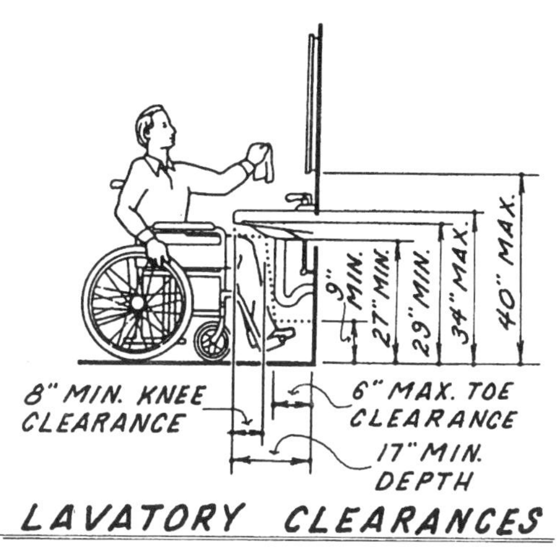

LAVATORY CLEARANCES

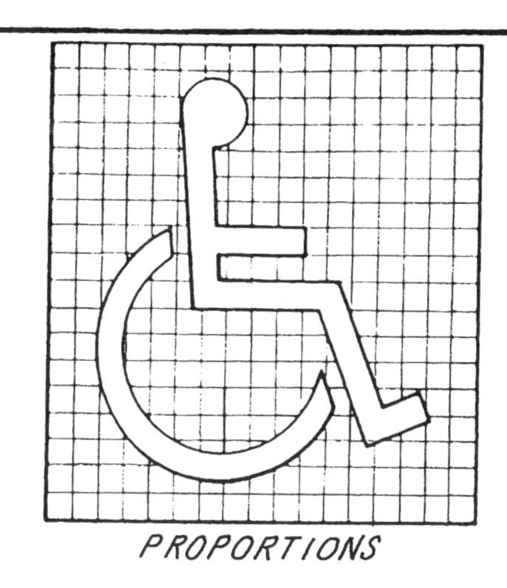

PROPORTIONS

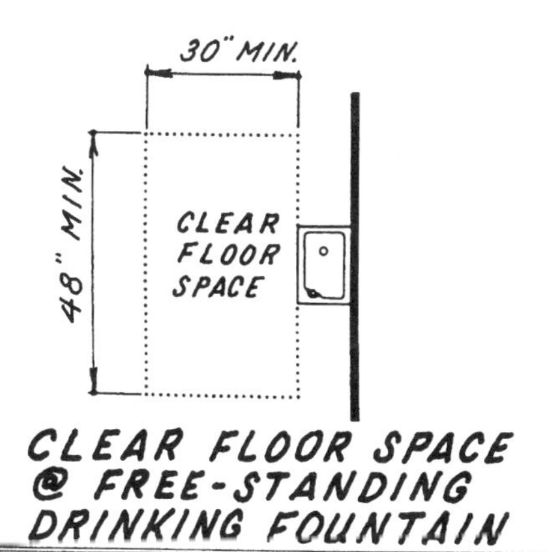

CLEAR FLOOR SPACE @ FREE-STANDING DRINKING FOUNTAIN

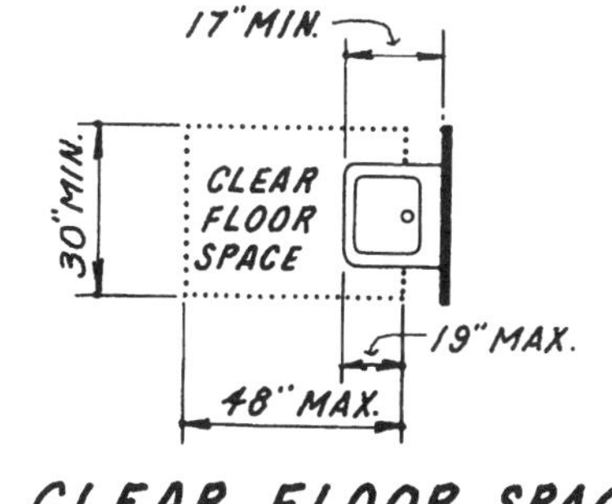

CLEAR FLOOR SPACE @ LAVATORIES

DISPLAY CONDITIONS

INTERNATIONAL SYMBOL OF ACCESSIBILITY

SIGNAGE DETAIL

Source: Morgan & Parmley, Ltd.

管道工程详图

Plumbing Layouts

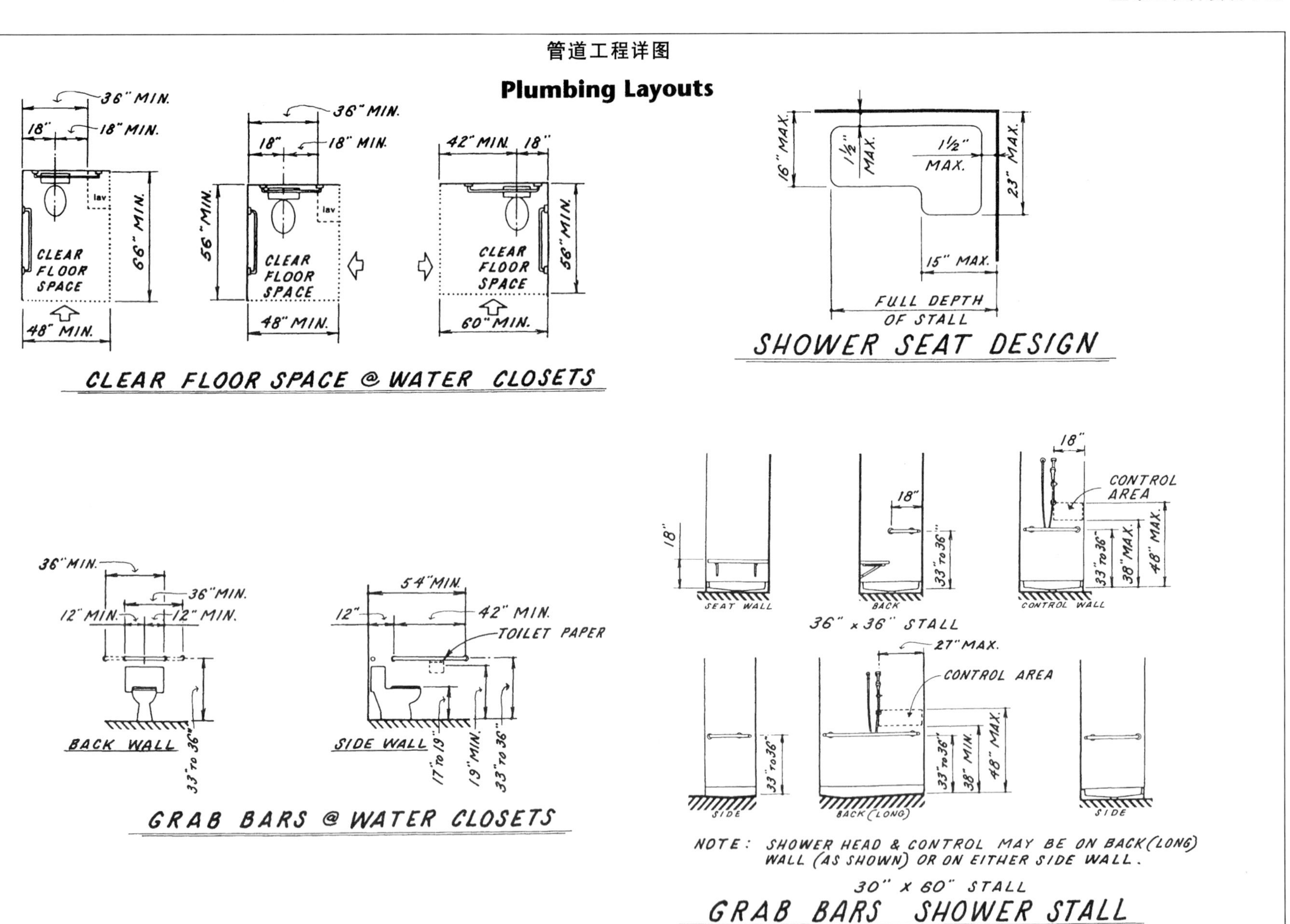

Source: Morgan & Parmley, Ltd.

设 计

第 22 节　技术说明

例　22

技术说明是对项目设计的书面说明，这些说明明确了设计图中的特定内容，而这些内容是不能很好的用绘图来表达的。

本节首先简要地介绍了由施工规范学会（Construction Specifications Institute，CSI）制定的 MasterFormat™。节选的技术说明以及一个实际项目的目录给读者提供了技术写作风格的一个实例。

节选的技术说明样张取自对现有饮用水供水设备进行更新的综合项目。

目　录

编号形式

下一节用到的编号和命名来自 MasterFormat™（1995年版），它是由建筑规格学会（CSI）和加拿大施工技术规范（Canada Construction Specifications，CSC）于2002年出版的。摘录这些材料得到了CSI的允许。

如果想要更深入的了解 MasterFormat™ 和它在建筑业中的应用，请联系：

The Construction Specifications Institute(CSI)

99 Canal Center Plaza，Suite 300

Alexandra，VA 223-14

800-689-2900；703-684-0300

CSINet URL：http://www.csinet.org

在过去的30年中，作为 MasterFormat 核心部分的编号和命名体系在建筑业中得到了越来越多的应用。现在，MasterFormat 是在美国和加拿大惟一得到广泛应用的施工规范组织体系。它将施工信息总结成“招投标必备条件”、“缔约必备条件”和16分项产品和行为是很通用的。

MasterFormat 得到美国国防部的采用，也得到州一级和市一级政府的采用。

CSC 在加拿大推动施工文件的标准化。它的努力已经得到加拿大皇家建筑学会、加拿大顾问工程师协会和加拿大建筑协会的支持。MasterFormat 也被省一级和市一级政府采用，用于组织主要的技术说明、数据文件和成本分析体系等。MasterFormat 是加拿大政府颁布的国家主控技术规范（National Master Specification，NMS）命名和编号体系的基础，它被所有加拿大联邦政府机构和大部分私营部门所采用。

当组织一份项目手册时，一件最重要的事情是记住技术说明节与其他内容没有层级关系。尽管 MasterFormat 对它的编号和命名建立了一套层级关系，但是技术说明节对其他内容和其他合同文件起补充作用。技术说明中的一节不能“控制”其他节，作为补充文件，所有的技术说明潜在地影响其他所有技术说明。

另外，重要的考虑是“概况”、“招投标必备条件”和“缔约必备条件”不是技术说明，通常也不是技术说明人员或专业设计人的职责。规范的法律实践倾向于仅仅依据名称来识别这些文件，而不是编号。MasterFormat 的编号在项目手册中可用于识别这些文件，但非常重要的是应避免采用“节”或者其他的叫法，否则容易将其与技术说明混淆。如果已经完成识别，也没有必要对已经印好的文件重新编号。这包括了多个专业协会发布的标准通用条件。

Courtesy：CSI

Level Two Numbers and Titles　第二层编号和标题

NOTE: See Master List of /Numbers 1995 Manual for complete system

Introductory Information

00001 Project Title Page
00005 Certifications Page
00007 Seals Page
00010 Table of Contents
00015 List of Drawings
00020 List of Schedules

Bidding Requirements

00100 Bid Solicitation
00200 Instructions To Bidders
00300 Information Available To Bidders
00400 Bid Forms and Supplements
00490 Bidding Addenda

Contracting Requirements

00500 Agreement
00600 Bonds and Certificates
00700 General Conditions
00800 Supplementary Conditions
00900 Addenda and Modifications

Facilities and Spaces

Facilities and Spaces

Systems and Assemblies

Systems and Assemblies

Construction Products and Activities

Division 1 General Requirements

01100 Summary
01200 Price and Payment Procedures
01300 Administrative Requirements
01400 Quality Requirements
01500 Temporary Facilities and Controls
01600 Product Requirements
01700 Execution Requirements
01800 Facility Operation
01900 Facility Decommissioning

Division 2 Site Construction

02050 Basic Site Materials and Methods
02100 Site Remediation
02200 Site Preparation
02300 Earthwork
02400 Tunneling, Boring, and Jacking
02450 Foundation and Load-Bearing Elements
02500 Utility Services
02600 Drainage and Containment
02700 Bases, Ballasts, Pavements, and Appurtenances
02800 Site Improvements and Amenities
02900 Planting
02950 Site Restoration and Rehabilitation

Divicion 3 Concrete

03050 Basic Concrete Materials and Methods
03100 Concrete Forms and Accessories
03200 Concrete Reinforcement
03300 Cast-in-Place Concrete
03400 Precast Concrete
03500 Cementitious Decks and Underlayment
03600 Grouts
03700 Mass Concrete
03900 Concrete Restoration and Cleaning

Division 4 Masonry

04050 Basic Masonry Materials and Methods
04200 Masonry Units
04400 Stone
04500 Refractories
04600 Corrosion-Resistant Masonry
04700 Simulated Masonry
04800 Masonry Assemblies
04900 Masonry Restoration and Cleaning

Division 5 Metals

05050 Basic Metal Materials and Methods
05100 Structural Metal Framing
05200 Metal Joists
05300 Metal Deck
05400 Cold-Formed Metal Framing
05500 Metal Fabrications
05600 Hydraulic Fabrications
05650 Railroad Track and Accessories
05700 Ornamental Metal
05800 Expansion Control
05900 Metal Restoration and Cleaning

Division 6 Wood and Plastics

00050 Basic Wood and Plastic Materials and Methods
06100 Rough Carpentry
06200 Finish Carpentry
06400 Architectural Woodwork
06500 Structural Plastics
06600 Plastic Fabrications
06900 Wood and Plastic Restoration and Cleaning

Courtesy: CSI

DIVISION 7 THERMAL AND MOISTURE PROTECTION

07050 BASIC THERMAL AND MOISTURE PROTECTION MATERIALS AND METHODS
07100 DAMPPROOFING AND WATERPROOFING
07200 THERMAL PROTECTION
07300 SHINGLES, ROOF TILES, AND ROOF COVERINGS
07400 ROOFING AND SIDING PANELS
07500 MEMBRANE ROOFING
07600 FLASHING AND SHEET METAL
07700 ROOF SPECIALTIES AND ACCESSORIES
07800 FIRE AND SMOKE PROTECTION
07900 JOINT SEALERS

DIVISION 8 DOORS AND WINDOWS

08050 BASIC DOOR AND WINDOW MATERIALS AND METHODS
08100 METAL DOORS AND FRAMES
08200 WOOD AND PLASTIC DOORS
08300 SPECIALTY DOORS
08400 ENTRANCES AND STOREFRONTS
08500 WINDOWS
08600 SKYLIGHTS
08700 HARDWARE
08800 GLAZING
08900 GLAZED CURTAIN WALL

DIVISION 9 FINISHES

09050 BASIC FINISH MATERIALS AND METHODS
09100 METAL SUPPORT ASSEMBLIES
09200 PLASTER AND GYPSUM BOARD
09300 TILE
09400 TERRAZZO
09500 CEILINGS
09600 FLOORING
09700 WALL FINISHES
09800 ACOUSTICAL TREATMENT
09900 PAINTS AND COATINGS

DIVISION 10 SPECIALTIES

10100 VISUAL DISPLAY BOARDS
10150 COMPARTMENTS AND CUBICLES
10200 LOUVERS AND VENTS
10240 GRILLES AND SCREENS
10250 SERVICE WALLS
10260 WALL AND CORNER GUARDS
10270 ACCESS FLOORING
10290 PEST CONTROL
10300 FIREPLACES AND STOVES
10340 MANUFACTURED EXTERIOR SPECIALTIES
10350 FLAGPOLES
10400 IDENTIFICATION DEVICES
10450 PEDESTRIAN CONTROL DEVICES
10500 LOCKERS
10520 FIRE PROTECTION SPECIALTIES
10530 PROTECTIVE COVERS
10550 POSTAL SPECIALTIES
10600 PARTITIONS
10670 STORAGE SHELVING
10700 EXTERIOR PROTECTION
10750 TELEPHONE SPECIALTIES
10800 TOILET, BATH, AND LAUNDRY ACCESSORIES
10880 SCALES
10900 WARDROBE AND CLOSET SPECIALTIES

DIVISION 11 EQUIPMENT

11010 MAINTENANCE EQUIPMENT
11020 SECURITY AND VAULT EQUIPMENT
11030 TELLER AND SERVICE EQUIPMENT
11040 ECCLESIASTICAL EQUIPMENT
11050 LIBRARY EQUIPMENT
11060 THEATER AND STAGE EQUIPMENT
11070 INSTRUMENTAL EQUIPMENT
11080 REGISTRATION EQUIPMENT
11090 CHECKROOM EQUIPMENT
11100 MERCANTILE EQUIPMENT
11110 COMMERCIAL LAUNDRY AND DRY CLEANING EQUIPMENT
11120 VENDING EQUIPMENT
11130 AUDIO-VISUAL EQUIPMENT
11140 VEHICLE SERVICE EQUIPMENT
11150 PARKING CONTROL EQUIPMENT
11160 LOADING DOCK EQUIPMENT
11170 SOLID WASTE HANDLING EQUIPMENT
11190 DETENTION EQUIPMENT
11200 WATER SUPPLY AND TREATMENT EQUIPMENT
11280 HYDRAULIC GATES AND VALVES
11300 FLUID WASTE TREATMENT AND DISPOSAL EQUIPMENT
11400 FOOD SERVICE EQUIPMENT
11450 RESIDENTIAL EQUIPMENT
11460 UNIT KITCHENS
11470 DARKROOM EQUIPMENT
11480 ATHLETIC, RECREATIONAL, AND THERAPEUTIC EQUIPMENT
11500 INDUSTRIAL AND PROCESS EQUIPMENT
11600 LABORATORY EQUIPMENT
11650 PLANETARIUM EQUIPMENT
11660 OBSERVATORY EQUIPMENT
11680 OFFICE EQUIPMENT
11700 MEDICAL EQUIPMENT
11780 MORTUARY EQUIPMENT
11850 NAVIGATION EQUIPMENT
11870 AGRICULTURAL EQUIPMENT
11900 EXHIBIT EQUIPMENT

DIVISION 12 FURNISHINGS

12050 FABRICS
12100 ART
12300 MANUFACTURED CASEWORK
12400 FURNISHINGS AND ACCESSORIES
12500 FURNITURE
12600 MULTIPLE SEATING
12700 SYSTEMS FURNITURE
12800 INTERIOR PLANTS AND PLANTERS
12900 FURNISHINGS RESTORATION AND REPAIR

DIVISION 13 SPECIAL CONSTRUCTION

13010 AIR-SUPPORTED STRUCTURES
13020 BUILDING MODULES
13030 SPECIAL PURPOSE ROOMS
13080 SOUND, VIBRATION, AND SEISMIC CONTROL
13090 RADIATION PROTECTION
13100 LIGHTNING PROTECTION
13110 CATHODIC PROTECTION
13120 PRE-ENGINEERED STRUCTURES
13150 SWIMMING POOLS
13160 AQUARIUMS
13165 AQUATIC PARK FACILITIES
13170 TUBS AND POOLS
13175 ICE RINKS
13185 KENNELS AND ANIMAL SHELTERS
13190 SITE-CONSTRUCTED INCINERATORS
13200 STORAGE TANKS
13220 FILTER UNDERDRAINS AND MEDIA
13230 DIGESTER COVERS AND APPURTENANCES
13240 OXYGENATION SYSTEMS
13260 SLUDGE CONDITIONING SYSTEMS
13280 HAZARDOUS MATERIAL REMEDIATION
13400 MEASUREMENT AND CONTROL INSTRUMENTATION
13500 RECORDING INSTRUMENTATION
13550 TRANSPORTATION CONTROL INSTRUMENTATION
13600 SOLAR AND WIND ENERGY EQUIPMENT
13700 SECURITY ACCESS AND SURVEILLANCE
13800 BUILDING AUTOMATION AND CONTROL
13850 DETECTION AND ALARM
13900 FIRE SUPPRESSION

DIVISION 14 CONVEYING SYSTEMS

14100 DUMBWAITERS
14200 ELEVATORS
14300 ESCALATORS AND MOVING WALKS
14400 LIFTS
14500 MATERIAL HANDLING
14600 HOISTS AND CRANES
14700 TURNTABLES
14800 SCAFFOLDING
14900 TRANSPORTATION

DIVISION 15 MECHANICAL

15050 BASIC MECHANICAL MATERIALS AND METHODS
15100 BUILDING SERVICES PIPING
15200 PROCESS PIPING
15300 FIRE PROTECTION PIPING
15400 PLUMBING FIXTURES AND EQUIPMENT
15500 HEAT-GENERATION EQUIPMENT
15600 REFRIGERATION EQUIPMENT
15700 HEATING, VENTILATING, AND AIR CONDITIONING EQUIPMENT
15800 AIR DISTRIBUTION
15900 HVAC INSTRUMENTATION AND CONTROLS
15950 TESTING, ADJUSTING, AND BALANCING

DIVISION 16 ELECTRICAL

16050 BASIC ELECTRICAL MATERIALS AND METHODS
16100 WIRING METHODS
16200 ELECTRICAL POWER
16300 TRANSMISSION AND DISTRIBUTION
16400 LOW-VOLTAGE DISTRIBUTION
16500 LIGHTING
16700 COMMUNICATIONS
16800 SOUND AND VIDEO

SPECIFICATIONS
-BIDDING DOCUMENTS-
for
WATER SUPPLY, PUMPHOUSE, DISTRIBUTION SYSTEM & ELEVATED STORAGE RESERVOIR

Sample Table of Contents from Specifications Book 技术手册上的目录示例

TITLE	PAGE

BIDDING DOCUMENTS:

i

TITLE	PAGE

TECHNICAL SPECIFICATIONS: **NOTE:** Some of the following section numbers may not match the current CSI System because these specifications pre-date the 2002 publication.

ii

* Represents specific specs that are reproduced.

Courtesy: Morgan & Parmley, Ltd.

iii

iv

v

NOTE: Specification Section Title

工程概况（包括特别规定）

SUMMARY OF WORK
(W/SPECIAL PROVISIONS)

PART ONE - GENERAL

1.1 BACKGROUND

1.1.1 History of Project: The High School Complex in Tony, is owned and operated by the Flambeau School District. Due to historically poor potable water and continuous maintenance problems, the School Board retained Morgan & Parmley, Ltd. to solve their problem. On May 23, 1977, construction was completed for their new water supply well located on Rusk County Airport property. The pumphouse, pumping equipment and transmission main were placed in operation early in 1979. This facility has been in continuous use, serving the High School Complex, since that time with no significant operational problems.

On several occasions, the Village Board attempted to establish a municipal water system without success. However, their recent efforts have proven successful by obtaining funding through the Rural Economic and Community Development, USDA (formally, FmHA).

An agreement between the Flambeau School District and the Village of Tony was recently executed. The Village will purchase the existing school well, pumphouse, transmission main and related appurtenances. These components will be incorporated into their proposed (total) water system; as detailed in the accompanying Plans and Specifications.

The Village has established their public water utility under the rules and regulations of the Wisconsin Public Service Commission.

The Well Site Survey and Engineering Report have been approved by DNR.

1.2 SCOPE OF WORK

1.2.1 Description: The proposed Project consists of the following major items:

1-Rehab Existing Pumphouse
2-Vertical Turbine Pump w/Controls
3-Interior Piping & Controls
4-Right Angle Drive Mechanism
5-Chlorination Facility
6-Electrical Upgrade
7-Watermain Distribution System
8-Elevated Water Tank & Tower
9-Services & Meters

NOTE: Page Number

NOTE: Section Number

NOTE: Project Number

93-159

01010-1

Courtesy: Morgan & Parmley, Ltd.

The foregoing list is only major items. Refer to the accompanying Plans, applicable sections of the Specifications and the Proposal for a total concept.

1.2.2 Proposals: There will be two (2) Proposals. The first will be for all items, except the elevated water tank, tower and related components which will be bid separately.

1.3 SCHEDULE

1.3.1 Anticipated Construction Schedule:

Commence Advertising 11-13-1995
Bid Opening 12-14-1995
Notice of Award 12-20-1995
Pre-Construction Conference 1-15-1996
Notice to Proceed 1-15-1996
Complete Construction 11-10-1996

1.3.2 Highway Construction: All construction on U.S. Highway 8 shall be substantially completed by June 30, 1996.

1.4 HYDRANT PATTERN

1.4.1 Description: Since the Village of Tony is served by the Ladysmith Fire Department, it is mandatory that Tony's fire hydrant pattern be compatable with their equipment.

PART TWO - SPECIAL PROVISIONS

2.1 WAGE RATES

2.1.1 Wages: The Contractors and Sub-Contractors shall meet the requirements of all applicable official State Wage Rate Determinations for this Project. A copy is attached to these Specifications.

2.2 EQUAL EMPLOYMENT OPPORTUNITY

2.2.1 Compliance: The Contractors and Sub-Contractors shall comply with all current and applicable rules and regulations governing equal employment opportunities.

2.3 APPROVALS

2.3.1 General: The Contractors and Sub-Contractors shall comply with all applicable items contained in the regulatory approvals and permits for this project. Approvals and permits from DNR, PSC, FAA, DOT, SHPO, RR, Rusk County, and others are attached to this

Specification Book under the heading; Supplemental Documents. It is your responsibility to understand and abide by the conditions contained in the documents.

2.4 EROSION CONTROL

2.4.1 Responsibility: It is the Contractor's responsibility to provide the specified erosion control for the project. Refer to Section 02770, sheet 15 of 15 of the Plans and DOT's Policy 96.55 and related sections.

2.5 TRAFFIC CONTROL

2.5.1 Responsibility: It is the Contractor's responsibility for traffic control. Refer to DOT's Policy 96.51 and related sections.

2.6 EXISTING UTILITIES

2.6.1 General: It is the responsibility of the Contractor to acquaint himself with the location of all underground structures which may be encountered or which may be affected by work under the Contract. The locations of any underground structures furnished, shown on the Plans or given on the site are based upon available records, but are not guaranteed to be complete or correct and are given only to assist the Contractor in making a determination of the location of all underground structures.

2.6.2 Utility Notification: The Contractor shall give notice in writing to all utilities that may be affected by Contractor's operations at least seventy-two (72) hours before starting work. The Contractor shall not hinder or interfere with any person in the protection of such work.

The Contractor is encouraged to utilize the "Call Before You Dig" telephone notification system.

2.6.3 Cable TV Notification: The Contractor shall notify Marcus Cable, 219 W. Miner Avenue, Ladysmtih, WI 54848,; Tele. No. 715/532-6040.

2.6.4 Electrical Utility Notification: The Contractor shall notify Northern States Power Co., 129 N. Lake Ave., Phillips, WI 54555; Atten: Joe Perkins; Tele. No. 715/836-1198.

2.6.5 Telephone Utility: The Contractor shall notify Ameritech @ 1-800-242-8511 (Diggers Hotline).

2.6.6 Village Sewer System: The Contractor shall notify the Village's Sanitary Sewer Utility Operator, Billy Mechelke, N5297 Little X, Tony, WI 54563, Tele. No. 715/532-7046.

2.6.7 Emergency Facilities: The Contractor shall give written notice to the Fire Department, Rusk Co. Sheriff's Office and Ambulance Service, at least twenty-four (24) hours before closing off or in any way affecting through vehicular traffic on any street.

Rusk Co. Sheriff's Office 715/532-2200

Ladysmith Rural Fire Department 715/532-2186

Ambulance Service 715-532-2100

2.7 CONSTRUCTION SIGN

2.7.1 Construction Sign: One (1) construction sign shall be provided and erected at a location to be specified by the Engineer at the Pre-Construction Conference. Refer to Supplemental Specifications, page A for its size and design. The construction sign will be a bid item in the Proposal.

2.8 SURVEY MARKERS

2.8.1 General: No survey markers (e.g. DOT, County, USGS, Property Corners, etc.) located in the project area shall be disturbed, unless prior approval has been obtained. Replacement will be at the expense of the Contractor.

2.9 STORAGE OF MATERIALS

2.9.1 General: Contractor shall confine his apparatus, storage of materials and operations of his workmen to limits indicated by law, ordinances, permits or directions of the Engineer.

All materials stored on the site shall be protected from the elements by watertight coverings or shed with wood floors raised above the ground and protected from damage by Contractor's operations by barriers. The Owner assumes no responsibility for stored materials.

All materials shall be delivered in original sealed containers, with seals unbroken and with labels plainly indicating materials and directions for storage. Directions for storage shall be complied with unless specified to the contrary in the detailed Contract Specifications.

Damaged materials shall be immediately removed from the site. Stored material will be paid for in accordance with standard practice.

2.10 PROTECTING WORK

2.10.1 General: The Contractor shall protect existing improvements and Work installed by him from damage. Improvements and Work damaged shall be replaced by the Contractor. All replacement methods shall be approved by the Engineer.

2.11 LABOR LAWS

2.11.1 General: All Contractors and Subcontractors employed upon the Work shall be required to conform to applicable Federal and State labor laws and the various acts amendatory and supplementary thereto and to all other laws, ordinances and legal requirements applicable thereto.

All labor shall be performed in the best and most workmanlike manner by mechanics skilled in their respective trades. The standard of the Work required throughout shall be such grade as will bring results of the first class only.

2.12 TEMPORARY FACILITIES

2.12.1 Description: The Contractor shall provide his own temporary facilities such as electricity, toilets and heat. He shall also furnish, erect and maintain a weatherproof bulletin board for displaying wage rates, Equal Opportunity Requirement, etc. All temporary facilities shall be removed prior to job completion.

2.13 COOPERATION

2.13.1 General: The Contractors and Subcontractors shall cooperate and coordinate their Work with adjacent Work and shall give due notice to other Contractors and Subcontractors of intersecting Work to assure that all items are installed at an agreeable time and to facilitate general progress of the Work.

2.14 PROGRESS SCHEDULE

2.14.1 Description: The Contractor shall, within ten (10) days after the effective date of Notice to Proceed, furnish three (3) copies of a preliminary progress schedule covering his operations for the first 30 days.

The preliminary progress schedule shall be a bar graph or an arrow diagram showing the times he intends to commence and complete the various Work stages, operations and contract items planned to be started during the first 30 days.

The Contractor shall submit for approval by the Engineer within 30 days after the effective Notice to Proceed three (3) copies of a detailed bar graph or a graphic network diagram. The graph or diagram shall be accompanied by a brief written explanation of the proposed schedule and a list of activities.

The bar graph or graphic network diagram shall clearly depict the order and interdependencies of activities planned by the Contractor as well as activities by others which affects the Contractor's planning. For those activities lasting more than 30 days, either the estimated time for 25-50 and 75% completion or other significant milestone in the course of the activity shall be shown. In addition to the actual construction operations, the schedule shall show such items as submittal of samples and shop drawings, delivery of materials and equipment, construction in the area by other forces and other significant items related to the progress of construction. The graph or network diagram shall be printed or neatly and legibly drawn to a time scale.

The list of activities shall show for each activity the estimated duration and anticipated starting and finishing dates. Activities which are critical to complete the project in shortest time shall be identified. The cost of each activity shall be shown so the schedule can be cost loaded.

The Contractor shall submit with each payment request a copy of the schedule showing the current status of the project and the activities for which payment is being requested. If the project is behind schedule, a detailed report shall accompany the submittal stating why the project is behind schedule and what means are being taken to get the project back on schedule, including, if appropriate, contract time extensions.

The Schedule shall be revised:

a) When a Change Order significantly affects the contract completion date or the sequence of activities.

b) When progress of any critical activity falls significantly behind the scheduled progress.

c) When delay on a noncritical activity is of such magnitude as to make it become critical.

d) At any time the Contractor elects to change any sequence of activities affecting the shortest completion time.

The revised analysis shall be made in the same form and detail as the original submittal and shall be accompanied by an explanation of the reasons for the revisions.

2.15 SHOP DRAWINGS

2.15.1 Description: Shop drawings and descriptive data shall be required on all manufactured or fabricated items. Seven (7) copies of drawings and descriptive data shall be submitted for approval. (Two of these will be returned to the Contractor after their approval.)

Shop drawings shall be scheduled to be submitted for approval within 4 weeks after the effective Notice to Proceed.

2.16 SUPERVISION OF ERECTION FOR ALL EQUIPMENT

2.16.1 General: All items of structural, mechanical or electrical equipment whose cost delivered to the Owner exceeds $1,000 shall have included in the price bid the services of a qualified representative of the manufacturer for one (1) day to inspect equipment in operation and train the operator. For items whose cost exceeds $2,500, the bid price shall include not less than one (1) day during erection and not less than one (1) day operation inspection. Over $5,000, the bid price shall include not less than three (3) days during erection and two (2) days operation inspection. After each trip, the manufacturer is to submit a report to the Engineer covering the findings made during the inspection including changes and/or repairs, if any. Lesser time than specified may be for "make up" trips during the twelve (12) month warranty period. Trips made for the purpose of correcting defective materials or workmanship are to be made without charge and are not to be credited to the required number.

2.17 SCHOOL WATER SUPPLY

2.17.1 Service: It is mandatory that the High School Complex be provided potable water on a continuous basis during school hours. Therefore, the Contractor shall schedule his Work to avoid any interruption of water during school operation or normal activity.

2.18 EXISTING SHRUBBERY

2.18.1 Protection: It shall be the responsibility of the Contractor to protect all existing private shrubbery, trees, etc. within the construction area. If they are damaged, or destroyed, they shall be replaced to the owner's satisfaction at the Contractor's expense.

2.19 SAFETY

2.19.1 Basic: All work shall be done in accordance with OSHA standards and other specified regulations as noted throughout the Specifications.

2.20 PROJECT RECORD DOCUMENTS

2.20.1 Contractor's Responsibility: At the conclusion of the project, prior to final payment, the Contractor shall supply the Engineer with a legible record of all work as installed. Example: a map detailing watermain installed with location, lengths, tie-offs, valves, hydrants and appurtenances accurately shown so an "as-built" layout can be developed. This information can be marked on the appropriate sheet of the Plans with a permanent color marker.

2.21 OPERATION & MAINTENANCE DATA

2.21.1 O & M Data: The Contractors shall furnish to the Engineer three (3) complete unbound copies of complete operation and maintenance data for all equipment items. This data shall include a list of parts and a recommendation of spare parts that should be kept on hand.

2.22 TESTING

2.22.1 Testing & Equipment: All mechanical and electrical equipment furnished for this project shall be field tested to insure full compliance with the Specifications. Testing shall be performed by the Contractor and/or equipment supplier in the presence of the Engineer. If the field test shows that the equipment does not meet the Specifications, it shall be replaced with acceptable equipment at Contractor's expense.

Full payment for equipment will not be made until it has been field tested and accepted.

2.23 FINAL CLEANUP

2.23.1 General: The Contractor shall be responsible for cleaning up this Work and his waste and rubbish. It is the General Contractor's responsibility to completely clean up the inside and outside of all buildings and the total construction site at the completion of the project.

2.24 PRIVATE SERVICES

2.24.1 Description: Per specified bid items, the Contractor shall furnish and install service laterals from the distribution main to (and including) the curb stop. The Contractor shall also furnish and install the water meter including fittings and appurtenances within each respective user's structure.

Please be advised that section of the service lateral from the curb stop to the meter is the responsibility of each user and is not part of this construction contract.

2.25 PAYMENT TO CONTRACTOR

2.25.1 General: The Owner will make progress payments to the Contractors in accordance with the General Conditions. It is understood that grant reimbursement payments to the Owner are dependent upon compliance by the Owner with all the grant conditions identified in the grant request and grant agreement. The Owner shall be responsible for progress payments to the Contractor even when failure to comply with said grant conditions delays grant payments from the funding agency.

2.25.2 Final Payment: Final payment to the Contractor will not be made until the Owner and Engineer are satisfied that the facility is fully operational and all aspects of the Plans, Specifications and Construction Contract have been satisfied.

2.26 FmHA NAME CHANGE

2.26.1 Clarification: Wherever the words "Farmers Home Administration (FmHA) or Rural Development Administration (RDA)" appear in the project manual, substitute the words "U.S. of America."

2.26 RAILROAD NOTIFICATION

2.26.1 General:

-END OF SECTION-

储水设施的消毒

DISINFECTION OF WATER STORAGE FACILITY

PART ONE - GENERAL

1.1 GENERAL

1.1.1 Work Included: The Wo-k described in this Section shall consist of the disinfection of the total elevated water storage facility so it can safely store potable water for use by the Tony Municipal Water system.

1.1.2 Related Work Described Elsewhere:

1) Summary of Work 01010
2) Field Engineering & Inspection 01050
3) Applicable Standards 01085
4) Submittals & Substitutions 01300
5) Testing Laboratory Services 01410
6) Water Systems 02713
7) Painting 09901
8) Surface Preparation Elev. Storage Tank 09910
9) Painting Elevated Water Storage Facility 09920
10) Elevated Water Storage Facility 13410

Applicable Standards & Codes

1.1.3 Timing: The newly painted water storage facility shall be allowed to dry for a minimum of 7 days before being subjected to this disinfection process.

1.1.4 Holding Period: Chlorine solutions will remain in contact with the surfaces to be disinfected for a minimum of 24 hours.

1.1.5 Forms of Chlorine: Three (3) forms of chlorine are acceptable for disinfecting water storage facilities. The CONTRACTOR will select the form of chlorine he intends to use and so notify the ENGINEER, three (3) days before beginning the disinfection process.

1) Liquid chlorine will meet the requirements of AWWA B-301.

2) Sodium Hypochlorite will meet the requirements AWWA B-300.

3) Calcium Hypochlorite will meet the requirements of AWWA B-300.

1.1.6 Disinfection Methods: The CONTRACTOR must use one of the three (3) alternative disinfection methods specified in AWWA Standard C-652.

CONTRACTOR shall notify the ENGINEER of the selected method at the same time he notifies him of the form of chlorine to be used.

1.1.7 Initial Fill: The initial filling of the elevated storage reservoir (tank) shall be accomplished by using water from the newly constructed distribution system. The expense for providing this water shall be borne by the Village of Tony.

1.1.8 Drainage: After the "holding period", the highly chlorinated water in the tank and riser shall be completely drained to waste, via the municipal sanitary sewer collection system at a rate that will not hydraulically and biologically upset the WWTF.

1.1.9 Refill: Following the complete draining of the tank and riser, as described in 1.1.8, the storage facility shall be refilled with potable water from the municipal distribution system at no expense to the Contractor. After refilling, samples of water should be taken from the tank and tested to demonstrate and record the good sanitary condition of the tank before it is placed into regular service.

The CONTRACTOR and ENGINEER will together drain, refill and sample the water in the tower. The ENGINEER will select the laboratory to analyze the collected water samples and the OWNER will pay the corresponding fee.

1.1.10 Unsuccessful Test: If the test for coliform organisms is positive, the tower and riser must be rechlorinated. The cost of rechlorination, water, chlorine, sample collection and sample analysis will be the responsibility of the CONTRACTOR. THE OWNER WILL CHARGE THE CONTRACTOR $1.15/1000 GALLONS OF WATER USED DURING ANY AND ALL RECHLORINATION WORK.

END OF SECTION

Courtesy: Morgan & Parmley, Ltd.

深井涡轮泵

DEEP WELL TURBINE PUMP

PART ONE - GENERAL

1.1 DESCRIPTION

1.1.1 Work included: This section of the Specifications applies to the furnishing and complete installation of one (1) deep well turbine water pump. Removal of the existing submersible pump and related appurtenances is also included in this Work.

1.1.2 Work location: The location of this pump installation is at Municipal Well No. 1, Rusk County Airport, as shown on the accompanying Plans.

1.1.3 Related work described elsewhere:

1) Summary of Work 01010
2) Field Engineering 01050
3) Submittals and Substitutions 01300
4) Testing & Laboratory Services 01410
5) Water Systems 02713
6) Right Angle Drive 11211
7) Chlorination Facility 11212
8) Internal Piping 11213
9) Pressure Switch & Surge Eliminator 11214
10) Water Supply Control System 13601
11) Electrical 16051

1.2 SUBMITTALS

1.2.1 Shop Drawings: Within thirty (30) days after Notice to Proceed, the Contractor shall submit tot he Engineer a complete manufacturer's specification and descriptive literature of the pump and related appurtenances.

1.2.2 DNR Form: The Contractor shall supply the Engineer with the necessary technical information required to complete Form 3300-226 *WELL PUMP SUBMITTAL CHECKLIST MUNICIPAL SYSTEMS*. A copy of this form is available to the Contractor upon request.

1.2.3 Operation & Maintenance Data: O & M data shall be supplied with illustrative literature for operator's Manual.

1.3 GUARANTEE

1.3.1 Manufacturer: The manufacturer shall guarantee all materials and workmanship for a period of one (1) year from date of acceptance. All costs of field repair and/or replacement of components or devices to enforce the warranty will be borne by the Contractor or his representative for warranty period.

1.3.2 Contractor: The Contractor shall guarantee the total installation for one (1) year from date of acceptance, free of any expense to the Owner.

1.4 ACCEPTANCE

The pump and its installation will be accepted when it has been tested and demonstrates that it complies with the Specifications and design.

PART TWO - PRODUCTS

2.1 DEEP WELL TURBINE PUMP

2.1.1 Operating Conditions: The capacity of the well pump (for bidding purposes) shall be 200 G.P.M. against a total (field) dynamic head of 230 feet. To this head the manufacturer shall add the column friction loss, the pump friction loss and the line shaft bearing loss, as well as the motor loss. All computations for loss shall be as stated in Standards of the Hydraulic Institute and in the Standards of the current Edition of the National Association of Vertical Turbine Pump Manufacturers. Calculations shall accompany the shop drawing submittal. The speed of the pump and motor shall not exceed 1800 rpm. The pump shall be water lubricated with open shaft and rubber bearing.

2.1.2 Efficiency: Each Bidder shall submit with his Proposal a guaranteed wire to water efficiency of the pump he proposes to furnish when operating at the above total head and capacity, and shall submit performance curves showing head, capacity, efficiency, and horsepower demand from no delivery to maximum. The guarantee shall be stated in kilowatt hours electrical consumption per 1000 gallons of water produced, and the pump may be throttled during the acceptance test to develop the guaranteed head.

2.1.3 Bowl Assembly: The bowl assembly shall consist of close grain iron bowls, stainless steel impeller shaft, bronze enclosed type impellers, and bronze renewable wear rings. A discharge nozzle shall be furnished with the bowl assembly, which shall contain an extra long main bearing, and shall be bolted or screwed to the top bowl. Pumps without a discharge nozzle will not be acceptable.

2.1.4 Discharge Column: The discharge column shall be of standard weight 6" I.D. nominal steel pipe with heavy screw type couplings located not less than every 10 feet along column. Sufficient column shall be furnished to set the top of the bowls at the top of the screen and adequate suction tail pipe shall be located to place the suction entrance at a point three feet

Courtesy: Morgan & Parmley, Ltd.

above the bottom of the screen. The line shaft shall be not less than 1" in diameter, and shall be composed of high carbon steel, and shall have adequately lubricated bearings located at least every 10 feet along the shaft. Water lubricated, Monel or Stainless Steel sleeves shall be located opposite rubber bearings fluted for water lubrication.

2.1.5 Pump Head: The pump head shall have a separate and removable cast iron base plate which can accommodate a one inch extension of the inner well casing above the top of the concrete pump foundation. The base plate shall be permanently grouted to the concrete pump base by the Contractor after the pump is installed and properly aligned. The pump head shall be large enough to accommodate a coupling in the motor drive shaft below the base of the motor and above the stuffing box. There shall be furnished and installed suitable protecting pipe guards to properly steady the protecting pipe and line shaft assembly. The Contractor shall furnish and install a pre-lubrication line, ½ inch solenoid valve for 110 volt operation, 90 second time delay relay and other electrical controls as needed to afford a 30 to 90 second prelubrication before each pump start-up. The pump shall have a 6-inch flanged above ground discharge drilled for 150 pound American Standard flanges.

The pump Contractor shall furnish and install all electrical equipment and controls required for the complete pump installation and chemical feed unit; commencing from the service entrance box as installed by the pumphouse subcontractor. Refer to technical Sections 11211, 11212, 11213, and 11214.

2.1.6 Electric Pump Motor: The Contractor shall furnish and install a vertical hollow shaft, part-winding, 240 volt, three phase, 60 Hz. drip-proof, 1800 rpm no load speed, motor of sufficient horsepower to operate the deep well pump at any head capacity condition, even at reduced heads, without overloading the name plate rating of the motor. The motor is to be manufactured by only a domestic manufacturer, such as G.E., Westinghouse, U.S., Allis-Chalmers, or equal.

2.1.7 Electrical Controls: This Contractor shall furnish with the pump at 240 volt, three phase, 60 Hz, part-winding magnetic starter in NEMA one enclosure equal to Square D, with Hand-Off-Automatic selector switch, together with fusible heavy duty disconnect switch. This Contractor shall also provide a time delay relay and pre-lubrication solenoid for operating the pre-lubrication line for up to 90 seconds prior to pump start. A Hand-Off-Automatic selector switch and fused disconnect switch shall be furnished and installed on the control circuit to the chemical feeder.

2.1.8 Pump Installation: Pump Sub-Contractor shall not sublet the installation of the pump; as it is intended that this Contractor shall be solely responsible for the satisfactory operation of the system. The Contractor shall provide an orifice and pressure gauge, and other testing equipment required to run a two-hour efficiency test upon completion of the installation. In the event the efficiency or performance of the pump does not meet his guarantee, the Contractor must make necessary modifications or replacements at his own expense.

2.1.9 Pump Control: The Pump Control System shall be the product of a manufacturer experienced, skilled and regularly engaged in the design and fabrication of this type of equipment, with similar installations within convenient inspection distance. Complete detail drawings and descriptive information shall be submitted to the Engineer for approval before fabrication. Refer to Section 13601.

2.1.10 Pressure Switch: The Contractor shall furnish and install pressure switch for sensing of main pressure for pump start/stop equipment. Refer to Specification 11214 for details.

2.1.11 Surge Eliminator: The Contractor shall furnish and install surge eliminator tank and related appurtenances. Refer to Specification 11214 for details.

2.1.12 Installation: The Contractor, through the manufacturer, shall provide the services of factory Engineers to inspect installation, supervise final adjustments and instruct the operating personnel in the use and maintenance of the equipment. This supervision shall be provided by the Contractor on the job site and is considered incidental to the contract.

2.1.13 Water Level Indicator: Contractor shall furnish and install a plastic airline with gage and tire pump to allow Waterworks Operator to measure water levels in well.

PART THREE - EXECUTION

3.1 GENERAL

3.1.1 Location: Refer to applicable drawings on the accompanying Plans for location and typical layouts.

3.1.2 Installation: All electrical Work shall be executed by a qualified electrician or under his direct supervision. All pump installation shall be executed by or under the direct supervision of fully qualified personnel.

3.1.3 Discrepancies: Where real discrepancies are found between the Plans and the Specifications they shall be brought to the attention of the Engineer as soon as found so that corrections can be made that will least affect the interests of the Owner and Contractor.

3.1.4 Final Connections: This Contractor shall be responsible for final connections of all electrical circuits at the control panel relative to pump installation and related appurtenances as defined.

3.1.5 Testing: This Contractor, together with a representative of the manufacturer, shall in cooperation with each other, check all circuitry before energizing.

3.1.6 Operation and Maintenance Manual Data: Spare parts Data and Shop Drawings requirements are specified in the Supplemental General Conditions. This Contractor is governed by those conditions.

As stated previously, workmanship and materials shall comply with Wisconsin and National Electric Codes, NEMA Standards and all equipment shall bear UL labels.

3.1.7 Labor for Testing: All labor, instruction, and equipment used for testing and fine tuning the electrical and mechanical systems shall be furnished by the Contractor and is considered incidental to the Contract.

3.2 ACCEPTANCE

3.2.1 Final Acceptance: By the Engineer and Owner, will follow the testing of the "total" pumping system and related equipment.

3.3 PAYMENT METHOD

3.3.1 Payment: Will be based upon the Proposal.

END OF SECTION

内部管道系统（泵房改造） INTERIOR PIPING

PART ONE - GENERAL

1.1 DESCRIPTION

1.1.1 Work included: This Section of the Specifications applies to the furnishing and installation of interior pumphouse piping and shall commence at the discharge of the Deep Well Turbine Pump assembly and terminate at the 6-inch riser pipe, as illustrated on the accompanying Plans. The existing interior piping and related components shall be removed by Contractor. Disposal is the Contractor's responsibility.

1.1.2 Related Work described elsewhere:

1) Summary of Work 01010
2) Field Engineering 01050
3) Submittals & Substitutions 01300
4) Water Systems 02713
5) Deep Well Turbine Pump 11210
6) Right Angle Gear Drive 11211
7) Pumphouse Rehab 15050

1.2 SUBMITTALS

1.2.1 Shop Drawings: Within thirty (30) days after Notice to Proceed, the Contractor shall submit to the Engineer a piping layout of interior piping.

1.3 GUARANTEE

Interior piping and equipment shall be fully guaranteed for one (1) year after final acceptance for all material, workmanship, and installation. All costs for field repair and/or replacement shall be borne by Contractor for duration of warranty.

1.4 ACCEPTANCE

Interior piping system shall be accepted when it has been fully tested and demonstrated in conjunction with the "total" pumping system.

PART TWO - PRODUCT

2.1 Description: The Contractor shall furnish and install the following list of materials and equipment to fully complete the interior piping for connection of pumping system to the existing 6-inch riser pipe, as shown on the accompanying Plans. Material shall be ductile iron pipe

Courtesy: Morgan & Parmley, Ltd.

with flanged joints, 6-inch I.D. with a minimum working pressure of 250 p.s.i.

Automatic Air Relief System
Check Valve (6-inch)
Smooth Sampling Tap (½-inch)
Sleeves & Couplings (6-inch) (as needed)
Meter (6-inch)
O.S. & Y Gate Valve (6-inch)
6" x 6" x 6" Tee W/Blind Flange
Screened Well Vent
Floor Pipe Supports

This Work, also includes the removal of existing interior piping and components not needed for new facility.

The foregoing list includes major components. However, the list is not necessarily complete. No claim for extras will be allowed.

Refer to accompanying Plans for basic layout and technical Section 02713 of the Specifications.

PART THREE - EXECUTION

3.1 GENERAL

3.1.1 Location: Refer to accompanying Plans for details and location.

3.1.2 Installation: This Contractor shall provide experienced personnel to properly install the equipment.

3.1.3 Testing: The Contractor shall test this system in conjunction with testing of Deep Well Turbine Pump.

3.2 ACCEPTANCE

3.2.1 Final Acceptance: Final acceptance by the Engineer and Owner will follow the successful testing of the "total" pump facility.

3.3 PAYMENT METHOD

3.3.1 Payment: Payment method is based upon a lump sum as bid for the segment as listed in the Proposal.

END OF SECTION

水塔

ELEVATED WATER STORAGE RESERVOIR

PART ONE - GENERAL

1.1 BASE BID

1.1.1 Description: The base bid shall be for a Single Pedestal Sphere-Shaped elevated water storage reservoir; including all applicable appurtenances.

1.2 ALTERNATE BID

1.2.1 Description: The alternate bid shall be for a Multi-Leg Ellipsoidal-Shaped elevated water storage reservoir; including all applicable appurtenances.

1.3 PROJECT DESCRIPTION

1.3.1 Scope of Work: The storage facility basically consists of furnishing all materials, supplies, equipment, tools, labor and related skills necessary for fabricating, delivering, erecting, painting, testing and certifying one (1) 50,000 gallon elevated steel storage reservoir, complete with the specified accessories; ready for service to the Tony Water Utility. The elevated reservoir (tank) and tower will be fabricated and erected in strict accordance with all applicable regulatory codes, plan details, these general specifications, NFPA and AWWA Standards so it will meet the Wisconsin Department of Natural Resources approval requirements.

1.3.2 Related Work described elsewhere:

1) Summary of Work 01010
2) Field Engineering & Inspection 01050
3) Applicable Standards 01085
4) Submittals & Substitutions 01300
5) Testing Laboratory Services 01410
6) Temporary Facilities & Controls 01500
7) Contract Closeout 01700
8) Clearing & Grubbing 02110
9) Earthwork 02200
10) Topsoil & Landscaping 02480
11) Water Systems 02713
12) Erosion Control 02770
13) Concrete Testing & Sampling 03050
14) Concrete Formwork 03100
15) Concrete Reinforcement 03200
16) Cast-in-Place Concrete 03300
17) Structural Steel 05120

Courtesy: Morgan & Parmley, Ltd.

18) Structural Steel Erection 05121
19) Metal Fabrication 05500
20) Flashing & Sheet Metal 07600
21) Sealants & Calking 07920
22) Painting 09901
23) Surface Preparation Elevated Tank 09910
24) Painting Elevated Storage Tank 09920
25) Water Supply Control System 13601
26) Electrical 16051
27) Standard Specifications: .... 5th Edition
Sewer & Water Construction in Wisconsin
28) Applicable Standards NFPA & AWWA
29) Refer to Soil Boring Report (Supplemental Document I)

1.4 LOCATION

1.4.1 Site: The elevated water storage reservoir shall be constructed on the 100' x 100' lot located westerly of the intersection of Cedar Street and Central Avenue; as shown on the accompanying Plans. The Village holds a long term lease on this property.

1.4.2 Access: The Owner (Village of Tony) will provide suitable access to the site, via a twenty foot (20') wide easement from the adjacent municipal street.

1.5 SUBMITTALS

1.5.1 Shop Drawings: Within thirty (30) days after executing the Notice to Proceed document, the Contractor shall submit to the Engineer five (5) complete sets of fabrication shop drawings detailing the total facility; including design of the concrete foundation.

All drawings and design submittals shall be stamped and signed by a Professional Engineer currently registered in the State of Wisconsin.

1.5.2 Structural Calculations Accompanying the previously noted shop drawings, shall be five (5) completed sets of structural calculations certifying the structural integrity of the concrete foundation, elevated water storage reservoir and tower. These calculations shall be stamped and signed by a Professional Engineer currently registered in the State of Wisconsin.

1.6 GUARANTEE

1.6.1 Manufacturer: The manufacturing/fabricator shall fully guarantee all materials and workmanship for a period of one (1) year from date of final acceptance by Owner. All costs of field repair and/or replacement of any component or device to enforce the warranty will be completely borne by the Contractor or his representative for the warranty period.

93-159 13410-2

1.6.2 Contractor: The Contractor shall fully guarantee the total installation for one (1) year from date of final acceptance by the Owner.

1.7 NOTIFICATION

1.7.1 Airport and General: The Contractor shall notify the manager of the Rusk County Airport, in writing, a minimum of five (5) working days, prior to commencing steel erection. The Contractor shall also, comply with any other applicable requirements specified in the regulatory approvals and permits; including FAA mandates.

PART TWO - PRODUCTS

2.1 GENERAL

2.1.1 Base-Bid Description: The reservoir (tank) and supporting structure will be of all-welded steel construction. The transition sections between the pedestal and tank, and between the pedestal and base shall be smooth curves.

2.1.2 Alternate - Bid Description: The reservoir (tank) and supporting structure will be of all-welded steel construction. The tank shall be supported on multiple tubular steel columns with a five (5) foot diameter dry riser. Access to the tank and valve vault shall be provided through the dry riser base, hereafter referred to as the bell. If the dry riser is to be used a support column, smooth curve transitions will be required between the dry riser and bell, and between the dry riser and tank.

2.2 DESIGN

2.2.1 Design Parameters: The following list of design parameters follows the basics outlined in the Forward of Section III, AWWA Standard D-100-84:

1) Capacity: 50,000 U.S. Gallons

2) Height to Overflow: 122 feet (nominal)

3) Type of Roof: Base Bid - Single Pedestal Sphere
Alternate Bid - Ellipsoidal

4) Head Range: 20 Feet (Nominal)

5) Riser Pipe: 8 inch diameter (steel)

6) Location: See Plans

93-159 13410-3

7) Time for Completion: See Bidding Documents

8) Nearest Town: Within corporate limits of the Village of Tony, Wisconsin

9) Nearest Railroad Siding: Wisconsin Central, Ltd. Siding adjacent to construction site

10) Type of Road Near Site: Municipal street (asphalt pavement) w/access road graveled

11) Electrical Power: Northern States Power (NSP) Company, local office in Ladysmith, WI (5 miles west of Tony)

12) Compressed Air: Not available at site

13) Single Pedestal Design Supplement: Not required

14) Welded Joints: Detailed drawings are required to be submitted. Welds must be continuous and on all sides of overlaps.

15) Copper-Bearing Steel: Not required.

16) Type of Pipe & Fittings: Steel

17) Snow Load: 25 PSF Minimum

18) Wind Load: 100 MPH Minimum

19) Earthquake Load: None required

20) Corrosion Allowance: None required

21) Inspection & Painting Balcony: Not applicable

22) Increased Wind Load: Required to accommodate future paint removal, dust control apparatus (tarps, curtains, etc.) and resistance to the overturning forces.

23) Manholes, Ladders, Additional Accessories: See Item No. 30

24) Pipe Connections: An 8 inch diameter steel pipe will carry water from the bottom of the tank to a base elbow 8 feet below grade. The riser pipe will have an Expansion joint immediately above the base elbow. The riser pipe will be insulated to protect against freezing to -40°F (outside temp.)

25) Removable Silt Stop: A removable silt stop is required, 12 inches high minimum on the riser pipe inside tank.

26) Overflow: A 6 inch diameter steel pipe over flow is required. The overflow piping will be fitted with an adequate anti-vortex entrance detail. The overflow piping will extend down the inside of the access tube and pedestal and discharge at a point perpendicular to and approximately 1 foot above ground level onto a concrete splash pad.

27) Roof Ladder: A roof ladder, handrail, or non-skid surface is not required for access to hatches or vents, provided however, that the access surface is sloped 2 inches in 12 inches or less.

28) Safety Devices: Safety devices will be installed on ladders in accordance with OSHA and the State of Wisconsin Safety codes.

29) Special Tank Vent: A special tank vent is required to handle pressure differential caused when pumping or withdrawing at a maximum rate of 750 GPM. The overflow piping will not be considered as a part of the venting system when designing this special vent. The vent will terminate in a "U"-Bend or mushroom cap constructed with the opening at lease 4 inches above the roof and covered with 4 mesh to the inch noncorrodible screen. The screen will be installed within the pipe or cap at a location protected from the environment that will insure fail-safe operation in the event that the insect screens frost over. The vent must be easily dismantled to remove the screens for cleaning.

30) Additional Accessories:

A) A 30" x 60" access door located in the supporting bell complete with locking device.

B) Ladders in bell, pedestal, access tube and tank, with safety devices to meet applicable State Safety Standards.

C) Steel condensate ceiling located at the junction of the pedestal cylinder and truncated cone complete with drain. The drain piping will extend down the inside of the supporting bell and discharge perpendicular to and approximately 1 foot above ground level onto the same concrete splash pad as the overflow piping.

D) A manhole giving access to two painter's rings located at the top of the pedestal. There will be a platform inside the pedestal at this point. All manholes will be rainproof and will have a 4" high curb. The cover will overlap the curb by at least 2 inches.

E) Eight interior water proof light sockets and bulbs with conduit, wiring, and switch will be provided inside of the pedestal and access tube. There will be one light located at the top of the access tube, one light near the top of the midpoint of the supporting pedestal, one light above the condensate ceiling, three lights in the bell and one light in the valve vault. Two (2) - 110 volt grounded double convenience outlets will also be installed as directed by the ENGINEER; one in the bell and the other near the top of the access tube.

F) An access tube 36" in diameter will be provided from the top of the pedestal to the tank roof. A manway will be provided at the base of the tube for access to the ladder on the exterior of the tube (interior of the tank)

G) All vents, overflows, or other openings in the tank will be screened with 4 mesh to the inch noncorrodible screen. Details of this screening will be submitted to the ENGINEER for approval.

H) Roof Lugs - Sufficient lugs will be furnished and installed in all sections of the tank to permit lashing of painters' scaffolding and boatswain's chairs for painting all portions of the tank.

I) Riser Insulation - The 8 inch piping will be insulated with precut, semi-circular sections of Extruded Polystyrene (2-inches minimum). Masking tape will be used to hold insulation in place until a .019" thick aluminum sheet is wrapped around the insulation and held in place with stainless steel bands and marmon clamps.

J) An aircraft warning obstruction light, enclosed in an aviation red obstruction light globe, complete with photo-electric cell, conduit, and wire to an electrical breaker panel located in the base, shall be provided. Obstruction light equipment shall meet approval of the Federal Aviation Administration (FAA).

K) Circulation Pump - Furnish and install one (1) complete circulation pump assembly inside the tank. The assembly will be securely mounted to the exterior of the access tube running up through the tank. The assembly will be located in plain view, with the pump inlet approximately one-third (1/3) of the way up the tank bowl. The pump power cable will be enclosed in rigid steel conduit attached securely to the exterior of the access tube. The circulation pump will be Red Jacket 50 (½ H.P.) CNW (2-wire), 1 (208/230 V), CN (Corrosion-Proof NEMA), 9 (number of stages) BC (10 GPM), or approved equal. The pump discharge will be fitted with 1-1/4"galvanized steel discharge piping as shown on the construction plans. The pump power cables will terminate in an electrical junction box located near the top of the inside of the access tube.

L) Electrical Service - The CONTRACTOR will make arrangements to provide 120/240 volt, single phase permanent underground electrical power service to the base of the tank. Temporary power requirements necessary for construction will also be the CONTRACTOR's responsibility.

The CONTRACTOR will provide a meter socket and disconnect switch in a NEMA 3R enclosure suitable for the permanent service. These items will be located on the outside of the tank base. The CONTRACTOR will also provide for all necessary conduit and conductors necessary to connect the electrical service to the 200 amp minimum load control circuit breaker panel located inside the tank base. Separate electrical circuits will be provided for as follows:

1. Circulation Pump No. 1
2. Circulation Pump No. 2
3. Aircraft obstruction lighting system
4. Interior lights/Convenience Outlets
5. Interior convenience outlets.
6. Valve pit sump pump.

M) Expansion Joint - A stainless steel, bellows type expansion joint will be installed on the vertical riser pipe.

N) Sediment Drain - A 4 inch drain will be installed at the bottom of the tank bowl for draining sediment. The drain will discharge to the 6" overflow pipe as shown on the plans. A water tight, removable cap will be provided on the 4 inch drain inside the tank.

31. Butt-Joint Welds: Continuous with 100% joint penetration required.

32. Mill or Shop Inspection: Mill or shop inspection is not required by the OWNER.

33. Certified Written Report: At the conclusion of the work, the CONTRACTOR will submit a written report prepared by the CONTRACTOR's qualified personnel certifying that the work was inspected as listed in Section 11.2.1 of AWWA, D-100-84.

34. Radiographic Film & Test Segments: These items will become the property of the OWNER.

35. Type of Inspection: Inspection of welded - shell butt joints will be by radiographic testing in accordance with Section 11 of AWWA D-100-84.

36. Cleaning & Painting: See detailed specifications, Sections 09900 and 09910 plus manufacturer's recommendations. The tank must be painted prior to water testing and filling of the tank.

37. Soil Investigation: See Detailed Soils Report, Supplemental Document I. The top of the foundation will be 6 inches above finish grade.

38. Piling Compensation adjustment: Refer to the Standard General Conditions of the construction contract section on extra work.

39. Buoyancy Effect: Buoyancy effect shall be considered in the foundation design.

40. Concrete requirements: All concrete used during construction will be in accordance specifications located herein and as stipulated in other specific conditions with the in Section 3000 of the Specifications.

41. Earth Cover: There will be 8 feet of cover from finished ground level to the crown of the inlet and outlet pipe.

42. Seismic Design Specification Sheet: Not required to be completed.

43. Vertical Acceleration: Not required to be considered.

44. Local Seismic Data: Not available.

2.3 FOUNDATION

2.3.1 Description: The concrete foundation is included in the elevated tank Contract and will be designed by the CONTRACTOR (utilizing the services of a Registered Professional Engineer), in accordance with allowable soil bearing values and minimum depths previously provided. The foundation design must include the additional future (temporary) load resulting from paint removal/dust control apparatus (tarps, curtains, etc.) A concrete pipe vault, dimensions as shown on the Plans, will be constructed inside the pedestal base (or beneath the dry riser base; if alternate selected). Water main, extending from centerline of the tower to (and including) the 8" x 6" tee is included as a part of the elevated tank Contract. Water main includes one hydrant package and valve. An 8-inch blind flange shall be furnished and installed on the 8" x 6" tee which is the end of this section of the Work.

PART THREE - QUALITY ASSURANCE

3.1 EXPERIENCE

3.1.1 Contractor's Experience: Bids will be accepted only from experienced tank Contractors who are skilled and have furnished and erected at least five (5) similar facilities of equal, or greater, capacity within recent years. A letter shall accompany the Bid listing the five (5) such examples and their location with relevant details.

3.2 CONCEPTUAL DESIGN

3.2.1 Included with Bid submittal: A drawing showing the dimensions of the tank and tower, including the tank diameter, the height to lower and upper capacity levels, sizes of principal members, and thickness of plates in all parts of the tank and tower shall be submitted with the Bid. Also, the maximum wind gross moment and shear on the foundation system shall be identified.

3.3 QUALIFICATIONS

3.3.1 Welders: All welders employed on this project shall be qualified by ASME requirement standards.

3.3.2 Electricians: All electrical work shall be executed by experienced electricians.

3.4 WELDING

3.4.1 Welding supervision: The Contractor shall employ the services of a welding supervisor independent of the tank/tower erection foreman's jurisdiction.

3.4.2 Inspection: Welding inspection shall be in strict accordance with Section 11 of AWWA Specifications and shall be provided by the tank manufacturer.

3.5 DISCREPANCIES

3.5.1 Definition: Where real discrepancies are found between the Plans and the Specifications, they shall be brought to the attention of the Engineer as soon as they are discovered so that corrections can be made that will least affect the interest of the Owner and Contractor.

3.6 FINAL CONNECTIONS

3.6.1 General: This Contractor shall be responsible for final connection of all applicable electrical circuits and watermain.

3.7 TESTING

3.7.1 Labor, Material and Equipment: This Contractor shall furnish all labor, material, supplies and equipment necessary to test the facility.

3.7.2 Chlorination: This Contractor shall furnish all labor, material, supplies and equipment necessary to successfully disinfect the total facility, pursuant to applicable codes and DNR's approval.

The OWNER will furnish and dispose of all water required for initial testing and sterilization purposes.

3.8 ACCEPTANCE

3.8.1 Final Acceptance: Final acceptance, by the Owner, will follow the successful testing of the total facility and assurance that all other specified requirements have been satisfied.

3.9 PAYMENT

3.9.1 Method: Payment will be based upon the Proposal's bid items for this phase of the Work. Periodic partial payments will be made as work progresses utilizing customary practice. Final payment will not be made until the total facility has been accepted by the Engineer and Owner.

END OF SECTION

93-159 13410-10

供水控制系统 WATER SUPPLY CONTROL SYSTEM

PART ONE - GENERAL

1.1 DESCRIPTION

1.1.1 Work Included: This section of the Specifications applies to the furnishing and installation of an integrated, solid state water supervisory control system with alarm monitoring of Tony's water system.

1.1.2 Water Location: Work sites are at the pumphouse and the elevated tank. Refer to the accompanying Plans for exact locations.

1.1.3 Related Work Described Elsewhere:

1) Summary of Work-Special Provisions 01010
2) Field Engineering 01050
3) Submittals & Substitutions 01300
4) Testing Laboratory Services 01410
5) Water Systems 02713
6) Deep Well Turbine Pump 11210
7) Elevated Water Storage Reservoir 13410

1.2 QUALITY ASSURANCE

1.2.1 General: The control system shall be the product of a manufacturer experienced, skilled, and regularly engaged in the design and fabrication of this type of equipment with similar installations within convenient inspection distance. The general design of the equipment shall be such that all working parts are readily accessible for inspection and repairs, easily duplicated and replaced, and each and every component suitable for the service required. Design, manufacture and testing of all components shall be in conformance with all requirements of governing and regulating agencies and all applicable industry codes.

1.2.2 Factory Supervision: The manufacturer shall provide the services of factory engineers to inspect the installation, supervise final adjustments, and instruct the operating personnel in the use and maintenance of the equipment. This supervision shall be incidental to the project and the cost included in the bid item.

1.2.3 Governing Codes & Inspections: The completed electrical installation shall meet all the requirements of the latest edition of the National Electric Code as well as state and local regulations as they may apply. This shall not be construed to permit a lower grade of construction where Plans and Specifications call for workmanship or materials in excess of code requirements.

93-159 13601-1

Courtesy: Morgan & Parmley, Ltd.

The work herein specified shall be subject to inspection and approved by authorized representatives of the National Board of Fire Underwriters, state and local governing authorities, and the Engineer.

The control system as specified herein, shall be constructed/manufactured by a UL certified #058 control manufacturing facility.

1.3 IDENTIFICATION

1.3.1 Labeling: All control devices and device enclosures shall be labeled with individual nameplates or legend plates.

Individual nameplates or legend plates shall be one of the following types:

1. Black laminated plastic or micarta with white cut letters.

2. Corrosion-resistant metal plates with engraved or raised letters and backfill.

1.4 ENVIRONMENTAL REQUIREMENTS

1.4.1 General: The Supervisory Control System shall function reliably within an operating temperature of -20^0 to +70^0 C (-40^0 to 158^0 F). Humidity shall be 0 to 95 percent non-condensing. Each supervisory or remote unit shall be furnished with integral lightning protection devices to protect against lightning induced transients on leased telephone and power lines.

The entire solid state Supervisory Control System shall provide a typical system accuracy and repeatability of better than ± ½% over a temperature range of 0^0 to +50^0 C (+32^0 to +122^0 F).

1.5 WARRANTY

1.5.1 General: The Contractor, thru the Manufacturer, shall warrant the Supervisory Control and Alarm System to be free of defective material and workmanship for a period of one (1) year from date of acceptance. The manufacturer shall be obligated to furnish and install, at no charge to the Village, replacement material items or unit proven defective within this warranty period. This warranty shall not be construed to cover lights, fuses, or other items normally consumed in service nor those items which have been damaged due to outside forces such as vandalism, lighting, etc. Service calls for Control System problems will not be considered part of the factory supervision.

1.6 SYSTEM OPERATION RESPONSIBILITY

1.6.1 General: The Contractor, thru the manufacturer, shall assume the total, undivided

responsibility for the correct operation of the Water Control System. This shall include everything required to make the system work, including additional control changes, if required.

The manufacturer shall maintain a service center with all necessary replacement parts/personnel. A certified letter shall be required to show compliance with the service center requirements.

PART TWO - PRODUCE

2.1 DESCRIPTION

2.1.1 Telemetry System: All control signals, status signals, alarm or variable analog data shall be transmitted and received between the tower and Well #1 utilizing an audio tone pulse frequency telemetry system operating on a low cost signal grade line or buried cable. The pulse frequency shall be directly proportional to the level in the tower and vary between 5-25 pulses per second. Each transmitter and receiver shall contain individual span and offset potentiometers for calibration and each unit shall contain a pulse indicating LED.

Audio tone telemetry system shall utilize 3 state FSK audio tone. Transmitter output shall be adjustable from -20 dbm to +3 dbm. Receiver sensitivity shall be adjustable to -40 dbm. Each transmitter/receiver unit shall be provided with a three stage surge suppression lighting protector consisting of gas tube arresters, varistors, and zener diodes.

2.1.2 LED Bargraph Indicator/Controller: The Control System shall sense the tower level over a calibrated range, display it on a 4" LED bar graph on the face of the controller, graphically display 8 level adjustments for automatic pump and alarm control in a coordinated arrangement with the level display and provide automatic operation of the pump and alarms as herein-after or otherwise described. Include:

A. 0-20 Ft. tower level range.
B. 40 Segment LED bar graph level display.
C. 40 Level adjustability of each 8 control levels utilizing gold plated adjusting pins.
D. "Raise-auto-lower" level simulation switch with spring return to "auto".
E. "Pump-up" control with eight level settings for four (4) demand levels.
F. Full-range adjustable dual settings for each control and alarm circuit, utilize gold plated adjusting pins to set levels.
G. LED indicator for each control and alarm circuit, or equal.
H. Power-up "wake-up" with pump and alarms off; held off for 18 seconds.
I. 20 Second time interval after 1 pump starts before another pump is allowed to start.
J. Rate of change limiting of analog level signal.
K. High and low Level alarm sensing.
L. Unit shall door mount on the outside of the enclosure.

The above controller format is desired for operating personnel because of the ease in comprehension.

2.1.3 Pump Control Module: The pump shall have a control module which combines the Hand-Off-Automatic switch with the pilot lights for "Required"-"Run"-"Failed". HOA switches shall be full size, rotary type units.

The pilot lights shall be LED type or equal and shall light when the pump is called for, and when it is running, or when it has failed.

2.1.4 Annunciator Panel:

A. An annunciator system shall be provided to announce a change of condition for all of the monitored status or alarm conditions.

B. The annunciator system shall consist of the required number of self-contained modules. The points shall be individually selectable for alarm annunciation or status indication. Individual points shall be grouped on the modules by location, category or function as designated by the Village to provide the operator with the easiest comprehension. Critical alarms shall be grouped on separate modules with an independent, easily identifiable audible alarm.

C. Alarm condition shall be annunciated by the ISA-1A sequence as follows:

1.

Condition	Light	Audible
Normal	Off	Off
Alarm	Flashing	On
Acknowledge	On	Off
Return to Normal	Off	Off

2. Silencing one alarm shall not inhibit sequence alarms.

3. All annunciator modules shall have a common acknowledge/silence push-button

4. A lamp test provision shall be included.

2.1.5 Elevated Tower Panel:

A. A solid state pressure responsive tower level/alarm transmitter system housed in a new NEMA 3R GALVANIZED enclosure with other components as hereinafter described shall be mounted on the Elevated Tower riser.

The enclosure shall incorporate a freezing protective dual electric heater with heavy duty adjustable thermostat to provide protection for the equipment located herein. ½" thick insulation of interior enclosure walls shall be provided. Provide the following major items of equipment:

1. Control breaker.
2. 120 VAC lighting protection as required.
3. Block and bleed valves.
4. 3½" Altitude gauge.
5. Solid state transducer to convert head range of the tower to an analog signal. Transducer to have stainless steel bellows and stainless steel internal parts to insure long life.
6. Surge quelling and suppression card. The process output of the transducer shall be fed into this card to remove electronically (not hydraulically) any irregularities in the process signal due to water pressure surges or friction losses. The rate of limit quelling or suppression shall be adjustable from 30 seconds to 180 seconds. Hydraulic pressure snubbers are not acceptable.
7. Telemetry transmitter as previously specified.

   Transmit:

   a. Elevated tower level data.
   b. Signal failure/Power failure.
   c. High water alarm.
   d. Low water alarm.
   e. Control panel low temperature.

8. Internal dual heaters to protect panel interior equipment at temperature of -40^0 Fahrenheit or wind chill of -85^0 Fahrenheit. Heaters to be screw-in type, Chromalox SCB150, or equal.
9. Low panel temperature alarm thermostat.
10. Enclosure shall mount on the elevated tower riser, with no tubing or piping exposed.

2.1.6 Supervisory Control Center at Well No. 1

A. The enclosure shall be a NEMA type 12 and all surfaces shall be phosphatized before painting and finished with rust-inhibiting base coat and a final coat of exterior grade baked enamel. All instrumentation and operator controls shall be mounted on the front panel.

B. The basic functions and components of the control center are to be as follows:

1. Provide a modular solid state type system with plug in type circuit cards.
2. Control breaker with 120 VAC lightning protection.
3. Telemetry receiver to receive tower level/alarm signals from the Elevated Tower (as previously specified).
4. Provide a seven-day circular chart electronic recorder (10") to record the water level in the Elevated Tower. Provide two year's supply of charts and ink.
5. Provide LED bargraph monitor/controller as previously specified for the Elevated Tower. This unit shall control the operation of the pumps. Control stages shall be as follows:

    1. High Level Alarm
    2. Lead pump on/off control
    3. Future pump on/off control
    4. Low Level Alarm

6. Provide a pump control module as previously specified with selector switch and pilot lights.
7. Provide an alarm annunciator as specified for the following alarms:

    A. Elevated Tower High Level
    B. Elevated Tower Low Level
    C. Elevated Tower Power/Signal Fail
    D. Elevated Tower Control Panel Low Temperature
    E. Well #1 Fail

Include audible horn/silence button. Any annunciator alarm shall activate an external 100 watt flashing red alarm light.

8. Provide space for future well pump control logic and future telephone dialer.

2.1.7 Discrepancies: Where real discrepancies are found between the Plans and Specifications they shall be brought to the attention of the Engineer as soon as found so that corrections can be made that will least affect the interests of the Owner and Contractor.

PART THREE - ACCEPTANCE

3.1 GENERAL

3.1.1 Testing: The Contractor, manufacturer's representative, Village personnel and Engineer shall be present during testing.

3.1.2 Instructions: The manufacturer's representative shall fully demonstrate, to Village personnel and Engineer, the correct operational procedures of the system.

3.1.3 O & M Data: The manufacturer shall provide copies of complete operation and maintenance data; plus complete parts catolog..

3.1.4. Final Acceptance: By the Engineer and Owner, will follow the testing of the new Well facility and Elevated Tower.

3.1.5 Payment: Will be based upon a lump sum as bid an listed in the Proposal.

-END OF SECTION-

招标程序

第23节 招标文件

例 23

招标程序是一个施工项目的关键步骤。由专业设计师直接检查整个过程是非常重要的，这样可以避免出现任何表达不清和理解混淆。招标程序的所有方面都应该是不受质疑的。

必须注意到，每个项目在很大范围上有自己的特殊要求。每个特定招标程序所适用的法规和规范必须符合当地、州和/或联邦政府的要求；也就是说，符合劳动标准、工资等级、非歧视、资金条件等。所以，接下来的内容仅反映了这个最重要程序的概览。在实施招标程序之前，我们提醒读者要完整地理解他们特定项目的所有条件和要求。联邦管理部门，例如环境保护局(EPA)、运输部(DOT)、美国农业部(USDA)乡村开发、经济开发署(EDA)等，对由他们投资的项目都有非常特别的条件，项目工程师必须完全遵循他们的特定规则。

工程师可能会选择由专业社团来准备招标文件，例如美国全国职业工程师协会(NSPE)或者美国建筑师学会(AIA)。无论是哪种情况，项目工程师在选择和准备招标文件时都必须进行自己的专业判断。

说明：用符号（*）标明的文件由下列单位共同准备并认可：

商务部经济开发署（Economic Development Administration，Department of Commerce）
环境保护局（Environment Protection Agency）
农业部农家信贷管理局（Farmers Home Administration，Department of Agriculture）
住房和城市发展部（Department of Housing and Urban Development）
美国顾问工程师委员会（American Consulting Engineer Council）
美国公共工程协会（American Public Works Association）
美国土木工程师学会（American Society of Civil Engineers）
美国总承包商联合会（Associated General Contractors of America）
美国职业工程师协会（National Society of Professional Engineers）
美国公共事业承包商协会（National Utility Contractor Association）

目 录

Advertisement for Bids 招标广告（示例）

Project No. 99-157

Drummond Sanitary District, Drummond, WI
(OWNER)

Sealed bids for Elevated Water Storage Tank Rehabilitation will be received by the Sanitary District at Drummond Fire Hall Meeting Room (North of Town Hall) until 2:00 o'clock., P.M., Local Time, Feb. 8, 2001 and then at said location publicly opened and read aloud.

MAJOR ITEMS OF BID

Removal and Disposal of Insulation
Blasting: Paint Removal & Disposal
Steel Tank Repair
Handrailing Installation
Circulation Pump System
Water Supply Control System
Related Appurtenances
Total Tank Painting

The Information for Bidders, Pre-Qualification Statement, Form of Bid, Form of Contract, Plans, Specifications, and Forms of Bid Bond, Performance and Payment Bond, and other contract documents may be examined at the following:

Eau Claire Builders Exchange, Wausau Builders Exchange, Duluth Builders Exchange, F.W. Dodge - Minneapolis, Morgan & Parmley, Ltd., Ladysmith, Wisconsin

Copies may be obtained at the office of Morgan & Parmley, Ltd. located at 115 W. 2nd Street, S., Ladysmith, WI 54848, (715) 532-3721 upon payment of a $ 25.00 non refundable handling fee.

The Owner reserves the right to waive any informalities or to reject any or all bids. Letting is subject to Section 61.54, 61.55, 61.56 and to Section 66.29, Wisconsin Statutes.

Each Bidder must deposit with their Bid, security in the amount, form and subject to the conditions provided in the Information for Bidders.

Attention of Bidders is particularly called to the requirements as to conditions of employment to be observed and minimum wage rates to be paid under the contract, Section 3, Segregated Facility, Section 109 and E.O. 11246.

No Bidder may withdraw their Bid within 60 days after the actual date of the opening thereof.

Date: Jan. 2, 2001 James Unseth, Commission President

Source: Morgan & Parmley, Ltd.

Advertisement for Bids 招标广告（示例）

____
Owner
____
Address
____

Separate sealed BIDS for the construction of (briefly describe nature, scope, and major elements of the work) ____________________

will be received by ____________________

at the office of ____________________

until __________, (Standard Time—Daylight Savings Time) __________, __________, and then at said office publicly opened and read aloud.

The CONTRACT DOCUMENTS may be examined at the following locations:

____

Copies of the CONTRACT DOCUMENTS may be obtained at the office of __________ located at __________

upon payment of $__________ for each set.

Any BIDDER, upon returning the CONTRACT DOCUMENTS promptly and in good condition, will be refunded his payment, and any non-bidder upon so returning the CONTRACT DOCUMENTS will be refunded $__________.

____________________ ____________________
Date

Source: ＊(See page 23-1)

Information for Bidders　给竞标者的通知

BIDS will be received by ____________________

(herein called the "OWNER"), at ____________________

until ____________, ________, and then at said office publicly opened and read aloud.

Each BID must be submitted in a sealed envelope, addressed to ____________ ____________ at ____________. Each sealed envelope containing a BID must be plainly marked on the outside as BID for ____________ and the envelope should bear on the outside the name of the BIDDER, his address, his license number if applicable and the name of the project for which the BID is submitted. If forwarded by mail, the sealed envelope containing the BID must be enclosed in another envelope addressed to the OWNER at ____________ ____________.

All BIDS must be made on the required BID form. All blank spaces for BID prices must be filled in, in ink or typewritten, and the BID form must be fully completed and executed when submitted. Only one copy of the BID form is required.

The OWNER may waive any informalities or minor defects or reject any and all BIDS. Any BID may be withdrawn prior to the above scheduled time for the opening of BIDS or authorized postponement thereof. Any BID received after the time and date specified shall not be considered. No BIDDER may withdraw a BID within 60 days after the actual date of the opening thereof. Should there be reasons why the contract cannot be awarded within the specified period, the time may be extended by mutual agreement between the OWNER and the BIDDER.

BIDDERS must satisfy themselves of the accuracy of the estimated quantities in the BID Schedule by examination of the site and a review of the drawings and specifications including ADDENDA. After BIDS have been submitted, the BIDDER shall not assert that there was a misunderstanding concerning the quantities of WORK or of the nature of the WORK to be done.

The OWNER shall provide to BIDDERS prior to BIDDING, all information which is pertinent to, and delineates and describes, the land owned and rights-of-way acquired or to be acquired.

The CONTRACT DOCUMENTS contain the provisions required for the construction of the PROJECT. Information obtained from an officer, agent, or employee of the OWNER or any other person shall not affect the risks or obligations assumed by the CONTRACTOR or relieve him from fulfilling any of the conditions of the contract.

Each BID must be accompanied by a BID bond payable to the OWNER for five percent of the total amount of the BID. As soon as the BID prices have been compared, the OWNER will return the BONDS of all except the three lowest responsible BIDDERS. When the Agreement is executed the bonds of the two remaining unsuccessful BIDDERS will be returned. The BID BOND of the successful BIDDER will be retained until the payment BOND and performance BOND have been executed and approved, after which it will be returned. A certified check may be used in lieu of a BID BOND.

A performance BOND and a payment BOND, each in the amount of 100 percent of the CONTRACT PRICE, with a corporate surety approved by the OWNER, will be required for the faithful performance of the contract.

Attorneys-in-fact who sign BID BONDS or payment BONDS and performance BONDS must file with each BOND a certified and effective dated copy of their power of attorney.

The party to whom the contract is awarded will be required to execute the Agreement and obtain the performance BOND and payment BOND within ten (10) calendar days from the date when NOTICE OF AWARD is delivered to the BIDDER. The NOTICE OF AWARD shall be accompanied by the necessary Agreement and BOND forms. In case of failure of the BIDDER to execute the Agreement, the OWNER may at his option consider the BIDDER in default, in which case the BID BOND accompanying the proposal shall become the property of the OWNER.

The OWNER within ten (10) days of receipt of acceptable performance BOND, payment BOND and Agreement signed by the party to whom the Agreement was awarded shall sign the Agreement and return to such party an executed duplicate of the Agreement. Should the OWNER not execute the Agreement within such period, the BIDDER may by WRITTEN NOTICE withdraw his signed Agreement. Such notice of withdrawal shall be effective upon receipt of the notice by the OWNER.

The NOTICE TO PROCEED shall be issued within ten (10) days of the execution of the Agreement by the OWNER. Should there be reasons why the NOTICE TO PROCEED cannot be issued within such period, the time may be extended by mutual agreement between the OWNER and CONTRACTOR. If the NOTICE TO PROCEED has not been issued within the ten (10) day period or within the period mutually agreed upon, the CONTRACTOR may terminate the Agreement without further liability on the part of either party.

The OWNER may make such investigations as he deems necessary to determine the ability of the BIDDER to perform the WORK, and the BIDDER shall furnish to the OWNER all such information and data for this purpose as the OWNER may request. The OWNER reserves the right to reject any BID if the evidence submitted by, or investigation of, such BIDDER fails to satisfy the OWNER that such BIDDER is properly qualified to carry out the obligations of the Agreement and to complete the WORK contemplated therein.

A conditional or qualified BID will not be accepted.

Award will be made to the lowest responsible BIDDER.

All applicable laws, ordinances, and the rules and regulations of all authorities having jurisdiction over construction of the PROJECT shall apply to the contract throughout.

Each BIDDER is responsible for inspecting the site and for reading and being thoroughly familiar with the CONTRACT DOCUMENTS. The failure or omission of any BIDDER to do any of the foregoing shall in no way relieve any BIDDER from any obligation in respect to his BID.

Further, the BIDDER agrees to abide by the requirements under Executive Order No. 11246, as amended, including specifically the provisions of the equal opportunity clause set forth in the SUPPLEMENTAL GENERAL CONDITIONS.

The low BIDDER shall supply the names and addresses of major material SUPPLIERS and SUBCONTRACTORS when requested to do so by the OWNER.

Inspection trips for prospective BIDDERS will leave from the office of the ____________________ at ____________

The ENGINEER is ____________________. His address is ____________________.

Source: * (See page 23-1)

Bid Bond 投标保单

KNOW ALL MEN BY THESE PRESENTS, that we, the undersigned, ________________

________________________________ as Principal, and

________________________________ as Surety, are hereby

held and firmly bound unto ________________________ as OWNER

in the penal sum of________________________________

for the payment of which, well and truly to be made, we hereby jointly and severally bind ourselves, successors and assigns.

Signed, this ______________day of______________, 20__________.

The Condition of the above obligation is such that whereas the Principal has submitted

to ________________________________a certain BID,

attached hereto and hereby made a part hereof to enter into a contract in writing, for the

____

____

____

NOW, THEREFORE,

(a) If said BID shall be rejected, or

(b) If said BID shall be accepted and the Principal shall execute and deliver a contract in the Form of Contract attached hereto (properly completed in accordance with said BID) and shall furnish a BOND for his faithful performance of said contract, and for the payment of all persons performing labor or furnishing materials in connection therewith, and shall in all other respects perform the agreement created by the acceptance of said BID,

then this obligation shall be void, otherwise the same shall remain in force and effect; it being expressly understood and agreed that the liability of the Surety for any and all claims hereunder shall, in no event, exceed the penal amount of this obligation as herein stated.

The Surety, for value received, hereby stipulates and agrees that the obligations of said Surety and its BOND shall be in no way impaired or affected by any extension of the time within which the OWNER may accept such BID; and said Surety does herby waive notice of any such extension.

IN WITNESS WHEREOF, the Principal and the Surety have hereunto set their hands and seals, and such of them as are corporations have caused their corporate seals to be hereto affixed and these presents to be signed by their proper officers, the day and year first set forth above.

________________________ (L.S.)
Principal

____
Surety

By: ________________________

IMPORTANT—Surety companies executing BONDS must appear on the Treasury Department's most current list (Circular 570 as amended) and be authorized to transact business in the state where the project is located.

Source: * (See page 23-1)

业主关于保单和保险的说明

Owner's Instructions for Bonds and Insurance

Project: ______________________ Date: ______________________
______________________ Project No.: ______________________
Owner: ______________________

The Contractor is hereby instructed that the following bonds and insurance are required, by the Owner, as a condition of this contract. Proof of coverage is not required for bidding but is required as a part of the agreement.

1) Bid Security:
Not Required

Required in the amount of $__________ OR ________ percent of Total Bid

2) Performance and Payment Bond:
Not Required

Required in the amount of $__________ OR ________ percent of Total Bid

3) Contractors Liability Insurance:
A. Workers Compensation

1)	State	Statutory
2)	Applicable Federal	Statutory
3)	Employers Liability	$__________
4)	Benefits Required by Union Labor Contracts	As Applicable

B: Commercial General Liability including: Premises-Operations, Independent Contractors' Protective, Products and Completed Operations, Broad Form Property Damage, Personal Injury and Contracted.

1) Bodily Injury & Property Damage Combined Single Limit
$__________ Each Occurrence
$__________ Aggregate Products and Completed Operations

2) Products and completed Operations Insurance shall be maintained after final payment for:

One Year

Two Years

3) Property Damage Liability Insurance shall include:

Explosion

Collapse

Underground

____

4) Contractual Liability (Hold Harmless Coverage) for bodily injury and property damage combined single limit:

$__________ Each Occurrence

5) Personal injury with Employment Exclusions deleted:

$__________ Aggregate

C: Commercial Automobile Liability; owned, non-owned, hired:
Bodily Injury & Property Damage $__________ Accident
Combined Single Limit $__________ Aggregate

D: Aircraft Liability; owned and non-owned:
Bodily Injury & Property Damage $__________ Each Accident
Combined Single Limit $__________ Aggregate

E: Watercraft Liability; owned and non-owned:
Bodily Injury & Property Damage $__________ Each Accident
Combined Single Limit $__________ Aggregate

F: Property Insurance: Contractor shall purchase all Risk, completed Value Builders Risk Insurance.
$ Full Value

G: Other Insurance:
(If none so state)

Description　　Amount

By: ______________________________ Date: ______________
Title: ______________________________

Source: Morgan & Parmley, Ltd.

General Conditions 一般条款

1. *DEFINITIONS*

1.1 Wherever used in the CONTRACT DOCUMENTS, the following terms shall have the meanings indicated which shall be applicable to both the singular and plural thereof:

1.2 ADDENDA—Written or graphic instruments issued prior to the execution of the Agreement which modify or interpret the CONTRACT DOCUMENTS, DRAWINGS and SPECIFICATIONS, by additions, deletions, clarifications or corrections.

1.3 BID—The offer or proposal of the BIDDER submitted on the prescribed form setting forth the prices for the WORK to be performed.

1.4 BIDDER—Any person, firm or corporation submitting a BID for the WORK.

1.5 BONDS—Bid, Performance, and Payment Bonds and other instruments of security, furnished by the CONTRACTOR and his surety in accordance with the CONTRACT DOCUMENTS.

1.6 CHANGE ORDER—A written order to the CONTRACTOR authorizing an addition, deletion or revision in the WORK within the general scope of the CONTRACT DOCUMENTS, or authorizing an adjustment in the CONTRACT PRICE or CONTRACT TIME.

1.7 CONTRACT DOCUMENTS—The contract, including Advertisement For Bids, Information For Bidders, BID, Bid Bond, Agreement, Payment Bond, Performance Bond, NOTICE OF AWARD, NOTICE TO PROCEED, CHANGE ORDER, DRAWINGS, SPECIFICATIONS, and ADDENDA.

1.8 CONTRACT PRICE—The total monies payable to the CONTRACTOR under the terms and conditions of the CONTRACT DOCUMENTS.

1.9 CONTRACT TIME—The number of calendar days stated in the CONTRACT DOCUMENTS for the completion of the WORK.

1.10 CONTRACTOR—The person, firm or corporation with whom the OWNER has executed the Agreement.

1.11 DRAWINGS—The part of the CONTRACT DOCUMENTS which show the characteristics and scope of the WORK to be performed and which have been prepared or approved by the ENGINEER.

1.12 ENGINEER—The person, firm or corporation named as such in the CONTRACT DOCUMENTS.

1.13 FIELD ORDER—A written order effecting a change in the WORK not involving an adjustment in the CONTRACT PRICE or an extension of the CONTRACT TIME, issued by the ENGINEER to the CONTRACTOR during construction.

1.14 NOTICE OF AWARD—The written notice of the acceptance of the BID from the OWNER to the successful BIDDER.

1.15 NOTICE TO PROCEED—Written communication issued by the OWNER to the CONTRACTOR authorizing him to proceed with the WORK and establishing the date of commencement of the WORK.

1.16 OWNER—A public or quasi-public body or authority, corporation, association, partnership, or individual for whom the WORK is to be performed.

1.17 PROJECT—The undertaking to be performed as provided in the CONTRACT DOCUMENTS.

1.18 RESIDENT PROJECT REPRESENTATIVE—The authorized representative of the OWNER who is assigned to the PROJECT site or any part thereof.

1.19 SHOP DRAWINGS—All drawings, diagrams, illustrations, brochures, schedules and other data which are prepared by the CONTRACTOR, a SUBCONTRACTOR, manufacturer, SUPPLIER or distributor, which illustrate how specific portions of the WORK shall be fabricated or installed.

1.20 SPECIFICATIONS—A part of the CONTRACT DOCUMENTS consisting of written descriptions of a technical nature of materials, equipment, construction systems, standards and workmanship.

1.21 SUBCONTRACTOR—An individual, firm or corporation having a direct contract with the CONTRACTOR or with any other SUBCONTRACTOR for the performance of a part of the WORK at the site.

1.22 SUBSTANTIAL COMPLETION—That date as certified by the ENGINEER when the construction of the PROJECT or a specified part thereof is sufficiently completed, in accordance with the CONTRACT DOCUMENTS, so that the PROJECT or specified part can be utilized for the purposes for which it is intended.

1.23 SUPPLEMENTAL GENERAL CONDITIONS—Modifications to General Conditions required by a Federal agency for participation in the PROJECT and approved by the agency in writing prior to inclusion in the CONTRACT DOCUMENTS, or such requirements that may be imposed by applicable state laws.

1.24 SUPPLIER—Any person or organization who supplies materials or equipment for the WORK, including that fabricated to a special design, but who does not perform labor at the site.

1.25 WORK—All labor necessary to produce the construction required by the CONTRACT DOCUMENTS, and all materials and equipment incorporated or to be incorporated in the PROJECT.

1.26 WRITTEN NOTICE—Any notice to any party of the Agreement relative to any part of this Agreement in writing and considered delivered and the service thereof completed, when posted by certified or registered mail to the said party at his last given address, or delivered in person to said party or his authorized representative on the WORK.

2. *ADDITIONAL INSTRUCTIONS AND DETAIL DRAWINGS*

2.1 The CONTRACTOR may be furnished additional instructions and detail drawings, by the ENGINEER, as necessary to carry out the WORK required by the CONTRACT DOCUMENTS.

2.2 The additional drawings and instruction thus supplied will become a part of the CONTRACT DOCUMENTS. The CONTRACTOR shall carry out the WORK in accordance with the additional detail drawings and instructions.

3. *SCHEDULES, REPORTS AND RECORDS*

3.1 The CONTRACTOR shall submit to the OWNER such schedule of quantities and costs, progress schedules, payrolls, reports, estimates, records and other data where applicable as are required by the CONTRACT DOCUMENTS for the WORK to be performed.

3.2 Prior to the first partial payment estimate the CONTRACTOR shall submit construction progress schedules showing the order in which he proposes to carry on the WORK, including dates at which he will start the various parts of the WORK, estimated date of completion of each part and, as applicable:

3.2.1. The dates at which special detail drawings will be required; and

3.2.2 Respective dates for submission of SHOP DRAWINGS, the beginning of manufacture, the testing and the installation of materials, supplies and equipment.

3.3 The CONTRACTOR shall also submit a schedule of payments that he anticipates he will earn during the course of the WORK.

4. *DRAWINGS AND SPECIFICATIONS*

4.1 The intent of the DRAWINGS and SPECIFICATIONS is that the CONTRACTOR shall furnish all labor, materials, tools, equipment, and transportation necessary for the proper execution of the WORK in accordance with the CONTRACT DOCUMENTS and all incidental work necessary to complete the PROJECT in an acceptable manner, ready for use, occupancy or operation by the OWNER.

4.2 In case of conflict between the DRAWINGS and SPECIFICATIONS, the SPECIFICATIONS shall govern. Figure dimensions on DRAWINGS shall govern over scale dimensions, and detailed DRAWINGS shall govern over general DRAWINGS.

4.3 Any discrepancies found between the DRAWINGS and SPECIFICATIONS and site conditions or any inconsistencies or ambiguities in the DRAWINGS or SPECIFICATIONS shall be immediately reported to the ENGINEER, in writing, who shall promptly correct such inconsistencies or ambiguities in writing. WORK done by the CONTRACTOR after his discovery of such discrepancies, inconsistencies or ambiguities shall be done at the CONTRACTOR'S risk.

5. *SHOP DRAWINGS*

5.1 The CONTRACTOR shall provide SHOP DRAWINGS as may be necessary for the prosecution of the WORK as required by the CONTRACT DOCUMENTS. The ENGINEER shall promptly review all SHOP DRAWINGS. The ENGINEER'S approval of any SHOP DRAWING shall not release the CONTRACTOR from responsibility for deviations from the CONTRACT DOCUMENTS. The approval of any SHOP DRAWING which substantially deviates from the requirement of the CONTRACT DOCUMENTS shall be evidenced by a CHANGE ORDER.

5.2 When submitted for the ENGINEER'S review, SHOP DRAWINGS shall bear the CONTRACTOR'S certification that he has reviewed, checked and approved the SHOP DRAWINGS and that they are in conformance with the requirements of the CONTRACT DOCUMENTS.

5.3 Portions of the WORK requiring a SHOP DRAWING or sample submission shall not begin until the SHOP DRAWING or submission has been approved by the ENGINEER. A copy of each approved SHOP DRAWING and each approved sample shall be kept in good order by the CONTRACTOR at the site and shall be available to the ENGINEER.

6. *MATERIALS, SERVICES AND FACILITIES*

6.1 It is understood that, except as otherwise specifically stated in the CONTRACT DOCUMENTS, the CONTRACTOR shall provide and pay for all materials, labor, tools, equipment, water, light, power, transportation, supervision, temporary construction of any nature, and all other services and facilities of any nature whatsoever necessary to execute, complete, and deliver the WORK within the specified time.

6.2 Materials and equipment shall be so stored as to insure the preservation of their quality and fitness for the WORK. Stored materials and equipment to be incorporated in the WORK shall be located so as to facilitate prompt inspection.

6.3 Manufactured articles, materials and equipment shall be applied, installed, connected, erected, used, cleaned and conditioned as directed by the manufacturer.

6.4 Materials, supplies and equipment shall be in accordance with samples submitted by the CONTRACTOR and approved by the ENGINEER.

6.5 Materials, supplies or equipment to be incorporated into the WORK shall not be purchased by the

Source: ＊(See page 23-1)

General Conditions *(continued)*

CONTRACTOR or the SUBCONTRACTOR subject to a chattel mortgage or under a conditional sale contract or other agreement by which an interest is retained by the seller.

7. *INSPECTION AND TESTING*

7.1 All materials and equipment used in the construction of the PROJECT shall be subject to adequate inspection and testing in accordance with generally accepted standards, as required and defined in the CONTRACT DOCUMENTS.

7.2 The OWNER shall provide all inspection and testing services not required by the CONTRACT DOCUMENTS.

7.3 The CONTRACTOR shall provide at his expense the testing and inspection services required by the CONTRACT DOCUMENTS.

7.4 If the CONTRACT DOCUMENTS, laws, ordinances, rules, regulations or orders of any public authority having jurisdiction require any WORK to specifically be inspected, tested, or approved by someone other than the CONTRACTOR, the CONTRACTOR will give the ENGINEER timely notice of readiness. The CONTRACTOR will then furnish the ENGINEER the required certificates of inspection, testing or approval.

7.5 Inspections, tests or approvals by the engineer or others shall not relieve the CONTRACTOR from his obligations to perform the WORK in accordance with the requirements of the CONTRACT DOCUMENTS.

7.6 The ENGINEER and his representatives will at all times have access to the WORK. In addition, authorized representatives and agents of any participating Federal or state agency shall be permitted to inspect all work, materials, payrolls, records of personnel, invoices of materials, and other relevant data and records. The CONTRACTOR will provide proper facilities for such access and observation of the WORK and also for any inspection, or testing thereof.

7.7 If any WORK is covered contrary to the written instructions of the ENGINEER it must, if requested by the ENGINEER, be uncovered for his observation and replaced at the CONTRACTOR'S expense.

7.8 If the ENGINEER considers it necessary or advisable that covered WORK be inspected or tested by others, the CONTRACTOR, at the ENGINEER'S request, will uncover, expose or otherwise make available for observation, inspection or testing as the ENGINEER may require, that portion of the WORK in question, furnishing all necessary labor, materials, tools, and equipment. If it is found that such WORK is defective, the CONTRACTOR will bear all the expenses of such uncovering, exposure, observation, inspection and testing and of satisfactory reconstruction. If, however, such WORK is not found to be defective, the CONTRACTOR will be allowed an increase in the CONTRACT PRICE or an extension of the CONTRACT TIME, or both, directly attributable to such uncovering, exposure, observation, inspection, testing and reconstruction and an appropriate CHANGE ORDER shall be issued.

8. *SUBSTITUTIONS*

8.1 Whenever a material, article or piece of equipment is identified on the DRAWINGS or SPECIFICATIONS by reference to brand name or catalogue number, it shall be understood that this is referenced for the purpose of defining the performance or other salient requirements and that other products of equal capacities, quality and function shall be considered. The CONTRACTOR may recommend the substitution of a material, article, or piece of equipment of equal substance and function for those referred to in the CONTRACT DOCUMENTS by reference to brand name or catalogue number, and if, in the opinion of the ENGINEER, such material, article, or piece of equipment is of equal substance and function to that specified, the ENGINEER may approve its substitution and use by the CONTRACTOR. Any cost differential shall be deductible from the CONTRACT PRICE and the CONTRACT DOCUMENTS shall be appropriately modified by CHANGE ORDER. The CONTRACTOR warrants that if substitutes are approved, no major changes in the function or general design of the PROJECT will result. Incidental changes or extra component parts required to accommodate the substitute will be made by the CONTRACTOR without a change in the CONTRACT PRICE or CONTRACT TIME.

9. *PATENTS*

9.1 The CONTRACTOR shall pay all applicable royalties and license fees. He shall defend all suits or claims for infringement of any patent rights and save the OWNER harmless from loss on account thereof, except that the OWNER shall be responsible for any such loss when a particular process, design, or the product of a particular manufacturer or manufacturers is specified, however if the CONTRACTOR has reason to believe that the design, process or product specified is an infringement of a patent, he shall be responsible for such loss unless he promptly gives such information to the ENGINEER.

10. *SURVEYS, PERMITS, REGULATIONS*

10.1 The OWNER shall furnish all boundary surveys and establish all base lines for locating the principal component parts of the WORK together with a suitable number of bench marks adjacent to the WORK as shown in the CONTRACT DOCUMENTS. From the information provided by the OWNER, unless otherwise specified in the CONTRACT DOCUMENTS, the CONTRACTOR shall develop and make all detail surveys needed for construction such as slope stakes, batter boards, stakes for pile locations and other working points, lines, elevations and cut sheets.

10.2 The CONTRACTOR shall carefully preserve bench marks, reference points and stakes and, in case of willful or careless destruction, he shall be charged with the resulting expense and shall be responsible for any mistakes that may be caused by their unnecessary loss or disturbance.

10.3 Permits and licenses of a temporary nature necessary for the prosecution of the WORK shall be secured and paid for by the CONTRACTOR unless otherwise stated in the SUPPLEMENTAL GENERAL CONDITIONS. Permits, licenses and easements for permanent structures or permanent changes in existing facilities shall be secured and paid for by the OWNER, unless otherwise specified. The CONTRACTOR shall give all notices and comply with all laws, ordinances, rules and regulations bearing on the conduct of the WORK as drawn and specified. If the CONTRACTOR observes that the CONTRACT DOCUMENTS are at variance therewith, he shall promptly notify the ENGINEER in writing, and any necessary changes shall be adjusted as provided in Section 13, CHANGES IN THE WORK.

11. *PROTECTION OF WORK, PROPERTY AND PERSONS*

11.1 The CONTRACTOR will be responsible for initiating, maintaining and supervising all safety precautions and programs in connection with the WORK. He will take all necessary precautions for the safety of, and will provide the necessary protection to prevent damage, injury or loss to all employees on the WORK and other persons who may be affected thereby, all the WORK and all materials or equipment to be incorporated therein, whether in storage on or off the site, and other property at the site or adjacent thereto, including trees, shrubs, lawns, walks, pavements, roadways, structures and utilities not designated for removal, relocation or replacement in the course of construction.

11.2 The CONTRACTOR will comply with all applicable laws, ordinances, rules, regulations and orders of any public body having jurisdiction. He will erect and maintain, as required by the conditions and progress of the WORK, all necessary safeguards for safety and protection. He will notify owners of adjacent utilities when prosecution of the WORK may affect them. The CONTRACTOR will remedy all damage, injury or loss to any property caused, directly or indirectly, in whole or in part, by the CONTRACTOR, any SUBCONTRACTOR or anyone directly or indirectly employed by any of them or anyone for whose acts any of them be liable, except damage or loss attributable to the fault of the CONTRACT DOCUMENTS or to the acts or omissions of the OWNER or the ENGINEER or anyone employed by either of them or anyone for whose acts either of them may be liable, and not attributable, directly or indirectly, in whole or in part, to the fault or negligence of the CONTRACTOR.

11.3 In emergencies affecting the safety of persons or the WORK or property at the site or adjacent thereto, the CONTRACTOR, without special instruction or authorization from the ENGINEER or OWNER, shall act to prevent threatened damage, injury or loss. He will give the ENGINEER prompt WRITTEN NOTICE of any significant changes in the WORK or deviations from the CONTRACT DOCUMENTS caused thereby, and a CHANGE ORDER shall thereupon be issued covering the changes and deviations involved.

12. *SUPERVISION BY CONTRACTOR*

12.1 The CONTRACTOR will supervise and direct the WORK. He will be solely responsible for the means, methods, techniques, sequences and procedures of construction. The CONTRACTOR will employ and maintain on the WORK a qualified supervisor or superintendent who shall have been designated in writing by the CONTRACTOR as the CONTRACTOR'S representative at the site. The supervisor shall have full authority to act on behalf of the CONTRACTOR and all communications given to the supervisor shall be as binding as if given to the CONTRACTOR. The supervisor shall be present on the site at all times as required to perform adequate supervision and coordination of the WORK.

13. *CHANGES IN THE WORK*

13.1 The OWNER may at any time, as the need arises, order changes within the scope of the WORK without invalidating the Agreement. If such changes increase or decrease the amount due under the CONTRACT DOCUMENTS, or in the time required for performance of the WORK, an equitable adjustment shall be authorized by CHANGE ORDER.

13.2 The ENGINEER, also, may at any time, by issuing a FIELD ORDER, make changes in the details of the WORK. The CONTRACTOR shall proceed with the performance of any changes in the WORK so ordered by the ENGINEER unless the CONTRACTOR believes that such FIELD ORDER entitles him to a change in CONTRACT PRICE or TIME, or both, in which event he shall give the ENGINEER WRITTEN NOTICE thereof within seven (7) days after the receipt of the ordered change. Thereafter the CONTRACTOR shall document the basis for the change in CONTRACT PRICE or TIME within thirty (30) days. The CONTRACTOR shall not execute such changes pending the receipt of an executed CHANGE ORDER or further instruction from the OWNER.

14. *CHANGES IN CONTRACT PRICE*

14.1 The CONTRACT PRICE may be changed only by a CHANGE ORDER. The value of any WORK covered by a CHANGE ORDER or of any claim for increase or decrease in the CONTRACT PRICE shall be determined by one or more of the following methods in the order of precedence listed below:

(a) Unit prices previously approved.

(b) An agreed lump sum.

(c) The actual cost for labor, direct overhead, materials, supplies, equipment, and other services necessary to complete the work. In addition there shall be added an amount to be agreed upon but not to exceed fifteen (15) percent of the actual cost of the WORK to cover the cost of general overhead and profit.

15. *TIME FOR COMPLETION AND LIQUIDATED DAMAGES*

15.1 The date of beginning and the time for completion of the WORK are essential conditions of the CONTRACT DOCUMENTS and the WORK embraced shall be commenced on a date specified in the NOTICE TO PROCEED.

15.2 The CONTRACTOR will proceed with the WORK at such rate of progress to insure full completion within the CONTRACT TIME. It is expressly understood and agreed, by and between the CONTRACTOR and the OWNER, that the CONTRACT TIME for the completion of the WORK described herein is a reasonable time, taking into consideration the average climatic and economic conditions and other factors prevailing in the locality of the WORK.

15.3 If the CONTRACTOR shall fail to complete the WORK within the CONTRACT TIME, or extension of time granted by the OWNER, then the CONTRACTOR will pay to the OWNER the amount for liquidated damages as specified in the BID for each calendar day that the CONTRACTOR shall be in default after the time stipulated in the CONTRACT DOCUMENTS.

15.4 The CONTRACTOR shall not be charged with liquidated damages or any excess cost when the delay in completion of the WORK is due to the following, and the CONTRACTOR has promptly given WRITTEN NOTICE of such delay to the OWNER or ENGINEER.

15.4.1 To any preference, priority or allocation

General Conditions *(continued)*

order duly issued by the OWNER.

15.4.2 To unforeseeable causes beyond the control and without the fault or negligence of the CONTRACTOR, including but not restricted to, acts of God, or of the public enemy, acts of the OWNER, acts of another CONTRACTOR in the performance of a contract with the OWNER, fires, floods, epidemics, quarantine restrictions, strikes, freight embargoes, and abnormal and unforeseeable weather; and

15.4.3 To any delays of SUBCONTRACTORS occasioned by any of the causes specified in paragraphs 15.4.1 and 15.4.2 of this article.

16. *CORRECTION OF WORK*

16.1 The CONTRACTOR shall promptly remove from the premises all WORK rejected by the ENGINEER for failure to comply with the CONTRACT DOCUMENTS, whether incorporated in the construction or not, and the CONTRACTOR shall promptly replace and re-execute the WORK in accordance with the CONTRACT DOCUMENTS and without expense to the OWNER and shall bear the expense of making good all WORK of other CONTRACTORS destroyed or damaged by such removal or replacement.

16.2 All removal and replacement WORK shall be done at the CONTRACTOR'S expense. If the CONTRACTOR does not take action to remove such rejected WORK within ten (10) days after receipt of WRITTEN NOTICE, the OWNER may remove such WORK and store the materials at the expense of the CONTRACTOR.

17. *SUBSURFACE CONDITIONS*

17.1 The CONTRACTOR shall promptly, and before such conditions are disturbed, except in the event of an emergency, notify the OWNER by WRITTEN NOTICE of:

17.1.1 Subsurface or latent physical conditions at the site differing materially from those indicated in the CONTRACT DOCUMENTS; or

17.1.2 Unknown physical conditions at the site, of an unusual nature, differing materially from those ordinarily encountered and generally recognized as inherent in WORK of the character provided for in the CONTRACT DOCUMENTS.

17.2 The OWNER shall promptly investigate the conditions, and if he finds that such conditions do so materially differ and cause an increase or decrease in the cost of, or in the time required for, performance of the WORK, an equitable adjustment shall be made and the CONTRACT DOCUMENTS shall be modified by a CHANGE ORDER. Any claim of the CONTRACTOR for adjustment hereunder shall not be allowed unless he has given the required WRITTEN NOTICE; provided that the OWNER may, if he determines the facts so justify, consider and adjust any such claims asserted before the date of final payment.

18. *SUSPENSION OF WORK, TERMINATION AND DELAY*

18.1 The OWNER may suspend the WORK or any portion thereof for a period of not more than ninety days or such further time as agreed upon by the CONTRACTOR, by WRITTEN NOTICE to the CONTRACTOR and the ENGINEER which notice shall fix the date on which WORK shall be resumed. The CONTRACTOR will resume that WORK on the date so fixed. The CONTRACTOR will be allowed an increase in the CONTRACT PRICE or an extension of the CONTRACT TIME, or both, directly attributable to any suspension.

18.2 If the CONTRACTOR is adjudged a bankrupt or insolvent, or if he makes a general assignment for the benefit of his creditors, or if a trustee or receiver is appointed for the CONTRACTOR or for any of his property, or if he files a petition to take advantage of any debtor's act, or to reorganize under the bankruptcy or applicable laws, or if he repeatedly fails to supply sufficient skilled workmen or suitable materials or equipment, or if he repeatedly fails to make prompt payments to SUBCONTRACTORS or for labor, materials or equipment or if he disregards laws, ordinances, rules, regulations or orders of any public body having jurisdiction of the WORK or if he disregards the authority of the ENGINEER, or if he otherwise violates any provision of the CONTRACT DOCUMENTS, then the OWNER may, without prejudice to any other right or remedy and after giving the CONTRACTOR and his surety a minimum of ten (10) days from delivery of a WRITTEN NOTICE, terminate the services of the CONTRACTOR and take possession of the PROJECT and of all materials, equipment, tools, construction equipment and machinery thereon owned by the CONTRACTOR, and finish the WORK by whatever method he may deem expedient. In such case the CONTRACTOR shall not be entitled to receive any further payment until the WORK is finished. If the unpaid balance of the CONTRACT PRICE exceeds the direct and indirect costs of completing the PROJECT, including compensation for additional professional services, such excess SHALL BE PAID TO THE CONTRACTOR. If such costs exceed such unpaid balance, the CONTRACTOR will pay the difference to the OWNER. Such costs incurred by the OWNER will be determined by the ENGINEER and incorporated in a CHANGE ORDER.

18.3 Where the CONTRACTOR'S services have been so terminated by the OWNER, said termination shall not affect any right of the OWNER against the CONTRACTOR then existing or which may thereafter accrue. Any retention or payment of monies by the OWNER due the CONTRACTOR will not release the CONTRACTOR from compliance with the CONTRACT DOCUMENTS.

18.4 After ten (10) days from delivery of a WRITTEN NOTICE to the CONTRACTOR and the ENGINEER, the OWNER may, without cause and without prejudice to any other right or remedy, elect to abandon the PROJECT and terminate the Contract. In such case, the CONTRACTOR shall be paid for all WORK executed and any expense sustained plus reasonable profit.

18.5 If, through no act or fault of the CONTRACTOR, the WORK is suspended for a period of more than ninety (90) days by the OWNER or under an order of court or other public authority, or the ENGINEER fails to act on any request for payment within thirty (30) days after it is submitted, or the OWNER fails to pay the CONTRACTOR substantially the sum approved by the ENGINEER or awarded by arbitrators within thirty (30) days of its approval and presentation, then the CONTRACTOR may, after ten (10) days from delivery of a WRITTEN NOTICE to the OWNER and the ENGINEER, terminate the CONTRACT and recover from the OWNER payment for all WORK executed and all expenses sustained. In addition and in lieu of terminating the CONTRACT, if the ENGINEER has failed to act on a request for payment or if the OWNER has failed to make any payment as aforesaid, the CONTRACTOR may upon ten (10) days written notice to the OWNER and the ENGINEER stop the WORK until he has been paid all amounts then due, in which event and upon resumption of the WORK, CHANGE ORDERS shall be issued for adjusting the CONTRACT PRICE or extending the CONTRACT TIME or both to compensate for the costs and delays attributable to the stoppage of the WORK.

18.6 If the performance of all or any portion of the WORK is suspended, delayed, or interrupted as a result of a failure of the OWNER or ENGINEER to act within the time specified in the CONTRACT DOCUMENTS, or if no time is specified, within a reasonable time, an adjustment in the CONTRACT PRICE or an extension of the CONTRACT TIME, or both, shall be made by CHANGE ORDER to compensate the CONTRACTOR for the costs and delays necessarily caused by the failure of the OWNER or ENGINEER.

19. *PAYMENTS TO CONTRACTOR*

19.1 At least ten (10) days before eacn progress payment falls due (but not more often than once a month), the CONTRACTOR will submit to the ENGINEER a partial payment estimate filled out and signed by the CONTRACTOR covering the WORK performed during the period covered by the partial payment estimate and supported by such data as the ENGINEER may reasonably require. If payment is requested on the basis of materials and equipment not incorporated in the WORK but delivered and suitably stored at or near the site, the partial payment estimate shall also be accompanied by such supporting data, satisfactory to the OWNER, as will establish the OWNER's title to the material and equipment and protect his interest therein, including applicable insurance. The ENGINEER will, within ten (10) days after receipt of each partial payment estimate, either indicate in writing his approval of payment and present the partial payment estimate to the OWNER, or return the partial payment estimate to the CONTRACTOR indicating in writing his reasons for refusing to approve payment. In the latter case, the CONTRACTOR may make the necessary corrections and resubmit the partial payment estimate. The OWNER will, within ten (10) days of presentation to him of an approved partial payment estimate, pay the CONTRACTOR a progress payment on the basis of the approved partial payment estimate. The OWNER shall retain ten (10) percent of the amount of each payment until final completion and acceptance of all work covered by the CONTRACT DOCUMENTS. The OWNER at any time, however, after fifty (50) percent of the WORK has been completed, if he finds that satisfactory progress is being made, shall reduce retainage to five (5%) percent on the current and remaining estimates. When the WORK is substantially complete (operational or beneficial occupancy), the retained amount may be further reduced below five (5) percent to only that amount necessary to assure completion. On completion and acceptance of a part of the WORK on which the price is stated separately in the CONTRACT DOCUMENTS, payment may be made in full, including retained percentages, less authorized deductions.

19.2 The request for payment may also include an allowance for the cost of such major materials and equipment which are suitably stored either at or near the site.

19.3 Prior to SUBSTANTIAL COMPLETION, the OWNER, with the approval of the ENGINEER and with the concurrence of the CONTRACTOR, may use any completed or substantially completed portions of the WORK. Such use shall not constitute an acceptance of such portions of the WORK.

19.4 The OWNER shall have the right to enter the premises for the purpose of doing work not covered by the CONTRACT DOCUMENTS. This provision shall not be construed as relieving the CONTRACTOR of the sole responsibility for the care and protection of the WORK, or the restoration of any damaged WORK except such as may be caused by agents or employees of the OWNER.

19.5 Upon completion and acceptance of the WORK, the ENGINEER shall issue a certificate attached to the final payment request that the WORK has been accepted by him under the conditions of the CONTRACT DOCUMENTS. The entire balance found to be due the CONTRACTOR, including the retained percentages, but except such sums as may be lawfully retained by the OWNER, shall be paid to the CONTRACTOR within thirty (30) days of completion and acceptance of the WORK.

19.6 The CONTRACTOR will indemnify and save the OWNER or the OWNER'S agents harmless from all claims growing out of the lawful demands of SUBCONTRACTORS, laborers, workmen, mechanics, materialmen, and furnishers of machinery and parts thereof, equipment, tools, and all supplies, incurred in the furtherance of the performance of the WORK. The CONTRACTOR shall, at the OWNER'S request, furnish satisfactory evidence that all obligations of the nature designated above have been paid, discharged, or waived. If the CONTRACTOR fails to do so the OWNER may, after having notified the CONTRACTOR, either pay unpaid bills or withhold from the CONTRACTOR'S unpaid compensation a sum of money deemed reasonably sufficient to pay any and all such lawful claims until satisfactory evidence is furnished that all liabilities have been fully discharged whereupon payment to the CONTRACTOR shall be resumed, in accordance with the terms of the CONTRACT DOCUMENTS, but in no event shall the provisions of this sentence be construed to impose any obligations upon the OWNER to either the CONTRACTOR, his Surety, or any third party. In paying any unpaid bills of the CONTRACTOR, any payment so made by the OWNER shall be considered as a payment made under the CONTRACT DOCUMENTS by the OWNER to the CONTRACTOR and the OWNER shall not be liable to the CONTRACTOR for any such payments made in good faith.

19.7 If the OWNER fails to make payment thirty (30) days after approval by the ENGINEER, in addition to other remedies available to the CONTRACTOR, there shall be added to each such payment interest at the maximum legal rate commencing on the first day after said payment is due and continuing until the payment is received by the CONTRACTOR.

General Conditions *(continued)*

20. *ACCEPTANCE OF FINAL PAYMENT AS RELEASE*

20.1 The acceptance by the CONTRACTOR of final payment shall be and shall operate as a release to the OWNER of all claims and all liability to the CONTRACTOR other than claims in stated amounts as may be specifically excepted by the CONTRACTOR for all things done or furnished in connection with this WORK and for every act and neglect of the OWNER and others relating to or arising out of this WORK. Any payment, however, final or otherwise, shall not release the CONTRACTOR or his sureties from any obligations under the CONTRACT DOCUMENTS or the Performance BOND and Payment BONDS.

21. *INSURANCE*

21.1 The CONTRACTOR shall purchase and maintain such insurance as will protect him from claims set forth below which may arise out of or result from the CONTRACTOR'S execution of the WORK, whether such execution be by himself or by any SUBCONTRACTOR or by anyone directly or indirectly employed by any of them, or by anyone for whose acts any of them may be liable:

21.1.1 Claims under workmen's compensation, disability benefit and other similar employee benefit acts;

21.1.2 Claims for damages because of bodily injury, occupational sickness or disease, or death of his employees;

21.1.3 Claims for damages because of bodily injury, sickness or disease, or death of any person other than his employees;

21.1.4 Claims for damages insured by usual personal injury liability coverage which are sustained (1) by any person as a result of an offense directly or indirectly related to the employment of such person by the CONTRACTOR, or (2) by any other person; and

21.1.5 Claims for damages because of injury to or destruction of tangible property, including loss of use resulting therefrom.

21.2 Certificates of Insurance acceptable to the OWNER shall be filed with the OWNER prior to commencement of the WORK. These Certificates shall contain a provision that coverages afforded under the policies will not be cancelled unless at least fifteen (15) days prior WRITTEN NOTICE has been given to the OWNER.

21.3 The CONTRACTOR shall procure and maintain at his own expense, during the CONTRACT TIME, liability insurance as hereinafter specified;

21.3.1 CONTRACTOR'S General Public Liability and Property Damage Insurance including vehicle coverage issued to the CONTRACTOR and protecting him from all claims for personal injury, including death, and all claims for destruction of or damage to property, arising out of or in connection with any operations under the CONTRACT DOCUMENTS, whether such operations be by himself or by any SUBCONTRACTOR under him, or anyone directly or indirectly employed by the CONTRACTOR or by a SUBCONTRACTOR under him. Insurance shall be written with a limit of liability of not less than $500,000 for all damages arising out of bodily injury, including death, at any time resulting therefrom, sustained by any one person in any one accident; and a limit of liability of not less than $500,000 aggregate for any such damages sustained by two or more persons in any one accident. Insurance shall be written with a limit of liability of not less than $200,000 for all property damage sustained by any one person in any one accident; and a limit of liability of not less than $200,000 aggregate for any such damage sustained by two or more persons in any one accident.

21.3.2 The CONTRACTOR shall acquire and maintain, if applicable, Fire and Extended Coverage insurance upon the PROJECT to the full insurable value thereof for the benefit of the OWNER, the CONTRACTOR, and SUBCONTRACTORS as their interest may appear. This provision shall in no way release the CONTRACTOR or CONTRACTOR'S surety from obligations under the CONTRACT DOCUMENTS to fully complete the PROJECT.

21.4 The CONTRACTOR shall procure and maintain, at his own expense, during the CONTRACT TIME, in accordance with the provisions of the laws of the state in which the work is performed, Workmen's Compensation Insurance, including occupational disease provisions, for all of his employees at the site of the PROJECT and in case any work is sublet, the CONTRACTOR shall require such SUBCONTRACTOR similarly to provide Workmen's Compensation Insurance, including occupational disease provisions for all of the latter's employees unless such employees are covered by the protection afforded by the CONTRACTOR. In case any class of employees engaged in hazardous work under this contract at the site of the PROJECT is not protected under Workmen's Compensation statute, the CONTRACTOR shall provide, and shall cause each SUBCONTRACTOR to provide, adequate and suitable insurance for the protection of his employees not otherwise protected.

21.5 The CONTRACTOR shall secure, if applicable, "All Risk" type Builder's Risk Insurance for WORK to be performed. Unless specifically authorized by the OWNER, the amount of such insurance shall not be less than the CONTRACT PRICE totaled in the BID. The policy shall cover not less than the losses due to fire, explosion, hail, lightning, vandalism, malicious mischief, wind, collapse, riot, aircraft, and smoke during the CONTRACT TIME, and until the WORK is accepted by the OWNER. The policy shall name as the insured the CONTRACTOR, the ENGINEER, and the OWNER.

22. *CONTRACT SECURITY*

22.1 The CONTRACTOR shall within ten (10) days after the receipt of the NOTICE OF AWARD furnish the OWNER with a Performance Bond and a Payment Bond in penal sums equal to the amount of the CONTRACT PRICE, conditioned upon the performance by the CONTRACTOR of all undertakings, covenants, terms, conditions and agreements of the CONTRACT DOCUMENTS, and upon the prompt payment by the CONTRACTOR to all persons supplying labor and materials in the prosecution of the WORK provided by the CONTRACT DOCUMENTS. Such BONDS shall be executed by the CONTRACTOR and a corporate bonding company licensed to transact such business in the state in which the WORK is to be performed and named on the current list of "Surety Companies Acceptable on Federal Bonds" as published in the Treasury Department Circular Number 570. The expense of these BONDS shall be borne by the CONTRACTOR. If at any time a surety on any such BOND is declared a bankrupt or loses its right to do business in the state in which the WORK is to be performed or is removed from the list of Surety Companies accepted on Federal BONDS, CONTRACTOR shall within ten (10) days after notice from the OWNER to do so, substitute an acceptable BOND (or BONDS) in such form and sum and signed by such other surety or sureties as may be satisfactory to the OWNER. The premiums on such BOND shall be paid by the CONTRACTOR. No further payments shall be deemed due nor shall be made until the new surety or sureties shall have furnished an acceptable BOND to the OWNER.

23. *ASSIGNMENTS*

23.1 Neither the CONTRACTOR nor the OWNER shall sell, transfer, assign or otherwise dispose of the Contract or any portion thereof, or of his right, title or interest therein, or his obligations thereunder, without written consent of the other party.

24. *INDEMNIFICATION*

24.1 The CONTRACTOR will indemnify and hold harmless the OWNER and the ENGINEER and their agents and employees from and against all claims, damages, losses and expenses including attorney's fees arising out of or resulting from the performance of the WORK, provided that any such claims, damage, loss or expense is attributable to bodily injury, sickness, disease or death, or to injury to or destruction of tangible property including the loss of use resulting therefrom; and is caused in whole or in part by any negligent or willful act or omission of the CONTRACTOR, and SUBCONTRACTOR, anyone directly or indirectly employed by any of them or anyone for whose acts any of them may be liable.

24.2 In any and all claims against the OWNER or the ENGINEER, or any of their agents or employees, by any employee of the CONTRACTOR, any SUBCONTRACTOR, anyone directly or indirectly employed by any of them, or anyone for whose acts any of them may be liable, the indemnification obligation shall not be limited in any way by any limitation on the amount or type of damages, compensation or benefits payable by or for the CONTRACTOR or any SUBCONTRACTOR under workmen's compensation acts, disability benefit acts or other employee benefits acts.

24.3 The obligation of the CONTRACTOR under this paragraph shall not extend to the liability of the ENGINEER, his agents or employees arising out of the preparation or approval of maps, DRAWINGS, opinions, reports, surveys, CHANGE ORDERS, designs or SPECIFICATIONS.

25. *SEPARATE CONTRACTS*

25.1 The OWNER reserves the right to let other contracts in connection with this PROJECT. The CONTRACTOR shall afford other CONTRACTORS reasonable opportunity for the introduction and storage of their materials and the execution of their WORK, and shall properly connect and coordinate his WORK with theirs. If the proper execution or results of any part of the CONTRACTOR'S WORK depends upon the WORK of any other CONTRACTOR, the CONTRACTOR shall inspect and promptly report to the ENGINEER any defects in such WORK that render it unsuitable for such proper execution and results.

25.2 The OWNER may perform additional WORK related to the PROJECT by himself, or he may let other contracts containing provisions similar to these. The CONTRACTOR will afford the other CONTRACTORS who are parties to such Contracts (or the OWNER, if he is performing the additional WORK himself), reasonable opportunity for the introduction and storage of materials and equipment and the execution of WORK, and shall properly connect and coordinate his WORK with theirs.

25.3 If the performance of additional WORK by other CONTRACTORS or the OWNER is not noted in the CONTRACT DOCUMENTS prior to the execution of the CONTRACT, written notice thereof shall be given to the CONTRACTOR prior to starting any such additional WORK. If the CONTRACTOR believes that the performance of such additional WORK by the OWNER or others involves him in additional expense or entities him to an extension of the CONTRACT TIME, he may make a claim therefor as provided in Sections 14 and 15.

26. *SUBCONTRACTING*

26.1 The CONTRACTOR may utilize the services of specialty SUBCONTRACTORS on those parts of the WORK which, under normal contracting practices, are performed by specialty SUBCONTRACTORS.

26.2 The CONTRACTOR shall not award WORK to SUBCONTRACTOR(s), in excess of fifty (50%) percent of the CONTRACT PRICE, without prior written approval of the OWNER.

26.3 The CONTRACTOR shall be fully responsible to the OWNER for the acts and omissions of his SUBCONTRACTORS, and of persons either directly or indirectly employed by them, as he is for the acts and omissions of persons directly employed by him.

26.4 The CONTRACTOR shall cause appropriate provisions to be inserted in all subcontracts relative to the WORK to bind SUBCONTRACTORS to the CONTRACTOR by the terms of the CONTRACT DOCUMENTS insofar as applicable to the WORK of SUBCONTRACTORS and to give the CONTRACTOR the same power as regards terminating any subcontract that the OWNER may exercise over the CONTRACTOR under any provision of the CONTRACT DOCUMENTS.

26.5 Nothing contained in this CONTRACT shall create any contractual relation between any SUBCONTRACTOR and the OWNER.

27. *ENGINEER'S AUTHORITY*

27.1 The ENGINEER shall act as the OWNER'S representative during the construction period. He shall decide questions which may arise as to quality and acceptability of materials furnished and WORK performed. He shall interpret the intent of the CONTRACT DOCUMENTS in a fair and unbiased manner. The

General Conditions *(continued)*

ENGINEER will make visits to the site and determine if the WORK is proceeding in accordance with the CONTRACT DOCUMENTS.

27.2 The CONTRACTOR will be held strictly to the intent of the CONTRACT DOCUMENTS in regard to the quality of materials, workmanship and execution of the WORK. Inspections may be made at the factory or fabrication plant of the source of material supply.

27.3 The ENGINEER will not be responsible for the construction means, controls, techniques, sequences, procedures, or construction safety.

27.4 The ENGINEER shall promptly make decisions relative to interpretation of the CONTRACT DOCUMENTS.

28. *LAND AND RIGHTS-OF-WAY*

28.1 Prior to issuance of NOTICE TO PROCEED, the OWNER shall obtain all land and rights-of-way necessary for carrying out and for the completion of the WORK to be performed pursuant to the CONTRACT DOCUMENTS, unless otherwise mutually agreed.

28.2 The OWNER shall provide to the CONTRACTOR information which delineates and describes the lands owned and rights-of-way acquired.

28.3 The CONTRACTOR shall provide at his own expense and without liability to the OWNER any additional land and access thereto that the CONTRACTOR may desire for temporary construction facilities, or for storage of materials.

29. *GUARANTY*

29.1 The CONTRACTOR shall guarantee all materials and equipment furnished and WORK performed for a period of one (1) year from the date of SUBSTANTIAL COMPLETION. The CONTRACTOR warrants and guarantees for a period of one (1) year from the date of SUBSTANTIAL COMPLETION of the system that the completed system is free from all defects due to faulty materials or workmanship and the CONTRACTOR shall promptly make such corrections as may be necessary by reason of such defects including the repairs of any damage to other parts of the system resulting from such defects. The OWNER will give notice of observed defects with reasonable promptness. In the event that the CONTRACTOR should fail to make such repairs, adjustments, or other WORK that may be made necessary by such defects, the OWNER may do so and charge the CONTRACTOR the cost thereby incurred. The Performance BOND shall remain in full force and effect through the guarantee period.

30. *ARBITRATION*

30.1 All claims, disputes and other matters in question arising out of, or relating to, the CONTRACT DOCUMENTS or the breach thereof, except for claims which have been waived by the making and acceptance of final payment as provided by Section 20, shall be decided by arbitration in accordance with the Construction Industry Arbitration Rules of the American Arbitration Association. This agreement to arbitrate shall be specifically enforceable under the prevailing arbitration law. The award rendered by the arbitrators shall be final, and judgment may be entered upon it in any court having jurisdiction thereof.

30.2 Notice of the demand for arbitration shall be filed in writing with the other party to the CONTRACT DOCUMENTS and with the American Arbitration Association, and a copy shall be filed with the ENGINEER. Demand for arbitration shall in no event be made on any claim, dispute or other matter in question which would be barred by the applicable statute of limitations.

30.3 The CONTRACTOR will carry on the WORK and maintain the progress schedule during any arbitration proceedings, unless otherwise mutually agreed in writing.

31. *TAXES*

31.1 The CONTRACTOR will pay all sales, consumer, use and other similar taxes required by the law of the place where the WORK is performed.

Supplemental General Conditions 补充一般条款

1. NOTICE TO UTILITIES. The Contractor shall give notice in writing to all utilities that may be affected by Contractor's operations at least seventy-two (72) hours before starting work. The Contractor shall not hinder or interfere with any person in the protection of such work.

   NOTICE TO FIRE AND POLICE. The Contractor shall give written notice to the Fire and Police Department of the municipality at least twenty-four (24) hours before closing off or in any way affecting through vehicular traffic on any street.

   NOTICE TO PUBLIC WORKS DEPARTMENT. The Contractor shall give notice in writing to the Village's Public Works Department at least seventy-two (72) hours before starting work.

2. PERMITS AND LICENSES. The Contractor shall procure all necessary permits and licenses, pay all charges and fees, and give all notices necessary and incidental to the due and lawful prosecution of the work unless otherwise specifically provided. Copies of all written notices and permits shall be submitted to the Engineer prior to the commencement of construction. All required right-of-way easements and property acquisition have been procured by the municipality.

3. LOCATION OF UNDERGROUND STRUCTURES. It is the responsibility of the Contractor to acquaint himself with the location of all underground structures which may be encountered or which may be affected by work under the Contract. The locations of any underground structures furnished, shown on the Plans or given on the site are based upon the available records, but are not guaranteed to be complete or correct and are given only to assist the Contractor in making a determination of the location of all underground structures.

4. ORDER OF PRIORITY OF DOCUMENTS. In addition to Section 4.2 of General Conditions, State Regulations govern over any other section followed by the Supplemental General Conditions, then General Conditions; then the technical Specifications in that order.

5. PROGRESS SCHEDULE. The Contractor shall, within ten (10) days after the effective date of Notice to Proceed, furnish three (3) copies of a preliminary progress schedule covering his operations for the first 30 days.

   The preliminary progress schedule shall be a bar graph or an arrow diagram showing the times he intends to commence and complete the various work stages, operations and contract items planned to be started during the first 30 days.

Source: Morgan & Parmley, Ltd.

The Contractor shall submit for approval by the Engineer within 30 days after the effective Notice to Proceed three (3) copies of a detailed bar graph or a graphic network diagram. The graph or diagram shall be accompanied by a brief written explanation of the proposed schedule and a list of activities.

The bar graph or graphic network diagram shall clearly depict the order and interdependencies of activities planned by the contractor as well as activities by others which affects the contractor's planning. For those activities lasting more than 30 days, either the estimated time for 25-50 and 75% completion or other significant milestone in the course of the activity shall be shown. In addition to the actual construction operations, the schedule shall show such items as submittal of samples and shop drawings, delivery of materials and equipment, construction in the area by other forces and other significant items related to the progress of construction. The graph or network diagram shall be printed or neatly and legibly drawn to a time scale.

The list of activities shall show for each activity the estimated duration and anticipated starting and finishing dates. Activities which are critical to complete the project in shortest time shall be identified. The cost of each activity shall be shown so the schedule can be cost loaded.

The Contractor shall submit with each payment request a copy of the schedule showing the current status of the project and the activities for which payment is being requested. If the project is behind schedule, a detailed report shall accompany the submittal stating why the project is behind schedule and what means are being taken to get the project back on schedule, including, if appropriate, contract time extensions.

The Schedule shall be revised when:

(a) When a Change Order significantly affects the contract completion date or the sequence of activities.

(b) When progress of any critical activity falls significantly behind the scheduled progress.

(c) When delay on a noncritical activity is of such magnitude as to make it become critical.

(d) At any time the Contractor elects to change any sequence of activities affecting the shortest completion time.

The revised analysis shall be made in the same form and detail as the original submittal and shall be accompanied by an explanation of the reasons for the revisions.

6. <u>SHOP DRAWINGS</u>. Shop drawings and descriptive data shall be required on all manufactured or fabricated items. Seven (7) copies of drawings and descriptive data shall be submitted for approval. (Two of these will be returned to the Contractor after their approval.)

Shop drawings shall be scheduled to be submitted for approval within 4 weeks after the effective Notice to Proceed.

7. <u>OPERATION AND MAINTENANCE DATA</u>. The Contractor shall furnish to the Engineer three (3) complete unbound copies of complete operation and maintenance data for all equipment items. This data shall include a list of parts and a recommendation of spare parts that should be kept on hand.

8. <u>SUPERVISION OF ERECTION FOR ALL EQUIPMENT</u>. All items of mechanical or electrical equipment whose cost delivered to the Owner exceeds $1,000.00 shall have included in the price bid the services of a qualified representative of the manufacturer for one (1) day to inspect equipment in operation and train the operator. For items whose cost exceeds $2,500.00, the bid price shall include not less than one (1) day during erection and not less than one (1) day operation inspection. Over $5,000.00, the bid price shall include not less than three (3) days during erection and two (2) days operation inspection. After each trip, the manufacturer is to submit a report to the Engineer covering the findings made during the inspection including changes and/or repairs, if any. Lesser time than specified may be needed. In this case, the Owner reserves the right to call for "make up" trips during the twelve (12) months following acceptance until time specified has been utilized. Unused time will lapse after the twelve (12) month period. Trips made for the purpose of correcting defective materials or workmansip are to be made without charge and are not to be credited to the required number.

9. <u>TESTING OF EQUIPMENT</u>. All mechanical and electrical equipment furnished for this project shall be field tested to insure full compliance with the Specifications. Testing shall be performed by the Contractor and/or equipment supplier in the presence of the Engineer. If the field test shows that the equipment does not meet the Specifications, it shall be replaced with acceptable equipment at Contractor's expense.

Full payment for equipment will not be made until it has been field tested and accepted.

Supplemental General Conditions *(continued)*

10. STORAGE OF MATERIALS. Contractor shall confine his apparatus, storage of materials and operations of his workmen to limits indicated by law, ordinances, permits or directions of the Engineer.

    All materials stored on the site shall be protected from the elements by watertight coverings or shed with wood floors raised above the ground and protected from damage by Contractor's operations by barriers. The Owner assumes no responsibility for stored materials.

    All materials shall be delivered in original sealed containers, with seals unbroken and with labels plainly indicating materials and directions for storage. Directions for storage shall be complied with unless specified to the contrary in the detailed Contract Specifications.

    Damaged materials shall be immediately removed from the site. Stored material will be paid for in accordance with Section 19 of the General Conditions.

11. PROTECTING WORK. The Contractor shall protect existing improvements and work installed by him from damage. Improvements and work damaged shall be replaced by the Contractor. All replacement methods shall be approved by the Engineer.

12. COOPERATION. The Contractors and Subcontractors shall cooperate and corrdinate their work with adjacent work and shall give due notice to other Contracotrs and Subcontractors of intersecting work to assure that all items are installed at an agreeable time and to facilitate general progress of the work.

13. LABOR LAWS. All Contractors and Subcontractors employed upon the work shall be required to conform to Federal and State labor laws and the various acts amendatory and supplementary thereto and to all other laws, ordinances and legal requirements applicable thereto.

    All labor shall be performed in the best and most workmanlike manner by mechanics skilled in their respective trades. The standard of the work required throughout shall be such grade as will bring results of the first class only.

14. TEMPORARY FACILITIES. The Contractor shall provide his own temporary facilities such as electricity, toilets and heat. He shall also furnish, erect and maintain a weatherproof bulletin board for displaying wage rates, Equal Opportunity Requirement, etc. All temporary facilities shall be removed prior to job completion.

15. FINAL CLEANUP. The Contractor shall be responsible for cleaning up his work and his waste and rubbish. It is the General Contractor's responsibility to completely clean up the inside and outside of all buildings and the construction site at the completion of the project.

16. PAYMENT TO CONTRACTOR. The Owner will make progress payment to the Contractors in accordance with Section 19 of the General Conditions. It is understood that grant reimbursement payment to the Owner are dependent upon compliance by the Owner with all the grant conditions identified in the grant request and grant agreement. The Owner shall be responsible for progress payments to the Contractor even when failure to comply with said grant conditions delays grant payments from the funding agency.

17. SAFETY. All work shall be done in accordance with OSHA standards.

18. AMERICAN-MADE PRODUCTS. Federally funded projects prohibit foreign-made products. Therefore, contractor(s) and sub-contractors shall buy/use American-made products on this project.

19. EASEMENTS. Some of the work will be done outside of existing right of ways. The Village has obtained needed easements. All disturbed areas within the construction limits shall be restored. Easements are available for review at the office of the Village Clerk.

20. MINORITY REQUIREMENTS. The goals for minority and female participation are 0.6% and 6.9% respectively. These goals are for both contractors and subcontractors. If these goals cannot be met, the contractor must be able to document a good faith effort in attempting to meet these goals. These requirements are explained elsewhere in the contract documents and the contractor is required to comply with them.

Summary of Work 工程概要（示例）

PART ONE - GENERAL (W/SPECIAL PROVISIONS)

1.1 BACKGROUND

1.1.1 History of Project: The High School Complex in Tony, is owned and operated by the Flambeau School District. Due to historically poor potable water and continuous maintenance problems, the School Board retained Morgan & Parmley, Ltd. to solve their problem. On May 23, 1977, construction was completed for their new water supply well located on Rusk County Airport property. The pumphouse, pumping equipment and transmission main were placed in operation early in 1979. This facility has been in continuous use, serving the High School Complex, since that time with no significant operational problems.

On several occasions, the Village Board attempted to establish a municipal water system without success. However, their recent efforts have proven successful by obtaining funding through the Rural Economic and Community Development, USDA (formally, FmHA).

An agreement between the Flambeau School District and the Village of Tony was recently executed. The Village will purchase the existing school well, pumphouse, transmission main and related appurtenances. These components will be incorporated into their proposed (total) water system; as detailed in the accompanying Plans and Specifications.

The Village has established their public water utility under the rules and regulations of the Wisconsin Public Service Commission.

The Well Site Survey and Engineering Report have been approved by DNR.

1.2 SCOPE OF WORK

1.2.1 Description: The proposed Project consists of the following major items:

1-Rehab Existing Pumphouse
2-Vertical Turbine Pump w/Controls
3-Interior Piping & Controls
4-Right Angle Drive Mechanism
5-Chlorination Facility
6-Electrical Upgrade
7-Watermain Distribution System
8-Elevated Water Tank & Tower
9-Services & Meters

93-159 01010-1

The foregoing list is only major items. Refer to the accompanying Plans, applicable sections of the Specifications and the Proposal for a total concept.

1.2.2 Proposals: There will be two (2) Proposals. The first will be for all items, except the elevated water tank, tower and related components which will be bid separately.

1.3 SCHEDULE

1.3.1 Anticipated Construction Schedule:

Commence Advertising 11-13-1995
Bid Opening 12-14-1995
Notice of Award 12-20-1995
Pre-Construction Conference 1-15-1996
Notice to Proceed 1-15-1996
Complete Construction 11-10-1996

1.3.2 Highway Construction: All construction on U.S. Highway 8 shall be substantially completed by June 30, 1996.

1.4 HYDRANT PATTERN

1.4.1 Description: Since the Village of Tony is served by the Ladysmith Fire Department, it is mandatory that Tony's fire hydrant pattern be compatable with their equipment.

PART TWO - SPECIAL PROVISIONS

2.1 WAGE RATES

2.1.1 Wages: The Contractors and Sub-Contractors shall meet the requirements of all applicable official State Wage Rate Determinations for this Project. A copy is attached to these Specifications.

2.2 EQUAL EMPLOYMENT OPPORTUNITY

2.2.1 Compliance: The Contractors and Sub-Contractors shall comply with all current and applicable rules and regulations governing equal employment opportunities.

2.3 APPROVALS

2.3.1 General: The Contractors and Sub-Contractors shall comply with all applicable items contained in the regulatory approvals and permits for this project. Approvals and permits from DNR, PSC, FAA, DOT, SHPO, RR, Rusk County, and others are attached to this

93-159 01010-2

Source: Morgan & Parmley, Ltd.

Summary of Work *(continued)*

Specification Book under the heading; Supplemental Documents. It is your responsibility to understand and abide by the conditions contained in the documents.

2.4 EROSION CONTROL

2.4.1 Responsibility: It is the Contractor's responsibility to provide the specified erosion control for the project. Refer to Section 02770, sheet 15 of 15 of the Plans and DOT's Policy 96.55 and related sections.

2.5 TRAFFIC CONTROL

2.5.1 Responsibility: It is the Contractor's responsibility for traffic control. Refer to DOT's Policy 96.51 and related sections.

2.6 EXISTING UTILITIES

2.6.1 General: It is the responsibility of the Contractor to acquaint himself with the location of all underground structures which may be encountered or which may be affected by work under the Contract. The locations of any underground structures furnished, shown on the Plans or given on the site are based upon available records, but are not guaranteed to be complete or correct and are given only to assist the Contractor in making a determination of the location of all underground structures.

2.6.2 Utility Notification: The Contractor shall give notice in writing to all utilities that may be affected by Contractor's operations at least seventy-two (72) hours before starting work. The Contractor shall not hinder or interfere with any person in the protection of such work.

The Contractor is encouraged to utilize the "Call Before You Dig" telephone notification system.

2.6.3 Cable TV Notification: The Contractor shall notify Marcus Cable, 219 W. Miner Avenue, Ladysmtih, WI 54848,; Tele. No. 715/532-6040.

2.6.4 Electrical Utility Notification: The Contractor shall notify Northern States Power Co., 129 N. Lake Ave., Phillips, WI 54555; Atten: Joe Perkins; Tele. No. 715/836-1198.

2.6.5 Telephone Utility: The Contractor shall notify Ameritech @ 1-800-242-8511 (Diggers Hotline).

2.6.6 Village Sewer System: The Contractor shall notify the Village's Sanitary Sewer Utility Operator, Billy Mechelke, N5297 Little X, Tony, WI 54563, Tele. No. 715/532-7046.

2.6.7 Emergency Facilities: The Contractor shall give written notice to the Fire Department, Rusk Co. Sheriff's Office and Ambulance Service, at least twenty-four (24) hours before closing off or in any way affecting through vehicular traffic on any street.

93-159 01010-3

Rusk Co. Sheriff's Office 715/532-2200

Ladysmith Rural Fire Department 715/532-2186

Ambulance Service 715-532-2100

2.7 CONSTRUCTION SIGN

2.7.1 Construction Sign: One (1) construction sign shall be provided and erected at a location to be specified by the Engineer at the Pre-Construction Conference. Refer to Supplemental Specifications, page A for its size and design. The construction sign will be a bid item in the Proposal.

2.8 SURVEY MARKERS

2.8.1 General: No survey markers (e.g. DOT, County, USGS, Property Corners, etc.) located in the project area shall be disturbed, unless prior approval has been obtained. Replacement will be at the expense of the Contractor.

2.9 STORAGE OF MATERIALS

2.9.1 General: Contractor shall confine his apparatus, storage of materials and operations of his workmen to limits indicated by law, ordinances, permits or directions of the Engineer.

All materials stored on the site shall be protected from the elements by watertight coverings or shed with wood floors raised above the ground and protected from damage by Contractor's operations by barriers. The Owner assumes no responsibility for stored materials.

All materials shall be delivered in original sealed containers, with seals unbroken and with labels plainly indicating materials and directions for storage. Directions for storage shall be complied with unless specified to the contrary in the detailed Contract Specifications.

Damaged materials shall be immediately removed from the site. Stored material will be paid for in accordance with standard practice.

2.10 PROTECTING WORK

2.10.1 General: The Contractor shall protect existing improvements and Work installed by him from damage. Improvements and Work damaged shall be replaced by the Contractor. All replacement methods shall be approved by the Engineer.

2.11 LABOR LAWS

93-159 01010-4

2.11.1 General: All Contractors and Subcontractors employed upon the Work shall be required to conform to applicable Federal and State labor laws and the various acts amendatory and supplementary thereto and to all other laws, ordinances and legal requirements applicable thereto.

All labor shall be performed in the best and most workmanlike manner by mechanics skilled in their respective trades. The standard of the Work required throughout shall be such grade as will bring results of the first class only.

2.12 TEMPORARY FACILITIES

2.12.1 Description: The Contractor shall provide his own temporary facilities such as electricity, toilets and heat. He shall also furnish, erect and maintain a weatherproof bulletin board for displaying wage rates, Equal Opportunity Requirement, etc. All temporary facilities shall be removed prior to job completion.

2.13 COOPERATION

2.13.1 General: The Contractors and Subcontractors shall cooperate and coordinate their Work with adjacent Work and shall give due notice to other Contractors and Subcontractors of intersecting Work to assure that all items are installed at an agreeable time and to facilitate general progress of the Work.

2.14 PROGRESS SCHEDULE

2.14.1 Description: The Contractor shall, within ten (10) days after the effective date of Notice to Proceed, furnish three (3) copies of a preliminary progress schedule covering his operations for the first 30 days.

The preliminary progress schedule shall be a bar graph or an arrow diagram showing the times he intends to commence and complete the various Work stages, operations and contract items planned to be started during the first 30 days.

The Contractor shall submit for approval by the Engineer within 30 days after the effective Notice to Proceed three (3) copies of a detailed bar graph or a graphic network diagram. The graph or diagram shall be accompanied by a brief written explanation of the proposed schedule and a list of activities.

The bar graph or graphic network diagram shall clearly depict the order and interdependencies of activities planned by the Contractor as well as activities by others which affects the Contractor's planning. For those activities lasting more than 30 days, either the estimated time for 25-50 and 75% completion or other significant milestone in the course of the activity shall be shown. In addition to the actual construction operations, the schedule shall show such items as submittal of samples and shop drawings, delivery of materials and equipment, construction in the area by other

forces and other significant items related to the progress of construction. The graph or network diagram shall be printed or neatly and legibly drawn to a time scale.

The list of activities shall show for each activity the estimated duration and anticipated starting and finishing dates. Activities which are critical to complete the project in shortest time shall be identified. The cost of each activity shall be shown so the schedule can be cost loaded.

The Contractor shall submit with each payment request a copy of the schedule showing the current status of the project and the activities for which payment is being requested. If the project is behind schedule, a detailed report shall accompany the submittal stating why the project is behind schedule and what means are being taken to get the project back on schedule, including, if appropriate, contract time extensions.

The Schedule shall be revised:

a) When a Change Order significantly affects the contract completion date or the sequence of activities.

b) When progress of any critical activity falls significantly behind the scheduled progress.

c) When delay on a noncritical activity is of such magnitude as to make it become critical.

d) At any time the Contractor elects to change any sequence of activities affecting the shortest completion time.

The revised analysis shall be made in the same form and detail as the original submittal and shall be accompanied by an explanation of the reasons for the revisions.

2.15 SHOP DRAWINGS

2.15.1 Description: Shop drawings and descriptive data shall be required on all manufactured or fabricated items. Seven (7) copies of drawings and descriptive data shall be submitted for approval. (Two of these will be returned to the Contractor after their approval.)

Shop drawings shall be scheduled to be submitted for approval within 4 weeks after the effective Notice to Proceed.

2.16 SUPERVISION OF ERECTION FOR ALL EQUIPMENT

2.16.1 General: All items of structural, mechanical or electrical equipment whose cost delivered to the Owner exceeds $1,000 shall have included in the price bid the services of a qualified representative of the manufacturer for one (1) day to inspect equipment in operation and train the operator. For items whose cost exceeds $2,500, the bid price shall include not less than one (1) day during erection and not less than one (1) day operation inspection. Over $5,000, the bid price shall include not less than three (3) days during erection and two (2) days operation inspection. After each trip, the manufacturer is to submit a report to the Engineer covering the findings made during the inspection including changes and/or repairs, if any. Lesser time than specified may be for "make up" trips during the twelve (12) month warranty period. Trips made for the purpose of correcting defective materials or workmanship are to be made without charge and are not to be credited to the required number.

2.17 SCHOOL WATER SUPPLY

2.17.1 Service: It is mandatory that the High School Complex be provided potable water on a continuous basis during school hours. Therefore, the Contractor shall schedule his Work to avoid any interruption of water during school operation or normal activity.

2.18 EXISTING SHRUBBERY

2.18.1 Protection: It shall be the responsibility of the Contractor to protect all existing private shrubbery, trees, etc. within the construction area. If they are damaged, or destroyed, they shall be replaced to the owner's satisfaction at the Contractor's expense.

2.19 SAFETY

2.19.1 Basic: All work shall be done in accordance with OSHA standards and other specified regulations as noted throughout the Specifications.

2.20 PROJECT RECORD DOCUMENTS

2.20.1 Contractor's Responsibility: At the conclusion of the project, prior to final payment, the Contractor shall supply the Engineer with a legible record of all work as installed. Example: a map detailing watermain installed with location, lengths, tie-offs, valves, hydrants and appurtenances accurately shown so an "as-built" layout can be developed. This information can be marked on the appropriate sheet of the Plans with a permanent color marker.

2.21 OPERATION & MAINTENANCE DATA

2.21.1 O & M Data: The Contractors shall furnish to the Engineer three (3) complete unbound copies of complete operation and maintenance data for all equipment items. This data shall include a list of parts and a recommendation of spare parts that should be kept on hand.

2.22 TESTING

2.22.1 Testing & Equipment: All mechanical and electrical equipment furnished for this project shall be field tested to insure full compliance with the Specifications. Testing shall be performed by the Contractor and/or equipment supplier in the presence of the Engineer. If the

field test shows that the equipment does not meet the Specifications, it shall be replaced with acceptable equipment at Contractor's expense.

Full payment for equipment will not be made until it has been field tested and accepted.

2.23 FINAL CLEANUP

2.23.1 General: The Contractor shall be responsible for cleaning up this Work and his waste and rubbish. It is the General Contractor's responsibility to completely clean up the inside and outside of all buildings and the total construction site at the completion of the project.

2.24 PRIVATE SERVICES

2.24.1 Description: Per specified bid items, the Contractor shall furnish and install service laterals from the distribution main to (and including) the curb stop. The Contractor shall also furnish and install the water meter including fittings and appurtenances within each respective user's structure.

Please be advised that section of the service lateral from the curb stop to the meter is the responsibility of each user and is not part of this construction contract.

2.25 PAYMENT TO CONTRACTOR

2.25.1 General: The Owner will make progress payments to the Contractors in accordance with the General Conditions. It is understood that grant reimbursement payments to the Owner are dependent upon compliance by the Owner with all the grant conditions identified in the grant request and grant agreement. The Owner shall be responsible for progress payments to the Contractor even when failure to comply with said grant conditions delays grant payments from the funding agency.

2.25.2 Final Payment: Final payment to the Contractor will not be made until the Owner and Engineer are satisfied that the facility is fully operational and all aspects of the Plans, Specifications and Construction Contract have been satisfied.

2.26 FmHA NAME CHANGE

2.26.1 Clarification: Wherever the words "Farmers Home Administration (FmHA) or Rural Development Administration (RDA)" appear in the project manual, substitute the words "U.S. of America."

Summary of Work *(continued)*

2.26　RAILROAD NOTIFICATION

2.26.1　General:

The Railroad is not a member of Diggers Hotline, therefore; before construction can begin, you should schedule signal wire location, a railroad flagger and the inspector at least three (3) working days in advance; to do so, please call the following people:

Construction Inspection
Cooper Engineering Company, Inc.
Maurice E. Smith
Telephone No. (715) 234-7008
Fax No. (715) 234-1025

Signal Wire Location
Wisconsin Central Ltd.
Jill Hamilton
Telephone No. (715) 345-2524
Fax No. (715) 345-2543

Railroad Flagger
Wisconsin Central Ltd.
Jerry Sernau
Telephone No. (715) 345-2511
Fax No. (715) 345-2507

-END OF SECTION-

93-159　　01010-9

Scope of Project　项目范围（示例）

WATER SUPPLY, PUMPHOUSE, DISTRIBUTION
SYSTEM & ELEVATED STORAGE RESERVOIR

Village
of
Tony, Wisconsin

The Village of Tony will exercise their option and purchase the existing well, pumphouse and transmission main from the Flabmeau School District and take title to the related easements.

The project proposes to rehab the existing pumphouse, located on the Rusk County Airport, replace the existing submersible pump with a 200 gpm vertical turbine pump and retrofit the interior piping. A right angle gear drive mechanism will be installed to provide emergency pumping via a farm tractor's PTO. A standby chlorination facility will be installed in the pumphouse.

A 6-inch watermain distribution system with adequate valves and fire hydrants will be installed within the Village's service area and all customers currently connected to the sanitary sewer collection system will be provided service from the new water system. All services will be furnished with water meters and billed accordingly, pursuant to PSC rate scheduling.

An elevated water storage reservoir will be erected to provide adequate pressure in the distribution system and have a reserve available for fire protection.

The work is fully described in the Specifications bound herein and on the following sheets of the accompanying Plans:

Source: Morgan & Parmley, Ltd.

Plan Sheet Index 设计图纸目录（示例）

SHEET NO.	DESCRIPTION
1	Title Sheet, Location Maps, Index & Notes
2	General Project Site Map w/Tower Site
3	Existing Well & Pumphouse Site Plan
4	Existing Well Log w/Construction Details
5	Pumphouse Rehab w/Piping Retrofit & Chlorination
6	Pumphouse Electrical w/H & V Layouts
7	Existing Transmission Main w/Easement - CSM
8	Village Map of Platted Area (w/watermain network)
9	Distribution System Layout (w/topography)
10	User Location Map w/Listing
11	Railroad & Highway Crossing: Oak St.
12	Railroad & Highway Crossing: Walnut St. (CTH "I")
13	Elevated Water Storage Reservoir
14	Typical Fire Hydrant & Service Details
15	Typical Erosion Control Details

投标书格式（示例）

Bid
Proposal "A" Format

Place:__________________

Date:__________________

Project: Water System

Project No.: 93-159

Proposal of ____

(Hereinafter called BIDDER), organized and existing under the laws of the State of ________

doing business as a ____________________________________.
(Corporation, Partnership, or Individual; delete non-applicable organization)

To the VILLAGE OF TONY, WISCONSIN:
(hereinafter called OWNER)

The BIDDER, in compliance with your Advertisement for Bids for the construction and installation of a MUNICIPAL WATER SYSTEM - CONTRACT A: Pumphouse Rehab, Pump, Right Angle Drive, Piping, Controls, Watermain Distribution System, Highway & R.R. Crossings, Meters, Services and Related Appurtenances; having examined the Plans and Specifications and related Documents prepared by Morgan & Parmley, Ltd., Consulting Engineers, and the site of the proposed Work and being familiar with all conditions surrounding the construction of the proposed project, including the availability of materials and labor, hereby proposes to furnish all labor, equipment, materials and supplies, and to construct the complete project in accordance with the Contract Documents within the time set forth therein and at the prices set forth below. Those prices are to cover all of the expenses incurred in performing the Work required under the Contract Documents, of which this Proposal is a part thereof.

The BIDDER hereby agrees to commence Work on the Contract on or before a date specified in the written "Notice to Proceed" of the Owner and to complete the project within 300 consecutive calendar days thereafter.

Source: Morgan & Parmley, Ltd.

The BIDDER agrees to pay, as liquidated damages, the sum of$350 for each consecutive calendar day thereafter as provided in Section 15 of the General Conditions.

Bidder acknowledges receipt of the following Addendum ______________________

____

____

BIDDER shall fill in Subcontractors and Material Supplier List and Similar Projects Lists attached hereto.

BIDDERS are advised that two (2) alternate pipe materials are acceptable for the distribution system (except where specified on the Plans). The alternatives are:

Ductile Iron - Class 52

PVC-DR-18

BIDDERS are required to complete all Bid Items.

-UNIT BID SCHEDULE-
for
PROPOSAL "A"
WATER SYSTEM - TONY, WISCONSIN

ITEM NO.	ITEM DESCRIPTION	UNIT	EST. QUAN.	UNIT PRICE	TOTAL
*1-A)	6" DISTRIBUTION MAIN: 6-Inch DI-Class 52 Watermain (150 psi W.P., 8'-0" Bury w/Cable Bond Connector @ all joints	L.F.	9,325	$	$
*1-B)	6" DISTRIBUTION MAIN: 6-Inch PVC-DR-18 Watermain (150 psi W.P., 8'-0" Bury)	L.F.	9,325		
*SPECIAL NOTE: BIDDER shall use the lowest priced alternate pipe material in Items 1 & 2 to determine the total Bid Price for Proposal A.					
*2-A)	8" DISTRIBUTION MAIN: 8-Inch DI-Class 52 Watermain (150 psi W.P., 8'-0" Bury w/Cable Bond Connector @ all joints	L.F.	625		
*2-B)	8" DISTRIBUTION MAIN: 8-Inch PVC-DR-18 Watermain (150 psi W.P., 8'-0" Bury)	L.F.	625		

ITEM NO.	ITEM DESCRIPTION	UNIT	EST. QUAN.	UNIT PRICE	TOTAL
3)	6" MAIN @ CROSSINGS: 6-Inch M.J.D.I., Class 52 Watermain (150 psi W.P., 8'-0" Bury) including sand bedding & sealing ends of casings	L.F.	285		
4)	8" MAIN @ CROSSINGS & TOWER: 8-Inch M.J.D.I., Class 52 Watermain (150 psi W.P., 8'-0" Bury) including sand bedding & sealing ends of casings	L.F.	390		
5)	6" VALVES: 6-Inch M.J. Gate Valves w/Three Piece Box @ 8'-0" Bury	Each	31		
6)	8" VALVES: 8-Inch M.J. Gate Valves w/Three Piece Box @ 8'-0" Bury	Each	4		
7)	6" X 6" CUT-IN-TEE: 6" x 6" x 6" M.J. Cut-In-Tee	Each	1		
8)	6" X 6" TEES: 6" x 6" x 6" M.J. Tees	Each	34		
9)	8" X 8" TEES: 8" x 8" x 8" M.J. Tees	Each	3		
10)	8" X 6" TEES: 8" X 8" X 6" M.J. Tees	Each	2		
11)	8" X 6" REDUCERS: 8" x 6" M.J. Reducers	Each	3		
12)	6" - 45° BENDS: 6-Inch M.J. ⅛ Bends	Each	2		
13)	8" - 45° BENDS: 8-Inch M.J. ⅛ Bends	Each	1		
14)	6" BLIND FLANGES: 6-Inch M.J. Blind Flanges	Each	10		
15)	8" BLIND FLANGES: 8-Inch M.J. Blind Flanges	Each	1		
16)	8" X 8" CROSS: 8-Inch x 8-Inch M.J. Cross	Each	1		
17)	HYDRANT GROUP: 6-Inch Hydrant (8'-0" Bury-Ladysmith Pattern) including Gate Valve & Box with 15 Ft. of 6" M.J.D.I. Lead (Does not include Tee)	Each	19		
18)	BEDDING-6" PIPE: Imported sand bedding, per Specs, for 6-Inch pipe	L.F.	500		

Bid Proposal "A" Format *(continued)*

ITEM NO.	ITEM DESCRIPTION	UNIT	EST. QUAN.	UNIT PRICE	TOTAL
19)	PIPE INSULATION: Extruded Polystyrene (4'-0" wide) Insulation, per Plans	L.F.	100		
20)	BEDDING - 8-INCH: Imported sand bedding, per Specs., for 8-Inch pipe.	L.F.	100		
21)	R.R. CASING - CTH "I": Bore & Jack 24" steel casing, per Plans & Permit (Watermain not included in this Item)	L.F.	92		
22)	R.R. CASING - OAK ST.: Bore & Jack 24" steel casing, per Plans & Permit (Watermain not included in this Item)	L.F.	65		
23)	HWY. CASING - CTH "I": Bore & Jack 24" steel casing, per Plans & Permit (Watermain not included in this Item)	L.F.	45		
24)	HWY. CASING - OAK ST.: Bore & Jack 24" steel casing, per Plans & Permit (Watermain not included in this Item)	L.F.	45		
25)	CONNECTION @ TANK: Connection of 8-Inch Watermain to Hydrant Tee @ End of Tank Yard Piping. (Remove & Salvage 8" Blind Flange)	L.S.	1		
26)	1" WATER SERVICE: 1-Inch Type K, Copper Service (Main to P/L) w/Corp., Saddle Clamp, Curb Stop and Box	Each	54		
27)	WATER METERS: ⅝" X ¾" Water Meter complete w/remote reader & connection fittings, per Specifications	Each	54		
28)	WATER METER (SCHOOL): 1½" Water Meter Complete w/connection fittings, per Specifications	Each	1		
29)	ACCESS DRIVEWAY: Construct and gravel Tower access driveway, per Plans	L.S.	1		
30)	CULVERT REPLACEMENT: 12-Inch CMCP, Complete with Gravel & Restoration	L.F.	180		

ITEM NO.	ITEM DESCRIPTION	UNIT	EST. QUAN.	TOTAL PRICE	TOTAL
31)	ABANDON CONNECTION: Disconnect School's existing 4-Inch cross-connection from abandoned well	L.S.	1		
32)	CURB & GUTTER REPLACEMENT: Replacement of concrete curb & gutter damaged during construction, including sub-base	L.F.	50		
33)	SIDEWALK REPLACEMENT: Replacement of concrete sidewalk damaged during construction; 4-Inch thick w/6 x 6 x 10/10 W.W.M. & 6-Inch sand lift	S.F.	800		
34)	BASE COURSE: Crushed aggregate base course; includes compaction and fine grading	C.Y.	800		
35)	SUB-BASE: Granular sub-base; includes compaction and fine grading	C.Y.	1,250		
36)	6" VALVE REPLACEMENT: Replacement of existing 6-Inch gate valve & box located in transmission main	Each	2		
37)	ASPHALT PAVING: Replacement of existing asphalt paving damaged during construction (2 - 1½" lifts); includes square cutting prior to excavation	S.F.	45,000		
38)	EROSION BALES: Includes stakes and installation, per Specifications	Each	50		
39)	SILT FENCE: Includes stakes and installation, per Specifications	L.F.	200		
40)	PUMPHOUSE REHAB: Complete rehabilitation of existing Pumphouse, per Plans & Specifications	L.S.	1		
41)	WELL PUMP: Deep Well Turbine Pump, complete per Plans & Specifications, including related appurtenances	L.S.	1		
42)	RIGHT ANGLE DRIVE: Right Angle Gear Drive mechanism; complete per Specifications including appurtenances and shafts (extension, combination and short)	L.S.	1		

Bid Proposal "A" Format *(continued)*

ITEM NO.	ITEM DESCRIPTION	UNIT	EST. QUAN.	UNIT PRICE	TOTAL
43)	INTERIOR PIPING: Complete interior piping and related components for Pumphouse, per Plans & Specifications	L.S.	1		
44)	CHLORINATION FACILITY: Chlorination Facility and related equipment, per Plans & Specifications	L.S.	1		
45)	ELECTRICAL: Complete electrical system for Pumphouse, per Plans & Specifications; including NSP application, entry, panels, general wiring, power roof ventilator, electrical heaters, lighting fixtures, control panels, and related Work not specifically Bid elsewhere	L.S.	1		
46)	SURGE ELIMINATOR: Surge eliminator and pressure switch; complete with related components (per Specifications)	L.S.	1		
47)	WATER CONTROL SYSTEM: Water Supply control system; complete per Specifications	L.S.	1		
48)	TRAFFIC CONTROL: Traffic control, per Specs. & Permit; including flagging personnel & signage	L.S.	1		
49)	MAIN DISINFECTION: Disinfection, flushing and testing of distribution system	L.S.	1		
50)	CONSTRUCTION SIGN: Provide one (1) project construction sign, per Specifications, and install at location designated by Engineer	L.S.	1		
51)	FIELD OFFICE: Provide construction field office at site during construction, per standard practice	L.S.	1		
52)	SITE RESTORATION: General restoration of construction site not specifically itemized elsewhere; including topsoil, seeding, fertilizing, sodding and mulching, as needed	L.S.	1		
53)	MOBILIZATION: Mobilization, bonds, insurance and other general project administration items	L.S.	1		
TOTAL BID: PROPOSAL A -- $					

The BIDDER understands that the OWNER reserves the right to reject any and all Bids, to waive any informalities in the Bidding process and to accept the BID most advantageous to the OWNER.

The undersigned further agrees to furnish surety in a penal sum of equal to or greater than the aggregate of the Contract and on the form hereto attached, to the OWNER, for carrying out the Work for which this PROPOSAL is submitted.

The undersigned agrees to execute the CONTRACT within ten (10) days from the date of notice from the OWNER that the CONTRACT has been awarded.

Accompanying this PROPOSAL is a check (certified) or Bidder's Bond in the amount of five (5%) percent of the bid amounting to ________________________________ Dollars ($____________) as required in the Advertisement for Bids.

In submitting this Bid it is understood that the right is reserved by the Owner to reject any or all Bids and to waive any informalities in the Bidding.

The Bidder agrees that this Bid shall be good and may not be withdrawn for a period of 90 calendar days after the scheduled closing for receiving Bids.

FIRM'S NAME	OFFICIAL ADDRESS
______________________________	______________________________
______________________________	______________________________
by ____________________________	______________________________
Title __________________________	______________________________
	Tele. No. (__)________________

(Seal: If Bid by Corporation)

Attest:

____

投标书格式（示例）

Bid
Proposal "B" Format

Place:____________

Date:____________

Project: Water System

Project No.: 93-159

Proposal of ______________________________

(Hereinafter called BIDDER), organized and existing under the laws of the State of ________

doing business as a ______________________________.

(Corporation, Partnership, or Individual; delete non-applicable organization)

To the VILLAGE OF TONY, WISCONSIN:
(hereinafter called OWNER)

The BIDDER, in compliance with your Advertisement for Bids for the construction and installation of a MUNICIPAL WATER SYSTEM - CONTRACT B: Elevated Steel Water Tank, Tower, Foundation, Valve Vault, Yard Piping, and Related Appurtenances; having examined the Plans and Specifications and related Documents prepared by Morgan & Parmley, Ltd., Consulting Engineers, and the site of the proposed Work and being familiar with all conditions surrounding the construction of the proposed project, including the availability of materials and labor, hereby proposes to furnish all labor, equipment, materials and supplies, and to construct the complete project in accordance with the Contract Documents within the time set forth therein and at the prices set forth below. Those prices are to cover all of the expenses incurred in performing the Work required under the Contract Documents, of which this Proposal is a part thereof.

The BIDDER hereby agrees to commence Work on the Contract on or before a date specified in the written "Notice to Proceed" of the Owner and to complete the project within 300 consecutive calendar days thereafter.

The BIDDER agrees to pay, as liquidated damages, the sum of$350 for each consecutive calendar day thereafter as provided in Section 15 of the General Conditions.

Bidder acknowledges receipt of the following Addendum ______________________

____

____

BIDDER shall fill in Subcontractors and Material Supplier List and Similar Projects List attached hereto.

BIDDERS are advised that two (2) types of Elevated Steel Water Storage Tanks are acceptable for use on this Project. The acceptable tank types are as follows:

BASE BID - Single Pedestal (Sphere) Elevated Storage Tank
ALTERNATE BID - Multi-Leg (Ellipsodal) Elevated Storage Tank

When evaluating Bids (Base Bid vs. Alternate Bid) a Cost Deduct Factor will be applied to the Total Amount Base Bid. This factor is a present worth cost of life-cycle paint costs for painting multi-leg tanks (Alternate Bid). The Cost Deduct Factor will be subtracted from the Base Bid Total Amount, and the resultant will be termed Equalized Base Bid.

COST DEDUCT FACTOR = $ 11,650

BIDDERS are required to complete the unit price schedule for both types of Elevated Storage Tanks.

Bid Proposal "B" Format *(continued)*

-UNIT BID SCHEDULE-
for
PROPOSAL "B"

WATER SYSTEM - TONY, WISCONSIN

SINGLE PEDESTAL (SPHERE) ELEVATED STORAGE TANK

ITEM NO.	ITEM DESCRIPTION	UNIT	EST. QUAN.	UNIT PRICE	TOTAL
1)	FOUNDATION: Concrete foundation & valve vault; complete w/related appurtenances	L.S.	1		$
2)	ELEVATED TANK: Fabricate & erect 50,000 gal. elevated steel water tank, tower & riser; complete w/accessories	L.S.	1		
3)	ELECTRICAL SYSTEM: Complete electrical system including service, panels, wiring, lighting, outlets, warning system and related components	L.S.	1		
4)	CIRCULATION PUMP: Duplex circulation pump system; complete per Specifications	L.S.	1		
5)	WELD TESTING: Radiographic weld testing with report	L.S.	1		
6)	PAINTING: Cleaning, surface preparation and painting of tank, tower and related components; including lettering of Village name on tank	L.S.	1		
7)	DISINFECTION: Complete disinfection of tank, riser & yard piping; including flushing	L.S.	1		
8)	YARD PIPING: 8-Inch, M.J.D.I. CL. 52 watermain connection from riser to hydrant tee; including hydrant, valve, 6-Inch lead, tee and blind flange	L.S.	1		
9)	SITE RESTORATION: Complete site restoration including grading, removal of construction debris, topsoil, seeding, fertilizing & mulching	L.S.	1		
10)	EROSION CONTROL: Erosion control of construction site, per Plans & Specifications	L.S.	1		
11)	MOBILIZATION: Mobilization, bonds, insurance and other general project administration items	L.S.	1		

TOTAL BASE BID (PROPOSAL B) -- $

COST DEDUCT FACTOR -- $ - (11,650)

EQUALIZED BASE BID -- $

-ALTERNATE BID-

ELLIPSOIDAL (MULTI-LEG) ELEVATED STORAGE TANK

ITEM NO.	ITEM DESCRIPTION	UNIT	EST. QUAN.	UNIT PRICE	TOTAL
1)	FOUNDATION: Concrete foundation & valve vault; complete w/related appurtenances	L.S.	1		
2)	ELEVATED TANK: Fabricate & erect 50,000 gal. elevated steel water tank, tower & riser; complete w/accessories	L.S.	1		
3)	ELECTRICAL SYSTEM: Complete electrical system including service, panels, wiring, lighting, outlets, warning system and related components	L.S.	1		
4)	CIRCULATION PUMP: Duplex circulation pump system; complete per Specifications	L.S.	1		
5)	WELD TESTING: Radiographic weld testing with report	L.S.	1		
6)	PAINTING: Cleaning, surface preparation and painting of tank, tower and related components; including lettering of Village name on tank	L.S.	1		
7)	DISINFECTION: Complete disinfection of tank, riser & yard piping; including flushing	L.S.	1		

Bid Proposal "B" Format *(continued)*

ITEM NO.	ITEM DESCRIPTION	UNIT	EST. QUAN.	UNIT PRICE	TOTAL
8)	YARD PIPING: 8-Inch, M.J.D.I. CL. 52 watermain connection from riser to hydrant tee; including hydrant, valve, 6-Inch lead, tee and blind flange	L.S.	1		
9)	SITE RESTORATION: Complete site restoration including grading, removal of construction debris, topsoil, seeding, fertilizing & mulching	L.S.	1		
10)	EROSION CONTROL: Erosion control of construction site, per Plans & Specifications	L.S.	1		
11)	MOBILIZATION: Mobilization, bonds, insurance and other general project administration items	L.S.	1		

TOTAL ALTERNATE BID (PROPOSAL B) ---------------------------- $

***SPECIAL NOTE:** BIDDER shall use the least costly Bid (i.e. Base or Alternate) total for proposed Base Price.

TOTAL BID: PROPOSAL B -- $

The BIDDER understands that the OWNER reserves the right to reject any and all Bids, to waive any informalities in the Bidding process and to accept the BID most advantageous to the OWNER.

The undersigned further agrees to furnish surety in a penal sum of equal to or greater than the aggregate of the Contract and on the form hereto attached, to the OWNER, for carrying out the Work for which this PROPOSAL is submitted.

The undersigned agrees to execute the CONTRACT within ten (10) days from the date of notice from the OWNER that the CONTRACT has been awarded.

Accompanying this PROPOSAL is a check (certified) or Bidder's Bond in the amount of five (5%) percent of the bid amounting to ______________________________ Dollars ($____________) as required in the Advertisement for Bids.

In submitting this Bid it is understood that the right is reserved by the Owner to reject any or all Bids and to waive any informalities in the Bidding.

The Bidder agrees that this Bid shall be good and may not be withdrawn for a period of 90 calendar days after the scheduled closing for receiving Bids.

FIRM'S NAME	OFFICIAL ADDRESS
______________________	______________________
______________________	______________________
by ____________________	______________________
Title __________________	______________________
	Tele. No. (___)________________

(Seal: If Bid by Corporation)

Attest:

____

Table of Contents　目录：招标文件（示例）

WATER SYSTEM
TONY, WISCONSIN

Source: Morgan & Parmley, Ltd.

Table of Contents (*continued*)

Table of Contents *(continued)*

招标程序

第 24 节 招标广告和开标

例 24

施工招标广告的法规和规范随州的不同以及投资机构的不同而不同。这样，我们建议读者按照管理机构和/或市政部门的要求来准备项目本阶段所需要的文件。

现今使用的很多官方许可文件和格式都不能包括进这本手册。所以在前一节，我们有选择性地给出了一般的介绍，用以在项目的这个阶段引导读者。

前一节（第 23 节）和本节有多个相互关联的部分和文件。所以，读者应当将两节一起研究以获得对招标程序的完全了解。

目 录

给出版商的提交信函

Submittal Letter to Publisher

Engineer's Letterhead

Date: ___________

Publication Name, Address, etc.

Attn: Legal Advertisement Editor

RE: Project Name: ___________
Project Location: ___________
Project Number: ___________

Dear *(Editor)*:

Please publish the enclosed ADVERTISEMENT for BIDS for subject project for *(Number)* consecutive weeks, commencing on *(Month/Day), (Year)* in your legal ads section. An affidavit of publication is required and requested to be submitted to our office within three (3) days prior to the Bid Opening which is *(Month/Day), (Year)*.

If you have any questions concerning this submittal, do not hesitate to contact me.

Sincerely,

____

Project Engineer

Enc. Ad for Bids

C: *(Owner/Client)*

Editor's Note: Refer to Section 23 for samples of Ads for Bids.

资格预审声明格式

Pre-Qualification Statement Format

PRE-QUALIFICATION STATEMENT

Submitted to ___________ Date Filed ___________

Project: ___________

NOTE: If the municipality, board, public body, or officer is not satisfied with the sufficiency of the answers to the questionnaire and financial statement, the bid may be rejected or disregarded or additional information may be required.

1. Name of Bidder ___________
2. Bidder's Address ___________
3. Direct any questions regarding information provided on this form to: ___________ at ___________
4. Type of organization(check one): Corporation_____ Partnership_____ Ind._____
5. When organized? ___________
6. If a corporation, when and where incorporated ___________
7. Attach a statement listing the corporate officers, partners or other principal members of your organization. Detail the background and experience of the principal members of your personnel, including the officers.
8. How many years has your organization been engaged in the contracting business under the present firm name? ___________
9. General description of work performed by your firm ___________
10. Attach a list of contracts on hand, for both public and private construction, including for each contract: the class of work; the contract amount; the percent completed; the estimated completion date; and the name and address of the owner or contracting officer.
11. Has your organization or any officer or partner acting either in their own name or as an officer or partner of some other organization ever defaulted on a contract or failed to complete a construction contract for any reason during the past 5 years?

Source: Morgan & Parmley, Ltd.

Pre Qualification Statement Format *(continued)*

12. Has your organization, any of its owners, a subsidiary or corporate parent, or any officer or director thereof, been subjected to any regulatory action or violation of law concerning construction projects or employees in the last 10 years? ____ If so, indicate:

    (a) The Date________________________ (b) Claimant __________________

    (c) Claimant's Mailing Address ________________________________ and

    (d) Attach a statement reciting the particulars of such violations(s).

13. Attach a list of the last 5 projects your organization has completed, including for each project: the class of work; the contract amount; the completion date; and the name, address and telephone number of the owner or contracting officer.

14. Attach a list of the major equipment which is available to your organization for the proposed work.

15. Attach a statement of your organization's experience in the construction or work similar in nature and importance to this project.

16. Credit available____.
    Attach a letter from your bank(s) or other financial institution(s) advising line of credit set up for your organization.

17. Name of Bonding Company and name, address and telephone number of agent:

    ____

    ____

18. Financial Statement (An official company Financial Statement may be attached in lieu of completing this Item)

    a. Cash $________________________

    b. Accounts Receivable $________________________

    c. Real estate equity $________________________

    d. Materials in stock $________________________

    e. Equipment, book value $________________________

    f. Furniture and fixtures, book value $________________________

    g. Other assets $________________________

Financial Statement (continued on next page)

TOTAL ASSETS ________________________

Liabilities

h. Accounts, notes & interest payable $________________________

i. Other liabilities ________________________

TOTAL LIABILITIES ________________________

NET WORTH $________________________

19. Additional information may be submitted if desired.

Dated at ______________________________ this ______ day of ________, ______.

Name of organization ____

By ____

Title ____

State of ______________________)
)
County of ______________________)

____ being duly sworn says that he/she is ____________________________________ of ______________________________
Title of Officer　　　　Name of Organization

and that the answers to the foregoing questions and all statements contained herein and in the attachments are true and correct.

Signed ____________________________________

Subscribed and sworn to before me this
______ day of ______________, ______.

____
Notary Public, State of ____________
My Commission Is/Expires _________

NOTE: This statement must be filed with the Owner or the Engineer not later than five (5) days prior to the opening of bids. This law further required that the information given by you in this statement shall be kept strictly confidential.

分包商和材料供应商名单

Sub-Contractors & Materials Suppliers List

All BIDDERS (Contractors) are required to list their Subcontractors and Material Suppliers including the phase of construction they will execute.

Failure to complete this list may be considered cause for rejection of PROPOSAL.

Name of Subcontractor and/or Material Supplier w/Address	Trade, Work or Material

Source: Morgan & Parmley, Ltd.

代理商或者分包商名单

List of Agents or Subcontractors Form

LIST OF AGENTS OR SUBCONTRACTORS

NAME ______________________ NAME ______________________
ADDRESS ______________________ ADDRESS ______________________
CITY, STATE, ZIP ______________________ CITY, STATE, ZIP ______________________
TELEPHONE # _(____)______________________ TELEPHONE # _(____)______________________

NAME ______________________ NAME ______________________
ADDRESS ______________________ ADDRESS ______________________
CITY, STATE, ZIP ______________________ CITY, STATE, ZIP ______________________
TELEPHONE # _(____)______________________ TELEPHONE # _(____)______________________

NAME ______________________ NAME ______________________
ADDRESS ______________________ ADDRESS ______________________
CITY, STATE, ZIP ______________________ CITY, STATE, ZIP ______________________
TELEPHONE # _(____)______________________ TELEPHONE # _(____)______________________

NAME ______________________ NAME ______________________
ADDRESS ______________________ ADDRESS ______________________
CITY, STATE, ZIP ______________________ CITY, STATE, ZIP ______________________
TELEPHONE # _(____)______________________ TELEPHONE # _(____)______________________

NAME ______________________ NAME ______________________
ADDRESS ______________________ ADDRESS ______________________
CITY, STATE, ZIP ______________________ CITY, STATE, ZIP ______________________
TELEPHONE # _(____)______________________ TELEPHONE # _(____)______________________

NAME ______________________ NAME ______________________
ADDRESS ______________________ ADDRESS ______________________
CITY, STATE, ZIP ______________________ CITY, STATE, ZIP ______________________
TELEPHONE # _(____)______________________ TELEPHONE # _(____)______________________

Source: Morgan & Parmley, Ltd.

Addendum Format　附录格式

ADDENDUM NO. 1
LADYSMITH SITE IMPROVEMENT PROJECT
LADYSMITH INDUSTRIAL PARK
RUSK COUNTY, WISCONSIN

Bid Date: July 10, 2000, 1:00 p.m.

Date of Addendum: June 21, 2000

CHANGES TO THE SPECIFICATIONS:

1 The bids received date on the INFORMATION TO BIDDERS form is revised to reflect the date on the ADVERTISEMENT FOR BIDS – July 10, 2000, 1:00 p.m.

2 Line 17 of the BID form is revised to require completion of the project within 160 days of the commencement date. The 1250 foot rail spur and crossing shall be completed within 60 consecutive calendar days of the commencement date. Line 14 of the AGREEMENT is revised to reflect the same change – 120 calendar days becomes 160 calendar days.

3 Revise line 24 of Section 13200-3, Pre-Engineered Metal Buildings, as follows:

Wall panels shall be 26 gauge (min.) galvanized,...

END OF ADDENDUM NO. 1

Morgan & Parmley, Ltd.
115 W. Second Street South
Ladysmith, WI 54848

Receipt of this Addendum is hereby acknowledged:

By:________________________ Date:______________

Contractor:______________________________________

PLEASE INCLUDE ADDENDUM IN YOUR PROPOSAL

By:________________________ Date:______________

Contractor:______________________________________

PLEASE INCLUDE ADDENDUM IN YOUR PROPOSAL

Source: Morgan & Parmley, Ltd.

Plan Holders List Form　设计图持有者名单

PLAN HOLDERS LIST

Page 1 of ___

Project: ____________

Bid Date:
Time:
Place:

Set No.	Contractor	Name & Address	Date Requested	Date Sent	Bid	Date Addendum Sent Out				Plan Deposit
						No. 1	No. 2	No. 3	No. 4	
1										
2										
3										
4										
5										
6										
7										
8										
9										
10										
11										
12										
13										
14										

Source: Morgan & Parmley, Ltd.

开标会出席签单

Bid Opening Attendance Sign-Up Sheet

ATTENDANCE SIGN-UP SHEET

BID OPENING

for

Project: ____________

Date: ____________

NAME	ADDRESS	REPRESENTING

Bid Tabulation Format 投标表格格式

Bid Date: ________

Time: ________

Place: ________

BID TABULATION

Name of Project

Page __ of __

No.	Item (NOTE: All items must include description)	CONTRACTOR	CONTRACTOR	CONTRACTOR
1				
2				
3				
4				
5				
6				
7				
8				
9				
10				
11				
12				
13				
14				
15				
16				
17				
18				
19				
20				
21				
22				
23				
24				
25				
26				
27				
28				
29				
30				
31				
32				
33				
34				
35				
	Project Total			

Source: Morgan & Parmley, Ltd.

Bid Tabulation Format *(continued)*

Bid Date: May 26, 2002
Time: 10:00 a.m.
Place: Ladysmith City Hall

BID TABULATION
River Avenue Street Reconstruction
Ladysmith, Wisconsin

No.	Item	Unit Price	Quantity	CONTRACTOR Baughman Trucking & Excavating, LLC	Unit Price	Quantity	CONTRACTOR Russ Thompson Excavting, Inc.	Unit Price	Quantity	CONTRACTOR A-1 Excavating, Inc.
1	Unclassified Excavation	6.00	2090	12540.00	4.55	2090	9509.50	5.20	2090	10868.00
2	Unclassified Excavtation - Above Base Bid	6.00	100	600.00	4.55	100	455.00	5.20	100	520.00
3	Clearing & Grubbing	1200.00	1	1200.00	700	1	700.00	23.50	1	23.50
4	Sawing Pavement	1.50	200	300.00	2.5	200	500.00	2.00	200	400.00
5	6" Valve	1600.00	2	3200.00	1163.01	2	2326.02	1100.00	2	2200.00
6	Hydrant Units	2300.00	1	2300.00	3000	1	3000.00	1900.00	1	1900.00
7	2" Polystyrene Insulation	1.25	320	400.00	2	320	640.00	1.60	320	512.00
8	1" Water Services	500.00	3	1500.00	800	3	2400.00	580.00	3	1740.00
9	4' Sanitary Sewer Manholes	2000.00	5	10000.00	1750	5	8750.00	1560.00	5	7800.00
10	8" PVC Sewermain	18.00	880	15840.00	35	880	30800.00	31.00	880	27280.00
11	4" Sewer Services	400.00	18	7200.00	650	18	11700.00	660.00	18	11880.00
12	6" Sewer Services	450.00	2	900.00	700	2	1400.00	740.00	2	1480.00
13	12" Storm Sewer CL III	35.00	30	1050.00	24	30	720.00	28.00	30	840.00
14	12" Storm Sewer CL V	38.00	65	2470.00	26.25	65	1706.25	28.00	65	1820.00
15	10" D.I. CL 52 Storm Sewer	32.00	190	6080.00	27.35	190	5196.50	29.50	190	5605.00
16	4' Storm Sewer Manhole	900.00	3	2700.00	1400	3	4200.00	1465.00	3	4395.00
17	2' Diameter Storm Sewer Inlets	300.00	2	600.00	500	2	1000.00	850.00	2	1700.00
18	Type 3 Inlet	750.00	6	4500.00	950	6	5700.00	955.00	6	5730.00
19	Rip Rap	50.00	2	100.00	100	2	200.00	88.00	2	176.00
20	Imported Sand Bedding - 8" Pipe	2.00	880	1760.00	1.5	880	1320.00	0.50	880	440.00
21	Imported Sand Bedding - 12" Pipe	2.00	95	190.00	1.5	95	142.50	0.50	95	47.50
22	Imported Sand Bedding - 10" Pipe	2.00	160	320.00	1	160	160.00	0.50	160	80.00
23	Crushed Aggregate Base Course	8.25	2000	16500.00	9	2000	18000.00	8.20	2000	16400.00
24	Curb & Gutter 30-inch	7.25	2200	15950.00	7	2200	15400.00	7.00	2200	15400.00
25	Topsoil Delivered & Placed	18.00	200	3600.00	18	200	3600.00	17.80	200	3560.00
26	Seed	3.00	50	150.00	5.5	50	275.00	8.50	50	425.00
27	Fertilizer	20.00	3	60.00	35	3	105.00	50.00	3	150.00
28	Mulch	500.00	1	500.00	350	1	350.00	200.00	1	200.00
29	Silt Fence Installation & Maintenance	2.00	300	600.00	2	300	600.00	2.20	300	660.00
30	Erosion Bales Installation & Maintenance	4.00	20	80.00	10	20	200.00	8.00	20	160.00
31	4" Concrete Sidewalk	10.00	20	200.00	5	20	100.00	5.00	20	100.00
32	Geotextile Fabric	1.25	500	625.00	1.1	500	550.00	1.00	500	500.00
33	Sediment Trap	500.00	1	500.00	1000	1	1000.00	800.00	1	800.00
34	Asphalt	40.00	800	32000.00	36.6	800	29280.00	38.60	800	30880.00
35	Cut in Type 3 Flat Grate	500.00	1	500.00	750	1	750.00	875.00	1	875.00
	Project Total			**$147,015.00**			**$162,735.77**			**$157,547.00**

Source: Morgan & Parmley, Ltd.

标书评审和中标办法
Bid Evaluation & Method of Award

After the bid opening, each proposal received will be reviewed, evaluated and checked for accuracy and for compliance with the bidding documents. Any non-responsive bids can be rejected by the Owner. A bid may be considered non-responsive for any of the following reasons:

1. Failure to fully complete each and every bid item listed in the BID SCHEDULE for the proposal submitted.

2. Failure to comply with the conditions of the ADVERTISEMENT FOR BIDS.

3. The Bidder imposes conditions or qualifies his bid in any way.

4. The Proposal is unsigned and/or not attested.

If a Contract is awarded, it will be awarded to the lowest responsible Bidder. A responsible Bidder is one who meets the following requirements.:

1. Has adequate financial resources to complete the Proposal for which the Bid was received.

2. Has the necessary experience, organization and technical qualifications and the necessary equipment to perform the work required.

3. Is able to comply with the required performance schedule to meet the completion date.

4. Has a satisfactory record of performance, integrity, judgment and skills.

5. Is qualified and eligible to receive the award under the applicable laws and regulations.

给业主（客户）的推荐信格式
Recommendation to Owner/Client Format

Engineer's Letterhead

Date: ______________

Owner/Client Name, Address, etc.

RE: Project Name: ______________________________
Project Number: ______________________________

Dear ______________:

Construction bids for subject project were opened *(Time & Date)* ______________.

Reference copies of the following documents are enclosed:

1) AFFIDAVIT of PUBLICATION
2) PLAN HOLDERS LIST
3) ATTENDANCE LIST @ BID OPENING
4) BID TABULATION
5) LOW BIDDER'S STATEMENT of QUALIFICATIONS

Based upon the foregoing data and the contractor's references, *(Name of Engineering Firm)* recommends that *(Name of Owner/Client)* award the construction contract to *(Low Bidder's Name)*; subject to concurrence from (if applicable) *(Grant Agency or similar funding source)*.

Please advise me when the award will be made so I can issue the Notice of Award to the contractor and schedule the Pre-Construction Conference.

Sincerely,

____
(Name of Project Engineer)

Enc.

C: *(Appropriate Officials)*

Source: Morgan & Parmley, Ltd.

招标程序

第 25 节　施工合同

例 25

施工合同（或协议）是业主和承包商之间签署的法律文件，它详细地规定了需要完成的工程和支付的价格。这份在法律意义上捆绑在一起的文件按惯例通常会包含在招标文件中。与大多数合同不同，施工合同有很多组成部分，包括设计图纸、技术规范、特殊要求、一般条款、补充一般条款，以及通常还包括的投资机构条款。

再一次向读者指出，管理机构或市政部门可能要求采用它们特定的合同文本。这样，聪明的做法是在规划阶段就取得这些文件，以避免将来的法律纠缠。本节后面给出的仅是一些常规材料，它们中的一部分已在本书主编的很多项目中得到了成功的使用。

还必须注意的是，很多专业协会都有被广泛使用的招标和施工合同文本。这样，专业设计师可能会希望采用由这些组织准备好的文本，例如美国职业工程师协会(NSPE)或者美国建筑师学会(AIA)等。

目　录

Notice of Award 中标通知书

To: ____________________ __________

____

____

____

____

PROJECT Description: ____

____

____

The OWNER has considered the BID submitted by you for the above described WORK in response to its Advertisement for Bids dated __________, __________, and Information for Bidders.

You are hereby notified that your BID has been accepted for items in the amount of $____________.

You are required by the Information for Bidders to execute the Agreement and furnish the required CONTRACTOR'S Performance BOND, Payment BOND and certificates of insurance within ten (10) calendar days from the date of this Notice to you.

If you fail to execute said Agreement and to furnish said BONDS within ten (10) days from the date of this Notice, said OWNER will be entitled to consider all your rights arising out of the OWNER'S acceptance of your BID as abandoned and as a forfeiture of your BID BOND. The OWNER will be entitled to such other rights as may be granted by law.

You are required to return an acknowledged copy of this NOTICE OF AWARD to the OWNER.

Dated this __________ day of ______________, __________.

____
Owner

By ____________________________

Title ____________________________

ACCEPTANCE OF NOTICE

Receipt of the above NOTICE OF AWARD is hereby acknowledged

by ____,

this the ____________________day of ________________________, _____

By ____________________________

Title ____________________________

Source: * (See page 23-1)

Agreement: Construction Contract 协议：施工合同

THIS AGREEMENT, made this __________day of__________, __________, by and between __________________________________, hereinafter called "OWNER"
(Name of Owner), (an Individual)

and __________________________________ doing business as (an individual,) or (a partnership,) or (a corporation) hereinafter called "CONTRACTOR".

WITNESSETH: That for and in consideration of the payments and agreements hereinafter mentioned:

1. The CONTRACTOR will commence and complete the construction of

____

2. The CONTRACTOR will furnish all of the material, supplies, tools, equipment, labor and other services necessary for the construction and completion of the PROJECT described herein.

3. The CONTRACTOR will commence the work required by the CONTRACT DOCUMENTS within __________ calendar days after the date of the NOTICE TO PROCEED and will complete the same within __________ calendar days unless the period for completion is extended otherwise by the CONTRACT DOCUMENTS.

4. The CONTRACTOR agrees to perform all of the WORK described in the CONTRACT DOCUMENTS and comply with the terms therein for the sum of $ __________, or as shown in the BID schedule.

5. The term "CONTRACT DOCUMENTS" means and includes the following:

(A) Advertisement For BIDS

(B) Information For BIDDERS

(C) BID

(D) BID BOND

(E) Agreement

Source: * (See page 23-1)

(F) General Conditions

(G) SUPPLEMENTAL GENERAL CONDITIONS

(H) Payment BOND

(I) Performance BOND

(J) NOTICE OF AWARD

(K) NOTICE TO PROCEED

(L) CHANGE ORDER

(M) DRAWINGS prepared by ____________________
numbered ________ through __________, and dated __________,
____

(N) SPECIFICATIONS prepared or issued by ______________
______________________________,
dated __________, ________

(O) ADDENDA:

No. ________, dated __________, ________

No. ________, dated __________, ________

No. ________, dated __________, ________

No. ________, dated __________, ________

No. ________, dated __________, ________

No. ________, dated __________, ________

6. The OWNER will pay to the CONTRACTOR in the manner and at such times as set forth in the General Conditions such amounts as required by the CONTRACT DOCUMENTS.

7. This Agreement shall be binding upon all parties hereto and their respective heirs, executors, administrators, successors, and assigns.

IN WITNESS WHEREOF, the parties hereto have executed, or caused to be executed by their duly authorized officials, this Agreement in (__________ (Number of Copies)) each of which shall be deemed an original on the date first above written.

OWNER:

____

BY ____________________

Name ____________________ (Please Type)

Title ____________________

(SEAL)

ATTEST:

____

Name ____________________ (Please Type)

Title ____________________

CONTRACTOR:

____

BY ____________________

Name ____________________ (Please Type)

Address ____________________

____

(SEAL)

ATTEST:

____

Name ____________________ (Please Type)

____

Performance Bond Form 履约保单

KNOW ALL MEN BY THESE PRESENTS: that

____
(Name of Contractor)

____
(Address of Contractor)

a ______________________________, hereinafter called Principal, and
(Corporation, Partnership, or Individual)

____
(Name of Surety)

____
(Address of Surety)

hereinafter called Surety, are held and firmly bound unto ______________________________

____
(Name of owner)

____
(Address of Owner)

hereinafter called OWNER, in the penal sum of ______________________________

______________________________Dollars, $(______________)

in lawful money of the United States, for the payment of which sum well and truly to be made, we bind ourselves, successors, and assigns, jointly and severally, firmly by these presents.

THE CONDITION OF THIS OBLIGATION is such that whereas, the Principal entered into a certain contract with the OWNER, dated the ______________day of______________, 19____, a copy of which is hereto attached and made a part hereof for the construction of:

____

____

____

____

NOW, THEREFORE, if the Principal shall well, truly and faithfully perform its duties, all the undertakings, covenants, terms, conditions, and agreements of said contract during the original term thereof, and any extensions thereof which may be granted by the OWNER, with or without notice to the Surety and during the one year guaranty period, and if he shall satisfy all claims and demands incurred under such contract, and shall fully indemnify and save harmless the OWNER from all costs and damages which it may suffer by reason of failure to do so, and shall reimburse and repay the OWNER all outlay and expense which the OWNER may incur in making good any default, then this obligation shall be void; otherwise to remain in full force and effect.

PROVIDED, FURTHER, that the said surety, for value received hereby stipulates and agrees that no change, extension of time, alteration or addition to the terms of the contract or to WORK to be performed thereunder or the SPECIFICATIONS accompanying the same shall in any wise affect its obligation on this BOND, and it does hereby waive notice of any such change, extension of time, alteration or addition to the terms of the contract or to the WORK or to the SPECIFICATIONS.

PROVIDED, FURTHER, that no final settlement between the OWNER and the CONTRACTOR shall abridge the right of any beneficiary hereunder, whose claim may be unsatisfied.

IN WITNESS WHEREOF, this instrument is executed in ________ counterparts, each
(Number)

one of which shall be deemed an original, this the ______________day of ______________ 19________.

ATTEST:

____
Principal

____
(Principal) Secretary

By ______________________________(s)

(SEAL)

____
(Witness as to Principal)

____
(Address)

____
(Address)

____

____

____
Surety

ATTEST:

____
(Surety) Secretary

(SEAL)

____
Witness as to Surety

By ______________________________
Attorney-in-Fact

____
(Address)

____
(Address)

____

____

NOTE: Date of BOND must not be prior to date of Contract.
If CONTRACTOR is Partnership, all partners should execute BOND.

IMPORTANT: Surety companies executing BONDS must appear on the Treasury Department's most current list (Circular 570 as amended) and be authorized to transact business in the state where the PROJECT is located.

Source: * (See page 23-1)

Payment Bond Form　付款保单

KNOW ALL MEN BY THESE PRESENTS: that

____
(Name of Contractor)

____
(Address of Contractor)

a ______________________________, hereinafter called Principal,
(Corporation, Partnership or Individual)

and ______________________________
(Name of Surety)

____
(Address of Surety)

hereinafter called Surety, are held and firmly bound unto ______________________________

____
(Name of Owner)

____
(Address of Owner)

hereinafter called OWNER, in the penal sum of________________Dollars, $(__________)

in lawful money of the United States, for the payment of which sum well and truly to be made, we bind ourselves, successors, and assigns, jointly and severally, firmly by these presents.

THE CONDITION OF THIS OBLIGATION is such that whereas, the Principal entered into a certain contract with the OWNER, dated the ________________day of __________ 19________, a copy of which is hereto attached and made a part hereof for the construction of:

____

____

____

____

NOW, THEREFORE, if the Principal shall promptly make payment to all persons, firms, SUBCONTRACTORS, and corporations furnishing materials for or performing labor in the prosecution of the WORK provided for in such contract, and any authorized extension or modification thereof, including all amounts due for materials, lubricants, oil, gasoline, coal and coke, repairs on machinery, equipment and tools, consumed or used in connection with the construction of such WORK, and all insurance premiums on said WORK, and for all labor, performed in such WORK whether by SUBCONTRACTOR or otherwise, then this obligation shall be void; otherwise to remain in full force and effect.

PROVIDED, FURTHER, that the said Surety for value received hereby stipulates and agrees that no change, extension of time, alteration or addition to the terms of the contract or to the WORK to be performed thereunder or the SPECIFICATIONS accompanying the same shall in any wise affect its obligation on this BOND, and it does hereby waive notice of any such change, extension of time, alteration or addition to the terms of the contract or to the WORK or to the SPECIFICATIONS.

PROVIDED, FURTHER, that no final settlement between the OWNER and the CONTRACTOR shall abridge the right of any beneficiary hereunder, whose claim may be unsatisfied.

IN WITNESS WHEREOF, this instrument is executed in ________counterparts, each
(number)

one of which shall be deemed an original, this the ____________ day of ______________ 19 __________.

ATTEST:

____
Principal

____
(Principal) Secretary

(SEAL)　　By ______________________________(s)

____
(Address)

____

____
tness as to Principal

____
(Address)

____

____
Surety

ATTEST:　　By ______________________________
Attorney-in-Fact

____
Witness as to Surety

____
(Address)

____
(Address)

____

NOTE: Date of BOND must not be prior to date of Contract.
If CONTRACTOR is Partnership, all partners should execute BOND.

IMPORTANT: Surety companies executing BONDS must appear on the Treasury Department's most current list (Circular 570 as amended) and be authorized to transact business in the State where the PROJECT is located.

Source: * (See page 23-1)

承包商的保险证明单

Contractor's Certificate of Insurance Form

Name of Contractor: ____________ Name of Owner: ____________

Address: ____________ Address: ____________

____________ ____________

KIND OF INSURANCE	POLICY NO.	EXPIRATION DATE	LIMITS
Workmen's Compensation Employers' Liability			$ ____ Statutory Workmen's Compensation $ ____ One Accident and Aggregate Disease
Commercial General Liab. Bodily Injury & Property Damage			$ ____ Each Occurrence $ ____ Aggregate $ ____ Aggregate Completed Operations and Products
Personal Injury			$ ____ Aggregate
Comprehensive Auto Liab. Bodily Injury & Property Damage			$ ____ Each Occurrence $ ____ Aggregate
Other Property Insurance			$

		Yes	No
1.	Does Property Damage Liability Insurance shown include coverage for XC and U Hazards?	__	__
2.	Is Occurrence Basis Coverage provided under Property Damage Liability?	__	__
3.	Is Broad Form Property Damage Coverage provided for this Project?	__	__
4.	Does Personal Injury Liability Insurance include coverage for personal injury sustained by any person as a result of an offense directly or indirectly related to the employment of such person by the insured?	__	__
5.	Is Coverage provided for Contractual Liability (including indemnification provision) assumed by Insured?	__	__
6.	Does Coverage above apply to non-owned and hired automobiles?	__	__
7.	Is Occurrence Basis Coverage provided under Automobile Damage Liability?	__	__

Notice: The above named Owner shall be notified 15 days prior to the cancellation of any of the above coverages.

____
NAME OF INSURANCE COMPANY

____
ADDRESS

____
SIGNATURE OF AUTHORIZED REPRESENTATIVE & DATE

Sample Letter to Contractor 给承包商的信函示例

Engineer's Letterhead

Date:

Contractor's Name, Address, etc.

Attn: *(Name and Title of C.E.O.)*

RE: Project Name: ____________

Project Number: ____________

Dear *(Name of C.E.O.)*:

You are hereby notified that the (*Owner/Client's Name*) has accepted your PROPOSAL for (*Name of Project*).

I have enclosed the following documents for you to execute:

1) NOTICE OF AWARD: (6 copies) Sign, date and return 5 copies to our office immediately. The extra copy is for your files.
2) AGREEMENT: (5 copies) Sign, date, seal and return all copies to our office.
3) PERFORMANCE BOND: (5 copies) Execute and return all copies to our office.
4) PAYMENT BOND: (5 copies) Execute and return all copies to our office.
5) CERTIFICATE OF INSURANCE: (5 copies) Execute and return all, copies to our office. (You may use your insurance company s form).
6) OWNER'S INSTRUCTIONS FOR BONDS & INSURANCE: (I copy for reference only)

After these documents have been received and reviewed, the Contract will be formally signed and sealed by the appropriate Officials. A copy of these documents, bound in the Specifications Book, will be delivered to you at the Preconstruction Conference.

If you have any questions concerning this material, please feel free to contact me.

Sincerely,

____
Project Engineer

Enc.

C: (*Owner/Client*)

Source: Morgan & Parmley, Ltd.

代理的证明单

Attorney's Certification Form

CERTIFICATE OF OWNER'S ATTORNEY

I, the undsigned, ______________________________, the duly authorized and acting legal representative of ____________________ ______________________, do hereby certify as follows:

I have examined the attached Contract(s) and Performance and Payment Bond(s) and the manner of execution thereof, and I am of the opinion that each-of the aforesaid Agreements are adequate and have has been duly executed by the proper parties thereto acting through their duly authorized representatives; that said representatives have full power and authority to execute said Agreements on behalf of the respective parties named thereon; and that the forgoing Agreements constitute valid and legally binding obligations upon the parties executing the same in accordance with terms, conditions, and provisions thereof.

Date: ______________

____
Attorney

NOTE: Delete phrase "performance and payment bonds" when not applicable.

Change Order Form　变更指令单

Order No. ____________________

Date: ____________________

Agreement Date: ____________________

NAME OF PROJECT: ____

____

OWNER: ____

CONTRACTOR: ____

The following changes are hereby made to the CONTRACT DOCUMENTS:

Justification:

Change to CONTRACT PRICE:

Original CONTRACT PRICE $______________

Current CONTRACT PRICE adjusted by previous CHANGE ORDER $______________

The CONTRACT PRICE due to this CHANGE ORDER will be (increased) (decreased)

by: $____________________

The new CONTRACT PRICE including this CHANGE ORDER will be $______________

Change to CONTRACT TIME:

The CONTRACT TIME will be (increased) (decreased) by__________calendar days.

The date for completion of all work will be ______________(Date).

Approvals Required:
To be effective this Order must be approved by the Federal agency if it changes the scope or objective of the PROJECT, or as may otherwise be required by the SUPPLEMENTAL GENERAL CONDITIONS.

Requested by: ____

Recommended by: ____

Ordered by: ____

Accepted by: ____

Federal Agency Approval (where applicable) ______________________________

Source: * (See page 23-1)

施　工

第26节　开工前会议

例　26

开工前会议是项目正式启动并推进到实际施工阶段的里程碑。施工合同正式签署，合同副本分发，而且一份正式的开工文件被送达给承包商。

在这个会议上，所有主要的参与者都将聚积在一起，讨论项目的重大问题，并回答相关的技术问题。所有特别规定都应进行充分的说明。大体上，所有的法规、规范、格式、进度计划和项目的相关参数都将经过彻底的回顾。无论如何强调施工前会议的重要性都是不过分的，它是项目从规划设计阶段过渡到最终的物质产品的桥梁。

对于高规格项目，很多情况下需要举行开工典礼。这个典礼可能会与施工前会议合并，但我们建议将开工典礼放在简要技术介绍后举行。

目　录

通知格式

Format of Notice

NOTICE
of
PRE-CONSTRUCTION CONFERENCE

Date: ______________

To: (*List Contractors, Utility Co. Representatives, Contractors, Regulatory Agency Staff, Municipal Officials, Engineering Staff and Other appropriate individuals who should be notified*)

RE: (*Name of Construction Project, Name of Owner and Project No.*)

Memo:

Please be advised that a PRE-CONSTRUCTION CONFERENCE for subject project will be held on ______________, ______________, ________,
(*Day Name*) (*Month*) (*Day*)

______________ @ ______________, in the ______________________,
(*Year*) (*Time*) (*Location*)

______________________, ________. (A location map is enclosed.)
(*City*) (*State*)

If the prime contractor wants their sub-contractors present at this meeting, it is their responsibility to notify them.

CONSTRUCTION CONTRACTS with a NOTICE TO PROCEED will be issued at this meeting.

Sincerely,

____

(*Name of Engineer with Title*)

Enc. Location Map

日程安排示例

Sample Agenda

Page 1 of 2

-AGENDA-
PRE-CONSTRUCTION CONFERENCE
for
Wastewater Collection & Treatment Facility
Catawba-Kennan Joint Sewage Commission
M & P Project No. 98-106
CDBG Project No. PF FY00-0089

DATE: January 30, 2002

TIME: 1:00 p.m.

LOCATION: Community Center, Kennan, Wisconsin

I. **REGISTRATION:**
  A. Sign Attendance List

II. **INTRODUCTION:**
  A. Welcome & General Overview of Project (M & P, Ltd.)
  B. Sewage Commission Officials
  C. Morgan & Parmley, Ltd. Staff
  D. Rural Development (USDA) Representatives
  E. Grant Administrator
  F. Contractors' Representatives
  G. Subcontractors' Representatives
  H. DNR Representatives
  I. WDOT Representatives
  J. Price Co. Highway Commission Representatives
  K. Town of Catawba Representatives
  L. Town of Kennan Representatives
  M. Utilities (Electric, Telephone, Etc.)
  N. Others

III. **CONSTRUCTION CONTRACTS:**
  A. Commission to Sign Contracts
  B. Rural Development to Approve Contracts
  C. Review: Notice to Proceed (Each Contract)
  D. Contractor (2) to Sign Notice to Proceed
  E. Distribute Contract (2)
    1-Contractor
    2-Commission
    3-Rural Development
    4-Grant Administrator
    5-Engineer

IV. **USDA-RURAL DEVELOPMENT:**
  A. Project Specialist
  B. Reference Pages 233 thru 250
  C. Project Sign

Source: Morgan & Parmley, Ltd.

Page 2 of 2

V. **CDBG REGULATIONS:**
A. Grant Administrator
B. Refer to Supplemental CDBG Agenda

VI. **SCOPE OF WORK:**
A. Review Section 01010 (Special Provisions)
B. Timetable (Construction Schedule)
C. Shop Drawing Process
D. Construction Area Video Taped (Fall 2001 by M & P, Ltd.)
E. Existing Utilities (Contact Diggers Hotline)
F. List of Key Contacts

VII. **PLANS & SPECIFICATIONS:**
A. General Review
B. Specific Items & Questions

VIII. **PAY REQUEST:**
A. RD Form (pg. 42 thru 44)
B. Submit to Engineer by 3rd of Month (3 copies)
C. Payroll Records (State & Federal)
D. Payments will be processed monthly, in accordance with the "Progress Schedule of Construction". These schedules, cumulatively, form the basis of the Sewage Commission's outlay management program.

IX. **APPROVALS:**
A. Contractors to comply with all applicable conditions of each respective approval/permit. Copies of these regulatory approvals and permits are contained in the Specifications & Contracts.

X. **GENERAL QUESTIONS & COMMENTS:**
A. Overview-Supplemental
B. Contractors' Representatives
C. Utilities Representatives
D. Ground Breaking Ceremony (Schedule)
E. Other Related Items
F. Closing Remarks

会议出席者签名表示例

Sample Attendance Signature List

Page __ of __

ATTENDANCE SIGNATURE LIST
for
PRE-CONSTRUCTION CONFERENCE

Wastewater Collection & Treatment Facility
Catawba-Kennan Joint Sewage Commission
M & P Project No. 98-106
CDBG Project No. PF FY00-0089
January 30, 2002

NAME	REPRESENTING

Source: Morgan & Parmley, Ltd.

一般的开工通知

Generic Notice to Proceed

NOTICE TO PROCEED

To: ______________________ Date: ______________________

______________________ Project: ______________________

______________________ ______________________

______________________ ______________________

______________________ ______________________

You are hereby notified to commence WORK in accordance with the Agreement dated __________, ______, on or before __________, ______, and you are to complete the WORK within ________ consecutive calendar days thereafter. The date of completion of all WORK is therefore __________, ______.

____
Owner

By ______________________

Title ______________________

ACCEPTANCE OF NOTICE

Receipt of the above NOTICE TO PROCEED is hereby acknowledged by ______ ______________________,

this the ______________________ day

of ______________________, ______

____

By ______________________

Title ______________________

Source: * (See page 23-1)

Generic Format for Construction Schedule　施工进度计划的一般格式

WASTEWATER COLLECTION & TREATMENT FACILITY

Project No. ______________________

January 30, 2002

BID ITEM	MAY 5-11	MAY 12-18	MAY 19-25	MAY/JUNE 26-1	JUNE 2-8	JUNE 9-15	JUNE 16-22	JUNE 23-29	JUNE/JULY 30-6	JULY 7-13	JULY 14-20	JULY 21-27	JULY/AUG 28-3	AUG 4-10	AUG 11-17
1) (*Name*)															
2) (*Name*)															
3) (*Name*)															
4) (*Name*)															
5) (*Name*)															
6) (*Name*)															
7) (*Name*)															
8) (*Name*)															
9) (*Name*)															

NOTE: 1) Restoration to be completed as installations progress.

Format for Application for Payment 付款申请格式

Page 1 of 2

APPLICATION FOR PAYMENT

Project: ______

Owner: ______ Contractor: ______

______ ______

______ ______

Date: ______ Pay Request No.: ______

From (Date): ______ To (Date): ______

1. Original Contract Amount .... $ ______
2. Net change by Change Orders .... $ ______

   Change Order No. Amount (+ or −)

   ______ ______

   ______ ______

   ______ ______

   ______ ______

   NET CHANGE $ ______

3. Contract Sum to Date .... $ ______
4. Total Completed and Stored to Date .... $ ______
   (Total from Page ___)
5. Retainage @ ____%. .... $ ______
6. Total Earned less Retainage .... $ ______
7. Less Previous Certificates for Payment .... $ ______
8. Current Payment Due .... $ ______
9. Balance to Finish, Plus Retainage .... $ ______

Contractor: ______ Date: ______

Certified By: ______ Date: ______

Accepted By: ______ Date: ______

Authorized By: ______ Date: ______
(When Applicable)

Page 2 of 2

SUMMARY OF COMPLETED WORK

Item No.	Description of Work	Contract Amount	Previous Application	This Application	Total Completed and Stored to Date	Balance to Finish
TOTAL						

Source: Morgan & Parmley, Ltd.

Payroll Record Format　工资单记录格式

PAYROLL

NAME OF CONTRACTOR____ or SUBCONTRACTOR____		ADDRESS	
PAYROLL NO.	FOR WEEK ENDING	PROJECT AND LOCATION	PROJECT NO.

(1) NAME, ADDRESS AND SOCIAL SECURITY NUMBER OF EMPLOYEE	(2) # OF EXEMP-TIONS	(3) WORK CLASSIFICATION	OT OR ST	(4) DAY AND DATE							(5) TOTAL HOURS	(6) RATE OF PAY	(7) Gross Amount Earned	(8) DEDUCTIONS						(9) NET WAGES PAID FOR WEEK
														FICA	WITH HOLDING TAX			OTHER	TOTAL DEDUCTIONS	
				Hours Worked Each Day																
			O																	
			S																	
			O																	
			S																	
			O																	
			S																	
			O																	
			S																	
			O																	
			S																	
			O																	
			S																	
			O																	
			S																	

施　工

第27节　加工图

例　27

设计图和技术说明描述了待建项目并详细说明了不同构件。然而，良好的实践是以这样一种方式来指定不同的组件，避免所谓的“冷门”，这样就有不止一个供应商能够满足技术要求，从而确保竞争性投标并满足预算规定。所以，工程师有责任审核项目所有的加工图，确保承包商通过它们的制造商/供应商符合设计图和技术说明的要求。

目　录

一般过程

项目的大量组件和设备都不是在项目现场制造的。这样在生产之前，这些特定单元的加工图就需经过准备并送给工程师审核。工程师有责任审查这些图纸，并在需要时标明修改，这样来保证最后的项目能够符合项目设计图和技术说明的要求。

至关重要的是，所有加工图的副本都应被保留作为工程师的永久文件。另外，项目监理也必须有这些副本，当组件到达现场时，用来检查并确认实际产品满足经过核准的标准。

保留一套完整的加工图有利于在施工完成后帮助准备运行和维护手册。

本节后面的加工图示例选自实际项目，在这里给出它们是为了让读者对它们的内容有一个大体的感觉，因为要想在本手册的有限篇幅中给出加工图的全部范畴是不可能的和不现实的。读者应当理解，电气组件和结构用钢的生产在它们特定图纸表示中是非常不同的。通常，生产有多种型号产品的制造商会提交一份它们的技术目录副本，目录的特定页做有标记用以表明所选产品满足项目设计图所要求的技术说明。一般来说，这被认为是一种可接收的提交。当然，这种方法与生产桥梁所需结构钢材的详细过程有很大不同。对于任何一种情况，读者都必须明了当今使用广泛的加工图技术。记住，加工图的目的是确保复杂的组件满足项目设计图的技术要求。

DEEP WELL TURBINE PUMP
ELECTRIC MOTOR DRIVEN

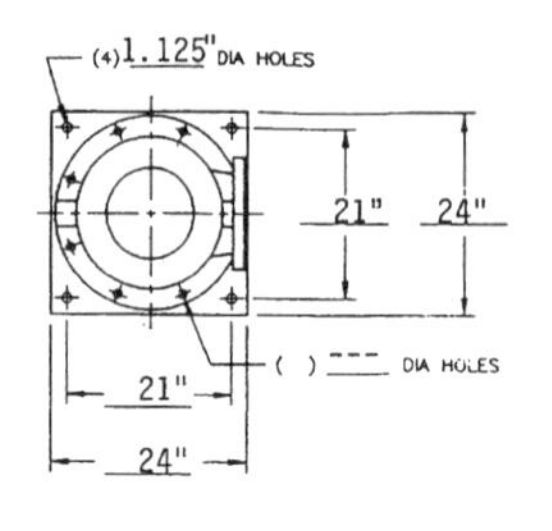

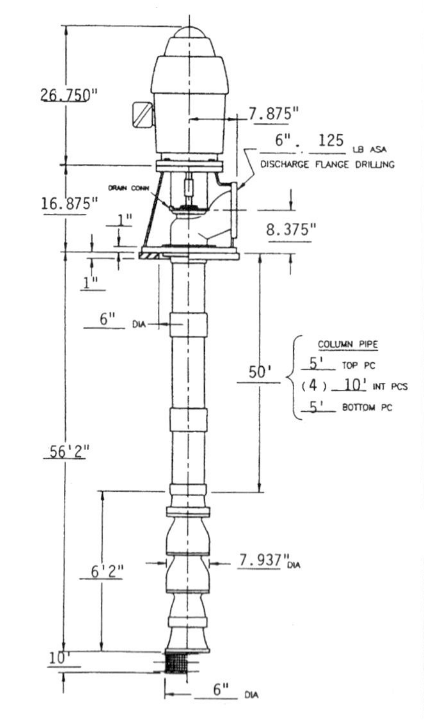

MATERIAL OF CONSTRUCTION			
Bowl	C.I. 30	Impeller	Bronze
Bowl Shaft	416	Shaft Coupling	C-1137
Bowl Bearings	Bronze	Shaft Bearings	Bronze
Strainer	N/A	Bowl W/R	Bronze
Impeller W/R	Bronze	Column Pipe	A120 B
Lineshaft	C-1045	Packing	Composition
Base Plate	Steel	Head	C.I. 30
PUMP			
Type	8 KC	Discharge Head	TR6B
Suction	6" N.P.T.	Discharge	6"
Lineshaft	C-1045	Column	6"
Lubrication	Product	Model	
Stage	9	GPM	200
TDH	221	Trim	6.182'
RPM	1770	BHP	14.36
MOTOR			
Make	U.S.	Type	AUE
Enclosure	W.P.I.	NRR	Yes
SRC	N/A	HP	15
RPM	1770	Phase	3
Hertz	60	Voltage	230/460
Frame No.	254 TP	Type Coupling	Bolted
OTHER SPECIFICATIONS			
Drawing No.		Serial No.	
Fluid	Well Water	Spec.Gravity	1
Viscosity		Temperature	
PH		No.Units Required	1

Customer			
Address			
City		ST	ZIP
Tel. ( )		Fax. ( )	
Rep.			
Supplier			
Salesman			

Source: Morgan & Parmley, Ltd.

示例：高架水箱加工图

Sample: Shop Drawings for Elevated Water Storage Tank

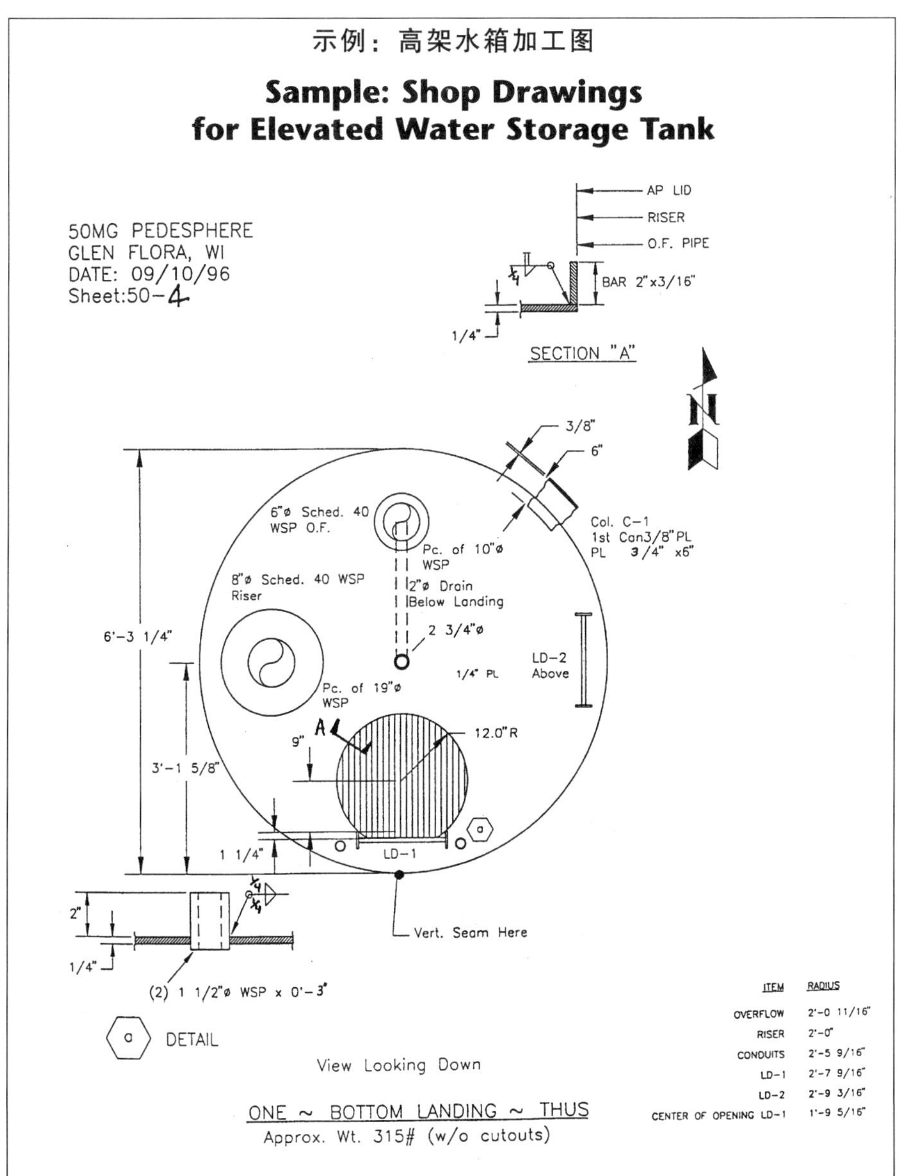

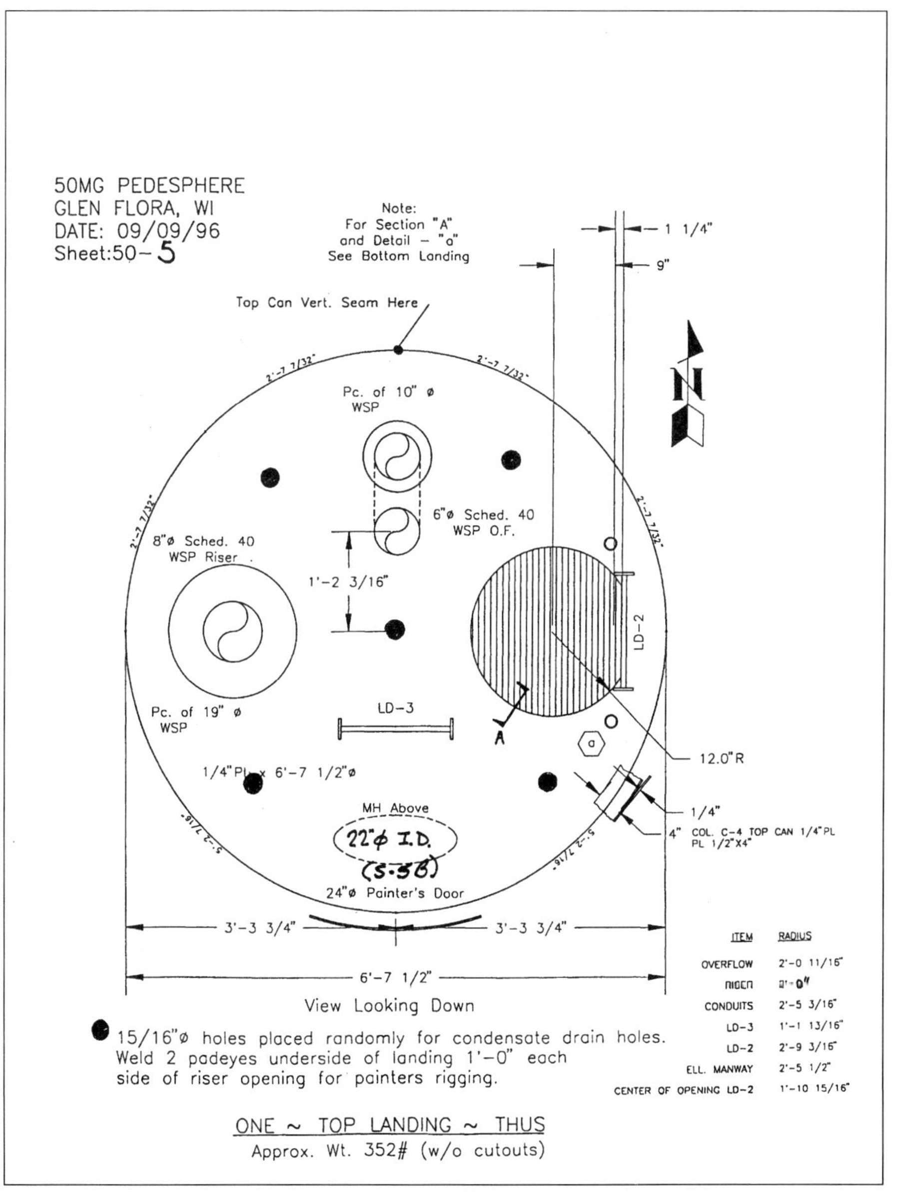

Source: Maguire Iron, Inc.

示例：高架水箱加工图

Sample: Shop Drawings for Elevated Water Storage Tank *(continued)*

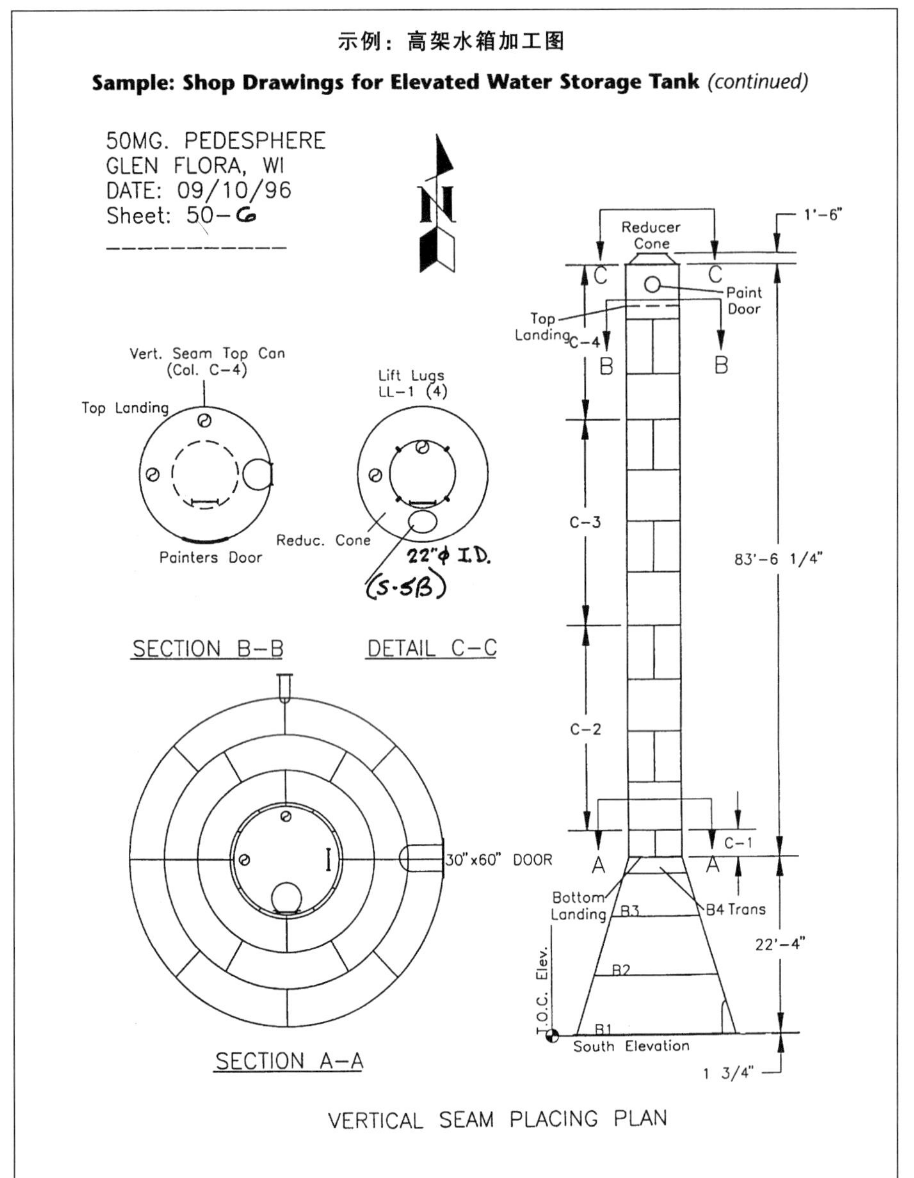

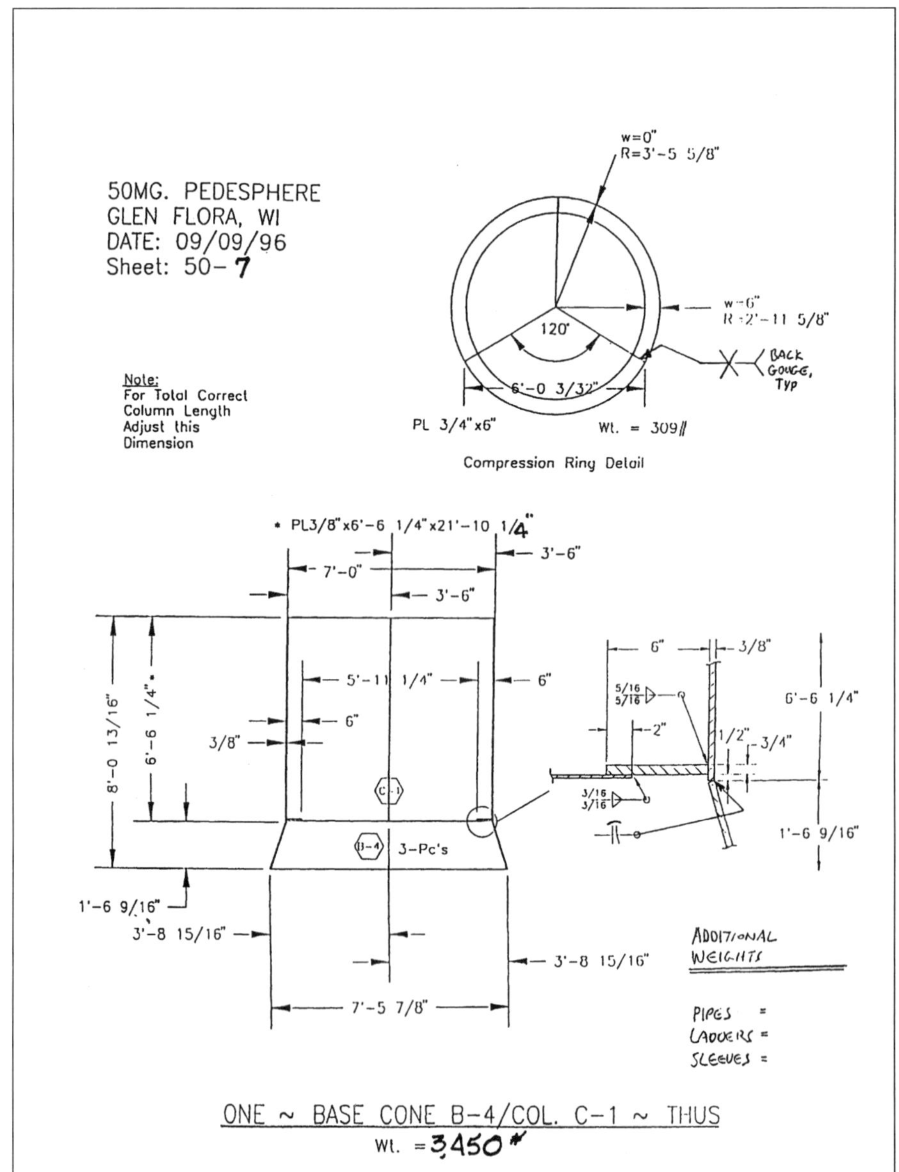

示例：高架水箱加工图

Sample: Shop Drawings for Elevated Water Storage Tank *(continued)*

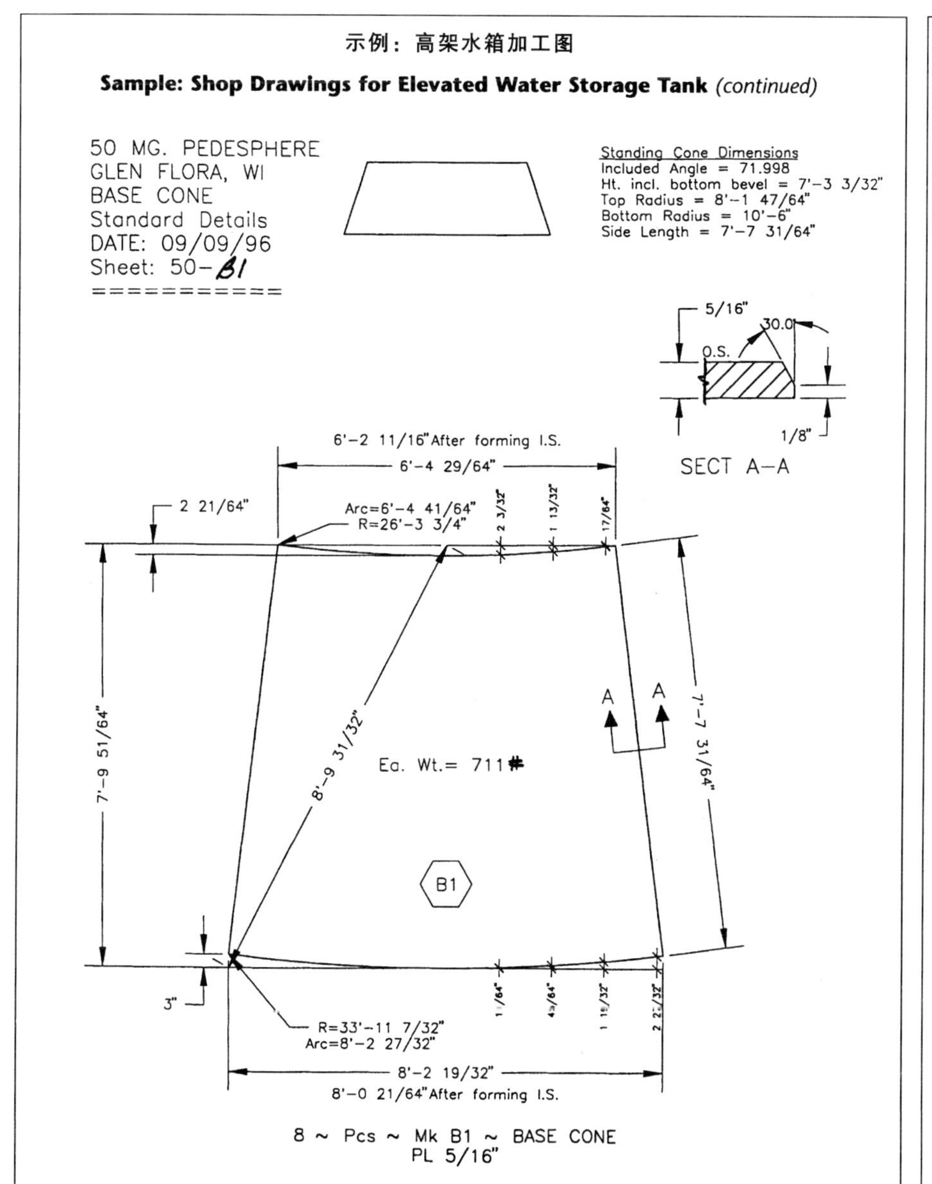

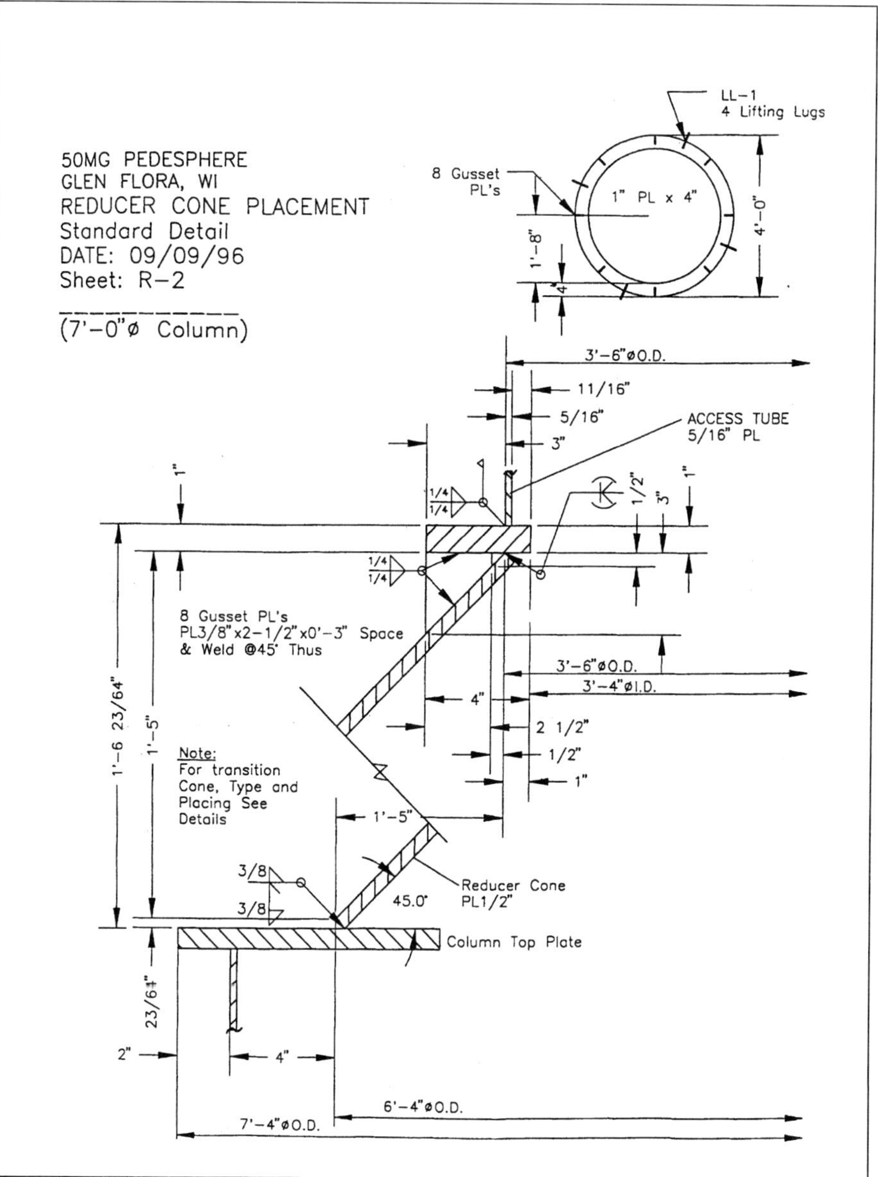

工程师的审核章

Engineer's Review Stamp

Review by the Engineer shall not be construed as a complete check of the submitted material nor shall it relieve the contractor from his responsibility to comply with the specifications.

Initials ________ Date ______ Reviewed: no exceptions noted

Initials ________ Date ______ Reviewed: exceptions noted

证实审核的一般方法

Typical Method of Certifying Review

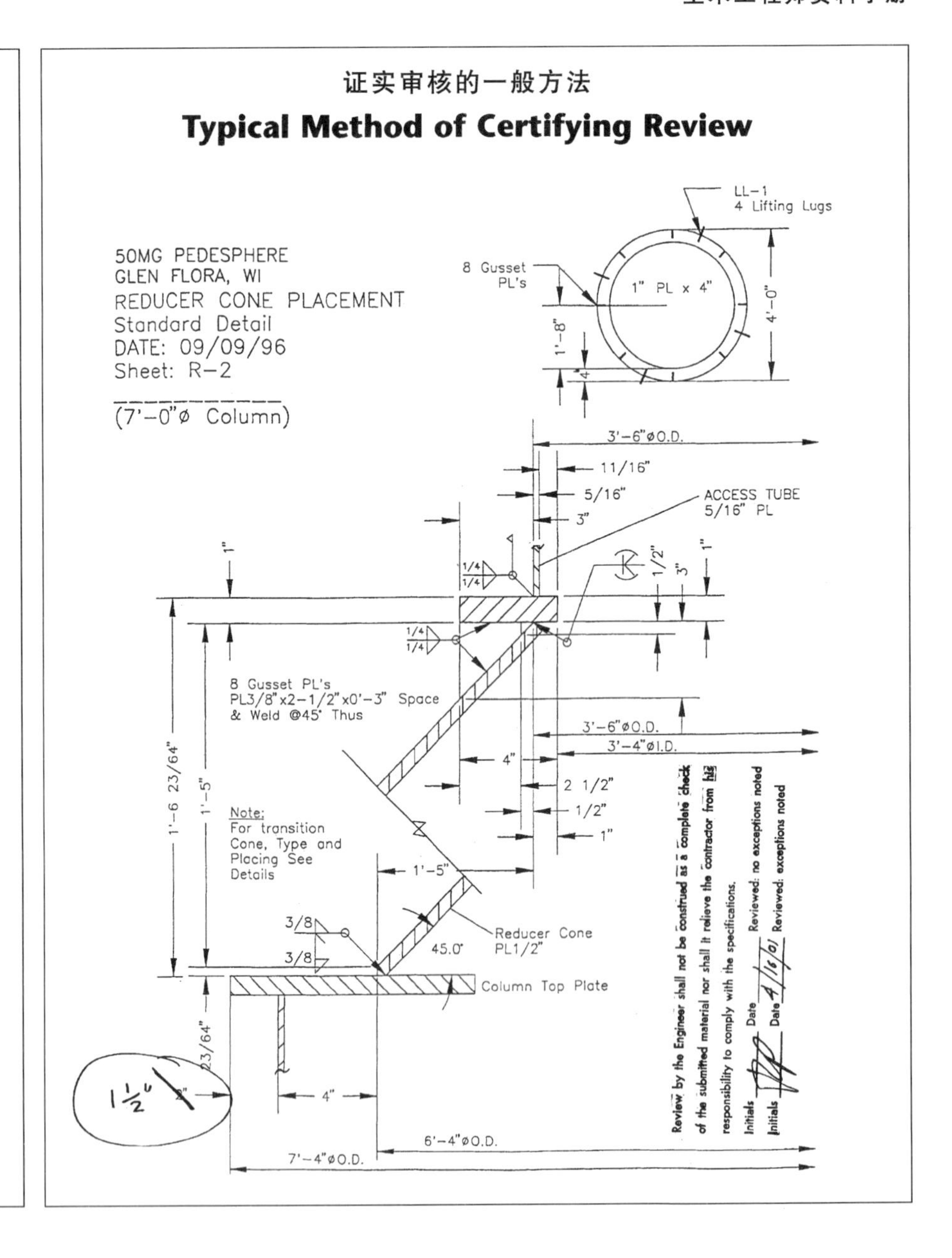

Source: Morgan & Parmley, Ltd. & Maguire Iron, Inc.

证实审核的一般方法

Typical Method of Certifying Review *(continued)*

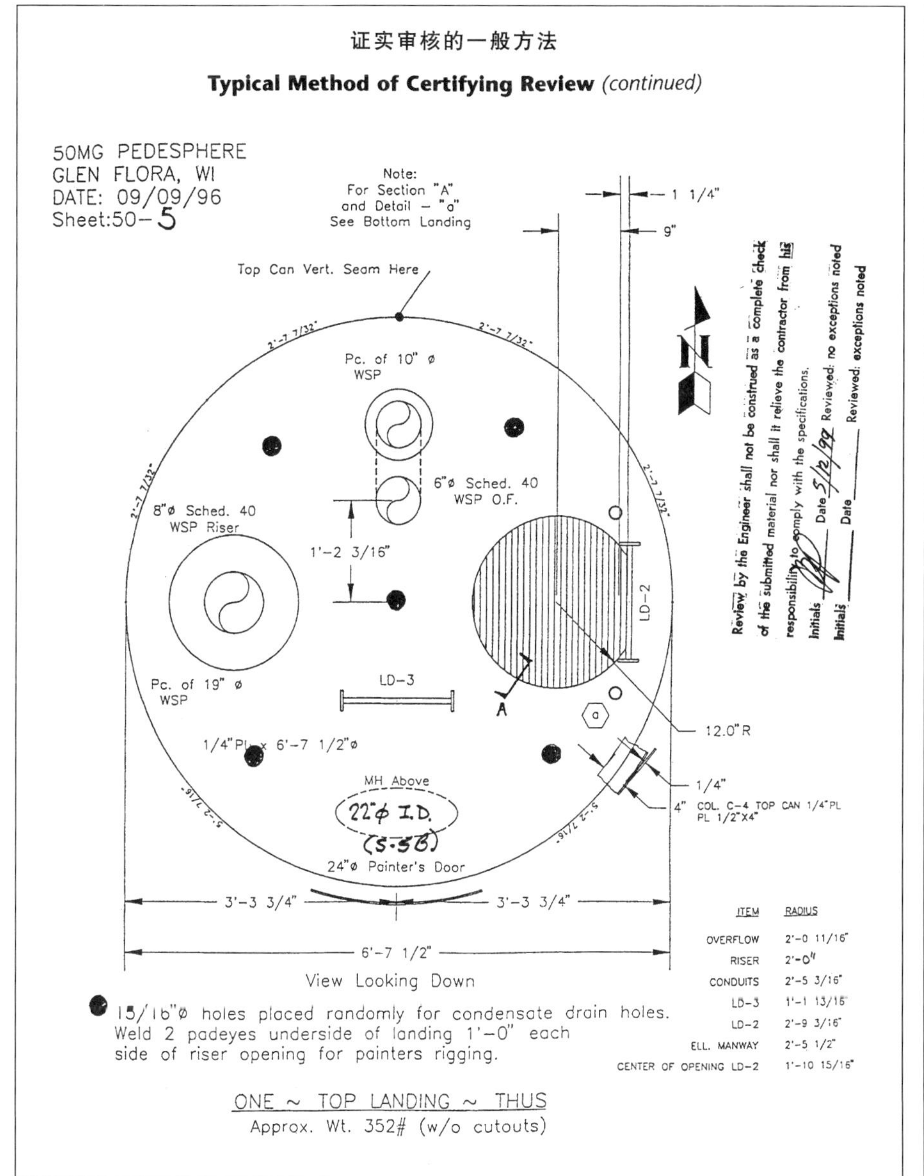

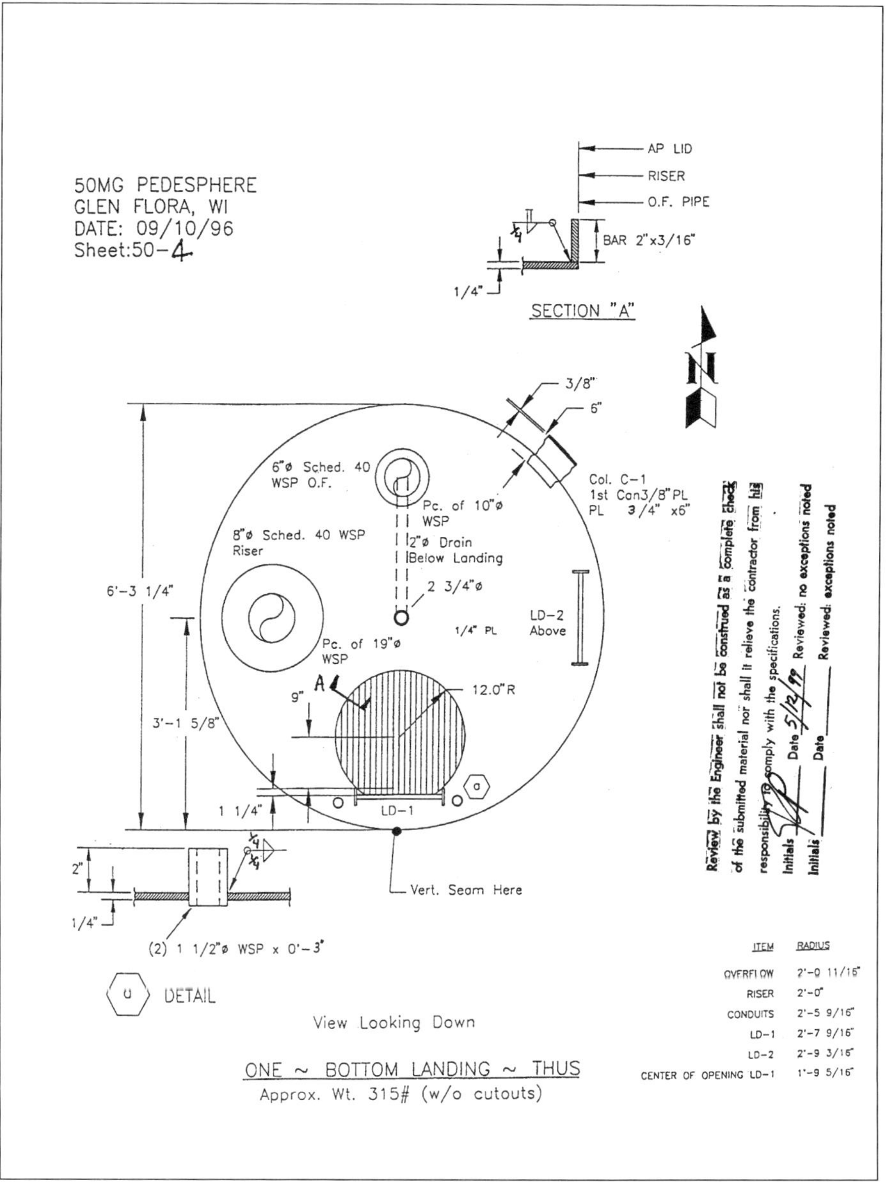

施　工

第28节　安全

例　28

千万不能低估施工现场的安全问题。在执行任何实际活动之前，非常重要的是在适当的位置采取了充分的安全措施。用于保护公众的恒久的警示和适当的栅栏以及对工作人员的保护是其中的关键措施。

本节的目的在于记录施工中这一重要因素，并将读者的注意力引导至一些基本原理上。我们包括了一些一般条件的参考信息。然而，这些材料远远不是无所不包，对于每个特定施工项目，我们建议读者咨询相关的管理机构、专业组织、地方、州和联邦法律和美国劳工部职业安全与健康署(OSHA)规范来了解相关的法规。一些在本质上是通用的，另一些则只适用于特定的施工类型。

安全语言必须能被所有人了解，不论他的经验、年龄、背景或者种族。施工现场的安全是每天都要关注的，并且需要持续的警惕。所以，安全是构成施工整体活动的一部分，必须被整合进项目执行的组织中来。

强烈建议召开针对所有人员的常规安全会议。对安全程序的经常性提醒能够避免很多问题。成功的事故防范需要团队协作。

对建筑工人的急救训练是从业的前提性条件，特别是对于某些比较危险的项目。

目　录

氧缺乏的潜在影响

Potential Effects of Oxygen-Deficient Atmospheres

Oxygen Content (% by Volume)	Effects and Symptoms (At Atmospheric Pressure)
19.5%	Minimum permissible oxygen level.
15-19%	Decreased ability to work strenuously. May impair coordination and may induce early symptoms in persons with coronary, pulmonary, or circulatory problems.
12-14%	Respiration increases in exertion, pulse up, impaired coordination, perception, judgment.
10-12%	Respiration further increases in rate and depth, poor judgment, lips blue.
8-10%	Mental failure, fainting, unconsciousness, ashen face, blueness of lips, nausea, and vomiting.
6-8%	8 minutes, 100% fatal; 6 minutes, 50% fatal; 4-5 minutes, recovery with treatment.
4-6%	Coma in 40 seconds, convulsions, respiration ceases, death.

These values are approximate and vary as to the individual's state of health and his physical activities.

Source: Scott Instruments

接触硫化氢的潜在影响

Potential Effects of Hydrogen Sulfide Exposure

PPM*	Effects and Symptoms	Time
10	Permissible Exposure Level	8 Hours
50-100	Mild Eye Irritation, Mild Respiratory Irritation	1 Hour
200-300	Marked Eye Irritation, Marked Respiratory Irritation	1 Hour
500-700	Unconsciousness, Death	½-1 Hour
1000 or More	Unconsciousness, Death	Minutes

These values are approximate and vary as to the individual's state of health and his physical activity.

Source: Scott Instruments

接触一氧化碳的潜在影响

Potential Effects of Carbon Monoxide Exposure

PPM*	Effects and Symptoms	Time
50	Permissible Exposure Level	8 Hours
200	Slight Headache, Discomfort	3 Hours
400	Headache, Discomfort	2 Hours
600	Headache, Discomfort	1 Hour
1000-2000	Confusion, Headache, Nausea	2 Hours
1000-2000	Tendency to Stagger	1½ Hours
1000-2000	Slight Palpitation of the Heart	30 Min.
2000-2500	Unconsciousness	30 Min.
4000	Fatal	Less Than 1 Hour

These values are approximate and vary as to the individual's state of health and his physical activity.

Source: Scott Instruments

潜在的危险区域

Potentially Dangerous Areas

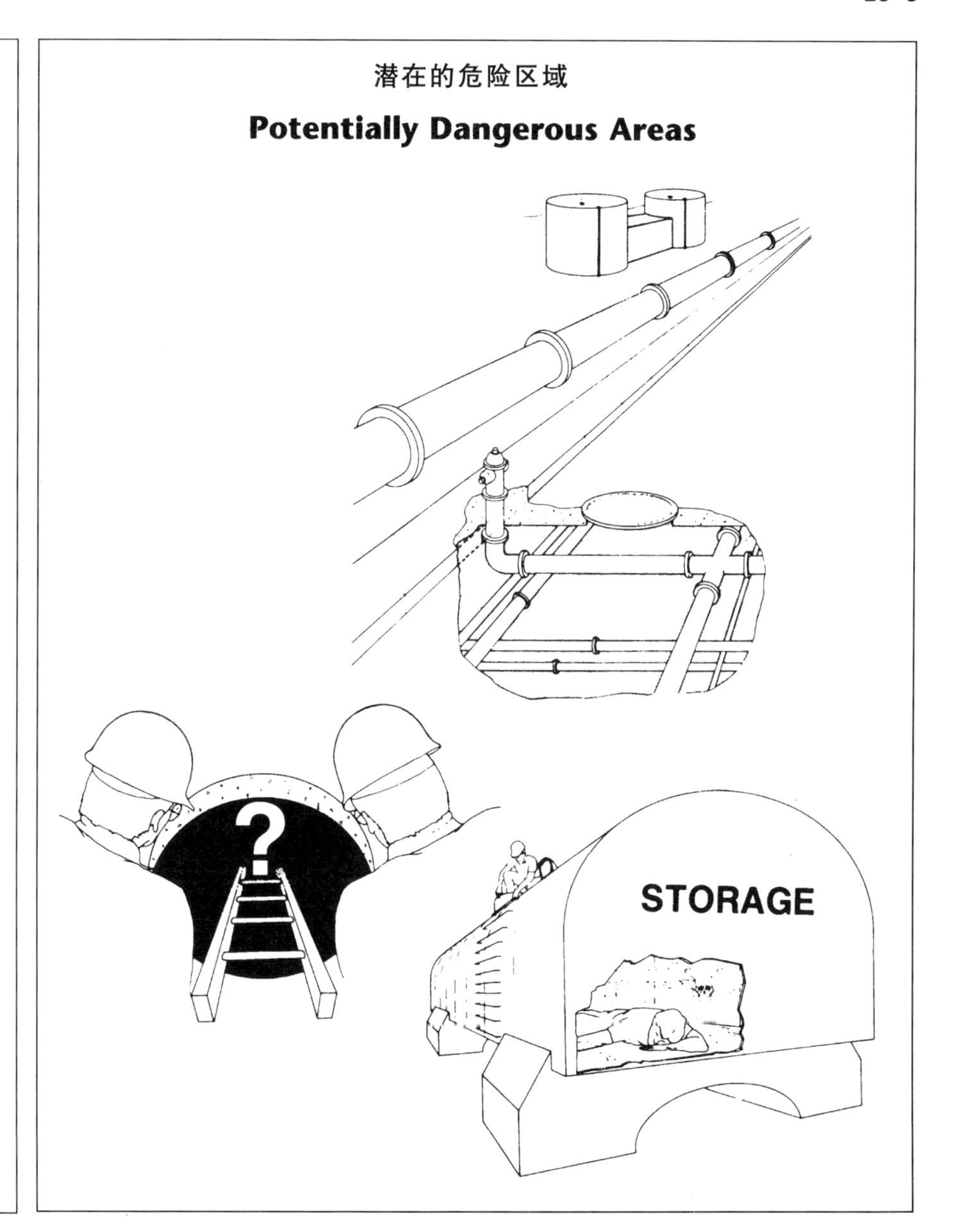

Source: Scott Instruments

潜在的危险区域

Potentially Dangerous Areas *Continued*

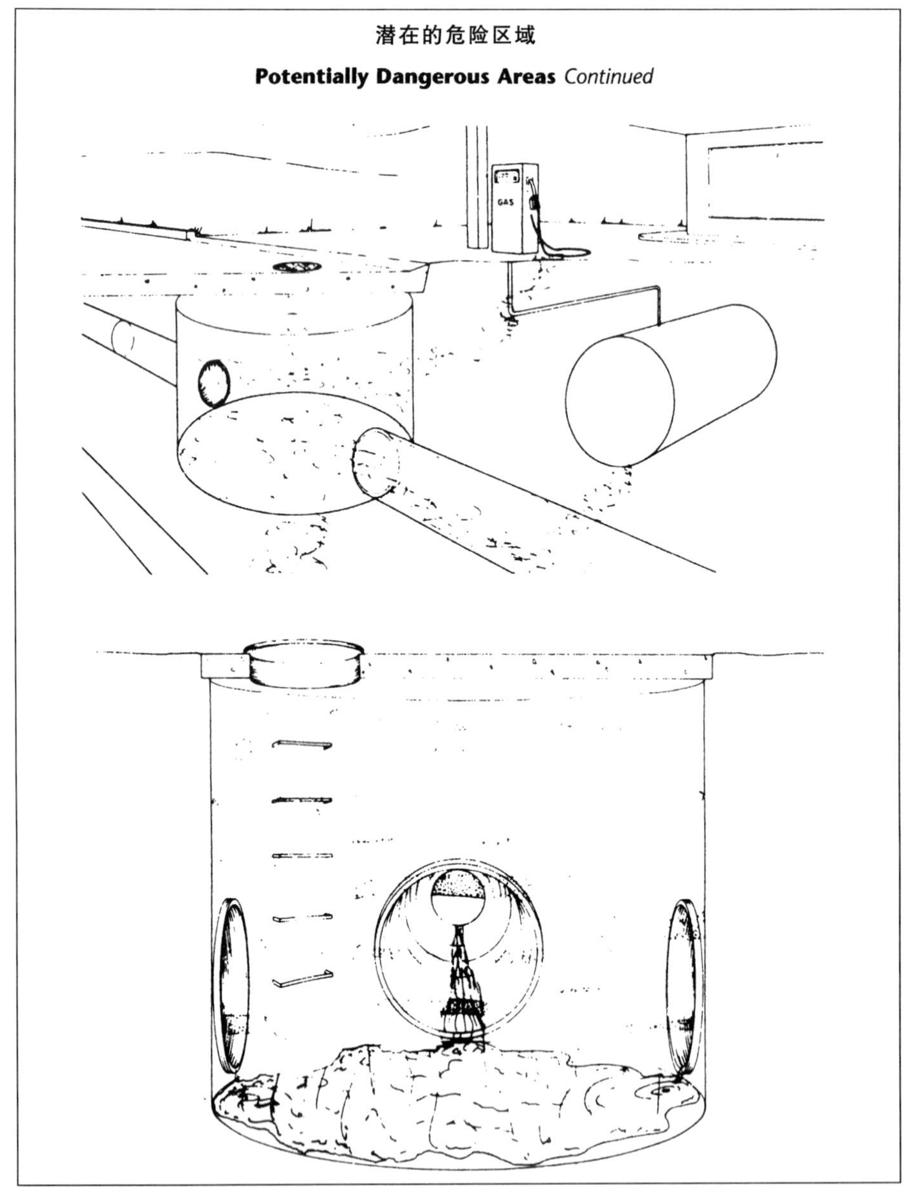

入口前的空气监测

Monitor Atmosphere Before Entry

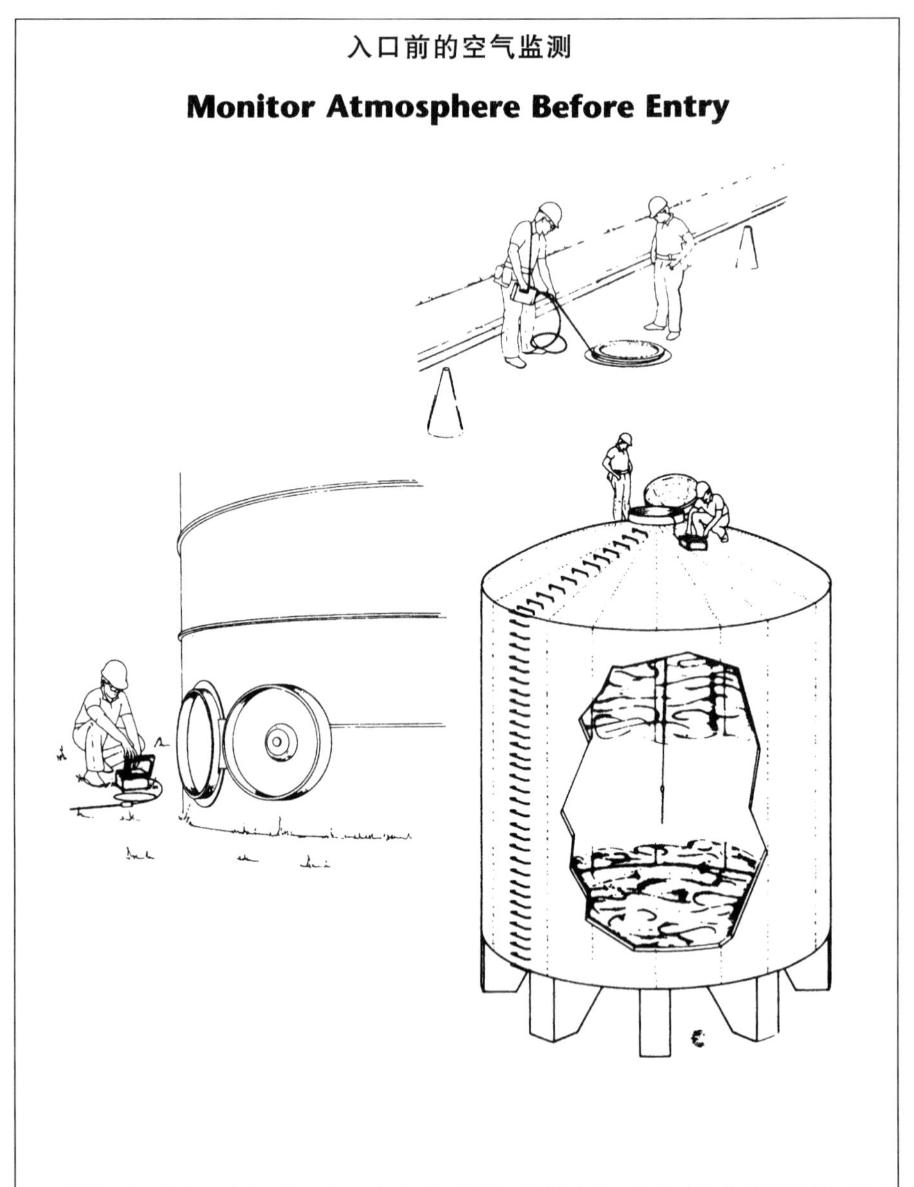

Source: Scott Instruments

施工期间对工作场所进行监测

Monitor Work Area During Construction

Source: Scott Instruments

限制入口的基本设备

Basic Equipment for Confined Entry

1 Gas meter
2 DO meter
3 PH meter
4 First-aid kit
5 Thermometers
6 Blower
7 Generator
8 Safety harness
9 Fire extinguishers
10 Tripod and winch
11 Lanyard with spreader bar
12 Airline respirator
13 Traffic cones
14 Safety vests
15 Air tank
16 Rope ladder
17 Cell phone
18 Protective coveralls
19 Rubber boots and gloves
20 Safety glasses or goggles

Source: R.O.Parmley, Field Engineer's Manual, 3/e © 2002 Published by McGraw-Hill and reproduced with permission of the McGraw-Hill Companies.

交通控制规划示例

Sample Traffic Control Plan

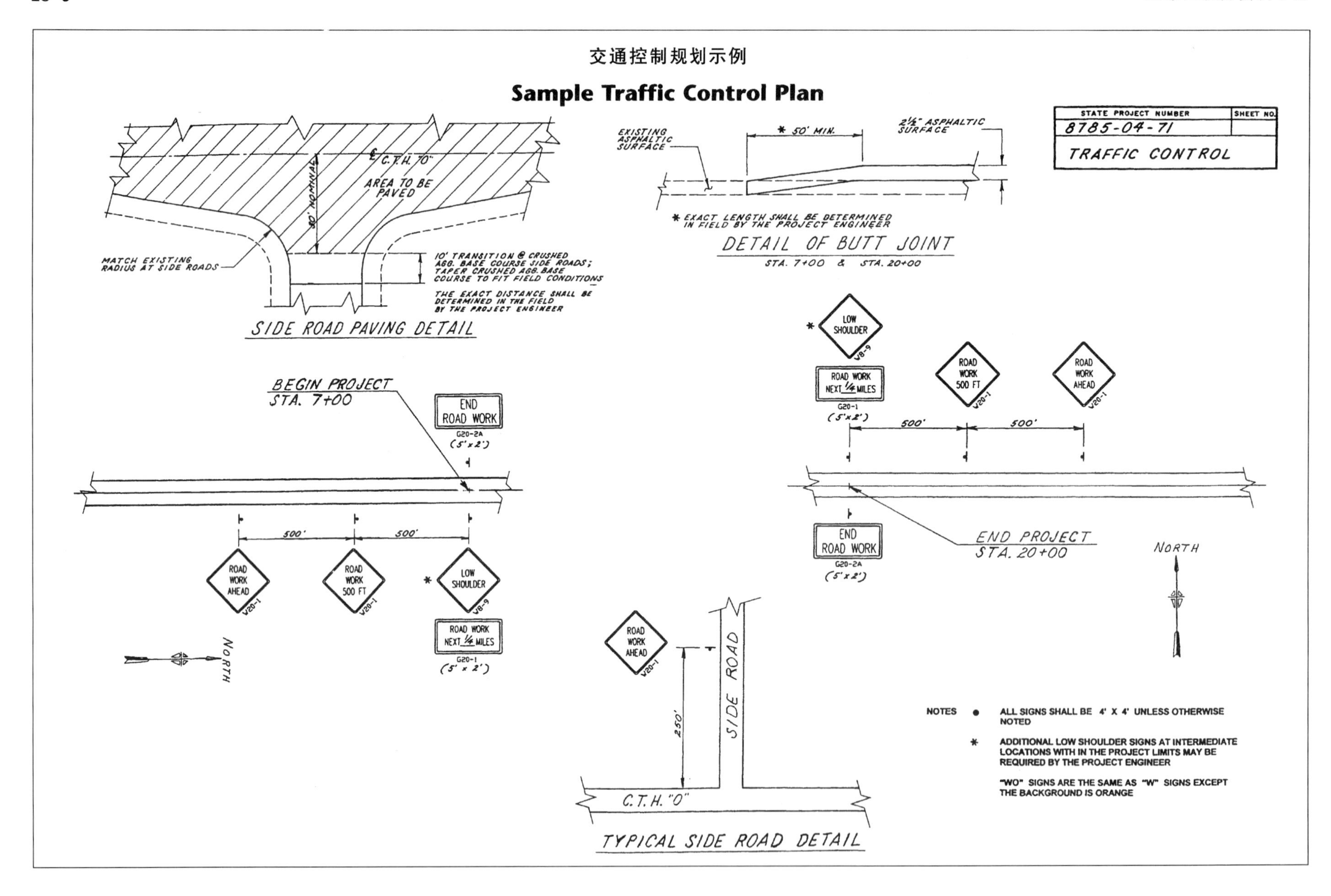

Source: Morgan & Parmley, Ltd.

Crane Signals　　起重机标志

From: Pittsburgh DesMoines Steel Co., Pittsburgh, PA

First Aid Suggestions　　急救建议

Wear a Med-A-Lert tag at all times if you have any allergies to a particular medication or have a disease, e.g., diabetes, that requires special attention.

Store the majority of items listed in the table below in a dust and waterproof container. Label the container "First Aid Kit." Place a list of contents on the inside of the first aid kit. Replenish items after each use. The kit should be carried in field vehicle; i.e., inspector's truck and survey crew's van.

Quantity	*Item*	*Use*
12	4 x 4 Compresses	Dressings for wounds and burns
12	2 x 2 Compresses	
1 box	Assorted Band-aids	
3	40″ Triangle bandages	Slings; to hold dressings in place
10 yards	1″ Roller gauze or flexible gauze	Slings, open wounds, or dry dressings for burns
10 yards	2″ Roller gauze or flexible gauze	
10 yards	4″ Roller gauze or flexible gauze	
1	2″ Roll elastic bandage	For sprains; to hold dressings in place
5 yards	1″ Adhesive tape	To hold dressings in place
5 yards	2″ Adhesive tape	
2	Small bath towels or large Pampers	Dressings for large wounds or burns
2	Eye pads	For eye injuries
1	Small bar of soap or bottle of liquid soap	Antiseptic for cleansing wounds
1	Single-edge razor blade, individually wrapped	For snake bite
1	Small container with needle, thread, and safety pins	To remove slivers; do emergency repair; fasten dressings
1	Flashlight with extra batteries	
1	Book matches	To sterilize needle (above)
1	Clean sheet, individually wrapped	For dressing large burns; splinting
1	Blanket	For transporting or shock
1	Cup and spoon	For giving fluids or for flushing wounds
1	First-Aid Book	
1	Tweezers	For removing foreign objects, insect stingers
1	Pocket knife	To cut splints
1	Scissors	To cut bandages
1	Small package salt	For shock, heat exhaustion (to be mixed with water)
1	Small package soda	
1	8″-long Tourniquet twist stick	As a last resort in bleeding; for snake bite
2	Small plastic bag	Container for ice or snow; for sealing chest wounds
1	1 Bottle antiseptic (Consult physician on selection and follow his or her recommendation.)	For general use

Note: The kit should include the latest publications of the American Safety & Health Institute (Emergency First Aid Information; CPR, Bleeding, Snake Bite, Etc.)

施　工

第29节 检查和测试

例 29

对于任何一项施工活动，监理都是其中的关键人员，而且是项目取得成功所必需的。他们是专业设计师、项目工程师和业主在工作现场的眼睛和耳朵。他们对细节的关注和监视能够确保项目按照设计图进行施工，并做精确的日志记录。

每个项目都将有其需要进行测试的特定领域。监理通常参与到这个过程中并见证这些事件。如果需要抽样，监理有责任进行记录和保管。

施工结束后，监理向项目工程师提交施工日志、记录、照片和取样，作为项目永久档案。

目 录

监理的技能

监理是专业设计师（建筑师或者工程师）在现场的“眼睛和耳朵”。监理的工作可能是施工中最艰巨的任务，他们从始至终被委以监视实际施工，确保按照设计图和技术说明生产最终产品。

理想的监理应该受过广泛的技术教育，而且是所监理施工领域经验丰富的技术人员。

监理的责任

每个项目都有其自身的特定法规和规范，取决于所涉及的组织和管理机构。一般来说，专业监理工程师或建筑师将给出方针，详细规定了监理的委托责任和权力范围。然而，有一些基本责任是所有监理都应被告知承担的。它们如下：

1. 了解监理的职责对于保证最高施工质量的重要性。
2. 拥有良好的交流技巧和技术知识，完全了解项目的设计图纸和技术说明。
3. 拥有合理、务实的判断力，并有良好写作和制图技巧。
4. 了解监理职责的范围。
5. 认可监理职责。
6. 不能逾越监理的权限。
7. 领会主管（建筑师和工程师）的意见和目标。
8. 知道将所有重大事件精确记录到监理日志的责任，包括所有相关的口头指令和合适的会谈。
9. 无论何时，监理都应以一种专业的方式处理问题，表现出礼貌、耐心并与各方协作确保团队的努力取得项目完成的成功。

监理的权力

监理有权和指令来向承包商指出与设计图和技术说明不同的地方。然而，监理却没有相关权限来允许这些变更。

在特定方针指导下开展工作，并受到规范中法律部分的约束，监理充当了施工阶段中的一个关键角色，因为他（或她）监视每天的工作；如实地记录所有活动作为档案。

监理的基本职责

基本职责通常适用于所有的监理，如下所述：

1. 在施工开始之前，全面地审视图纸。向项目建筑师和工程师汇报所有错误、疏漏、矛盾和缺失。所有需要进一步解释的部分要在施工之前由建筑师和工程师来澄清。
2. 保存一套完整的建筑施工图、技术说明和合同文件作为个人在工作时使用。在设计图纸上用“红”笔记录所有相关的现场记录，这样就能保证在工作完成时也就准备好了所谓的“竣工图”。
3. 获取所有加工图的完整档案，并在安装之前将它们与运送到工作现场的实际设备和物资进行对比。
4. 保存一本整洁和组织良好的日志簿，准确地记录所有遇到的相关项目活动。
5. 总是对可能的索赔和未来的纠纷情形保持警惕。当可能的索赔或者纠纷正在酝酿时，通报给项目建筑师和工程师并将有关的事实准确地记录进日志簿。

6. 保持与所有可能延误正式进度计划的情况齐头并进。定期向项目建筑师和工程师汇报任何可能影响施工按时完成的情况。
7. 如果可能，监管劳动标准并核实官方的工资率、工作时间和工资单记录。
8. 保留现场材料清单的副本，并监视储藏设施，注意其安全措施、供暖和通风，以及应对恶劣天气的保护措施。
9. 协助项目建筑师和工程师审核承包商的付款请求，确保付款请求所包含的工程已经准确的按照认可的设计图和技术说明完成了。
10. 在身边保留所有确保项目顺利实施所需要的项目许可、审核批准、通行权、切片图纸和其他档案。
11. 在所有关键阶段拍摄相关照片。用拍立得照相机确保照片记录了关键区域。做一个摄影日志，并在必要时可采用带录音功能的摄像机。应该意识到，带有文字/草图补充的照片对于将来使用具有不可估量的价值。如果拿不准时，千万不要犹豫，拍照片总是有益的。
12. 大量地使用录音机。在随后的时间里将这些口头资料转记入日志簿。
13. 将总是对项目保持公平和公正作为你的首要目标。
14. 千万不要做任何假设。如果拿不准时，查阅设计图、技术说明、招标文件、合同、加工图、档案、审查批准、许可或者其他相关文件；随时在头脑中记得，如果问题不能在项目文件中找到就必须咨询项目建筑师和工程师。

监理指南

有几本优秀的关于施工监理的手册已出版，建议所有参与监理的工作人员拥有一本。本手册篇幅有限，不能完整叙述和详细介绍施工监理的完整而复杂的范畴。所以，本书不包括具体的指导意见。

安　全

要尽可能保持项目现场施工区域的有序和安全，从而保障工人的安全。良好的现场管理总是非常重要，这也是监理的观察和总管能发挥巨大作用的地方。

对于大型的施工工作，安全不是监理的职责。一般来说，一个安全工程师和足够的职员被指定来承担这些职责，强制执行美国劳工部职业安全与健康署(OSHA)标准和类似管理法规。然而，对于任何潜在的危险情况，监理在警告相关责任人时仍然充当了关键角色。大量的人员受伤事故都是由粗心地处理碎屑引起的，而警觉的监理的警告和压力能够改善工作区域的条件。

另外，经验丰富的监理总能留意到这样的要求：在任何实际开挖活动之前，对埋置在地下的设施做好合适的标记。

Event Calendar – (Basic Engineering Tasks Summary) 工作日程表

Item No.	Description	Projected Dates
1	Define project scope	
2	Client authorizes engineers to begin	
3	Initial conference: client/engineer	
4	Research records and files	
5	Field survey and reconnaissance	
6	Video photograph project site	
7	Buried utilities located and flagged	
8	Research property survey records	
9	Soil borings, test drilling, and subsurface investigation	
10	Committee meeting—refine project scope	
11	Reduce field notes	
12	Preliminary budget—construction cost estimates	
13	Committee meeting—review and approve preliminary concept	
14	Preliminary budget—construction cost estimates	
15	Preliminary plans and specifications	
16	Client meeting—review and approve preliminary P & S	
17	Easement survey and legal documents (if required)	
18	Final plans and specifications	
19	Engineering report (summary of work)	
20	Proposal and bidding documents	
21	Obtain concurrence from client for final P & S	
22	Submit plans and specifications to regulatory agencies	

Item No.	Description	Projected Dates
23	Obtain approvals from regulatory agencies	
24	Department of Natural Resources	
25	Public Service Commission	
26	Department of Transportation	
27	Corps of Engineers (if required)	
28	Private utility company approvals	
29	Water utility	
30	Sewer utility	
31	Railroad company	
32	Gas company	
33	Cable TV company	
34	Electrical company	
35	Telephone company	
36	Public hearing (assessments to property owners) (if required)	
37	Organize bidding process (commence)	
38	Obtain wage rates (state and/or federal)	
39	Advertise for bids	
40	Supply plans and specs to Builders Exchange	
41	Send plans and specifications to contractors	
42	Prebid meeting	
43	Bid opening	
44	Minutes, bid tabulation, and analysis of proposals	

工作日程表

Event Calendar – (Basic Engineering Tasks Summary) *Continued*

Item No.	Description	Projected Dates
45	Recommendation to client	
46	Client gives notice of award	
47	Prepare construction contract documents	
48	Secure client attorney's certification of construction contract	
49	Obtain client resolution to sign construction contract	
50	Supervise signing of construction contract	
51	Preconstruction conference	
52	Client signs notice to proceed	
53	Shop drawing review	
54	Construction staking	
55	General inspection services, photos, daily log, etc.	
56	Construction records	
57	Contractor payment requests review and certification	
58	Status reports (periodic)	
59	Inspection and testing of installation	
60	Prepare punch list	
61	Final inspection and certification	
62	Preparation of construction record drawings (as built)	
63	Final payment request review and certification	

Note: The projected dates are targets and may vary as a result of conditions beyond the control of the client, engineer, and contractor(s).

Inspector's Equipment List 监理的设备清单

Briefcase
Plans and specifications with contract documents
Project log book
Video camera
35-mm camera with spare film
Polaroid camera with spare film
Project file
List of key project personnel
Event calendar
Project milestone list
Proper personal ID
A/E scales
Magnetic locator
Divining rods
Magnifying glass
Keys for project locks on restricted area
Clipboard
Table, pencils, markers, pen, etc.
Tape recorder
Carpenter's hand level
6-ft folding ruler (fractional/decimal)
Cloth tape (100-ft English/metric)
Electronic calculator (slide rule design)
Measuring wheel (metric/English)
First aid kit
Watch with second hand
Safety glasses
Safety cones
Specific testing equipment
Rain gear
Orange vest
Hard hat
Duct tape
Paper toweling
Tool kit
Coveralls
Rubber boots
Rubber gloves
Surveying equipment (if required)
Cellular telephone
Spray paint (red, green, blue, and white)
Confined entry safety equipment (if applicable)
Containers for collecting samples
Traffic cones
Flags
Surveyor's ribbon
Field vehicle
Two-way radios

Miscellaneous Nomenclature 其他术语

A/E	Architect/engineer
BB	Distance back-to-back
BC	Bottom chord
BF	Board feet
BM	Bench mark
BOD	Biochemical oxygen demand
BS	Back sight
C	Celsius
CC	Distance center-to-center
CDF	Creosoted Douglas fir (pressure-treated)
CF	Cubic feet
CFS	Cubic feet per second
CL	Centerline
COD	Chemical oxygen demand
CY	Cubic yard
D	Diameter
EL	Elevation
F	Fahrenheit
GPC/D	Gallons per capita per day
GPD	Gallons per day
GPM	Gallons per minutes
HI	Height of instrument
HL	Head loss or heat loss
HP	Horsepower
HWM	High-water mark
IE	Invert elevation
J	Joule
KIP	1000 lb
L	Liter
M	Meter
MGD	Millions of gallons per day
MH	Manhole
PC	Point of curve
PI	Point of intersection
PL	Property line
P & S	Plans and specifications
PT	Point of tangent
Q	Quantity
R	Radius
RC	Reinforced concrete
RW	Redwood
R/W	Right of way
S	Slope (grade)
SF	Square feet
SS	Suspended solids
STA	Station
SY	Square yards
T	Tons
TC	Top chord
TSS	Total suspended solids
UF	Untreated fir
V	Volt
VC	Vertical clearance
W	Watt
WL	Water level

Source: R.O.Parmley, Field Engineer's Manual, 3/e © 2002 Published by McGraw-Hill and reproduced with permission of the McGraw-Hill Companies.

收集混凝土试样

基本要求：如果试样是从模板中铲取的，那么不要收集混凝土堆表面的水面上混凝土。如果试样取自混凝土罐车或者搅拌器卸出的混凝土注流，除接近开始和结束时外，则在三个不同时间间隔取样。

试样重量应在100磅左右。将其用两个桶提到圆柱试件制作和存放的地方。将试样在手推车或金属盘中用铲子混合并搅拌，确保灌注进模具之前搅拌均匀。

圆柱试件浇注：使用带底板的直径为6英寸，高12英寸的钢制、塑料或者石蜡纸模具。将模具放置在光滑、坚固、表面平整的表面，随后三层基本等厚的混凝土，每层用直径为5/8英寸的钢筋插捣25次。确保轻微地刺透到前一层。在插捣后，敲击模具的侧面以填充任何空隙。第三层应该有一定的余量，在插捣后才能保持水平表面。通常需要制作三个圆柱试件用于测试。

填写数据表格，完整地描述混合物和位置。然后，将一个带有数据表格副本的信封贴在每个圆柱模具的侧面，并用塑料袋封口以防止水分挥发。

圆柱试件的处理：24小时内不能移动和扰动圆柱试件。将它们保存在温度为60－80℉的受保护的区域。然后将它们包裹在湿润的锯屑中，防止失水并提供保护，再运送到实验室。

坍落度试验：在不能吸水表面重新混合混凝土。将圆锥容器用三层等体积混凝土填充，并用直径为5/8英寸的钢筋对每层插捣25次。在提起圆锥容器之前，去除底部周围多余的混凝土。要非常小心，确保坍落度圆锥被竖直提起。在圆锥被提起后，混凝土的下陷量（单位为英寸）就叫“坍落度”。用直尺测量这个数据。如果模具中的混凝土完全跌落，撤消这次测试，重新开始。

使用的截头圆锥顶部直径4英寸，底部直径8英寸，高12英寸，用16级金属板做成。确保坍落度圆锥经过清理和预湿。

Source: R.O.Parmley, Field Engineer's Manual, 3/e © 2002 Published by McGraw-Hill and reproduced with permission of the McGraw-Hill Companies.

细骨料和粗骨料混合物的筛分析

Sieve Analysis for Mixture of Fine & Coarse Aggregates

Project and/or Project No. Morgan & Parmley / 02LAB

Deposit Identification Filter Sand

Contract | County Rusk

Specifications

Contractor and/or Producer

Sample No. 1

☐ Crushed Stone ☐ Base Course ☐ Grade 2

☐ Crushed Gravel X Other X Other

☐ Blend Filter Sand Special

Materials accepted at: ☐ Belt Stockpile ☐ Roadway

Date: 7/23/02

Sampled at Location (pit, grid, station, etc.)

Time

Comments

MOISTURE CONTENT

Weight of Sample (moist): 1041

Weight of Sample (dry): 1011

Moisture Loss: 30

% Moisture: 3.0%

Weight of Total Sample (dry, unwashed) 1011

Weight of R-4 (dry, Unwashed) 1011 = 1.000

Weight of P-4 (dry, unwashed) = 0.000

R-4 MATERIAL — Washed: ☑ Yes ☐ No

P-4 MATERIAL — Washed: ☐ Yes ☐ No — Wt. = (minimum 500 grams)

TOTAL MATERIALS (% Passing)

Sieve	Weight Retained	% Retained	% Pass (C)	Weight Retained	% Retained	% Pass (D)	R-4 (A*C)	P-4 (B*D)	Dry Sieved	Corr. Factor	Corr. or Washed Results	Spec.
3/8"	0	0.0	100.0								100.0	100
4	246	24.3	75.7								75.7*	77-100
8	574	56.8	43.2								43.2*	53-100
16	847	83.8	16.2								16.2	15-80
30	946	93.6	6.4								6.4	3-50
50	989	97.8	2.2								2.2*	0-1
100	1003	99.2	0.8								0.8	0-1
200	1006	99.5	0.5								0.5	0-1
pan	2											

R-4 FRACTURE COUNT

Fracture Particles:

Total Particles:

% Fracture:

PLASTICITY CHECK

Can P-40 be rolled into 1/8" thread when moist?

☐ Yes ☐ No

Weight/c.y. =

TEST RESULTS

☐ Passed X Failed

Morgan & Parmley

Sampled by:

Project Engineer: Morgan & Parmley

Tested by: B.Walker

Date Tested: 7/24/02

Source: Morgan & Parmley, Ltd.

混凝土抗压强度报表

Report Form of Concrete Compressive Strength

REPORT OF CONCRETE COMPRESSIVE STRENGTH

Project:

Contractor:

Client:

Date:

CEC File No.:

FIELD DATA

Location of Placement:	______	Specified Strength:	______
		Supplier:	______
Method of Placement:	______	Mix Number:	______
Date Placed:	______	Cement (lbs):	______
Time of Tests:	______	Fly Ash (lbs):	______
Slump (in):	______	Water (lbs):	______
Air Content (%):	______	Fine Aggregate (lbs):	______
Concrete Temperature:	______	Coarse Aggregate #1 (lbs):	______
Weather Conditions:	______	Coarse Aggregate #2 (lbs):	______
Date Received in Lab:	______	Admixture #1 (oz):	______
Field Data Submitted by:	______	Admixture #2 (oz):	______

LABORATORY RESULTS

Test Method: ASTM C39-86

Set No.	Break Type	Age (days)	Load (lbs)	Strength (psi)

COMMENTS

☐ Cylinder cast by CEC

☐ Cylinders picked up by CEC

☐ Cylinders meet project specifications

☐ Cylinders cast by Contractors or Architects Representative

☐ Cylinders delivered to CEC laboratory

☐ Cylinders do not meet project specifications

Source: Cooper Engineering Co., Inc.

核试验现场数据表

Nuclear Testing Field Data Sheet

CLIENT:	PROJECT NAME:
CLIENTS PROJECT #:	COOPER PROJECT #:
CONTRACTOR:	
TESTING METHOD:	GAUGE #:
DATE:	TEST PERFORMED BY:

Test Number	Station, Offset Northing, Easting Grid Lines-x,y Other (circle one)	Approx. Elevation	Dry Density (pcf)	Moisture (percent)	Wet Density (pcf)	Maximum Density (pcf)	Proctor Sample Number	Relative Compaction (percent)	Spec's Met?

Source: Cooper Engineering Co., Inc.

核试验电子表格

Nuclear Testing Data Spreadsheet

CLIENT:	PROJECT NAME:
CLIENT'S PROJECT #:	COOPER PROJECT #:
CONTRACTOR:	DATE:
TESTING METHOD:	GAUGE #:
LOT #'s:	TESTS PERFORMED BY:

Date Placed	Date Tested	Test Number	Station, Offset	Lower or Upper Course	Mainline or Shoulder	Maximum Density (pcf)	Density Count	Wet Density (pcf)	% Maximum Density

Lot #: Average % Maximum Density = Specified Density =

☐ This lot meets project specifications. ☐ This lot does not meet project specifications.

Date Placed	Date Tested	Test Number	Station, Offset	Lower or Upper Course	Mainline or Shoulder	Maximum Density (pcf)	Density Count	Wet Density (pcf)	% Maximum Density

Lot #: Average % Maximum Density = Specified Density =

☐ This lot meets project specifications. ☐ This lot does not meet project specifications.

Date Placed	Date Tested	Test Number	Station, Offset	Lower or Upper Course	Mainline or Shoulder	Maximum Density (pcf)	Density Count	Wet Density (pcf)	% Maximum Density

Lot #: Average % Maximum Density = Specified Density =

☐ This lot meets project specifications. ☐ This lot does not meet project specifications.

Date Placed	Date Tested	Test Number	Station, Offset	Lower or Upper Course	Mainline or Shoulder	Maximum Density (pcf)	Density Count	Wet Density (pcf)	% Maximum Density

Lot #: Average % Maximum Density = Specified Density =

☐ This lot meets project specifications. ☐ This lot does not meet project specifications.

Source: Cooper Engineering Co., Inc.

起重机和吊车词汇表

A frame 人字架：参见 Gantry 和 Jib mast。有时它也配合塔架使用，特别是装备在驳船上，用于支承起重臂根部和顶部台升。

Anti two-block device 滑车防撞装置：一种机械装置，当启用时，它可以切断所有会导致下部承载滑车撞击上部承载滑车或者起重臂端部滑轮的起重机运行。

Articulated jib 铰接臂：塔吊起重臂，通常在其中部有一个销接点；也称为回转动臂。

Axis of rotation 旋转轴：起重机或吊车围绕旋转的竖直线；也称为转动中心和回转轴。

Back-hitch gantry 门架：参见 Gantry。

Ballast 压重：添加到起重机底部的重物以产生额外的稳定性；当起重机旋转时，压重并不转动。

Barrel 筒壁：卷扬机绳索卷筒的套筒或主体部分。

Base mounting 底座：构成起重机或吊车最下部分的结构；它将荷载传递到地面或其他支承面。对于移动式起重机，底座就是托架或履带装置。对于塔式吊车，这个词包括了移动底座、拐弯式构架底座或者固定底座（基础）。

Base section 底部节段：伸缩式起重臂最下面的节段；它不能伸缩，但包括了起重臂根部的销栓装置和起重臂液压油缸超级末端装置。

Basic boom 基本起重臂：可组合格构式起重臂能够装配和运行的最小长度，通常仅由起重臂底部和顶端节段组成。

Bogie 转向架：由两个或两个以上的车轴组成的装置，既允许竖向轮位移，同时均分作用在轮上的荷载。

Boom 起重臂：起重机或吊车部件，用于向前支撑上端的卷扬机滑车在径向移动或高度与径向的组合移动；在欧洲称为 jib。

Boom angle 起重臂角：水平线与起重臂中心轴线的夹角。

Boom base 起重臂底部：可组合格构式起重臂最下面的部件，在下端包括连接件或起重臂根部的销栓装置；也称为起重臂粗端或粗端节段。

Boom butt 起重臂粗端：参见 Boom base。

Boom foot mast 起重臂根立柱：某些移动式起重机伸臂悬挂系统的部件。它由在起重臂根部或附近铰接的构架组成，有助于增加固定伸臂悬吊索内侧端的高度，从而加大这些绳索与起重臂之间的夹角。它的作用是减小作用在起重臂上的轴压力。

Boom guy line 起重臂吊索：参见 Pendant。

Boom head 起重臂头部：起重臂上安装负载滑轮的部位。

Boom hoist 起重臂伸降机：控制起重臂的俯仰运动的绳索卷筒、动力和绳索穿绕。

Boom-hoist cylinder 起重臂伸降机液压油缸：取代绳子悬吊作用的液压油缸；这是改变伸缩式起重臂倾角的最常见方法。

Booming in(out) 变幅：参见 Derricking。

Boom inserts 中间起重臂：可组合格构式起重臂的中间节段，通常包括四条平行弦。

Boom point 起重臂顶点：参见 Boom tip section。

Boom stay 起重臂吊索：参见 Pendant。

Boom stop 起重臂止块：用于限制起重臂最大倾角的装置。

Boom suspension 起重臂悬挂：由固定长度或可变长度的绳索和配件组成的系统，支撑起重臂并控制起重臂角度。

Boom tip section 起重臂顶部节段：可组合格构式起重臂最上面的部件，通常包括安装了上部承载滑轮的焊接件作为整体部件；也称为起重臂点，头部节段，或者锥形顶部。

Bridle 束带：参见 Floating harness。

Bull pole 调控杆：通常由钢管制成的杆，安装在起重机立柱底座上，侧向伸出。用于手动转动吊车架。

Bull wheel 起重机转盘：水平安装的圆形框架，固定在起重机立柱底座上，用于容纳和引导摆动绳。

Source: H.I.Shapiro et al, Cranes and Derricks, 2d ed., McGraw-Hill, New York, © 1991. Reprinted with permission of the McGraw-Hill Companies.

Butt section 粗端节段：参见 Boom base。

Carbody 车体：履带吊车底部装置部分，承载上部结构，也是履带侧框架用于连接的部位。

Carrier 托架：带轮子的底盘，汽车式吊车和粗糙场地吊车的底部装置。

Center of rotation 转动中心：参见 Axis of rotation。

Cheek weight 滑车的外壳重量：下部负载滑车边板上的绳索放松重量。

Climbing frame 爬升套架：塔吊立柱上的附加结构，形成环绕的套架；它被用于起重机的顶部爬升。

Climbing ladder 爬升阶梯：从爬升套架伸出的带有横臂（成对使用）的钢制构件，作为塔吊爬升时的顶车支承点。

Climbing schedule 爬升进度：随着工程建设，建筑结构高度不断增加，用于协调塔吊周期性抬升（爬升）的图表。

Counter jib 平衡臂：塔吊起重臂的水平构件，其上安装有配重，通常还安装有提升机；又称为配重臂。

Counterweight 配重：加在起重机上部结构上的重量，用于增加额外的稳定性。它们随着起重机一起转动。

Counterweight jib 配重臂：参见 Counter jib。

Crawler frames 履带构架：履带式起重机底部装置的一部分；与车体连接并支承履带、轨道滚轴、主动链轮和空转链轮。履带构架将起重机重量和运行荷载传到地面；也称为侧构架。

Cribbing 垫板：设置在移动式起重机轨道下或外伸支架承载板下的叠木、钢板或其他结构构件，用于降低其下承载面上的单位承载压力。

Crossover points 跨越点：在绳索卷盘上，本层绳索跨越上一层绳索的接触点。

Dead end 固定端：在动绳系统中，绳索张拉紧的那一端，而另一端（活动端）在绳索卷盘上拉紧。

Deadman 临时支撑物：本身存在或有意修建的物体或结构，用作吊索的锚固。

Derricking 变幅：通过改变起重臂悬挂索的长度来改变起重臂倾角；也称为起重机臂上下俯仰运动。

Dog 止块：与设置在绳索卷盘一个轮缘上的棘齿配合工作的倒齿，用于锁定卷盘，防止反卷方向的转动；也是支承塔吊重量的一套凸耳中的一个。

Dogged off 止转：绳索卷盘的止块啮合时的状况。

Drift 伸展距离：吊重顶和处于最高位置时的起重机吊钩之间的间距；它是对载荷进行吊装或处理的可能距离的一种量度。

Drifting 平移：侧向牵引被悬挂的载荷，改变它的水平位置。

Drum hoist 绞盘：由一个或多个绳索卷盘构成的提升机构；也称为提升引擎。

Duty cycle work 周期工作：对于一个或多个日倒班，在相对不变的载荷等级和相对不变的短周期内的稳定的工作。

EOT crane 电动桥式起重机：电动桥式起重机（Electric overhead traveling crane）。

Expendable base 一次性底座：对于静力装配塔吊，一种浇筑在混凝土基础中的底部立柱节段。这个立柱节段的全部或者部分不能再在今后的装置中使用。

Extension cylinder 伸展油缸：用于伸展伸缩式起重臂的液压油缸；它是最常用的起重机部件动力伸展方式，但不是惟一的。

Fall 吊绳：参见 Parts of line。

Flange point 轮缘点：绳索与卷盘轮缘的接触点，绳索在此改变其索层。

Fleet angle 绳索偏角：绳索在卷盘上缠绕时与卷盘转动轴的垂线形成的夹角。

Fleeting sheave 跟踪滑轮：安装在与绳索卷盘轴相平行的轴上的滑轮，它可以在缠绕绳索时侧向滑动，允许近距离的滑轮布置而不造成过大的绳索偏角。

Float 承载板：外伸支架脚板，将外伸支架传来的荷载分散到支承面或其下的垫板上；它是外伸支承系统的一部分。

Floating harness 吊具：组成起重臂悬挂系统的一部分，支承滑轮活动吊索并连接到固定吊索；也称为束带。

Fly section　套筒节段：伸缩臂上最外层的电动伸缩节段。

Footblock　底滑车：钢制焊接件或装配件，用做缆索式桅杆起重机、起重把杆和芝加哥动臂式起重机的底座装备。

Free fall　自由下落：通过重力降低吊钩（载荷）；降落速度仅由减速装置（如制动器）控制。

Frequency of vibration　振动频率：单位时间内振动发生的周次，单位通常采用赫兹（每秒周次）；也称为频率。

Front end attachments　前端配件：用在移动式起重机上的选装承载构件，例如锥形起重臂、锤头起重臂、吊索起重机配件和塔架配件等。

Gantry　门架：一种高度固定或可调的结构，构成了起重机上部结构的一部分，吊具（承担起重臂活动吊索）锚固在它的上面；也称为人字形门架，人字架。

Gate block　开口滑车：参见 Snatch block。

Gooseneck boom　折臂：上部起重臂与下部起重臂的纵向中心线成一定夹角。

Gudgeon pin　活塞销轴：起重机立柱顶部的销轴，为星形轮或刚性腿塔架提供枢轴。

Guy rope　牵索：固定长度的支承索，有意保持两个连接点之间的标称固定长度；也称为吊索。

Hammerhead boom　锤头起重臂：起重臂端部装置，其中的起重臂悬吊和起升绳都偏移起重臂纵向中心线很远，提供更大的负载空间。

Head section　头部节段：参见 Boom tip section。

Hog line　吊索：参见 Pendant，Intermediate suspension。

Hoist　提升，升降机：提升物品；用于提升的机器设备。

Horse　起重臂脚架：参见 Jib mast。

Impact　冲击：因为动力原因，增大竖向荷载作用。

In-service wind　工作状态风荷载：起重机工作时遇到的风荷载；通常用于定义起重机必须停止工作前现场允许的最高风压和风速等级。

Intermediate suspension　中间悬吊：附加的一套起重臂悬吊绳索，与起重臂连接在主悬吊连接点和起重臂根部的中间某个位置上。对于移动式起重机，中间悬吊用于安装时减小起重臂的弹性挠度；对于水平起重臂塔吊，它用作主要支承的一部分；也称为中点悬吊、中点索结、中间索结或者中间吊索。

Jib　副臂：在美国习惯中，起重臂在端部的延伸称为副臂，它与起重臂纵轴线同线或有所偏移。它装备了自己的悬索，与起重臂端部紧密连接，起重臂端部反过来由牵索或变幅系统支承（在欧洲称它为副臂，jib 是指 boom）。

Jib mast　起重臂顶杆：安装在起重臂头部的短压杆或构架，为起重臂支承索提供了连接方法；也称为人字架。

Jib strut　起重臂支杆：参见 Jib mast。

Jumping　抬升：与建筑物完成一层楼板相对应，将塔吊或牵索起重机从一个工作标高抬升到下一个标高。

Lagging　套筒：抽取式套筒（选装件），用在绳索卷筒上用于改变索拉力和索速度。

Latticed boom　格构式起重臂：由四根纵向的角部构件（称为弦杆）和横向对角构件（称为缀条）组成的起重臂，在两个方向形成桁架结构。弦杆承受起重臂的轴向力和弯矩，缀条抵抗剪力。

Layer　索层：一系列围绕绳索卷筒的钢缆，从一个轮缘到另一个轮缘。

Level luffing　水平变幅：一种自动装备，起重机或吊车通过它可以在起重臂俯仰时并不显著改变吊钩的高度。

Line pull　索拉力：绳索在卷筒或某个特定的层心直径（索层数）套筒上能够获得的拉力。

Line speed　索速度：绳索在卷筒或某个特定的层心直径（索层数）套筒上能够获得的速度。

Live spreader　吊具：参见 Floating harness。

Load jib　承载臂：参见 Saddle jib。

Load radius　荷载半径：参见 Radius。

Loading 荷载：能够在结构构件中引起内力的外部作用。它可能是一种直接物理量的形式（例如加在结构上的重力或者风压力）或者是一种较抽象的形式（例如与运动相关的惯性力）。

Lower spreader 吊具：组成起重臂悬挂系统的一部分，支承滑轮活动吊索并连接到起重机架或上部结构。

Luffing 起重机臂上下俯仰运动：见 Derricking。

Luffing jib 变幅臂：塔吊起重臂，通过围绕一个销轴抬高或降低来沿径向移动吊钩，也就是改变工作半径。

Lumped masses 集中质量：一种数学简化分析概念，将分布质量或者离散质量用集中在点上的质量块来代替，其由质量引起的惯性力是等效的。

Machine resisting moment 机器抵抗矩：起重机或吊车的自重减去起重臂的重量后关于倾覆支点的弯矩；也就是抗倾覆弯矩；也称为机器矩或稳定矩。

Manual insert 手动插入节段：伸缩式起重臂上选装的无动力节段，构成最外层的起重臂节段，使用时要么完全伸展，要么完全缩进。

Mast 立柱：起重机或吊车上基本竖直的承载构件；塔吊的塔架。

Mast cap 立柱帽：参见 Spider。

Midpoint suspension 中点悬吊：参见 Intermediate suspension。

Midsection 中间节段：伸缩臂中安装在底座和套筒节段之间的中间电动伸缩节段。

OET crane 电动桥式起重机：参见 EOT crane。

Offset angle 偏移角：副臂纵向中心线与安装它的起重臂纵向中心线的夹角。

Operating radius 工作半径：参见 Radius。

Operating sectors 工作扇面：围绕移动式起重机转动轴的圆的一部分，它划定了各种区域的边界，包括可用的侧面、后面和前面工作范围。

Out-of-service wind 非工作状态风荷载：处于非工作状态的起重机能够抵抗的风速或风压上限；通常用于定义处于非工作状态的起重机能够安全承受的设计最大风载等级。

Outriggers 外伸支架：连接到起重机底座上的伸缩臂，能够分担起重机车轮（履带）上的重量；用于增加稳定性。

Overhauling weight 绳索放松重量：需要施加在承载吊绳上的重量，用于克服阻力，当没有支承活载时允许绳索卷盘不反绕；也请参见 Cheek weights。

Overturning moment 倾覆弯矩：负载和起重臂自重之和关于倾覆支点的弯矩。需要时可以考虑风荷载和动力影响。

Parking track 泊车轨道：对于装有轨道的移动式起重机，一段经过支撑的轨道，它能够承受暴风引起的转向架荷载；需要时可以采用防风锚。

Parts of line 吊绳：支承载荷或力的若干动绳；也称为提桶。

Paying out 放线：通过放松（卷出）绳子，加大绳子的松弛度或者减小绳上的荷载。

Pendant 吊索：固定长度的绳索，构成伸臂悬挂系统的一部分；也称为起重臂牵索、静索。

Pitch diameter 层心直径：从绳索中心线开始量测的滑轮或绳索卷筒的直径；等于支撑面直径加上绳索直径。

Pivoted luffing jib 回转动臂：参见 Articulated jib。

Preventer 辅助索：在装配索具过程中，一种通常包括但不止于钢丝绳的措施，防止出现不利的运动或事故，或者在锚固和连接失效后作为备用设施。

Radius, load(operating) 载荷（工作）半径：转动轴到吊重重心之间的名义水平距离。对于移动式起重机，它更常定义为加载前的转动轴线在地面的投影到加载后竖直张拉索中心的距离。

Range diagram 范围图：起重机的立面图，用圆弧表示所有起重臂长度下的端点的变幅路径，径向线表示起重臂倾角。竖向刻度表示地面以上的高度，水平刻度表示工作半径。该图可用于确定起吊高度、载荷底部空隙，以及障碍物上的吊装间隙。

Reach　伸臂长度：距离起重机或吊车转动轴的长度；有时与半径同义。

Recurrence period　再现周期：发生重复事件之间的时间间隔；事件（例如特定级别的暴风）重复发生之间的统计预期时间间隔。

Reeving diagram　绳索缠绕图：表示绳索通过滑轮（滑车）系统缠绕路径的图解。

Revolving superstructure　上部旋转结构：参见 Upperstructure。

Rooster　冠架：起重臂和立柱顶端的一根或多根压杆的行业术语，例如起重臂压杆，塔吊的塔顶或者移动式起重机塔架附件的立柱顶端压杆。

Root diameter　根直径：参见 Tread diameter。

Rope drum　绳索卷筒：卷扬机的部件，由侧边带轮缘的转动圆筒组成，起重索在其上缠绕（卷绕）。

Rotation-resistant rope　防转索：由编绕方向相反的内层索股和外层索股组成的钢丝绳。

Running line　动绳：在滑轮或卷筒上运动的绳索。

Saddle jib　鞍型臂：锤头型塔吊的水平活荷载支承构件，其载荷支承索由在起重臂上来回移动的台车支承；也叫承载臂。

Sheave　滑轮：带有圆周形凹槽的滑轮或滑车，凹槽是为特定粗细的钢丝绳设计的；用于改变绳索的运动方向。

Side frames　侧构架：参见 Crawler frames。

Side guys　侧向索：支撑起重臂或立柱侧面的绳索，防止发生侧向运动或侧向失稳。

Side loading　侧向荷载：与起重臂的竖直平面成任意角度的荷载。

Sill　下横梁：刚性腿起重机的一种水平稳定构件，它保护竖向交叉的刚性腿。

Slewing　回转：参见 Swing。

Snatch block　开口滑车：将单滑轮或双滑轮滑车装配成一个或两个护板都能打开，在不使用绳索自由端头的情况下就能将滑车缚住。

Spider　星形轮：安装在起重机立柱顶部枢轴（耳轴销轴）上的装置，为牵索提供连接点；也称为立柱帽。

Spreader　吊具：参见 Floating harness，Lower spreader。

Spreader bar　吊具：参见 Floating harness。

Stabilizers　稳定器：增加起重机稳定性的装置；它们与起重机的底座相连，但它们并不能减轻轮（履带）式吊车的轮压。

Stabilizing moments　稳定矩：参见 Machine resisting moment。

Standing line　静索：不通过卷筒进行缠绕而支承荷载的固定长度的绳索；绳索两端都是固定的；也称为牵索，吊索。

Static base　静力座：塔式起重机的支座（底座装置），起重机立柱就固定在它的上面或者进入基础。

Stay rope　静索：参见 Guy rope，Pendant。

Strand　索股：一组钢丝螺旋形拧在一起，形成一股钢丝绳或者一根钢丝绳的部分。

Strength factor　强度系数：破坏荷载（或应力）被允许工作荷载（或应力）除。

Structural competence　结构能力：设备及其部件支承运行荷载引起的应力的能力，使应力不超过给定的限值。

Superstructure　上部结构：参见 Upperstructure。

Swing　旋转：起重机或吊车的功能，其中起重臂或承载构件围绕竖向轴（旋转轴）转动；也称为回转。

Swing axis　回转轴：参见 Axis of rotation。

Tackle　滑车：由绳索和滑轮组成的装置，用于牵引。

Tagline　跟随索：与负载相连的绳索（通常为纤维质），用于从地面控制负载转动和对齐。同时，对于抓斗操作，钢丝绳用于防止料斗转动和摇摆。

Tailing crane　尾部控制起重机：在将一个长物件从起始水平位置竖立起来的多机联合作业中，尾部控制起重机负责控制物件的底部。

Taking up　收线：消除绳索中松弛的过程，或者收紧（绞紧）绳索；通过张紧，给绳索施加荷载。

Tapered tip　锥形端：参见 Boom tip section。

Telescoping　伸缩：通过在外塔增加节段随后提升内塔的方法，增加移动式或独立式塔吊的高度的过程。也有吊车是通过在下部增加内塔节段的方式来伸缩。

Tipping fulcrum　倾覆支点：当起重机或吊车倾覆时将要围绕其旋转的水平线；倾覆时，起重机或吊车的全部重量将施加到这些点上。

Tipping load　倾覆荷载：对于一特定操作半径，使起重机或吊车开始发生倾覆的荷载。

Top climbing　顶部爬升：通过增加立柱节段来抬升塔吊的一种方法；当吊车处于平衡时，爬升框架将吊车上部结构抬起，在转盘下插入一段新的塔吊节段。这个过程可按需求重复。

Top tower　顶塔：一些塔吊在起重臂上面安装的塔，为吊索提供连接；也称为塔顶。

Topping　变幅：参见 Derricking。

Topping lift　起重臂俯仰：参见 Boom hoist。

Tower　塔架：塔吊立柱；一部分塔吊设有顶塔，用于支承起重臂和平衡臂吊索。

Tower attachment　塔架附件：竖直立柱和俯仰起重臂的组合，安装在汽车式吊车的前端，用于替代传统的起重臂。

Tower head　塔顶：参见 Top tower。

Transit　转运：起重机从一个工作地点到另一个工作地点的转移和运输。

Travel　行程：汽车式或轮式起重机在工作场地内依靠自身动力进行的移动。

Travel base　行程底座：装备有轮子（可移动）塔吊的底座装置。

Tread diameter　支撑面直径：滑轮直径或者从槽底量测的带凹槽绳索卷筒的直径；绳索卷筒的光滑筒壁的直径。

Trolley　台车：带着吊钩沿塔吊水平安装的伸臂做径向运动的小车；有的塔吊上的配重小车允许配重沿径向移动，从而调节它们的反弯矩与负载吊钩半径成正比。

Two blocking　滑轮碰撞：过量张拉滑车绳索，使得两个滑轮碰到一起。

Upper　上部结构：参见 Upperstructure。

Upper spreader　吊具：参见 Floating harness。

Upperstructure　上部结构：对于移动式起重机，去掉前端附件的整个转动结构；也称为上部转动结构。

Vangs(vang lines)　支索：缚在起重臂上的边索，用于转动起重臂。

Walking beam　摆动梁：转向架构件，它的独立轴标称水平而且平行于运动方向；摆动梁中心装有枢轴，在其两端装有轮、轮对、轮轴或者另一摆动梁的中轴。它的目的是允许轮子摆动，在不规则行程路线中等分轮压。

Weathervane　风标：随风摆动，停下来时迎风面最小。

Whip line　辅助起升索：第二根或者辅助起升索。

Winch　绞盘：参见 Drum hoist。

Wrap　卷盘：将钢缆卷在圆周形的筒上。

施　工

第30节　施工立桩定线

例　30

因为激光的普遍使用使得施工立桩定线变得更加容易，但是当然不是更加不重要了。

在过去，使用定斜板、平行线和测高杆将地坪传递给污水管道安装队。现在，设置到准确角度的（冷）激光束被用于人孔与人孔之间。除此之外，一条旋转激光束可用于公路、建筑场地和广泛的施工项目。然而，大部分项目确实需要用实际的立桩来确定转角、中心线、基线、人孔位置、路缘和水沟对齐以及类似的控制点。

本节后面包含一些施工立桩定线的典型示例，而且其中包括在激光普及之前采用的历史上的方法。

如前所述，本手册只能包括一些基本材料。这样，读者应参考一些完全针对测量、平整土地和施工立桩定线的指南和手册。

目　录

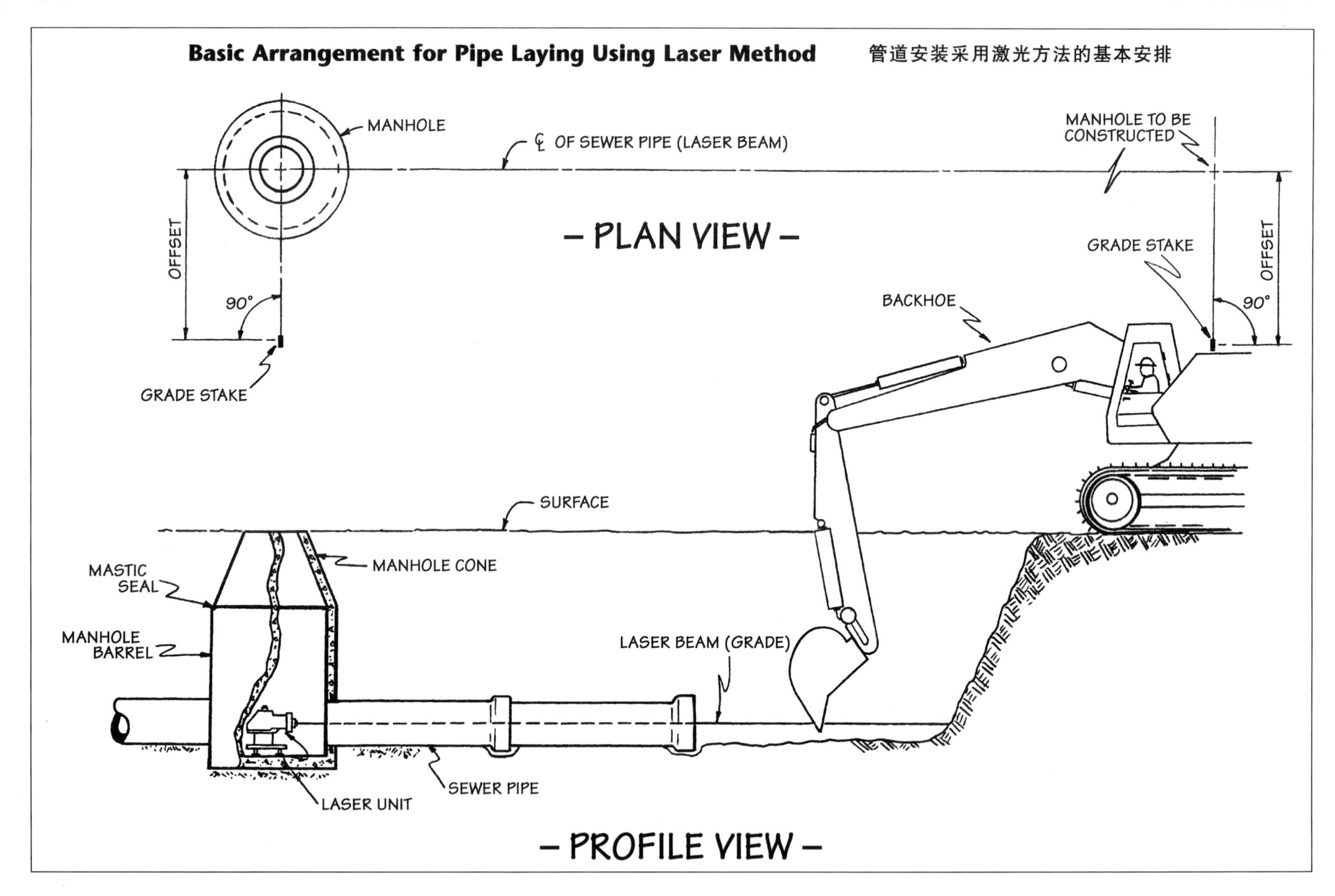
Basic Arrangement for Pipe Laying Using Laser Method
管道安装采用激光方法的基本安排
MANHOLE
℄ OF SEWER PIPE (LASER BEAM)
MANHOLE TO BE CONSTRUCTED
OFFSET
90°
GRADE STAKE
– PLAN VIEW –
GRADE STAKE
OFFSET
90°
BACKHOE
SURFACE
MASTIC SEAL
MANHOLE CONE
MANHOLE BARREL
LASER BEAM (GRADE)
LASER UNIT
SEWER PIPE
– PROFILE VIEW –

Typical Cut Sheet Form　典型的切片图纸格式

PROJECT No. -												
PROJECT NAME -												
STA.	MARK No.	(INLET / E.W. / M.H.)	STAKE ELEV.	FLOW LINE	CUT	PIPE INVERT	CUT	OFFSET	FINISHED CENTER LINE	CUT/ FILL	LOCATION	NOTES

Source: Morgan & Parmley, Ltd.

Batter Board Method for Pipe Laying 管道安装的定斜板法

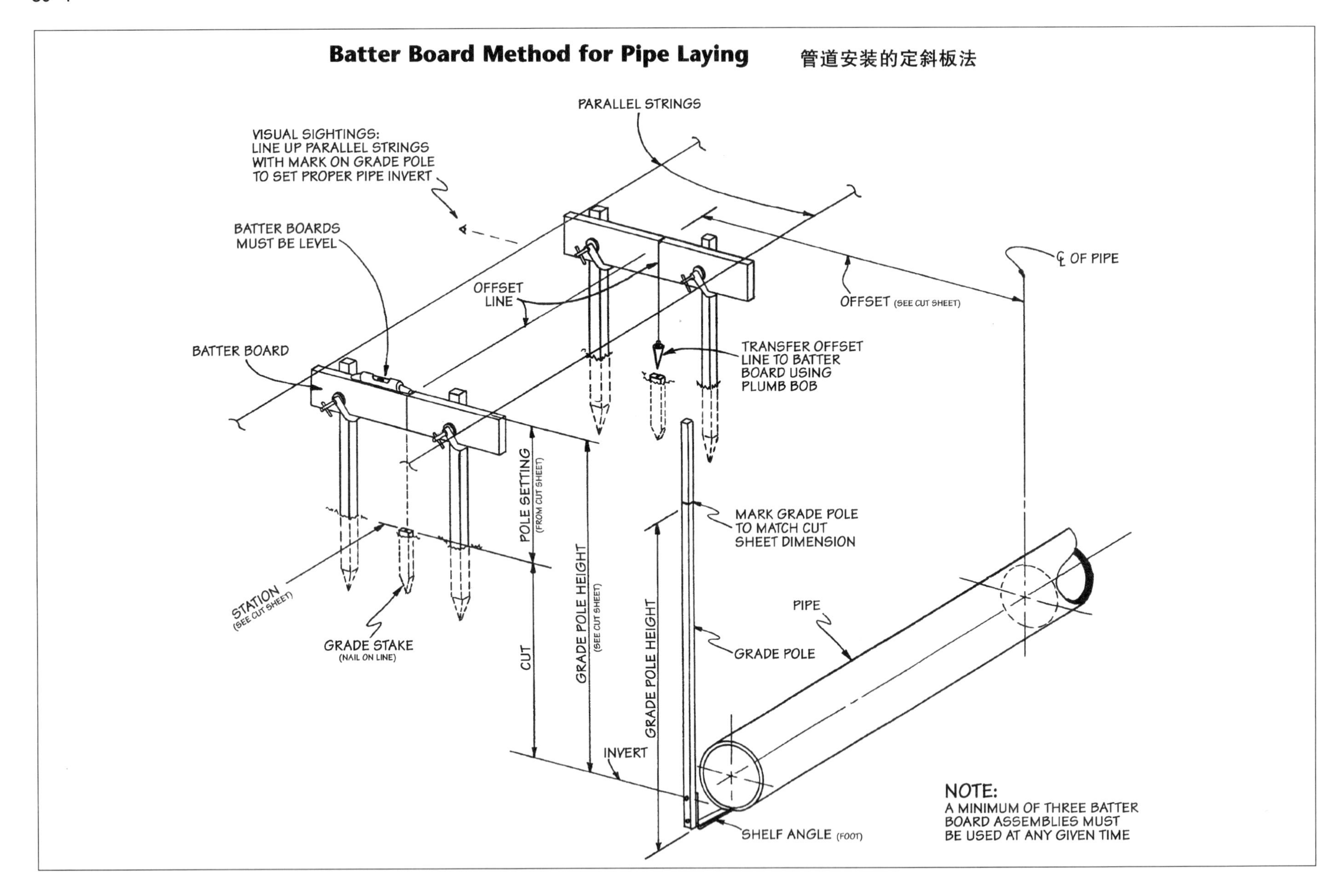

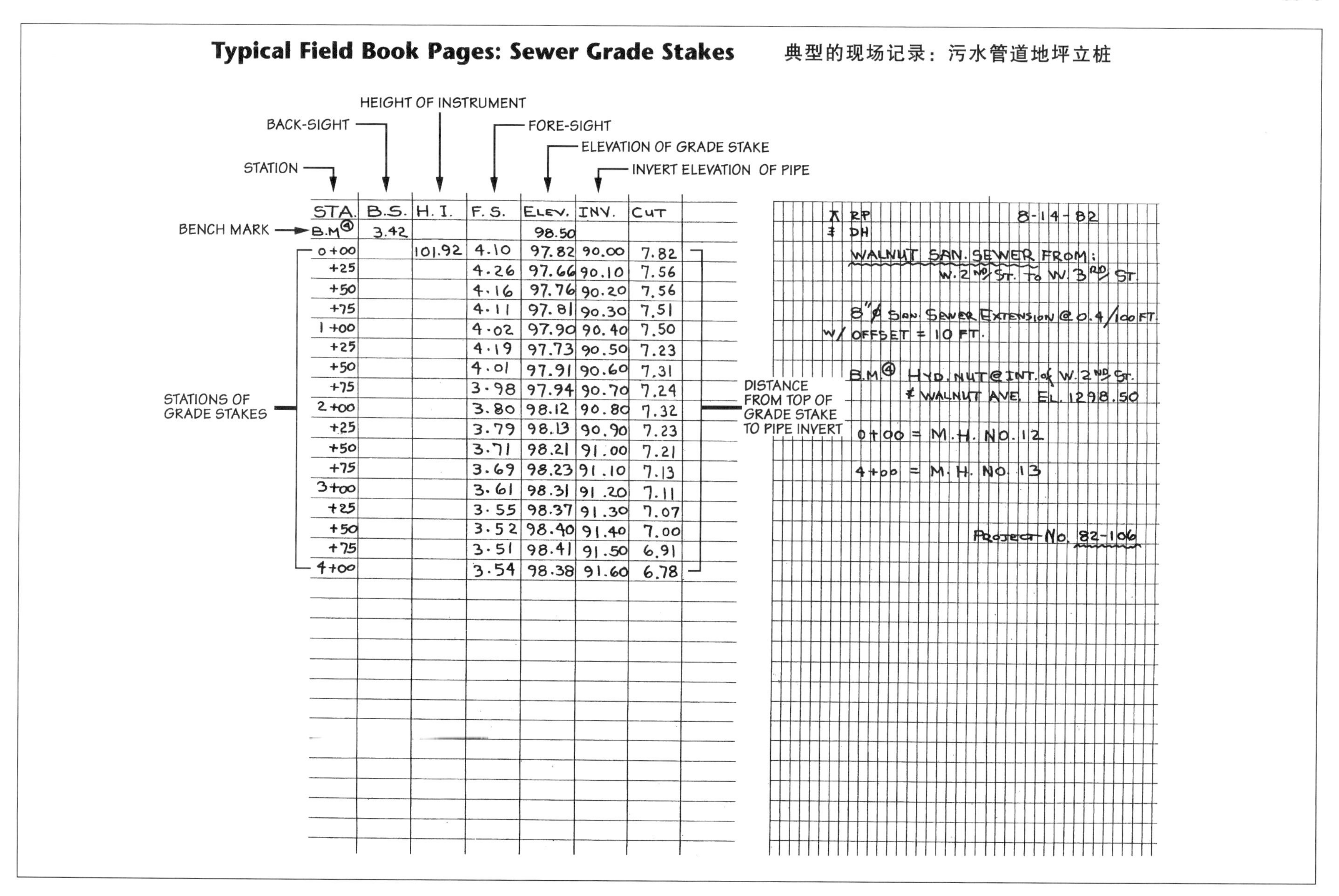

STA.	B.S.	H.I.	F.S.	ELEV.	INV.	CUT
B.M.④	3.42			98.50		
0+00		101.92	4.10	97.82	90.00	7.82
+25			4.26	97.66	90.10	7.56
+50			4.16	97.76	90.20	7.56
+75			4.11	97.81	90.30	7.51
1+00			4.02	97.90	90.40	7.50
+25			4.19	97.73	90.50	7.23
+50			4.01	97.91	90.60	7.31
+75			3.98	97.94	90.70	7.24
2+00			3.80	98.12	90.80	7.32
+25			3.79	98.13	90.90	7.23
+50			3.71	98.21	91.00	7.21
+75			3.69	98.23	91.10	7.13
3+00			3.61	98.31	91.20	7.11
+25			3.55	98.37	91.30	7.07
+50			3.52	98.40	91.40	7.00
+75			3.51	98.41	91.50	6.91
4+00			3.54	98.38	91.60	6.78

Typical Grade Sheet for Batter Board Method 定斜板法的典型地坪表

GRADE SHEET

DATE: August 14, 1982 PAGE 1 OF 1

PROJECT: SANITARY SEWER EXTENSION NO.82-106 (8-INCH)

LOCATION: WALNUT AVENUE

FROM: W. 2ND STREET TO: W. 3RD STREET

SLOPE: 0.40'/100 FT. OFFSET: 10 FT.

BENCH MARK ELEVATION: 1298.50 LOCATION: HYD. NUT@WALNUT & W.2ND

STATION	STAKE ELEVATION	GRADE ELEVATION	FILL	CUT	GRADE POLE 9 FT.	NOTES
0+00	97.82	90.00		7.82	1.18	M. H. NO.12
+25	97.66	90.10		7.56	1.44	
+50	97.76	90.20		7.56	1.44	
+75	97.81	90.30		7.51	1.49	
1+00	97.90	90.40		7.50	1.50	
+25	97.73	90.50		7.23	1.77	
+50	97.91	90.60		7.31	1.69	
+75	97.94	90.70		7.24	1.76	
2+00	98.12	90.80		7.32	1.68	
+25	98.13	90.90		7.23	1.77	
+50	98.21	91.00		7.21	1.79	
+75	98.23	91.10		7.13	1.87	
3+00	98.31	91.20		7.11	1.89	
+25	98.37	91.30		7.07	1.93	
+50	98.40	91.40		7.00	2.00	
+75	98.41	91.50		6.91	2.09	
4+00	98.38	91.60		6.78	2.22	M.H. 13

GRADE SHEET

DATE: PAGE ___ OF ___

PROJECT:

LOCATION:

FROM: TO:

SLOPE: OFFSET:

BENCH MARK ELEVATION: LOCATION:

STATION	STAKE ELEVATION	GRADE ELEVATION	FILL	CUT	GRADE POLE	NOTES

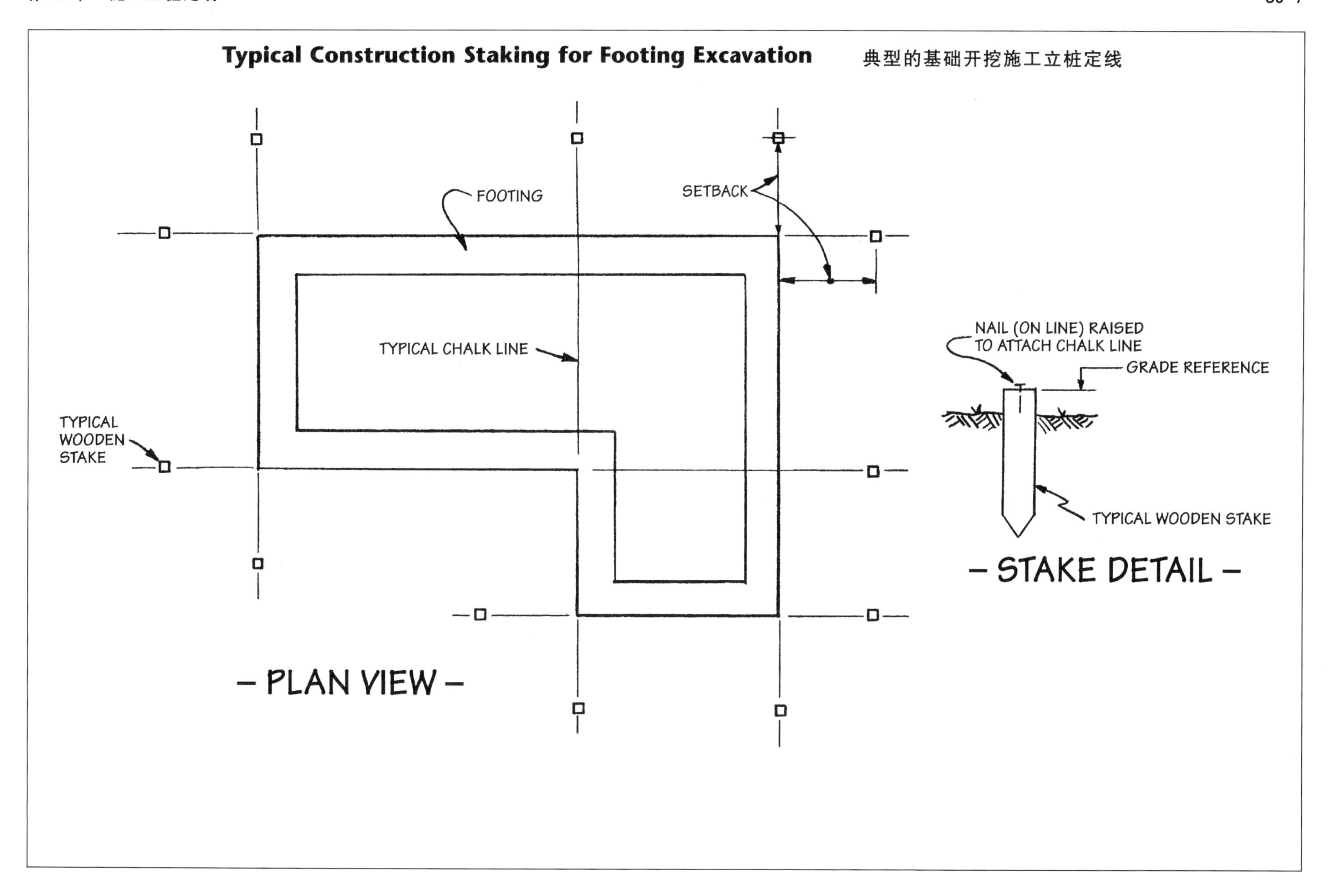
Typical Construction Staking for Footing Excavation
典型的基础开挖施工立桩定线
FOOTING
SETBACK
TYPICAL CHALK LINE
TYPICAL WOODEN STAKE
– PLAN VIEW –
NAIL (ON LINE) RAISED TO ATTACH CHALK LINE
GRADE REFERENCE
TYPICAL WOODEN STAKE
– STAKE DETAIL –

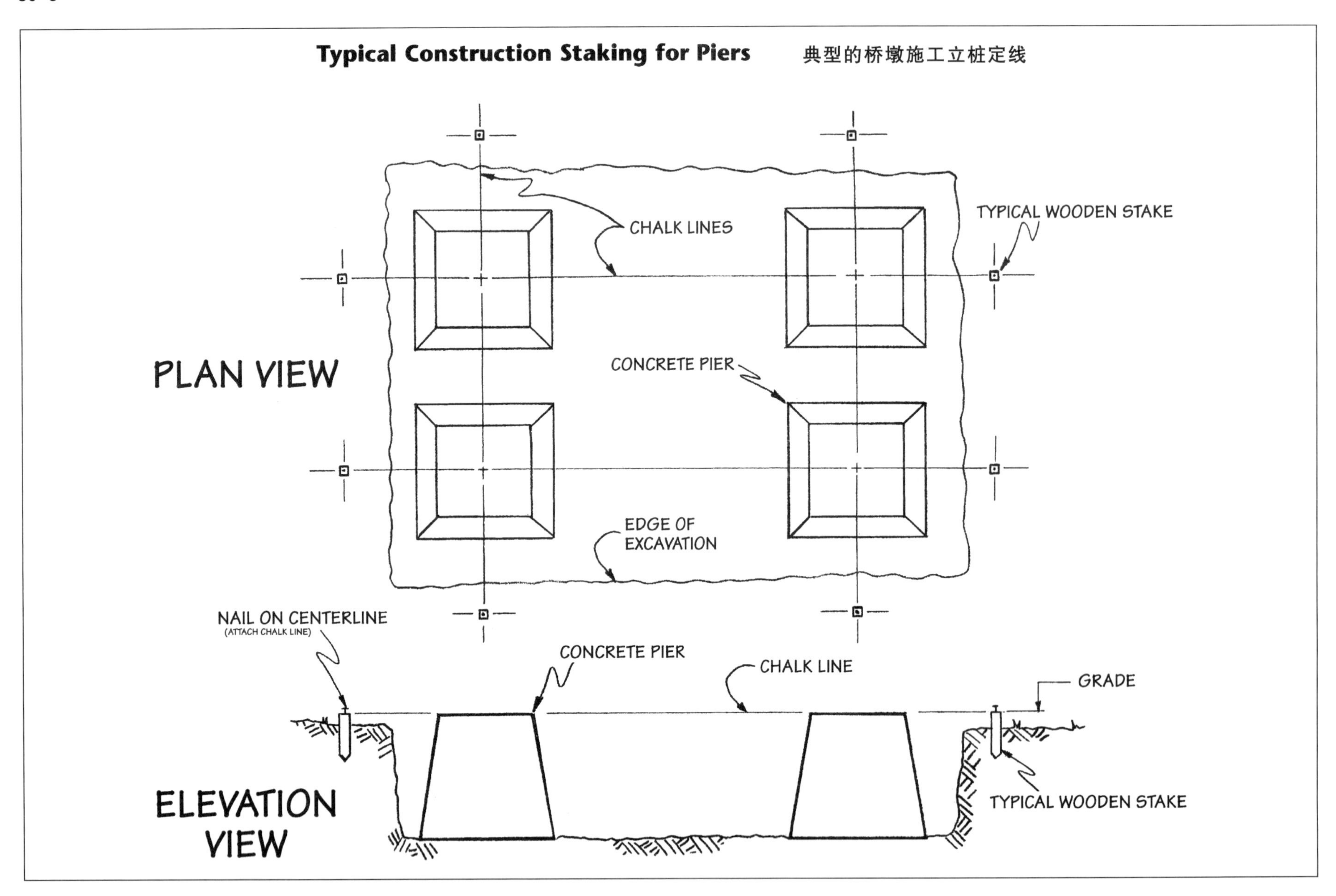
Typical Construction Staking for Piers 典型的桥墩施工立桩定线
CHALK LINES
TYPICAL WOODEN STAKE
PLAN VIEW
CONCRETE PIER
EDGE OF EXCAVATION
NAIL ON CENTERLINE
(ATTACH CHALK LINE)
CONCRETE PIER
CHALK LINE
GRADE
ELEVATION VIEW
TYPICAL WOODEN STAKE

施　工

第 31 节　竣工

例　31

最后 5% 的施工有时是一个项目最容易遇到挫折的时期。虽然施工大部分已经完成而且设备也已能够运行，但是仍然有大量细小的项目有待完成以满足设计图和技术说明的要求。承包商的施工人员以及分包商几乎都已撤离施工现场。监理不断重复催促工头完成建设缺陷清单上所列的项目，但都没能成功。这时就成为项目工程师的职责去正式发布截止日期，并代表业主利益确保所有项目细节都得到相应的完成。大部分项目都以一种有序的方式完成，如期竣工。然而，罕见的情形是业主通常有权将前期付款中提留的保证金用于完善工程，倘若这一行为是依据相关法律进行的。

目　录

建设缺陷清单格式（示例）

Punch List Format

(Sample)

MORGAN and PARMLEY, Ltd.
Professional Consulting Engineers
115 West 2nd Street, S.
LADYSMITH, WI 54848
Phone 715-532-3721
FAX 715-532-5305

DATE:

-MEMO-

TO: (Name of Contractor)

RE: (Name & Number of Project)

PUNCH LIST:

1) ISCO Representative 2nd Training Session (3.3 – 15960)

2) Supply 3 hole paper punch (2.2.2 – 11600)

3) Supply 3 drawer, metal file cabinet (2.2.3 – 11600)

4) Supply UV spare parts (2.3.10 – 16601)

5) Complete telemetry @ lift station (2.8 – 02606) w/demo & testing

6) Furnish & install cap for emergency riser @ lift station

7) Install neutral (4th wire) from CT cabinet to control panel in service building (per spec & Xcel Energy)

8) Replace faulty control for Blower #2 and adjust properly.

9) Resolve timer sequence problem for effluent pump to insure UV bulbs submergence in weir channel and replace check valve & piping (alignment may be problem).

10) Install small rip rap @ outfall for balance & clean channel to river

11) Furnish 3 complete sets of O & M material for:

A-Lift Station; pumps, controls & electrical panel
B-Telemetry System @ Lift Station
C-Air Conditioner @ Service Building
D-Electrical equipment @ Service Building
  *Toe Heaters
  *Base Board Heaters
  *Power Ventilators
  *Suspended Heaters
E-UV (Ultraviolet Disinfection System)

Page 2

12) Furnish executed forms:

A-Consent of Surety Co. to final payment (Pg. 57)

B-Contractor's Affidavit of Release of Liens (Pg. 58)

C-Contractor's Affidavit of Debts & Claims (Pg. 59)

13) Written document for (3 yr.) extended warranty of structural integrity of service building.

14) Add gravel around air valves (flush w/surface) at south end of cell No. 1; i.e. north of parking lot.

15) Fill (seed, fert. & mulch) four (4) mud puddles at fence corners.

16) Shim exit door (east: Blower Room) Does not swing free.

17) Replace all window screens: all screen frames are bent.

18) Seal/Fix wall plate & clean-out in rest room.

19) Ceiling cove trim has shrunk: tighten and caulk to keep insulation from leaking; entire building.

20) Fix insulation leak @ existing fan in rest room.

21) Fix broken handle on lid to Valve Vault chamber "D" (effluent/UV pump area).

22) Replace all UV bulbs or guarantee their longevity: much damage was evidenced during failure of channel to keep bulbs submerged.

23) East fascia on Service Building wrinkled; must be fixed, but could wait until spring.

24) Provide staff gate for UV unit; reference specs 16601, Section 2.3.6, Item 5.

25) Provide 2-inch Gate Valve (for throttling) on effluent pump discharge, per Plans.

Sincerely,

Robert O. Parmley, P.E.
Project Engineer

Source: Morgan & Parmley, Ltd.

最终付款的担保公司承诺

Consent of Surety Company to Final Payment

PROJECT: ______________________

____

Owner: ______________________

Contractor: ______________________

Address: ______________________

____

Contract Date: ______________________

In accordance with the provisions of the Contract between the Owner and the Contractor.

______________________, SURETY COMPANY,

on bond of

______________________, CONTRACTOR,

hereby approves of the final payment to the Contractor, and agrees that final payment to the Contractor shall not relieve the Surety Company of any of its obligations to

______________________, OWNER,

as set forth in the said Surety Company's bond.

IN WITNESS WHEREOF,
the Surety Company has hereunto set its hand this _______ day of ______________, 20___.

____
Surety Company

____
Signature of Authorized Representative

Attest:
(Seal):

____
Title

Source: Morgan & Parmley, Ltd.

承包商的债务和权利主张书

Contractor's Affidavit of Debts and Claims

PROJECT: ______________________

____

Owner: ______________________

Contractor: ______________________

Address: ______________________

____

Contract Date: ______________________

State:

County of:

The undersigned, pursuant to Section 19.6 of the General Conditions of the Contract, hereby certifieS that, except as listed below, he has paid in full or has otherwise satisfied all obligations for all materials and equipment furnished, for all work, labor, and services performed, and for all known indebtedness and claims against the Contractor for damages arising in any manner in connection with the performance of the Contract referenced above for which the Owner or his property might in any way be held responsible.

EXCEPTIONS: (If none, write "None".)

CONTRACTOR:

Address:

BY:

Subscribed and sworn to before me

this ____ day of __________, 20__

Notary Public:

My Commission Expires:

Source: Morgan & Parmley, Ltd.

承包商留置权释放书

Contractor's Affidavit of Release of Liens

PROJECT: ____________________

____

Owner: ____________________

Contractor: ____________________

Address: ____________________

____

Contract Date: ____________________

State of:

County of:

The undersigned, pursuant to Section 19.6 of the General Conditions of the Contract, hereby certifies that to the best of his knowledge, information and belief, except as listed below, the Releases or Waivers of Lien attached hereto include the Contractor, all Subcontractors, all suppliers of materials and equipment, and all performers of Work, labor or services who have or may have liens against any property of the Owner arising in any manner out of the performance of the Contract referenced above.

EXCEPTIONS: (if none, write "None".)

ATTACHMENTS:

CONTRACTOR:

Address:

BY:

Subscribed and sworn before me this ____ day of ________________, 20__

Notary Public:

My Commission Expires:

Source: Morgan & Parmley, Ltd.

实质完工证明

Certificate of Substantial Completion

OWNER'S Project No. ____________ ENGINEER'S Project No. __________

Project Name: ________________________________

CONTRACTOR ________________________________

Contract Amount ____________________ Contract Date ________________

This Certificate of Substantial Completion applies to all Work under the Contract Documents or to the following specified parts thereof:

To: ________________________________
OWNER

And To ________________________________
CONTRACTOR

The Work to which this Certificate applies has been inspected by authorized representatives of OWNER, CONTRACTOR and ENGINEER, and that Work is hereby declared to be substantially complete in accordance with the Contract Documents on:

____
Date of Substantial Completion

A Punch List of items to be completed or corrected is attached hereto. This list may not be all-inclusive, and the failure to include an item does not alter the responsibility of CONTRACTOR to complete all the Work in accordance with the Contract Documents. The items in the Punch List shall be completed or corrected by CONTRACTOR within _____ consecutive days of the above date of Substantial Completion.

Source: Morgan & Parmley, Ltd.

运行和维护手册（前序材料和目录）

O & M Manual

(Sample Front Material and Table of Contents)

M & P Project No. 98-109

VOLUME I

-OPERATION & MAINTENANCE MANUAL-

WASTEWATER TREATMENT

FACILITY

Village

of

Sheldon, Wisconsin

Prepared by:

MORGAN & PARMLEY, LTD.
Professional Consulting Engineers
115 West 2nd Street, South
Ladysmith, Wisconsin 54848

PREFACE

The Sheldon Wastewater Treatment Facility and Sewer Collection System is owned and operated by the Village of Sheldon. The Village Board provides all necessary administrative and supervisory controls, including all monies to insure proper operation and maintenance (O & M) of the complete sewerage facility.

The following is a suggested list of management responsibilities:

1. Establish staff requirements, prepare job descriptions, develop organizational charts and assign personnel as required.
2. Provide operational personnel with sufficient funds to properly operate and maintain the treatment facility and collection system.
3. Ensure that operational personnel are paid a salary to commensurate with their level of responsibility.
4. Provide good working conditions, safety equipment and proper tools for the operational personnel.
5. Establish operator training programs.
6. Make periodic inspections of the treatment facility to discuss mutual problems with Operators and to observe operational practices.
7. Maintain good public relations and create an atmosphere that will make the Operators feel that they can bring all problems to management's attention.
8. Plan for future expansions and modifications to the sewage collection system and treatment facility.

-i-

Source: Morgan & Parmley, Ltd.

(Continued)

Sound water management goals, relative to wastewater treatment, should have the following primary objectives:

1. Provide an adequate level of waste treatment to meet applicable water quality standards and BPWTT (Best Practicable Wastewater Treatment Technology) requirements.
2. Provide adequate capacity in both conveyance and treatment facilities to handle both present and anticipated future flows.
3. Allow for expansion of treatment facilities in the future, should growth necessitate such expansion.

These three primary objectives have been addressed by the Village Board and achieved by the study, analysis, planning and construction of their new Wastewater Treatment Facility. Financing for Facility Planning, Design and Construction was supported by a grant-loan package from the USDA-Rural Development.

Volume I of the O & M Manual contains a detailed description of the total facility with related technical data for successful operation and maintenance. Volume II is devoted to suppliers' literature and manufacturers' technical material for all individual components and equipment.

It is recommended that the WWTP Operator insert additional material and add notes as he judges appropriate to assist him in fully personalize this manual.

We wish you GOOD LUCK and if you need any assistance, please do not hesitate to contact us: Morgan & Parmley, Ltd. Our telephone number is 532-3721 and our FAX number is 532-5305.

-ii-

VOLUME I

Sheldon WWTF O & M Manual

-TABLE OF CONTENTS-

-iii-

(Continued)

(Continued)

-vi-

*These publications are supplied separately.

LIST OF DRAWINGS, TABLES, FORMS AND CHARTS

-vii-

(Continued)

Typical Construction Record Drawing　　典型的施工记录图

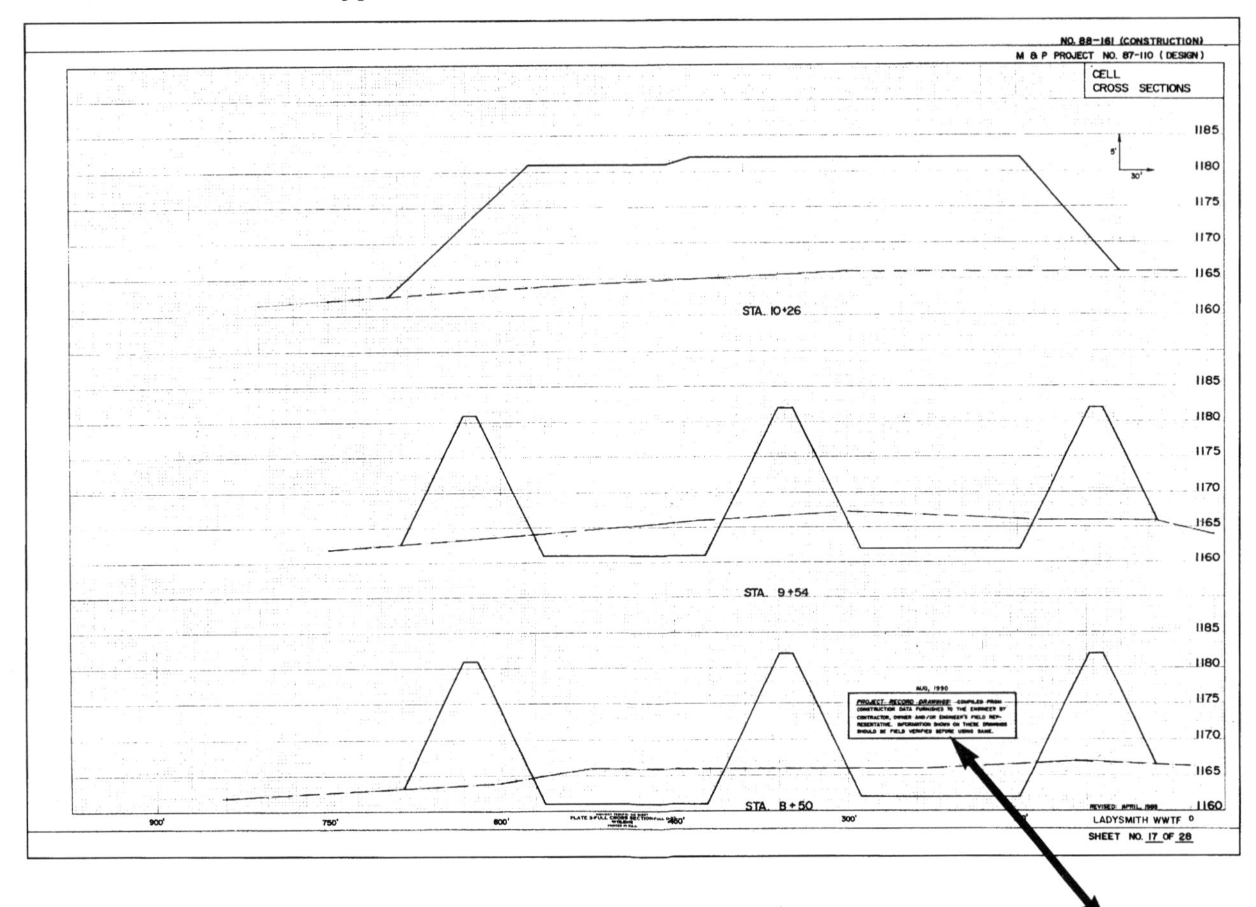

PROJECT RECORD DRAWINGS: COMPILED FROM CONSTRUCTION DATA FURNISHED TO THE ENGINEER BY CONTRACTOR, OWNER AND/OR ENGINEER'S FIELD REPRESENTATIVE. INFORMATION SHOWN ON THESE DRAWINGS SHOULD BE FIELD VERIFIED BEFORE USING SAME.

Source: Morgan & Parmley, Ltd.

工程师的认可证明

Engineer's Certificate of Approval

PROJECT NAME: __________

PROJECT NO: __________

I, ___________________________, a registered Professional Engineer in the State of ___________, and project engineer representing _____________________, do hereby certify that I have inspected _______________________________________ and find the same accomplished according to the Specifications and/or duly authorized Change Orders to the prime contract of ____ .

I do approve of the above referred to improvements and recommend acceptance of this work.

Signature: __________________________________
Engineer

Title: ______________________________________

Date: ______________________________________

....

As an authorized representative of the Owner ___________________________,
I do hereby accept the improvements referred herein this ____day of ______________, _____.

The warranty period begins ___________________ and ends __________________

Signature:___________________________________
Representative

Title: ______________________________________

Date: ______________________________________

Source: Morgan & Parmley, Ltd.

补充材料

补充材料

第 32 节　技术参考

例　32

每一个熟练工程师都有大量的参考文献，适合他们感兴趣和专业学科的特定领域。然而，本书的最后一节并不打算用来替代或者重复这些文献，主编感兴趣的是一些较难找到或者可能晦涩的材料，读者在将来可能用得到。本节还讨论了公制计量的基本内容和格式，后附一份详细的英制到公制的转换系数列表。

目　录

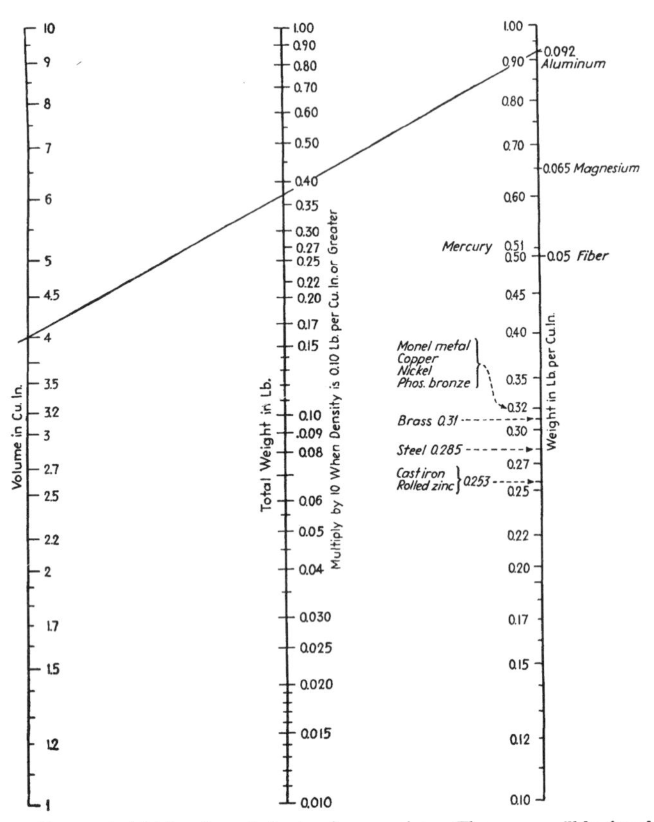

Draw a straight line through the two known points. The answer will be found at the intersection of this line with the third scale.

Example: 4 cu. in. of aluminum weighs 0.37 lb.

水平放置平头圆形容器中的体积

Volumes in Horizontal Round Tanks with Flat Ends

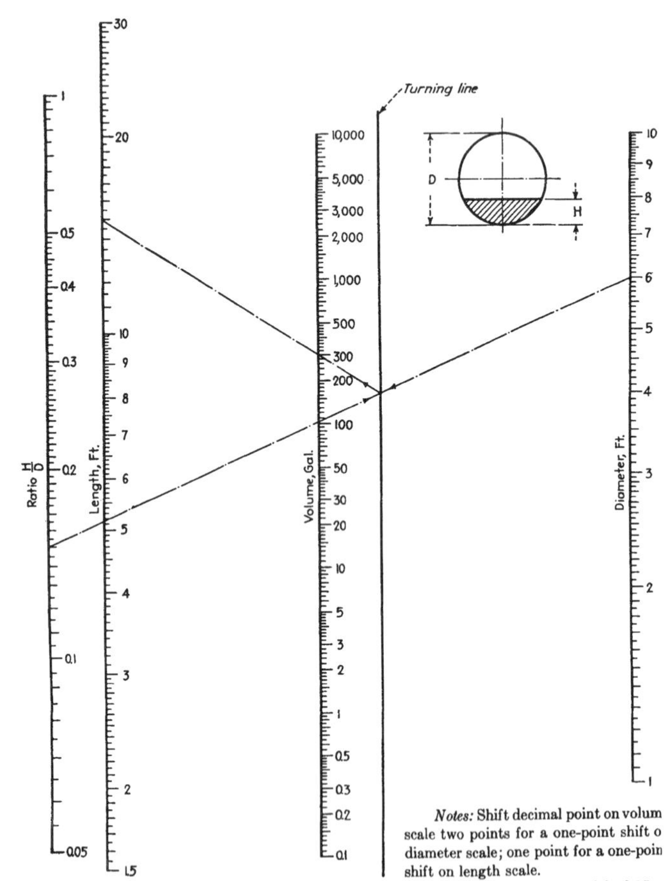

Notes: Shift decimal point on volum scale two points for a one-point shift o diameter scale; one point for a one-poin shift on length scale.

Example: Tank is 6 ft. in diameter and 15 ft. long. $H = 0.9$ ft. $H/D = 0.15$. Join 0.15 o H/D scale with 6 on diameter scale. From point of intersection with turning line, draw line t 15 ft. on the length scale. The volume scale shows 300 gal. If D had been 0.6 ft., H 0.09 ft and length the same, the answer would be 3.00 gal.

计算泵送液体所需的马力

Find How Much Horsepower to Pump Liquids

The charts on these two pages will make it easier to design pumps and other equipment for handling liquid flowing in pipes.

First, it is often helpful to know the theoretical horsepower required to raise the liquid to various heads. This is obtained from the horsepower-gpm charts. They are plotted for a 10-ft head of water: For other heads, multiply hp by $H/10$ where H is the revised head; for other liquids, multiply by the corresponding specific gravity.

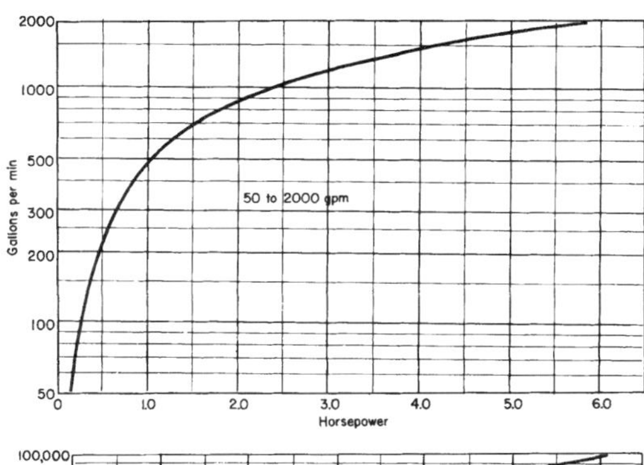

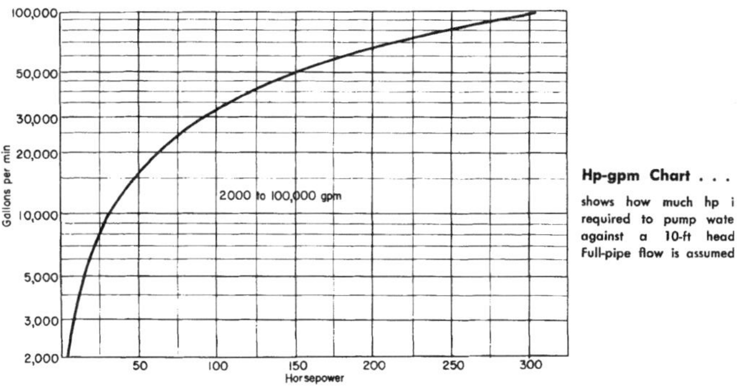

Hp-gpm Chart . . .

shows how much hp i required to pump wate against a 10-ft head Full-pipe flow is assumed

SYMBOLS

G = flow rate, gpm
H_1 = head loss, ft
L = length of pipe, ft
P = pressure, psi
R = Reynold's number $= 0.0833\ Vd/\nu$
V = velocity, fps
d = pipe dia, in.
f = friction factor
w = fluid density, lb per cu ft
r = relative roughness of pipe $= \epsilon/d$
ϵ = effective height of roughness particle, in.
ν = kinematic viscosity, ft^2/sec

Next, for practical results friction losses must be accounted for. These vary and should be known for each individual case.

Much used in liquid-flow calculations is the Darcy formula

$$H_1 = f\frac{6LV^2}{d32.16}$$

which can be modified, if velocity is in gpm, to

$$H_1 = 0.0312 f\frac{LG^2}{d^5}w$$

or for head loss in psi units

$$P = 0.000217 f\frac{LG^2}{d^5}w$$

Practical values of f vary from about 0.01 to 0.06 depending on pipe smoothness and dia. For laminar flow, $f = 64.4/R$. The flow chart gives f for various values of R and pipe roughness. Values ϵ for various pipes are: 0.00006 in. for smooth drawn tubing; 0.0018 in. for wrought iron; 0.01 in. for cast iron. Curves for relative roughness values of 0.0005 to 0.01 are plotted. Most of these lie in the transition zone between laminar flow and complete turbulence.

Fluid-flow Chart . . .

gives friction for various pipe conditions and values of Reynold's Number.

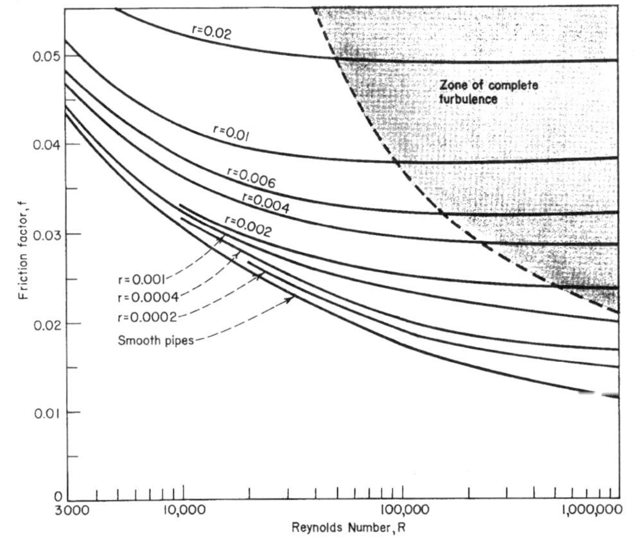

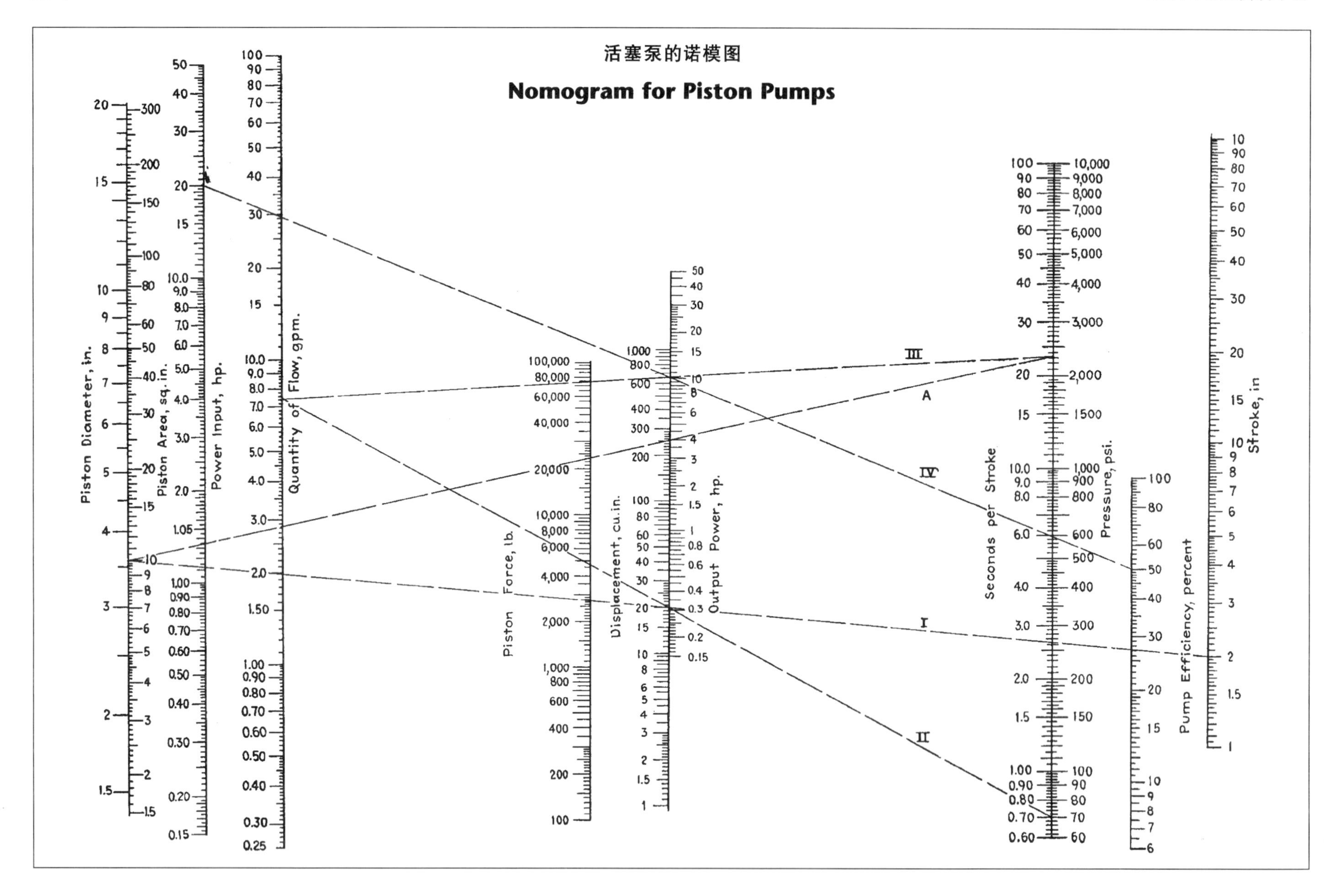

活塞泵的诺模图
Nomogram for Piston Pumps
Piston Diameter, in.
Piston Area, sq.in.
Power Input, hp.
Quantity of Flow, gpm.
Piston Force, lb.
Displacement, cu.in.
Output Power, hp.
Seconds per Stroke
Pressure, psi.
Pump Efficiency, percent
Stroke, in
I
II
III
IV
A

计算开口和封闭皮带的长度

Find the Length of Open & Closed Belts

The following formulas give the answers (see the illustrations for notation):
Open length, $L = \pi D + (\tan\theta - \theta)(D - d)$
Closed length, $L = (D + d)[\pi + (\tan\theta - \theta)]$
You can find θ from (for open belts): $\cos\theta = (D - d)/2C$; (for closed belts) $\cos\theta = (D + d)/2C$.

When you want to find the **center distance of belt drives**, however, it is much quicker if you have a table that gives you $y = \cos\theta$ in terms of $x = (\tan\theta) - \theta$. **Sidney Kravitz,** of Picatinny Arsenal, has compiled such a table. Now, all you need do to find C is first calculate $x = [L/(D + d)] - \pi$ for open drives.

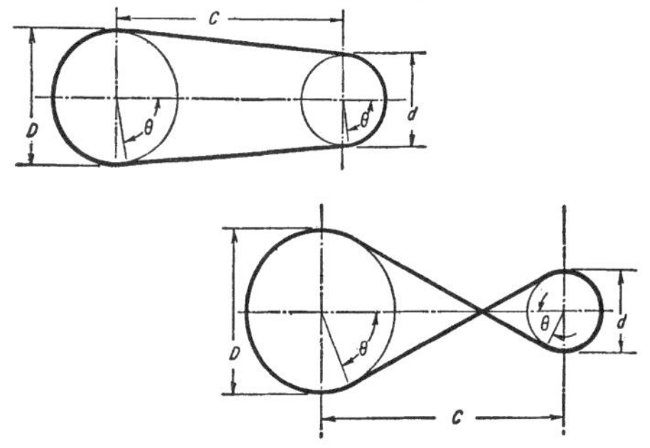

y values[1]

x	0.00	0.01	0.02	0.03	0.04	0.05	0.06	0.07	0.08	0.09
0.0	1.00000	0.95332	0.92731	0.90626	0.88804	0.87175	0.85690	0.84318	0.83039	0.81839
0.1	.80705	.79630	.78606	.77629	.76693	.75795	.74931	.74098	.73295	.72518
0.2	.71767	.71038	.70332	.69646	.68980	.68332	.67701	.67086	.66487	.65902
0.3	.65331	.64774	.64230	.63698	.63177	.62668	.62169	.61681	.61202	.60733
0.4	.60274	.59822	.59380	.58946	.58520	.58101	.57690	.57286	.56889	.56499
0.5	0.56116	0.55738	0.55367	0.55002	0.54643	0.54289	0.53941	0.53598	0.53260	0.52927
0.6	.52600	.52277	.51958	.51645	.51336	.51031	.50730	.50433	.50141	.49852
0.7	.49567	.49286	.49009	.48735	.48465	.48198	.47935	.47675	.47417	.47164
0.8	.46913	.46665	.46420	.46179	.45940	.45703	.45470	.45239	.45011	.44785
0.9	.44562	.44342	.44123	.43908	.43694	.43483	.43274	.43068	.42863	.42661
	.0	.1	.2	.3	.4	.5	.6	.7	.8	.9
1	0.42461	0.40568	0.38850	0.37284	0.35848	0.34526	0.33304	0.32170	0.31115	0.30130
2	.29208	.28344	.27531	.26766	.26043	.25359	.24712	.24098	.23515	.22960
3	.22431	.21926	.21445	.20984	.20544	.20121	.19717	.19328	.18955	.18596
4	.18251	.17918	.17598	.17289	.16991	.16703	.16424	.16156	.15895	.15644
5	0.15400	0.15163	0.14935	0.14712	0.14497	0.14287	0.14084	0.13886	0.13694	0.13508
6	.13326	.13149	.12977	.12810	.12646	.12487	.12332	.12181	.12033	.11889
7	.11748	.11611	.11477	.11346	.11217	.11092	.10970	.10850	.10733	.10618
8	.10506	.10396	.10289	.10183	.10080	.09979	.09880	.09783	.09688	.09594
9	.09503	.09413	.09325	.09238	.09153	.09070	.08988	.08908	.08829	.08751
	0	1	2	3	4	5	6	7	8	9
10	0.08675	0.07980	0.07389	0.06879	0.06436	0.06046	0.05701	0.05393	0.05116	0.04867
20	.04641	.04435	.04246	.04073	.03914	.03766	.03629	.03502	.03384	.03273
30	.03160	.03072	.02980	.02894	.02812	.02735	.02663	.02594	.02528	.02466
40	.02406	.02350	.02296	.02244	.02195	.02148	.02103	.02059	.02018	.01978
50	0.01939	0.01903	0.01867	0.01833	0.01800	0.01768	0.01737	0.01708	0.01679	0.01651
60	.01624	.01598	.01573	.01549	.01525	.01502	.01480	.01459	.01438	.01417
70	.01397	.01378	.01359	.01341	.01323	.01310	.01289	.01273	.01257	.01241
80	.01226	.01211	.01197	.01183	.01169	.01155	.01142	.01129	.01117	.01104
90	.01092	.01080	.01069	.01057	.01046	.01036	.01025	.01015	.01004	.00994
100	0.00985 (see note below for x > 100)									

[1] If $x = (\tan\psi) - \psi$; then $y = \cos\psi$.

If $x > 100$, calculate C from $C = \frac{L}{2} - \frac{\pi}{4}(D + d)$ for both open and closed belts.

$$x = \frac{L}{D + d} \quad \text{for closed drives}$$

Then

$$C = \frac{D - d}{2y} \quad \text{for open drives}$$

$$C = \frac{D + d}{2y} \quad \text{for closed drives}$$

Example: $L = 60.0$, $D = 15.0$, $d = 10.0$, $x = (L - \pi D)/(D - d) = 2.575$, $y = 0.24874$ by linear interpolation in the table. $C = (D - d)/2y = 10.051$.

• • • •

Morton P. Matthew's letter on fractional derivatives (*PE*—July 22 '63, p 105) drew several interesting comments from readers. Here's what Professor Komkov of the University of Utah had to say on the subject. He pointed out that the question raised by Mr Matthew is well known in mathematics, but very little publicized.

"The definition of fractional derivatives goes back to Abel, who developed around 1840 this fascinating little formula:

$$D^s(f) = \frac{d^s f(x)}{dx^s} = \frac{1}{\Gamma(-s)} \int_0^x (\xi - t)^{(-s-1)} f(t)\, dt$$

($\Gamma(n)$ is the Euler's Gamma Function).

An elementary proof of this formula is given for example in Courant's *Differential and Integral Calculus,* Part II, page 340. Abel claimed that the formula works for all real values of *S*, although there is no guarantee that the range of values obtained can be bounded. For a negative *S* Abel's operator D^s becomes an integral operator:

$$\iota^s(f(x)) = D^{-s}(f(x)) = \frac{1}{\Gamma(s)} \int_0^x (\xi - t)^{s-1} f(t)\, dt.$$

All results quoted by Mr Matthew may be easily obtained by application of Abel's Formula.

"There exists a generalization to partial differential equations of the fractional derivative. This is the so-called Riesz Operator. In one dimensional case it becomes Abel's derivative of fractional order.

"Details of the Riesz technique are explained, for example, in Chapter 10 of *Partial Differential Equations* by Duff. Unfortunately I know of no textbook which devotes more than a few pages to the subject of fractional derivatives. However, there exists a large number of papers on the subject in mathematical journals. I remember reading one by Professor John Barrett in the Pacific Journal of Mathematics (I think it was 1947) which discussed the equation:

$$\frac{d^s y}{dx^s} + \iota y = 0 \quad \text{where } 1 \leq s \leq 2$$

"There are some interesting applications in engineering and science for this theory. I was interested some years ago in formulation of elasticity equations for some plastics. I have never completed that investigation but I have established that in some cases, the behavior of plastics may be better simulated by assuming stress-strain relationship to be of the type:

$$\iota_{ij} = C_{ijkl} \frac{d^s \epsilon^{kl}}{dt^s}$$

where *s* is some number between 0 and 1, than by the usual assumption of linear superposition of Hooke's law and Newtonian Fluid properties. In case of some rubbers *s* worked to be close to 0.7."

90°弯头所需材料长度

Length of Material for 90 Degree Bends

As shown in Fig. 1, when a sheet or flat bar is bent, the position of the neutral plane with respect to the outer and inner surfaces will depend on the ratio of the radius of bend to the thickness of the bar or sheet. For a sharp corner, the neutral plane will lie one-third the distance from the inner to the outer surface. As the radius of the bend is increased, the neutral plane shifts until it reaches a position midway between the inner and outer surfaces. This factor should be taken into consideration when calculating the developed length of material required for formed pieces.

The table on the following pages gives the developed length of the material in the 90-deg. bend. The following formulas were used to calculate the quantities given in the table, the radius of the bend being measured as the distance from the center of curvature to the inner surface of the bend.

1. For a sharp corner and for any radius of bend up to T, the thickness of the sheet, the developed length L for a 90-deg. bend will be

$$L = 1.5708\left(R + \frac{T}{3}\right)$$

2. For any radius of bend greater than $2T$, the length L for a 90-deg. bend will be

$$L = 1.5708\left(R + \frac{T}{2}\right)$$

3. For any radius of bend between $1T$ and $2T$, the value of L as given in the table was found by interpolation.

The developed length L of the material in any bend other than 90 deg. can be obtained from the following formulas:

1. For a sharp corner or a radius up to T:

$$L = 0.0175\left(R + \frac{T}{3}\right) \times \text{degrees of bend}$$

2. For a radius of $2T$ or more:

$$L = 0.0175\left(R + \frac{T}{2}\right) \times \text{degrees of bend}$$

R = Inside radius T = Stock thickness

Sharp corner R = T or less R = 1T to 2T R = 2T or more

Fig. 1.

For double bends as shown in Fig. 2, if $R_1 + R_2$ is greater than B:

$$X = \sqrt{2B(R_1 + R_2 - B/2)}$$

With R_1, R_2, and B known:

$$\cos A = \frac{R_1 + R_2 - B}{R_1 + R_2}$$

$$L = 0.0175(R_1 + R_2)A$$

where A is in degrees and L is the developed length.

If $R_1 + R_2$ is less than B, as in Fig. 3,

$$Y = B \operatorname{cosec} A - (R_1 + R_2)(\operatorname{cosec} A - \operatorname{cotan} A)$$

Fig. 2. Fig. 3.

The value of X when B is greater than $R_1 + R_2$ will be

$$X = B \cot A + (R_1 + R_2)(\operatorname{cosec} A - \operatorname{cotan} A)$$

The total developed length L required for the material in the straight section plus that in the two arcs will be

$$L = Y + 0.0175(R_1 + R_2)A$$

To simplify the calculations, the table on this page gives the equations for X, Y, and the developed length for various common angles of bend. The table on following pages gives L for values of R and T for 90-deg. bends.

EQUATIONS FOR X, Y, AND DEVELOPED LENGTHS

Angle A, deg.	X	Y	Developed length
15	$3.732B + 0.132(R_1 + R_2)$	$3.864B - 0.132(R_1 + R_2)$	$3.864B + 0.130(R_1 + R_2)$
22½	$2.414B + 0.199(R_1 + R_2)$	$2.613B - 0.199(R_1 + R_2)$	$2.613B + 0.194(R_1 + R_2)$
30	$1.732B + 0.268(R_1 + R_2)$	$2.000B - 0.268(R_1 + R_2)$	$2.000B + 0.256(R_1 + R_2)$
45	$B + 0.414(R_1 + R_2)$	$1.414B - 0.414(R_1 + R_2)$	$1.414B + 0.371(R_1 + R_2)$
60	$0.577(B + R_1 + R_2)$	$1.155B - 0.577(R_1 + R_2)$	$1.155B + 0.470(R_1 + R_2)$
67½	$0.414B + 0.668(R_1 + R_2)$	$1.082B - 0.668(R_1 + R_2)$	$1.082B + 0.510(R_1 + R_2)$
75	$0.268B + 0.767(R_1 + R_2)$	$1.035B - 0.767(R_1 + R_2)$	$1.035B + 0.542(R_1 + R_2)$
90	$R_1 + R_2$	$B - R_1 - R_2$	$B + 0.571(R_1 + R_2)$

矢高和弦长

Chordal Height & Length of Chord

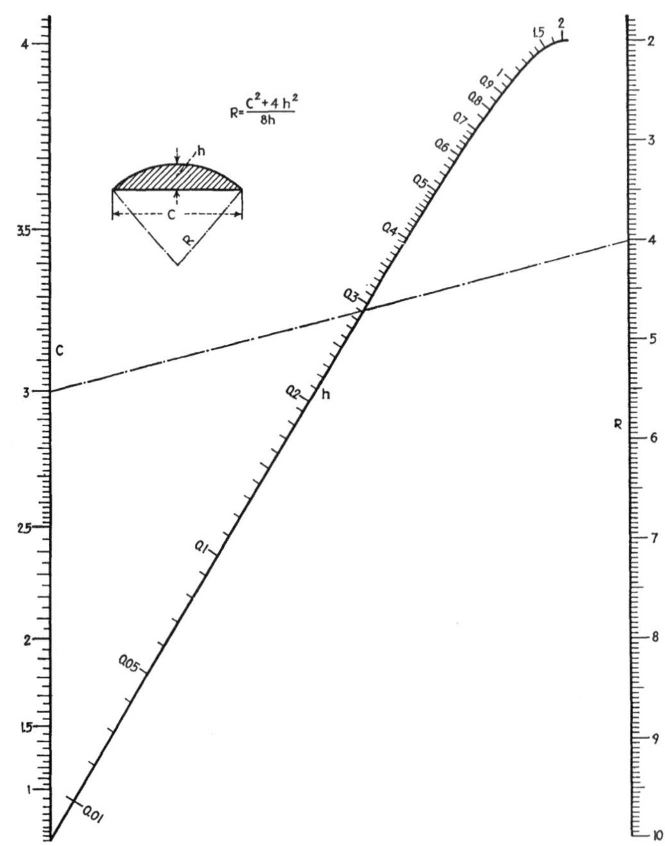

Draw a straight line through the two known points. The answer will be found at the intersection of this line with the third scale.

Example: Length of chord is 3 in., and radius of circle is 4 in. The height h of the chord is 0.29 in.

等臂肘接的内力

Forces in Toggle Joint with Equal Arms

$$\frac{P}{F}=\frac{S}{4h}$$

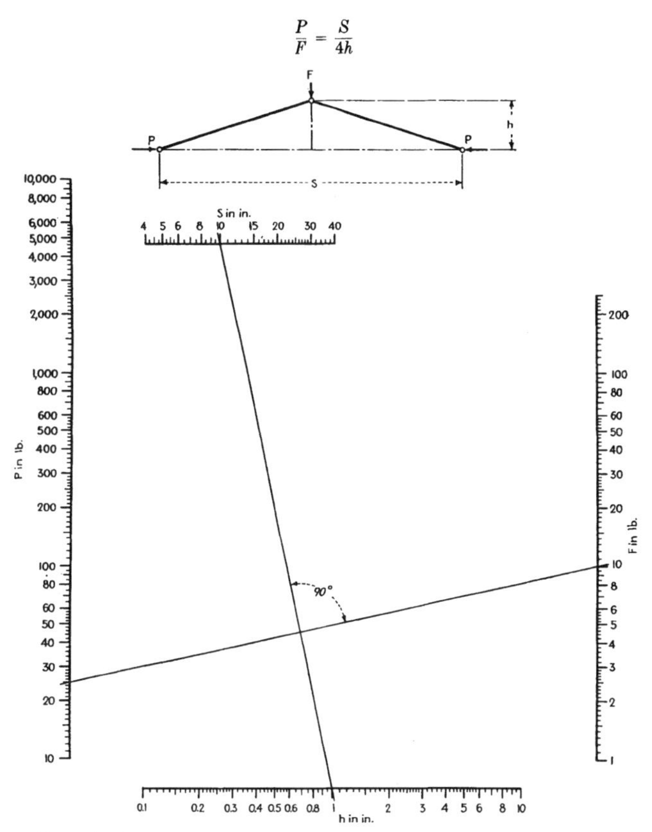

Example: Use mutually perpendicular lines drawn on tracing cloth or celluloid In the example given for S = 10 in. and h = 1 in., a force F of 10 lb. exerts pressures P of 25 lb. each.

棱柱关于 aa 轴的转动惯性矩

Moment of Inertia of a Prism About the Axis aa

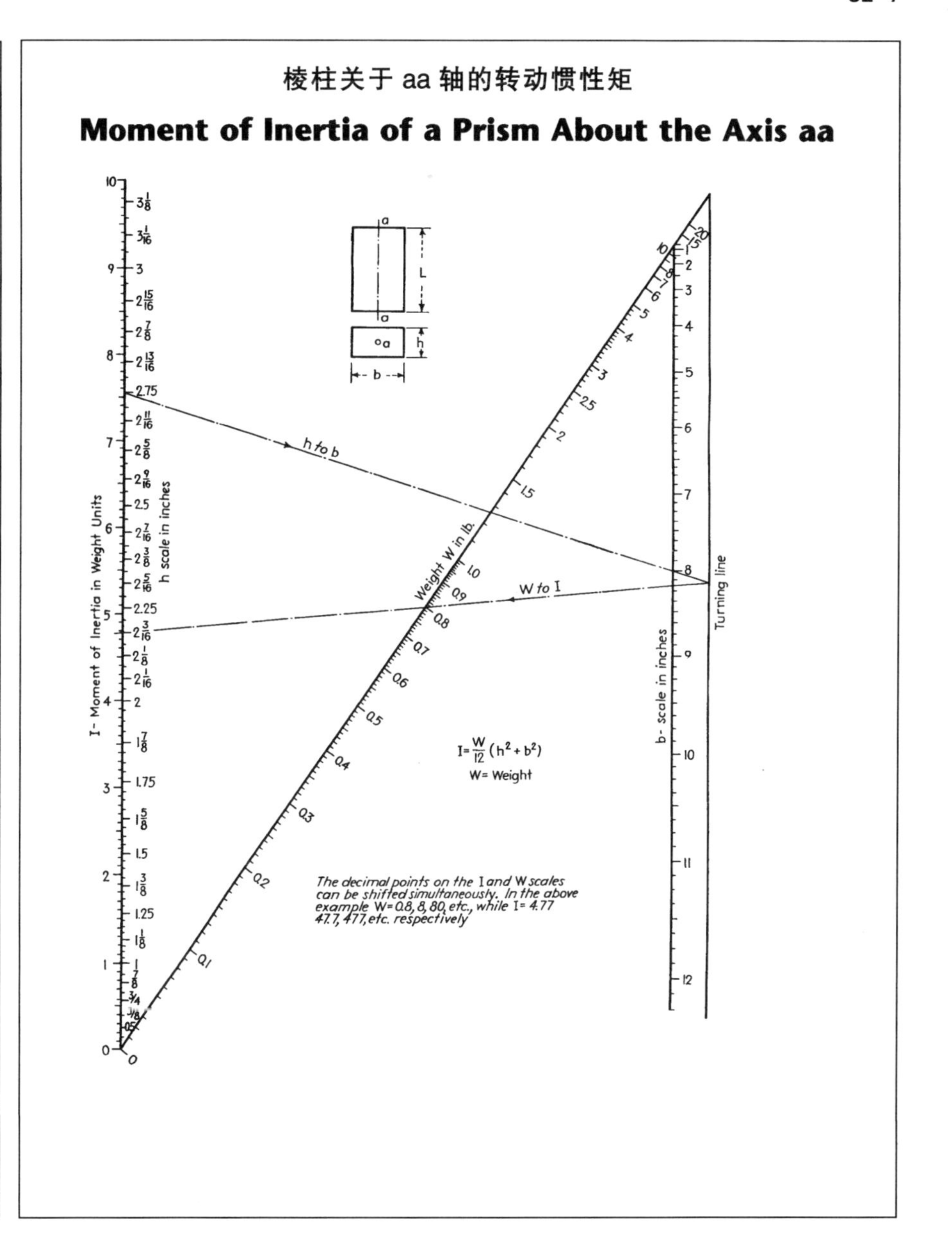

惯性矩的平移图表

Chart for Transferring Moment of Inertia

$$I = I_0 + WX^2$$

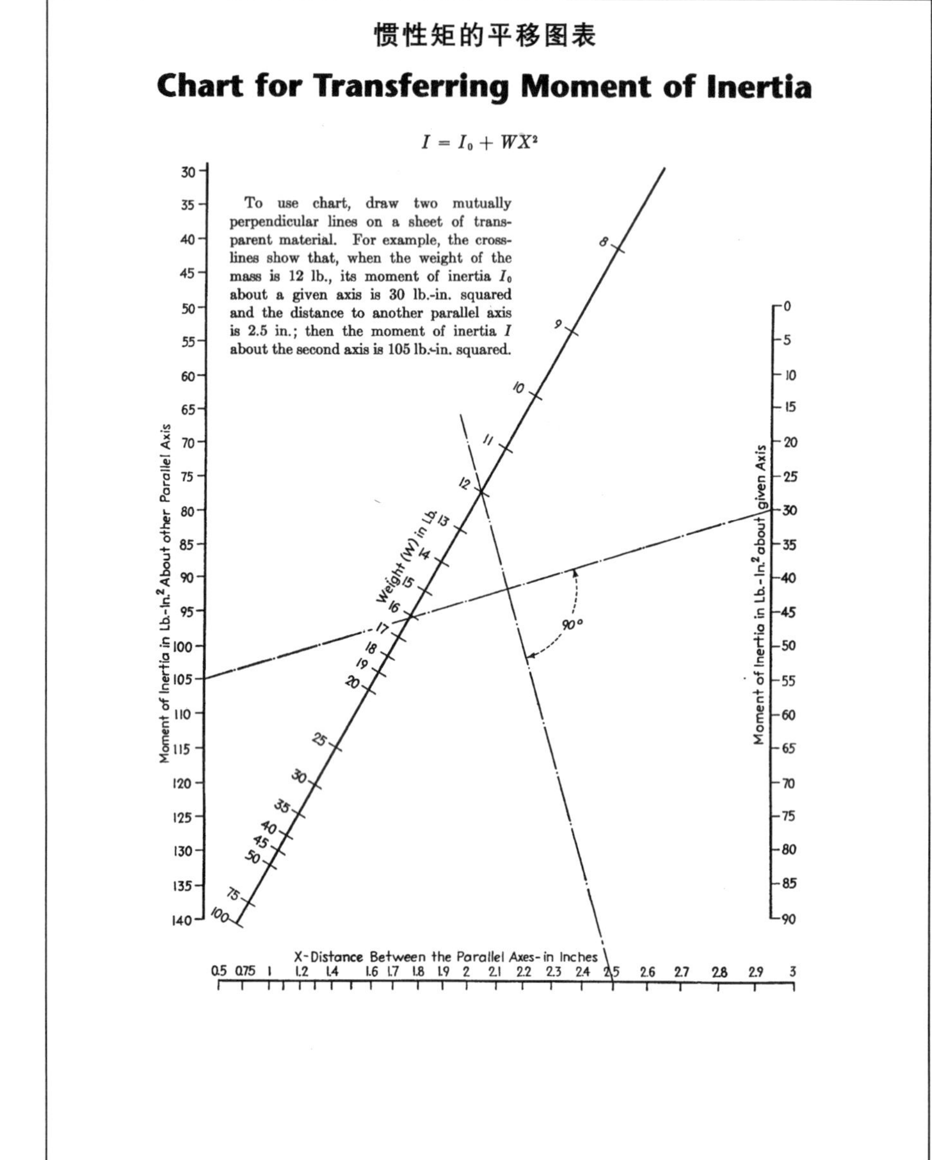

To use chart, draw two mutually perpendicular lines on a sheet of transparent material. For example, the cross-lines show that, when the weight of the mass is 12 lb., its moment of inertia I_0 about a given axis is 30 lb.-in. squared and the distance to another parallel axis is 2.5 in.; then the moment of inertia I about the second axis is 105 lb.-in. squared.

加速直线运动

Accelerated Linear Motion

$$\frac{2S}{T^2} = \frac{V}{2S} = \frac{V}{T} = \frac{32.16F}{W} = G$$

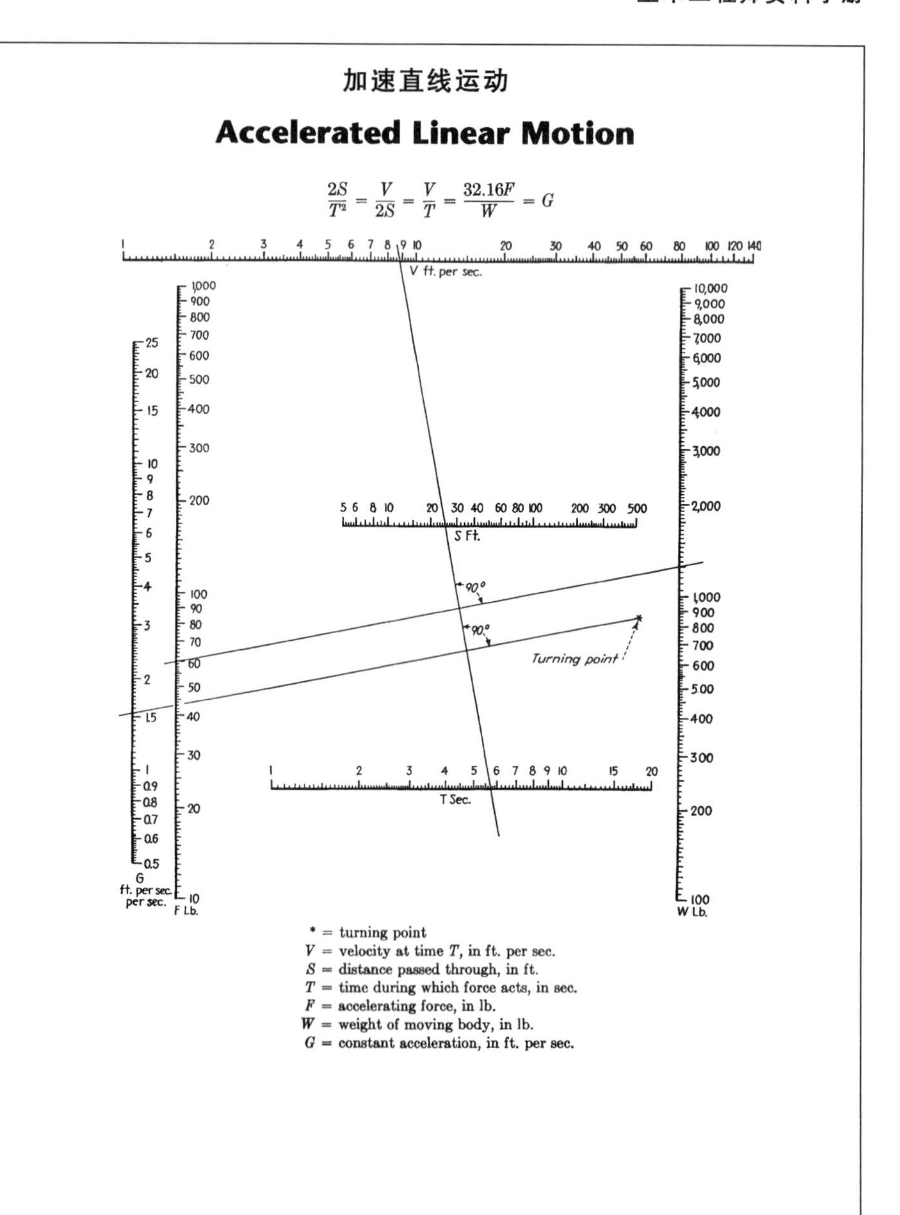

* = turning point
V = velocity at time T, in ft. per sec.
S = distance passed through, in ft.
T = time during which force acts, in sec.
F = accelerating force, in lb.
W = weight of moving body, in lb.
G = constant acceleration, in ft. per sec.

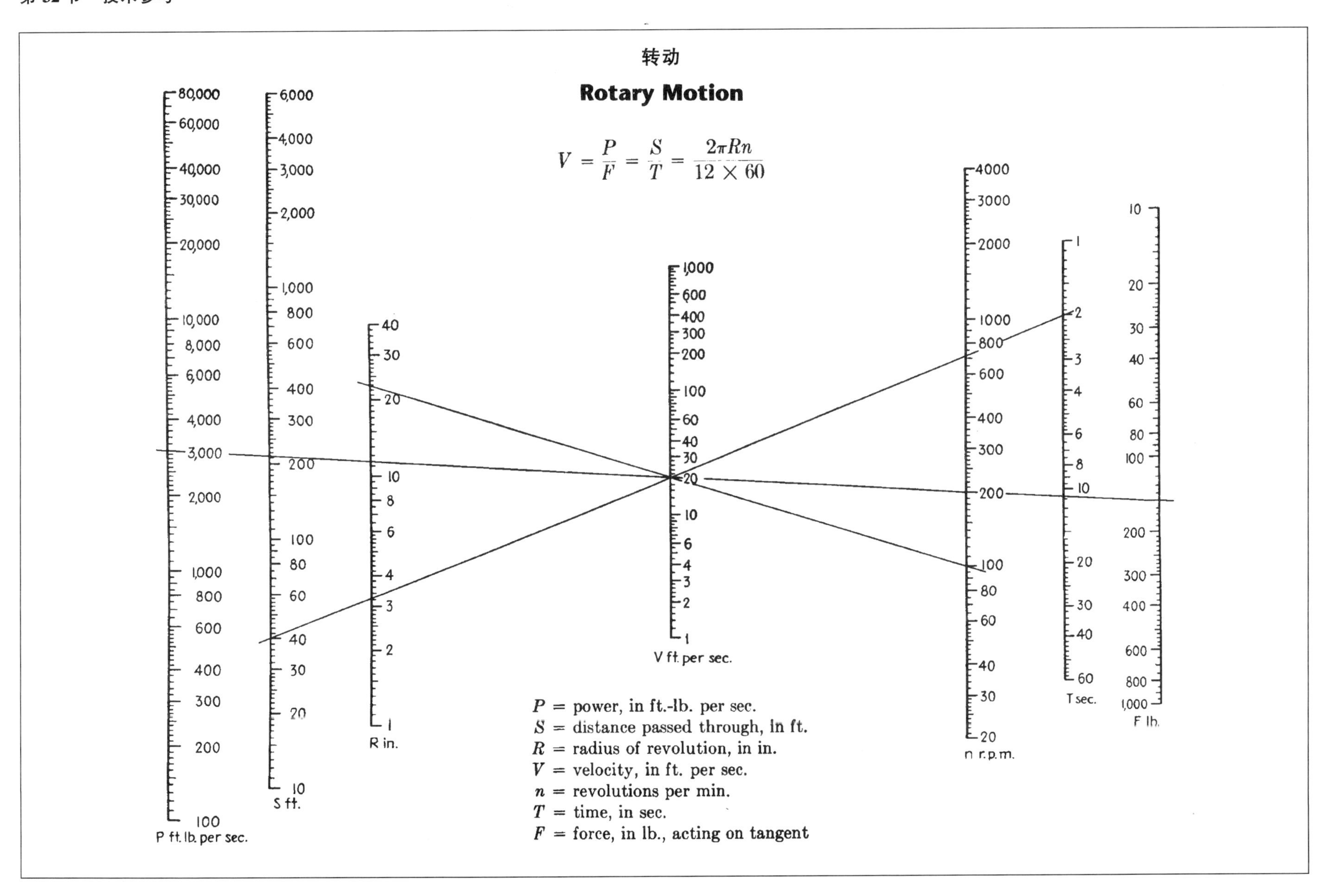
转动
Rotary Motion
$V = \frac{P}{F} = \frac{S}{T} = \frac{2\pi Rn}{12 \times 60}$
P ft. lb. per sec.
S ft.
R in.
V ft. per sec.
n r.p.m.
T sec.
F lb.
P = power, in ft.-lb. per sec.
S = distance passed through, in ft.
R = radius of revolution, in in.
V = velocity, in ft. per sec.
n = revolutions per min.
T = time, in sec.
F = force, in lb., acting on tangent

转动体的回转半径

Radii of Gyration for Rotating Bodies

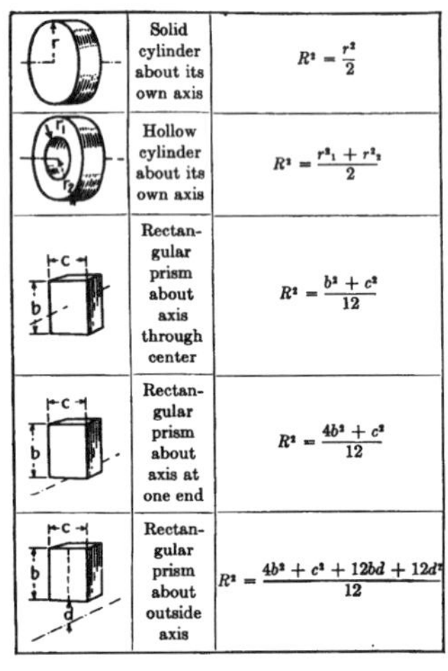

	Solid cylinder about its own axis	$R^2 = \frac{r^2}{2}$
	Hollow cylinder about its own axis	$R^2 = \frac{r_1^2 + r_2^2}{2}$
	Rectangular prism about axis through center	$R^2 = \frac{b^2 + c^2}{12}$
	Rectangular prism about axis at one end	$R^2 = \frac{4b^2 + c^2}{12}$
	Rectangular prism about outside axis	$R^2 = \frac{4b^2 + c^2 + 12bd + 12d^2}{12}$

	Cylinder about axis through center	$R^2 = \frac{l + 3r^2}{12}$
	Cylinder about axis at one end	$R^2 = \frac{4l^2 + 3r^2}{12}$
	Cylinder about outside axis	$R^2 = \frac{4l^2 + 3r^2 + 12dl + 12d^2}{12}$
Center of gravity; Center of rotation	Any body about axis outside its center of gravity $R_1^2 = R_0^2 + d^2$ where R_0 = radius of gyration about axis through center of gravity; R_1 = radius of gyration about any other parallel axis; d = distance between center of gravity and axis of rotation	

APPROXIMATIONS FOR CALCULATING MOMENTS OF INERTIA

Name of Part	Moment of Inertia
Flywheels (not applicable to belt pulleys)	Moment of inertia equal to 1.08 to 1.15 times that of rim alone
Flywheel (based on total weight and outside diameter)	Moment of inertia equal to two-thirds of that of total weight concentrated at the outer circumference
Spur or helical gears (teeth alone)	Moment of inertia of teeth equal to 40 per cent of that of a hollow cylinder of the limiting dimensions
Spur or helical gears (rim alone)	Figured as a hollow cylinder of same limiting dimensions
Spur or helical gears (total moment of inertia)	Equal to 1.25 times the sum of that of teeth plus rim
Spur or helical gears (with only weight and pitch diameter known)	Moment of inertia considered equal to 0.60 times the moment of inertia of the total weight concentrated at the pitch circle
Motor armature (based on total weight and outside diameter)	Multiply outer radius of armature by following factors to obtain radius of gyration:
	Large slow-speed motor 0.75–0.85
	Medium speed d-c or induction motor 0.70–0.80
	Mill-type motor 0.60–0.65

WR^2 OF SYMMETRICAL BODIES

For computing WR^2 of rotating masses of weight per unit volume ρ, by resolving the body into elemental shapes.

Note: ρ in pounds per cubic inch and dimensions in inches give WR^2 in lb.-in. squared.

1. Weights per Unit Volume of Materials.

Material	Weight, Lb. per Cu. In.
Cast iron	0.260
Cast-iron castings of heavy section i.e., flywheel rims	0.250
Steel	0.283
Bronze	0.319
Lead	0.410
Copper	0.318

2. Cylinder, about Axis Lengthwise through the Center of Gravity.

$$\text{Volume} = \frac{\pi}{4}L(D_1^2 - D_2^2)$$

(a) For any material:

$$WR^2 = \frac{\pi}{32}\rho L(D_1^4 - D_2^4)$$

where ρ is the weight per unit volume.

(b) For cast iron:

$$WR^2 = \frac{L(D_1^4 - D_2^4)}{39.2}$$

(c) For cast iron (heavy sections):

$$WR^2 = \frac{L(D_1^4 - D_2^4)}{40.75}$$

(d) For steel:

$$WR^2 = \frac{L(D_1^4 - D_2^4)}{36.0}$$

3. Cylinder, about an Axis Parallel to the Axis through Center of Gravity.

$$\text{Volume} = \frac{\pi}{4}L(D_1^2 - D_2^2)$$

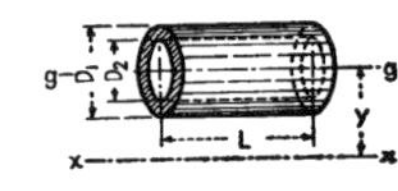

(a) For any material:

$$WR^2_{x-x} = \frac{\pi}{4}\rho L(D_1^2 - D_2^2)\left(\frac{D_1^2 + D_2^2}{8} + y^2\right)$$

(b) For steel:

$$WR^2_{x-x} = \frac{(D_1^2 - D_2^2)L}{4.50}\left(\frac{D_1^2 + D_2^2}{8} + y^2\right)$$

4. Solid Cylinder, Rotated about an Axis Parallel to a Line that Passes through the Center of Gravity and Is Perpendicular to the Center Line.

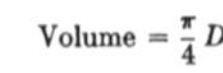

$$\text{Volume} = \frac{\pi}{4}D^2L$$

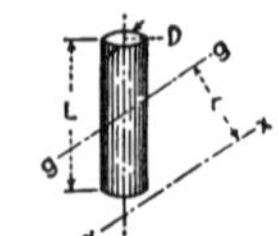

(a) For any material:

$$WR^2_{x-x} = \frac{\pi}{4}D^2L\rho\left(\frac{L^2}{12} + \frac{D^2}{16} + r^2\right)$$

(b) For steel:

$$WR^2_{x-x} = \frac{D^2L}{4.50}\left(\frac{L^2}{12} + \frac{D^2}{16} + r^2\right)$$

转动体的回转半径

Radii of Gyration for Rotating Bodies *Continued*

5. Rod of Rectangular or Elliptical Section, Rotated about an Axis Perpendicular to and Passing through the Center Line.

For rectangular cross sections:

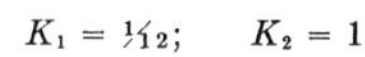

$$K_1 = 1/12; \qquad K_2 = 1$$

For elliptical cross sections:

$$K_1 = \frac{\pi}{64}; \qquad K_2 = \frac{\pi}{4}$$

$$\text{Volume} = K_2 abL$$

(*a*) For any material:

$$WR^2_{x'-x'} = \rho abL\left\{K_2\left[\frac{L^2}{3} + r_1(r_1 + L)\right] + K_1 a^2\right\}$$

(*b*) For a cast-iron rod of elliptical section ($\rho = 0.260$):

$$WR^2_{x'-x'} = \frac{abL}{4.90}\left[\frac{L^2}{3} + r_1(r_1 + L) + \frac{a^2}{16}\right]$$

6. Elliptical Cylinder, about an Axis Parallel to the Axis through the Center of Gravity.

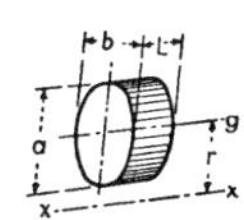

$$\text{Volume} = \frac{\pi}{4} abL$$

(*a*) For any material:

$$WR^2_{x-x} = \rho\frac{\pi}{4} abL\left(\frac{a^2 + b^2}{16} + r^2\right)$$

(*b*) For steel:

$$WR^2_{x-x} = \frac{abL}{4.50}\left(\frac{a^2 + b^2}{16} + r^2\right)$$

7. Cylinder with Frustum of a Cone Removed.

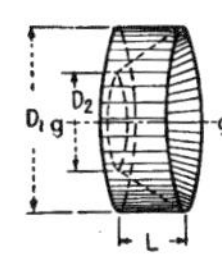

$$\text{Volume} = \frac{\pi L}{2(D_1 - D_2)}\left[\frac{1}{3}(D^3{}_1 - D^3{}_2) - \frac{D^2}{2}(D^2{}_1 - D^2{}_2)\right]$$

$$WR^2_{g-g} = \frac{\pi\rho L}{8(D_1 - D_2)}\left[\frac{1}{5}(D^5{}_1 - D^5{}_2) - \frac{D_2}{4}(D^4{}_1 - D^4{}_2)\right]$$

8. Frustum of a Cone with a Cylinder Removed.

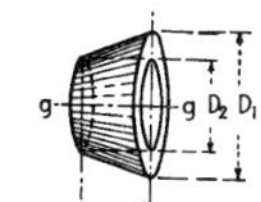

$$\text{Volume} = \frac{\pi L}{2(D_1 - D_2)}\left[\frac{D_1}{2}(D^2{}_1 - D^2{}_2) - \frac{1}{3}(D^3{}_1 - D^3{}_2)\right]$$

$$WR^2_{g-g} = \frac{\pi\rho L}{8(D_1 - D_2)}\left[\frac{D_1}{4}(D^4{}_1 - D^4{}_2) - \frac{1}{5}(D^5{}_1 - D^5{}_2)\right]$$

9. Solid Frustum of a Cone.

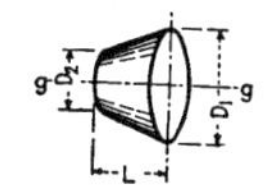

$$\text{Volume} = \frac{\pi L}{12}\frac{(D^3{}_1 - D^3{}_2)}{(D_1 - D_2)}$$

$$WR^2_{g-g} = \frac{\pi\rho L}{160}\frac{(D^5{}_1 - D^5{}_2)}{(D_1 - D_2)}$$

10. Chamfer Cut from Rectangular Prism Having One End Turned about a Center.

Distance to center of gravity, where $A = R_2/R_1$ and $B = C/2R_1$

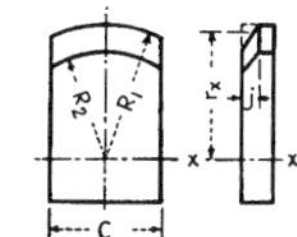

$$r_x = \frac{jR^3{}_1B}{\text{volume} \times (1 - A)}\left[\frac{1}{3}(A^3 - 3A + 2) + \frac{B^2}{3}\left(1 - A - A\log_e\frac{1}{A}\right) + \frac{3}{40}\frac{B^4}{A}(A^2 - 2A + 1) + \frac{5}{672}\frac{B^6}{A^3}(3A^4{}_1 - 4A^3 + 1)\cdots\right]$$

$$\text{Volume} = \frac{jR^2{}_1B}{(1 - A)}\left\{(A^2 - 2A + 1) + \frac{B^2}{3}\left[\log_e\frac{1}{A} - (1 - A)\right] + \frac{1}{40}\frac{B^4}{A^2}(2A^3 - 3A + 1) + \frac{1}{224}\frac{B^6}{A^4}(4A^5 - 5A^4 + 1) + \cdots\right\}$$

$$WR^2_{x-x} = -\frac{\rho jR^4{}_1B}{6(1 - A)}\left\{(A^4 - 4A + 3) + B^2(A^2 - 2A + 1) + \frac{9}{10}B^4\left[\log_e\frac{1}{A} - (1 - A)\right] + \frac{5}{56}\frac{B^6}{A^2}(2A^3 - 3A^2 + 1) + \cdots\right\}$$

11. Complete Torus.

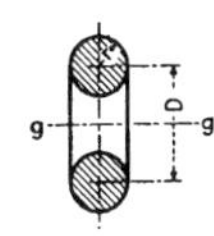

$$\text{Volume} = \pi^2 Dr^2$$

$$WR^2_{g-g} = \frac{\pi^2\rho Dr^2}{4}(D^2 + 3r^2)$$

12. Outside Part of a Torus.

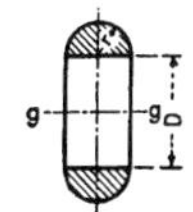

$$\text{Volume} = 2\pi r^2\left(\frac{\pi D}{4} + \frac{2}{3}r\right)$$

$$WR^2_{g-g} = \pi\rho r^2\left[\frac{D^2}{4}\left(\frac{\pi D}{2} + 4r\right) + r^2\left(\frac{3\pi}{8}D + \frac{8}{15}r\right)\right]$$

转动体的回转半径

Radii of Gyration for Rotating Bodies *Continued*

13. Inside Part of a Torus.

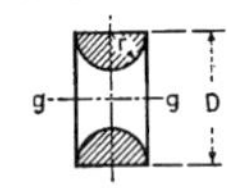

$$\text{Volume} = 2\pi r^2\left(\frac{\pi D}{4} - \frac{2}{3}r\right)$$

$$WR^2_{g-g} = \pi\rho r^2\left[\frac{D^2}{4}\left(\frac{\pi D}{2} - 4r\right) + r^2\left(\frac{3\pi}{8}D - \frac{8}{15}r\right)\right]$$

14. Circular Segment about an Axis through Center of Circle.

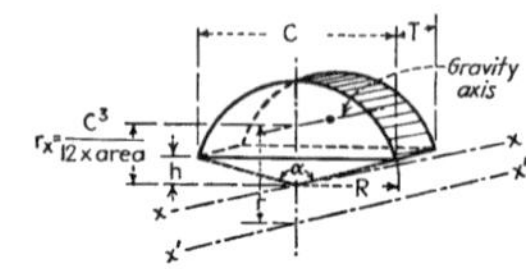

$$\alpha = 2\sin^{-1}\frac{C}{2R}\ \text{deg.}$$

$$\text{Area} = \frac{R^2\alpha}{114.59} - \frac{C}{2}\sqrt{R^2 - \frac{C^2}{4}}$$

(*a*) Any material:

$$WR^2_{x-x} = \rho T\left[\frac{R^4\alpha}{229.2} - \frac{1}{6}\left(3R^2 - \frac{C^2}{2}\right)\frac{C}{2}\sqrt{R^2 - \frac{C^2}{4}}\right]$$

(*b*) For steel:

$$WR^2_{x-x} = \frac{T}{3.534}\left[\frac{R^4\alpha}{229.2} - \frac{1}{6}\left(3R^2 - \frac{C^2}{2}\right)\frac{C}{2}\sqrt{R^2 - \frac{C^2}{4}}\right]$$

15. Circular Segment about Any Axis Parallel to an Axis through the Center of the Circles. (Refer to 14 for Figure.)

$$WR^2_{x'-x'} = WR^2_{x-x} + \text{weight}\,(r^2 - r^2_x)$$

16. Rectangular Prism about an Axis Parallel to the Axis through the Center of Gravity.

$$\text{Volume} = WLT$$

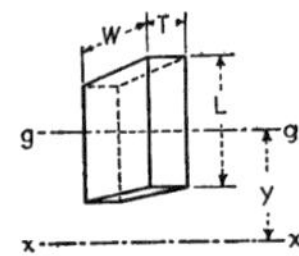

(*a*) For any material:

$$WR^2_{x-x} = \rho WLT\left(\frac{W^2 + L^2}{12} + y^2\right)$$

(*b*) For steel:

$$WR^2_{x-x} = \frac{WLT}{3.534}\left(\frac{W^2 + L^2}{12} + y^2\right)$$

17. Isosceles Triangular Prism, Rotated about an Axis through Its Vertex.

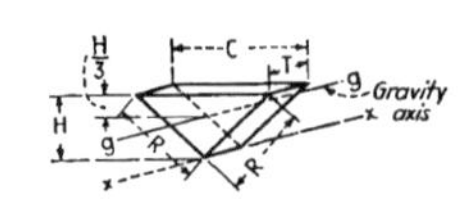

$$\text{Volume} = \frac{CHT}{2}$$

$$WR^2_{x-x} = \frac{\rho CHT}{2}\left(\frac{R^2}{2} - \frac{C^2}{12}\right)$$

18. Isosceles Triangular Prism, Rotated about Any Axis Parallel to an Axis through the Vertex.

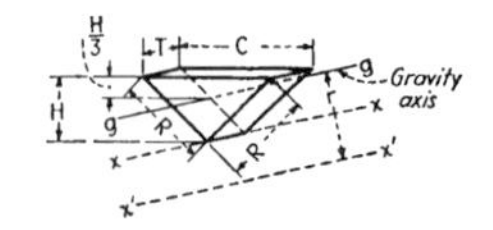

$$\text{Volume} = \frac{CHT}{2}$$

$$WR^2_{x'-x'} = \frac{\rho CHT}{2}\left(\frac{R^2}{2} - \frac{C^2}{12} - \frac{4}{9}H^2 + r^2\right)$$

19. Prism with Square Cross Section and Cylinder Removed, along Axis through Center of Gravity of Square.

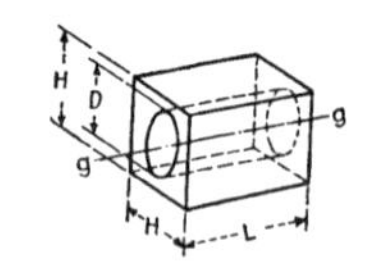

$$\text{Volume} = L\left(H^2 - \frac{\pi D^2}{4}\right)$$

$$WR^2_{g-g} = \frac{\pi\rho L}{32}(1.697H^4 - D^4)$$

20. Any Body about an Axis Parallel to the Gravity Axis, When WR^2 about the Gravity Axis Is Known.

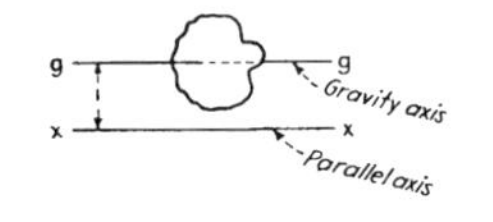

$$WR^2_{x-x} = WR^2_{g-g} + \text{weight} \times r^2$$

21. WR^2 of a Piston, Effective at the Cylinder Center Line, about the Crankshaft Center Line.

$$WR^2 = r^2W_p\left(\frac{1}{2} + \frac{r^2}{8L^2}\right)$$

where r = crank radius
L = center-to-center length of connecting rod
W_p = weight of complete piston, rings, and pin

转动体的回转半径

Radii of Gyration for Rotating Bodies *Continued*

22. WR^2 of a Connecting Rod, Effective at the Cylinder Center Line, about the Crankshaft Center Line.

$$WR^2 = r^2\left[W_1 + W_2\left(\frac{1}{2} + \frac{r^2}{8L^2}\right)\right]$$

where r = crank radius
L = center-to-center length of connecting rod
W_1 = weight of the lower or rotating part of the rod = $[W_R(L - L_1)]/L$
W_2 = weight of the upper or reciprocating part of the rod = $W_R L_1/L$
$W_R = W_1 + W_2$, the weight of the complete rod
L_1 = distance from the center line of the crankpin to the center of gravity of the connecting rod

23. Mass Geared to a Shaft.—The equivalent flywheel effect at the shaft in question is

$$WR^2 = h^2(WR^2)'$$

where h = gear ratio $= \dfrac{\text{r.p.m. of mass geared to shaft}}{\text{r.p.m. of shaft}}$
$(WR^2)'$ = flywheel effect of the body in question about its own axis of rotation

24. Mass Geared to Main Shaft and Connected by a Flexible Shaft.—The effect of the mass $(WR^2)'$ at the position of the driving gear on the main shaft is

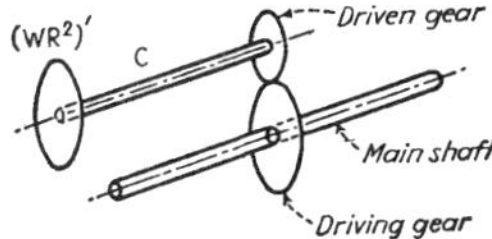

$$WR^2 = \frac{h^2(WR^2)'}{1 - \dfrac{(WR^2)'f^2}{9.775C}}$$

where h = gear ratio $= \dfrac{\text{r.p.m. of driven gear}}{\text{r.p.m. of driving gear}}$
$(WR^2)'$ = flywheel effect of geared-on mass
f = natural torsional frequency of the shafting system, in vibrations per sec.
C = torsional rigidity of flexible connecting shaft, in pound-inches per radian

25. Belted Drives.—The equivalent flywheel effect of the driven mass at the driving shaft is

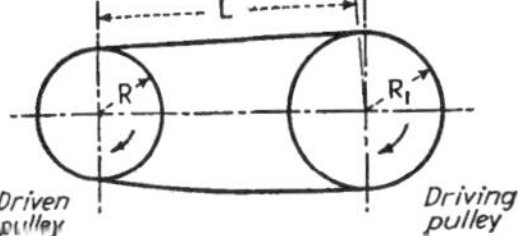

$$WR^2 = \frac{h^2(WR^2)'}{1 - \dfrac{(WR^2)'f^2}{9.775C}}$$

where $h = R_1/R$ $= \dfrac{\text{r.p.m. of pulley belted to shaft}}{\text{r.p.m. of shaft}}$
$(WR^2)'$ = flywheel effect of the driven body about its own axis of rotation
f = natural torsional frequency of the system, in vibrations per sec.
$C = R^2AE/L$
A = cross-sectional area of belt, in sq. in.
E = modulus of elasticity of belt material in tension, in lb. per sq. in.
R = radius of driven pulley, in in.
L = length of tight part of belt which is clear of the pulley, in in.

26. Effect of the Flexibility of Flywheel Spokes on WR^2 of Rim.—The effective WR^2 of the rim is

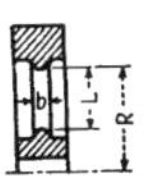

$$WR^2 = \frac{(WR^2)'}{1 - \dfrac{(WR^2)'f^2}{9.775C}}$$

where $(WR^2)'$ = flywheel effect of the rim
f = natural torsional frequency of the system of which the flywheel is a member, in vibrations per sec.
C = torque required to move the rim through one radian relative to the hub

$$C = \frac{12_g Eka^3bR}{L^2}\left(\frac{L}{3R} + \frac{R}{L} - 1\right)$$

where g = number of spokes
E = bending modulus of elasticity of the spoke material
$k = \pi/64$ for elliptical, and $k = 1/12$ for rectangular section spokes

All dimensions are in inches.

For cast-iron spokes of elliptical section:

$$E = 15 \times 10^6 \text{ lb. per sq. in.}$$

$$C = \frac{ga^3bR \times 10^6}{0.1132L^2}\left(\frac{L}{3R} + \frac{R}{L} - 1\right)\frac{\text{lb.-in.}}{\text{radians}}.$$

Note: It is found by comparative calculations that with spokes of moderate taper very little error is involved in assuming the spoke to be straight and using cross section at mid-point for area calculation.

TYPICAL EXAMPLE

The flywheel shown below is used in a Diesel engine installation. It is required to determine effective WR^2 for calculation of one of the natural frequencies of torsional vibration. The anticipated natural frequency of the system is 56.4 vibrations per sec.

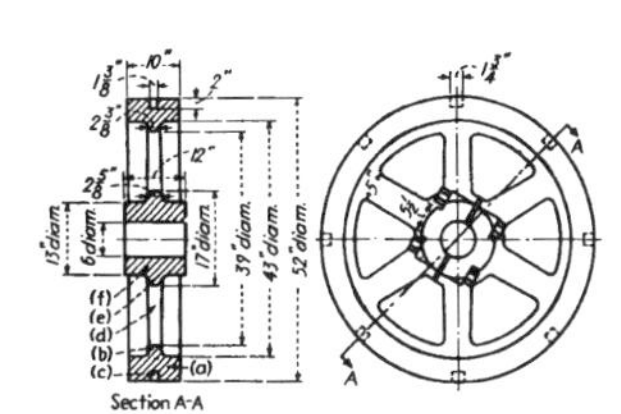

Note: Since the beads at the ends of the spokes comprise but a small part of the flywheel WR^2, very little error will result in assuming them to be of rectangular cross section. Also, because of the effect of the clamping bolts, the outer hub will be considered a square equal to the diameter. The spokes will be assumed straight and of mid-point cross section.

Part of fly wheel	Formula	WR^2
(a)	2c	$\dfrac{10[(52)^4 - (43)^4]}{40.75} = 955{,}300$
(b)	2b	$\dfrac{2.375[(43)^4 - (39)^4]}{39.2} = 67{,}000$
(c)	16a neglecting $\left(\dfrac{W^2 + L^2}{12}\right)$	$-0.250 \times 1.75 \times 2 \times 1.375(25)^2 \times 8 = -6{,}000$
		Total for rim = 1,016,300 lb.-in. squared
(d)	5b	$6 \times \dfrac{5.25 \times 2.5 \times 11}{4.90}\left[\dfrac{(11)^2}{3} + 8.5(8.5+11) + \dfrac{(5.25)^2}{16}\right] = 36{,}800$
(e)	2b	$\dfrac{2.625[(17)^4 - (13)^4]}{39.2} = 3{,}700$
(f)	19	$\dfrac{\pi \times 0.250 \times 12}{32}[1.697 \times (13)^4 - (6)^4] = 13{,}900$
		Total for remainder of flywheel = 54,400 lb.-in. squared

From formula (26)

$$C = \frac{6 \times (5.25)^3 \times 2.5 \times 19.5 \times 10^6}{0.1132 \times (11)^2}\left(\frac{11}{3 \times 19.5} + \frac{19.5}{11} - 1\right) = 2{,}970 \times 10^6 \frac{\text{lb.-in.}}{\text{radians}}$$

$$\text{and } WR^2 = \frac{1{,}016{,}300}{1 - \dfrac{1{,}016{,}300 \times (56.4)^2}{9.775 \times 2{,}970 \times 10^6}} + 54{,}400$$

$$= \underline{1{,}197{,}000} \text{ lb.-in. squared}$$

国际单位制基本单位

Basic SI Units

Quantity	*Unit*
length	meter (m)
mass	kilogram (kg)
time	second (s)
electric current	ampere (A)
temperature (thermodynamic)	kelvin (K)
amount of substance	mole (mol)
luminous intensity	candela (cd)

Source: R.O.Parmley, Field Engineer's Manual, 3/e © 2002
Published by McGraw-Hill and reproduced with
permission of the McGraw-Hill Companies.

国际单位制词头

Prefixes for SI Units

Multiple and submultiple	*Prefix*	*Symbol*
$1,000,000,000,000 = 10^{12}$	tera	T
$1,000,000,000 = 10^{9}$	giga	G
$1,000,000 = 10^{6}$	mega	M
$1,000 = 10^{3}$	kilo	k
$100 = 10^{2}$	hecto	h
$10 = 10$	deka	da
$0.1 = 10^{-1}$	deci	d
$0.01 = 10^{-2}$	centi	c
$0.001 = 10^{-3}$	milli	m
$0.000\ 001 = 10^{-6}$	micro	μ
$0.000\ 000\ 001 = 10^{-9}$	nano	n
$0.000\ 000\ 000\ 001 = 10^{-12}$	pico	p
$0.000\ 000\ 000\ 000\ 001 = 10^{-15}$	femto	f
$0.000\ 000\ 000\ 000\ 000\ 001 = 10^{-18}$	atto	a

Source: R.O.Parmley, Field Engineer's Manual, 3/e © 2002
Published by McGraw-Hill and reproduced with
permission of the McGraw-Hill Companies.

土木工程和力学工程中采用的国际单位制导出单位

SI Derived Units in Civil & Mechanical Engineering

Quantity	*Usual units*	*Symbol*
water usage	liter per person day	l/person·d
runoff	cubic meter per square kilometer	m^3/km^2
precipitation	millimeter per hour	mm/h
river flow	cubic meter per second	m^3/s
discharge	cubic meter per day	m^3/d
hydraulic load per unit area, e.g., filtration rates	cubic meter per square meter day	$m^3/m^2\cdot d$
concentration	gram per cubic meter	g/m^3
	milligram per liter	mg/l
BOD produced	kilogram per day	kg/d
BOD loading	kilogram per cubic meter day	$kg/m^3\cdot d$
hydraulic load per unit volume, e.g., biological filters	cubic meter per cubic meter day	$m^3/m^3\cdot d$
sludge treatment	square meter per person	m^2/person
flow in pipes, channels, etc., peak demands; surface water runoff	cubic meters per second	m^3/s
flow in small pipes; demands of sanitary fittings; pumping rates	liters per second	l/s
water demand; water supply	cubic meters per day	m^3/d
force	kilonewtons	kN
moment of force	newton meter	Nm
	kilonewton meter	kNm

Source: *McGraw-Hill Metrication Manual*. Copyright © 1971 by McGraw-Hill, Inc. New York. (Used with permission of the McGraw-Hill Co.)

国际体系的导出单位

Derived Units of the International System

Quantity	*Name of unit*	*Unit symbol or abbreviation, where differing from basic form*	*Unit expressed in terms of basic or supplementary units**
area	square meter		m^2
volume	cubic meter		m^3
frequency	hertz, cycle per second†	Hz	s^{-1}
density	kilogram per cubic meter		kg/m^3
velocity	meter per second		m/s
angular velocity	radian per second		rad/s
acceleration	meter per second squared		m/s^2
angular acceleration	radian per second squared		rad/s^2
volumetric flow rate	cubic meter per second		m^3/s
force	newton	N	$kg\cdot m/s^2$
surface tension	newton per meter, joule per square meter	N/m, J/m^2	kg/s^2
pressure	newton per square meter, pascal†	N/m^2, Pa†	$kg/m\cdot s^2$
viscosity, dynamic	newton-second per square meter, poiseuille†	N s/m^2, Pl†	kg/m·s
viscosity, kinematic	meter squared per second		m^2/s
work, torque, energy, quantity of heat	joule, newton-meter, watt-second	J, N·m, W·s	$kg\cdot m^2/s^2$
power, heat flux	watt, joule per second	W, J/s	$kg\cdot m^2/s^3$
heat flux density	watt per square meter	W/m^2	kg/s^3
volumetric heat release rate	watt per cubic meter	W/m^3	$kg/m\cdot s^3$
heat transfer coefficient	watt per square meter degree	$W/m^2\cdot deg$	$kg/s^3\cdot deg$
heat capacity (specific)	joule per kilogram degree	J/kg·deg	$m^2/s^2\cdot deg$
capacity rate	watt per degree	W/deg	$kg\cdot m^2/s^3\cdot deg$
thermal conductivity	watt per meter degree	W/m·deg, $\frac{Jm}{s\cdot m^2\cdot deg}$	$kg\cdot m/s^3\cdot deg$
quantity of electricity	coulomb	C	A·s
electromotive force	volt	V, W/A	$kg\cdot m^2/A\cdot s^3$
electric field strength	volt per meter		V/m
electric resistance	ohm	Ω, V/A	$kg\cdot m^2/A^2\cdot s^3$
electric conductivity	ampere per volt meter	A/V·m	$A^2 s^3/kg\cdot m^3$
electric capacitance	farad	F, A·s/V	$A^3 s^1/kg\cdot m^2$
magnetic flux	weber	Wb, V·s	$kg\cdot m^2/A\cdot s^2$
inductance	henry	H, V·s/A	$kg\cdot m^2/A^2 s^2$
magnetic permeability	henry per meter	H/m	$kg\cdot m/A^2 s^2$
magnetic flux density	tesla, weber per square meter	T, Wb/m^2	$kg/A\cdot s^2$
magnetic field strength	ampere per meter		A/m
magnetomotive force	ampere		A
luminous flux	lumen	lm	cd sr
luminance	candela per square meter		cd/m^2
illumination	lux, lumen per square meter	lx, lm/m^2	$cd\cdot sr/m^2$

*Supplementary units are plane angle, radian (rad); solid angle, steradian (sr).
†Not used in all countries.

Source: *McGraw-Hill Metrication Manual*. Copyright © 1971 by McGraw-Hill, Inc. New York. (Used with permission of the McGraw-Hill Co.)

Conversion Factors as Exact Numerical Multiples of SI Units 国际单位制的转换系数

The following tables express the definitions of various units of measure as exact numerical multiples of coherent SI units, and provide multiplying factors for converting numbers and miscellaneous units to corresponding new numbers and SI units.

The first two digits of each numerical entry represent a power of 10. An asterisk follows each number which expresses an exact definition. For example, the entry "–02 2.54*" expresses the fact that 1 inch = 2.54×10^{-2} meter, exactly, by definition. Most of the definitions are extracted from National Bureau of Standards documents. Numbers not followed by an asterisk are only approximate representations of definitions or are the results of physical measurements. The conversion factors are listed alphabetically and by physical quantity.

The Listing by Physical Quantity includes only relationships which are frequently encountered and deliberately omits the great multiplicity of combinations of units which are used for more specialized purposes. Conversion factors for combinations of units are easily generated from numbers given in the Alphabetical Listing by the technique of direct substitution or by other well-known rules for manipulating units. These rules are adequately discussed in many science and engineering textbooks and are not repeated here.

ALPHABETICAL LISTING

To convert from	*to*	*multiply by*
abampere	ampere	+01 1.00*
abcoulomb	coulomb	+01 1.00*
abfarad	farad	+09 1.00*
abhenry	henry	–09 1.00*
abmho	mho	+09 1.00*
abohm	ohm	–09 1.00*
abvolt	volt	–08 1.00*
acre	meter²	+03 4.046 856 422 4*
ampere (international of 1948)	ampere	–01 9.998 35
angstrom	meter	–10 1.00*
are	meter²	+02 1.00*
astronomical unit	meter	+11 1.495 978 9
atmosphere	newton/meter²	+05 1.013 25*
bar	newton/meter²	+05 1.00*
barn	meter²	–28 1.00*
barrel (petroleum, 42 gallons)	meter³	–01 1.589 873
barye	newton/meter²	–01 1.00*
British thermal unit (ISO/TC 12)	joule	+03 1.055 06
British thermal unit (International Steam Table)	joule	+03 1.055 04
British thermal unit (mean)	joule	+03 1.055 87
British thermal unit (thermochemical)	joule	+03 1.054 350 264 488
British thermal unit (39° F)	joule	+03 1.059 67
British thermal unit (60° F)	joule	+03 1.054 68
bushel (U.S.)	meter³	–02 3.523 907 016 688*
cable	meter	+02 2.194 56*
caliber	meter	–04 2.54*
calorie (International Steam Table)	joule	+00 4.1868
calorie (mean)	joule	+00 4.190 02
calorie (thermochemical)	joule	+00 4.184*
calorie (15° C)	joule	+00 4.185 80
calorie (20° C)	joule	+00 4.181 90
calorie (kilogram, International Steam Table)	joule	+03 4.1868
calorie (kilogram, mean)	joule	+03 4.190 02
calorie (kilogram, thermochemical)	joule	+03 4.184*
carat (metric)	kilogram	–04 2.00*
Celsius (temperature)	kelvin	$t_K = t_C + 273.15$
centimeter of mercury (0° C)	newton/meter²	+03 1.333 22
centimeter of water (4° C)	newton/meter²	+01 9.806 38
chain (engineer or ramden)	meter	+01 3.048*
chain (surveyor or gunter)	meter	+01 2.011 68*
circular mil	meter²	–10 5.067 074 8
cord	meter³	+00 3.624 556 3
coulomb (international of 1948)	coulomb	–01 9.998 35
cubit	meter	–01 4.572*
cup	meter³	–04 2.365 882 365*
curie	disintegration/second	+10 3.70*
day (mean solar)	second (mean solar)	+04 8.64*
day (sidereal)	second (mean solar)	+04 8.616 409 0
degree (angle)	radian	–02 1.745 329 251 994 3
denier (international)	kilogram/meter	–07 1.00*
dram (avoirdupois)	kilogram	–03 1.771 845 195 312 5*
dram (troy or apothecary)	kilogram	–03 3.887 934 6*
dram (U.S. fluid)	meter³	–06 3.696 691 195 312 5*
dyne	newton	–05 1.00*
electron volt	joule	–19 1.602 10
erg	joule	–07 1.00*
Fahrenheit (temperature)	kelvin	$t_K = (5/9)\ (t_F + 459.67)$
Fahrenheit (temperature)	Celsius	$t_C = (5/9)\ (t_F - 32)$

Source: *McGraw-Hill Metrication Manual*. Copyright © 1971 by McGraw-Hill, Inc. New York. (Used with permission of the McGraw-Hill Co.)

Conversion Factors as Exact Numerical Multiples of SI Units *(Continued)*　国际单位制的转换系数

To convert from	*to*	*multiply by*
farad (international of 1948)	farad	−01 9.995 05
faraday (based on carbon 12)	coulomb	+04 9.648 70
faraday (chemical)	coulomb	+04 9.649 57
faraday (physical)	coulomb	+04 9.652 19
fathom	meter	+00 1.828 8*
fermi (femtometer)	meter	−15 1.00*
fluid ounce (U.S.)	meter^3	−05 2.957 352 956 25*
foot	meter	−01 3.048*
foot (U.S. survey)	meter	+00 1200/3937*
foot (U.S. survey)	meter	−01 3.048 006 096
foot of water (39.2° F)	newton/meter^2	+03 2.988 98
foot-candle	lumen/meter^2	+01 1.076 391 0
foot-lambert	candela/meter^2	+00 3.426 259
furlong	meter	+02 2.011 68*
gal (galileo)	meter/second^2	−02 1.00*
gallon (U.K. liquid)	meter^3	−03 4.546 087
gallon (U.S. dry)	meter^3	−03 4.404 883 770 86*
gallon (U.S. liquid)	meter^3	−03 3.785 411 784*
gamma	tesla	−09 1.00*
gauss	tesla	−04 1.00*
gilbert	ampere turn	−01 7.957 747 2
gill (U.K.)	meter^3	−04 1.420 652
gill (U.S.)	meter^3	−04 1.182 941 2
grad	degree (angular)	−01 9.00*
grad	radian	−02 1.570 796 3
grain	kilogram	−05 6.479 891*
gram	kilogram	−03 1.00*
hand	meter	−01 1.016*
hectare	meter^2	+04 1.00*
henry (international of 1948)	henry	+00 1.000 495
hogshead (U.S.)	meter^3	−01 2.384 809 423 92*
horsepower (550 foot lbf/second)	watt	+02 7.456 998 7
horsepower (boiler)	watt	+03 9.809 50
horsepower (electric)	watt	+02 7.46*
horsepower (metric)	watt	+02 7.354 99
horsepower (U.K.)	watt	+02 7.457
horsepower (water)	watt	+02 7.460 43
hour (mean solar)	second (mean solar)	+03 3.60*
hour (sidereal)	second (mean solar)	+03 3.590 170 4
hundredweight (long)	kilogram	+01 5.080 234 544*
hundredweight (short)	kilogram	+01 4.535 923 7*
inch	meter	−02 2.54*
inch of mercury (32° F)	newton/meter^2	+03 3.386 389
inch of mercury (60° F)	newton/meter^2	+03 3.376 85
inch of water (39.2° F)	newton/meter^2	+02 2.490 82
inch of water (60° F)	newton/meter^2	+02 2.4884
joule (international of 1948)	joule	+00 1.000 165
kayser	1/meter	+02 1.00*
kilocalorie (International Steam Table)	joule	+03 4.186 74
kilocalorie (mean)	joule	+03 4.190 02
kilocalorie (thermochemical)	joule	+03 4.184*
kilogram mass	kilogram	+00 1.00*
kilogram force (kgf)	newton	+00 9.806 65*
kilopond force	newton	+00 9.806 65*
kip	newton	+03 4.448 221 615 260 5*
knot (international)	meter/second	−01 5.144 444 444
lambert	candela/meter^2	+04 1/π*
lambert	candela/meter^2	+03 3.183 098 8
langley	joule/meter^2	+04 4.184*
lbf (pound force, avoirdupois)	newton	+00 4.448 221 615 260 5*
lbm (pound mass, avoirdupois)	kilogram	−01 4.535 923 7*
league (British nautical)	meter	+03 5.559 552*
league (international nautical)	meter	+03 5.556*
league (statute)	meter	+03 4.828 032*
light year	meter	+15 9.460 55
link (engineer or ramden)	meter	−01 3.048*
link (surveyor or gunter)	meter	−01 2.011 68*
liter	meter^3	−03 1.00*
lux	lumen/meter^2	+00 1.00*
maxwell	weber	−08 1.00*
meter	wavelengths Kr 86	+06 1.650 763 73*
micron	meter	−06 1.00*
mil	meter	−05 2.54*
mile (U.S. statute)	meter	+03 1.609 344*
mile (U.K. nautical)	meter	+03 1.853 184*
mile (international nautical)	meter	+03 1.852*
mile (U.S. nautical)	meter	+03 1.852*
millibar	newton/meter^2	+02 1.00*
millimeter of mercury (0° C)	newton/meter^2	+02 1.333 224
minute (angle)	radian	−04 2.908 882 086 66
minute (mean solar)	second (mean solar)	+01 6.00*
minute (sidereal)	second (mean solar)	+01 5.983 617 4
month (mean calendar)	second (mean solar)	+06 2.628*
nautical mile (international)	meter	+03 1.852*
nautical mile (U.S.)	meter	+03 1.852*
nautical mile (U.K.)	meter	+03 1.853 184*
oersted	ampere/meter	+01 7.957 747 2
ohm (international of 1948)	ohm	+00 1.000 495
ounce force (avoirdupois)	newton	−01 2.780 138 5
ounce mass (avoirdupois)	kilogram	−02 2.834 952 312 5*
ounce mass (troy or apothecary)	kilogram	−02 3.110 347 68*
ounce (U.S. fluid)	meter^3	−05 2.957 352 956 25*
pace	meter	−01 7.62*
parsec	meter	+16 3.083 74
pascal	newton/meter^2	+00 1.00*
peck (U.S.)	meter^3	−03 8.809 767 541 72*
pennyweight	kilogram	−03 1.555 173 84*
perch	meter	+00 5.0292*
phot	lumen/meter^2	+04 1.00
pica (printers)	meter	−03 4.217 517 6*
pint (U.S. dry)	meter^3	−04 5.506 104 713 575*
pint (U.S. liquid)	meter^3	−04 4.731 764 73*
point (printers)	meter	−04 3.514 598*
poise	newton second/meter^2	−01 1.00*
pole	meter	+00 5.0292*
pound force (lbf avoirdupois)	newton	+00 4.448 221 615 260 5*

Conversion Factors as Exact Numerical Multiples of SI Units *(Continued)* 国际单位制的转换系数

To convert from	to	multiply by
pound mass (lbm avoirdupois)	kilogram	−01 4.535 923 7*
pound mass (troy or apothecary)	kilogram	−01 3.732 417 216*
poundal	newton	−01 1.382 549 543 76*
quart (U.S. dry)	meter^3	−03 1.101 220 942 715*
quart (U.S. liquid)	meter^3	−04 9.463 529 5
rad (radiation dose absorbed)	joule/kilogram	−02 1.00*
Rankine (temperature)	kelvin	$t_K=(5/9)t_R$
rayleigh (rate of photon emission)	1/second meter^2	+10 1.00*
rhe	meter^2/newton second	+01 1.00*
rod	meter	+00 5.0292*
roentgen	coulomb/kilogram	−04 2.579 76*
rutherford	disintegration/second	+06 1.00*
second (angle)	radian	−06 4.848 136 811
second (ephemeris)	second	+00 1.000 000 000
second (mean solar)	second (ephemeris)	Consult American Ephemeris and Nautical Almanac
second (sidereal)	second (mean solar)	−01 9.972 695 7
section	meter^2	+06 2.589 988 110 336*
scruple (apothecary)	kilogram	−03 1.295 978 2*
shake	second	−08 1.00
skein	meter	+02 1.097 28*
slug	kilogram	+01 1.459 390 29
span	meter	−01 2.286*
statampere	ampere	−10 3.335 640
statcoulomb	coulomb	−10 3.335 640
statfarad	farad	−12 1.112 650
stathenry	henry	+11 8.987 554
statmho	mho	−12 1.112 650
statohm	ohm	+11 8.987 554
statute mile (U.S.)	meter	+03 1.609 344*
statvolt	volt	+02 2.997 925
stere	meter^3	+00 1.00*
stilb	candela/meter^2	+04 1.00
stoke	meter^2/second	−04 1.00*
tablespoon	meter^3	−05 1.478 676 478 125*
teaspoon	meter^3	−06 4.928 921 593 75*
ton (assay)	kilogram	−02 2.916 666 6
ton (long)	kilogram	+03 1.016 046 908 8*
ton (metric)	kilogram	+03 1.00*
ton (nuclear equivalent of TNT)	joule	+09 4.20
ton (register)	meter^3	+00 2.831 684 659 2*
ton (short, 2000 pound)	kilogram	+02 9.071 847 4*
tonne	kilogram	+03 1.00*
torr (0° C)	newton/meter^2	+02 1.333 22
township	meter^2	+07 9.323 957 2
unit pole	weber	−07 1.256 637
volt (international of 1948)	volt	+00 1.000 330
watt (international of 1948)	watt	+00 1.000 165
yard	meter	−01 9.144*
year (calendar)	second (mean solar)	+07 3.1536*
year (sidereal)	second (mean solar)	+07 3.155 815 0
year (tropical)	second (mean solar)	+07 3.155 692 6
year 1900, tropical, Jan., day 0, hour 12	second (ephemeris)	+07 3.155 692 597 47*
year 1900, tropical, Jan., day 0, hour 12	second	+07 3.155 692 597 47

Listing of Conversion Factors by Physical Quantity　按物理量列表的换算系数

ACCELERATION

To convert from	to	multiply by
foot/second^2	meter/second^2	−01 3.048*
free fall, standard	meter/second^2	+00 9.806 65*
gal (galileo)	meter/second^2	−02 1.00*
inch/second^2	meter/second^2	−02 2.54*

AREA

To convert from	to	multiply by
acre	meter^2	+03 4.046 856 422 4*
are	meter^2	+02 1.00*
barn	meter^2	−28 1.00*
circular mil	meter^2	−10 5.067 074 8
foot^2	meter^2	−02 9.290 304*
hectare	meter^2	+04 1.00*
inch^2	meter^2	−04 6.4516*
mile^2 (U.S. statute)	meter^2	+06 2.589 988 110 336*
section	meter^2	+06 2.589 988 110 336*
township	meter^2	+07 9.323 957 2
yard^2	meter^2	−01 8.361 273 6*

DENSITY

To convert from	to	multiply by
gram/centimeter^3	kilogram/meter^3	+03 1.00*
lbm/inch^3	kilogram/meter^3	+04 2.767 990 5
lbm/foot^3	kilogram/meter^3	+01 1.601 846 3
slug/foot^3	kilogram/meter^3	+02 5.153 79

ENERGY

To convert from	to	multiply by
British thermal unit (ISO/TC 12)	joule	+03 1.055 06
British thermal unit (International Steam Table)	joule	+03 1.055 04
British thermal unit (mean)	joule	+03 1.055 87
British thermal unit (thermochemical)	joule	+03 1.054 350 264 488
British thermal unit (39° F)	joule	+03 1.059 67
British thermal unit (60° F)	joule	+03 1.054 68
calorie (International Steam Table)	joule	+00 4.1868
calorie (mean)	joule	+00 4.190 02
calorie (thermochemical)	joule	+00 4.184*
calorie (15° C)	joule	+00 4.185 80
calorie (20° C)	joule	+00 4.181 90
calorie (kilogram, International Steam Table)	joule	+03 4.1868
calorie (kilogram, mean)	joule	+03 4.190 02
calorie (kilogram, thermochemical)	joule	+03 4.184*
electron volt	joule	−19 1.602 10
erg	joule	−07 1.00*
foot lbf	joule	+00 1.355 817 9
foot poundal	joule	−02 4.214 011 0
joule (international of 1948)	joule	+00 1.000 165
kilocalorie (International Steam Table)	joule	+03 4.1868
kilocalorie (mean)	joule	+03 4.190 02
kilocalorie (thermochemical)	joule	+03 4.184*
kilowatt hour	joule	+06 3.60*
kilowatt hour (international of 1948)	joule	+06 3.600 59
ton (nuclear equivalent of TNT)	joule	+09 4.20
watt hour	joule	+03 3.60*

ENERGY/AREA TIME

To convert from	to	multiply by
Btu (thermochemical)/foot^2 second	watt/meter^2	+04 1.134 893 1
Btu (thermochemical)/foot^2 minute	watt/meter^2	+02 1.891 488 5
Btu (thermochemical)/foot^2 hour	watt/meter^2	+00 3.152 480 8
Btu (thermochemical)/inch^2 second	watt/meter^2	+06 1.634 246 2
calorie (thermochemical)/cm^2 minute	watt/meter^2	+02 6.973 333 3
erg/centimeter^2 second	watt/meter^2	−03 1.00*
watt/centimeter^2	watt/meter^2	+04 1.00*

FORCE

To convert from	to	multiply by
dyne	newton	−05 1.00*
kilogram force (kgf)	newton	+00 9.806 65*
kilopond force	newton	+00 9.806 65*
kip	newton	+03 4.448 221 615 260 5*
lbf (pound force, avoirdupois)	newton	+00 4.448 221 615 260 5*
ounce force (avoirdupois)	newton	−01 2.780 138 5
pound force, lbf (avoirdupois)	newton	+00 4.448 221 615 260 5*
poundal	newton	−01 1.382 549 543 76*

LENGTH

To convert from	to	multiply by
angstrom	meter	−10 1.00*
astronomical unit	meter	+11 1.495 978 9
cable	meter	+02 2.194 56*
caliber	meter	−04 2.54*
chain (surveyor or gunter)	meter	+01 2.011 68*
chain (engineer or ramden)	meter	+01 3.048*
cubit	meter	−01 4.572*
fathom	meter	+00 1.8288*
fermi (femtometer)	meter	−15 1.00*
foot	meter	−01 3.048*
foot (U.S. survey)	meter	+00 1200/3937*
foot (U.S. survey)	meter	−01 3.048 006 096
furlong	meter	+02 2.011 68*
hand	meter	−01 1.016*
inch	meter	−02 2.54*
league (U.K. nautical)	meter	+03 5.559 552*
league (international nautical)	meter	+03 5.556*
league (statute)	meter	+03 4.828 032*
light year	meter	+15 9.460 55
link (engineer or ramden)	meter	−01 3.048*
link (surveyor or gunter)	meter	−01 2.011 68*
meter	wavelengths Kr 86	+06 1.650 763 73*
micron	meter	−06 1.00*
mil	meter	−05 2.54*
mile (U.S. statute)	meter	+03 1.609 344*
mile (U.K. nautical)	meter	+03 1.853 184*
mile (international nautical)	meter	+03 1.852*
mile (U.S. nautical)	meter	+03 1.852*
nautical mile (U.K.)	meter	+03 1.853 184*
nautical mile (international)	meter	+03 1.852*
nautical mile (U.S.)	meter	+03 1.852*

Listing of Conversion Factors by Physical Quantity *(Continued)* 按物理量列表的换算系数

To convert from	*to*	*multiply by*
pace	meter	−01 7.62*
parsec	meter	+16 3.083 74
perch	meter	+00 5.0292*
pica (printers)	meter	−03 4.217 517 6*
point (printers)	meter	−04 3.514 598*
pole	meter	+00 5.0292*
rod	meter	+00 5.0292*
skein	meter	+02 1.097 28*
span	meter	−01 2.286*
statute mile (U.S.)	meter	+03 1.609 344*
yard	meter	−01 9.144*

MASS

To convert from	*to*	*multiply by*
carat (metric)	kilogram	−04 2.00*
dram (avoirdupois)	kilogram	−03 1.771 845 195 312 5*
dram (troy or apothecary)	kilogram	−03 3.887 934 6*
grain	kilogram	−05 6.479 891*
gram	kilogram	−03 1.00*
hundredweight (long)	kilogram	+01 5.080 234 544*
hundredweight (short)	kilogram	+01 4.535 923 7*
kgf second^2 meter (mass)	kilogram	+00 9.806 65*
kilogram mass	kilogram	+00 1.00*
lbm (pound mass, avoirdupois)	kilogram	−01 4.535 923 7*
ounce mass (avoirdupois)	kilogram	−02 2.834 952 312 5*
ounce mass (troy or apothecary)	kilogram	−02 3.110 347 68*
pennyweight	kilogram	−03 1.555 173 84*
pound mass, lbm (avoirdupois)	kilogram	−01 4.535 923 7*
pound mass (troy or apothecary)	kilogram	−01 3.732 417 216*
scruple (apothecary)	kilogram	−03 1.295 978 2*
slug	kilogram	+01 1.459 390 29
ton (assay)	kilogram	−02 2.916 666 6
ton (long)	kilogram	+03 1.016 046 908 8*
ton (metric)	kilogram	+03 1.00*
ton (short, 2000 pound)	kilogram	+02 9.071 847 4*
tonne	kilogram	+03 1.00*

POWER

To convert from	*to*	*multiply by*
Btu (thermochemical)/second	watt	+03 1.054 350 264 488
Btu (thermochemical)/minute	watt	+01 1.757 250 4
calorie (thermochemical)/second	watt	+00 4.184*
calorie (thermochemical)/minute	watt	−02 6.973 333 3
foot lbf/hour	watt	−04 3.766 161 0
foot lbf/minute	watt	−02 2.259 696 6
foot lbf/second	watt	+00 1.355 817 9
horsepower (550 foot lbf/second)	watt	+02 7.456 998 7
horsepower (boiler)	watt	+03 9.809 50
horsepower (electric)	watt	+02 7.46*
horsepower (metric)	watt	+02 7.354 99
horsepower (U.K.)	watt	+02 7.457
horsepower (water)	watt	+02 7.460 43
kilocalorie (thermochemical)/minute	watt	+01 6.973 333 3
kilocalorie (thermochemical)/second	watt	+03 4.184*
watt (international of 1948)	watt	+00 1.000 165

PRESSURE

To convert from	*to*	*multiply by*
atmosphere	newton/meter^2	+05 1.013 25*
bar	newton/meter^2	+05 1.00*
barye	newton/meter^2	−01 1.00*
centimeter of mercury (0° C)	newton/meter^2	+03 1.333 22
centimeter of water (4° C)	newton/meter^2	+01 9.806 38
dyne/centimeter^2	newton/meter^2	−01 1.00*
foot of water (39.2° F)	newton/meter^2	+03 2.988 98
inch of mercury (32° F)	newton/meter^2	+03 3.386 389
inch of mercury (60° F)	newton/meter^2	+03 3.376 85
inch of water (39.2° F)	newton/meter^2	+02 2.490 82
inch of water (60° F)	newton/meter^2	+02 2.4884
kgf/centimeter^2	newton/meter^2	+04 9.806 65*
kgf/meter^2	newton/meter^2	+00 9.806 65*
lbf/foot^2	newton/meter^2	+01 4.788 025 8
lbf/inch^2 (psi)	newton/meter^2	+03 6.894 757 2
millibar	newton/meter^2	+02 1.00*
millimeter of mercury (0° C)	newton/meter^2	+02 1.333 224
pascal	newton/meter^2	+00 1.00*
psi (lbf/inch^2)	newton/meter^2	+03 6.894 757 2
torr (0° C)	newton/meter^2	+02 1.333 22

SPEED

To convert from	*to*	*multiply by*
foot/hour	meter/second	−05 8.466 666 6
foot/minute	meter/second	−03 5.08*
foot/second	meter/second	−01 3.048*
inch/second	meter/second	−02 2.54*
kilometer/hour	meter/second	−01 2.777 777 8
knot (international)	meter/second	−01 5.144 444 444
mile/hour (U.S. statute)	meter/second	−01 4.4704*
mile/minute (U.S. statute)	meter/second	+01 2.682 24*
mile/second (U.S. statute)	meter/second	+03 1.609 344*

TEMPERATURE

To convert from	*to*	*multiply by*
Celsius	kelvin	$t_K = t_C + 273.15$
Fahrenheit	kelvin	$t_K = (5/9)(t_F + 459.67)$
Fahrenheit	Celsius	$t_C = (5/9)(t_F - 32)$
Rankine	kelvin	$t_K = (5/9)t_R$

TIME

To convert from	*to*	*multiply by*
day (mean solar)	second (mean solar)	+04 8.64*
day (sidereal)	second (mean solar)	+04 8.616 409 0
hour (mean solar)	second (mean solar)	+03 3.60*
hour (sidereal)	second (mean solar)	+03 3.590 170 4
minute (mean solar)	second (mean solar)	+01 6.00*
minute (sidereal)	second (mean solar)	+01 5.983 617 4
month (mean calendar)	second (mean solar)	+06 2.628*
second (ephemeris)	second	+00 1.000 000 000
second (mean solar)	second (ephemeris)	Consult American Ephemeris and Nautical Almanac
second (sidereal)	second (mean solar)	−01 9.972 695 7
year (calendar)	second (mean solar)	+07 3.1536*
year (sidereal)	second (mean solar)	+07 3.155 815 0
year (tropical)	second (mean solar)	+07 3.155 692 6
year 1900, tropical, Jan., day 0, hour 12	second (ephemeris)	+07 3.155 692 597 47*
year 1900, tropical, Jan., day 0, hour 12	second	+07 3.155 692 597 47

Listing of Conversion Factors by Physical Quantity *(Continued)*　按物理量列表的换算系数

To convert from	*to*	*multiply by*
	VISCOSITY	
centistoke	$meter^2$/second	−06 1.00*
stoke	$meter^2$/second	−04 1.00*
$foot^2$/second	$meter^2$/second	−02 9.290 304*
centipoise	newton second/$meter^2$	−03 1.00*
lbm/foot second	newton second/$meter^2$	+00 1.488 163 9
lbf second/$foot^2$	newton second/$meter^2$	+01 4.788 025 8
poise	newton second/$meter^2$	−01 1.00*
poundal second/$foot^2$	newton second/$meter^2$	+00 1.488 163 9
slug/foot second	newton second/$meter^2$	+01 4.788 025 8
rhe	$meter^2$/newton second	+01 1.00*
VOLUME		
acre foot	$meter^3$	+03 1.233 481 9
barrel (petroleum, 42 gallons)	$meter^3$	−01 1.589 873
board foot	$meter^3$	−03 2.359 737 216*
bushel (U.S.)	$meter^3$	−02 3.523 907 016 688*
cord	$meter^3$	+00 3.624 556 3
cup	$meter^3$	−04 2.365 882 365*
dram (U.S. fluid)	$meter^3$	−06 3.696 691 195 312 5*
fluid ounce (U.S.)	$meter^3$	−05 2.957 352 956 25*
$foot^3$	$meter^3$	−02 2.831 684 659 2*
gallon (U.K. liquid)	$meter^3$	−03 4.546 087
gallon (U.S. dry)	$meter^3$	−03 4.404 883 770 86*
gallon (U.S. liquid)	$meter^3$	−03 3.785 411 784*
gill (U.K.)	$meter^3$	−04 1.420 652
gill (U.S.)	$meter^3$	−04 1.182 941 2
hogshead (U.S.)	$meter^3$	−01 2.384 809 423 92*
$inch^3$	$meter^3$	−05 1.638 706 4*
liter	$meter^3$	−03 1.00*
ounce (U.S. fluid)	$meter^3$	−05 2.957 352 956 25*
peck (U.S.)	$meter^3$	−03 8.809 767 541 72*
pint (U.S. dry)	$meter^3$	−04 5.506 104 713 575*
pint (U.S. liquid)	$meter^3$	−04 4.731 764 73*
quart (U.S. dry)	$meter^3$	−03 1.101 220 942 715*
quart (U.S. liquid)	$meter^3$	−04 9.463 529 5
stere	$meter^3$	+00 1.00*
tablespoon	$meter^3$	−05 1.478 676 478 125*
teaspoon	$meter^3$	−06 4.928 921 593 75*
ton (register)	$meter^3$	+00 2.831 684 659 2*
$yard^3$	$meter^3$	−01 7.645 548 579 84*

英汉词汇对照

A

Abutments，桥墩
Accelerated linear motion，加速直线运动
Addendum format，补遗格式
Advertisement for Bids，招标公告
Affidavits，书面陈述
Aggregates，骨料
Agreement，协议

Air Conditioning，空调
Airports，机场（见第 16 节）
  clearing plan，净场平面图
  fencing，栅栏
  markings，标记
  pavement controls，道路控制
  pavement joints，道路分缝
  pavement markings，道路标记
  runway，飞机跑道
  safety plan，安全计划
  taxiway，滑行道
  terminal area，候机厅区域
  terminal plan，候机厅平面
  typical section，典型剖面
Annuity，年金
  amount，总计
  given capita，给定资本
  present worth，现值
  sinking fund，偿债基金
Approvals，审核批准（见第 5 节）
Ash collection，粉尘收集
Ash storage，粉尘储藏
Assessment hearing，评估听证会
Athletic facilities，体育设施（见第 17 节）：
  ball field，球场
  field lighting，场地照明
  hockey rink，曲棍球
  quarter－mile track，四分之一英里跑道
  racquetball court，短网拍墙球场
  running track，跑道
  tennis court，网球场
Atmosphere monitoring，大气监测
Attendance list，出席者名单

B

Ball fields，球场
Batter boards，定斜板
Belts，皮带
Bends，弯头
Bid （投标）：
  bond，投标保单
  evaluation，标书评审
  opening，开标
  proposal，投标书
  tabulation，投标表格
Bidding documents （招标文件，见第 23、24 和 25 节）：
  ad for bids，招标公告
  agreement，协议
  bid bond，投标保单
  change order，变更指令
  information for bidders，给竞标者的通知
  general conditions，一般条款
  notice of award，中标通知
  notice to proceed，开工通知
  payment bond，付款保单
  performance bond，履约保单
  process，招标程序（见第 24 节）
  proposal，投标书
Boiler installation，锅炉安装
Bollard，系缆柱
Bonds，保单：
  bid，投标保单
  payment，付款保单
  performance，履约保单
Boundary mapping，红线图
Bridges，桥梁（见第 15 节）
Buildings，建筑（见第 8 节）
Buttress，扶壁

C

Campgrounds，露营场所
Casing，套管：
  common，普通套管
Catch basins，收集地面污水的贮水池
Ceremonies，典礼
Certification，证明：
  attorney's，代理机构证明
  engineer's，工程师证明
  insurance，保险证明
  submittal completion，实质完工证明
Certified survey maps，经核准的测量图
Chain link fence，链条栅栏
Chalk line，墨线
Change order，变更指令
Chordal，弦：
  height，矢高
  length，弦长
Church，教堂
Civil engineer，土木工程师
Clearing plan，净场平面图
Close-out，竣工（见第 31 节）：
  affidavit of debts & claims，债务和权利主张书
  affidavit of release of liens，留置权释放书
  certification of substantial completion，实质完工证明
  consent of surety，担保承诺
  construction record drawings，施工记录图
  O & M manual，运行和维护手册
  punch list，建设缺陷清单
Collection system，收集系统
Committee meetings，委员会会议
Common trench，普通管沟

D

E

F

G

H

I

J

K

L

M

T

U

V

W

Y

Z

译后记

在中国加入世界贸易组织后，土木工程行业面临着巨大的机遇和挑战，了解和掌握以美国为代表的国外土木工程实践领域的规则和程序是我国土木工程行业从业者走向世界的第一步。本书的作者是美国的职业工程师，拥有多年的工程实践经验，书中大量实用的图表、表格、图纸和其他数据大都选自美国的实际工程项目，具有较大的参考价值。

本书的组织结构完全是按照建设项目的全过程循序渐进展开的，从规划、设计到招标和施工。值得说明的是，本书对原始的图纸、图表并未翻译，而仅是翻译每节开篇的文字说明和目录，这样既保留了资料的原貌，同时也便于国内读者对照。附录中的技术参考提供了大量的技术数据信息，文末的英汉对照便于国内读者阅读时参考。

姚燕女士为本书早日出版付出了辛勤的劳动，在此表示诚挚的谢意。由于译者水平及时间有限，译文中一定存在不少错误和欠妥之处，请读者批评指正。

清华大学
杨军
2004年8月